Surface and Ground Water, Weathering, and Soils

Citations
Please use the following example for citations:
Amundson R. (2003) Soil formation, pp. 1–35. In *Surface and Ground Water, Weathering, and Soils* (ed. J.I. Drever) Vol. 5
Treatise on Geochemistry (eds. H.D. Holland and K.K. Turekian), Elsevier–Pergamon, Oxford.

Surface and Ground Water, Weathering, and Soils

Edited by

J. I. Drever
University of Wyoming, WY, USA

TREATISE ON GEOCHEMISTRY
Volume 5

Executive Editors

H. D. Holland
Harvard University, Cambridge, MA, USA

and

K. K. Turekian
Yale University, New Haven, CT, USA

ELSEVIER

2005

AMSTERDAM – BOSTON – HEIDELBERG – LONDON – NEW YORK – OXFORD
PARIS – SAN DIEGO – SAN FRANCISCO – SINGAPORE – SYDNEY – TOKYO

ELSEVIER B.V.
Radarweg 29
P.O. Box 211, 1000 AE Amsterdam
The Netherlands

ELSEVIER Inc.
525 B Street, Suite 1900
San Diego, CA 92101-4495
USA

ELSEVIER Ltd
The Boulevard, Langford Lane
Kidlington, Oxford OX5 1GB
UK

ELSEVIER Ltd
84 Theobalds Road
London WC1X 8RR
UK

First edition 2005

Library of Congress Cataloging in Publication Data
A catalog record is available from the Library of Congress.

British Library Cataloguing in Publication Data
A catalogue record is available from the British Library.

ISBN: 0-08-044719-8 (Paperback)

The following chapters are US Government works in the public domain and not subject to copyright:
    Geochemistry of Saline Lakes
    Stable Isotope Applications in Hydrologic Studies
    Natural Weathering Rates of Silicate Minerals
    Modeling Low-temperature Geochemical Processes
    Deep Fluids in the Continents: I. Sedimentary Basins
    Geochemistry of Groundwater
    Mass-balance Approach to Interpreting Weathering Reactions in Watershed Systems

⊗ The paper used in this publication meets the requirements of ANSI/NISO Z39.48-1992 (Permanence of Paper).

Printed and bound in the United Kingdom

Transferred to Digital Print 2011

DEDICATED
TO

WERNER STUMM
(1924–1999)

Photograph provided by James I. Drever

Contents

Executive Editors' Foreword

H. D. Holland and

Harvard University, Cambridge, MA, USA

and

K. K. Turekian

Yale University, New Haven, CT, USA

Geochemistry has deep roots. Its beginnings can be traced back to antiquity, but many of the discoveries that are basic to the science were made between 1800 and 1910. The periodic table of elements was assembled, radioactivity was discovered, and the thermodynamics of heterogeneous systems was developed. The solar spectrum was used to determine the composition of the Sun. This information, together with chemical analyses of meteorites, provided an entry to a larger view of the universe.

During the first half of the twentieth century, a large number of scientists used a variety of methods to determine the major-element composition of the Earth's crust, and the geochemistries of many of the minor elements were defined by V. M. Goldschmidt and his associates using the then new technique of emission spectrography. V. I. Vernadsky founded biogeochemistry. The crystal structures of most minerals were determined by X-ray diffraction techniques. Isotope geochemistry was born, and age determinations based on radiometric techniques began to define the absolute geologic timescale. The intense scientific efforts during World War II yielded new analytical tools and a group of people who trained a new generation of geochemists at a number of universities. But the field grew slowly. In the 1950s, a few journals were able to report all of the important developments in trace-element geochemistry, isotopic geochronometry, the exploration of paleoclimatology and biogeochemistry with light stable isotopes, and studies of phase equilibria. At the meetings of the American Geophysical Union, geochemical sessions were few, none were concurrent, and they all ranged across the entire field.

Since then the developments in instrumentation and the increases in computing power have been spectacular. The education of geochemists has been broadened beyond the old, rather narrowly defined areas. Atmospheric and marine geochemistry have become integrated into solid Earth geochemistry; cosmochemistry and biogeochemistry have contributed greatly to our understanding of the history of our planet. The study of Earth has evolved into "Earth System Science," whose progress since the 1940s has been truly dramatic.

Major ocean expeditions have shown how and how fast the oceans mix; they have demonstrated the connections between the biologic pump, marine biology, physical oceanography, and marine sedimentation. The discovery of hydrothermal vents has shown how oceanography is related to economic geology. It has revealed formerly unknown oceanic biotas, and has clarified the factors that today control, and in the past have controlled the composition of seawater.

Seafloor spreading, continental drift and plate tectonics have permeated geochemistry. We finally understand the fate of sediments and oceanic crust in subduction zones, their burial and their exhumation. New experimental techniques at temperatures and pressures of the deep Earth interior have clarified the three-dimensional structure of the mantle and the generation of magmas.

Moon rocks, the treasure trove of photographs of the planets and their moons, and the successful search for planets in other solar systems have all revolutionized our understanding of Earth and the universe in which we are embedded.

Geochemistry has also been propelled into the arena of local, regional, and global anthropogenic problems. The discovery of the ozone hole came as a great, unpleasant surprise, an object lesson for optimists and a source of major new insights into the photochemistry and dynamics of the atmosphere. The rise of the CO_2 content of the atmosphere due to the burning of fossil fuels and deforestation has been and will continue to be at the center of the global change controversy, and will yield new insights into the coupling of atmospheric chemistry to the biosphere, the crust, and the oceans.

The rush of scientific progress in geochemistry since World War II has been matched by organizational innovations. The first issue of *Geochimica et Cosmochimica Acta* appeared in June 1950. The Geochemical Society was founded in 1955 and adopted *Geochimica et Cosmochimica Acta* as its official publication in 1957. The International Association of Geochemistry and Cosmochemistry was founded in 1966, and its journal, *Applied Geochemistry*, began publication in 1986. *Chemical Geology* became the journal of the European Association for Geochemistry.

The Goldschmidt Conferences were inaugurated in 1991 and have become large international meetings. Geochemistry has become a major force in the Geological Society of America and in the American Geophysical Union. Needless to say, medals and other awards now recognize outstanding achievements in geochemistry in a number of scientific societies.

During the phenomenal growth of the science since the end of World War II an admirable number of books on various aspects of geochemistry were published. Of these only three attempted to cover the whole field. The excellent *Geochemistry* by K. Rankama and Th.G. Sahama was published in 1950. V. M. Goldschmidt's book with the same title was started by the author in the 1940s. Sadly, his health suffered during the German occupation of his native Norway, and he died in England before the book was completed. Alex Muir and several of Goldschmidt's friends wrote the missing chapters of this classic volume, which was finally published in 1954.

Between 1969 and 1978 K. H. Wedepohl together with a board of editors (C. W. Correns, D. M. Shaw, K. K. Turekian and J. Zeman) and a large number of individual authors assembled the *Handbook of Geochemistry*. This and the other two major works on geochemistry begin with integrating chapters followed by chapters devoted to the geochemistry of one or a small group of elements. All three are now out of date, because major innovations in instrumentation and the expansion of the number of practitioners in the field have produced valuable sets of high-quality data, which have led to many new insights into fundamental geochemical problems.

At the Goldschmidt Conference at Harvard in 1999, Elsevier proposed to the Executive Editors that it was time to prepare a new, reasonably comprehensive, integrated summary of geochemistry. We decided to approach our task somewhat differently from our predecessors. We divided geochemistry into nine parts. As shown below, each part was assigned a volume, and a distinguished editor was chosen for each volume. A tenth volume was reserved for a comprehensive index:

(i) *Meteorites, Comets, and Planets*: Andrew M. Davis

(ii) *Geochemistry of the Mantle and Core*: Richard Carlson

(iii) *The Earth's Crust*: Roberta L. Rudnick

(iv) *Atmospheric Geochemistry*: Ralph F. Keeling

(v) *Freshwater Geochemistry, Weathering, and Soils*: James I. Drever

(vi) *The Oceans and Marine Geochemistry*: Harry Elderfield

(vii) *Sediments, Diagenesis, and Sedimentary Rocks*: Fred T. Mackenzie

(viii) *Biogeochemistry*: William H. Schlesinger

(ix) *Environmental Geochemistry*: Barbara Sherwood Lollar

(x) *Indexes*

The editor of each volume was asked to assemble a group of authors to write a series of chapters that together summarize the part of the field covered by the volume. The volume editors and chapter authors joined the team enthusiastically. Altogether there are 155 chapters and 9 introductory essays in the Treatise. Naming the work proved to be somewhat problematic. It is clearly not meant to be an encyclopedia. The titles *Comprehensive Geochemistry* and *Handbook of Geochemistry* were finally abandoned in favor of *Treatise on Geochemistry*.

The major features of the Treatise were shaped at a meeting in Edinburgh during a conference on Earth System Processes sponsored by the Geological Society of America and the Geological Society of London in June 2001. The fact that the Treatise is being published in 2003 is due to a great deal of hard work on the part of the editors, the authors, Mabel Peterson (the Managing Editor), Angela Greenwell (the former Head of Major Reference Works), Diana Calvert (Developmental Editor, Major Reference Works),

Bob Donaldson (Developmental Manager), Jerome Michalczyk and Rob Webb (Production Editors), and Friso Veenstra (Senior Publishing Editor). We extend our warm thanks to all of them. May their efforts be rewarded by a distinguished journey for the Treatise.

Finally, we would like to express our thanks to J. Laurence Kulp, our advisor as graduate students at Columbia University. He introduced us to the excitement of doing science and convinced us that all of the sciences are really subdivisions of geochemistry.

Contributors to Volume 5

R. Amundson
University of California, Berkeley, CA, USA

E. K. Berner
Yale University, New Haven, CT, USA

R. A. Berner
Yale University, New Haven, CT, USA

R. Blomqvist
Geological Survey of Finland, Espoo, Finland

J. D. Blum
The University of Michigan, Ann Arbor, MI, USA

A. Blyth
University of Waterloo, ON, Canada

C. J. Bowser
University of Wisconsin, Madison, WI, USA

S. L. Brantley
Pennsylvania State University, University Park, PA, USA

O. P. Bricker
USGS Water Resources Division, Reston, VA, USA

M. C. Castro
University of Michigan, Ann Arbor, USA

F. H. Chapelle
US Geological Survey, Columbia, SC, USA

D. M. Deocampo
US Geological Survey, Reston, VA, USA

D. H. Doctor
United States Geological Survey, Menlo Park, CA, USA

B. Dupré
Laboratoire des Mécanismes et Transferts en Géologie, Toulouse, France

Y. Erel
The Hebrew University, Jerusalem, Israel

S. K. Frape
University of Waterloo, ON, Canada

J. Gaillardet
Institut de Physique du Globe de Paris, France

M. Gascoyne
Gascoyne GeoProjects Inc., Pinawa, MB, Canada

J. S. Hanor
Louisiana State University, Baton Rouge, LA, USA

B. F. Jones
USGS Water Resources Division, Reston, VA, USA

C. Kendall
United States Geological Survey, Menlo Park, CA, USA

Y. K. Kharaka
US Geological Survey, Menlo Park, CA, USA

R. H. McNutt
University of Toronto, ON, Canada

M. Meybeck
University of Paris VI, CNRS, Paris, France

[†]K. L. Moulton
Kent State University, OH, USA

D. K. Nordstrom
US Geological Survey, Boulder, CO, USA

E. M. Perdue
Georgia Institute of Technology, Atlanta, GA, USA

F. M. Phillips
New Mexico Tech, Socorro, NM, USA

G. J. Retallack
University of Oregon, Eugene, OR, USA

J. D. Ritchie
Georgia Institute of Technology, Atlanta, GA, USA

M. Tranter
University of Bristol, UK

J. Viers
Laboratoire des Mécanismes et Transferts en Géologie, Toulouse, France

A. F. White
US Geological Survey, Menlo Park, CA, USA

Volume Editor's Introduction

J. I. Drever

University of Wyoming, WY, USA

It is impossible to pick a point in time that represents the start of modern approaches to understanding the chemistry of surface- and groundwaters. The most influential papers were probably those of Garrels and Mackenzie (1967) and Garrels (1967). They proposed two concepts: (i) the concept of mass balance—that the composition of a water could be explained by a series of mineral dissolution and precipitation reactions, and (ii) the concept that chemical equilibria between minerals and water were an important control on both the composition of the water and the identity of minerals precipitated from those waters. A further concept that evolved at about the same time (e.g., Mackenzie and Garrels, 1966; Holland, 1968, 1978) was the importance of understanding weathering and diagenetic reactions in order to understand the carbon dioxide balance of the atmosphere and hence, through the greenhouse effect, global temperature (following the ideas of Urey (1956)). To a large extent, the chapters in this volume represent the evolution of these concepts over the intervening years. An additional impetus for research in geochemistry has been the necessity of predicting the effects of human activities such as mining and waste disposal (including radioactive waste) on water quality. This has led to various models for predicting surface water chemistry (see Chapter 5.02). The reliability of various modeling approaches for predicting future water quality is subject to considerable debate.

1 MASS BALANCE AS A MEANS OF CONSTRAINING CHEMICAL REACTIONS

The relatively simple approach of Garrels and Mackenzie has been systematized and extended into spreadsheet programs (see Chapter 5.04) and computer codes (Chapters 5.02 and 5.14). Probably the most important developments have been the inclusion of isotopic species (both stable and radiogenic: Chapters 5.11 and 5.12) in the calculations, and the inclusion of redox species (compounds of, e.g., carbon, oxygen, sulfur, and iron) in the overall balance. One result has been a vastly improved understanding of microbial redox processes as controls on groundwater composition; however, the identification of silicate weathering reactions is still subject to considerable ambiguity. It is clear that the more isotopic species that can be brought in, the better constrained the overall balance. It is also clear that uptake and storage of elements in plants can have a major influence on the composition of surface waters (see Chapter 5.06). Mass-balance approaches have also been applied to the solid phases transported by rivers (Chapter 5.09). The integration of information on solid phases with information on solutes provides important additional insights into weathering processes and the denudation of the continents.

2 CHEMICAL EQUILIBRIA AND KINETICS

The early works on weathering stressed the importance of equilibria between natural waters and secondary products (primarily clay minerals) formed during weathering. This topic has received relatively little attention in subsequent years. The general consensus is that when a mineral is newly formed, i.e., it does not inherit its structure from another mineral, the assumption of equilibrium between the newly formed phase and solution is a reasonable first approximation (Drever, 1997). It is, however, only an approximation. The free energies of formation of most secondary silicates are poorly defined—kaolin minerals occur with varying degrees of structural order and hence free energy; smectites are highly variable in composition—so it is not really possible to evaluate precise departures from equilibrium. Secondary minerals that inherit part of their structure from a primary mineral (e.g., vermiculite or smectite formed from biotite) are typically nowhere near equilibrium with the solutions in which they form.

The focus in recent years has been to try to relate the composition of natural waters to the mechanism and rate of dissolution of primary minerals. Werner Stumm, to whom this volume is dedicated, has been a leader in this field, particularly in applying the concepts of coordination chemistry to mineral dissolution. This has led to the idea of surface speciation (adsorption or loss of protons; adsorption of ligands) to form precursor complexes as the key to understanding dissolution rates far from equilibrium (see Chapter 5.03). This represents only one of Stumm's many contributions to aquatic chemistry (see also Volume 9). Dissolution of silicates such as feldspars, even in the laboratory, has turned out to be a complex subject that is still not fully understood. A particular problem is in understanding dissolution rates of minerals in solutions that are not "far from equilibrium." Rates decrease as equilibrium is approached—but what causes this? Is it simply because of the degree of undersaturation (Burch *et al.*, 1993), or is it related to adsorption of a species such as aluminum, which has a strong inhibiting effect (Oelkers *et al.*, 1994)? The next fundamental question is the relationship between dissolution rates in lab experiments and the weathering rates of minerals in the field (see Chapters 5.03 and 5.05). Rates in the field are typically one-to-three orders of magnitude slower than would be predicted by simple extrapolation of laboratory results—why? Is the reason "aging" of mineral surfaces? Approach to saturation? Unrealistic

assumptions in the comparison? It appears that all these effects are important (Chapter 5.05).

3 CHEMISTRY OF DEEP SUBSURFACE WATERS ON THE CONTINENTS

Our understanding of subsurface waters has greatly advanced by the use of an increasing number of radiogenic and cosmogenic isotopes to determine flow rates and subsurface processes (see Chapters 5.15–5.17). It has been known for a long time that saline waters were present at depth in many sedimentary basins, but the origin of the salinity was controversial. These processes are becoming better understood through the application of a range of tracers (Chapter 5.16). Deep fluids in crystalline rocks (Chapter 5.17) have always been something of a mystery. The chemistry of these waters is quite diverse and many of them are highly saline. There appear to be several processes responsible for the high salinities: e.g., seawater incursion, uptake of water by hydration reactions, upward movement of fluids released by metamorphic reactions at greater depth. This is a field where our knowledge is still very limited.

4 GLOBAL FLUXES AND ATMOSPHERIC CARBON DIOXIDE

During the last few decades there has been tremendous interest in understanding the Earth as a system (see Volume 8). The challenge here has been to extrapolate from laboratory experiments and detailed studies of small catchments to the global scale. How do we calculate CO_2 consumption by weathering on a global scale? How does it vary regionally as a function of climate, lithology, and topography? What is the precise functional relationship between atmospheric CO_2 content, CO_2 consumption by weathering, and global climate? Can we use these concepts to understand past climates? Can we predict the effect on climate of anthropogenic inputs of greenhouse gases? The first major synthesis was that of Holland (1978); there have been many compilations and models since. The most influential modeling approach is that of Berner (Berner *et al.*, 1983; Berner, 1991, 1994; Berner and Kothavala, 2001; Chapters 5.06 and 5.18). A focal point for some of these ideas has been the question of whether uplift of the Himalayas caused drawdown of atmospheric CO_2 and hence global cooling through accelerated chemical weathering of silicates (Raymo and Ruddiman, 1992). The strontium isotopic curve for seawater and the general plausibility of the idea that finely ground minerals (from glacial action) should weather rapidly favored the idea;

subsequent studies (e.g., Blum *et al.*, 1998) have suggested that silicate weathering rates were in fact rather low in the high Himalayas, primarily as a consequence of low temperatures and the absence of vegetation. The highly radiogenic $^{87}Sr/^{86}Sr$ ratio was a consequence of weathering of carbonate minerals containing unusually radiogenic strontium. The argument continues: it is interesting because it integrates many of the different approaches used to understand river-water chemistry and it illustrates the importance of weathering and erosion in controlling global climate.

5 BIOLOGICAL PROCESSES

Perhaps the most important trend in recent years has been the integration of biology into our understanding of geochemical processes. If one looks back at papers from the 1960s, biology received only a passing mention. Yes, it was a source of CO_2 in soils, and yes, microorganisms served as catalysts for redox reactions, but otherwise geochemists thought largely in terms of inorganic processes. If we look at papers today on weathering (Chapter 5.06), the chemistry of rivers (Chapter 5.08), of groundwater (Chapter 5.14), or even of subglacial waters (Chapter 5.07), we see the overwhelming influence of biological processes. Geomicrobiology is currently the fastest-growing subfield of geochemistry (e.g., Banfield and Nealson, 1997).

Finally, when talking about the role of organisms in geological processes, we cannot ignore the role of humans. Most of the discussion of human influences is in Volume 9, but we must recognize that the chemistry of rivers and the fluxes of elements transported by rivers to the ocean have been greatly influenced by human activities (see Chapters 5.08 and 5.09). These human influences are not simply the result of direct pollution, but are also a consequence of changes in land use, such as agriculture and deforestation.

REFERENCES

Banfield J. F. and Nealson K. H. (1997) *Geomicrobiology: Interactions between Microbes and Minerals.* Reviews in Mineralogy, Mineralogical Society of America, vol. 35, 448pp.

Berner R. A. (1991) A model for atmospheric CO_2 over Phanerozoic time. *Am. J. Sci.* **291**, 339–376.

Berner R. A. (1994) GEOCARB: II. A revised model of atmospheric CO_2 over Phanerozoic time. *Am. J. Sci.* **294**, 56–91.

Berner R. A. and Kothavala Z. (2001) Geocarb: III. A revised model of atmospheric CO_2 over Phanerozoic time. *Am. J. Sci.* **301**, 182–204.

Berner R. A., Lasaga A. C., and Garrels R. M. (1983) The carbonate–silicate geochemical cycle and its effects on atmospheric carbon dioxide and climate. *Am. J. Sci.* **283**, 641–683.

Blum J. D., Gazis C. A., Jacobson A. D., and Chamberlain C. P. (1998) Carbonate versus silicate weathering in the Raikot Watershed within the High Himalayan Crystalline Series. *Geology* **26**, 411–414.

Burch T. E., Nagy K. L., and Lasaga A. C. (1993) Free energy dependence of albite dissolution kinetics at 80 °C and pH 8.8. *Chem. Geol.* **105**, 137–162.

Drever J. I. (1997) *The Geochemistry of Natural Waters: Surface and Groundwater Environments* (3rd edn.). Prentice-Hall, Upper Saddle River, NJ, 436pp.

Garrels R. M. (1967) Genesis of some ground waters from igneous rocks. In *Researches in Geochemistry* (ed. P. H. Abelson). Wiley, New York, vol. 2, pp. 405–420.

Garrels R. M. and Mackenzie F. T. (1967) Origin of the chemical compositions of some springs and lakes. In *Equilibrium Concepts in Natural Water Systems*, Advances in Chemistry Series 67 (ed. R. F. Gould). American Chemical Society, Washington, DC, pp. 222–242.

Holland H. D. (1968) The abundance of CO_2 in the Earth's atmosphere through geologic time. In *Origin and Distribution of the Elements* (ed. L. H. Ahrens). Pergamon, New York, pp. 949–954.

Holland H. D. (1978) *The Chemistry of the Atmosphere and Oceans.* Wiley, New York, p. 351.

Mackenzie F. T. and Garrels R. M. (1966) Chemical mass balance between rivers and oceans. *Am. J. Sci.* **264**, 507–525.

Oelkers E. H., Schott J., and Devidal J.-L. (1994) The effect of aluminum, pH, and chemical affinity on the rates of aluminosilicate dissolution reactions. *Geochim. Cosmochim. Acta* **58**, 2011–2024.

Raymo M. E. and Ruddiman W. F. (1992) Tectonic forcing of late Cenozoic climate. *Nature* **359**, 117–122.

Urey H. C. (1956) Regarding the early history of the earth's atmosphere. *Geol. Soc. Am. Bull.* **67**, 1125–1128.

5.01
Soil Formation

R. Amundson

University of California, Berkeley, CA, USA

...that the Earth has not always been here—that it came into being at a finite point in the past and that everything here, from the birds and fishes to the loamy soil underfoot, was once part of a star. I found this amazing, and still do.

Timothy Ferris (1998)

5.01.1 INTRODUCTION

Soil is the biogeochemically altered material that lies at the interface between the lithosphere (Volume 3) and the atmosphere (Volume 4). *Pedology* is the branch of the natural sciences that concerns itself, in part, with the biogeochemical processes (Volume 8) that form and distribute soil across the globe. Pedology originated during the scientific renaissance of the nineteenth century as a result of conceptual breakthroughs by the Russian scientist Vassali Dochuchaev (Krupenikov, 1992; Vil'yams, 1967) and conceptual

and administrative efforts by the American scientist Eugene Hilgard (Jenny, 1961; Amundson and Yaalon, 1995).

Soil is the object of study in pedology, and while the science of pedology has a definition that commands some general agreement, there is no precise definition for soil, nor is there likely ever to be one. The reason for this paradox is that soil is a part of a continuum of materials at the Earth's surface (Jenny, 1941). At the soil's base, the exact line of demarcation between "soil" and "nonsoil" will forever elude general agreement, and horizontal changes in soil properties may occur so gradually that similar problems exist in delineating the boundary between one soil "type" and another. The scientific path out of this conundrum is to divide the soil continuum, albeit arbitrarily, into *systems* that suit the need of the investigator. Soil systems are necessarily open to their surroundings, and through them pass matter and energy which measurably alter the properties of the system over timescales from seconds to millennia. It was the recognition by Dokuchaev (1880), and later the American scientist Hans Jenny (1941), that the properties of the soil system are controlled by *state factors* that ultimately formed the framework of the fundamental paradigm of pedology.

The purpose of this chapter is to present an abridged overview of the factors and processes that control soil formation, and to provide, where possible, some general statements of soil formation processes that apply broadly and commonly.

5.01.2 FACTORS OF SOIL FORMATION

Jenny (1941) applied principles from the physical sciences to the study of soil formation. Briefly, Jenny recognized that soil systems (or if the aboveground flora and fauna are considered, *ecosystems*) exchange mass and energy with their surroundings and that their properties can be defined by a limited set of *independent variables*. From comparisons with other sciences, Jenny's *state factor model* of soil formation states that

$$\underbrace{\text{Soils/ecosystems}}_{\text{dependent variables}} = f \begin{pmatrix} \text{initial state of system,} \\ \text{surrounding environment,} \\ \text{elapsed time} \end{pmatrix}$$
$$\underbrace{\phantom{f \begin{pmatrix} \text{initial state of system,} \\ \text{surrounding environment,} \\ \text{elapsed time} \end{pmatrix}}}_{\text{independent variables}}$$

$$(1)$$

From field observations and the conceptual work of Dokuchaev, a set of more specific environmental factors have been identified which encompass the controls listed above:

$$\underbrace{\text{Soils/ecosystems}}$$
$$= f(\underbrace{\text{climate, organisms}}_{\text{surrounding environment}},$$
$$\underbrace{\text{topography, parent material}}_{\text{initial state of system}}, \text{time}, \ldots) \quad (2)$$

These so-called "state factors of soil formation" have the following important characteristics: (i) they are independent of the system being studied and (ii) in many parts of the Earth, the state factors vary independently of each other (though, of course, not always). As a result, through judicious site (system) selection, the influence of a single factor can be observed and quantified in nature.

Table 1 provides a brief definition of the state factors of soil formation. A field study designed to observe the influence of one state factor on soil properties or processes is referred to as a *sequence*, e.g., a series of sites which have similar state factor values except climate is referred to as a *climosequence*. Similar sequences can, and have been, established to examine the effect of other state factors on soils. An excellent review of soil state factor studies is presented by Birkeland (1999). An informative set of papers discussing the impact of Jenny's state factor model on advances in pedology, geology, ecology, and related sciences is presented in Amundson *et al.* (1994a,b).

The state factor approach to studying soil formation has been, and continues to be, a powerful quantitative means of linking soil properties to important variables (Amundson and Jenny, 1997). As an example, possibly the best characterized soil versus factor relationship is the relationship of soil organic carbon and nitrogen storage to climate (mean annual temperature and precipitation) (Figure 1). The pattern—increasing carbon storage with decreasing temperature and increasing precipitation—illustrated in Figure 1 is the result of nearly six decades of work, and is based on thousands of soil observations (Miller *et al.*, 2002). This climatic relationship is important in global change research and in predicting the response of soil carbon storage to climate change (Schlesinger and Andrews, 2000). However, the relationship, no matter how valid, provides no insight into the rates at which soils achieve their carbon storage, nor the mechanisms involved in the accumulation. Thus, other approaches, again amenable to systems studies, have been applied in pedology to quantify soil formation. These are discussed in later sections.

Table 1 The major state factors of soil and ecosystem formation, and a brief outline of their characteristics.

State factor	Definition and characteristics
Climate	Regional climate commonly characterized by mean annual temperature and precipitation
Organisms	Potential biotic flux into system (as opposed to what is present at any time)
Topography	Slope, aspect, and landscape configuration at time $t = 0$
Parent material	Chemical and physical characteristics of soil system at $t = 0$
Time	Elapsed time since system was formed or rejuvenated
Humans	A special biotic factor due to magnitude of human alteration of Earth's and humans' possession of variable cultural practices and attitudes that alter landscapes

Sources: Jenny (1941) and Amundson and Jenny (1997).

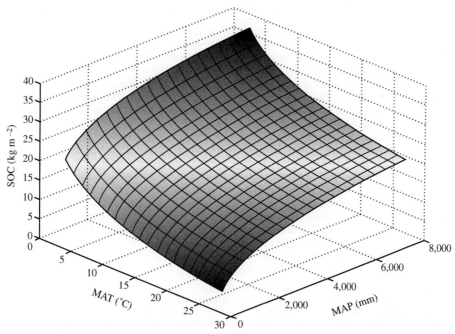

Figure 1 The distribution of global soil C in relation to variations in mean average temperature (MAT) and precipitaiton (MAP). The curve is derived from a multiple regression model of published soil C data versus climate (Amundson, 2001) (reproduced by permission of Annual Reviews from *Ann. Rev. Earth Planet. Sci.* **2001**, *29*, 535–562).

5.01.3 SOIL MORPHOLOGY

A trend in present-day pedology is to incorporate ever more sophisticated chemical and mathematical tools into our understanding of soil and their formation. Yet, an examination of soils *in situ* is required, in order to develop the appropriate models or to even logically collect samples for study.

Soil *profiles* are two-dimensional, vertical exposures of the layering of soils. The net result of the transport of matter and energy is a vertical differentiation of visible, distinctive layers called soil *horizons*. Soil horizons reflect the fact that soil formation is a depth-dependent process. They are layers that are readily identified by visual and tactile procedures (field based) that have been developed over the years (Soil Survey Staff, 1993). A nomenclature has developed over the past century, first started by the pioneering Russian scientists in the nineteenth century, that involves the "naming"of soil horizons on the basis of *how they differ from the starting parent material*. Therefore, horizon naming requires data acquisition and hypothesis development. Soil horizon names are commonly assigned from field-based data, and may ultimately be modified as a result of subsequent laboratory investigations.

The present US horizon nomenclature has two components: (i) an upper case, "master horizon" symbol and (ii) a lower case "modifier" that provides more information on the horizon characteristics or the processes that formed it.

Tables 2(a) and (b) provide definitions of both the common master and modifier symbols. The detailed rules for their use can be found in the *Soil Survey Manual* (Soil Survey Staff, 1993).

Most soil process models are (roughly) continuous with depth. However, during the observation of many soil profiles, it is apparent that horizons do not always, or even commonly, grade gradually into one another. Sharp or abrupt horizon boundaries are common in soils around the world. This indicates that our concepts and models of soil formation capture only a part of the long-term trajectory of soil development. Some processes are not continuous with depth (the formation of carbonate horizons for example), while some may be continuous for some time period and then, due to feedbacks, change their character (the formation of clay-rich horizons which, if they reach a critical clay content, restrict further water and clay transport. This causes an abrupt buildup of additional clay at the top of the horizon). In the following sections, the author examines various approaches to understanding soil formation, and examines some of their attributes and limitations.

5.01.4 MASS BALANCE MODELS OF SOIL FORMATION

Detailed chemical analyses of soils and the interpretation of that data relative to the composition of the parent material have been

Table 2(a) A listing, and brief definitions, of the nomenclature used to identify master soil horizons.

Master horizons	Definition and examples of lower case modifiers
O	Layers dominated by organic matter. State of decomposition determines type: highly (Oa), moderately (Oe), or slightly (Oi)[a] decomposed
A	Mineral horizons that have formed at the surface of the mineral portion of the soil or below an O horizon. Show one of the following: (i) an accumulation of humified organic matter closely mixed with minerals or (ii) properties resulting from cultivation, pasturing, or other human-caused disturbance (Ap)
E	Mineral horizons in which the main feature is loss of silicate clay, iron, aluminum, or some combination of these, leaving a concentration of sand and silt particles
B	Horizons formed below A, E, or O horizons. Show one or more of the following: (i) illuvial[b] concentration of silicate clay (Bt), iron (Bs), humus (Bh), carbonates (Bk), gypsum (By), or silica (Bq) alone or in combination; (ii) removal of carbonates (Bw); (iii) residual concentration of oxides (Bo); (iv) coatings of sesquioxides[c] that make horizon higher in chroma or redder in hue (Bw); (v) brittleness (Bx); or (vi) gleying[d] (Bg).
C	Horizons little affected by pedogenic processes. May include soft sedimentary material (C) or partially weathered bedrock (Cr)
R	Strongly indurated[e] bedrock
W	Water layers within or underlying soil

Source: Soil Survey Staff (1999).
[a] The symbols in parentheses illustrate the appropriate lower case modifiers used to describe specific features of master horizons. [b] The term illuvial refers to material transported into a horizon from layers above it. [c] The term sesquioxide refers to accumulations of secondary iron and/or aluminum oxides. [d] Gleying is a process of reduction (caused by prolonged high water content and low oxygen concentrations) that results in soil colors characterized by low chromas and gray or blueish hues. [e] The term indurated means strongly consolidated and impenetrable to plant roots.

Table 2(b) Definitions used to identify the subordinate characteristics of soil horizons.

Lower case modifiers of master horizons	Definitions (relative to soil parent material)
a	Highly decomposed organic matter (O horizon)
b	Buried soil horizon
c	Concretions or nodules of iron, aluminum, manganese, or titanium
d	Noncemented, root restricting natural or human-made (plow layers, etc.) root restrictive layers
e	Intermediate decomposition of organic matter (O horizon)
f	Indication of presence of permafrost
g	Strong gleying present in the form of reduction or loss of Fe and resulting color changes
h	Accumulation of illuvial complexes of organic matter which coat sand and silt particles
i	Slightly decomposed organic matter (O horizon)
j	Presence of jarosite (iron sulfate mineral) due to oxidation of pyrite in previously reduced soils
k	Accumulation of calcium carbonate due to pedogenic processes
m	Nearly continuously cemented horizons (by various pedogenic minerals)
n	Accumulation of exchangeable sodium
o	Residual accumulation of oxides due to long-term chemical weathering
p	Horizon altered by human-related activities
q	Accumulation of silica (as opal)
r	Partially weathered bedrock
s	Illuvial accumulation of sesquioxides
ss	Presence of features (called slickensides) caused by expansion and contraction of high clay soils
t	Accumulation of silicate clay by weathering and/or illuviation
v	Presence of plinthite (iron rich, reddish soil material)
w	Indicates initial development of oxidized (or other) colors and/or soil structure
x	Indicates horizon of high firmness and brittleness
y	Accumulation of gypsum
z	Accumulation of salts more soluble than gypsum (e.g., Na_2CO_3)

Source: Soil Survey Staff (1999).

performed since nearly the origins of pedology (Hilgard, 1860). Yet, quantitative estimates of total chemical denudation, and associated physical changes that occur during soil formation, were not rigorously performed until the late 1980s when Brimhall and co-workers (Brimhall and Dietrich, 1987; Brimhall *et al.*, 1991) began applying a mass balance model originally derived for ore body studies to the soil environment. Here the author presents the key components of this model, and reports the results of its application to two issues: (i) the behavior of many of the chemical elements in soil formation and (ii) general trends of soil physical and chemical behavior as a function of time during soil formation.

A representation of a soil system during soil formation is shown in Figure 2. While the figure illustrates a loss of volume during weathering, volumetric increases can also occur, as will be shown later. The basic expression, describing mass gains or losses of a given chemical element (j), in the transition from parent material (p) to soil (s) in terms of volume (V), bulk density (ρ), and chemical composition (C) is

$$m_{j,\text{flux}} = m_{j,\text{s}} - m_{j,\text{p}} \qquad (3)$$

where m is the mass of element j added/lost (flux) in the soil (s) or parent material (p). Incorporating volume, density, and concentration (in percent) into the model gives

$$\underbrace{m_{j,\text{flux}}}_{\substack{\text{mass of element}(j)\text{ into/out}\\\text{of parent material volume}}} = \underbrace{\frac{V_{\text{s}}\rho_{\text{s}}C_{j,\text{s}}}{100}}_{\substack{\text{mass of element}(j)\text{ in soil}\\\text{volume of interest}}}$$
$$- \underbrace{\frac{V_{\text{p}}\rho_{\text{p}}C_{j,\text{p}}}{100}}_{\substack{\text{mass of element}(j)\text{ in}\\\text{parent material volume}}} \qquad (4)$$

Definitions of all terms used in these mass balance equations are given in Table 3. The 100 in the denominator is needed only if concentrations are in percent.

During soil development, volumetric collapse (ΔV, Figure 2) may occur through weathering losses while expansion may occur through biological or physical processes. Volumetric change is defined in terms of strain (ε):

$$\varepsilon_{i,\text{s}} = \frac{\Delta V}{V_{\text{p}}} = \left(\frac{V_{\text{s}}}{V_{\text{p}}} - 1\right) = \left(\frac{\rho_{\text{p}}C_{i,\text{p}}}{\rho_{\text{s}}C_{i,\text{s}}} - 1\right) \qquad (5)$$

where the subscript i refers to an immobile, index element. Commonly zirconium, titanium, or other members of the titanium or rare earth groups of the periodic table are used as index elements. The fractional mass gain or loss of an element j relative to the mass in the parent material (τ) is

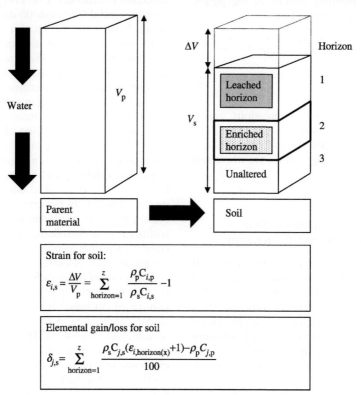

Figure 2 Diagram illustrated a mass balance perspective of soil formation (based on similar figures in Brimhall and Dietrich, 1987).

Table 3 Definition of parameters used in mass balance model.

Parameter	Definition
V_p (cm^3)	Volume of parent material
V_s (cm^3)	Volume of soil
ρ_p (g cm^{-3})	Parent material bulk density
ρ_s (g cm^{-3})	Soil bulk density
$C_{j,p}$ (%,ppm)	Concentration of mobile element j in parent material
$C_{j,s}$ (%,ppm)	Concentration of mobile element j in soil
$C_{i,p}$ (%,ppm)	Concentration of immobile element i in parent material
$C_{i,s}$ (%,ppm)	Concentration of immobile element i in soil
$m_{j,\text{flux}}$ (g cm^{-3})	Mass of element j added or removed via soil formation
$\varepsilon_{i,s}$	Net strain determined using element i
τ	Fractional mass gain or loss of element j relative to immobile element i
$\delta_{j,s}$ (g cm^{-3})	Mass gain or loss per unit volume of element j relative to immobile element i.

Sources: Brimhall and Dietrich (1987) and Brimhall *et al.* (1992).

defined by combining Equations (3)–(5):

$$\tau = \frac{m_{j,\text{flux}}}{m_{j,p}} = \left(\frac{\rho_s C_{j,s}}{\rho_p C_{j,p}}(\varepsilon_{i,s} + 1) - 1 \right) \quad (6)$$

Through substitution, Equation (6) reduces to

$$\tau = \frac{R_s}{R_p} - 1 \quad (7)$$

where $R_s = C_{j,s}/C_{i,s}$ and $R_p = C_{j,p}/C_{i,p}$. Thus, τ can be calculated readily from commonly available chemical data and does not require bulk density data. Absolute gains or losses of an element in mass per unit volume of the parent material ($\delta_{j,s}$) can be expressed as

$$\delta_{j,s} = \frac{m_{j,\text{flux}}}{V_p} = \frac{\rho_s C_{j,s}(\varepsilon_{i,s} + 1) - \rho_p C_{j,p}}{100}$$

$$= \frac{\tau C_{j,p}\rho_p}{100} \quad (8)$$

In applying the mass balance expressions, analyses are commonly performed by soil horizon, and total gains or losses (or collapse or expansion) can be plotted by depth or integrated for the whole soil profile (Figure 2).

5.01.4.1 Mass Balance Evaluation of the Biogeochemistry of Soil Formation

The chemical composition of soils is the result of a series of processes that ultimately link the soil to the history of the universe (Volume 1), with the principal processes of chemical differentiation being: (i) chemical evolution of universe/solar system; (ii) chemical differentiation of Earth from the solar system components; and (iii) the biogeochemical effects of soil formation on crustal chemistry.

The chemical composition of the solar system (Figure 3) has been widely discussed (Greenwood and Earnshaw, 1997; Chiappini, 2001). Today,

99% of the universe is comprised of hydrogen and helium, which were formed during the first few minutes following the big bang. The production of elements of greater atomic number requires a series of nuclear processes that occur during star formation and destruction. Thus, the relative elemental abundance versus atomic number is a function of the age of the universe and the number of cycles of star formation/termination that have occurred (e.g., Allègre, 1992).

The chemical composition of average crustal rock (Taylor and McLennan, 1985; Bowen, 1979) relative to the solar system reveals systematic differences (Brimhall, 1987) (Figure 4) that result from elemental fractionation during: (i) accretion of the Earth (and the interior planets) (Allègre, 1992) and (ii) differentiation of the core, mantle, and crust (Brimhall, 1987) and possibly unique starting materials (Drake and Righter, 2002). In general, the crust is depleted in the noble gases (group VIIIA) and hydrogen, carbon, and nitrogen, while it is enriched in many of the remaining elements. For the remaining elements, there is a trend toward decreasing enrichment with increasing atomic number within a given period, due to increasing volatility with increasing atomic number (Brimhall, 1987). The depletion of the siderophile elements in the crust relative to the solar system has been attributed to their concentration within the core (Brimhall, 1987), though the crust composition may also reflect late-stage accretionary processes (Delsemme, 2001).

The result of these various processes is that the Earth's crust, the parent material for soils, is dominated (in mass) by eight elements (oxygen, silicon, aluminum, iron, calcium, sodium, magnesium, and potassium). These elements, with the exception of oxygen, are not the dominant elements of the solar system. Thus, soils on Earth form in a matrix dominated by oxygen and silicon, the elements which form the backbone of the silicate minerals that dominate both the primary and secondary minerals found in soils.

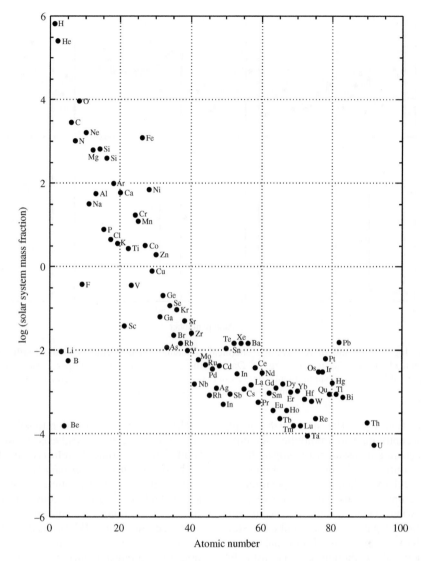

Figure 3 The log of the mass fraction (mg kg^{-1}) of elements in the solar system arranged by atomic number (source Anders and Grevesses, 1989, table 1).

There are a variety of compilations of the concentrations of many of the chemical elements for both crustal rocks (see above and Volume 3) and for soils (Bowen, 1979; Shacklette and Boerngen, 1984). In the case of soils, the samples analyzed are usually from a standard surface sampling depth, or from the uppermost horizon. Thus, these samples give a somewhat skewed view of the overall process of soil formation because, as will be discussed, soil formation is a depth-dependent process. Nonetheless, the data do provide a general overview of soil biogeochemistry that is applicable across broad geographical gradients.

When analyzing large chemical data sets, it is common to evaluate the behavior of elements in soils, and how they change during soil formation, by dividing the mass concentration of the elements in soils by that in crustal rocks, with

the resulting ratio being termed the *enrichment factor*—values less than 1 indicating loss, more than 1 indicating gains. A disconcerting artifact of this analysis is that some mobile elements, particularly silicon, commonly show enrichment factors greater than 1. Silicon is one of the major elements lost via chemical weathering, having an annual flux to the ocean of 6.1×10^{12} mol Si yr^{-1} (Tréguer *et al.*, 1995), so that there is a large net loss of the element from landscapes. The reason for the apparent enrichment is that although silicon is lost via weathering, the concentration of chemically resistant silicates (e.g., quartz) leads to a relative retention of the element. These discrepancies can be avoided by relating soil and parent material concentrations to immobile index elements such as zirconium.

The present analysis uses τ, the fractional elemental enrichment factor relative to the parent

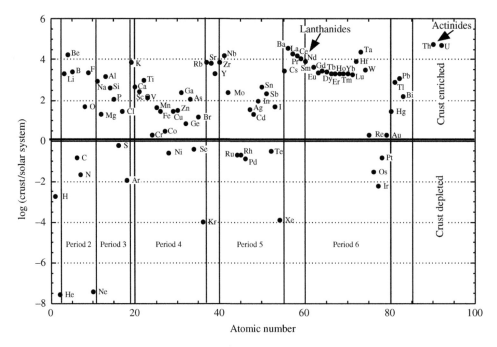

Figure 4 The log of the ratio of the average chemical concentration of elements in the Earth's upper crust (mg kg^{-1}) to that of the solar system (mg kg^{-1}). Data on the geochemistry of the upper crust from Taylor and McLennan (1985, table 2.15) with supplemental data from Bowen (1979, table 3.3). The chemistry of the solar system from Anders and Grevesse (1989).

material (Equation (7)). The normalization of elemental concentrations to immobile elements provides an accurate assessment of biogeochemical behavior during soil formation. Figure 5 illustrates the relative chemical composition of soil surface samples versus that of crustal rock (log($\tau + 1$)), where positive values indicate soil enrichment and negative values indicate soil depletion, relative to the crust. Elemental losses are due to chemical weathering, and the ultimate removal of weathering products to oceans. Elemental gains are due primarily to biological processes—the addition of elements to soils, primarily by land plants.

The comparison (soils to average continental crust) indicates that soils are: (i) particularly depleted, due to aqueous weathering losses, in the alkali metals and alkaline earths (particularly magnesium, sodium, calcium, potassium, and beryllium) and some of the halides; (ii) depleted, to a lesser degree, in silicon, iron, and aluminum; and (iii) enriched in carbon, nitrogen, and sulfur. The losses are clearly due to chemical weathering, as the chemical composition of surface waters illustrates an enrichment of these same elements relative to that of the crust (Figure 6). Plants directly assimilate elements from soil water (though they exhibit elemental selectivity across the root interface (Clarkson, 1974)), and are therefore enriched, relative to the crust, in elements derived from chemical weathering.

The key elemental addition to soils by plants is carbon, because photosynthesis greatly increases plant carbon content relative to the crust. Globally, *net primary production* (NPP) (gross photosynthetic carbon fixation–plant respiration) is ~60 Gt C yr^{-1}, an enormous carbon flux rate that nearly equals ocean/atmosphere carbon exchange (Sundquist, 1993). In addition to enriching the soil in carbon, the variety of organic molecules produced during the cycling of this organic material, coupled with the CO_2 generated in the soil by the decomposition of the organic compounds by heterotrophic microorganisms, greatly accelerate rates of chemical weathering (see Chapter 5.05). As a result, plants are responsible not only for enrichments of soil carbon, but also for enhanced rates of chemical weathering.

Second only to carbon inputs, nitrogen fixation by both symbiotic and nonsymbiotic organisms comprises an enormous biologically driven elemental influx to soils. Biological nitrogen fixation occurs via the following reaction (Allen *et al.*, 1994):

$$N_{2(atmosphere)} + 10H^+ + nMgATP + 8e^-$$
$$= 2NH_4^+ + H_2 + nMgADP + nP_i (n \geq 16)$$

where P_i is inorganic P. The breakage of the triple bonds in N_2 is a highly energy demanding process (thus the consumption of ATP), and in

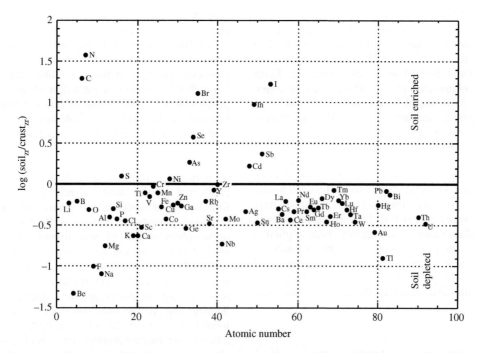

Figure 5 The log of the ratio of the mass fraction of an element in soil (relative to Zr) to that in the crust (relative to Zr). Soil data from Bowen (1979, table 4.4) and the crust data from Taylor and McLennan (1985, table 2.25) with supplemental data from Bowen (1979, table 3.3).

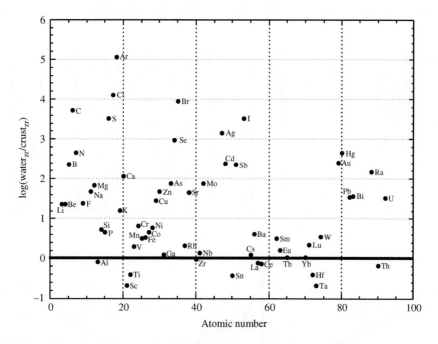

Figure 6 The log of the ratio of the mass fraction of an element in freshwater (rivers) (relative to Zr) to that of the crust (relative to Zr) as a function of atomic number. Water data from Bowen (1979, table 2.3) and crust data from Taylor and McLennan (1985, table 2.25) with supplemental data from Bowen (1979, table 3.3).

nature microorganisms have developed symbiotic relations with certain host plants (particularily legumes), deriving carbon sources from the host plant, and in turn enzymatically reducing atmospheric N_2 to NH_4^+, a form which is plant available and becomes part of plant proteins.

Globally, it is estimated that, prior to extensive human activity, biological nitrogen fixation was

~$(90-140) \times 10^{12}$ g N yr^{-1} (Vitousek *et al.*, 1997a). This rate is increasing because of the agriculturally induced nitrogen fixation. The ability to fix nitrogen is one of the most fundamental biological developments on Earth (Navarro-Gonzalez *et al.*, 2001), since nitrogen availability is one of the key limiting elements to plant growth, and hence to virtually all biogeochemical processes.

The summary of this brief discussion is that weathering losses plus plant additions characterize soil formation. This model, while capturing some important themes, neglects one of the key characteristics of soils—the distinctive and widely varying ways in which their properties vary with depth. Mass balance analyses have been applied to complete soil profiles along gradients of landform age, giving us a general perspective on the rates and directions of physical and chemical changes of soils with time. This is discussed in the next section.

5.01.4.2 Mass Balance of Soil Formation versus Time

5.01.4.2.1 *Temperate climate*

The main conclusions that can be summarized by mass balance analyses of soil formation over

time in nonarid environments are that: in early phases of soil formation, the soil experiences volumetric dilation due to physical and biological processes; the later stages of soil formation are characterized by volumetric collapse caused by large chemical losses of the major elements that, given sufficient time, result in nutrient impoverishment of the landscape. The key studies that contribute to this understanding are summarized below.

On a time series of Quaternary marine terraces in northern California, Brimhall *et al.* (1992) conducted the first mass balance analysis of soil formation over geologic time spans. This analysis provided quantitative data on well-known qualitative observations of soil formation: (i) the earliest stages of soil formation (on timescales of 10^1–10^3 yr) are visually characterized by loss of sedimentary/rock structure, the accumulation of roots and organic matter, and the reduction of bulk density; and (ii) the later stages of soil development ($>10^3$ yr) are characterized by the accumulation of weathering products (iron oxides, silicate clays, and carbonates) and the loss of many products of weathering.

Figure 7 shows the trend in ε, volumetric strain (Equation (5)), over ~2.40×10^5 yr. The data show the following physical changes: (i) large volumetric expansion ($\varepsilon > 0$) occurred in the

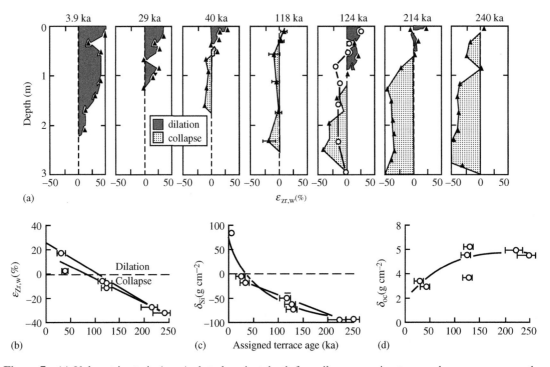

Figure 7 (a) Volumetric strain ($\varepsilon_{Zr,w}$) plotted against depth for soils on a marine terrace chronosequence on the Mendocino Coast of northern California; (b) average strain for entire profiles versus time (integrated strain to sampling depth divided by sampling depth); (c) integrated flux of Si (δ_{Si}) for entire profiles versus time; and (d) integrated flux of organic C versus time (Brimhall *et al.*, 1992) (reproduced by permission of the American Association from the Advancement of Science from *Science* 1992, *255*, 695).

young soil (Figure 7(a)); (ii) integrated expansion for the whole soil declined with age (Figure 7(b)); and (iii) the cross-over point between expansion and collapse ($\varepsilon < 0$) moved progressively toward the soil surface with increasing age (Figure 7(a)).

Biological processes, along with abiotic mixing mechanisms, drive the distinctive first phases of soil formation. The large positive strain (expansion) measured in the young soil on the California coast was due to an influx of silicon-rich beach sand (Figure 7(c)) and the accumulation of organic matter from plants (Figure 7(d)). In many cases, there is a positive relationship between the mass influx of carbon to soil (δ_{oc}) and strain; Jersak *et al.* (1995)). Second, in addition to adding carbon mass relative to the parent material, the plants roots (and other subterranean organisms) expand the soil, create porosity, and generally assist in both mixing and expansion. Pressures created by growing roots can reach 15 bar (Russell, 1977), providing adequate forces to expand soil material. Brimhall *et al.* (1992) conducted an elegant lab experiment showing the rapid manner in which roots can effectively mix soil, and incorporate material derived from external sources. Over several

hundred "root growth cycles" using an expandable/collapsible tube in a sand mixture (Figure 8(a)), they demonstrated considerable expansion and depth of mixing (Figure 8(b)), with an almost linear relation between expansion and depth of translocation of externally added materials (Figures 8(c) and (d)).

The rate of physical mixing and volumetric expansion caused by carbon additions declines quickly with time. Soil carbon accumulation with time (Figure 7(d)) can be described by the following first-order decay model (Jenny *et al.*, 1949):

$$\frac{\mathrm{d}C}{\mathrm{d}t} = I - kC \qquad (9)$$

where I is plant carbon inputs (kg m^{-2} yr^{-1}), C the soil carbon storage (kg m^{-2}), and k the decay constant (yr^{-1}). Measured and modeled values of k for soil organic carbon (Jenkinson *et al.*, 1991; Raich and Schlesinger, 1992) indicate that steady state should be reached for most soils within $\sim 10^2 - 10^3$ yr. Thus, as rates of volumetric expansion decline, the integrated effects of mineral weathering and the leaching of silicon (Figure 7(c)), calcium, magnesium, sodium,

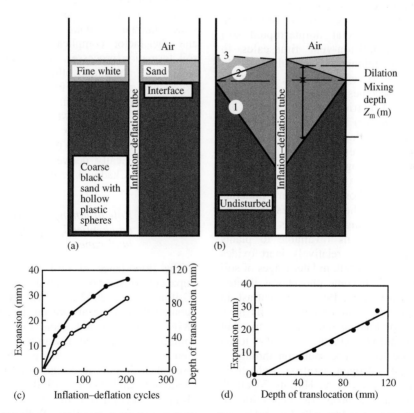

Figure 8 (a) Initial state of a cyclical dilation mixing experiment, with a surgical rubber tube embedded in a sandy matrix; (b) features after mixing: line 1 is depth of mixing after mixing, line 2 is the dilated surface, and line 3 is the top of the overlying fine sand lense; (c) expansion (o) and depth of mixing (●) as a function of mixing cycles; and (d) relationship of soil expansion to mixing depth (Brimhall *et al.*, 1992) (reproduced by permission of the American Association from the Advancement of Science from *Science* **1992**, *255*, 695).

potassium, and other elements begin to become measurable, and over time tend to eliminate the measured expansion not only near the surface (Figure 7(a)), but also for the whole profile (Figure 7(b)).

5.01.4.2.2 Cool tropical climate

The integrated mass losses of elements over time are affected by parent material mineralogy, climate, topography, etc. The mass balance analysis of soil formation of the temperate California coast (Brimhall *et al.*, 1991; Chadwick *et al.*, 1990; Merritts *et al.*, 1992) is complemented by an even longer time frame on the Hawaiian Islands (Vitousek *et al.*, 1997b; Chadwick *et al.*, 1999). The Hawaiian chronosequence encompasses ~4 Myr in a relatively cool, but wet, tropical setting. Because of both steady and cyclic processes of erosion and deposition, few geomorphic surfaces in temperate settings on Earth are older than Pleistocene age. Yet the exceptions to this rule: the Hawaiian Island chronosequence, river terrace/glacial outwash sequence in the San Joaquin Valley of California (e.g., Harden, 1987; White *et al.*, 1996), and possibly others provide glimpses into the chemical fate of the Earth's surface in the absence of geological rejuvenation.

The work by Vitousek and co-workers on Hawaii demonstrates that uninterrupted soil development on million-year timescales in those humid conditions depletes the soil in elements essential to vegetation and, ultimately, the ecosystem becomes dependent on atmospheric sources of nutrients (Chadwick *et al.*, 1999). Figures 9(a)–(g) illustrate: (i) silicon and alkali and alkaline earth metals are progressively depleted, and nearly removed from the upper 1 m; (ii) soil mineralogy shifts from primary minerals to secondary iron and aluminum oxides with time; (iii) phosphorus in primary minerals is rapidly depleted in the early stages of weathering, and the remaining phosphorus is sequestered into organic forms (available to plants through biocycling) and relatively inert oxides and hydroxides. As a result, in later stages of soil formation, the soils become phosphorus limited to plants (Vitousek *et al.*, 1997b). In contrast, in very early phases of soil formation, soils have adequate phosphorus in mineral forms, but generally lack nitrogen (Figure 9(g)) due to an inadequate time for its accumulation through a combination of nitrogen fixation (which is relatively a minor process on Hawaii; Vitousek *et al.* (1997b) and atmospheric deposition of NO_3^-, NH_4^+, and organic N (at rates of ~5–50 kg N yr^{-1}; Heath and Huebert (1999)). The rate of nitrogen accumulation in soil and the model describing it generally parallel the case of organic carbon, because carbon storage hinges on the availability of nitrogen (e.g., Parton *et al.*, 1987).

In summary, soils in the early stages of their development contain most of the essential elements for plant growth with the exception of nitrogen. Soil nitrogen, like carbon, reaches maximum steady-state values in periods on the order of thousands of years and, as a result, NPP of the ecosystems reach maximum values at this stage of soil development (Figure 10). Due to progressive removal of phosphorus and other plant essential elements (particularly calcium), plant productivity declines (Figure 10), carbon inputs to the soil decline, and both soil carbon and nitrogen storage begin a slow decline (Figure 9). This trend, because of erosive rejuvenating processes, is rarely observed in climatically and tectonically active parts of the Earth. Alternatively, low latitude, tectonically stable continental regions may reflect these long-term processes. Brimhall and Dietrich (1987) and Brimhall *et al.* (1991, 1992) discuss the pervasive weathering and elemental losses from cratonal regions such as Australia and West Africa. These regions, characterized by an absence of tectonic activity and glaciation, and by warm (and sometimes humid) climates, have extensive landscapes subjected to weathering on timescales of millions of years. However, over such immense timescales, known and unknown changes in climate and other factors complicate interpretation of the soil formation processes. An emerging perspective is that these and other areas of the Earth experience atmospheric inputs of dust and dissolved components that wholly or partially compensate chemical weathering losses, ultimately creating complex soil profiles and ecosystems which subsist on the steady but slow flux of atmospherically derived elements (Kennedy *et al.*, 1998; Chadwick *et al.*, 1999).

5.01.4.2.3 Role of atmospheric inputs on chemically depleted landscapes

The importance of dust deposition, and its impact on soils, is not entirely a recent observation (Griffin *et al.*, 2002). Darwin complained of Saharan dust while aboard the Beagle (Darwin, 1846), and presented some discussion of its composition and the research on the phenomenon at the time. In the pedological realm, researchers in both arid regions (Peterson, 1980; Chadwick and Davis, 1990) and in humid climates recognized the impact of aerosol inputs on soil properties. With respect to the Hawaiian Islands—a remarkably isolated volcanic archipelago—Jackson *et al.* (1971) recognized the presence of prodigous quantities of quartz in the basaltic soils of Hawaii. Oxygen-isotope

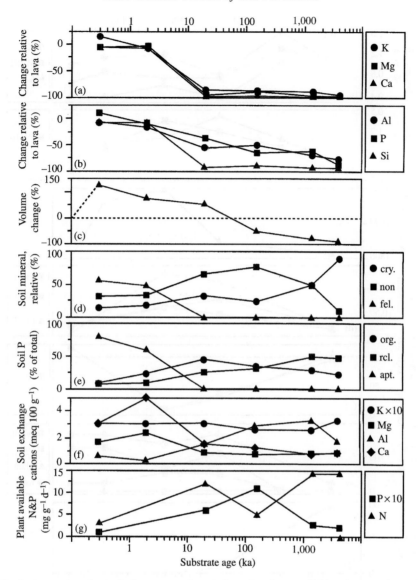

Figure 9 Weathering, mineralogical changes, and variations in plant-available elements in soils (to 1 m in depth) as a function of time in Hawaii: (a) total soil K, Mg, and Ca; (b) total soil Al, P, and Si; (c) volumetric change; (d) soil feldspar, crystalline and noncrystalline secondary minerals; (e) soil P pools (organic, recalcitrant, apatite); (f) exchangeable K, Mg, Al, and Ca; and (g) changes in resin-extractable (biologically available) inorganic N and P (Vitousek *et al.*, 1997b) (reproduced by permission of the Geological Society of America from *GSA Today*, **1997b**, *7*, 1–8).

analyses of these quartz grains showed that the quartz was derived from continental dust from the northern hemisphere, a source now well constrained by atmospheric observations (Nakai *et al.*, 1993) and the analysis of Pacific Ocean sediment cores (Rea, 1994).

The work by Chadwick *et al.* (1999) has demonstrated that the calcium and phosphorus nutrition of the older Hawaiian Island ecosystems depends almost entirely on atmospheric sources. With respect to calcium, strontium-isotope analyses of soils and plants indicates that atmospherically derived calcium (from marine sources) increases from less than 20% to more

than 80% of total plant calcium with increasing soil age. With respect to phosphorus, the use of rare earth elements and isotopes of neodymium all indicated that from ~0.5 to more than 1.0 mg P m^{-2} yr^{-1} is delivered in the form of dust each year and, in the old soils, the atmospheric inputs approach 100% of the total available phosphorus at the sites. Brimhall *et al.* (1988, 1991) demonstrated that much of the zirconium in the upper part of the soils in Australia and presumably other chemical constituents are derived from atmospheric inputs. In summary, it is clear that the ultimate fate of soils in the absence of geological rejuvenation,

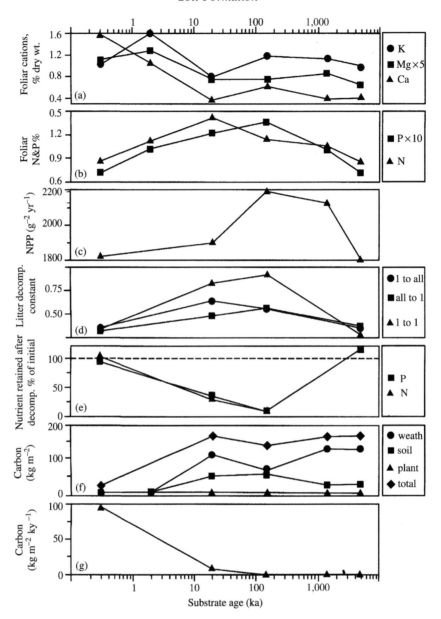

Figure 10 Plant nutrients, production and decomposition, and carbon sinks during soil development in Hawaii: (a) K, Mg, and Ca in canopy leaves of dominant tree (*Metrosideros*); (b) N and P in leaves of dominant tree; (c) changes in NPP of forests; (d) decomposition rate constant (*k*) of *metrosideros* leaves decomposed in the site collected (1-to-1), collected at each site and transferred to common site (all-to-1), and collected from one site and transferred to all sites (1-to-all); (e) fraction of N and P in original leaves remaining after 2 yr of decomposition; (f) carbon storage in the ecosystems in the form of plant biomass, soil organic matter, and as CO_2 consumed during silicate weathering; and (g) instantaneous rate of total ecosystem carbon storage (Vitousek *et al.*, 1997b) (reproduced by permission of the Geological Society of America from *GSA Today*, **1997b**, *7*, 1–8).

in humid climates, is a subsistence on atmospheric elemental sources.

5.01.4.3 Mass Balance Evaluation of Soil Formation versus Climate

Hyperarid regions offer a unique view of the importance of atmospheric elemental inputs to soils, because in these areas, the inputs are not removed by leaching. The western coasts of southern Africa and America lie in hyperarid climates that have likely persisted since the Tertiary (Alpers and Brimhall, 1988). These regions, particularly the Atacama desert of Chile, are known for their commercial-grade deposits of sulfates, iodates, bromates, and particularly nitrates (Ericksen, 1981; Böhlke *et al.*, 1997). These deposits may form due to several processes and salt sources, including deflation around playas, spring deposits (Rech *et al.*, 2003), and other settings. However, many of the deposits are simply heavily concentrated soil horizons, formed by the long accumulation of soluble constituents over long time spans (Erickson, 1981). The ultimate origin of these salts (and their elements) is in some cases obscure, but it is clear that they arrive at the soils via the atmosphere. In Chile, sources of elements may be from fog, marine spray, reworking of playa crusts, and more general long-distance atmospheric sources (Ericksen, 1981; Böhlke *et al.*, 1997).

Recent novel research by Theimens and co-workers has convincingly demonstrated the atmospheric origin (as opposed to sea spray, playa reworking, etc.) of sulfates in Africa and Antarctica (Bao *et al.*, 2000a,b, 2001). Briefly, these researchers have shown that as sulfur undergoes chemical reactions in the stratosphere and troposphere, a mass-independent fractionation of oxygen isotopes in the sulfur oxides occurs. The mechanism for these fractionations is obscure (Thiemens, 1999), but the presence of mass-independent ratios of ^{17}O and ^{18}O in soil sulfate accumulations is a positive indicator of an atmospheric origin. In both the Nambian desert (Bao *et al.*, 2001) and Antarctica (Bao *et al.*, 2000a), there was an increase in the observed ^{17}O isotopic anomaly with distance from the ocean, possibly due to a decrease in the ratio of sea salt-derived sulfate in atmospheric sulfate.

There have been few, if any, systematic pedological studies of the soils of these hyperarid regions. Here, we present a preliminary mass balance analysis of soil formation along a precipitation gradient in the presently hyperarid region of the Atacama desert, northern Chile (Sutter *et al.*, 2002). Along a south-to-north gradient, precipitation decreases from ~15 mm yr^{-1} to ~2 mm yr^{-1} (http://www.worldclimate.com/worldclimate/index.htm). For the study, three sites were chosen ~50 km inland on the oldest (probably Mid- to Early Pleistocene) observable fluvial landform (stream terrace or alluvial fan) in the region. Depths of observation were restricted by the presence of salt-cemented soil horizons.

Using the mass balance equations presented earlier and titanium as an immobile index element, the volumetric and major element changes with precipiation were calculated (Figure 11). The calculations for ε show progressive increases in volumetric expansion with decreasing precipitation (approaching 400% in some horizons at Yungay, the driest site) (Figure 11(a)). These measured expansions are due primarily to the accumulation and retention of NaCl and CaSO$_4$ minerals. For example, Yungay has large expansions near the surface and below

120 cm. Figure 11(c) shows that sulfur (in the form of CaSO$_4$) is responsible for the upper expansion, while chlorine (in the form of NaCl) is responsible for the lower horizon expansion. As the chemical data indicate, the type and depth of salt movement is climatically related: the mass of salt changes from (Cl, S) to (S, CaCO$_3$) to (CaCO$_3$) with increasing rainfall (north to south). Additionally, the depth of S and CaCO$_3$ accumulations increase with rainfall.

These data suggest that for the driest end-member of the transect, the virtual absence of chemical weathering (for possibly millions of years), and the pervasive input and retention of atmospherically derived chemical constituents (due to a lack of leaching), drives the long-term trajectory of these soils toward continued volumetric expansion (due to inputs) and the accretion, as opposed to the loss, of plant-essential elements. Therefore, it might be hypothesized that there is a critical water balance for soil formation (precipitation–evapotranspiration) at which the long-term accumulation of atmospherically derived elements exceeds weathering losses, and landscapes undergo continual dilation as opposed to collapse. The critical climatic cutoff point is likely to be quite arid. In the Atacama desert, the crossover point between the accretion versus the loss of soluble atmospheric inputs such as nitrate and sulfate is somewhere between 5 mm and 20 mm of precipitation per year. These Pleistocene (or older) landscapes have likely experienced changes in climate (Betancourt *et al.*, 2000; Latorre *et al.*, 2002), so the true climatic barrier to salt accumulations is unknown. Nonetheless, it is clear that the effect of climate drives strongly contrasting fates of soil formation (collapse and nutrient impoverishment versus dilation and nutrient accumulation) over geological time spans.

5.01.5 PROCESSES OF MATTER AND ENERGY TRANSFER IN SOILS

Chadwick *et al.* (1990) wrote that depth-oriented mass balance analyses change the study of soil formation from a "black to gray box." The "grayness" of the mass balance approach is due to the fact that it does not directly provide insight into mechanisms of mass transfer, and it does not address the transport of heat, water, and gases. The mechanistic modeling and quantification of these fluxes in field settings is truly in its infancy, but some general principles along with some notable success stories have emerged on the more mechanistic front of soil formation. In this section, the author discusses the general models that describe mass transfer in soils, and examines in some detail how

Soil Formation

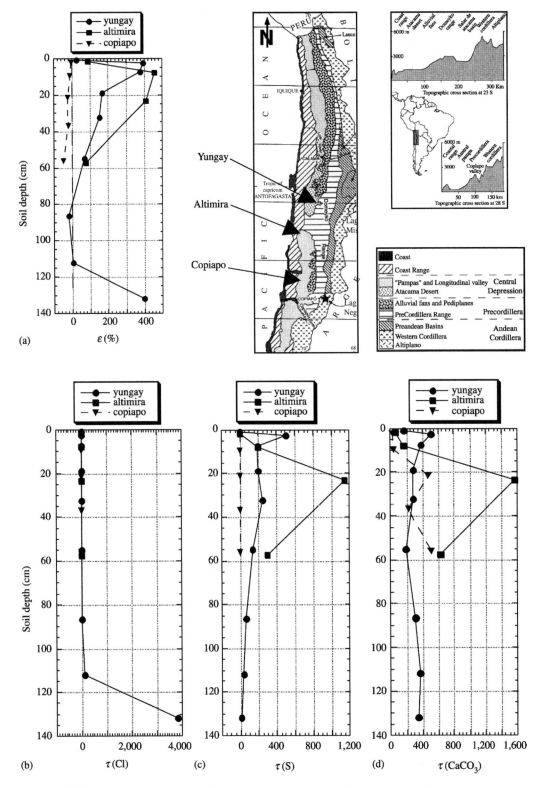

Figure 11 (a) Volumetric change along climate gradient in Atacama desert, northern Chile; (b) fractional mass gains of Cl; (c) fractional mass gains of S; and (d) fractional mass gains of CaCO₃ for three sites sampled along a precipitation gradient (Sutter *et al.*, 2002).

these models have been successfully used to describe observed patterns in soil gas and organic matter concentrations, and their isotopic composition.

The movement of most constituents in soils can ultimately be described as variants of diffusive (or in certain cases, advective) processes. The study of these processes, and their modeling, has long been the domain of *soil physics*, an experimental branch of the soil sciences. There are several good textbooks in soil physics that provide introductions to these processes (Jury *et al.*, 1991; Hillel, 1998, 1980a,b). However, it is fair to say that the application of this work to natural soil processes, and to natural soils, has been minimal given the focus on laboratory or highly controlled field experiments. However, notable exceptions to this trend exist, exceptions initiated by biogeochemists who adapted or modified these principles to illuminate the soil "black box."

There are few, if any, cases where these various models have been fully coupled to provide an integrated view of soil formation. However, one group of soil processes that has been extensively studied and modeled comprise the soil carbon cycle. We review the mechanisms of this cycle and the modeling approaches that have received reasonably wide acceptance in describing the processes.

5.01.5.1 Mechanistic Modeling of the Organic and Inorganic Carbon Cycle in Soils

The processing of carbon in soils has long received attention due to its importance to agriculture (in the form of organic matter) and the marked visual impact it imparts to soil profiles. A schematic perspective of the flow of carbon through soils is given in Figure 12. Carbon is fixed from atmospheric CO_2 by plants, enters soil in organic forms, undergoes decomposition, and is cycled back to the atmosphere as CO_2. In semi-arid to arid regions, a fraction of the CO_2 may ultimately become locked in pedogenic $CaCO_3$, whereas in humid regions a portion may be leached out as dissolved organic and inorganic carbon with groundwater. On hillslopes, a portion of the organic carbon may be removed by erosion (Stallard, 1998). Most studies of the organic part of the soil carbon assume the latter three mechanisms to be of minor importance (to be discussed more fully below) and consider the respiratory loss of CO_2 as the main avenue of soil carbon loss.

Jenny *et al.* (1949) were among the first to apply a mathematical framework to the soil carbon cycle. For the organic layer at the surface of forested soils, Jenny applied, solved, and

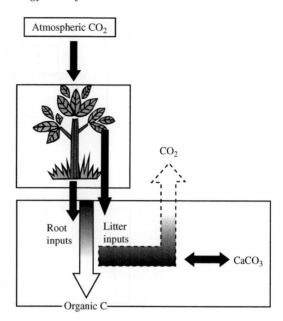

Figure 12 Schematic diagram illustrating C flow through terrestrial ecosystems.

evaluated the mass balance model given by Equation (9) discussed earlier. This approach can be applied to the soil organic carbon pool as a whole. This has proved useful in evaluating the response of soil carbon to climate and environmental change (Jenkinson *et al.*, 1991).

The deficiency of the above model of soil carbon is that it ignores the interesting and important variations in soil carbon content with soil depth. The depth variations of organic carbon in soils vary widely, suggesting a complex set of processes that vary from one environment to another. Figure 13 illustrates just three commonly observed soil carbon (and nitrogen) trends with depth: (i) exponentially declining carbon with depth (common in grassland soils or Mollisols); (ii) randomly varying carbon with depth (common on young fluvial deposits or, as here, in highly stratified desert soils or Aridisols); and (iii) a subsurface accumulation of carbon below a thick plant litter layer on the soil surface (sandy, northern forest soils or Spodosols). The mechanisms controlling these distributions will be discussed, in reverse order, culminating with a discussion of transport models used to describe the distribution of carbon in grasslands.

Spodosols are one of the 12 soil orders in the USDA Soil Taxonomy. The soil orders, and their key properties, are listed in Table 4. The distribution of the soil orders in the world is illustrated in Figure 14. With respect to Spodosols, which are common to the NE USA, Canada, Scandinavia, and Russia, the key

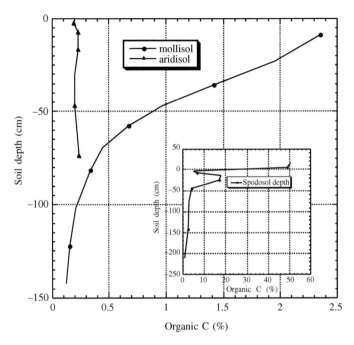

Figure 13 Soil organic C depth in three contrasting soil orders (source Soil Survey Staff, 1975).

Table 4 The soil orders of the US Soil Taxonomy and a brief definition of their characteristics.

Order	Characteristics
Alfisols	Soils possessing Bt horizons with >35% base saturation; commonly Pleistocene aged
Andisols	Soils possessing properties derived from weathered volcanic ash, such as low bulk density, high extractable Al, and high P retention
Aridisols	Soils of arid climates that possess some degree of subsurface horizon development; commonly Pleistocene aged
Entisols	Soils lacking subsurface horizon development due to young age, resistant parent materials, or high rates of erosion
Gelisols	Soils possessing permafrost and/or evidence of cryoturbation
Histosols	Soils dominated by organic matter in 50% or more of profile
Inceptisols	Soils exhibiting "incipient" stages of subsurface horizon development due to age, topographic position, etc.
Mollisols	Soils possessing a relatively dark and high C and base saturation surface horizon; commonly occur in grasslands
Oxisols	Soils possessing highly weathered profiles characterized by low-cation-exchange-capacity clays (kaolinite, gibbsite, etc.), few remaining weatherable minerals, and high clay; most common on stable, tropical landforms
Spodosols	Soils of northern temperate forests characterized by intense (but commonly shallow) biogeochemical downward transport of humic compounds, Fe, and Al; commonly Holocene aged
Ultisols	Soils possessing Bt horizons with <35% base saturation; commonly Pleistocene aged
Vertisols	Soils composed of >35% expandable clay, possessing evidence of shrink/swell in form of cracks, structure, or surface topography

After Soil Survey Staff (1999).

characteristics that lead to their formation are: sandy or coarse sediment (commonly glacial outwash or till), deciduous or coniferous forest cover, and humid, cool to cold, climates. Organic matter added by leaf and branch litter at the surface accumulates to a steady-state thickness of organic horizons (Figure 13). During the decomposition of this surface material, soluble organic molecules are released which move downward with water. The organic molecules have reactive functional groups which complex with iron and aluminum, stripping the

Global soil regions

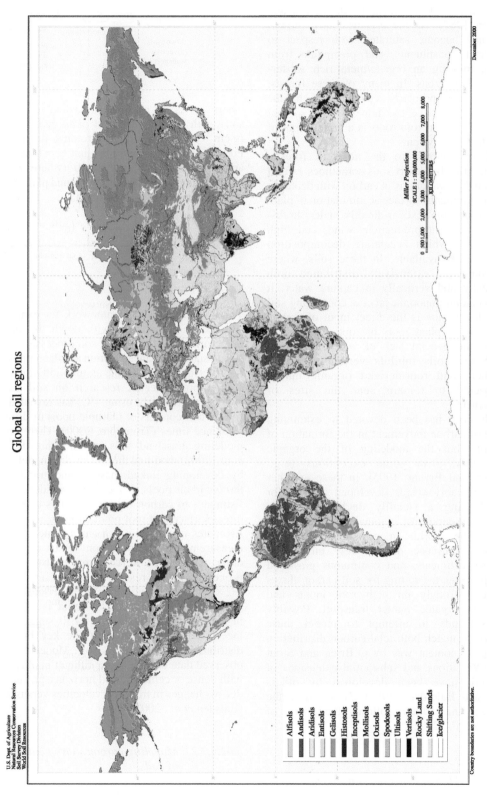

Alfisols
Andisols
Aridisols
Entisols
Gelisols
Histosols
Inceptisols
Mollisols
Oxisols
Spodosols
Ultisols
Vertisols
Rocky Land
Shifting Sands
Ice/glacier

Miller Projection
SCALE 1 : 100,000,000

0 500 1,000 2,000 3,000 4,000 5,000 6,000 7,000 8,000
KILOMETERS

December 2000

Figure 14 The global distribution of soil orders (source http://www.nrcs.usda.gov/technical/worldsoils/mapindx/).

upper layers of the mineral soils that contain these elements and leaving a white, bleached layer devoid, or depleted in iron, aluminum, and organic matter (De Coninck, 1980; Ugolini and Dahlgren, 1987). As the organics move downward, they become saturated with respect to their metal constituents, and precipitate from solution, forming an organo/metal-rich subsurface set of horizons. In many of these forests, tree roots are primarily concentrated in the organic layers above the mineral soil, so that addition of carbon from roots is a minor input of carbon to these systems.

In sharp contrast to the northern forests, stratified, gravelly desert soils sometimes exhibit almost random variations in carbon with depth. In these environments, surface accumulation of plant litter is negligible (except directly under shrubs) due to low plant production, wind, and high temperatures which accelerate decomposition when water is available. In these soils, where plant roots are extensively distributed both horizontally and vertically to capture water, it appears that an important process controlling soil carbon distribution is the direct input of carbon from decaying plant roots or root exudates. In addition, the general lack of water movement through the soils inhibits vertical transport of carbon, and root-derived organic matter is expected to remain near the sites of emplacement.

Much work has been devoted to examining the role of carbon movement in the formation of Spodosols, but the modeling of the organic carbon flux, with some exceptions (e.g., Hoosbeek and Bryant, 1995), in these or other soils is arguably not as developed as it is for soils showing a steadily declining carbon content with depth that is found in the grassland soils of the world. In grassland soils, the common occurrence of relatively unstratified Holocene sediments, and continuous grassland cover, provides the setting for soil carbon fluxes dependent strongly on both root inputs and subsequent organic matter transport. Possibly the first study to attempt to model these processes to match both total carbon distribution and its ^{14}C content was by O'Brien and Stout (1978). Variations and substantial extensions of this work have been developed by others (Elzein and Balesdent, 1995). To illustrate the approach, the author follows the work of Baisden *et al.* (2002).

5.01.5.1.1 *Modeling carbon movement into soils*

The soil carbon mass balance is hypothesized to be, for grassland soils, a function of plant inputs (both surface and root), transport, and decomposition:

$$\frac{dC}{dt} = \underbrace{-v\frac{dC}{dz}}_{\substack{\text{downward} \\ \text{advective} \\ \text{transport}}} - \underbrace{kC}_{\text{decomposition}} + \underbrace{\frac{F}{L}e^{-z/L}}_{\substack{\text{plant inputs} \\ \text{distributed} \\ \text{exponentially}}}$$

(10)

where $-v$ is the advection rate (cm yr^{-1}), z the soil depth (cm), F the total plant carbon inputs (g cm^{-2}yr^{-1}), and L the e-folding depth (cm). For the boundary conditions that $C = 0$ at $z = \infty$ and $-v(dC/dz) = F_A$ at $z = 0$ (where F_A are aboveground and F_B the belowground plant carbon inputs), the steady-state solution is

$$C(z) = \underbrace{\frac{F_A}{v}e^{-kz/v}}_{\substack{\text{above–ground} \\ \text{input/transport}}} + \underbrace{\frac{F_B}{kL-v}e^{-kz/v}\left(e^{z(kL-v)/vL} - 1\right)}_{\text{root input/transport}}$$

(11)

This model forms the framework for examining soil carbon distribution with depth. It contains numerous simplifications of soil processes such as steady state, constant advection and decomposition rates versus depth, and the assumption of one soil carbon pool. Recent research on soil carbon cycling, particularly using ^{14}C, has revealed that soil carbon consists of multiple pools of differing residence times (Trumbore, 2000). Therefore, in modeling grassland soils in California, Baisden *et al.* (2002) modified the soil carbon model above by developing linked mass balance models for three carbon pools of increasing residence time. Estimates of carbon input parameters came from direct surface and root production measurements. Estimates of transport velocities came from ^{14}C measurements of soil carbon versus depth, and other parameters were estimated by iterative processes. The result of this effort for a ~2×10^5 yr old soil (granitic alluvium) in the San Joaquin Valley of California is illustrated in Figure 15. The goodness of fit suggests that the model captures at least the key processes distributing carbon in this soil. Model fitting to observed data became more difficult in older soils with dense or cemented soil horizons, presumably due to changes in transport velocities versus depth (Baisden *et al.*, 2002).

5.01.5.1.2 *Modeling carbon movement out of soils*

The example above illustrates that long-observed soil characteristics are amenable to analytical or numerical modeling, and it illustrates the importance of transport in the vertical

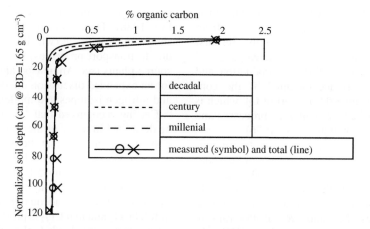

Figure 15 Measured total organic C versus depth and modeled amounts of three fractions of differing residence time ($\sim 10^0$ yr, 10^2 yr, and 10^3 yr) for a $\sim$600 ka soil formed on granitic alluvium in the San Jaoquin Valley of California (source Baisden, 2000).

distribution of soil properties, in this case organic carbon. As the model and observations indicate, the primary pathway for carbon loss from soil is the production of CO_2 from the decomposition of the organic carbon. It has long been recognized that CO_2 leaves the soil via diffusion, and various forms of Fick's law have been applied to describing the transport of CO_2 and other gases in soils (e.g., Jury *et al.*, 1991; Hillel, 1980b). However, the application of these models to natural processes and to the issue of stable isotopes in the soil gases is largely attributable to the work of Thorstenson *et al.* (1983) and, in particular, of Cerling (1984). Cerling's (1984) main interest was in describing the carbon-isotope composition of soil CO_2, which he recognized ultimately controls the isotope composition of pedogenic carbonate. We begin with the model describing total CO_2 diffusion, then follow that with the extension of the model to soil carbon isotopes.

Measurements of soil CO_2 concentrations versus depth commonly reveal an increase in CO_2 content with depth. The profiles and the maximum CO_2 levels found at a given depth are climatically controlled (Amundson and Davidson, 1990) due to rates of C inputs from plants, decomposition rates, etc. Given that most plant roots and soil C are concentrated near the surface, the production rates of CO_2 would be expected to decline with depth. Cerling developed a production/diffusion model to describe steady-state soil CO_2 concentrations:

$$\varepsilon \frac{dCO_2}{dt} = 0 = \underbrace{D\frac{d^2CO_2}{dz^2}}_{\text{net diffusion}} + \underbrace{\phi}_{\substack{\text{biological} \\ \text{production}}} \tag{12}$$

where ε is free air porosity in soil, CO_2 the concentration of CO_2 (mol cm^{-3}), and D the effective diffusion coefficient of CO_2 in soil

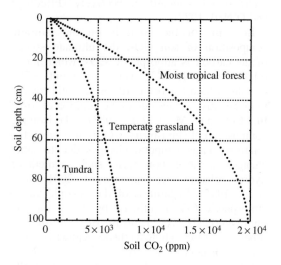

Figure 16 Calculated soil CO_2 concentrations versus depth for three contrasting ecosystems using Equation (13) and the following data: total soil respiration rates (tundra = 60 g C m^{-2}yr^{-1}, grassland = 442 g C m^{-2} yr^{-1}, and tropical forest = 1,260 g C m^{-2}yr^{-1}), $D_s = 0.021$ cm^{2}s^{-1}, atmospheric CO_2 = 350 ppm, and $L = 100$ cm (respiration data from Raich and Schlesinger, 1992).

(cm^2 s^{-1}), z the soil depth (cm), and $\phi = CO_2$ production (mol cm^{-3} s^{-1}). Using reported soil respiration rates, and reasonable parameter values for Equation (11), the CO_2 concentration profiles for three strongly contrasting ecosystems are illustrated in Figure 16.

For the boundary conditions of an impermeable lower layer and $CO_2(0) = CO_2$(atm), the solution to the model (with exponentially decreasing CO_2 production with depth) is

$$CO_2(z) = \frac{\phi_{z=0}z_0^2}{D}\left(1 - e^{-z/z_0}\right) + C_{\text{atm}} \tag{13}$$

where z_0 is the depth at which the production is $\phi_{z=0}/e$. At steady state, the flux of CO_2 from the soil to the atmosphere is simply the first derivative of Equation (13) evaluated at $z = 0$ (e.g., Amundson *et al.*, 1998). The production, and transport of CO_2, is accompanied by the consumption and downward transport of O_2, which is driven and described by analogous processes and models (e.g., Severinghaus *et al.*, 1996).

Cerling's primary objective was the identification of the processes controlling the carbon-isotope composition of soil CO_2, and a quantitative means of describing the process. In terms of notation, carbon isotopes in compounds are evaluated as the ratio (R) of the rare to common stable isotope of carbon ($^{13}C/^{12}C$) and are reported in delta notation: $\delta^{13}C(\permil)$ $= (R_s/R_{std} - 1)1,000$, where R_s and R_{std} refer to the carbon-isotope ratios of the sample and the international standard, respectively (Friedman and O'Neil, 1977).

In terms of the controls on the isotopic composition of soil CO_2, the ultimate source of the carbon is atmospheric CO_2, which has a relatively steady $\delta^{13}C$ value of about $-7\permil$ (a value which has been drifting recently toward more negative values due to the addition of fossil fuel CO_2 (e.g., Mook *et al.*, 1983)). The isotopic composition of atmospheric CO_2 is also subject to relatively large temporal changes due to other changes in the carbon cycle, such as methane hydrate releases, etc. (Jahren *et al.*, 2001; Koch *et al.*, 1995). Regardless of the isotopic value, atmospheric CO_2 is utilized by plants through photosynthesis. As a result of evolutionary processes, three photosynthetic pathways have evolved in land plants: (i) C_3: $\delta^{13}C = \sim -27\permil$, (ii) C_4: $\sim -12\permil$, and (iii) CAM: isotopically intermediate between C_3 and C_4, though it is commonly close to C_3. It is believed that C_3 photosynthesis is an ancient mechanism, whereas C_4 photosynthesis (mainly restricted to tropical grasses) is a Cenozoic adaptation to decreasing CO_2 levels (Cerling *et al.*, 1997) or to strong seasonality and water stress (e.g., Farquhar *et al.*, 1988). Once atmospheric carbon is fixed by photosynthesis, it may eventually be added to soil as dead organic matter through surface litter, root detritus, or as soluble organics secreted by living roots. This material is then subjected to microbial decomposition (see Volume 8), and partially converted to CO_2 which then diffuses back to the overlying atmosphere. An additional source of CO_2 production is the direct respiration of living roots, which is believed to account for $\sim 50\%$ of the total CO_2 flux out of soils (i.e., soil respiration; Hanson *et al.* (2000)). During decomposition of organic matter, there is a small ($\sim 2\permil$ or more) discrimination of carbon isotopes, whereby ^{12}C is preferentially lost as CO_2 and ^{13}C

remains as humic substances (Nadelhoffer and Fry, 1988). Thus, soil organic matter commonly shows an enrichment (which increases with depth due to transport processes) of ^{13}C relative to the source plants (see Amundson and Baisden (2000) for an expanded discussion of soil organic carbon and nitrogen isotopes and modeling.

Cerling recognized that $^{12}CO_2$ and $^{13}CO_2$ can be described in terms of their own production and tranport models, and that the isotope ratio of CO_2 at any soil depth is described simply by the ratio of the ^{13}C and the ^{12}C models. For the purposes of illustration here, if we assume that the concentration of ^{12}C can be adequately described by that of total CO_2 and that CO_2 is produced at a constant rate over a given depth L, then the model describing the steady-state isotopic ratio of CO_2 at depth z is (see Cerling and Quade (1993) for the solution where the above assumptions are not applied)

$$R_s^{13} = \frac{(\phi R_p^{13}/D_s^{13})(Lz - z^2/2) + C_{atm}R_{atm}^{13}}{(\phi/D_s)(Lz - z^2/2) + C_{atm}} \quad (14)$$

where R_s, R_p, and R_{atm} refer to the isotopic ratios of soil CO_2, plant carbon, and atmospheric CO_2, respectively, and D^{13} is the diffusion coefficient of $^{13}CO_2$ which is $D_s/1.0044$. The $\delta^{13}C$ value of the CO_2 can be calculated by inserting R_s into the equation given above.

Quade *et al.* (1989a) examined the $\delta^{13}C$ value of soil CO_2 and pedogenic carbonate along elevation (climate) gradients in the Mojave Desert/Great Basin and utilized the models described above to analyze the data. Quade *et al.* found that there were systematic trends in soil CO_2 concentrations, and $\delta^{13}C$ values, due to changes in CO_2 production rates with increasing elevation, and soil depth (Figure 17). The primary achievement of this work was that the observations were fully explainable using the mechanistic model represented here by Equation (13), a result that opened the door for the use of pedogenic carbonates in paleoenvironmental (e.g., Quade *et al.*, 1989b) and atmospheric p_{CO_2} (Cerling, 1991) studies.

5.01.5.1.3 Processes and isotope composition of pedogenic carbonate formation

In arid and semi-arid regions of the world, where precipitation is exceeded by potential evapotranspiration, soils are incompletely leached and $CaCO_3$ accumulates in significant quantities. Figure 18 illustrates the global distribution of carbonate in the upper meter of soils. As the figure illustrates, there is a sharp boundary between calcareous and noncalcareous soil in the USA at about the 100th meridian. This long-recognized boundary reflects the soil water balance. Jenny and Leonard (1939)

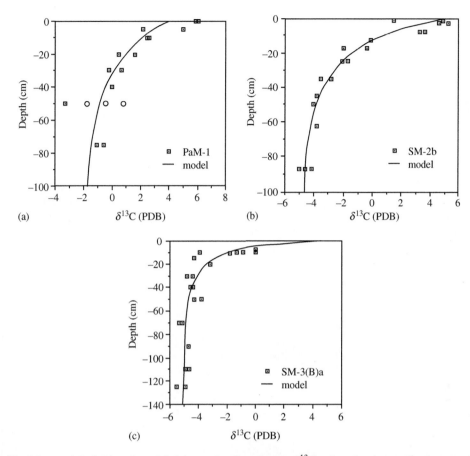

Figure 17 Measured (points) and modeled (curves) soil carbonate $\delta^{13}C$ values for three soils along an elevation (climate) gradient in the Mojave desert/Great Basin of California and Nevada. Modeled carbonate values based on Equation (14) plus the fractionation between CO_2 and carbonate (~10‰). The elevations (and modeled soil respiration rates that drive the curve fit) are: (a) 330 m (0.18 mmol CO_2 m^{-2} h^{-1}); (b) 1,550 m (0.4 mmol CO_2 m^{-2} h^{-1}); and (c) 1,900 m (1.3 mmol CO_2 m^{-2} h^{-1}). The $\delta^{13}C$ value of soil organic matter (CO_2 source) was about -21‰ at all sites. Note that depth of atmospheric CO_2 isotopic signal decreases with increasing elevation and biological CO_2 production (Quade *et al.*, 1989a) (reproduced by permission of Geological Society of America from *Geol. Soc. Am. Bull.* **1989**, *101*, 464–475).

examined the depth to the top of the carbonate-bearing layer in soils by establishing a climo-sequence (precipitation gradient) along an east to west transect of the Great Plains (Figure 19). They observed that at constant mean average temperature (MAT) below 100 cm of mean average precipitation (MAP), carbonate appeared in the soils, and the depth to the top of the carbonate layer decreased with decreasing precipitation. An analysis has been made of the depth to carbonate versus precipitation relation for the entire USA (Royer, 1999). She found that, in general, the relation exists broadly but as the control on other variables between sites (temperature, soil texture, etc.) is relaxed, the strength of the relationship declines greatly.

In addition to the depth versus climate trend, there is a predictable and repeatable trend of carbonate amount and morphology with time (Gile *et al.* (1966); Figure 20) due to the

progressive accumulation of carbonate over time, and the ultimate infilling of soil porosity with carbonate cement, which restricts further downward movement of water and carbonate.

The controls underlying the depth and amount of soil carbonate hinge on the water balance, Ca^{+2} availability, soil CO_2 partial pressures, etc. Arkley (1963) was the first to characterize these processes mathematically. His work has been greatly expanded by McFadden and Tinsley (1985), Marian *et al.* (1985), and others to include numerical models. Figure 21 illustrates the general concepts of McFadden and Tinsley's numerical model, and Figure 22 illustrates the results of model predictions for a hot, semi-arid soil (see the figure heading for model parameter values). These predictions generally mimic observations of carbonate distribution in desert soils, indicating that many of the key processes have been identified.

Soil inorganic carbon

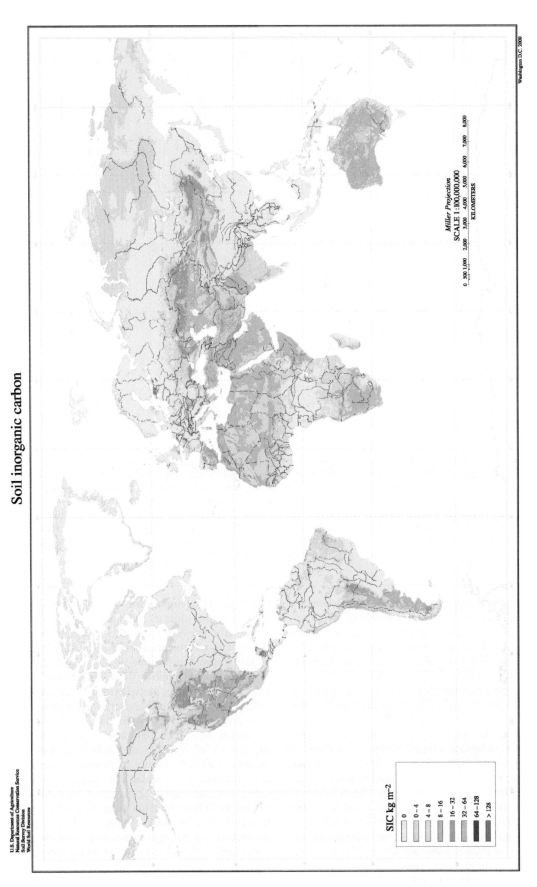

Washington D.C. 2000

Miller Projection
SCALE 1:100,000,000

KILOMETERS

0 500 1,000 2,000 3,000 4,000 5,000 6,000 7,000 8,000

SIC kg m^{-2}

0
0 – 4
4 – 8
8 – 16
16 – 32
32 – 64
64 – 128
> 128

Figure 18 Global distribution of pedogenic carbonate (source http://www.nrcs.usda.gov/technical/worldsoils/mapindx/).

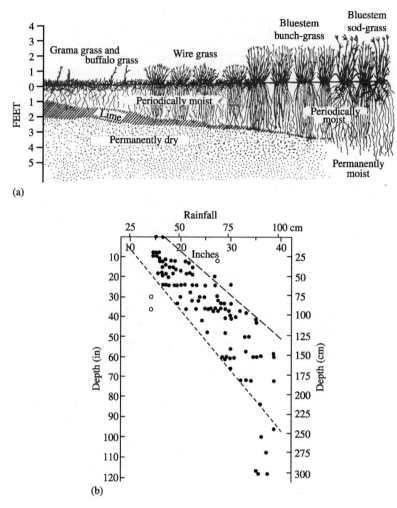

(a)

(b)

Figure 19   (a) Schematic view of plant and soil profile changes on an east (right) to west (left) transect of the Great Plains and (b) measured depth to top of pedogenic carbonate along the same gradient (source Jenny, 1941).

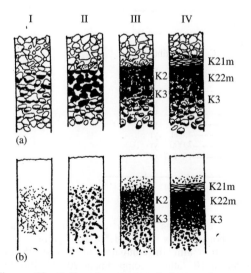

(a)

(b)

Figure 20   Soil carbonate morphology and amount versus time for: (a) gravelly and (b) fine-grained soils (Gile *et al.*, 1966) (reproduced by permission of Williams and Wilkins from *Soil Sci.* **1966**, *101*, 347–360).

The general equation describing the formation of carbonate in soils is illustrated by the reaction

$$CO_2 + H_2O + Ca^{+2} = CaCO_3 + 2H^+$$

From an isotopic perspective, in unsaturated soils, soil CO_2 represents an infinite reservoir of carbon and soil water an infinite reservoir of oxygen, and the $\delta^{13}C$ and $\delta^{18}O$ values of the pedogenic carbonate (regardless of whether its calcium is derived from silicate weathering, atmospheric sources, or limestone) are entirely set by the isotopic composition of soil CO_2 and H_2O. Here we focus mainly on the carbon isotopes. However, briefly for completeness, we outline the oxygen-isotope processes in soils. The source of soil H_2O is precipitation, whose oxygen-isotope composition is controlled by a complex set of physical processes (Hendricks *et al.*, 2000), but which commonly shows a positive correlation with MAT (Rozanski *et al.*, 1993). Once this water enters the soil, it is subject to transpirational

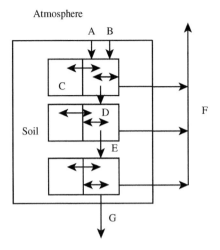

Figure 21 Diagram of compartment model for the numerical simulation of calcic soil development. Compartments on left represent solid phases and compartments on right represent aqueous phases. Line A represents precipitation, line B represents dust influx, line C represents the transfer of components due to dissolution and precipitation, line D represents transfer between aqueous phases, line E represents downward movement of solutes due to gravitational flow of soil water, line F represents evapotranspirational water loss, and line G represents leaching losses of solutes (McFadden *et al.*, 1991) (reproduced by permission of Soil Science Society of America from *Occurrence, Characteristics, and Genesis of Carbonate, Gypsum and Silica Accumulations in Soils*, **1991**).

(largely unfractionating) and evaporative (highly fractionating) losses. Barnes and Allison (1983) presented an evaporative soil water model that consists of processes for an: (i) upper, vapor transport zone and (ii) a liquid water zone with an upper evaporating front. The model describes the complex variations observed in soil water versus depth following periods of extensive evaporation (Barnes and Allison, 1983; Stern *et al.*, 1999). Pedogenic carbonate that forms in soils generally mirrors these soil water patterns (e.g., Cerling and Quade, 1993). In general, in all but hyperarid, poorly vegetated sites (where the evaporation/ transpiration ratio is high), soil $CaCO_3$ $\delta^{18}O$ values roughly reflect those of precipitation (Amundson *et al.*, 1996).

The carbon-isotope model and its variants have, it is fair to say, revolutionized the use of soils and paleosols (see Chapter 5.18) in paleobotany and climatology. Some of the major achievements and uses of the model include the following.

- The model adequately describes the observed increases in the $\delta^{13}C$ value of both soil CO_2 and $CaCO_3$ with depth (Figure 17).
- The model clearly provides a mechanistic understanding of why soil CO_2 is enriched in ^{13}C relative to plant inputs (steady-state diffusional enrichment of ^{13}C).
- The model indicates that for reasonable rates of CO_2 production in soils, the $\delta^{13}C$ value of soil

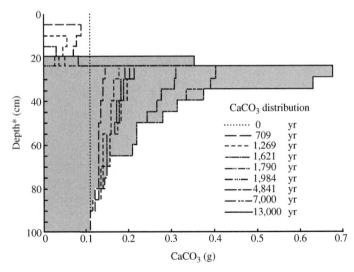

Figure 22 Predicted pattern of Holocene pedogenic carbonate accumulation in a cm^2 column in a semi-arid, thermic climate (leaching index = 3.5 cm). External carbonate flux rate = 1.5×10^{-4} g cm^{-2} yr^{-1}, $p_{CO_2} = 1.5 \times 10^{-3.3}$ atm in compartment 1 increasing to $10^{-2.5}$ atm in compartment 5 (20–25 cm). Below compartment 5, the p_{CO_2} decreases to a minimum value of 10^{-4} atm in compartment 20 (95–100 cm). Dotted line shows carbonate distribution at $t = 0$. Gray area indicates final simulated distribution. Depth* = absolute infiltration depth in <2 mm fraction (McFadden *et al.*, 1991) (reproduced by permission of Soil Science Society of America from *Occurrence, Characteristics, and Genesis of Carbonate, Gypsum and Silica Accumulations in Soils*, **1991**).

CO_2 should, at a depth within 100 cm of the surface, represent the $\delta^{13}C$ value of the standing biomass plus 4.4‰. The $\delta^{13}C$ of $CaCO_3$ will also reflect this value, plus an equilibrium fractionation of ~10‰ (depending on temperature). Therefore, if paleosols are sampled below the "atmospheric mixing zone," whose thickness depends on CO_2 production rates (Figure 17), the $\delta^{13}C$ value of the carbonate will provide a guide to the past vegetation (Cerling *et al.*, 1989).

Cerling (1991) recognized that Equation (14), if rearranged and solved for C_{atm}, could provide a means of utilizing paleosol carbonates for reconstructing past atmospheric CO_2 partial pressures. To do so, the values of the other variables (including the $\delta^{13}C$ value of the atmospheric CO_2—see Jahren *et al.* (2001)) must be known, which for soils of the distant past is not necessarily a trivial problem. Nonetheless, an active research field has developed using this method, and a compilation of calculated atmospheric CO_2 levels is emerging (Ekart *et al.*,1999; Mora *et al.*, 1996), with estimates that correlate well with model calculations by Berner (1992).

Cerling's (1984) approach to modeling stable carbon isotopes in soil CO_2 has been expanded and adapted to other isotopes in soil CO_2: (i) $^{14}CO_2$—for pedogenic carbonate dating (Wang *et al.*, 1994; Amundson *et al.*, 1994a,b) and soil carbon turnover studies (Wang *et al.*, 2000) and (ii) $C^{18}O^{16}O$—for hydrological tracer applications and, more importantly, as a means to constrain the controls on global atmospheric CO_2–^{18}O budgets (Hesterberg and Siegenthaler, 1991; Amundson *et al.*, 1998; Stern *et al.*, 1999, 2001; Tans, 1998). The processes controlling the isotopes, and the complexity of the models, greatly increase from ^{14}C to ^{18}O (see Amundson *et al.* (1998) for a detailed account of all soil CO_2 isotopic models).

5.01.5.2 Lateral Transport of Soil Material by Erosion

The previous section, devoted to the soil carbon cycle, emphasized the conceptual and mathematical focus on the vertical transport of materials. Models emphasizing vertical transport dominate present pedological modeling efforts. However, for soils on hillslopes or basins and floodplains, the lateral transport of soil material via erosive processes exerts an overwhelming control on a variety of soil properties. The study of the mechanisms by which soil is physically moved— even on level terrain—is an evolving aspect of pedology. Here we focus on new developments in the linkage between geomorphology and pedology in quantifying the effect of sediment transport on

soil formation. The focus here is on natural, undisturbed landscapes as opposed to agricultural landscapes, where different erosion models may be equally or more important.

This section begins by considering soils on divergent, or convex, portions of the landscape. On divergent hillslopes, slope increases with distance downslope such that (for a two-dimensional view of a hillslope)

$$\frac{\Delta\text{slope}}{\Delta\text{distance}} = \left(\frac{\partial}{\partial x}\right)\left(\frac{\partial z}{\partial x}\right)$$
$$= \text{negative quantity} \quad (15)$$

where z and x are distances in a vertical and horizontal direction, respectively. On convex slopes, there is an ongoing removal of soil material due to transport yr processes. In portions of the landscape where the derivative of slope versus distance is positive (i.e., convergent landscapes), there is a net accumulation of sediment, which will be discussed next.

In many divergent landscapes (those not subject to overland flow or landslides), the ongoing soil movement may be almost imperceptible on human timescales. Research by geomorphologists (and some early naturalists such as Darwin) has shed light on both the rate and mechanisms of this process. In general, a combination of physical and biological processes, aided by gravity, can drive the downslope movement of soil material. These processes are sometimes viewed as roughly diffusive in that soil is randomly moved in many directions, but if a slope gradient is present, the net transport is down the slope. Kirkby (1971), in laboratory experiments, showed that wetting and drying cycles caused soil sensors to move in all directions, but in a net downslope direction. Black and Montgomery (1991) examined pocket gopher activity on sloping landscapes in California, discussing their underground burrowing and transport mechanisms, and the rate at which upthrown material then moves downslope. Tree throw also contributes to diffusive-like movement (Johnson *et al.*, 1987). In a truly far-sighted study, Darwin (1881) quantified both the mass and the volume of soil thrown up by earthworms, and the net amount moved downslope by wind, water, and gravity.

The general transport model for diffusive-like soil transport is (Kirkby, 1971)

$$Q = -\rho_s K \frac{dz}{dx} \quad (16)$$

where Q is soil flux per unit length of a contour line (ρ_s) (g cm^{-1} yr^{-1}) and K is transport coefficient (cm^2 yr^{-1}). The concepts behind this expression are generally attributed to Davis (1892) and Gilbert (1909). However,

Darwin (1881) clearly recognized this principle in relation to sediment transport by earthworms on slopes. For several sites in England (based on a few short-term measurements), Darwin reported Q and slope, allowing us to calculate a K based solely on earthworms. The values of ~ 0.02 cm^2 yr^{-1}, while significant, are about two orders of magnitude lower than overall K values calculated for parts of the western USA that capture the integrated biological and abiotic mixing/transport mechanisms (McKean et al., 1993; Heimsath et al., 1999).

At this point, it is worth commenting further on the effect and role of bioturbation in soil formation processes whether it results in net downslope movement or not. Beginning with Darwin, there has been a continued, but largely unappreciated, series of studies concerning the effects of biological mixing on soil profile features (Johnson et al., 1987; Johnson, 1990). These studies concluded that most soils (those maintaining at least some plant biomass to support a food chain) have a "biomantle" (a highly mixed and sorted zone) that encompasses the upper portions of the soil profile (Johnson, 1990). The agents are varied, but include the commonly distributed earthworms, gophers, ground squirrels, ants, termites, wombats, etc. In many cases, the biomantle is equivalent to the soil's A horizons. The rapidity at which these agents can mix and sort the biomantle is impressive. The invasion of earthworms into the Canadian prairie resulted in complete homogenization of the upper 10 cm in 3 yr (Langmaid, 1964). Johnson (1990) reports that Borst (1968) estimated that the upper 75 cm of soils in the southern San Joaquin Valley of California are mixed in 360 yr by ground squirrels. Darwin reported that in England, $\sim 1.05 \times 10^4$ kg of soil per hectare are passed through earthworms yearly, resulting in a turnover time for the upper 50 cm of the soil of just over 700 yr. The deep burial of Roman structures and artifacts in England by earthworm casts supports this estimated cycling rate.

Bioturbation on level ground may involve a large gross flux of materials (as indicated above), but no net movement in any direction due to the absence of a slope gradient. Nonetheless, the mixing has important physical and chemical implications for soil development: (i) rapid mixing of soil and loss of stratification at initial stages of soil formation; (ii) rapid incorporation of organic matter (and the subsequent slowed decomposition of this material below ground); (iii) periodic cycling of soil structure which prevents soil horizonation within the biomantle; and (iv) in certain locations, striking surficial expressions of biosediment movement caused by nonrandom sediment transport. The famed "mima mounds" of the Plio-Pleistocene fluvial terraces

and fans of California's Great Valley (Figure 23(a)) are a series of well-drained sandy loam to clay loam materials that overlie a relatively level, impermeable layer (either a dense, clay-rich horizon or a silica cemented hardpan)(Figure 23(b)). The seasonally water-logged conditions that develop over the impermeable layers have probably caused gophers to move soil preferentially into better drained landscape segments, thereby producing this unusual landscape. These landform features are now relatively rare due to the expansion of agriculture in the state, but the original extent of this surface feature is believed to have been more than 3×10^5 ha (Holland, 1996).

The conceptual model for diffusive soil transport down a hillslope is shown in Figure 24 (Heimsath et al., 1997, 1999). In any given section of the landscape, the mass of soil present is the balance of transport in, transport out, and soil production (the conversion of rock or sediment to soil). If it is assumed that the processes have been operating for a sufficiently long period of time, then the soil thickness is at steady state. The model describing this condition is

$$\rho_s \frac{\partial h}{\partial t} = 0 = \underbrace{(\rho_r \phi)}_{\text{soil production}} + \underbrace{\left(K\rho_s \frac{\partial^2 z}{\partial x^2} \right)}_{\substack{\text{balance between} \\ \text{diffusive inputs/losses}}} \quad (17)$$

where h is the soil thickness (cm), ρ_s and ρ_r the soil and rock bulk density (g cm^{-3}), respectively, and ϕ the soil production rate (cm yr^{-1}).

Soil production is a function of soil depth (Heimsath et al., 1997), parent material, and environmental conditions (Heimsath et al., 1999). As soil thickens, the rate of the conversion of the underlying rock or sediment to soil decreases. This has been shown using field observations of the relation between soil thickness and the abundance of cosmogenic nuclides (^{10}Be and ^{26}Al) in the quartz grains at the rock–soil interface (Figure 24). From this work, soil production can be described by

$$\phi = \rho_r \phi_0 e^{-\alpha h} \quad (18)$$

where ϕ and ϕ_0 are soil production for a given soil thickness and no soil cover, respectively; α the constant (cm^{-1}), and h the soil thickness (cm). The soil production model described by Equation (18) is applicable where physical disruption of the bedrock is the major soil production mechanism. By inserting Equation (18) into Equation (17), and rearranging, one can solve for soil thickness at any position on a hillslope:

$$h = \frac{1}{\alpha} \left(-\ln \left(-\frac{K\rho_s \partial^2 z}{\phi_0 \rho_r \partial x^2} \right) \right) \quad (19)$$

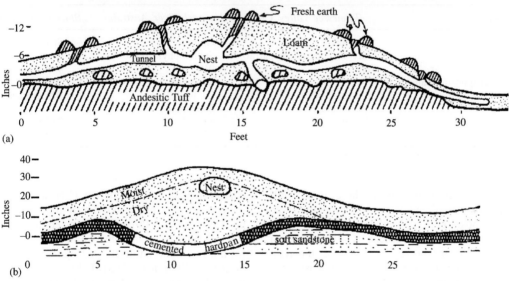

Figure 23 (a) Mima mounds (grass-covered areas) and vernal pools (gravel veneered low areas) on a Plio-Pleistocene aged fluvial terrace of the Merced river, California. Observations indicate that the landscape is underlain by a dense, clay pan (formed by long intervals of soil Bt horizon formation) capped by a gravel lense (the vernal pool gravels extend laterally under the Mima mounds). (b) Schematic diagram of "pocket gopher theory" of mima mound formation, illustrating preferential nesting and soil movement in/toward the well-drained mound areas and away from the seasonally wet vernal pools (Arkley and Brown, 1954) (reproduced by permission of Soil Science Society of America from *Soil Sci. Soc. Am. Proc.* **1954**, *18*, 195–199).

As this equation indicates, the key variables controlling soil thickness on hillslopes are slope curvature, the transport coefficient, and the soil production rate. The importance of curvature is intuitive in that changes in slope gradient drive the diffusive process. The value of the transport coefficient K varies greatly from one location to another (Fernandes and Dietrich, 1997; Heimsath *et al.*,1999), and seems to increase with increasing humidity and decreasing rock competence (Table 5). Production, like transport, is likely dependent on climate and rock composition (Heimsath *et al.*,1999).

The production/transport model of sediment transport provides a quantitative and mechanistic insight into soil profile thickness on hillslopes, an area of research that has received limited, but in some cases insightful, attention in the pedological literature. The study of the distribution of soil along hillslope gradients are called *toposequences*

Soil Formation

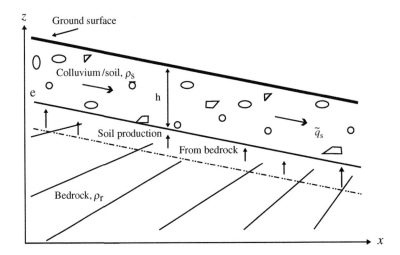

Figure 24 Schematic diagram of soil production and downslope transport on diffusion-dominated hillslopes (Heimsath *et al.*, 1997) (reproduced by permission of Nature Publishing Group from *Nature* **1997**, *388*, 358–361).

Table 5 Soil production (P_0), erosion, and diffusivity values for three watersheds in contrasting climates and geology.

Location	Tennessee Valley, CA	Nunnock River, Australia	Black Diamond, CA
Bedrock	Sandstone	Granitic	Shale
Vegetation	Grass and coastal chaparral	Schlerophyll forest	Grass
Precipitation (mm yr^{-1})	760	910	450
Major transport mechanism	Gophers	Wombats, tree throw	Soil creep
Erosion rate (g m^{-2} yr^{-1})	50: backslope	20: backslope	625 average
	130: shoulder	60: shoulder	
P_0 (mm kyr^{-1})	77 ± 9	53 ± 3	2,078
K (cm^2 yr^{-1})	25	40	360

Sources: Heimsath (1999) and McKean *et al.* (1993).

or *catenas*. In many areas, there is a well-known relationship between soil properties and hillslope position (e.g., Nettleton *et al.* (1968) for toposequences in southern California, for example), but the processes governing the relations were rather poorly known—at least on a quantitative level.

Beyond the effect of slope position (and environmental conditions) on soil thickness, a key factor is the amount of time that a soil on a given hillslope has to form. Time is a key variable that determines the amount of weathering and horizon formation that has occurred (Equation (2)). Yet, constraining "soil age" on erosional slopes has historically been a challenging problem. Here the author takes a simple approach and views soil age on slopes in terms of residence time (τ), where

$$\tau = \frac{\text{soil mass/area}}{\rho_r \phi_0 e^{-\alpha h}} = \frac{\text{soil mass/area}}{\rho_s K (\partial^2 z/\partial x^2)} \quad (20)$$

This is a pedologically meaningful measure that defines the rate at which soil is physically moved through a soil "box." As the expression indicates, soil residence time increases with decreasing

curvature (and from Equation (19) above, with increasing soil thickness). From measured soil production rates in three contrasting environments, soil residence times on convex hillslopes may vary from a few hundred to 10^5 yr (Figure 25). These large time differences clearly help to explain many of the differences in soil profile development long observed on hillslope gradients.

On depositional landforms, sloping areas with concave slopes (convergent curvature) or level areas on floodplains, the concept of residence time can also be applied and quantitative models of soil formation can be derived. In these settings, residence time can be viewed as the amount of time required to fill a predetermined volume (or thickness) of soil with incoming sediment.

On active floodplains of major rivers, the soil residence time can be in most cases no longer than a few hundred years. Soil profiles here exhibit stratification, buried, weakly developed horizons, and little if any measurable chemical weathering.

In a landmark study of soils on depositional settings, Jenny (1962) examined soils on the

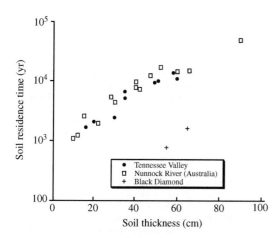

Figure 25 Calculated soil residence time versus soil thickness in three contrasting watersheds using Equation (20) and data from Table 5.

floodplain of the Nile River, Eygpt. Available data suggested that the sedimentation rate prior to the construction of the Aswan Dam was 1 mm yr^{-1}, giving a residence time of 1,500 yr for a soil 150 cm thick. Jenny observed that in these soils, used for agriculture for millennia, nitrogen (the most crop-limiting element in agriculture) decreased with depth. This trend was not due to plant/atmospheric nitrogen inputs and crop cycling which creates the standard decrease in carbon and nitrogen with depth (see previous section on soil organic carbon). Instead, the continual deposition of nitrogen-rich sediment derived from soils in the Ethopian highlands was the likely source.

Jenny demonstrated that the observed nitrogen decrease with depth was a result of microbial degradation of the nitrogen-bearing organic matter in the sediment. The mineralized nitrogen was used by the crops and was then removed from the site. Stated mathematically

$$N(z) = N(0)e^{-kz} \tag{21}$$

where $N(z)$ and $N(0)$ are the soil nitrogen at depth z and 0, respectively, and k the decomposition rate constant. Jenny's analysis showed that on the Nile floodplain, in the absence of nitrogen fertilization, long-term high rates of crop production could only be maintained by continued flood deposition of Nile sediments.

More generally, on concave (or convergent) slopes, where sedimentation is caused by long-term diffusional soil movement, one can calculate rates of sedimentation (or, of course, measure them directly) using a version of the geomorphic models discussed above. In deposition settings, soil production from bedrock or sediment can be viewed as being zero, and the

change in soil height with time is (Fernandes and Dietrich, 1997)

$$\frac{\partial h}{\partial t} = K \frac{\partial^2 z}{\partial x^2} \tag{22}$$

Soils in these depositional settings are sometimes referred to as "cumulic" soils in the US Soil Taxonomy, and commonly exhibit high concentrations of organic carbon and nitrogen to depths of several meters. This organic matter is partially due to the burial of organic matter produced *in situ*, but also reflects the influx of organic-matter-rich soil from upslope positions. This process of organic carbon burial has attracted the attention of the global carbon cycle community, because geomorphic removal and burial of organic carbon may represent a large, neglected, global carbon flux (Stallard, 1998; Rosenblum *et al.*, 2001). Beyond the carbon burial, the relatively short residence time of any soil profile thickness inhibits the accumulation of chemical weathering products, vertical transport, and horizon development.

In summary, an under-appreciated fact is that all soils—regardless of landscape position—are subject to slow but measurable physical and (especially) biological turbation. On level terrain, the soil/sediment flux is relatively random. It results in extensive soil surface mixing and, only in certain circumstances, striking directional movement. The addition of a slope gradient, in combination with this ongoing soil turbation, results in a net downslope movement dependent on slope curvature and the characteristics of the geological and climatic settings. The thickness of a soil, and its residence time, is directly related to the slope curvature. These conditions also determine the expression of chemical weathering and soil horizonation that can occur on a given landscape position.

5.01.6 SOIL DATA COMPILATIONS

Soils, and their properties, are now being used in many regional to global biogeochemical analyses. The scientific and education community is fortunate to have a wealth of valuable, high-quality, data on soil properties available on the worldwide web and in other electronic avenues. Here the author identifies some of these data sources.

During soil survey operations, numerous soil profile descriptions, and large amounts of laboratory data, are generated for the soils that are being mapped. Generally, the soils being described, sampled, and analyzed are representatives of *soil series*, the most detailed (and restrictive) classification of soils in the USA. There are $\sim 2.1 \times 10^4$ soil series that have been identified and mapped in the USA. The locations, soil descriptions, and lab data for many of these series are now available on the worldwide web via the USDA-NRCS (Natural Resource Conservation Service) National Soil Survey Center web site (http://www.statlab. iastate.edu/soils/ssl/natch_data.html). The complete data set is archived at the National Data Center, and is available to researchers.

The "characterization" data sheet is likely to be of interest to most investigators. In general, these data sheets are somewhat regionally oriented, reporting different chemical analyses for different climatic regions. For example, salt chemistry is reported for arid regions, whereas a variety of iron and aluminum oxide analyses are reported for humid, northern regions. Some major analyses of each soil horizon that are commonly reported include: particle size data, organic carbon and nitrogen, chemically extractable metals, cation exchange capacity and base saturation, bulk density, water retention, extractable bases and acidity, calcium carbonate, pH, chemistry of water extracted from saturated pastes, clay mineralogy, and sand and silt mineralogy (to list some of the major analyses). A complete and thorough discussion of the types of soil analyses, and the methods used, are presented by the Soil Survey Staff (1995), and on the web.

5.01.7 CONCLUDING COMMENTS

The scientific study of soils is just entering its second century. An impressive understanding of soil geography and, to a lesser degree, soil processes has evolved. There is growing interest in soils among scientists outside the agricultural sector, particularly geochemists and ecologists. Soils are central in the present attention to the global carbon cycle and its management, and as the human impact on the planet continues to increase, soils and their properties and services are being considered from a geobiodiversity perspective (Amundson, 1998; Amundson *et al.*, 2003). In many ways, the study of soils is at a critical and exciting point, and interdisciplinary cross-fertilization of the field is sure to lead to exciting new advances that bring pedology back to its multidisciplinary origins.

REFERENCES

Allègre C. (1992) *From Stone to Star: A View of Modern Geology*. Harvard University Press, Cambridge.
Allen R. M., Chatterjee R., Madden M. S., Ludden P. W., and Shar V. K. (1994) Biosynthesis of the iron–molybdenum co-factor of nitrogenase. *Crit. Rev. Biotech.* **14**, 225–249.
Alpers C. N. and Brimhall G. H. (1988) Middle Miocene climatic change in the Atacama desert, northern Chile: evidence for supergene mineralization at La Escondida. *Geol. Soc. Am. Bull.* **100**, 1640–1656.

Amundson R. (1998) Do soils need our protection? *Geotimes* March, 16–20.
Amundson R. (2001) The carbon budget in soils. *Ann. Rev. Earth Planet. Sci.* **29**, 535–562.
Amundson R. and Baisden W. T. (2000) Stable isotope tracers and mathematical models in soil organic matter studies. In *Methods in Ecosystem Science* (eds. O. E. Sala, R. B. Jackson, H. A. Mooney, and R. W. Howarth). Springer, New York, pp. 117–137.
Amundson R. and Davidson E. (1990) Carbon dioxide and nitrogenous gases in the soil atmosphere. *J. Geochem. Explor.* **38**, 13–41.
Amundson R. and Jenny H. (1997) On a state factor model of ecosystems. *Bioscience* **47**, 536–543.
Amundson R. and Yaalon D. (1995) E. W. Hilgard and John Wesley Powell: efforts for a joint agricultural and geological survey. *Soil Sci. Soc. Am. J.* **59**, 4–13.
Amundson R., Harden J., and Singer M. (1994a) *Factors of Soil Formation: A Fiftieth Anniversary Retrospective*, SSSA Special Publication No. 33. Soil Science Society of America, Madison, WI.
Amundson R., Wang Y., Chadwick O., Trumbore S., McFadden L., McDonald E., Wells S., and DeNiro M. (1994b) Factors and processes governing the ^{14}C content of carbonate in desert soils. *Earth Planet Sci. Lett.* **125**, 385–405.
Amundson R., Chadwick O. A., Kendall C., Wang Y., and DeNiro M. (1996) Isotopic evidence for shifts in atmospheric circulation patterns during the late Quaternary in mid-North America. *Geology* **24**, 23–26.
Amundson R., Stern L., Baisden T., and Wang Y. (1998) The isotopic composition of soil and soil-respired CO_2. *Geoderma* **82**, 83–114.
Amundson R., Guo Y., and Gong P. (2003) Soil diversity and landuse in the United States. *Ecosystems*, doi: 10.1007/s10021-002-0160-2.
Anders E. and Grevesse N. (1989) Abundances of the elements: meteoric and solar. *Geochim. Cosmochim. Acta* **53**, 197–214.
Arkley R. J. (1963) Calculations of carbonate and water movement in soil from climatic data. *Soil Sci.* **96**, 239–248.
Arkley R. J. and Brown H. C. (1954) The origin of mima mound (hog wallow) microrelief in the far western states. *Soil Sci. Soc. Am. Proc.* **18**, 195–199.
Baisden W. T. (2000) Soil organic matter turnover and storage in a California annual grassland chronosequence. PhD Dissertation, University of California, Berkeley.
Baisden W. T., Amundson R., Brenner D. L., Cook A. C., Kendall C., and Harden J. W. (2002) A multi-isotope C and N modeling analysis of soil organic matter turnover and transport as a function of soil depth in a California annual grassland chronosequence. *Glob. Biogeochem. Cycles*, **16**(4), 1135, doi: 10.1029/2001/GB001823, 2002.
Bao H., Campbell D. A., Bockheim J. G., and Theimens M. H. (2000a) Origins of sulphate in Antarctic dry-valley soils as deduced fro anomalous ^{17}O compositions. *Nature* **407**, 499–502.
Bao H., Thiemens M. H., Farquhar J., Campbell D. A., Lee C. C.-W., Heine K., and Loope D. B. (2000b) Anomalous ^{17}O compositions in massive sulphate deposits on the Earth. *Nature* **406**, 176–178.
Bao H., Thiemens M. H., and Heine K. (2001) Oxygen-17 excesses of the Central Namib gypcretes: spatial distribution. *Earth Planet. Sci. Lett.* **192**, 125–135.
Barnes C. J. and Allison G. B. (1983) The distribution of deuterium and ^{18}O in dry soils: I. Theory. *J. Hydrol.* **60**, 141–156.
Betancourt J. L., Latorre C., Rech J. A., Quade J., and Rlander K. A. (2000) A 22,000-year record of monsoonal precipitation from Chile's Atacama desert. *Science* **289**, 1542–1546.
Black T. A. and Montgomery D. R. (1991) Sediment transport by burrowing animals, Marin county, California. *Earth Surf. Proc. Landforms* **16**, 163–172.

Berner R. A. (1992) Palaeo-CO_2 and climate. *Nature* **358**, 114.

Birkeland P. W. (1999) *Soils and Geomorphology*, 3rd edn. Oxford University Press, New York.

Böhlke J. K., Erickson G. E., and Revesz K. (1997) Stable isotope evidence for an atmospheric origin of desert nitrate deposits in northern Chile and southern California, USA. *Chem. Geol.* **136**, 135–152.

Borst G. (1968) The occurrence of crotovinas in some southern California soils. *Trans. 9th Int. Congr. Soil Sci., Adelaide* **2**, 19–22.

Bowen H. J. M. (1979) *Environmental Chemistry of the Elements*. Academic Press, London.

Brimhall G. (1987) Preliminary fractionation patterns of ore metals through Earth history. *Chem. Geol.* **64**, 1–16.

Brimhall G. H. and Dietrich W. E. (1987) Constitutive mass balance relations between chemical composition, volume, density, porosity, and strain in metasomatic hydrochemical systems: results on weathering and pedogenesis. *Geochim. Cosmochim. Acta* **51**, 567–587.

Brimhall G. H., Lewis C. J., Ague J. J., Dietrich W. E., Hampel J., Teague T., and Rix P. (1988) Metal enrichment in bauxite by deposition of chemically-mature eolian dust. *Nature* **333**, 819–824.

Brimhall G. H., Lewis C. J., Ford C., Bratt J., Taylor G., and Warren O. (1991) Quantitative geochemical approach to pedogenesis: importance of parent material reduction, volumetric expansion, and eolian influx in laterization. *Geoderma* **51**, 51–91.

Brimhall G., Chadwick O. A., Lewis C. J., Compston W., Williams I. S., Danti K. J., Dietrich W. E., Power M. E., Hendricks D., and Bratt J. (1992) Deformational mass balance transport and invasive processes in soil evolution. *Science* **255**, 695–702.

Cerling T. E. (1984) The stable isotopic composition of modern soil carbonate and its relationship to climate. *Earth Planet. Sci. Lett.* **71**, 229–240.

Cerling T. E. (1991) Carbon dioxide in the atmosphere: evidence from Cenozoic and Mesozoic paleosols. *Am. J. Sci.* **291**, 377–400.

Cerling T. E. and Quade J. (1993) Stable carbon and oxygen isotopes in soil carbonates. In *Climate Change in Continental Isotopic Records* (eds. P. K. Swart, K. C. Lohmann, J. McKenzie, and S. Sabin). Geophysical Monograph 78. American Geophysical Union, Washington, DC, pp. 217–231.

Cerling T. E., Quade J., Wang Y., and Bowman J. R. (1989) Carbon isotopes in soils and palaeosols as paleoecologic indicators. *Nature* **341**, 138–139.

Cerling T. E., Harris J. M., MacFadden B. J., Leakey M. G., Quade J., Eisenmann V., and Ehleringer J. R. (1997) Global vegetation change through the Miocene/Pliocene boundary. *Nature* **389**, 153–158.

Chadwick O. A. and Davis J. O. (1990) Soil-forming intervals caused by eoloian sediment pulses in the Lahontan basin, northwestern Nevada. *Geology* **18**, 243–246.

Chadwick O. A., Brimhall G. H., and Hendricks D. M. (1990) From a black to a gray box—a mass balance interpretation of pedogenesis. *Geomorphology* **3**, 369–390.

Chadwick O. A., Derry L. A., Vitousek P. M., Huebert B. J., and Hedin L. O. (1999) Changing sources of nutrients during four million years of ecosystem development. *Nature* **397**, 491–497.

Chiappini C. (2001) The formation and evolution of the Milky Way. *Am. Sci.* **89**, 506–515.

Clarkson D. T. (1974) *Ion Transport and Cell Structure in Plants*. McGraw-Hill, New York, 350pp.

Darwin C. (1846) An account of the fine dust which often falls on vessels in the Atlantic ocean. *Quat. J. Geol. Soc. London* **2**, 26–30.

Darwin C. (1881) *The Formation of Vegetable Mould through the Action of Worms*. Appleton, New York.

Davis W. M. (1892) The convex profile of badland divides. *Science* **20**, 245.

De Coninck F. (1980) Major mechanisms in formation of spodic horizons. *Geoderma* **24**, 101–128.

Delsemme A. H. (2001) An argument for the cometary origin of the biosphere. *Am. Sci.* **89**, 432–442.

Dokuchaev V. V. (1880) Protocol of the meeting of the branch of geology and mineralogy of the St. Petersburg Society of Naturalists. *Trans. St. Petersburg Soc. Nat.* **XII**, 65–97. (Translated by the Department of Soils and Plant Nutrition, University of California, Berkeley).

Drake M. J. and Righter K. (2002) Determining the composition of the earth. *Nature* **416**, 39–51.

Ekart D. D., Cerling T. E., Montañáñez I. P., and Tabor N. J. (1999) A 400 million year carbon-isotope record of pedogenic carbonate: implications for paleoatmospheric carbon dioxide. *Am. J. Sci.* **299**, 805–827.

Elzein A. and Balesdent J. (1995) Mechanistic simulation of vertical distribution of carbon concentrations and residence times in soils. *Soil Sci. Soc. Am. J.* **59**, 1328–1335.

Ericksen G. E. (1981) Geology and origin of the Chilean nitrate deposits. Geological Survey Professional Paper 1188, US Geological Survey, 37pp.

Farquhar G. D., Hubick K. T., Condon A. G., and Richards R. A. (1988) Carbon-isotope fractionation and plant water-use efficiency. In *Stable Isotopes in Ecological Research* (eds. P. W. Rundell, J. R. Ehleringer, and K. A. Nagy). Springer, New York, pp. 21–40.

Fernandes N. F. and Dietrich W. E. (1997) Hillslope evolution by diffusive processes: the timescale for equilibrium adjustments. *Water Resour. Res.* **33**, 1307–1318.

Ferris T. (1998) Seeing in the dark. *The New Yorker* August 10, 55–61.

Friedman I. and O'Neil J. R. (1977) Compilation of stable isotope fractionation factors of geochemical interest. In *Data of Geochemistry*, 6th edn., US Geological Survey Professional Paper 440-KK (ed. M. Fleischer). chap. KK.

Gilbert G. K. (1909) The convexity of hilltops. *J. Geol.* **17**, 344–350.

Gile L. H., Peterson F. F., and Grossman R. B. (1966) Morphological and genetic sequences of carbonate accumulation in desert soils. *Soil Sci.* **101**, 347–360.

Greenwood N. N. and Earnshaw A. (1997) *Chemistry of the Elements*, 2nd edn. Heinemann, Oxford.

Griffin D. W., Kellogg C. A., Garrison V. H., and Shinn E. A. (2002) The global transport of dust. *Am. Sci.* **90**, 228–235.

Hanson P. J., Edwards N. T., Garten C. T., and Andrews J. A. (2000) Separating root and soil microbial contributions to soil respiration: a review of methods and observations. *Biogeochemistry* **48**, 115–146.

Harden J. (1987) Soils developed in granitic alluvium near Merced, California. *US Geol. Surv. Bull.* **1590-A**, pp. A1–A65.

Heath J. A. and Huebert B. J. (1999) Cloudwater deposition as a source of fixed nitrogen in a Hawaiian montane forest. *Biogeochemistry* **44**, 119–134.

Hendricks M. B., DePaolo D. J., and Cohen R. C. (2000) Space and time variation of delta O-18 and delta D in precipitation: can paleotemperature be estimated from ice cores? *Global Biogeochem. Cycles* **14**, 851–861.

Heimsath A. M., Dietrich W. E., Nishiizumi K., and Finkel R. C. (1997) The soil production function and landscape equilibrium. *Nature* **388**, 358–361.

Heimsath A. M., Dietrich W. E., Nishiizumi K., and Finkel R. C. (1999) Cosmogenic nuclides, topography, and the spatial variation of soil depth. *Geomorphology* **27**, 151–172.

Hesterberg R. and Siegenthaler U. (1991) Production and stable isotopic composition of CO_2 in a soil near Bern, Switzerland. *Tellus, Ser. B* **43**, 197–205.

Hilgard E. W. (1860) Report on the geology and agriculture of the state of Mississippi, Jackson, MS.

Hillel D. (1980a) *Fundamentals of Soil Physics*. Academic Press, New York.

Hillel D. (1980b) *Applications of Soil Physics*. Academic Press, New York.

Hillel D. (1998) *Environmental Soil Physics*. Academic Press, San Diego.

Holland R. F. (1996) Great Valley vernal pool distribution, Photorevised 1996. In *Ecology, Conservation, and Management of Vernal Pool Ecosystems,* Proceedings of a 1996 Conference, California Native Plant Society, Sacramento, CA (eds. C. W. Witham, E. T. Bauder, D. Belk, W. R. Ferren, Jr., and R. Ornduff), pp. 71–75.

Hoosbeek M. R. and Bryant R. B. (1995) Modeling the dynamics of organic carbon in a Typic Haplorthod. In *Soils and Global Change* (eds. R. Lal, J. M. Kimble, E. R. Levine, and B. A. Stuart). CRC Lewis Publishers, Chelsea, MI, pp. 415–431.

Jackson M. L., Levelt T. W. M., Syers J. K., Rex R. W., Clayton R. N., Sherman G. D., and Uehara G. (1971) Geomorphological relationships of tropospherically derived quartz in the soils of the Hawaiian Islands. *Soil Sci. Soc. Am. Proc.* **35**, 515–525.

Jahren A. H., Arens N. C., Sarmiento G., Guerrero J., and Amundson R. (2001) Terrestrial record of methane hydrate dissociation in the early Cretaceous. *Geology* **29**, 159–162.

Jenkinson D. S., Adams D. E., and Wild A. (1991) Model estimates of CO_2 emissions from soil in response to global warming. *Nature* **351**, 304–306.

Jenny H. (1941) *Factors of Soil Formation: A System of Quantitative Pedology*. McGraw-Hill, New York.

Jenny H. (1961) *E. W. Hilgard and the Birth of Modern Soil Science*. Collan della Rivista "Agrochimica", Pisa, Italy.

Jenny H. (1962) Model of a rising nitrogen profile in Nile Valley alluvium, and its agronomic and pedogenic implications. *Soil Sci. Soc. Am. Proc.* **27**, 273–277.

Jenny H. and Leonard C. D. (1939) Functional relationships between soil properties and rainfall. *Soil Sci.* **38**, 363–381.

Jenny H., Gessel S. P., and Bingham F. T. (1949) Comparative study of decomposition rates of organic matter in temperate and tropical regions. *Soil Sci.* **68**, 419–432.

Jersak J., Amundson R., and Brimhall G. (1995) A mass balance analysis of podzolization: examples from the northeastern United States. *Geoderma* **66**, 15–42.

Johnson D. L. (1990) Biomantle evolution and the redistribution of earth materials and artifacts. *Soil Sci.* **149**, 84–102.

Johnson D. L., Watson-Stegner D., Johnson D. N., and Schaetzl R. J. (1987) Proisotopic and proanisotropic processes of pedoturbation. *Soil Sci.* **143**, 278–292.

Jury W. A., Gardner W. R., and Gardner W. H. (1991) *Soil Physics*, 5th edn. Wiley, New York.

Kirkby M. J. (1971) Hillslope process-response models based on the continuity equation. *Inst. Prof. Geogr. Spec. Publ.* **3**, 15–30.

Koch P. L., Zachos J. C., and Dettman D. L. (1995) Stable isotope stratigraphy and paleoclimatology of the Paleogene Bighorn Basin. *Palaeogeogr. Palaeoclimat. Palaeoecol.* **115**, 61–89.

Kennedy M. J., Chadwick O. A., Vitousek P. M., Derry L. A., and Hendricks D. M. (1998) Changing sources of base cations during ecosystem development, Hawaiian Islands. *Geology* **26**, 1015–1018.

Krupenikov I. A. (1992) *History of Soil Science: From its Inception to the Present*. Amerind Publishing, New Delhi.

Langmaid K. K. (1964) Some effects of earthworm inversion in virgin podzols. *Can. J. Soil Sci.* **44**, 34–37.

Latorre C., Betancourt J. L., Rylander K. A., and Quade J. (2002) Vegetation invasions into absolute desert: a 45,000 yr rodent midden record from the Calama-Salar de Atacama basins, northern Chile (lat. 22°–24° S). *Geol. Soc. Am. Bull.* **114**, 349–366.

Marian G. M., Schlesinger W. H., and Fonteyn P. J. (1985) Caldep: a regional model for soil $CaCO_3$ (calcite) deposition in southwestern deserts. *Soil Sci.* **139**, 468–481.

McFadden L. D. and Tinsley J. C. (1985) Rate and depth of pedogenic-carbonate accumulation in soils: formation and testing of a compartment model. In *Quaternary Soils and Geomorphology of the Southwestern United States.* (ed. D. L. Weide). Geological Society of America Special Paper 203. Geological Society of America, Boulder, pp. 23–42.

McFadden L. D., Amundson R. G., and Chadwick O. A. (1991) Numerical modeling, chemical, and isotopic studies of carbonate accumulation in soils of arid regions. In *Occurrence, Characteristics, and Genesis of Carboante, Gypsum, and Silica Accumulations in Soils* (ed. W. D. Nettleon). Soil Science Society of America, Madison, WI, pp. 17–35.

McKean J. A., Dietrich W. E., Finkel R. C., Southon J. R., and Caffee M. W. (1993) Quantification of soil production and downslope creep rates from cosmogenic ^{10}Be accumulations on a hillslope profile. *Geology* **21**, 343–346.

Merritts D. J., Chadwick O. A., Hendricks D. M., Brimhall G. H., and Lewis C. J. (1992) The mass balance of soil evolution on late Quaternary marine terraces, northern California. *Geol. Soc. Am. Bull.* **104**, 1456–1470.

Miller A. J., Amundson R., Burke I. C., and Yonker C. (2003) The effect of climate and cultivation on soil organic C and N. *Biogeochemistry* (in press).

Mook W. G., Koopmans M., Carter A. F., and Keeling C. D. (1983) Seasonal, latitudinal, and secular variations in the abundance and isotopic ratios of atmospheric carbon dioxide: 1. Results from land stations. *J. Geophys. Res.* **88**(C15), 10915–10933.

Mora C. I., Driese S. G., and Colarusso L. A. (1996) Middle to late Paleozoic atmospheric CO_2 levels from soil carbonate and organic matter. *Science* **271**, 1105–1107.

Nadelhoffer K. and Fry B. (1988) Controls on the nitrogen-15 and carbon-13 abundances in forest soil organic matter. *Soil Sci. Soc. Am. J.* **52**, 1633–1640.

Nakai S., Halliday A. N., and Rea D. K. (1993) Provenance of dust in the Pacific Ocean. *Earth Planet. Sci. Lett.* **119**, 143–157.

Navarro-Gonzalez R., McKay C. P., and Mvondo D. N. (2001) A possible nitrogen crisis for Archaean life due to reduced nitrogen fixation by lightning. *Nature* **412**, 61–64.

Nettleton W. D., Flach K. W., and Borst G. (1968) *A Toposequence of Soils on Grus in the Southern California Peninsular Range*. US Dept. Agric. Soil Cons. Serv., Soil Surv. Invest. Rep. No. 21, 41p.

O'Brien J. B. and Stout J. D. (1978) Movement and turnover of soil organic matter as indicated by carbon isotope measurements. *Soil Biol. Biochem.* **10**, 309–317.

Parton W. J., Schimel D. S., Cole C. V., and Ojima D. S. (1987) Analysis of factors controlling soil organic matter levels in Great Plains grasslands. *Soil Sci. Soc. Am. J.* **5**, 1173–1179.

Peterson F. D. (1980) Holocene desert soil formation under sodium salt influence in a playa-margin environment. *Quat. Res.* **13**, 172–186.

Quade J., Cerling T. E., and Bowman J. R. (1989a) Systematic variations in the carbon and oxygen isotopic composition of pedogenic carbonate along elevation transects in the southern Great Basin, United States. *Geol. Soc. Am. Bull.* **101**, 464–475.

Quade J., Cerling T. E., and Bowman J. R. (1989b) Development of the Asian monsoon revealed by marked ecological shift during the latest Miocene in northern Pakistan. *Nature* **342**, 163–166.

Raich J. W. and Schlesinger W. H. (1992) The global carbon dioxide flux in soil respiration and its relationship to vegetation and climate. *Tellus* **B44**, 48–51.

Rea D. K. (1994) The paleoclimatic record provided by eolian deposition in the deep sea: the geological history of wind. *Rev. Geophys.* **5**, 193–259.

Rech J. A., Quade J., and Hart B. (2002) Isotopic evidence for the source of Ca and S in soil gypsum, anhyrite and calcite in the Atacama desert, Chile. *Geochim. Cosmochim. Acta* **67**(4), 575–586.

Rosenblum N. A., Doney S. C., and Schimel D. S. (2001) Geomorphic evolution of soil texture and organic matter in eroding landscapes. *Global Biogeochem. Cycles* **15**, 365–381.

Royer D. L. (1999) Depth to pedogenic carbonate horizon as a paleoprecipitation indicator? *Geology* **27**, 1123–1126.

Rozanski K., Araguás- Araguás L., and Gonfiantini R. (1993) Isotopic patterns in modern global precipitation. In *Climate Change in Continental Isotopic Records*. American Geophysical Union Monograph 78 (ed. P. K. Swart, *et al.*). American Geophysical Union, Washington, DC, pp. 1–36.

Russell R. S. (1977) *Plant Root Systems: Their Function and Interaction with the Soil*. McGraw-Hill, London.

Schlesinger W. H. and Andrews J. A. (2000) Soil respiration and the global carbon cycle. *Biogeochemistry* **48**, 7–20.

Severinghaus J. P., Bender M. L., Keeling R. F., and Broecker W. S. (1996) Fractionation of soil gases by diffusion of water vapor, gravitational settling, and thermal diffusion. *Geochim. Cosmochim. Acta* **60**, 1005–1018.

Shacklette H. T. and Boerngen J. G. (1984) Element concentrations in soils and other surficial materials of the conterminous United States. US Geological Survey Professional Paper 1270.

Soil Survey Staff (1975) *Soil Taxonomy: A Basic System of Soil Classification for Making and Interpreting Soil Surveys*, Agr. Handbook 426. US Government Printing Office, Washington, DC.

Soil Survey Staff (1993) *Soil Survey Manual*. USDA Handbook No. 18. US Government Printing Office, Washington, DC.

Soil Survey Staff (1995) *Soil Survey Laboratory Information Manual*. Soil Survey Investigations Report No. 45, Version 1.0, US Government Printing Office, Washington, DC.

Soil Survey Staff (1999) *Keys to Soil Taxonomy*, 8th edn. Pocahontas Press, Blacksburg, VA.

Stallard R. F. (1998) Terrestrial sedimentation and the carbon cycle: coupling weathering and erosion to the carbon cycle. *Global Biogeochem. Cycles* **12**, 231–257.

Stern L. A., Baisden W. T., and Amundson R. (1999) Processes controlling the oxygen-isotope ratio of soil CO_2: analytic and numerical modeling. *Geochim. Cosmochim. Acta* **63**, 799–814.

Stern L. A., Amundson R., and Baisden W. T. (2001) Influence of soils on oxygen-isotope ratio of atmospheric CO_2. *Global Biogeochem. Cycles* **15**, 753–759.

Sundquist E. T. (1993) The global carbon dioxide budget. *Science* **259**, 934–941.

Sutter B., Amundson R., Ewing S., Rhodes K. W., and McKay C. W. (2002) The chemistry and mineralogy of Atacama Desert soils: a possible analog for Mars soils. American Geophysical Union Fall Meeting, San Francisco, pp. 71A–0443.

Tans P. P. (1998) Oxygen isotopic equilibrium between carbon dioxide and water in soils. *Tellus* **B50**, 163–178.

Taylor S. R. and McLennan S. M. (1985) *The Continental Crust: Its Composition and Evolution*. Blackwell, Oxford.

Thiemens M. H. (1999) Mass-independent isotope effects in planetary atmospheres and the early solar system. *Science* **283**, 341–345.

Thorstenson D. C., Weeks E. P., Haas H., and Fisher D. W. (1983) Distribution of gaseous $^{12}CO_2$, $^{13}CO_2$, and $^{14}CO_2$ in sub-soil unsaturated zone of the western US Great Plains. *Radiocarbon* **25**, 315–346.

Tréguer P., Nelson D. M., Van Bennekom A. J., DeMaster D. J., Leynaert A., and Quéguiner (1995) The silica balance in the world ocean: a reestimate. *Science* **268**, 375–379.

Trumbore S. E. (2000) Age of soil organic matter and soil respiration: radiocarbon constraints on belowground dynamics. *Ecol. Appl.* **10**, 399–411.

Ugolini F. C. and Dahlgren R. (1987) The mechanism of podzolization as revealed by soil solution studies. In *Podzols et Podolization* (eds. D. Fighiand and A. Chavell). Assoc. Fr. Estude Sol, Plassier, France, pp. 195–203.

Vil'yams V. R. (1967) V. V. Dokuchaev's role in the development of soil science. In *Russian Chernozem*. (translated by Israel Program for Scientific Translations, Jerusalem).

Vitousek P. M., Aber J. D., Howarth R. W., Likens G. E., Matson P. A., Schindler D. W., Schlesinger W. H., and Tilman D. G. (1997a) Human alterations of the global nitrogen cycle: sources and consequences. *Ecol. Appl.* **7**, 737–750.

Vitousek P. M., Chadwick O. A., Crews T. E., Fownes J. H., Hendricks D. M., and Herbert D. (1997b) Soil and ecosystem development across the Hawaiian Islands. *GSA Today* **7**, 1–8.

Wang Y., Amundson R., and Trumbore S. (1994) A model for soil $^{14}CO_2$ and its implications for using ^{14}C to date pedogenic carbonate. *Geochim. Cosmochim. Acta* **58**, 393–399.

Wang Y., Amundson R., and Niu X.-F. (2000) Seasonal and altitudinal variation in decomposition of soil organic matter inferred from radiocarbon measurements of soil CO_2 flux. *Global Biogeochem. Cycles* **14**, 199–211.

White A. F., Blum A. E., Schulz M. S., Bullen T. D., Harden J. W., and Peterson M. L. (1996) Chemical weathering rates of a soil chronosequence on granitic alluvium: I. Quantification of mineralogical and surface area changes and calculation of primary silicate reaction rates. *Geochim. Cosmochim. Acta* **60**, 2533–2550.

5.02
Modeling Low-temperature Geochemical Processes

D. K. Nordstrom

US Geological Survey, Boulder, CO, USA

A model takes on the quality of theory when it abstracts from raw data the facts that its inventor perceives to be fundamental and controlling, and puts these into relation to each other in ways that were not understood before—thereby generating predictions of surprising new facts.

H. F. Judson (1980), *The Search for Solutions*

5.02.1 INTRODUCTION

Geochemical modeling has become a popular and useful tool for a wide number of applications from research on the fundamental processes of water–rock interactions to regulatory requirements and decisions regarding permits for industrial and hazardous wastes. In low-temperature environments, generally thought of as those in the temperature range of 0–100 °C and close to atmospheric pressure (1 atm = 1.01325 bar = 101,325 Pa), complex hydrobio-geochemical reactions participate in an array of interconnected processes that affect us, and that, in turn, we affect. Understanding these complex processes often requires tools that are sufficiently sophisticated to portray multicomponent, multiphase chemical reactions yet transparent enough to reveal the main driving forces. Geochemical models are such tools. The major processes that they are required to model include mineral dissolution and precipitation; aqueous inorganic speciation and complexation; solute adsorption and desorption; ion exchange; oxidation–reduction; or redox; transformations; gas uptake or production; organic matter speciation and complexation; evaporation; dilution; water mixing; reaction during fluid flow; reaction involving biotic interactions; and photoreaction. These processes occur in rain, snow, fog, dry atmosphere, soils, bedrock weathering, streams, rivers, lakes, groundwaters, estuaries, brines, and diagenetic environments. Geochemical modeling attempts to understand the redistribution of elements and compounds, through anthropogenic and natural means, for a large range of scale from nanometer to global. "Aqueous geochemistry" and "environmental geochemistry" are often used interchangeably with "low-temperature geochemistry" to emphasize hydrologic or environmental objectives.

Recognition of the strategy or philosophy behind the use of geochemical modeling is not often discussed or explicitly described. Plummer (1984, 1992) and Parkhurst and Plummer (1993) compare and contrast two approaches for modeling groundwater chemistry: (i) "forward modeling," which predicts water compositions from hypothesized reactions and user assumptions and (ii) "inverse modeling," which uses water, mineral, and isotopic compositions to constrain hypothesized reactions. These approaches simply reflect the amount of information one has to work with. With minimal information on a site, a modeler is forced to rely on forward modeling. Optimal information would include detailed mineralogy on drill cores or well cuttings combined with detailed water analyses at varying depths and sufficient spatial distribution to follow geochemical reactions and mixing of waters along defined flow paths. With optimal information, a modeler will depend on inverse modeling.

This chapter outlines the main concepts and key developments in the field of geochemical modeling for low-temperature environments and illustrates their use with examples. It proceeds with a short discussion of what modeling is, continues with concepts and definitions commonly used, and follows with a short history of geochemical models, a discussion of databases, the codes that embody models, and recent examples of how these codes have been used in water-rock interactions. An important new stage of development seems to have been reached in this field with questions of reliability and validity of models. Future work will be obligated to document ranges of certainty and sources of uncertainty, sensitivity of models and codes to parameter errors and assumptions, propagation of errors, and delineation of the range of applicability.

5.02.1.1 What is a Model?

A "model" has a relation to "reality," but reality is experiential, and the moment we attempt to convey the experience we are seriously constrained by three limitations: our own ability to communicate, the capability of the communication medium to portray the experience, and the ability of the person receiving the communication to understand it. Communication requires the use of language and the use of percepts and concepts, and the process of perceiving, conceiving, and communicating is an abstracting process that removes us from the immediate experience. Any attempt to transform an experience into language suffers from this abstraction process so that we lose the immediate experience to gain some understanding of it. The problem is to transform existential knowledge to communicated or processed knowledge. To communicate knowledge, we must use a simplified and abstract symbolism (words, mathematics, pictures, diagrams, analogues, allegories, three-dimensional physical constructs, etc.) to describe a material object or phenomenon, i.e., a model. Models take on many forms but they are all characterized by being a simplification or idealization of reality. "... models include only properties and relationships that are needed to understand those aspects of the real system that we are interested in..." (Derry, 1999). Assumptions, both explicit and implicit, are made in model construction, because either we do not know all the necessary properties or we do not need to know them. Hopefully, models have caught the essential properties of that which they are attempting to portray so that they are useful and lead to new insights and to better answers for existing questions and problems.

Scientific models begin as ideas and opinions that are formalized into a language, often, but not necessarily, mathematical language. Furthermore, they must have testable consequences. The emphasis on testability, or falsifiability, is a fundamental attribute (Popper, 1934).

Greenwood (1989) stated that the word model should be reserved "… for well-constrained logical propositions, not necessarily mathematical, that have necessary and testable consequences, and avoid the use of the word if we are merely constructing a scenario of possibilities." A scientific model is a testable idea, hypothesis, theory, or combination of theories that provide new insight or a new interpretation of an existing problem. An additional quality often attributed to a model or theory is its ability to explain a large number of observations while maintaining simplicity (Occam's razor). The simplest model that explains the most observations is the one that will have the most appeal and applicability.

Models are applied to a "system," or a portion of the observable universe separated by well-defined boundaries for the purpose of investigation. A chemical model is a theoretical construct that permits the calculation of chemical properties and processes, such as the thermodynamic, kinetic, or quantum mechanical properties of a system. A geochemical model is a chemical model developed for geologic systems. Geochemical models often incorporate chemical models such as ion association and aqueous speciation together with mineralogical data and assumptions about mass transfer to study water–rock interactions.

A computer code is obviously not a model. A computer code that incorporates a geochemical model is one of several possible tools for interpreting water–rock interactions in low-temperature geochemistry. The computer codes in common use and examples of their application will be the main focus of this chapter. It is unfortunate that one commonly finds, in the literature, reference to the MINTEQ model or the PHREEQE model or the EQ3/6 model when these are not models but computer codes. Some of the models used by these codes are the same so that a different code name does not necessarily mean a different model is being used.

5.02.2 MODELING CONCEPTS AND DEFINITIONS

5.02.2.1 Modeling Concepts

Many different forms of models are utilized, usually dictated by the objectives of research. *Conceptual models* are the most fundamental. All of us have some kind of concept of water–rock interactions. For a groundwater interacting with the aquifer minerals during its evolution, one might conceive that most minerals would be undersaturated in the area of recharge but that some minerals (those that dissolve fastest) would become saturated at some point down gradient, having reached their equilibrium solubility

(a state of *partial equilibrium*). The conceptual model can be formalized into a set of mathematical equations using chemical principles, the *mathematical model*, entered into a computer program, the *code*, and predictions made to test the assumptions (and the databases) against the results from real field data. This exercise helps to quantify and constrain the possible reactions that might occur in the subsurface. Mathematical equations for complex interacting variables are not always solved exactly and, therefore, systems of numerical approximations, or *numerical models*, have been developed. Alternatively, an experiment could be set up in the laboratory with columns made of aquifer material with groundwater flowing through to simulate the reactions, the *experimental model* or *scale model*. Having obtained results from a mathematical or scale model, some unexpected results often occur which force us to change our original conceptual model. This example demonstrates how science works; it is an ongoing process of approximation that iterates between idea, theory, observational testing of theory, and back to modifications of theory or development of new theories.

5.02.2.2 Modeling Definitions

In low-temperature geochemistry and geochemical modeling, it is helpful to define several words and phrases in common use.

Aqueous speciation. The distribution of dissolved components among free ions, ion pairs, and complexes. For example, dissolved iron in acid mine drainage (AMD) can be present as $Fe_{(aq)}^{2+}$ (free ferrous iron), $FeSO_{4(aq)}^{0}$ (ion pair), $Fe_{(aq)}^{3+}$ (free ferric iron), $Fe(OH)_{(aq)}^{2+}$, and $FeSO_{4(aq)}^{+}$ species. These species are present in a single phase, aqueous solution. Aqueous speciation is not uniquely defined but depends on the theoretical formulation of mass action equilibria and activity coefficients, i.e. it is model dependent. Some aqueous speciation can be determined analytically but operational definitions and assumptions are still unavoidable.

Phase. Commonly defined as a uniform, homogeneous, physically distinct, and mechanically separable part of a system. Unfortunately, mineral phases are often not uniform, homogeneous, or mechanically separable except in theory. Sophisticated microscopic and spectroscopic techniques and operational definitions are needed to define some mineral phases.

Phase speciation. The distribution of components among two or more phases. In a wet soil, iron can be present in at least three phases: as the X-ray amorphous oxyhydroxide, ferrihydrite; as goethite; and as dissolved aqueous iron.

Mass transfer. The transfer of mass between two or more phases, for example, the precipitation and dissolution of minerals in a groundwater.

Reaction-path calculation. A sequence of mass transfer calculations that follows defined phase (or reaction) boundaries during incremental steps of reaction progress.

Mass transport. Solute movement by mass flow of a fluid (could be liquid, gas, or mixture of solid and liquid and/or gas).

Reactive transport. Mass transfer combined with mass transport; commonly refers to geo-chemical reactions during stream flow or groundwater flow.

Forward geochemical modeling. Given an initial water of known composition and a rock of known mineralogy and composition, the rock and water are computationally reacted under a given set of conditions (constant or variable temperature, pressure, and water composition) to produce rock and water (or set of rocks and waters). In forward modeling the products are inferred from an assumed set of conditions (equilibrium or not, phases allowed to precipitate or not, etc.) and thermodynamic and/or kinetic data are necessary.

Inverse geochemical modeling. Given a set of two or more actual water analyses along a flow path that have already reacted with a rock of known mineralogy, the reactions are inferred. In inverse modeling the reactions are computed from the known field conditions (known water chemistry evolution and known mineralogy) and assumed mineral reactivity. Inverse modeling is based on mass or mole balancing, and thermodynamic and kinetic data are not necessary although saturation indices provide useful constraints. Relativistic kinetics (i.e. relative mineral reaction rates) are deduced as part of the results.

5.02.2.3 Inverse Modeling, Mass Balancing, and Mole Balancing

A key concept to interpreting water–rock interactions is mass balancing between water chemistry data, mineral/gas transformations, and biological uptake or release. The mass-balance concept is a means of keeping track of reacting phases that transfer mass during fluid flow. It is an integral part of the continuity equation for steady-state and transient-state flow conditions. Formalizing mass balances for solutes during fluid flow is simple if the solutes are conservative but more complex when reactions take place. For applications to the environment, different types of mass balances have been used and these require some explanation.

Three types of mass-balance approaches will be discussed. In its simplest form, mass balancing is utilizing the law of conservation of mass that reflects that any partitioning of mass through a system must sum to the total. The first type of mass-balance approach, catchment mass balances (or mass fluxes), use "input–output" accounting methods to identify overall gains and losses of solutes during the flow of water from meteoric inputs to stream outflow at a specified point (Bormann and Likens, 1967; Likens *et al.*, 1977). Pačes (1983, 1984) called this the integral mass-balance approach:

$$[\text{Mass rate}_{in} - \text{mass rate}_{out}] \pm \text{mass rate}_{internal}$$
$$= \text{rate of accumulation/depletion} \qquad (1)$$

In difference form

$$\frac{\Delta m_{trans}}{\Delta t} + \frac{\Delta m_{int}}{\Delta t} = \frac{\Delta m_{tot}}{\Delta t} \qquad (2)$$

where $\Delta m_{trans}/\Delta t$ is the difference in rate of mass (of some component) transported into and out of the system via fluid flow, $\Delta m_{int}/\Delta t$ is the rate of mass produced or removed from the fluid by internal chemical reaction, and $\Delta m_{tot}/\Delta t$ is the total rate of change in mass for a component in the fluid. Steady-state conditions are often assumed so that the accumulation/depletion rate becomes zero and the difference between input and output for a component becomes equal to the internal processes that produce or remove that component. These mass rates of change are usually obtained as integral mass fluxes (computed as discharge times concentration).

For catchments in which input from precipitation and dry deposition can be measured and the output through a weir or similar confining outflow can be measured, the difference in solute flow averaged over an appropriate period of time provides an accounting of solutes gained or lost. Overviews and summaries of essential concepts of catchment mass balances have been presented by Drever (1997a,b), Bassett (1997), and Bricker *et al.* (Chapter 5.04). The input–output mass balance treats the catchment as a black box, but knowledge of biogeochemical processes, use of lab experiments, additional field data, and well-controlled small-scale field-plot studies can help to identify the dominant processes causing the gain or loss of a particular solute. Load (or mass flux) calculations for a river or stream use the same type of accounting procedure.

A second type of mass-balance approach is quantitative incorporation of mass balances within a reactive-transport model and could be applied to groundwaters, surface waters, and surface-water–groundwater interactions. Pačes (1983, 1984) calls this the local mass-balance approach. There are numerous examples and explanations of this approach (e.g., Freeze and Cherry, 1979; Domenico and Schwartz, 1990).

A third type of mass-balance approach is to leave the physical flow conditions implicit, assume steady-state flow along a flow line in an aquifer, and account for observed chemical changes that occur as the groundwater flows from recharge to discharge (Plummer and Back, 1980). This approach was introduced formally by the classic paper of Garrels and Mackenzie (1967), who calculated mass balances derived from springwater data in the Sierra Nevada Mountains of California. The modeler begins with aqueous chemical data along a flow path in an aquifer (or catchment) of known mineralogy and accounts for changes in solute concentrations by specified reaction sources and sinks. The analytical solution is achieved by solving a set of simultaneous linear equations known as mole balancing:

$$\sum_{j=1}^{j} b_p \alpha_p^i = \Delta m_i = m_{i(\text{final})} - m_{i(\text{initial})} \quad (3)$$

in which m_i is the number of moles of element/component i per kilogram of water, α_p^i is the stoichiometric coefficient of element i in phase p, b_p is the mass transfer coefficient of the pth phase, and j is the number of phases (Plummer *et al.*, 1983). Because groundwaters are frequently mixtures of different water quality types, mixing fractions can also be employed in the matrix array. Redox chemistry can be included through conservation of electrons and water balance can be used to simulate evaporation. Plummer (1984) has called this type of modeling the inverse method, because it proceeds in a reverse manner to that of forward modeling, i.e., it backs out the probable reactions from known data on water chemistry, isotopes, and mineralogy. Parkhurst (1997) simply refers to it as mole balancing. Semantics aside, this model is developed for water–rock interactions from "a set of mixing fractions of initial aqueous solutions and mole transfers of minerals and gases that quantitatively account for the chemical composition of the final solution" (Parkhurst, 1997).

Two papers point out the fundamental nature of mole balancing for interpreting groundwater chemistry. Plummer (1984) describes attempts to model the Madison Limestone aquifer by both forward and inverse methods. Two valuable conclusions from this paper were that information gained from mole balancing was needed to provide relativistic reaction rates for forward modeling and that even with these rates, the well-water compositions could not be closely matched, whereas they can be exactly matched with inverse modeling. Furthermore forward modeling required much more effort and contained more uncertainty. Similar conclusions were reached by Glynn and Brown (1996) in their detailed study of the acidic plume moving in the Pinal Creek Basin. They found that the best approach was to pursue a series of mole-balance calculations, improving their modeling with each re-examination of the phases and reactions considered. Then they took their refined mole-balance results and made further improvements by forward modeling. Inverse modeling, however, was a key to their successful interpretation. The complex nature of interpreting a dispersed contaminant plume and the associated mineral dissolution front required the use of reactive-transport modeling. The reactive-transport modeling, however, would have been much more uncertain without the inverse modeling at the front end.

It has been pointed out that inverse modeling assumes advective, steady-state flow and that "reaction inversion" does not occur (Steefel and Van Cappellen, 1998). Although these are important issues, they are often not serious limitations. These assumptions are not a firm requirement for the system being studied, it is only essential that the consequences of assuming them for non-steady-state flow (with or without dispersion) do not have a significant effect on the results of the calculation. For example, it can be argued that over a long enough period of time or for a large enough aquifer, steady-state conditions never truly apply. Indeed, steady state is an approximation of the physical and chemical state of an aquifer system that works well for many, if not most, aquifers. As Zhu and Anderson (2002) pointed out, the most important criterion is that the same parcel of water has traveled from point A to point B. They also state that chemical steady state is sufficient but not necessary "because the underlying mathematical equations to be solved... are the integrated forms of mass conservation (Lichtner, 1996)." At the Iron Mountain mines (Alpers *et al.*, 1992), high concentrations of sulfate (tens of grams per liter) and metals (grams per liter) discharge from two major portals. The groundwater has advective and nonadvective (dispersive and convective) properties, and the variable effluent flows would indicate transient state. The chemistry, however, is still dominated by oxidation of sulfide minerals and acid dissolution of aluminosilicate minerals, and a mass-balance calculation will still reflect these geochemical processes. A mass balance going from rainwater to portal effluent will delineate the key minerals that dissolve and precipitate along the flow path. Steefel and Van Cappellen (1998) correctly state that inverse modeling cannot be applied to contaminant plumes or reaction fronts unless the spatial delineation of the reaction front is clearly defined. "Reaction inversion," such as the dissolution of a particular mineral changing to precipitation between two points along a flow path such that no overall reaction appears to have taken place, is an interesting problem. Without additional information it would not be possible to distinguish between no reaction and reaction inversion. In such instances, it makes no difference to the modeling or the conclusions, because the net result is the same. It is as if the element of concern was conserved in the solid phase during that increment of fluid flow, a safe assumption for silica in quartz but not for silica in weathering feldspars. It might only matter if these small-scale processes affect other processes such as the unidirectional release or attenuation of trace elements (released upon mineral dissolution but not co-precipitated or adsorbed upon mineral precipitation or vice versa) or similar irreversible processes involving nutrients. Further discussion of inverse modeling is found in Sections 5.02.7.1 and 5.02.8.3.

5.02.3 SOLVING THE CHEMICAL EQUILIBRIUM PROBLEM

Although both kinetic and equilibrium expressions can be used in geochemical modeling computations, the "equilibrium problem" is the foundation of most calculations of this type. Simply stated, the chemical equilibrium problem is the determination of the most stable state of a system for a given set of pressure (P), temperature (T), and compositional constraints (X_i). These variables, P, T, and X_i, need not be fixed, but they must be defined for the given system. The chemical state of a system is given by the total Gibbs free energy, G, and its differential change with the progress variable, ξ, denotes the state of the system in mathematical terms. For a system at equilibrium

$$\left(\frac{\partial G}{\partial \xi}\right)_{P,T} = 0 \qquad (4)$$

and any perturbation from equilibrium will cause this differential to be other than 0. The progress variable is a singular quantity that expresses the extent to which a reaction has taken place. It is equal to the change in the number of moles, n_i, of a reactant (or product) normalized to the stoichiometric coefficient, ν_i, for that component, element, or species. In differential form,

$$\partial \xi = \frac{\partial n_i}{\nu_i} \qquad (5)$$

Solving the equilibrium problem means finding the minimum in the free energy curve for the defined system. Two general approaches have been used: the equilibrium constant approach and the free-energy minimization approach. As the names suggest they primarily use equilibrium constants to solve the problem or they use free energies. Of course, the approaches are considered the same because they are related by $\Delta_r G^\circ = RT \ln K$, where R is the universal gas constant, T is the absolute temperature in kelvin, and K is the equilibrium constant, but the logarithmic conversion leads to different numerical techniques. Both approaches have to employ mass-balance and mass-action equations. Upon substitution of mass-action into mass-balance expressions (or vice versa), a set of nonlinear equations is derived that are readily solved by a numerical method coded for a computer. The mathematical formulation of the chemical equilibrium problem is explained in Zeleznik and Gordon (1968), Van Zeggeren and Storey (1970), Wolery (1979), and Smith and Missen (1982). Harvie and Weare (1980) and Harvie *et al.* (1987) have improved the free-energy minimization algorithm for solving the general chemical equilibrium problem and applied it to finding parameters for the Pitzer method. Rubin (1983, 1990) provides more formalism for dealing with reactive transport and nonequilibrium conditions. Further discussion of the numerical techniques can be found in the books by Smith and Missen (1982), Nordstrom and Munoz (1994), and many other textbooks and papers. A simple problem involving speciation of an aqueous solution at equilibrium with gypsum is solved by both the Newton–Raphson method and the "continued fraction" method in Nordstrom and Munoz (1994).

5.02.4 HISTORICAL BACKGROUND TO GEOCHEMICAL MODELING

The founders of geochemistry, F. W. Clarke, G. M. Goldschmidt, V. I. Vernadsky, and A. E. Fersman, clearly made major contributions to our "models" of geochemical processes. Today, however, we think of geochemical models in terms of high-speed quantitative computations of water–rock interactions made possible with fast processors and disks with large memories. This direction of research began with the principles of physical chemistry adapted for solving aqueous low-temperature geochemical problems by Hem (1959), Garrels (1960), Sillén (1961), Garrels and Christ (1965), and Krauskopf (1967). The scholarly structures that these authors built would not have been possible without the foundations laid by those who came before and developed the science of physical chemistry: Wilhelm Ostwald, Jacobius Henricus van't Hoff, and Svante Arrhenius (Servos, 1990). These chemists, along with numerous others they influenced, established the principles upon which aqueous low-temperature geochemistry and geochemical modeling is based. Van't Hoff himself was an early developer of geochemical models in his efforts to apply physical chemistry to the interpretation of the equilibrium factors determining the stability of gypsum and anhydrite (van't Hoff *et al.*, 1903) and the equilibrium interpretation of the Permian Zechstein evaporite deposits (van't Hoff (1905, 1909), 1912; Eugster, 1971).

Probably the earliest paper to apply a speciation calculation to natural water was Goldberg's (1954) estimate of the ionic species form of copper in seawater based on Niels Bjerrum's theory of ion-pair formation and Jannik Bjerrum's data (see Bjerrum, 1950). He found that 65% was in the free ion (Cu^{2+}) form and another 33% was present as either a cation or neutral species form (copper-chloride complexes). This calculation, however, did not take into account simultaneous competitive complexing from all the other ions in seawater. Speciation was central to Krauskopf's (1951) study on gold solubility, his discussion of trace-metal enrichment in sedimentary deposits (Krauskopf, 1955), and his examination of the factors controlling trace-metal concentrations in seawater (Krauskopf, 1956). In the US Geological Survey (USGS), Hem and Cropper (1959) and Hem (1961) were estimating distribution of ionic species and activities of ions at chemical equilibrium in natural waters. Sillén's (1961) first paper on seawater considered hydrolysis and complex formation. At this time he and his colleagues were pioneering the application of computers to solve complex aqueous solution equilibria (Sillén, 1962; Ingri and Sillén, 1962; Ingri *et al.*, 1967; Dyrssen *et al.*, 1968). The next paper to apply aqueous speciation to natural water was that of Garrels and Thompson (1962) in their "Chemical Model of Seawater," which contained unique features such as the mean-salt method for estimating activity coefficients.

The adaptation of water–rock geochemical modeling to computers was pioneered by Helgeson and his colleagues, who focused on ore deposits and high-temperature, high-pressure reactions (Helgeson, 1964, 1967, 1969) but also considered low-temperature processes (Helgeson *et al.*, 1969, 1970). Helgeson's (1971) research continued with the application of computers to the calculation of geochemical kinetics. Theoretical approaches were developed for geochemical kinetics (Aagard and Helgeson, 1982; Helgeson and Murphy, 1983) and computations compared to measured reaction rates (Helgeson *et al.*, 1984; Murphy and Helgeson, 1987) and coupled with diffusion (Murphy *et al.*, 1989). These approaches provide insight to the mechanisms of controlled laboratory studies on mineral dissolution rates, but difficulties are encountered when applications are made to the complex conditions of the natural environment. Experimental and theoretical approaches to mineral dissolution fail to predict

the actual dissolution rates in catchments, because experiments do not reproduce the composition and structure of natural mineral surfaces, they do not account for adsorbed inhibitors, they do not consider the effect of variable climate and hydrologic flow rates, and they do not account for biological activity. These factors may have taken anywhere from decades to millenia to develop and they can change over short periods of time. Computer codes that incorporate reaction kinetics into groundwater or catchment codes usually do so with some type of generalized first-order rate law.

Beginning in the 1960s and expanding considerably during the 1970s, a series of computer codes were developed that could perform a wide variety of aqueous geochemical calculations (Bassett and Melchior, 1990; Mangold and Tsang, 1991) and these evolved considerably during the 1980s and 1990s to include simulation of more processes, coupling of processes (especially reaction with transport), and the availability of more options (Alpers and Nordstrom, 1999). The advances in geochemical modeling are apparent in the evolution of Barnes' "*Geochemistry of Hydrothermal Ore Deposits*" from first edition (Barnes, 1967) to third edition (Barnes, 1997), in the American Chemical Society (ACS) books *Chemical Modeling in Aqueous Systems* that have shown significant advances from the first (Jenne, 1979) to the second (Melchior and Bassett, 1990), and in the plethora of papers on this subject since the 1980s. Nevertheless, advanced sophistication for geochemical codes does not imply a parallel advance in our understanding of geochemical processes. The sophistication of software has outdistanced our capacity to evaluate the software over a range of conditions and it has outdistanced our ability to obtain the field data to constrain and test the software.

5.02.5 THE PROBLEM OF ACTIVITY COEFFICIENTS

5.02.5.1 Activity Coefficients

Aqueous electrolytes and the equilibrium constants that define various reactions in low-temperature geochemistry are inexorably linked with the problem of activity coefficients, or the problem of nonideality for aqueous electrolyte solutions. Thermodynamic equilibrium constants, defined by an extrapolation to infinite dilution for the standard state condition (not the only standard state), require the use of activity coefficients. Unfortunately, there is neither a simple nor universal nonideality method that works for all electrolytes under all conditions. This section provides a brief overview of a major subject still undergoing research and development but for which several satisfactory approaches are available.

If the activity of a solute or ion were ideal, it could be taken as equivalent to the molal concentration, m_i, of the i ion or solute. However, interactions with other ions and with the solvent water are strong enough to cause nonideal behavior and the characteristic property relating concentration to chemical potential is the activity coefficient, γ_i:

$$a_i = \gamma_i \, m_i \qquad (6)$$

With the accumulation of activity coefficient measurements and the search for a theoretical expression ensued, it was discovered that in dilute solutions the logarithm of the activity coefficient was an approximate function of the square root of the molality:

$$\log \gamma_i = \alpha_i m_i^{1/2} \qquad (7)$$

where α_i was simply a constant. Brønsted (1922) introduced a linear term to Equation (7) in which the coefficient, β, is a "specific-ion interaction parameter,"

$$\log \gamma_i = \alpha_i m_i^{1/2} + \beta m_i \qquad (8)$$

Later modifications of this general approach became known as the specific-ion interaction theory (SIT) because of the explicit dependence on the solute or ions being considered.

An important concept that aided in the development of electrolyte theory was the ionic strength, I, introduced by Lewis and Randall (1921):

$$I = \tfrac{1}{2} \sum m_i z_i^2 \qquad (9)$$

where z_i is the ionic charge. Thereafter, the ionic strength was used as a parameter in activity coefficient equations. Debye and Hückel (1923) derived an equation from electrostatic theory which, in the limit of infinite dilution (hence, the Debye–Hückel limiting law), becomes

$$\log \gamma_i = -A z_i^2 I^{1/2} \qquad (10)$$

where A is a Debye–Hückel solvent parameter (dependent on the properties of the solvent).

The extended Debye–Hückel equation is

$$\log \gamma_i = -\frac{A z_i^2 I^{1/2}}{1 + B a_i I^{1/2}} \qquad (11)$$

where B is another Debye–Hückel solvent parameter and a_i is an ion-size diameter derived by empirical fit but approximates the hydrated ion diameter.

The use of the extended Debye–Hückel equation with the appropriate equilibrium constants for mass action expressions to solve a complex chemical equilibrium problem is known as the ion-association (IA) method.

Several modifications of these equations were tried over the next several decades. Contributions by Guggenheim (1935) and Scatchard (1936) and the KTH or Swedish group (Grenthe *et al.*, 1992, 1997) became the Brønsted–Guggenheim–Scatchard or SIT method:

$$\log \gamma_i = -\frac{A z_i^2 I^{1/2}}{1 + 1.5 I^{1/2}} + \sum_k \varepsilon(i, k, I) m_k \qquad (12)$$

in which the linear term is a function of molality summed for all the other ions in solutions, k, and $\varepsilon(i, k)$ tends to be constant at higher molalities but solute or ion specific. The last term in Equation (12) represents the deviation of the experimentally measured activity coefficient from the prediction of the Debye–Hückel equation. Other groups, such as Truesdell and Jones (1974), kept the ionic strength dependence throughout the equation, and the coefficient, b, became the difference factor but still a constant for a given solute or ion:

$$\log \gamma_i = -\frac{A z_i^2 I^{1/2}}{1 + B a_i I^{1/2}} + b_i I \qquad (13)$$

The IA method using Equation (13) for major ions in natural water compares well with other more precise methods (see below) up to an ionic strength of seawater (0.7) but not much higher.

The next better approximation is to allow the linear coefficient to be a function of the ionic strength (Pitzer and Brewer, 1961) and was utilized by Helgeson and Kirkham (1974a,b, 1976), Helgeson *et al.* (1981), and Tanger and Helgeson (1988) in the development of the Helgeson–Kirkham–Flowers (HKF) equations, which include consideration of the Born function and ion hydration.

Pitzer (1973) re-examined the statistical mechanics of aqueous electrolytes in water and derived a different but semi-empirical method for activity coefficients, commonly termed the Pitzer specific-ion-interaction model. He fitted a slightly different function for behavior at low concentrations and used a virial coefficient formulation for high concentrations. The results have proved extremely fruitful for modeling activity coefficients over a very large range of molality. The general equation is

$$\ln \gamma_\pm = -\frac{A}{3} |z_+ z_-| f(I) + \frac{2 v_+ v_-}{v} B(I) m + \frac{2 (v_+ v_-)^{3/2}}{v} C m^2 \qquad (14)$$

where

$$f(I) = \frac{I^{1/2}}{1 + 1.2 I^{1/2}} + 1.67 \ln (1 + 1.2 I^{1/2})$$

and

$$B(I) = 2\beta°$$

$$+ \frac{2\beta'}{\alpha^2 I}\left[1 - \left(1 + \alpha I^{1/2} - \frac{1}{2}\alpha^2 I\right)e^{-\alpha I^{1/2}} \right]$$

and where $\beta°$ and β' are specific-ion parameters, α is a constant for a similarly charged class of electrolytes and C is a specific-ion parameter independent of ionic strength. The B and C parameters are second and third virial coefficients. The parameter $v = v_+ + v_-$, is the sum of the stoichiometric coefficients for the cation, v_+, and the anion, v_-, of the solute. The Pitzer parameters have been fitted for a wide range of solutes and have been used for mixed electrolyte solutions to model the mineral solubility behavior in brines (Harvie and Weare, 1980; Harvie *et al.*, 1980). Grenthe *et al.* (1997) have compared the SIT method and the Pitzer method in detail and came to the following conclusions: (i) the more extensively parametrized Pitzer model allows the most precise modeling of activity coefficients and equilibrium constants provided that all the inter-action coefficients are known; (ii) when Pitzer parameters are missing (not available from experimental data), they must be estimated and the precision and accuracy of the activity coefficients can be significantly compromised; (iii) the parameters in the SIT model can be related to those in the Pitzer model and provide another means of estimating Pitzer parameters; (iv) the SIT model is in good agreement with the Pitzer model for the range $m = 0.1–4$ mol kg^{-1}; and (v) the Pitzer model is the preferred formalism for solutions or brines of high ionic strength. An extensive discussion of the development of the Pitzer model and several hybrid approaches can be found in Pitzer (1991a) and Millero (2001). It appears that hybrid approaches provide the best promise in the future, because they combine the extensive data available on equilibrium constants with a better formulation of activity coefficients.

The difference between the extended Debye–Hückel equation and the Pitzer equations has to do with how much of the nonideality of electrostatic interactions is incorporated into mass action expressions and how much into the activity coefficient expression. It is important to remember that the expression for activity coefficients is inexorably bound up with equilibrium constants and they must be consistent with each other in a chemical model. Ion-pair interactions can be quantified in two ways, explicitly through stability constants (IA method) or implicitly through empirical fits with activity coefficient parameters (Pitzer method). Both approaches can be successful with enough effort to achieve consistency. At the present, the Pitzer method works much better for brines, and the IA method works better for dilute waters because of the greater number of components and species for which basic data exist. When the effort is made to compare both approaches for the same set of high-quality data, they appear to be comparable (Felmy *et al.*, 1990). The primary challenge for the future will be to insure that consistency is maintained between the thermodynamic data, the expressions for non-ideality, and the mass-action expressions in geochemical modeling codes and to incorporate trace elements and redox species within the same formulation.

5.02.5.2 Saturation Indices

After speciation and activities have been calculated for all the free ions, ion pairs, triplets, etc., a mineral saturation index can be computed. The saturation index, SI, is defined as the logarithm of the ratio of the ion-activity product, IAP, to the solubility product constant, K_{sp},

$$SI = \log\left[\frac{IAP}{K_{sp}} \right] \qquad (15)$$

If the solution is at equilibrium, the IAP $= K_{sp}$ and the SI $= 0$. If the SI > 0, then the solution is supersaturated and the mineral would tend to precipitate; if the SI < 0, the solution is under-saturated and the mineral would tend to dissolve, if present. Because the SI is affected by the stoichiometry of the mineral formula, it is best to normalize the SI to the total mineral stoichiometry as pointed out by Zhang and Nancollas (1990):

$$SI = \log\left[\frac{IAP}{K_{sp}} \right]^{1/v} \qquad (16)$$

where $v = v^+ + v^-$, the sum of the stoichio-metries of the positive and negative components in the mineral formula.

These computations describe the "tendency" of a water sample to be saturated, but they do not necessarily demonstrate whether mineral dissolu-tion or precipitation is taking place. For dissolu-tion to take place, the mineral must be present and it must dissolve at a rate that is fast enough relative to the flow rate of the water to affect the water chemistry (Berner, 1978). Likewise for a mineral to precipitate it must do so at a fast enough rate. The kinetics of precipitation and dissolution reactions must be applied to get a realistic interpretation of water–rock interactions.

5.02.6 GEOCHEMICAL DATABASES

Input for aqueous geochemical codes consists of field data (geology, petrology, mineralogy, and water analyses), thermodynamic and aqueous electrolyte data, and possibly kinetic and sorption data. Thermodynamic data are fundamental to most geochemical computations and several major compilations are available. Although the guidelines and necessary relations for thermodynamic consistency are well established (Rossini *et al.*, 1952; Helgeson, 1968, 1969; Haas and Fisher, 1976; Nordstrom *et al.*, 1990; Nordstrom and Munoz, 1994), it has been difficult to employ them in the development of databases. Two general approaches have been recognized, serial networks and simultaneous regression. With serial networks, an evaluator begins with a single starting point, such as the standard state properties for an element (or all elements) and gradually builds in compound properties through the appropriate reaction combinations. Serial networks, such as the National Bureau of Standards (NBS) database (Wagman *et al.*, 1982), achieve continuity for a large data set but lose thermodynamic constraints, whereas simultaneous regression, which preserves thermodynamic relationships, can only be done on a limited subset of data, and the weighting of the regression fit depends on the judgment of the evaluator (Archer and Nordstrom, 2003). Computer codes have been developed for the purpose of correlating and evaluating a diversity of thermodynamic data (Haas, 1974; Ball *et al.*, 1988), but they have not been widely used. Serious discrepancies in thermodynamic properties do appear among these compilations and some of these discrepancies have been incorporated into the databases of geochemical codes. Many users of these codes are not familiar with these databases and the possible uncertainties propagated through the computations from thermodynamic errors. Such error propagation may or may not be important, depending on the specific objectives of the geochemical problem being addressed.

5.02.6.1 Thermodynamic Databases

Since the early compilations of thermodynamic data (e.g., Lewis and Randall, 1923), numerous measurements were made during the twentieth century leading to the well-known compilations of Latimer (1952) and Rossini *et al.* (1952). Measurements and compilations continued to expand through the latter part of the twentieth century, and comprehensive inventories such as Sillén and Martell (1964), Martell and Smith (1974–1976), Wagman *et al.* (1982), Chase (1998), Robie and Hemingway (1995), and Gurvich *et al.* (1993) summarized and organized a considerable amount of data. Unfortunately, the methods used to evaluate the data are not always transparent and occasionally the source references are not known.

Notable exceptions were the publication of the CODATA tables (Garvin *et al.*, 1987; Cox *et al.*, 1989), the Organization Economic Co-Operation and Development/Nuclear Energy Agency Thermochemical Data Base (OECD/NEA TDB) publications (Grenthe *et al.*, 1992; Silva *et al.*, 1995; Rard *et al.*, 1999; Lemire, 2001), and the International Union of Pure and Applied Chemistry (IUPAC) compilations (e.g., Lambert and Clever, 1992; Scharlin, 1996). The OECD/NEA TDB, although focusing on radionuclides of most concern to safe disposal of high-level radioactive wastes and military wastes, contains a considerable amount of auxiliary data on other common aqueous and solid species that had to be evaluated along with the actinides. For aqueous geochemical modeling the equilibrium constant tables of Nordstrom *et al.* (1990) have proven useful and are found in USGS and US Environmental Protection Agency (USEPA) codes described below and in popular water-chemistry/aqueous-geochemistry textbooks (Appelo and Postma, 1993; Stumm and Morgan, 1996; Langmuir, 1997; Drever, 1997a). These tables have been recompiled, expanded, and tabulated in terms of thermodynamic properties (G, H, S, C_P, and log K for reaction and individual species (Nordstrom and Munoz, 1994)). The tabulation of Nordstrom and Munoz (1994) was organized in such a way that individual species can be compared with reactions measured independently. Consequently, it is easy to check thermodynamic relations to see if the properties for individual species are consistent with tabulated reaction equilibria. Only values that agreed within a close range of the stated uncertainties were included in the tables.

These tabulations represent a portion of the data needed for geochemical model computations and there have been some obvious inconsistencies. By the mid-1990s critical evaluations of thermodynamic data were tapering off and many errors and inconsistencies had not been resolved. As recently as 1994, serious inconsistencies still existed between calorimetrically derived and solubility-derived properties for such common minerals as celestine and barite, and a less common but important phase such as radium sulfate (Nordstrom and Munoz, 1994). This set of discrepancies has been resolved for celestine and barite by new measurements using high-temperature oxide-melt and differential-scanning calorimetries (Majzlan *et al.*, 2002). Listing of sources of thermodynamic data compilations can be found in Nordstrom and Munoz (1994) and in Grenthe and Puigdomenech (1997).

5.02.6.2 Electrolyte Databases

The classic books by Harned and Owen (1958) and Robinson and Stokes (1959) put electrolyte theory on a firm basis and provided important tabulations of electrolyte data for calculations of activity coefficients and related properties. Advances in electrolyte theory, especially the use of the Pitzer method (Pitzer, 1973, 1979), led to alternate methods of calculating activities and speciation. The books edited first by Pytkowicz (1979) and later by Pitzer (1991b) provided a general overview of the different approaches and hybrid approaches to these computations. The more recent reference providing a thorough overview of the subject with tables of molal properties, activity coefficients and parameters, and dissociation/association constants is Millero (2001). These references explain the general use of the IA method with extended forms of Debye–Hückel equations, the Pitzer specific-ion interaction method, and hybrid approaches that take advantage of the best aspects of both approaches.

5.02.7 GEOCHEMICAL CODES

This listing and description of geochemical codes is not meant to be either exhaustive or complete. It is meant to be a brief overview of several codes in current and common use. Reviews by Yeh and Tripathi (1989a), Mangold and Tsang (1991), Parkhurst and Plummer (1993), and Alpers and Nordstrom (1999) provide information on more codes, their evolution, and references to earlier reviews on codes and geochemical modeling. Loeppert et al. (1995) also contains useful information on geochemical codes and modeling, especially for soil interactions. It is important to remember that most codes in active use often undergo enhancements, database updates, and other improvements. The descriptions given in this section may not apply to some of these same codes five years from now.

5.02.7.1 USGS Codes

The USGS has developed several codes that are useful for the interpretation of water chemistry data and for simulating water–rock interactions. Two similar aqueous speciation codes were developed in parallel, WATEQ (Truesdell and Jones, 1974) and SOLMNEQ (Kharaka and Barnes, 1973). The primary aim of these programs was to aid in the interpretation of water quality data. SOLMNEQ, however, had a different subroutine for calculating temperature and pressure dependence and could calculate reaction equilibria above 100 °C. WATEQ was intended for temperatures of 0–100 °C. Both of these programs have been updated. WATEQ4F v. 2 (Ball and Nordstrom, 1991, with database updates to 2002) uses the IA method with an expanded

form of the extended Debye–Hückel equation for major ions (the Truesdell–Jones formulation or hybrid activity coefficient; Truesdell and Jones (1974) and Nordstrom and Munoz (1994)), includes independent redox speciation computations that assume redox disequilibrium, and has database updates for uranium (Grenthe et al., 1992), chromium (Ball and Nordstrom, 1998), and arsenic redox species (Archer and Nordstrom, 2003; Nordstrom and Archer, 2003). SOLMINEQ.88 (Kharaka et al., 1988; Perkins et al., 1990) covers the temperature range of 0–350 °C and 1–1,000 bar pressure, includes both the IA method and the Pitzer method, and has mass-transfer options such as boiling, fluid mixing, gas partitioning, mineral precipitation, mineral dissolution, ion exchange, and sorption. It has been found to be particularly applicable to deep sedimentary basins, especially those containing oil and gas deposits. The latest version, SOLMINEQ.GW is explained in an introductory text on groundwater geochemistry (Hitchon et al., 1996).

Parkhurst et al. (1980) developed the PHREEQE code to compute, in addition to aqueous speciation, mass transfer, and reaction paths. A separate but similar code, PHRQPITZ, uses the Pitzer method for brine calculations (Plummer et al., 1988; Plummer and Parkhurst, 1990). The PHREEQE code has been regularly enhanced and the recent version, PHREEQC v. 2, includes ion exchange, evaporation, fluid mixing, sorption, solid-solution equilibria, kinetics, one-dimensional transport (advection, dispersion, and diffusion into stagnant zones or dual porosity), and inverse modeling (Parkhurst and Appelo, 1999). An interface (PHREEQCI) was developed for interactive modification of the input files by Charlton et al. (1997). The latest development (β-test version) is PHAST, which includes three-dimensional transport. The program PHAST uses the solute-transport simulator HST3D (Kipp, 1987, 1998) and iterates at every time step with PHREEQC. Thorstenson and Parkhurst (2002) have developed the theory needed to calculate individual isotope equilibrium constants for use in geochemical models and utilized them in PHREEQC to calculate carbon-isotope compositions in unsaturated zone with seasonally varying CO_2 production (Parkhurst et al., 2001).

Geochemical modeling of reactants in flowing mountainous stream systems can be done with the USGS codes OTIS (Runkel, 1998) and OTEQ (Runkel et al., 1996, 1999) that model solute transport and reactive transport, respectively. OTIS, or one-dimensional transport with inflow and storage, is based on the earlier work of Bencala (1983) and Bencala and Walters (1983). The OTEQ code combines the OTIS code with MINTEQA2 for chemical reaction at each incremental step of transport. These codes are calibrated with constant-flow tracer injection studies and testable assumptions regarding the solubility product constant of the precipitating phases (Kimball *et al.*, 2003). They have been tested in numerous settings, primarily with the objective of quantifying and predicting the sources, transport, and fate of acid drainage from mined environments in the western US.

Codes for inverse modeling began with the program BALANCE (Parkhurst *et al.*, 1982) from which the interactive code NETPATH (Plummer *et al.*, 1991) evolved. NETPATH, in addition to mass balances, does database management for a suite of wells and can compute speciation and saturation indices with WATEQ. NETPATH has now been incorporated into PHREEQC along with uncertainty propagation (Parkhurst, 1997). Bowser and Jones (1990, 2002) have incorporated the mass-balance approach into a spreadsheet format that allows graphical output and a quick reconnaissance of ranges of mineral compositions that are permissible models for silicate solid-solution series (see Chapter 5.04).

USGS codes and manuals can be downloaded free of charge at http://water.usgs.gov/nrp/gwsoftware/. The more current version and additional bibliographic and information files for PHREEQC are available at http://wwwbrr.cr.usgs.gov/projects/GWC_coupled/. Updates on revised thermodynamic data and WATEQ4F can be found at http://wwwbrr.cr.usgs.gov/projects/GWC_chemtherm/ with links to software and thermodynamic property evaluation. The OTIS code can be accessed at http://co.water.usgs.gov/otis (see Chapter 5.04).

5.02.7.2 LLNL Codes

A set of computer codes known as EQ3/6 was originally developed by Wolery (1979) to model rock–water interactions in hydrothermal systems for the temperature range of 0–300 °C. Software development was later sponsored by the US Department of Energy at Lawrence Livermore National Laboratories (LLNL) to model geochemical processes anticipated in high-level nuclear waste repositories. This geochemical code has become one of the most sophisticated and most applicable for a wide range of conditions and processes. In addition to speciation and mass transfer computations, it allows for equilibrium and nonequilibrium reactions, solid-solution reactions, kinetics, IA and Pitzer methods, and both inorganic and organic species. The program has been used for several municipal and industrial waste situations and has been used to assess natural and engineered remediation processes. It has five supporting thermodynamic data files

and the thermodynamic data are evaluated and updated with the SUPCRT92 software (Johnson *et al.*, 1992), based on Helgeson's formulation for activity coefficients and aqueous thermodynamic properties over a wide range of temperature and pressure (Helgeson and Kirkham, 1974a,b, 1976; Helgeson *et al.*, 1981) and solid-phase thermodynamic properties (Helgeson *et al.*, 1978). Thermodynamic data are updated with the availability of published research (e.g., Shock and Helgeson, 1988, 1990; Shock *et al.*, 1989). Several manuals for operation and general information are available at http://geosciences.llnl.gov/esd/geochem/eq36.html (last accessed February 17, 2003).

5.02.7.3 Miami Codes

It would be difficult to find more comprehensive or more detailed studies on the physical chemistry of seawater than those done at the University of Miami (Millero, 2001). Several programs were developed for calculation of activity coefficients and speciation of both major ions and trace elements in seawater. The activity coefficient models have been influenced strongly by the Pitzer method but are best described as hybrid because of the need to use ion-pair formation constants (Millero and Schreiber, 1982). The current model is based on Quick Basic; computes activity coefficients for 12 major cations and anions, 7 neutral solutes, and more than 36 minor or trace ions. At 25 °C the ionic strength range is 0–6 m. For major components, the temperature range has been extended to 0–50 °C, and in many cases the temperature dependence is reasonably estimated to 75 °C. Details of the model and the parameters and their sources can be found in Millero and Roy (1997) and Millero and Pierrot (1998). Comparison of some individual-ion activity coefficients and some speciation for seawater computed with the Miami model is shown in Section 5.02.8.6 on model reliability.

5.02.7.4 The Geochemist's Workbench™

A set of five programs known as The Geochemist's Workbench™ or GWB was developed by Bethke (1994) with a wide range of capabilities similar to EQ3/6 and PHREEQC v. 2. GWB performs speciation, mass transfer, reaction-path calculations, isotopic calculations, temperature dependence for 0–300 °C, independent redox calculations, and sorption calculations. Several electrolyte databases are available including ion association with Debye–Hückel activity coefficients, the Pitzer formulation, the Harvie–Møller–Weare formulation, and a

PHRQPITZ-compatible formulation. The program X2t allows the coupling of two-dimensional transport with geochemical reaction. Basin3 is a basin-modeling program that can be linked to GWB. Another advantageous feature is the plotting capability that can produce $p\varepsilon$–pH or activity–activity or fugacity–fugacity diagrams from GWB output. Bethke (1996) has published a textbook on *Geochemical Reaction Modeling* that guides the reader through a variety of geochemical computations using GWB and provides the basis for a course in modeling. The GWB is a registered code and must be obtained from Dr. Bethke and the University of Illinois.

5.02.7.5 REDEQL-MINTEQ Codes

One of the first speciation programs that included mass transfer reactions at equilibrium was REDEQL (Morel and Morgan, 1972). The primary aim of this set of codes was to compute the equilibrium chemistry of dilute aqueous solutions in the laboratory and was among the first to include sorption. It also has been widely used to interpret water quality in environmental systems. This FORTRAN code has evolved through several versions, parallel with advances in computer hardware and software. Incorporation of the WATEQ3 database (Ball *et al.*, 1981) with the MINEQL program (Westall *et al.*, 1976) produced MINTEQ (Felmy *et al.*, 1984) which became the USEPA-supported code, MINTEQA2 (Allison *et al.*, 1991). The more recent upgrade is MINTEQA2/PRODEFA2 v. 4 (USEPA, 1998, 1999 (revised)), and contains code revisions, updates in thermodynamic data, and modifications to minimize nonconvergence problems, to improve titration modeling, to minimize phase rule violations, to enhance execution speed, and to allow output of selected results to spreadsheets. PRODEFA2 is an ancillary program that produces MINTEQA2 input files using an interactive preprocessor. Thermodynamic data and computational abilities for Be, Co(II/III), Mo(IV), and Sn (II, IV) species have been added to the program. Unfortunately, the numerous inconsistencies, lack of regular upgrades, and tendency for numerous independent researchers to make their own modifications to the program have led to several versions that differ in reliability. Examples of test cases to demonstrate the input setup, capabilities of the code, and comparisons of standard test cases have not been included in the documentation as has been done for other programs. The more recent upgrade information can be obtained at http://epa.gov/ceampubl/mmedia/minteq/supplei. pdf (last accessed February 17, 2003).

Perhaps the most noteworthy aspect of the upgraded MINTEQA2 code is the Gaussian model for interactions of dissolved organic matter with cations. The model formulation is based on the statistical treatment of proton binding (Posner, 1964) that was developed by Perdue and Lytle (1983), Perdue *et al.* (1984), and Dobbs *et al.* (1989). This approach uses a continuous distribution of sites as opposed to discrete site binding. It is equivalent to a collection of monoprotic ligands, each able to bind protons and metal cations, with the variations in log K described by a single Gaussian distribution for each class of sites. Another code, MODELm (Huber *et al.*, 2002), uses a linear differential equilibrium function to account for trace-metal complexation with natural organic matter. Computations were compared to those done by MINTEQA2. MODELm predicted greater amounts of metal-organic complexing than MINTEQA2, but no conclusions were drawn as to the cause of this difference. MODELm can be used in MINTEQA2 instead of the Gaussian distribution model. Although cation-organic binding in natural waters is a highly complex and challenging subject to quantify for modeling purposes, advances since the early 1990s are leading us much closer to practical approaches that can be incorporated into computerized equilibrium and nonequilibrium chemical codes. Geochemistry will benefit from continued research in this area because organic matter exerts such a strong control on the behavior of trace elements in most aquatic systems. More testing and evaluation of a range of natural waters with different trace-element concentration and organic matter types is needed in the future. This can be accomplished by comparing analytical speciation with computed speciation.

A Windows version of MINTEQA2 v. 4.0, known as Visual MINTEQ, is available at no cost from http://www.lwr.kth.se/English/OurSoftware/vminteq/ at The Royal Institute of Technology, Sweden (last accessed February 17, 2003). It is supported by two Swedish research councils, VR and MISTRA. The code includes the National Institute of Standards and Technology (NIST) database, adsorption with five surface complexation models, ion-exchange, and metal-humate complexation with either the Gaussian DOM Model or the Stockholm Humic Model. Input data are accepted from Excel spreadsheets and output is exported to Excel. A major update was completed in January 2003. The Royal Institute of Technology also has produced HYDRA and MEDUSA for creating a database for a given system and creating activity–activity and $p\varepsilon$–pH diagrams. Links to other similar programs can be found online.

A Windows version of MINTEQA2, known as MINEQL + v. 4.5, has several enhancements that make it an attractive alternative to other versions.

The user interface with a relational database technique for scanning thermodynamic data, a utility for creating a personal database, a relational spreadsheet editor for modifying chemical species, a multi-run manager, special reports for convenient data extraction or additional calculations, and a tutorial with several examples provide a much more flexible and practical code for modelers. It also provides graphic output for log C–pH plots, ion-fraction diagrams, solubility plots, titration curves, and sensitivity plots. It is limited in temperature to 0–50 °C and ionic strength <0.5 M. The program is available through Environmental Research Software, http://www.mineql.com/ (last accessed February 21, 2003).

5.02.7.6 Waterloo Codes

A considerable history of hydrogeochemical modeling and its application has evolved at the University of Waterloo, Ontario, Canada, and a series of reactive-transport codes have evolved with it. These codes have been primarily applied to mine tailings piles. A finite element transport module, PLUME2D, was utilized with the MINTEQA2 module in an efficient two-step sequential solution algorithm to produce MINTRAN (Walter *et al.*, 1994). Leachates from heterogeneous mine overburden spoil piles that made a contaminant front were modeled with MINTRAN and other codes at open-pit lignite mines in Germany (Gerke *et al.*, 1998). When a numerical model that coupled oxygen diffusion and sulfide-mineral oxidation, PYROX, was added to MINTRAN, it became MINTOX (Wunderly *et al.*, 1996). MINTOX proved to be capable of simulating 35 yr of contaminant transport at the Nickel Rim mine tailings impoundment (Bain *et al.*, 2000). The more recent version is MIN3P (Mayer *et al.*, 1999), which is a general reactive transport code for variably saturated media. It has been applied to the Nickel Rim impoundment and to the contaminated groundwater downgradient of the Königstein uranium mine in Saxony, Germany (Bain *et al.*, 2001). At Königstein, reactions involving iron, uranium, sulfate, cadmium, chromium, nickel, lead, and zinc were all modeled. A short review of similar codes and an update on those codes can be found in Mayer *et al.*, (2003).

5.02.7.7 Harvie–Møller–Weare Code

With the arrival of the Pitzer method for calculating activity coefficients at high ionic strengths (≥ 1 m), research by Harvie and Weare (1980) led to computations of equilibrium mineral solubilities for brines. They could calculate solubility data from gypsum ($I < 0.06$ m) to bischofite saturation (>20 m). This capability made it possible to more accurately calculate the mineral sequences during seawater evaporation and quantitatively solve a problem that had puzzled van't Hoff (Harvie *et al.*, 1980). The original model included the components Na, K, Mg, Ca, Cl, SO_4, and H_2O and was applied to the simpler salt systems and to mineral equilibria in seawater (Eugster *et al.*, 1980). With further revision of the parameters the model was expanded to include more complex electrolyte mixtures and their salts for seawater evaporation (Harvie *et al.*, 1982). Later the number of components was expanded to include H, OH, HCO_3, CO_3, and CO_2 (Harvie *et al.*, 1984). Incorporating the temperature dependence to allow calculations from 25–250 °C was achieved by Møller (1988) and for lower temperatures (0–250 °C) by Greenberg and Møller (1989) and Spencer *et al.* (1990). Marion and Farren (1999) extended the model of Spencer *et al.* (1990) for more sulfate minerals during evaporation and freezing of seawater down to temperatures of -37 °C. Pressure dependence for aqueous solutes and minerals in the $Na–Ca–Cl–SO_4–H_2O$ system to 200 °C and 1 kbar were obtained by Monnin (1990) and applied to deep Red Sea brines and sediment pore waters (Monnin and Ramboz, 1996). Møller *et al.* (1998) developed the TEQUIL code for geothermal brines. Ptacek and Blowes (2000) have reviewed the status of the Pitzer method applications for sulfate mineral solubilities and demonstrated the applicability of a modified Harvie–Møller–Weare (HMW) to mineral solubilities in mine tailings piles.

5.02.7.8 WHAM Models

Tipping (1994) has proposed an alternative and practical model for humic acid–metal binding embodied in the speciation code Windemere Humic Aqueous Model (WHAM) that has two versions, one for water and one for soils (Tipping, 1998). This aqueous metal-organic model is based on the concept of electrostatic interactions at discrete sites. Hence, it is readily amenable to use with inorganic chemical speciation programs for aqueous solutions and includes ionic strength and temperature effects. Tipping *et al.* (2002) provide a thorough evaluation of humic substance binding with aluminum and iron in freshwaters. They demonstrate that 60–70% of the dissolved organic carbon (DOC) in their samples was humic substances involved in metal complexation: after accounting for organic complexing, the inorganic speciation is consistent with control of dissolved aluminum and iron concentration by their hydrous oxides. A full explanation of the developments that led up to the WHAM model and descriptions of many other models (multiple versus single discrete models, continuous models, competitive versus noncompetitive models, empirical models, site heterogeneity/polyelectrolyte models, and Gaussian distribution models) can be found in Tipping (2002). The latest WHAM code, version 6, is available at http://www.ife.ac.uk/aquatic_processes/wham/WHAMtitlebar.htm (last accessed February 16, 2003). An example of a continuous distribution for binding sites can be found in the development of the Natural Organic Anion Equilibrium Model, or NOAEM (Gryzb, 1995) which also can be used in conjunction with ionic components and electrostatic theory.

5.02.7.9 Additional Codes

The coupled code developed by Steefel and Lasaga (1994) for multicomponent reactive transport with kinetics of precipitation and dissolution of minerals has been developed further into the OS3D/GIMRT code (Steefel and Yabusaki, 1996). This model has been applied to reaction fronts in fracture-dominated flow systems (Steefel and Lichtner, 1998). Further developments for nonuniform velocity fields by Yabusaki *et al.* (1998) required the use of massively parallel processing computers, although "… the accuracy of the numerical formulation coupling the nonlinear processes becomes difficult to verify."

The model BIOKEMOD has been developed to simulate geochemical and microbiological reactions in batch aqueous solutions (Salvage and Yeh, 1998). It has been tested and found to simulate a range of processes that include complexation, adsorption, ion-exchange, precipitation/dissolution, biomass growth, degradation of chemicals by metabolism of substrates, metabolism of nutrients, and redox. The code has been coupled to HYDROGEOCHEM (Yeh and Tripathi, 1989b) for simulation of reactive transport modeling with biogeochemical transformation of pollutants in groundwaters. FEREACT was developed for two-dimensional steady-state flow with equilibrium and kinetically controlled reactions (Tebes-Stevens *et al.*, 1998). Another code, that was developed for biogeochemical transport and interactions of oxidative decay of organics with oxygen, iron, manganese, and sulfur redox species, has been introduced by Hunter *et al.* (1998).

Lichtner (2001) developed the computer code FLOTRAN, with coupled thermal–hydrologic–chemical (THC) processes in variably saturated, nonisothermal, porous media in 1, 2, or 3 spatial dimensions. Chemical reactions included in FLOTRAN consist of homogeneous gaseous reactions, mineral precipitation/dissolution, ion exchange, and adsorption. Kinetic rate laws and redox disequilibrium are allowed with this code. Debye–Hückel and Pitzer options are both available for computing activity coefficients, and thermodynamic data are based on the EQ3/6 database or user defined databases.

Several options are available in FLOTRAN for representing fractured media. The equivalent continuum model (ECM) formulation represents fracture and matrix continua as an equivalent single continuum. Two distinct forms of dual continuum models also are available, defined in terms of connectivity of the matrix. These models are the dual continuum connected matrix (DCCM) and the dual continuum disconnected matrix (DCDM) options. A parallel version of the code, PFLOTRAN, has been developed based on the PETSC parallel library at Argonne National Laboratory.

Park and Ortoleva (2003) have developed WRIS.TEQ, a comprehensive reaction-transport-mechanical simulator that includes kinetic and thermodynamic properties with mass transport (advection and diffusion). A unique property of this code is a dynamic compositional and textural model specifically designed for sediment alteration during diagenesis.

Further research on reactive transport theory, modeling, and codes can be found in the *Reviews in Mineralogy* volume edited by Lichtner *et al.* (1996) and the special issue of *Journal of Hydrology*, vol. 209 (1998).

5.02.8 WATER–ROCK INTERACTIONS

About a dozen or so major hydrogeochemical processes dominate the compositions of most surface and groundwaters. These processes include calcite dissolution and precipitation, gypsum dissolution and precipitation, pyrite oxidation and formation of hydrous ferric oxide, silicate mineral dissolution (feldspars, micas, chlorites, amphiboles, olivines, and pyroxenes) and clay mineral formation (kaolinitization, laterization, and illitization), dolomite dissolution and calcite precipitation (dedolomitization), dolomite formation (dolomitization), sulfate reduction and pyrite formation, silica precipitation, evaporation, and cation exchange. They are explained in several available textbooks (e.g., Appelo and Postma, 1993; Langmuir, 1997; Drever, 1997a), and only a few examples are given here to demonstrate how complex geochemistry can be easily computed with available codes. Aqueous speciation is discussed first, because it is required for more complex computations. For geochemical modeling, aqueous speciation occurs so quickly that it can safely be assumed to be at equilibrium. This assumption is valid for the vast majority of aqueous reactions but not for redox reactions.

5.02.8.1 Aqueous Speciation

Some brief examples of aqueous speciation are given here to demonstrate the large decreases in free ion concentration or activity that occur as a result of ion-pair formation or complexing. Polyvalent ions have a greater tendency to associate with other ions of opposite charge and among polyvalent cations that are found commonly in waters, iron and aluminum are prime examples. Alpers and Nordstrom (1999) tabulate analyses of four acid mine waters with pH in the range 4.9–0.48 and show the effects of speciation for copper, sulfate, aluminum, and iron using the WATEQ4F code. Table 1 shows the total dissolved concentrations and the main species for aluminum and sulfate. Several conclusions are readily apparent. First, to approximate the free ion concentrations by the total dissolved concentrations is fair, as for sulfate in AMD-A (83%), to poor, as for aluminum in AMD-D (4.1%). Furthermore, for a wide range of total dissolved sulfate concentration, the $AlSO_4^+$ ion pair is always important (50–70% of the dissolved aluminum) but with increasing sulfate concentration and decreasing pH, the $Al(SO_4)_2^-$ ion triplet becomes increasingly important. This ion triplet has an association constant that is not well established and some codes might not include it in their database.

Another example of aqueous speciation that includes redox can be shown with the arsenic $p\varepsilon$–pH diagram shown in Figure 1. Arsenic can exist in several oxidation states including As(-III) as in arsine gas (AsH_3), As(0) as in elemental arsenic, As(II) as in realgar (AsS), As(III) as in orpiment (As_2S_3) and dissolved arsenite, and As(V) as in dissolved arsenate. Figure 1 shows the dominant dissolved species, arsenate and arsenite, and their hydrolysis products as a function of redox potential and pH based on the thermodynamic evaluation of Nordstrom and Archer (2003). These results show the dominance of hydrolysis for arsenate species, but it is of minor consequence for the arsenite species.

Hydrolysis is of major importance in understanding mineral reactions, kinetics of reactions, and sorption behavior. At neutral to high pH, the adsorption of arsenate onto hydrous ferric oxides is weak to nonexistent, and the high anionic charge on the dissolved arsenate developed through hydrolysis combined with negatively charged surfaces helps to account for this lack of adsorption. The lack of significant arsenite hydrolysis helps explain the more competitive adsorption of arsenate relative to arsenite.

5.02.8.2 Modeling Sorption Reactions

Although sorption modeling should be included in our discussion, this extensively researched area warrants an entirely separate chapter. Several books cover the subject well, including Dzombak and Morel (1990), Davis and Hayes (1986), Stumm (1987), and Hochella and White (1990).

5.02.8.3 Model Simulations of Mineral Reactions

One of the most ubiquitous geochemical processes is the dissolution of calcite. Dissolution of calcite in the environment can be the dominant source of dissolved calcium in many waters, but dissolved inorganic carbon (DIC) can have at least two sources: calcite and CO_2 produced from organic decay. Calcite dissolution has been modeled with PHREEQCI for a range of carbon dioxide partial pressure, P_{CO_2}. Figure 2(a) is a plot of calcite dissolution in terms of calcium and DIC concentrations at P_{CO_2} values ranging from atmospheric ($10^{-3.5}$) to 10^{-1}. Computationally, increments of calcite were dissolved in water at fixed partial pressure until solubility equilibrium was reached. The solid line represents the equilibrium solubility of calcite for this range of carbon dioxide partial pressure and the

Table 1 Example of aqueous aluminum and sulfate speciation for acid mine waters covering a range of pH and composition.

Sample	AMD-A	AMD-B	AMD-C	AMD-D
Temperature (°C)	16.0	19.5	24.0	34.8
pH	4.9	3.25	1.10	0.48
Total dissolved Al (mg L^{-1})	5.06	19.8	1,410	2,210
% as Al^{3+}	29	26	10	4.1
% as $AlSO_4^+$	51	66	57	61
% as $Al(SO_4)_2^-$	2	4.5	32	32
Total dissolved SO_4 (mg L^{-1})	206	483	50,000	118,000
% as SO_4^{2-}	83	71	18	8
% as HSO_4^-	0	2	32	53
% as $AlSO_4^+$ and $Al(SO_4)_2^-$	4.5	11	12	8
% as Fe(II/III)–SO_4 ions	0	2	29	26

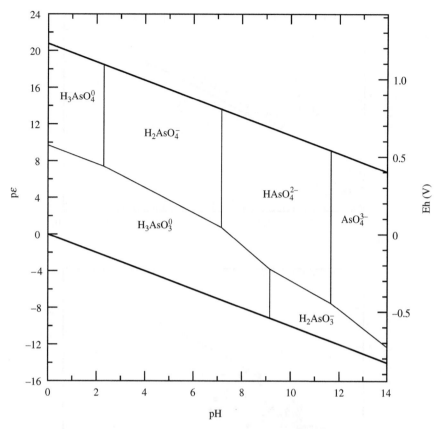

Figure 1 Species predominance diagram for dissolved arsenic at 25 °C and 1 bar (source Nordstrom and Archer, 2003).

equilibrium pH values at each P_{CO_2} from 8.3 to 6.7 are shown in parentheses.

Many shallow groundwaters reflect calcite dissolution as the dominant control on water quality. Groundwaters incorporate higher P_{CO_2} than that of the atmosphere because of carbon dioxide production in the soil zone and organic matter decomposition in the groundwater. A range of P_{CO_2} from 10^{-2} to 10^{-1} and pH values from 7 to 8 are common for most groundwaters. Water analyses in a carbonate terrain of Pennsylvania (Langmuir, 1971) are compared in Figure 2(b) to the predictions in Figure 2(a). Although this plot simplifies the water chemistry and does not take into account dissolution of other minerals, it does show the dominant control by a relatively simple reaction. The P_{CO_2} fall within the expected range and calcite solubility equilibrium provides an upper boundary. The saturation index plot in Figure 3 for the same samples takes into account temperature and ionic strength effects on the activities and also shows that saturation with respect to calcite is reached and provides an upper limit to calcium and DIC concentrations. As the pH increases, there is a greater proportion of carbonate ions relative to bicarbonate that increases the saturation with respect to calcite.

At pH values below ~7, waters are nearly always undersaturated with respect to calcite (and most other carbonate minerals).

Pyrite oxidation is a complex hydrobiogeochemical processes that accounts for 11% of the sulfate found in river drainages (E. K. Berner and R. A. Berner, 1996). Mining activities have increased the rate of pyrite oxidation and caused severe contamination of many waterways with acid and metals. When pyrite oxidizes, the sulfur rapidly converts to sulfate but the oxidation of the iron proceeds more slowly, depending on pH. Simulating the oxidation of pyrite is instructive in summarizing the chemistry of this complex process (Nordstrom, 2000). In Figure 4, PHREEQCI was used to simulate pyrite oxidation by adding increments of oxygen to pyrite in water. The reaction also could be simulated by adding increments of pyrite to an excess of oxygen. The results vary with the amount of oxidation allowed and are represented in the figure by a solution pH as a function of the amount of pyrite oxidized. First, pyrite oxidation to an acid ferrous sulfate solution only, is shown by the solid line. Second, the same reaction occurs but the ferrous iron is allowed to oxidize without forming a precipitate, shown by the dashed line with a crossover

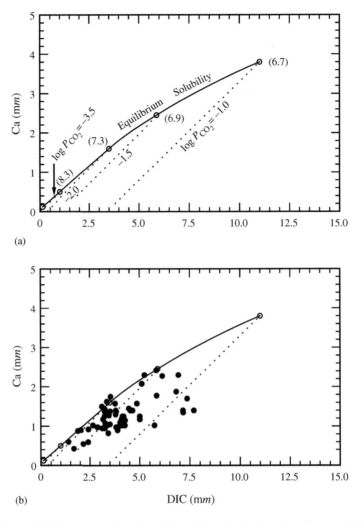

(a)

(b)

DIC (m*m*)

Figure 2 (a) Calcium and DIC concentrations plotted for calcite dissolution at log P_{CO_2} = −3.5, −2, −1.5, and −1 (dotted lines) up to equilibrium calcite solubility (solid line). The pH values for each equilibrium solubility at the specified log P_{CO_2} are shown in parentheses. Plot was computed with PHREEQCI for 25 °C and 1 bar. (b) Calcium and DIC concentrations plotted from data of Langmuir (1971) for groundwaters taken from a limestone aquifer in Pennsylvania. Dotted and dashed lines are the same as those in (a).

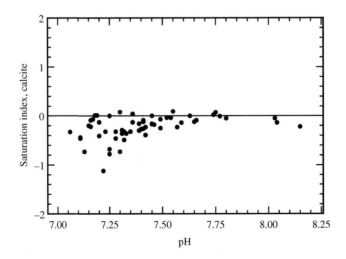

pH

Figure 3 Saturation indices for calcite from the same groundwaters as in Figure 2, plotted as a function of pH.

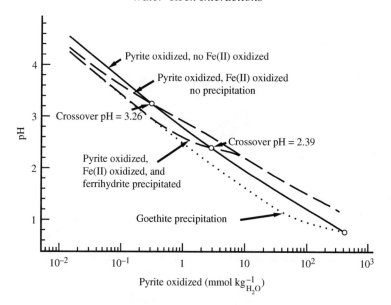

Figure 4 Change in pH as a function of the amount of pyrite oxidized under four scenarios: (i) pyrite oxidizes to an acid ferrous sulfate solution without any further oxidation (solid line); (ii) pyrite oxidizes and the resultant ferrous sulfate solution is allowed to oxidize, but no precipitation is allowed (upper dashed line); (iii) pyrite oxidizes, the ferrous sulfate solution oxidizes and precipitates ferrihydrite ($pK_{sp} = 4.89$, lower dashed line); and (iv) pyrite oxidizes, ferrous sulfate solution oxidizes, and goethite precipitates (dotted line). Computed with PHREEQCI at 25 °C and 1 bar, thermodynamic data from Nordstrom *et al.* (1990).

pH of 3.26. Third, a precipitate such as ferrihydrite is allowed to form, which lowers the crossover pH to 2.39. If the K_{sp} of the precipitating phase is lower, such as that for crystalline goethite, the crossover pH is much lower as shown by the dotted line. However, crystalline goethite is not stable at pH values below 2 and jarosite would precipitate instead. The reason for showing the goethite curve is to demonstrate the effect of lowering the K_{sp} for a precipitating phase of the same stoichiometry. The dashed lines cross over the original line, because the oxidation of ferrous iron involves both proton-consuming and proton-producing reactions. The oxidation of Fe^{2+} to Fe^{3+} consumes protons:

$$Fe^{2+} + \tfrac{1}{4}O_2 + H^+ \rightarrow Fe^{3+} + \tfrac{1}{2}H_2O \quad (17)$$

which happens at all pH values. The hydrolysis and precipitation of a ferric hydroxide phase produces protons:

$$Fe^{3+} + H_2O \rightarrow Fe(OH)^{2+} + H^+ \quad (18)$$

$$Fe^{3+} + 3H_2O \rightarrow Fe(OH)_3 \downarrow + 3H^+ \quad (19)$$

which only happens if the pH has reached the point of hydrolysis, i.e., near or above a $pH = pK_1 = 2.2$ for Fe^{3+} hydrolysis. Consequently, at lower pH values there is no hydrolysis and the pH can only increase on oxidation. The crossover pH reflects the balance between the proton-consuming and the proton-producing

reactions, a small buffering process represented by the small plateau in the curves. This diagram, although explaining some of the complexities of the chemistry of acid rock drainage, does not include the consequences of acid dissolution involving calcite and aluminosilicate minerals. Nevertheless, some water analyses are available from mine sites that are predominantly affected by pyrite oxidation and little else. These samples, collected from the Leviathan mine area, California, and Iron Mountain, California (Ball and Nordstrom, 1989; Nordstrom, 1977), include a pH measured *in situ* and a pH measured some weeks later after the ferrous iron had oxidized and precipitated. Figure 5 reproduces Figure 4 with only the two lines shown for pyrite oxidation but no ferrous iron oxidation (solid line) and pyrite oxidation with precipitation of ferrihydrite (or the most soluble hydrous ferric oxide, as a dashed line). Because there are low concentrations of cations in these samples, the initial pH values (open circles) can be assumed to be caused by pyrite oxidation. The final pH after oxidation and precipitation is shown by the closed circles. The final pH can be seen to closely approximate the dashed line in agreement with the simulation. In this comparison of a simulation with actual field data, two aspects are noteworthy. First is the good agreement but since this simulation is sensitive to the chosen K_{sp} of the precipitating phase, the agreement indicates that the solubility product constant for freshly

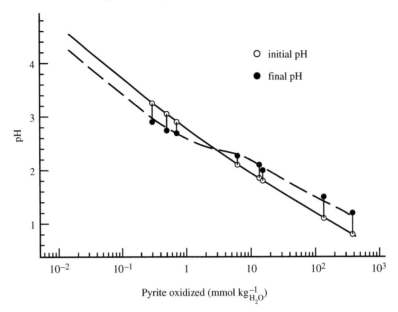

Figure 5 Change in pH as a function of the amount of pyrite oxidized under scenarios (i) and (iii) from Figure 4. Points are actual values measured from acid mine waters. Initial pH is based on field measurements taken on site and the amount of pyrite oxidized is derived from stoichiometry assuming that all the dissolved sulfate is from pyrite oxidation. Final pH was measured after the sample had been allowed to oxidize for at least two weeks in an unfiltered, unpreserved bottle.

precipitating hydrous ferric oxide is a reasonable model choice.

Waters not only undergo geochemical reactions but they commonly mix. One of the classic examples of mixing is seawater intrusion into coastal aquifers, often enhanced because of groundwater withdrawal. When seawater and fresh groundwater mix, both at saturation with respect to calcite, the result can lead to undersaturation with respect to calcite, i.e. calcite dissolution. The possibility of this scenario and other combinations (some leading to mineral precipitation from the mixing of waters that are both initially undersaturated) was outlined by Runnells (1969). Computations made by Plummer (1975) showed the proportion of mixing over which the undersaturation effect, and consequent dissolution of coastal limestone, was operative. Figures 6(a)–(d) are examples from Plummer (1975) and demonstrate that for a large range of seawater in the mixture, calcite undersaturation can occur, depending on carbon dioxide partial pressure and temperature. These calculations were done over a range of temperatures, partial pressures, and pH values. They also were conducted with actual carbonate groundwater compositions and coastal seawater for several locations along the Florida coast.

The inverse or mass-balance modeling approach provides additional constraints on reactant and product mineral phases when the mineral mass transfers are plotted as a function of the range of solid solution compositions. Bowser and

Jones (2002) discovered this application when investigating the effect of compositional ranges of feldspars, phyllosilicates, and amphiboles on mass balances for catchments and groundwaters dominated by silicate hydrolysis. They investigated nine drainages in six areas of widely varying lithology and climatic settings in the US. One of these is shown below in Figure 7 for the Wyman Creek drainage in the Inyo Mountains, California. Mineral mass transfers for dissolution (positive values in $mol\,kg^{-1}$) and precipitation (negative values in $mol\,kg^{-1}$) are plotted as a function of the percent montmorillonite solid solution. The gray band shows the restricted range of solid solution composition (27–45%) determined by the crossover points for K-feldspar and goethite, respectively. These restrictions apply, because K-feldspar cannot precipitate in these dilute waters and goethite is a weathering product being formed, not dissolved. Similar arguments show that there are restrictions on the range of plagioclase subject to weathering. Indeed, it is surprising to find how closely tied the fluid compositions for major cations are to the Na/Ca ratios of plagioclase, the Ca/Mg ratio of ferromagnesian silicates, and to the Fe/Mg ratio of ferromagnesian silicates.

The final example in this set is from the Madison regional aquifer study by Plummer *et al.* (1990). The Madison Limestone aquifer occurs in Wyoming, Montana, and South Dakota. Plummer *et al.* (1990) utilized a combination of saturation index constraints, inverse modeling, and carbon

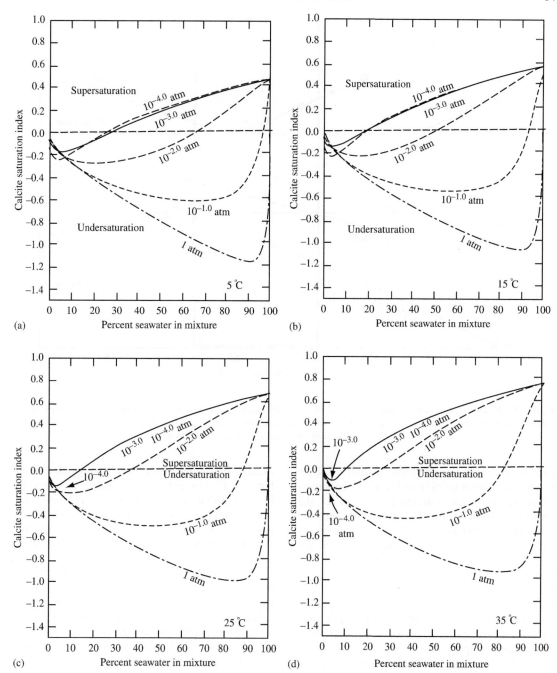

Figure 6 Calcite saturation indices plotted as a function of percent seawater mixing with a carbonate aquifer water with contours representing different levels of P_{CO_2} (source Plummer, 1975).

and sulfur isotopes to delineate geochemical reaction models for the flow paths. The models indicated that the major reaction is dedolomitization, i.e., dolomite dissolution and calcite precipitation driven by anhydrite dissolution, sulfate reduction, $[Ca^{2+} + Mg^{2+}]/Na^+$ cation exchange, with some local halite dissolution. Sulfur isotopes were treated as an isotope dilution problem and carbon isotopes were treated as Rayleigh distillations (see Chapter 5.11). Corroboration of

the isotopic modeling was achieved by predicting the isotopic compositions of the dolomite and the anhydrite. Actual $\delta^{34}S$ values for anhydrite fit the values assumed in the model calculations. Further consistency was found when the adjusted ^{14}C ages combined with Darcy's law resulted in groundwater flow velocities that agreed within a factor of five of those calculated by a digital flow model. Figure 8 portrays the saturation indices of calcite, dolomite, and gypsum (surrogate for anhydrite)

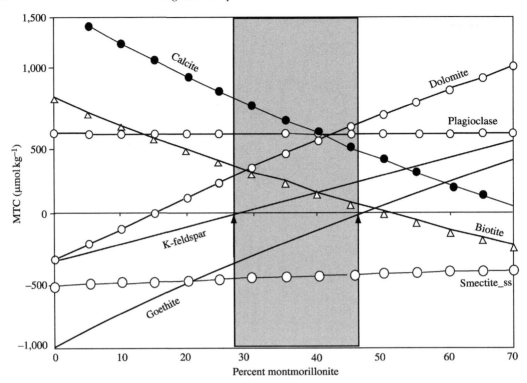

Figure 7 Mineral mass transfer coefficients versus smectite solid solution composition for Wyman Creek mass balances, Inyo Mountains (Bowser and Jones, 2002). Upper and lower bounds on possible smectite compositions are where goethite and K-feldspar mass transfers are equal to zero.

for waters from the Madison aquifer. Calcite reaches saturation quickly and tends to be supersaturated. This supersaturation may reflect the pressure effect on the ion activities and the solubility product constant (pressure corrections were not modeled), it may be caused by pH changes on pumping pressurized water to the surface, it may result from calcite that has some substituted elements displacing the SI values, it may be caused by an inhibition on calcite precipitation kinetics by magnesium, or it may be caused by gypsum dissolving faster than calcite can precipitate.

A subset of the mass-balance results for the Madison aquifer study is shown in Table 2 covering the range of parameters encountered from recharge to discharge although these selected samples are not along the same flow path. The general trend in chemistry is indicated here. Samples near recharge are low in sulfate and the amount of mass transfer is low. As the water moves down gradient and evolves chemically, it increases in sulfate from anhydrite dissolution. Increased anhydrite dissolution leads to increased dissolution of dolomite and increased calcite precipitation. With increasing age there is more organic matter (represented by CH_2O in Table 2) available for decomposition which results in greater amounts of hydrous ferric

oxides dissolving (reducing) and pyrite being formed. Cation exhange in Table 2 refers to $(Ca + Mg)/Na$, i.e., exchange of calcium and magnesium for sodium. The measured sulfur isotopic compositions of the H_2S and of the anhydrite match nicely with the model simulation and the stable carbon isotopes.

Mazor *et al.* (1993) pointed out that some of the samples from the Madison aquifer had high tritium contents when the ^{14}C results indicated dates too old for tritium and suggested that significant mixing of younger and older waters may have occurred. Mixing and dilution trends can have similar chemical appearances to hydrochemical evolution, and caution must be used to distinguish evolutionary trends from mixing trends by age-dating techniques.

5.02.8.4 Reactive-Transport Modeling in Streams

Although biological reactions, mostly dissolved oxygen degradation or nutrient uptake, have been modeled in flowing streams and rivers for a long time, trace-metal reactions have not been modeled until relatively recently. Bencala (1983) considered solute transport in pool-and-riffle streams using a kinetic term for sorption; Bencala and Walters (1983) introduced transient storage into

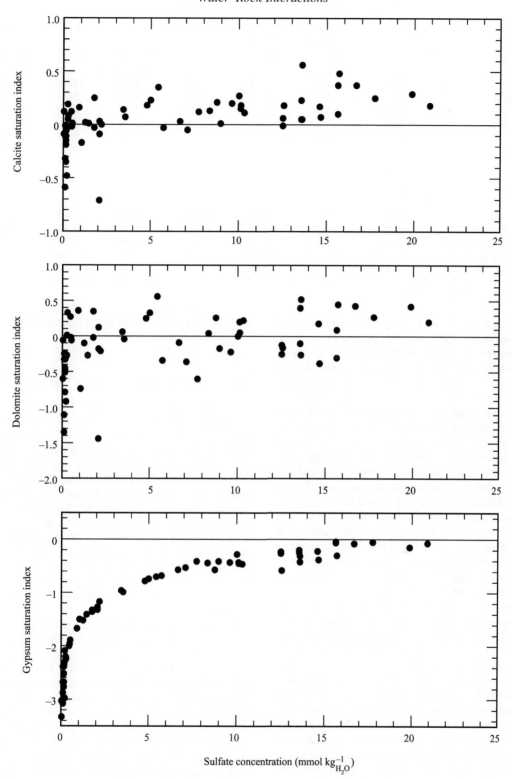

Figure 8   Saturation indices for calcite, dolomite, and gypsum for groundwaters from the Madison limestone aquifer
(source Plummer *et al.*, 1990).

their modeling of trace-metal transport. One of the most obvious inputs of trace metals to surface waters is from mining and mineral-processing wastes. Techniques for modeling these acid mine water reactions continued to develop during the 1990s from empirical rate constants and partition coefficients to mechanistic sorption models such as surface complexation and equilibrium

Table 2 Selected mass transfer results from Plummer *et al.* (1990) for the Madison Limestone aquifer, units in millimoles per kilogram water.

Well no.	F6-19	F6-17	F6-13	F7-10	F8-25	F8-21	F3-20
SO_4	0.18	0.52	2.19	5.73	8.97	13.56	19.86
Dolomite	0.14	0.14	0.60	0.74	1.87	2.90	3.54
Calcite	−0.18	−0.41	−1.62	−2.64	−4.70	−7.15	−5.33
Anhydrite	0.15	0.51	2.27	5.93	9.19	13.56	20.15
CH_2O	0.01	0.02	0.20	0.44	0.71	0.30	0.87
FeOOH	0.00	0.01	0.05	0.12	0.21	0.11	0.09
Pyrite	−0.00	−0.01	−0.05	−0.12	−0.19	−0.06	−0.09
Ion exch.	0.01	0.03	0.04	0.44	0.19	0.14	8.28
NaCl	−0.01	−0.02	0.03	2.83	1.38	0.97	15.31
KCl	0.01	0.02	0.02	0.27	0.28	0.33	2.52
CO_2	−0.07	0.17	−0.21	−0.42	−1.18	−0.40	−0.04
$\delta^{13}C$ (‰) calculated	−8.86	−9.33	−7.79	−9.67	−5.52	−3.59	−2.21
$\delta^{13}C$ (‰) measured	−7.82	−10.0	−6.80	−9.70	−5.50	−3.50	−2.34
Apparent age (yr)	Modern	Modern	Modern	2,386	14,461	13,310	22,588

Positive values signify dissolution, negative values precipitation. Samples, although not along the same flow path, are in the general direction of recharge to discharge from left to right.

precipitation of mineral phases (Brown and Hosseinipour, 1991; Kimball *et al.*, 1994, 1995; Runkel *et al.*, 1996, 1999). Reaction–transport models such as these require more accurate stream discharge measurements than can be obtained from current-meter measurements. These are obtained from constant-flow tracer injection studies along with synoptic sampling, including all possible inflows, so that the transport model can be reliably calibrated. Not only has this approach been used to define sources and sinks of trace metals in mountainous streams but it also has been helpful in predicting remediation scenarios for mine sites (Runkel and Kimball, 2002).

An example of modeling reactive transport of acid mine waters with OTEQ is shown in Figure 9 (Ball *et al.*, 2003) for the drainage released from the Summitville mine in the San Juan Mountains of southwestern Colorado into the Alamosa River. The model was calibrated with tracer injection and synoptic sampling techniques including measurements of Fe(II/III) that helped constrain precipitation of hydrous ferric oxides in the model. Sodium chloride was used as the tracer. Figure 9(a) shows measured pH and simulated pH with reaction. The main in-stream reactive chemistry in this system is the oxidation of iron, the precipitation of hydrous ferric oxides, the precipitation of hydrous aluminum oxides, the adsorption of trace metals, and pH changes (controlled by the oxidation, hydrolysis, and precipitation reactions and by the neutralization of inflows). Simulated and measured dissolved and total aluminum and iron concentrations are shown in Figures 9(b) and (c), respectively. Iron and aluminum concentrations decrease rapidly because of precipitation during downstream transport. Figure 9(d) shows the copper concentrations; when adsorption is invoked in the model the agreement between

measured and simulated copper concentrations is within analytical error.

5.02.8.5 Geochemical Modeling of Catchments

Attempts to model chemical weathering of catchments have used a variety of approaches and were originally designed to understand acidification processes. The BIRKENES code (Christophersen *et al.*, 1982) was one of the first developed to model catchment stream chemistry. It used cation–anion charge balance, a gibbsite equilibrium solubility control for aluminum concentrations, a Gapon ion exchange for metals sorption, and rates for sulfate adsorption/desorption in a two-reservoir model. The model was calibrated by input mass fluxes and output mass fluxes for the Birkenes catchment in Norway to provide the water flux information and to fit empirical parameters.

The ILWAS code (Integrated Lake-Watershed Acidification Study; Chen *et al.*, 1983; Gherini *et al.*, 1985) also contains a semi-empirical model but is much more detailed than the BIRKENES code with respect to its hydrologic and geochemical compartments. Alkalinity was one of the key chemical components in the model. Because the code was originally calibrated on three watersheds receiving acid rain in upstate New York, it was programmed with mineral rate dissolution data, gibbsite solubility equilibrium, and other parameters pertinent to those environments. The considerable quantity of detailed data needed to calibrate the ILWAS code limits its general usefulness.

The MAGIC code (Model of Acidification of Groundwater In Catchments; Cosby *et al.*, 1985a,b) is similar in many respects to the BIRKENES code, but parameters for soil-water

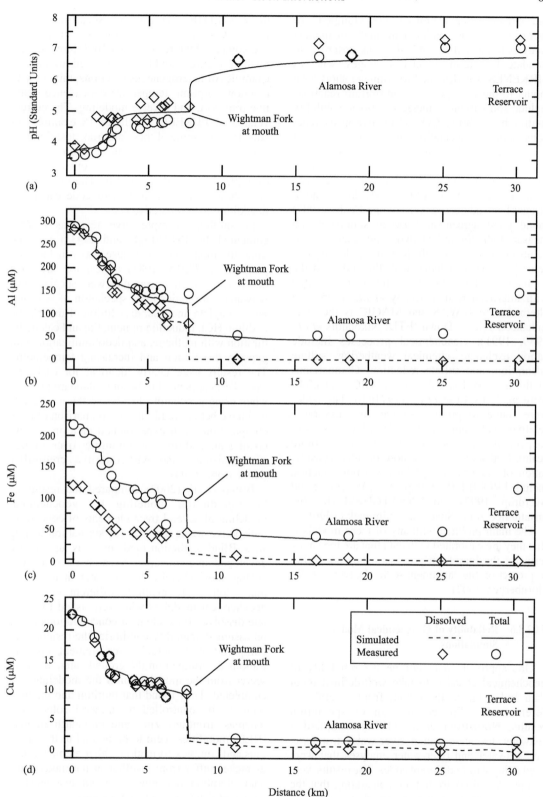

Figure 9 Downstream profiles in the Wightman Fork/Alamosa River system of: (a) pH measured and simulated; (b) dissolved aluminum concentrations measured and simulated; (c) dissolved iron concentrations measured and simulated; and (d) copper concentrations measured and simulated. Simulations were obtained with the OTEQ code after calibration on tracer-injection data and Fe(II/III) determinations.

and stream-water chemistry are "lumped" or averaged over the spatial scale that can include many heterogeneities. MAGIC was designed to simulate long-term (annual) averages, whereas BIRKENES was designed to simulate short-term (hours to days) responses. Equilibrium reactions include soil cation-exchange reactions, solubility control by gibbsite, CO_2-H_2O hydrolysis reactions, aqueous speciation of aluminum among sulfate and fluoride complexes, water dissociation, ion balance, and temperature dependence. Activity coefficients appear not to have been used and the sensitivity of the model to this factor has not been addressed. Wright and Cosby (1987) found good agreement between simulated and measured alkalinities for two manipulated catchments in Norway using the MAGIC code. Organic acids have been included with the code and the simulations of the Norway catchments revisited with improved results (Cosby et al., 1995). An alternative to ILWAS and MAGIC is the code Enhanced Trickle Down (ETD; Nikolaidis et al., 1989, 1991). Geochemical processes include cation exchange, chemical weathering, sulfate adsorption, and sulfate reduction. Comparisons of the three codes, ILWAS, MAGIC, and ETD, were made by Rose et al. (1991a,b). The codes were found to provide similar forecasts on a relative scale but substantial differences with respect to specific concentrations (absolute scale). Codes that contain more detailed consideration of chemistry, especially reaction kinetics, include PROFILE (Sverdrup, 1990; Warfvinge and Sverdrup, 1992) and UNSATCHEM (for the unsaturated zone, Suarez and Simunek (1996)).

The main problem with any of these models is that they are calibrated with data that are too short-term for the long-term processes they are trying to predict on the catchment scale (Drever, 1997b; Hornberger, 2002).

5.02.8.6 Reliability of Geochemical Model Simulations

One approach to determine the reliability of geochemical codes is to take well-defined input data and compare the output from several different codes. For comparison of speciation results, Nordstrom et al. (1979) compiled a seawater test case and a river-water test case, i.e., seawater and river-water analyses that were used as input to 14 different codes. The results were compared and contrasted, demonstrating that the thermodynamic databases, the number of ion pairs and complexes, the form of the activity coefficients, the assumptions made for redox species, and the assumptions made for equilibrium solubilities of mineral phases were prominent factors in the results. Additional arsenic, selenium, and uranium redox test cases were designed for testing of the WATEQ4F code (Ball and Nordstrom, 1991). Broyd et al. (1985) used a groundwater test case and compared the results from 10 different codes. Such test cases could be expanded to include examples of heterogeneous reactions, mixing with reaction, sorption, temperture dependence, reactive transport, and inverse modeling. The current version of PHREEQC includes 18 examples of code testing. Some of these are true "tests" in that they can be compared with measurements or independent computations, while others are just examples of the code capability. The EQ3/6 code also shows several examples of code capability in addition to comparative tests.

In another example, five test cases were computed by PHREEQE and EQ3/6 and the same thermodynamic database was run for each program (INTERA, 1983) to test for any code differences. The five examples were speciation of seawater with major ions, speciation of seawater with complete analysis, dissolution of microcline in dilute HCl, reduction of hematite and calcite by titration with methane, and dedolomitization with gypsum dissolution and increasing temperature. The results were nearly identical for each test case. Test cases need to become standard practice when using geochemical codes so that the results will have better credibility. A comparison of code computations with experimental data on activity coefficients and mineral solubilities over a range of conditions also will improve credibility (Nordstrom, 1994).

Bruno et al. (2002) compared the results of blind-prediction modeling for geochemical modeling of several trace elements of concern to radioactive waste disposal for six natural analog sites. Blind-prediction modeling was an exercise developed in radioactive research on repository analog sites whereby actual water analyses for major ions were given to different groups of geochemical modelers who were asked to simulate dissolved trace-element concentrations based on assumed solubility equilibria; the results were compared with actual trace-element concentrations determined on the groundwaters but kept secret from the modelers until the modeling was completed. They found that thorium and uranium seem to be controlled by mineral solubilities, whereas strontium, zinc, and rare-earth element (REE) mobilities seem to be related to the major ions and complex formation. Other elements such as nickel suffer from insufficient thermodynamic data. Sorption reactions were not examined in sufficient detail to draw conclusions.

Another approach to determining model reliability is to perform sensitivity or uncertainty analyses. One of the first examples is the examination of equilibrium aluminum computations for surface waters using the Monte Carlo method of randomizing sources of error in the water analysis and thermodynamic data (Schecher and Driscoll, 1987, 1988). The results showed that uncertainties in the input data did not contribute significantly to computational speciation errors. Nordstrom and Ball (1989) used two different sets of water analyses, groundwater analyses from a granite in Sweden, and surface-water analyses from a creek receiving AMD in California, to evaluate uncertainties in the water analyses and uncertainties in the thermodynamic data as possible sources of consistent super-saturation effects for calcite, fluorite, barite, ferric oxyhydroxide, and aluminum hydroxide. Instead of Monte Carlo methods, they used a brute force approach and recomputed the speciation assuming reasonable errors in the input analytical data and in the thermodynamic data. The conclusion was that supersaturation for these minerals could not be accounted for by errors in the water analyses nor in the thermodynamic data. For ferric oxyhydroxide, iron colloids were probably getting through the filter and contributing to the apparent dissolved Fe(III) concentrations and the supersaturation effect. For aluminum hydroxide, the saturation indices did not indicate super-saturation with respect to amorphous aluminum hydroxide and hence they were probably reasonable. For the remaining minerals it was concluded that the supersaturation may have been realistic, because it could not be accounted for by propagation of these errors. Other causes of supersaturation include impure mineral phases (solid substitution or defects), disordered or fine-grained nature, inhibition of precipitation rates, and possible unaccounted for pressure effects on equilibria.

Criscenti *et al.* (1996) determined overall uncertainty from geochemical computations by a combination of propagating Monte-Carlo-generated analytical and thermodynamic uncertainties through a geochemical code and applying the "generalized sensitivity analysis" (Spear and Hornberger, 1980) to the output. One of the results of this study is that the aqueous speciation scheme used in many geochemical codes is not necessarily consistent with the speciation scheme used to define standard pH buffers by the National Bureau of Standards. This conclusion raises the possibility that geochemical computations introduce an error when speciating a natural water based on a field pH measurement calibrated with one of these buffers. Cabaniss (1999) investigated methods of uncertainty propagation, comparing the derivative-based method (assumes a linear approximation) with the Monte Carlo technique for solubility equilibria computations of gibbsite, calcite, and jarosite. He found that derivative methods and the assumption of Gaussian uncertainty can misrepresent the propagated uncertainty.

Computational speciation can be compared to analytical speciation for some species. There is always the problem that analytical methods also suffer from operational definitions, interferences, limits of detection, and associated assumptions. Nevertheless, there is no better method of determining accuracy of speciation than by comparing analytical results with computational results (Nordstrom, 1996). In the few instances where this has been done, the comparison ranges from excellent to poor. Examples of studies of this type can be found in Leppard (1983), Batley (1989), and Nordstrom (1996, 2001). Sometimes comparison of two analytical methods for the same speciation can give spurious results. In Table 3, measured and calculated ionic activity coefficients for seawater at 25 °C and 35‰ salinity are compared, after adjusting to a reference value of $\gamma_{Cl} = 0.666$ (Millero, 2001). These values would indicate that for a complex saline solution such as seawater, the activity coefficients can be

Table 3 Comparison of measured and calculated ion activity coefficients in seawater at 25 °C and 35 ‰ salinity, referenced to $\gamma_{Cl} = 0.666$ (Millero, 2001).

Ion	Measured	Calculated
H^+	0.590	0.592
Na^+	0.668	0.674
	0.670	
	0.678	
K^+	0.625	0.619
NH_4^+	0.616	0.624
	0.592	
Mg^{2+}	0.242	0.211
	0.22	
Ca^{2+}	0.203	0.205
	0.180	
F^-	0.296	0.297
Cl^-	0.666	0.666
OH^-	0.242	0.263
	0.254	
	0.254	
HS^-	0.681	0.688
	0.673	
	0.550	
HCO_3^-	0.576	0.574
	0.592	
	0.528	
$B(OH)_4^-$	0.419	0.384
	0.398	
	0.351	
SO_4^{2-}	0.104	0.110
	0.112	
	0.121	
	0.121	
CO_3^{2-}	0.040	0.041
	0.041	
	0.035	
	0.035	

calculated with a high degree of confidence. Much more effort along these lines is needed to better determine the errors and uncertainties for speciation calculations and to identify reaction equilibria and species for which thermodynamic data need to be measured.

Iron redox speciation was evaluated by comparing the Eh values computed from a speciation code that accepts Fe(II) and Fe(III) concentrations determined analytically with Eh values measured electrometrically with a platinum electrode. An example of this comparison is shown in Figure 10 (Nordstrom, 2000). For iron concentrations greater than 10^{-6} m, the comparison is generally excellent (usually to within 30 mV), indicating that the analytical data, speciation calculations, and redox measurements are all consistent with equilibrium as predicted by the Nernst equation. This comparison is one of two redox speciation computations that can be tested electrometrically. Deviations from equilibrium are apparent at low iron concentrations, because the electroactivity of iron is low enough that other electron acceptors, such as dissolved oxygen, begin to interfere. The other redox condition that permits testing with the Nernst equation is sulfide. Berner (1963) showed that the platinum electrode gives Nernstian behavior in anoxic marine sediments in response to dissolved sulfide activities. All other redox species in natural waters are not sufficiently electroactive to establish Nernstian equilibrium.

Nordstrom (2001) demonstrated that if one compares computed free-fluoride ion activities or concentrations (in this example, with WATEQ4F) with measured values obtained with a fluoride ion-selective electrode, the results are in good agreement down to $\sim\!10^{-6}$ m. This agreement between measurement and calculation corroborates the fluoride speciation by the IA method.

5.02.9 FINAL COMMENTS

Numerous models and codes have been developed over the last century for interpreting and portraying low-temperature geochemical processes. They have been applied to a great variety of conditions and processes and have enhanced our ability to understand how low-temperature earth systems work. However, many of these models have a dangerous sophistication for computing almost any type of possibility without adequately constraining what is probable. Perhaps this affair is little different from many centuries ago when we had fewer scientific tools and more imagination to play with. The danger in greater model sophistication is the difficulty in testing and refuting it. If we cannot test a model, then we have no means of determining its reliability. And yet if we do not strive for greater sophistication, then the model lacks representativeness (Oreskes, 2000).

What we must remember is that modeling is a tool—a useful tool to be sure, but not something that can ever replace the experience gained from directly working on a hydrogeochemical problem for several years. Expert judgment, developed over long time periods and involving many mistakes, along with carefully acquired empirical observations in the field and in the laboratory, will ultimately guide our models from possibility to probability. "… even the most mathematically and computationallly sophisticated model will not absolve us of the need for judgment, nor of the need to justify our judgment in human terms" (Oreskes, 2000). Expert judgment is particularly important in identifying the appropriateness of assumptions in applying a model and constitutes a bigger problem than that of model formulation.

Model reliability is a very important aspect of communicating computational results to

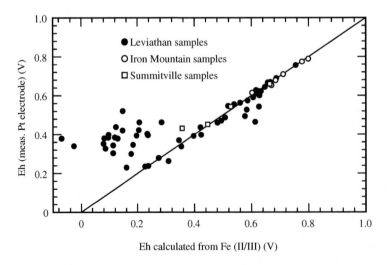

Figure 10   Eh measured with a Pt electrode on site compared to Eh calculated from Fe(II/III) determinations and speciated with WATEQ4F (source Nordstrom, 2000).

managers, risk assessors, stakeholders, politicians, and the public. Unfortunately, sophisticated model computations are not easy to interpret, model results are often nonunique, and modeling is often state-of-the-art science whose reliability has not been adequately tested. Because reliability has often been couched in terms of "verification" and "validation" for the convenience of regulatory requirements (Jenne and Krupka, 1984; Mattigod, 1995; OECD/NEA, 1994; Kirchner *et al.*, 1996; Huguet, 2001; Freedman and Ibaraki, 2003; Celia *et al.*, 1992), confusion about the limitations of models and even misunderstandings about how science works have been propagated. Words such as verification and validation might make sense in a regulatory or legal setting but are inappropriate and even incompatible with scientific research and the scientific method (Konikow and Bredehoeft, 1994; Oreskes *et al.*, 1994). Validation is a matter of legitimacy and has a different context and meaning than what is sought in science. Science progresses by testing hypotheses. Success is measured by consistency between observations and calculations (which is not proof of validity), logical structure, simplicity combined with wide applicability, and consensus through independent peer review (also not proof of validity). Oreskes (2000) quotes Richard Feynman, "Doubt is the essence of understanding." Yet doubt is exactly what regulatory agencies are trying to minimize or eliminate. Oreskes (2000) quotes Richard Feynman, "Doubt is the essence of understanding". Yet doubt is exactly what regulatory agencies are trying to minimize or eliminate. Oreskes (2000) offers a solution that seems obvious the earth scientists: our computational process has far exceeded our observational data on natural systems. More field data and related empirical observations are needed. Field data will provide the necessary constraints to achieve the legitimacy that is being sought by the public. Sophisticated computations, especially in the hands of the unskilled, have the possibility of achieving any pre-conceived result unless adequately constrained by empirical data.

Future efforts should be directed toward developing standardized test cases for a wide variety of processes against which code performance can be compared and tested, incorporation of reliable methods of accounting for metal-organic complexing involving humic substances, recognition of artifacts of sample collection in the determination of trace elements in natural waters, more comparisons of analytical versus computed speciation to obtain accuracy estimates of aqueous speciation, development of routine techniques for estimating uncertainties in model calculations, and more detailed studies of fine-grained mineralogy that are reactive phases in geochemical systems.

ACKNOWLEDGMENTS

This paper was accomplished through the support of the National Research Program and it has benefited substantially from the hosts of geochemical modelers, who have brought the state-of-the-art to its present status. The author is grateful for the reviews and editing of Rob Lee and Tim Drever, who properly questioned writing misdemeanors and awkward styles.

REFERENCES

Aagard P. and Helgeson H. C. (1982) Thermodynamic and kinetic constraints on reaction rates among minerals and aqueous solutions: I. Theoretical considerations. *Am. J. Sci.* **282**, 237–285.

Allison J. D., Brown D. S., and Novo-Gradac K. J. (1991) *MINTEQA2/PRODEFA2, A Geochemical Assessment Model for Environmental Systems, Version 3.0 User's Manual.* US Environ. Prot. Agency (EPA/600/3-91/021).

Alpers C. N. and Nordstrom D. K. (1999) Geochemical modelling of water-rock interactions in mining environments. In *Reviews in Economic Geology, vol. 6A, The Environmental Geochemistry of Mineral Deposits: Part A. Processes, Methods and Health Issues.* (eds. G. S. Plumlee and M. J. Logsdon). Soc. Econ. Geol., Littleton, CO, pp. 289–324.

Alpers C. N., Nordstrom D. K., and Burchard J. M. (1992) *Compilation and Interpretation of Water Quality and Discharge Data for Acid Mine Waters at Iron Mountain, Shasta County, California, 1940–91.* US Geol. Surv. Water-Resour. Invest. Report 91-4160, 173pp.

Appelo C. A. J. and Postma D. (1993) *Geochemistry, Groundwater, and Pollution.* A. A. Balkema, Rotterdam, 536pp.

Archer D. G. and Nordstrom D. K. (2003) Thermodynamic properties of some arsenic compounds of import to groundwater and other applications. *J. Chem. Eng. Data* (accepted but in suspension).

Bain J. G., Blowes D. W., Robertson W. D., and Frind E. O. (2000) Modelling of sulfide oxidation with reactive transport at a mine drainage site. *J. Contamin. Hydrol.* **41**, 23–47.

Bain J. G., Mayer K. U., Blowes D. W., Frind E. O., Molson J. W. H., Kahnt R., and Jenk U. (2001) Modelling the closure-related geochemical evolution of groundwater at a former uranium mine. *J. Contamin. Hydrol.* **52**, 109–135.

Ball J. W. and Nordstrom D. K. (1989) *Final Revised Analyses of Major and Trace Elements from Acid Mine Waters in the Leviathan Mine Drainage Basin, California and Nevada—October, 1981 to October, 1982.* US Geol. Surv. Water-Resour. Invest. Report 89-4138, 49pp.

Ball J. W. and Nordstrom D. K. (1991) *User's Manual for WATEQ4F, with Revised Thermodynamic Data Base and Test Cases for Calculating Speciation of Major, Trace, and Redox Elements in Natural Waters.* US Geol. Surv. Open-File Report 91-183, 189pp.

Ball J. W. and Nordstrom D. K. (1998) Critical evaluation and selection of standard state thermodynamic properties for chromium metal and its aqueous ions, hydrolysis species, oxides, and hydroxides. *J. Chem. Eng. Data* **43**, 895–918.

Ball J. W., Jenne E. A., and Cantrell M. W. (1981) *WATEQ3—A Geochemical Model with Uranium added.* US Geol. Surv. Open-File Report 81-1183, 81pp.

Ball J. W., Parks G. A., Haas J. L., Jr., and Nordstrom D. K. (1988) *A Personal Computer Version of PHAS20, for the Simultaneous Multiple Regression of Thermochemical Data.* US Geol. Surv. Open-File Report 88-489-A, 119pp.

Ball J. W., Runkel R. L., and Nordstrom D. K. (2003) Evaluating remedial alternatives for the Alamosa River and Wightman Fork, Summitville, Colorado: application of a reactive-transport model to low- and high-flow synoptic studies. In *Environ. Sci. Environ. Comput.* (ed. P. Zanetti). EnviroComp Institute, Fremont, CA, vol. 2 (in press).

Barnes H. L. (ed.) (1967) *Geochemistry of Hydrothermal Ore Deposits* Holt, Rinehart and Winston, NY, 670pp.

Barnes H. L. (ed.) (1997) *Geochemistry of Hydrothermal Ore Deposits* 3rd edn. Wiley, NY, 972pp.

Bassett R. L. (1997) Chemical modelling on the bare-rock or forested watershed scale. *Hydrol. Proc.* **11**, 695–718.

Bassett R. L. and Melchior D. C. (1990) Chemical modelling of aqueous systems: an overview. In *Chemical Modelling of Aqueous Systems II*, Symp. Ser. 416 (eds. D. C. Melchior and R. L. Bassett). Am. Chem. Soc., Washington, DC, pp. 1–14.

Batley G. E. (ed.) (1989) *Trace Element Speciation: Analytical Methods and Problems.* CRC Press, Boca Raton, Florida, 350pp.

Bencala K. E. (1983) Simulation of solute transport in a mountain pool-and-riffle stream with a kinetic mass transfer model for sorption. *Water Resour. Res.* **19**, 732–738.

Bencala K. E. and Walters R. A. (1983) Simulation of solute transport in a mountain pool-and-riffle stream: a transient storage model. *Water Resour. Res.* **19**, 718–724.

Berner R. A. (1963) Electrode studies of hydrogen sulfide in marine sediments. *Geochim. Cosmochim. Acta* **27**, 563–575.

Berner R. A. (1978) Rate control of mineral dissolution under earth surface conditions. *Am. J. Sci.* **278**, 1235–1252.

Berner E. K. and Berner R. A. (1996) *Global Environment: Water, Air, and Geochemical Cycles.* Prentice Hall, NJ, 376pp.

Bethke C. M. (1994) *The Geochemist's Workbench™ Version 2.0: A User's Guide to Rxn, Act2. Tact, React, and Gtplot.* University of Illinois, 213pp.

Bethke C. M. (1996) *Geochemical Reaction Modelling: Concepts and Applications.* Oxford University Press, NY, 397pp.

Bjerrum J. (1950) On the tendency of metal ions toward complex formation. *Chem. Rev.* **46**, 381–402.

Bormann F. H. and Likens G. E. (1967) Nutrient cycling. *Science* **155**, 424–429.

Bowser C. J. and Jones B. F. (1990) Geochemical constraints on ground waters dominated by silicate hydrolysis: an interactive spreadsheet, mass balance approach. *Chem. Geol.* **84**, 33–35.

Bowser C. J. and Jones B. F. (2002) Mineralogic controls on the composition of natural waters dominated by silicate hydrolysis. *Am. J. Sci.* **302**, 582–662.

Brown K. P. and Hosseinipour E. Z. (1991) New methods for modelling the transport of metals from mineral processing wastes into surface waters. *J. Environ. Sci. Health A* **26**, 157–203.

Broyd T. W., Grant M. M., and Cross J. E. (1985) A report on intercomparison studies of computer programs which respectively model: (i) Radionuclide migration (ii) Equilibrium chemistry of groundwater. EUR 10231 EN, Commission of the European Communities, Luxembourg.

Brønsted J. N. (1922) Studies on solubility: IV. The principle of specific interaction of ions. *J. Am. Chem. Soc.* **44**, 877–898.

Bruno J., Duro L., and Grivé M. (2002) The applicability and limitations of thermodynamic geochemical models to simulate trace element behaviour in natural waters: lessons learned from natural analogue studies. *Chem. Geol.* **190**, 371–393.

Cabaniss S. E. (1999) Uncertainty propagation in geochemical calculations: non-linearity in solubility equilibria. *Appl. Geochim.* **14**, 255–262.

Celia M. A., Gray W. G., Hassanizadeh S. M., and Carrera J. (eds.) (1992) Validation of geo-hydrological models: Part I. *Adv. Water Resour.* **15**, 1–274.

Charlton S. R., Macklin C. L., and Parkhurst D. L. (1997) *PHREEQCI—A Graphical User Interface for the Geochemical Computer Program PHREEQC.* US Geol. Surv. Water-Resour. Invest. Report 97-4222, 9pp.

Chase M. W., Jr. (ed.) (1998) *NIST-JANAF Thermochemical Tables*, 4th edn. *J. Phys. Chem. Ref. Data*, Monogr. 9, Parts I and II. American Chemical Society and American Institute of Physics, NY, 1963pp.

Chen C. W., Gherini S. A., and Goldstein R. A. (1983) *The Integrated Lake-watershed Acidification Study, Vol. 1: Model Principles and Application Procedures.* EPRI Report EA-3221.

Christophersen N., Seip H. M., and Wright R. F. (1982) A model for streamwater chemistry at Birkenes, Norway. *Water Resour. Res.* **18**, 977–996.

Cosby B. J., Hornberger G. M., Galloway J. N., and Wright R. F. (1985a) Modelling the effects of acid deposition: assessment of a lumped-parameter model of soil water and streamwater chemistry. *Water Resour. Res.* **21**, 51–63.

Cosby B. J., Hornberger G. M., Galloway J. N., and Wright R. F. (1985b) Time scales of catchment acidification. *Environ. Sci. Technol.* **19**, 1144–1149.

Cosby B. J., Wright R. F., and Gjessing E. (1995) An acidification model (MAGIC) with organic acids evaluated using whole-catchment manipulations in Norway. *J. Hydrol.* **170**, 101–122.

Cox J. D., Wagman D. D., and Medvedev V. A. (1989) *CODATA Key Values for Thermodynamics.* Hemisphere Publishing Corp., 271pp.

Criscenti L. J., Laniak G. F., and Erikson R. L. (1996) Propagation of uncertainty through geochemical code calculations. *Geochim. Cosmochim. Acta* **60**, 3551–3568.

Davis J. A. and Hayes K. F. (eds.) (1986) *Geochemical Processes at Mineral Surfaces*, Am. Chem. Soc. Symp. Series 323. American Chemical Society, Washington, DC, 683pp.

Debye P. and Hückel E. (1923) On the theory of electrolytes. *Phys. Z.* **24** 185–208, 305–325.

Derry G. N. (1999) *What Science is and How it Works.* Princeton University Press, NJ, 311pp.

Dobbs J. C., Susetyo W., Carreira L. A., and Azarraga L. V. (1989) Competitive binding of protons and metal ions in humic substances by lanthanide ion probe spectroscopy. *Anal. Chem.* **61**, 1519–1524.

Domenico P. A. and Schwartz F. W. (1990) *Physical and Chemical Hydrogeology.* Wiley, NY, 824pp.

Drever J. I. (1997a) *The Geochemistry of Natural Waters: Surface and Groundwater Environments*, 3rd edn. Prentice Hall, NJ, 436pp.

Drever J. I. (1997b) Catchment mass balance. In *Geochemical Processes, Weathering, and Groundwater Recharge in Catchments* (eds. O. Saether and P. de Cariat). A. A. Balkema, Rotterdam, pp. 241–261.

Dyrssen D., Jagner D., and Wengelin F. (1968) *Computer Calculations of Ionic Equilibria and Titration Procedures.* Almqvist and Wiksell, Stockholm, 250pp.

Dzombak D. A. and Morel F. M. M. (1990) *Surface Complexation Modelling: Hydrous Ferric Oxide.* Wiley, NY, 393pp.

Eugster H. P. (1971) The beginnings of experimental petrology. *Science* **173**, 481–489.

Eugster H. P., Harvie C. E., and Weare J. H. (1980) Mineral equilibria in the six-component seawater system, Na–K–Mg–Ca–SO_4–Cl–H_2O, at 25°C. *Geochim. Cosmochim. Acta* **44**, 1335–1347.

Felmy A. R., Girvin D. C., and Jenne E. A. (1984) *MINTEQ— A Computer Program for Calculating Aqueous Geochemical Equilibria.* US Environ. Prot. Agency (EPA-600/3-84-032).

Felmy A. R., Rai D., and Amonette J. E. (1990) The solubility of barite and celestite in sodium sulfate: evaluation of thermodynamic data. *J. Sol. Chem.* **19**, 175–185.

Freedman V. L. and Ibaraki M. (2003) Coupled reactive mass transport and fluid flow: issues in model verification. *Adv. Water Resour.* **26**, 117–127.

Freeze R. A. and Cherry J. A. (1979) *Groundwater*. Prentice-Hall, Englewood Cliffs, NJ, 604pp.

Garrels R. M. (1960) *Mineral Equilibria*. Harper and Brothers, NY, 254pp.

Garrels R. M. and Christ C. L. (1965) *Solutions, Minerals, and Equilibria*. Harper and Row, NY, 450pp.

Garrels R. M. and Mackenzie F. T. (1967) Origin of the chemical composition of some springs and lakes. In *Equilibrium Concepts in Natural Water Systems*. Adv. Chem. Series 67, Am. Chem. Soc., Washington, DC, pp. 222–242.

Garrels R. M. and Thompson M. E. (1962) A chemical model for seawater at 25°C and one atmosphere total pressure. *Am. J. Sci.* **260**, 57–66.

Garvin D., Parker V. B., and White H. J. (1987) *CODATA Thermodynamic Tables—Selections for Some Compounds of Calcium and Related Mixtures: A Prototype Set of Tables*. Hemisphere Publishing Corp., Washington, DC, 356pp.

Gerke H. H., Molson J. W., and Frind E. O. (1998) Modelling the effect of chemical heterogeneity on acidification and solute leaching in overburden mine spoils. *J. Hydrol.* **209**, 166–185.

Gherini S. A., Mok L., Hudson R. J. M., Davis G. F., Chen C. W., and Goldstein R. A. (1985) The ILWAS model: formulation and application. *Water Air Soil Pollut.* **26**, 425–459.

Glynn P. and Brown J. (1996) Reactive transport modelling of acidic metal-contaminated ground water at a site with sparse spatial information. In *Reactive Transport in Porous Media*, Rev. Mineral. (eds. P. C. Lichtner, C. I. Steefel, and E. H. Oelkers). Mineralogical Society of America and Geological Society, Washington, DC, vol. 34, pp. 377–438.

Goldberg E. D. (1954) Marine geochemistry: 1. Chemical scavengers of the sea. *J. Geol.* **62**, 249–265.

Greenberg J. P. and Møller N. (1989) The prediction of mineral solubilities in natural waters: a chemical equilibrium model for the Na–K–Ca–Cl–SO$_4$–H$_2$O system to high concentrations from 0 to 250°C. *Geochim. Cosmochim. Acta* **53**, 2503–2518.

Greenwood H. (1989) On models and modelling. *Can. Mineral.* **27**, 1–14.

Grenthe I. and Puigdomenech I. (eds.) (1997) *Modelling in Aquatic Chemistry*. Organization for Economic Co-operation and Development/Nuclear Energy Agency Publications, Paris, France, 742pp.

Grenthe I., Fuger J., Konings R. J. M., Lemire R. J., Muller A. B., Nguyen-Trung C., and Wanner H. (1992) *Chemical Thermodynamics of Uranium*. Elsevier, Amsterdam, 715pp.

Grenthe I., Plyasunov A. V., and Spahiu K. (1997) Estimations of medium effects on thermodynamic data. In *Modelling in Aquatic Chemistry* (eds. I. Grenthe and I. Puigdomenech). Organization for Economic Co-operation and Development/ Nuclear energy Agency Publications, Paris, France, pp. 343–444.

Gryzb K. R. (1995) NOAEM (natural organic anion equilibrium model): a data analysis algorithm for estimating functional properties of dissolved organic matter in aqueous environments: Part I. Ionic component speciation and metal association. *Org. Geochem.* **23**, 379–390.

Guggenheim G. A. (1935) The specific thermodynamic properties of aqueous solutions of strong electrolytes. *Phil. Mag.* **19**, 588–643.

Gurvich L. V., Veyts I. V., Alcock C. B., and Iorish V. S. (1993) *Thermodynamic Properties of Individual Substances, Parts 1 and 2*. CRC Press, Boca Raton, FL, vol. 3, 1120pp.

Haas J. L., Jr. (1974) *PHAS20, a Program for Simultaneous Multiple Regression of a Mathematical Model to Thermochemical Data*. Natl. Tech. Info. Service Report AD-780 301, 158pp.

Haas J. L. and Fisher J. R. (1976) Simultaneous evaluation and correlation of thermodynamic data. *Am. J. Sci.* **276**, 525–545.

Harned H. S. and Owen B. B. (1958) *The Physical Chemistry of Electrolyte Solutions*, 3rd edn. Reinhold, NY, 354pp.

Harvie C. E. and Weare J. H. (1980) The prediction of mineral solubilities in natural waters: the Na–K–Mg–Ca–Cl–SO$_4$–H$_2$O system from zero to high concentration at 25°C. *Geochim. Cosmochim. Acta* **44**, 981–997.

Harvie C. E., Weare J. H., Hardie L. A., and Eugster H. P. (1980) Evaporation of seawater: calculated mineral sequences. *Science* **208**, 498–500.

Harvie C. E., Eugster H. P., and Weare J. H. (1982) Mineral equilibria in the six-component seawater system, Na–K–Mg–Ca–SO$_4$–Cl–H$_2$O at 25°C: II. composition of the saturated solutions. *Geochim. Cosmochim. Acta* **46**, 1603–1618.

Harvie C. E., Møller N., and Weare J. H. (1984) The prediction of mineral solubilities in natural waters: the Na–K–Mg–Ca–H–Cl–SO$_4$–OH–HCO$_3$–CO$_3$–CO$_2$–H$_2$O system to high ionic strengths at 25°C. *Geochim. Cosmochim. Acta* **48**, 723–751.

Harvie C. E., Greenberg J. P., and Weare J. H. (1987) A chemical equilibrium algorithm for highly non-ideal multiphase systems: free energy minimization. *Geochim. Cosmochim. Acta* **51**, 1045–1057.

Helgeson H. C. (1964) *Complexing and Hydrothermal Ore Deposition*. Pergamon, NY, 128pp.

Helgeson H. C. (1967) Thermodynamics of complex dissociation in aqueous solution at elevated temperatures. *J. Phys. Chem.* **71**, 3121–3136.

Helgeson H. C. (1968) Evaluation of irreversible reactions in geochemical processes involving minerals and aqueous solutions: I. Thermodynamic relations. *Geochim. Cosmochim. Acta* **32**, 853–877.

Helgeson H. C. (1969) Thermodynamics of hydrothermal systems at elevated temperatures and pressures. *Am. J. Sci.* **167**, 729–804.

Helgeson H. C. (1971) Kinetics of mass transfer among silicates and aqueous solutions. *Geochim. Cosmochim. Acta* **35**, 421–469.

Helgeson H. C. and Kirkham D. H. (1974a) Theoretical prediction of the thermodynamic behavior of aqueous electrolytes at high pressures and temperatures: I. Summary of the thermodynamic/electrostatic properties of the solvent. *Am. J. Sci.* **274**, 1089–1198.

Helgeson H. C. and Kirkham D. H. (1974b) Theoretical prediction of the thermodynamic behavior of aqueous electrolytes at high pressures and temperatures: II. Debye-Hückel parameters for activity coefficients and relative partial molal properties. *Am. J. Sci.* **274**, 1199–1261.

Helgeson H. C. and Kirkham D. H. (1976) Theoretical prediction of the thermodynamic behavior of aqueous electrolytes at high pressures and temperatures: III. Equation of state for aqueous species at infinite dilution. *Am. J. Sci.* **276**, 97–240.

Helgeson H. C. and Murphy W. M. (1983) Calculation of mass transfer as a function of time and surface area in geochemical processes: I. Computational approach. *Math. Geol.* **15**, 109–130.

Helgeson H. C., Garrels R. M., and Mackenzie F. T. (1969) Evaluation of irreversible reactions in geochemical processes involving minerals and aqueous solutions: II. Applications. *Geochim. Cosmochim. Acta* **33**, 455–481.

Helgeson H. C., Brown T. H., Nigrini A., and Jones T. A. (1970) Calculation of mass transfer in geochemical processes involving aqueous solution. *Geochim. Cosmochim. Acta* **34**, 569–592.

Helgeson H. C., Delany J. M., Nesbitt H. W., and Bird D. W. (1978) Summary and critique of the thermodynamic properties of rock-forming minerals. *Am. J. Sci.* **278-A**, 1–229.

Helgeson H. C., Kirkham D. H., and Flowers G. C. (1981) Theoretical prediction of the thermodynamic behavior of aqueous electrolytes at high pressures and temperatures: IV. Calculation of activity coefficients and relative partial moral properties to 600°C and 5 kb. *Am. J. Sci.* **281**, 1249–1516.

Helgeson H. C., Murphy W. M., and Aagard P. (1984) Thermodynamic and kinetic constraints on reaction rates among minerals and aqueous solutions: II. Rate constants, effective surface area, and the hydrolysis of feldspar. *Geochim. Cosmochim. Acta* **48**, 2405–2432.

Hem J. D. (1959) *Study and Interpretation of the Chemical Characteristics of Natural Water*. US Geol. Surv. Water-Supply Paper 1473, 269pp.

Hem J. D. (1961) *Calculation and Use of Ion Activity*. US Geological Survey Water-Supply Paper 1535-C.

Hem J. D. and Cropper W. H. (1959) *Survey of Ferrous-ferric Chemical Equilibria and Redox Potentials*. US Geological Survey Water-Supply Paper 1459-145A.

Hitchon B., Perkins E. H., and Gunter W. D. (1996) *Introduction of Ground Water Geochemistry and SOLMINEQ.GW*. Geoscience Publishing Ltd., Alberta.

Hochella M. F., Jr. and White A. F. (eds.) (1990) *Mineral–Water Interface Geochemistry, Rev. Mineral.* Mineralogical Society of America, Washington, DC, vol. **23**, 603pp.

Hornberger G. M. (2002) Impacts of acidic atmospheric deposition on the chemical composition of stream water and soil water. In *Environmental Foresight and Models: A Manifesto* (ed. M. B. Beck). Elsevier, NY, pp. 131–145.

Huber C., Filella M., and Town R. M. (2002) Computer modelling of trace metal ion speciation: practical implementation of a linear continuous function for complexation by natural organic matter. *Comput. Geosci.* **28**, 587–596.

Huguet J. M. (2001) Testing and validation of numerical models of groundwater flow, solute transport and chemical reactions in fractured granites: a quantitative study of the hydrogeological and hydrochemical impact produced. ENRESA Tech. Publ. 06/2001, 253pp.

Hunter K. S., Wang Y., and Van Cappellen P. (1998) Kinetic modelling of microbially-driven redox chemistry of subsurface environments: coupling transport, microbial metabolism and geochemistry. *J. Hydrol.* **209**, 53–80.

Ingri N. and Sillén L. G. (1962) High-speed computers as a supplement to graphical methods: II. Some computer programs for studies of complex formation equilibria. *Acta Chem. Scand.* **16**, 173–191.

Ingri N., Kakolowicz W., Sillén L. G., and Warnqvist B. (1967) HALTAFALL, a general program for calculating the composition of equilibrium mixtures. *Talanta* **14**, 1261–1286.

INTERA (1983) Geochemical models suitable for performance assessment of nuclear waste storage: comparison of PHREEQE and EQ3/EQ6. INTERA Environmental Consultants, Inc., ONWI-473, 114pp.

Jenne E. A. (ed.) (1979) *Chemical Modelling of Aqueous Systems*. Am. Chem. Soc. Symp. Series 93, Washington, DC, 914pp.

Jenne E. A. and Krupka K. M. (1984) *Validation of Geochemical Models*. USNRC Report PNL-SA-12442, 9pp.

Johnson J. W., Oelkers E. H., and Helgeson H. C. (1992) SUPCRT92: a software package for calculating the standard moral thermodynamic properties of minerals, gases, aqueous species, and reactions from 1 to 5,000 bars and 0° to 1,000°C. *Comp. Geosci.* **18**, 899–947.

Judson H. F. (1980) *The Search for Solutions*. Holt, Rinehart and Winston, NY, 211pp.

Kharaka Y. K. and Barnes I. (1973) *SOLMNEQ: Solution-mineral Equilibrium Computations*. Natl. Tech. Info. Serv. Report PB214-899, 82pp.

Kharaka Y. K., Gunter W. D., Aggarwal P. K., Perkins E. H., and Debraal J. D. (1988) *SOLMINEQ.88: A Computer Program for Geochemical Modelling of Water-rock Interactions*. US Geol. Surv. Water-Resour. Invest. Report 88-4227, 420pp.

Kimball B. A., Broshears R. E., Bencala K. E., and McKnight D. M. (1994) Coupling of hydrologic transport and chemical reactions in a stream affected by acid mine drainage. *Environ. Sci. Tech.* **28**, 2065–2073.

Kimball B. A., Callender E., and Axtmann E. V. (1995) Effects of colloids on metal transport in a river receiving acid mine drainage, Upper Arkansas River, Colorado USA. *Appl. Geochem.* **10**, 285–306.

Kimball B. A., Runkel R. L., and Walton-Day K. (2003) Use of field-scale experiments and reactive transport modelling to evaluate remediation alternatives in streams affected by acid mine drainage. In *Environmental Aspects of Mine Wastes*, Mineral. Assoc. Canada Short Course Series 31 (eds. J. L. Jambor, D. W. Blowes, and A. I. M. Ritchie). Mineralogical Association of Canada, Ottawa, pp. 261–282.

Kipp K. L., Jr. (1987) *HST3D—A Computer Code for Simulation of Heat and Solute Transport in Three-dimensional Ground-water Flow Systems*. US Geol. Surv. Water-Resour. Invest. Report 86-4095, 517pp.

Kipp K. L., Jr. (1998) *Guide to the Revised Heat and Solute Transport Simulator: HST3D*. US Geol. Surv. water-Resour. Report 97-4157, 149pp.

Kirchner J. W., Hooper R. P., Kendall C., Neal C., and Leavesley G. (1996) Testing and validating environmental models. *Sci. Tot. Environ.* **183**, 33–47.

Konikow L. F. and Bredehoeft J. D. (1992) Ground-water models cannot be validated. *Adv. Water Resour.* **15**, 75–83.

Krauskopf K. B. (1951) The solubility of gold. *Econ. Geol.* **46**, 858–870.

Krauskopf K. B. (1955) Sedimentary deposits of rare metals. *Econ. Geol. Fiftieth Anniv. Part I*, pp. 411–463.

Krauskopf K. B. (1956) Factors controlling the concentrations of thirteen rare metals in sea water. *Geochim. Cosmochim. Acta* **9**, 1–32.

Krauskopf K. B. (1967) *Introduction to Geochemistry*. McGraw-Hill, NY, 721pp.

Lambert I. and Clever H. L. (eds.) (1992) *Alkaline Earth Hydroxides in Water and Aqueous Solutions*. IUPAC Solubility Data Series. Pergamon, Oxford, UK, vol. 52, 365pp.

Langmuir D. (1971) The geochemistry of some carbonate ground waters in central Pennsylvania. *Geochim. Cosmochim. Acta* **35**, 1023–1045.

Langmuir D. (1997) *Aqueous Environmental Geochemistry*. Prentice Hall, NJ, 600pp.

Latimer W. M. (1952) *The Oxidation State of the Elements and their Potentials in Aqueous Solution*. Prentice-Hall, NY, 392pp.

Lemire R. J. (2001) *Chemical Thermodynamics of Neptunium and Plutonium*. North-Holland/Elsevier, Amsterdam, 836pp.

Leppard G. G. (1983) *Trace Element Speciation in Surface Waters and Its Ecological Implications*, NATO Conf. Series. Plenum, NY, vol. 6, 320pp.

Lewis G. N. and Randall M. (1921) The activity coefficient of strong electrolytes. *J. Am. Chem Soc.* **43**, 1112–1153.

Lewis G. N. and Randall M. (1923) *Thermodynamics and the Free Energy of Chemical Substances*. McGraw-Hill, NY, 653pp.

Lichtner P. C. (1996) Continuum formulation of multi-component-multiphase reactive transport. In *Reactive Transport in Porous Media* (eds. P. C. Lichtner, C. I. Steefel, and E. H. Oelkers). Rev. Mineral., Mineralogical Society of America, Washington, DC, vol. 34, pp. 1–81.

Lichtner P. C. (2001) *FLOTRAN User's Manual*. Los Alamos National Laboratory Report LA-UR-01-2349, Los Alamos, NM.

Lichtner P. C., Steefel C. I., and Oelkers E. H. (eds.) (1996) *Reactive Transport in Porous Media*. Rev. Mineral., Mineralogical Society of America, Washington, DC, vol. 34, 438pp.

Likens G. E., Bormann F. H., Pierce R. S., Eaton J. S., and Johnson N. M. (1977) *Biogeochemistry of a Forested Ecosystem*. Springer, NY, 146pp.

Loeppert R. H. Schwab. A. P., and Goldberg S. (eds.) (1995) *Chemical Equilibrium and Reaction Models*, Spec. Publ. 42. Soil Science Society of America, Madison, WI, 422pp.

Mangold D. C. and Tsang C.-F. (1991) A summary of subsurface hydrological and hydrochemical models. *Rev. Geophys.* **29**, 51–79.

Majzlan J., Navrotsky A., and Neil J. M. (2002) Energetics of anhydrite, barite, celestine, and anglesite: a high-temperature and differential scanning calorimetry study. *Geochim. Cosmochim. Acta* **66**, 1839–1850.

Marion G. M. and Farren R. E. (1999) Mineral solubilities in the Na–K–Mg–Ca–Cl–SO₄–H₂O system: a re-evaluation of the sulfate chemistry in the Spencer–Møller–Weare model. *Geochim. Cosmochim. Acta* **63**, 1305–1318.

Martell A. E. and Smith R. M. (1974–1976) *Critical Stability Constants*: Vol. 1 Amino Acids, Vol. 2 Amines, Vol. 3 Other Organic Ligands, Vol. 4 Inorganic Compounds, Plenum, NY, 1636pp.

Mattigod S. V. (1995) Validation of geochemical models. In *Chemical Equilibrium and Reaction Models*, Spec. Publ. 42 (eds. R. H. Loeppert, A. P. Schwab, and S. Goldberg). Soil Science Society of America, Madison, WI, pp. 201–218.

Mayer K. U., Benner S. G., and Blowes D. W. (1999) The reactive transport model MIN3P: application to acid-mine drainage generation and treatment—Nickel Rim mine site, Sudbury, Ontario. In *Mining and Environment 1* (eds. D. Goldsack, N. Bezlie, P. Yearwood, and G. Hall). Laurentian University, Sudbury, Ontario, Canada, pp. 145–154.

Mayer K. U., Blowes D. W., and Frind E. O. (2003) Advances in reactive-transport modelling of contaminant release and attenuation from mine-waste deposits. In *Environmental Aspects of Mine Wastes*, Mineralogical Association of Canada Short Course Series (eds. J. L. Jambor, D. W. Blowes, and A. I. M. Ritchie). Mineralogical Association of Canada, Ottawa, Canada, vol. 31, pp. 283–302.

Mazor E., Drever J. I., Finlay J., Huntoon P. W., and Lundy D. A. (1993) Hydrochemical implications of groundwater mixing: an example from the Southern Laramie Basin, Wyoming. *Water Resour. Res.* **29**, 193–205.

Melchior D. C. and Bassett R. L. (eds.) (1990) *Chemical Modelling of Aqueous Systems: II*. Symp. Ser Am. Chem. Soc., 416, Washington, DC, 556pp.

Millero F. J. (2001) *The Physical Chemistry of Natural Waters*. Wiley, NY, 654pp.

Millero F. J. and Pierrot D. (1998) A chemical model for natural waters. *Aqua. Geochem.* **4**, 153–199.

Millero F. J. and Roy R. (1997) A chemical model for the carbonate system in natural waters. *Croat. Chem. Acta* **70**, 1–38.

Millero F. J. and Schreiber D. R. (1982) Use of the ion pairing model to estimate activity coefficients of the ionic components of natural waters. *Am. J. Sci.* **282**, 1508–1540.

Monnin C. (1990) The influence of pressure on the activity coefficients of the solutes and on the solubility of minerals in the system Na–Ca–Cl–SO₄–H₂O to 200°C and 1 kbar, and to high NaCl concentration. *Geochim Cosmochim. Acta* **52**, 821–834.

Monnin C. and Ramboz C. (1996) The anhydrite saturation index of the ponded brines and the sediment pore waters of the Red Sea deeps. *Chem. Geol.* **127**, 141–159.

Morel F. and Morgan J. J. (1972) A numerical method for computing equilibria in aqueous systems. *Environ. Sci. Tech.* **6**, 58–67.

Murphy W. M. and Helgeson H. C. (1987) Thermodynamic and kinetic constraints on reaction rates among minerals and aqueous solutions: III. Activated complexes and the pH dependence of the rates of feldspar, pyroxene, wollastonite, and olivine hydrolysis. *Geochim. Cosmochim. Acta* **51**, 3137–3153.

Murphy W. M., Oelkers E. H., and Lichtner P. C. (1989) Surface reaction versus diffusion control of mineral dissolution and growth rates in geochemical processes. *Chem. Geol.* **78**, 357–380.

Møller N. (1988) The prediction of mineral solubilities in natural waters: a chemical equilibrium model for the Na–Ca–Cl–SO₄–H₂O system, to high temperature and concentration. *Geochim. Cosmochim. Acta* **52**, 821–837.

Møller N., Greenberg J. P., and Weare J. H. (1998) Computer modelling for geothermal systems: predicting carbonate and silica scale formation, CO₂ breakout and H₂S exchange. *Transport Porous Media* **33**, 173–204.

Nikolaidis N. P., Schnoor J. L., and Geogakakos K. P. (1989) Modelling of long-term lake alkalinity responses to acid deposition. *J. Water Pollut. Control Feder.* **61**, 188–199.

Nikolaidis N. P., Muller P. K., Schnoor J. L., and Hu H. L. (1991) Modelling the hydrogeochemical response of a stream to acid deposition using the enhanced trickle-down model. *Res. J. Water Pollut. Control Feder.* **63**, 220–227.

Nordstrom D. K. (1977) Hydrogeochemical and microbiological factors affecting the heavy metal chemistry of an acid mine drainage system. PhD Dissertation, Stanford University, 210pp.

Nordstrom D. K. (1994) On the evaluation and application of geochemical models, Appendix 2. In *Proc. 5th CEC Natural Analogue Working Group and Alligator Rivers Analogue Project, Toledo, Spain, October 5–19, 1992*. EUR 15176 EN, pp. 375–385.

Nordstrom D. K. (1996) Trace metal speciation in natural waters: computational vs. analytical. *Water, Air, Soil Pollut.* **90**, 257–267.

Nordstrom D. K. (2000) Advances in the hydrogeochemistry and microbiology of acid mine waters. *Int. Geol. Rev.* **42**, 499–515.

Nordstrom D. K. (2001) A test of aqueous speciation: measured vs. calculated free fluoride ion activity. In *Proc. 10th Int. Symp. Water–Rock Interaction, WRI-10, Villasimius, Sardinia, July 10–15, 2001* (ed. R. Cidu), pp. 317–320.

Nordstrom D. K. and Archer D. G. (2003) Arsenic thermodynamic data and environmental geochemistry. In *Arsenic in Ground Water* (eds. A. H. Welch and K. G. Stollenwerk). Kluwer, Boston, MA, pp. 1–25.

Nordstrom D. K. and Ball J. W. (1989) Mineral saturation states in natural waters and their sensitivity to thermodynamic and analytic errors. *Sci. Géol. Bull.* **42**, 269–280.

Nordstrom D. K. and Munoz J. L. (1994) *Geochemical Thermodynamics*, 2nd edn. Blackwell, Boston, 493pp.

Nordstrom D. K., Plummer L. N., Wigley T. M. L., Wolery T. J., Ball J. W., Jenne E. A., Bassett R. L., Crerar D. A., Florence T. M., Fritz B., Hoffman M., Holdren G. R., Jr., Lafon G. M., Mattigod S. V., McDuff R. E., Morel F., Reddy M. M., Sposito G., and Thrailkill J. (1979) A comparison of computerized chemical models for equilibrium calculations in aqueous systems. In *Chemical Modelling of Aqueous Systems*, Symp. Ser. 93. American Chemical Society, Washington, DC, pp. 857–892.

Nordstrom D. K., Plummer L. N., Langmuir D., Busenberg E., May H. M., Jones B. F., and Parkhurst D. L. (1990) Revised chemical equilibrium data for major water-mineral reactions and their limitations. In *Chemical Modelling of Aqueous Systems II*, Symp. Ser. 416 (eds. D. C. Melchior and R. L. Bassett). American Chemical Society, Washington, DC, pp. 398–413.

OECD/NEA (1994) *In situ* Experiments at the Stripa Mine. In *Proc. 4th Int. NEA/SKB Symp., Stockholm, Sweden, Oct. 14–16, 1992*, 468pp.

Oreskes N. (2000) Why believe a computer? Models, measures, and meaning in the natural world. In *The Earth Around Us* (ed. Jill Schneiderman). W. H. Freeman, NY, pp. 70–82.

Oreskes N., Shrader-Frechette K., and Belitz K. (1994) Verification, validation, and confirmation of numerical models in the earth sciences. *Science* **263**, 641–646.

Pačes T. (1983) Rate constants of dissolution derived from the measurements of mass balance in hydrological catchments. *Geochim. Cosmochim. Acta* **47**, 1855–1863.

Pačes T. (1984) Mass-balance approach to the understanding of geochemical processes in aqueous systems. In *Hydrochemical Balances of Freshwater Systems*,

Proc. Uppsala Symp., IAHS Publ. No. 150 (ed. E. Eriksson). International Association of Hydrological Sciences, Washington, DC, pp. 223–235.

Park A. J. and Ortoleva P. J. (2003) WRIS.TEQ: multimineralic water-rock interaction, mass-transfer and textural dynamics simulator. *Comput. Geosci.* **29**, 277–290.

Parkhurst D. L. (1997) Geochemical mole-balancing with uncertain data. *Water Resour. Res.* **33**, 1957–1970.

Parkhurst D. L. and Appelo C. A. J. (1999) *User's Guide to PHREEQC (version 2)—A Computer Program for Speciation, Batch-reaction, One-dimensional Transport, and Inverse Geochemical Calculations.* US Geol. Surv. Water-Resour. Invest. Report 99-4259, 312pp.

Parkhurst D. L. and Plummer L. N. (1993) Geochemical models. In *Regional Ground-water Quality* (ed. W. M. Alley). Van Nostrand Reinhold, NY, pp. 199–226.

Parkhurst D. L., Plummer L. N., and Thorstenson D. C. (1980) *PHREEQE—A Computer Program for Geochemical Calculations.* US Geol. Surv. Water-Resour. Invest. Report 80-96, 195pp.

Parkhurst D. L., Plummer L. N., and Thorstenson D. C. (1982) *BALANCE—A Computer Program for Calculating Mass Transfer for Geochemical Reactions in Ground Water.* US Geol. Surv. Water-Resour. Invest. Report 82-14, 29pp.

Parkhurst D. L., Thorstenson D. C., and Kipp K. L. (2001) Calculating carbon-isotope compositions in an unsaturated zone with seasonally varying CO_2 production. In *Hydrogeology and the Environment* (ed. Yanxin Wang). China Environmental Science Press, Wuhan, China, pp. 220–224.

Perdue E. M. and Lytle C. R. (1983) Distribution model for binding of protons and metal ions by humic substances. *Environ. Sci. Tech.* **17**, 654–660.

Perdue E. M., Reuter J. H., and Parrish R. S. (1984) A statistical model of proton binding by humus. *Geochim. Cosmochim. Acta* **48**, 1257–1263.

Perkins E. H., Kharaka Y. K., Gunter W. D., and Debraal J. D. (1990) Geochemical modelling of water-rock interactions using SOLMINEQ.88. In *Chemical Modelling of Aqueous Systems II D*, Symp. Ser. 416 (eds. C. Melchior and R. L. Bassett). American Chemical Society, Washington, DC, pp. 117–127.

Pitzer K. S. (1973) Thermodynamics of electrolytes: I. Theoretical basis and general equations. *J. Phys. Chem.* **77**, 268–277.

Pitzer K. S. (1979) Theory: ion interaction approach. In *Activity Coefficients in Electrolyte Solutions* (ed. R. M. Pytkowicz). CRC Press, Boca Raton, FL, vol. 1, pp. 157–208.

Pitzer K. S. (ed.) (1991a) *Activity Coefficients in Electrolyte Solutions,* 2nd edn. CRC Press, Boca Raton, FL, 542pp.

Pitzer K. S. (1991b) Theory: ion interaction approach. In *Activity Coefficients in Electrolyte Solutions,* 2nd edn. CRC Press, Boca Raton, FL, pp. 75–153.

Pitzer K. S. and Brewer L. (revision of Lewis and Randal) (1961) *Thermodynamics.* McGraw-Hill, NY, 723pp.

Plummer L. N. (1975) Mixing of sea water with calcium carbonate ground water. *Geol. Soc. Am. Mem.* **142**, 219–236.

Plummer L. N. (1984) Geochemical modelling: a comparison of forward and inverse methods, In First Canadian/American Conf Hydrogeol. In *Practical Applications of Ground Water Geochemistry* (eds. B. Hitchon and E. I. Wallick). National Well Water Association, Worthington, OH, pp. 149–177.

Plummer L. N. (1992) Geochemical modelling of water-rock interaction: past, present, future. In *Proc. 7th Intl. Symp. Water–Rock Interaction, Park City, Utah* (eds. Y. K. Kharaka and A. S. Maest). A. A. Balkema, Rotterdam, pp. 23–33.

Plummer L. N. and Back W. (1980) The mass balance approach: applications to interpreting chemical evolution of hydrologic systems. *Am. J. Sci.* **280**, 130–142.

Plummer L. N. and Parkhurst D. L. (1990) Application of the Pitzer equations to the PHREEQE geochemical model. In *Chemical Modelling of Aqueous Systems II*, Symp. Ser. 416 (eds. D. C. Melchior and R. L. Bassett). American Chemical Society, Washington, DC, pp. 128–137.

Plummer L. N., Parkhurst D. L., and Thorstenson D. C. (1983) Development of reaction models for groundwater systems. *Geochim. Cosmochim. Acta* **47**, 665–685.

Plummer L. N., Parkhurst D. L., Fleming G. W., and Dunkle S. A. (1988) *A Computer Program Incorporating Pitzer's Equations for Calculation of Geochemical Reactions in Brines.* US Geol. Surv. Water-Resour. Invest. Report 88-4153, 310pp.

Plummer L. N., Busby J. F., Lee R. W., and Hanshaw B. B. (1990) Geochemical modelling of the Madison aquifer in parts of Montana, Wyoming, and South Dakota. *Water Resour. Res.* **26**, 1981–2014.

Plummer L. N., Prestemon E. C., and Parkhurst D. L. (1991) *An Interactive Code (NETPATH) for Modelling Net Geochemical Reactions along a Flow Path.* US Geol. Surv. Water-Resour. Invest. Report 91-4078, 227pp.

Popper K. R. (1934) *Logik der Forschung (The Logic of Scientific Discovery).* Springer, Berlin.

Posner A. M. (1964) Titration curves of humic acid. In *Proc. 8th Int. Congr. Soil Sci.*, Part II, Bucharest, Romania.

Ptacek C. and Blowes D. (2000) Predicting sulfate-mineral solubility in concentrated waters. In *Sulfate Minerals: Crystallography, Geochemistry, and Environmental Significance*, Rev. Mineral., Mineralogical Society of America and Geochemical Society, Washington, DC, vol. 40, pp. 513–540.

Pytkowicz R. M. (ed.) (1979). *Activity Coefficients in Electrolyte Solutions*, vol. 1 and 2, CRC Press, Boca Raton, FL, 228pp. and 330pp.

Rard J. A., Rand M. H., Anderegg G., and Wanner H. (1999) *Chemical Thermodynamics of Technetium.* North-Holland/Elsevier, Amsterdam, 568pp.

Robie R. A. and Hemingway B. S. (1995) *Thermodynamic Properties of Minerals and Related Substances at 298.15 K and 1 bar (10^5 Pascals) Pressure and at Higher Temperatures.* US Geol. Surv. Bull. 2131, 461pp.

Robinson R. A. and Stokes R. H. (1959) *Electrolyte Solutions*, 2nd edn. Academic Press, Butterworth, London, 571pp.

Rose K. A., Cook R. B., Brenkert A. L., Gardner R. H., and Hettelingh J. P. (1991a) Systematic comparison of ILWAS, MAGIC, and ETD watershed acidification models: 1. Mapping among model inputs and deterministic results. *Water Resour. Res.* **27**, 2577–2589.

Rose K. A., Cook R. B., Brenkert A. L., Gardner R. H., and Hettelingh J. P. (1991b) Systematic comparison of ILWAS, MAGIC, and ETD watershed acidification models: 2. Monte Carlo analysis under regional variability. *Water Resour. Res.* **27**, 2591–2603.

Rossini F. D., Wagman D. D., Evans W. H., Levine S., and Jaffe I. (1952) *Selected Values of Chemical Thermodynamic Properties.* Natl. Bur. Stds. Circular 500, 1268pp.

Rubin J. (1983) Transport of reacting solutes in porous media: relation between mathematical nature of problem formulation and chemical nature of reactions. *Water Resour. Res.* **19**, 1231–1252.

Rubin J. (1990) Solute transport with multisegment, equilibrium-controlled reactions: a feed forward simulation method. *Water Resour. Res.* **26**, 2029–2055.

Runkel R. L. (1998) *One-dimensional Transport with Inflow and Storage (OTIS): A Solute Transport Model for Streams and Rivers.* US Geol. Surv. Water-Resour. Invest. Report 98-4018.

Runkel R. L. and Kimball B. A. (2002) Evaluating remedial alternatives for an acid mine drainage stream: application of a reactive transport model. *Environ. Sci. Technol.* **36**, 1093–1101.

Runkel R. L., Bencala K. E., Broshears R. E., and Chapra S. C. (1996) Reactive solute transport in streams: 1. Development of an equilibrium-based model. *Water Resour. Res.* 409–418.

Runkel R. L., Kimball B. A., McKnight D. M., and Bencala K. E. (1999) Reactive solute transport in streams: a surface complexation approach for trace metal sorption. *Water Resour. Res.* **35**, 3829–3840.

Runnells D. D. (1969) Diagenesis, chemical sediments, and the mixing of natural waters. *J. Sedim. Petrol.* **39**, 1188–1201.

Salvage K. M. and Yeh G.-T. (1998) Development and application of a numerical model of kinetic and equilibrium microbiological and geochemical reactions (BIOKEMOD). *J. Hydrol.* **209**, 27–52.

Scatchard G. (1936) Concentrated solutions of electrolytes. *Chem. Rev.* **19**, 309–327.

Scharlin P. (ed.) (1996) *Carbon Dioxide in Water and Aqueous Electrolyte Solutions*, IUPAC Solubility Data Series, vol. 62, Oxford University Press, UK, 383pp.

Schecher W. D. and Driscoll C. T. (1987) An evaluation of uncertainty associated with aluminum equilibrium calculations. *Water Resour. Res.* **23**, 525–534.

Schecher W. D. and Driscoll C. T. (1988) An evaluation of the equilibrium calculations within acidification models: the effect of uncertainty in measured chemical components. *Water Resour. Res.* **24**, 533–540.

Servos J. W. (1990) *Physical Chemistry from Ostwald to Pauling: The Making of a Science in America.* Princeton University Press, NY, 402pp.

Shock E. L. and Helgeson H. C. (1988) Calculation of the thermodynamic and transport properties of aqueous species at high pressures and temperatures: correlation algorithms for ionic species and equation of state predictions to 5 kb and 1000°C. *Geochim. Cosmochim. Acta* **52**, 2009–2036.

Shock E. L. and Helgeson H. C. (1990) Calculation of the thermodynamic and transport properties of aqueous species at high pressures and temperatures: standard partial molal properties of organic species. *Geochim. Cosmochim. Acta* **54**, 915–946.

Shock E. L., Helgeson H. C., and Sverjensky D. A. (1989) Calculation of the thermodynamic and transport properties of aqueous species at high pressures and temperatures: standard partial moral properties of inorganic neutral species. *Geochim. Cosmochim. Acta* **53**, 2157–2183.

Sillén L. G. (1961) *The Physical Chemistry of Seawater.* Am. Assoc. Adv. Sci. Publ. 67, pp. 549–581.

Sillén L. G. (1962) High-speed computers as a supplement to graphical methods: I. The functional behavior of the error square sum. *Acta Chem. Scand.* **16**, 159–172.

Sillén L. G. and Martell A. E. (1964) *Stability Constants of Metal-ion Complexes.* Spec. Publ. 17. Chem. Soc., London, 754pp.

Silva R. J., Bidoglio G., Rand M. H., Robouch P. B., Wanner H., and Puigdomenech I. (1995) *Chemical Thermodynamics of Americium.* North-Holland/Elsevier, Amsterdam, 392pp.

Smith W. R. and Missen R. W. (1982) *Chemical Reaction Equilibrium Analysis.* Wiley, NY, 364pp.

Spear R. C. and Hornberger G. M. (1980) Eutrophication in Peel Inlet: II. Identification of critical uncertainties via Generalized Sensitivity Analysis. *Water Res.* **14**, 43–59.

Spencer R. J., Møller N., and Weare J. H. (1990) The prediction of mineral solubilities in natural waters: a chemical equilibrium model for the Na–K–Ca–Mg–Cl–SO$_4$–H$_2$O system at temperatures below 25°C. *Geochim. Cosmochim. Acta* **54**, 575–590.

Steefel C. I. and Lasaga A. C. (1994) A coupled model for transport of multiple chemical species and kinetic precipitation/dissolution reactions with application to reactive flow in single phase hydrothermal systems. *Am. J. Sci.* **294**, 529–592.

Steefel C. I. and Lichtner P. C. (1998) Multicomponent reactive transport in discrete fractures: I. Controls on reaction front geometry. *J. Hydrol.* **209**, 186–199.

Steefel C. I. and Van Cappellen P. (1998) Reactive transport modelling of natural systems. *J. Hydrol.* **209**, 1–7.

Steefel C. I. and Yabusaki S. B. (1996) OS3D/GIMRT, Software for multicomponent-multidimensional reactive transport. User manual and programmer's guide, PNL-11166, Richland.

Stumm W. (1987) *Aquatic Surface Chemistry.* Wiley, NY, 519pp.

Stumm W. and Morgan J. J. (1996) *Aquatic Chemistry,* 3rd edn. Wiley, NY, 1022pp.

Suarez D. L. and Simunek J. (1996) Solute transport modelling under variably saturated water-flow conditions. In *Reactive Transport in Porous Media* (eds. P. C. Lichtner, C. I. Steefel, and E. H. Oelkers). Rev. Mineral., Mineralogical Society of America, Washington, DC, vol. 34, pp. 229–268.

Sverdrup H. U. (1990) *The Kinetics of Base Cation Release due to Chemical Weathering.* Lund University Press, Lund, Sweden, 246pp.

Tanger J. C. and Helgeson H. C. (1988) Calculation of the thermodynamics and transport properties of aqueous species at high pressures and temperatures: revised equations of state for the standard partial moral properties of ions and electrolytes. *Am. J. Sci.* **288**, 19–98.

Tebes-Stevens C., Valocchi A. J., VanBriesen J. M., and Rittmann B. E. (1998) Multicomponent transport with coupled geochemical and microbiological reactions: model description and example simulations. *J. Hydrol.* **209**, 8–26.

Tipping E. (1994) WHAM—a chemical equilibrium model and computer code for waters, sediments and soils incorporating a discrete site/electrostatic model of ion-binding by humic substances. *Comput. Geosci.* **20**, 973–1023.

Tipping E. (1998) Modelling the properties and behavior of dissolved organic matter in soils. *Mitteil. Deutsh. Boden. Gesell.* **87**, 237–252.

Tipping E. (2002) *Cation Binding by Humic Substances.* Cambridge University Press, UK, 434pp.

Tipping E., Rey-Castro C., Bryan S. E., and Hamilton-Taylor J. (2002) Al(III) and Fe(III) binding by humic substances in freshwaters, and implications for trace metal speciation. *Geochim. Cosmochim. Acta* **66**, 3211–3224.

Thorstenson D. C. and Parkhurst D. L. (2002) *Calculation of Individual Isotope Equilibrium Constants for Implementation in Geochemical Models.* Water-Resour. Invest. Report 02-4172, 129pp.

Truesdell A. H. and Jones B. F. (1974) WATEQ, a computer program for calculating chemical equilibria of natural waters. *J. Res. US Geol. Surv.* **2**, 233–248.

USEPA (1998, 1999) MINTEQA2/PRODEFA2, a geochemical assessment model for environmental systems. User manual supplement for version 4.0, online at epa.gov/ceampubl/mmedia/minteq/supplei.pdf, 76pp.

Van Zeggeren F. and Storey S. H. (1970) *The Computation of Chemical Equilibria.* Cambridge University Press, NJ, 176pp.

van't Hoff J. H. (1905, 1909) Zur Bildung der ozeanischen Salzlagerstätten, Vieweg, Braunschweig.

van't Hoff J. H. (1912) Untersuchungen über die Bildung-sterhältnisse der ozeanischen Salzablagerungen, insobesondere das Stassfürter Salzlagers, Leipzig.

van't Hoff J. H., Armstrong E. F., Hinrichsen W., Weigert F., and Just G. (1903) Gypsum and anhydrite. *Z. Physik. Chem.* **45**, 257–306.

Wagman D. D., Evans W. H., Parker V. B., Schumm R. H., Halow I., and Bailey S. M. (1982) The NBS tables of chemical thermodynamic properties. *J. Phys. Chem. Ref. Data II* Suppl. 2, 1–392, Washington, DC.

Walter A. L., Frind E. O., Blowes D. W., Ptacek C. J., and Molson J. W. (1994) Modelling of multicomponent reactive transport in groundwater: 1. Model development and evaluation. *Water Resour. Res.* **30**, 3137–3148.

Warfvinge P. and Sverdrup H. U. (1992) Calculating critical loads of acid deposition with PROFILE—A steady state soil chemistry model. *Water Air Soil Pollut.* **63**, 119–143.

Westall J., Zachary J. L., and Morel F. M. M. (1976) MINEQL, a computer program for the calculation of chemical equilibrium composition of aqueous systems. Mass. Inst. Tech. Dept. Civil Eng. Tech. Note 18, 91pp.

Wolery T. J. (1979) Calculation of chemical equilibrium between aqueous solution and minerals: the EQ3/6 software package, Lawrence Livermore Laboratory, UCRL-52658, 41pp.

Wright R. F. and Cosby B. J. (1987) Use of a process-oriented model to predict acidification at manipulated catchments in Norway. *Atmos. Environ.* **21**, 727–730.

Wunderly M. D., Blowes D. W., Frind E. O., and Ptacek C. J. (1996) Sulfide mineral oxidation and subsequent reactive transport of oxidation products in mine tailings impoundments: a numerical model. *Water Resour. Res.* **32**, 3173–3187.

Yabusaki S. B., Steefel C. I., and Wood B. D. (1998) Multidimensional, multicomponent, subsurface reactive transport in nonuniform velocity fields: code verification using advective reactive streamtube approach. *J. Contamin. Hydrol.* **30**, 299–331.

Yeh G. T. and Tripathi V. S. (1989a) A critical evaluation of recent developments in hydrogeochemical transport models of reactive multichemical components. *Water Resour. Res.* **25**, 93–108.

Yeh G. T. and Tripathi V. S. (1989b) *HYDROGEOCHEM, A Coupled Model of Hydrologic Transport and Geochemical Equilibria of Reactive Multicomponent Systems.* Oak Ridge National Laboratory Report ORNL-6371, Oak Ridge, TN.

Zhang J.-W. and Nancollas G. H. (1990) Mechanisms of growth and dissolution of sparingly soluble salts. In *Mineral-water Interface Geochemistry* (eds. M. F. Hochella, Jr. and A. F. White). Rev. Mineral., Mineralogical Society of America, Washington, DC, vol. 23, pp. 365–396.

Zeleznik F. J. and Gordon S. (1968) Calculation of complex chemical equilibria. *Ind. Eng. Chem.* **60**, 27–57.

Zhu C. and Anderson G. M. (2002) *Environmental Applications of Geochemical Modelling.* Cambridge University Press, Cambridge, UK, 284pp.

5.03

Reaction Kinetics of Primary Rock-forming Minerals under Ambient Conditions

S. L. Brantley

Pennsylvania State University, University Park, PA, USA

NOMENCLATURE

a_i	activity of species i in solution
a, a'	geometric constant (Equations (26) and (27))
A	specific surface area (surface area per unit mass)
A_i	ith aqueous species in reaction (22)
A'	pre-exponential factor in Arrhenius equation
A''	constant in dissolution rate law (Equation (64))
A^T	total surface area
A_{ads}	specific surface area measured by gas adsorption
A_{geo}	specific geometric surface area
A_{geo}^T	total geometric surface area
A_{int}	specific surface area of internal surface
b	constant for prediction of surface area (Equation (29))
B	constant in dissolution rate law (Equation (64))
c	concentration
c_i	inlet concentration
c_o	outlet concentration
C	connectedness
d	dimensional constant
D	grain diameter
$f(\Delta G)$	affinity function in Equation (65)
E_a	activation energy
E_a'	apparent activation energy measured at a given pH value
$H_2CO_3^*$	$CO_{2(aq)} + H_2CO_{3(aq)}$
$[i]$	concentration of species i
k	rate constant for dissolution
k_0	rate constant at 25 °C
k_H	rate constant for proton-promoted dissolution
k_{H_2O}	rate constant for dissolution in neutral pH
k_{min}	rate constant measured far from equilibrium for Equation (63)
k_{OH}	rate constant for hydroxyl-promoted dissolution
k_{L_i}	rate constant for ligand-promoted dissolution
k_{SiOH}	rate constant for reaction at $\equiv$SiOH
k_{SiO^-}	rate constant for reaction at $\equiv$SiO$^-$ and $\equiv$SiONa
k_{solv}	rate constant for exchange of water molecules around a cation in solution
k_1, k_2, k_3	rate constants in the mechanism of dissolution of calcite
k_5, k_6	rate constants in Equation (62)
k_1'	rate constant for reaction (11)
k_4	rate constant for precipitation of calcite
k_+	forward rate constant
K_H^{ads}	equilibrium constant for adsorption of proton onto a mineral surface (Equation (52))
$K_{M_i}^{ads}$	equilibrium constant for adsorption of cation M_i onto a mineral surface
K_{Al}	equilibrium constant for exchange of Al^{3+} and H^+ at the mineral surface (Equation (24))
K_{ex}	equilibrium constant for exchange of four protons for two magnesium atoms on forsterite surface (Equation (52))
K_2	dissociation constant for bicarbonate
K_c	solubility product constant for calcite
K_i	equilibrium constant for exchange of cation M_i and a proton at mineral surface (Equation (25))
$K^\bullet$	equilibrium constant for formation of precursor species by surface protonation (Equation (24))
l	dimension of a geometric solid used to model a powder grain for surface area (Equations (26) and (27))
L_i	ligand i
m	order with respect to hydroxyl in hydroxyl-promoted dissolution equation
m_1, m_2	constants used in a rate equation with an affinity term
m_j	order with respect to species appearing in the rate limiting step of a reaction (Equation (57))
M	mass
M_i	cation
n	order with respect to a reactant in Equations (9) or (42); elsewhere, order with respect to H^+ in Equations (17) and (32)
n_1	constant in a rate equation using an affinity term
n_o	order with respect to H^+ in Equation (17) as measured at the reference temperature
p	order with respect to OH^-
pH_{ppzc}	pH of pristine point of zero charge
$P^\bullet$	precursor species (Equation (22))
q	release rate of component to solution
q_1, q_2, q_3	order with respect to surface species in Equations (20) and (54)
q_A, q_B	charge on a species
Q	flow rate
Q^i	a tetrahedrally coordinated atom surrounded by i bridging oxygens

r	surface-area normalized rate of reaction
r_{net}	net rate of reaction (forward rate−backward rate)
R	gas constant
s	number of Al^{3+} ions exchanged to create a precursor species in reaction (22)
$\equiv S$	metal cation at a surface site on a mineral
$\equiv SL_i$	surface metal−ligand complex
$\equiv SO^-$	deprotonated surface hydroxyl
$\equiv SOH$	surface hydroxyl species
$\equiv SOH_2^+$	protonated surface hydroxyl
$\equiv SiOAl\equiv$	bridging oxygen linkage between Si and Al cations
$\equiv SiOSi\equiv$	bridging oxygen linkage between two Si cations
$\equiv SiONa$	surface silanol with adsorbed Na ion
t	time
T	temperature
T_0	reference temperature
v	rate of interface advance during dissolution or growth (m s^{-1})
X	number of cations in nontetrahedral sites in a structure
X/Si	ratio of number of cations in nontetrahedral sites to cations in tetrahedral sites in a structure
V	volume
V_o	pore volume of a PFR or total volume of a CST reactor
$\bar{V}$	molar volume of a mineral
X/Si	ratio of the number of non-tetrahedrally coordinated to tetrahedrally coordinated cations in a silicate
α	rate of change of pH dependence with temperature (Equation (48))
β	# of precursors formed during adsorption of 3 protons (Equation (24))
Θ_{SiOH}	fraction of the silica surface covered by silanol groups
$\Theta_{SiO^-_{tot}}$	sum of the fractions of total sites existing as deprotonated surface hydroxyls and as sites with adsorbed Na^+ on a silica surface
λ	surface roughness
λ_{ext}	external surface roughness
ν_i	stoichiometric coefficient for species i
ρ	density
Ω	ratio of reaction quotient to equilibrium constant for a reaction
ΔG	Gibbs free energy of reaction
ΔG_{crit}	critical Gibbs free energy for spontaneous nucleation of etch pits at dislocations

5.03.1 INTRODUCTION

Mineral dissolution kinetics influence such phenomena as development of soil fertility, amelioration of the effects of acid rain, formation of karst, acid mine drainage, transport and sequestration of contaminants, sequestration of carbon dioxide at depth in the earth, ore deposition, and metamorphism. On a global basis, mineral weathering kinetics are also involved in the long-term sink for CO_2 in the atmosphere:

$$CaSiO_3 + CO_2 = CaCO_3 + SiO_2 \quad (1)$$

$$MgSiO_3 + CO_2 = MgCO_3 + SiO_2 \quad (2)$$

These reactions (Urey, 1952) describe the processes that balance the volcanic and metamorphic CO_2 production to maintain relatively constant levels of atmospheric CO_2 over 10^5-10^6 yr timescales. In these equations, Ca- and $MgSiO_3$ represent all calcium- and magnesium-containing silicates. Calcium- and magnesium-silicates at the Earth's surface are predominantly plagioclase feldspars, Ca−Mg-pyroxenes, amphiboles, and phyllosilicates, Ca−Mg orthosilicates. Although dissolution of the other main rock−forming mineral class, carbonate minerals, does not draw down CO_2 from the atmosphere over geologic timescales, carbonate dissolution is globally important in controlling river and ground water chemistry.

Despite the importance of mineral dissolution, field weathering rates are generally observed to be up to five orders of magnitude slower than laboratory dissolution rates (White, 1995), and the reason for this discrepancy remains a puzzle. For example, mean lifetimes of 1 mm spheres of rock-forming minerals calculated from measured rate data following Lasaga (1984) are much smaller than the mean half-life of sedimentary rocks (600 My, Garrels and Mackenzie, 1971). As pointed out by others (Velbel, 1993a), the order of stability of minerals calculated from measured dissolution kinetics (Table 1) generally follow weathering trends observed in the field (e.g., Goldich, 1938) with some exceptions. Some have suggested that quantitative prediction of field rates will be near-impossible, although such rate trends may be predictable (Casey et al., 1993a). For studies with mineral substrates identical between laboratory and field, however, the discrepancy between field and laboratory rate estimate is generally on the order of one to two orders of magnitude (e.g., Schnoor, 1990; Swoboda-Colberg and Drever, 1993; White and Brantley, in press). Some of the discrepancy may be related to factors in the field that have not been well mimicked in laboratory systems

Table 1 Mean lifetime (t) of a 1 mm crystal at pH 5.

Mineral	Log (dissolution rate) (mol m^{-2} s^{-1})	Dissolution rate (mol m^{-2} s^{-1})	t (yr)	References
Quartz		4.1×10^{-14}	34,000,000	Rimstidt and Barnes, 1980
K-feldspar	-12.4	5.0×10^{-13}	740,000	Table 2
Muscovite	-12.5	3.2×10^{-13}	720,000	Table 3
Phlogopite	-12.5	3.2×10^{-13}	670,000	Table 3
Albite	-12.2	6.3×10^{-13}	500,000	Table 2
Diopside	-11.4	3.6×10^{-12}	140,000	Table 3
Anorthite	-11.4	4.0×10^{-12}	80,000	Table 2
Enstatite	-10.5	3.2×10^{-11}	16,000	Table 3
Tremolite	-11.0	1.1×10^{-11}	10,000	Table 3
Forsterite	-9.4	3.6×10^{-10}	2,000	Table 3
Fayalite	-9.4	3.6×10^{-10}	1,900	Table 3

Revised from Lasaga (1984).

(White (in press) see Chapter 5.05). For example, to extrapolate mineral reaction rates from one system to another, the following variables must be understood: (i) mechanism of dissolution, (ii) reactive surface area, (iii) mineral composition, (iv) temperature of dissolution, (v) chemistry of dissolving solutions, (vi) chemical affinity of dissolving solutions, (vii) duration of dissolution, (viii) hydrologic parameters, and (ix) biological factors. In this chapter, general techniques of measurement of dissolution and precipitation rates of rock-forming silicates and carbonates are discussed, and then, seven of these nine factors are discussed sequentially. A full discussion of the biological effects (discussed by Berner *et al.* in Chapter 5.06) and hydrological parameters are outside the scope of this chapter.

Empirical models predicting the rates of mineral-specific dissolution as a function of pH are summarized within the section on mineral composition in an attempt to provide a useful database for predicting dissolution rates for both laboratory and field systems. Equations describing near-equilibrium mineral dissolution and precipitation rates are summarized in the section on chemical affinity.

5.03.2 EXPERIMENTAL TECHNIQUES FOR MINERAL DISSOLUTION

5.03.2.1 Chemical Reactors

In general for a dissolution reaction such as,

$$A_{\nu_A}B_{\nu_B(s)} \rightarrow \nu_A A^{q_A}_{(aq)} + \nu_B B^{q_B}_{(aq)} \qquad (3)$$

the rate of reaction, r, is expressed as

$$r = -\frac{d[A_{\nu_A}B_{\nu_B}]}{dt} = \frac{1}{\nu_A}\frac{d[A^{q_A}]}{dt} = \frac{1}{\nu_B}\frac{d[B^{q_B}]}{dt} \qquad (4)$$

where $[i]$ refers to the concentration of species i, ν_i is the stoichiometric coefficient of the reaction, and q_i is the charge of the aqueous species. Because most dissolution rates are expressed on an $ML^{-2}T^{-1}$ basis, and because the rate of disappearance of a mineral is more difficult to monitor than the rate of increase of solute in solution, the rate is usually defined by an expression such as

$$r = \frac{V_o}{\nu_B AM}\frac{d[B^{q_B}]}{dt} \qquad (5)$$

where V_o is the volume of water, A is the specific surface area of the mineral (L^2M^{-1}, discussed below), and M is the mass of the mineral.

To determine r, investigators have dissolved mineral separates under controlled temperature, pH, and solution composition conditions. Batch, continuously stirred flow-through, plug flow, and fluidized bed reactors have all been used (e.g., Hill, 1977; van Grinsven and van Riemsdijk, 1992). Typically, batch reactors are stirred tank reactors run without flow. Simple to use, these reactors allow estimation of the progress of reaction for systems from the change in solution chemistry with time. Batch reactors may be open or closed to the atmosphere and may be run in constant (pH-stat) or changing pH mode. By monitoring the concentration of dissolution products as a function of time, and correcting for removal of sample during monitoring, the reaction rate can be calculated from an equation such as (5) (Posey-Dowty *et al.*, 1986; Laidler, 1987). The reaction rate is expressed as either the release rate of a given component, or as the rate of dissolution of moles of the phase per surface area per unit time. Interpretation of reaction rates for a batch reactor is complicated by changing solution chemistry over time and the effects of back-reactions including precipitation of secondary phases (see, e.g., Oelkers *et al.*, 2001).

Interpretation of reaction rates using stirred flow-through reactors is more straightforward than for batch reactors because solution chemistry remains constant during dissolution. In a continuously stirred tank reactor (CSTR) or a mixed flow reactor (Rimstidt and Dove, 1986) a mineral sample is placed in a reactor of volume V_o and fluid is pumped through at flow rate Q ($L^3 T^{-1}$). Fluid is stirred by a propeller or by agitation. The rate of reaction, r (mol m^{-2} s^{-1}), is calculated from the inlet (c_i) and outlet concentrations (c_o) of a component released during dissolution of the mineral:

$$r = \frac{Q(c_o - c_i)}{v_i AM} \tag{6}$$

During dissolution, mass of the mineral decreases and specific surface area generally increases. Most researchers use the initial mineral mass and surface area to normalize reaction rate, but for experiments where the extent of reaction is large, the final surface area may be used to normalize the rate (Stillings and Brantley, 1995). Reactors are run until outlet concentration reaches a constant steady-state value. Dissolution rates are then reported with respect to solution chemistry as measured in the effluent. For example, measured rate is reported with respect to the outlet rather than inlet pH.

The fluidized bed reactor (FBR) is a stirred tank reactor that utilizes a slow single-pass flow and a second faster recirculating flow that stirs the mineral powder by suspending the particles in the reactor. The values of c_i and c_o are analyzed in the single pass flow to determine the chemical reaction rate (e.g., Chou and Wollast, 1985).

Some researchers use plug-flow reactors (PFRs), also known as packed bed reactors or column reactors (if run vertically) to model natural systems. In an ideal plug-flow or column reactor, fluid is pumped or drained through a packed bed of mineral grains and every fluid packet is assumed to have the same residence or contact time (Hill, 1977). The residence time equals the ratio of the pore volume of the reactor (V_o) divided by flow rate Q. With no volume change in the reaction, radial flow, or pooling of fluid in the reactor (Laidler, 1987), the outlet concentration varies from the inlet concentration according to:

$$\frac{1}{1-n}\left[\frac{1}{c_o^{n-1}} - \frac{1}{c_i^{n-1}}\right] = \frac{kAM}{Q} \tag{7}$$

for $n \neq 1$; and

$$c_o = c_i \exp\left(\frac{-kAM}{Q}\right) \tag{8}$$

for $n = 1$. Here we assume that $v_i = 1$ and that the reaction rate r can be described by an nth order reaction:

$$r = kc^n \tag{9}$$

For such a rate equation, n represents the order of the reaction with respect to the component whose concentration is depicted by c. PFRs mimic geologic systems more closely than CSTRs; however, because the chemical conditions change along the length of the reactor, determination of the dependence of rate on affinity or on individual solute species can become numerically complex (e.g., Taylor *et al.*, 2000). In addition, a packed bed reactor may become transport-controlled (see Section 5.03.2.1), and measured kinetics may represent kinetics of transport instead of interface-limited dissolution. For example, van Grinsven and van Riemsdijk (1992) have observed that weathering rates of soils in column reactors increase with the square root of the percolation rate. In contrast, a fluidized bed reactor suspends particles to accelerate transport to and from the mineral surface maintaining constant chemistry throughout the reactor.

A few researchers have investigated the rate of dissolution through microscopic examination of mineral surfaces (e.g., Hellmann *et al.*, 1992; MacInnis and Brantley, 1992, 1993; Dove and Platt, 1996; Mellott *et al.*, 2002). For such studies, the dissolution rate (mol m^{-2} s^{-1}) is generally obtained from the expression

$$v = r\bar{V} \tag{10}$$

where $\bar{V}$ is the molar volume (m^3 mol^{-1}) of the dissolving mineral and v is the rate of advance of the mineral surface (m s^{-1}). Early on, geochemists discovered that by shielding the dissolving mineral surface from dissolution, the overall surface retreat could be measured as the difference between dissolution of shielded and unshielded areas on the mineral surface (MacInnis and Brantley, 1992). On unshielded areas, etch pit nucleation and growth was investigated and a dissolution model based upon the distribution of etch pit sizes was developed (MacInnis and Brantley, 1993). Use of Equation (10) implicitly assumes a geometric surface area estimated at the scale of the microscopic technique utilized (Brantley *et al.*, 1999), despite arguments to the contrary (Luttge *et al.*, 1999). Comparisons of dissolution rates derived from microscopic measurements to powder dissolution measurements by many of these authors have been found to be useful in determining the relative contributions of surface sites to dissolution (Figure 1(a)). Figure 1(a) shows that at the scale of observation of the vertical scanning interferometer, dissolution can be conceptualized as occurring uniformly across broad planes of the surface (surface retreat)

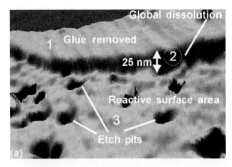

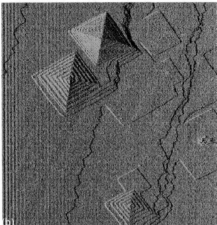

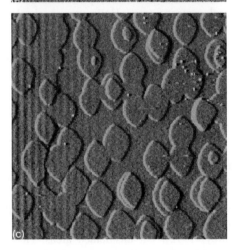

Figure 1 (a) Vertical scanning interferometric (VSI) image of a dolomite cleavage plane documenting dissolution during 1 h of dissolution at pH 3 and 25 °C (reproduced by permission of E. Schweizerbart'sche Verlagsbuchhandlung from *Euro. J. Mineral.*, in press; © E. Schweizerbart'sche Verlagsbuchhandlung) collected from experiments in Luttge *et al.* (2003)). Glue was used to block dissolution of the reference surface, shown here after glue removal as point (1). The extent of retreat of the overall surface is shown as the difference in levels at point (2), and dissolution at etch pits is documented as shown at (3). (b) Fluid Cell Atomic Force Microscope (AFM) image depicting growth mechanisms on a 15 μm × 15 μm area of the $(10\bar{1}4)$ face of calcite. Growth in this image occurred at $\ln \Omega = 0.4$ within minutes at spiral dislocations (three large features) and on monomolecular layers. (c) Similar 15 μm × 15 μm image

and more intensively at high surface free energy sites (etch pits). Generally, these microscopic techniques are easiest to implement for faster dissolving phases (Dove and Platt, 1996).

5.03.2.2 Interpreting Dissolution Rates

When a reaction rate is measured in a chemical reactor, the reaction is generally a composite reaction comprised of a sequence of elementary reactions. An elementary reaction is a reaction that occurs at the molecular level exactly as written (Laidler, 1987). The mechanism of the reaction is the sequence of elementary reactions that comprise the overall or composite reaction. For example, mineral dissolution reactions generally include transport of reactant to the surface, adsorption of reactant, surface diffusion of the adsorbate, reaction of the surface complex and release into solution, and transport of product species away from the surface. These reactions occur as sequential steps. Reaction of surface complexes and release to solution may happen simultaneously at many sites on a surface, and each site can react at a different rate depending upon its free energy (e.g., Schott *et al.*, 1989). Simultaneous reactions occurring at different rates are known as parallel reactions. In a series of sequential reactions, the rate-determining step is the step which occurs most slowly at the onset of the reaction, whereas for parallel steps, the rate-determining step is the fastest reaction.

Some investigators have used a rotating disk of dissolving substrate to investigate the dissolution rate of fast-dissolving phases and to elucidate transport versus surface reaction control (Sjoberg, 1976; Sjoberg and Rickard, 1983; MacInnis and Brantley, 1992; Alkattan *et al.*, 1998; Gautelier *et al.*, 1999). The stirring rate of the dissolving disk or powder can be varied to determine whether transport limits dissolution (e.g., Shiraki and Brantley, 1995). However, several researchers have criticized stirring of reactors containing powders for causing continuous abrasion and surface area changes during dissolution. In effect, one chooses either the problem of changing solution chemistry and possible transport control

as in (b), except growth occurred at $\ln \Omega = 1.6$ within tens of seconds of growth. In contrast to (b), dominant growth under these conditions occured by randomly distributed two-dimensional nucleation of ~100 nm scale nuclei on the surface. Some of the imaged nuclei have coalesced into larger nuclei. Continuous surface nucleation has also been imaged in several locations. Images (b) and (c) after Teng *et al.* (2000).

(PFR) or abrasion of particles and continuous production of new surface area (CSTRs, FBRs). Such problems inherent in each technique contribute to differences in rates when measured in different types of reactors (van Grinsven and van Riemsdijk, 1992; Clow and Drever, 1996). For example, van Grinsven and van Riemsdijk observed that dissolution rates measured with stirred batch reactors yielded higher rates than column experiments. Furthermore, where pH changes within the reactor are substantial, some investigators have used buffer solutions to maintain constant pH; however, the presence of cations and anions other than H^+ or OH^- influence the rate and mechanism of dissolution (e.g., Dove and Crerar, 1990; Wogelius and Walther, 1991; Stillings and Brantley, 1995).

Many silicates are observed to dissolve nonstoichiometrically either due to precipitation of secondary minerals or to preferential leaching of elements (see Section 5.03.3.3.1). To distinguish these possibilities, more than one element must be analyzed during dissolution, precipitates must be investigated, and chemical affinity must be calculated for reacting solutions. Furthermore, surface chemistry measurements can be completed (e.g., Farquhar *et al.*, 1999a). Apparent nonstoichiometric dissolution due to dissolution of impurity minerals, even in hand-picked and carefully cleaned samples, also complicates the analysis of laboratory rates (Zhang *et al.*, 1996; Kalinowski *et al.*, 1998; Brantley *et al.*, 1998; Rosso and Rimstidt, 2000) and field rates (White *et al.*, 1999). Preferential dissolution of exsolved phases can also cause apparent nonstoichiometric dissolution in experiments with natural materials (e.g., Inskeep *et al.*, 1991; Stillings and Brantley, 1995; Chen and Brantley, 1998). One promising method to avoid problematic inhomogeneities in starting materials is to dissolve glasses of composition similar to the minerals of interest: Hamilton *et al.* (2000) have observed similar rates of dissolution for albite crystal and glass.

For minerals that dissolve incongruently, the determination of reaction rate depends upon which component released to solution is used in Equation (5). Due to preferential release of cations such as calcium and magnesium during inosilicate dissolution, for example, dissolution rates for these phases are usually calculated from observed silicon release (Brantley and Chen, 1995). Here, we report silicate dissolution rates based upon silicon release, but we normalize by the stoichiometry of the mineral and report as mol mineral per unit surface area per unit time. It is important to note that dissolution rates reported on this basis depend upon both the formula unit and the monitored solute.

5.03.3 MECHANISMS OF DISSOLUTION

5.03.3.1 Rate-limiting Step and the Effect of Dislocations

To extrapolate a rate confidently from one system to another, the rate of an elementary reaction must be known and the same reaction must control the rate in the new system. Perhaps the most important end-member cases for rate control are transport and interface limitation. Where the interface reaction is rate-limiting, an increase in fluid flow or diffusion across a boundary layer will not change the rate of reaction because no concentration gradients exist within the solution or at the mineral–water interface and the rate of reaction is therefore entirely determined by the dissolution rate of the mineral. In contrast, where diffusion in solution is rate-limiting, an increase in fluid flow will generally enhance dissolution because a concentration gradient exists at the mineral–water interface and the flow rate can change this gradient. For the transport-limited case, the dissolution rate constant no longer controls the net rate of reaction: instead, the rate is controlled either by rate of diffusion or advection. For dissolution or precipitation of silicates under ambient conditions, many authors assume that the interface reaction is rate-limiting in both the laboratory and in the field (see discussion in Kump *et al.*, 2000). However, others have suggested that transport control related to differences in hydrology may explain slower rates observed in the field (Swoboda-Colberg and Drever, 1993; Velbel, 1993a). For many fast-dissolving phases in natural systems or for high-temperature reactions, either diffusive or advective transport is probably rate limiting (Murphy *et al.*, 1989; Steefel and Lasaga, 1992).

One distinguishing difference (Lasaga, 1984) between transport and interface control of dissolution is the activation energy of reaction (see Section 5.03.6.1), E_a. For transport in solution, E_a ($\sim$5 kcal mol^{-1}) $\ll$ E_a for the interface reaction ($\sim$15 kcal mol^{-1}). In addition, where a reaction is rate-limited by the interface reaction, ion detachment is slow, and portions of the mineral surface may selectively dissolve (Berner, 1978). The resulting etch pits (e.g., Figure 1(a)) are considered by some to document an interface-controlled reaction (Berner *et al.*, 1980). For such a condition, it is suggested that the concentration of solution at the mineral–solution interface is equal to that in the bulk solution because transport is fast compared to the interface reaction. In contrast, where reactions are rate-controlled by diffusion, mineral surfaces are expected to be rounded and devoid of etch pits, because the concentration at the mineral–solution interface is expected to approach equilibrium and only the most highly energetic sites (e.g., corners, edges) should dissolve. Although many authors have inferred interface control from the presence of etch pits, etch pits can form on minerals even when dissolved under conditions where rates of diffusion affect the rate of dissolution (see, e.g., Alkattan *et al.*, 1998). The presence or absence of etch pits may therefore not prove the rate-limiting step of dissolution.

Far from equilibrium, etch pits can nucleate everywhere on a mineral surface, while close to equilibrium etch pits may not form extensively (Brantley *et al.*, 1986; Blum and Lasaga, 1987; Schott *et al.*, 1989; Lasaga and Luttge, 2001). For most situations, the change from dissolution far from equilibrium to dissolution close to equilibrium may coincide with a change in rate limitation from the interface reaction to transport. The critical saturation index above which spontaneous opening of dislocation etch pits does not occur thus may mark a change in mechanism for dissolution of minerals from nucleation and growth of etch pits at dislocations (far from equilibrium) to dissolution only at more energetic defects such as edges and corners (close to equilibrium). Such a mechanism change, causing a uniform rounding of the surface for dissolution near equilibrium, may in turn be marked by a change in slope of the rate versus ΔG curve (Burch *et al.*, 1993; Lasaga and Luttge, 2001), contributing to the existence of a dissolution plateau (see Section 5.03.8).

Despite the importance of etching at dislocation outcrops, increases in dislocation density generally have relatively minor effects on dissolution of a variety of minerals far from equilibrium. Rates are observed to increase by factors less than 14, and generally by factors less than 3, for increases in dislocation density up to four orders of magnitude (Casey, 1988a,b; Schott *et al.*, 1989; Murphy, 1989; Blum *et al.*, 1990; MacInnis and Brantley, 1992). Lee *et al.* (1998) point out, however, that in comparison to laboratory dissolution, etching at dislocation outcrops will be extremely important early in natural weathering due to the fact that natural solutions are generally closer to saturation than laboratory solutions. Mechanical breakage of heavily etched minerals may also expose new surface area to solution in natural systems, enhancing dissolution in late stages of weathering (Lee *et al.*, 1998).

5.03.3.2 Carbonate Dissolution Mechanism

The kinetics of calcite dissolution have generally been studied more than those of any other mineral except perhaps quartz (for more extensive reference lists see Alkattan *et al.* (1998) and Wollast (1990)). Here we summarize the classic work of Plummer *et al.* (1978) who suggested that dissolution occurred by the following reactions

that occur simultaneously at the calcite surface:

$$CaCO_{3(s)} + H^+_{(aq)} = Ca^{2+}_{(aq)} + HCO^-_{3(aq)} \quad (11)$$

$$CaCO_{3(s)} + H_2CO^0_{3(aq)} = Ca^{2+}_{(aq)} + 2HCO^-_{3(aq)} \quad (12)$$

$$CaCO_{3(s)} + H_2O_{(aq)}$$
$$= Ca^{2+}_{(aq)} + HCO^-_{3(aq)} + OH^-_{(aq)} \quad (13)$$

Rates of these reactions were posited to be described by the following equations:

$$r = k_1 a_{H^+} \quad (14)$$

$$r = k_2 a_{H_2CO^*_3} \quad (15)$$

$$r = k_3 a_{H_2O} \quad (16)$$

In each case, a surface site is thought to react with an aqueous ion at the mineral surface. Plummer *et al.* (1978) observed three regions of dissolution: region 1 was transport-controlled while the rates of dissolution in regions 2 and 3 were slower than rates controlled by transport. While the boundaries of these regions depended upon solution chemistry and P_{CO_2}, the demarcation between region 1 and region 2 occurred generally below pH 3.5. Rate equations and models that explicitly incorporate the concentration of complexes at the carbonate surface yield more mechanistic rate equations, and workers have recently been suggesting such models for carbonates (e.g., Pokrovsky *et al.*, 1999a,b, 2000; Pokrovsky and Schott, 1999).

5.03.3.3 Silicate and Oxide Dissolution Mechanisms

5.03.3.3.1 Nonstoichiometric dissolution

Early investigators of silicate dissolution identified parabolic kinetics ($r = kt^{-1/2}$) for solute release rates (Luce *et al.*, 1972; Paces, 1973). Such parabolic kinetics have been identified for many reacting phases and were generally attributed to diffusion through a surface layer (e.g., Doremus, 1983). However, leached layers on dissolving mineral surfaces were observed by X-ray photoelectron spectroscopy (XPS) to be only a few angstroms thick (Petrovic *et al.*, 1976; Berner and Holdren, 1979; Holdren and Berner, 1979); parabolic kinetics for minerals were soon attributed to sample preparation artifacts (Holdren and Berner, 1979). Thorough reviews of this early work have been presented by Lasaga (1984) and by Velbel (1986).

Although silicate dissolution is now *not* generally thought to be rate limited by diffusion through an armoring alteration layer, much evidence has accumulated documenting the development of silicon-rich cation-depleted layers of varying thickness on many dissolving silicates, especially at low pH. After long duration dissolution in acid solutions, it is generally thought (see, however, Nesbitt *et al.*, 1991) that the rate of diffusion of leaching cations through an altered surface layer equals the rate of release of silicon to solution from the exterior of the altered layer, and that dissolution rates of minerals such as feldspar become stoichiometric (Schweda, 1990; Stillings and Brantley, 1995). The early transient period of dissolution of many silicates may thus document the formation of a steady-state altered layer, including proton exchange and diffusion of alkalis and alkaline earths out of the leached layer, hydrolysis and release to solution of aluminum and silicon, and condensation of silanols to siloxanes in the layer (e.g., Schweda *et al.*, 1997). Diffusion rates in alteration layers of feldspars have been quantified in some cases (Chou and Wollast, 1984; Nesbitt *et al.*, 1991; Blum, 1994; Hellmann, 1997a; Chen *et al.*, 2000). Where samples do not dissolve stoichiometrically, even after long durations of reaction, exsolution lamellae or impurity phases may explain the incongruent dissolution (e.g., Inskeep *et al.*, 1991; Oxburgh *et al.*, 1994; Stillings and Brantley, 1995). It has also been suggested that the alteration layers on some phases may not be spatially homogeneous in composition, structure, or thickness, and may be spatially related to etch pits on the surface (e.g., Gout *et al.*, 1997). In contrast, others have suggested that leached layers are uniform over the mineral surface (e.g., Schweda *et al.*, 1997).

Altered surfaces have been inferred from solution chemistry measurements (e.g., Chou and Wollast, 1984, 1985) and from spectroscopic measurements of altered surfaces, using such techniques as secondary ion mass spectrometry (for altered layers that are several tens of nm thick (e.g., Schweda *et al.*, 1997), Auger electron spectroscopy (layers <10 nm thick (e.g., Hochella, 1988), XPS (layers <10 nm thick (e.g., Hochella, 1988; Muir *et al.*, 1990), transmission electron microscopy (TEM, e.g., Casey *et al.*, 1989b), Raman spectroscopy (e.g., Gout *et al.*, 1997), Fourier transform infrared spectroscopy (e.g., Hamilton *et al.*, 2001), *in situ* high-resolution X-ray reflectivity (Farquhar *et al.*, 1999b; Fenter *et al.*, 2003), nuclear magnetic resonance (Tsomaia *et al.*, 2003), and other spectroscopies (e.g., Hellmann *et al.*, 1997).

Alteration layers have been described for feldspars dissolved at subneutral pH (Chou and Wollast, 1984, 1985; Casey *et al.*, 1988, 1989a; Nesbitt and Muir, 1988; Muir *et al.*, 1989, 1990; Hellmann *et al.*, 1989, 1990a,b, 1991; Shotyk *et al.*, 1990; Hochella, 1990; Nesbitt *et al.*, 1991; Hellmann, 1994, 1995, 1997a,b; Stillings and Brantley, 1995, 1996; Gout *et al.*, 1997),

for glasses (Guy and Schott, 1989; Doremus, 1994; Hamilton *et al.*, 2000, 2001), for inosilicates (Brantley and Chen, 1995), and for some phyllosilicates (Nagy, 1995; Kalinowski and Schweda, 1996). For example, preferential leaching of cations in the M2, and to a lesser extent, M1 sites in pyroxenes have been reported and attributed to differences in Madelung site energy (e.g., Schott and Berner, 1983, Schott and Berner, 1985; Sverdrup, 1990). Preferential leaching of the M4 as compared to the M1, M2, and M3 sites in amphiboles has been reported (Schott and Berner, 1983, 1985; Brantley and Chen, 1995; Chen and Brantley, 1998). In fact, some of the deepest leaching of cations (several thousand angstroms) has been reported for the inosilicate mineral wollastonite dissolved at pH 2 (Casey *et al.*, 1993c). Extensive leaching of sodium and aluminum has also been reported for a glass of inosilicate composition (jadeite; Hamilton *et al.* (2001)). Thick alteration layers on dissolved feldspars or glasses (see Figure 2) often demonstrate characteristics of amorphous silica (Casey *et al.*, 1989b; Hamilton *et al.*, 2000, 2001; Hellmann *et al.* 2003).

Thickness of altered layers on feldspars generally decreases with increasing dissolved cation content of the leaching solution (Nesbitt *et al.*, 1991) and increases with decreasing pH

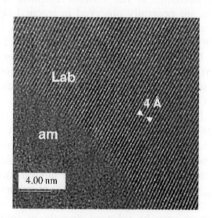

Figure 2 A high-resolution TEM photomicrograph of the amorphous altered layer (lower left) developed on crystalline labradorite (bulk material, upper right) after dissolution at pH 1. The blurry lattice fringes at the interface reflect the varying boundary orientation with respect to the ultrathin section. Interface thickness is ~0.5–2 nm. Energy filtered (EF) TEM was also used to chemically characterize the alteration zone, which was found to be depleted in Ca, Na, K, and Al, and enriched in H, O, and Si. The sharp structural interface shown here and the sharp chemical interface observed with EFTEM are interpreted by the authors to indicate that the alteration layer is formed by dissolution-precipitation. Such amorphous altered layers are often high in porosity and yield high BET surface areas (reproduced by permission of Springer from *Phys. Chem. Min.*, 2003, *30*, 192–197).

(Schweda, 1990). In fact, above pH 3, the presence of a true leached layer on dissolving feldspars is not well documented (Blum, 1994). The thickness of the altered layer on plagioclase feldspars dissolved at pH 3.5 also varies as a function of the Al/Si ratio of the feldspar (Muir *et al.*, 1990; Stillings and Brantley, 1995). Labradorite (Al/Si = 0.66–0.8) and bytownite (Al/Si = 0.8–0.9) crystals show much thicker altered layers than albite (Al/Si < 0.4), oligoclase (Al/Si = 0.4–0.5) and andesine (Al/Si = 0.5–0.66) crystals. In contrast, dissolution of anorthite (Al/Si = 1.0) is almost stoichiometric with respect to aluminum at low pH (Amrhein and Suarez, 1988). Similarly, aluminum depletion is observed after acid leaching of albite and jadeite glasses (Al/Si < 1), but not on nepheline glass (Al/Si = 1) (Hamilton *et al.*, 2001).

Within the alteration layer of feldspars, hydrolysis of bridging oxygens has been presumed to reduce the connectedness of network atoms from 4 (Q^4) to 3 (Q^3) to 2 (Q^2) to 1 (Q^1) before hydrolysis releases the atom to solution (Hellmann *et al.*, 1990a, 2001; Brantley and Stillings, 1997; Tsomaia *et al.*, 2003). Here, connectedness refers to the number of bridging oxygens around the tetrahedrally coordinated atom (Liebau, 1985). However, alteration layers on silicates also contain high proton concentrations, as alteration layers form during proton-cation exchange at the surface. Because fewer protons are generally observed in leached layers than would be predicted based on 1:1 cation exchange (Petit *et al.*, 1987; Schott and Petit, 1987; Casey *et al.*, 1988, 1989a; Casey *et al.*, 1989b), condensation of silanols to siloxanes in the alteration layer of phases such as feldspars (e.g., Casey and Bunker 1990; Hellmann *et al.*, 1990a,b; Stillings and Brantley, 1995) and chain silicates (Casey *et al.*, 1993c; Weissbart and Rimstidt, 2000) has been inferred. Furthermore, it has been suggested that connectedness of silicon may not be reduced to zero at the altered surface during dissolution of many silicates at subneutral pH, because silica polymers rather than silica monomers may be released directly to solution (Dietzel, 2000; Weissbart and Rimstidt, 2000). Reconstruction of alteration layers may also lead to formation of clays and amorphous phases directly at the surface, without a solution step (Casey *et al.*, 1993c). In fact, six-coordinate aluminum has been observed on the albite surface dissolved under acid conditions, whereas bulk albite exhibits only four-coordinate aluminum (Tsomaia *et al.*, 2003). This six-coordinate aluminum may result from reconstruction within the altered layer, as suggested by Tsomaia *et al.*, or may represent precipitation or adsorption of aqueous aluminum onto the surface.

Distinguishing the difference between these two mechanisms of formation of altered layers (reconstruction/leaching versus back-reactions) is not trivial. A few workers have suggested that alteration layers on some feldspars formed at low pH (Figure 2) may result from solution–precipitation reactions rather than leaching reactions *per se*. Interestingly, these reactions are proposed to occur well below solubility limits for secondary phases; and if they occur, they do so only at the dissolving mineral surface. For example, *in situ* atomic force microscopy and high-resolution X-ray reflectivity have documented that under acidic conditions the altered layer on dissolving orthoclase has a thickness of only one unit cell (Teng *et al.*, 2001; Fenter *et al.*, 2003). The observation of formation of coatings on dissolving orthoclase on terraces under some conditions (under slow flow rates but not at faster flow rates) (Teng *et al.*, 2001) has been cited as evidence for precipitation or polymerization of silica at the alteration layer. Very fast dissolution and altered-layer formation at the dissolving feldspar interface observed by atomic force microscopy at low pH and high temperature (Hellmann *et al.*, 1992; Jordan *et al.*, 1999) may also be related to such fast reconstruction or reprecipitation processes instead of leaching of cations. The chemical and structural sharpness of the boundary between leached surface and bulk mineral on labradorite crystals dissolved at low pH and low temperature has been interpreted to reflect a solution–precipitation process rather than a diffusional leaching process (Figure 2; Hellmann *et al.*, 2003). Clearly, back-reactions of aqueous silicon and aluminum with the dissolving feldspar surface occur, and determination of the relative contributions from hydrolysis, diffusion, and back-reactions remains to be elucidated. Distinguishing differences between back-reactions occurring at or within the altered layer from reconstruction reactions within a gel-like alteration layer may be difficult.

Development of alteration layers on minerals dissolved under neutral and alkaline conditions has not been thoroughly investigated, but some work has been completed, especially on feldspar compositions (Chou and Wollast, 1984; Hellmann *et al.*, 1989, 1990a; Muir *et al.*, 1990; Nesbitt *et al.*, 1991; Hellmann, 1995, Hamilton *et al.*, 2000). Under neutral conditions, the leached layer thickness (tens of angstroms to a few hundred angstroms) is generally less than that observed for more acid dissolution, with variability reported in the composition and thickness of the layer; i.e., sodium depletion is generally observed, but both aluminum depletion and enrichment (with respect to silicon) have been reported. Variations in solution chemistry (see Section 5.03.7) and feldspar composition may explain some of these differences for chemistry and thickness of leached layers reported in the literature.

Release of trace elements such as strontium from feldspar is also observed to be nonstoichiometric (Brantley *et al.*, 1998). At pH 3, bytownite, microcline, and albite all release strontium at an initially fast rate that slows to near stoichiometric values at steady state. In addition, aqueous strontium is enriched in [87]Sr compared to the bulk mineral early in dissolution. All feldspars studied eventually released strontium in isotopic abundance roughly equal to that of the bulk mineral. Nonstoichiometric release of strontium was explained by the presence of defects or accessory phases in the minerals. Taylor *et al.* (2000) also reported that the initial dissolution of labradorite was nonstoichiometric during dissolution in column reactors with inlet solution pH 3, but that the mineral dissolved and released strontium stoichiometrically at steady state. In contrast to the earlier work, however, [87]Sr/[86]Sr in solution did not differ from that of the bulk labradorite during dissolution in the column experiments.

5.03.3.3.2 Surface complexation model

Despite the complications of layer formation on many silicates, dissolution is generally assumed to be interface-limited and to be accelerated by the presence of protons or hydroxyl ions. Rates are thus often described by the following empirical equation:

$$r = k_H a_{H^+}^n + k_{OH} a_{OH^-}^m \qquad (17)$$

where k_H and k_{OH} are the rate constants for proton- and hydroxyl-promoted dissolution, respectively, a_i is the activity of species i in solution, and n and m are the reaction orders. Values of n and m vary as a function of mineral composition (Tables 2 and 3). Some authors include a third rate constant, k_{H_2O}, to describe dissolution at neutral pH. Values of m have not been as well documented for most minerals. Brady and Walther (1992) suggested that m equals 0.3 for all aluminosilicates. Several models have been proposed to explain the pH-dependence of mineral dissolution and the value of the exponent, n.

In the surface complexation model, Stumm and co-workers (Furrer and Stumm, 1983, 1986; Stumm and Furrer, 1987; Stumm and Wieland, 1990) suggested that adsorption or desorption of protons on an oxide surface polarizes the metal–oxygen bonds, weakening the bonding between the cation and the underlying lattice and explaining the pH-dependence of rates. Surface complexation reactions for an oxide mineral can be written as follows (Schindler, 1981):

$$\equiv SO^- + H^+ \Leftrightarrow \equiv SOH \qquad (18)$$

$$\equiv SOH + H^+ \Leftrightarrow \equiv SOH_2^+ \qquad (19)$$

where $\equiv SOH_2^+$, $\equiv SOH$, and $\equiv SO^-$ represent the positively charged, neutral, and negatively charged surface complexes, respectively. According to this model, the dissolution rate, r, in solutions without reactive ligands is related to surface speciation:

$$r = k_H[\equiv SOH_2^+]^{q_1} + k_{OH}[\equiv SO^-]^{q_2} \qquad (20)$$

Here, $[\equiv SOH_2^+]$ or $[\equiv SO^-]$ represents the concentration of protonated or deprotonated surface sites, respectively, on the mineral surface, and the exponents are constants for each mineral. According to this model, the rate of dissolution of most oxides is slowest in solutions where pH = pH_{ppzc}, the pH of the pristine point of zero charge where the surface charge of the mineral of interest equals zero (Figures 3 and 4). Some authors include a separate rate term describing dissolution at near-neutral pH (= $k_{H_2O}[\equiv SOH]$). Above and below the pH_{ppzc}, oxides are predicted to show enhanced dissolution due to protonated and deprotonated surface sites, respectively.

The proton-promoted surface complexation model was applied to silicates by several research groups in the late 1980s (e.g., Blum and Lasaga, 1988, 1991; Brady and Walther, 1989, 1992; Schott, 1990). For example, Blum and Lasaga (1988, 1991) performed dissolution and titration experiments for albite and found that the surface charge under acid conditions,

Table 2 Feldspar rate parameters at 25 °C for $r = k_H(a_{H^+})^n + k_{H_2O} + k_{OH}(a_{OH^-})^m$.

	$\log k_H$ (mol m^{-2} s^{-1})	$\log k_{H_2O}$ (mol m^{-2} s^{-1})	$\log k_{OH}$ (mol m^{-2} s^{-1})	n	m	pH_{ppzc}[a]
Microcline	−9.9 to −9.4		−10.4 to −9.2[b]	0.4 − 0.5[b]	0.3 − 0.7[b]	6.1
Albite	−9.7 to −9.5[b]	−12.2 to −11.8[b]	−9.9[a]	0.5[b]	0.3[b]	5.2
Labradorite	−9.3 to −8.3[b]			0.4 − 0.5[b]		
Anorthite	−5.9 to −4.5[b]			0.9 − 1.1[b]		5.6

[a] Values of the pristine point of zero charge are from calculations summarized by Sverjensky (1994). Values of pH_{ppzc} for feldspars may be significantly lower than these quoted values, depending upon the model used for calculation (e.g. Parks (1967) and upon the degree of proton-exchange of the surface. [b] Blum and Stillings (1995). Note that where a range of values is presented, the slowest rate constant is most likely to represent the steady state rate of dissolution.

Table 3 Summary of dissolution rates for selected silicates (Equation (17)).

Phase	log k_H	n	pH range	Data used	Formula used
Forsterite[a,b]	−7.0	0.49	<5.7	Blum and Lasaga, 1988; Wogelius and Walther, 1991; as compiled by Chen and Brantley, 2000	O_4
Forsterite[a,b]		0.5	1 < pH ≤ 8	Pokrovsky and Schott, 2000b	
Forsterite[a,b]		0.1	9 ≤ pH ≤ 12	Pokrovsky and Schott, 2000b	
Fayalite	−6.0	0.69		Wogelius and Walther, 1992	O_4
Epidote	−10.7	0.26	2–4	Kalinowski et al., 1998	$O_{12}(OH)_2$
Enstatite[b]	−8.6	0.11	2–7	Ferruzzi, 1993	O_6
Enstatite[b]	−9.3	0.24	1–13	Oelkers and Schott, 2001	O_6
Bronzite[a,b]	−9.5	0.36	≤5	Grandstaff, 1977; Schott and Berner, 1983	O_6
Diopside[b]	−9.4	0.15	2–10	Knauss et al., 1993	O_6
Diopside[b]	−9.0	0.12	2–6	Knauss et al., 1993	O_6
Diopside[c]	−10.5	0.19	1–4	Chen and Brantley, 1998	O_6
Augite[a,b]	−6.7	0.85	≤6	Siegel and Pfannkuch, 1984; Sverdrup, 1990	O_6
Wollastonite[a,b]	−7.8	0.2	≤7.2	Xie, 1994; Xie and Walther, 1994	O_6
Wollastonite[a,b]	−11	0.27	≥7	Xie, 1994; Xie and Walther, 1994	O_6
Wollastonite[a,b]	−9	0	2–6	Weissbart and Rimstidt, 2000	O_6
Rhodonite[b]	−9	0.27	2.1–7.1	Banfield et al., 1995	O_6
Spodumene[a,b]	−4.2	0.64	3–7	Sverdrup, 1990	O_6
Jadeite[a,b]	−7	0.18	3–6	Sverdrup, 1990	O_6
Anthophyllite[b]	−11.85	−0.05	2–5	Mast and Drever, 1987	$O_{22}(OH)_2$
Anthophyllite[c]	−12.5	0.24	1–4	Chen and Brantley, 1998	$O_{22}(OH)_2$
Tremolite[a,b]	−11.5	−0.11	1,6	Schott et al., 1981	$O_{22}(OH)_2$
Hornblende[b]	−11.2	0.33	1–5	Frogner and Schweda, 1998	$O_{22}(OH)_2$
Hornblende[b]	−10.4	0.53	3–5	Frogner and Schweda, 1998	$O_{22}(OH)_2$
Hornblende[b]	−10.4	0.36	1–5.7	Brantley et al., unpublished data	$O_{22}(OH)_2$
Hornblende[b]	−10.9	0.34	1–5.7	Frogner and Schweda, 1998; as reviewed by Brantley et al., in prep.	$O_{22}(OH)_2$
Glaucophane[a,b]	−6.1	0.64	3–7	Sverdrup, 1990	$O_{22}(OH)_2$
Muscovite[b]	−11.8	0.14	1–4	Kalinowski and Schweda, 1996	$O_{20}(OH)_4$
Phlogopite[a,b]	−10.5	0.40	1–4	Kalinowski and Schweda, 1996	$O_{20}(OH)_4$
Biotite[b]	−9.49	0.61	1–4	Kalinowski and Schweda, 1996	$O_{20}(OH)_4$
Chrysotile[a]	−12.2	0.24		Bales and Morgan, 1985	$O_5(OH)_4$
Talc	−12.5	0		Lin and Clemency, 1981	$O_{10}(OH)_2$

All rates expressed as mol mineral m^{-2} s^{-1}, where formula units are given and where rates derive from Si release.
[a] Indicates dissolution measured for less than 1000 h.　[b] Indicates normalized by initial surface area.　[c] Indicates normalized by final surface area.

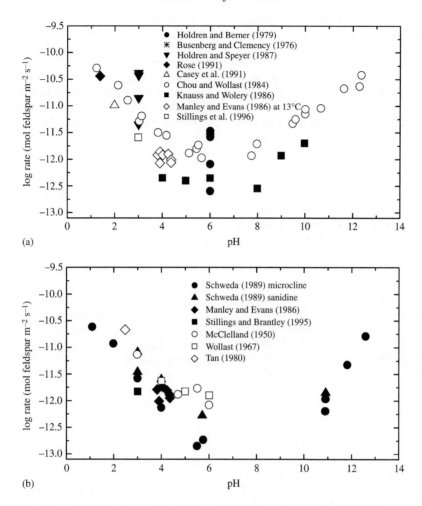

Figure 3 Log (dissolution rate) versus pH for (a) albite, and (b) K-feldspar. Data for (a) from published sources Busenberg and Clemency (1976), Holdren and Berner (1979), Chou and Wollast (1984), Knauss and Wolery (1986), Manley and Evans (1986), Casey *et al.* (1991), Rose (1991), and Stillings *et al.* (1996), and for (b) McClelland (1950), Wollast (1967), Tan (1980), Manley and Evans (1986), Schweda (1989), Stillings and Brantley (1995) (after Blum and Stillings, 1995).

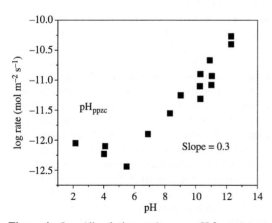

Figure 4 Log (dissolution rate) versus pH for quartz at 25 °C measured in various pH buffers in agitated batch reactors. The slope of the log rate − pH curve equals 0.3 above the pristine point of zero charge (Brady and Walther, 1992) (source Brady and Walther, 1990).

attributed to $[\equiv SOH_2^+]$, is proportional to $a_{H^+}^{0.5}$. Given that Chou and Wollast (1984) measured the rate of albite dissolution and showed that n (Equation (17)) equals ~0.5, Blum and Lasaga proposed that the dissolution rate for albite at 25 °C was directly proportional to $[\equiv SOH_2^+]$ ($q_1 = 1$ for Equation (20)). On the basis of these and other observations at low pH, Schott (1990), Blum and Lasaga (1991), Brady and Walther (1992), and Walther (1996) suggested that protonation of terminal aluminum hydroxyl sites on feldspars controls dissolution. At high pH, Brady and Walther (1992) and Blum and Lasaga (1991) hypothesized that deprotonation of terminal silicon or aluminum sites, respectively controls dissolution. For many minerals, however, the value of the exponents in Equation (20) is not unity. Several workers have suggested that q_1 corresponds to the number of protonation steps required to release a cation from the surface, which corresponds to the oxidation number of the central metal ion in the crystalline lattice (Furrer and Stumm, 1986; Zinder *et al.*, 1986; Guy and Schott, 1989; Schott, 1990): e.g., $q_1 = 2(BeO)$, $= 3(\delta\text{-}Al_2O_3)$, $= 3(\alpha\text{-}Fe_2O_3)$, and $= 4(SiO_2)$.

Despite these attempts to predict the order with respect to protons, q_1 cannot be easily related to an activated complex for mixed oxides (Casey and Ludwig, 1996). Researchers have thus attempted to use proton-promoted models developed specifically for mixed oxides (Hiemstra *et al.*, 1989a; Hiemstra *et al.*, 1989b; Hiemstra and van Riemsdijk, 1990). However, the proton-promoted surface complexation model suffers from many limitations when applied to some silicates (Brantley and Stillings, 1996, Brantley and Stillings, 1997). First, the high surface charge accumulation on feldspars is most likely related to penetration of protons into the altered surface layer, and not due to simple protonation at the water-mineral interface, as assumed in the proton-promoted surface complexation model. Second, for many mixed oxides such as feldspars, the theoretical value of the pH where dissolution should be at a minimum is not well-defined (see e.g., Walther, 1996; Mukhopadhyay and Walther, 2001), and for some minerals, values of the pH_{ppzc} do not correspond to the pH of minimum dissolution (e.g., enstatite, Oelkers and Schott, 2001). Third, application of the proton-promoted model relies upon a comparison of the slope of the log (dissolution rate) versus pH curve to the slope of the log (protonated surface site density) versus pH curve for a given mineral. The slopes derived from titration-based plots (Blum and Lasaga, 1988, 1991; Wollast and Chou, 1992) are strongly dependent upon the value of the pH_{ppzc} chosen, and the pH range over which the slope is determined. Of related interest is the problem that no value for the pH_{ppzc} for feldspars is generally accepted (Parks, 1967; Sverjensky, 1994; Stumm and Morgan, 1996; Brantley and Stillings, 1996, 1997; Walther, 1996; Hellmann, 1999; Mukhopadhyay and Walther, 2001). Fourth, below the pH_{ppzc} of a mineral, proton-promoted dissolution should be enhanced by increasing dissolved salts because of the compression of the diffuse layer and subsequent increase in surface protonation: instead, feldspar dissolution rates decrease with increasing salt concentration (Stillings and Brantley, 1995). Fifth, *ab initio* calculations suggest that protonation of the *Al–O–Si* bridging oxygen during acid dissolution may explain the increase in dissolution rate with decreasing pH (Brand *et al.*, 1993; Xiao and Lasaga, 1995); the proton-promoted model can only be consistent with this inference if the number of protons adsorbed to the mineral surface

at subneutral pH is a constant function of the number adsorbed to bridging oxygens. Furthermore, *ab initio* calculations also document that protonation of terminal $\equiv SiOH$ and $\equiv AlOH$ groups strengthens rather than weakens the bonds connecting these groups to the underlying mineral (Kubicki *et al.*, 1996), suggesting that protonation of terminal hydroxyl groups should inhibit rather than accelerate dissolution. The *ab initio* results do suggest, however, that deprotonation of these groups weakens the bonds, suggesting that, in contrast to low pH, the surface complexation model may be useful for high pH dissolution.

5.03.3.3.3 *Cation exchange mechanism*

To overcome these shortcomings, researchers have suggested surface exchange models that explicitly incorporate proton-metal exchange reactions into the model for acid dissolution (see Oelkers *et al.*, 1994; Oelkers and Schott, 1995a; Hellmann, 1995, Hellmann, 1997b; Brantley and Stillings, 1996, 1997; Pokrovsky and Schott, 2000b; Oelkers, 2001b). In effect, these models now attempt to incorporate both leached layer development and surface reaction. For example, Brantley and Stillings (1996, 1997) proposed that protonation of Al–O–Si bridging oxygens after alkali or alkaline earth cation removal throughout the leached layer promotes dissolution of feldspar at low pH. The importance of the Al–O–Si bond had been emphasized by previous workers (Hellmann *et al.*, 1990b; Oxburgh *et al.*, 1994; Xiao and Lasaga, 1995). Of the cation exchange models, the model developed by Oelkers and co-workers is discussed more extensively here because that model has been applied to a wide range of minerals (Oelkers, 2001b).

Oelkers (2001b) suggests that, for mixed oxides, metal-proton exchange reactions occur for each component metal at different rates, comprising a series of steps, the slowest of which control dissolution. For silicates, the slow step is generally hydrolysis of Si–O network bonds. Acceleration of network hydrolysis occurs when metals bonded to the Si–O bridging bonds are hydrolyzed and leached from the mineral surface. For minerals requiring removal of these metals to form rate-controlling precursors, dissolution rate will depend on the activity of the metal in solution. Examples of such minerals include alkali feldspars, plagioclase feldspars with An < 80%, and inosilicates. Minerals such as anorthite and forsterite, wherein removal of the first metals (e.g., calcium, aluminum, and magnesium, respectively) leaves behind silicon tetrahedra without any bridging oxygens, dissolve according to a proton-promoted mechanism.

The model of Oelkers (2001b) was originally developed and applied to alkali feldspar dissolution at 150 °C and pH 9 (Gautier *et al.*, 1994; Oelkers *et al.*, 1994):

$$r = k \frac{\left(a_{H^+}^{3s} / a_{Al^{3+}}^{s}\right) \prod_i a_{A_i}^{\nu_i}}{1 + K_{Al}\left(a_{H^+}^{3s} / a_{Al^{3+}}^{s}\right) \prod_i a_{A_i}^{\nu_i}} \qquad (21)$$

Here, k is the effective dissolution rate constant. K_{Al} is the equilibrium constant for the exchange of Al^{3+} and H^+ at the mineral surface and formation of $P^\bullet$, the precursor species:

$$M - sAl + 3sH^+ + \sum_i \nu_i A_i \Leftrightarrow P^\bullet + sAl^{3+} \quad (22)$$

The activities, a_i, of the species i in Equation (21) are raised to the powers ν_i, the stoichiometric coefficient of the species in Equation (22). Here, $M - sAl$, is an aluminum-filled Si–O–Si-linked surface site, s equals the number of aluminum ions exchanged to create a surface precursor site, and A_i stands for the ith aqueous species involved in the formation of the precursor. The rate model described in Equation (21) predicts two types of behavior (Oelkers and Schott, 1995a): at low aluminum activity, rates are at a maximum and are independent of aqueous aluminum activity (or concentration), and at higher aluminum activity, fewer precursors are found on the mineral surface and the rate is slower and dependent on dissolved aluminum activity (or concentrations). As discussed later, these authors further argue that a region exists in which the rate is dependent also on an affinity term (Equation (66)).

If the activity of $M - sAl \gg$ activity of $P^\bullet$, and if $\nu_i = 0$ for all species other than H_2O in reaction (22), Equation (21) can be simplified to

$$r = k\left(\frac{a_{H^+}^{3s}}{a_{Al^{3+}}^{s}}\right) \qquad (23)$$

Equation (23) suggests that feldspars that dissolve according to this mechanism should show a dependence upon aqueous Al^{3+} activity. However, not all feldspars dissolve according to such a mechanism. For example, the increase in dissolution rate observed above a threshold concentration of An_{70-80} for plagioclase compositions has been attributed by Blum and Stillings (1995) to the lack of Si–O–Si linkages in compositions with high aluminum content. A mineral with a value of Al/Si approaching 1 is therefore hypothesized to dissolve by a different mechanism because aluminum is completely removed from the dissolving surface during dissolution, leaving only silicon atoms surrounded by nonbridging oxygens (Oelkers and Schott, 1995b). Oelkers and Schott (1995b) suggest that reaction at the anorthite surface

forms a precursor, $(H_{4/\beta}Al_{1/\beta}Si_{1/\beta}O_{4/\beta}^{+3/\beta})^\bullet$ as protons adsorb to the mineral surface. Here, β refers to the number of precursors formed as three hydrogens adsorb, and $(H_{1/\beta}Al_{1/\beta}Si_{1/\beta}O_{4/\beta})$ designates a surface site. $K^\bullet$ is defined to be the equilibrium constant for adsorption of these hydrogens. Oelkers and Schott (1995b) derive the rate law:

$$r = k \frac{a_{H^+}^{3/\beta}}{1 + K^\bullet a_{H^+}^{3/\beta}} \qquad (24)$$

Here, if $K^\bullet a_{H^+}^{3/\beta} \ll 1$, then the rate dependence simplifies such that $r \propto a_{H^+}^{3/\beta}$. For such a condition, the dissolution rate far from equilibrium is dependent only on pH. The value of β, determined from measurement of the pH dependence of anorthite dissolution rates, used by Oelkers and Schott (1995b) was 2 (note that other values could be inferred based on data of Amrhein and Suarez (1988) ($\beta = 1$), and Sverdrup (1990) ($\beta = 3$).

The alkali feldspar model of Equation (21) (Oelkers *et al.*, 1994) differs from the anorthite model of Equation (24) (Oelkers and Schott, 1995b) in the assumed chemistry of the precursor: for alkali feldspars, the precursor contains no aluminum (and thus dissolution depends on aluminum activity in solution) while for anorthite, the precursor contains aluminum (and therefore exhibits no dependence on aluminum activity in solution). In the first model, Al–H^+ exchange occurs at the mineral surface, while in the second model H^+ adsorption is the precursor to dissolution. These models have also been compared to glass dissolution for compositions ranging from albite to nepheline (Hamilton *et al.*, 2001).

Building upon these exchange models for feldspars, Oelkers (2001b) proposed a general kinetic mechanism for multioxides. His general rate equation is expressed as

$$r = k \prod_{i=1, i \neq k}^{i} \left[\frac{K_i\left(a_{H^+}^{\nu_i} / a_{M_i}^{\nu_i}\right)^s}{\left(1 + K_i\left(a_{H^+}^{\nu_i} / a_{M_i}^{\nu_i}\right)^s\right)} \right] \qquad (25)$$

Here k is the rate constant, and K_i is the equilibrium constant for an exchange reaction between protons and the metal M_i at the surface. Oelkers argues that when the term $K_i(a_{H^+}^{\nu_i} / a_{M_i}^{\nu_i})^s$ is small, significant M_i remains in the surface leached layer, and the rate equation simplifies in that the denominator becomes unity. For such a case, the logarithm of the far-from-equilibrium rate becomes linearly related to the logarithm of the activity of the aqueous species M_i and is dependent only upon pH and activity of M_i. Oelkers (2001b) has used this simplified rate equation to describe dissolution of basalt glass

(25 °C, pH 3), kaolinite (pH 2, 150 °C), enstatite (70 °C, pH 2), and muscovite (150 °C, pH 2). In contrast, for forsterite (25 °C, pH 2), and for anorthite (60 °C, pH 2.5), $K_i(a_{H^+}^{\nu_i}/a_{M_i}^{\nu_i})^s$, is large, exchange of Mg^{2+} or Al^{3+} goes to completion, and dissolution rates become independent of Mg^{2+} and Al^{3+} aqueous activities, respectively.

While the preceding model focused upon the pH-dependence of dissolution, the value of k for Equation (17) has also been observed to vary smoothly as a function of the An content of plagioclase (Table 2). The mechanism for such changes has been investigated using *ab initio* techniques (Hamilton *et al.*, 2001). The average bond length of Si–O bonds within Al–O–Si linkages increase from 1.58 Å to 1.60 Å to 1.62 Å as the Al/Si ratio increases in the tetrahedral unit, suggesting that hydrolysis of Al–O–Si bonds across the plagioclase join becomes easier as the Al/Si ratio increases. This observation is consistent with the increase in dissolution rate of plagioclase minerals as An content increases (Blum and Stillings, 1995).

5.03.3.3.4 Mechanism of dissolution of redox-sensitive silicates

The presence of metals of variable redox state in many silicates allows a complex set of oxidation-reduction reactions to occur in parallel with dissolution. Siever and Woodford (1979) investigated the effects of oxygen on the rate of weathering of hypersthene and other iron-containing minerals. They concluded that the dissolution rate of hypersthene at pH 4.5 decreased in the presence of oxygen, perhaps because of precipitation of iron oxyhydroxides onto the dissolving mineral grains. Schott and Berner (1983, 1985) similarly reported several lines of evidence that the increased Fe/Si ratio of the bronzite surface after dissolution at pH 6 under oxic conditions is due to a hydrous ferric oxide precipitate. They argued that this ferric silicate surface should be less reactive than the original ferrous silicate and that the increase in thickness of the ferric layer with time could explain the observed parabolic dissolution behavior of bronzite. Rates of dissolution for oxidized mineral surfaces dominated by ferric ions are predicted to be slower than dissolution rates of comparable ferrous minerals according to the ligand exchange model (Casey *et al.*, 1993a). In contrast to these observations and predictions Zhang (1990) suggested that oxidation of Fe(II) to Fe(III) in hornblende might increase dissolution rates, and Hoch *et al.* (1996) concluded that the dissolution rate of augite was about three times faster in oxygenated compared to sealed experiments.

At least one other study concluded that the presence of oxygen might enhance the weathering rates of iron-containing silicates. White and Yee (1985) investigated oxidation and dissolution reactions of augite and hornblende at ambient temperature in aqueous solutions between pH 1 and pH 9. As observed previously by Schott and Berner (1983, 1985) for bronzite, they concluded that surface iron on these two minerals was dominated by ferric ions. While they reported seeing linear release rates of silicon, potassium, calcium, and magnesium over month-long periods, they observed that release rates of ferrous iron to solution were rapid at first and then decreased as iron oxides precipitated (>pH 3.5).

They also reported ample evidence for a coupled surface-solution oxidation–reduction mechanism whereby aqueous ferric iron can be reduced by ferrous iron on the mineral surface. The oxidation of the surface iron is accompanied by release of a lattice cation to maintain charge balance. Ferrous iron from several angstroms to tens of angstroms depth can be oxidized by these reactions. Hydrolysis of the silicate is thought to occur either simultaneously or subsequent to oxidation. Therefore, cations can be released from the hornblende or augite surface either during hydrolysis reactions (with uptake of protons), or coupled with oxidation of the iron at the surface with no pH change. When Fe^{3+} is released during hydrolysis, it is reduced in solution by ferrous ions deep in the mineral, regenerating the oxidized iron surface layer. Oxidation of ferrous iron in biotite by ferric ions in solution has also been observed by other workers (Kalinowski and Schweda, 1996). For pre-weathered augite, White and Yee (1985) noticed that surface oxidation preceded silicate hydrolysis, but for freshly ground augite the two reactions occurred in parallel. The rate of oxidation of surface ferrous ions by dissolved oxygen in both augite and hornblende was observed to increase with decreasing pH.

5.03.4 SURFACE AREA

5.03.4.1 BET and Geometric Surface Area

To report dissolution rates, elemental release rates are generally normalized either by geometric or BET mineral surface area (Brantley and Mellott, 2000). The total geometric surface area, A_{geo}^T, of a dissolving particle can be calculated from

$$A_{geo}^T = al^d \tag{26}$$

For a cube, a equals 6 and l is the side dimension of the cube. For Euclidean geometries, the term, a is a constant (e.g., for spheres of radius l, $a = 4\pi$), and $d = 2$. The geometric surface area per gram of solid, A_{geo} (the specific surface area) can be expressed as

$$A_{geo} = a' \rho^{-1} l^{d-3} \qquad (27)$$

Here the density of the solid is ρ and a' is another geometric constant. Geometric surface areas have often been used to estimate rates of weathering in field systems (White and Peterson, 1990). In addition, when volumetric rates of dissolution are estimated from rates of interface advance by Equation (10), these rates are implicitly normalized by geometric surface area (e.g., Brantley et al., 1993; Brantley and Chen, 1995).

Surface areas of naturally weathered minerals deviate from A_{geo} due to surface roughness and porosity (Figures 1 and 2). Adsorption of gas is generally used to measure mineral powder surface area (Brantley et al., 1999; Brantley and Mellott, 2000). One complication of such measurements arises because the surface areas measured using different adsorbates differ (for example, N_2 surface areas are generally larger than areas measured with krypton). Reproducibility of values of specific surface area measured by BET on laboratory-ground silicate powders range from poor ($\pm 70\%$ for specific surface areas $<1,000 \text{ cm}^2 \text{ g}^{-1}$) to excellent ($\pm 5\%$ for specific surface areas $>4,000 \text{ cm}^2 \text{ g}^{-1}$) (Brantley and Mellott, 2000). Several researchers have also tried to estimate surface area using atomic force microscopy (e.g., Brantley et al., 1999; Mellott et al., 2002); these surface area estimates generally differ from those based on BET or on geometric calculation.

The ratio of specific surface area measured by adsorption, A_{ads}, to the geometric surface area defines the surface roughness (Helgeson et al., 1984), λ:

$$\lambda = A_{ads}/A_{geo} \qquad (28)$$

The roughness of some minerals correlates with microtexture: for example, λ correlates with microtexture for some alkali feldspars (Hodson et al., 1997).

The measured surface area consists of both external and internal area where internal surface area includes all cracks or connected pores that are deeper than they are wide, varying from subatomic defects to pores of extreme size (Gregg and Sing, 1982). For example, micropores are defined as pores with radius <2 nm, mesopores as pores with radius from 2 nm to 50 nm, and macropores as those with pores of diameter >50 nm. The main distinction between internal and external surface is that advection can control transport to and away from external surface while diffusion must control transport for internal pore space (Hochella and Banfield, 1995). Porosity may be related to crystallization or replacement processes (Putnis, 2002).

The BET specific surface area for a given particle size of laboratory-ground monomineralic samples generally follows the trend of quartz $\approx$ olivine $\approx$ albite $<$ oligoclase $\approx$ bytownite $<$ hornblende $\approx$ diopside (Brantley and Mellott, 2000). For good examples of nonporous silicates with insignificant internal surface area such as laboratory-ground Amelia albite and San Carlos olivine, the log (A_{ads}, $\text{m}^2 \text{g}^{-1}$) can be calculated from

$$\log(A_{ads}) = b + d \log(D) \qquad (29)$$

where $D =$ grain diameter (μm), $b = 1.2 \pm 0.2$ and $d = -1.0 \pm 0.1$ (Brantley and Mellott, 2000). In cases where mineral powders show specific surface areas greater than predicted by Equation (29), these high values may be attributed to the presence of second phase particulates or to the presence of meso- or micropores. Porosity that contributes significantly to BET surface area has been inferred for examples of laboratory-ground diopside, hornblende, potassium feldspar, and all compositions of plagioclase except albite and for naturally weathered quartz, plagioclase, potassium feldspar, and hornblende. However, researchers have suggested that micropores observed on some laboratory-ground alkali feldspars may be related to cracks in the mineral surface generated during grinding rather than due to grain-crosscutting pores (Hodson, 1999).

5.03.4.2 Reactive Surface Area

The reactive surface area (Helgeson et al., 1984), estimated from the dissolution rate of a solid powder, can also differ from the BET surface area (Avnir et al., 1985; Holdren and Speyer, 1985, 1986, 1987; White and Peterson, 1990; Adamson, 1992; Gautier et al., 2001). For example, the adsorption surface area does not equal the reactive surface area for some inosilicates (Xie and Walther, 1994) and for some phyllosilicates (Turpault and Trotignon, 1994; Kalinowski and Schweda, 1996). The latter case may be generally true for phyllosilicates due to differences in reactivity of edge sites and sites on the basal planes. The discrepancy between reactive and adsorption surface area can thus be related to differences in free energy of surface sites (Lee and Parsons, 1995). Spectacular examples of etching due to differences in surface energy have been documented by Lee and Parsons (1995) and others for alkali feldspars: the most reactive sites on some alkali feldspars consist of the surface outcrops of edge dislocations along semi-coherent exsolution lamellae.

Holdren and Speyer (1985, 1986, 1987) observed that dissolution rates of feldspars ground to different grain sizes did not scale with surface area, suggesting that the predominant reactive sites for dissolution are not uniformly distributed within the mineral at the scale of tens of microns. Such differences among grain sizes may be related to the microtextural controls on fracturing of feldspars (e.g., Hodson *et al.*, 1997). Similarly, Acker and Bricker (1992) observed different rates for dissolution of biotite as a function of grain size. Dissolution rates of the mineral component of the B horizon of a granitic podzol were also observed to differ as a function of grain size after removal of organic matter and precipitates, despite similar mineralogy and composition (Hodson, 2002).

Reactive surface area may also change with time due to aging, possibly related to condensation of surface silanols to form siloxanes (Eggleston *et al.*, 1989; Stillings *et al.*, 1995). During dissolution, etching of reactive surface sites may also change reactive surface area. As an example of the latter phenomenon, several researchers have emphasized that naturally weathered feldspars may lose much of their highly reactive surface area, possibly to the formation of etch pits, while laboratory samples may still retain these highly reactive sites (Velbel, 1990; Anbeek, 1992a,b, 1993; Anbeek *et al.*, 1994a,b; Lee and Parsons, 1995).

Discrepancies between reactive and adsorption surface area may also be related to the presence of deep etch pits or pore outcrops which can constitute transport-limited micro-environments for dissolution (Jeschke and Dreybrodt, 2002). Much of the BET surface area for some alkali feldspars used for dissolution in the laboratory has been attributed to grinding-induced microporosity (Hodson *et al.*, 1999), and such pore outcrops are candidates for transport limitation. If such induced surfaces react differently than surfaces of weathered samples, then the BET surface area may be an inappropriate parameter to use for extrapolating interface-limited kinetics from laboratory to field (Lee *et al.*, 1998; Brantley and Mellott, 2000; Jeschke and Dreybrodt, 2002) and consideration may need to be given to length and extent of grinding for laboratory samples (Hodson, 1999). It may be more appropriate to use geometric rather than BET surface area to extrapolate kinetics for samples where etch pits or pore outcrops are important contributors to BET surface area (Gautier *et al.*, 2001; Jeschke and Dreybrodt, 2002; Mellott *et al.*, 2002).

While some researchers have used the initial surface area to normalize dissolution rates, others, recognizing that natural weathering rates are generally normalized by the final rather than initial surface area, have used final surface areas of laboratory-dissolved samples (Brantley and Chen, 1995). For example, Stillings and Brantley (1995) noticed that elemental release rates increased for some dissolving feldspars at pH 3 at the same time that surface area of the grains increased. They concluded that steady-state dissolution rates could only be inferred from rates normalized by final rather than initial surface areas (see also, Schweda, 1990). However, others have observed that dissolution rates decrease or remain constant with time while surface area increases, and have noticed that final surfaces of dissolved feldspars, glasses, and wollastonite are dominated by pores of spalled, altered layers (Casey *et al.*, 1989a; Hamilton *et al.*, 2000; Weissbart and Rimstidt, 2000). Mass-normalized dissolution rates of quartz dissolved at high temperature have also been observed to remain constant while the BET surface area of the quartz increased, mainly due to growth of etch pit walls (Gautier *et al.*, 2001). Observations such as these have led many to suggest that final surface area is not all reactive. For such cases, initial surface area is thought best for use in normalizing dissolution.

One approach to quantifying the reactive surface area is to define the reactive site density on the mineral surface. Site densities, ranging up to $\sim$40 sites nm^{-2} (Davis and Kent, 1990; Koretsky *et al.*, 1998), have been summarized by several workers for representative phases. Such site densities generally must be multiplied by BET surface area to predict the quantity of sites dissolving in any experimental system.

An additional problem in defining reactive surface area on natural samples is the observation that coatings containing organic material, iron, aluminum, and/or silicon on natural mineral surfaces may dominate some surface properties (Davis, 1982; Dove, 1995). Nugent *et al.* (1998) has shown that development of surface coatings on silicates happens very quickly in temperate soils, and that the coatings are often impossible to image using standard techniques of secondary electron or optical microscopy. Protective mineral coatings may form especially on iron-containing minerals in oxidized environments (Velbel, 1993b) and may serve to increase specific surface area of naturally weathered phases (Hodson *et al.*, 1998). If such coatings decrease the reactivity of the surface, then quantification of the chemistry and coverage of such coatings is necessary to predict natural rates of mineral weathering (Dove, 1995).

5.03.5 RATE CONSTANTS AS A FUNCTION OF MINERAL COMPOSITION

Dissolution rates for silicates are generally only reproducible to within $\pm$0.25 log units within one laboratory, and to within two orders of magnitude among different laboratories (see figures 2 and 3 in Brantley and Chen, 1995). However, the database of rate constants for dissolution of primary silicate minerals (White and Brantley, 1995) has led to general models to predict kinetic parameters (e.g., Sverjensky, 1992), and these models should yield better tools toward the assessment of published rate data. Because the magnitudes of dissolution rates, surface areas and surface site densities are not constrained precisely, empirical rate equations have been commonly used for rock-forming minerals. Here, we review empirical rate equations for silicates, and we generate a general rate equation for *ortho-* and inosilicates. Dissolution kinetics of carbonates are also reviewed. The reader is referred to other discussions of dissolution kinetics of secondary silicates, oxides, and sulfides for rates of those processes (e.g., Nordstrom, 1982; Stumm and Wieland, 1990; Sverjensky, 1992; Rimstidt and Newcomb, 1993; Rimstidt *et al.*, 1994; Nagy, 1995).

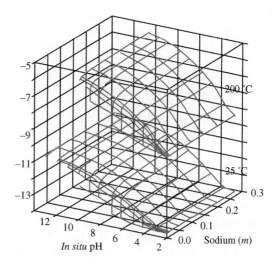

Figure 5 Log (dissolution rate) of quartz $(mol\ m^{-2}\ s^{-1})$ plotted as a function of solution pH and dissolved sodium concentration at 25 °C and 200 °C as predicted by Equation (30) (reproduced by permission of *Am. J. Sci.* **1994**, *294*, 665–712).

5.03.5.1 Silica

The most comprehensive dataset for a silicate has been compiled for quartz dissolution over a range of temperature and pressure (e.g., Rimstidt and Barnes, 1980; Knauss and Wolery, 1988; Wollast and Chou, 1988; Brady and Walther, 1990; Dove and Crerar, 1990; Hiemstra and van Riemsdijk, 1990; Dove and Elston, 1992; Dove and Rimstidt, 1994; Tester *et al.*, 1994; Dove, 1995). Dove (1994, 1995) has modelled the rate of dissolution of quartz as a function of temperature by the rate equation,

$$r = k_{\text{SiOH}}(T)\theta_{\equiv\text{SiOH}} + k_{\text{SiO}^-}(T)\theta_{\equiv\text{SiO}^-_{\text{tot}}} \quad (30)$$

where each rate constant describes the reaction at the respective surface sites, $\theta_{\equiv\text{SiOH}}$ is the fraction of total surface sites with a proton, and $\theta_{\equiv\text{SiO}^-_{\text{tot}}}$ is the sum of the fractions of total sites existing as deprotonated surface hydroxyls and as sites with adsorbed Na^+ ($\equiv\text{SiONa}^+$). Dove argues that this equation describes quartz dissolution over a span of a factor of 10^{11}, from 25 °C to 300 °C for pH 2–12 in variable ionic strength and sodium concentrations to 0.5 m (Figure 5). The first term in the rate equation describes dissolution under acidic conditions near pH = pH$_{\text{ppzc}}$ while the second term describes dissolution at higher pH and NaCl concentration. She points out that both rate constants follow an Arrhenius temperature dependence, with an activation energy of 15.8 kcal mol^{-1} and 19.8 kcal mol^{-1}, respectively (see Section 5.03.6).

Rates of dissolution for amorphous silica into solutions with and without NaCl are compiled by Icenhower and Dove (2000) for the rate model,

$$r = k_+ \left(a_{\text{SiO}_{2(\text{aq})}}\right)\left(a_{\text{H}_2\text{O}}\right)^2 (1 - \Omega) \quad (31)$$

as a function of temperature. Here, k_+ is the forward rate constant and Ω ($=\exp(\Delta G/RT)$) is a measure of the departure from equilibrium. Rates were observed to be ~10 times faster than comparable dissolution rates measured for quartz, but two types of amorphous silica (fused purified quartz and pyrolyzed $SiCl_4$) dissolved at equivalent rates.

5.03.5.2 Feldspars

The most commonly used rate equation for feldspar dissolution under ambient conditions can be written as

$$r = k_{\text{H}}a_{\text{H}^+}^n + k_{\text{H}_2\text{O}} + k_{\text{OH}}a_{\text{OH}^-}^m \quad (32)$$

Here, $k_{\text{H}_2\text{O}}$ is the rate constant for dissolution under neutral pH, and the rate constant k_{H} can be conceptualized as the rate extrapolated to pH 0, while k_{OH} is the rate extrapolated to pOH 0.

As discussed earlier, Equation (32) generally predicts that log dissolution rate versus pH has a V shape (Figure 3), where the trough of low dissolution rate occurs approximately at pH = pH$_{\text{ppzc}}$ (however, see discussion in Brantley and Stillings, 1996; Walther, 1996; Mukhopadhyay and Walther, 2001; Oelkers, 2001b). Parameters for Equation (32) for feldspar minerals are summarized in Table 2; however, the value of the rate constants appropriate at neutral and basic pH ($k_{\text{H}_2\text{O}}$, k_{OH}) have generally not been well-constrained. Under neutral pH, dissolution is so slow that rates may be unmeasurable in laboratory timeframes, and many workers have set the second term equal to zero in Equation (32).

As reviewed by Blum and Stillings (1995), at pH 2–3 at ambient temperature, the log (dissolution rate) for plagioclase feldspar increases linearly with increasing anorthite content from An0 to An80. The rate of dissolution of An100 is significantly faster, and lies off these trends (Fleer, 1982; Chou and Wollast, 1985; Holdren and Speyer, 1987; Mast and Drever, 1987; Amrhein and Suarez, 1988; Sjoberg, 1989; Sverdrup, 1990; Casey *et al.*, 1991; Amrhein and Suarez, 1992; Oxburgh *et al.*, 1994; Stillings and Brantley, 1995). At pH 5, reproducibility of rates in the literature is poor, but a similar dependence on aluminum content may be observed. According to Blum and Stillings (using data from papers cited above), the value of *n* equals 0.5 from An0 to An70, but increases to ~0.75 at An76 and ~1.0 at An100, suggesting that a threshold aluminum content exists such that dissolution behavior changes drastically. Several workers have suggested that log k_H (or *n*) in Equation (32) can be expressed as an empirical function of the Al/(Al + Si) ratio in the feldspar (Casey *et al.*, 1991; Welch and Ullman, 1996).

5.03.5.3 Nonframework Silicates

Dissolution rates for silicates other than feldspars are summarized in Table 3, based upon the first term of rate Equation (32). Except for a few studies, only rates under acid conditions for longer durations are summarized, and, where multiple researchers have published rates, only those rates are summarized that were measured under controlled conditions (usually flow reactors).

Although correlations between these compiled data have not always been successful due to factors such as differences in solution chemistry or reactor characteristics, Casey and Westrich (1992) and Westrich *et al.* (1993) found that the dissolution rates of orthosilicates at pH 2 and 25 °C measured in their laboratory are correlated to the rate constants of solvent exchange (k_{solv}, the rate constant of exchange of H_2O molecules complexed to a metal ion in solution) around the corresponding hydrated, divalent cations. Using data and predictions from those workers and summarized in other references (Blum and Lasaga, 1988; Wogelius and Walther, 1991, 1992; Casey and Westrich, 1992; Casey *et al.*, 1993b; Westrich *et al.*, 1993; Chen and Brantley, 2000; Pokrovsky and Schott, 2000b; Rosso and Rimstidt, 2000) the correlation can be expressed as

$$\log r(\text{pH } 2) = 1.22(\pm 0.08)\log k_{solv} - 14.7(\pm 0.5) \qquad (33)$$

where *r* is rate in mol silicate $m^{-2} s^{-1}$ (formula unit M_2SiO_4). It is expected that a similar correlation exists between the rate constant of solvent exchange (k_{solv}) and the value of the rate constant k_H for orthosilicate dissolution. Using data from the above mentioned sources, the rate constants for the orthosilicates are plotted as log k_H and log(rate, pH 2) versus log k_{solv} (Figure 6). As expected, the correlation between log k_H and log k_{solv} is not as strong as that

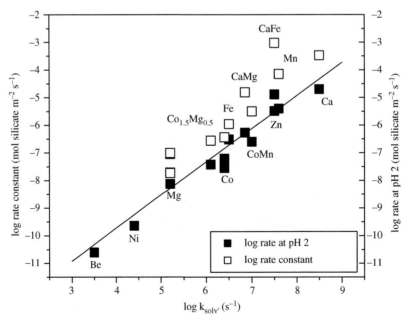

Figure 6 Log (rate constant for dissolution) plotted versus log (solvation constant). Dissolution rates at pH 2 or rate constants at pH 0 are plotted for different compositions of orthosilicate as indicated. References cited in text.

expressed in Equation (33). Assuming an identical slope for the two plots, however, the correlation between k_H and k_{solv} is expressed as:

$$\log k_H = 1.22 \log k_{solv} - 13.6(\pm 0.18) \quad (34)$$

Using the same approach, the relationship between dissolution rates of the inosilicates and the rate constants of solvent exchange for the corresponding divalent cations was also investigated (Banfield *et al.*, 1995). Their results show that the log (rate) for inosilicates may also

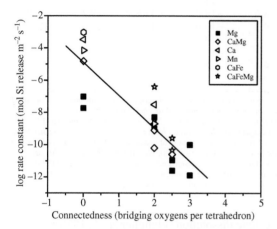

Figure 7 Log (rate constant for dissolution) versus the connectedness, or average number of bridging oxygens per tetrahedrally coordinated atom. A connectedness of 0 reflects an orthosilicate, 2 reflects a pyroxene, 2.5 reflects an amphibole, and 3 reflects a phyllosilicate. Data are reviewed in the text.

correlate with $\log k_{solv}$, although the correlation is not as strong as for orthosilicates.

In general, values of k_H measured for an end-member inosilicate are slower than k_H measured for the end-member orthosilicate, documenting that dissolution rate decreases with increasing polymerization. Brantley and Chen (1995) plotted the log rate constant (mol Si cm^{-2} s^{-1}) of *ortho-*, single-chain, and double-chain silicates against the number of bridging oxygens per tetrahedron (or connectedness, Liebau, 1985) of each mineral (Figure 7). A similar plot is introduced here, wherein the log rate (mol Si cm^{-2} s^{-1}) is plotted versus the ratio of nontetrahedral to tetrahedrally coordinated cations, X/Si, in Figure 8. X/Si equals 2 for orthosilicates, 1 for pyroxenes and pyroxenoids, 0.875 for amphiboles, 1.5 for chrysotile, and 0.75 for talc. The linear correlation here also shows that the rate constant (mol Si m^{-2} s^{-1}) increases with increasing X/Si, following this expression:

$$\log k_H = -5.18 + 13.51 \log(0.5 \, X/Si) \quad (35)$$

Using Equation (32) and the relations shown in Figures 6 and 8, the rate constants for dissolution of the *ortho-*, ino-, and phyllosilicates can be estimated in terms of the X/Si ratio:

$$k_H = 10^{13.51 \log(0.5 \, X/Si) + 1.22 \log k_{solv} - 13.6} / \nu_{Si} \quad (36)$$

where k_H is the rate constant at 25 °C for the rate model expressed in Equation (32) (in mol silicate m^{-2} s^{-1}), and ν_{Si} is the number of silicon in the mineral formula, for example, $\nu_{Si} = 2$ for pyroxene and 8 for amphibole.

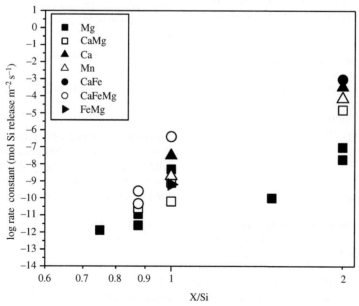

Figure 8 Log (rate of release of Si) versus X/Si for the dissolving mineral. X/Si equals 2 for orthosilicates, 1 for pyroxenes and pyroxenoids, 0.875 for amphiboles, 1.5 for chrysotile, and 0.75 for talc. Data sources are discussed in text.

Table 4 Values of k_{solv} and predicted dissolution rate constants for selected ortho-, ino-, and phyllosilicates using Equation (36).[a]

M	$\log k_{solv}$[b] (s^{-1})	$\log k_H$[c] (mol silicate $m^{-2} s^{-1}$)							
		ortho-[d]		single-chain[e]		double-chain[f]		phyllosilicates	
		Predicted	Measured	Predicted	Measured	Predicted	Measured	Predicted	Measured
Ca	12.5	−3.3	−3.5	−7.7	−7.8 −6.0	−6.0			
Mn	7.6	−4.4	−4.2	−8.7	−9.0	−10.1			
Zn	7.5	−4.5		−8.9		−10.2			
Fe	6.5	−5.7	−6.0	−10.1		−11.4			
Co	6.4	−5.8	−6.4	−10.2		−11.5			
Mg	5.2	−7.2	−7.0	−11.6	−8.6	−13.0	−11.8	−9.2	−9.9[g]
	5.2		−7.7	−11.6	−9.3	−13.0	−12.2	−13.6	−12.5[h]
Ni	4.4	−8.2		−12.6		−14.0			
Be	3.5	−9.3		−13.6		−15.1			
CaMg	6.85	−5.3	−4.8	−9.6	−9.4 −10.0	−11.0			

[a] Values listed are log rates of *ortho-* and inosilicate hypothetical end-members. Some end-members may not exist in nature. Rate constants for mixed composition silicates can be calculated as weighted averages of log rate constants of the end members. Rate equation is expressed: $k_H = 10^{13.51 \log(0.5 \; X/Si)+1.22 \log k_{solv}-13.6}/\nu_{Si}$. [b] Data from Hewkin and Prince (1970). [c] Log rate constant at 25 °C at pH 0 for rate Equation (17). [d] Orthosilicates of composition M_2SiO_4. Data for measured rates from Blum and Lasaga (1988), Wogelius and Walther (1991, 1992), Westrich *et al.* (1993), Casey *et al.* (1993b), Chen and Brantley (2000), and Rosso and Rimstidt (2000). [e] Single-chain inosilicates of composition $M_2Si_2O_6$. Data for measured rates from Ferruzzi (1993), Knauss *et al.* (1993), Xie (1994), Banfield *et al.* (1995), Chen and Brantley (1998), Weissbart and Rimstidt (2000), and Oelkers and Schott (2001). [f] Double-chain inosilicates of composition $M_7Si_8O_{22}(OH)_2$. Data for measured rates from Chen and Brantley (1998), and Mast and Drever (1987). [g] Chrysotile of composition $Mg_3Si_2O_5(OH)_4$. Rate was assumed to be constant with pH (Bales and Morgan, 1985). [h] Talc of composition $Mg_3Si_4O_{10}(OH)_2$. Rate was assumed to be constant with pH (Lin and Clemency, 1981, and Nagy, 1995).

This general rate equation predicts orthosilicate dissolution accurately (Table 4). In addition, dissolution rates for diopside, measured by two independent research groups, are also predicted well. In contrast, based on Figure 6, we expect k_H(enstatite) $< k_H$(diopside), contradicting the data compiled in Table 4. Equation (36) is thus consistent with the conclusion that published values of enstatite dissolution rates may have been measured before steady-state dissolution was achieved. Similarly, measured dissolution rates for talc are also more than an order of magnitude faster than predicted by Equation (36). Other researchers have also tried to predict dissolution rates as a function of composition and structure, and such correlations should become increasingly useful in terms of culling experimental data for artifacts (Sverjensky, 1992).

5.03.5.4 Carbonates

Carbonate dissolution has been well-investigated and models based upon elementary reactions are summarized here, while other models for carbonate dissolution based on affinity are summarized in Section 5.03.8. The rates of forward reaction for the three reactions summarized earlier (Equations (11)–(13)), occurring in parallel, suggested the following equation describing r_{net}, the net rate of dissolution minus precipitation, for calcite dissolution (Plummer *et al.*, 1978):

$$r_{net} = k_1 a_{H^+} + k_2 a_{H_2CO_3^*} + k_3 a_{H_2O} - k_4 a_{Ca^{2+}} a_{HCO_3^-} \qquad (37)$$

Here k_1, k_2, k_3 are rate constants for reactions 11, 12, and 13, and k_4 represents the rate constant for the precipitation of calcite. The values for these rate constants (mmol $cm^{-2} s^{-1}$) are summarized as a function of temperature, T (K), in the following reactions:

$$\log k_1 = 0.198 - 444/T \qquad (38)$$

$$\log k_2 = 2.84 - 2177/T \qquad (39)$$

$$\log k_3 = -1.10 - 1737/T \qquad (40)$$

Equations (38) and (39) were found to describe dissolution over the temperature range 5–48 °C, while Equation (40) only described dissolution from 25 °C to 48 °C. The rate constant for the precipitation reaction is described by

$$k_4 = \frac{K_2}{K_c}\left(k_1' + \frac{1}{a_{H^+_{(s)}}}\right)\left(k_2 a_{H_2CO_{3(s)}^*} + k_3 a_{H_2O}\right) \qquad (41)$$

Here, K_2 is the dissociation constant for bicarbonate, K_c is the solubility product constant for calcite, k_1' is the rate constant for reaction (11), and the subscript (s) refers to concentrations in the surface adsorption layer. Chou *et al.* (1989) suggested a similar set of equations to describe calcite dissolution over a range of conditions at ambient temperature.

Busenberg and Plummer (1982) have similarly described the dissolution of dolomite using the equation

$$r_{net} = k_1 a_{H^+}^n + k_2 a_{H_2CO_3^*}^n + k_3 a_{H_2O} - k_4 a_{HCO_3^-}$$

$$(42)$$

The value of the rate constants (mmol cm^{-2} s^{-1}) are given by

$$\log k_1 = 2.12 - 1880/T \qquad (43)$$

$$\log k_2 = -0.07 - 1800/T \qquad (44)$$

$$\log k_3 = 0.53 - 2700/T \qquad (45)$$

$$\log k_4 = 3.16 - 2300/T \qquad (46)$$

for low iron, sedimentary dolomites. Significant differences in rates of dissolution were observed for sedimentary and hydrothermal dolomites. The value of the exponent n in Equation (42) has been observed to equal 0.5 (Busenberg and Plummer, 1982) or 0.75 (Chou et al., 1989). This rate mechanism has been related to surface speciation for dolomite (Pokrovsky et al., 1999a; Pokrovsky et al., 2000) wherein, under acid conditions, dissolution was controlled by protonated carbonate surface species, and under neutral to alkaline conditions, dissolution was controlled by surface alkaline earth hydroxyl groups. The importance of transport control for dolomite dissolution has also been investigated (Gautelier et al., 1999).

5.03.6 TEMPERATURE DEPENDENCE

5.03.6.1 Activation Energy

The literature data for ortho-, soro-, ino-, and phyllosilicate dissolution at 25 °C derived from long duration dissolution experiments ($>$ a month except for wollastonite and forsterite, Tables 3 and 5) bracket the value of the order with respect to H^+, n (see Equation (17)), between 0 and 0.85 at 25 °C. The higher values of n from the literature tend to be for silicates containing iron. Extrapolating the rate constant for the proton-promoted dissolution rate constant to other temperatures, $k_H(T)$, can be accomplished with the Arrhenius equation:

$$k_H(T) = k_H(T_0)e^{-E_a/RT} \qquad (47)$$

where $k_H(T_0)$ (mol silicate cm^{-2} s^{-1}) is the temperature-independent pre-exponential factor,

E_a is the apparent activation energy, R is the gas constant, and T is absolute temperature (Laidler, 1987).

Measured values of E_a are reported in Table 6 for a range of mineral compositions for dissolution far from equilibrium using Equation (17). Also included are activation energies for the rate constant, k, for dissolution of quartz and amorphous silica in pure water assuming rate Equation (31) (Icenhower and Dove, 2000). Significantly, the measured activation energies do not show a discernible pattern as a function of the connectedness of the mineral (Figure 9). Indeed, Wood and Walther (1983) predicted that all mineral dissolution reactions could be modeled over a large temperature range satisfactorily by an activation energy of 13 kcal mol^{-1}. For comparison, ab initio calculations have shown that the activation energy of hydrolysis of an Si–O$_{br}$ bond (where the subscript refers to a bridging bond) decreases with the connectedness of the silicon atom from 49 kcal mol^{-1} (connectedness of 4) to 33 (3) to 22 (2) to 17 (1) (Pelmenschikov et al., 2000, 2001). The Pelmenschikov et al. approach assumes no relaxation of the molecular geometry at the mineral surface and thus presumably overestimates the activation energy of hydrolysis. A comparison of these calculated values to the measured values summarized in Table 6 or the predicted value of Wood and Walther suggests that the connectedness of the silicon atom of the precursor molecule for silicate dissolution may be equal to or less than 2, even for feldspars and quartz. Such a conclusion may thus be in agreement with predictions made for silicates and aluminosilicates (Kubicki et al., 1996; Fenter et al., 2003). If Q^3, Q^2, and Q^1 sites (where Q^i refers to tetrahedrally coordinated atoms surrounded by i bridging oxygens) are equivalent to edge, layer, and adatom sites, respectively (Hellmann et al., 1990a), then this suggests that the rate-limiting step of dissolution might relate to layer retreat on the mineral surface. Such concepts may not be fully useful, however, when applied to leached layer surfaces where the coordination of atoms could vary throughout the layer. Furthermore, interpretation of activation energies may be complex because apparent activation energies of dissolution controlled by surface complexation may include heats of adsorption (Casey and Sposito, 1992).

5.03.6.2 Effect of Solution Chemistry

Some of the discrepancies among activation energies in the literature may be related to the fact that the apparent activation energy of dissolution may change as solution chemistry changes. For example, apparent activation energies for dissolution of silicates have been observed to decrease in

Table 5 pH dependence for selected silicates as a function of temperature (pH < 7).

Phase	n	Run duration	Solution	References
Orthosilicates				
Forsterite	0.5[a]	5,000–25,000 min	$HNO_3 + H_2O$	Rosso and Rimstidt (2000)
	0.48 (25 °C)			
	0.47 (35 °C)			
	0.53 (45 °C)			
Forsterite	0.49 (25 °C)			Blum and Lasaga (1988), Wogelius and Walther (1991), Chen and Brantley (2000)
	0.70 (65 °C)			
Sorosilicates				
Epidote	0.2[a]			Rose (1991)
	0.20 (90 °C)	1,000–4,700 h	$H_2O \pm H_2SO_4 \pm HCl$	Kalinowski *et al.* (1998)
Epidote	0.15 to 0.62 (25 °C)			
Single chain silicates	0.2[a]			
Enstatite	0.25	1,500–2,000 h	0.1 M buffer solutions	Ferruzzi (1993)
Enstatite	0.24 (25 °C)		0.0001 m Mg, 0.01 m ionic strength	Oelkers and Schott (2001)
	0.24 (50 °C)			
	0.23 (150 °C)			
Diopside[b]	0.2 (25, 70, 90 °C)	40–60 d	buffer solutions pH 2–10	Knauss *et al.* (1993)
Diopside[c]	0.19 (25 °C)	3,400 h	$HCl–H_2O$	Chen and Brantley (1998)
	0.76 (90 °C)			
Wollastonite	0.24	80–220 h	$HCl \pm H_2O \pm KCl$	Xie (1994)
Rhodonite/pyroxmangite and synthetic rhodonite	0.27	2,000–3,500 h	buffer solutions	Banfield *et al.* (1995)
Double chain silicates				
Anthophyllite[b]	0.2 (w/o Fe)[a]			
Anthophyllite[c]	0.4 (w/ Fe)[a]			
	0	1,000–2,000 h	$HCl + H_2O$	Mast and Drever (1987)
	0.24 (25 °C)	3,400 h	$HCl + H_2O$	Chen and Brantley (1998)
	0.63 (90 °C)			
Hornblende	0.47 (25 °C)	1 yr	$HCl–H_2O$	Frogner and Schweda (1998)
Hornblende	0.4 (25 °C)			Brantley *et al.* unpublished data
Phyllosilicates	0.14[a]			
Muscovite	0.14 (25 °C)	3,000 h	$H_2SO_4 + HCl + H_2O$	Kalinowski and Schweda (1996)
Muscovite	0.37 (70 °C)		buffer solutions	Knauss and Wolery (1989)

[a] Suggested value for groups of silicates at 25 °C. [b] Indicates normalized by initial surface area. [c] Indicates normalized by final surface area.

Table 6 pH-independent E_a for dissolution of selected silicates far from equilibrium.

Phase	E_a (kcal mol^{-1})	Temperatures	References
Carbonates			
Calcite	2	5–48 °C	Equation (11), Plummer *et al.* (1978)
	10	5–48 °C	Equation (12), Plummer *et al.* (1978)
	8	25–48 °C	Equation (13), Plummer *et al.* (1978)
Orthosilicates			
Tephroite	22	25–45 °C	Casey *et al.* (1993a)
Monticellite	17	25–45 °C	Westrich *et al.* (1993)
Forsterite	19	25–65 °C	Wogelius and Walther (1991, 1992)
Forsterite	30	25–65 °C	Chen and Brantley (2000), with data from literature
Forsterite	10	26–45 °C	Rosso and Rimstidt (2000)
Forsterite	15	26–65 °C	Oelkers (2001a), measured at pH 2
Pyroxenes			
Enstatite	12	28–168 °C	Oelkers and Schott (2001), measured at pH 2
Diopside	11–15	35–65 °C	Sanemasa and Kataura (1973)
Diopside	10	25–70 °C	Knauss *et al.* (1993)
Diopside	23	25–90 °C	Chen and Brantley (1998)
Amphiboles			
Anthophyllite	19	25–90 °C	Chen and Brantley (1998)
Phyllosilicates			
Kaolinite	16	25–80 °C	Carroll and Walther (1990)
Kaolinite	13	0–50 °C	Kline and Fogler (1981)
Illite	13	0–45 °C	Kline and Fogler (1981)
Pyrophyllite	13	25–70 °C	Kline and Fogler (1981)
Muscovite	13	38–70 °C	Kline and Fogler (1981)
Talc, Phlogopite	10	25–60 °C	Kline and Fogler (1981)
Tectosilicates			
Am. Silica	18	25–250 °C	Rimstidt and Barnes (1980), Mazer and Walther (1994), Icenhower and Dove (2000)
Quartz[a]	16–18	25–300 °C	Rimstidt and Barnes (1980)
Quartz[a]	21	25–625 °C	Tester *et al.* (1994)
Quartz[a]	19	100–300 °C	Polster (1994)
Quartz[a]	17	175–290 °C	Dove (1999)
Microcline	12	5–100 °C	Blum and Stillings (1995) (E_a for basic pH = 14 kcal mol^{-1})
Albite	14	5–300 °C	Blum and Stillings (1995) (E_a for neutral and basic pH = 16 and 12 kcal mol^{-1}, respectively)
Albite	16	5–300 °C	Chen and Brantley (1997)
Labradorite	14	8–70 °C	Blum and Stillings (1995)
Anorthite	19	25–95 °C	Blum and Stillings (1995)

[a] Data only compiled for dissolution in distilled deionized water.

the presence of organic anions (Welch and Ullman, 2000), and in the presence of metabolizing bacteria (Welch and Ullman, 1999). Furthermore, the temperature-dependence of the order of dissolution with respect to $H^+(n)$ and $OH^-(m)$ must also be known for accurate extrapolation, as these values can change with temperature (Table 5). According to model predictions (Brady and Walther, 1992; Casey and Sposito, 1992), for pH < pH_{ppzc}, an increase in temperature will result in an increase in n in Equation (17). The increased value of n at higher temperature has been observed by experiments for some of the silicate minerals (e.g., kaolinite (Carroll and Walther, 1990) and inosilicates (Chen and Brantley, 1998). However, for other minerals, controversy exists as to whether n varies or

remains constant with temperature (e.g., albite (Hellmann, 1994; Chen and Brantley, 1997) and olivine (Casey *et al.*, 1993b; Chen and Brantley, 2000; Rosso and Rimstidt, 2000)).

Where n varies with temperature, the apparent activation energy of dissolution will vary with pH of the system. As a consequence, in the literature, two kinds of E_a are discussed: a pH-dependent and a pH-independent activation energy (Chen and Brantley, 1998). The pH-dependent (apparent) activation energy, E_a', is reported by investigators who plot ln (*rate*) versus $1/T$, and is valid only at the pH of measurement. The pH-independent E_a is determined from a plot of $\ln(k_H)$ versus $1/T$. Where n is independent of temperature, $E_a' = E_a$. For phases where n increases with temperature, activation energies reported in Table 6 are larger

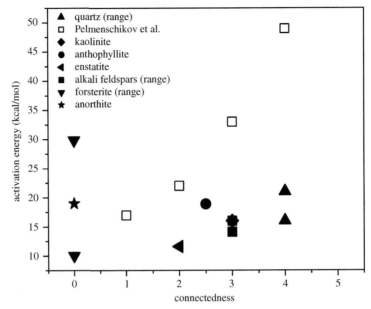

Figure 9 Estimates of the pH-independent activation energy for dissolution at low pH for selected minerals plotted versus connectedness for the dissolving phase. The maximum and minimum values of the E_a are plotted for quartz, alkali feldspars, and forsterite. Connectedness of anorthite is assumed during dissolution to equal 0 because Al is leached first, leaving Si tetrahedra without any bridging oxygens. Similarly, alkali feldspars are plotted at a connectedness of 3. Estimates of activation energy calculated by Pelmenschikov *et al.* (2000, 2001) are included (see text). Experimental data are summarized in Table 6.

than the apparent values that would describe temperature dependence for systems at near-neutral pH. In fact, Arrhenius activation energies for mineral dissolution are theoretically predicted to be lowest at pH values near the pH_{znpc} (Casey and Cheney, 1993). Table 6 summarizes estimates of the pH-independent activation energy for selected silicates. Values of E_a even for low pH dissolution are still relatively poorly defined, and little is known about E_a for pH > neutral.

For those silicates where n varies with temperature, the rate of change in n with respect to temperature (Table 7) has been modeled with a linear dependence on temperature (Kump *et al.*, 2000):

$$\alpha = (n - n_o)/(T - T_0) \qquad (48)$$

where n_o is n at temperature T_0. Combining Equations (17), (47), and (48) and setting $T_0 = 298.15$ K, a general rate equation for silicates can be obtained:

$$r = k_0 e^{(E_a/R)(1/298.15 - 1/T)}(a_{H^+})^{\alpha(T-298.15)+n_o} \qquad (49)$$

or

$$\log r = \log k_0 + [E_a/(2.303R)](1/298.15 - 1/T) - [\alpha(T - 298.15) + n_o]pH \qquad (50)$$

where k_0 is the rate constant at 25 °C and n_o is the reaction order with respect to H^+ at 25 °C. This rate

equation, where we have assumed that contributions due to k_{H_2O} and k_{OH} are absent, is valid for values of pH < pH_{ppzc} for silicate minerals.

Accepting the theoretical prediction that n increases with temperature (Brady and Walther, 1992; Casey and Sposito, 1992), and using the data compiled for orthosilicates (Casey *et al.*, 1993b; Westrich *et al.*, 1993; Chen and Brantley, 2000) and inosilicates (Chen and Brantley, 1998), the values of 0.0062 K^{-1}, 0.0088 K^{-1}, and 0.0060 K^{-1} were chosen as representative values of α (Equations (49) and (50)) for orthosilicate, pyroxene, and amphibole dissolution, respectively (Kump *et al.*, 2000). For such low values of α, changes in pH dependence will only be important for large changes in temperature, and setting $\alpha = 0$ will generally be appropriate for temperatures near ambient.

5.03.7 CHEMISTRY OF DISSOLVING SOLUTIONS

5.03.7.1 Cationic Species

Groundwater or soil pore-water chemistry generally differs from the chemistry of laboratory solutions due to both inorganic and organic solutes which can act as catalysts or inhibitors to dissolution (e.g., Ganor and Lasaga, 1998). For example, much effort has been expended to

Table 7 pH-dependence as a function of temperature.

Phase	α (1/K)	Temperatures	References
Orthosilicates			
Monticellite	0.0055	25–45 °C	Westrich *et al.* (1993)
Tephroite	0.0075	25–45 °C	Casey *et al.* (1993b)
Forsterite	0	25–65 °C	Wogelius and Walther (1991, 1992)
Forsterite	0.0055	25–65 °C	Chen and Brantley (2000)
Forsterite	0	25–45 °C	Rosso and Rimstidt (2000)
Pyroxenes			
Diopside	0	25–70 °C	Knauss *et al.* (1993)
Diopside	0.0088	25–90 °C	Chen and Brantley (1998)
Amphiboles			
Anthophyllite	0.0060	25–90 °C	Chen and Brantley (1998)

understand the inhibitory effect of both dissolved magnesium and phosphate on calcite dissolution (see brief review by Alkattan *et al.*, 1998 and references therein). Similarly, much research has focused upon the catalytic effects of cations on quartz dissolution. The presence of alkali or alkaline earth cations at low concentrations in solution has been observed to increase the dissolution rate of quartz and amorphous silica by factors of up to 100 in low and high-temperature solutions (Dove and Crerar, 1990; Bennett, 1991; Gratz and Bird, 1993; Dove, 1994, 1995; Dove and Nix, 1997; Icenhower and Dove, 2000). Near neutral pH, the trend of rate enhancement follows the order, water $<$ Mg^{2+} $<$ Ca^{2+} $\approx$ Li^{+} $\approx$ Na^{+} $\approx$ K^{+} $<$ Ba^{2+} (Dove and Nix, 1997). Rate enhancements in single salt solutions can be predicted by equations that follow the form of Langmuir-type isotherms, whereas in mixed salt solutions, behavior follows a competitive cation-surface interaction model (Dove, 1999). Dove and co-workers argue that the alkali and alkaline earth cations enhance dissolution by modifying characteristics of the solvent at the mineral–solution interface.

In contrast, the presence of these cations has been observed to decrease the rate of dissolution of several feldspars (Sjoberg, 1989; Nesbitt *et al.*, 1991; Stillings and Brantley, 1995) and to decrease the thickness of the leached layers formed on the feldspar surface. Schweda (1990), Nesbitt *et al.* (1991), and Brantley and Stillings (1996) proposed that feldspar dissolution in acidic solution far from equilibrium is related to competitive adsorption of H^{+} and other cations on the surface of feldspar. Brantley and Stillings derived a rate equation based on a leached layer and Langmuir adsorption model:

$$r = k\left[\frac{K_{\mathrm{H}}^{\mathrm{ads}}a_{\mathrm{H}^+}}{1 + K_{\mathrm{H}}^{\mathrm{ads}}a_{\mathrm{H}^+} + \sum_i K_{M_i}^{\mathrm{ads}}a_{M_i}}\right]^{0.5} \quad (51)$$

where k is the apparent rate constant, M_i refers to adsorbing cations, and $K_{\mathrm{H}}^{\mathrm{ads}}$ and $K_{M_i}^{\mathrm{ads}}$ refer to the

equilibrium constants for adsorption of H^{+} and M_i on the $\equiv$SiOAl$\equiv$ surface site, respectively. Here, $\equiv$SiOAl$\equiv$ denotes a surface bridging oxygen between silicon and aluminum (Brantley and Stillings, 1996, 1997).

Dove (1995) further summarizes evidence suggesting that adsorption of both Al^{3+} and Fe^{3+} onto quartz surfaces inhibits reactivity of that phase. Inhibition of feldspar dissolution also occurs when Al^{3+} is present in solution (Chou and Wollast, 1985; Nesbitt *et al.*, 1991; Chen and Brantley, 1997). For example, Nesbitt *et al.* (1991) argued that adsorption of Al^{3+} retarded the rate of dissolution of labradorite more than other cations. Furthermore, the effect of aqueous Al^{3+} on dissolution of albite may increase with increasing temperature due to the enhanced adsorption of cations with temperature (Machesky, 1989; Chen and Brantley, 1997; note however that Oelkers (2001b) disputes this trend). In contrast, the addition of aqueous aluminum was not observed to affect the rate of forsterite dissolution at pH 3 and 65 °C (Chen and Brantley, 2000). It may be that aqueous aluminum becomes incorporated into surfaces and affects dissolution wherever the connectedness of surface silicon atoms is $>$0. Brantley and Stillings (1996, 1997) and Chen and Brantley (1997) suggest that Equation (51) can be used to model aluminum inhibition on feldspars. Sverdrup (1990) has reviewed the effects of aqueous Al^{3+} on many minerals and incorporated these effects into rate equations.

Equation (51) represents a competitive Langmuir model for cation adsorption to the mineral surface. A similar competitive Langmuir model has been proposed to describe the effect of magnesium on forsterite dissolution (Pokrovsky and Schott, 2000b):

$$r = \frac{kK_{\mathrm{H}}^{\mathrm{ads}}a_{\mathrm{H}^+}^{0.5}}{1 + K_{\mathrm{H}}^{\mathrm{ads}}a_{\mathrm{H}^+}^{0.5}\left(a_{\mathrm{Mg}^{2+}}^2/\left(K_{\mathrm{ex}}a_{\mathrm{H}^+}^4\right)\right)} \quad (52)$$

where K_H^{ads} represents the equilibrium constant for adsorption of a proton and K_{ex} represents the equilibrium constant for exchange of four protons for two magnesium atoms at the forsterite surface. Pokrovsky and Schott (2000b) have suggested that such a rate equation may apply to any silicate whose dissolution is promoted by ion-exchange followed by rate-controlling proton adsorption on exchange sites.

In general, the effect of differences in solution chemistry other than pH on rates of dissolution of ino- and orthosilicates has been reported to be small (see Brantley and Chen, 1995, Chen and Brantley, 2000, and references therein). For example, the effect of dissolved magnesium on forsterite dissolution at pH 3 to pH 6 is insignificant (Pokrovsky and Schott, 2000a,b), presumably because the equilibrium constant for Mg–H exchange reaction (K_{ex} in Equation (52)) is very large. In contrast, forsterite dissolution decreases by factors up to 5 at pH > 8.8 and 25 °C with the addition of silica in solution (Pokrovsky and Schott, 2000a,b). This effect has been attributed to surface complexation reactions at the magnesium-rich surface layer formed in alkaline solutions. In contrast, Oelkers (2001a) reports that forsterite dissolution is independent of both dissolved magnesium and silicon at pH 2 from 25 °C to 65 °C.

As summarized earlier, Oelkers (2001b) has proposed that the effect of dissolved cations derived from the dissolving silicate can be modeled using equations such as (25). For example, Oelkers and Schott (2001) reported that the rate of enstatite dissolution at pH 2 was independent of silica activity but decreased with increasing magnesium activity according to the following equation:

$$r = A' \exp\left(\frac{-E_a}{RT}\right)\left(\frac{a_{H^+}^2}{a_{Mg^{2+}}}\right)^{1/8} \quad (53)$$

where A' is the pre-exponential factor and $E_a = 12.0$ kcal mol^{-1}. Most of the data supporting this rate equation were measured at above ambient temperature.

5.03.7.2   Dissolved Carbon Dioxide

As a consequence of the realization that dissolution of magnesium- and calcium-containing silicates controls carbon dioxide concentrations in the atmosphere over geologic time (e.g., Equations (1)–(2)) and that such reactions may be important for subsurface carbon sequestration over human timescales, researchers have also become interested in the effect of dissolved inorganic carbon on dissolution. The indirect effect of CO_2 species on dissolution of

rock-forming minerals can be predicted based upon the effect of this dissolved gas on pH. However, adsorption of carbon dioxide to surfaces is also known to occur for iron oxides (e.g., goethite (Russell *et al.*, 1975; Zeltner and Anderson, 1988; Van Geen *et al.*, 1994), and such surface complexation may also affect dissolution rates. Sverdrup (1990) models the effect of P_{CO_2} on dissolution rates of several minerals. Carbonate complexation at the surface may explain the effect of P_{CO_2} on hematite dissolution (Bruno *et al.*, 1992). Similarly, a two orders of magnitude decrease in release rate of silica from wollastonite, observed in the presence of ambient CO_2 as compared to CO_2-free NaOH solution at high pH, has been related to formation of surface calcium carbonate complexes (Xie, 1994; Xie and Walther, 1994). The rate of dissolution of forsterite has also been observed to decrease with increasing P_{CO_2} under alkaline conditions (Wogelius and Walther, 1991; Pokrovsky and Schott, 2000b). Surface complexation reactions have also been used to explain the observation that the rate of aluminum release from anorthite varies as a function of $[CO_3^{2-}]^{0.24}$ for $5.5 <$ pH < 8.5 under ambient conditions (Berg and Banwart, 2000). In contrast, under alkaline conditions, the release of calcium and magnesium from diopside during dissolution decreased slightly or not at all for $P_{CO_2} >$ atmospheric (Knauss *et al.*, 1993). Similarly, the dissolution rates of augite and anorthite at pH 4 do not vary measurably as a function of P_{CO_2} in organic-rich solutions (Brady and Carroll, 1994). Malmström and Banwart (1997) report that release of iron, magnesium, and aluminum during dissolution of biotite is similar between CO_2-containing and CO_2-free experiments, but that silicon release is lower in the former system. Despite some interesting observations such as the reaction of carbon dioxide at the fresh albite surface (Wollast and Chou, 1992), the importance of carbonate complexation at the surface of silicate minerals remains to be investigated.

5.03.7.3   Anionic Species

5.03.7.3.1   *Models of ligand-promoted dissolution*

The effect of other inorganic anions on dissolution of silicates is generally considered to be minimal (Sjoberg, 1989; Nesbitt *et al.*, 1991). Incorporation of such anions into the leached layer of feldspar has been observed (Hellmann *et al.*, 1989, 1990a). However, few workers have investigated the effects of inorganic anions on dissolution of minerals. In contrast, many workers have investigated the effects of organic compounds. Results from much of the research investigating the effects of organic anions on silicate dissolution show small but often contradictory effects (Drever and Stillings, 1997). For example, citrate and oxalate have been observed to increase the solubility and rate of dissolution of quartz at 25 °C between pH 5.5 and pH 7 (Bennett *et al.*, 1988; Bennett, 1991) or at other pH values (Welch and Ullman, 1996) even though little evidence for oxalate complexation with aqueous silicon nor adsorption at the quartz surface has been established under ambient conditions (Iler, 1979; Poulson *et al.*, 1997; Pokrovski and Schott, 1998; Kubicki *et al.*, 1999). Increased dissolution of quartz in organic compound-containing solutions may be related to the presence of Na^+ or other aqueous cations in organic salt solutions (Poulson *et al.*, 1997). Some of the disagreement in the literature may also be related to the fact that generally, the rate effects of organic anions are relatively small, only becoming significant at millimolar organic concentrations (Drever, 1994). For example, the effect of inorganic and organic anions on dissolution rates is generally less than a factor of 10 for ortho- and inosilicates in millimolar solutions of organic ligands (Wogelius and Walther, 1991; Brantley and Chen, 1995), and the increase in dissolution rate of feldspar by millimolar concentrations of oxalic acid is less than a factor of 15 (Drever and Stillings, 1997). Such effects are generally pH- and ligand-dependent (Wogelius and Walther, 1991; Welch and Ullman, 1996), and differences in pH and ligand content of experimental solutions may contribute to discrepancies in published rate values.

Given the lack of agreement in experimental data for dissolution in the presence of anions, no model predicting the effect of anions on silicate dissolution has been generally accepted (Drever and Stillings, 1997). Many have suggested that organic ligands have a direct effect on dissolution through surface complexation reactions (Zinder *et al.*, 1986; Amrhein and Suarez, 1988; Chin and Mills, 1991; Wogelius and Walther, 1991; Wieland and Stumm, 1992; Welch and Ullman, 1996; Stillings *et al.*, 1998) where the strength of the ligand-cation interaction is thought to polarize and weaken the bonds between the cation and mineral lattice (Furrer and Stumm, 1986). It has been proposed that ligand-promoted dissolution occurs in parallel with proton promoted dissolution and that the rates of each are additive (Furrer and Stumm, 1986):

$$r = k_H[\equiv SOH_2^+]^{q_1} + k_{OH}[\equiv SO^-]^{q_2} + \sum_i k_{L_i}[\equiv SL_i]^{q_3} \qquad (54)$$

Here L_i refers to a ligand, k_{L_i} refers to the rate constant for ligand-promoted dissolution, and

$\equiv SL_i$ represents a metal–ligand surface complex. Values of k_{L_i} for a given mineral may follow predictable trends. For example, the rate of dissolution of Al_2O_3 was observed to decrease in the presence of organic ligands (at constant concentration and pH) in the order oxalate > malonate > succinate for dicarboxylic acids, and salicylate > phthalate > benzoate for aromatic acids (Furrer and Stumm, 1986). Dicarboxylic ligands were observed to promote dissolution to a greater extent than monocarboxylates. Ludwig *et al.* (1995) suggested that rate constants for ligand-promoted dissolution can be predicted from the equilibrium constants for aqueous metal-organic complexes, and they attribute this to the similarity of activated surface complexes and the corresponding aqueous complexes. Others have concluded that there is no unique stability sequence of the silicate minerals in different organic acids (Barman *et al.*, 1992).

Other researchers argue that the effect of organic ligands on aluminosilicate dissolution is indirect, i.e., due to complexation of ions such as aluminum in solution (Oelkers and Schott, 1998). Complexation is thought to either decrease inhibition or to change the affinity of dissolution (Drever and Stillings, 1997). Oelkers and Schott (1998) use Equation (21) to predict that the log of the dissolution rate of feldspar (composition of An_x where $x = 80\%$ or less) should vary linearly with $\log(a_{H^+}^3/a_{Al}^{3+})$, yielding a slope of 0.33. According to their calculated values of the slope (0.33 for bytownite and labradorite (Welch and Ullman, 1993) and 0.2 for labradorite and andesine (Stillings *et al.*, 1996) dissolved in the presence of organic acids, observations agree with model predictions within error. Oelkers and Schott (1998) argue that the effect of organic ligands on dissolution of feldspars at low pH cannot be related to aluminum surface complexation since most aluminum is leached from the feldspar surface under acid conditions (Brantley and Stillings, 1996). They point out that aluminum complexation will also change the affinity of dissolution, indirectly changing the dissolution rate (Drever *et al.*, 1996).

In agreement with an indirect effect for organic ligands, Kubicki *et al.* (1999) report the formation of inner-sphere monodentate or bidentate bridging surface complexes for oxalic acid on albite, illite, and kaolinite, but *ab initio* calculations of such model molecules indicate that the strengths of these interactions are not enough to enhance the release of the cation to solution. They also report little strong inner-sphere adsorption for acetic, citric, benzoic, salicylic, and phthalic acids at pH 3 and pH 6 on quartz and albite, especially in comparison to adsorption onto iron hydroxide-containing illite. Kubicki *et al.* (1999) argue that enhancement of feldspar and phyllosilicate dissolution in the presence of organic ligands may be related to changes in the ΔG of dissolution due to aqueous complexation of aluminum or to complexation of charge-balancing alkali or alkaline earth cations opening up sites for proton attack.

5.03.7.3.2 Ligand-promoted dissolution of individual minerals

The effect of organic ligands on dissolution of feldspar at ambient temperature has been observed by several groups to follow the same order as that observed by Furrer and Stumm (1986) for alumina: citrate, oxalate > salicylate, tartrate > aspartate, acetate (Drever and Stillings, 1997). High-temperature studies of aluminosilicate dissolution in the presence of organic acids have also been completed, but are not reviewed here (see, however the review by Oelkers and Schott, 1998).

Stillings *et al.* (1996) reported experiments with a range of feldspar compositions, pH, and concentrations of oxalic acid. In the presence of 1 mM oxalic acid, the dissolution rate of feldspar increased by a factor of 2–15 over rates in inorganic solutions at the same pH. The composition of the feldspar did not affect the magnitude of the enhancement of dissolution rate in the presence of oxalic acid over that in an inorganic solution at pH 3, but the relative enhancement may increase with increasing An content for dissolution at higher pH (Mast and Drever, 1987; Amrhein and Suarez, 1988; Drever and Stillings, 1997). Welch and Ullman (1996) also report dissolution rates for a range of feldspars and for quartz from pH 3 to pH 10, with and without oxalate and catechol. The organic compounds enhanced dissolution rates compared to inorganic solutions by up to a factor of 10, and the effect increased with increasing aluminum content.

Rate equations for the dissolution of feldspar in the presence of organic ligands have been proposed and have been related to surface complexation at the feldspar surface (Stillings *et al.*, 1996; Welch and Ullman, 2000). These latter workers also report that the apparent activation energy of silica release from bytownite decreases from $\sim 10 \text{ kcal mol}^{-1}$ to $\sim 7 \text{ kcal mol}^{-1}$ in the presence of oxalate and gluconate in neutral solutions (Welch and Ullman, 2000).

The effect of organic anions on silicates other than quartz and feldspar has also been investigated. Dissolution rates of kaolinite are enhanced more by oxalate than salicylate, while malonate and phthalate show little effect (Carroll-Webb and Walther, 1988; Chin and Mills, 1991; Wieland and Stumm, 1992). Dissolution of hornblende is accelerated at pH 4 in the presence of organic acids at 2.5 mM: rate enhancement was observed to follow the trend oxalic > citric > tannic > polygalacturonic acids (Zhang and Bloom, 1999). The effect of other organic acids on inosilicates was reviewed by Brantley and Chen (1995).

5.03.7.3.3 Complex ligands

Only a few investigators have dissolved minerals in the presence of fulvic and humic acids (reviewed by Drever and Stillings, 1997; van Hees *et al.*, 2002; Zhang and Bloom, 1999). Most interestingly, both inhibition and acceleration of dissolution have been reported for naturally occurring high molecular weight organic acids (Lundström and Öhman, 1990; Ochs *et al.*, 1993). For example, adsorbed humic molecules were observed to both inhibit dissolution (pH 4, 4.5) and accelerate dissolution (pH 3; Ochs *et al.*, 1993), a contradictory observation that has been attributed to the difference in protonation and nature of complexation of the organic molecule at the mineral surface (Ochs, 1996). Lundström and Öhman (1990) demonstrated that the reactivity of organic acids diminished after incubation with live bacteria. The presence of extracellular polysaccharides has also been observed to both accelerate and inhibit dissolution of minerals (Welch *et al.*, 1999).

In recent years, progress has also been made quantifying the effects of bacteria on mineral dissolution (Banfield and Hamers, 1997; Maurice *et al.*, 2001a,b), and interest has grown in the effect of siderophores, low molecular weight iron chelates secreted by bacteria and fungi, on dissolution of iron-containing minerals in terrestrial and marine systems (Hersman, 2000; Kalinowski *et al.*, 2000b; Yoshida *et al.*, 2002). Experiments utilizing purified siderophores or siderophore analogues show enhanced iron release and dissolution of goethite (Watteau and Berthelin, 1994; Holmen and Casey, 1996; Kraemer *et al.*, 1999), hematite (Hersman *et al.*, 1995) and hornblende (Kalinowski *et al.*, 2000a; Kalinowski *et al.*, 2000b; Liermann *et al.*, 2000). Reduction of iron minerals by siderophore-producing bacteria has also been documented (Hersman *et al.*, 2000). Most of these authors have attributed enhanced dissolution to adsorption of siderophores at the mineral surface (Holmen *et al.*, 1997; Kalinowski *et al.*, 2000a). However, differences in dissolution caused by a hydroxamate siderophore and its derivative at temperatures $\leq 40\,^{\circ}\text{C}$ were found to disappear at $55\,^{\circ}\text{C}$, while temperature was found to have no effect on adsorption to the mineral surface (Holmen *et al.*, 1997). More recently, Cheah *et al.* (2003) found that oxalate decreased adsorption of the hydroxamate desferrioxamine B onto goethite, but the hydroxamate did not affect adsorption of oxalate. These workers also observed that the dissolution rate of goethite in the presence of desferrioxamine B alone increased with ligand concentration much faster than in the presence of oxalate alone. Furthermore, it was observed that dissolution of the mineral in the presence of desferrioxamine B is doubled by the addition of oxalate, whereas dissolution in the presence of oxalate is increased by an order of magnitude when the hydroxamate is added, suggesting that the siderophore removes Fe(III) from the oxalate complex such that the oxalate is free to react with the goethite repeatedly. Maurice *et al.* (1996) have observed heterogeneous surface features on hematite and goethite incubated with *Pseudomonas* sp., and Buss *et al.* (2002, 2003) have used atomic force microscope (AFM), XPS, and vertical scanning interferometry to document etch pits on hornblende glass surfaces incubated with *Bacillus* sp. and the commercially available hydroxamate siderophore salt, desferrioxamine mesylate. The importance of high affinity ligands in affecting mineral dissolution in nature remains to be documented.

5.03.8 CHEMICAL AFFINITY

5.03.8.1 Linear Rate Laws

The net rate of dissolution and precipitation of a given phase is a function of ΔG, the driving force for reaction. For example, for albite dissolution, ΔG is defined for the reaction

$$NaAlSi_3O_8 + 4H^+ + 4H_2O$$
$$\rightarrow Na^+ + Al^{3+} + 3H_4SiO_4 \qquad (55)$$

as

$$\Delta G = RT \ln(\Omega) \qquad (56)$$

where Ω is the ratio of the reaction activity quotient and the equilibrium constant for the reaction at the specified temperature and pressure. The chemical affinity of the reaction, defined as $-\Delta G$, equals 0 at equilibrium. Few researchers have measured the rates of dissolution of rock-forming silicates under near-equilibrium conditions and ambient temperatures because of the slow kinetics of reaction; however, many researchers have investigated such equations for carbonates. Examples from both carbonates and silicates are described below.

A general definition for the net rate of precipitation minus dissolution, r_{net}, derived from transition state theory (Lasaga, 1981; Aagaard and Helgeson, 1982; Helgeson *et al.*, 1984; Lasaga, 1984), yields a rate law that is linear with respect to ΔG near equilibrium:

$$r_{net} = -k_+ \prod a_j^{m_j} \left(1 - \exp\left(\frac{n_1 \Delta G}{RT}\right)\right) \qquad (57)$$

Here k_+ is the forward rate constant, a_j is the activity of species j in the rate-determining reaction, m_j and n_1 are constants, and R and T are the gas constant and absolute temperature, respectively. The sign of the rate indicates whether the reaction goes forward or backward. The relationship of this equation to transition state theory and irreversible kinetics has been discussed in the literature (Lasaga, 1995; Alekseyev *et al.*, 1997; Lichtner, 1998; Oelkers, 2001b). The use of this equation with $n_1 = 1$ is generally associated with a composite reaction in which all the elementary reactions are near equilibrium except for one step which is rate-determining. This step must be shared by both dissolution and precipitation.

Rate expressions of this form were derived for calcite precipitation with $n_1 = 1$ (Nancollas and Reddy, 1971; Reddy and Nancollas, 1971), and with $m_j = 0$ and $n_1 = 0.5$ (Sjoberg, 1976; Kazmierczak *et al.*, 1982; Rickard and Sjoberg, 1983; Sjoberg and Rickard, 1983). Rate equations such as (57) wherein rates are linearly proportional to ΔG close to equilibrium have been attributed to adsorption-controlled growth (Nielsen, 1983; Shiraki and Brantley, 1995). Such rate models have been used by some researchers to model dissolution and precipitation of quartz over a wide range in temperature and pressure (Rimstidt and Barnes, 1980); however, it has been pointed out that this has only been confirmed with experiments at high temperature (Dove, 1995).

5.03.8.2 Nonlinear Rate Laws

Many rate data for growth and dissolution of minerals cannot be fit by equations that are linear in ΔG near equilibrium (Nagy and Lasaga, 1990, 1992, 1993; Nagy *et al.*, 1991; Devidal *et al.*, 1992; Murphy *et al.*, 1992; Burch *et al.*, 1993; Alekseyev *et al.*, 1997; Berger *et al.*, 1994; Cama *et al.*, 1994; Gautier *et al.*, 1994; Schott and Oelkers, 1995; Taylor *et al.*, 2000; Oelkers, 2001b). For example, a rate model nonlinear in ΔG was derived for calcite dissolution (Morse, 1978):

$$r_{net} = k(1 - \Omega)^{n_1} \qquad (58)$$

In this equation, the rate constant k and the exponent n_1 were observed to vary for different samples of natural calcite. Here, $\Omega = \exp(\Delta G/RT)$ and the equation can be rewritten as

$$r_{net} = k\left(1 - \exp\left(\frac{\Delta G}{RT}\right)\right)^{n_1} \qquad (59)$$

Equations (58)–(59) are often used to describe dissolution and precipitation rates controlled by crystal defects (Lasaga, 1981). These equations yield nonlinear relationships between r_{net} and ΔG for all values of $n_1 \neq 1$. It has been shown that spiral growth at screw dislocations may be well-described by such a rate law with $n_1 = 2$ (e.g., Burton *et al.*, 1951; Blum and Lasaga, 1987; Teng *et al.*, 2000). Such a nonlinear rate law has also been used to model dissolution and growth of kaolinite (Nagy *et al.*, 1990, 1991), gibbsite (Nagy and Lasaga, 1992), quartz (Berger *et al.*, 1994), and analcime (Murphy *et al.*, 1992) over limited ranges of chemical affinity.

Validation of the mechanisms controlling dissolution or precipitation described by such rate laws is difficult. Teng *et al.* (2000) used *in situ* atomic force microscopy to point out that interpretation of the value of the exponent in an equation such as (58) or (59) in terms of mechanism of precipitation is problematic: for example, the value depends on the supersaturation range and the sample itself (Figures 1(b) and (c)). The presence of ions such as Mg^{2+} also inhibit and affect the mechanism of crystal growth as imaged by AFM (Davis *et al.*, 2000). However, Teng *et al.* demonstrated that a rate order of 2 is obtained when growth occurs on calcite very close to equilibrium at simple, single-sourced dislocation spirals (Figure 1(b)). At higher supersaturation, overall growth is controlled by a different mechanism (the sum of spiral growth and two-dimensional nucleation controlled growth) and is described by a different rate law. Two-dimension nucleation controlled growth (Figure 1(c)) yields an exponential dependence on supersaturation and dominates growth at very high supersaturation (Nielsen, 1983; Shiraki and Brantley, 1995; Teng *et al.*, 2000). Teng *et al.* (2000) point out that rate laws such as those described in Equations (57)–(59) represent composites of multiple mechanisms and that no single rate-determining step may control reaction. Similarly, Shiraki and Brantley (1995) documented that no one rate equation could fit all of the available data over the full range of ΔG and rather suggested a family of curves to describe dissolution and precipitation of calcite at 100 °C as a function of pH, P_{CO_2}, and total inorganic carbon.

A good example of a composite equation that has been successfully used for several minerals including for precipitation of calcite (Inskeep and Bloom, 1985) can be written as

$$r_{net} = k\left(\exp\left(\frac{n_1 \Delta G}{RT}\right) - 1\right)^{m_1} \quad (60)$$

which is equivalent to

$$r_{net} = k(\Omega^{n_1} - 1)^{m_1} \quad (61)$$

In this case, the equation is used to describe precipitation, so it is written so that the rate will be positive for the precipitation reaction. Inskeep and Bloom suggested $n_1 = 0.5$ and $m_1 = 2$ to describe

calcite precipitation under ambient conditions. This rate equation, with different values of the exponents, has been incorporated by many researchers (e.g., Alekseyev *et al.*, 1997) to model dissolution and growth of earth materials, and is used in many geochemical reactive transport codes (as written, $r_{net} > 0$ refers to precipitation and $r_{net} < 0$ refers to dissolution).

Equations (58)–(61) predict that net reaction occurs as a nonlinear function of ΔG. As m_1 increases, the reaction rate becomes a more sigmoidal function of ΔG, defining three behaviors during dissolution (Figure 10): the "dissolution plateau" where rate is constant and is far from equilibrium (Burch *et al.*, 1993), an intermediate region where rate changes rapidly as a function of ΔG, and a region of very slow rate found very close to equilibrium (see review by Alekseyev *et al.*, 1997). Alekseyev *et al.* use rate Equation (61) to describe sanidine dissolution at 300 °C at pH 9. These workers also discuss the complexities of predicting primary mineral dissolution when secondary phases also precipitate.

In some cases, even a single equation such as (60) or (61) is inadequate to describe dissolution over a range of chemical affinity. For example, Burch *et al.* (1993) suggested that the rate of albite dissolution at 80 °C near pH 8.8 (Figure 10) can be written as a sum of parallel rates:

$$r = k_5\left[1 - \exp\left(-\left(\frac{n_1|\Delta G|}{RT}\right)^{m_1}\right)\right]$$
$$+ k_6\left[1 - \exp\left(-\left(\frac{|\Delta G|}{RT}\right)^{m_2}\right)\right] \quad (62)$$

where k_5 and k_6 are rate constants, and n_1, m_1, and m_2 are fitting parameters. Burch *et al.* (1993) suggested that if the ΔG for albite dissolution was greater than ~ -25 kJ mol^{-1} (closer to equilibrium) at 80 °C, then albite dissolution showed a measurable decrease in dissolution rate. Burch *et al.* suggest that dissolution where ΔG is less than -25 kJ mol^{-1} (farther from equilibrium) occurs on the "dissolution plateau" where the rate is constant as a function of affinity (Figure 10). Burch *et al.* argued that the transition from the dissolution plateau to the affinity-dependent region may be related to the critical saturation index for etch pit nucleation at dislocations. For values of the saturation index representing solutions closer to equilibrium than the critical saturation index (Brantley *et al.*, 1986; Blum and Lasaga, 1987; Lasaga and Luttge, 2001), theory predicts that etch pits will not open at either perfect surface or dislocation outcrops.

Far-from-equilibrium dissolution on the dissolution plateau ($\Delta G < -42$ kJ mol^{-1}) was also attributed to etch pit formation at defects for labradorite dissolved in column reactors under ambient conditions with inlet solution pH ~ 3

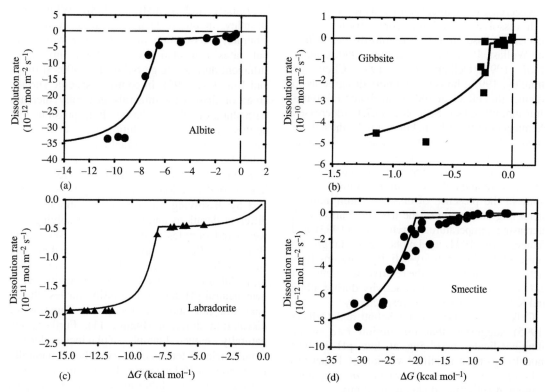

Figure 10 Fits of Equation (64) to data for (a) albite dissolving at pH 8.8, 80 °C (Burch *et al.*, 1993), (b) gibbsite dissolving at pH 3, 80 °C (Mogollon *et al.*, 1996; Nagy and Lasaga, 1992), (c) labradorite dissolving at pH 3, 25 °C (Taylor *et al.*, 2000), and (d) smectite dissolving at pH 3, 80 °C (Cama *et al.*, 2000). In all cases, dissolution rate (plotted for convenience as a negative value) is plotted versus Gibbs free energy of reaction. Far from equilibrium, the rates increase until they approach a dissolution plateau (Lasaga and Luttge, 2001) (Reprinted (abstracted excerpted) with permission from American Association for the Advancement of Science from *Science* **2001**, *291*, 2400–2404).

(Taylor *et al.*, 2000). As equilibrium was approached for labradorite dissolution (Figure 10), etch pit formation was interpreted to be less significant. Taylor *et al.* (2000) calculated the critical value of ΔG for etch pit nucleation on labradorite to be between -26 kJ mol^{-1} and -11 kJ mol^{-1}, roughly in agreement with the transition between the dissolution plateau and the affinity-dependent region. Taylor *et al.* used an inverse model to calculate ΔG in their column reactor, and suggested the following dependence for labradorite dissolution rate on ΔG:

$$r = -k_{min}\left\{0.76\left[1-\exp\left(-1.3\times10^{-17}\left(\frac{\Delta G}{RT}\right)^{14}\right)\right]\right.$$
$$\left. - 0.24\left[1-\exp\left(\frac{-0.35|\Delta G|}{RT}\right)\right]\right\} \quad (63)$$

where k_{min} is the far-from-equilibrium rate constant. The two terms in each of the rate models of Burch *et al.* and Taylor *et al.* represent parallel reactions at etch pits and uniform reaction over the mineral surface (see Figure 1(a), where a similar

pattern of etch pits and surface retreat is imaged for dissolving dolomite).

A generalized model for dissolution incorporating dissolution stepwaves generated by etch pits has been applied to several minerals to explain nonlinearities as a function of chemical affinity (Figure 10) (Lasaga and Luttge, 2001):

$$r = A''(1 - \exp(\Delta G/RT))\tanh\left[\frac{B}{f(\Delta G)}\right]f(\Delta G) \quad (64)$$

where $f(\Delta G)$ is defined as

$$f(\Delta G) = 1 - \left(\frac{1 - \exp(\Delta G_{crit}/RT)}{1 - \exp(\Delta G/RT)}\right) \quad (65)$$

Here, the parameters A'' and B are determined from dissolution far from and near equilibrium, respectively, and ΔG_{crit} is the critical value for the driving force for dissolution (ΔG) such that when $|\Delta G| > |\Delta G_{crit}|$ etch pits spontaneously nucleate at dislocations. Fits of Equation (64) to several mineral dissolution datasets are shown in Figure 10.

A decrease in chemical affinity in the presence of high dissolved aluminum concentrations may decelerate dissolution of aluminosilicates (Chou and Wollast, 1984, 1985; Sverdrup, 1990; Gautier *et al.*, 1994; Oelkers *et al.*, 1994; Chen and Brantley, 1997); however, authors disagree as to whether this effect should be attributed to aluminum inhibition (see Section 5.03.7.1), decreased chemical affinity, or both. For example, the data of Chou and Wollast (1985) for albite dissolution at 25 °C can be modelled as aluminum inhibition or as a decreased driving force. Chen and Brantley (1997) attributed a decrease in dissolution rate of albite in the presence of dissolved aluminum to increasing adsorption of the trivalent cation with increasing temperature. Burch *et al.* (1993) and Gautier *et al.* (1994) used plots of constant growth rate contoured on activity–activity diagrams (isotach plots) to determine the relative importance of ΔG and aqueous aluminum activity during albite and K-feldspar dissolution, respectively.

Oelkers *et al.* (1994) and Schott and Oelkers (1995) suggested that dissolution rate of an aluminosilicate dissolving at low pH is best modeled using a modified version of Equations (21) or (25) that incorporates *both* aqueous aluminum activity and chemical affinity:

$$r = k_+ \left(\frac{\left(a_{H^+}^{3s}/a_{Al^{3+}}^s\right)\prod_i a_{A_i}^{v_i}}{1 + K_{Al}\left(a_{H^+}^{3s}/a_{Al^{3+}}^s\right)\prod_i a_{A_i}^{v_i}} \right)$$
$$\times \left(1 - \exp\left(\frac{\Delta G}{v_{Si}RT}\right)\right) \qquad (66)$$

Variables are defined earlier for Equations (21) and (25). Here, v_{Si} signifies the stoichiometric number of moles of silicon in one mole of dissolving aluminosilicate. Oelkers (2001b) has generalized this model for minerals that dissolve either nonstoichiometrically or stoichiometrically and has discussed the concept of chemical affinity for a mineral surface chemistry that differs from the bulk mineral.

5.03.9 DURATION OF DISSOLUTION

Several authors have noted that dissolution rates appear to decrease with time in both laboratory and field systems (e.g., Colman, 1981; Locke, 1986; Bain *et al.*, 1993; White, 1995; Brantley and Chen, 1995; Gislason *et al.*, 1996; Blum and Erel, 1997; Dahms *et al.*, 1997; Hodson *et al.*, 1999; White and Brantley, in press). In fact, many researchers have pointed out that experiments must run over long periods to attain steady-state kinetics, and that steady-state kinetics may not be attainable (e.g., Brantley and Chen, 1995; Frogner and Schweda, 1998;

Weissbart and Rimstidt, 2000; White and Brantley, in press). Generally, mass-normalized laboratory dissolution rates decrease with time, but rates can also increase with time after sufficient duration (e.g., Schweda, 1990; Stillings and Brantley, 1995). In contrast, specific surface areas of dissolving minerals generally increase with duration of dissolution. It is possible that neither steady-state dissolution nor a steady-state surface ever develops (MacInnis and Brantley, 1992; Weissbart and Rimstidt, 2000).

To investigate some of these ideas, researchers have varied duration of weathering in two ways: first by dissolving samples in the laboratory for different durations and second by dissolving pre-weathered samples in the laboratory. White and Brantley (2003) compiled literature data and showed that literature dissolution rates measured in the laboratory on freshly ground plagioclase at near-neutral pH during dissolution for less than a year did not show a trend as a function of dissolution duration (Figure 11). Furthermore, naturally weathered samples of plagioclase derived from glaciated terrains (i.e., weathered naturally for periods of 10 kyr) were observed to dissolve at similar rates to laboratory ground samples

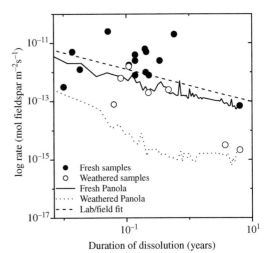

Figure 11 Measured dissolution rate for plagioclase from Panola granite, Georgia, USA compared to published dissolution rates of other plagioclase samples. Panola plagioclase was dissolved either as fresh unweathered samples, or as naturally pre-weathered samples (see text). Also plotted are dissolution rates for plagioclase under ambient conditions in the laboratory at near-neutral pH for freshly ground samples from other localities (solid symbols) and for samples weathered naturally in other field localities and then dissolved in the laboratory (open symbols). All rates were normalized by BET surface area. Dashed line is a fit to all field and laboratory data, including field data from systems weathering for periods of time $\gg 1$ yr (data not shown). Figure adapted from White and Brantley (2003), and all data are attributed in that paper.

(Schnoor, 1990; Swoboda-Colberg and Drever, 1993; Clow and Drever, 1996). Rates of dissolution of laboratory-ground alkali feldspar (Schweda, 1989) are also similar to rates of dissolution at pH 2 and 25 °C of naturally weathered feldspar from Scotland (another glaciated terrain) ground to similar grain size (Lee *et al.*, 1998). These observations are consistent with the conclusion that the intrinsic differences in surfaces between naturally and laboratory-weathered samples are not large and that the duration of reaction is not significant over tens of thousands of years. Futhermore, rates for laboratory-ground samples are significantly faster than the rate measured over 4 yr in the laboratory for naturally weathered plagioclase from a California spodsol (Suarez and Wood, 1995), suggesting that either the longer duration of dissolution (4 yr) or the longer duration of natural weathering (500 kyr) for the latter study, or both, is responsible for slower rates.

White and Brantley (in press) analyzed the effects of both duration of dissolution and the effect of pre-weathering on dissolution rates. The initial rate of dissolution of laboratory-ground plagioclase from Panola granite was similar to the rates of dissolution of laboratory-ground samples published in the literature and to pre-weathered glaciated samples compiled from the literature (Figure 11). However, the final dissolution rate of the freshly ground Panola plagioclase ($10^{-13.2}$ mol m^{-2} s^{-1}) was slower than all the literature dissolution rates except that of the samples from the CA spodsol. In addition, the final dissolution rate of naturally weathered Panola plagioclase measured after 6 yr ($10^{-14.6}$ mol m^{-2} s^{-1}) was slower still. Furthermore, this slower rate was similar to the rate measured over 4 yr in the laboratory for the plagioclase from the California spodsol (Suarez and Wood, 1995). Both of these rates were still faster, however, than the rate of weathering of Panola granite in the field ($10^{-15.6}$ mol m^{-2} s^{-1}). In both the case of Panola and the CA spodsol, the duration of natural weathering *in situ* prior to collection was ~500 kyr. The slow rates of dissolution of naturally weathered Panola and the plagioclase from the CA spodsol in the laboratory are therefore best explained by both the extensive natural weathering experienced by each sample and by the length of duration of weathering in the laboratory.

White and Brantley (in press) suggest several factors to explain these observations. Whereas the chemical affinity of the laboratory weathering solutions are similar and relatively undersaturated ($\Delta G = -13.3$ kcal mol^{-1} and -15.6 kcal mol^{-1}), groundwaters associated with natural weathering of the Panola granite were only slightly undersaturated (-2.6 kcal mol^{-1}) with

respect to albite (White *et al.*, 2001). The order of magnitude difference in dissolution rate between laboratory and field in this setting is therefore largely attributed to the difference in chemical affinity of leaching solutions. In contrast, the more than one order of magnitude difference in dissolution rate between freshly ground and weathered Panola plagioclase must be due to factors intrinsic to the mineral grains themselves, including such factors as armoring of the reactive surface by leached layers, etching of dislocation cores on the mineral surface after long-term natural weathering, or increased dislocation density caused by grinding of fresh surfaces. Suarez and Wood (1995) suggest that even sonication and sieving can enhance weathering rates of minerals for as long as 100 d.

5.03.10 CONCLUSION

Our understanding of mineral dissolution kinetics has advanced significantly over the last several decades as documented in the series of volumes on geochemical kinetics (Lasaga, 1981, 1998; Stumm, 1990; Sverdrup, 1990; White and Brantley, 1995). The widest dataset for minerals, summarized in this chapter, is available for a simple model of dissolution at sub-neutral pH: $r = k_H a_{H^+}^n$. Similar terms have been parametrized for some minerals at neutral and above neutral pH. More specific rate models have also been parametrized for some minerals to incorporate not only the effects of pH, but also dissolved species, surface chemistry, surface topography, and chemical affinity. The emerging field of theoretical molecular geochemistry is now yielding insights into the mechanisms of mineral reaction, and these predictions are amenable to experimental tests through new spectroscopies or through bulk measurements.

The general form of mineral reaction rate equations are better constrained than the rate constants themselves, probably because the dissolution rates of minerals vary as a function of solution chemistry, and as a function of mineral structure, composition, and preparation. These unconstrained variables contribute to the explanation for why rates measured in different laboratories even on similar specimens do not always agree. In general, however, the relative rates of mineral weathering in the laboratory are established for many minerals. Some minerals, such as quartz and calcite, have been particularly thoroughly investigated. Activation energies of dissolution for silicates generally vary between 10 kcal mol^{-1} and 30 kcal mol^{-1}, and show no trend as a function of connectedness. Lack of reproducibility in activation energies measured for a given mineral are probably related to differences in mineral preparation/chemistry and solution chemistry.

Newly parametrized models allow the comparison of reaction rates measured at different scales, ranging from the nanoscale using microscopic observation to the bulk scale using solution chemistry to the field scale using natural observations. However, extrapolating from one scale to another (scaling up) is often not quantitatively successful. For example, quantitative extrapolation of laboratory rates to field systems remains difficult, and we now recognize that multiple factors contribute to this discrepancy. For example, intrinsic factors related to differences in mineral samples prepared in the laboratory or weathered in the field contribute to the laboratory-field discrepancy. Of particular importance is the reactive surface area of dissolving minerals: this term must be investigated and understood more thoroughly. In addition, laboratory rates are generally measured far from equilibrium, whereas natural weathering often occurs closer to equilibrium where dissolution is slower. This difference in chemical affinity is a consequence of the hydrological complexity of natural systems wherein fluids at mineral interfaces may approach equilibrium. Thus, dissolution of laboratory samples may generally be rate-limited by the interface reaction while dissolution in field systems may at times become partially rate-limited by transport. The scaling-up problem inherent in the laboratory-field discrepancy may well be solved through increased understanding developed as we bridge scales of analysis from the nano- to global scale. As reaction transport modeling advances in scope and utility, it should become increasingly possible to elucidate the complexities of mineral reaction kinetics across these scales.

ACKNOWLEDGMENTS

SLB acknowledges Y. Chen, B. Alexander, A. White, J. Schulz, C. Steefel, L. Criscenti, E. Oelkers, R. Hellmann, L. Liermann, and T. Drever for discussions and reviews. Figures provided by A. Luttge, P. Dove, R. Hellmann, A.F. White, H. Teng, and R. Arvidson were also appreciated. Much of this chapter was written while SLB was a Visiting Scientist at the US Geological Survey at Menlo Park CA.

REFERENCES

Aagaard P. and Helgeson H. C. (1982) Thermodynamic and kinetic constraints on reaction rates among minerals and aqueous solutions: I. Theoretical considerations. *Am. J. Sci.* **282**, 237–285.

Acker J. G. and Bricker O. P. (1992) The influence of pH on biotite dissolution and alteration kinetics at low temperature. *Geochim. Cosmochim. Acta* **56**, 3073–3092.

Adamson A. W. (1992) *Physical Chemistry of Surfaces*. Wiley, New York.

Alekseyev V. A., Medvedeva L. S., Prisyagina N. I., Meshalkin S. S., and Balabin A. (1997) Change in the dissolution rates of alkali feldspars as a result of secondary mineral precipitation and approach to equilibrium. *Geochim. Cosmochim. Acta* **59**, 19–31.

Alkattan M., Oelkers E. H., Dandurand J. L., and Schott J. (1998) An experimental study of calcite and limestone dissolution rates as a function of pH from -1 to 3 and temperature from 25 to 80°C. *Chem. Geol.* **151**(1–4), 199–214.

Amrhein C. and Suarez D. L. (1988) The use of a surface complexation model to describe the kinetics of ligand-promoted dissolution of anorthite. *Geochim. Cosmochim. Acta* **52**, 2785–2793.

Amrhein C. and Suarez D. L. (1992) Some factors affecting the dissolution kinetics of anorthite at 25 °C. *Geochim. Cosmochim. Acta* **56**, 1815–1826.

Anbeek C. (1992a) The dependence of dissolution rates on grain size for some fresh and weathered feldspars. *Geochim. Cosmochim. Acta* **56**, 3957–3970.

Anbeek C. (1992b) Surface roughness of minerals and implications for dissolution studies. *Geochim. Cosmochim. Acta* **56**, 1461–1469.

Anbeek C. (1993) The effect of natural weathering on dissolution rates. *Geochim. Cosmochim. Acta* **57**, 4963–4975.

Anbeek C., Bloom P. R., Nater E. A., and Zhang H. (1994a) Comment on Zhang et al. (1995). *Geochim. Cosmochim. Acta* **58**, 4601–4613.

Anbeek C., van Breemen N., Meijer E. L., and van der Plas L. (1994b) The dissolution of naturally weathered feldspar and quartz. *Geochim. Cosmochim. Acta* **58**, 4601–4614.

Avnir D., Farin D., and Pfeifer P. J. (1985) Surface geometric irregularity of particulate materials: the fractal approach. *J. Coll. Interface Sci.* **103**, 112–113.

Bain D. C., Mellor A., Roberston-Rintoul M. S. E., Buckland S. T. (1993) Variations in weathering processes and rates with time in a chronosequence of soils from Glen Feshie, Scotland. *Geoderma* **57**, 275–293.

Bales R. C. and Morgan J. J. (1985) Dissolution kinetics of chrysotile at pH 7 to 10. *Geochim. Cosmochim. Acta* **49**(11), 2281–2288.

Banfield J. F. and Hamers R. J. (1997) Processes at minerals and surfaces with relevance to microorganisms and prebiotic synthesis. In *Geomicrobiology: Interactions between Microbes and Minerals* (eds. J. F. Banfield and K. H. Nealson). Mineralogical Society of America, Washington, DC, vol. 35, pp. 81–122.

Banfield J. F., Ferruzzi C. G., Casey W. H., and Westrich H. R. (1995) HRTEM study comparing naturally and experimentally weathered pyroxenoids. *Geochim. Cosmochim. Acta* **59**(1), 19–31.

Barman A. K., Varadachari C., and Ghosh K. (1992) Weathering of silicate minerals by organic acids: I. Nature of cation solubilisation. *Geoderma* **53**, 45–63.

Bennett P. C. (1991) Quartz dissolution in organic-rich aqueous systems. *Geochim. Cosmochim. Acta* **55**, 1781–1797.

Bennett P. C., Melcer M. E., Siegel D. I., and Hassett J. P. (1988) The dissolution of quartz in dilute aqueous solutions of organic acids at 25 °C. *Geochim. Cosmochim. Acta* **52**, 1521–1530.

Berg A. and Banwart S. A. (2000) Carbon dioxide mediated dissolution of Ca-feldspar: implications for silicate weathering. *Chem. Geol.* **163**, 25–42.

Berger G., Cadore E., Schott J., and Dove P. M. (1994) Dissolution rate of quartz in lead and sodium electrolyte solutions between 25 and 300 °C: effect of the nature of surface complexes and reaction affinity. *Geochim. Cosmochim. Acta* **58**, 541–551.

Berner R. A. (1978) Rate control of mineral dissolution under earth surface conditions. *Am. J. Sci.* **278**(9), 1235–1252.

Berner R. A. and Holdren G. R. (1979) Mechanism of feldspar weathering: II. Observation of feldspars from soils. *Geochim. Cosmochim. Acta* **43**(8), 1173–1186.

Berner R. A., Sjoberg E. L., Velbel M. A., and Krom M. D. (1980) Dissolution of pyroxenes and amphiboles during weathering. *Science* **207**(4436), 1205–1206.

Blum A. E. (1994) Feldspars in weathering. In *Feldspars and their Reactions* (ed. I. Parsons). Kluwer Academic, Dordrecht, Boston, pp. 595–630.

Blum A. E. and Lasaga A. C. (1987) Monte Carlo simulations of surface reaction rate laws. In *Aquatic Surface Chemistry: Chemical Processes at the Particle-water Interface* (ed. W. Stumm). Wiley, New York, pp. 255–291.

Blum A. E. and Lasaga A. C. (1988) Role of surface speciation in the low-temperature dissolution of minerals. *Nature* **331**, 431–433.

Blum A. E. and Lasaga A. C. (1991) The role of surface speciation in the dissolution of albite. *Geochim. Cosmochim. Acta* **55**, 2193–2201.

Blum A. E. and Stillings L. L. (1995) Feldspar dissolution kinetics. In *Chemical Weathering Rates of Silicate Minerals* (eds. A. F. White and S. L. Brantley). Mineralogical Society of America, Washington, DC, vol. 31, pp. 291–351.

Blum A. E., Yund R. A., and Lasaga A. C. (1990) The effect of dislocation density on the dissolution rate of quartz. *Geochim. Cosmochim. Acta* **54**, 283–297.

Blum J. D. and Erel Y. (1997) Rb–Sr isotope systematics of a granitic soil chronosequence: the importance of biotite weathering. *Geochim. Cosmochim. Acta* **61**, 3193–3204.

Brady P. V. and Carroll S. A. (1994) Direct effects of CO_2 and temperature on silicate weathering: possible implications for climate control. *Geochim. Cosmochim. Acta* **58**, 1853–1856.

Brady P. V. and Walther J. V. (1989) Controls on silicate dissolution rates in neutral and basic pH solutions at 25°C. *Geochim. Cosmochim. Acta* **53**, 2823–2830.

Brady P. V. and Walther J. V. (1990) Kinetics of quartz dissolution at low temperatures. *Chem. Geol.* **82**, 253–264.

Brady P. V. and Walther J. V. (1992) Surface chemistry and silicate dissolution at elevated temperatures. *Am. J. Sci.* **292**, 639–658.

Brand H. V., Curtiss L. A., and Iton L. E. (1993) *Ab initio* molecular orbital cluster studies of the zeolite AZM-5: 1. Proton affinities. *J. Phys. Chem.* **97**, 12773–12782.

Brantley S. L. and Chen Y. (1995) Chemical weathering rates of pyroxenes and amphiboles. In *Chemical Weathering Rates of Silicate Minerals* (eds. A. F. White and S. L. Brantley). Mineralogical Society of America, Washington, DC, vol. 31, pp. 119–172.

Brantley S. L. and Mellott N. (2000) Specific surface area and porosity of primary silicate minerals. *Am. Mineral.* **85**, 1767–1783.

Brantley S. L. and Stillings L. L. (1996) Feldspar dissolution at 25°C and low pH. *Am. J. Sci.* **296**, 101–127.

Brantley S. L. and Stillings L. L. (1997) Reply to comment: feldspar dissolution at 25°C and low pH. *Am. J. Sci.* **297**, 1021–1032.

Brantley S. L., Crane S. R., Crerar D., Hellmann R., and Stallard R. (1986) Dissolution at dislocation etch pits in quartz. *Geochim. Cosmochim. Acta* **50**, 2349–2361.

Brantley S. L., Blai A., MacInnis I., Cremeens D., and Darmody D. (1993) Natural etching rates of hornblende and feldspar. *Aquat. Sci.* **55**, 262–272.

Brantley S. L., Chesley J. T., and Stillings L. L. (1998) Isotopic ratios and release rates of Sr measured from weathering feldspars. *Geochim. Cosmochim. Acta* **62**, 1492–1500.

Brantley S. L., White A. F., and Hodson M. (1999) Specific surface area of primary silicates. In *Growth and Dissolution in Geosystems* (eds. B. Jamtveit and P. Meakin). Kluwer Academic, Dordrecht, Boston, pp. 291–326.

Bruno J., Stumm W., Wersin P., and Brandberg F. (1992) On the influence of carbonate in mineral dissolution: I. The thermodynamics and kinetics of hematite dissolution in bicarbonate solutions at T = 25°C. *Geochim. Cosmochim. Acta* **56**, 1139–1147.

Burch T. E., Nagy K. L., and Lasaga A. C. (1993) Free energy dependence of albite dissolution kinetics at 80°C and pH 8.8. *Chem. Geol.* **105**, 137–162.

Burton W. K., Cabrera N., and Frank F. C. (1951) The growth of crystals and the equilibrium structure of their surfaces. *Phil. Trans. Roy. Soc. London* **243**, 299–358.

Busenberg E. and Clemency C. V. (1976) The dissolution kinetics of feldspars at 25°C and 1 atm. CO_2 partial pressure. *Geochim. Cosmochim. Acta* **40**(1), 41–49.

Busenberg E. and Plummer L. N. (1982) The kinetics of dissolution of dolomite in CO_2–H_2O systems at 1.5 to 65 °C and 0 to 1 atm pCO_{22}. *Am. J. Sci.* **282**, 45–78.

Buss H. L., Luttge A., and Brantley S. L. (2002) Etch pits and leached layers on iron-silicate surfaces during siderophore-promoted dissolution. *Geochim. Cosmochim. Acta* **66**(15A Suppl. 1), A113–A113.

Buss H. L., Brantley S. L., and Liermann L. J. (2003) Nondestructive methods for removal of bacteria from silicate surfaces. *Geomicrobiol. J.* **20**, 25–42.

Cama J., Ganor J., and Lasaga A. C. (1994) The kinetics of smectite dissolution. *Min. Mag.* **58A**, 140–141.

Cama J., Ganor J., Ayora C., and Lasaga A. C. (2000) Smectite dissolution kinetics at 80 degrees C and pH 8.8. *Geochim. Cosmochim. Acta* **64**, 2701–2717.

Carroll S. A. and Walther J. V. (1990) Kaolinite dissolution at 25°, 60°, and 80°C. *Am. J. Sci.* **290**, 797–810.

Carroll-Webb S. A. and Walther J. V. (1988) A surface complex model for the pH-dependence of corundum and kaolinite dissolution rates. *Geochim. Cosmochim. Acta* **52**, 2609–2623.

Casey W. H. and Bunker B. C. (1990) Leaching of mineral and glass surfaces during dissolution. In *Mineral–Water Interface Geochemistry* (eds. M. F. Hochella Jr. and A. F. White). Mineralogical Society of America, Washington, DC, vol. 23, pp. 397–494.

Casey W. H. and Cheney M. A. (1993) Bronsted reactions on oxide mineral surfaces and the temperature-dependence of their dissolution rates. *Aquat. Sci.* **55**, 304–313.

Casey W. H. and Ludwig C. (1996) The mechanism of dissolution of oxide minerals. *Nature* **381**, 506–509.

Casey W. H. and Sposito G. (1992) On the temperature dependence of mineral dissolution rates. *Geochim. Cosmochim. Acta* **56**, 3825–3830.

Casey W. H. and Westrich H. R. (1992) Control of dissolution rates of orthosilicate minerals by divalent metal–oxygen bonds. *Nature* **355**, 157–159.

Casey W. H., Carr M. J., and Graham R. A. (1988a) Crystal defects and the dissolution kinetics of rutile. *Geochim. Cosmochim. Acta* **52**, 1545–1556.

Casey W. H., Westrich H. R., and Arnold G. W. (1988b) Surface chemistry of labradorite feldspar reacted with aqueous solutions at pH = 2, 3 and 12. *Geochim. Cosmochim. Acta* **52**, 2795–2807.

Casey W. H., Westrich H. R., Arnold G. W., and Banfield J. F. (1989a) The surface chemistry of dissolving labradorite feldspar. *Geochim. Cosmochim. Acta* **53**, 821–832.

Casey W. H., Westrich H. R., Massis T., Banfield J. F., and Arnold G. W. (1989b) The surface chemistry of labradorite feldspar after acid hydrolysis. *Chem. Geol.* **78**, 205–218.

Casey W. H., Westrich H. R., and Holdren G. R. (1991) Dissolution rates of plagioclase at pH = 2 and 3. *Am. Mineral.* **76**, 211–217.

Casey W. H., Banfield J. F., Westrich H. R., and McLaughlin L. (1993a) What do dissolution experiments tell us about natural weathering? *Chem. Geol.* **105**(1–3), 1–15.

Casey W. H., Hochella M. F., Jr. and Westrich H. R. (1993b) The surface-chemistry of manganiferous silicate minerals as inferred from experiments on tephroite (Mn_2SiO_4). *Geochim. Cosmochim. Acta* **57**(4), 785–793.

Casey W. H., Westrich H. R., Banfield J. F., Ferruzzi G., and Arnold G. W. (1993c) Leaching and reconstruction at the surfaces of dissolving chain-silicate minerals. *Nature* **366**(6452), 253–256.

Cheah S.-F., Kraemer S. M., Cervini-Silva J., and Sposito G. (2003) Steady-state dissolution kinetics of goethite in the presence of desferrioxamine B and oxalate ligands: implications for the microbial acquisition of iron. *Chem. Geol.* **198**, 63–75.

Chen Y. and Brantley S. L. (1997) Temperature- and pH-dependence of albite dissolution rate at acid pH. *Chem. Geol.* **135**, 275–292.

Chen Y. and Brantley S. L. (1998) Diopside and anthophyllite dissolution at 25°C and 90°C and acid pH. *Chem. Geol.* **147**(3–4), 233–248.

Chen Y. and Brantley S. L. (2000) Dissolution of forsteritic olivine at 65°C and 2 < pH < 5. *Chem. Geol.* **165**(3–4), 267–281.

Chen Y., Brantley S. L., and Ilton E. S. (2000) X-ray photoelectron spectroscopic measurement of the temperature dependence of leaching of cations from the albite surface. *Chem. Geol.* **163**(1–4), 115–128.

Chin P. K. F. and Mills G. L. (1991) Kinetics and mechanism of kaolinite dissolution—effects of organic ligands. *Chem. Geol.* **90**, 307–317.

Chou L. and Wollast R. (1984) Study of the weathering of albite at room temperature and pressure with a fluidized bed reactor. *Geochim. Cosmochim. Acta* **48**, 2205–2217.

Chou L. and Wollast R. (1985) Steady-state kinetics and dissolution mechanisms of albite. *Am. J. Sci.* **285**, 963–993.

Chou L., Garrels R. M., and Wollast R. (1989) Comparative study of the kinetics and mechanisms of dissolution of carbonate minerals. *Chem. Geol.* **78**, 269–282.

Clow D. W. and Drever J. I. (1996) Weathering rates as a function of flow through an alpine soil. *Chem. Geol.* **132**, 131–141.

Cocozza C., Tsao C. C. G., Cheah S.-F., Kraemer S. M., Raymond K. N., Miano T. M., and Sposito G. (2002) Temperature dependence of goethite dissolution promoted by trihydroxamate siderophores. *Geochim. Cosmochim. Acta* **66**, 431–438.

Colman S. M. (1981) Rock-weathering rates as functions of time. *Quat. Res.* **15**(3), 250–264.

Dahms D. E., *et al.* (1997) Relation between soil age and silicate weathering rates determined from the chemical evolution of a glacial chronosequence: comment and reply. *Geology*, 381–383.

Davis J. A. (1982) Adsorption of natural dissolved organic matter at the oxide/water interface. *Geochim. Cosmochim. Acta* **46**(11), 2381–2393.

Davis J. A. and Kent D. B. (1990) Surface complexation modelling in aqueous geochemistry. In *Mineral-water Interface Geochemistry* (eds. M. F. Hochella, Jr. and A. F. White). Mineralogical Society of America, Washington, DC, vol. 23, pp. 177–248.

Davis K. J., Dove P. M., and De Yoreo J. J. (2000) The role of Mg_2^+ as an impurity in calcite growth. *Science* **290**, 1134–1137.

Devidal J.-L., Dandurand J. L., and Schott J. (1992) Dissolution and precipitation kinetics of kaolinite as a function of chemical affinity (T = 150°C, pH = 2 and 7.8). *Water–Rock Interact.* **7**, 93–96.

Dietzel M. (2000) Dissolution of silicates and the stability of polysilicic acid. *Geochim. Cosmochim. Acta* **64**, 3275–3281.

Doremus R. H. (1983) Diffusion-controlled reaction of water with glass. *J. Non-crystall. Solids* **55**, 143–147.

Doremus R. H. (1994) Chemical durability: reaction of water with glass. In *Glass Science*. Wiley, New York, pp. 215–240.

Dove P. M. (1994) The dissolution kinetics of quartz in sodium chloride solutions at 25° to 300°C. *Am. J. Sci.* **294**, 665–712.

Dove P. M. (1995) Kinetic and thermodynamic controls on silica reactivity in weathering environments. In *Chemical Weathering Rates of Silicate Minerals* (eds. A. F. White and S. L. Brantley). Mineralogical Society of America, Washington, DC, Short Course, pp. 236–290.

Dove P. M. (1999) The dissolution kinetics of quartz in aqueous mixed cation solutions. *Geochim. Cosmochim. Acta* **63**, 3715–3727.

Dove P. M. and Crerar D. A. (1990) Kinetics of quartz dissolution in electrolyte solutions using a hydrothermal mixed flow reactor. *Geochim. Cosmochim. Acta* **54**, 955–969.

Dove P. M. and Elston S. F. (1992) Dissolution kinetics of quartz in sodium chloride solutions: analysis of existing data and a rate model for 25°C. *Geochim. Cosmochim. Acta* **56**, 4147–4156.

Dove P. M. and Nix C. J. (1997) The influence of the alkaline earth cations, magnesium, calcium, and barium on the dissolution kinetics of quartz. *Geochim. Cosmochim. Acta* **61**, 3329–3340.

Dove P. M. and Platt F. M. (1996) Compatible real-time rates of mineral dissolution by Atomic Force Microscopy (AFM). *Chem. Geol.* **127**, 331–338.

Dove P. M. and Rimstidt J. D. (1994) Silica-water interactions. In *Silica: Physical Behavior, Geochemistry and Materials Applications* (eds. P. J. Heaney, C. T. Prewitt, and G. V. Gibbs). Mineralogical Society of America, Washington, DC, vol. 29, pp. 259–308.

Drever J. I. (1994) The effect of land plants on weathering rates of silicate minerals. *Geochim. Cosmochim. Acta* **58**, 2325–2332.

Drever J. I. and Stillings L. (1997) The role of organic acids in mineral weathering. *Coll. Surf.* **120**, 167–181.

Drever J. I., Poulson S. R., Stillings L. L., and Sun Y. (1996) The effect of oxalate on the dissolution rate of quartz and plagioclase feldspars at 20–25°C. *Geochemistry of Crustal Fluids: Water/Rock Interaction during Natural Processes*, 39.

Eggleston C. M., Hochella M. F., and Parks G. A. (1989) Sample preparation and aging effects on the dissolution rate and surface composition of diopside. *Geochim. Cosmochim. Acta* **53**, 797–804.

Farquhar M. L., Wogelius R. A., and Tang C. C. (1999a) *In situ* synchrotron x-ray reflectivity study of the oligoclase feldspar mineral-fluid interface. *Geochim. Cosmochim. Acta* **63**, 1587–1594.

Farquhar M. L., Wogelius R. A., and Tang C. C. (1999b) *In situ* synchrotron x-ray reflectivity study of the oligoclase feldspar mineral-fluid interface. *Geochim. Cosmochim. Acta* **63**, 1587–1594.

Fenter P., Park C., Zhang Z., Krekeler M. P. S., and Sturchio N. C. (2003) Orthoclase dissolution kinetics probed by *in situ* X-ray reflectivity: effects of temperature, pH and crystal orientation. *Geochim. Cosmochim. Acta* **67**, 197–211.

Ferruzzi G. G. (1993) The character and rates of dissolution of pyroxenes and pyroxenoids. MS, University of California, Davis.

Fleer V. N. (1982) The dissolution kinetics of anorthite $(CaAl_2Si_2O_8)$ and synthetic strontium feldspar (Sr $Al_2Si_2O_8$) in aqueous solutions at temperatures below 100°C: applications to the geological disposal of radioactive nuclear wastes. PhD, Pennsylvania State University.

Frogner P. and Schweda P. (1998) Hornblende dissolution kinetics at 25°C. *Chem. Geol.* **151**(1–4), 169–179.

Furrer G. and Stumm W. (1983) The role of surface coordination in the dissolution of γ-Al_2O_3 in dilute acids. *Chimie* **37**, 338–341.

Furrer G. and Stumm W. (1986) The coordination chemistry of weathering: I. Dissolution kinetics of delta-Al_2O_3 and BeO. *Geochim. Cosmochim. Acta* **50**, 1847–1860.

Ganor J. and Lasaga A. C. (1998) Simple mechanistic models for inhibition of a dissolution reaction. *Geochim. Cosmochim. Acta* **62**, 1295–1306.

Garrels R. M. and Mackenzie F. T. (1971) *Evolution of Sedimentary Rocks.* Norton, New York.

Gautelier M., Oelkers E. H., and Schott J. (1999) An experimental study of dolomite dissolution rates as a function of pH from −0.5 to 5 and temperature from 25 to 80°C. *Chem. Geol.* **157**, 13–26.

Gautier J.-M., Oelkers E. H., and Schott J. (1994) Experimental study of K-feldspar dissolution rates as a function of chemical affinity at 150°C and pH 9. *Geochim. Cosmochim. Acta* **58**(21), 4549–4560.

Gautier J.-M., Oelkers E. H., and Schott J. (2001) Are quartz dissolution rates proportional to BET surface areas? *Geochim. Cosmochim. Acta* **65**(7), 1059–1070.

Gislason S. R., Arnorsson S., and Armannsson H. (1996) Chemical weathering of basalt in southwest Iceland: effects of runoff, age of rocks and vegetative/glacial cover. *Am. J. Sci.* **296**(8), 837–907.

Goldich S. (1938) A study of rock weathering. *J. Geol.* **46**, 17–58.

Gout R., Oelkers E. H., Schott J., and Zwick A. (1997) The surface chemistry and structure of acid-leached albite: new insights on the dissolution mechanisms of the alkali feldspars. *Geochim. Cosmochim. Acta* **61**(14), 3013–3018.

Grandstaff D. E. (1977) Some kinetics of bronzite orthopyroxene dissolution. *Geochim. Cosmochim. Acta* **41**(8), 1097–1103.

Gratz A. J. and Bird P. (1993) Quartz dissolution: negative crystal experiments and a rate law. *Geochim. Cosmochim. Acta* **57**, 965–976.

Gregg S. J. and Sing K. S. W. (1982) *Adsorption, Surface Area, and Porosity.* Academic Press, London, New York.

Guy C. and Schott J. (1989) Multisite surface reaction versus transport control during the hydrolysis of a complex oxide. *Chem. Geol.* **78**, London, New York, 181–204.

Hamilton J. P., Pantano C. G., and Brantley S. L. (2000) Dissolution of albite glass and crystal. *Geochim. Cosmochim. Acta* **64**, 2603–2615.

Hamilton J. P., Brantley S. L., Pantano C. G., Criscenti L. J., and Kubicki J. D. (2001) Dissolution of nepheline, jadeite and albite glasses: toward better models for aluminosilicate dissolution. *Geochim. Cosmochim. Acta* **65**(21), 3683–3702.

Helgeson H. C., Murphy W. M., and Aagard P. (1984) Thermodynamic and kinetic constraints on reaction rates among minerals and aqueous solutions: II. Rate constants, effective surface area, and the hydrolysis of feldspar. *Geochim. Cosmochim. Acta* **48**, 2405–2432.

Hellmann R. (1994) The albite-water system: Part I. The kinetics of dissolution as a function of pH at 100, 200, and 300°C. *Geochim. Cosmochim. Acta* **58**, 595–611.

Hellmann R. (1995) The albite-water system: Part II. The time-evolution of the stoichiometry of dissolution as a function of pH at 100, 200, and 300°C. *Geochim. Cosmochim. Acta* **59**, 1669–1697.

Hellmann R. (1997a) The albite-water system: Part IV. Diffusion modelling of leached and hydrogen-enriched layers. *Geochim. Cosmochim. Acta* **61**, 1595–1612.

Hellmann R. (1997b) Hydrogen penetration of minerals and glasses during hydrolysis at hydrothermal conditions. *Lunar Planet. Inst. Contrib.* **921**, 91.

Hellmann R. (1999) The dissolution behavior of albite feldspar at elevated temperatures and pressures: the role of surface charge and speciation. *Mitteilungen der Osterreichischen Mineralogischen Gesellschaft* **144**, 69–100.

Hellmann R., Eggleston C. M., Hochella M. F., Jr. and Crerar D. A. (1989) Altered layers on dissolving albite: I. Results. *Sixth International Symposium on Water–Rock Interaction, WRI-6*, 293–296.

Hellmann R., Eggleston C. M., Hochella M. F., Jr. and Crerar D. A. (1990a) The formation of leached layers on albite surfaces during dissolution under hydrothermal conditions. *Geochim. Cosmochim. Acta* **54**, 1267–1281.

Hellmann R., Schott J., Dran J.-C., Petit J.-C., and Della Mea G. (1990b) A comparison of the dissolution behavior of albite and albite glass under hydrothermal conditions. *Geol. Soc. Am.: Ann. Meet. Abstr.* A292.

Hellmann R., Schott J., Dran J.-C., Petit J.-C., and Della Mea G. (1991) The formation of leached layers on hydrothermally altered albite and albite glass. *Terra Abstr.* **3**(1), 468.

Hellmann R., Drake B., and Kjoller K. (1992) Using atomic force microscopy to study the structure, topography, and dissolution of albite surfaces. *7th Int. Symp. Water–Rock Interact.* 149–152.

Hellmann R., Dran J.-C., and Della Mea G. (1997) The albite water system: Part III. Characterization of leached and hydrogen-enriched layers formed at 300c using MeV ion beam techniques. *Geochim. Cosmochim. Acta* **61**, 1575–1594.

Hellmann R., Penisson J.-M., Hervig R., Thomassin J. H., and Abrioux M.-F. (2002) Solution-reprecipitation responsible for altered near-surface zones during feldspar dissolution: do leached layers really exist? *Goldschmidt Conference Abstracts*, A320.

Hellmann R., Penisson J.-M., Hervig R. L., Thomassin J.-H., and Abrioux M.-F. (2003) An EFTEM/HRTEM high-resolution study of the near surface of labradorite feldspar altered at acid pH: evidence for interfacial dissolution–reprecipitation. *Phys. Chem. Min.* **30**, 192–197.

Hersman L., Lloyd T., and Sposito G. (1995) Siderophore-promoted dissolution of hematite. *Geochim. Cosmochim. Acta* **59**(16), 3327–3330.

Hersman L. E. (2000) The role of siderophores in iron oxide dissolution. In *Environmental Microbe-metal Interactions* (ed. D. Lovley). ASM Press, Washington, DC, pp. 145–157.

Hersman L. E., Huang A., Maurice P. A., and Forsythe J. H. (2000) Siderophore production and iron reduction by pseudomonas mendocina in response to iron deprivation. *Geomicrobiol. J.* **17**, 261–273.

Hewkin D. J. and Prince R. H. (1970) The mechanism of octahedral complex formation by labile metal ions. *Coordinat. Chem. Rev.* **5**, 45–73.

Hiemstra T. and van Riemsdijk W. H. (1990) Multiple activated complex dissolution of metal (hydr)oxides: a thermodynamic approach to quartz. *J. Coll. Interface Sci.* **136**, 132–150.

Hiemstra T., van Riemsdijk W. H., and Bolt G. H. (1989a) Multisite proton adsorption modelling at the solid/solution interface of (hydr)oxides: a new approach: I. Model description and evaluation of intrinsic reaction constants. *J. Coll. Interface Sci.* **133**, 91–104.

Hiemstra T., van Riemsdijk W. H., and Bolt G. H. (1989b) Multisite proton adsorption modelling at the solid/solution interface of (hydr)oxides: a new approach: II. Application to various important (hydr)oxides. *J. Coll. Interface Sci.* **133**, 105–117.

Hill C. G. (1977) *An Introduction to Chemical Engineering Kinetics and Reactor Design.* Wiley, New York.

Hoch A. R., Reddy M. M., and Drever J. I. (1996) The effect of iron content and dissolved O_2 on dissolution fates of clinopyroxene at pH 5.8 and 25°C: preliminary results. *Chem. Geol.* **132**(1–4), 151–156.

Hochella M. F., Jr. (1988) Auger electron and X-ray photoelectron spectroscopies. In *Spectroscopic Methods in Mineralogy and Geology* (ed. F. C. Hawthorne). Mineralogical Society of America, Washington, DC, vol. 18, pp. 573–637.

Hochella M. F. J. (1990) Atomic structure, microtopography, composition, and reactivity of mineral surfaces. In *Mineral–Water Interface Geochemistry* (eds. M. F. Hochella, Jr. and A. F. White). Mineralogical Society of America, Washington, DC, vol. 23, pp. 87–132.

Hochella M. F. J. and Banfield J. F. (1995) Chemical weathering of silicates in nature: a microscopic perspective with theoretical considerations. In *Chemical Weathering Rates of Silicate Minerals* (eds. A. F. White and S. L. Brantley). Mineralogical Society of America, Washington, DC, vol. 31, pp. 353–406.

Hodson M. (2002) Variation in element release rate from different mineral size fractions from the B horizon of a granitic podzol. *Chem. Geol.* **190**, 91–112.

Hodson M. E. (1999) Micropore surface area variation with grain size in unweathered alkali feldspars: implications for surface roughness and dissolution studies. *Geochim. Cosmochim. Acta* **62**(21–22), 3429–3435.

Hodson M. E., Lee M. R., and Parsons I. (1997) Origins of the surface roughness of unweathered alkali feldspar grains. *Geochim. Cosmochim. Acta* **61**(18), 3885–3896.

Hodson M. E., Langan S. J., Kennedy F. M., and Bain D. C. (1998) Variation in soil surface area in a chronosequence of soils from Glen Feshie, Scotland and its implications for mineral weathering rate calculations. *Geoderma* **85**, 1–18.

Hodson M. E., Langan S. J., and Wilson M. J. (1999) The influence of soil age on calculated mineral weathering rates. *Appl. Geochem.* **14**, 387–394.

Holdren G. R. and Berner R. A. (1979) Mechanism of feldspar weathering: I. Experimental studies. *Geochim. Cosmochim. Acta* **43**(8), 1161–1172.

Holdren G. R. and Speyer P. M. (1985) Reaction rate-surface area relationships during the early stages of weathering: I. Initial observations. *Geochim. Cosmochim. Acta* **49**, 675–681.

Holdren G. R. and Speyer P. M. (1986) Stoichiometry of alkali feldspar dissolution at room temperature and various pH values. In *Rates of Chemical Weathering of Rocks and Minerals* (eds. S. L. Colman and D. P. Dethier). Academic Press, Orlando, pp. 61–81.

Holdren G. R. and Speyer P. M. (1987) Reaction rate-surface area relationships during the early stages of weathering: II. Data on eight additional feldspars. *Geochim. Cosmochim. Acta* **51**, 2311–2318.

Holmen B. A. and Casey W. H. (1996) Hydroxymate ligands, surface chemistry, and the mechanism of ligand-promoted dissolution of goethite [α-FeOOH(s)]. *Geochim. Cosmochim. Acta* **60**, 4403–4416.

Holmen B. A., Tejedor-Tejedor M. I., and Casey W. H. (1997) Hydroxamate complexes in solution and at the goethite-water interface: a cylindrical internal reflection Fourier transform infrared spectroscopy study. *Langmuir* **13**, 2197–2206.

Icenhower J. and Dove P. M. (2000) The dissolution kinetics of amorphous silica into sodium chloride solutions: effects of temperature and ionic strength. *Geochim. Cosmochim. Acta* **64**, 4193–4203.

Iler R. K. (1979) *The Chemistry of Silica*. Wiley, New York.

Inskeep W. P. and Bloom P. R. (1985) An evaluation of rate equations for calcite precipitation kinetics at P_{CO_2} less than 0.01 atm and pH greater than 8. *Geochim. Cosmochim. Acta* **49**, 2165–2180.

Inskeep W. P., Nater E. A., Bloom P. R., Vandervoort D. S., and Erich M. S. (1991) Characterization of laboratory weathered labradorite surfaces using X-ray photoelectron spectroscopy and transmission electron microscopy. *Geochim. Cosmochim. Acta* **55**, 787–800.

Inskeep W. P., Nater E. A., Bloom P. R., Vandervoort D. S., and Erich M. S. (1991) Characterization of laboratory weathered labradorite surfaces using X-ray photoelectron spectroscopy and transmission electron microscopy. *Geochim. Cosmochim. Acta* **55**, 787–800.

Jeschke A. A. and Dreybrodt W. (2002) Dissolution rates of minerals and their relation to surface morphology. *Geochim. Cosmochim. Acta* **66**, 3055–3062.

Jordan G., Higgins S. R., Eggleston C. M., Swapp S. M., Janney D. E., and Knauss K. G. (1999) Acidic dissolution of plagioclase: *in-situ* observations by hydrothermal atomic force microscopy. *Geochim. Cosmochim. Acta* **63**, 3183–3191.

Kalinowski B., Faith-Ell C., and Schweda P. (1998) Dissolution kinetics and alteration of epidote in acidic solutions at 25°C. *Chem. Geol.* **151**, 181–197.

Kalinowski B. E. and Schweda P. (1996) Kinetics of muscovite, phiogopite and biotite dissolution and alteration at pH 1–4, room temperature. *Geochim. Cosmochim. Acta* **60**, 367–385.

Kalinowski B. E., Liermann L. J., Brantley S. L., Barnes A., and Pantano C. G. (2000a) X-ray photoelectron evidence for bacteria-enhanced dissolution of hornblende. *Geochim. Cosmochim. Acta* **64**(8), 1331–1343.

Kalinowski B. E., Liermann L. J., Givens S., and Brantley S. L. (2000b) Rates of bacteria-promoted solubilization of Fe from minerals: a review of problems and approaches. *Chem. Geol.* **169**, 357–370.

Kazmierczak T. F., Tomson M. G., and Nancollas G. H. (1982) Crystal growth of calcium carbonate: a controlled composition kinetic study. *J. Phys. Chem.* **86**, 103–107.

Kline W. E. and Fogler H. S. (1981) Dissolution kinetics: the nature of the particle attack of layered silicates in HF. *Chem. Eng. Sci.* **36**, 871–884.

Knauss K. G. and Wolery T. J. (1986) Dependence of albite dissolution kinetics on pH and time at 25°C and 70°C. *Geochim. Cosmochim. Acta* **50**, 2481–2497.

Knauss K. G. and Wolery T. J. (1988) The dissolution kinetics of quartz as a function of pH and time at 70°C. *Geochim. Cosmochim. Acta* **52**, 43–53.

Knauss K. G. and Wolery T. J. (1989) Muscovite dissolution kinetics as a function of pH and time at 70°C. *Geochim. Cosmochim. Acta* **53**, 1493–1501.

Knauss K. G., Nguyen S. N., and Weed H. C. (1993) Diopside dissolution kinetics as a function of pH, CO_2, temperature, and time. *Geochim. Cosmochim. Acta* **57**(2), 285–294.

Koretsky C. M., Sverjensky D. A., and Sahai N. (1998) A model of surface site types on oxide and silicate minerals based on crystal chemistry: implications for site types and densities, multi-site adsorption, surface infrared spectroscopy, and dissolution kinetics. *Am. J. Sci.* **298**(5), 349–438.

Kraemer S. M., Cheah S.-F., Zapf R., Xu J., Raymond K. N., and Sposito G. (1999) Effect of hydroxamate siderophores on Fe release and Pb(II) adsorption by goethite. *Geochim. Cosmochim. Acta* **63**, 3003–3008.

Kubicki J. D., Blake G. A., and Apitz S. E. (1996) *Ab initio* calculations on aluminosilicate Q^3 species: implications for atomic structures of mineral surfaces and dissolution mechanisms of feldspars. *Am. Mineral.* **81**, 789–799.

Kubicki J. D., Schroeter L. M., Itoh M. J., Nguyen B. N., and Apitz S. E. (1999) Attenuated total reflectance Fourier-transform infrared spectroscopy of carboxylic acids adsorbed onto mineral surfaces. *Geochim. Cosmochim. Acta* **63**, 2709–2725.

Kump L., Brantley S. L., and Arthur M. A. (2000) Chemical weathering, atmospheric CO_2 and climate. *Earth Planet. Sci. Rev.* **28**, 611–667.

Laidler K. J. (1987) *Chemical Kinetics*. Harper and Row, New York.

Lasaga A. C. (1981) Transition state theory. In *Kinetics of Geochemical Processes* (eds. A. C. Lasaga and R. J. Kirkpatrick). Mineralogical Society of America, Washington, DC, vol. 8, pp. 135–169.

Lasaga A. C. (1984) Chemical kinetics of water-rock interactions. *J. Geophys. Res.* **89**, 4009–4025.

Lasaga A. C. (1995) Fundamental approaches in describing mineral dissolution and precipitation rates. In *Chemical Weathering Rates of Silicate Minerals* (eds. A. F. White and S. L. Brantley). Mineralogical Society of America, Washington, DC, vol. 31, pp. 23–81.

Lasaga A. C. (1998) *Kinetic Theory in the Earth Sciences*. Princeton University Press.

Lasaga A. C. and Luttge A. (2001) Variation of crystal dissolution rate based on a dissolution stepwise model. *Science* **291**, 2400–2404.

Lasaga A. C. and Lüttge A. (2003) A model for crystal dissolution. *Euro. J. Mineral.*, **15**, 603–615.

Lee M. R. and Parsons I. (1995) Microtextural controls of weathering of perthitic alkali feldspars. *Geochim. Cosmochim. Acta* **59**(21), 4465–4488.

Lee M. R., Hodson M. E., and Parsons I. (1998) The role of intragranular microtextures and microstructures in chemical and mechanical weathering: direct comparisons of experimentally and naturally weathered alkali feldspars. *Geochim. Cosmochim. Acta* **62**, 2771–2788.

Lichtner P. C. (1998) Modelling reactive flow and transport in natural systems. *Proceedings of the Rome Seminar on Environmental Geochemistry*, 5–72.

Liebau F. (1985) *Structural Chemistry of Silicates*. Springer, Berlin, New York.

Liermann L. J., Kalinowski B. E., Brantley S. L., and Ferry J. G. (2000) Role of bacterial siderophores in dissolution of hornblende. *Geochim. Cosmochim. Acta* **64**(4), 587–602.

Lin C. L. and Clemency C. V. (1981) The dissolution kinetics of brucite, antigorite, talc, and phlogopite at room temperature and pressure. *Am. Mineral.* **66**(7–8), 801–806.

Locke W. W. (1986) Rates of hornblende etching in soils on glacial deposits, Baffin Island, Canada. In *Rates of Chemical Weathering of Rocks and Minerals* (eds. S. M. Colman and D. P. Dethier). Academic Press, pp. 129–145.

Luce R. W., Bartlett R. W., and Parks G. A. (1972) Dissolution kinetics of magnesium silicates. *Geochim. Cosmochim. Acta* **36**(1), 35–50.

Ludwig C., Casey W. H., and Rock P. A. (1995) Prediction of ligand-promoted dissolution rates from the reactivities of aqueous complexes. *Nature* **375**, 44–47.

Lundström U. and Öhman L.-O. (1990) Dissolution of feldspars in the presence of natural organic solutes. *J. Soil. Sci.* **41**, 359–369.

Luttge A., Bolton E. W., and Lasaga A. C. (1999) An interferometric study of the dissolution kinetics of anorthite: the role of reactive surface area. *Am. J. Sci.* **299**, 652–678.

Luttge A., Winkler U., and Lasaga A. C. (2003) An interferometric study of the dolomite dissolution: a new conceptual model for mineral dissolution. *Geochim. Cosmochim. Acta* **67**, 1099–1116.

Machesky M. L. (1989) Influence of temperature on ion adsorption by hydrous metal oxides. In *Chemical Modeling of Aqueous System II* (eds. D. C. Melchior and R. L. Bassett). American Chemical Society, Washington, DC, vol. 416, pp. 282–292.

MacInnis I. N. and Brantley S. L. (1992) The role of dislocations and surface morphology in calcite dissolution. *Geochim. Cosmochim. Acta* **56**(3), 1113–1126.

MacInnis I. N. and Brantley S. L. (1993) Development of etch pit size distributions (PSD) on dissolving minerals. *Chem. Geol.* **105**(1–3), 31–49.

Malmström M. and Banwart S. (1997) Biotite dissolution at 25°C: the pH dependence of dissolution rate and stoichiometry. *Geochim. Cosmochim. Acta* **61**(14), 2779–2799.

Manley E. P. and Evans L. J. (1986) Dissolution of feldspars by low-molecular weight aliphatic and aromatic acids. *Soil Sci.* **141**(2), 106–112.

Mast M. A. and Drever J. I. (1987) The effect of oxalate on the dissolution rates of oligoclase and tremolite. *Geochim. Cosmochim. Acta* **51**(9), 2559–2568.

Maurice P., Forsythe J., Hersman L., and Sposito G. (1996) Application of atomic-force microscopy to studies of microbial interactions with hydrous Fe(III)-oxides. *Chem. Geol.* **132**(1–4), 33–43.

Maurice P. A., Vierkorn M. A., Hersman L. E., and Fulghum J. E. (2001a) Dissolution of well and poorly ordered kaolinites by an aerobic bacterium. *Chem. Geol.* **180**, 81–97.

Maurice P. A., Vierkorn M. A., Hersman L. E., Fulghum J. E., and Ferryman A. (2001b) Enhancement of kaolinite dissolution by an aerobic *Pseudomonas mendocina* bacterium. *Geomicrobiol. J.* **18**, 21–35.

Mazer J. J. and Walther J. V. (1994) Dissolution kinetics of silica glass as a function of pH between 40 and 85°C. *J. Noncrystall. Solids* **170**(1), 32–45.

McClelland J. E. (1950) The effect of time, temperature, and particle size on the release of bases from some common soil-forming minerals of different crystal structure. *Soil Sci. Soc. Proc.* 301–307.

Mellott N. P., Brantley S. L., and Pantano C. G. (2002) Topography of polished plates of albite crystal and glass during dissolution. In *Water–Rock Interactions, Ore Deposits, and Environmental Geochemistry, A Tribute to David A. Crerar*, vol. Spec. Pub. No. 7 (eds. R. Hellmann and S. Wood). The Geochemical Society, St. Louis, MO, pp. 83–96.

Mogollon J. L., Ganor J., Soler J. M., and Lasaga A. C. (1996) Column experiments and the full dissolution rate law of gibbsite. *Am. J. Sci.* **296**, 729–765.

Morse J. W. (1978) Dissolution kinetics on calcium carbonate in sea water: VI. The near equilibrium dissolution kinetics of calcium carbonate-rich deep sea sediments. *Am. J. Sci.* **278**, 344–353.

Muir I. J., Bancroft G. M., and Nesbitt H. W. (1989) Characteristics of altered labradorite surface by SIMS and XPS. *Geochim. Cosmochim. Acta* **53**(6), 1235–1241.

Muir I. J., Bancroft M., Shotyk W., and Nesbitt H. W. (1990) A SIMS and XPS study of dissolving plagioclase. *Geochim. Cosmochim. Acta* **54**, 2247–2256.

Mukhopadhyay B. and Walther J. V. (2001) Acid-base chemistry of albite surfaces in aqueous solutions at standard temperature and pressure. *Chem. Geol.* **174**, 415–443.

Murphy W. M. (1989) Dislocations and feldspar dissolution. *Euro. J. Mineral.* **1**, 315–326.

Murphy W. M., Oelkers E. H., and Lichtner P. C. (1989) Surface reaction versus diffusion control of mineral dissolution and growth rates in geochemical processes. *Chem. Geol.* **78**(3–4), 357–380.

Murphy W. M., Pabalan R. T., Prikryl J. D., and Goulet C. J. (1992) Dissolution rate and solubility of analcime at 25°C. *Water–Rock Interact.* **7**, 107–110.

Nagy K. L. (1995) Dissolution and precipitation kinetics of sheet silicates. In *Chemical Weathering Rates of Silicate Minerals* (eds. A. F. White and S. L. Brantley). Mineralogical Society of America, Washington, DC, vol. 31, pp. 173–225.

Nagy K. L. and Lasaga A. C. (1990) The effect of deviation from equilibrium on the kinetics of dissolution and precipitation of kaolinite and gibbsite. *Chem. Geol.* **84**, 283–285.

Nagy K. L. and Lasaga A. C. (1992) Dissolution and precipitation kinetics of gibbsite at 80°C and pH 3: the dependence on solution saturation state. *Geochim. Cosmochim. Acta* **56**, 3093–3111.

Nagy K. L. and Lasaga A. C. (1993) Simultaneous precipitation kinetics of kaolinite and gibbsite at 80°C and pH 3. *Geochim. Cosmochim. Acta* **57**, 4329–4335.

Nagy K. L., Steefel C. I., Blum A. E., and Lasaga A. C. (1990) Dissolution and precipitation kinetics of kaolinite: initial results at 80°C with application to porosity evolution in a sandstone. *Am. Assoc. Petrol. Geol. Mem.* **49**, 85–101.

Nagy K. L., Blum A. E., and Lasaga A. C. (1991) Dissolution and precipitation kinetics of kaolinite at 80°C and pH 3: the dependence on solution saturation state. *Am. J. Sci.* **291**, 649–686.

Nancollas G. H. and Reddy M. M. (1971) The crystallization of calcium carbonate: II. Calcite growth mechanism. *J. Coll. Interface Sci.* **37**, 824–830.

Nesbitt H. W. and Muir I. J. (1988) SIMS depth profiles of weathered plagioclase and processes affecting dissolved Al and Si in some acidic soils. *Nature* **334**(6180), 336–338.

Nesbitt H. W., Macrae N. D., and Shotyk W. (1991) Congruent and incongruent dissolution of labradorite in dilute acidic salt solutions. *J. Geol.* **99**, 429–442.

Nielsen A. (1983) Precipitates: formation, coprecipitation, and aging. In *Treatise on Analytical Chemistry* (eds. I. M. Kolthoff and P. J. Elving). Wiley, New York, pp. 269–374.

Nordstrom D. K. (1982) Aqueous pyrite oxidation and the consequent formation of secondary iron minerals. In *Acid Sulfate Weathering* (ed. D. M. Kral). Soil Science Society of America, Madison, WI, vol. 10, pp. 37–56.

Nugent M. A., Brantley S. L., Pantano C. G., and Maurice P. A. (1998) The influence of natural mineral coatings on feldspar weathering. *Nature* **395**(6702), 588–591.

Ochs M. (1996) Influence of humified and non-humified natural organic compounds on mineral dissolution. *Chem. Geol.* **132**, 119–124.

Ochs M., Brunner I., Stumm W., and Cosovic B. (1993) Effects of root exudates and humic substances on weathering kinetics. *Water Air Soil Pollut.* **68**(1–2), 213–229.

Oelkers E. H. (2001a) An experimental study of forsterite dissolution rates as a function of temperature and aqueous Mg and Si concentrations. *Chem. Geol.* **175**(3–4), 485–494.

Oelkers E. H. (2001b) General kinetic description of multioxide silicate mineral and glass dissolution. *Geochim. Cosmochim. Acta* **65**(21), 3703–3719.

Oelkers E. H. and Schott J. (1995a) Dissolution and crystallization rates of silicate minerals as a function of chemical affinity. *Pure Appl. Chem.* **67**, 903–910.

Oelkers E. H. and Schott J. (1995b) Experimental study of anorthite dissolution and the relative mechanism of feldspar hydrolysis. *Geochim. Cosmochim. Acta* **59**(24), 5039–5053.

Oelkers E. H. and Schott J. (1998) Does organic acid adsorption affect alkali-feldspar dissolution rates? *Chem. Geol.* **151**(1–4), 235–245.

Oelkers E. H. and Schott J. (2001) An experimental study of enstatite dissolution rates as a function of pH, temperature, and aqueous Mg and Si concentration, and the mechanism of pyroxene/pyroxenoid dissolution. *Geochim. Cosmochim. Acta* **65**(8), 1219–1231.

Oelkers E. H., Schott J., and Devidal J.-L. (1994) The effect of aluminum, pH, and chemical affinity on the rates of aluminosilicate dissolution reactions. *Geochim. Cosmochim. Acta* **58**(9), 2011–2024.

Oelkers E. H., Schott J., and Devidal J.-L. (2001) On the interpretation of closed system mineral dissolution experiments: Comment on "Mechanism of kaolinite dissolution at room temperature and pressure: Part II. Kinetic study" by Huertas *et al.* (1999). *Geochim. Cosmochim. Acta* **65**, 4429–4432.

Oxburgh R., Drever J. I., and Sun Y. (1994) Mechanism of plagioclase dissolution in acid solution at 25°C. *Geochim. Cosmochim. Acta* **58**(2), 661–669.

Paces T. (1973) Steady-state kinetics and equilibrium between ground water and granitic rock. *Geochim. Cosmochim. Acta* **37**(12), 2641–2663.

Parks G. A. (1967) Aqueous surface chemistry of oxides and complex oxide minerals. In *Equilibrium Concepts in Natural Water Systems*. American Chemical Society, Washington, DC, vol. 67, pp. 121–160.

Pelmenschikov A., Strandh H., Pettersson L. G. M., and Leszczyynski J. (2000) Lattice resistance to hydrolysis of Si–O–Si bonds of silicate minerals: *Ab initio* calculations of a single water attack onto the (001) and (111) beta-quistobalite surfaces. *J. Phys. Chem. B* **104**, 5779–5783.

Pelmenschikov A., Leszczynski J., and Pettersson G. M. (2001) Mechanism of dissolution of neutral silica surfaces: including effect of self-healing. *J. Phys. Chem. A* **105**, 9528–9532.

Petit J. C., Della Mea G., Dran J.-C., and Schott J. (1987) Mechanism of diopside dissolution from hydrogen depth profiling. *Nature* **325**, 705–707.

Petrovic R., Berner R. A., and Goldhaber M. B. (1976) Rate control in dissolution of alkali feldspars: I. Study of residual feldspar grains by X-ray photoelectron spectroscopy. *Geochim. Cosmochim. Acta* **40**(5), 537–548.

Plummer L. N., Wigley T. M. L., and Parkhurst D. L. (1978) The kinetics of calcite dissolution in CO_2—water systems at 5 to 60°C and 0.0 to 1.0 atm CO_2. *Am. J. Sci.* **278**, 179–216.

Pokrovski G. S. and Schott J. (1998) Experimental study of the complexation of silicon and germanium with aqueous species: implications for germanium and silicon transport and Ge/Si ratio in natural waters. *Geochim. Cosmochim. Acta* **62**(21–22), 3413–3428.

Pokrovsky O. S. and Schott J. (1999) Processes at the magnesium-bearing carbonates/solution interface: II. Kinetics and mechanism of magnesite dissolution. *Geochim. Cosmochim. Acta* **63**, 881–897.

Pokrovsky O. S. and Schott J. (2000a) Forsterite surface composition in aqueous solutions: a combined potentiometric, electrokinetic, and spectroscopic approach. *Geochim. Cosmochim. Acta* **64**, 3299–3312.

Pokrovsky O. S. and Schott J. (2000b) Kinetics and mechanism of forsterite dissolution at 25°C and pH from 1 to 12. *Geochim. Cosmochim. Acta* **64**(19), 3313–3325.

Pokrovsky O. S., Schott J., and Thomas F. (1999a) Dolomite surface speciation and reactivity in aquatic systems. *Geochim. Cosmochim. Acta* **63**, 3133–3143.

Pokrovsky O. S., Schott J., and Thomas F. (1999b) Processes at the magnesium-bearing carbonates/solution interface: I. A surface speciation model for magnesite. *Geochim. Cosmochim. Acta* **63**, 863–880.

Pokrovsky O. S., Mielczarski J. A., Barres O., and Schott J. (2000) Surface speciation models of calcite and dolomite/aqueous solution interfaces and their spectroscopic evaluation. *Langmuir* **16**, 2677–2688.

Polster W. (1994) Hydrothermal precipitation and dissolution of silica: Part I. Conditions in geothermal fields and sedimentary basins: Part 2. Experimental evaluation of kinetics. PhD, Pennsylvania State University.

Posey-Dowty J., Crerar D., Hellmann R., and Chang C. D. (1986) Kinetics of mineral-water reactions: theory, design and application of circulating hydrothermal equipment. *Am. Mineral.* **71**, 85–94.

Poulson S. R., Drever J. I., and Stillings L. L. (1997) Aqueous Si-oxalate complexing, oxalate adsorption onto quartz, and the effect of oxalate upon quartz dissolution rates. *Chem. Geol.* **140**(1–2), 1–7.

Putnis A. (2002) Mineral replacement reactions: from macroscopic observations to microscopic mechanisms. *Mineral. Mag.* **66**(5), 689–708.

Reddy M. M. and Nancollas G. H. (1971) The crystallization of calcium carbonate: I. Isotopic exchange and kinetics. *J. Coll. Interface Sci.* **36**, 166–172.

Rickard D. and Sjoberg E. L. (1983) Mixed kinetic control of calcite dissolution rates. *Am. J. Sci.* **283**, 815–830.

Rimstidt J. D. and Barnes H. L. (1980) The kinetics of silica-water reactions. *Geochim. Cosmochim. Acta* **44**(11), 1683–1700.

Rimstidt J. D. and Dove P. M. (1986) Mineral solution reaction rates in a mixed flow reactor: wollastonite hydrolysis. *Geochim. Cosmochim. Acta* **50**(11), 2509–2516.

Rimstidt J. D. and Newcomb W. D. (1993) Measurement and analysis of rate data: the rate of reaction of ferric iron with pyrite. *Geochim. Cosmochim. Acta* **57**(9), 1919–1934.

Rimstidt J. D., Chermak J. A., and Gagen P. M. (1994) Rates of reaction of galena, sphalerite, chalcopyrite, and arsenopyrite with Fe(III) in acidic solutions. In *Environmental Geochemistry of Sulfide Oxidation* (eds. C. N. Alpers and D. W. Blowes). American Chemical Society, Washington, DC, vol. 550, pp. 2–13.

Rose N. M. (1991) Dissolution rates of prehnite, epidote, and albite. *Geochim. Cosmochim. Acta* **55**(11), 3273–3286.

Rosso J. J. and Rimstidt J. D. (2000) A high resolution study of forsterite dissolution rates. *Geochim. Cosmochim. Acta* **64**(5), 797–811.

Russell J. D., Paterson E., Fraser A. R., and Farmer V. C. (1975) Adsorption of carbon dioxide on goethite surfaces, and its implications for anion adsorption. *Chem. Soc. Faraday Trans. I* **71**, 1623–1630.

Sanemasa I. and Kataura T. (1973) The dissolution of $CaMg(SiO_3)_2$ in acid solutions. *Bull. Chem. Soc. Japan* **46**, 3416–3422.

Schindler P. W. (1981) Surface complexes at oxide-water interfaces. In *Adsorption of Inorganics at Solid–Liquid Interfaces* (eds. M. A. Anderson and A. J. Rubin). Ann. Arbor Sci. Publ. Ann Arbor, MI, pp. 1–150.

Schnoor J. L. (1990) Kinetics of chemical weathering: a comparison of laboratory and field weathering rates. In *Aquatic Chemical Kinetics* (ed. W. Stumm). Wiley, New York, pp. 475–504.

Schott J. (1990) Modelling of the dissolution of strained and unstrained multiple oxides: the surface speciation approach. In *Aquatic Chemical Kinetics* (ed. W. Stumm). Wiley, New York, pp. 337–365.

Schott J. and Berner R. A. (1983) X-ray photoelectron studies of the mechanism of iron silicate dissolution during weathering. *Geochim. Cosmochim. Acta* **47**, 2233–2240.

Schott J. and Berner R. A. (1985) Dissolution mechanisms of pyroxenes and olivines during weathering. In *The Chemistry of Weathering* (ed. J. I. Drever). D. Reidel Publishing Co., Dordrecht, Boston, vol. 149.

Schott J. and Oelkers E. H. (1995) Dissolution and crystallization rates of silicate minerals as a function of chemical affinity. *Pure Appl. Chem.* **67**(6), 903–910.

Schott J. and Petit J. C. (1987) New evidence for the mechanisms of dissolution of silicate minerals. In *Aquatic Surface Chemistry: Chemical Processes at the Particle-water Interface* (ed. W. Stumm). Wiley, New York, pp. 293–315.

Schott J., Berner R. A., and Sjoberg E. L. (1981) Mechanism of pyroxene and amphibole weathering: I. Experimental studies of iron-free minerals. *Geochim. Cosmochim. Acta* **45**(11), 2123–2135.

Schott J., Brantley S. L., Crerar D., Guy C., Borcsik M., and Willaime C. (1989) Dissolution kinetics of strained calcite. *Geochim. Cosmochim. Acta* **53**(2), 373–382.

Schweda P. (1989) Kinetics of alkali feldspar dissolution at low temperature. In *Proc. 6th Int. Symp. Water/Rock Interact.* (ed. D. L. Miles). A. A. Balkema, Rotterdam, Brookfield, Vt., pp. 609–612.

Schweda P. (1990) Kinetics and mechanisms of alkali feldspar dissolution at low temperatures. PhD, Stockholm University.

Schweda P., Sjoberg L., and Sodervall U. (1997) Near-surface composition of acid-leached labradorite investigated by SIMS. *Geochim. Cosmochim. Acta* **61**(10), 1985–1994.

Shiraki R. and Brantley S. L. (1995) Kinetics of near-equilibrium calcite precipitation at 100°C: an evaluation of elementary reaction-based and affinity-based rate laws. *Geochim. Cosmochim. Acta* **59**(8), 1457–1471.

Shotyk W., Nesbitt H. W., and Fyfe W. S. (1990) The behavior of major and trace elements in complete vertical peat profiles from three *Sphagnum* bogs. *Int. J. Coal Geol.* **15**(3), 163–190.

Siegel D. I. and Pfannkuch H. O. (1984) Silicate mineral dissolution at pH 4 and near standard temperature and pressure. *Geochim. Cosmochim. Acta* **48**, 197–201.

Siever R. and Woodford N. (1979) Dissolution kinetics and the weathering of mafic minerals. *Geochim. Cosmochim. Acta* **43**(5), 717–724.

Sjoberg E. L. (1976) A fundamental equation for calcite dissolution kinetics. *Geochim. Cosmochim. Acta* **40**, 441–447.

Sjoberg E. L. and Rickard D. (1983) Calcite dissolution kinetics: surface speciation and the origin of the variable pH dependence. *Chem. Geol.* **42**, 119–136.

Sjoberg L. (1989) Kinetics and non-stoichiometry of labradorite dissolution. *6th Int. Conf. Water–Rock Interact.*

Steefel C. I. and Lasaga A. C. (1992) Putting transport into water–rock interaction models. *Geology* **20**, 680–684.

Stillings L., Drever J. I., and Poulson S. R. (1998) Oxalate adsorption at a plagioclase (An(47)) surface and models for ligand-promoted dissolution. *Environ. Sci. Technol.* **32**, 2856–2864.

Stillings L. L. and Brantley S. L. (1995) Feldspar dissolution at 25°C and pH 3: reaction stoichiometry and the effect of cations. *Geochim. Cosmochim. Acta* **59**, 1483–1496.

Stillings L. L., Brantley S. L., and Machesky M. (1995) Proton adsorption at an adularia feldspar surface. *Geochim. Cosmochim. Acta* **59**(8), 1473–1482.

Stillings L. L., Drever J. I., Brantley S. L., Sun Y., and Oxburgh R. (1996) Rates of feldspar dissolution at pH 3–7 with 0–8 mM oxalic acid. *Chem. Geol.* **132**(1–4), 79–89.

Stumm W. (1990) *Aquatic Chemical Kinetics*. Wiley, New York.

Stumm W. and Furrer G. (1987) The dissolution of oxides and aluminum silicates: examples of surface-coordination-controlled kinetics. In *Aquatic Surface Chemistry: Chemical Processes at the Particle-water Interface* (ed. W. Stumm). Wiley, New York, pp. 197–220.

Stumm W. and Morgan J. J. (1996) *Aquatic Chemistry: Chemical Equilibria and Rates in Natural Waters*. Wiley, New York.

Stumm W. and Wieland E. (1990) Dissolution of oxide and silicate minerals: rates depend on surface speciation. In *Aquatic Chemical Kinetics* (ed. W. Stumm). Wiley, New York, pp. 367–400.

Suarez D. L. and Wood J. D. (1995) Short-term and long-term weathering rates of a feldspar fraction isolated from an arid zone soil. *Chem. Geol.* **132**, 143–150.

Sverdrup H. U. (1990) *The Kinetics of Base Cation Release due to Chemical Weathering*. Lund University Press.

Sverjensky D. A. (1992) Linear free energy relations for predicting dissolution rates of solids. *Nature* **358**, 310–313.

Sverjensky D. A. (1994) Zero-point-of-charge prediction from crystal chemistry and solvation theory. *Geochim. Cosmochim. Acta* **58**(14), 3123–3129.

Swoboda-Colberg N. G. and Drever J. I. (1993) Mineral dissolution rates in plot-scale field and laboratory experiments. *Chem. Geol.* **105**(1–3), 51–69.

Tan K. H. (1980) The release of silicon, aluminum, and potassium during decomposition of soil minerals by humic acid. *Soil Science* **129**, 5–11.

Taylor A. S., Blum J. D., and Lasaga A. C. (2000) The dependence of labradorite dissolution and Sr isotope release rates on solution saturation state. *Geochim. Cosmochim. Acta* **64**(14), 2389–2400.

Teng H. H., Dove P. M., and DeYoreo J. J. (2000) Kinetics of calcite growth: surface processes and relationships to macroscopic rate laws. *Geochim. Cosmochim. Acta* **64**, 2255–2266.

Teng H. H., Fenter P., Cheng L., and Sturchio N. C. (2001) Resolving orthoclase dissolution processes with atomic force microscopy and X-ray reflectivity. *Geochim. Cosmochim. Acta* **65**, 3459–3474.

Tester J. W., Worley W. G., Robinson B. A., Grigsby C. O., and Feerer J. L. (1994) Correlating quartz dissolution kinetics in pure water from 25 to 625°C. *Geochim. Cosmochim. Acta* **58**, 2407–2420.

Tsomaia N., Brantley S. L., Hamilton J. P., Pantano C. G., and Mueller K. T. (2003) NMR evidence for formation of octahedral and tetrahedral Al and repolymerization of the Si network during dissolution of aluminosilicate glass and crystal. *Am. Mineral.* **88**, 54–67.

Turpault M. P. and Trotignon L. (1994) The dissolution of biotite single crystals in dilute HNO_3 at 24°C: evidence of an anisotropic corrosion process of micas in acidic solutions. *Geochim. Cosmochim. Acta* **58**, 2761–2775.

Urey H. C. (1952) *The Planets, their Origin and Development.* Yale University Press, New Haven.

Van Geen A., Robertson A. P., and Leckie J. O. (1994) Complexation of carbonate species at the goethite surface: implications for adsorption of metal ions in natural waters. *Geochim. Cosmochem. Acta* **58**, 2073–2086.

van Grinsven J. J. M. and van Riemsdijk W. H. (1992) Evaluation of batch and column techniques to measure weathering rates in soils. *Geoderma* **52**, 41–57.

van Hees P. A. W., Lundstrom U. S., and Morth C.-M. (2002) Dissolution of microcline and labradorite in a forest O horizon extract: the effect of naturally occurring organic acids. *Chem. Geol.* **189**, 199–211.

Velbel M. A. (1986) Influence of surface area, surface characteristics, and solution composition on feldspar weathering rates. In *Geochemical Processes at Mineral Surfaces* (eds. J. A. Davis and K. F. Hayes). American Chemical Society, Washington, DC, pp. 618–634.

Velbel M. A. (1990) Influence of temperature and mineral surface characteristics on feldspar weathering rates in natural and artificial systems: a first approximation. *Water Resour. Res.* **26**(12), 3049–3053.

Velbel M. A. (1993a) Constancy of silicate-mineral weathering-rate ratios between natural and experimental weathering: implications for hydrologic control of differences in absolute rates. *Chem. Geol.* **105**(1–3), 89–99.

Velbel M. A. (1993b) Formation of protective surface layers during silicate-mineral weathering under well-leached, oxidizing conditions. *Am. Mineral.* **78**(3–4), 405–414.

Walther J. V. (1996) Relation between rates of aluminosilicate mineral dissolution, pH, temperature, and surface charge. *Am. J. Sci.* **296**, 693–728.

Watteau F. and Berthelin J. (1994) Microbial dissolution of iron and aluminum from soil minerals: efficiency and specificity of hydroxamate siderophores compared to aliphatic acids. *Euro. J. Soil Biol.* **30**(1), 1–9.

Weissbart E. J. and Rimstidt J. D. (2000) Wollastonite: incongruent dissolution and leached layer formation. *Geochim. Cosmochim. Acta* **64**(23), 4007–4016.

Welch S. A. and Ullman W. J. (1993) The effect of organic acids on plagioclase dissolution rates and stoichiometry. *Geochim. Cosmochim. Acta* **57**(12), 2725–2736.

Welch S. A. and Ullman W. J. (1996) Feldspar dissolution in acidic and organic solutions: compositional and pH dependence of dissolution rate. *Geochim. Cosmochim. Acta* **60**, 2939–2948.

Welch S. A. and Ullman W. J. (1999) The effect of microbial glucose metabolism on bytownite feldspar dissolution rates between 5 degrees and 35 degrees C. *Geochim. Cosmochim. Acta* **63**, 3247–3259.

Welch S. A. and Ullman W. J. (2000) The temperature dependence of bytownite feldspar dissolution in neutral aqueous solutions of inorganic and organic ligands at low temperature (5–35 °C). *Chem. Geol.* **167**, 337–354.

Welch S. A., Barker W. W., and Banfield J. F. (1999) Microbial extracellular polysaccharides and plagioclase dissolution. *Geochim. Cosmochim. Acta* **63**(9), 1405–1419.

Westrich H. R., Cygan R. T., Casey W. H., Zemitis C., and Arbold G. W. (1993) The dissolution kinetics of mixed-cation orthosilicate minerals. *Am. J. Sci.* **293**(9), 869–893.

White A. F. (1995) Chemical weathering rates of silicate minerals in soils. In *Chemical Weathering Rates of Silicate Minerals* (eds. A. F. White and S. L. Brantley). Mineralogical Society of America, Washington, DC, vol. 31, pp. 407–461.

White A. F. and Brantley S. L. (eds.) (1995) *Chemical Weathering Rates of Silicate Minerals.* Mineralogical Society of America, vol. 31, Washington, DC.

White A. F. and Brantley S. L. The effect of time on the experimental and natural weathering rates of silicate minerals. *Chem. Geol.* (in press).

White A. F. and Peterson M. L. (1990) Role of reactive-surface-area characterization in geochemical kinetic models. In *Chemical Modelling of Aqueous Systems II* (eds. D. C. Melchior and R. L. Bassett). American Chemical Society, Washington, DC, vol. 416, pp. 461–475.

White A. F. and Yee A. (1985) Aqueous oxidation-reduction kinetics associated with coupled electron-cation transfer from iron-containing silicates at 25°C. *Geochim. Cosmochim. Acta* **49**, 1263–1275.

White A. F., Bullen T. D., Vivit D. V., Schulz M. S., and Clow D. W. (1999) The role of disseminated calcite in the chemical weathering of granitoid rocks. *Geochim. Cosmochim. Acta* **63**(13–14), 1939–1953.

White A. F., Bullen T. D., Schulz M. S., Blum A. E., Huntington T. G., and Peters N. E. (2001) Differential rates of feldspar weathering in granitic regoliths. *Geochim. Cosmochim. Acta* **65**(6), 847–869.

Wieland E. and Stumm W. (1992) Dissolution kinetics of kaolinite in acidic aqueous solutions at 25°C. *Geochim. Cosmochim. Acta* **56**(9), 3339–3355.

Wogelius R. A. and Walther J. V. (1991) Olivine dissolution at 25°C: effects of pH, CO_2, and organic acids. *Geochim. Cosmochim. Acta* **55**(4), 943–954.

Wogelius R. A. and Walther J. V. (1992) Olivine dissolution kinetics at near-surface conditions. *Chem. Geol.* **97**(1–2), 101–112.

Wollast R. (1967) Kinetics of the alteration of K-feldspar in buffered solutions at low temperature. *Geochim. Cosmochim. Acta* **31**, 635–648.

Wollast R. (1990) Rate and mechanism of dissolution of carbonates in the system $CaCO_3$–$MgCO_3$. In *Aquatic Chemical Kinetics* (ed. W. Stumm). Wiley-Interscience, New York, pp. 431–445.

Wollast R. and Chou L. (1988) Rate control of weathering of siliate minerals at room temperature and pressure. In *Physical and Chemical Weathering in Geochemical Cycles* (eds. A. Lerman and M. Meybeck). Kluwer Academic Publishers, Dordrecht, Boston, vol. 251, pp. 11–31.

Wollast R. and Chou L. (1992) Surface reactions during the early stages of weathering of albite. *Geochim. Cosmochim. Acta* **56**, 3113–3121.

Wood B. J. and Walther J. V. (1983) Rates of hydrothermal reactions. *Science* **222**, 413–415.

Xiao Y. and Lasaga A. C. (1995) *Ab initio* quantum mechanical studies of the kinetics and mechanisms of silicate dissolution: H^+ (H_3O^+) catalysis. *Geochim. Cosmochim. Acta* **58**, 5379–5400.

Xie Z. (1994) Surface properties of silicates, their solubility and dissolution kinetics. PhD, Northwestern University.

Xie Z. and Walther J. V. (1994) Dissolution stoichiometry and adsorption of alkali and alkaline earth elements to the acid-reacted wollastonite surface at 25°C. *Geochim. Cosmochim. Acta* **58**(12), 2587–2598.

Yoshida T., Hayashi K., and Ohmoto H. (2002) Dissolution of iron hydroxides by marine bacterial siderophore. *Chem. Geol.* **184**(1–2), 1–9.

Zeltner W. A. and Anderson M. A. (1988) Surface charge development at the goethite/aqueous solution interface: effects of CO_2 adsorption. *Langmuir* **4**, 469–474.

Zhang H. (1990) Factors determining the rate and stoichiometry of hornblende dissolution. PhD, University of Minnesota.

Zhang H. and Bloom P. R. (1999) Dissolution kinetics of hornblende in organic acid solutions. *Soil Sci. Soc. Am. J.* **63**, 815–822.

Zhang H. L., Bloom P. R., Nater E. A., and Erich M. S. (1996) Rates and stoichiometry of hornblende dissolution over 115 days of laboratory weathering at pH 3.6–4.0 and 25°C in 0.01 M lithium acetate. *Geochim. Cosmochim. Acta* **60**(6), 941–950.

Zinder B., Furrer G., and Stumm W. (1986) The coordination chemistry of weathering: II. Dissolution of Fe(III) oxides. *Geochim. Cosmochim. Acta* **50**(9), 1861–1869.

5.04

Mass-balance Approach to Interpreting Weathering Reactions in Watershed Systems

O. P. Bricker and B. F. Jones

USGS Water Resources Division, Reston, VA, USA

and

C. J. Bowser

University of Wisconsin, Madison, WI, USA

5.04.1 INTRODUCTION

The mass-balance approach is conceptually simple and has found widespread applications in many fields over the years. For example, chemists use mass balance (Stumm and Morgan, 1996) to sum the various species containing an element in order to determine the total amount of that element in the system (free ion, complexes). Glaciologists use mass balance to determine the changes in mass of glaciers (Mayo *et al.*, 1972 and references therein). Groundwater hydrologists use this method to interpret changes in water balance in groundwater systems (Rasmussen and Andreasen, 1959; Bredehoeft *et al.*, 1982; Heath, 1983; Konikow and Mercer, 1988; Freeze and Cherry, 1979; Ingebritsen and Sanford, 1998). This method has also been used to determine changes in chemistry along a flow path (Plummer *et al.*, 1983; Bowser and Jones, 1990) and to quantify lake hydrologic budgets using stable isotopes (Krabbenhoft *et al.*, 1994). Blum and Erel

(see Chapter 5.12) discuss the use of strontium isotopes, Chapelle (see Chapter 5.14) treats carbon isotopes in groundwater, and Kendall and Doctor (see Chapter 5.11) and Kendall and McDonnell (1998) discuss the use of stable isotopes in mass balance. Although the method is conceptually simple, the parameters that define a mass balance are not always easy to measure. Watershed investigators use mass balance to determine physical and chemical changes in watersheds (Garrels and Mackenzie, 1967; Plummer *et al.*, 1991; O'Brien *et al.*, 1997; Drever, 1997). Here we focus on describing the mass-balance approach to interpret weathering reactions in watershed systems including shallow groundwater.

Because mass balance is simply an accounting of the flux of material into a system minus the flux of material out of the system, the geochemical mass-balance approach is well suited to interpreting weathering reactions in watersheds (catchments) and in other environmental settings

(Drever, 1997). It is, perhaps, the most accurate and reliable way of defining weathering reactions in natural systems. The usefulness of the mass-balance approach to interpreting weathering reactions was first brought to the attention of the geochemical fraternity by the classic work of Garrels (1967) and Garrels and Mackenzie (1967) in their investigations of the origin of the composition of Sierra Nevada Springs.

The primary objectives of mass-balance studies are: (i) quantify the mass fluxes into and out of watershed systems; (ii) interpret the reactions and processes occurring in the watershed that cause the observed changes in composition and flux; (iii) determine weathering rates of the various minerals constituting the bedrock, regolith, and soils of the watershed; and (iv) evaluate which mineral phases are critically involved in controlling water chemistry to help develop models of more general applicability (i.e., transfer value).

5.04.2 METHODS OF MASS BALANCE

Several variations of mass balance have been employed. Some of the most commonly used methods are: (i) comparing the changes in bulk chemistry with depth in a weathering profile. Goldich (1938) was one of the first to use this approach; (ii) indexing losses of weatherable minerals in the soil, regolith, and weathered rock against minerals in the fresh bedrock known not to weather easily such as zircon, titanite, etc. (April *et al.*, 1986); and (iii) input–output of water and dissolved substances from a well-characterized watershed. This approach was developed by Garrels and Mackenzie (1967) in their classic investigation of Sierra Nevada springwaters. All of these mass-balance approaches provide information on changes in chemistry due to weathering.

In order to use any mass-balance method to best advantage, a number of criteria must be met. In method (i), the purely chemical method, changes in bulk chemistry through a profile from fresh bedrock to the soil provide the information needed to calculate chemical losses from the system, but this method tells nothing about the mineralogical transformations that occur during the weathering process. To determine the weathering rate, it must be assumed that the time when weathering began is known, that there has been no removal of weathered material by physical erosion since that time, or that losses due to erosion can be compensated. The Brimhall–Dietrich calculations to address this problem are discussed in detail by Amundson (see Chapter 5.01) and White (see Chapter 5.05). The use of an inert tracer such as zircon or rutile to correct for losses due to erosion has been discussed by Bain *et al.* (1990) and by

Nesbitt and Wilson (1992). Chemical losses due to weathering and overall weathering rates can be calculated from this information.

In method (ii), the mineralogical method, exemplified by the work of April *et al.* (1986), the mineralogy of the fresh bedrock and the weathered material must be known and, for the indexing method, there must be a significant amount of a refractory mineral such as zircon against which to measure relative losses of weatherable minerals. If the time of inception of weathering is known, information on the weathering rate can be derived. An advantage of methods (i) and (ii) is that they represent true weathering rates unaffected by biomass uptake or ion-exchange processes. A potential disadvantage is that they deal only with very long term rates (e.g., decades to millennia). Short-term rates (e.g., months to several years) cannot be addressed.

An example of the mineral depletion method (method (ii) above) is illustrated in a study by April *et al.* (1986) of Woods and Panther lakes in the Adirondack Mountains of New York. These investigators examined the decrease in weatherable minerals from fresh glacial till at depth to the present-day soil surface to determine which minerals were reacting and to obtain the long-term rates of weathering in the system. It was assumed that weathering began at the end of glaciation 1.4×10^4 yr ago and that there was no mechanical erosion to truncate the profile. Since soil formation is commonly not isovolumetric, changes in the content of the refractory minerals, magnetite and ilmenite, were used to correct for volume changes. Provided that these assumptions are valid, the depletion curves can then be used to calculate the amount and rate of weathering in the system. Because this method relies only on changes in mineral abundances, the rate is not confounded by ion-exchange processes or the biomass uptake. In the Adirondack region, the long-term rates determined by mineral depletion or by bulk chemistry (methods (i) and (ii)) differ significantly from the rate determined from the flux of solutes method (method (iii)).

A more powerful and flexible method, and one that is widely used in watershed weathering studies, relates changes in the chemistry of input–output waters to reactions between the waters and solid phases in the system. This approach, termed chemical mass balance (method (iii)), is considered the best way of making quantitative estimates of rates of elemental transfers in the Earth's surface environment (Clayton, 1979). In method (iii), the hydrochemical method developed by Garrels and Mackenzie (1967), mineralogy and composition of the primary and secondary phases must be well characterized. The composition and volume of the initial water and the final water must be known, and it must be assumed that the basin is hydrologically "tight," i.e., there is no loss or gain of water and its solute load through groundwater entering or leaving the catchment (in cases where the composition of the groundwater is well known, it may be possible to account for groundwater contributions to the stream). In watersheds that do not suffer from road salt applications or where buried evaporite minerals are not present, chloride commonly is used as a conservative tracer to check for gain or loss of water by pathways other than the stream that drains the watershed. If there are other pathways, the system is not hydrologically closed. If the input and output of chloride balance, then the flux of other dissolved material carried out of the watershed in the runoff water reflects reactions for the weathering of primary to secondary minerals. The amounts of dissolved ions leaving the watershed can be used to quantify the loss of weatherable phases, the gain of weathering products, and the reactions determining the composition of the resulting water. To use the method to interpret chemical weathering along a groundwater flow path, changes in these same chemical parameters between wells along the flow path must be known.

The mass-balance approach has also been used in groundwater studies to interpret physical water balances and chemical changes along a flow path. Chapelle (see Chapter 5.14) discusses mass-balance calculations in groundwater extensively. In using this application one must be sure that the wells are along a flow path and do not intersect two or more different groundwater reservoirs that are independent of each other.

Mass-balance relations are used to define mass-transfer changes in the chemical evolution of one water to another. This relation was generalized mathematically by Plummer and Back (1980). If the compositions of the initial and final waters and the reactants and products are known, a chemical mass-balance for the change in total molality of the cth constituent along the reaction path can be calculated. This requires n equations, one for each constituent, in the form (Equation (1))

$$\sum_{j=1}^{\Phi} \alpha_j \beta_{cj} = \Delta m_c \big|_{c=1,n} \qquad (1)$$

where Φ is the total number of reactants or products (minerals and gases) in the reaction, n is the minimum number of constituents needed to define the compositions of the appropriate minerals and gases, α is the stoichiometric coefficient of the jth component, β_{cj} is the stoichiometric coefficient of the cth constituent in the jth component, and Δm_c is the change in concentration of the cth constituent between the initial and final waters. If the values of Δm_c are known

and $n = \Phi$, the simultaneous solution of the n equations will define the values of n of the reaction coefficients and provide a mass balance for the system.

This mass-balance expression has been simplified by (Bowser and Jones, 2002) to

$$\sum_{p=1}^{j} \text{MTC}_p \alpha_p^i = \Delta m_i = m_{i(\text{final})} - m_{i(\text{initial})} \qquad (2)$$

where j is the number of phases, p are the plausible reactant and product phases used in the mass-balance model, $\text{MTC}_p =$ mass-transfer coefficient for any phase (p) in moles, α is the stoichiometric coefficient of element i in phase p, m_i is the total moles of element I, and i represents the chemical element used in mass balance.

The solution of mass-balance systems requires the simultaneous solution of a set of equations in which the number of constraints (solutes) equals the number of mineral phases. This can be represented in the form of a matrix:

$$\begin{bmatrix} \alpha_{p1,1} & \alpha_{p1,2} & \cdots & \alpha_{p1,i} \\ \alpha_{p2,1} & \alpha_{p2,2} & \cdots & \alpha_{p2,i} \\ \cdots & \cdots & \cdots & \cdots \\ \alpha_{pj,1} & \alpha_{pj,2} & \cdots & \alpha_{pj,i} \end{bmatrix}^{-1} \begin{bmatrix} \Delta m_{i1} \\ \Delta m_{i2} \\ \cdots \\ \Delta m_{ii} \end{bmatrix}$$

$$= \begin{bmatrix} \text{MTC}_{p1} \\ \text{MTC}_{p2} \\ \cdots \\ \text{MTC}_{pj} \end{bmatrix} \qquad (3)$$

This can be represented in simplified form by

$$[p]^{-1}[\Delta W] = [\text{MTC}] \qquad (4)$$

where $[p]^{-1}$ is the inverse matrix of mineral phase stoichiometric coefficients, $[\Delta W]$ is the vector of solution composition changes, and [MTC] is the resultant vector of mass-transfer coefficients for each of the plausible mineral phases involved in the solute evolution of the water.

In weathering studies the system is usually a watershed (catchment) with well-defined boundaries. Equation (5) is a simple summary of the water balance, the hydrologic input and output terms usually measured in mass-balance studies (Jenkins *et al.*, 1994). Winter (1981) evaluated the hydrologic methodology used in a number of water-balance studies of lakes in the US and found that one or more terms in the budget are commonly calculated as residuals. Comparisons of several lake-water balances show that the residual, if interpreted as groundwater, can differ from independent estimates of groundwater by over 100%. Thus, it is important in chemical mass-balance studies to have a complete and accurate water budget in which error terms are not lumped into one residual term:

$$R = P - \text{ET} + dS \qquad (5)$$

where R is runoff, P is precipitation, ET is evapotranspiration, and dS is change in the storage of water in the system.

Equation (6) summarizes the chemical terms (Bricker *et al.*, 1994):

$$N_i = W_i + P_i + A_i - R_i - M_i - B_i - S_i \qquad (6)$$

where N_i is the net accumulation or depletion of component i in the system, W_i is the total weathering input of component i in the system, P_i is the atmospheric deposition input of component i in the system (wet precipitation, dry deposition, gaseous deposition), A_i is the anthropogenic input of component i to the system, R_i is the chemical erosion of component i in the system (dissolved components transported out of watershed), M_i is the mechanical erosion of component i in the system (particulate components transported out of watershed), B_i is the biomass uptake of component i in the system (harvest), and S_i is the storage of component i in watershed.

If it is assumed that the biomass is at steady state and there is no change in storage of water or in the exchange pool in the system, Equation (6) reduces to (Drever, 1997)

$$N_i = P_i + W_i - R_i - M_i \qquad (7)$$

Commonly, the term M_i is not addressed or quantified, because particulates are not directly involved in chemical weathering and do not add or subtract dissolved species to or from the waters (Langmuir, 1997); then the above Equation (7) is reduced to

$$N_i = P_i + W_i - R_i \qquad (8)$$

Pitfalls exist in making the above assumptions/simplifications (Drever, 1997). If one or all of the reservoirs (the exchange pool, the vegetation, or the water storage) were not at steady state during the period of measurement, the output from the watershed would not represent reactions over the long term. Only those immediate reactions representing short-term measurements of the current state of the system would be reflected. In the case of water storage, if measurements were made either during a period of drought or a period of unusually high rainfall, the long-term state of the system would not be represented. Drever (1997) states, "The question of timescale is very important in catchment-scale studies: it is rarely possible to construct a meaningful catchment budget for a timescale of less than a year." A growing appreciation for interannual variations in climate and hydrology over decadal or longer times emphasizes that even data from long-term

observation periods may not always be adequate. Interannual mass transfers must be correlated with the precipitation record. The annual flux for each year should be used to provide a mass balance for that year. Then the annual mass balances can be averaged to provide the long-term mass balance for the system. Averaging all the years of flux measurements first and then doing a mass balance on the average can be misleading (Bowser and Jones, 2002).

April *et al.* (1986) found that input–output calculations (short term) did not compare well with their mineral depletion determinations (long term). They speculated that the input–output balances reflected recent acid-rain effects, while the mineral-depletion calculations represented the long-term state of the system prior to changes in atmospheric deposition composition. Over long time periods, a system's exchange complex will equilibrate with the composition of the incoming precipitation. If substantial changes in the chemistry of the input waters occur (e.g., in atmospheric deposition composition due to acid rain), the exchange complex will no longer be in equilibrium with the deposition and will readjust to the new conditions. This readjustment process may take decades to stabilize, or if the deposition composition continues to change, a new equilibrium may never be established (Drever, 1997).

There are some watersheds (catchments), common in the western US (Baron, 1992) and in the Scandinavian countries (Christophersen *et al.*, 1982) that have little or no vegetative biomass. In such cases the influence of biomass on the mass balance can be safely ignored. In forested regions, particularly in the eastern US, it is sometimes assumed that the biomass is at steady state in the system being investigated. In that case the same amount of material is returned to the system by decomposition of biomass as is being taken up (for discussions of biomass effect, see Taylor and Velbel (1991), Berner and Cochran (1998), Moulton and Berner (1998), Likens *et al.* (1977), Cleaves *et al.* (1970), and Bowser and Jones (2002)). However, forests are seldom at steady state with respect to biomass unless there is a continuum of ages and species of trees over the long term. Each age class and species requires different nutrition; therefore, there will be a differential uptake of nutrients depending on the age and species distribution of trees in the forest. For example, Likens *et al.* (1977) suggest that as much as 40% of the calcium released by weathering of bedrock at Mirror Lake is taken up by the biomass and does not appear in the runoff waters.

There is some question about how tightly the upper soil is coupled with the lower zone where more intense weathering is occurring. There has been speculation that the vegetation derives most or all of its nutrient needs from the upper soil zone and is completely decoupled from the weathering zone below (Ugolini *et al.*, 1977; Schulz *et al.*, 2002; Bullen *et al.*, 2002). There may be a tightly closed cycling of nutrients between the upper soil and the biota. If this is the case, stream-water composition faithfully reflects only the weathering reactions in the system and is independent of biomass effects. Therefore, the biomass can be ignored in mass-balance calculations; but each watershed must be evaluated to determine the role of biomass.

Garrels and Mackenzie (1967) proposed that the composition of water in the Sierra Nevada was controlled by weathering reactions between incoming CO_2-charged precipitation and minerals in the bedrock. The precipitation, which was further enriched in CO_2 while passing through the soil, reacted with bedrock minerals to produce secondary phases and to contribute solutes to the water. The composition of the waters emanating from the bedrock was assumed to reflect the weathering reactions. The differences in composition between the incoming precipitation and the springwaters were attributed to weathering reactions, and the reactions could thus be quantified. It was assumed that the solutes were all derived from weathering, and that there were no gains or losses due to other processes. If this were the case, appropriate amounts of each of the solutes could be "back-reacted" with the secondary minerals to reconstitute the primary minerals, and at the end of the calculation no residual solutes should remain (Table 1). In the calculation for Sierra Nevada waters (Table 1) a small amount of silica was left over, but this was well within the error of the analyses and was not considered significant. (Garrels and Mackenzie's work has been discussed in detail in Drever (1997).)

No vegetative biomass was involved, and ion-exchange was unnecessary to explain the results. Thus, mineral weathering was considered responsible for the changes in chemistry between the input and output waters. The calculation worked quite well for the Sierra Nevada waters, confirming that weathering was the only significant process active in creating the changes observed in the system (Table 1). Garrels and Mackenzie based their balance on differences in the concentrations of solutes in the incoming and outgoing waters and reaction with granitic bedrock containing relatively few reactive minerals. There was little biomass in this system to confound the calculations. The rock consists primarily of quartz and feldspar (nearly equal amounts of plagioclase and orthoclase) with accessory biotite. Based on the disparate Ca/Na ratio of the parent plagioclase and the deeper springwaters, the authors proposed the presence of calcite in the deeper, less weathered rock (it had been essentially removed by weathering to

Table 1 Garrels and Mackenzie (1967) mass-balance calculations for Sierra Nevada ephemeral springs.

Reaction (coefficients × 10⁴)	Concentrations in mol L⁻¹ × 10⁻⁴								Products (mol L⁻¹ × 10⁴)
	Na^+	Ca^{2+}	Mg^{2+}	K^+	HCO_3^-	SO_4^{2-}	Cl^-	SiO_2	
Initial concentrations in springwater	1.34	0.78	0.29	0.28	3.28	0.10	0.14	2.73	
Minus concentrations in snow water	1.10	0.68	0.22	0.20	3.10			2.70	Derived from rock
• *Change kaolinite back into plagioclase*					Minus plagioclase				
$1.23Al_2Si_2O_5(OH)_4 + 1.10Na^+ + 0.62Ca^{2+} + 2.44HCO_3^- + 2.20SiO_2$ Kaolinite	0.00	0.00	0.22	0.20	0.64	0.00	0.00	0.50	$1.77Na_{0.62}Ca_{0.38}$ feldspar
$= 1.77Na_{0.62}Ca_{0.38}Al_{1.38}Si_{2.62}O_8 + 2.44CO_2 + 3.67H_2O$ plagioclase									
• *Change kaolinite into biotite*					Minus biotite				
$0.037Al_2Si_2O_5(OH)_4 + 0.073K^+ + 0.22Mg^{2+} + 0.15SiO_2 + 0.51HCO_3^-$ Kaolinite	0.00	0.00	0.00	0.13	0.13	0.00	0.00	0.35	0.073 biotite
$= 0.073KMg_3AlSi_3O_{10}(OH)_2 + 0.51CO_2 + 0.26H_2O$ biotite									
• *Change kaolinite back into K-feldspar*					Minus K-feldspar				
$0.065Al_2Si_2O_5(OH)_4 + 0.13K^+ + 0.13HCO_3^- + 0.26SiO_2$ Kaolinite	0.00	0.00	0.00	0.00	0.00	0.00	0.00	0.12	0.13 K-feldspar
$= 0.13KAlSi_3O_8 + 0.13CO_2 + 0.195H_2O$ K-feldspar									

Reproduced by permission of American Chemical Society from *Equilibrium Concepts in Natural Water Systems*, **1967**, pp. 222–242.

the depth that waters from the ephemeral springs circulated). The major alteration mineral in ephemeral springs was kaolinite and, in the perennial springs, a combination of kaolinite and Ca-smectite. The mass balance worked best for the ephemeral springs, but was satisfactory for the perennial springwaters that contained excess calcium contributed by dissolution of small amounts of calcite. The balances were well within the error of the chemical analyses and led to some important and far-reaching conclusions.

(i) Although the rock consisted of nearly equal amounts of plagioclase and orthoclase, virtually all of the solutes were contributed by weathering of the plagioclase. Minor amounts of potassium and silica came from orthoclase and the magnesium, the remaining potassium, and silica were derived from biotite. Although quartz was a major component of the rock, virtually no silica was derived from the dissolution of quartz.

(ii) The above leads to the conclusion that the reactivity of a mineral is much more important than its quantity in contributing solutes to weathering solutions. This is confirmed by the large amount of calcium contributed to the solute load of the perennial springs by the miniscule content of calcite in the rock. Similar observations have been made about the contribution of large amounts of calcium to solute loads of weathering solutions by trace amounts of calcite in bedrock (Bricker *et al.*, 1968; Katz *et al.*, 1985; Mast *et al.*, 1990; White *et al.*, 1999a; Bowser and Jones, 1993).

(iii) All of the solutes could be accounted for by weathering of primary minerals in the bedrock; ion exchange was therefore considered to be insignificant in this system. It has been pointed out that the Garrels and Mackenzie solution is not unique, and that a mass-balance calculation incorporating ion exchange could also explain the Sierra Nevada spring compositions. We note that this is a theoretical possibility, but favor the argument expressed by Velbel (1993): "the minimal contribution of cation exchange to watershed solute budgets cannot be proven, but it is strongly suggested by the fact that rates calculated by the geochemical mass-balance method give relative rates (rate ratios) which compare most favorably with relative rates determined in laboratory kinetic studies of mineral weathering—in other words, cation exchange is a negligible influence on solute output—." Two fundamental differences between cation exchange and weathering reactions, rate and reversibility, are discussed in detail by Bowser and Jones (2002).

The work of Garrels and Mackenzie employed concentrations rather than fluxes of the input and output solutions. These were values at the time of sampling. The balance worked out well, because the compositions they used were close to long-term average concentrations. Solute concentrations alone, although useful for identifying weathering reactions, do not provide the information necessary to calculate the penetration rate of the weathering front, denudation rates, or absolute dissolution rates of individual minerals. Studies using the mass flux of solutes instead of concentration were conducted on gneiss (Bricker *et al.*, 1968; Cleaves *et al.*, 1970); serpentinite (Cleaves *et al.*, 1974); greenstone (Katz *et al.*, 1985); schists and gneisses (Velbel, 1985); amphibolite (Velbel, 1992); and granodioritic bedrock (Mast *et al.*, 1990). These studies employed flux instead of concentration and therefore permitted rate calculations.

In the earlier studies, the idealized composition of the minerals involved in the weathering reactions was used. This generally gave satisfactory balances. However, where measured compositions of primary minerals and their reaction products were used, the balances were improved and provided a closer approximation to reality (Bowser and Jones, 2002). Differences between the results obtained by using idealized versus actual mineral compositions in mass-balance studies of Sierra Nevada perennial springwaters are illustrated in Table 2 (Bowser and Jones, 2002).

Table 2 Comparison of MTCs using theoretical, pure phlogophite (Garrels and Mackenzie, 1967) versus Sierra Nevada average biotite (Dodge *et al.*, 1969) for perennial springwaters.

Biotite composition →	Phlogopite	Sierran avg.	
MTC ($\mu mol\ L^{-1}$)			
Plagioclase	175.8	175.8	
K-feldspar	8.7	8.2	
Biotite	16.1	42.3	←
CO$_2$ gas	445.2	450.6	
Silica	0.0	0.0	
Calcite	121.8	116.4	
Smectite solid solution	−111.9	−130.8	
Goethite	0.0	−54.6	←
Halite	16.0	16.0	
Gypsum	15.0	15.0	
Mineral composition			
Plagioclase An no.	38.0	38.0	
Pct. montmorillonite	11.120	12.527	
Interlayer-K	0.114	0.252	
Interlayer charge	0.500	0.500	
Ratios			
Pag/K-feldspar	20.1	21.5	
K-feldspar/biotite	0.543	0.193	
Fe/Mg (biotite)	0.000	1.088	

After Bowser and Jones (2002).
Arrows indicate mineral mass transfers most significantly affected by using analyzed biotites. Mineral composition data and mineral mass-transfer coefficient ratios are also shown.

Although actual mineral composition information is not always easy to obtain, the use of specific mineral chemistry can substantially improve mass-balance analyses (Bowser and Jones, 2002). As can be seen in Table 1, the original idealized mineral compositions used by Garrels and Mackenzie account well for mineral reactions and product amounts. However, the use of actual mineral compositions changes the amounts slightly. The "biotite" composition that Garrels and Mackenzie used is that of the ideal magnesium end-member phlogopite, containing no iron, rather than the actual composition (analyzed later). The relative mass-transfers of K-feldspar and biotite are dependent on the Fe/Mg ratio of the biotite in the system. The Mg/K ratio of solutions derived from the dissolution of biotite is determined by its Fe/Mg ratio, which in turn determines how much K has come from biotite rather than K-feldspar. Bowser and Jones used an actual analyzed biotite composition with an Fe/Mg ratio close to 1 (Sierra Nevada biotite analyzed by Dodge *et al.* (1969)). Using this actual biotite composition rather than the idealized composition led to changes in the amounts of K-feldspar and biotite dissolved and the amount of goethite precipitated. Likewise, using the actual composition of smectite solid solution rather than that of the theoretical Ca-smectite changed the amount of calcite dissolved and the amount of smectite solid solution precipitated (Table 2). Although the changes are not large, and do not invalidate the original calculations of Garrels and Mackenzie, they provide a closer approximation to reality in the system. Similar calculations for Loch Vale using analyzed mineral compositions reduced the residual errors associated with the use of theoretical mineral compositions (Mast *et al.*, 1990). Unfortunately, there are not many systems for which analyzed mineral compositions are available (see Bowser and Jones (2002), for a detailed discussion of the use of actual versus theoretical mineral compositions).

Previous studies that have used fluxes rather than concentrations, and a few studies that have involved specific, well-characterized mineral compositions (Mast *et al.*, 1990; Bowser and Jones, 2002) confirm the results of Garrels and Mackenzie and show that mineral weathering processes dominate the export of solutes from watersheds in a variety of lithologic settings.

The export of solutes released by weathering from watersheds exhibits a strong relationship to rock type (Bricker and Rice, 1989). The same relationship had been demonstrated by Velbel (1985) in a study of weathering at the Coweeta Hydrologic Laboratory in North Carolina. He employed the mass-balance formalization of Plummer and Back (1980) to determine the mass flux from the watersheds, to interpret the mineral

weathering reactions that were responsible for the observed changes in water chemistry on several different types of bedrock, to calculate the rate of weathering of biotite, garnet, and plagioclase, and to calculate the rate of penetration of the weathering front into the fresh rock (saprolitazation). He was able to represent the system with four equations (the mass balances for potassium, calcium, magnesium, and sodium) in four unknowns (the number of moles of plagioclase, biotite, and garnet weathered, and the kilograms of biomass growth). The equations were solved algebraically, and the results were verified with a standard linear algebra matrix solving routine. Bricker *et al.* (1968) used a 4×4 matrix in their mass-balance calculations at Pond Branch, and Katz *et al.* (1985) used a 6×6 matrix in their Catoctin studies. Bowser and Jones (2002) expanded this to a 10×10 matrix to include all of the major reactants and products in the system. These authors found that it was possible to increase the number of phases that could be considered in the matrix to beyond 10 by using the concept of "fictive phases" and mean reactive compositions in some cases where there were more mineral phases than components in a system. (A fictive phase is defined as the proportional composition of two or more similar phases that have fixed dissolution rate ratios, e.g., biotite and chlorite. Mean reactive composition was used specifically for mixtures of plagioclase feldspar where the "An" number is defined by the Ca/Na ratio of the water that would be derived from the mixture.)

The reactivity of feldspars differs widely in weathering environments. Orthoclase is quite resistant to weathering relative to plagioclase. This was noted by Goldich (1938) in his classic work, and by many subsequent investigators (e.g., White *et al.*, 2001; Garrels and Mackenzie, 1967). The solubility and rate of dissolution of plagioclase feldspar makes it a major contributor of sodium and calcium ions to weathering solutions. For many rock types, plagioclase is the only silicate source of these ions. It is to be expected that the weathering solutions would have a Ca/Na ratio that reflects congruent dissolution of plagioclase, and this is observed in some watershed systems. More frequently, however, the Ca/Na ratio is larger than the ratio in the plagioclase, implying either the presence of another source of calcium or a sink for sodium. Clayton (1986, 1988) observed that the calcium-rich cores of zoned plagioclase and the more calcium-rich lamellae of structurally evolved plagioclase (also see Inskeep *et al.*, 1991) weathered more rapidly than the sodic rims, and cited this as one possible explanation for the "excess calcium" problem. Clayton pointed out, however, that the Ca/Na ratio in the water could be no greater than the most calcic portion of the plagioclase. Often, the ratio is higher than can be accounted for by dissolution of plagioclase and this has been called the "calcium problem" (Bowser and Jones, 1993 and references therein; 2002). Sverdrup (1990) attributed the excess calcium to dissolution of epidote, but there is no observed epidote in many of the rocks modeled by him. Epidote also is not nearly as soluble and dissolves much more slowly than calcite. The "excess calcium" effect has now been attributed, in most part, to dissolution of the calcite that occurs in small amounts in most rocks (Mast *et al.*, 1990; White *et al.*, 1999a; also see below), although other contributory sources (e.g., amphibole, pyroxene) may need consideration in some geologic settings.

5.04.3 MASS-BALANCE MODELING

A number of models have been developed to quantitatively define chemical weathering and watershed (catchment) mass balance, particularly during the period that acid rain was considered an environmental threat. Some of the more prominent models developed to predict the effects of acid rain on watersheds are ILWAS (Chen *et al.*, 1983), Birkenes (Christophersen *et al.*, 1982), MAGIC (Cosby *et al.*, 1985; Wright, 1987), the Trickle Down model (Schnoor *et al.*, 1984, 1986) and a mixing model "EMMA" (Hooper *et al.*, 1990). All of these models suffer from a lack of attention to soil mineralogy and texture. Each of them was developed for a specific watershed or region, and were calibrated using data from that system. Most of them perform reasonably well in describing weathering chemistry and mass-balance relations in the system for which they were developed; however, they have little transfer value for other watersheds.

The models listed above are, for the most part, site specific, although they were claimed by their developers to have broad applicability for predicting the effects of acid rain on watershed systems. Parkhurst *et al.* (1980) developed PHREEQE, a general computer program, to model geochemical reactions. It is based on an ion pairing aqueous model and can calculate pH, redox potential, and mass transfer as a function of reaction progress. This was followed in 1991 by the NETPATH program for use in modeling net geochemical reactions along a flow path (Plummer *et al.*, 1991, 1994). NETPATH was subsequently imbedded as a subroutine in PHREEQC, version I, which was written in the C programming language to implement the capabilities of PHREEQ. It added ion-exchange equilibria, surface complexation equilibria, fixed-pressure gas equilibria, and advective transport. Parkhurst (1995) published a user's guide to PHREEQC,

version I, for interpreting speciation, reaction path and advective transport modeling (forward modeling), as well as inverse geochemical calculations. Forward modeling has been applied successfully to systems dominated by carbonate and evaporite mineral geochemistry, for which reliable, internally consistent thermodynamic data exist. Applications to systems dominated by silicate hydrolysis have been severely hampered by a lack of internally consistent thermodynamic data for common silicate minerals (Nordstrom *et al.*, 1991; Melchior and Bassett, 1990). For these systems, inverse modeling, or mass-balance modeling, based on the principles of mass conservation during mineral–water reaction, has been used successfully. Inverse modeling is unconstrained by thermodynamic considerations and was adapted for general hydrochemical computation by Plummer *et al.* (1983). Geochemical modeling is discussed in Nordstrom's chapter (see Chapter 5.02).

An updated and expanded version of PHREEQC (version II) was published by Parkhurst and Appelo (1999). Version II has all of the capabilities of version I, and includes new routines for kinetically controlled reactions, solid-solution equilibria, fixed-volume gas-phase equilibria, variation of the number of exchange or surface sites in proportion to a mineral or kinetic reactant, diffusion or dispersion in one-dimensional (1D) transport, 1D transport coupled with diffusion into stagnant zones, and isotope mole balance in inverse modeling.

The above programs calculate the saturation state of a large number of minerals based on thermodynamic data to aid in identifying which mineral phases are likely to dissolve or to precipitate, but the mass-balance calculations are based solely on the amounts of material that must enter or leave the aqueous phase (through dissolution or precipitation) to match the observed water composition. It is important to recognize that mass-balance analysis requires no thermodynamic constraints, whereas programs such as PHREEQE and PHREEQC are highly dependent on thermodynamics. This presents a difficulty in silicate systems owing to the uncertainties that exist for free energy data (and their temperature derivatives) not only of primary silicate phases of complex chemical composition (e.g., amphiboles, pyroxenes, and even feldspars) but also of product phases (e.g., clays). For any system, there are usually a number of model solutions found by PHREEQE or PHREEQC that satisfy the criteria, and it is up to the investigator to pick the most likely one based on geological and geochemical knowledge. All of these programs provide models strictly in the form of numerical tables (Table 3) rather than graphical representation. None of the programs gives a unique solution to the

Table 3 Typical output of NETPATH (mmol L^{-1}) of water. All of the models meet the mass-transfer requirements, but only model 1 is reasonable from a geochemical standpoint.

	Model				
	1	*2*	*3*	*4*	*5*
NaCl	0.01600	0.01600	0.01600	0.01600	0.01600
Gypsum	0.01500	0.01500	0.01500	0.01500	0.01500
Kaolinite	−0.03355	−0.82741	2.07993	−0.12833	
Ca-montmorillonite	−0.08136	0.60007	−1.89550		−0.11015
CO_2 gas	0.30296	0.41676		0.31655	0.29815
Calcite	0.11380		0.41676	0.10021	0.11861
Biotite	0.01400	0.01400	0.01400	0.01400	0.01400
SiO_2		−0.91311	2.43096	−0.10902	0.03859
Plagioclase38	0.17584	0.17584	0.17584	0.17584	0.17584

Source: Plummer *et al.* (1994).

mass-balance problem. For instance, each of the five models in Table 3 satisfies the water chemistry constraints. The thermodynamic saturation calculation, which is integral to the NETPATH model, can help to eliminate models in which one or more of the product minerals is not saturated in the system, or is so grossly oversaturated as to be unreasonable. Looking at the model output, model 1 is reasonable. In this model, kaolinite and Ca-montmorillonite, two possible weathering products are forming, and the other minerals are dissolving. Model 2 is not likely from a geologic standpoint, because Ca-montmorillonite must dissolve to match the composition of the output water. This is not reasonable, since montmorillonite is a mineral that is formed during the weathering process and is not part of the primary mineral assemblage. Model 3 requires dissolution of a large amount of kaolinite, a mineral observed to form as a product in the system. The model is therefore questionable. Model 4, in which kaolinite and silica are precipitated during the weathering process, is not reasonable, because kaolinite plus silica should form smectite rather than existing as separate phases, although crystallization of smectite may present a kinetic problem. The last one, model 5, is not reasonable. Kaolinite is not precipitated in the weathering process, although it is known to be a major alteration product in the system.

Sverdrup (1990) and Sverdrup and Warfvinge (1995) developed the PROFILE model to calculate mineral weathering rates by means of a geochemical mass-balance procedure. This model differs from the others in that it uses dissolution constants, which were, for the most part, determined in the laboratory. Empirical fitting parameters, such as surface area of mineral exposed, are used to adjust the model to the real system being described. The model appears to work satisfactorily in many catchments if the fitting parameters are chosen judiciously. This requires a considerable amount of knowledge about the system and the mineral–water reactions that take place there.

The rate at which minerals dissolve plays a critical role in watershed weathering reactions and in the chemistry of water in the watershed. Several factors influence the dissolution rates of minerals: higher temperatures in many field sites increase the dissolution rates (White *et al.*, 1999b; Velbel, 1990); inability to determine surface area in the natural environment, coupled with an inability to quantify the wetted surface area, is one reason for the differences between field and laboratory dissolution rates (Velbel, 1986; Clow and Drever, 1996); the saturation state of the solution phase relative to the mineral(s) being weathered. The "affinity effect" has been cited as a governing factor in dissolution rates (Clow and Drever, 1996; Burch *et al.*, 1993; Taylor *et al.*, 2000; Aagaard and Helgeson, 1982). The closer to saturation a weathering solution is, the slower the dissolution rate of a mineral. Most work on the "affinity effect" has been done in the laboratory; examples of this effect in natural systems can be found in Gislason *et al.* (1996) and in Bowser and Jones (2002). None of these explanations, however, completely accounts for the differences in reaction rates in field sites and the laboratory experiments.

A recent contribution to mass-balance modeling of weathering (Bowser and Jones, 2002) utilizes a spreadsheet graphical method to interpret mass balance in watershed systems in place of the strictly numerical solution methods. Key to the approach is to solve the mass-balance equation for a 10×10 matrix as a function of mineral composition (specifically in terms of plagioclase feldspar and smectite compositions). Exploration by means of models that cover mineral compositional space limited the range of possible compositions and restricted the range of possible mass-balance solutions. Figure 1 (after Bowser and Jones, 2002) is an example of the spreadsheet approach applied to the Sierra Nevada ephemeral

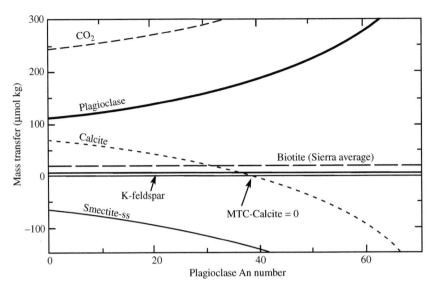

Figure 1 Plot of MTCs versus plagioclase composition for ephemeral springwaters of the Sierra Nevada Mountains, CA (Garrels and Mackenzie, 1967). Positive values are for mineral dissolution, and negative for precipitation. The average plagioclase composition (An-38) for Sierran granodiorites is near the point where the MTC for calcite changes from positive to negative values (after Bowser and Jones, 2002).

springs of Garrels and Mackenzie (1967). Mass balances and mass transfers are plotted as a function of the assumed plagioclase composition.

With the spreadsheet model, it is possible to make a detailed and rapid assessment of the effects of silicate mineral composition on natural water chemistry and to test the response of mass-balance results to variations in multiple phase assemblages and related water chemistry. As an example, Garrels and Mackenzie argued that calcite dissolution was necessary to account for what they considered Ca in ephemeral springs in excess of that contributed by the dissolution of plagioclase. As shown in Figure 1, the MTC for calcite is positive (dissolves) at low An numbers and negative (precipitates) at higher values. The point where calcite MTC is zero (the crossover point) is at a plagioclase composition of An-38, the average composition of Sierra Nevada plagioclase that Garrels and Mackenzie (1967) used. There is, therefore, no need to call on the dissolution of calcite. In perennial springs, however, there is more calcium than can be derived from plagioclase. Some Ca must therefore be supplied from the solution of calcite present in the deeper rocks. Using the spreadsheet approach, multiple mass-balance comparisons of variable watershed inputs can be made. In addition, causal delineation of hydrochemical changes between wells along flow lines can be explored, as Bowser and Jones (2002) have done for groundwater systems in the Basin and Range of southern Arizona (Vekol Valley) and shallow glacial aquifers of Northern Wisconsin (Trout Lake area).

5.04.4 SUMMARY

Mass-balance studies are widely considered to be the most reliable means of making quantative determinations of elemental transfer rates in natural systems. Garrels (1967) and Garrels and Mackenzie (1967) pioneered the use of mass-balance calculations for mineral weathering in their classic study of Sierra Nevada springwaters. These waters were chosen because a careful set of water analyses and associated primary igneous rock minerals and the soil mineral alteration products were known. Since the actual compositions of the minerals were not known, Garrels and Mackenzie used the theoretical formulas for the minerals.

This work was conducted using the concentration of elements in the output solutions. It produced some very interesting observations. In both the ephemeral and perennial springs most of the dissolved components in the output water were derived from the dissolution of plagioclase feldspar by attack of carbonic acid-rich waters, although K-feldspar and quartz are abundant constituents of the rock. Thus, bicarbonate is the major anion balancing the cations. Minor amounts of silica and potassium were derived from K-feldspar, and all of the magnesium, and some potassium, came from biotite. Virtually no silica came from quartz dissolution, although quartz is the most abundant silicate mineral in the rock. In the perennial springs, the effluent waters contained more calcium than could be derived from the plagioclase. The deeper rocks from which the perennial springs emanate presumably contain trace amounts of calcite. Dissolution of a small amount of this mineral was necessary to provide a calcium balance. This conclusion emphasizes that the reactivity of minerals is generally more important than their abundance in determining their contribution to the composition of weathering solutions. This was confirmed by the overwhelming effect of minor amounts of calcite on the calcium content of waters draining many other areas (Mast *et al.*, 1990; Katz *et al.*, 1985). Since all of the solutes in the Sierra Nevada waters could be accounted for by the weathering of primary minerals in the bedrock, neither ion exchange nor biotic uptake is important there.

The use of concentrations precludes the calculation of weathering rates for landscapes or for individual minerals. Bricker *et al.* (1968) and Cleaves *et al.* (1970, 1974) modeled changes in solute fluxes, enabling them to calculate weathering rates for the catchment and rates for individual mineral reactions. Velbel extended this type of calculation to a variety of rocks and minerals of the southern Blue Ridge mountains (Velbel, 1985, 1992), and other investigators have used this method to examine rock and mineral weathering rates in the field (Paces, 1983; Berner, 1981; Sverdrup, 1990; Creasey *et al.*, 1986; Dethier, 1986; Marchand, 1971; Yuretich and Batchelder, 1988; White *et al.*, 2001).

Rates determined in field studies are commonly one to three orders of magnitude slower than rates determined in the laboratory. A number of factors have been cited as responsible for this discrepancy (Bowser and Jones, 2002). A few of the more realistic suggestions ranked in decreasing order of importance are:

- difficulty in quantifying the surface area and wetted surface (Brantley and Stillings, 1996; Brantley *et al.*, 1999; Clow and Drever, 1996; Drever, 1997; White and Blum, 1995; White *et al.*, 1996);
- degree of saturation (affinity) (Clow and Drever, 1996; Gislason *et al.*, 1996); and
- temperature differences between experimental and natural systems (White *et al.*, 1999b; Velbel, 1990).

Some less important factors are:
- surface coatings and biological effects (Velbel, 1993, 1995);
- brief duration of experiments relative to that in natural systems (Swoboda-Coberg and Drever, 1993; Blum, 1994);
- disturbed layer at surface (more soluble) due to grinding or other preparation procedures (Eggleston *et al.*, 1989); and
- lower pH values used in laboratory experiments than found in natural settings (Casey *et al.*, 1993).

None of these factors alone completely explains the observed discrepancies in reaction rates, but mineral dissolution rates play a significant role in determining water composition in watersheds. The difference between the rate of dissolution of a primary phase and the precipitation of a secondary product (alteration phase) may be an important factor in determining whether or not a balance can be achieved. This, in turn, impacts mass-balance relations in natural systems. Mass-balance modeling of the effects of mineral weathering on water chemistry in catchment systems has progressed from pencil and paper exercises to sophisticated computer calculations. Although all of the answers concerning the kinetics of mineral weathering reactions are not in, the mass-balance approach appears to hold the most promise for quantifying and understanding elemental transfer rates in the Earth's surface environment. The spreadsheet approach provides advantages over previous numerical models in that it has the ability to evaluate output as a function of one or more mineral solid-solution compositions, and to view the results in graphical form. Mass-balance models are providing an improved understanding of the contribution of rock weathering to the dynamics and total ion concentrations of natural waters.

REFERENCES

Aagaard P. and Helgeson H. C. (1982) Thermodynamic and kinetic constraints on reaction rates among minerals and aqueous solutions. *Am. J. Sci.* **282**, 237–285.

April R., Newton R., and Coles L. T. (1986) Chemical weathering in two Adirondack watersheds: past and present-day rates. *Geol. Soc. Am. Bull.* **97**, 1232–1238.

Bain D. C., Mellor A., and Wilson M. (1990) Nature and origin of an aluminous vermiculite weathering product in acid soils from upland catchments in Scotland. *Clay Minerals* **25**, 467–475.

Baron J. (1992) *Biogeochemistry of a Subalpine Ecosystem: Loch Vale Watershed.* Ecological Series 90, Springer, New York.

Berner R. A. (1981) Kinetics of weathering and diagenesis. In *Kinetics of Geochemical Processes,* Reviews in Mineralogy (eds. A. C. Lasaga and R. S. Kirkpatrick). Mineralogical Society of America, Washington, DC, pp. 111–134.

Berner R. A. and Cochran M. F. (1998) Plant-induced weathering of Hawaiian basalts. *J. Sedim. Res.* **68**, 723–728.

Blum A. E. (1994) Feldspars in weathering. In *Feldspars and their Reactions* (ed. I. Parson). Kluwer Academic, Dordrecht, The Netherlands, pp. 595–630.

Bowser C. J. and Jones B. F. (1990) Geochemical constraints on groundwaters dominated by silicate hydrolysis: an interactive spreadsheet mass balance approach. *Chem. Geol.* **84**, 33–35.

Bowser C. J. and Jones B. F. (1993) Mass balances of natural water: silicate dissolution, clays, and the calcium problem. In *Biogeomon Symposium on Ecosystem Behavior: Evaluation of Integrated Monitoring in Small Catchments* (ed. J. Cerny). Czech Geological Survey, Prague, pp. 30–31.

Bowser C. J. and Jones B. F. (2002) Mineralogic controls on the composition of natural waters dominated by silicate hydrolysis. *Am. J. Sci.* **302**, 582–662.

Brantley S. L. and Stillings L. L. (1996) Feldspar dissolution at 25 °C and low pH. *Am. J. Sci.* **296**, 101–127.

Brantley S. L., White A. F., and Hodson M. E. (1999) Surface area of primary silicate minerals. In *Growth, Dissolution, and Pattern Formation in Geosystems* (eds. B. Jamveit and P. Meakin). Kluwer Academic, Dordrecht, The Netherlands, pp. 291–326.

Bredehoeft J. D., Papadopulos S. S., and Cooper H. H., Jr. (1982) Groundwater—the water-budget myth. In *Scientific Bases of Water-resources Management*. National Academy Press, Washington, DC, pp. 51–57.

Bricker O. P. and Rice K. C. (1989) Acid deposition to streams: a geology based method predicts their sensitivity. *Environ. Sci. Technol.* **23**, 379–385.

Bricker O. P., Godfrey A. E., and Cleaves E. T. (1968) Mineral–water interactions during the chemical weathering of silicates. In *Trace Inorganics in Water*, Advances in Chemistry Series 73 (ed. R. F. Gould). American Chemical Society, Washington, DC, pp. 128–142.

Bricker O. P., Pačes T., Johnson C. E., and Sverdrup H. (1994) Weathering and erosion aspects of small catchment research. In *Biogeochemistry of Small Catchments: A Tool for Environmental Research* (eds. B. Moldan and J. Cerny). Wiley, Chichester, England, pp. 85–105.

Bullen T., Wiegand B., Chadwick O., Vitousek P., Bailey S., and Creed I. (2002) New approaches to understanding calcium budges in watersheds: Sr isotopes meet Ca isotopes. In *Biogeomon, 4th International Symposium on Ecosystem Behavior*.

Burch T. E., Nagy K. L., and Lasaga A. C. (1993) Free energy dependence of albite dissolution kinetics at 80 °C, pH 8.8. *Chem. Geol.* **105**, 137–162.

Casey W. H., Banfield J. F., Westreich H. R., and McLaughlin L. (1993) What do dissolution experiments tell us about natural weathering? *Chem. Geol.* **105**, 1–15.

Chen C. W., Gehrini J. D., Hudson J. M., and Dean J. D. (1983) *The Integrated Lake–Watershed Acidification Study*. Final Report EPRI EA-3221, Electrical Power Research Institute, Palo Alto, California.

Christopheren N., Seip H. M., and Wright R. F. (1982) A model for streamwater chemistry at Birkenes, Norway. *Water Resour. Res.* **18**, 977–996.

Clayton J. L. (1979) Nutrient supply to soil by rock weathering. In *Impact of Intensive Harvesting on Forest Nutrient Cycling*,. College of Environmental Science and Forestry. State University of New York, Syracuse, NY, pp. 75–96.

Clayton J. L. (1986) An estimate of plagioclase weathering rate in the Idaho batholith based upon geochemical transport rates. In *Rates of Chemical Weathering of Rocks and Minerals* (eds. S. Coleman and D. Dethier). Academy Press, Orlando, FL, pp. 453–467.

Clayton J. L. (1988) Some observations of the stoichiometry of feldspar hydrolysis in granitic soils. *J. Environ. Qual.* **17**, 153–157.

Cleaves E. T., Godfrey A. E., and Bricker O. P. (1970) Geochemistry of a small watershed and its geomorphic implications. *Geol. Soc. Am. Bull.* **81**, 3015–3032.

Cleaves E. T., Fisher D. W., and Bricker O. P. (1974) Chemical weathering of serpentinite in the eastern Piedmont of Maryland. *Geol. Soc. Am. Bull.* **85**, 437–444.

Clow D. and Drever J. I. (1996) Weathering rates as a function of flow through an alpine soil. *Chem. Geol.* **132**, 131–141.

Cosby B. J., Wright R. F., Hornberger G. M., and Galloway J. N. (1985) Assessment of a lumped parameter model for soil water and stream water chemistry. *Water Resour. Res.* **21**, 51–63.

Creasey J., Edwards A. C., Reid J. M., Macleod D. A., and Cresser M. S. (1986) The use of catchment studies for assessing chemical weathering rates in two contrasting upland areas of Northeast Scotland. In *Rates of Chemical Weathering of Rocks and Minerals* (eds. S. M. Colman and D. P. Dethier). Academic Press, Orlando, FL, pp. 467–501.

Dethier D. P. (1986) Weathering rates and chemical fluxes from catchments in the Pacific Northwest USA. In *Rates of Chemical Weathering of Rocks and Minerals* (eds. S. M. Colman and D. P. Dethier). Academic Press, Orlando, FL, pp. 503–528.

Dodge F. D. W., Smith V. C., and Mays R. E. (1969) Biotites from granitic rocks of the central Sierra Nevada batholith, California. *J. Petrol.* **10**, 250–271.

Drever J. I. (1997) *The Geochemistry of Natural Waters*. Prentice Hall, Englewood Cliffs, NJ, 436pp.

Eggleston C. M., Hochella M. F., Jr., and Parks G. A. (1989) Sample preparation and ageing effects on the dissolution rate and surface composition of diopside. *Geochim. Cosmochim. Acta* **53**, 797–804.

Freeze R. A. and Cherry J. A. (1979) *Groundwater*. Prentice Hall, Englewood Cliffs, NJ.

Garrels R. M. (1967) Genesis of some ground waters from igneous rocks. In *Researches in Geochemistry* (ed. P. H. Ableson). Wiley, New York, pp. 405–420.

Garrels R. M. and Mackenzie F. T. (1967) Origin of the chemical composition of some springs and lakes. In *Equilibrium Concepts in Natural Water Systems*, Advances in Chemistry Series, No. 67 (ed. R. F. Gould). American Chemical Society, Washington, DC, pp. 222–242.

Gislason S. R., Arnorsson S., and Armannsson H. (1996) Chemical weathering of basalt in southwest Iceland: effects of runoff, age of rocks, and vegetative/glacial cover. *Am. J. Sci.* **296**, 837–907.

Goldich S. (1938) A study in rock weathering. *J. Geol.* **46**, 17–38.

Heath R. C. (1983) Basic ground-water hydrology. *US Geol. Surv., Water-Supply Pap.* **2220**, 84.

Hooper R. P., Christophersen N., and Peters N. E. (1990) Modelling streamwater chemistry as a mixture of soilwater end-members—an application to the Panola Mountain catchment, Georgia, USA. *J. Hydrol.* **116**, 321–343.

Ingebritsen S. E. and Sanford W. E. (1998) *Groundwater in Geologic Processes*. Cambridge University Press, New York.

Inskeep W. P., Nater E. A., Bloom P. R., Vandervoort D. S., and Erich M. S. (1991) Characterization of laboratory weathered laboradorite surfaces using X-ray photoelectron spectroscopy and transmission electron microscopy. *Geochim. Cosmochim. Acta* **55**, 787–800.

Jenkins A., Peters N. E., and Rodhe A. (1994) Hydrology. In *Biogeochemistry of Small Catchments: A Tool for Environmental Research* (eds. B. Moldan and J. Cerny). Wiley, Chichester, England, pp. 31–54.

Katz B. G., Bricker O. P., and Kennedy M. M. (1985) Geochemical mass-balance relationships for selected ions in precipitation and stream water, Catoctin Mountains, Maryland. *Am. J. Sci.* **285**, 931–962.

Kendall C. and McDonnell J. J. (1998) *Isotope Tracers in Catchment Hydrology*. Elsevier, Amsterdam, The Netherlands, p. 837.

Konikow L. F. and Mercer J. W. (1988) Groundwater flow and transport modeling. *J. Hydrol.* **100**(2), 379–409.

Krabbenhoft D. P., Bowser C. J., Kendall C., and Gat J. R. (1994) Use of oxygen-18 and deuterium to assess the hydrology of groundwater-lake systems. In *Environmental Chemistry of Lake and Reservoirs* (ed. L. A. Baker). American Chemical Society, Washington, DC, pp. 67–90.

Langmuir D. (1997) *Aqueous Environmental Geochemistry*. Prentice Hall, Upper Saddle River, NJ.

Likens G. E., Bormann R. S., Pierce J. S., and Johnson N. M. (1977) *Biogeochemistry of a Forested Ecosystem*. Springer, New York.

Marchand D. E. (1971) Rates and modes of denudation, White Mountains, Eastern California. *Am. J. Sci.* **270**, 109–135.

Mast M. A., Drever J. I., and Baron J. (1990) Chemical weathering in the Loch Vale watershed, Rocky Mountain National Park, Colorado. *Water Resour. Res.* **26**, 2971–2978.

Mayo L. R., Meir M. F., and Tangborn W. V. (1972) A system to combine stratigraphic and annual mass-balance systems: a contribution to the International Hydrological Decade. *J. Glaciol.* **61**, 3–14.

Melchior D. C. and Bassett R. L. (1990) *Chemical Modeling of Aqueous Systems II*. American Chemical Society, Washington, DC, vol. 416, p. 556.

Moulton K. L. and Berner R. A. (1998) Quantification of the effect of plants on weathering: studies in Iceland. *Geology* **26**, 895–898.

Nesbitt H. W. and Wilson R. E. (1992) Recent chemical weathering of basalts. *Am. J. Sci.* **292**, 740–777.

Nordstrom D. K., Plummer L. N., Wigley T. M. L., Wolery T. J., Ball J. W., Jenne E. A., Bassett R. L., Crerar D. A., Florence T. M., Fritz B., Hoffman M., Holdren G. R., Jr., Lafon G. M., Mattigod S. V., McDuff R. E., Morel R., Reddy M. M., Sposito G., and Thraikill J. (1991) Comparison of computerized chemical models for equilibrium calculations in aqueous solutions. In *Chemical Modeling in Aqueous Systems* (ed. E. A. Jenne). American Chemical Society, Washington, DC, pp. 857–894.

O'Brien A. K., Rice K. C., Bricker O. P., Kennedy M. M., and Anderson R. T. (1997) Use of geochemical mass balance modelling to evaluate the role of weathering in determining stream chemistry in five Mid-Atlantic watersheds on different lithologies. In *Water Quality Trends and Geochemical Mass Balances: Advances in Hydrologic Processes* (eds. N. E. Peters, O. P. Bricker, and M. M. Kennedy). Wiley, New York, pp. 309–334.

Pačes T. (1983) Rate constants of dissolution derived from the measurements of mass balances in hydrological catchments. *Geochim. Cosmochim. Acta* **47**, 1855–1863.

Parkhurst D. L. (1995) *User's Guide to PHREEQC—A Computer Program for Speciation, Reaction-path Advective-transport, and Inverse Geochemical Calculations*. US Geol. Surv., Water Resour. Invest. Report. 95-4227.

Parkhurst D. L. and Appelo C. A. J. (1999) *User's Guide to PHREEQC (Version 2)—A Computer Program for Speciation, Batch-reaction, One-dimensional Transport, and Inverse Geochemical Calculations*. US Geol. Surv., Water Resour. Invest. Report 99-4259.

Parkhurst D. L., Thorstenson D. C., and Plummer L. N. (1980) *PHREEQE—A Computer Program for Geochemical Calculations*. US Geol. Surv., Water Resour. Invest. Report, pp. 80–96.

Plummer L. N. and Back W. (1980) The mass balance approach: application to interpreting the chemical evolution of hydrologic systems. *Am. J. Sci.* **280**, 130–142.

Plummer L. N., Parkhurst D. L., and Thorstenson D. C. (1983) Development of reaction models for groundwater systems. *Geochim. Cosmochim. Acta* **47**, 665–685.

Plummer L. N., Prestemon E. C., and Parkhurst D. L. (1991) *An Interactive Code (NETPATH) for Modeling Net Geochemical Reactions along a Flow Path*. US Geol. Surv., Water Resour. Invest. Report 91-4087, p. 227.

Plummer L. N., Prestemon E. C., and Parkhurst D. L. (1994) *An Interactive Code (NETPATH) for Modeling Net Geochemical Reactions along a Flow Path Version 2.0*. US Geol. Surv., Water Resour. Invest. Report 94-4169, p. 130.

Rasmussen W. C. and Andreasen G. E. (1959) Hydrologic budget of the Beaverdam Creek Basin, Maryland. *US Geol. Surv., Water-Supply Pap.* **1472**, 106.

Schnoor J. L., Palmer W. D., Jr., and Glass G. E. (1984) Modeling impacts of acid precipitation for Northeastern Minnesota. In *Modeling of Total Acid Precipitation Impacts* (ed. J. L. Schnoor). Butterworths, London.

Schnoor J. L., Lee S., Nickolaidis N. P., and Nair D. R. (1986) Lake resources at risk to acid deposition in eastern United States. *Water Air Soil Pollut.* **31**, 1091–1101.

Schulz J. S., White A. F., Harden J., Stonesdstrom D., Anderson S., and Vivit D. (2002) Silicate weathering of marine terraces north of Santa Cruz, California. *Trans. Am. Geophys. Union* **83**, H61C-0809.

Stumm W. and Morgan J. J. (1996) *Aquatic Chemistry: Chemical Equilibria and Rates in Natural Waters*. Wiley-Interscience, New York, NY, 1022pp.

Sverdrup H. U. (1990) *The Kinetics of Base Cation Release due to Chemical Weathering*. Lund University Press, Lund, Sweden, 246pp.

Sverdrup H. U. and Warfvinge P. (1995) Estimating field weathering rates using laboratory kinetics. In *Chemical Weathering Rates of Silicate Minerals*, Reviews in Mineralogy (eds. A. F. White and S. L. Brantley). Mineralogical Society of America, Washington, DC, pp. 485–541.

Swoboda-Coberg N. G. and Drever J. I. (1993) Mineral dissolution rates in plot-scale field and laboratory experiments. *Chem. Geol.* **105**, 51–69.

Taylor A. B. and Velbel M. A. (1991) Geochemical mass balances and weathering rates in forested watersheds of the southern Blue Ridge. *Am. J. Sci.* **285**, 904–930.

Taylor A. S., Blum J. D., and Lasaga A. C. (2000) The dependence of laboradorite dissolution and Sr isotope release rate on solution saturation state. *Geochim. Cosmochim. Acta* **64**, 2389–2400.

Ugolini F. C., Minden R., Dawson H., and Zachara J. (1977) An example of soil processes in the *abies amabilis* zone of central Cascades, Washington. *Soil Sci.* **124**, 291–302.

Velbel M. A. (1985) Geochemical mass balances and weathering rates in forested watersheds of the southern Blue Ridge. *Am. J. Sci.* **285**, 904–930.

Velbel M. A. (1986) The mathematical basis for determining the rates of geochemical and geomorphic processes in small forested watersheds by mass balance: examples and implications. In *Rates of Chemical Weathering of Rocks and Minerals* (eds. S. M. Coleman and D. P. Dethier). Academic Press, Orlando, FL, pp. 439–451.

Velbel M. A. (1990) Influence of temperature and mineral surface characteristics on feldspar weathering rates in natural and artificial systems: a first approximation. *Water Resour. Res.* **26**, 3049–3053.

Velbel M. A. (1992) Geochemical mass balances and weathering rates in forested watersheds of the southern Blue Ridge: III. Cation budgets and the weathering rate of amphibole. *Am. J. Sci.* **292**, 58–78.

Velbel M. A. (1993) Formation of protective surface layers during silicate weathering under well-leached oxidizing conditions. *Am. Mineral.* **78**, 405–414.

Velbel M. A. (1995) Interaction of ecosystem processes and weathering processes. In *Solute Modeling in Catchment Systems* (ed. S. T. Trudgill). Wiley, Chichester, England, pp. 193–209.

White A. F. and Blum A. E. (1995) Effects of climate on chemical weathering in watersheds. *Geochim. Cosmochim. Acta* **59**, 1729–1747.

White A. F., Blum A. E., Schulz M. S., Bullen T. D., Harden J. W., and Peterson M. L. (1996) Chemical weathering of a soil chromosequence on granitic alluvium: 1. Quantification of mineralogical and surface area changes and calculation of primary silicate reaction rates. *Geochim. Cosmochim. Acta* **60**, 2533–2556.

White A. F., Blum A. E., Bullen T. D., Vivit D. V., Schulz M. W., and Clow D. W. (1999a) The role of disseminated calcite in the chemical weathering of granitoid rocks. *Geochim. Cosmochim. Acta* **63**, 1939–1954.

White A. F., Blum A. E., Bullen T. D., Vivit D. V., Schulz M., and Fitzpatrick J. (1999b) The effect of temperature on experimental and natural chemical weathering rates of granitoid rocks. *Geochim. Cosmochim. Acta* **63**, 3277–3291.

White A. F., Bullen T. D., Schulz M. S., Blum A. E., Huntington T. G., and Peters N. E. (2001) Differential rates of feldspar weathering of granitic regoliths. *Geochim. Cosmochim. Acta* **65**, 847–870.

Winter T. C. (1981) Uncertainties in estimating the water balance of lakes. *Water Res. Bull.* **17**(1), 82–114.

Wright R. F. (1987) Influence of acid rain on weathering rates. In *Physical and Chemical Weathering in Geochemical Cycles* (eds. M. Meybeck and A. Lerman). D. Reidel, Dordrecht, The Netherlands, pp. 181–196.

Yuretich R. F. and Batchelder G. L. (1988) Hydrogeochemical cycling and chemical denudation in the Fort River watershed, central Massachussets: an appraisal of mass-balance studies. *Water Resour. Res.* **24**, 105–114.

5.05
Natural Weathering Rates of Silicate Minerals

A. F. White

US Geological Survey, Menlo Park, CA, USA

NOMENCLATURE

b_{solid}	solid weathering gradient (m kg mol^{-1})
b_{solute}	solute weathering gradient (m L mol^{-1})
c	concentration of weatherable component in solute (mol)
$C_{j,w}$	final concentration of weatherable component j in regolith (mol m^{-3})
$C_{j,p}$	initial concentration of weatherable component j in protolith (mol m^{-3})
$C_{i,w}$	final concentration of inert component i in regolith (mol m^{-3})
$C_{i,p}$	initial concentration of inert component i in protolith (mol m^{-3})
d	mineral grain diameter (m)
D_j	diffusion coefficient (m^2 s^{-1})
E_a	activation energy (kJ mol^{-1})
G	pit growth rate (m s^{-1})
ΔG	excess free energy of reaction (kJ mol^{-1})
h_g	gravitational head (m m^{-1})
h_p	pressure head (m m^{-1})
∇H	hydraulic gradient (m m^{-1})
IAP	ionic activity product
k	intrinsic rate constant
K_m	hydraulic conductivity (m s^{-1})
K_s	mineral solubility product
m_j	atomic weight of component j (g)
ΔM	change in mass due to weathering (mol)
n	number of etch pits
P	annual precipitation (m)
q_h	flux density of water (m s^{-1})
R_{solid}	reaction rate based on solid concentrations (mol m^{-2} s^{-1})
R_{solute}	reaction rate based on solute concentrations (mol m^{-2} s^{-1})
R'	gas constant (J mol^{-1} K^{-1})
S	surface area (m^2)
T	temperature (K)
t	duration of weathering (s)
v_x	half wide opening rate (m s^{-1})
V_m	molar volume (m^3)
V_s	volume of solute (m^3)
W	etch diameter (m)
z	distance (m)
ε	volumetric strain
ϕ	porosity (m^3 m^{-3})
λ	surface roughness (m^2 m^{-2})
v	solute weathering velocity (m s^{-1})
θ	pit wall slope (m m^{-1})
ρ	specific mineral density (m^3 g^{-1})
ρ_p	density of regolith (cm^3 g^{-1})
ρ_w	density of protolith (cm^3 g^{-1})
τ_j	mass transfer coefficient
τ_s	sheer modulus
ω	solid-state weathering velocity (m s^{-1})
Ω	thermodynamic saturation state

5.05.1 INTRODUCTION

Silicates constitute more than 90% of the rocks exposed at Earth's land surface (Garrels and Mackenzie, 1971). Most primary minerals comprising these rocks are thermodynamically unstable at surface pressure/temperature conditions and are therefore susceptible to chemical weathering. Such weathering has long been of interest in the natural sciences. Hartt (1853) correctly attributed chemical weathering to "the efficacy of water containing carbonic acid in promoting the decomposition of igneous rocks." Antecedent to the recent interest in the role of vegetation on chemical weathering, Belt (1874) observed that the most intense weathering of rocks in tropical Nicaragua was confined to forested regions. He attributed this effect to "the percolation through rocks of rain water charged with a little acid from decomposing vegetation." Chamberlin (1899) proposed that the enhanced rates of chemical weathering associated with major mountain building episodes in Earth's history resulted in a drawdown of atmospheric CO_2 that led to periods of global cooling. Many of the major characteristics of chemical weathering had been described when Merrill (1906) published the groundbreaking volume *Rocks, Rock Weathering, and Soils*.

The major advances since that time, particularly during the last several decades, have centered on understanding the fundamental chemical, hydrologic, and biologic processes that control weathering and in establishing quantitative weathering rates. This research has been driven by the importance of chemical weathering to a number environmentally and economically important issues. Undoubtedly, the most significant aspect of chemical weathering is the breakdown of rocks to form soils, a process that makes life possible on the surface of the Earth. The availability of many soil macronutrients such as magnesium, calcium, potassium, and PO_4 is directly related to the rate at which primary minerals weather. Often such nutrient balances are upset by anthropogenic activities. For example, Huntington *et al.* (2000) show that extensive timber harvesting in the southeastern forests of the United States, which are underlain by intensely weathered saprolites, produces net calcium exports that exceed inputs from weathering, thus creating a long-term regional problem in forest management.

The role of chemical weathering has long been recognized in economic geology. Tropical bauxites, which account for most of world's aluminum ores, are typical examples of residual concentration of silicate rocks by chemical weathering over long time periods (Samma, 1986). Weathering of ultramafic silicates such as peridotites forms residual lateritic deposits that contain significant deposits of nickel and cobalt. Ores generated by chemical mobilization include uranium deposits that are produced by weathering of granitic rocks under oxic conditions and subsequent concentration by sorption and precipitation (Misra, 2000).

Over the last several decades, estimating rates of silicate weathering has become important in addressing new environmental issues. Acidification of soils, rivers, and lakes has become a major concern in many parts of North America and Europe. Areas at particular risk are uplands where silicate bedrock, resistant to chemical weathering, is overlain by thin organic-rich soils (Driscoll *et al.*, 1989). Although atmospheric deposition is the most important factor in watershed acidification, land use practices, such as conifer reforestation, also create acidification problems (Farley and Werritty, 1989). In such environments, silicate hydrolysis reactions are the principal buffer against acidification. As pointed out by Drever and Clow (1995), a reasonable environmental objective is to decrease the inputs of acidity such that they are equal to or less than the rate of neutralization by weathering in sensitive watersheds.

The intensive interest in past and present global climate change has renewed efforts to understand quantitatively feedback mechanisms between climate and chemical weathering. On timescales longer than a million years, atmospheric CO_2 levels have been primarily controlled by the balance between the rate of volcanic inputs from the Earth's interior and the rate of uptake through chemical weathering of silicates at the Earth's surface (Ruddiman, 1997). Weathering is proposed as the principal moderator in controlling large increases and decreases in global temperature and precipitation through the greenhouse effects of CO_2 over geologic time (R. A. Berner and E. K. Berner, 1997). Weathering processes observed in paleosols, discussed elsewhere in this volume (see Chapter 5.18), have also been proposed as indicating changes in Archean atmospheric CO_2 and O_2 levels (Ohmoto, 1996; Rye and Holland, 1998).

5.05.2 DEFINING NATURAL WEATHERING RATES

Chemical weathering is characterized in either qualitative or quantitative terms. Qualitative approaches entail comparing the relative weathering of different minerals or the weathering of the same mineral in different environments. Early observations by Goldich (1938) showed that the weathering sequence of igneous rocks in the field was the reverse of Bowen's reaction series that ranked minerals in the order of crystallization from magmas. Thermodynamic and kinetic considerations have added new dimensions to characterizing the relative weatherability of various minerals and rocks (Nesbitt and Markovics, 1997). Silicate weathering has long been recognized as exhibiting consistent trends with time and climate (Jenny, 1941; Oillier, 1984). Such studies provide valuable insights into weathering processes and into the influence of environmental conditions such as climate, vegetation, and geomorphology. A number of these issues are discussed elsewhere in this volume (see Chapters 5.01 and 5.06).

The primary focus of this chapter is on the development of rates that quantitatively describe silicate mineral and rock weathering. The advantages of this approach are that such rates are related to reaction mechanisms and can be used as predictive tools in estimating how weathering will behave under various environmental conditions.

The weathering rate R ($mol\ m^{-2}\ s^{-1}$) of a primary silicate mineral is commonly defined by the relationship

$$R = \frac{\Delta M}{S \cdot t} \qquad (1)$$

where ΔM (mol) is the mass change due to weathering, S (m^2) is the total surface area involved and t (s) is the duration of the reaction. The weathering rate is therefore equivalent to a chemical flux. Much of the following discussion will center on how the terms on the right side of Equation (1) are determined and quantified. This discussion is applicable in varying degrees to other types of minerals, such as carbonates, sulfates, and sulfides that commonly weather at significantly faster rates.

An important feature of silicate weathering is the extremely large spatial and temporal scales over which the parameters describing weathering are measured (Equation (1)). Natural weathering can be characterized by changes in microscopic surface morphologies of mineral grains, by solid and solute changes in soil profiles, from solute fluxes in small watersheds and large river basins and by continental and global-scale element cycles. Clearly, the degree to which individual weathering parameters are characterized depends strongly on the magnitude of these scales. Several other chapters in this volume deal in greater detail with some of these specific weathering environments, i.e., soils (see Chapter 5.01), glacial environments (see Chapter 5.07), watersheds (see Chapter 5.08), and river systems (see Chapter 5.09).

5.05.3 MASS CHANGES RELATED TO CHEMICAL WEATHERING

Silicate weathering involves hydrolysis reactions that consume reactant species, i.e., primary minerals and protons and form weathering products, i.e., solute species and secondary minerals. A typical hydrolysis reaction is the weathering of albite feldspar to form kaolinite

$$2NaAlSi_3O_8 + 2H^+ + H_2O$$
$$\rightarrow Al_2Si_2O_5(OH)_4 + 2Na^+ + 4SiO_2 \quad (2)$$

which consumes hydrogen ions and water and produces kaolinite and solute sodium and SiO_2. For a more detailed discussion of the reaction mechanisms of silicate weathering, the reader is referred to White and Brantley (1995 and references therein), Lasaga (1998), and Brantley (see Chapter 5.03).

The rate of weathering is reflected by the mass change ΔM with time of any of the individual components in a reaction if the overall stoichiometry is known. The consumption of hydrogen ions can, in principle, determine the rate of plagioclase reaction (Equation (2)), However, such measurements are difficult due to the many sources and sinks of hydrogen ion in the weathering environment, e.g., CO_2, organic acids and ion exchange reactions. Use of secondary minerals to define weathering rates is also difficult due to compositional heterogeneities and nonconservative behavior. In Equation (2), aluminum is completely immobilized in kaolinite during plagioclase weathering. However, in low-pH and/or organic-rich weathering environments, aluminum exhibits significant mobility and the quantity of kaolinite produced no longer directly reflects the amount of plagioclase that has reacted.

Most weathering rate studies have focused on measuring decreases in the solid-state reactant and/or increases in one or more of the mobile solute products. Examples of such changes due to weathering are shown in Figure 1 for solid-state and pore-water sodium distributions with depth in a 160 kyr old marine terrace in California. Mineralogical analyses indicate that significant sodium is contained only in plagioclase. Therefore, both the decrease in solid-state sodium with decreasing depth and the increase in solute sodium with increasing depth can be used to define the rate of plagioclase weathering (Equation (2)).

The above composition trends (Figure 1) can be conceptualized as solid and solute weathering gradients describing mineral weathering (White, 2002). Under simple steady-state conditions, as shown in Figure 2, gradients, b_{solid} and b_{solute}, are defined in terms the respective weathering rates R_{solid} and R_{solute} and weathering velocities ω and v (m s^{-1}). The state-state weathering velocity is the rate at which the weathering front propagates through the regolith and the solute weathering velocity is the rate of solute transport or fluid flux.

Increasing the weathering rate produces a shallower weathering gradient in which a measured solid or solid component, such as

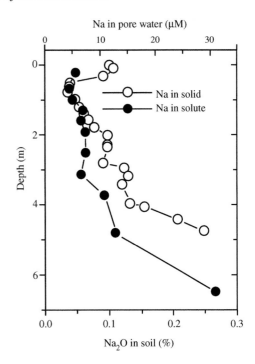

Figure 1 Distributions of sodium in the solids and pore waters of a Santa Cruz marine terrace, California, USA.

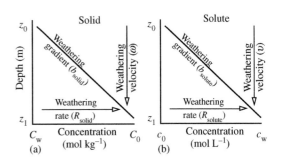

Figure 2 Schematic showing (a) solid and (b) solute distributions of a mobile element in a weathering regolith. C_w is the concentration in the most weathered material at shallow depth z_0 and C_p is the initial protolith concentration at depth z_1. c_w defines the corresponding solute weathered concentration and c_0 is the initial solute concentration. The weathering gradients b_{solid} and b_{solute}, which describe the solid and solute elemental distributions, are defined in terms of respective weathering rates (R_{solid} and R_{solute}) and velocities (ω and v) (after White, 2002).

sodium, increases more rapidly with depth in the regolith. Increasing weathering velocity, due to the rate at which weathering front or the pore water moves through the regolith steepens the weathering gradient. The slope of these gradients, measured under field conditions (Figure 1), can be converted to weathering rates in combination with additional physical and chemical parameters (White, 2002).

Defining the stoichiometry of the weathering reaction is often more difficult than suggested by plagioclase weathering (Equation (2)). The first stage of biotite weathering, for example, involves the oxidation of ferrous atoms and the concurrent release of potassium to form oxybiotite (Amonette *et al.*, 1988). Depending on the weathering environment, biotite subsequently undergoes direct transformation to kaolinite with congruent release of magnesium (Murphy *et al.*, 1998) or it can weather to vermiculite with magnesium partially retained in the solid (Fordham, 1990). Therefore, knowledge of the specific reaction pathway is required in order to define the rate of mineral weathering.

5.05.3.1 Bulk Compositional Changes in Regoliths

Determining weathering rates from solid-state mass changes (ΔM in Equation (1)) involves comparing concentration differences of a mineral, element or isotope in a weathered regolith with the corresponding concentration in the initial protolith (Merrill, 1906; Barth, 1961). For a component j, such as sodium in Figure 1, regolith concentrations are defined respectively as $C_{j,w}$ and $C_{j,p}$. A comparison of these concentrations is relatively straightforward for regoliths such as saprolites or laterites that are developed *in situ* on crystalline bedrock. The determinations of the parent concentrations are more difficult for sedimentary deposits, due to heterogeneities produced during deposition, e.g., the Santa Cruz marine terraces (Figure 1). In such cases, the protolith is commonly represented by the deepest, least weathered portion of the profile (Harden, 1987).

The measured ratio $C_{j,w}/C_{j,p}$ is dependent not only the weatherability of component j but also on gains and losses of other components in the regolith, as well as external factors such as compaction or dilation of the soil or regolith. The most common method for overcoming such effects is to compare the ratio of the mobile component j to the ratio of an additional component i, which is chemically inert during weathering. Such a comparison is commonly defined in terms of the mass transfer coefficient τ_j such that (Brimhall and Dietrich, 1987)

$$\tau_j = \frac{C_{j,w}/C_{j,p}}{C_{i,w}/C_{i,p}} - 1 \qquad (3)$$

When $\tau_j = 1$, no mobilization of j occurs, if $\tau_j = 0$, j has been completely mobilized. The concurrent volume change due to compaction or dilation by weathering and bioturbation is defined by the volumetric strain ε such that

$$\varepsilon = \frac{\rho_p C_{i,p}}{\rho_w C_{i,w}} - 1 \qquad (4)$$

where ρ_p and ρ_w are respective densities of the protolith and regolith. Positive values of ε denote regolith expansion and negative values indicate collapse. A value of $\varepsilon = 0$ denotes isovolumetric weathering. The reader is referred to Amundson (see Chapter 5.01) for a more detailed discussion of mass transfer coefficients and volumetric strain.

Conservative elements (C_i) include zirconium (Harden, 1987), titanium (Johnsson *et al.*, 1993), rare earth elements, and niobium (Brimhall and Dietrich, 1987). Considerable disagreement occurs in the literature as to the relative mobility of these elements under differing weathering conditions (Hodson, 2002). Also, these elements are concentrated in the heavy mineral fractions and are often not suitable for describing weathering ratios in depositional environments where selective concentration and winnowing occurs. For such conditions, relatively inert minerals such as quartz can be considered (Sverdrup, 1990; White, 1995).

The applicability of the above approach (Equations (3) and (4)) in describing elemental mobilities is shown in Figure 3 for a 350 kyr old regolith at Panola, Georgia, USA (White *et al.*, 2001; White *et al.*, 2002). This profile consists of a dense unstructured kaolinitic soil, overlying a porous saprolite resting on granodiorite bedrock. Saprolites, common in many subtropical to tropical weathering environments, are clay-rich regoliths that retain the original bedrock texture. Volume changes in the Panola saprolite and bedrock, calculated from zirconium, titanium, and niobium concentrations, center close to zero, i.e., $\varepsilon = 0$ (Equation (4)), indicating that weathering is isovolumetric (Figure 3(a)). In contrast, the shallowest soils have undergone volume increases due to bioturbation and the introduction of organic matter.

Values for τ_j (Equation (3)) cluster near zero in the Panola bedrock below 10.5 m indicating no elemental mobility (Figure 3(b)). Calcium and sodium are almost completely mobilized at depths of between 9.5 m and 4.5 m, where silicon is moderately depleted. Bedrock weathering is highly selective, with other elements remaining relatively immobile. Potassium and magnesium are mobilized in the overlying saprolite and iron and aluminum are strongly depleted in the upper soil and enriched in the lower soil horizons, which is indicative of downward iron and aluminum mobilization and subsequent re-precipitation.

The total elemental mass loss or gain occurring in the Panola regolith ($\Delta M_{j,\text{solid}}$) is determined by integrating the mass transfer coefficient τ_j over the regolith depth z (m) and unit surface area (1 m^2) such that (Chadwick *et al.*, 1990; White *et al.*, 1996)

$$\Delta M_{j,\text{solid}} = \left(\rho_p \frac{C_{j,p}}{m_j} 10^4 \right) \int_0^z - (\tau_j) \, dz \qquad (5)$$

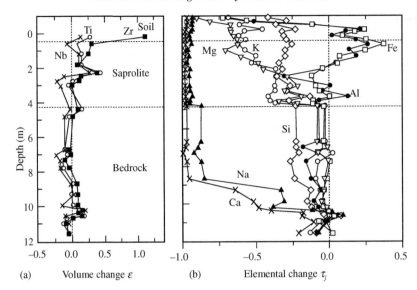

Figure 3 (a) Volume and (b) elemental changes in the Panola regolith, Georgia, USA. Volume changes are calculated separately assuming conservancy of Zr, Nb, and Ti. Positive values of ε (Equation (4)) indicate dilation and negative values indicate compaction. Positive values of τ_j (Equation (3)) indicate elemental enrichment and negative values denote elemental mobilization and loss (after White *et al.*, 2001).

Table 1 Elemental mobility ($kmol\,m^3$) and total % change in the Panola regolith. Negative values indicate element increases.

	Bedrock	Saprolite	Soil	%loss
Ca	4,550	40	−10	77
Na	8,190	670	−90	67
Si	15,030	5,520	−160	16
K	−33	3,360	10	18
Mg	−229	630	−20	16
Fe	−550	−130	540	−5
Al	1,520	−2,030	1,780	0

where m_j is the atomic weight of element j. The elemental losses in each horizon of the Panola profile are calculated from Equation (5) and tabulated in Table 1.

Element mobilities in the Panola regolith can be converted to changes in primary and secondary mineral abundances (Figure 4) using a series of linear equations describing mineral stochiometries (see Chapter 5.04). Table 1 indicates that initial bedrock weathering removes most of the plagioclase. K-feldspar weathering predominates within the overlying saprolite and biotite weathering occurs both in the saprolite and shallower soil. Secondary kaolinite is produced within the bedrock from plagioclase weathering and in the saprolite and deeper soils from K-feldspar and biotite weathering. The loss of kaolinite in the shallowest soils implies either dissolution or physical translocation to deeper soil horizons. Iron oxides are formed by oxidation of iron released from biotite weathering in the saprolite

(Figure 4). The abundance of iron-oxides, like that of kaolinite, decreases in the shallow soils.

Mass balance calculations produce important insights into element and mineral mobilities in regoliths in different weathering environments, in addition to providing quantitative estimates of mass changes required in calculating chemical weathering rates (ΔM, Equation (1)). Rates of plagioclase and biotite weathering in the Panola regolith, based on this approach, are listed in Table 2 (nos. 6 and 38). Additional examples of weathering mass balances are contained in Merritts *et al.* (1992), Stewart *et al.* (2001) and Amundson (see Chapter 5.01).

5.05.3.2 Small-scale Changes in Mineral and Rock Compositions

In addition to bulk changes in regolith composition, chemical weathering is also characterized by smaller-scale physical and chemical changes in individual mineral grains and rock surfaces. The formation of weathering and hydration rinds is commonly observed on volcanic rocks. Rind thicknesses are measured using optical and scanning electron microscopy (SEM). The most pervasive use of hydration rinds has been in the study of natural glasses in which thicknesses are used to date cultural artifacts (Friedman *et al.*, 1994). Weathering rinds developed on volcanic clasts have also been evaluated in terms of weathering duration and used as a tool for dating glacial deposits and other geomorphologic

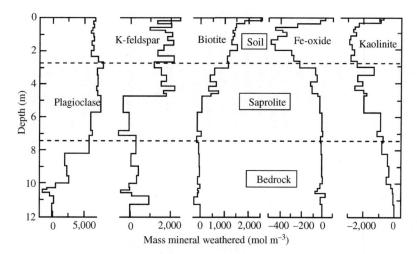

Figure 4 Primary minerals lost (positive values) and secondary minerals produced (negative values) during weathering of the Panola regolith. Horizontal lines denote regolith horizons (after White *et al.*, 2001).

features (Colman, 1981). The weathering reactions producing rinds are considered to be transport-controlled and limited by diffusion of reactants and products through the altered surfaces of parent material.

Changes in mineral surface morphologies have also been extensively correlated with silicate weathering. Dorn (1995) used the digital processing of backscatter electron imagery to determine void spaces produced by weathering of plagioclase in Hawaiian basalts and to determine climatic effects on weathering rates.

Of all the morphological changes associated with weathering, the development of etch pits has received the most attention. Examples of etch pits developed on microcline and hornblende during soil weathering are shown by SEM images in Figure 5. Recently atomic force microscopy and interferometry have contributed higher resolution and greater depth sensitivities to etch pit imaging (Luttge and Lasaga, 1999; Maurice *et al.*, 2002).

Increases in etch pit formation have been translated into weathering rates by MacInnis and Brantley (1993) using a pit-size-distribution model (PSD) that considers the number of pits n as an inverse exponential function of the pit width W (cm) divided by the growth rate G (m s^{-1}) such that

$$n = n^0 \exp\left(\frac{-W}{\tau_s G}\right) \qquad (6)$$

where τ_s is the shear modulus and n^0 is the initial pit nucleation density. Based on this equation, a plot of log n versus W will produce a straight line with a slope of $-1/\tau_s G$. This relationship was derived for several soil profiles (Figure 6) using hornblende and K-feldspar data from Cremeens *et al.* (1992).

The contribution of etch pits to the bulk weathering rate of a mineral depends on the number, density and rate of dissolution of each pit. By integrating over the PSD and assuming a pyramidal geometry for each pit, MacInnis and Brantley (1993) derived a simple expression for the weathering rate

$$R_{\text{solid}} = \frac{2n^0 v_x \tan \theta}{V_m} (\tau_s G)^3 \qquad (7)$$

where V_m is the molar volume (m^3 mol^{-1}), θ is the average slope of the wall of the etch pits and v_x is the half-width opening rate ($=0.5G$). Etch pit distributions produced quantitative weathering rates for hornblende and K-feldspar that are comparable to natural rates estimated by other techniques (Table 2, nos. 22 and 34).

In addition to characterizing mass losses, textural features can be used to discern the mechanisms and chemical environment associated with silicate dissolution. It is widely accepted that pitted surfaces, such as that shown in Figure 5, indicate surface reaction whereas smooth rounded surfaces result from diffusion-controlled dissolution (Lasaga, 1998; see Chapter 5.03).

5.05.3.3 Changes Based on Solute Compositions

The alternative method for calculating rates is based on the mass of solutes produced during weathering (Equation (1)). This approach commonly involves comparing changes between initial and final solute concentrations Δc_j (M) in a known volume of water V_s (L) such that

$$\Delta M_{j,\text{solute}} = \Delta c_j V_s \qquad (8)$$

The above equation is analogous to Equation (5), which describes mass changes in the solid state.

Table 2 Specific weathering rates of common silicates (rates in parentheses are BET surface area-normalized rates).

	Weathering environment	log rate (mol m^{-2}s^{-1})	Long time (yr)	pH	Surface measurement	Surface area S (m^2 g^{-1})	Grain size (μm)	Roughness (λ)	References
	Plagioclase								
1	Davis Run saprolite, VA, USA	−16.4	5.93	ND	BET	1.0	ND	ND	White et al. (2001)
2	China Hat Soil, Merced CA, USA	−16.3	6.48	6.0–7.0	BET	1.5	500	331	White et al. (1996)
3	Turlock Lake Soil, Merced, CA, USA	−16.0	5.78	6.0–7.0	BET	1.48	730	477	White et al. (1996)
4	Modesto Soil, Merced, CA, USA	−15.8	4.00	6.0–7.0	BET	0.26	320	37	White et al. (1996)
5	Riverbank Soil, Merced, CA, USA	−15.7	5.40	6.0–7.0	BET	0.46	650	132	White et al. (1996)
6	Panola bedrock, GA, USA	−15.7	5.70	6.5	BET	1.0*	500	221	White et al. (2001)
7	Modesto Soil, Merced, CA, USA	−15.1	4.00	6.0–7.0	BET	0.26	320	37	White et al. (1996)
8	Filson Ck. Watershed, MN, USA	−14.7 (−15.3)	4.00	5.0	GEO	0.50	63–125	21	Siegal and Pfannkuch (1984)
9	Bear Brook Watershed, ME, USA	−14.5 (15.2)	4.00	5.8	GEO	0.50	63–125	15	Schnoor (1990)
10	Crystal Lake Aquifer, WI, USA	−13.7 (15.3)	4.00	6.0–7.5	GEO	0.12	1–100	2	Kenoyer and Bowser (1992)
11	Bear Brook watershed soil, ME, USA	−13.3 (−15.2)	4.00	3.0–4.5	GEO	0.020	1–4,000	1	Swoboda-Colberg and Drever (1992)
12	Gardsjon watershed soil, Sweden	−13.1	4.00	6.0	BET	1.53	NA	ND	Sverdrup (1990)
13	Loch Vale Nano-Catchment, CO, USA	−12.8	4.00	7.0	BET	0.21	53–208	9	Clow and Drever (1996)
14	Coweeta Watershed, NC, USA	−12.5 (−14.9)	5.50	6.8	GEO	0.003	1,000	1	Velbel (1985)
15	Trnavka River watershed, Czech.	−12.1 (−14.0)	4.00	5.0	GEO	0.057	ND	1	Paces (1986)
16	Plastic Lake, Ontario, Canada	−11.8 (−13.7)	4.00	ND	GEO	0.030	100	1	Kirkwood and Nesbitt (1991)
	K-feldspar								
17	Davis Run Saprolite, VA, USA	−16.8	5.93	ND	BET	1.0	ND	ND	White et al. (2001)
18	China Hat Soil, Merced, CA, USA	−16.6	6.48	6.0–7.0	BET	0.81	500	179	White et al. (1996)
19	Riverbank Soil, Merced, CA, USA	−16.4	5.52	6.0–7.0	BET	0.94	650	270	White et al. (1996)
20	Turlock Lake Soil, Merced, CA, USA	−16.3	5.78	6.0–7.0	BET	0.81	730	261	White et al. (1996)
21	Upper Modesto Soil, Merced, CA, USA	−15.3	4.00	6.0–7.0	BET	0.26	170	20	White et al. (1996)
22	Adams County, IL	−14.7 (−16.6)	4.10	6.0–7.0	GEO	0.02	53–100	1	Brantley et al. (1993)
23	Loch Vale Nano-Catchment, CO, USA	−13.8	4.00	7.0	BET	NA	NA	ND	Clow and Drever (1996)

24	Bear Brook watershed soil, ME, USA	−13.3 (−15.2)	4.00	4.5	GEO	0.020	1–4,000	1	Swoboda-Colberg and Drever (1992)
25	Gardsjon watershed soil, Sweden	−13.3	4.08	5.6–6.1	BET	1.53	ND	ND	Sverdrup (1990)
26	Surface exposures of Shap Granite	−12.9 (−14.8)	3.70	ND	GEO	ND	NA	1	Lee et al. (1998)
27	Plastic Lake, Ontario, Canada	−11.8 (−13.6)	4.00	ND	GEO	0.03	100	1	Kirkwood and Nesbitt (1991)
Hornblende									
28	China Hat Soil, Merced CA, USA	−17.0	6.48	6.0–7.0	BET	0.67	500	179	White et al. (1996)
29	Turlock Lake Soil, Merced, CA, USA	−16.4	5.78	6.0–7.0	BET	0.67	730	261	White et al. (1996)
30	Riverbank Soil, Merced, CA, USA	−16.0	5.40	6.0–7.0	BET	0.72	650	250	White et al. (1996)
31	Riverbank Soil, Merced, CA, USA	−15.9	5.11	6.0–7.0	BET	0.72	650	250	White et al. (1996)
32	Lower Modesto Soil, Merced, CA, USA	−15.7	4.60	6.0–7.0	BET	0.34	320	58	White et al. (1996)
33	Bear Brook watershed soil ME, USA	−14.5 (−16.4)	4.00	4.5	GEO	0.020	1–4,000	1	Swoboda-Colberg and Drever (1993)
34	Adams County, IL	−14.1 (−16.0)	4.10	ND	GEO	0.02	53–100	1	Brantley et al. (1993)
35	Lake Gardsjon	−13.6	4.00	4.0	BET	1.53	NA	ND	Sverdrup (1990)
36	Plastic Lake, Ontario, Canada	−12.3 (−14.2)	4.00	ND	GEO	0.03	100	1	Kirkwood and Nesbitt (1991)
Biotite									
38	Panola, GA, USA	−16.5	5.70	5.0–5.5	BET	5	100–500	775	White (2002)
39	Crystal Lake Aquifer, WI, USA	−15.5 (−17.4)	4.00	6.0–7.5	GEO	12	1–100	1	Kenoyer and Bowser (1992)
40	Rio Icacos, PR	−15.4	5.48	4.5–5.5	BET	8.1	200–1,200	2,900	White (2002)
41	Rio Icacos, PR	−15.0	5.48	4.5–5.5	BET	8.1	200–1,200	2,900	Murphy et al. (1998)
42	Loch Vale Nano-Catchment, CO, USA	−14.1	4.00	7.0	BET	3.2	NA	ND	Clow and Drever (1996)
43	Bear Brook watershed soil, ME, USA	−14.0 (−15.9)	4.00	4.5	GEO	0.0800	1–4,000	1	Swoboda-Colberg and Drever (1992)
44	Coweeta, NC, USA	−12.9 (−14.8)	5.48	6.0	GEO	0.006	1	1	Velbel (1985)
Quartz									
45	Rio Icacos, PR	−15.1	5.48	4.5–5.5	BET	0.3	450		Schulz and White (1999)

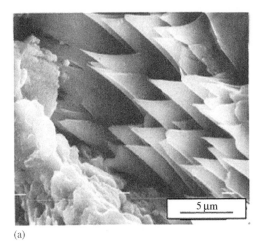
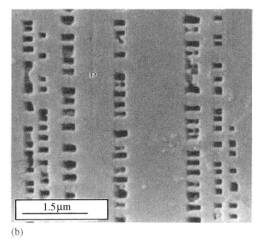

(a)   (b)

Figure 5 Etch pit morphologies: (a) denticulated margins on weathered hornblende formed by side-by-side coalescence of lenticular etch pits, Blue Ridge soil, North Carolina, USA (source Velbel, 1989) and (b) etch pits along a microperthite lamellae, weathered Shap Granite, England (source Lee and Parsons, 1995).

The calculation of mineral weathering rates based on solute concentrations is complicated by the fact that individual solute species are commonly produced by more than one weathering reaction. In granite weathering, for example, aqueous silicon is produced not only from plagioclase weathering (Equation (2)), but also by K-feldspar, biotite, hornblende and quartz dissolution. In addition, secondary minerals such as kaolinite take up aqueous silicon (Equation (2)).

The problem of multi-mineral sources for solutes was initially addressed using a spread sheet approach popularized by Garrels and Mackenzie (1967) in a study of weathering contributions from springs in the Sierra Nevada. This approach has since been incorporated into geochemical computer codes which simultaneously calculate mineral masses. Depending on the number of minerals and solutes considered, nonunique results are commonly generated that require independent confirmation of the actual weathering reactions (Parkhurst and Plummer, 1993). A detailed discussion of the various aspects of mass balance calculations is presented by Bricker *et al.* (see Chapter 5.04).

5.05.3.3.1 *Characterization of fluid transport*

Water is the medium in which weathering reactants and products are transported to and from the minerals undergoing reaction. Clearly, a detailed discussion of hydrologic principles is beyond the scope of this chapter, but several aspects critical to understanding natural weathering will be briefly mentioned.

In the case of diffusive transport, bulk water is immobile and movement of solutes occurs by Brownian motion. Under such conditions, the flux

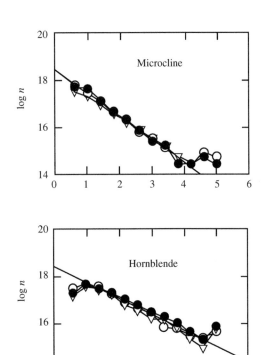

Etch pit diameter W (μm)

Figure 6 Etch pit distributions (PSD) of naturally weathered microcline and hornblende in soils from Illinois, USA (after Brantley *et al.*, 1993).

Q_j of a solute reactant or product is determined by the product of the diffusion coefficient D (m^2 s^{-1}) and the concentration gradient as given by Fick's First Law (Allen and Cherry, 1979)

$$Q_j = -D_j \frac{\partial c_j}{\partial z} \qquad (9)$$

The diffusion coefficients of solutes are commonly on the order of $10^{-9} \, m^2 \, s^{-1}$ making diffusive transport effective only at or near the mineral surface. On a nanometer-scale, solutes diffuse along and across water layers structurally influenced by the silicate surface. Microscopic diffusion occurs through stagnant water contained within internal pore spaces and etch pits or along grain boundaries.

Solutes also moved through the weathering environment along with the advective flow of water. In this case, the solute flux is the product of the solute concentration and the flux density of water q_h, i.e.,

$$Q_j = c_j q_h \qquad (10)$$

where q_h (m s^{-1}) is defined as the product of the hydraulic conductivity K_m (m s^{-1}), and the hydraulic gradient ∇H (m m^{-1}) (Hillel, 1982)

$$q_h = -K_m \nabla H \qquad (11)$$

The above relationships (Equations (9)–(11)) define the basic processes by which solutes are transported during chemical weathering.

5.05.3.3.2 Weathering based on solutes in soils

The number of studies characterizing weathering rates based on soil/regolith pore-water solutes is relatively limited, e.g., Sverdrup (1990) and White et al. (2002). This deficiency is somewhat surprising considering the general accessibility of the soil/regolith environment and the relative ease with which pore waters can be sampled using lysimeters and suction water devices. An example of solute distributions generated by such sampling is shown in Figure 7

for a deep weathering profile in the tropical Luquillo Mountains of Puerto Rico (White et al., 1998).

The upper 0.5 m of this regolith is a dense kaolinitic clay overlying ~10 m of saprolite resting on a quartz diorite bedrock consisting predominately of quartz, plagioclase, hornblende, and biotite. As indicated, significant variations in solute potassium, magnesium, and silicon occur in the upper clay zone due to the effects of evapotranspiration and biological cycling of mineral nutrients. At greater depths (>2 m), this variability is damped with solute species exhibiting consistent increases with depth. Such distributions reflect both the rate at which reaction products are contributed from chemical weathering and the rate of pore-water movement through the soil (Figure 2). Most soil pore-water studies assume one-dimensional vertical advective transport in which progressive increases in solute concentrations with depth reflect increasing reaction times along a single flow path. Flow paths in soils with significant topographic relief or macropore flow are more complex.

Another complexity is that pore-water flow in soils commonly occurs under unsaturated or vadose conditions. This is shown in Figure 8(a) for the Luquillo regolith in which fluid saturation varies between 65% and 95% over a depth of 7 m (White et al., 1998). In such a case, the hydraulic conductivity K_m (Equation (11)) is strongly dependent on the soil moisture content (Stonestrom et al., 1998). Experimental conductivities produced for cores taken from the Luquillo regolith (Figure 8(c)) decrease between 2 and 4 orders of magnitude with less than a 30% decrease in moisture saturation. This is caused by the entrapment of air within the pore spaces, which physically obstructs the movement of water.

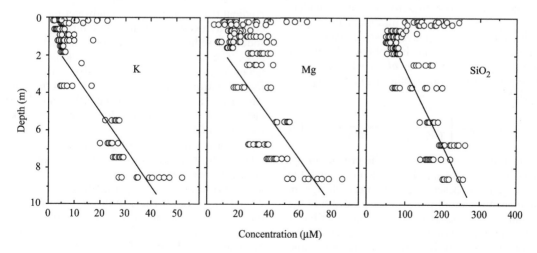

Figure 7 Selected pore-water concentrations as functions of depth in a deeply weathered regolith in the Luquillo Mountains of Puerto Rico. Diagonal solid lines are linear fits to the weathering gradients used by Murphy et al. (1998) to calculate biotite weathering rates (Equation (13)) (after White et al., 1998).

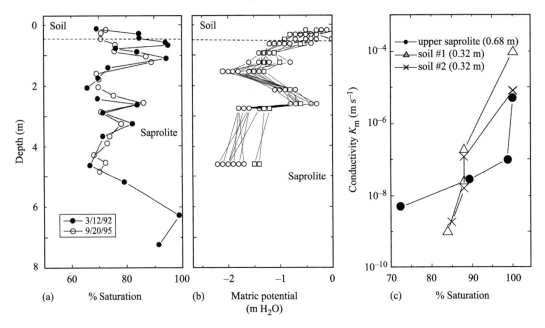

Figure 8 Hydrologic parameters defining pore-water flow in the Luquillo regolith: (a) moisture saturation, (b) matric potential describing capillary tension, and (c) experimental hydrologic conductivities as functions of moisture saturation at several regolith depths (after White *et al.*, 1998).

The hydraulic gradient ∇H in Equation (11) also becomes dependent on water content during unsaturated flow such that (Hillel, 1982)

$$\frac{dH}{dz} = \frac{dh_g}{dz} + \frac{dh_p}{dz} \qquad (12)$$

where dh_g/dz is the gravitational potential and dh_p/dz is the matric potential describing the capillary tension on soil mineral surfaces. The inverse relationship between soil-water saturation and the matric or capillary tension is shown in Figures 8(a) and (b) for the Luquillo regolith. Pore waters close to saturation at shallow depths have pressure heads close to zero. The most under-saturated pore waters at intermediate depths exhibit the greatest negative pressure heads.

Based on the solute and hydrologic information outlined above, Murphy *et al.* (1998) derived a rate equation for biotite weathering in the Luquillo regolith

$$R_{biotite} = \left[\frac{q_h}{\beta V_{biotite} \rho (1 - \phi) S} \right] \frac{dc}{dz} \qquad (13)$$

in which the weathering rate is the product of the water flux density q_h (Equation (11)) and the solute gradient describing the linear change in solute magnesium or potassium with depth in the regolith dc/dz. $V_{biotite}$ is the volume fraction of biotite in the regolith, β is the stiochometric coefficient describing the magnesium and potassium content of the biotite. ρ, ϕ, and S are the respective density, porosity, and BET surface area terms. The resulting biotite weathering rate in

the Luquillo regolith, based on magnesium and potassium pore-water solute distributions, is reported in Table 2 (no. 41).

5.05.3.3.3 Weathering based on solutes in groundwater

Several characteristics distinguish chemical weathering in aquifers from soils. Groundwater weathering occurs under phreatic conditions in which fluid flow is governed by Darcy's law and is proportional to the product of the saturated hydraulic conductivity and the gravitational gradient (Equation (11)). Groundwater movement commonly occurs as matrix flow through uncon-solidated sediments and bedrock or as fracture flow. The saturated hydraulic conductivities are dependent on porosity, tortuosity, and other factors intrinsic to the geologic material and vary over a wide range from 10^{-4} m s^{-1} for clean sand to 10^{-12} m s^{-1} for crystalline rocks (Allen and Cherry, 1979).

The physical isolation of groundwater aquifers relative to that of soils is advantageous for characterizing weathering rates because biologic and short-term climatic perturbations are dimi-nished. However, because of the greater isolation, the characterization of mineral distributions, aquifer heterogeneities and mineral surface areas becomes more difficult. The earliest and simplest characterization of chemical weathering in groundwater was based on spring or seep discharge (Garrels and Mackenzie, 1967). In such cases, the changes in solute concentration due to weathering are assumed to be the difference between the discharge compositions and atmospheric contributions in precipitation (Equation (8)). The fluid flux is the measured spring or seep discharge. The disadvantage of such an approach is that no information on the spatial extent of the weathering environment is known, making normalization of the weathering rate to a surface area impossible.

A more rigorous and intrusive approach to groundwater weathering is to characterize changes in solute concentrations measured in an array of wells in an aquifer. Determination of chemical weathering rates based on this approach requires that changes in solute concentrations occur along defined flow paths and that the rate of groundwater flow is known. This approach is commonly referred to as an inverse problem in which weathering reactions are computationally fitted to account for the observed chemical changes in the groundwater (Parkhurst and Plummer, 1993). Such studies tie the rates of chemical weathering closely to a detailed understanding of fluid flow in an aquifer, which is commonly described by groundwater flow models.

The above approach is illustrated in the determination of chemical weathering rates in groundwaters in glacial sediments in northern Wisconsin, USA (Kenoyer and Bowser, 1992; Bullen *et al.*, 1996). The flow path is defined by subaerial recharge from Crystal Lake moving downgradient across a narrow isthmus and discharging into Big Muskellunge Lake (cross-sections in Figure 9). Water levels and *in situ* slug tests, measured in an array of piezometers, determined the hydraulic conductivities and gradients plotted in Figures 9(a) and (b). Groundwater velocities were determined by application of Darcy's law to these measurements and were checked using tracer tests. The resulting fluid residences in groundwater at various positions in the aquifer were then calculated based on the distance along the flow path (Figure 9(c)).

Groundwater solutes, sampled from the piezometers, exhibited significant trends as shown by silicon and calcium distributions plotted in Figure 10. The core of the groundwater at the upgradient end of the cross-section is recharged directly from Crystal Lake. Significant changes in the chemistry of this plume take place along the flow path due to the dissolution of well-characterized silicate minerals that compose the aquifer. Various proportions of minerals phases were computationally dissolved using the PHREEQE code (Parkhurst and Plummer, 1993) until the best fit to the solute trends along the flow path was obtained. The resulting mineral masses were than divided by the fluid residence time and

estimates of mineral surface area to produce the plagioclase and biotite weathering rates tabulated in Table 2 (nos. 10 and 39).

5.05.3.3.4 Weathering based on surface-water solutes

The composition and concentrations of solutes in surface waters in catchments and rivers are used extensively to characterize silicate weathering rates. As in the case of soils and aquifers, the determination of such rates in watersheds is closely associated with an understanding of the hydrology. Commonly, solute compositions and fluid fluxes are determined at a gauged position on a stream or river. Fluid flow, expressed in terms of stage, is continuously measured, and water samples are periodically taken. Since a strong inverse correlation commonly exists between stage and solute concentration, a number of approaches were developed to produce a continuous estimate of discharge-weighted solute compositions (Likens *et al.*, 1977; Zeman and Slaymaker, 1978). Integrating both the discharge and solute concentrations over a given time interval, commonly on an annual basis, and dividing this output by the geographical area of the watershed produces a solute flux $Q_{watershed}$ (Equation (14)), commonly expressed as $mol\ ha^{-1}\ yr^{-1}$.

The weathering components associated with watershed discharge must be corrected for other potential sources and sinks within the watershed including atmospheric and anthropogenic inputs, biological cycling, and changes in ion exchange processes in the soil. Such a relationship can be represented by the expression (Paces, 1986)

$$Q_{weathering} = Q_{watershed\ output} - Q_{precipitation}$$
$$- Q_{anthropedgenic} \pm Q_{biology}$$
$$\pm Q_{exchange} \qquad (14)$$

Usually, for pristine watersheds, biologic and exchange reactions are assumed to be at steady state and the weathering flux is the difference between the watershed output and the precipitation input. In perturbed watersheds, the calculations become more complex due to agricultural inputs or changes in the exchangeable ions due to watershed acidification.

The measurement of solute fluxes in surface-water discharge is an indirect approach to estimating chemical weathering rates. Due to low mineral-to-fluid ratios and short residence times, minimal silicate weathering occurs in the streambed and the hyporheic environment. Rather, surface-water solutes represent discharges from other weathering environments that are spatially and temporally integrated by the watershed flux.

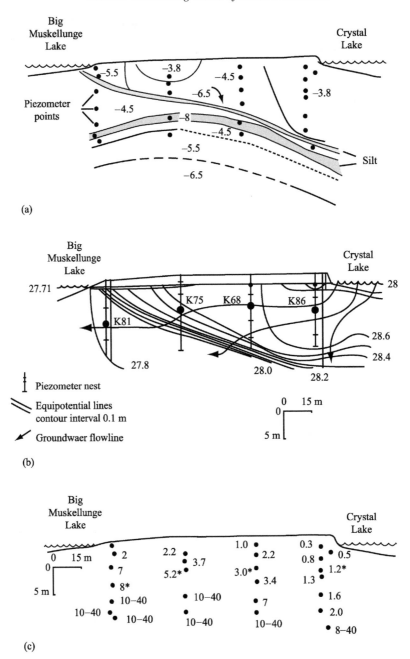

Figure 9 Hydrologic properties of a sandy silicate aquifer in northern Wisconsin: (a) hydrologic conductivities (log K_m, m s^{-1}), (b) groundwater potentials (m), and (c) groundwater residence times (yr) (reproduced by permission of American Geophysical Union from *Water Resources Research*, **1992**, *28*, 579–589).

The presence of multiple solute sources was demonstrated for variations in discharge from the Panola watershed in central Georgia, USA (Hooper *et al.*, 1990). As shown in Figure 11, variations in solute silicon and magnesium, are explained as a mixture of three end-member sources; a groundwater component contained in fractured granite, a hill-slope component representing waters from soils and saprolites and an organic component consisting of near-surface runoff in the shallow soils. To further complicate the interpretation of the discharge flux, the relative proportions of these components vary seasonally. Hill-slope waters dominate during the wet season (January to May), groundwater dominates at low flow during seasonally dry periods, and the organic component is present only during intense storm events. Different weathering processes and rates are associated with each of these environments.

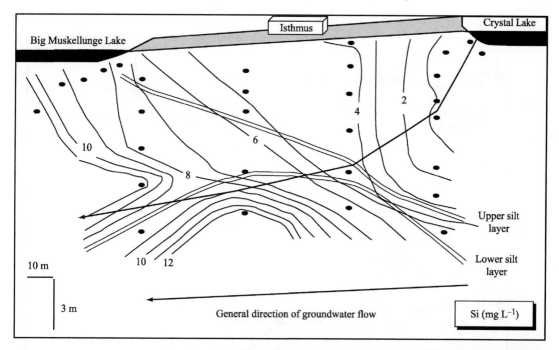

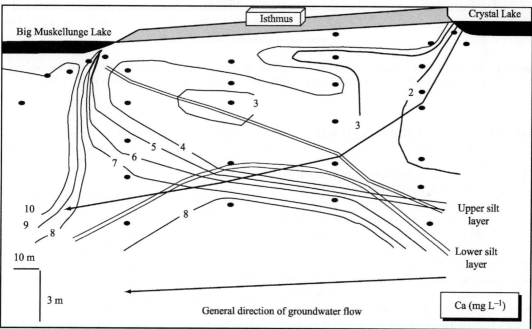

Figure 10 Cross-sections showing the vertical distribution of Si and Ca in groundwater in the North Wisconsin aquifer (after Bullen *et al.*, 1996).

The above example serves as a cautionary note to the complexities involved in interpreting solute discharge fluxes and calculating weathering rates even in an intensely studied watershed. However, significant efforts, such as those contained in Trudgill (1995), are directed to resolving the various hydrologic and geochemical sources of watershed discharge using a number of computational approaches. The compilation of watershed solute fluxes is effective in establishing the impact of rock type, precipitation, temperature, and other parameters on chemical weathering rates, e.g., Dethier (1986), Bluth and Kump (1994), and White and Blum (1995). Such fluxes are also used to calculate silicate weathering rates (Table 2, nos. 9, 10, and 14).

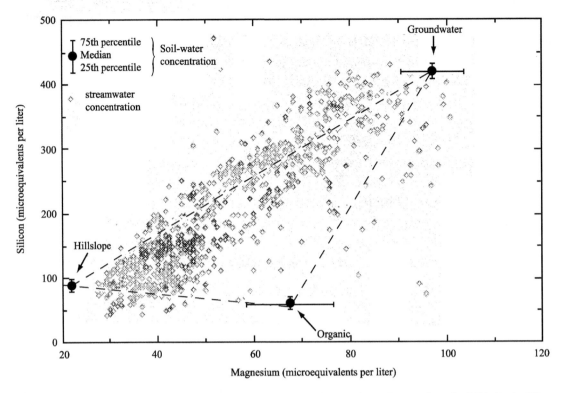

Figure 11 Dissolved silicon and magnesium in discharge from the Panola catchment, Georgia, USA. Dashed lines encompass the hillslope, groundwater, and shallow soil (organic) inputs (after Hooper *et al.*, 1990).

5.05.4 TIME AS A FACTOR IN NATURAL WEATHERING

The weathering rate is inversely proportional to duration of chemical weathering (Equation (1)), i.e., the longer the time required to produce a given change in mass, the slower is the reaction rate. The duration of chemical weathering in natural systems involves very different timescales depending on whether changes in solid-state or solute concentrations are measured. Solid-state mass differences commonly reflect weathering over geologic timescales of thousands to millions of years (t_{solid}). The duration of chemical weathering reflected in solute compositions (t_{solute}) is equivalent to the residence time of the water in the regolith, commonly the time elapsed since initial recharge as precipitation. For most weathering environments such as soils, watersheds, and shallow groundwaters, solute residence times range from days to decades.

Recent advances have significantly increased our ability to quantitatively establish the duration of weathering associated with both solid-state and solute mass changes. Of particular importance is the use of cosmogenic isotopes such as ^{36}Cl, ^{26}Al, and ^{10}Be in age dating geomorphic surfaces and the use of ^{36}Cl, $^{3}He/^{4}He$, and chloroflurocarbons

to establish fluid residence times over time spans of years to tens of thousands of years (see Chapter 5.15).

5.05.4.1 Comparison of Contemporary and Geologic Rates

Rates based on solute compositions are measurements of present-day weathering that correlate with contemporary environmental conditions including hydrology, climate, and vegetation. Weathering rates based on solid-state composition changes reflect average rates integrated over geologic time during which conditions influencing weathering can only be indirectly estimated. However, under steady-state conditions, assuming constant surface area, the respective weathering rates R_{solid} and R_{solute} must be equal and proportional to the respective mass changes with time, i.e.,

$$R_{solid} = R_{solute} = \frac{\Delta M_{solid}}{t_{solid}} = \frac{\Delta M_{solute}}{t_{solute}} \quad (15)$$

Some authors have documented that present-day base cation fluxes measured in North America and Northern Europe watersheds are significantly faster than long-term past weathering rates due to the impacts of acidic precipitation

(April *et al.*, 1986; Kirkwood and Nesbitt, 1991; Land *et al.*, 1999). Likewise, Cleaves (1993) found watershed solute discharge in the Piedmont of the eastern USA, to be 2–5 times faster than for past periglacial periods represented by long-term saprolite weathering. He attributed these weathering rate differences to past periods of lower precipitation, colder temperatures and lower soil gas CO_2.

In contrast, Pavich (1986) concluded that long-term weathering rates of saprolites in the Virginia Piedmont are comparable to current-day weathering rates based on stream solute fluxes. Rates of saprolite formation are also found to be similar to current weathering rates in the Luquillo Mountains of Puerto Rico (White *et al.*, 1998) and in the Panola watershed in northern Georgia, USA (White *et al.*, 2002). As the literature indicates, considerable debate exists regarding the extent to which contemporary fluxes correspond to increased weathering or result from other contributions such as enhanced cation exchange rates, deforestation and other parameters contained in Equation (14).

5.05.4.2 Utilization of Soil Chronosequences

A particularly valuable approach to investigating the effect of time on chemical weathering has been the use of soil chronosequences which are defined as a group of soils that differ in age and, therefore, in duration of weathering. These soils have similar parent materials and have formed under similar climatic and geomorphic conditions (Jenny, 1941). Individual soils, therefore, provide "snapshots" of the progressive nature of chemical weathering with time.

Chronsequences include marine and alluvial deposits and lava flows. Studies of chronosequences (e.g., Bockheim, 1980; Birkland, 1984; Harden, 1987; Jacobson *et al.*, 2002) have quantified systematic changes in weathering properties of soils with ages ranging from 1 kyr to 4,000 kyr. Mass balance studies have characterized mineralogical changes (Mokma *et al.*, 1973; Mahaney and Halvorson, 1986), chemical changes (Brimhall and Dietrich, 1987; Merritts *et al.*, 1992), and development of etch pitting (Hall and Horn, 1993) as functions of soil age. An important conclusion of these chronosequence studies is that the rates of primary silicate weathering generally decrease with duration of weathering.

5.05.5 NORMALIZATION OF WEATHERING TO SURFACE AREA

The mass losses or gains resulting from weathering are normalized to the surface area in the rate equation (Equation (1)). On a fundamental level, the rate of reaction is directly proportional to the reactive surface area, which defines the density of reactive sites on a silicate surface at which hydrolysis reactions occur. While such site distributions are characterized in terms of ligand exchange sites and dislocation densities under laboratory conditions (Blum and Lasaga, 1991; see Chapter 5.03), natural weathering studies generally equate reactive surface area with the physical surface area of the weathering environment.

5.05.5.1 Definitions of Natural Surface Areas

Surface areas in natural weathering studies are geographic, volumetric, or mineral specific. The weathering rate based on geographic surface area is defined as the mass of silicate mineral weathered per unit area of Earth's surface. Generally, this surface area is not corrected for aspect or slope of the terrain. Geographic surface areas are used extensively to describe weathering rates based on solute discharge in watersheds, which are normalized to basin areas ranging from small experimental catchments to the world's rivers (see Chapters 5.09 and 5.08).

Rates normalized to geographic surface area contain no information on the vertical dimension of the weathering environment, i.e., there is no distinction between weathering in a soil that is one meter or ten meters thick. One approach in overcoming this problem is to employ a volumetric surface area, which is commonly done for groundwater systems. Weathering rates are defined either in terms of the volume of the aquifer or of the groundwater. Commonly, surface areas contained in this volume are estimated based on the distributions of fracture surfaces or on sizes and distributions of pore spaces (Paces, 1973; Gislason and Eugster, 1987).

The more fundamental approach to addressing the physical dimensions involved in weathering is to characterize the surface areas of the individual minerals, i.e., the specific mineral surface area S ($m^2 g^{-1}$). The extent to which this specific surface area scales directly with the reactive surface areas in natural environments is a matter of considerable debate, particularly in regard to the accessibility of water. For unsaturated environments, such as those in most soils, the wetted surface area may be considerably less than the physical surface area of contained mineral grains (Drever and Clow, 1995). In addition, surface areas of microscopic features such as external pits and internal pores may be associated with stagnant water that is thermodynamically saturated and not actively involved in weathering reactions (Oelkers, 2001).

5.05.5.2 Measurements of Specific Surface Areas

The scale of the measurement technique operationally defines the specific surface area of a mineral. Geometric estimates of surface area S_{Geo} involve grain size analyses with physical dimensions commonly on the scale of millimeters to centimeters. Brunauer, Emmett, and Teller (BET) surface measurements S_{BET} utilize isotherms describing low temperature sorption of N_2 or argon on mineral surfaces. The scale of BET measurements, defined by atomic dimensions, is on the order of nanometers. As expected, specific surface areas determined by geometric techniques produce significantly smaller values of physical surface area than do BET measurements because the former technique does not consider microscopic irregularities on the external surface nor the internal porosity of mineral grains that have undergone significant weathering. Recently developed techniques such as atomic force microscopy (Maurice *et al.*, 2002) and interferometry (Luttge *et al.*, 1999) have the potential to span the gap between the scales of geometric and BET surface area measurements.

BET surface areas are almost universally measured in experimental weathering studies using freshly ground and prepared minerals (see Chapter 5.03). BET surface areas are also commonly measured on samples of natural unconsolidated soils, saprolites, and sediments. Alteration and destruction of the physical fabric of such materials by disaggregation is required for such measurements but produces uncertainties in the reported data. In addition, secondary clay and oxyhydroxides commonly coat natural silicate surfaces and contribute to the formation of mineral-organic aggregates (Tisdall, 1982; Brantley *et al.*, 1999; Nugent *et al.*, 1998). When not removed, these coatings can produce erroneously high surface areas. Alternatively, removing these phases, using mechanical and chemical methods, may expose additional silicate surfaces that are not present in the natural weathering environment (White *et al.*, 1996).

In spite of the above issues, consistent trends of increasing surface area with increasing intensity of natural weathering are observed (Brantley *et al.*, 1999). An example of this increase is shown in Figure 12(a) in which BET surfaces of primary minerals increase with increasing age of soils in the Merced chronosequence (White *et al.*, 1996). The extent of this increase depends on the specific mineral phase. The more readily weathered aluminosilicates exhibit greater surface area increases than quartz. Application of BET measurements to characterize fractures and porosity in consolidated rocks has remained generally untested; these surface area estimates have been

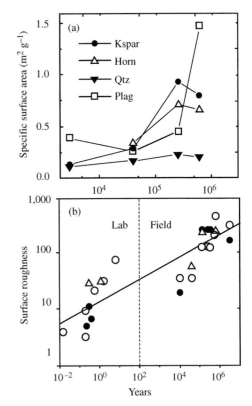

Figure 12 Surface area parameters: (a) surface area of primary silicate minerals contained in soils of the Merced chronosequence (after White *et al.*, 1996). (b) Surface roughness of silicate minerals as functions of the duration of experimental and natural weathering (source White and Brantley, 2003).

confined to geometric approaches (Paces, 1973; Gislason and Eugster, 1987).

5.05.5.3 Surface Roughness

Surface roughness λ is defined as the ratio of the geometric (S_{Geo}) to the BET (S_{BET}) surface areas (Helgeson, 1971). For a perfectly smooth surface, without internal porosity, the two surface area measurements should be the same, i.e., $\lambda = 1$. For nonideal surfaces of geometric spheres, the roughness can be related directly to the particle diameter d and the mineral density ρ such that

$$\lambda = \frac{S_{BET}}{S_{GEO}} = \frac{\rho \, d s_{BET}}{6} \quad (16)$$

White and Peterson (1990) calculated an average roughness factor of $\lambda = 7$ for a wide size range of fresh silicate surfaces. Surface roughness generally increases during weathering as shown in Figure 12(b) for a compilation of surface roughnesses measured for experimentally and naturally reacted silicate minerals (after White and Brantley, 2003).

5.05.6 TABULATIONS OF WEATHERING RATES OF SOME COMMON SILICATE MINERALS

The diverse approaches described above have produced a large number of published weathering rates that are not always comparable in terms of units and dimensions. In general, weathering rates can be classified as being element or mineral specific. In the following sections, weathering rates based on geographic and volumetric surface areas are referred to as weathering fluxes, and weathering based on specific surface areas are referred to as mineral specific weathering rates.

5.05.6.1 Elemental Fluxes

The most prominent examples of element-based rates are those reported for annual stream or river discharge and normalized to the watershed area. The relatively straightforward nature of these measurements, coupled with a significant amount of watershed research related to land management issues, has produced a large database describing such fluxes. A tabulation of these data is beyond the scope of the present paper and the reader is referred to previous compilations (Dethier, 1986; Meybeck, 1979; Bluth and Kump, 1994; White and Blum, 1995).

Watershed fluxes are equivalent to chemical denudation commonly defined as the sum of rock-derived base cations solubilized per unit area of geographic surface (Bluth and Kump, 1994; Drever and Clow, 1995). These rates serve as a means to evaluate the importance of a number of environmental controls on chemical weathering including precipitation, temperature, vegetation, and rock type. These topics are addressed briefly in later sections of this chapter as well as elsewhere in this volume (see Chapters 5.06 and 5.08).

5.05.6.2 Mineral Fluxes

The alternative approach is to report silicate weathering fluxes on a mineral specific basis. As previously discussed, this requires distributing elemental fluxes among mineral phases using mass balance approaches described in this chapter and elsewhere in this volume (see Chapter 5.04).

Mineral flux data in the literature are less common than element-based data. Table 3 contains a tabulation of weathering rates for the common silicate minerals based on average annual watershed fluxes and geographic areas. Significantly more data are available for plagioclase feldspar than for other mineral phases, which partly reflects its common occurrence in crystalline rocks. In addition, the rate of plagioclase weathering is the easiest to calculate, because of the conventional assumption that it scales directly with the sodium discharge flux after correction for

Table 3 Weathering rates for selected silicate minerals based on average annual watershed discharge.

Watershed	Rate[a]	An[b]	References
Plagioclase			
Emerald Lake, CA, USA	44	0.24	Williams *et al.* (1993)
Loch Vale Watershed, CO, USA	86	0.27	Mast *et al.* (1990)
Gigndal Watershed, Norway	100	0.30	Frogner (1990)
Silver Creek, ID, USA	102	0.19	Clayton (1986)
Lapptrasket Basin, Sweden	133	0.00	Andersson-Calles and Ericksson (1979)
Pond Branch, MD, USA	148	0.22	Cleaves *et al.* (1970)
Velen Basin, Sweden	148	0.00	Andersson-Calles and Ericksson (1979)
Rio Parana, Brazil	174	0.49	Benedetti *et al.* (1994)
Kassjoan Basin, Sweden	181	0.00	Andersson-Calles and Ericksson (1979)
Plastic Lake, Canada	185	0.34	Kirkwood and Nesbitt (1991)
Filson Creek, MN, USA	235	0.64	Siegal and Pfannkuch (1984)
Catoctin Mtn., MI, USA	316	0.00	Katz (1989)
K-feldspar			
Pond Branch, MD, USA	11	na	Cleaves *et al.* (1970)
Silver Creek, ID, USA	44	na	Clayton (1986)
Plastic Lake, Canada	69	na	Kirkwood and Nesbitt (1991)
Biotite			
Lapptrasket Basin, Sweden	3	na	Andersson-Calles and Ericksson (1979)
Loch Vale Watershed, CO, USA	26	na	Mast *et al.* (1990)
Hornblende			
Lapptrasket Basin, Sweden	2	na	Andersson-Calles and Ericksson (1979)
Plastic Lake, Canada	8	na	Kirkwood and Nesbitt (1991)
Emerald Lake, CA, USA	24	na	Williams *et al.* (1993)

[a] $mol\,ha^{-1}\,yr^{-1}$.　[b] Mole fraction anorthite.

precipitation inputs. Weathering rates of other silicates are more difficult to determine from elemental fluxes because their major components are commonly derived from more than one mineral.

The reported weathering rates in Table 3 are normalized to the geographic surface area of the watershed but not to the relative proportions or surface areas of the specific mineral present. The presence of relatively large amounts of plagioclase in most igneous rocks accounts for the fact that the rates of plagioclase weathering, calculated on a watershed basis, are more rapid than those of other silicates such as hornblende, which is counter to Goldichs's order of weatherability (Goldich, 1938).

5.05.6.3 Specific Mineral Rates

Natural weathering rates, normalized to specific mineral surface areas, are reported in Table 2 for a number of common silicate minerals (after White and Brantley, 2003). Also included in the Table 2 are the approximate pH ranges, ages of the weathering environment and parameters defining the physical surface area of the minerals.

Large variations are apparent in the reported weathering rates for each of the minerals. For example, reported plagioclase rates vary from 4×10^{-17} mol m^{-2} s^{-1} in the Davis Run saprolite in Virginia (White *et al.*, 2001) to 1.5×10^{-12} mol m^{-2} s^{-1} in the Plastic Lake watershed in Ontario, Canada (Kirkwood and Nesbitt, 1991). Rates based on geometric surface areas are more rapid than those based on BET measurements. This correlation, in part, explains age trends in the rate data. Rates for younger weathering environments are based primarily on watershed fluxes and geometric surface area estimates. In nearly all cases, these watersheds are developed on glaciated topographies developed over the last 12 kyr. In contrast, weathering rates in older soils, varying in ages from 10 kyr to 3,000 kyr, are based on BET surface measurement techniques.

5.05.6.4 Normalizing Rate Data

One approach for overcoming the effect of different methods of surface area measurement is to normalize all the weathering rates by a surface roughness factor λ (Equation (16)), which converts geometric-based rates to their BET equivalent. This is done by statistically describing the relationship between increasing surface roughness and time for data presented in Figure 12(b) such that (White and Brantley, 2003)

$$\lambda = 1.13t^{0.182} \qquad (17)$$

The resulting roughness-normalized weathering rates are in parentheses to the right of reported rates based on geometric surface areas (Table 2).

This approach is an obvious simplification, because surface morphologies and roughnesses developed during weathering are mineral specific (Figure 5). In addition, controversy exists whether weathering rates are directly dependent on surface roughness. However, considering the large discrepancies between rates calculated by geometric versus BET derived surface areas, normalizing to an estimated surface roughness factor is warranted. It is required in any comparison with experimental weathering rates, which are almost exclusively normalized to BET surface areas (see Chapter 5.03).

After correction for surface roughness, a significant trend persists between decreasing weathering rates and increasing weathering duration for the rate data in Table 2. This is shown for the plagioclase and K-feldspar data plotted on log–log scales in Figure 13. Also included on the

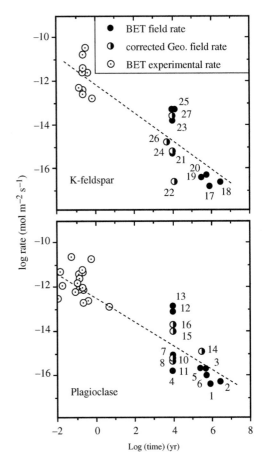

Figure 13 Relationship between weathering rate and reaction time for plagioclase and alkali feldspars. Labels for field rates correspond to those in Table 2. Experimental rates are tabulated in White and Brantley (2003).

plots are experimental dissolution rates of fresh feldspars summarized by White and Brantley (2003). Experimental rates are orders of magnitude faster than natural rates, an observation that has been frequently made in the weathering literature (Schnoor, 1990; Brantley, 1992; Velbel, 1993; Drever and Clow, 1995).

Regression fits to the plagioclase and K-feldspar data (dashed lines in Figure 13) produce the respective power rate laws

$$R_{plag} = 12.4t^{-0.54}, \quad r = 0.01$$
$$R_{K-spar} = 12.1t^{-0.54}, \quad r = 0.83$$ (18)

that describe the progressive decrease in plagioclase and K-feldspar weathering with time. Equation (18) is similar to a relationship developed by Taylor and Blum (1995) to describe the decrease in chemical fluxes in a suite of soils with time and is comparable to parabolic kinetics describing transport control in the hydration of obsidian and the development of rock weathering rinds (Coleman, 1981; Friedman et al., 1994).

5.05.7 FACTORS INFLUENCING NATURAL WEATHERING RATES

Weathering rates are dependent on a number of factors that can be classified as either intrinsic or extrinsic to a specific mineral (White and Brantley, 2003). Intrinsic properties are physical or chemical characteristics such as mineral composition, surface area, and defect densities. If intrinsic properties dominate weathering, such characteristics should be transferable between environments, e.g., laboratory and field rates of the same mineral should be comparable. Extrinsic features reflect environmental conditions external to the silicate phase that impact chemical weathering such as solution composition, climate, and biological activity. These processes are dependent on external environmental conditions that are difficult to recreate fully under laboratory simulations.

5.05.7.1 Mineral Weatherability

Mineral composition and structure are the primary intrinsic factors controlling weathering rates. Based on early weathering studies, Goldich (1938) observed that the weathering sequence for common igneous rocks in the field was the reverse of Bowen's reaction series that ranked minerals in the order of crystallization from magma. Amphiboles and pyroxenes are expected to weather faster than feldspars which weather faster than

quartz. In addition, field data suggest that volcanic glasses weather an order of magnitude faster than crystalline minerals of comparable composition (Gislason et al., 1996).

Although no fundamental method exists for predicting the weathering order of all silicate minerals, certain trends in rates between minerals are apparent. Silicate weathering is commonly viewed as a ligand exchange process with the metal ions bonded in the mineral structure. The rates are dependent on the relative strengths of coordinated metal ions within the mineral structure relative to the strength of the metal ligand bond. Casey and Westrich (1992) showed that for simple compositional series, such as olivines, the dissolution rate decreased with the increase in the metal valence state. Evidence also indicates that the relative weathering rates of multi-oxides such as feldspars approximate the relative dissolution rates of the single oxide components (Oelkers, 2001). A more detailed discussion of the relationship between weathering mechanisms and rates is presented elsewhere in this volume (see Chapter 5.03).

For structurally complex minerals undergoing incongruent or stepwise weathering in the natural environment, the relative rates become very dependent on specific reaction pathways. In the case of micas, for example, biotite and muscovite are structurally similar but weather at very different rates as shown by residual biotite and muscovite compositions in the saprolite developed on the Panola Granite (White et al., 2002). Biotite exhibits a progressive loss of potassium and increase in the Al/Si ratio that correlates with increasing weathering intensity and decreasing depth in the profile (Figure 14(a)). These changes are driven by the relatively rapid oxidation of ferrous iron in biotite (Amonette et al., 1988). As shown by the electron backscatter image in Figure 14(b), this reaction produces epitaxial replacement of biotite by kaolinite. In contrast, the muscovite, which is structurally similar, but contains low concentrations of iron, weathers very slowly, exhibiting no potassium loss (Figure 14(a)). The dramatic differences in the behavior of mica minerals emphasize the importance of mineral-specific reactions in the weathering environment.

Mineral surfaces are heterogeneous substrates possessing compositional and structural features that may differ from those of the bulk mineral phase (Hochella et al., 1991). Compositional differences within single mineral grains due to zonation and exsolution can influence reaction rates. As predicted by the Bowen reaction sequence, the crystallization of plagioclase phenocrysts often produces calcic cores surrounded by more sodic rims. During the reverse process of chemical weathering, the plagioclase cores weather more rapidly, producing preferential

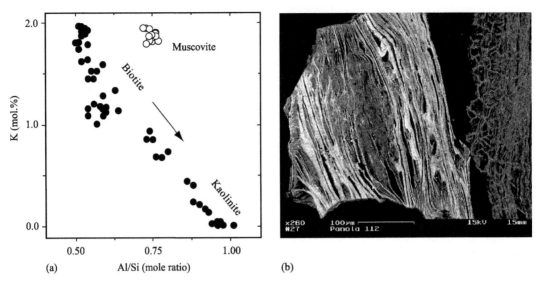

Figure 14 Mica weathering in a saprolite profile from central Georgia, USA. (a) changes in K and Al/Si mole ratios in coexisting biotite and muscovite and (b) SEM backscatter photomicrograph of residual biotite grains at 2.8 m depth. Bright layers in the left grain are unaltered biotite and gray areas are kaolinite replacement. Right grain is completely kaolinized biotite with surficial boxwork structure (after White *et al.*, 2002).

release of solute calcium relative to sodium (Clayton, 1986). Such incongruent dissolution is a process that has been invoked to explain the commonly observed excess solute calcium to sodium ratios in watershed fluxes relative to that predicted from the bulk plagioclase stoichiometry (Stauffer, 1990).

Incongruent weathering also occurs at exsolution features. Detailed SEM studies of perthitic textures of naturally weathered feldspars by Lee and Parsons (1995) showed that the etch density was strongly dependent on the composition of these exsolution features, i.e., the sodium component dissolved more rapidly than the potassium component (Figure 5(b)). The highest densities of defect structures are also created along the strained lamellae. Weathering rates of specific minerals may therefore be dependent on the crystallization history of igneous rocks as well as the effects of tectonics and deformation on defect formation.

Weathering acts on mineral surfaces by decreasing the overall surface free energy by selectively dissolving more soluble components and attacking structural defects and dislocations. With continued weathering, the overall surface energy and the corresponding weathering rate may decrease, as observed for the feldspar data plotted in Figure 13. The progressive occlusion of the reactive mineral substrate by secondary clays and iron and aluminium oxides, the formation of depleted leached layers and the adsorption of organic compounds, have also been proposed as mechanisms to decrease surface reactivity with time (Banfield and Barker, 1994; Nugent *et al.*, 1998).

Whether these secondary coatings represent effective barriers to the transport of reactants and products from the reactive silicate surface is still actively debated.

5.05.7.2 Solute Chemistry and Saturation States

Of the extrinsic factors influencing natural weathering rates, solute compositions have the most direct impact. Chemical weathering is ultimately dependent on the concentration of reactants that complex with and detach the oxygen-bonded metal species in the silicate structure (Casey and Ludwig, 1995). As documented by numerous experimental studies, the principal species involved in this reaction are hydrogen ions, although other complexing agents such as organic anions can participate in this process (see Chapter 5.03). In contrast, some solute species, such as aluminum and sodium ions, inhibit experimental weathering rates by interfering with and competing with the ligand exchange processes (Oelkers and Schott, 1995; Stillings *et al.*, 1996; see Chapter 5.03).

The direct effect of aqueous species on natural weathering is difficult to demonstrate. For example, a summary of experimental findings indicates that feldspar weathering rates decrease about two orders of magnitude over a pH range of 2–6 (Blum and Stillings, 1995; also see Chapter 5.03). However, when this effect is looked for in natural systems, the evidence is ambiguous. Over a 25 yr period, the pH of solute discharge from Hubbard Brook has increased from pH 4.1 to 4.4 due to a gradual recovery from acid deposition (Driscoll *et al.*, 1989). However, solute silicon concentrations, a good indicator of silicate weathering, have remained essentially unchanged (Figure 15). Although, these results do not disprove the effect of hydrogen ion on weathering rates, as demonstrated experimentally, it does serve to show that other processes sometimes overwhelm expected results for natural systems.

Experimental rates of silicate dissolution decrease as solutions approach thermodynamic equilibrium (Burch *et al.*, 1993; Taylor *et al.*, 2000). The saturation state Ω is defined as the product of the solute activities (IAP) divided by the saturation constant of the specific mineral K_s and is related to the net free energy of reaction ΔG (kJ mol^{-1}) by the relationship

$$\Omega = \exp\left(\frac{\Delta G}{R'T}\right) \quad (19)$$

where T is the temperature and R' is the gas constant. When the system is undersaturated, i.e., Ω is <1, ΔG is negative, leading to dissolution. At a $\Omega = 1$, saturation is achieved and $\Delta G = 0$. If $\Omega > 1$, the system is supersaturated, ΔG is positive and precipitation may occur.

Based on transition-state theory, the relationship between the reaction rate and the solute saturation state is represented as (Nagy *et al.*, 1991)

$$R = k[1 - \Omega^m]^n \quad (20)$$

where k is the intrinsic rate constant and m and n are orders of reaction. In strongly under-saturated solutions (large negative values of ΔG), net detachment reactions dominate over attachment reactions, the dissolution rate is independent of saturation state and R is directly proportional to the intrinsic rate constant. At near saturation (small values of ΔG), Equation (20) reduces to a form in which R becomes linearly dependent on ΔG. At equilibrium ($\Delta G = 0$) the weathering rate is zero. Although the relationship between

decreasing reaction rates and approach to equilibrium (Equation (20)) has been verified for a few minerals at low pH and elevated temperature, no comparable studies have been carried out under conditions more representative of natural weathering.

Calculations involving mineral saturation state (Ω in Equations (19) and (20)) are dependent on accurate characterization of the thermodynamic states of the reactants and products and are commonly calculated using speciation codes (see Chapter 5.02 for additional information). Although the equilibrium constants for most simple primary silicates have been determined, thermodynamic data do not exist for many complex silicates or for solid solutions.

White (1995) found that the apparent thermodynamic supersaturation of silicate minerals in most soil pore waters resulted from excessive values for total dissolved aluminum. In reality, much of this aluminum is complexed with dissolved organics in shallow soils and does not contribute to the thermodynamic saturation state of silicate minerals. Solubility calculations involving low dissolved organic concentrations in deeper soil horizons and in groundwater appear to produce much clearer equilibrium relationships (Paces, 1972; Stefansson and Arnorsson, 2000; Stefansson, 2001).

Natural weathering involves much longer times and smaller solute/solid ratios than occur in experimental studies. Such weathering may therefore occur much closer to thermodynamic equilibrium. This difference in saturation states may explain two common discrepancies observed when comparing experimental and natural weathering rates. As has been frequently noted by others and documented in detail in the Figure 13, natural weathering rates are commonly several orders magnitude slower than experimental rates. Further, the relative weatherability of minerals in the natural environment often differs from that predicted on the basis of experimental dissolution rates. For example, in a review of experimental feldspar dissolution studies, Blum and Stillings (1995) concluded that under neutral to acidic pH conditions, the rates of sodic plagioclase and K-feldspar dissolution were essentially the same. However, K-feldspar is commonly much more resistant to weathering than is plagioclase during natural weathering (Nesbitt *et al.*, 1997).

Weathering of the Panola Granite is an extreme example of the relative weathering rates of feldspars (White *et al.*, 2001). As shown by the mass balance calculations (Figures 3 and 4), kaolinization of plagioclase occurs to depths of 10 m in the granite regolith (Figure 4). This is evident in SEM images (Figure 16(a)) in which, except for residual rims, plagioclase grains are

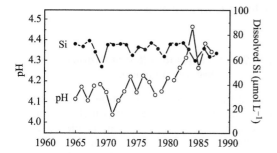

Figure 15 Variations in mean annual pH and dissolved silica concentration in Hubbard Brook, New Hampshire, USA (after Driscoll *et al.*, 1989).

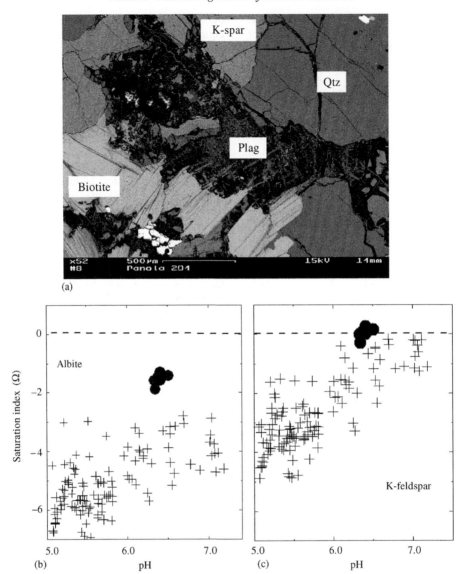

(a)

(b)   (c)

Figure 16 Primary silicate weathering in the Panola regolith: (a) backscatter SEM micrograph of weathered granite at a depth 7.5 m. The center grain is a plagioclase phenocryst almost completely replaced by kaolinite (dark region). Minor residual plagioclase remains on outside rims of grain. Saturation states (log Ω) of soil pore water (+) and groundwater (●) relative to (b) plagioclase (albite) and (c) K-feldspar (after White *et al.*, 2001).

completely replaced by kaolinite. In contrast, K-feldspar in the immediate proximity of plagioclase grains remains pristine and unaffected by weathering.

The significant difference in feldspar weathering rates at depth in the Panola was explained by White *et al.* (2001) by the difference in their respective thermodynamic saturation states (Figures 16(b) and (c)). Soil pore waters are generally undersaturated with respect to both plagioclase and K-feldspar (Figures 16(b) and (c)). In contrast, groundwater in the underlying granite bedrock is saturated with K-feldspar but undersaturated with more soluble plagioclase. Equation (20) predicts that plagioclase will

weather in contact with groundwater while K-feldspar will remain stable.

5.05.7.3 Coupling the Effect of Hydrology and Chemical Weathering

The thermodynamic saturation state is dependent on solute concentrations, which in turn, are controlled by the volume and residence times of fluids moving through the weathering environment. The roles of hydrology and chemical weathering are commonly coupled in a weathering regolith and can be viewed as co-evolving as the intensity of the weathering increases. This effect is evident both in the initial phases of chemical weathering in bedrock environments and development of argillic horizons in older soils.

5.05.7.3.1 Initial stages of weathering

The kaolinization of plagioclase at depth in the Panola granite is clearly an example of the initial stage of chemical weathering in which other primary mineral phases remain unreacted (Figure 16(a)). This process was modeled by White *et al.* (2001) as a coupled relationship between weathering rates and the development of secondary permeability as shown in Figure 17. At the initiation of weathering, the permeability of the fresh granite is extremely low, placing severe constraints on the fluid flux q_p and the mass of feldspar that can dissolve before becoming thermodynamically saturated. Under such conditions, weathering is limited by the availability of water and not by the kinetic rate of feldspar weathering.

Over long times, slow rates of transport-limited weathering occur, resulting in mass loss from the granite (Figure 3(b)). Based on density estimates, the conversion of plagioclase to kaolinite (Equation (2)) produces a porosity increase of ~50%. This change in porosity slowly increases the flux density of water (Equation (11)), plotted as the ratio of q_s/q_p in Figure 17. Increased

pore-water flow accelerates saturation-limited weathering and produces greater porosity and even higher fluid fluxes. This coupled feedback accelerates plagioclase weathering, which gradually shifts from a transport limited to a kinetic limited reaction (Figure 17). At a q_s/q_p ratios >150, plagioclase weathering becomes completely controlled by kinetics and no longer reflects additional increases in conductivity.

The rate of K-feldspar weathering shows a comparable transition from transport to kinetic control but at significant higher flux ratios due to its lower solubility and not to slightly slower reaction kinetics (Figure 17). Concurrent plagioclase dissolution enhances this effect by producing solutes, principally silicon, which further suppresses K-feldspar dissolution by increasing the saturation state. Nahon and Merino (1997) and Soler and Lasaga (1996) present additional discussions of the role of solute transport and solubility on textures and mineral distributions in saprolites and other regoliths.

5.05.7.3.2 Late-stage weathering

Weathering in older soils produces decreasing permeabilities due to *in situ* secondary mineral formation and the development of hard pans and argillic horizons. Such zones of secondary clay and iron-oxides are clearly evident in the increased aluminum and iron at a depth of a meter in the Panola regolith (Figure 3). Low permeabilities in such features are commonly related to the absence of continuous pores due to the formation of thick cutans of clay (O'Brien and Buol, 1984). This process is enhanced by physical translocation and collapse of saprolite structures (Torrent and Nettleton, 1978).

Argillic horizons often correlate with the maximum depth of effective evapotranspiration, commonly 1–2 m. Water loss initiates the precipitation of secondary clays and oxides from solutes. Such precipitation may also be related to the loss of dissolved organic species and the de-complexation of soluble aluminum. The resulting low permeabilities further retard the downward percolation of pore water, commonly creating transient perched water tables directly above the hardpans. Periodic drying in this zone focuses addition secondary mineral precipitation in the vicinity of the hardpan, which then leads to a lower permeability and more clay formation.

The progressive development of such argillic features with increasing weathering intensity is commonly documented in soil chronosequence studies (Harden, 1990). An example of the development of argillic horizons is shown for the Merced chronosequenence in Figure 18 (White *et al.*, 2003). Over a time span of

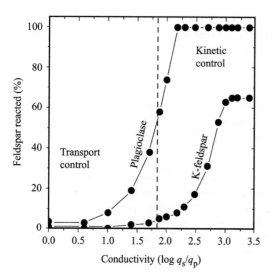

Figure 17 The percent of plagioclase and K-feldspar reacted as a function of the ratio of secondary to primary flux densities of water (q_s/q_p) in the Panola Granite. The vertical dashed line is the approximate flux ratio at which plagioclase is completely weathered and the K-feldspar weathering has not yet commenced. This condition evident in thin section in Figure 16 (after White *et al.*, 2001).

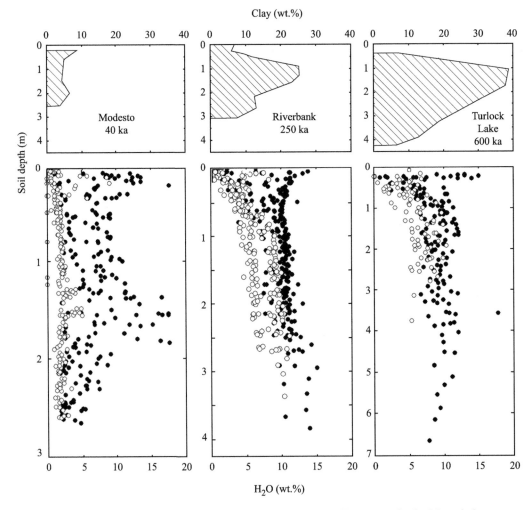

Figure 18 Correlation between clay and water contents of soils of different ages in the Merced chronosequence, California. The closed circles are soil moistures measured during the wet season (December to May) and open circles are data from the dry season (June to November) (after White *et al.*, 2003).

40–600 kyr, the maximum clay content in these horizons increases from ~5% to ~40%. As expected, the development of these intense argillic horizons has a profound effect on the water content within and below these features.

The Merced site has a Mediterranean climate with pronounced semiannual wet and dry precipitation cycles. This seasonal variability is delineated in the soil moisture contents by the open and closed symbols in Figure 18. In the youngest soil, with a poorly developed argillic horizon, the climate signal is propagated to the base of the profile. In the older soils, with increasingly more developed argillic zones, this seasonal signal becomes progressively lost, reflecting decreases in the infiltration rate of pore waters through soil. In these older soils, solubility calculations indicate that relatively long pore-water residence times result in solutes that are thermodynamically saturated with

respect to the primary feldspars, a situation leading to progressively slower weathering as a function of soil age (Table 2).

The preceding examples indicate that regolith development is coupled in terms of weathering and hydrology. Ultimately, the impact of chemical weathering on hydrology is related to changes in secondary porosity and permeability, which, in turn, increase or decrease the degree of thermodynamic saturation and the weathering rate.

5.05.7.4 Role of Climate on Chemical Weathering

The term "weathering" implies that chemical weathering is related to climate. This relationship is important both from the standpoint of the potential long-term feedback during much of

Earth's history as well as the current impacts on changes in nutrient cycling produced by present-day CO_2 inputs to the atmosphere. The importance of climate relative to other controls on weathering, in particular, topography and tectonics remains controversial (Ruddiman, 1997, and references therein). Examples of contrasting results comparing weathering rates in small-scale catchments are contained in White and Blum (1995), who found that climate was the most important control, whereas Riebe *et al.* (2001) determined that topography and physical erosion were the dominant factors.

A definitive study would involve a weathering environment that has undergone consistent sustained climate change. However, the long-term data required for such a study are not available and surrogate weathering studies comparing spatially separated climatic regimes are used. The utility of such comparisons depends on the ability to isolate the effects of climate from other variables influencing weathering including lithology, geomorphology, and vegetation. Some of these parameters often correlate with climate, making the isolation of individual variables difficult. Physical erosion generally increases with precipitation, exposing fresher mineral surfaces and increasing weathering rates. Likewise, higher rainfall may produce greater plant productivity, increased soil CO_2 and higher dissolved organic concentration, all of which tend to vary systematically with climate.

The ability to isolate climate effects decreases as the scale of the weathering process increases. For example, a number of studies comparing weathering rates in soils and small catchments have found a significant climate effect (Velbel, 1993; White and Blum, 1995; Dessert *et al.*, 2001). In contrast, comparison of solute concentrations and fluxes originating from large scale river systems commonly fail to detect a climate signature (Edmond *et al.*, 1995; Huh *et al.*, 1998).

5.05.7.4.1 Temperature

The effect of temperature on weathering is the easiest climate parameter to predict on a fundamental basis. The rates of most chemical reactions, including silicate hydrolysis, increase exponentially with temperature according to the Arrhenius expression. This relationship can be represented as the ratio of reaction rates R/R_0 at different temperatures T and T_0 (K) (Brady and Carroll, 1994):

$$\frac{R}{R_0} = \exp\left[\frac{E_a}{R'}\left(\frac{1}{T_0} - \frac{1}{T}\right)\right] \quad (21)$$

The activation energy E_a (kJ mol^{-1}K^{-1}) of most common silicate minerals ranges between 50 kJ

and 120 kJ (Brady and Carroll, 1994; Casey and Sposito, 1992; White *et al.*, 1999). Equation (21) predicts that rates should increase by about an order of magnitude between 0 °C and 25 °C, the temperature range encountered in most natural weathering. If temperature were the only variable in weathering, the effect should be readily observable when comparing weathering environments at substantially different temperatures.

Temperature effects on the rate of chemical weathering are observed in well-characterized environments. Velbel (1993) estimated elevation-dependent temperature differences in the Coweeta watershed in North Carolina, USA (10.6–11.7 °C) and calculated an activation energy of 77 kJ mol^{-1} for plagioclase weathering. Dorn and Brady (1995) used plagioclase porosity formed by etch pitting in Hawaiian basalt flows at different elevations and temperatures (12.5–23.3 °C) to calculate an activation energy of 109 kJ mol^{-1} (Figure 19(a)). Dessert *et al.* (2001) calculated an E_a of 42 kJ mol^{-1} for basalt weathering (2–27 °C) from a compilation of river fluxes (Figure 19(b)). Studies of larger river systems have tended to discount the importance of temperature as a significant control on weathering (Edmond *et al.*, 1995).

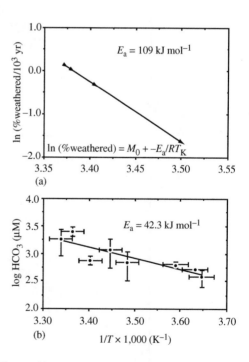

Figure 19 Arrhenius relationship between temperature and rates for basalt weathering: (a) rates based on porosity increases in plagioclase from Hawaiian basalt flows (Dorn and Brady, 1995) and (b) rates based on solute bicarbonate fluxes from a global distribution of watersheds underlain by basalt (Dessert *et al.*, 2001).

5.05.7.4.2 *Precipitation and recharge*

A number of studies have observed a linear correlation between precipitation or runoff and solute fluxes or chemical denudation rates (Dunne, 1978; Dethier, 1986; Stewart *et al.*, 2001). The relationship between solute silicon fluxes and runoff of the large watershed data set plotted in Figure 20 shows comparable relationships for granitic and basaltic rock types (Bluth and Kump, 1994).

As pointed out by Drever and Clow (1995), mineral surfaces undergo weathering only in the presence of liquid water. Increases in precipitation increase regolith recharge and the wetted surface areas of minerals, thereby promoting increased weathering. In addition, increased water flow increases the rates at which reactants, such as dissolved CO_2 and organic acids, are flushed through the system. Finally, increased precipitation dilutes the solute concentrations of pore

waters in the regolith, decreasing thermodynamic saturation.

5.05.7.4.3 *Coupling climate effects*

In order to explain anomalously rapid rates of chemical weathering in upland topical watersheds such as in Puerto Rico and Malaysia, White and Blum (1995) and White *et al.* (1999) proposed that a coupled climate effect in which the solute fluxes were proportional to the product of a linear precipitation function and an exponential temperature function such that

$$Q = (aP)\exp\left[-\frac{E_a}{R}\left(\frac{1}{T} - \frac{1}{T_0}\right)\right] \quad (22)$$

Equation (22) was fitted to data for well-characterized upland catchments underlain by granitoid rocks. A three-dimensional plot

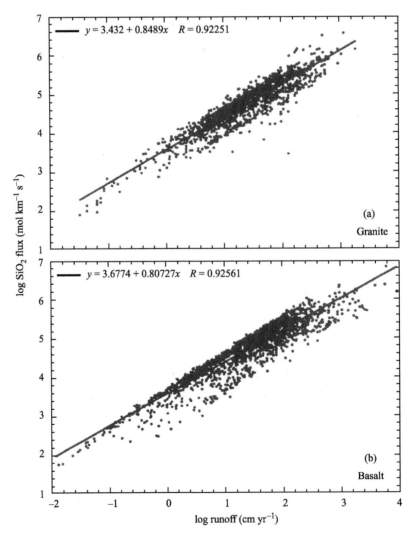

Figure 20 Relationship between Si fluxes and runoff from watersheds draining: (a) granitic and (b) basaltic watershed (after Bluth and Kump, 1994).

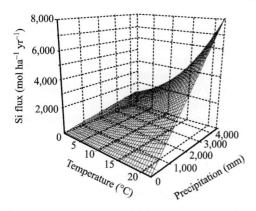

Figure 21 Three-dimensional surface representing the optimized fit of watershed Si fluxes to Equation (22) as functions of precipitation and temperature (after White and Blum, 1995).

generated for silicon fluxes is shown in Figure 21. The net predicted effect is to reinforce weathering in watersheds with both high temperatures and precipitation and to decrease weathering fluxes in watersheds with low temperature and precipitation. This conclusion is in agreement with Meybeck (1994), who has shown that maximum concentrations of dissolved species from chemical weathering occur in rivers draining humid mountainous terrains and that minimum fluxes occur in arid and arctic rivers.

5.05.7.5 Role of Physical Weathering

Chemical weathering, in combination with physical erosion, are the processes that produce global element cycling. R. A. Berner and E. K. Berner (1997) estimated that the combined denudation rate of the continents is 252 km yr^{-1} while the average chemical denudation rate is ~20% of that value. As expected, the absolute rates and relative ratios of physical to chemical weathering are strongly related to differences in topography. For example, the chemical denudation rate is less than 10% of the physical denudation rate for watersheds of high relief such as the Ganges River, which drains the Himalayas, whereas it approaches 45% of the physical denudation rate for lowland rivers such as the Congo.

5.05.7.5.1 Transport versus chemical weathering regimes

A important relationship between physical and chemical weathering was proposed by Carson and Kirby (1972) and Stallard and Edmond (1983), who differentiated mineral selectivity in

terms of weathering-limited and transported-limited weathering regimes. In the weathering-limited case, mechanical erosion is faster than chemical weathering. Therefore the most reactive phases will be available for weathering. An example is the oxidation of ferrous-containing biotite to oxybiotite with the release of interlayer potassium. This rapid weathering reaction is documented by high potassium fluxes in present-day glacial watersheds in which large amounts of fresh rock are exposed by physical denudation (Andersson et al., 1997; Blum and Erel, 1997; also see Chapter 5.07). Such excess potassium is not observed in watersheds that are geomorphically older.

Under transport-limited weathering, the amount of fresh rock available to weathering is limited. Chemical weathering is faster then physical weathering and available minerals ultimately contribute to the solute load in proportion to their abundance in the bedrock. Such is the case for the nearly complete destruction of alumino-silicates from old laterites and saprolites in which base cations are all effectively removed from the regolith.

5.05.7.5.2 Physical development of regoliths

In addition to controlling the relative proportions of minerals available for weathering, physical erosion also has a profound effect on the physical development of weathering environments. Regoliths can be viewed as aggrading, steady-state, or degrading systems depending on whether their weathering thicknesses increase, remain constant or decrease with time due to physical erosion. These conditions control the position of the weathering gradient shown in Figure 2.

An example of an aggrading regolith is one for which the present topographic surface represents the original depositional surface. A lack of physical erosion on flat alluvial or marine terraces can be verified by the concentrations of cosmogenic isotopes (Pavich et al., 1986; Perg et al., 2001) and is an important basis for using soil chronosequences in weathering studies. In terms of the definitions of Stallard and Edmond (1983) such as system would shift progressively from a chemically-limited to a transport-limited regime as primary minerals become more isolated from the surficial weathering environment.

A degrading weathering environment is one in which rates of physical erosion exceed the rate of chemical weathering and the regolith thickness decreases. Such decreases may result from natural catastrophic events such as landslides in tropical terrains or as a result of human influences such as deforestation, which result in increased rates of erosion (Larsen and Concepicon, 1998). Under such conditions, the weathering environment will shift from weathering limited to transport limited.

As discussed by Stallard (1995), given enough time, a dynamic equilibrium may be reached between the chemical and physical weathering rates. This situation has been documented for thick saprolite sequences, as developed at Panola and elsewhere in the southeastern USA. In these cases, the physical and chemical denudation rates are comparable (Pavich, 1986). Under such a "conveyor belt" scenario, the regolith depth will remain constant as additional rock is chemically weathered at depth and soils are physically removed at the surface. The position of the weathering gradient, defining the change in chemistry and mineralogy, will remain constant with depth (Figure 2).

5.05.8 WEATHERING RATES IN GEOCHEMCIAL MODELS

As discussed by Warfvinge (1995), the incorporation of weathering kinetics into geochemical codes can be discussed in terms of the degree to which rates are coupled to other geochemical and hydrologic processes. A significant number of models have been developed to explain and predict the impact of acid precipitation on watershed solutes. The reaction rates in such models are commonly coupled to hydrogen ion concentration by the relationship

$$R = k \cdot [H^+]^n \qquad (23)$$

where the rate constant k is a fitted parameter or else is determined from empirical or experimental data. With the exponential term $n = 0$, the weathering rate is independent of pH, as is assumed in the MAGIC code (Cosby *et al.*, 1985). At other powers of n, the rate becomes exponentially dependent on pH, as in the SMART code (De Vries and Kros, 1989) where $n = 0.5$. Such models have been applied in a number of detailed studies of watershed fluxes and have served as a basis for environmental regulation of acid emissions.

Another class of models relevant to chemical weathering is based on the reaction path approach originally developed by Helgeson *et al.* (1970). EQ3/EQ6 (Wolery *et al.* (1990), PHREEQC (Parkhurst and Plummer, 1993), and PATHARC.94 (Gunter *et al.*, 2000) are some codes currently used to describe the progressive reaction of primary silicates and the precipitation of secondary phases as a function of time and mass. These codes are discussed in Nordstrom (see Chapter 5.02). They commonly permit the introduction of user-defined silicate reaction rates. Such models also commonly consider solubility controls on reaction kinetics as defined by

Equation (20). Such models generally assume that reactions occur across a constant surface area and in a given volume of water.

A final group of more computationally complex codes consider additional factors pertinent to rates of chemical weathering. The PROFILE and SAFE codes (Sverdrup and Warfvinge, 1995) describe the evolution of solute compositions during weathering in soils and catchments and require detailed site-specific data including the number and thickness of soil horizons, bulk density, moisture content, and surface areas. These codes use an internal database describing the kinetic weathering rates of common silicate minerals. The rate coefficients are dependent on base cation and aluminum concentrations and the degree of saturation based on transition state theory. Fluid transport is one dimensional with flow dependent on relative inputs and outputs from precipitation and evapotranspiration unless otherwise specified by independent means. Surface areas vary with the degree of surface wetting depending on moisture saturation. A critical evaluation of the PROFILE code is presented by Hodson *et al.* (1997).

Other models directly couple chemical reaction with mass transport and fluid flow. The UNSATCHEM model (Suarez and Simunek, 1996) describes the chemical evolution of solutes in soils and includes kinetic expressions for a limited number of silicate phases. The model mathematically combines one- and two-dimensional chemical transport with saturated and unsaturated pore-water flow based on optimization of water retention, pressure head, and saturated conductivity. Heat transport is also considered in the model. The IDREAT and GIMRT codes (Steefel and Lasaga, 1994) and Geochemist's Workbench (Bethke, 2001) also contain coupled chemical reaction and fluid transport with input parameters including diffusion, advection, and dispersivity. These models also consider the coupled effects of chemical reaction and changes in porosity and permeability due to mass transport.

An example of the results generated by the last type of coupled code is that of Soler and Lasaga (1996), who used the IDREAT to model bauxite formation from a granite protolith. As mentioned in the introduction, bauxite formation is an example of an important economic resource produced by long-term chemical weathering. The main objective of the study was to reproduce the mineral succession, and especially the thickness of the transition zone containing gibbsite and kaolinite by making use of kinetic reaction rates for the primary phases albite, microcline quartz, and phlogopite. The calculations also made use of laboratory-based pH and saturation-dependent rate laws as well as field-based estimates of dispersion, diffusion, and hydraulic conductivity.

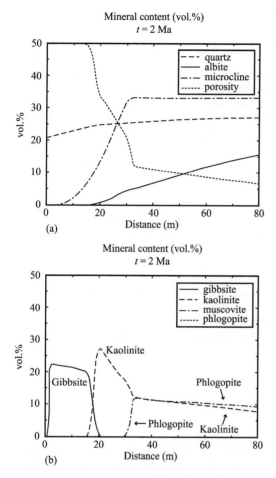

Mineral content (vol.%)
$t = 2$ Ma

(a)

Mineral content (vol.%)
$t = 2$ Ma

(b)

Figure 22 A 1DREACT computer simulation of: (a) residual primary silicate distributions and secondary porosity and (b) secondary mineral distributions with depth for Los Pijiguaos bauxite deposit, Venezuela after 2 Ma years of weathering (after Soler and Lasaga, 1998).

A sample of the calculation (Figure 22) shows the resulting primary and secondary mineral distributions in the Los Pijiguaos bauxite deposit, Venezuela after 2 Myr of weathering. The simulation was run with a nucleation barrier for muscovite precipitation. Due to solubility constraints, albite is selectively lost and microcline preserved in the profile, which is consistent with results observed by White *et al.* (2001) at Panola (Figures 16 and 17). The succession and relative thickness of the gibbsite and kaolinite were shown to be dependent on the kinetic rates at which the various mineral phases dissolve and precipitate. The increase in regolith porosity, as the granite is converted to saprolite, is also demonstrated.

5.05.9 SUMMARY

This paper reviews the chemical, physical, and hydrologic processes that control silicate mineral weathering rates at the Earth's surface. Quantitative rates of weathering are important in understanding reaction mechanisms and in addressing a number of economic and environment issues. Silicate weathering rates are fluxes that describe mass transport across an interface over a given time interval.

Mass change, defined in terms of elements, isotopes, or mineral abundances is determined from either solid-state or solute compositions. Solid-state mass changes in a bulk regolith must be corrected for the mobility of all components as well as volume and density changes due to compaction and bioturbation. This correction is commonly done by normalizing the concentrations of reactive species against an element or mineral considered to be resistant to chemical weathering. Solid-state weathering is also characterized on microscopic levels based on weathering rind thickness and mass losses due to porosity and etch pit development.

The calculation of rates based on changes in solute species concentrations in soils, aquifers, and watersheds requires partitioning the reactant between sources produced by primary mineral dissolution and sinks created by secondary mineral precipitation. Calculation of weathering rates based on solute transport requires knowing the nature and rate of fluid flow through soils, aquifers, and watersheds.

The time-period over which weathering is being evaluated is very different depending upon whether changes in solid-state or solute concentrations are measured. Solid-state mass differences reflect weathering over geologic timescales, whereas solute compositions reflect the residence time of the water. These mass losses or gains are normalized to surface area. Rates of reaction are directly proportional to the density of reactive sites on a silicate surface. Natural rates equate this density with the physical surface area of the weathering environment, generally defined on a geographic, volumetric, or mineral-specific basis. Specific surface areas are either geometric estimates based on particle size dimensions or are based on BET gas sorption isotherms, which include surface areas associated with microscopic roughness and internal porosity. The difference in the scales of these measurements produces major differences in surface-normalized mineral weathering rates.

The majority of rate studies are based on watershed solute fluxes normalized to catchment area and are equivalent to rates of chemical denudation. Previous efforts have tabulated these rates, which have proved to be valuable in evaluating the importance of a number of environmental controls on chemical weathering including precipitation, temperature, vegetation, and rock type. The present chapter summarizes literature data for the rates of weathering of several common silicate minerals based on geographic and specific surface areas. In the latter case, differences between geometric and BET-based surface areas are removed by normalizing the rates against an estimated time-dependent surface area roughness factor. Results confirm that a significant correlation exists between decreases in weathering rates and increases in weathering duration.

Factors affecting weathering rates can be categorized as either intrinsic or extrinsic to a specific silicate mineral. The primary intrinsic factors controlling weathering rates are mineral composition and structure. Reaction rates are also affected by compositional differences within single-mineral grains, such as zonation and exsolution features, and by secondary surface coatings and leached layers.

Of the extrinsic factors influencing natural weathering rates, solute composition has the most direct impact. Solutions provide reactants that complex with and detach the oxygen-bonded metal species in the silicate structure. In addition, reaction rates depend on the thermodynamic saturation state of the dissolving phase. Unlike most experimental studies, natural weathering commonly involves long times and small solute/solid ratios, producing solutes close to thermodynamic saturation. This difference in saturation states explains, in part, why natural weathering rates are commonly orders magnitude slower than experimental rates and why the relative weatherability of minerals in the natural environment is often different from that predicted by experimental dissolution rates.

The origin of the term "weathering" implies that chemical weathering is related to climate. However, the exact nature of this relationship remains an area of significant controversy. Studies comparing weathering rates in soils and small catchments often find a significant climate effect, whereas solute concentrations and fluxes originating from large river systems often fail to detect a climate signature. This discrepancy is related to the difficulty of isolating the effects of climate from other often co-dependent parameters, a difficulty that increases as the scale of the weathering processes increases. An exponential relationship between reaction rate and temperature is observed in some weathering environments. Several studies have also documented linear increases in weathering with precipitation due to increases in wetted surface areas and thermodynamic reaction affinities.

As expected, the ratio of physical to chemical weathering is strongly related to differences in topography. An important relationship between physical weathering and chemical weathering is summarized in terms of weathering-limited and transport-limited weathering regimes. In addition to controlling the relative proportions of minerals available for weathering, physical erosion also has a profound effect on the physical development of the weathering environment. Regoliths can be viewed as aggrading, steady-state, or degrading systems, depending on whether their thickness increases, remains constant, or decreases with time due to physical erosion.

The incorporation of weathering kinetics into geochemical codes is classified in terms of the level at which the rates are coupled to other geochemical and hydrologic processes. A significant number of models explain and predict the impact of acid precipitation on watershed solutes. Another class of models is based on a reaction-path approach, which models the progressive loss of primary silicates and the precipitation of secondary phases as functions of time and mass. A final group of more computationally complex codes considers additional factors pertinent to rates of chemical weathering, including fluid and heat transport and the coupling of chemical reaction to changes in porosity and permeability.

REFERENCES

Allen R. and Cherry J. A. (1979) *Groundwater*. Prentice Hall, Englewood Cliffs, NJ.

Amonette J., Ismail F. T., and Scott A. D. (1988) Oxidation of biotite by different oxidizing solutions at room temperature. *Soil Sci. Am. J.* **49**, 772–777.

Andersson-Calles U. M. and Ericksson E. (1979) Mass balance of dissolved inorganic substances in three representative basins in Sweden. *Nordic Hydrol.* **28**, 99–114.

Andersson S. P., Drever J. I., and Humphrey N. F. (1997) Chemical weathering in glacial environments. *Geology* **25**, 399–402.

April R., Newton R., and Coles L. T. (1986) Chemical weathering in two Adirondack watersheds: past and present-day rates. *Geol. Soc. Am. Bull.* **97**, 1232–1238.

Banfield J. F. and Barker W. W. (1994) Direct observation of reactant–product interfaces formed in natural weathering of exsolved, defective amphibole to smectite: evidence for episodic, isovolumetric reactions involving structural inheritance. *Geochim. Cosmochim. Acta* **58**, 1419–1429.

Barth T. F. (1961) Abundance of the elements, aerial averages and geochemical cycles. *Geochim. Cosmochim. Acta* **23**, 1–8.

Belt T. (1874) *The Naturalist in Nicaragua*. University of Chicago press, Chicago.

Benedetti M. F., Menard O., Noack Y., Caralho A., and Nahon D. (1994) Water–rock interactions in tropical catchments: field rates of weathering and biomass impact. *Chem. Geol.* **118**, 203–220.

Berner R. A. and Berner E. K. (1997) Silicate weathering and climate. In *Tectonic Uplift and Climate Change* (ed. W. F. Ruddiman). Plenum, NY, pp. 353–364.

Bethke C. M. (2001) The GEOCHEMIST WORKBENCH: a users guide to Rxn, ACT2, Tact, React and Gtplot. University of Illinois, Urbana.

Birkland P. W. (1984) Holocene soil chronofunctions, Southern Alps, New Zealand. *Geoderma* **34**, 115–134.

Blum A. E. and Lasaga A. C. (1991) The role of surface speciation in the dissolution of albite. *Geochim. Cosmochim. Acta* **55**, 2193–2201.

Blum A. E. and Stillings L. L. (1995) Feldspar dissolution kinetics. In *Chemical Weathering Rates of Silicate Minerals*. Reviews in Mineralogy, 31 (eds. A. F. White and S. L. Brantley). Mineralogical Society of America, Washington, DC, pp. 176–195.

Blum J. D. and Erel Y. (1997) Rb–Sr isotope systematics of a granitic soil chronosequence: the importance of biotite weathering. *Geochim. Cosmochim. Acta* **61**, 3193–3204.

Bluth G. S. and Kump L. R. (1994) Lithologic and climatic controls of river chemistry. *Geochim. Cosmochim. Acta* **58**, 2341–2359.

Bockheim J. G. (1980) Solution and use of chronofunctions in studying soil development. *Geoderma* **24**, 71–85.

Brady P. V. and Carroll S. A. (1994) Direct effects of CO_2 and temperature on silicate weathering: possible implications for climate control. *Geochim. Cosmochim. Acta* **58**, 1853–1863.

Brantley S. L. (1992) Kinetics of dissolution and precipitation—experimental and field results. In *Water–Rock Interaction WRI7* (eds. Y. K. Kharaka and A. S. Maest). A. A. Balkema, Rotterdam, pp. 3–6.

Brantley S. L., Blai A. C., Cremens D. L., MacInnis I., and Darmody R. G. (1993) Natural etching rates of feldspar and hornblende. *Aquat. Sci.* **55**, 262–272.

Brantley S. L., White A. F., and Hodson M. E. (1999) Surface areas of primary silicate minerals. In *Growth, Dissolution, and Pattern Formation in Geosystems* (eds. B. Jamtveit and P. Meakin). Kluwer Academic, Dordrecht, pp. 291–326.

Brimhall G. H. and Dietrich W. E. (1987) Constitutive mass balance relations between chemical composition, volume, density, porosity, and strain in metasomatic hydrochemical systems: results on weathering and pedogenesis. *Geochim. Cosmochim. Acta* **51**, 567–587.

Bullen T. D., Krabbenhoft D. P., and Kendal C. (1996) Kinetic and mineralogic controls on the evolution of groundwater chemistry and $^{87}Sr/^{86}Sr$ in a sandy silicate aquifer, northern Wisconsin, USA. *Geochim. Cosmochim. Acta* **60**, 1807–1821.

Burch T. E., Nagy K. L., and Lasaga A. C. (1993) Free energy dependence of albite dissolution kinetics at 80°C, and pH 8.8. *Chem. Geol.* **105**, 137–162.

Carson M. A. and Kirby M. J. (1972) *Hillslope, Form, and Process*. Cambridge University Press, Cambridge.

Casey W. H. and Ludwig C. (1995) Silicate mineral dissolution as a ligand exchange reaction. In *Chemical Weathering Rates of Silicate Minerals*, Reviews in Mineralogy, 31 (eds. A. F. White and S. L. Brantley), Mineralogical Society of America, Washington, DC, pp. 241–294.

Casey W. H. and Sposito G. (1992) On the temperature dependence of mineral dissolution rates. *Geochim. Cosmochim. Acta* **56**, 3825–3830.

Casey W. H. and Westrich H. R. (1992) Control of dissolution rates of orthosilicate minerals by divalent metal-oxygen bonds. *Nature* **355**, 157–159.

Chadwick O. A., Brimhall G. H., and Hendricks D. M. (1990) From black box to a grey box: a mass balance interpretation of pedogenesis. *Geomorphology* **3**, 369–390.

Chamberlin T. C. (1899) An attempt to frame a working hypothesis of the cause of glacial periods on an atmospheric basis. *J. Geol.* **7**, 545–584.

Clayton J. L. (1986) An estimate of plagioclase weathering rate in the Idaho batholith based upon geochemical transport rates. In *Rates of Chemical Weathering of Rocks and Minerals* (eds. S. Coleman and D. Dethier). Academic Press, Orlando, pp. 453–466.

Cleaves A. T., Godfrey A. E., and Bricker O. P. (1970) Geochemical balance of a small watershed and its geomorphic implications. *Geol. Soc. Am. Bull.* **81**, 3015–3032.

Cleaves E. T. (1993) Climatic impact on isovolumetric weathering of coarse-grained schist in the northern Piedmont province of the central Atlantic states. *Geomophology* **8**, 191–198.

Clow D. W. and Drever J. I. (1996) Weathering rates as a function of flow through an alpine soil. *Chem. Geol.* **132**, 131–141.

Coleman S. M. (1981) Rock-weathering rates as functions of time. *Quat. Res.* **15**, 250–264.

Cosby B. J., Wright R. F., Hornberger G. M., and Galloway J. N. (1985) Modelling the effects of acid deposition: assessment of a lumped parameter model for soil water and stream chemistry. *Water Resour. Res.* **21**, 51–63.

Cremeens D. L., Darmody R. G., and Norton L. D. (1992) Etch-pit size and shape distribution on orthoclase and pyriboles in a loess catena. *Geochim. Cosmochim. Acta* **56**, 3423–3434.

De Vries W. and Kros J. (1989) Simulation of the long term soil response to acid deposition in various buffer ranges. *Water Air Soil Pollut.* **48**, 349–390.

Dessent C., Dupre B., Francois L., Schott J., Gaillard J., Chakrapani G., and Bajpai S. (2001) Erosion of Deccan traps determined by river geochemistry: impact on global climate and the $^{87}Sr^{86}Sr$ ratio of seawater. *Earth Planet. Sci. Lett.* **188**, 459–474.

Dethier D. P. (1986) Weathering rates and the chemical flux from catchment in the Pacific northwest USA. In *Rates of Chemical Weathering of Rocks and Minerals* (eds. S. Coleman and D. Dethier). Academic Press, Orlando, pp. 503–530.

Dorn R. (1995) Digital processing of back-scatter electron imagery: a microscopic approach to quantifying chemical weathering. *Geol. Soc. Am. Bull.* **107**, 725–741.

Dorn R. I. and Brady P. V. (1995) Rock-based measurement of temperature-dependent plagioclase weathering. *Geochim. Cosmochim. Acta* **59**, 2847–2852.

Drever J. I. and Clow D. W. (1995) Weathering rates in catchments. In *Chemical Weathering Rates of Silicate Minerals*, Reviews in Mineralogy (eds. A. F. White and S. L. Brantley), Mineralogical Society of America, Washington, DC, pp. 463–481.

Driscoll C. T., Likens G. E., Hedlin L. O., Eaton J. S., and Bormann F. H. (1989) Changes in the chemistry of surface waters. *Environ. Sci. Technol.* **23**, 137–142.

Dunne T. (1978) Rates of chemical denudation of silicate rocks in tropical catchments. *Nature* **274**, 244–246.

Edmond J. M., Palmer M. R., Measures C. I., Grant B., and Stallard R. F. (1995) The fluvial geochemistry and denudation rate of the Guayana shield in Venezuela, Colombia, and Brazil. *Geochim. Cosmochim. Acta* **59**, 3301–3325.

Farley D. A. and Werritty A. (1989) Hydrochemical budgets for the Loch Dee experimental catchments, southwest Scotland (1981–1985). *J. Hydrol.* **109**, 351–368.

Fordham A. W. (1990) Weathering of biotite into dioctahedral clay minerals. *Clay Min.* **25**, 51–63.

Friedman I., Trembour F. W., Smith F. L., and Smith G. I. (1994) Is obsidian hydration dating affected by relative humidity? *Quat. Res.* 185–190.

Frogner T. (1990) The effect of acid deposition on cation fluxes in artificially acidified catchments in western Norway. *Geochim. Cosmochim. Acta* **54**, 769–780.

Garrels R. M. and Mackenzie F. T. (1967) Origin of the chemical composition of some springs and lakes. In *Equilibrium Concepts in Natural Water Systems*, Adv. Chem. Series 67 (ed. W. Stumm). American Chemical Society, Washington, DC, pp. 222–242.

Garrels R. M. and Mackenzie F. T. (1971) *Evolution of Sedimentary Rocks*. W. W. Norton, NY.

Gislason S. R. and Eugster H. P. (1987) Meteoric water-basalt interactions: II. A field study in NE Iceland. *Geochim. Cosmochim. Acta* **51**, 2841–2855.

Gislason S. R., Arnorsson S., and Armannsson H. (1996) Chemical weathering of basalt in southwest Iceland: effects of runoff, age of rocks and vegetative/glacial cover. *Am. J. Sci.* **296**, 837–907.

Goldich S. S. (1938) A study of rock weathering. *J. Geol.* **46**, 17–58.

Gunter W. D., Perkins E. H., and Hutcheon I. D. (2000) Aquifer disposal of acid gases: modeling of water–rock reactions for trapping of acid wastes. *Appl. Geochem.* **15**, 1085–1095.

Hall R. D. and Horn L. L. (1993) Rates of hornblende etching in soils in glacial deposits of the northern Rocky mountains (Wyoming-montana, USA): influence of climate and characteristics of the parent material. *Chem. Geol.* **105**, 17–19.

Harden J. W. (1987) Soils developed in granitic aavium near Merced, California. *US Geol. Surv. Bull.* **1590-A**, 121p.

Harden J. W. (1990) Soil development on stable landforms and implications for landscape studies. *Geomorphology* **3**, 391–398.

Hartt C. F. (1853) *Geologia e geografia fidsca do Brasil.* Companhia ditoria Nacional.

Helgeson H. C. (1971) Kinetics of mass transfer among silicates and aqueous solutions. *Geochim. Cosmochim. Acta* **35**, 421–469.

Helgeson H. C., Brown T. H., Nigrini A., and Jones T. A. (1970) Calculation of mass transfer in geochemical processes involving aqueous solutions. *Geochim. Cosmochim. Acta* **24**, 569–592.

Hillel D. (1982) *Introduction to Soil Physics.* Academic Press, Orlando.

Hochella M. F., Jr., Eggleston C. M., Johnsson P. A., Stipp S. L., Tingle T. N., Blum A. E., White A. F., and Fawcett J. J. C. (1991) Examples of mineral surface structure, composition, and reactivity observed at the molecular and atomic levels. In *Program with Abstracts—Geological Association of Canada, Mineralogical Association of Canada, Canadian Geophysical Union, Joint Annual Meeting* **16**, 56, Ontario.

Hodson M. E. (2002) Experimental evidence for the mobility of Zr and other trace elements in soils. *Geochim. Cosmochim. Acta* **66**, 819–828.

Hodson M. E., Langan S. J., and Wilson M. J. (1997) A critical evaluation of the use of the PROFILE model in calculating mineral weathering rates. *Water Air Soil Pollut.* **98**, 79–104.

Hooper R. P., Christophersen N., and Peters N. E. (1990) Modelling stream water chemistry as a mixture of soil water end-members-an application to the Panola mountain catchment, Georgia, USA. *J. Hydrol.* **116**, 321–343.

Huntington T. G., Hooper R. P., Johnson C. E., Aulenbach B. T., Cappellato R., and Blum A. E. (2000) Calcium depletion in forest ecosystems of the southeastern United States. *Soil Sci. Soc. Am. J.* **64**, 1845–1858.

Huh Y., Panteleyev G., Babich D., Zaitsev A., and Edmond J. (1998) The fluvial geochemistry of the rivers of eastern Siberia: II. Tributaries of the Lena, Omloy, Yana, Indigirka, Kolyma, and Anadyr draining the collisional/accretionary zone of the Verkhoyansk and Cherskiy ranges. *Geochim. Cosmochim. Acta* **62**, 2053–2075.

Jacobson A. D., Blum J. D., Chanberlain C. P., Poage M. A., and Sloan V. F. (2002) Ca/Sr and Sr isotope systematics of a Himalayan glacial chronosequence: carbonate versus silicate weathering rates as a function of landscape surface age. *Geochim. Cosmochim. Acta* **66**, 13–27.

Jenny H. (1941) *Factors of Soil Formation.* McGraw-Hill, NY.

Johnsson M. J., Ellen S. D., and McKittrick M. A. (1993) Intensity and duration of chemical weathering: an example from soil clays of the southeastern Koolau mountains, Oahu, Hawaii. *Geol. Soc. Am. Spec. Publ.* **284**, 147–170.

Katz B. G. (1989) Influence of mineral weathering reactions on the chemical composition of soil water, springs, and groundwater, Catoctin mountains, Maryland. *Hydrol. Process.* **3**, 185–202.

Kenoyer G. J. and Bowser C. J. (1992) Groundwater evolution in a sandy silicate aquifer in northern Wisconsin: 1. Patterns and rates of change. *Water Resour. Res.* **28**, 579–589.

Kirkwood S. E. and Nesbitt H. W. (1991) Formation and evolution of soils from an acidified watershed: Plastic Lake, Ontario, Canada. *Geochim. Cosmochim. Acta* **55**, 1295–1308.

Land M., Ingri J., and Ohlander B. (1999) Past and present weathering rates in northern Sweden. *Appl. Geochem.* **14**, 761–774.

Larsen M. C. and Concepicon I. M. (1998) Water budgets of forested and agriculturally developed watersheds in Puerto Rico: proceedings, tropical hydrology and Caribbean water resources. In *Proc. American Water Resources Association Annual Meeting San Juan Puerto Rico* (ed. R. I. Segarra-Garcia), pp. 199–204.

Lasaga A. C. (1998) *Kinetic Theory in the Earth Sciences.* Princeton University Press, Princeton.

Lee M. R. and Parsons I. (1995) Microtextural controls of weathering of perthitic alkali feldspars. *Geochim. Cosmochim. Acta* **59**, 4465–4492.

Lee M. R., Hodson M. E., and Parsons I. (1998) The role of intragranular microtextures and microstructures in chemical and mechanical weathering: direct comparisons of experimentally and naturally weathered alkali feldspars. *Geochim. Cosmochim. Acta* **62**, 2771–2788.

Likens G. E., Bormann F. H., Pierce R. S., Eaton J. S., and Johnson N. M. (1977) *Biogeochemistry of a Forested Ecosystem.* Springer-Verlag, Berlin.

Luttge A. B., Bolten E. W., and Lasaga A. C. (1999) An interferometric study of the dissolution kinetics of anorthite: the role of reactive surface area. *Am. J. Sci.* **299**, 652–678.

MacInnis I. N. and Brantley S. L. (1993) Development of etch pit size distributions on dissolving minerals. *Chem. Geol.* **105**, 31–49.

Mahaney W. C. and Halvorson D. L. (1986) Rates of mineral weathering in the wind river mountains, western Wyoming. In *Rates of Chemical Weathering of Rocks and Minerals* (eds. S. Coleman and D. Dethier). Academic Press, Orlando, pp. 147–168.

Mast M. A., Drever J. I., and Barron J. (1990) Chemical weathering in the Loch Vale watershed, Rocky Mountain National Park, Colorado. *Water Resour. Res.* **26**, 2971–2978.

Maurice P. A., McKnight D. M., Leff L., Fulghum J. E., and Gooseff M. (2002) Direct observations of aluminosilicate weathering in the hyporheic zone of an Antarctic dry valley stream. *Geochim. Cosmochim. Acta* **66**, 1335–1347.

Merino E., Nahon D., and Wang Y. (1993) Kinetic and mass transfer of pseudomorphic replacement: application to replacement of parent minerals and kaolinite by Al, Fe, and Mn oxides during weathering. *Am. J. Sci.* **293**, 135–155.

Merrill G. P. (1906) *A Treatise on Rocks, Rock Weathering and Soils.* McMillian, NY.

Merritts D. J., Chadwick O. A., Hendricks D. M., Brimhall G. H., and Lewis C. J. (1992) The mass balance of soil evolution on late quaternary marine terraces, northern California. *Geol. Soc. Am. Bull.* **104**, 1456–1470.

Meybeck M. (1979) Concentrations des eaux fluviales en elements majeurs et apports en solution aux oceans. *Revue De Geologie Dynamique et De Geographie Physique* **21**, 215–246.

Meyback M. (1994) *Material Fluxes on the Surface of the Earth.* National Academy Press, Washington.

Misra K. C. (2000) *Understanding Mineral Deposits.* Kluwer Academic, Rotterdam.

Mokma D. L., Jackson M. L., Syers J. K., and Steens P. R. (1973) Mineralogy of a chronosequence of soils from greywacke and mica-schist alluvium. *NZ J. Sci.* **16**, 769–797.

Murphy S. F., Brantley S. L., Blum A. E., White A. F., and Dong H. (1998) Chemical weathering in a tropical watershed, Luquillo mountains, Puerto Rico: II. Rate and mechanism of biotite weathering. *Geochim. Cosmochim. Acta* **62**, 227–243.

Nagy K. L., Blum A. E., and Lasaga A. C. (1991) Dissolution and precipitation kinetics of kaolinite at 80°C and pH 3:

the dependence on the saturation state. *Am. J. Sci.* **291**, 649–686.

Nahon D. and Merino E. (1997) Pseudomorphic replacement in tropical weathering: evidence, geochemical consequences, and kinetic-rheological origin. *Am. J. Sci.* **297**, 393–417.

Nesbitt H. W. and Markovics G. (1997) Weathering of granodioritic crust, long-term storage of elements in weathering profiles, and petrogenesis of siliciclastic sediments. *Geochim. Cosmochim. Acta* **61**, 1653–1670.

Nesbitt H. W., Fedo C. M., and Young G. M. (1997) Quartz and feldspar stability, steady and non steady state weathering and petrogenesis of siliciclastic sands and muds. *J. Geol.* **105**, 173–191.

Nugent M. A., Brantley S. L., Pantano C. G., and Maurice P. A. (1998) The influence of natural mineral coatings on feldspar weathering. *Nature* **396**, 527–622.

O'Brien E. L. and Buol S. W. (1984) Physical transformations in a vertical soil-saprolite sequence. *Soil Sci. Am. J.* **48**, 354–357.

Oelkers E. H. (2001) General kinetic description of multioxide silicate mineral and glass dissolution. *Geochim. Cosmochim. Acta* **65**, 3703–3719.

Oelkers E. H. and Schott J. (1995) Experimental study of anorthite dissolution and the relative mechanism of feldspar hydrolysis. *Geochim. Cosmochim. Acta* **59**, 5039–5053.

Ohmoto H. (1996) Evidence in pre-2.2 Ga paleosols for the early evolution of atmospheric oxygen and terrestrial biota. *Geology* **24**, 1135–1138.

Oillier C. (1984) *Weathering.* Longman, London.

Paces T. (1972) Chemical characteristics and equilibrium between groundwater and granitic rocks. *Geochim. Cosmochim. Acta* **36**, 217–240.

Paces T. (1973) Steady-state kinetics and equilibrium between ground water and granitic rock. *Geochim. Cosmochim. Acta* **37**, 2641–2663.

Paces T. (1986) Rates of weathering and erosin derived from mass balance in small drainage basins. In *Rates of Chemical Weathering of Rocks and Minerals* (eds. S. Coleman and Dethier). Academic Press, Orlando, pp. 531–550.

Parkhurst D. L. and Plummer L. N. (1993) Geochemical models. In *Regional Ground–Water Quality* (ed. W. M. Alley). Van Nostrand Reinhold, pp. 199–225.

Pavich M. J. (1986) Processes and rates of saprolite production and erosion on a foliated granitic rock of the Virgina piedmont. In *Rates of Chemical Weathering of Rocks and Minerals* (eds. S. Coleman and D. Dethier). Academic Press, Orlando, pp. 551–590.

Pavich M. J., Brown L., Harden J., Klein J., and Middleton R. (1986) [10]Be distribution in soils from Merced River terraces, California. *Geochim. Cosmochim. Acta* **50**, 1727–1735.

Perg L. A., Andersson R. S., and Finkel R. C. (2001) Use of a new [10]Be and [26]Al inventory method to date marine terraces, Santa Cruz, California, USA. *Geology* **29**, 879–882.

Riebe C. S., Kirchner J. W., Granger D. E., and Finkel R. C. (2001) Strong tectonic and weak climatic control of long-term chemical weathering rates. *Geology* **29**, 511–514.

Ruddiman W. F. (1997) *Tectonic Uplift and Climate Change.* Plenum, NY.

Rye R. and Holland H. D. (1998) Palosols and the evolution of atmospheric oxygen: a critical review. *Am. J. Sci.* **298**, 621–672.

Samma J. C. (1986) *Ore Fields and Continental Weathering.* Van Nostrand, NY.

Schnoor J. L. (1990) Kinetics of chemical weathering: a comparison of laboratory and field rates. In *Aquatic Chemical Kinetics* (ed. W. Stumm). Wiley, NY, pp. 475–504.

Schulz M. S. and White A. F. (1999) Chemical weathering in a tropical watershed, Luquillo mountains, Puerto Rico: III. Quartz dissolution rates. *Geochim. Cosmochim. Acta* **63**, 337–350.

Siegal D. and Pfannkuch H. O. (1984) Silicate mineral dissolution at pH 4 and near standard temperature and pressure. *Geochim. Cosmochim. Acta* **48**, 197–201.

Soler J. M. and Lasaga A. C. (1996) A mass transfer model of bauxite formation. *Geochim. Cosmochim. Acta* **60**, 4913–4931.

Stallard R. F. (1995) Tectonic, environmental, and human aspects of weathering and erosion: a global review using a steady-state perspective. *Ann. Rev. Earth Planet. Sci.* **23**, 11–39.

Stallard R. F. and Edmond J. M. (1983) Geochemistry of the Amazon: 2. The influence of geology and weathering environment on dissolved load. *J. Geophys. Res.* **88**, 9671–9688.

Stauffer R. E. (1990) Granite weathering and the sensitivity of Alpine Lakes to acid depositon. *Limnol. Oceangr.* **35**, 1112–1134.

Stefansson A. (2001) Dissolution of primary minerals of basalt in natural waters: I. Calculation of mineral solubilities from 0°C to 350°C. *Chem. Geol.* **172**, 225–250.

Stefansson A. and Arnorsson S. (2000) Feldspar saturation state in natural waters. *Geochim. Cosmochim. Acta* **64**, 2567–2584.

Steefel C. I. and Lasaga A. C. (1994) A coupled model for the transport of multiple chemical species and kinetic precipitation/dissolution reactions with application to reactive flow in a single phase hydrothermal system. *Am. J. Sci.* **294**, 529–592.

Stewart B. W., Capo R. C., and Chadwick O. A. (2001) Effects of rainfall on weathering rate, base cation provenance, and Sr isotope composition of Hawaiian soils. *Geochim. Cosmochim. Acta* **65**, 1087–1099.

Stillings L. L., Drever J. I., Brantley S., Sun Y., and Oxburgh R. (1996) Rates of feldspar dissolution at pH 3–7 with 0–8 M oxalic acid. *Chem. Geol.* **132**, 79–89.

Stonestrom D. A., White A. F., and Akstin K. C. (1998) Determining rates of chemical weathering in soils-solute transport versus profile evolution. *J. Hydrol.* **209**, 331–345.

Suarez D. L. and Simunek J. (1996) Solute transport modelling under variably saturated water flow conditions. In *Reactive Transport in Porous Media* (eds. P. C. Lichtner, C. I. Steefel, and E. H. Oelkers). Mineralogical Society of America, Washington DC, pp. 230–268.

Sverdrup H. U. (1990) *The Kinetics of Base Cation Release due to Chemical Weathering.* Lund University, Lund.

Sverdrup K. and Warfvinge P. (1995) Estimating field weathering rates using laboratory kinetics. In *Chemical Weathering Rates of Silicate Minerals.* Reviews in Mineralogy 31 (eds. A. F. White and S. L. Brantley). Mineralogical Society of America, Washington DC, pp. 485–541.

Swoboda-Colberg N. G. and Drever J. I. (1992) Mineral dissolution rates: a comparison of laboratory and field studies. In *Water–Rock Interaction WRI 7* (eds. Y. K. Kharaka and A. S. Maest). Balkema, Rotterdam pp. 115–117.

Taylor A. and Blum J. D. (1995) Relation between soil age and silicate weathering rates determined from the chemical evolution of a glacial chronosequence. *Geology* **23**, 979–982.

Taylor A. S., Blum J. D., and Lasaga A. C. (2000) The dependence of labradorite dissolution and Sr isotope release rates on solution saturation state. *Geochim. Cosmochim. Acta* **64**(14), 2389–2400.

Tisdall J. M. (1982) Organic matter and water-stable aggregates in soils. *J. Soil Sci.* **33**, 141–163.

Torrent J. and Nettleton W. D. (1978) Feedback processes in soil genesis. *Geoderma* **20**, 281–287.

Trudgill S. T. (1995) *Solute Modelling in Catchment System.* Wiley, NY.

Velbel M. A. (1985) Geochemical mass balances and weathering rates in forested watersheds of the southern Blue ridge. *Am. J. Sci.* **285**, 904–930.

Velbel M. A. (1989) Weathering of hornblende to ferruginous products by dissolution-reprecipitation mechanism: petrography and stoichiometry. *Clays Clay Min.* **37**, 515–524.

Velbel M. C. (1993) Temperature dependence of silicate weathering in nature: how strong a feedback on long-term accumulation of atmospheric CO_2 and global greenhouse warming. *Geology* **21**, 1059–1062.

Warfvinge P. (1995) Basic principals of frequently used models. In *Solute Modeling in Catchment Systems* (ed. S. T. Trudgill). Wiley, NY, pp. 57–71.

White A. F. (1995) Chemical weathering rates in soils. In *Chemical Weathering Rates of Silicate Minerals* (eds. A. F. White and S. L. Brantley). Mineralogical Society of America, Washington, DC, pp. 407–458.

White A. F. (2002) Determining mineral weathering rates based on solid and solute weathering gradients and velocities: application to biotite weathering in saprolites. *Chem. Geol.* **190**, 69–89.

White A. F. and Blum A. E. (1995) Effects of climate on chemical weathering rates in watersheds. *Geochim. Cosmochim. Acta* **59**, 1729–1747.

White A. F. and Brantley S. L. (1995) Chemical weathering rates of silicate minerals. In *Reviews in Mineralogy.* Mineralogical Society of America, Washington DC, vol. 31, 584pp.

White A. F. and Brantley S. L. (2003) The effect of time on the weathering of silicate minerals: why to weathering rates differ in the laboratory and field? *Chem. Geol.* (in press).

White A. F. and Peterson M. L. (1990) Role of reactive surface area characterization in geochemical models. In *Chemical Modelling of Aqueous Systems II.* Am. Chem. Soc. Symp. Ser. 416 (eds. D. C. Melchior and R. L. Bassett), Am. Chem. Soc., Washington DC, pp. 461–475.

White A. F., Blum A. E., Schulz M. S., Bullen T. D., Harden J. W., and Peterson M. L. (1996) Chemical weathering of a soil chronosequnece on granitic alluvium: 1. Reaction rates based on changes in soil mineralogy. *Geochim. Cosmochim. Acta* **60**, 2533–2550.

White A. F., Blum A. E., Schulz M. S., Vivit D. V., Larsen M., and Murphy S. F. (1998) Chemical weathering in a tropical watershed, Luquillo mountains, Puerto Rico: I. Long-term versus short-term chemical fluxes. *Geochim. Cosmochim. Acta* **62**, 209–226.

White A. F., Blum A. E., Bullen T. D., Vivit D. V., Schulz M., and Fitzpatrick J. (1999) The effect of temperature on experimental and natural weathering rates of granitoid rocks. *Geochim. Cosmochim. Acta* **63**, 3277–3291.

White A. F., Blum A. E., Stonestrom D. A., Bullen T. D., Schulz M. S., Huntington T. G., and Peters N. E. (2001) Differential rates of feldspar weathering in granitic regolths. *Geochim. Cosmochim. Acta* **65**, 847–869.

White A. F., Blum A. E., Schulz M. S., Huntington T. G., Peters N. E., and Stonestrom D. A. (2002) Chemical weathering of the Panola granite: solute and regolith elemental fluxes and the dissolution rate of biotite. In *Water–Rock Interaction, Ore Deposits, and Environmental Geochemisty: a Tribute to David A. Crerar*, Spec. Publ. 7 (eds. R. Hellmann and S. A. Wood). The Geochemical Society, St Louis, pp. 37–59.

White A. F., Schulz M. J., Davison V., Blum A. E., Stonestrom D. A., and Harden J. W. (2003) Chemical weathering rates of a soil chronosequence on granitic alluvium: III. The effects of hydrology, biology, climate on pore water compositions and weathering fluxes. *Geochim. Cosmochim. Acta* (in press).

Williams M. W., Brown A. D., and Melack J. M. (1993) Geochemical and hydrologic controls on the composition of surface water in a high-elevation basin, Sierra Nevada, California. *Limnol. Oceanogr.* **38**, 775–797.

Wolery T. J., Jackson K. J., Bourcier W. L., Bruton C. J., and Viani B. E. (1990) Current status of the EQ3/6 software package for geochemical modelling. In *Chemical Modelling of Aqueous Systems II* (eds. D. C. Melchior and R. L. Bassett). American Chemical Society, Washington, DC, pp. 104–116.

Zeman L. J. and Slaymaker O. (1978) Mass balance model for calculation of ionic input loads in atmospheric fallout and discharge from a mountainous basin. *Hydrol. Sci.* **23**, 103–117.

5.06
Plants and Mineral Weathering: Present and Past

E. K. Berner and R. A. Berner

Yale University, New Haven, CT, USA

and

[†]K. L. Moulton

Kent State University, OH, USA

5.06.1 INTRODUCTION

Geological studies of weathering have traditionally focused either on bulk chemical changes at the outcrop or on the minerals undergoing dissolution and those formed as weathering products. Laboratory studies have focused on the rates and mechanisms of mineral-water interaction. In both cases weathering is treated as a simple inorganic example of rock–water interaction. However, there is a third aspect that is very important to weathering and that is the effect of biology. Only recently, geologists have paid appreciable attention to how plants and microbiota affect weathering. To try to remedy this, in the present review, the effects of higher plants on

---

[†] Deceased.

weathering will be discussed; the effects of microbes alone are left to another review.

Possibly the earliest study, not only of the chemistry of weathering itself, but also of the effects of plants on weathering, was that of Ebelmen (1845). He suggested that roots and organic acids secreted by plants were promoting the weathering of rocks, and that silicate rock weathering was a key factor in the control of atmospheric carbon dioxide over geological time. We hope to show here that Ebelmen was correct (and prescient) on both accounts.

5.06.2 STUDIES OF MODERN WEATHERING

Rooted vascular plants (and their associated microbiota—bacteria and fungi) accelerate mineral weathering for a number of reasons. These effects can be grouped into chemical processes and physical processes.

5.06.2.1 Chemical Processes

The effects of plants on chemical weathering have been studied both in the field and in the laboratory. From these studies general principles have been elucidated as to how plants affect the chemistry of the soil and the dissolution of minerals.

5.06.2.1.1 General principles

Forest ecosystems made up of vascular plants cycle nutrient elements between various compartments—atmosphere, trees, soil and soil solutions, and the bedrock. Outside inputs to the system include: (i) weathering of primary and secondary soil minerals; (ii) precipitation and direct depositional input of dust and marine aerosols; and (iii) gaseous inputs (Figure 1). The major source of the nutrient elements calcium, magnesium, potassium, sodium, silicon, and iron is mineral weathering and the major input of nitrogen, sulfur, phosphorus, and chlorine comes from precipitation and dust (Likens *et al.*, 1977; Knoll and James, 1987). The largest part (>90%) of nutrients taken up by trees on an annual basis is returned to forest litter, where they are recycled back into plants. However, the recycling of plant nutrients is not perfect. Two major sinks for nutrients are storage in the woody parts of plants and removal to secondary minerals in the soil. There is also an annual net loss of nutrients from ecosystems in runoff, stream flow, and groundwater—particularly calcium, magnesium, sodium, potassium, and

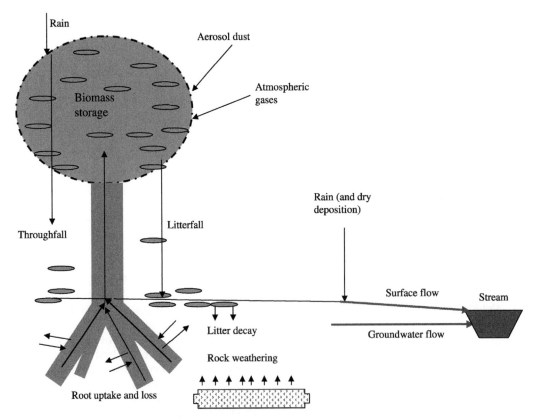

Figure 1 Biogeochemical cycling in forests.

silicon. These nutrients, which must be replaced in order to support plant growth, at most localities come primarily from rock weathering (Likens *et al.*, 1977).

When the concentration of relatively insoluble elements in soil solutions reaches saturation with secondary minerals, they precipitate in soils, particularly tropical soils. Lucas (2001) points out that precipitation of silicon, aluminum, and iron in tropical soils occurs because of the low solubility of clay minerals and oxides and the high concentrations of these elements in the topsoil horizons produced by plant cycling of these elements.

Vegetation alters the chemistry of soil solutions primarily in the upper soil horizons and around plant roots. When bacteria decompose litterfall from trees in the upper part of the soil, carbon dioxide is produced, forming carbonic acid, which acts as a weathering agent for minerals. Due to ion exchange and the secretion of organic acids by roots and associated microflora, the soil solution around fine plant roots usually has a lower pH (as low as 3) than surrounding soil solutions (pH 5–7) (Arthur and Fahey, 1993). This zone is known as the rhizosphere; it extends out only a few millimeters from the roots. Lower rhizosphere pH values, with a difference of up to two units from that of the bulk soil, have also been found by Marschner and Romheld (1983). However, the pH of the root–soil interface varies with the plant species and nitrogen source. If the nitrogen source is ammonium or nitrogen fixation, the pH at the soil–root interface is acidic (pH 4); when the source is nitrate, it is alkaline (pH 7.5) (Marschner and Romheld, 1983; Marschner *et al.*, 1991). Fine roots excrete H^+ in exchange for the cationic nutrients (Mg^{2+}, Ca^{2+}, NH_4^+, and K^+), which they take up, in order to maintain charge balance (Lucas, 2001). Roots also release carbon dioxide in respiration, forming carbonic acid in the root zone.

Plant roots and the symbiotic organisms such as mycorrhizal fungi associated with them secrete organic acids and chelates which attack primary minerals to release nutrient cations needed by the roots (Griffiths *et al.*, 1994). Organic acids (such as oxalic acid) increase the availability of PO_4^{3-}, an important soil nutrient. Oxalic acid also forms soluble complexes with aluminum and iron, which are less toxic to higher plants than the free metal ions. This process enhances the dissolution of minerals containing aluminum and iron by removing normally insoluble aluminum and iron in solution (Drever and Vance, 1994; Drever, 1994; Cochran and Berner, 1996; Berner and Cochran, 1998). Symbiotic rhizosphere bacteria associated with red cedar, which has higher calcium accumulations than white pine, solubilize calcium silicate minerals more effectively than those associated with white pine (Jackson and Voight, 1971).

Plants through their roots obtain nutrients such as calcium, magnesium, and potassium from the primary minerals that undergo dissolution. Mineral grains in soil adjacent to roots are affected mechanically, chemically, and mineralogically by invading roots (April and Keller, 1990). The grains are fractured parallel to the roots, which increases the amount of mineral surface area available for dissolution. In this way mineral grains adjoining roots are preferentially dissolved. Chemical interaction between the roots and mineral grains can bring about the precipitation of hydrous aluminum oxides, opaline, and amorphous silica and calcium oxalate in root cells.

Studies of young Hawaiian basalts (Cochran and Berner, 1996; Berner and Cochran, 1998) show similar evidence of root interaction with minerals. Even 150-year-old rocks show evidence of selective plagioclase dissolution by invading fungal hyphae and root hairs. After 3,000 years of weathering, basalt beneath higher plant stands shows removal of plagioclase from surface rocks down to a depth of several centimeters or more, particularly along fractures. Plagioclase grains are dissolved out completely by invading plant roots resulting in voids in the rock. The aluminum and iron are largely removed as soluble oxalic acid chelates. The resulting porosity was more than 30–40%. Transpiration causes fine roots to pull water out of the rocks before it becomes saturated with dissolved products, which would otherwise fill the voids and inhibit further chemical dissolution. The water is then replaced by unsaturated downward percolating rainwater, resulting in continued dissolution and chemical weathering (Cochran and Berner, 1996).

Van Breemen *et al.* (2000) and Jongmans *et al.* (1997) found that symbiotic ectomycorrhizal mycelia from fungi are able to create micropores by the production of aluminum-complexing organic acids at their hyphal tips. These hyphae penetrate directly into silicate minerals in granitic rock to remove calcium and phosphorus from apatite (calcium phosphate) inclusions, in isolation from soil solutions. Dissolved material can thus be transferred directly to the host (in this case conifer) plant roots, bypassing the bulk soil solution and competition from other organisms for nutrients. Phosphorus is the limiting nutrient in these soils and is present in feldspars as apatite inclusions, which photomicrographs show to be dissolved. This provision of nutrients, directly to conifers from inside mineral grains by symbiotic fungi, involves direct control of weathering by plants. Soluble aluminum and silicon which are not taken up by roots are exuded from the hyphae or stored in place. These "rock-eating" mycorrhizal fungi play a role in the formation of podzol soils. The soils under European conifers have been depleted in cations by acid deposition. However, direct nutrient uptake from inside minerals by microrrhizal fungi associated with conifers bypasses the soil solutions and somewhat insulates trees from the effects of acid deposition. Also, organically complexed aluminum, which is produced during pore formation by the fungi, is less phytotoxic than inorganic Al^{3+} in soils, which results from acid deposition. These effects may explain why European forest productivity has not been strongly affected by excess soil acidification.

Blum *et al.* (2002) and Probst *et al.* (2000) have found, in areas similarly cation-depleted by acid deposition, that a considerable proportion of calcium released by weathering came from apatite dissolution. This apatite was utilized directly by ectomycorrhizal tree species (spruce and fir) bypassing the soil exchange complex. In the Blum *et al.* (2002) study, ectomycorrhizal fungi associated with the roots of conifers provided 95% of the calcium found in the foliage of the trees and 35% of the Ca^{2+} leaving the mixed conifer hardwood watershed in stream water.

5.06.2.1.2 *Experimental studies*

Robert and Berthelin (1986) discuss experimental studies of the weathering of minerals by root systems of plants and their associated rhizospheric microorganisms—mycorrhizal symbiotic organisms and nonsymbiotic bacteria and fungi. The studies, in agreement with field observations, found that rhizospheric microorganisms solubilize plant nutrients—phosphorus from insoluble phosphates and potassium, magnesium, and iron from insoluble silicates such as biotite and feldspar. During the experimental transformation of biotite to kaolinite (Spyridakis *et al.*, 1967) potassium, magnesium, and iron are removed from biotite, and H_2O is added to form kaolinite. These reactions occur as a consequence of the production of low molecular weight organic acids by organisms: oxalic, citric, and tartaric acids among others. The organic acids cause hydrolysis, attack by hydrogen ion, and complexing to release silicon and aluminum from silicates. Mica is destroyed or transformed, with preservation of the crystal structure, into vermiculite and smectite by various organic acids (Mortland *et al.,* 1956; Boyle and Voight, 1973; Hinsinger *et al.*, 1992). Organic acids destroy mica by dissolution at low pH ($<$3). In the pH range 3–5, protons enter interlayer space in mica in exchange for cations (potassium, calcium, and magnesium), and complexes are formed with aluminum and iron. The processes vary with different organic acids and different silicate minerals.

Drever (1994) and Drever and Stillings (1997) have summarized laboratory experiments which indicate the effect of pH on silicate mineral dissolution. Dissolution rates of silicates are generally independent of pH at neutral pH; below about pH 4–5 dissolution rates increase with decreasing pH in acid solutions, and above about pH 8, they increase with pH in alkaline solutions. Besides organic acids and carbonic acid formed from soil CO_2, nitric and sulfuric acids are also produced as part of the biological cycle. Nitrogen fixation, for example, can result in high nitric acid concentrations (Homann *et al.*, 1992). Other sources of low pH are anthropogenic acid rain resulting in high sulfuric and nitric acid depositions, and high sulfuric acid concentrations from pyrite weathering.

Dissolution by organic acids has been studied extensively in the laboratory (Drever, 1994; Drever and Stillings, 1997). Organic acids, in addition to decreasing the solution pH, form complexes with cations at mineral surfaces, change the saturation state of the solution with respect to minerals, chelate aluminum in solution decreasing its chemical activity and thereby increase mineral dissolution rate. Oxalate, a common exudate of mycorrhizal fungi, is particularly effective because it complexes aluminum (Amrhein and Suarez, 1988). Laboratory measurements of the effect of oxalate on silicate dissolution show an increase by a factor of up to 15 for a concentration of 1mM oxalate as summarized by Drever (1994) and Drever and Stillings (1997) depending on the pH and aluminum content of the mineral. Levels of oxalate this high are not ordinarily found in soil solutions, but can occur in the rhizosphere around roots and fungal hyphae. High molecular weight humic acids have been found in the laboratory to dissolve aluminous minerals but not nearly as extensively as low molecular weight organic acids such as oxalate (Ochs *et al.*, 1993).

Welch and Ullman (2000) conducted laboratory experiments comparing differences between the dissolution rates of calcic plagioclase in neutral pH solutions by organic and inorganic anions. They found enhanced dissolution by gluconate and oxalate, relative to nitrate and distilled water. This correlates with the strong aluminum complexing ability of both gluconate and oxalate. The effect of change in temperature on dissolution rate was found to be less in the presence of these organics than with nitrate or distilled water. This agrees with the results of Brady *et al.* (1999), who, from field studies in Hawaii, showed a similar reduced temperature effect on silicate dissolution by lichens (which are notable secreters of oxalate) compared to adjacent lichen-free areas.

Experimental evidence of the importance of higher plants in weathering of basaltic rocks was obtained in the laboratory by Hinsinger *et al.* (2001). Diverse plant species were used including white lupin and oilseed rape, which release protons plus citrate ion in their rhizosphere, and also banana and maize. Plants enhanced the release rate of silicon, calcium, magnesium, and sodium over leaching by a dilute salt solution by factors of 1–5. Calcium and sodium were preferentially released presumably from plagioclase weathering. The amount of iron released from basalt by banana and maize, which as phytosiderophores produce strong iron chelators, was 100–500 times larger than that with a salt solution. The experiments suggest that the dissolution of basalt is increased partially by the uptake of elements by plants and also by the release of H^+ ions by roots.

In general, the effects of organic acids on mineral weathering rates in laboratory studies are considerably less pronounced than in soils, although the experiments are normally conducted at the same pH and organic acid content as bulk soil water. Bulk soil solution, as pointed out earlier, contains considerably lower concentrations of low molecular weight chelating acids than solutions within the rhizosphere where lower pH and much weathering occurs. Thus, experiments that reproduce the pH and organic acid concentrations of bulk soil water are not representative of actual weathering by plants. The difference between the chemical composition of bulk soil water and rhizosphere water may be due to inhibited transport of soil solutions between rhizosphere micropores and bulk soil macropores (Drever and Stillings, 1997). The presence of clays, oxides, or hydroxides could reduce the mobility of organic ligands, because they are easily absorbed on these substances (Lucas, 2001).

5.06.2.2 Physical Processes

Plants affect the circulation of water through the soil and atmosphere both on a regional and on a small scale. Evapotranspiration is the combination of transpiration, the release of water through stomata in plant leaves, and the evaporation of water from the ground surface. Plants, by increasing evapotranspiration, increase the amount of water vapor available locally to produce rain (Shukla and Mintz, 1982). In the Amazon River Basin, for example, rain entering from the Atlantic undergoes several evaporation and precipitation cycles before running off in rivers (Stallard and Edmond, 1981). By binding fine particles in the soil, plant roots cause retention of water between rain events. This retained water results in continued chemical weathering (Drever, 1994). Increased rainfall also means that saturated soil

solutions are replaced by freshwater, increasing the weathering rate.

Plant roots anchor soil and slow down physical erosion in areas of steep topography. This allows longer water contact times with primary minerals undergoing dissolution, thereby increasing weathering in such weathering-limited regimes (Stallard, 1985; Drever, 1994). Plant roots also reduce deep drainage by their uptake of water. This is particularly important in the humid tropics where the upper soil solutions reach high concentrations of silicon, aluminum, and iron, reducing mineral dissolution (Lucas, 2001). In low relief areas with low physical erosion, chemical weathering produces a thick clayey soil cover. This thick layer protects the primary minerals from weathering, because most plant activity occurs near the surface, where nutrients are recycled, and because the deep bedrock becomes isolated from precipitation.

However, even in such areas tree roots can penetrate great distances (25–30 ft (1 ft = 0.3048 m)) below the surface (Nepstad *et al.*, 1994) and reach bedrock (see also Figure 2). If there is a change in the plant cover due to drought, flood, disease, storms, or forest fires, so that the depth of root penetration changes, rate of weathering in low relief areas can change. Also, there

Figure 2 Deep rooting of trees in tropical weathering (in Hawaii). The depth of the roots, based on the figure for scale, is ~7 m (photo by C. Bowser).

can be upturned roots from windthrow (Bormann *et al.*, 1998). However, a rapid increase of runoff occurs in such cases resulting in accelerated loss of forest nutrients originally stored in trees or organic debris (Bormann and Likens, 1979). The resulting increase in physical erosion tends to expose new bedrock to the chemical weathering necessary to produce more nutrients in order to sustain new forest growth (Knoll and James, 1987). Also, on long geological timescales, tectonic uplift can convert clay-covered lowland areas to areas of high relief resulting in enhanced erosion, landsliding, and exposure of primary minerals to weathering. Volcanic eruptions and glaciation can also expose fresh rock to weathering.

Plants exert additional effects on weathering. The presence of forest trees tends to lower soil surface temperatures by shading the soil surface and reducing surface albedo (Kelly *et al.*, 1998). Also, roots fracture mineral grains and increase soil porosity and permeability, which allows greater contact between soil solutions and minerals (April and Keller, 1990; Colin *et al.*, 1992).

5.06.2.3 Field Studies Evaluating the Role of Large Land Plants on Weathering

Weathering rates in the field are as much as one to two orders of magnitude slower than dissolution rates measured in the laboratory (Benedetti *et al.*, 1994; see Chapter 5.05). The difference is due to a number of factors: (i) there are differences in surface area between laboratory minerals and natural minerals; (ii) secondary precipitates may protect primary mineral surfaces in the field; (iii) in soils, most flow is through macropores and not all mineral surfaces are continually exposed to flowing solutions as they are in laboratory experiments; and (iv) most

field studies integrate over longer time intervals than laboratory experiments (White *et al.*, 1996; Benedetti *et al.*, 1994). Thus, field studies evaluating weathering rates in natural systems provide valuable quantitative information not available from laboratory studies.

5.06.2.3.1 High-latitude and high-altitude weathering

An exhaustive field study by Moulton *et al.* (2000) (see also Moulton and Berner, 1998) quantified plant effects on weathering in western Iceland. The study area includes several adjacent basaltic areas one of which was barren (covered only by sporadic moss and lichens) and the others populated by birches and conifers. All of the areas have similar slopes and microclimate and do not have either acid rain (a problem in the northeastern USA, Europe, and elsewhere) or hydrothermal activity. Also, calcium carbonate is absent which, because it weathers rapidly, could confuse the results.

As shown in Table 1, the weathering release of calcium and magnesium is ~4 times as great in tree-covered areas as in barren areas. These weathering fluxes include storage in trees and soils as well as removal in stream water. Storage in stable secondary weathering products in soils for calcium and magnesium is comparable to tree storage. Soil storage prevents the transfer of calcium and magnesium in streams to the ocean, where they could be removed as carbonate resulting in atmospheric CO_2 removal. Mineral mass balance studies of tree-covered versus barren areas showed that plagioclase weathering was increased by a factor of 2 and pyroxene weathering by a factor of 10 in the forested areas.

In Iceland the major rock-forming elements stored in trees are calcium and magnesium, but the streamwater removal flux of these elements is far

Table 1 Ratio of weathering fluxes from vegetated areas versus unvegetated areas (numbers in parentheses in doubt).

	HCO_3	Na	Mg	Ca	Si	References
Iceland[a]						
Birch/bare	3	1	4	3	2	Moulton *et al.* (2000)
Conifer/bare	3	2	4	3	3	Moulton *et al.* (2000)
Iceland[b]						
Birch/bare		1.5	5	3	2	Moulton *et al.* (2000)
Conifer/bare		2	5	3	2.5	Moulton *et al.* (2000)
S. Swiss Alps[c,d]	8				8	Drever and Zobrist (1992)
Colo. Rocky Mts.		Na + K + Mg + Ca = 4				Arthur and Fahey (1993)
Hubbard Brk, NH[b]			(18)	(10)		Bormann *et al.* (1998)
Hubbard Brk, NH[e]			2	2		Bormann *et al.* (1998)
Hubbard Brk, NH[c]	2–10					Berner and Rao (1997)

After Moulton *et al.* (2000).
[a] Tree storage plus runoff. [b] Tree storage + runoff + soil storage. [c] Runoff only. [d] Corrected for temperature by elevation.
[e] Tree storage only.

greater than tree storage (by a factor of 7 for calcium and 22 for magnesium). By contrast, the storage of potassium in trees is much larger relative to that in the streamwater flux. (Basalt is a poor source of potassium; in fact the largest source of potassium is marine rain.) At Hubbard Brook, New Hampshire, an aggrading northern temperate forest, the streamwater flux is greater than the tree storage flux by a factor of 1.5 for calcium and 3.5 for magnesium. However, the streamwater flux is only one-third of the tree storage flux for potassium (Likens *et al.*, 1977). A large percentage (70%) of the input of potassium to the ecosystem is stored in the vegetation. In other areas, such as the Cascade Mountains (Homann *et al.*, 1992), where forests are growing much faster and the trees are much bigger, the amount of tree-storage is a factor of 100 times greater than that in Iceland.

A high-elevation watershed in Rocky Mountain Park in north-central Colorado was studied by Arthur and Fahey (1993). Although the area covered by subalpine spruce and fir forest comprises only 6% of the watershed with 82% bare rock, and the remainder permanent snowfields plus 1% subalpine meadows, the surface water chemistry in the watershed is controlled primarily by mineral weathering from under the forest. The rate of cation (sodium, potassium, magnesium, and calcium) denudation per unit area ($131 \ \mu mol \ m^{-2} \ yr^{-1}$) under the forest plus cation storage in trees ($29 \ \mu mol \ m^{-2} \ yr^{-1}$) amounts to ~4 times the cation denudation of the whole watershed ($38.2 \ \mu mol \ m^{-2} \ yr^{-1}$) and also 4 times the bare rock cation denudation rate (calculated as $37 \ \mu mol \ m^{-2} \ yr^{-1}$) (see Table 1). Arthur and Fahey attribute this increased weathering under forests to H^+ exuded from roots during nutrient uptake and production of organic acids in forest soils.

Chemical weathering in young glacial sediments has been studied in the foreland of a retreating glacier in south-central Alaska by Anderson *et al.* (2000). The sediment is metagraywacke and the climate is very cold (average $-4 \ °C$) and wet ($2.7 \ m \ yr^{-1}$). In the youngest sediments, carbonate dissolution and sulfide oxidation account for 90% of the solute formation and biotite alteration to vermiculite, by interlayer cation loss, accounts for 5–11%. Silicate weathering is dominant only in the oldest ($10^4 \ yr$) sediments and increases with distance from the glacier along with increasing extent of vegetation. The authors suggest that establishment of vegetative cover of alpine tundra vegetation (lichens, mosses, small flowering plants, and miniature trees) might be responsible for a greater silicate weathering flux. These field results were simulated by laboratory experiments in young glacial sediments where, initially, carbonate

dissolution dominates the solute flux, and after carbonate dissolution declines, there is still no appreciable silica flux.

5.06.2.3.2 Temperate weathering

In young, actively growing forests the nutrient input and output may involve increased biomass storage and they may not maintain steady state. Likens *et al.* (1977) and Bormann *et al.* (1998) found that failure to consider the uptake of elements by growing trees and other biota would cause weathering rates to be underestimated by a factor of 2 in New Hampshire forested watersheds. In Coweeta, NC, in the southern Appalachians, neglecting this process would cause an underestimate of mineral weathering and soil formation rates by up to a factor of 4 (Velbel, 1985; Taylor and Velbel, 1991). Benedetti *et al.* (1994) have estimated that including the biomass effects increases the calculated weathering rates by a factor of 1.3–5 in the humid subtropical conditions of the Paraná Basin in Brazil.

Non-steady-state storage of nutrients in biomass implies an input of nutrients from weathering. However, a large portion of the nutrients for net primary production in forests comes from the decay of organic litter, not mineral weathering. Likens *et al.* (1977) found that 85–90% of the mineral nutrients potassium, calcium, and magnesium at Hubbard Brook, NH, are recycled, and at Coweeta, NC, 50–90% of these nutrients are recycled (Taylor and Velbel, 1991). The remaining nutrient input comes from mineral weathering and/or from atmospheric input.

Taylor and Velbel point out that biomass storage may also decrease—i.e., the biomass does not always cause a net uptake of nutrients. If organic decay returns more nutrients to the soil solution than are taken up by plants, then a balance does not require mineral weathering, and nutrients are being lost to streams. Biomass decay can exceed biomass uptake when there is, e.g., insect defoliation of trees, storm losses of trees, timber cutting, or fire (Taylor and Velbel, 1991).

A study of chemical weathering of silicate rocks as a function of elevation in the southern Swiss Alps found that cation denudation rate (as a measure of chemical weathering) decreased by a factor of 25, more or less exponentially with elevation (Drever and Zobrist, 1992). Increasing elevation is a surrogate for a number of variables: decreasing temperature, changes in vegetation, decreasing soil thickness, and increasing physical erosion rate, which makes it difficult to see how much effect vegetation has on the change in chemical weathering rate. Drever and Zobrist (1992) assumed that the decrease of mineral weathering rate due to lower temperature at higher elevation was by a factor of 3. This number was derived from the work of Velbel *et al.* (1990) and Velbel (1993a,b). The temperature drop expected from increased elevation is ~13.5 °C based on the lapse rate of 6.5 °C/1,000 m. Combining this with data from White and Blum (1995) for the temperature dependence of silicate weathering rate, the decrease in silicate weathering (measured by silica flux) should be about a factor of 3.

With increasing elevation, the vegetation changes from deciduous forest to conifers to alpine pasture to bare rock and talus. If the only other major change is temperature, then the total drop in chemical weathering rate with elevation, a factor of 25, can be divided by 3 to obtain a factor ~8 for the effect of vegetation changes. The whole area has acid precipitation (pH 4.3–4.9) and steep relief, which would tend to increase the chemical weathering rate, but not affect the overall decrease in weathering rate with elevation.

Bormann *et al.* (1998) attempted to determine whether rates of weathering of primary minerals were being underestimated due to failure to include weathering products accumulating in the biomass and in soil. They compared two sandbox ecosystems (large monitored lysimeters) at Hubbard Brook, NH, one with red pine and the other relatively nonvegetated (containing sporadic lichens and mosses).

Their results showed a factor of 2 faster weathering of calcium and magnesium in the forested compared to the nonforested sandbox, when only the drainage flux and plant biomass storage (Table 1) were considered. When they also included soil weathering products, the vegetated weathering flux of magnesium was 18 times greater and that of calcium, 10 times greater than the unvegetated flux. However, we have found that their method of measuring calcium and magnesium in weathering products is problematical and nonreproducible (parenthetical entries in Table 1).

Berner and Rao (1997) found that at the very beginning of spring snowmelt the dissolved HCO_3^- concentration in outflow from the red-pine sandbox was 2–10 times greater than the concentration in outflow from the nonvegetated sandbox. Bicarbonate ion generally indicates the uptake of CO_2 via silicate weathering (in the absence of carbonate). Hubbard Brook has acid precipitation (pH of ~4.2). Because of the acid rain, HCO_3^- is observable only during winter weathering under the snow, when soil solutions are protected from acid precipitation by the snow cover.

When the snow begins to melt in the spring the bicarbonate-rich water under the pine trees is pushed out by the downward percolating acid snowmelt. As a result, there is alkaline water flow only for a few weeks before the acid snowmelt begins to dominate and neutralizes the alkaline soil solutions. It is during this short period that

the ratio of 2–10 for the weathering rate in the vegetated versus unvegetated plots is observed.

Weathering of primary minerals as a source of nutrients for trees results eventually in almost complete depletion of these minerals near the soil surface. In soils of the Appalachian area in the southeastern USA, layers of saprolite (undisturbed weathered rock), up to tens of meters in thickness, develop below most of the biologically active soil and thus are relatively protected from disturbance by bioturbation. However, even in saprolites weathering is not complete, and tree roots are still able to get enough nutrients from the saprolite for growth (Velbel, 1985).

5.06.2.3.3 Tropical weathering

In a humid subtropical area such as the Paraná Basin in Brazil, the weathering profile is 30 m thick and has a lower yellow part which is a saprolite. Above this the upper red soil called "Terra Rossa" consists of kaolinite and iron oxides (Benedetti *et al.*, 1994). The rock being weathered in this area is a 140-million-year-old basalt. Benedetti *et al.* (1994) found that the calculated weathering rate is increased by a factor of 1.3–5 by including biomass uptake and release of calcium, magnesium, and potassium in their weathering model.

The biota consists of pasture, secondary forest, and crops, and there is some anthropogenic effect. The effect of the biota on weathering rate may be due to the release of organic acids and the complexing of aluminum in solution, thus increasing the apparent solubility of aluminum and indirectly increasing the dissolution rate of gibbsite, which is dissolving. The increase in release rate of aluminum along with faster silicon release rates, owing to faster dissolution rates of augite and labradorite, brings about kaolinite saturation in the water and precipitation of kaolinite. Thus, the major process affecting the weathering profile is kaolinitization probably due to the impact of the biota.

In the tropics thick layers of weathered soil make access to fresh rock more difficult, but tree roots can penetrate very deeply. Nepstad *et al.* (1994) found root penetration up to 10 m in the Amazon Basin, and deep penetration is also evident in Hawaii (see Figure 2).

5.06.2.3.4 Biogenic nutrient cycling and biogenic mineral production

Plants can cause the precipitation of minerals in their roots or in their aboveground parts. Calcite is precipitated in or outside the roots of plants in alternate dry and wet (Mediterranean) climates.

Calcite precipitation can result in the formation of *calcrete*, soil horizons cemented by calcite which can be used as climatic indicators (Jaillard *et al.*, 1991; Wright *et al.*, 1995).

Biological pumping is the recycling of inorganic constituents by plants (Lucas, 2001). The major nutrients used by plants are nitrogen, phosphorus, sulfur, potassium, magnesium, and calcium. Minor nutrients include iron, manganese, zinc, copper, boron, molybdenum, chlorine, and nickel. The elements silicon, aluminum, and to a lesser extent iron are major constituents of both primary and secondary minerals and are strongly involved in weathering. However, these elements are also plant nutrients. Silicon is only essential for some plants but is helpful to many others (Epstein, 1994). Aluminum can be toxic to plants in high concentrations in soil solutions, but it does occur in certain plants in low concentrations (Lucas, 2001).

There are three main types of plant cycling of elements as they affect secondary mineral formation (Lucas, 2001).

(i) An element, e.g., nitrogen and sulfur, which is not part of a soil mineral. This cycle only involves the plant, soil solutions, and the atmosphere.

(ii) An element which is a constituent of soil minerals, and whose concentration in the soil solutions is below saturation with a secondary mineral. Minerals containing the element dissolve, e.g., calcium, magnesium, and potassium.

(iii) An element which is a constituent of soil minerals and whose concentration in soil solutions is at saturation with a secondary mineral. In this case, the plant cycle interacts with secondary mineral formation, e.g., silicon, aluminum, and iron.

Biogenic silica cycling keeps silica at higher concentrations in soil solution than would be the case if silica could all escape in runoff. Soluble biogenic silica can be lost from the soil by leaching in runoff. It can also form silicate minerals such as kaolinite, it can precipitate as soil opaline minerals (laminar opaline silica), or it can remain in the vegetation as *phytoliths* (plant opalline silica) (Kelly *et al.*, 1998). Silica-accumulator plants include grasses, bamboo, palms, corn, many tropical timber trees, and some temperate hardwoods (beech, walnut) (Lovering, 1959).

Plants take up silica as soluble monosilicic acid (H_4SiO_4) and deposit it inside the cell, in cell walls, and near evaporating surfaces. When a plant dies and decays, these microscopic opaline silica phytoliths are returned to the soil. Soil phytolith accumulations are associated with certain plant types such as bamboo (Meunier *et al.*, 1999). Some phytoliths persist in soils for thousands of years (Lucas, 2001).

Alexandre *et al.* (1997) found that the biogenic silica input into the biogeochemical silica cycle from the dissolution of phytoliths is twice as large as silica input from primary silicate mineral weathering in the tropical Congo rainforest. Biogenic (opaline) silica dissolves faster than silicate minerals. While most of the phytoliths dissolve rapidly with a mean residence time of ~ 6 months (Alexandre *et al.*, 1994), and the silica is recycled by the forest, a small part (7.5%) does not dissolve and is preserved in the soil.

A plant-controlled weathering system exists in the central Amazon Basin (Lucas, 2001) which has high temperatures and high rainfall. Typical ferrasol (laterite or bauxite), clayey, acid soil in the humid tropical Central Amazonian rainforest consists of a high silica upper layer containing of kaolinitic clay (8 m thick), underlain by a 3 m thick aluminous lower layer of more gibbsitic nodular clay, followed by an 8 m transition to saprolite. The primary mineral, quartz, which is being dissolved, is also present in the sandy-clay parent soil, along with iron oxides and hydroxides. The pH is 3.9–4.8. The soil sequence of kaolinite overlying gibbsite is the opposite of what geochemical soil models predict. Lucas *et al.* (1993) found that kaolinite is stabilized against dissolution and conversion to gibbsite by high dissolved silica concentrations in the shallow soil solutions due to silica recycling by trees. Iron and aluminum organocomplexes are adsorbed on clay minerals until they are biologically decomposed, releasing aluminum and iron and further helping to stabilize kaolinite (Lucas, 2001). The Amazonian forest litterfall contains large quantities of silicon and aluminum, which are removed by leaching and carried by soil water percolating through the upper soil. The amount of silica dissolved from the litterfall is 4 times the amount provided by weathering from below.

At places in the Amazon the ferrasol is transformed over time to a podzol, a highly weathered, bleached, sandy soil capped by a thick, peat-like accumulation of litter on the top. Everything except quartz is rapidly leached out of the podzol soil including aluminum and iron. The latter are present initially as constituents of organic complexes, but they are not immobilized, because there is a relative lack of clay minerals for adsorption.

5.06.2.3.5 Angiosperm versus gymnosperm weathering

Moulton *et al.* (2000) found that in Iceland, the weathering of basalt under the same microclimate was more rapid under angiosperms (dwarf birch) than under gymnosperms (conifers) by a factor of 1.5 for calcium and magnesium and a factor of 2 for potassium. The weathering rate was normalized to biomass, because the birch trees were much smaller than the conifers (Moulton *et al.*, 2000). They attribute the greater angiosperm weathering rates to more efficient nutrient scavenging by gymnosperms, which are favored in low nutrient environments because they contain nutrient-scavenging mycorrhizae.

In addition, gymnosperms lose their leaves (needles) throughout the year rather than in the fall as do angiosperms. Gymnosperm needles decay more slowly and release nutrients more gradually. Loss in runoff due to leaching of litter during the nongrowing season results in greater nutrient loss from soil covered by angiosperms (Knoll and James, 1987).

In comparisons of weathering data from angiosperm and gymnosperm watersheds other variables must be controlled, particularly lithology. Knoll and James (1987), using data from Likens *et al.* (1977) for temperate forested ecosystems, conclude that the rate of angiosperm-assisted weathering exceeds that of gymnosperm-assisted weathering. However, the failure to exclude data derived from carbonate weathering severely compromises the results, because carbonate weathering is much faster than silicate weathering (Robinson, 1991; Berner and Kothavala, 2001).

Red alder (angiosperm) forest in the western Washington Cascade Mountain foothills loses 8 times more calcium and magnesium and 35 times more potassium than adjacent Douglas fir forest (gymnosperms) of similar age and biomass (Homann *et al.*, 1992). Because the exchangeable soil cations are not being depleted in the red alder soil, most of the removed cations come from mineral weathering. The unusually high rates of weathering and cation loss under the symbiotic nitrogen-fixing alder are probably due to nitric acid from nitrification. Homann *et al.* (1992) also summarize net cation weathering rates from various types of trees and soils at other localities.

Data from Johnson and Todd (1990) quoted by Homann *et al.* (1992), concerning the Walker Branch Watershed in Tennessee, show considerable variations in weathering rates for various angiosperms (yellow poplar, chestnut oak, oak, and hickory) relative to that for pine. These are summarized in Table 2. The difference between weathering rates measured under angiosperms and those measured under pine varies depending on which cation is used to calculate the weathering rate. For example, as shown in Table 2, oak and hickory are 1.6 times as effective in weathering calcium as pine. The reverse is true for magnesium and potassium; these ions are more effectively weathered under pines.

Quideau *et al.* (1996) studied two experimental ecosystems in southern California and concluded that a more rapid release of calcium and magnesium from primary minerals occurred under gymnosperms (pine) than under angiosperms (scrub oak). They did not have data for loss of cations in runoff, and based their results solely on chemical analyses of the soil and the biota. They attributed the greater weathering rate by gymnosperms to the greater acidity found under the pine plot (pH 4.9) versus the oak plot (pH 5.8). However, the measured concentration changes in soil cations, and silica, over a 40-year period were very small and subject to large statistical error, especially for silica. Analyses of biomass show that the oaks store cations more efficiently than pine. This suggests that more cations are lost to output by the pines. Vermiculite appears to be converted to mica by the fixation of potassium in soil under oak trees, because the oak litter contains higher concentrations of potassium than litter found under pines.

With similar parent material (monzonitic granite) in Mont Lozère, France, the calcium flux was found to be almost 4 times greater from spruce forest (11 kg ha^{-1} yr^{-1}) than from beech thicket (3 kg ha^{-1} yr^{-1}) (Lelong *et al.*, 1990). Grassland

Table 2 Ratios of weathering rates under angiosperms to rates under gymnosperms.

Location	Ca	Mg	K	References
Iceland				
Birch/conifers	1.5	1.5	2	Moulton *et al.* (2000) (basalt)
Cascade Mts., USA				
Red alder/Douglas fir	8	8	35	Homann *et al.* (1992) (alder is N-fixer) (spodsol/inceptisol)
S. California, USA				
Scrub oak/pine	0.8	0.8	0.9	Quideau *et al.* (1996) (diorite)
Mt. Lozère, France				
Beech/spruce	0.27			Lelong *et al.* (1990) (monzonitic granite)
Walker Branch Watershed, TN, USA				
Yellow poplar/ pine	0.9	1.2	1.3	Johnson and Todd (1990) quoted by
Chestnut oak/pine	1.3	0.9	1.4	Homann *et al.* (1992) (utisol)
Oak and hickory/pine	1.6	0.4	0.8	

was found to be intermediate in the rate of calcium release (7 kg ha^{-1} yr^{-1}).

The studies discussed so far are based on chemical analyses of soils and waters draining watersheds. Comparisons of weathering rates based on direct studies of weathered minerals also exist. Bouabid *et al.* (1995) in an area of northern Minnesota found that plagioclase of fixed calcium/sodium composition exhibited approximately equal degrees of surface etch pitting in well-drained soils underneath stands of pine and oak-basswood, whereas under more acidic and less well-drained white cedar etching was greater. Augusto *et al.* (2000) inserted weighed mineral samples under a variety of stands of conifers and hardwoods (Norway spruce, Scots pine, sessile oak, pedunculated oak, and European beech) in northern France and found that after nine years there was a distinctly greater mass loss of plagioclase relative to quartz (used as a control) under the conifers than under the hardwoods.

The results summarized here offer no consistent answer to the question whether weathering rates under angiosperms are faster or slower than under gymnosperms. There are variations in the reported rate of release by weathering among different types of angiosperms and gymnosperms depending upon plant species, rock type, soil type, age of stand, climate, and methods of study. Temperature differences may also affect the results, because the normal progression of trees with increasing elevation and decreasing chemical weathering is from angiosperms at lower elevations to gymnosperms at higher elevations. These variations make it difficult to arrive at a valid generalization regarding the rate of weathering under angiosperms relative to that under gymnosperms. The ratio of the weathering rate under angiosperms to that under gymnosperms is usually between 0.5 and 1.5 (Table 2). As pointed out by Kothavala and Berner (2001), more carefully designed studies are needed before more precise conclusions can be reached.

5.06.2.3.6 *Effect of lichens and bryophytes on weathering*

Lichens are symbiotic associations of algae and fungi characterized by low growth rates and nutrient requirements that enable them to play the role of pioneer vegetation in the colonization of fresh rocks (Chapin, 1980). Bryophytes, which include mosses and liverworts, are small green land plants that lack vascular tissues, leaves, stems, and roots; they live only in moist habitats. Both lichens and bryophytes were early colonizers of the land, and there has been considerable interest in and discussion of the role of lichens

in enhancing rates of mineral weathering (Schwatrzman and Volk, 1989). The organization of the lichen–mineral system has been used by Banfield *et al.* (1999) as a model of mineral weathering in the rhizosphere (root area) of trees, which have symbiotic fungi and bacteria. There is an area of direct contact between the fungal part of the lichen and mineral surfaces; beyond this area contact is only with microbial products such as organic acids. In the contact area fungal hyphae, acting together with physical weathering process such as freeze–thaw, penetrate mineral cleavages and grain boundaries, and accumulate fragments of minerals within the thallus of the fungi. Cells are attached to mineral surfaces, and high molecular weight polymers are produced that cause dissolution, transport, and recrystallization on the polymer nucleus. These processes accelerate chemical weathering relative to that on bare rock. There is also enhanced dissolution in areas away from the contact zones due to the effect of organic acids. A similar description is given by Robert and Berthelin (1986). These authors also found calcium oxalate within lichens; oxalic acid is a common lichen exudate. Most of the lichen weathering products are either poorly crystallized or amorphous compounds (e.g., Jackson and Keller, 1970) such as silica gels and ferrihydrite.

In a study across a transect in Hawaii where variation in rainfall and temperature is systematic, Brady *et al.* (1999) studied the effect of abiotic weathering versus lichen weathering. They compared the porosity developed in plagioclase and olivine crystals by weathering under lichens versus that developed on totally bare rocks. Using a model that calculates temperature and rainfall, they found that lichen weathering was more sensitive to rainfall than abiotic weathering and that the weathering rate under lichens was 2–18 times as great as under abiotic conditions. However, it is very difficult to interpret their results quantitatively, because the actual temperatures and the degree of wetting of the rock surfaces that they studied were not measured.

Lichens, and other nonvascular small ground cover such as mosses, cannot accelerate weathering to as a great an extent as large vascular plants such as trees. This is due to a number of factors. (i) The interfacial area between organisms like lichens and minerals is very small compared to that between the huge mass of root hairs associated with vascular plants. For example, the area of roots and root hairs for a variety of small vascular (crop) plants is over 1,000 times greater than the actual spatial area occupied by the plants (Wild, 1993). (ii) The rate of growth of most vascular plants, especially trees, is far greater than that of lichens or algae. Thus, the rate of uptake of nutrients from the weathering of minerals is much greater for trees. (iii) Naturally occurring lichen acid–metal complexes are rare, have limited solubility in water, and are located on the upper thallus, which is not in contact with the rock. This suggests that they are not important in mineral weathering (Barker *et al.*, 1997). The rarity of soils developed under lichens, as compared to those under trees, further attests to the greater efficacy of trees in accelerating weathering.

5.06.2.3.7 Carbon cycling

Certain implications for the global carbon cycle follow from the results of Moulton *et al.* (2000) in Iceland. In addition to storage of carbon via growth, forests, because they accelerate silicate weathering, cause increased losses of HCO_3^- to streams and ultimately to the ocean thus reducing atmospheric CO_2 over the short term. By reacting with silicate minerals during weathering, CO_2 is converted to dissolved HCO_3^- (see Section 5.06.3.1). In Iceland the yearly rate of carbon storage in trees is only ~3 times greater than the rate of HCO_3^--C release to streams, indicating that, in this environment, the rate of CO_2 removal as bicarbonate in streams is by no means negligible compared to removal by plant uptake (see Table 3).

Chadwick *et al.* (1994) also found a considerable rate of loss of carbon due to silicate weathering as compared to carbon fixation in plants and soil at a number of places (Table 3). Kelly *et al.* (1998) found that the yearly rate of plant carbon storage along a transect in Kohala,

Hawaii, increased with increasing precipitation (18–350 cm yr^{-1}) and that this was accompanied by a factor of 6 increase in the long-term flux of HCO_3^--C due to weathering. The HCO_3^--C flux in areas of both low and high rainfall was greater than the rate of storage of organic carbon in soil, thus providing an important removal mechanism for atmospheric CO_2.

The effect of an increase in ambient atmospheric CO_2 concentration on forest growth and soil CO_2 dynamics has been studied by Andrews and Schlesinger (2001) in the Duke Forest Free-Air CO_2 Enrichment Experiment (FACE). Because the concentration of soil CO_2 and soil acidity increases with greater atmospheric CO_2 concentrations, the rate of silicate weathering increases in the soil. Chemical weathering of calcium and magnesian silicates in the soil converts soil CO_2 (ultimately derived from the atmosphere) to dissolved bicarbonate which is carried by streams to the sea, where half of the carbon is precipitated as carbonate minerals. This exerts a major control on atmospheric CO_2 over long periods of geological time (Section 5.06.3.1). For an increase in ambient atmospheric CO_2 of 55% (+200 ppmv) over two years in FACE, soil respiration increased by 27%, mainly due to increased root and rhizosphere respiration. The resulting increase in carbonic acid weathering of silicate minerals caused a 33% greater flux of bicarbonate to groundwater; this provides a negative feedback for increases in atmospheric CO_2 (see Table 3). From the increase in the HCO_3^- flux to groundwater with elevated CO_2, Andrews and Schlesinger (2001) estimated the net global

Table 3 Carbon storage and carbon loss from forests as bicarbonate due to silicate weathering (in g C m^{-2} yr^{-1}).

Location/conditions	C storage	C loss	References
Iceland			Moulton *et al.* (2000)
S. Birch	1.68	0.66	
N. Birch	2.31	0.75	
Conifer	3.57	0.76	
Temperate grassland			Chadwick *et al.* (1994)
Young soil (3.9 kyr)	7.4	0.11	
Old soil (240 kyr)	0.6	0.06	
Moist temperate New England deciduous forest			Chadwick *et al.* (1994)
Young soil (14 kyr)	2.5	0.31	
Young soil (14 kyr)	0.93	0.11	
Young soil (14 kyr)	1.93	0.44	
Desert scrubland			Chadwick *et al.* (1994)
Young soil (12 kyr)	0.19	0.004[a]	
Old soil (1,700 kyr)	0.005	0.002[a]	
Tropical soil, Kohala, Hawaii on basalt/tephra			Kelly *et al.* (1998)
18 cm yr^{-1} rain	0.02	0.08	
350 cm yr^{-1} rain	0.27	0.46	
FACE[b] *in NC*			Andrews and Schlesinger (2001)
Ambient CO_2		16.8	
570/360 × CO_2		21.5	

[a] Accumulation in soil as carbonate from silicate weathering.   [b] Duke Forest Free-Air CO_2 Enrichment Experiment.

sink for CO$_2$ due to greater plant activity, which increases silicate weathering, to be $(0.12-0.23) \times 10^{15}$ g C yr^{-1}.

5.06.3 GEOLOGICAL HISTORY OF PLANTS, WEATHERING, AND ATMOSPHERIC CO$_2$

The principles and observations made in modern plant/soil/weathering studies can be applied to estimate the effect of plants on weathering over geological time. In this section the focus is on the Phanerozoic eon (the past 550 million years), a time during which major plant evolution occurred.

5.06.3.1 Weathering Reactions and CO$_2$

As discussed in detail in earlier sections, large vascular plants, such as trees, accelerate the weathering of silicate minerals. This is especially important for calcium and magnesium silicates, because the removal of atmospheric CO$_2$ via calcium and magnesium silicate weathering is permanent, on a million-year timescale, whereas the weathering of carbonates and potassium and sodium silicates is not. The weathering of calcium and magnesium silicates and carbonates can be represented by the reactions:

Ca–Mg silicates:

$$2CO_2 + H_2O + CaSiO_3 \rightarrow Ca^{2+} + 2HCO_3^- + SiO_2 \quad (1)$$

$$2CO_2 + H_2O + MgSiO_3 \rightarrow Mg^{2+} + 2HCO_3^- + SiO_2 \quad (2)$$

Ca–Mg carbonates:

$$CO_2 + H_2O + CaCO_3 \rightarrow Ca^{2+} + 2HCO_3^- \quad (3)$$

$$2CO_2 + 2H_2O + CaMg(CO_3)_2 \rightarrow Ca^{2+} + Mg^{2+} + 4HCO_3^- \quad (4)$$

The cations and bicarbonate released by calcium and magnesium silicate and carbonate weathering are carried rapidly by rivers to the oceans where they are removed as Ca–Mg carbonates:

$$Ca^{2+} + 2HCO_3^- \rightarrow CO_2 + H_2O + CaCO_3 \quad (5)$$

$$Ca^{2+} + Mg^{2+} + 4HCO_3^- \rightarrow 2CO_2 + 2H_2O + CaMg(CO_3)_2 \quad (6)$$

Most dolomite forms by the reaction of Mg^{2+} with CaCO$_3$, but when combined with reaction (5)

the overall reaction is the same as reaction (6). Note that reactions (5) and (6) are the reverse of reactions (3) and (4). Thus, the weathering of calcium and magnesium carbonates, with the eventual formation of new calcium and magnesium carbonates, results in no net removal of atmospheric CO$_2$. By contrast, the overall reactions, resulting from combining reactions (1) and (2) with (5) and (6), are

$$CO_2 + CaSiO_3 \rightarrow CaCO_3 + SiO_2 \quad (7)$$

$$CO_2 + MgSiO_3 \rightarrow MgCO_3 + SiO_2 \quad (8)$$

These together with the burial of organic matter are the fundamental weathering reactions that control atmospheric CO$_2$ over geological time (Ebelmen, 1845; Urey, 1952). Carbon dioxide is transferred from the atmosphere to carbonate rocks. The weathering of potassium and sodium silicates does not have a direct effect on CO$_2$, because sodium and potassium are not removed from seawater as carbonates, but rather as silicates (Mackenzie and Garrels, 1966). These reactions return the CO$_2$ to the atmosphere.

It should be noted that reactions (1) and (2) represent the sum of many intermediate reactions. These include the photosynthetic fixation of CO$_2$ as biomass, the secretion of organic and carbonic acids by roots and associated microflora, the decomposition of litter and soil organic matter to form soil CO$_2$, the reaction of CO$_2$ with water to form carbonic acid, the reaction of organic and carbonic acids with calcium and magnesium silicates, and the formation of dissolved calcium, magnesium, and bicarbonate in soil and groundwater.

5.06.3.2 The GEOCARB Model

Based partly on reactions (7) and (8) and the reasoning presented in the previous sections, a theoretical model, GEOCARB (Berner, 1991, 1994; Berner and Kothavala, 2001), has been constructed that considers factors affecting the carbon cycle and atmospheric CO$_2$ on a multi-million year, i.e., geological timescale. Besides calcium and magnesium silicate weathering, this includes the diagenetic + metamorphic + volcanic degassing of CO$_2$ and the weathering and burial of sedimentary organic matter. Factors affecting the rate of weathering that are quantified over time are: uplift of mountains as they affect the rate of erosion and exposure of Ca–Mg silicate minerals to weathering, the evolution of land plants, continental drift as it affects the temperature and hydrology of the continents, the constant slow increase in solar radiation as it affects land temperature and hydrology, the feedback effect of elevated CO_2 as it affects global warming and an accelerated hydrological cycle via the atmospheric greenhouse effect, and the feedback effect of elevated CO_2 on plant growth and plant-assisted weathering.

Further details of the modeling are omitted here, except where they relate to plants and their effect on weathering. The model considers three major effects of the evolution of plants on weathering and CO_2. These are the rise of large vascular plants in the Devonian, the rise of angiosperms in the Cretaceous, and the effect of CO_2 fertilization on plant growth. The relationship between past atmospheric CO_2 levels and paleosols is also discussed by Retallack (see Chapter 5.18).

5.06.3.3 Prevascular Plants and Weathering

There is little doubt (e.g., see Gensel and Edwards (2001) and Willis and McElwain (2002) for summaries) that there was some biological land cover prior to the rise of large vascular plants in the Devonian. In the Early Paleozoic fossil evidence of tetrad spores and bryophyte remains attest to this. In fact, the surface of the land may have been occupied even earlier, during the Precambrian by some sort of photosynthetic microbial mats (Horodyski and Knauth, 1994). The effect of these organisms on the rate of mineral weathering cannot be quantitatively deduced, although using lichens as a modern analogue, Schwartzman and Volk (1989) calculated that nonvascular plants should have had a large effect on global weathering and atmospheric CO_2 during the Precambrian. There is solid evidence that etching of minerals via organic acids secreted by lichens does occur (see Section 5.06.2.3.6). However, the study cited by Schwartzman and Volk, as the basis of their calculation, was of lichens that trap windblown dust, and not those that actively weather rock (Cochran and Berner, 1996), casting some doubt on their conclusions.

It has been shown in Section 5.06.2.3.6 that lichens and mosses do not accelerate weathering to the same degree as trees, because the efficiency of nonvascular plants in accelerating weathering is much lower. GEOCARB modeling of the period prior to the rise of large vascular plants in the Devonian is treated solely in terms of the reaction of natural acid rain falling from a high-CO_2 atmosphere, directly on calcium and magnesium silicate minerals on the ground. For this purpose the results of mineral dissolution experiments and soil hydrological arguments, as they affect the pH of soil solutions, are utilized (Berner, 1991, 1994).

5.06.3.4 The Rise of Large Vascular Plants and their Effect on Weathering and CO_2

During the Devonian, from about 390 Ma to 360 Ma, large vascular plants with deep and extensive rooting systems arose on the continents (e.g., Niklas, 1997; Gensel and Edwards, 2001; Willis and McElwain, 2002). Of special interest was the spread of these plants to well-drained upland terrains where the plants could accelerate mineral weathering because of adequate drainage (Retallack, 1997; Algeo and Scheckler, 1998). Previously, vascular plants were small and confined to lowland water courses (Willis and McElwain, 2002).

Quantifying the effect of the rise of the large plants in the Devonian is difficult but can be estimated from studies of modern plants and weathering. From the studies summarized in Table 1 we may conclude that the effects of modern trees have resulted an increase in the weathering rates of calcium and magnesium silicates by factor of 2–10. Application of this range of values to the GEOCARB model results in big differences for the level of CO_2 prior to the rise of large plants (Figure 3). We believe that the only modern study that had the best control of factors other than plants on weathering rate is that of Moulton *et al.* (2001). On the basis of this study, a value for the acceleration of the rate of CO_2 removal by plant-assisted weathering of a factor of 4 is used as the "best estimate" in Figure 3. The resulting curve is in good agreement with calculated concentrations of CO_2 (Berner, 1998, 2003) based on independent studies of the carbon isotopic composition of carbon in paleosols (Mora *et al.*, 1996; Mora and Driese, 1999; Cox *et al.*, 2001) and the stomatal density of fossil plants (McElwain and Chaloner, 1995). Use of a plant accelerating factor of 10 or more in the GEOCARB model results in Early Paleozoic levels of CO_2 that are excessively high (Figure 3), suggesting that lower values for this factor are more correct for this period of time.

Figure 3 shows that the rise of large vascular plants in the Devonian (390–360 Ma) had a very major effect on atmospheric CO_2. In fact, no other process in the GEOCARB model was found to have such a large effect. The great drop in CO_2 was not simply due to its removal from the atmosphere or even to a large increase in the overall global weathering flux. This is because of the necessity of balancing inputs and outputs of CO_2 to/from the atmosphere. The carbon dioxide supplied to the atmosphere by degassing from volcanoes, metamorphism, and diagenesis must be consumed by weathering at an essentially equal rate to avoid physically impossible variations in CO_2 level (Berner and Caldeira, 1997). Thus, for

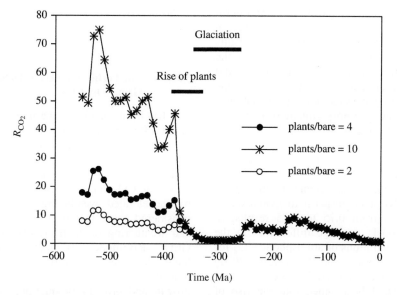

Figure 3 Plot of atmospheric CO$_2$ versus time for the Phanerozoic, as calculated via the GEOCARB model, showing sensitivity to the rise of large vascular land plants. R_{CO_2} is the ratio of the mass of CO$_2$ at some past time to that for the pre-industrial present. The ratios plants/bare refer to the relative rates of CO$_2$ consumption via weathering in the presence of vascular plants versus prevascular ground cover ("bare").

constant degassing and weathering fluxes, the accelerating effect of the plants in the Devonian was balanced by the decelerating effect of a drop in CO$_2$. Less CO$_2$, via the greenhouse effect, meant a cooler and drier earth. The large drop in CO$_2$ shown in Figure 3 occurred slowly over several million years and represents only a tiny imbalance between almost constant rates of both degassing and uptake via weathering. If the volcanic plus metamorphic input of CO$_2$ were to remain constant and the weathering rate simply increased by a factor of 4 without any greenhouse or other compensating effect, all CO$_2$ would be removed from the atmosphere in less than a million years (Berner and Caldeira, 1997).

A further drop in CO$_2$ later in the Carboniferous (~350 Ma) was also brought about by plants. Woody plants are rich in lignin, which is relatively nonbiodegradable. The production of lignin and other plant degradation-resistant materials by Carboniferous forests, followed by their burial and preservation in sediments, led to the further removal of CO$_2$ from the atmosphere. The burial was both on land (and where sufficiently abundant as coal) and in the sea after transport there by rivers. This increased burial of organic carbon, along with enhanced weathering starting in the Devonian, helps to explain the overall negative excursion in CO$_2$ shown in Figure 3 for the Late Paleozoic.

The drop in CO$_2$ going into the Carboniferous and the low CO$_2$ concentrations maintained into the Permian (285–245 Ma) correspond closely

with large-scale continental glaciation. The longest and most areally extensive glaciation of the entire Phanerozoic occurred at the time of low values of CO$_2$ calculated by the modeling (see also Crowley and Berner, 2001; Beerling *et al.*, 2002) (Figure 3). Thus, the rise of large vascular plants, because of their effect on CO$_2$ and the atmospheric greenhouse, played a major role in bringing about global climate change.

5.06.3.5 The Rise of Angiosperms

From the Late Permian to the Jurassic (270–140 Ma) the biomass on land was dominated by gymnosperms. During the Cretaceous, ~130 Ma, flowering plants began to populate the continents (Willis and McElwain, 2002). From that time on, their land coverage increased until angiosperms became the dominant occupants of the continents at ~80 Ma. This dominance has continued to the present, especially at low latitudes. Weathering at low latitudes is enhanced by high temperatures and high rainfall (providing there is sufficient relief); thus, the changeover from gymnosperms to angiosperms may have resulted in a change in global rates of plant-assisted weathering (Volk, 1989). The question is: did the replacement of gymnosperms by angiosperms accelerate weathering?

Studies of modern ecosystems reviewed in Section 6.01.2.3.5 indicate that there is no consensus over their relative effects on weathering rates. The data in Table 2 show a total range from 0.25 to 8 for the angiosperm/gymnosperm flux ratio for the release of dissolved calcium and magnesium. However, most of the results are between 0.5 and 1.5. It is obvious that the weathering effect varies between species and within each group due to such factors as the ability to fix nitrogen. As a first approximation it is likely that the gymnosperm/angiosperm rate ratio is relatively close to 1.

Unfortunately, for purposes of carbon cycle modeling it is critical to have a reasonably accurate estimate of the relative effects on weathering of angiosperms versus gymnosperms. Variation of the ratio of angiosperm to gymnosperm weathering rates from 0.75 to 1.25 results, in GEOCARB models, in large differences in the calculated concentrations of atmospheric CO_2 for the Mesozoic (Berner and Kothavala, 2001) (Figure 4). For the "standard" or best estimate case (Figure 4), it is assumed that the rise of angiosperms brought about a linear increase with time of the plant weathering factor from 0.875 at 130 Ma to 1.00 at 80 Ma and thereafter. The standard situation is indirectly reproduced from a crude fit to obtain independent CO_2 estimates, also shown in the diagram. This fit, however, should not be considered a justification for the use of a pre-angiosperm weathering factor of 0.875, because additional factors, especially volcanic degassing, are important in controlling atmospheric CO_2 over this time period and these factors are also not well defined (e.g., see Wallmann, 2001; Tajika, 1998). There is a great need for further research on the relative roles of angiosperms and gymnosperms as they affect the rate of weathering.

5.06.3.6 The Fertilization Over Time of Plant Growth by CO_2

It has been found in experimental studies that plants often grow faster and store more carbon in roots at elevated atmospheric CO_2 concentrations (e.g., Godbold *et al.*, 1997). As CO_2 varied over the Phanerozoic, how could variations in CO_2 have affected growth and plant root-accelerated weathering? Volk (1989) has considered the effect of CO_2 on plant growth and weathering on a theoretical basis, and the GEOCARB model has incorporated a negative feedback against CO_2 variation based on similar reasoning. The idea is that with increased growth and primary production, plants take up nutrients at a higher rate and thus weather rocks more rapidly. Actual experimental demonstration of this effect is shown by the work of Andrews and Schlesinger (2001), who subjected a forest plot to elevated CO_2 and measured, among many other things, the flux of dissolved bicarbonate from the plots (for a further discussion see Section 5.06.2.3.7). They found a 33% increase in the bicarbonate flux accompanying an increase in ambient CO_2 from 360 ppm to 570 ppm. This is in reasonable agreement with the value (23%) calculated for the same change in CO_2 by the GEOCARB model (Berner and Kothavala, 2001).

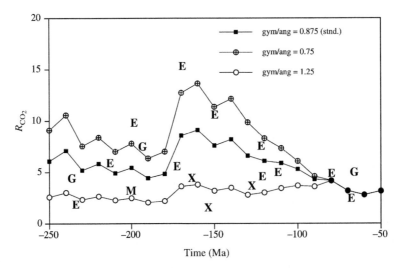

Figure 4 Plot of atmospheric CO_2 versus time for the Mesozoic as calculated via the GEOCARB model showing sensitivity to the rise of angiosperms. R_{CO_2} is the ratio of the mass of CO_2 at some past time to that for the pre-industrial present. The ratios gym/ang refer to the relative rates of CO_2 consumption via weathering. The superimposed letters refer to independent estimates of R_{CO_2} from studies of paleosols (Ekart *et al.*, 1999; Chen *et al.*, 2001; Ghosh *et al.*, 2001) and from the study of the stomatal density of fossil plants (McElwain and Chaloner, 1995). E: Ekart *et al.*—soil carbonate; G: Ghosh—soil carbonate; X: Chen *et al.*—stomata; and M: McElwain Tr–J—stomata.

There are two problems in applying the results of studies, such as that of Andrews and Schlesinger, to models of the carbon cycle. One is that their experiment lasted only for a few years. Would the effect continue? The other is that the proportion of plants on a global basis that respond similarly is not known. It is well established that plant growth is limited at many places by light, nutrients, or water availability (e.g., Bazzaz, 1990) and, as a result, will not respond to changes in atmospheric CO_2. In the GEOCARB model the dimensionless plant growth fertilization weathering effect is formulated as

$$f(\text{plant growth fertilization})$$
$$= [2R_{CO_2}/(1 + R_{CO_2})]^{\text{FERT}}$$

where R_{CO_2} is the ratio of CO_2 mass in an ancient atmosphere to that of the pre-industrial period and FERT is an exponent representing the global response of plant-assisted weathering to CO_2. The expression is constructed so that as CO_2 rises, the weathering effect approaches a maximum, of factor of 2, in the limit of FERT = 1. This accounts for the fact that the land can accommodate just so much total biomass. The real problem is determining the value of FERT. Figure 5 shows the effect of varying FERT from 0 to 1. Rather arbitrarily, an intermediate value of FERT = 0.4 is assumed as a standard value. The value of 0.4 is equivalent to assuming that only ~35% of plants globally are influenced by changes in CO_2, whereas values of 0 and 1 are for the situation or no plants or all plants being influenced by CO_2. Note the considerable variability shown in calculated values of CO_2 over the past 250 million years due to a lack of knowledge of the value for FERT. This shows the need for more studies like that of Andrews and Schlesinger but applied to areas where plant growth may be more severely limited by light, nutrients, or water.

5.06.4 SUMMARY

Weathering of minerals ultimately reduces simply to their solubilization by soil solutions. Plant roots and their associated microbiota, by their ability to secrete organic acids, greatly accelerate the rate of dissolution. This is accomplished by: (i) the organic acids which provide additional hydrogen ions for attack on the silicate framework and (ii) the formation of metal organic complexes, chelates, that solubilize the less soluble elements such as aluminum and iron. Also, plant roots exchange hydrogen ions directly with cations in soil waters. Plants have evolved processes that accelerate weathering, probably because of the need to pick up nutrients for growth with principal ones being potassium, calcium, magnesium, and phosphorus. Accelerated weathering is also accomplished by the recirculation of water by evapotranspiration, by roots holding soil against removal by erosion, and by fracturing of rocks by roots, thereby allowing greater water–mineral contact area.

Field studies have shown that the rate of silicate weathering can be increased by plants by a factor ~(2–10). Use of these values in modeling of the long-term geological carbon cycle shows that the

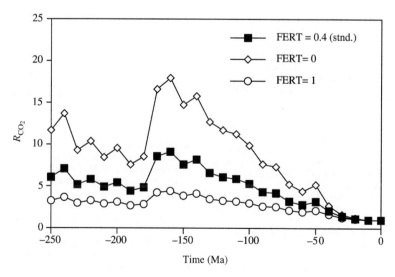

Figure 5 Plot of atmospheric CO_2 versus time for the Mesozoic and Cenozoic as calculated via the GEOCARB model showing sensitivity to the proportion of land plants whose growth and weathering rates are accelerated by rises in atmospheric CO_2. R_{CO_2} is the ratio of the mass of CO_2 at some past time to that for the pre-industrial present. The values of FERT = 0, 0.4, and 1 refer to the proportion of land plants whose growth and weathering are/were globally accelerated by CO_2.

population of the continents by large, deeply rooted land plants in the Devonian led to a very large drop in atmospheric CO_2. This was caused by plant-enhanced weathering of calcium and magnesium silicate minerals with carbon transferred from the atmosphere ultimately to carbonate sediments deposited on the ocean floor. Later, a further decrease in CO_2 occurred due to enhanced organic matter burial in sediments derived from plants. The decrease in CO_2, as a result of the atmospheric greenhouse effect, led to global cooling and the development of vast continental glaciers in the Carboniferous and Permian. This illustrates that plant evolution has affected weathering, and has been a major factor in the evolution of the atmosphere and of climate over geological time.

ACKNOWLEDGMENTS

The research of Robert Berner is supported by DOE Grant DE-FG02-01ER15173 and NSF Grant EAR 0104797. Katherine Moulton was intended to be the principal author originally, but she was killed in a tragic accident on November 9, 2001. This chapter is dedicated to her.

REFERENCES

Alexandre A., Colin F., and Meunier J.-D. (1994) Les phytoliths, indicateurs du cycle biogéochimique du silicium en forêt equatoriale. *C. R. Acad. Sci. Paris (II)* **319**, 453–458.

Alexandre A., Meunier J.-D., Colin F., and Koud J.-M. (1997) Plant impact on the biogeochemical cycle of silicon and related weathering processes. *Geochim. Cosmochim. Acta* **61**, 677–682.

Algeo T. J. and Scheckler S. E. (1998) Terrestrial–marine teleconnections in the Devonian: links between the evolution of land plants, weathering processes, and marine anoxic events. *Roy. Soc. London Phil. Trans. B* **353**, 113–130.

Amrhein C. and Suarez D. L. (1988) The use of surface complexation model to describe the kinetics of ligand-promoted dissolution of anorthite. *Geochim. Cosmochim. Acta* **52**, 2785–2793.

Anderson S., Drever J. I., Frost C. D., and Holden P. (2000) Chemical weathering in the foreland of a retreating glacier. *Geochim. Cosmochim. Acta* **64**, 1173–1189.

Andrews J. H. and Schlesinger W. H. (2001) Soil CO_2 dynamics, acidification, and chemical weathering in a temperate forest with experimental CO_2 enrichment. *Global Bigeochem. Cycles* **15**(1), 149–162.

April R. and Keller D. (1990) Mineralogy of the rhizosphere in forest soils of the eastern United States. *Biogeochemistry* **9**, 1–18.

Arthur M. A. and Fahey T. J. (1993) Controls on soil solution chemistry in a subalpine forest in north-central Colorado. *Soil Sci. Soc. Am. J.* **57**, 1123–1130.

Augusto L., Turpault M. P., and Ranger J. (2000) Impact of forest tree species on feldspar weathering rates. *Geoderma* **96**, 215–237.

Banfield J. F., Barker W. W., Welch S. A., and Taunton A. (1999) Biological impact on mineral dissolution: application of the lichen model to understanding mineral weathering in the rhizosphere. *Proc. Natl. Acad. Sci. USA* **96**, 3404–3411.

Barker W. W., Welch S. A., and Banfield J. F. (1997) Biogeochemical weathering of silicate minerals. In *Geomicrobiology: Interactions between Microbes and Minerals*, Reviews in Mineralogy (eds. J. F. Banfield and K. H. Nealson), Mineralogical Society of America, vol. 35, 448pp.

Bazzaz F. A. (1990) The response of natural ecosystems to the rising global CO_2 levels. *Ann. Rev. Evol. Syst.* **21**, 167–196.

Beerling D. J. (2002) Low atmospheric CO_2 levels during the Permo-carboniferous glaciation inferred from fossil lycopsids. *Proc. Natl. Acad. Sci. USA* **99**, 12567–12571.

Benedetti M. F., Menard O., Noack Y., Carvalho A., and Nahon D. (1994) Water–rock interactions in tropical catchments: field rates of weathering and biomass impact. *Chem. Geol.* **118**, 203–220.

Berner R. A. (1991) A model for atmospheric CO_2 over Phanerozoic time. *Am. J. Sci.* **291**, 339–376.

Berner R. A. (1994) Geocarb II: a revised model of atmospheric CO_2 over Phanerozoic time. *Am. J. Sci.* **294**, 56–91.

Berner R. A. (1998) The carbon cycle and CO_2 over Phanerozoic time: the role of land plants. *Phil. Trans. Roy. Soc. B* **353**, 75–82.

Berner R. A. (2003) The rise of trees and their effects on Paleozoic atmospheric CO_2, climate and geology. In *A History of Atmospheric CO_2 and its Effect on Plants, Animals, and Ecosystems* (eds. T. E. Cerling, M. D. Dearing and J. R. Ehleringer). Springer (in press).

Berner R. A. and Caldeira K. (1997) The need for mass balance and feedback in the geochemical carbon cycle. *Geology* **25**, 955–956.

Berner R. A. and Cochran M. F. (1998) Plant-induced weathering of Hawaiian basalts. *J. Sedim. Res.* **68**(5), 723–726.

Berner R. A. and Kothavala Z. (2001) Geocarb III: a revised model of atmospheric CO_2 over Phanerozoic time. *Am. J. Sci.* **301**, 182–204.

Berner R. A. and Rao J.-L. (1997) Alkalinity buildup during silicate weathering under a snow cover. *Aquat. Geochim.* **2**, 301–312.

Blum J. D., Klaue A., Nezat C. A., Driscoll C. T., Johnson C. E., Siccama T. G., Eagar C., Fahey T. J., and Likens G. E. (2002) Mycorrhizal weathering of apatite as an important calcium source in base-poor forest ecosystems. *Nature* **417**, 729–731.

Bormann B. T., Wang D., Bormann F. H., Benoit G., April R., and Snyder M. C. (1998) Rapid plant induced weathering in an aggrading experimental ecosystem. *Biogeochemistry* **43**, 129–155.

Bormann F. H. and Likens G. E. (1979) *Pattern and Process in a Forested Ecosystem*. Springer, New York, 253pp.

Bouabid R., Nater E. A., and Bloom P. R. (1995) Characterization of the weathering status of feldspar minerals in sandy soils of Minnesota using SEM and EDX. *Geoderma* **66**, 137–149.

Boyle J. R. and Voight G. K. (1973) Biological weathering of silicate minerals: implications for tree nutrition and soil genesis. *Plant Soil* **38**, 191–201.

Brady P. V., Dorn R. I., Brazel A. J., Moore R. B., and Glidewell T. (1999) Direct measurement of the combined effects of lichen, rainfall, and temperature on silicate weathering. *Geochim. Cosmochim. Acta* **63**, 3293–3300.

Chadwick O. A., Kelly E. F., Merritts D. M., and Amundson R. G. (1994) Carbon dioxide consumption during soil development. *Biogeochemistry* **24**, 115–127.

Chapin F. S., III (1980) The mineral nutrition of wild plants. *Ann. Rev. Ecol. Syst.* **11**, 233–260.

Chen L. Q., Li C. S., Chaloner W. G., Beerlin D. J., Su Q. G., Collinson M. E., and Mitchell P. L. (2001) Assessing the potential for the stomatal characters of extant and fossil Ginkgo leaves to signal atmospheric CO_2 change. *Am. J. Bot.* **88**, 1309–1315.

Cochran M. F. and Berner R. A. (1996) Promotion of chemical weathering by higher plants: field observations on Hawaiian basalts. *Chem. Geol.* **132**, 71–77.

Colin F., Brimhall G. H., Nahon D., Lewis C. J., Baronnet A., and Danti K. (1992) Equatorial rain forest lateritic mantles: a geomembrane filter. *Geology* **20**, 523–526.

Cox J. E., Railsback L. B., and Gordon E. A. (2001) Evidence from Catskill pedogenic carbonates for a rapid large Devonian decrease in atmospheric carbon dioxide concentrations. *Northeastern Geol. Environ. Sci.* **23**, 91–102.

Crowley T. J. and Berner R. A. (2001) CO_2 and climate change. *Science* **292**, 870–872.

Drever J. I. (1994) The effect of land plants on weathering rates of silicate minerals. *Geochim. Cosmochim. Acta* **58**, 2325–2332.

Drever J. I. and Stillings L. L. (1997) The role of organic acids in mineral weathering. *Coll. Surf. A: Physicochem. Eng. Asp.* **120**, 167–181.

Drever J. I. and Vance G. F. (1994) Role of soil organic acids in mineral weathering processes. In *Role of Soil Organic Acids in Geological Processes* (eds. M. D. Lewan and E. D. Pittman). Springer, New York, pp. 138–161.

Drever J. I. and Zobrist J. (1992) Chemical weathering of silicate rocks as a function of elevation in the southern Swiss Alps. *Geochim. Cosmochim. Acta* **56**, 3209–3216.

Ebelmen J. J. (1845) Sur les produits de la decomposition des especes minerales de la familie des silicates. *Ann. des Mines* **7**, 3–66.

Ekart D., Cerling T. E., Montanez I. P., and Tabor N. J. (1999) A 400 million year carbon isotope record of pedogenic carbonate: implications for paleoatmospheric carbon dioxide. *Am. J. Sci.* **299**, 805–827.

Epstein E. (1994) The anomaly of silicon in plant biology. *Proc. Natl. Acad. Sci. USA* **91**, 11–17.

Gensel P. G. and Edwards D. (2001) *Early Land Plants and their Environments.* Columbia University Press, New York.

Ghosh P., Ghosh P., and Bhattacharya S. K. (2001) CO_2 levels in the Late Palaeozoic and Mesozoic atmosphere from soil carbonate and organic matter, Satpura basin, Central India. *Paleogeogr. Paleoclimatol. Paleoecol.* **170**, 219–236.

Godbold D. L., Berntson G. M., and Bazzaz F. A. (1997) Growth and mycorrhizal colonization of three North American tree species under elevated atmospheric CO_2. *New Phytologist* **137**, 433–440.

Griffiths R. P., Baham J. E., and Caldwell B. A. (1994) Soil solution chemistry of ectomycorrhizal mats in forest soil. *Soil Biol. Biochem.* **26**, 331–337.

Hinsinger P., Jaillard B., and Dufey J. E. (1992) Rapid weathering of a trioctahedral mica by the roots of ryegrass. *Soil Sci. Am. J.* **56**, 977–982.

Hinsinger P., Barros O. N. F., Benedetti M. F., Noack Y., and Callot G. (2001) Plant-induced weathering of a basaltic rock: experimental evidence. *Geochim. Cosmochim. Acta* **65**(1), 137–152.

Homann P. S., Van Miegroet H., Cole D. W., and Wolfe G. V. (1992) Cation distribution, cycling, and removal from mineral soil in Douglas fir and red alder forests. *Biogeochemistry* **16**, 121–150.

Horodyski R. J. and Knauth L. P. (1994) Life on land in the Precambrian. *Science* **263**, 494–498.

Jackson T. A. and Keller W. D. (1970) Comparative study of the role of lichens and inorganic processes in the chemical weathering of recent Hawaiian lava flows. *Am. J. Sci.* **269**, 446–466.

Jackson T. A. and Voight G. K. (1971) Biochemical weathering of calcium-bearing minerals by rhizosphere microorganisms, and its influence on calcium accumulation in trees. *Plant Soil* **35**, 655–658.

Jaillard B., Guyon A., and Maurin A. S. (1991) Structure and composition of calcified roots, and their identification in calcareous soils. *Geoderma* **50**, 197–210.

Johnson D. W. and Todd D. E. (1990) Nutrient cycling in forests of Walker Branch Watershed, Tennessee: roles of uptake and leaching in causing soil changes. *J. Environ. Qual.* **19**, 97–104.

Jongmans A. G., Van Breeman N., Lundstrom U., van Hees P. A. W., Finlay R. D., Srinivasan M., Unestam T., Giesle R., Melkerud P.-A., and Olsson M. (1997) Rock-eating fungi. *Nature* **389**, 682–683.

Kelly E. F., Chadwick O. A., and Hill C. A. (1998) The effect of plants on mineral weathering. *Biogeochemistry* **42**, 21–43.

Knoll M. A. and James W. C. (1987) Effect of the advent and diversification of vascular land plants on mineral weathering through geologic time. *Geology* **15**, 1099–1102.

Lelong F., Dupreaz C., Durand P., and Didon-Lescat J. F. (1990) Effects of vegetation type on the biogeochemistry of small catchments (Mont Lozère, France). *J. Hydrol.* **166**, 125–145.

Likens G. E., Bormann F. H., Pierce R. S., Eaton J. S., and Johnson N. M. (1977) *Biogeochemistry of a Forested Ecosystem.* Springer, New York, 146pp.

Lovering T. S. (1959) Significance of accumulator plants in rock weathering. *Geol. Soc. Am. Bull.* **70**, 781–800.

Lucas Y. (2001) The role of plants in weathering. *Ann. Rev. Earth Planet. Sci.* **29**, 135–163.

Lucas Y., Luizao F. J., Chauvel A., Rouiller J., and Nahon D. (1993) The relation between biological activity of the rainforest and mineral composition of the soils. *Science* **260**, 521–523.

Mackenzie F. T. and Garrels R. M. (1966) Chemical mass balance between rivers and oceans. *Am. J. Sci.* **264**, 507–525.

Marschner H. and Romheld V. (1983) *In vivo* measurement of root-induced pH changes at the soil–root interface; effect of plant species and nitrogen source. *Z. Pflanzenphysiol.* **111**, S.241–S.251.

Marschner H., Hussling M., and George E. (1991) Ammonium and nitrate uptake rates and rhizosphere pH in non-mycorrhizal roots of Norway spruce (*Picea abies* (L.) Karst.). *Trees* **5**, 14–21.

McElwain J. C. and Chaloner W. G. (1995) Stomatal density and index of fossil plants track atmospheric carbon dioxide in the Paleozoic. *Ann. Bot.* **76**, 389–395.

Meunier J. C., Colin F., and Alarcon C. (1999) Biogenic storage of silica in soils. *Geology* **27**, 835–838.

Mora C. I. and Driese S. G. (1999) Palaeoenvironment, palaeoclimate and stable carbon isotopes of Palaeozoic red-bed palaeosols, Appalachian Basin, USA and Canada. *Spec. Pub. Intl. Assoc. Sedimentologists* **27**, 61–84.

Mora C. I., Driese S. G., and Colarusso L. A. (1996) Middle and Late Paleozoic atmospheric CO_2 levels from soil carbonate and organic matter. *Science* **271**, 1105–1107.

Mortland M. M., Lawton K., and Uehhara G. (1956) Alteration of biotite to vermiculite by plant growth. *Soil Sci.* **82**, 477–481.

Moulton K. L. and Berner R. A. (1998) Quantification of the effect of plants on weathering: studies in Iceland. *Geology* **26**(10), 895–898.

Moulton K. L., West J., and Berner R. A. (2000) Solute flux and mineral mass balance approaches to the quantification of plant effects on silicate weathering. *Am. J. Sci.* **300**, 539–570.

Nepstad A. C., de Carvalho C. R., Davidson E. A., Jipp P. H., Lefebvre P. A., Negreiros G. H., da Silva E. D., Stone T. A., Trumbore S. E., and Vieira S. (1994) The role of deep roots in the hydrological and carbon cycles of Amazonian forests and pastures. *Nature* **372**, 666–669.

Niklas K. J. (1997) *The Evolutionary Biology of Plants.* University of Chicago Press, Chicago.

Ochs M., Brunner I., Stumm W., and Cosovic B. (1993) Effects of root exudates and humic substances on weathering kinetics. *Water Air Soil Pollut.* **68**, 213–229.

Probst A., El Gh'mari A., Aubert D., Fritz B., and McNutt R. (2000) Strontium as a tracer of weathering processes in a silicate catchment polluted by acid atmospheric inputs, Strengbach, France. *Chem. Geol.* **170**, 203–219.

Quideau S. A., Chadwick O. A., Graham R. C., and Wood H. B. (1996) Base cation biogeochemistry and weathering under oak and pine: a controlled long-term experiment. *Biogeochemistry* **35**, 377–398.

Retallack G. J. (1997) Early forest soils and their role in Devonian global change. *Science* **276**, 583–585.

Robert H. and Berthelin J. (1986) Role of biological and biochemical factors in soil mineral weathering. In *Interactions of Soil Minerals with Natural Organics and Microbes*, Soil Science Society of America Special Publication 17 (eds. P. M. Huang and M. Schnitzer). Soil Science Society of America, Madison, Wisconsin, pp. 453–496.

Robinson J. M. (1991) Land plants and weathering. *Science* **252**, 860.

Schwartzman D. W. and Volk T. (1989) Biotic enhancement of weathering and the habitability of Earth. *Nature* **340**, 457–460.

Shukla J. and Mintz Y. (1982) Influence of land–surface evapotranspiration on the Earth's climate. *Science* **215**, 1498–1501.

Spyridakis D. E., Chesters G., and Wilde S. A. (1967) Kaolinization of biotite as a result of coniferous and deciduous seedling growth. *Soil Sci. Soc. Am.* **31**, 203–210.

Stallard R. F. (1985) River chemistry, geology, geomorphology, and soils in the Amazon and Orinoco Basins. In *The Chemistry of Weathering* (ed. J. I. Drever). Reidel, Dordrecht, pp. 293–316.

Stallard R. F. and Edmond J. M. (1981) Geochemistry of the Amazon: 2. The influence of the geology and weathering environment on the dissolved load. *J. Geophys. Res.* **86**, 9844–9858.

Tajika F. (1998) Climate change during the last 150 million years: reconstruction from a carbon cycle model. *Earth Planet. Sci. Lett.* **160**, 695–707.

Taylor A. B. and Velbel M. A. (1991) Geochemical mass balances and weathering rates in forested watersheds of the southern Blue Ridge: II. Effects of botanical uptake terms. *Geoderma* **51**, 29–50.

Urey H. C. (1952) *The Planets: Their Origin and Development*. Yale University Press, New Haven.

Van Breemen N., Finlay R., Lundstrom U., Jongmans A. G., Giesler R., and Olsson M. (2000) Mycorrhizal weathering: a true case of mineral plant nutrition? *Biogeochemistry* **49**, 53–67.

Velbel M. A. (1993a) Temperature dependence of silicate weathering in nature: how strong of a negative feedback on long term accumulation of atmospheric CO_2 and global greenhouse warming? *Geology* **21**, 1059–1062.

Velbel M. A. (1993b) Constancy of silicate-mineral weathering-rate ratios between natural and experimental weathering: implication for hydrologic control of differences in absolute rates. *Chem. Geol.* **105**, 89–99.

Velbel M. A. (1985) Geochemical mass balances and weathering rates in forested watersheds of the southern Blue Ridge. *Am. J. Sci.* **285**, 904–930.

Velbel M. A., Taylor A. B., and Romero N. L. (1990) Effect of temperature on feldspar weathering in alpine and non-alpine watersheds. *Geol. Soc. Am. Abstr. Prog.* **22**, 49.

Volk T. (1989) Rise of angiosperms as a factor in long-term climate cooling. *Geology* **17**, 107–110.

Wallmann K. (2001) Controls on the Cretaceous and Cenozoic evolution of seawater composition, atmospheric CO_2 and climate. *Geochim. Cosmochim. Acta* **65**, 3005–3025.

Welch S. and Ullman W. J. (2000) The temperature dependence of bytownite feldspar dissolution in neutral aqueous solutions of inorganic and organic ligands at low temperature (5–35 °C). *Chem. Geol.* **167**, 337–354.

White A. F. and Blum A. E. (1995) Effects of climate on chemical weathering in watersheds. *Geochim. Cosmochim. Acta* **59**, 1729–1747.

White A. F., Blum A. E., Schultz M. S., Bullen T. D., Harden J. W., and Peterson M. L. (1996) Chemical weathering rates of a soil chronosequence on granitic alluvium: I. Quantification of mineralogical and surface area changes and calculation of primary silicate reaction rates. *Geochim. Cosmochim. Acta* **60**, 2533–2550.

Wild A. (1993) *Soils and the Environment: An Introduction*. Cambridge University Press, Cambridge, 287pp.

Willis K. J. and McElwain J. C. (2002) *The Evolution of Plants*. Oxford University Press, New York, 378pp.

Wright V. P., Platt N. H., Mariott S. B., and Beck V. H. (1995) A classification of rhizogenic (root-formed) calcretes, with examples from the Upper Jurassic–Lower Cretaceous of Spain and Upper Cretaceous of southern France. *Sediment. Geol.* **100**, 143–158.

5.07

Geochemical Weathering in Glacial and Proglacial Environments

M. Tranter

University of Bristol, UK

5.07.1 INTRODUCTION

It seems counterintuitive that chemical erosion in glaciated regions proceeds at rates comparable to those of temperate catchments with comparable specific runoff (Anderson *et al.*, 1997). All the usual factors that are associated with elevated rates of chemical weathering (Drever, 1988, 1994), such as water, soil, and vegetation, are either entirely absent or absent for much of the year. For example, glaciated regions are largely frozen for significant periods each year, the residence time of liquid water in the catchment is low (Knight, 1999), there are thin, skeletal soils at best, and vegetation is either absent or limited (French, 1997). Other chapters in this volume have highlighted how these factors are important in other, more temperate and tropical environments. Even so, chemical erosion rates in glaciated terrain are usually near to or greater than the continental average (Sharp *et al.*, 1995; Wadham *et al.*, 1997; Hodson *et al.*, 2000). This is because glaciated catchments usually have high specific runoff, there are high concentrations of freshly comminuted rock flour, which is typically silt sized and coated with microparticles, and adsorbed organic matter or surface precipitates that may hinder water–rock interactions are largely absent (Tranter, 1982). In short, the rapid flow of water over fine-grained, recently crushed, reactive mineral surfaces maximizes both the potential rates of chemical weathering and chemical erosion.

A range of both lab- and field-based studies of glacial chemical weathering have been undertaken, mainly on the smaller glaciers of Continental Europe (e.g., Brown *et al.*, 1993a,b), Svalbard (e.g., Hodson *et al.*, 2002), and North America (e.g., Anderson *et al.*, 2000). The field-based studies typically generate hydrographs of glacier runoff, which show a characteristic diurnal cycle during summer in low latitudes (Figure 1), and more subdued diurnal cycles at high latitudes (Figures 2 and 3). The concentration of ions in solution, typically monitored by electrical conductivity, is often inverse with discharge on both a diurnal and a seasonal basis at lower latitudes, but is more complex at higher latitudes (Figures 1–3). Figures 1–3 also show that the total flux of glacial solutes is usually dominated by fluxes associated with high discharge, dilute waters. The chemical weathering reactions that are inferred to occur from the field studies have been supported, in part, by controlled laboratory studies (e.g., Brown *et al.*, 1993a). Recent stable-isotope studies have

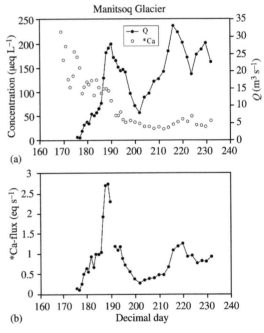

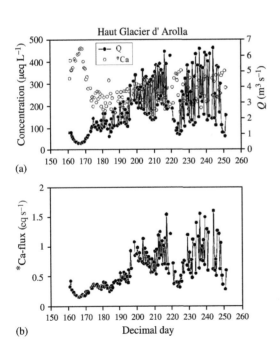

Figure 1 (a) The temporal variation of discharge and in non-sea-salt calcium ($^*Ca^{2+}$) concentration in runoff from Haut Glacier d'Arolla, a small, warm-based, valley glacier in the Swiss Alps, during 1989 (Brown *et al.*, 1993a). (b) The temporal variation in $^*Ca^{2+}$ flux from Haut Glacier d'Arolla during 1989. Maximum fluxes are associated with higher discharge waters.

Figure 2 (a) The temporal variation of discharge (Q) and non-sea-salt $^*Ca^{2+}$ concentration in runoff from Manitsoq Glacier, a small outlet glacier on the SW margin of the Greenland Ice Sheet, during 1999. The glacier is warm based, but has a cold-based margin during the winter and early ablation season, so displays polythermal-based hydrological features (Skidmore *et al.*, in preparation). (b) The temporal variation in $^*Ca^{2+}$ flux from Manitsoq Glacier during 1999. Maximum flux is associated with an early season "outburst" event, where longer stored subglacial water first exits the glacier. Otherwise, maximum fluxes are associated with higher discharge waters.

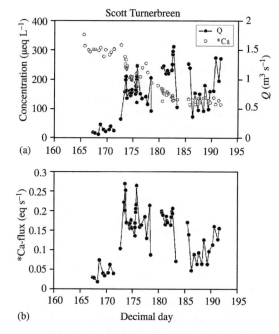

(a)

(b) Decimal day

Figure 3 (a) The temporal variation of discharge and $^*Ca^{2+}$ concentration in runoff from Scott Turnerbreen, a small, cold-based, valley glacier on Svalbard, during 1994 (after Hodgkins *et al.*, 1997). (b) The temporal variation in $^*Ca^{2+}$ flux from Scott Turnerbreen during 1994. Maximum fluxes are associated with higher discharge waters, although the precise associations are complex.

reported the key involvement of microbial processes in certain regions of the glacier bed (Bottrell and Tranter, 2002), and these processes are yet to be incorporated in lab-based chemical weathering studies.

This chapter will expand on these themes, and endeavor to integrate ongoing research into a state-of-the-science review, as well as indicating the areas into which the next generation of studies are likely to proceed. It is first necessary to understand a little basic glaciology and glacier hydrology to appreciate the principal features of the different chemical weathering environments that will be described below. The following sections summarize the types of glacial environments in which water flows, their typical debris content, the relative residence time of water, and the typical reactions that occur within them.

5.07.2 BASIC GLACIOLOGY AND GLACIER HYDROLOGY

Approximately 10% of the Earth's surface is presently covered by glaciers ($\sim 15.9 \times 10^6$ km^2 glacierized cf. 148.8×10^6 km^2 of total land surface; Knight, 1999). Some 91% of the Earth's land ice covers Antarctica, a further 8% covers Greenland, and glaciers in other regions

Table 1 Area of land surface covered by glaciers in different regions of the world, together with estimates of volume and the equivalent sea-level rise that the volume implies.

Region	Area (km^2)	Volume (km^3)	Sea level equivalent (m)
Antarctica	13,600,000	25,600,000	64
Greenland	1,730,000	2,600,000	6
North America	276,000		
Asia	185,000		
Europe	54,000	200,000	0.5
South America	25,900		
Australasia	860		
Africa	10		
Total	15,900,000	28,400,000	70.5

After Knight (1999).

Table 2 Approximate areas and volumes of the ice sheets at the last glacial maximum.

Ice sheet	Area (10^6 km^2)	Volume (10^6 km^3)	Sea level equivalent (m)
Antarctic	13.8	26	66
Greenland	2.3	3.5	11
Laurentide	13.4	30	74
Cordilleran	2.4	3.6	9
Scandinavian	6.7	13	34
Other smaller ice masses	5.2	1.1	3
Total	43.8	77.2	197

After Knight (1999).

contribute $\sim 1\%$ (Table 1). Estimates of the current annual global glacial runoff range from 0.3×10^{12} m^3 yr^{-1} to 1×10^{12} m^3 yr^{-1} (Jones *et al.*, 2002), of which $\sim 0.3 \times 10^{12}$ m^3 yr^{-1} is from Greenland (Oerlemans, 1993) and $(0.04-5) \times 10^{10}$ m^3 yr^{-1} is from Antarctica (Jacobs *et al.*, 1992). Hence, glaciers are estimated to contribute $\sim 0.6-1.0\%$ to the global annual runoff, which is believed to be $\sim 4.6 \times 10^{13}$ m^3 yr^{-1} (Holland, 1978).

The area of ice cover approximately doubled or trebled at the last glacial maximum (LGM; ~ 21 ka). Some $(33.4-43.7) \times 10^6$ km^2 was covered by glaciers, or $\sim 22-29\%$ of the present-day land surface. Table 2 shows that much of the additional ice cover was in the two great ice sheets of the northern hemisphere: the Laurentide Ice Sheet which grew over North America, and the Scandinavian Ice Sheet, which grew over Europe.

Chemical weathering can occur in a number of glacial environments. These include the margins of glaciers (in lateral moraine and in debris covering the sidewall slopes), on top of the glacier, in so-called *supraglacial* environments, and within the glacier, in *englacial* environments. However, most occurs in the sediments and rock flour beneath the glacier, in *subglacial* environments. In addition, the *proglacial* zone, the mosaic of moraine in front of the glacier, is also an important geochemical weathering environment.

5.07.2.1 Glaciers, Ice Caps, and Ice Sheets

Crudely, there are two main types of glaciers (Sugden and John, 1976), those that are unconstrained and blanket the topography, and those that are constrained by topography. The former type include ice sheets and ice caps, and the latter include valley glaciers (which contain most of the mass in this category), ice fields, and cirques. Ice sheets cover areas which are typically $>5 \times 10^4 \, km^2$, whereas ice caps cover smaller areas, $<5 \times 10^4 \, km^2$. Much of the Earth's ice cover is contained in ice sheets on Antarctica and Greenland (Table 1). Ice masses constrained by topography have areas which are typically $\sim 1-100 \, km^2$. Most research on the geochemical weathering of glaciers has been conducted on systems of this size, mainly because of the pragmatics of logistics and costs. Fortunately, it seems that, to a first approximation, the biogeochemical processes inferred from the small systems are similar to those that occur in the large systems, although the literature on large systems is currently sparse. The biogeochemical processes are similar, because the hydrology and erosional processes near the margins of larger ice masses have similarities with those that occur in smaller system.

5.07.2.1.1 Cold or warm ice at the bed?

The basal thermal regime defines whether or not liquid water can reside at the ice–bedrock interface (Paterson, 1994; Fountain and Walder, 1998). It has a first-order control on the type of subglacial drainage system that is present, which in turn defines the geochemical weathering environment (see below) and the area over which subglacial chemical weathering can occur. The basal thermal regime is a balance between factors that heat basal ice, most notably the geothermal heat flux, energy produced by basal friction and energy released by freezing water, and conduction into cold overlying ice or the frozen substrate beneath. If the basal ice is colder than the pressure melting point, the glacier is largely frozen to the bed and is termed *cold based*. This means that water has little opportunity to make contact with freshly comminuted bedrock at the glacier bed, and geochemical weathering is limited to ice marginal

environments. If the basal ice is at the pressure melting point, at least a thin layer of water is present at the interface between the ice and the bedrock, and the glacier is said to be *warm based*. Warm-based glaciers usually develop complex and seasonally variable basal drainage systems, in which water can flow through or access comminuted bedrock and basal debris. Consequently, there is a great potential for geochemical weathering to occur.

Ice sheets and ice caps usually contain both cold and warm basal sectors (Siegert, 2001). Most melt water is generated by surface melting within 10 or so kilometer of the ice margin, it is likely that the glacier bed in this region is warm, and that there is some penetration of water to the bed (Zwally *et al.*, 2002). The very margin of the ice mass is likely to be frozen to the bed during winter, since the winter cold wave penetrates through the snow and ice to the bed. This results in a barrier to water flow, which enhances water–rock contact times during the early melt season, promoting enhanced chemical weathering at the bed. Water breaking through the cold barrier is more enriched in solutes than waters with similar discharge later in the season (Skidmore and Sharp, 1999). Most valley glaciers in mid-latitudes are warm based, and rock–water reactions occur predominately at the bed (Collins, 1979; Tranter *et al.*, 2002a). Valley glaciers at higher latitudes are polythermal based, having warm-based centers and cold-based margins (Wadham *et al.*, 2000). Rock–water interaction occurs both at the bed and along the margins of the glacier. The cold-based margin can act as a barrier to water flow in the early melt season, so enhancing water–rock contact times and elevating the extent of chemical weathering, in a similar manner as in larger systems. Thin valley glaciers at high latitude are cold based, and so chemical weathering is limited to ice-margin environments (Hodgkins *et al.*, 1997).

5.07.2.1.2 Sources of water and flow paths

The source of water and the flow paths that it follows through glacial systems influence two crucial factors for geochemical weathering, namely, the rock–water contact time and the rock : water ratio (Raiswell, 1984). In addition, the flow path determines whether or not waters access freshly comminuted glacial debris, which usually has more reactive surfaces (those with strained and amorphous zones, and covered with microparticles; Petrovic *et al.*, 1976) and higher concentrations of trace reactive minerals, such as carbonates, kerogen, fluid inclusions, and sulfides. Finally, the flow path determines the ease of access to atmospheric gases or those derived from gas bubbles or gas clathrates or hydrates in melting ice.

The main source of water in most glacier systems is snow and/or ice melt. Some water is also derived from rain, and a little is derived from geothermal melting and internal deformation (Paterson, 1994). Water flows rapidly over impermeable ice, often in surface or supraglacial channels. These water courses either drain to the glacier margins, to surface depressions to form ponds or lakes, or they terminate at crevasses or moulins, from where the water descends to the glacier bed or into an englacial drainage system. Supraglacial ice melt is usually quite dilute ($I < 40\ \mu mol\ L^{-1}$). Two exceptions are due to flow through surficial moraine, in which chemical weathering can occur, or during early snow melt, when much of the solute in the snow cover is eluted by the first melt water (Tranter and Jones, 2001). The suspended sediment concentration of supraglacial melt water is typically $<1\ mg\ L^{-1}$ (unless there is access to surficial moraine or wind-blow debris) and the residence time of most of the water at the glacier surface is of the order of hours.

The continued capture of ice melt by crevasses and the flow of ice downstream eventually leads to the formation of moulins, cylindrical tubes that drain water from the surface to depth. Moulins are the main drains of surface water to the glacier bed. The melt water then flows towards the glacier terminus in large arterial tunnels or channels, which are believed to be ~1 m high and 2–5 m wide. This so-called channelized drainage system is often referred to as a low-pressure drainage system. It is water full at maximum diurnal discharge. At lower diurnal discharges, the arterial channels allow ingress of atmospheric gases into localized areas of the glacier bed (Fountain and Walder, 1998). Sediment can be acquired from the bed and from the margins of the channels. Rock : water ratios are ~1 g L^{-1}, and the dominant grain-size of suspended sediment is silt. Elevated concentrations of suspended sediment is may occur when there are major changes in the configuration of the channelized drainage system. These occur seasonally and are a consequence of the changing locus and magnitude of surface melt production.

By contrast to ice melt, snowmelt may be retained for days in snow cover, and may even

refreeze in deeper, colder packs to form super-imposed ice (Paterson, 1994). Early season snowmelt may be concentrated, since the snow-pack solute is preferentially eluted by melt water. Additionally, dust may weather chemically in wet snowpack. Later season snowmelt is as dilute as ice melt. The suspended sediment concentration of snowmelt is typically $<1\ mg\ L^{-1}$. Snowpacks are often underlain by crevasses and moulins, and snowmelt slowly draining through these to the bed feeds a high-pressure or distributed drainage system. This type of system flanks and feeds into the channelized drainage system (see below). It consists, for example, of linked cavities—lenses of water at the bed, which are poorly connected by small pipes—thin films of water at the ice–water interface, and slow flow through subglacial debris (Fountain and Walder, 1998). The residence time of water in these types of environments is of the order of hours to months, and rock : water ratios may be up to ~10 kg L^{-1} (i.e., the equivalent of water-saturated basal sediment).

5.07.2.1.3 Rock : water ratios and rock–water contact times

It is now possible to define six main types of glacial environment on the basis of their rock : water ratios and rock–water contact times. These are shown in Table 3. The distributed drainage system often produces the most concen-trated glacial melt waters (Figure 4), but dilute waters transported through the channelized and lateral drainage systems in large volumes are responsible for the largest fluxes of solute from glaciers (Figures 1–3).

5.07.2.2 The Proglacial Zone

5.07.2.2.1 Broad definition: zone of ice advance and retreat

Most glaciers fluctuate in length as a result of climatic forcing on a variety of timescales. One consequence is that the area in front of the glacier terminus is usually a complex mosaic of moraine and sediment reworked by runoff (Wadham *et al.*, 2001). This loosely defined proglacial zone may

Table 3 Crude representation of typical rock : water ratios and rock–water contact times in different glacial geochemical environments.

Rock–water contact times	*Rock : water ratio*	
	High (order: kg L^{-1})	*Low* (order: mg L^{-1})
High (order: days to months)	Distributed drainage system	Wet snowpack
Low (order: hours)	Channelized drainage system	Supraglacial drainage system
	Lateral channels	Englacial drainage system

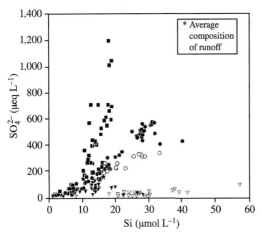

Figure 4 A scatterplot of SO_4^{2-} versus Si for waters sampled directly from the bed of Haut Glacier d'Arolla in 1993 and 1994. Many of these waters are more concentrated than those in contemporaneous bulk runoff (denoted by the star). The other symbols refer to the several types of waters that are found at the bed (after Tranter *et al.*, 2002a) (reproduced by permission of John Wiley & Sons from *Hydrol. Process.* **2002**, *16*, 959–993). The maximum SO_4^{2-} concentration that can be obtained from sulfide oxidation using surface waters saturated with oxygen is $\sim$400 μeq L^{-1}. Some of the subglacial waters sampled have $\sim$3$\times$ this concentration.

be seasonally flooded by glacial melt water, and hence the glacial debris is water saturated for at least part of the ablation season. Some of the chemical reactions that occur within the zone are similar to those that occur in glacial environments, and hence geochemical weathering within the proglacial zone is also included in this chapter.

5.07.2.2.2 Permafrost

A characteristic of the proglacial zone, particularly in high latitudes, is that winter cold often penetrates the snow cover and freezes the ground (French, 1996). Permafrost occurs when summer melting does not completely thaw the deeper ground. Hence, the thawed surface, or active layer, which is often saturated with snowmelt during the early thaw, is underlain by an impermeable seal of frozen ground. Sectors of the active layer remain water saturated for much of the ablation season, particularly near stream channels, where there is exchange of water on a variety of timescales, and in preferential water flow zones. Hence, rock : water ratios are high, rock–water contact times are high, and the extent of chemical weathering is maximized.

5.07.2.2.3 Seasonal freezing

The seasonal freezing serves to concentrate solute in residual waters, to such an extent that

secondary evaporite-like minerals are precipitated at depth in the proglacial zone. Additionally, biological cells may be ruptured or lysed during freezing, and subsequent thaws may release elevated concentrations of DOC and nutrients to melt waters.

5.07.2.2.4 Evapoconcentration

Elevated surfaces within the proglacial zone may dry during the ablation season. The waters drawn to the surface by capillary action contain solutes, and there is progressive concentration of solutes in surface-capillary waters. Eventually, secondary evaporitic salts are formed. These salts can be recycled following rainfall, or may remain over winter and are scavenged by early season snowmelt. These processes impart unusual chemistries to groundwaters flowing within the proglacial zone.

5.07.3 COMPOSITION OF GLACIAL RUNOFF

5.07.3.1 General Features in Comparison with Global Riverine Runoff

The chemical composition of glacial runoff from ice sheets, ice caps, and glaciers around the world is shown in Table 4 (after Brown, 2002), which also includes the composition of global mean river water for comparative purposes. Glacial runoff is a dilute Ca^{2+}–HCO_3^-–SO_4^{2-} solution, with variable Na^+–Cl^-. The sum of cation equivalents ranges from $\sim$10 μeq L^{-1} to 3,500 μeq L^{-1}. Glacial runoff is usually more dilute than global mean river water, and usually contains more K^+ and less Si for a given specific runoff (Anderson *et al.*, 1997). Figure 5 plots the Ca^{2+} : Si ratio of the glacial meltwaters in Figures 1–3 versus the corresponding HCO_3^- : SO_4^{2-} ratio. Also included in Figure 5 are the principal world river waters (E. K. Berner and R. A. Berner, 1996). It can be seen that glacial runoff is an end-member of global riverine water. Generally, Ca^{2+} : Si ratios are high and HCO_3^- : SO_4^{2-} ratios are low. The reasons for these features will become apparent from the glacial chemical weathering scenarios that are outlined below.

5.07.3.2 Relation to Lithology

Raiswell (1984) showed that the base-cation composition of glacial melt waters does not reflect that of the lithology of the bedrock. The predominant cation is always Ca^{2+}, even on acid igneous and metamorphic bedrocks. This is because the dissolution kinetics of Ca^{2+} from trace carbonates, which are ubiquitous in most

Table 4 The concentration of major ions in glacial runoff from different regions of the world. Concentrations are reported in μeq L^{-1}.

Region	Ca^{2+}	Mg^{2+}	Na^+	K^+	HCO_3^-	SO_4^{2-}	Cl^-	Source
Greenland	130–170	68–98	78–110	5.0–9.0	220–340	90–200	16–30	1
Antarctica	72–1,300	120–336	360–1,400	0.8–110	91–1,600	34–1,200	0.6–1,000	2
Iceland	110–350	30–120	30–480	2.8–12	190–570	26–130	30–87	3,4
Alaska	550	36	25	61	430	260	2.0	5
Canadian high Arctic	260–2,600	21–640	1.0–190	0.1–39	210–690	59–3,900		6
Canadian rockies	960–1,100	290–310	3.7–36	5.8–9.2	890–920	380–520	1.7–25	7
Cascades	35–80	8.3–20	2.5–17	9.7–37	83–100	7.9–29		8
European alps	20–640	6.0–140	4.9–92	5.9–33	11–400	10–240	0.9–92	9, 10, 11
Himalayas	75–590	6.6–230	25–65	22–51	200–730	160–410	1.0–22	12
Norway	8.8–623	1.6–66	8.3–210	1.0–29	1.4–680	7.0–140	0.9–190	13
Svalbard	120–1,000	99–540	110–270	5.1–41	110–940	96–760	5.0–310	14, 15, 16
Global mean runoff	670	280	220	33	850	170	160	17

After Brown (2002).
Sources: (1) Rasch *et al.* (2000); (2) De Mora *et al.* (1994); (3) Raiswell and Thomas (1984); (4) Steinporsson and Oskarsson (1983); (5) Anderson *et al.* (2000); (6) Skidmore and Sharp (1999); (7) Sharp *et al.* (2002); (8) Axtmann and Stallard (1995); (9) Brown (2002); (10) Collins (1979); (11) Thomas and Raiswell (1984); (12) Hasnain and Thayyan (1999); (13) Brown (2002); (14) Hodgkins *et al.* (1997); (15) Hodson *et al.* (2000); (16) Wadham *et al.* (1997); and (17) Livingstone (1963).

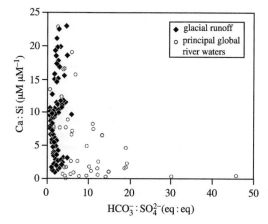

Figure 5 A scatterplot of the ratio of Ca^{2+}: Si versus the ratio of HCO$_3^-$: SO$_4^{2-}$ for glacial meltwaters sampled at Scott Turnerbreen (Hodgkins *et al.*, 1997) (reproduced by permission of John Wiley & Sons from *Hydrol. Process.* **1997**, *11*, 1813–1832), Haut Glacier d'Arolla (Brown *et al.*, 1993a), and Manitsoq Glacier (Skidmore *et al.*, in preparation), in comparison to those of principal world rivers (E. K. Berner and R. A. Berner, 1996). Glacial meltwaters typically have high Ca^{2+}: Si ratios and low HCO$_3^-$: SO$_4^{2-}$ ratios than river waters.

bedrocks, and from aluminosilicates are more rapid than those of monovalent ions. Hence, Ca^{2+} may be a relatively minor base cation in the bedrock, but becomes the dominant base cation in solution (White *et al.*, 2001).

5.07.3.3 Chemical Erosion Rates

Holland (1978) showed that specific annual discharge is the most significant control upon chemical erosion in temperate catchments,

and the same is true in glacierized basins (Anderson *et al.*, 1997; Hodson *et al.*, 2000). The lithology of the catchment is an important secondary control on chemical erosion rates, with carbonate-rich and basaltic lithologies exhibiting the highest chemical weathering rates (Table 5 and Figure 6). Data in Table 5 (after Hodson *et al.*, 2000) also show that there is year by year variability in specific runoff, and therefore on cationic denudation rate. Currently, there are more studies of chemical erosion rates in the glacierized basins of Svalbard than in other regions. Crustally derived solute fluxes in these 10 basins are equivalent to a mean cationic denudation rate of 350 meq m^{-2} yr^{-1} (range: 160–560 meq m^{-2} yr^{-1}), which lies within the global range of 94–1,650 meq m^{-2} yr^{-1} for the other 15 glacier basins in the northern hemisphere (Hodson *et al.*, 2000). The mean value for Svalbard is close to the continental average of 390 meq m^{-2} yr^{-1} (Livingstone, 1963).

5.07.3.4 pH, P_{CO_2}, and P_{O_2}

The pH of glacial runoff is usually in the range of ~7–10. The partial pressure of CO$_2$ in glacial runoff is seldom in equilibrium with the ambient atmosphere (Tranter *et al.*, 1993). Crudely, dilute waters at maximum daily runoff can be expected to have P_{CO_2} of an order of magnitude below that of equilibrium with the atmosphere, while those at low daily runoff will be closer to, if not greater than the equilibrium P_{CO_2}. The reasons for non-equilibrium P_{CO_2} will be outlined below.

Supraglacial waters are often saturated with respect to ambient atmospheric-O$_2$ concentrations,

Table 5 Catchment characteristics, runoff and solute fluxes for glaciers on Svalbard and continental and maritime glaciers at lower latitudes.

Basin	Specific runoff (m yr^{-1})	Catchment area km^{-2} (and % ice cover)	Total solute flux (kg km^{-2} yr^{-1}) ($\times 10^3$)	Cationic denudation rate (meq m^{-2})	Lithology	Source
Svalbard						
Kvikåa	0.56	5.0 (35)	17		Plutonic-metamorphic	1
Kvikåa	0.48	5.0 (35)	20		Plutonic-metamorphic	1
Beinbekken	0.24	5.0 (0)	14		Sedimentary-mixed	1
Beinbekken	0.32	5.0 (0)	20		Sedimentary-mixed	1
Glopbreen	0.41	5.6 (52)	30		Plutonic-metamorphic	1
Erikbreen	0.51	12.4 (75)	31	320	Carbonate-rich	1
Austre Brøggerbreen	0.84	9.9 (71)	23	240	Sedimentary-mixed	2
Austre Brøggerbreen	1.3	9.9 (71)	28	270	Sedimentary-mixed	2
Brøggerbreen Lower site	0.82	32 (53)	41	470	Carbonate-rich	2
Brøggerbreen Lower site	0.99	32 (53)	48	500	Carbonate-rich	2
Hannabreen	0.80	13.3 (69)	30	320	Carbonate-rich	2
Scott Turnerbreen	0.52	13 (32)	16	160	Sedimentary-mixed	3
Finsterwalderbreen	0.84	68 (65)		440	Sedimentary-mixed	4
Finsterwalderbreen	0.35	68 (65)		210	Sedimentary-mixed	4
Erdmannbreen	0.81	16 (70)	16	190	Sedimentary-mixed	2
Midtre Lovénbreen	1.5	5 (80)	41	450	Plutonic-metamorphic	2
Midtre Lovénbreen	1.3	7.4 (80)	47	560	Plutonic-metamorphic	2
Alpine glaciers						
Haut Glacier d'Arolla	1.7	12 (54)	50	640	Plutonic-metamorphic	5
Haut Glacier d'Arolla	2.3	12 (54)	61	685	Plutonic-metamorphic	5
Gornergletscher	1.4	82 (83)		450	Plutonic-metamorphic	6
Glacier d'Tsidjiore Nouve		4.8 (67)		510	Plutonic-metamorphic	7
North American glaciers						
South Cascade Glacier	3.3	6.1 (50)		930	Plutonic-metamorphic	8
South Cascade Glacier	3.9	6.1 (34)		676	Plutonic-metamorphic	9
Worthington Glacier	7.7	13 (83)		1600	Plutonic-metamorphic	10
Berendon Glacier	3.7	53 (62)		947	Plutonic-metamorphic	11
Lewis River	0.71	205 (89)		94	Plutonic-metamorphic	12
Icelandic glaciers						
Tungufljót	2.1	720 (35)	98	720	Basalt	13
Hvítá-south, Gullfoss	2.1	2,000 (19)	123	1100	Basalt	13
Hvítá-west, Kljáfoss	1.8	1,700 (21)	80	650	Basalt	13
Asian glaciers						
Batura Glacier	2.0	608 (60)		1600	Carbonate-rich	14
Chhota-Shigri Glacier	3.5	40 (25)		750	Plutonic-metamorphic	15
Dokriani Glacier	1.1	23 (45)		462	Plutonic-metamorphic	16
Other						
Olenek plateau, Siberia	0.15	237,000 (0)			Mixed	17
Continental average				390		18

After Hodson *et al.* (2000).
Sources: (1) Barsch *et al.* (1994); (2) Hodson *et al.* (2002); (3) Hodgkins *et al.* (1997); (4) Wadham *et al.* (1997); (5) Sharp *et al.* (1995); (6) Collins (1983); (7) Souchez and Lemmens (1987); (8) Reynolds and Johnson (1972); (9) Axtmann and Stallard (1995); (10) Anderson *et al.* (1997); (11) Eyles *et al.* (1982); (12) Church (1974); (13) Gislason *et al.* (1996); (14) Collins *et al.* (1996); (15) Hasnain *et al.* (1989); (16) Hasnain and Thayyan (1999); (17) Huh *et al.* (1998); and (18) Livingstone (1963).

but some supraglacial waters have oxygen saturations of ~50%. The few studies of oxygen saturation in glacial runoff to date usually show that the oxygen saturation is <100% (Brown *et al.*, 1993b).

5.07.3.5 ^{87}Sr : ^{86}Sr Ratios

It has been suggested that there is a link between the marine ^{87}Sr/^{86}Sr record and the extent of continental glaciation (Hodell *et al.*, 1990; Zachos *et al.*, 1999), since continental shields with high ^{87}Sr/^{86}Sr ratios could be extensively chemically weathered due to glacial comminution. In particular, crushed biotite could produce high ^{87}Sr/^{86}Sr ratios in subglacial environments (Sharp *et al.*, 2002). Additionally, Ca/Sr and ^{87}Sr/^{86}Sr ratios can be used to partition solutes between carbonate and silicate sources (Blum *et al.*, 1998). There is only one intensive field study of strontium isotopes in

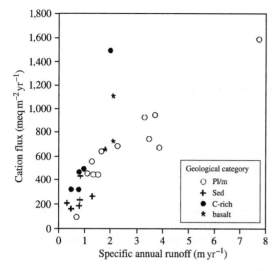

Figure 6 Glacial chemical-erosion rates, as measured by the cationic fluxes, for a range of glaciers world-wide as a function of specific annual runoff. Higher cationic fluxes for a given specific runoff are associated with carbonate/carbonate–rich (C-rich) and basaltic bedrock lithologies. Lower fluxes are associated with other sedimentary (sed) and plutonic/metamorphic bedrocks (Pl/m) (after Hodson *et al.*, 2000) (reproduced by permission of John Wiley & Sons from *Earth Surf. Process. Landforms* **2000**, *25*, 1447–1471).

glacial runoff to date, at Robertson Glacier, Alberta, which has carbonate bedrock. The study does show that enhanced weathering of muscovite occurs in subglacial environments, but in this small system, the strontium flux and the ratio of $^{87}Sr/^{86}Sr$ in the runoff are dominated by carbonate sources. It was concluded that there is unlikely to be a link between the marine $^{87}Sr/^{86}Sr$ record and the extent of continental glaciation, but the extrapolation of one site to all glacierized catchments leads to an element of uncertainty in this observation.

$^{87}Sr/^{86}Sr$ ratios in several samples of glacial runoff in the Himalayas are consistent with carbonate being the main source of calcium and strontium, but Ca/Sr ratios are relatively low. This suggests that these waters may precipitate calcite or aragonite when they warm (Jacobson *et al.*, 2002).

5.07.3.6 Ge/Si Ratios

The Ge/Si ratio preserved in marine opals was formerly thought to be an index of the intensity of terrestrial chemical weathering (Froelich *et al.*, 1992). Germanium is preferentially retained in secondary weathering products, such as clays. Hence, low-opaline Ge/Si ratios are associated with periods of low-chemical weathering intensity (primary minerals supply most solute to the oceans), and high Ge/Si ratios are associated with high-chemical weathering intensity (secondary minerals supply significant quantities of solute to the oceans). Glacial chemical weathering largely attacks the surfaces of aluminosilicate minerals, and there is insufficient time to form more complex secondary minerals such as clays. Consequently, the Ge/Si ratio of glacial runoff reflects more faithfully the ratio in the primary minerals, and is high (Jones *et al.*, 2002). Model estimates of potential glacial Ge/Si ratios and fluxes suggest that ~10% of the variation in the marine-opaline Ge/Si can be attributed to glacial runoff (Jones *et al.*, 2002).

5.07.3.7 $\delta^{18}O$, $\delta^{13}C_{DIC}$, $\delta^{34}S$, and $\delta^{18}O_{SO_4^{2-}}$

There are relatively few studies of the above combination of stable-isotope compositions in glacial runoff, but those that are appearing offer great potential in determining redox conditions and reaction pathways in subglacial environments. Most stable-isotope studies usually are concerned with $\delta^{18}O$ variations, since these give some indication of water sources and flow paths (Theakstone, 1988; Theakstone and Knudsen, 1996). However, recent studies of $\delta^{13}C_{DIC}$ variations have demonstrated a need to understand new fractionation effects during the initial hydrolysis of carbonates (Skidmore *et al.*, in preparation). This initial fractionation confounds attempts to use this isotope as an unequivocal fingerprint of dissolved inorganic carbon (DIC) sources, which include rock carbonate (usually enriched in ^{13}C), the atmosphere (usually enriched in ^{12}C), and organic matter (more highly enriched in ^{12}C). However, it is possible to show that source information can be obtained provided wholesale dissolution of carbonate grains occurs, rather than just the surface. Initial results point to microbial utilization of organic matter in subglacial environments (Bottrell *et al.*, submitted).

Studies of $\delta^{34}S$ are equally few. They have been used to show that sulfide is the main source of SO_4^{2-} in metamorphic and sedimentary catchments. There may be some kinetic effects on isotope fractionation if only the surfaces of the sulfides are being oxidized, and the SO_4^{2-} may become enriched in ^{34}S if sulfate reduction occurs (Bottrell *et al.*, submitted).

Variations in $\delta^{18}O$-SO_4 can give unequivocal information on the redox conditions of the chemical weathering environment at the glacier bed, since the $\delta^{18}O$ of atmospheric oxygen and glacier water are very different. $\delta^{18}O$-SO_4 becomes enriched in ^{16}O as oxygen becomes limited (Bottrell and Tranter, 2002). By contrast, when there is sulfate reduction, $\delta^{18}O$-SO_4 becomes very enriched in ^{18}O. Recent work has shown that sulfide oxidation takes place in anoxic conditions, and that sulfate reduction can also occur in certain subglacial environments (Bottrell *et al.*, submitted).

5.07.3.8 Nutrients

Sulfide oxidation and sulfate reduction are reactions that are usually microbially mediated in Earth-surface environments. It is likely that this is also true of subglacial environments. If this is the case, there is a requirement for nutrients, such as nitrogen and phosphorus. Snow- and ice melt provide limited quantities of nitrogen, mainly as NO_3^- and NH_4^+, and it is likely that phosphorus is derived from comminuted rock debris. However, there may well be a rock source of NH_4^+ from mica and feldspar dissolution (Holloway *et al.*, 1998), and some may also be obtained from the oxidation of organic matter. The concentration of NO_3^- in glacial runoff is usually $<30\ \mu eq\ L^{-1}$, and often between $0\ \mu eq\ L^{-1}$ and $2\ \mu eq\ L^{-1}$. On occasion, NO_3^- concentrations are below the detection limit, which may be evidence for microbial uptake in subglacial environments.

5.07.4 GEOCHEMICAL WEATHERING REACTIONS IN GLACIATED TERRAIN

The chemical composition of glacial melt waters described above is the result of a series of reactions that are controlled first by reaction kinetics, and then by microbial activity. The following is a summary of the principal reactions in glaciated terrain, first on bedrock that is primarily composed of silicates and aluminosilicates.

5.07.4.1 Trace-reactive Bedrock Components are Solubilized

Glacial comminution crushes bedrock and exposes the trace-reactive components more rapidly than in temperate and tropical soils, where new minerals are ultimately accessed via solubilization of aluminosilicate minerals. Hence, glaciers are effective at promoting the solubilization of trace-reactive components in the bedrock, which include carbonates, sulfides, and fluid inclusions.

5.07.4.2 Carbonate and Silicate Hydrolysis

Laboratory experiments and direct sampling of waters from the glacier bed (Tranter *et al.*, 1997, 2002a) show that the initial reactions that occur when dilute snow- and ice melt access glacial flour are carbonate and silicate hydrolysis (Equations (1) and (2)). These reactions raise the pH, to high values (>9), lower the P_{CO_2} (to $\sim 10^{-6}$ atm), and maximize the waters potential to adsorb CO_2.

$$Ca_{(1-x)}Mg_xCO_3(s) + H_2O(l) = (1-x)Ca^{2+}(aq)$$
calcite

$$+\ xMg^{2+}(aq) + HCO_3^-(aq) + OH^-(aq) \qquad (1)$$

$$KAlSi_3O_8(s) + H_2O(l) = HAlSi_3O_8(s) + K^+(aq)$$
K-feldspar

$$+\ OH^-(aq) \qquad (2)$$

5.07.4.3 Cation Exchange

The relatively dilute melt water in contact with fine-grained glacial flour favors the exchange of divalent ions from solution for monovalent ions on surface-exchange sites. Hence, some of the Ca^{2+} and Mg^{2+} released from carbonate and silicate hydrolysis is exchanged for Na^+ and K^+ (Equation (3)):

$$Na_{(2-z)}K_z\text{-glacial flour}(s) + (1-x)Ca^{2+}(aq)$$

$$+\ xMg^{2+}(aq) = Ca_{(1-x)}, Mg_x\text{-glacial flour}(s)$$

$$+\ (2-z)Na^+(aq) + zK^+(aq) \qquad (3)$$

5.07.4.4 Carbonate and Silicate Dissolution: Sources of CO_2 and Strong Acids

The high pH derived from hydrolysis enhances the dissolution of aluminosilicate minerals, since aluminum and silicon become more soluble at pH >9. Hydrolysis of carbonates results in a solution that is near saturation with calcite and aragonite. It is only in these types of waters that aluminosilicate dissolution is greater than carbonate dissolution. The influx of gases (including CO_2 and O_2), either from the atmosphere or from basal ice, and CO_2 produced by microbial respiration (see below) both lower the pH and the saturation with respect to carbonates. In addition, sulfide oxidation produces acidity (see below). Hence, almost all subglacial melt waters are undersaturated with respect to carbonates. Carbonate dissolution has a large impact on melt-water chemistry, despite the fact that carbonates are often present in only trace concentrations in the bedrock.

The acid hydrolysis of silicates and carbonates that arises from the dissociation of CO_2 in solution is known as carbonation (Equations (4) and (5)). Carbonation only occurs in a restricted number of subglacial environments, because ingress of atmospheric gases to these water-filled environments is restricted. It largely occurs in the major arterial channels at low flow, particularly near the terminus, and at the bottom of crevasses and moulins that reach the bed. Sediment is flushed rapidly from these environments, and there is little time for the formation of secondary weathering products, such as clays, to form. Hence, silicates dissolve incongruently, as crudely represented by Equation (4):

$$CaAl_2Si_2O_8(s) + 2CO_2(aq) + 2H_2O(l)$$
anorthite

$$= Ca^{2+}(aq) + 2HCO_3^-(aq)$$
$$+ H_2Al_2Si_2O_8(s) \qquad (4)$$
weathered feldspar surfaces

$$Ca_{(1-x)}Mg_xCO_3(s) + CO_2(aq) + H_2O(l)$$
calcite

$$= (1-x)Ca^{2+}(aq) + xMg^{2+}(aq)$$
$$+ 2HCO_3^-(aq) \qquad (5)$$

There is a limited body of evidence which suggests that microbial oxidation of bedrock kerogen occurs (Equation (6)). If this is the case, carbonation as a consequence of microbial respiration may occur in debris-rich environments, such as the distributed drainage system and the channel marginal zone:

$$C_{org}(s) + O_2(aq) + H_2O(l)$$
organic carbon

$$= H^+(aq) + HCO_3^-(aq) \qquad (6)$$

5.07.4.5 Sulfide Oxidation Using O_2 and Fe(III)

The dominant reaction in subglacial environments is sulfide oxidation. Following hydrolysis, this is the major reaction which provides protons to solution, lowering the pH, decreasing the saturation index of carbonates, and allowing more carbonate dissolution (Equation (7)). Sulfide oxidation occurs predominantly in debris-rich environments, where comminuted bedrock is first in contact with water. It is microbially mediated, occurring several orders of magnitude faster than in sterile systems. It consumes oxygen, driving down the P_{O_2} of the water:

$$4FeS_2(s) + 16Ca_{(1-x)}Mg_xCO_3(s) + 15O_2(aq)$$
pyrite

$$+ 14H_2O(l) = 16(1-x)Ca^{2+}(aq)$$
$$+ 16xMg^{2+}(aq) + 16HCO_3^-(aq)$$
$$+ 8SO_4^{2-}(aq) + 4Fe(OH)_3(s) \qquad (7)$$
ferric
oxyhydroxides

The oxidation of sulfides preferentially dissolves carbonates, rather than silicates, because the rate of carbonate dissolution is orders of magnitude faster. For example, Haut Glacier d'Arolla has a bedrock which is composed of metamorphic silicate rocks. Carbonates and sulfides are present in trace quantities in bedrock samples (0.00–0.58% and <0.005–0.71%, respectively). There are also occasional carbonate veins present in the schistose granite. Despite the bedrock being

dominated by silicates, sulfide oxidation in subglacial environments dissolves carbonate to silicate in a ratio of ~5 : 1 (Tranter *et al.*, 2002a), compared to the global average of ~1.3 : 1 (Holland, 1978).

Earlier studies suggested that the limit on sulfide oxidation was the oxygen content of supraglacial melt, since subglacial supplies of oxygen are limited to that released from bubbles in the ice during regelation, the process of basal ice melting and refreezing as it flows around bedrock obstacles. However, studies of water samples from boreholes drilled to the glacier bed show that the SO_4^{2-} concentrations may be 2 or 3 times that allowed by the oxygen content of supraglacial meltwaters (Figure 4; Tranter *et al.*, 2002a). This suggests that oxidizing agents other than oxygen are present at the glacier bed. It seems very likely that microbially mediated sulfide oxidation drives certain sectors of the bed towards anoxia, and that in these anoxic conditions, Fe(III), rather than O_2, is used as an oxidizing agent (Equation (8)). Sources of Fe(III) include the products of the oxidation of pyrite and other Fe(II) silicates in a previous oxic environment, as well as that found in magnetite and hematite:

$$FeS_2(s) + 14Fe^{3+}(aq) + 8H_2O(l) = 15Fe^{2+}(aq)$$
$$+ 2SO_4^{2-}(aq) + 16H^+(aq) \qquad (8)$$

Support for anoxia within subglacial environments comes from the $\delta^{18}O-SO_4$, which is enriched in ^{16}O when sulfide is oxidized in the absence of oxygen (Bottrell and Tranter, 2002).

5.07.4.6 Oxidation of Kerogen, Sulfate Reduction and Onwards to Methanogenesis?

The realization that there is microbial mediation of certain chemical-weathering reactions in subglacial environments (Sharp *et al.*, 1999; Skidmore *et al.*, 2000; Bottrell and Tranter, 2002) has resulted in a paradigm shift, since the types of reactions that may occur in anoxic sectors of the bed include the common redox reactions that occur, for instance, in lake or marine sediments. A key difference in glacial systems is that the supply of new or recent organic matter is limited to that inwashed from the glacier surface, such as algae, insects, and animal feces, or overridden soils during glacier advance. By contrast, the supply of old organic matter from comminuted rocks is plentiful. Given the thermodynamic instability of organic matter in the presence of O_2 or SO_4^{2-}, it seems likely that microbes will have evolved to colonize subglacial environments and utilize kerogen as an energy source. The first data to support this assertion is stable-isotope analysis from Finsterwalderbreen, a small polythermal-based glacier on Svalbard which has shale as a significant component of its bedrock (Bottrell *et al.*, submitted). The $\delta^{18}O$-SO_4 of waters upwelling from subglacial sediments are very enriched in ^{34}S, which suggests that cyclical sulfate reduction and oxidation has been occurring. The $\delta^{13}C$ of DIC is negative, consistent with the assertion that organic matter has been oxidized. Mass-balance calculations suggest that the most-probable source of organic matter is kerogen. Hence, sectors of the bed at Finsterwalderbreen appear to be anoxic, in which a significant geochemical weathering reaction is sulfate reduction linked to oxidation of kerogen (Equation (9)):

$$2CH_2O(s) + SO_4^{2-}(aq)$$
$$\text{organic carbon}$$
$$= 2HCO_3^-(aq) + H_2S(aq) \qquad (9)$$

It is possible that methanogenesis occurs under certain ice masses, since methanogens have been isolated from subglacial debris (Skidmore *et al.*, 2000). The low $\delta^{13}C$-CH_4 and high concentration of methane found in gas bubbles within the basal ice of the Greenland Ice Sheet are consistent with methanogenesis within the basal organic-rich paleosols. Skidmore *et al.* (2000) speculate that there could have been methanogenesis in organic matter trapped beneath the Laurentian Ice Sheet, resulting in a small source of CH_4 during deglaciation.

5.07.4.7 Nutrients from Glacial Flour

The colonization of subglacial environments by microbes suggests that both energy and nutrient sources are readily available. Energy sources, such as sulfides and kerogen, have been discussed above. Comminuted bedrock may also provide a source of nutrients. Average crustal rock contains 1,050 ppm of phosphorus. Typically, this is contained in sparingly soluble minerals such as apatite and in calcium, aluminium, and ferrous phosphates (O'Neill, 1985). Comminuted bedrock and basal debris provide a renewable source of phosphorus on mineral surfaces, and it is likely that uptake of phosphorus by microbes maximizes the extraction of phosphorus from these activated surfaces. Hodson *et al.* (submitted) suggest that $1-23\ \mu g\ P\ g^{-1}$ is present as readily extractable phosphorus on the surface of glacial flour. Sources of nitrogen are potentially more problematic, since the nitrogen content of rocks is typically 20 ppm (Krauskopf, 1967). However, bedrock has been shown to be a source of NH_4^+ derived from schists in the Sierra Nevadas, California (Holloway *et al.*, 1998), and there may be appreciable concentrations of NH_4^+, which substitutes for K^+, in biotite, muscovite, K-feldspar, and pagioclase (Mingram and Brauer, 2001). It follows that glacial comminution of bedrock and

basal debris maximizes the likelihood that nitrogen-producing surfaces are exposed to melt waters and microbes, and, given that bedrock in the Sierra Nevadas can act as an nitrogen source, it is likely that comminuted glacial debris is also a potential source of nitrogen.

5.07.4.8 Other Lithologies

The dominance of carbonate hydrolysis, carbonation, and sulfide oxidation in subglacial weathering reactions on aluminosilicate/silicate bedrock is also found on carbonate bedrock. However, the balance between carbonate dissolution and sulfide oxidation depends on the spatial distribution of sulfides in the bedrock and basal debris (Fairchild *et al.*, 1999). Noncongruent dissolution of strontium and magnesium from carbonate is also observed in high rock : water weathering environments, such as the distributed drainage systems, in which water flow is also low (Fairchild *et al.*, 1999).

There have been few studies on glacial chemical weathering on bedrock with a significant evaporitic content to date. Work at John Evans Glacier in the Canadian High Arctic has shown that gypsum is dissolved in some areas of the bed, and that mixing of relatively concentrated Ca^{2+}–SO_4^{2-} waters with more dilute Ca^{2+}–HCO_3^-–SO_4^{2-} waters results in $CaCO_3$ precipitation due to the common-ion effect (Skidmore, personal communication). Kennicott Glacier, Alaska, is underlain by a sabkha facies limestone, which contains trace quantities of halite. Waters accessing sites of active erosion readily acquire Na^+ and Cl^- (Anderson *et al.*, in press).

5.07.4.9 Little Necessity for Atmospheric CO_2

A key feature of the above chemical-weathering scenarios is that relatively little atmospheric or biogenic CO_2 is involved. Hence, whereas ~23% and ~77% of solutes, excluding recycled sea salt, found in global mean river water are derived from the atmosphere and rock, respectively (Holland, 1978), atmospheric sources account for a maximum of 3–11% of solute in glacial runoff (after Hodson *et al.*, 2000).

5.07.5 GEOCHEMICAL WEATHERING REACTIONS IN THE PROGLACIAL ZONE

5.07.5.1 Similarities with Subglacial Environments

The proglacial zone has been identified as a potential zone of high geochemical activity, because it contains a variety of comminuted glacial debris, is subject to reworking by glaciofluvial activity, and can be colonized by vegetation (Anderson *et al.*, 2000; Cooper *et al.*, 2002). It is likely that similar reactions to those identified above are the source of solute in environments which have little plant colonization. However, significant differences between subglacial and proglacial environments are that the surface of the proglacial zone freezes, thaws, and dries on an annual basis, and that it is a deposition site for snow and rain. These processes perturb the composition of groundwaters within the proglacial zone. Further, there may be ingress of atmospheric gases through the dry surface sediment, resulting in the surface groundwaters remaining oxic during unsaturated conditions. Water flowing through the sediments mainly acquire new solutes via sulfide oxidation and carbonate dissolution in relatively oxic areas (Anderson *et al.*, 2000; Cooper *et al.*, 2002), particularly those containing new or recently deposited glacial debris. There is no literature to date on the waters becoming anoxic, but it is likely that the reactions which occur in anoxic sections of glacier beds are also found in anoxic areas of the active layer. In addition, there is concentration and recycling of salts via evaporoconcentration and freeze concentration.

5.07.5.2 Evapoconcentration and Freeze Concentration

The surfaces of proglacial zones dry during the summer, and solutes contained in the water drawn to the surface by capillary attraction can become so concentrated that salts precipitate. It is common to see white evaporitic salts on dry proglacial surfaces in Svalbard. Carbonates precipitate first, followed by sulfates and chlorides/nitrates. The salts redissolve in the reverse order by rain or snowmelt, and such waters may then become concentrated solutions with SO_4^{2-} and Cl^- as the dominant anions and Ca^{2+} and Na^+ as the dominant cations. Mixing of these waters with pre-existing groundwaters which have the more common $Ca^{2+}-HCO_3^--SO_4^{2-}$ composition results in local precipitation of $CaCO_3$ (Figure 7).

Freezing of waters in the active zone over winter results in the progressive concentration of the residual solution, until first carbonates, then sulfates and chlorides/nitrates precipitate out of solution. These salts can be scavenged by the first waters of the thaw, and these may also become concentrated solutions of chlorides/nitrates and sulfates in the first instance. Thereafter, both secondary and primary carbonate can be scavenged, since, as Ca^{2+} concentrations decline, the saturation index of calcium carbonate becomes negative. Hence, groundwaters in proglacial areas may span the spectrum from relatively concentrated solutions of $Ca^{2+}-SO_4^{2-}-Cl^-$

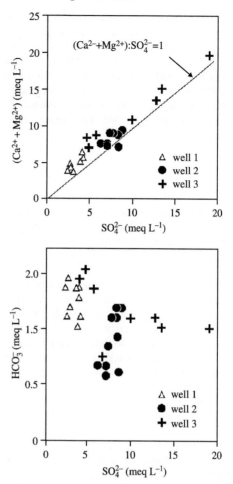

Figure 7 Scatterplots of $(Ca^{2+} + Mg^{2+})$ and HCO_3^- versus SO_4^{2-} for groundwaters in the proglacial zone of Finsterwalderbreen, a polythermal-based valley glacier on Svalbard. The groundwaters were sampled from three wells, and show high concentrations of the divalent cations and SO_4^{2-} due to the dissolution of secondary sulfate salts and freeze concentration effects. By contrast, HCO_3^- concentrations are depressed because of the high concentrations of Ca^{2+}, so that there is an inverse association between HCO_3^- and SO_4^{2-}. Full details can be found in Cooper *et al.* (2002) (reproduced by permission of Elsevier from *J. Hydrol.* **2002**, *269*, 208–223).

$(-NO_3^-)$ to more dilute solutions of $Ca^{2+}-HCO_3^-$ $-SO_4^{2-}$ (Cooper *et al.*, 2002).

5.07.5.3 Ingress of Water from Channels into the Proglacial Zone

Waters invading the proglacial zone from rising discharge in glaciofluvial channels interact with sediment in a similar manner to the reactions in the channel marginal zone in subglacial environments. Readily solubilized minerals, such as sulfides, sulfates, and carbonates, will be weathered first, and become depleted following repeated cycles of water ingress and drainage. Hence, silicate weathering becomes more dominant over time. The type of effect is seen on the larger scale in deglaciated terrain at Bench Glacier, Alaska, where sulfide oxidation and carbonate dissolution are dominant reactions in recently deglaciated terrain, and silicate dissolution dominates in progressively older terrain (Anderson *et al.*, 2000).

5.07.5.4 Enhancement of Glacial Solute Fluxes

The first field studies that have attempted to measure the enhancement of chemical weathering rates in the proglacial zone was conducted on recently deglaciated terrain at Bench Glacier, Alaska (Anderson *et al.*, 2000) and Finsterwalderbreen, Svalbard (Wadham *et al.*, 2001). The lithology of the former was metasedimentary, and that of the latter was a mixture of carbonates, sandstone, and black shale. Both studies noted an enhancement of chemical weathering rate in the proglacial zone relative to that of the glacier by a factor of 3–4. Carbonate dissolution and sulfide oxidation are dominant reactions initially, but as carbonates and sulfides become exhausted, silicate weathering increases in prominence. Colonization of the proglacial zone by plants is likely to further increase the rate of chemical weathering of silicates (Anderson *et al.*, 2000).

The enhancement of solute fluxes, predominantly Ca^{2+}, HCO_3^-, and SO_4^{2-}, was ~20–30% over a distance of ~1 km at Finsterwalderbreen (Wadham *et al.*, 2001). These values are likely to be high in comparison with other lithologies, since carbonate, sulfide, and kerogen are present in relatively high concentrations.

5.07.6 SUMMATION: IMPACT ON LOCAL AND GLOBAL GEOCHEMICAL CYCLES

5.07.6.1 Local Effects

5.07.6.1.1 *Redox conditions and pH may range widely within subglacial environments*

Microbial mediation of chemical weathering reactions in subglacial environments produces redox conditions that range from oxic to anoxic. The comminution of bedrock that accompanies glacial erosion produces potential reducing agents, sulfides and kerogen, whose oxidation can be catalyzed by microbial activity. Comminution of bedrock also maximizes the potential supply of the nutrients, nitrogen and phosphorus, from rock-forming minerals. It follows that the seasonal hydrology of the glacier, which affects key factors such as the water-flow path, the

rock : water ratio, and the rock–water contact time, has a fundamental impact both on the nature of the subglacial weathering environment and the types of microbes that colonize the environment.

5.07.6.1.2 *Groundwaters in the proglacial zone*

The concentration of solutes in waters of the proglacial zone is generally higher than those found within glacial runoff. The dominant reactions which occur within the zone are carbonate dissolution and sulfide oxidation, similar to the dominant reactions within the subglacial zone, in new or recently deposited sediment. Silicate weathering becomes progressively dominant in older sediment, as carbonates and sulfides are exhausted. The chemical weathering rate is 3–4 times that of the glaciated catchment, and the solute flux out of the proglacial zone may be enhanced by up to 30% of that leaving the glacier.

5.07.6.2 Global Effects

5.07.6.2.1 *Impact on atmospheric CO_2 concentrations over time*

Glacial chemical weathering does not remove large quantities of CO_2 from the atmosphere. This is because rock components are mostly solubilized in the absence of free contact with the atmosphere. Instead, glacial chemical weathering may be a source of CO_2 to the atmosphere. The Ca^{2+} and HCO_3^- dissolved by sulfide oxidation (Equation (7)) is ultimately deposited as carbonate in the oceans, and this gives rise to a net release of CO_2 to the atmosphere (Equation (10))

$$Ca^{2+}(aq) + 2HCO_3^-(aq) = CaCO_3(s)$$
$$+ H_2O(l) + CO_2(g) \qquad (10)$$

To date modeling that attempts to determine the impact of glacial chemical weathering on atmospheric CO_2 concentrations has not predicted any significant perturbations related to glacial or proglacial environments (Ludwig *et al.*, 1998; Jones *et al.*, 2002; Tranter *et al.*, 2002b). However, linkage of terrestrial chemical-erosion models to ocean carbon-cycle models is still at an early stage of development. It may well be the case that the next generation of ocean carbon-cycle models will be able to explore in greater detail the change in DIC species at depth that arise from changes in ocean circulation (Broecker, 1995; Dokken and Jansen, 1999). Atmospheric CO_2 perturbations from changing terrestrial chemical erosion may be amplified as a consequence, since it is anticipated that enhanced terrestrial chemical erosion during times of reduced sea level is less well buffered by slower deep-water turnover (Jones *et al.*, 2002).

The impact of sediment deposited by retreating glaciers and ice sheets, and the deposition of glacio-fluvial sediment in downstream environments, has yet to be fully evaluated in the context of global chemical erosion rates. In principle, the high silt-sized fraction of the sediment should enhance the geochemical reactivity. Hence, under given favorable climatic conditions and vegetative cover, glacially derived sediments and consequent soils might potentially give rise to elevated rates of chemical erosion. This might be the case particularly in the Alps and the Himalayas, where silt-sized sediment is deposited in warmer environments at lower elevations (Drever, personal communications). The available literature is insufficient to form a firm opinion.

5.07.6.2.2 Impact on global silicon flux and marine Ge/Si ratio

It is unlikely that glacial chemical weathering has a large impact on the global silicon cycle, since a typical concentration of silicon in glacial runoff is $\sim 6\ \mu eq\ L^{-1}$, and glacial runoff during the last glacial cycle was estimated to have been equivalent to current global riverine runoff for only a few relatively short periods during deglaciation. Global glacial runoff is usually <10% of current global riverine runoff for much of the last glacial cycle (Tranter *et al.*, 2002b). It is possible to perturb the riverine Ge/Si ratio significantly (Jones *et al.*, 2002), on the assumption that the glacial Ge/Si ratio is high, but the germanium flux calculated is a factor of ~ 5 too small to give rise to the observed increase in marine opals (Froelich *et al.*, 1992; Mortlock *et al.*, 1991). Further work is required to explore whether or not glacial runoff from the Great Ice Sheets had sufficiently high Ge/Si ratios to perturb the marine Ge/Si ratio.

5.07.6.2.3 Impact on hafnium-isotope composition of seawater

van de Flierdt *et al.* (2002) hypothesize that zircons are preferentially weathered during periods of intense glaciation, and release relatively unradiogenic hafnium to the oceans. They tested this hypothesis by examining neodymium- and hafnium-isotope time series in ferromanganese nodules in the NW Atlantic, and find that glaciation in the northern hemisphere does indeed coincide with periods when unradiogenic hafnium may be released to the oceans. Hence, it may be that hafnium isotopes in ferromanganese nodules offer the best record of the intensity of past global glacial chemical weathering.

REFERENCES

Anderson S. P., Drever J. I., and Humphrey N. F. (1997) Chemical weathering in glacial environments. *Geology* **25**, 399–402.

Anderson S. P., Drever J. I., Frost C. D., and Holden P. (2000) Chemical weathering in the foreland of a retreating glacier. *Geochim. Cosmochim. Acta* **64**, 1173–1189.

Anderson S. P., Longacre S. A., and Kraal E. R. (xxxx) Patterns of water chemistry and discharge in the glacier-fed Kennicott river, Alaska: evidence for subglacial water storage cycles. *Chem. Geol.* (in press).

Axtmann E. V. and Stallard R. F. (1995) Chemical weathering in the South Cascade Glacier Basin, comparison of subglacial and extra-glacial weathering. *IAHS Publication No.* **228**, 431–439.

Barsch D., Gude M., Mäusbascher R., Schukraft G., and Schulte A. (1994) Recent fluvial sediment budgets in glacial and periglacial environments, NW Spitsbergen. *Z. Geomorph. Suppl. Bd.* **97**, 111–122.

Berner E. K. and Berner R. A. (1996) *Global Environment: Water, Air, and Goechemical Cycles.* Prentice-Hall, Englewood, NJ, 376pp.

Blum J. D., Gazis C. A., Jacobson A. D., and Chamberlain C. P. (1998) Carbonate versus silicate weathering in the Raikhot watershed within the high Himalayan crystalline series. *Geology* **26**, 411–414.

Bottrell S. H. and Tranter M. (2002) Sulphide oxidation under partially anoxic conditions at the bed of Haut Glacier d'Arolla, Switzerland. *Hydrol. Process.* **16**, 2363–2368.

Bottrell S. H., Tranter M., Wadham J. L., and Raiswell R. Microbes reduce sulphate with kerogen in anoxic subglacial environments. *Earth Planet. Sci. Lett.* (submitted).

Broecker W. S. (1995) *The Glacial World According to Wally*, 2nd edn. Eldigio Press, Palisades, NY.

Brown G. H. (2002) Glacier meltwater hydrochemistry. *Appl. Geochem.* **17**, 855–883.

Brown G. H., Tranter M., Sharp M. J., and Gurnell A. M. (1993a) The impact of post-mixing chemical reactions on the major ion chemistry of bulk meltwaters draining the Haut Glacier d'Arolla, Valais, Switzerland. *Hydrol. Process.* **8**, 465–480.

Brown G. H., Tranter M., Sharp M. J., Davies T. D., and Tsiouris S. (1993b) Dissolved oxygen variations in Alpine glacial meltwaters. *Earth Surf. Process. Landforms* **19**, 247–253.

Church M. (1974) On the quality of some waters on Baffin Island, Northwest Territories. *Can. J. Earth Stud.* **11**, 1676–1688.

Collins D. N. (1979) Hydrochemistry of meltwaters draining from an alpine glacier. *Arctic Alpine Res.* **11**, 307–324.

Collins D. N. (1983) Solute yield from a glacierised high mountain basin. *IAHS Publ. No.* **141**, 41–50.

Collins D. N., Lowe A. T., and Boult S. (1996) Solute fluxes in meltwaters draining from glacierised high mountain basins. In *Fourth International Symposium on the Geochemistry of the Earth's Surface* (ed. S. H. Bottrell). University of Leeds, Ilkley, UK, pp. 728–732.

Cooper R. J., Wadham J. L., Tranter M., Hodgkins R., and Peters N. E. (2002) Groundwater hydrochemistry in the active layer of the proglacial zone, Finsterwalderbreen, Svalbard. *J. Hydrol.* **269**, 208–223.

De Mora S. J., Whitehead R. F., and Gregory M. (1994) The chemical composition of glacial meltwater ponds and streams on the McMurdo Ice Shelf, Antarctica. *Antarct. Sci.* **6**, 17–27.

Dokken T. E. and Jansen E. (1999) Rapid changes in the mechanism of ocean convection during the last glacial period. *Nature* **401**, 451–458.

Drever J. I. (1988) *The Geochemistry of Natural Waters*, 2nd edn. Prentice-Hall, Englewood Cliff, NJ, 437pp.

Drever J. I. (1994) The effect of land plants on weathering rates of silicate minerals. *Geochim. Cosmochim. Acta* **58**, 2325–2332.

Eyles N., Sasseville D. R., Slatt R. M., and Rogerson R. J. (1982) Geochemical denudation rates and solute transport mechanisms in a maritime temperate glacier basin. *Can. J. Earth Sci.* **19**, 1570–1581.

Fairchild I. J., Killawee J. A., Sharp M. J., Hubbard B., Lorrain R. D., and Tison J. L. (1999) Solute generation and transfer from a chemically reactive alpine glacial–proglacial system. *Earth Surf. Process. Landforms* **4**, 1189–1211.

Fountain A. G. and Walder J. S. (1998) Water flow through temperate glaciers. *Rev. Geophys.* **36**, 299–328.

French H. M. (1996) *The Periglacial Environment*, 2nd edn. Addison Wesley Longman, Harlow, 341pp.

Froelich P. N., Blanc V., Mortlock R. A., Chillrud S. N., Dunstan W., Udomkit A., and Peng T.-H. (1992) River fluxes of dissolved silica to the ocean were higher during glacials: Ge/Si in diatoms, rivers, and oceans. *Paleoceanography* **7**, 739–767.

Gíslason S., Arnorsson S., and Armannsson H. (1996) Chemical weathering in southwest Iceland: effects of runoff, age of rocks and vegetative/glacial cover. *Am. J. Sci.* **296**, 837–907.

Hasnain S. I. and Thayyan R. J. (1999) Sediment transport and solute variation in meltwaters of Dokriani Glacier (Bamak), Garwhal Himalaya. *J. Geol. Soc. India* **47**, 731–739.

Hasnain S. I., Subramanian V., and Dhanpal K. (1989) Chemical characteristics and suspended sediment load of meltwaters from a Himalayan glacier in India. *J. Hydrol.* **106**, 99–108.

Hodell D. A., Mead G. A., and Mueller P. A. (1990) Variations in the strontium isotope composition of seawater (8 Ma to present): implications for chemical weathering rates and dissolved fluxes to the oceans. *Chem. Geol.* **80**, 291–307.

Hodgkins R., Tranter M., and Dowdeswell J. A. (1997) Solute provenance, transport and denudation in a high Arctic glacierised catchment. *Hydrol. Process.* **11**, 1813–1832.

Hodson A. J., Tranter M., and Vatne G. (2000) Contemporary rates of chemical weathering and atmospheric CO_2 sequestration in glaciated catchments: an Arctic perspective. *Earth Surf. Process. Landforms* **25**, 1447–1471.

Hodson A. J., Tranter M., Gurnell A., Clark M., and Hagen O. J. (2002) The hydrochemistry of Bayelva, a high Arctic proglacial stream in Svalbard. *J. Hydrol.* **257**, 91–114.

Hodson A. J., Mumford P., and Lister D. Suspended sediment and phosphorus in proglacial rivers: bioavailability and potential impacts upon the P status of ice-marginal receiving waters. *Hydrol. Process.* (in press).

Holland H. D. (1978) *The Chemistry of the Atmosphere and Oceans*. Wiley, New York, 351pp.

Holloway J. M., Dahlgren R. A., Hansen B., and Casey W. H. (1998) Contribution of bedrock nitrogen to high nitrate concentrations in stream water. *Nature* **395**, 785–788.

Huh Y., Tsoi M.-Y., Zaitsev A., and Edmond J. M. (1998) The fluvial geochemistry of the rivers of Eastern Siberia: I. Tributaries of the Lena river draining the sedimentary platform of the Siberian Craton. *Geochim. Cosmochim. Acta* **62**, 1657–1676.

Jacobs S. S., Helmer H. H., Doake C. S. M., Jenkins A., and Frolich R. M. (1992) Melting of ice shelves and the mass balance of Antarctica. *J. Glaciol.* **38**, 375–387.

Jacobson A. D., Blum J. D., and Walter L. M. (2002) Reconciling the elemental and Sr isotope composition of Himalayan weathering fluxes: insights from the carbonate geochemistry of stream waters. *Geochim. Cosmochim. Acta* **66**, 3417–3429.

Jones I. W., Munhoven G., Tranter M., Huybrechts P., and Sharp M. J. (2002) Modelled glacial and non-glacial HCO_3^-, Si, and Ge fluxes since the LGM: little potential for impact on atmospheric CO_2 concentrations and the marine Ge:Si ratio. *Global Planet. Change* **33**, 139–153.

Knight P. G. (1999) *Glaciers*. Cheltenham, Stanley Thornes, 261pp.

Krauskopf K. B. (1967) *Introduction to Geochemistry*. McGraw-Hill, New York, 721pp.

Livingstone D. A. (1963) *Chemical Compositions of Rivers and Lakes*. US Geol. Surv. Prof. Pap., vol. 440-G, 64pp.

Ludwig W., Amiotte-Suchet P., Munhoven G., and Probst J.-L. (1998) Atmospheric CO_2 comsumption by continental erosion: present-day controls and implications for the last glacial maximum. *Global Planet. Change* **16–17**, 107–120.

Mingram B. and Brauer K. (2001) Ammonium concentration and nitrogen isotope composition in metasedimentary rocks from different tectonometamorphic units of the European Variscan Belt. *Geochim. Cosmochim. Acta* **65**, 273–287.

Mortlock R. A., Charles C. D., Froelich P. N., Zibello M. A., Saltzman J., Hays J. D., and Burkle L. H. (1991) Evidence for lower productivity in the Antarctic Ocean during the last glaciation. *Nature* **351**, 220–223.

Oerlemans J. (1993) Evaluating the role of climate cooling in iceberg production and Heinrich events. *Nature* **364**, 783–786.

O'Neill P. O. (1985) *Environmental Chemistry*. George Allen and Unwin, London, 232pp.

Paterson W. S. B. (1994) *The Physics of Glaciers*, 3rd edn. Pergamon, Oxford, 480pp.

Petrovic R., Berner R. A., and Goldhaber M. B. (1976) Rate control in dissolution of alkali feldspars: I. Study of residual grains by X-ray photoelectron spectroscopy. *Geochim. Cosmochim. Acta* **40**, 537–548.

Raiswell R. (1984) Chemical models of solute acquisition in glacial meltwaters. *J. Glaciol.* **30**, 49–57.

Raiswell R. and Thomas A. G. (1984) Solute acquisition in glacial meltwaters: I. Fjällsjökull (South-East Iceland): bulk meltwaters with closed system characteristics. *J. Glaciol.* **30**, 35–43.

Rasch M., Elbering B., Jakobsen B. H., and Hasholt B. (2000) High-resolution measurements of water discharge, sediment, and solute transport in the River Zackenbergelven, Northeast Greenland. *Arctic Antarct. Alpine Res.* **32**, 336–345.

Reynolds R. C. and Johnson N. M. (1972) Chemical weathering in the temperate glacial environment of the Northern Cascade Mountains. *Geochim. Cosmochim. Acta* **36**, 537–544.

Sharp M., Tranter M., Brown G. H., and Skidmore M. (1995) Rates of chemical denudation and CO_2 drawdown in a glacier-covered alpine catchment. *Geology* **23**, 61–64.

Sharp M., Parkes J., Cragg B., Fairchild I. J., Lamb H., and Tranter M. (1999) Bacterial populations at glacier beds and their relationship to rock weathering and carbon cycling. *Geology* **27**, 107–110.

Sharp M., Creaser R. A., and Skidmore M. (2002) Strontium isotope composition of runoff from a glaciated carbonate terrain. *Geochim. Cosmochim. Acta* **66**, 595–614.

Siegert M. J. (2001) *Ice sheets and Quaternary Environmental Change*. Wiley, Chichester, 231pp.

Skidmore M. L. and Sharp M. J. (1999) Drainage system behaviour of a high-Arctic polythermal glacier. *Ann. Glaciol.* **28**, 209–215.

Skidmore M. L., Foght J. M., and Sharp M. J. (2000) Microbial life beneath a high Arctic glacier. *Appl. Environ. Microbiol.* **66**, 3214–3220.

Skidmore M. L., Sharp M. J., and Tranter M. Fractionation of carbon isotopes during the weathering of carbonates in glaciated catchments: I. Kinetic effects during the initial phases of dissolution in laboratory experiments. *Geochim. Cosmochim. Acta* (in press).

Skidmore M. L., Tranter M., Grust K., Jackson A. C., Jones I. W., Bottrell S. H., Nienow P. W., Raiswell R. W., Sharp M. J., Siegert M. J., and Statham P. J. Subglacial drainage system structure at the margin of the Greenland Ice Sheet (in preparation).

Souchez R. A. and Lemmens M. M. (1987) Solutes. In *Glacio-fluvial Sediment Tranfer: An Alpine Perspective* (eds. A. M. Gurnell and M. J. Clark). Wiley, Chichester, pp. 285–303.

Steinporsson S. and Oskarsson N. (1983) Chemical monitoring of Jokulhaup water in Skeidara and the geothermal system in Grimsvotn, Iceland. *Jokull* **33**, 73–86.

Sugden D. E. and John B. S. (1976) *Glaciers and Landscape.* Arnold, London, 376pp.

Theakstone W. H. (1988) Temporal variations of isotopic composition of glacier river water during the summer: observations at Austre Okstindbreen, Okstindan, Norway. *J. Glaciol.* **34**, 309–317.

Theakstone W. H. and Knudsen N. T. (1996) Isotopic and ionic variations in glacier river water during three contrasting ablation seasons. *Hydrol. Process.* **10**, 523–540.

Thomas A. G. and Raiswell R. (1984) Solute acquisition in glacial meltwaters: II. Argentière (French Alps): bulk meltwaters with open system characteristics. *J. Glaciol.* **30**, 44–48.

Tranter M. (1982) *Controls on the Chemical Composition of Alpine Glacial Meltwaters.* PhD Thesis, University of East Anglia.

Tranter M. and Jones H. G. (2001) The chemistry of snow: processes and nutrient cycling. In *The Ecology of Snow* (eds. H. G. Jones, J. W. Pomeroy, D. A. Walker, and R. Hoham). Cambridge University Press, Cambridge, pp. 127–167.

Tranter M., Brown G. H., Raiswell R., Sharp M. J., and Gurnell A. M. (1993) A conceptual model of solute acquisition by Alpine glacial meltwaters. *J. Glaciol.* **39**, 573–581.

Tranter M., Sharp M. J., Brown G. H., Willis I. C., Hubbard B. P., Nielsen M. K., Smart C. C., Gordon S., Tulley M., and Lamb H. R. (1997) Variability in the chemical composition of *in situ* subglacial meltwaters. *Hydrol. Process.* **11**, 59–77.

Tranter M., Sharp M. J., Lamb H. R., Brown G. H., Hubbard B. P., and Willis I. C. (2002a) Geochemical weathering at the bed of Haut Glacier d'Arolla, Switzerland–a new model. *Hydrol. Process.* **16**, 959–993.

Tranter M., Huybrechts P., Munhoven G., Sharp M. J., Brown G. H., Jones I. W., Hodson A. J., Hodgkins R., and Wadham J. L. (2002b) Glacial bicarbonate, sulphate and base cation fluxes during the last glacial cycle, and their potential impact on atmospheric CO_2. *Chem. Geol.* **190**, 33–44.

van de Flierdt T., Frank M., Lee D.-C., and Halliday A. N. (2002) Glacial weathering and the hafnium isotope composition of seawater. *Earth Planet. Sci. Lett.* **201**, 639–647.

Wadham J. L., Hodson A. J., Tranter M., and Dowdeswell J. A. (1997) The rate of chemical weathering beneath a quiescent, surge-type, polythermal based glacier, southern Spitsbergen. *Ann. Glaciol.* **24**, 27–31.

Wadham J. L., Tranter M., and Dowdeswell J. A. (2000) The hydrochemistry of meltwaters draining a polythermal-based, high Arctic glacier, Svalbard: II. Winter and early Spring. *Hydrol. Process.* **14**, 1767–1786.

Wadham J. L., Cooper R. J., Tranter M., and Hodgkins R. (2001) Enhancement of glacial solute fluxes in the proglacial zone of a polythermal glacier. *J. Glaciol.* **47**, 378–386.

White A. F., Bullen T. D., Schulz M. S., Blum A. E., Huntington T. G., and Peters N. E. (2001) Differential rates of feldspar weathering in granitic regoliths. *Geochim. Cosmochim. Acta* **65**, 847–869.

Zachos J. C., Opdyke B. N., Quinn T. M., Jones E. C., and Halliday A. N. (1999) Early Cenozoic glaciation. Antarctic weathering and seawater $^{87}Sr/^{86}Sr$: is there a link? *Chem. Geol.* **161**, 165–180.

Zwally H. J., Abdalati W., Herring T., Larson K., Saba J., and Steffen K. (2002) Surface melt-induced acceleration of Greenland ice-sheet flow. *Science* **297**, 218–222.

5.08
Global Occurrence of Major Elements in Rivers

M. Meybeck

University of Paris VI, CNRS, Paris, France

5.08.1 INTRODUCTION

Major dissolved ions (Ca^{2+}, Mg^{2+}, Na^+, K^+, Cl^-, SO_4^{2-}, HCO_3^-, and CO_3^{2-}) and dissolved silica (SiO_2) in rivers have been studied for more than a hundred years for multiple reasons: (i) geochemists focus on the origins of elements and control processes, and on the partitioning between dissolved and particulate forms; (ii) physical geographers use river chemistry to determine chemical denudation rates and their spatial distribution; (iii) biogeochemists are concerned with the use of carbon, nitrogen, phosphorus, silica species, and other nutrients by terrestrial and aquatic biota; (iv) oceanographers need to know the dissolved inputs to the coastal zones, for which rivers play the dominant role; (v) hydrobiologists and ecologists are interested in the temporal and spatial distribution of ions, nutrients, organic carbon, and pH in various water bodies; (vi) water users need to know if waters comply with their standards for potable water, irrigation, and industrial uses.

The concentrations of the major ions are commonly expressed in $mg\,L^{-1}$; they are also reported in $meq\,L^{-1}$ or $\mu eq\,L^{-1}$, which permits a check of the ionic balance of an analysis: the sum of cations (Σ^+ in $eq\,L^{-1}$) should equal the sum of anions (Σ^- in $eq\,L^{-1}$). Dissolved silica is generally not ionized at pH values commonly found in rivers; its concentration is usually expressed in $mg\,L^{-1}$ or in $\mu mol\,L^{-1}$. Ionic contents can also be expressed as percent of Σ^+ or Σ^- ($\%C_i$), which simplifies the determination of ionic types. Ionic ratios (C_i/C_j) in $eq\,eq^{-1}$ are also often tabulated (Na^+/Cl^-, Ca^{2+}/Mg^{2+}, Cl^-/SO_4^{2-}, etc.). As a significant fraction of sodium can be derived from atmospheric sea salt and from sedimentary halite, a chloride-corrected sodium concentration is commonly reported ($Na^{\#} = Na^+ - Cl^-$ (in $meq\,L^{-1}$)). The export rate of ions and silica, or the yield ($Y_{Ca^{2+}}$, Y_{SiO_2}) at a given station is the average mass transported per year divided by the drainage area: it is expressed in units of $t\,km^{-2}\,yr^{-1}$ (equal to $g\,m^{-2}\,y^{-1}$) or in $eq\,m^{-2}\,yr^{-1}$.

This chapter covers the distribution of riverine major ions, carbon species, both organic and inorganic, and silica over the continents, including internal regions such as Central Asia, and also the major factors such as lithology and climate that control their distribution and yields.

Based on an unpublished compilation of water analyses in 1,200 pristine and subpristine basins, I am presenting here an idealized model of global river chemistry. It is somewhat different from the model proposed by Gibbs (1970), in that it includes a dozen major ionic types. I also illustrate the enormous range of the chemical composition of rivers—over three orders of magnitude for concentrations and yields—and provide two global average river compositions: the median composition and the discharge-weighted composition for both internally and externally draining regions of the world.

A final section draws attention to the human alteration of river chemistry during the past hundred years, particularly for Na^+, K^+, Cl^-, and SO_4^{2-}; it is important to differentiate anthropogenic from natural inputs. Trace element occurrence is covered by Gaillardet (see Chapter 5.09).

5.08.2 SOURCES OF DATA

Natural controls of riverine chemistry at the global scale have been studied by geochemists since Clarke (1924) and the Russian geochemists Alekin and Brazhnikova (1964). Regional studies performed prior to industrialization and/or in remote areas with very limited human impacts are rare (Kobayashi, 1959, 1960) and were collected by Livingstone (1963). Even some of his river water data are affected by mining, industries, and the effluents from large cities. These impacts are obvious when the evolution of rivers at different periods is compared; Meybeck and Ragu (1996, 1997) attempted this comparison for all the rivers flowing to the oceans from basins with areas exceeding 10^4 km². More recently interest in the carbon cycle and its riverine component has been the impetus for a new set of field studies (Stallard and Edmond, 1981, 1983; Degens *et al.*, 1991; Gaillardet *et al.*, 1997, 1999; Huh *et al.*, 1998/1999; Millot *et al.*, 2002; Guyot, 1993) that build on observations made since the 1970s (Reeder *et al.*, 1972; Stallard and Edmond, 1981, 1983). These data were used to construct the first global budgets of river dissolved loads (Meybeck, 1979) and their controls (Holland, 1978; Meybeck, 1987, 1994; Bluth and Kump, 1994; Berner, 1995; Stallard, 1995a,b).

Most of the annual means derived from these data have been collected into a global set of pristine or subpristine rivers and tributaries (PRISRI, $n = 1,200$) encompassing all the continents, including exorheic and endorheic (internal

drainage) runoff. The largest basins, such as the Amazon, Mackenzie, Lena, Yenisei, and Mekong, have been subdivided into several smaller sub-basins. In some regions (e.g., Indonesia, Japan) the size of the river basins included in these summaries may be less than 1,000 km². In the northern temperate regions only the most reliable historic analyses prior to 1920 have been included. Examples are analyses performed by the US Geological Survey (1909–1965) in the western and southwestern United States and in Alaska.

In order to study the influence of climate, total cationic contents (Σ^+) of PRISRI rivers have been split into classes based on annual runoff (q in mm yr^{-1}) and Σ^+ (meq L^{-1}). A medium-sized subset has also been used (basin area from 3,200 km² to 200,000 km², $n = 700$). The PRISRI data base covers the whole globe but has poor to very poor coverage of western Europe, Australia, South Africa, China, and India due to the lack of data for pre-impact river chemistry in these regions. PRISRI includes rivers that flow permanently ($q > 30$ mm yr^{-1}) and seasonally to occasionally ($3 < q < 30$ mm yr^{-1}). Lake outlets may be included in PRISRI.

5.08.3 GLOBAL RANGE OF PRISTINE RIVER CHEMISTRY

The most striking observation in global river chemistry is the enormous range in concentrations (C_i), ionic ratios (C_i/C_j), and the proportions of ions in cation and anion sums ($\%C_i$) as illustrated by the 1% and 99% quantiles (Q_1 and Q_{99}) of their distribution (Table 1). The Q_{99}/Q_1 ratio of solute concentrations is the lowest for potassium and silica, about two orders of magnitude, and it is very high for chloride and sulfate, exceeding three orders of magnitude. The concept of "global average river chemistry," calculated from the total input of rivers to oceans divided by the total volume of water, can only be applied for global ocean chemistry and elemental cycling; it is not a useful reference in either weathering studies, river ecology, or water quality.

Ionic ratios and ionic proportion distributions also show clearly that all major ions, except potassium, can dominate in multiple combinations: ionic ratios also range over two to three orders of magnitude, and they can be greater or less than unity for all except the Na^+/K^+ ratio, in which sodium generally dominates. River compositions found in less than 1% of analyses can be termed *rare*; for analyses from Q_1 to Q_{10} and Q_{90} to Q_{99}, I propose the term *uncommon*, from Q_{10} to Q_{25} and Q_{75} to Q_{90}, *common*, and between Q_{25} and Q_{75}, *very common*. An example of this terminology is shown in the next section (Figure 2) for dissolved inorganic carbon (DIC).

Table 1 Global range of pristine river chemistry (medium-sized basins).

	Ca^{2+}	Mg^{2+}	Na^+	K^+	Cl^-	SO_4^{2-}	HCO_3^-	SiO_2	Σ^+
Ionic contents (C_i)									
Q_{99}	9,300	5,900	14,500	505	17,000	14,500	5,950	680	32,000
Q_1	32	10	18	3.9	3.7	5	47	3.3	128
Q_{99}/Q_1	290	590	805	129	4,600	2,900	126	206	250
Ionic proportions (%C_i)									
Q_{99}	84	48	72	19.5	69	67	96		
Q_1	11	0.1	1	0.1	0.1	0.1	9		

	$\dfrac{Ca^{2+}}{Mg^{2+}}$	$\dfrac{Ca^{2+}}{Na^+}$	$\dfrac{Mg^{2+}}{Na^+}$	$\dfrac{Na^+}{K^+}$	$\dfrac{Na^+}{Cl^-}$	$\dfrac{Ca^{2+}}{SO_4^{2-}}$	$\dfrac{SO_4^{2-}}{HCO_3^-}$	$\dfrac{Cl^-}{SO_4^{2-}}$	$\dfrac{SiO_2}{\Sigma^+}$
Ionic ratios (C_i/C_j)									
Q_{99}	20.3	56	20	164	29	51	3.5	8.5	1.3
Q_1	0.01	0.14	0.01	0.95	0.33	0.19	0.01	0.01	0.0

C_i: ionic contents (μeq L^{-1} and μmol L^{-1} for silica); %C_i: proportion of ions in the sum of total cations or anions; C_i/C_j: ionic ratios; and Q_1 and Q_{99}: lowest and highest percentiles of distribution.

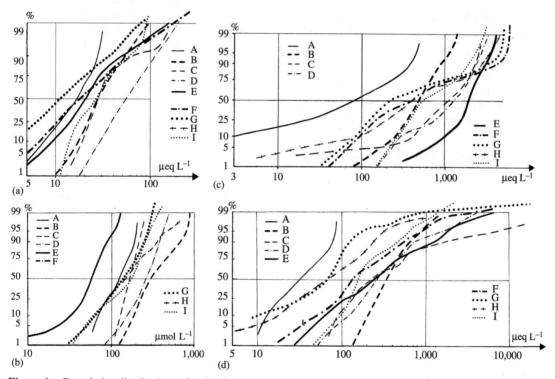

Figure 1 Cumulative distributions of major dissolved elements in pristine regions: (a) K^+, (b) SiO_2, (c) HCO_3^-, and (d) Na^+. A: Central and lower Amazon, B: Japan, C: Andean Amazon Basin, D: Thailand, E: Mackenzie Basin, F: French streams, G: temperate stream model, H: monolithologic miscellaneous French streams, I: major world rivers (source Meybeck, 1994).

When plotted on a log-probability scale (Henry's law diagram) the distributions of elemental concentrations show four patterns (Figure 1): (i) lognormal distribution as for potassium in Thailand (a, D), that can be interpreted as a single source of the element and limited control on its concentration; (ii) retention at lower concentrations (there is a significant break in the distribution as for silica in the Mackenzie Basin (b, E)); (iii) retention at higher concentrations (there is a significant break in the distribution as for bicarbonate (DIC) (c, C, E, F, G, I)); and (iv) additional source at higher concentrations (the break in the distribution is in the opposite direction, as for chloride, suggesting another source of material than the one observed between Q_{10} and Q_{90} (d, all except central Amazon)).

5.08.4 SOURCES, SINKS, AND CONTROLS OF RIVER-DISSOLVED MATERIAL

The observed distribution patterns have been interpreted by many authors (Likens *et al.*, 1977; Holland, 1978; Drever, 1988; Hem, 1989; E. K. Berner and R. A. Berner, 1996) in terms of multiple sources, sinks, and controls on riverine chemistry (Table 2). The sources of ions may be multiple: rainfall inputs, generally of oceanic origin, (rich in NaCl and also in $MgSO_4$), differential weathering of silicate minerals and carbonate minerals, dissolution of evaporitic minerals contained in some sedimentary rocks (gypsum and anhydrite, halite) or leached during rainstorms from surficial soils of semi-arid regions (Garrels and Mackenzie, 1971; Drever, 1988; Stallard, 1995a,b). Sinks are also multiple: silica may be retained in lakes and wetlands due to uptake by aquatic biota. Carbonate minerals precipitate in some eutrophic lakes and also precipitate when the total dissolved solids increases, generally above $\Sigma^+ = 6$ meq L^{-1}, due to evaporation.

In semi-arid regions ($30 < q < 140$ mm yr^{-1}) and arid regions (chosen here as $3 < q < 30$ mm yr^{-1})

surface waters gradually evaporate and become concentrated, reaching saturation levels of calcium and magnesium carbonates first, and then of calcium sulfate (Holland, 1978): these minerals are deposited in soils and in river beds during their drying stage, a process termed evaporation/ crystallization by Gibbs (1970).

There is also growing evidence of active recycling of most major elements, including silica, in terrestrial vegetation, particularly in forested areas (Likens *et al.*, 1977). Each element may, therefore, have multiple natural sources and sinks. The main controlling factors for each element are also multiple (Table 2). They can be climatic: higher temperature generally increases mineral dissolution; water runoff increases all weathering rates. Conversely, soil retention increases at lower runoff. The weathering of silicate and carbonate minerals is also facilitated by organic acids generated by terrestrial vegetation (see Chapter 5.06). Higher lake residence times ($\tau > 6$ months) favor biogenic silica retention and the precipitation of calcite (e.g., 30 g SiO_2 m^{-2} yr^{-1} and 300 g $CaCO_3$ m^{-2} yr^{-1} in Lake Geneva). The distance to the oceans is the key factor controlling the inputs of sea salts.

Tectonics also exercises important regional controls: active tectonics in the form of volcanism and uplift is associated with the occurrence at the Earth's surface of fresh rock that is more

T0010 **Table 2** Dominant sources, sinks, and controls of major ions in present day rivers.

	Major ions								Controls						
	SiO_2	K^+	Na^+	Cl^-	SO_4^{2-}	Mg^{2+}	Ca^{2+}	DIC	$T°$	τ	q	d	QH	V	T
Natural sources															
Atmosphere		←-----			-----→		←→					−			+
Silicate weathering	←------		→		←-----		-----→		+		+		×	+	+
Pyrite	←-----		-----→	←-----			-----→			+					
Carbonate					←-----		-----→			+				+	
Gypsum				←-----			-----→			+					
Halite		←-----	-----→							+					
Deep waters	←------							-----→							+
Natural sinks															
Terrestrial vegetation	←------		---					-----→	+		+			+	−
Soils	←------							-----→		−			×		−
Lakes	←→									+			×		
									prod	pop	fert	treat	irrig	nb	
Anthropogenic sources															
Mines		←-----			-----→				+		+				
Industries		←-----			-----→				+		+				
Cities			←-----		-----→					+					
Agriculture	←------			-----→							+				
Anthropogenic sinks															
Reservoirs	←→				←-----		-----→							+	
Irrigated soils	←------							-----→					+		

$+$, $-$: increase and decrease with related control; ×: complex relation; DIC: dissolved inorganic carbon.
$T°$: temperature; q: runoff; d: distance to ocean; QH: Quaternary history; V: terrestrial biomass; τ: water residence time; T: volcanism, tectonic uplift and rifting.
Prod: production; pop: urban population; fert: fertilization rate; treat: wastewater treatment and recycling; irrig: water loss through irrigation; nb: volume of reservoirs and eutrophication.

easily weathered than surficial rocks that have been exposed to weathering for many millions of years on a stable craton (Stallard, 1995b): there, the most soluble minerals have been dissolved and replaced by the least soluble ones such as quartz, aluminum and iron oxides, and clays. Active tectonics also limits the retention of elements by terrestrial vegetation, and retention in arid soils due to high mechanical erosion rates. Rifting is generally associated with inputs of saline deep waters that can have a major influence on surface water chemistry at the local scale. In formerly glaciated shields as in Canada and Scandinavia, glacial abrasion slows the development of a weathered soil layer. This limits silicate mineral weathering even under high runoff. These low relief glaciated areas are also characterized by a very high lake density, an order of magnitude higher than in most other regions of the world; lakes are sinks for silica and nutrients. The effects of Quaternary history in semi-arid and arid regions can be considerable, due to the inheritance of minerals precipitated under past climatic conditions. Due to the effect of local lithology, distance to the ocean, regional climate and tectonics, and the occurrence of lakes, river chemistry in a given area may be very variable.

5.08.4.1 Influence of Lithology on River Chemistry

Lithology is an essential factor in determining river chemistry (Garrels and Mackenzie, 1971; Drever, 1988, 1994), especially at the local scale (Strahler stream orders 1–3) (Miller, 1961; Meybeck, 1986). On more regional scales, there is generally a mixture of rock types, although some large river basins (area >0.1 Mkm^2) may contain one major rock type such as a granitic shield or a sedimentary platform. When selecting nonimpacted monolithologic river basins in a given region such as France (Table 3, A–E), the influence of climate, tectonics, and distance from the ocean can be minimized, revealing the dominant control of lithology, which in turn depends on (i) the relative abundance of specific minerals and (ii) the sensitivity of each mineral to weathering. The weathering scale most commonly adopted is that of Stallard (1995b) for mineral stability in tropical soils: quartz $\gg$ K-feldspar, micas $\gg$ Na-feldspar $>$ Ca-feldspar, amphiboles $>$ pyroxenes $>$ dolomite $>$ calcite $\gg$ pyrite, gypsum, anhydrite $\gg$ halite (least stable).

Table 3F provides examples of stream and river chemistry under peculiar conditions: highly weathered quartz sand (Rio Negro), peridotite (Dumbea), hydrothermal inputs (Semliki and Tokaanu), evaporated (Saoura),

black shales (Powder and Redwater), sedimentary salt deposits (Salt). They illustrate the enormous range of natural river chemistry (outlets of acidic volcanic lakes are omitted here).

Table 3 lists examples of more than a dozen different chemical types of river water. Although Ca^{2+} and HCO_3^- are generally dominant, Mg^{2+} dominance over Ca^{2+} can be found in rivers draining various lithologies such as basalt, peridotite, serpentinite, dolomite, coal, or where hydrothermal influence is important (Semliki). Sodium may dominate in sandstone basins, in black shales (Powder, Redwater in Montana), in evaporitic sedimentary basins (Salt), in evaporated basins (Saoura), and where hydrothermal and volcanic influence is important (Semliki, Tokaanu). K^+ rarely exceeds 4% of cations, except in some clayey sands, mica schists, and trachyandesite; it exceeds 15% in extremely dilute waters of Central Amazonia and in highly mineralized waters of rift lake outlets (Semliki, Ruzizi).

The occurrence of highly reactive minerals, such as evaporitic minerals, pyrite and even calcite, in low proportions—a percent or less— in a given rock, e.g., calcareous sandstone, pyritic shale, marl with traces of anhydrite, granite with traces of calcite, may determine the chemical character of stream water (Miller, 1961; Drever, 1988). In a study of 200 streams from monolithologic catchments underlain by various rock types under similar climatic conditions in France, the relative weathering rate based on the cation sum (Meybeck, 1986) ranges from 1 for quartz sandstone to 160 for gypsiferous marl.

When the lithology in a given region is fairly uniform, the distribution of major-element concentrations is relatively homogeneous with quantile ratios Q_{90}/Q_{10} well under 10 as, for example, in Japan (mostly volcanic), the Central Amazon Basin (detrital sand and shield) (Figure 1, distributions B and A); when a region is highly heterogeneous with regards to lithology, as in the Mackenzie Basin and in France, the river chemistry is much more heterogeneous (Figure 1, distributions E and F) and quantile ratios Q_{90}/Q_{10}, Q_{99}/Q_1 may reach those observed at the global scale (Figure 1, distributions G) (Meybeck, 1994). In the first set, regional geochemical background compositions can be easily defined, but not in the second set.

5.08.4.2 Carbon Species Carried by Rivers

The carbon cycle and its long-term influence on climate through the weathering of fresh silicate rocks (Berner *et al.*, 1983) has created a new interest in the river transfer of carbon. Bicarbonate (HCO_3^-) is the dominant form of DIC in the pH range of most world rivers

Table 3 A–E: composition of pristine waters draining single rock types in France (medians of analyses corrected for atmospheric inputs) (from 3 to 26 analyses in each class, except for estimates that are based on one analysis only). F: other river chemistry from various origins uncorrected for atmospheric inputs (Meybeck, 1986). Cation and anion proportions in percent of their respective sums (Σ^+ and Σ^-).

	SiO_2 ($\mu mol^{-1} L^{-1}$)	Σ^+ ($\mu eq^{-1} L^{-1}$)	Ca^{2+} (%)	Mg^{2+} (%)	Na^+ (%)	K^+ (%)	Cl^- (%)	SO_4^{2-} (%)	HCO_3^- (%)
A. *Noncarbonate detrital rocks*									
Quartz sand and sandstones	170	170	30	20	45	5	0	(40)	(60)
Clayey sands	135	300	53	17	13	17	0	(30)	(70)
Arkosic sands	200	400	48	35	12	5	0	(20)	(80)
Graywacke	90	350	58	22	18	2	0	(20)	(80)
Coal-bearing formations	150	5,000	30	55	14	1	0	20–90	80–10
B. *Carbonate-containing detrital rocks*									
Shales	90	500	60	30	8	1.5	0	25	75
Permian shales	175	2,200	53	35	8	4	(0–20)	10	90–70
Molasse	280	2,500	80	17	3	0.3	0	2	98
Flysch	50	2,200	79	19	1.5	0.5	0	2	98
Marl	90	3,000	83	14	2.5	0.5	0	2	98
C. *Limestones*									
Limestones	60	4,500	95	3	0.6	0.4	0	2.5	97.5
Dolomitic limestones	60	4,500	72	26	0.6	0.4	0	3.5	96.5
Chalk[a]	200	4,500	95	2.5	2.2	0.5	0	2.5	97.5
Dolomite[a]	67	5,900	54	46	0.1	0.1	0	8.5	91.5
D. *Evaporites*									
Gypsum marl[a]	160	22,000	77	22	0.2	0.2	0	83	17
Salt and gypsum marl[a]	133	27,500	34	32	33	(0.8)	36	44	20
E. *Plutonic, metamorphic, and volcanic rocks*									
Alkaline granite, gneiss, mica schists	140	130	15	15	65	5	0	(30)	(70)
Calc-alkaline granite, gneiss, mica schists	100	300	54	25	17	4	0	(15)	(85)
Serpentinite	225	1,500	38	60	7	1	0	(7)	93
Peridotite	180	600	5	93	1	1	0	(15)	(85)
Amphibolite	65	1,600	85	17	2	1	0	(7)	(93)
Marble[a]	150	3,400	86	11	1.8	0.6	0	(12)	88
Basalt	200	500	42	38	17	2.5	0	(2)	98
Trachyandesite	190	220	32	25	35	8	0	(2)	98
Rhyolite	190	550	53	15	25	2	0	(2)	98
Anorthosite	260	400	45	25	26	4	0	(2)	98
F. *Miscellaneous river waters (not rain corrected)*									
Rio Negro tributaries[b]	75	18.1	10.5	16.5	51.9	20.9			
Dumbea (New Caledonia)	232	1,175	3.9	84.7	10.9	0.3	14.2	7.1	78.5
Cusson (Landes, France)[b]	251	1,463	13.8	21.2	61.5	3.4	69	12.3	18.7
Semliki (Uganda)[c]	213	8,736	6.4	36.2	39.6	17.5	14.6	23.3	62.0
Powder (Montana)	148	20,200	27.2	20.3	51.5	0.9	15.0	62.8	22.2
Saoura (Marocco)[d]		26,150	23.3	16.8	59.2	0.7	62.8	27.7	9.6
Redwater (Montana)	116	40,700	10.7	24.5	64.1	0.6	1.0	72.2	26.7
Tokaanu (New Zeal.)[c]	4,760	41,600	14.4	3.0	79.5	3.1	90.6	5.8	3.5
Salt (NWT, Canada)	20	312,000	9.7	1.8	88.4	0.1	89.7	9.3	1.0

[a] Estimates. [b] Rain dominated. [c] Hydrothermal inputs. [d] Evaporated.

(6 < pH < 8.2); carbonate (CO_3^{2-}) is significant only at higher pH, which occurs in a few eutrophic rivers such as the Loire River, where pH exceeds 9.2 during summer algal blooms, and in waters that have undergone evaporation. Undissociated dissolved CO_2 is significant only in very acidic waters rich in humic substances such as the Rio Negro (Amazonia), but this is unimportant at the global scale. In terms of fluxes, bicarbonate DIC dominates (Table 4). It has two different sources: (i) carbonic acid weathering of noncarbonate minerals, particularly of silicates such as feldspars, micas, and olivine, (ii) dissolution of carbonate minerals such as calcite and dolomite, in which half of the resulting DIC originates from soil and/or atmospheric

Table 4 Riverine carbon transfer and global change.

Sources		Age (yr)	Flux[a] (10^{12} g C yr^{-1})	A	B	C	D	E	F	
PIC	Geologic	10^4–10^8	170	•					•	
DIC	Geologic	10^4–10^8	140		•	•			•	
	Atmospheric	0–10^2	245		•	•			•	
DOC	Soils	10^0–10^3	200			•			•	
	Pollution	10^{-2}–10^{-1}	(15) ?					•		TAC
CO_2	Atmospheric	0	(20–80)		•	•	•		•	
POC	Soils	10^0–10^3	(100)	•					•	
	Algal	10^{-2}	(<10)				•		•	
	Pollution	10^{-2}–10^0	(15) ?					•		
	Geologic	10^4–10^8	(80)	•					•	

A = land erosion, B = chemical weathering, C = global warming and UV changes, D = eutrophication, E = organic pollution, F = basin management, TAC = Total atmospheric carbon.
[a] Present global flux to oceans mostly based on Meybeck (1993), 10^{12} g C yr^{-1}.

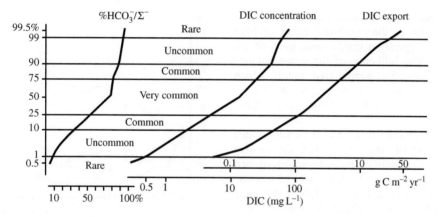

Figure 2 Global distribution of DIC concentration, ratio and export rate (yield) in medium-sized basins (3,500–200,000 km^2).

CO_2, and half from the weathered rock. Other forms of riverine carbon include particulate inorganic carbon (PIC) due to mechanical erosion in carbonate terrains, and dissolved and particulate organic carbon (DOC and POC) that are largely due to soil leaching and erosion; fossil POC in loess and shale may also contribute to river POC; organic pollution and algal growth in eutrophic lakes and rivers contribute minor fluxes (Meybeck, 1993). The "ages" of these carbon species, i.e., the time since their original carbon fixation fall into two categories: (i) those from 0 to ~1,000 yr, representing the fast cycling external part of the cycle are termed total atmospheric carbon (TAC); (ii) those from 50 kyr (Chinese Loess Plateau) to 100 Myr representing "old" carbon from sedimentary rocks. The sensitivity of these transfers to global change is complex (Table 4).

River carbon transfers are also very variable at the global scale: DIC concentrations range from 0.06 mg DIC L^{-1} (Q_1) to 71 mg DIC L^{-1}

(Q_{99}); DIC export, or yield, ranges from 0.16 g C m^{-2} yr^{-1} (Q_1) to 33.8 g C m^{-2} yr^{-1} (Q_{99}). In 50% of river basins bicarbonate makes up more than 80% of the anionic charge. The distribution quantiles for these variables are shown in Figure 2.

Organic acids, particularly those present in wetland-rich basins, also play an important role in weathering (Drever, 1994; Viers *et al.*, 1997; see also Chapters 5.06 and 5.10).

5.08.4.3 Influence of Climate on River Chemistry

The influence of air temperature is nearly impossible to measure for large basins due to its spatial heterogeneity and its absence in most databases. Some studies of mountain streams have considered the effect of air temperature (Drever and Zobrist, 1992). It is generally agreed that silicate weathering is more rapid

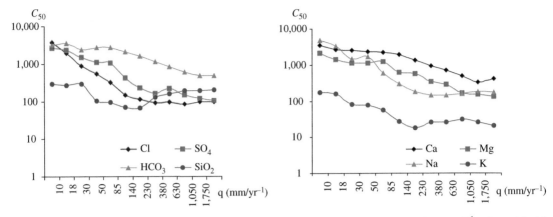

Figure 3 Median concentrations C_{50} in 12 classes of river runoff (q_1: <10 mm yr^{-1}, then 10–18, 18–30, 30–50, 50–85, 85–140, 140–230, 230–380, 380–630, 630–1,050, 1,050–1,750 and q_{12}: >1,750 mm yr^{-1}) (ions in μeq L^{-1}, silica in μmol L^{-1}) (total sample, $n = 1,091$).

in wet tropical regions (other than areas of very low relief) due to the combined effect of high temperature, vegetation impact and, most of all, high runoff. The influence of the water balance can be studied easily; annual river runoff, which is generally documented, integrates entire river basins (Meybeck, 1994; White and Blum, 1995).

The set of PRISRI rivers has been subdivided into 12 classes of runoff for which the median elemental concentrations have been determined (Figure 3). The Q_{25} and Q_{75} quantiles and, to some extend the Q_{10} to Q_{90} quantiles follow the same patterns for all elements. On the basis of Na$^+$, K$^+$, Cl$^-$, and SO$_4^{2-}$ two different clusters can be distinguished: (i) above 140 mm yr^{-1} the influence of runoff is not significant, (ii) below 140 mm yr^{-1} there is a gradual increase in elemental concentrations as runoff decreases. This pattern is interpreted as due to evaporation. For other elements there is a gradual increase of Ca^{2+}, Mg^{2+}, and HCO$_3^-$ from the highest runoff to 85 mm yr^{-1} or 50 mm yr^{-1}, then a stabilization of these concentrations at lower runoff values ($q < 50$ mm yr^{-1}). The second part of this pattern is most probably due to the precipitation of carbonate minerals as a result of evaporative concentration. The decrease in Ca^{2+}, Mg^{2+}, and HCO$_3^-$ concentrations at very high runoff, while the silicate-weathering related Na$^+$ and K$^+$ are stable, could be due to a lower occurrence of carbonate rocks in wetter regions. This does not imply a climate control of these rock types so much as the dominance of the Amazon and Congo basins, which are underlain by silicate rocks, and volcanic islands in the population of high-runoff rivers. The silica pattern is even more complex: the increase in silica concentration with runoff from 140 mm yr^{-1} to >1,750 mm yr^{-1} could be linked with a greater occurrence of crystalline

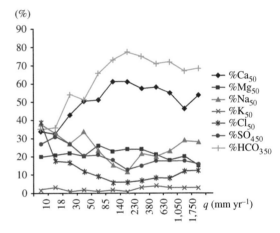

Figure 4 Median proportions (in percent of cations or anions sum) in 12 classes of river runoff (q_1: <10 mm yr^{-1}, then 10–18, 18–30, 30–50, 50–85, 85–140, 140–230, 230–380, 380–630, 630–1,050, 1,050–1,750 and q_{12}: >1,750 mm yr^{-1}, sample $n = 1,091$.

rocks, particularly of volcanic rocks, in wetter regions as in volcanic islands (Iceland, Japan, Indonesia) or coasts (Washington, Oregon), which are particularly abundant in the PRISRI database. These interactions between climate and lithology are possible but need to be verified. The lower silica value from 30 mm yr^{-1} to 140 mm yr^{-1} could also be attributed to the importance of the Canadian and Siberian rivers in these classes, where SiO$_2$ retention in lakes is important.

The relative median proportions of ions in the runoff classes of the PRISRI rivers confirm this pattern (Figure 4): below 140 mm yr^{-1} the %HCO$_3^-$ and %Ca^{2+} drop from 75% and 60%, respectively, to ~33%, while those of

Cl^- and SO_4^{2-} increase from 6% to 39% and from 13% to 27%, respectively: the proportion of K^+ does not vary significantly. There is a significant increase in the median proportions of Na^+ and Cl^- above 1,050 mm yr^{-1} that could correspond with the increased marine influence in these categories. The median proportion of Mg^{2+} is the most stable of all ions: from 15% to 26%.

5.08.5 IDEALIZED MODEL OF RIVER CHEMISTRY

Gibbs (1970) defined three main categories of surface continental waters: rain dominated, with Na^+ and Cl^- as the major ions, weathering dominated, with Ca^{2+} and HCO_3^- as the major ions, and evaporation/crystallization dominated, with Na^+ and Cl^- as the major ions. This typology can still be used, but it is too simplified. According to the PRISRI database there are many more types of water and controlling mechanisms, although the ones described by Gibbs may account for ~80% of the observed chemical types. The sum of cations (Σ^+), which is a good indicator of weathering, ranges from 50 μeq L^{-1} to 50,000 μeq L^{-1}, and the corresponding types of water, characterized by the dominant ions may reach as many as a dozen (Table 3). The true rain dominance type is found near the edges of continents facing oceanic aerosol inputs, as in many small islands and in Western Europe: it is effectively of the Na^+-Cl^- type and Σ^+ may be as high as 1 meq L^{-1}, for example, the Cusson R. (Landes, France) draining arkosic sands (Table 3, F). In continental interiors, in some rain forests, this water type occurs with much lower ionic concentrations ($\Sigma^+ < 0.1$ meq L^{-1}) in areas with highly weathered soils and very low mechanical erosion rates, such as the Central Amazon Basin (Rio Negro tributaries, Table 3F) or Cameroon: the very limited cationic inputs in rainfall are actively utilized and recycled by the forest and may even be stored (Likens *et al.*, 1977; Viers *et al.*, 1997). If the DOC level is high enough (~10 mg L^{-1}) the pH is often so low, that HCO_3^- is insignificant: the dominant anions are SO_4^{2-} and organic anions or Cl^-. This water chemistry is actually controlled by the terrestrial vegetation and can have a variety of ionic assemblages. The silica generated by chemical weathering is exported as dissolved SiO_2 and also as particulate biogenic SiO_2 from phytoliths or sponge spicules, as in the Rio Negro Basin.

Numerous water types reflect weathering control; these include $Mg^{2+}-HCO_3^-$, $Ca^{2+}-HCO_3^-$, $Na^+-HCO_3^-$, $Mg^{2+}-SO_4^{2-}$, $Ca^{2+}-SO_4^{2-}$, $Na^+-SO_4^{2-}$, Na^+-Cl^- types (Table 3). The evaporation–crystallization control found in semi-arid and arid

regions such as Central Asia (Alekin and Brazhnikova, 1964) also gives rise to multiple water types: $Mg^{2+}-SO_4^{2-}$, $Ca^{2+}-SO_4^{2-}$, $Na^+-SO_4^{2-}$, Na^+-Cl^-, $Mg^{2+}-HCO_3^-$, and even $Mg^{2+}-Cl^-$ (see Chapter 5.13). It is difficult to document the precipitation–redissolution processes (either during rare rain storms for occasional streams or in allochtonous rivers flowing from wetter headwaters such as the Saoura in Marocco, Table 3F), which are likely to have $\Sigma^+ > 6$ meq L^{-1} and/or runoff below 140–85 mm yr^{-1}. In some Canadian and Siberian basins, very low runoff (<85 mm yr^{-1}) is not associated with very high evaporation but with very low precipitation: weathering processes still dominate.

In rift and/or volcanic regions and in recent mountain ranges such as the Caucasus, hydrothermal inputs may add significant quantities of dissolved material (Na^+, K^+, Cl^-, SO_4^{2-}, SiO_2) to surface waters. The Semliki River, outlet of Lake Edward, is particularly enriched in K^+ (Table 3, F); the Tokaanu River (New Zealand) drains a hydrothermal field with record values of silica, Na^+ and Cl^- (Table 3, F).

Overall, on the basis of ionic proportions and total concentrations, only 8.2% of the rivers (in number) in the PRISRI database can be described as evaporation controlled, 2.6% as rain dominated and vegetation controlled, and 89.2% as weathering dominated, including rivers affected by large water inputs.

A tentative reclassification of the major ionic types is presented on Figure 5 showing the occurrence of major ion sources for different

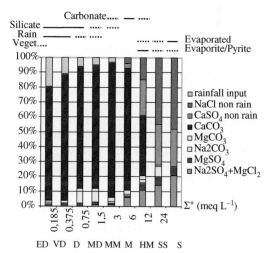

Figure 5 Idealized occurrence of water types (dominant ions and dominant control factors) per classes of increasing cationic content (Σ^+ in meq L^{-1}). ED: extremely dilute waters, VD: very dilute, D: dilute, MD: medium dilute, MM: medium mineralized, HM: highly mineralized, SS: subsaline, S: saline.

classes of Σ^+. The rainfall dominance types also include rivers with vegetation control and correspond to multiple ionic types ($Ca^{2+}-SO_4^{2-}$, Na^+-Cl^-, $Ca^{2+}-Cl^-$, $Na^+-HCO_3^-$). Other water types correspond to various rock weathering modes, including $Mg^{2+}-HCO_3^-$ waters observed in many volcanic regions. The most common type, $Ca^{2+}-HCO_3^-$, dominates from $\Sigma^+ < 0.185$ meq L^{-1} to 6 meq L^{-1}: in the most dilute waters it originates from silicate rock weathering, above 1.5 meq L^{-1} from weathering of carbonate minerals. This water type virtually disappears above 12 meq L^{-1}. "Non-rain" NaCl and $CaSO_4$ water types appear gradually above 6 meq L^{-1}. However, with these water types it is impossible to differenciate between evaporite rock dissolution, which can be observed even in the wet tropics (Stallard and Edmond, 1981) or in the Mackenzie (Salt R., Table 3F), from the leaching of salinized soils very common in Central Asia (Alekin and Brazhnikova, 1964). $Mg^{2+}-SO_4^{2-}$, $Na^+-SO_4^{2-}$, and $Mg^{2+}-Cl^-$ types are occasionally found in streams with Σ^+ values below 3 meq L^{-1}; they are commonly the result of pyrite weathering that can lead to very high Σ^+ as in the Powder and Redwater Rivers (Table 3, F).

The global proportions of the different water types depend on the global representativeness of the database. The exact occurrence of water types can only be estimated indirectly through modelling on the basis of lithologic maps, water balance, and oceanic fallout. Moreover, it will depend on spatial resolution: a very fine scale gives more importance to the smallest, rain-dominated coastal basins. In PRISRI, 70% of the basins have areas exceeding 3,200 km², thus limiting the appearance of oceanic influence.

A river salinity scale based on Σ^+ is also proposed (Figure 5). The least mineralized waters ($\Sigma^+ < 0.185$ meq L^{-1}), termed here "extremely dilute" correspond to a concentration of total dissolved solids of ~ 10 mg L^{-1} in NaCl equivalent. The most mineralized waters ($\Sigma^+ > 24$ meq L^{-1}) are here termed "saline" up to 1.4 g L^{-1} NaCl equivalent, a value slightly less than the conventional limit of 3 g L^{-1} NaCl adopted for "saline" lakes.

5.08.6 DISTRIBUTION OF WEATHERING INTENSITIES AT THE GLOBAL SCALE

Present-day weathering intensities can theoretically be assessed by the export of dissolved material by rivers. Yet many assumptions and corrections have to be made: (i) human impacts (additional sources and sinks) should be negligible, (ii) atmospheric inputs should be subtracted, (iii) products of chemical weathering should not be carried as particulates (e.g., phytoliths), nor (iv) accumulated within river basins in lakes or soils. In addition, it must be remembered that 100% of the HCO_3^- may originate from the atmosphere in noncarbonate river basins (calcite is only found in trace amounts in granites) and $\sim 50\%$ in carbonate terrains (see Chapter 5.11). In basins of mixed lithologies the proportions range between 50% and 100%. The total cation export or yield Y^+ (which excludes silica) is used to express the weathering intensity. Y^+, expressed in eq m^{-2} yr^{-1}, the product of annual runoff (m yr^{-1}) and Σ^+ (meq L^{-1}), is extremely variable at the Earth's surface since it combines both runoff variability and river chemistry variability (Table 5).

The concentration of elements that are derived from rock weathering (Ca^{2+}, Mg^{2+}, Na^+, K^+), are less variable than runoff even in the driest conditions. The opposite is observed for Cl^- and SO_4^{2-}, which are characterized by very low Q_1 quantiles. The retention of silica, particularly under the driest conditions, makes its yield the most variable. The lowest yearly average runoff in this data set (3.1 mm yr^{-1} for Q_1) actually corresponds to the conventional limit for occasional river flow (3 mm yr^{-1}). Under such extremely arid conditions, flow may occur only few times per hundred years, as for some tributaries of Lake Eyre in Central Australia. The other runoff quantile Q_{99} corresponds to the wettest regions of the planet bordering the coastal zone.

Since the variability of runoff, represented by the percentile ratio Q_{99}/Q_1, is generally much greater than the variability of concentration (Tables 1 and 5), ionic yields primarily depend on runoff. With a given rock type, there is a strong correlation between yield and runoff, that corresponds to clusters of similar concentrations as for Ca^{2+} and silica (Figure 6). This influence

Table 5 Global distribution of ionic (eq m^{-2} yr^{-1}) and silica (mol m^{-2} yr^{-1}) yields and annual runoff (q in mm yr^{-1}) in medium-sized basins (3,200–200,000 km², $n = 685$).

Y_i	Ca^{2+}	Mg^{2+}	Na^+	K^+	Cl^-	SO_4^{2-}	HCO_3^-	Σ^+	SiO_2	q
Q_{99}	2.77	0.95	1.3	0.115	1.05	1.25	2.82	3.0	0.82	3,040
Q_1	0.0045	0.002	0.002	0.0002	0.0003	0.0004	0.0046	0.0005	0.0001	3.1
Q_{99}/Q_1	615	475	650	575	3500	3100	613	600	8200	980

Q_1 and Q_{99}: lowest and highest percentiles.

was used by Meybeck (1994), Bluth and Kump (1994), and Ludwig *et al.* (1998) to model ionic yields and inputs of carbon species to the ocean. If all PRISRI basins are considered, this runoff control on chemical yields is still observed, although the relationship is complex. The data set is subdivided into 12 classes of runoff from $q < 10$ mm yr^{-1} to $q > 1,750$ mm yr^{-1} (see Figure 3), for which the median yields of major ions and silica have been determined (Figure 7). In this log–log diagram, domains of equal concentration are parallel to the diagonal (1 : 1) line. Several types of evolution can be observed:

(i) sulfate and chloride yields are fairly constant below 140 mm yr^{-1}, suggesting that lower runoff is compensated by a concentration increase through evaporation (except for SO_4^{2-} below 10 mm yr^{-1});

(ii) above 140 mm yr^{-1} all ion and silica yields are primarily linked to runoff;

(iii) above 140 mm yr^{-1} the median concentrations of Ca^{2+}, Mg^{2+}, and HCO_3^- decrease gradually, whereas silica and potassium concentrations increase gradually, suggesting a greater influence of silicate weathering relative to carbonate weathering;

(iv) below 140 mm yr^{-1} median Ca^{2+} and Mg^{2+} concentrations are constant and median HCO_3^- concentration decreases slightly, which suggest a regulation mechanism through precipitation of carbonate minerals in semi-arid and arid regions;

(v) potassium yield is stable between 30 mm yr^{-1} and 140 mm yr^{-1}; below 30 mm yr^{-1} it decreases, suggesting retention; and

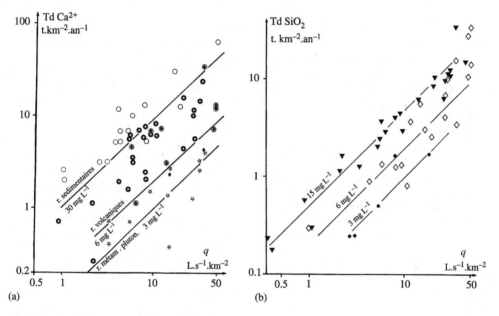

Figure 6 Relationship between calcium yield (a) and silica yield (b) versus river runoff for world rivers. Diagonals represent lines of equal concentrations. (a) Calcium yield for carbonated basins (average 30 mg Ca^{2+}L^{-1}), volcanic basins (6 mg L^{-1}), shield and plutonic (3 mg L^{-1}). (b) Silica yield for tropical regions (average 15 mg SiO$_2$ L^{-1}), temperate regions (6 mg L^{-1}), and cold regions (3 mg L^{-1}) (source Meybeck, 1994).

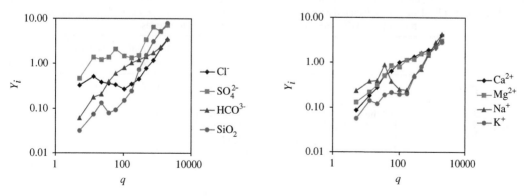

Figure 7 Median yields of ions and silica (Y_i in meq m^{-2} yr^{-1}) for 12 classes of river runoff ($n = 1,091$, all basins). Yields $= C_{i50} \times q_{50}$ in each class.

(vi) median silica yield is the most complex: there is certainly a retention in the arid regions. These observed trends need to be confirmed particularly below 30 mm runoff, where there are fewer silica analyses in the PRISRI base than analyses of major ions.

Detailed weathering controls, particularly for silicate rocks, can be found in Holland (1978), Drever and Clow (1995), Drever and Zobrist (1992), Gallard *et al.* (1999), Stallard (1995a,b), and White and Blum (1995).

5.08.7 GLOBAL BUDGET OF RIVERINE FLUXES

The unglaciated land surface of the present Earth amounts to $\sim$133 Mkm^2. Excluding land below a 3 mm yr^{-1} runoff threshold (at a 0.5° resolution), $\sim$50 Mkm^2 can be considered as nonexposed to surface water weathering (arheic, where $q < 3$ mm yr^{-1}), whether draining to the ocean (exorheic) or draining internally such as the Caspian Sea basin (endorheic) (Table 6). The land area effectively exposed to weathering by meteoric water is estimated to be $\sim$82.8 Mkm^2 (rheic regions); this, in turn, has to be divided into exorheic regions (76.1 Mkm^2) and endorheic regions (6.7 Mkm^2). Since the weathering intensity in the rheic regions is highly variable (see above), three main groupings have been made on the basis of weathering intensity or runoff: the least active or oligorheic regions (36.4 Mkm^2), the regions of medium activity or mesorheic (42.3 Mkm^2), and the most active or hyperrheic regions (4.1 Mkm^2). The corresponding ionic fluxes can be computed on the basis of the PRISRI database. The hyperrheic regions, which are exclusively found in land that drains externally, represent 37% of all DIC fluxes for only 2.75% of the land area; the mesorheic regions (28.3% of

land area) contribute 55% of the DIC fluxes, and the oligorheic regions (24.4% of land area) contribute only 8% of these fluxes (Table 6). If a finer spatial distribution of these fluxes is extracted from the PRISRI data, 1% of the land surface (carbonate rocks in very wet regions) contributes 10% of the river DIC fluxes. Similar figures are found for all other ions.

These budgets can also be broken down into classes on the basis of ionic contents (Figure 8), assuming the PRISRI database to be fully representative at the global scale. The extremely dilute waters correspond in PRISRI to $\sim$3% of the land area, but they are areas of very high runoff, therefore, their weight in terms of water volume is $\sim$7.5%. However, their influence on the global ionic fluxes is very low, less than 1%. At the other end of the salinity scale the subsaline and saline waters, mostly found in semi-arid and arid areas, correspond to $\sim$5% of the PRISRI data set; they contribute much less than 1% of runoff but $\sim$7% of the ionic fluxes.

Although we have drawn attention to the extreme variability of ionic and silica concentrations, ionic proportions, ionic ratios, ionic and silica yields throughout this chapter, two sets of global averages are proposed here (Table 7). The world spatial median values (WSM) of the medium-sized PRISRI data set (3,200–200,000 km^2 endorheic and exorheic basins) correspond to the river water chemistry most commonly found on continents at this resolution. The world weighted average (WWA) has been computed by summing the individual ionic fluxes of the largest 680 basins, including endorheic basins (Aral, Caspian, Titicaca, Great Basin, Chad basins), in the PRISRI set. The runoff values in both averages are different, although very close considering the global range. The WSM lists higher concentrations than the WWA because the dry and very dry regions are more common in the database than the very humid regions. Ionic ratios are more similar between

Table 6 Distribution of global land area (Mkm^2) exposed to chemical weathering and to river transfer of soluble material. A: percent of land area (nonglaciated area also contains alpine glaciers). B: percent of weathering generated fluxes (e.g., DIC flux). C: percent of river fluxes to oceans.

					A % land area	B % weath. flux	C % flux to ocean
Total land 149 Mkm^2 {	Glaciated 16 Mkm^2				10.7	0.1?	0.1?
	Nonglaciated 133 Mkm^2 {	Arheic 50.2 Mkm^2 {	Endorheic[a]		33.7	0	0
			Exorheic[a]				
		Rheic 82.8 Mkm^2 {	Oligorheic {	Endorheic	3.4	1	0
				Exorheic	21	7	7
			Mesorheic {	Endorheic	1.1	2	0
				Exorheic	27.3	53	55
			Hyperrheic	Exorheic	2.75	37	38

[a] Potentially.

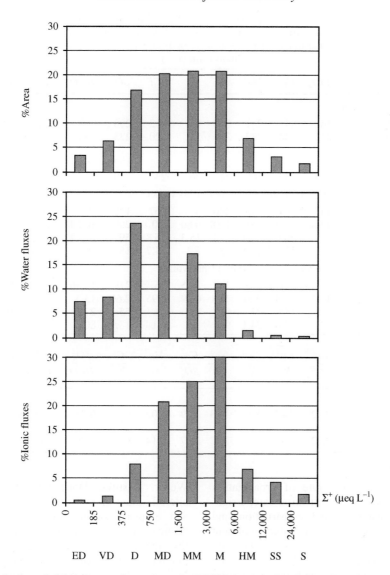

Figure 8 Distribution of global river attributes from the PRISRI database for classes of increasing total cations (Σ^+) based on PRISRI rivers (rheic regions). ED: extremely dilute waters, VD: very dilute, D: dilute, MD: medium dilute, MM: medium mineralized, HM: highly mineralized, SS: subsaline, S: saline.

the two averages. The composition of river inputs to the oceans is not stable: (i) it has varied since the Late Glacial Maximum and during the geological past (Kump and Arthur, 1997); (ii) present-day river chemistry is much altered by human activities. Four estimates of average exorheic rivers are presented in Table 7: considering the natural variations of river chemistry, these averages are relatively close to each other and to WWA and WSM, the differences are due to the nature of the data sets and to the inclusion or exclusion of presently altered rivers. A re-estimation of the pristine inputs (under present-day climatic conditions) should now be done with the new data, although human impacts are difficult to quantify.

5.08.8 HUMAN ALTERATION OF RIVER CHEMISTRY

River chemistry is very sensitive to alteration by many human activities, particularly mining and the chemical industries, but also to urbanization as urban wastewaters are much more concentrated than rural streams, and to agriculture through the use of fertilizers (Table 2) (Meybeck *et al.*, 1989; Meybeck, 1996; Flintrop *et al.*, 1996). New sinks are also created such as reservoirs (calcite and silica trapping; enhanced evaporation) and irrigated soils, which may retain soluble elements if they are poorly drained. New controls correspond to these anthropogenic influences, such as mining

Table 7   WWA ionic concentrations (C_i^* in μeq L^{-1}, and μmol L^{-1} for silica) and yields (Y_d^* in g m^{-2} yr^{-1}, Y_i^* in meq m^{-2} yr^{-1}, and mmol m^{-2} yr^{-1} for silica), ionic ratios (C_i/C_j^* in eq eq^{-1}) and relative ionic proportion ($\%C_i^*$). Same variables for the WSM determined for mesobasins (3,000–200,000 km^2) $Na^{\#} = Na-Cl$; q = global average runoff (mm yr^{-1}).

	SiO_2	Ca^{2+}	Mg^{2+}	Na^+	K^+	Cl^-	SO_4^{2-}	HCO_3^-	Σ^+	Σ^-	q
World river inputs to oceans											
Clarke (1924) C_i	161	843	231	208	45	132	208	969	1,327	1,309	(320)
Livingstone (1963) C_i	217	748	337	274	59	220	233	958	1,418	1,411	314
Meybeck (1979)											
pristine C_i	174	669	276	224	33	162	172	853	1,202	1,187	374
1970 C_i	174	733	300	313	36	233	239	869	1,382	1,341	374
World weighted average (WWA) (endorheic + exorheic) (this work)											340
C_i^*	145	594	245	240	44	167	175	798	1,125	1,139	
Y_d^*	2.98	4.04	1.01	1.88	0.60	2.0	2.87	16.54			
Y_i^*		202	83.3	81.8	15.3	56.8	59.7	271.3	382	388	
$\%C_i^*$		52.2	21.5	23	3.3	14.7	15.4	69.9	100	100	
World spatial median (WSM) (endorheic + exorheic) (this work)											222
C_{i50}	134	1,000	375	148	25.5	96	219	1,256	1,548	1,571	
Y_{d50}	1.82	3.78	0.91	1.31	0.29	0.95	2.66	13.3			
Y_{i50}		188	75	57	7.4	22	55	278	327	300	
$\%C_{i50}$		64.6	24.1	9.6	1.7	6.1	13.9	80			

	$\dfrac{Ca^{2+}}{Mg^{2+}}$	$\dfrac{Ca^{2+}}{Na^+}$	$\dfrac{Na^+}{K^+}$	$\dfrac{Na^+}{Cl^-}$	$\dfrac{Ca^{2+}}{SO_4^{2-}}$	$\dfrac{Mg^{2+}}{Na^+}$	$\dfrac{Cl^-}{SO_4^{2-}}$	$\dfrac{SO_4^{2-}}{HCO_3^-}$	$\dfrac{Na^{\#}}{K^+}$	$\dfrac{Ca^{2+}}{Na^{\#}}$	$\dfrac{Mg^{2+}}{Na^{\#}}$	$\dfrac{SiO_2}{\Sigma^+}$
WWA $(C_i/C_j)^*$	2.42	2.47	5.5	1.43	3.39	1.02	0.95	0.22	1.65	8.1	3.35	0.13
WSM $(C_i/C_j)_{50}$	2.32	2.54	6.55	2.27	2.97	1.14	0.48	0.26	3.0	4.3	1.9	0.12

and industrial production, rate of urbanization and population density, fertilization rate, irrigation rate and practices, and construction and operation of reservoirs. Wastewater treatment and/or recycling can be effective as a control on major ions originating from mines (petroleum and gas exploitation, coal and lignite, pyritic ores, potash and salt mines) and industries, yet their effects are seldom documented. Urban wastewater treatment does not generally affect the major ions. Human impacts on silica are still poorly studied apart from retention in reservoirs. At the pH values common in surface waters, silica concentration is limited and evidence of marked excess silica in rivers due to urban or industrial wastes has not been observed by this author. The gradual alteration of river chemistry was noted very early, for example, for SO_4^{2-} (Berner, 1971) and Cl^- (Weiler and Chawla, 1969) in the Mississippi and the Saint Lawrence systems. Regular surveys made since the 1960's and comparisons with river water analyses performed a hundred years ago reveal a worldwide increase in Na^+, Cl^-, and SO_4^{2-} concentrations, whereas Ca^{2+}, Mg^{2+}, and HCO_3^- concentrations are more stable (Meybeck *et al.*, 1989; Kimstach *et al.*, 1998). Some rivers affected by mining (Rhine,

Weser, Vistula, Don) may be much more altered than rivers affected by urbanization and industrialization only (Mississippi, Volga, Seine) (Figure 9). When river water is diverted and used for irrigation, there is a gradual increase of ionic content, particularly Na^+, K^+, Cl^-, and SO_4^{2-} as for the Colorado, Murray, Amu Darya (Figure 9). Salinization may also result from agriculture as for the Neman River (Figure 9). Salinization is discussed in detail by Vengosh (see Chapter 9.09)

It is difficult to assess the different anthropogenic sources of major ions which depend on the factors mentioned above. These human factors also vary in time for a given society and reflect the different environmental concerns of these societies, resulting in multiple types of river–society relationships (Meybeck, 2002). In the developed regions of the northern temperate zone it is now difficult to find a medium-sized basin that is not significantly impacted by human activities. In industrialized countries, each person generates dissolved salt loadings that eventually reach river systems (Table 8).

These anthropogenic loads are higher for Na^+, Cl^-, and SO_4^{2-} relative to natural loads. This partially explains the higher sensitivity of river chemistry to human development. At a

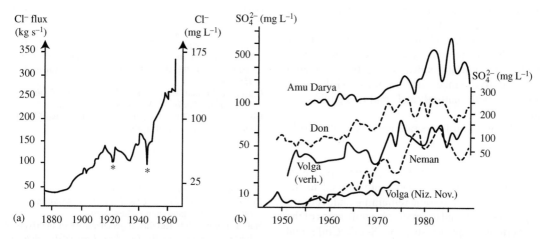

Figure 9 (a) Chloride evolution (fluxes, concentrations) in the Rhine River at mouth from 1,875 to 1,970 (ICPR, 1975). (b) Sulfate evolution in selected rivers of the former USSR, Volga at Nizhny Novgorod and Neman at Sovetsk (lower scale), Don at Aksai, and Volga at Verhnelebyazhye (medium scale), Amu Darya at Kzyl Djar (upper scale) (Tsikurnov in Kimstach *et al.*, 1998). * reduced mining activity in the Rhine Basin due to crisis and conflicts.

Table 8 Excess loads of major elements normalized to basin population in industrialized regions (kg cap^{-1} yr^{-1}).

	SiO_2	Ca^{2+}	Mg^{2+}	Na^+	K^+	Cl^-	SO_4^{2-}	HCO_3^-	Period
Per capita loads in residential urban sewage (kg cap^{-1} yr^{-1})									
Montreal	ND	3.2	0.65	6.6	1.0	8.2	13.5	24	1970s
US sewer	2.4	3	1.5	14	2	15	6	20	1960s
Brussels	1.2	2.6	ND	9.5	1.6	8.4	5.8	14.7	1980s
Paris	0.2	1.2	0.7	6.4	1.5	6.3	11.0	14.5	1990s
Evolution of rivers[a] (kg cap^{-1} yr^{-1})									
	ND	51	17	85	6	100	136	ND	1900–1970
Estimated anthropogenic input to ocean (10^6 t yr^{-1})									
	ND	47	10.5	78	5	93	124	100	1970

Sources: Meybeck (1979) and Meybeck *et al.* (1998).
[a] Mississippi, St. Lawrence, Seine, Rhine, Odra, Wisla.

certain population density in impacted river basins the anthropogenic loads equal (Na^+, K^+, Cl^-, SO_4^{2-}) or greatly exceed the natural ones (NO_3^-, PO_4^{2-}), defining a new era, the Anthropocene (Meybeck, 2002) when humans control geochemical cycles.

The silica trend in world rivers has recently attracted attention: dissolved silica is decreasing in impounded and/or eutrophied rivers, at the same time as nitrate is increasing in agricultural basins. As a result the Si : N ratio, which was generally well above 10 a hundred years ago, has dropped below 1.0 g g^{-1} in rivers such as the Mississippi, resulting in a major shift in the coastal algal assemblages (Rabalais and Turner, 2001), leading to dystrophic coastal areas. This trend, which is also observed for other river systems (Seine, Danube), could expand in the next decades due to increased fertilizer use and to increased reservoir construction.

5.08.9 CONCLUSIONS

Riverine chemistry is naturally highly variable at the global scale, which confirms the observation made by (the former) Soviet geochemists on a 20 Mkm^2 subset of land area (Alekin and Brazhnikova, 1964) and the conclusions of Clarke (1924), Livingstone (1963), Holland (1978), and Drever (1988). Ionic concentrations and yields (weathering rates) commonly vary over two to three orders of magnitude: only a fraction of the surface of the continents is actively exposed to weathering by meteoric water. The related river water types are multiple. There are at least a dozen types depending on surficial rock exposed to weathering, the water balance, and atmospheric inputs. Gibbs' (1970) global scheme for water chemistry holds for ~80% of river waters but is oversimplified for the remaining 20%. In very dilute waters (cation sum <0.185 μeq L^{-1}) water chemistry is probably controlled by vegetation. In these areas water chemistry is likely to be very sensitive to climate change and to forest cutting, although the related ionic fluxes are small. Each major ion and silica should be considered individually, since its sources, sinks and controls, both natural and anthropogenic, are different.

In the northern temperate regions of the world from North America to East Asia, the human impacts should be carefully filtered from the natural influence: the study of pristine river geochemistry will be more and more limited to the subarctic regions, to some remaining undeveloped tropical regions, and to small temperate areas of the southern hemisphere such as South Chile and New Zealand. There is a global-scale increase of riverine Na^+, K^+, Cl^-, and SO_4^{2-}. HCO_3^- is still very stable. Impacts of riverine changes on the Earth System (Li, 1981) should be further addressed.

REFERENCES

Alekin O. A. and Brazhnikova L. V. (1964) *Runoff of Dissolved Substances from the USSR Territory* (in Russian). Nauka, Moscow.

Berner E. K. and Berner R. A. (1996) *Global Environment, Water, Air, and Geochemical Cycles*. Prentice Hall, Englewoods Cliff, NJ.

Berner R. A. (1971) Worldwide sulfur pollution of rivers. *J. Geophys. Res.* **76**, 6597–6600.

Berner R. A. (1996) Chemical weathering and its effect on atmospheric CO_2 and climate. In *Chemical Weathering Rates of Silicate Minerals*, Reviews in Mineralogy (eds. A. F. White and S. L. Brantley). Mineralogical Society of America, Washington, DC, vol. 31, pp. 565–583.

Berner R. A., Lasaga A. C., and Garrels R. M. (1983) The carbonate–silicate geochemical cycle and its effect on atmospheric carbon dioxide over the past 100 millions years. *Am. J. Sci.* **301**, 182–204.

Bluth G. J. S. and Kump L. R. (1994) Lithologic and climatologic controls of river chemistry. *Geochim. Cosmochim. Acta* **58**, 2341–2359.

Clarke F. W. (1924) *The Data of Geochemistry*, 5th edn. US Geol. Surv. Bull. vol. 770, USGS, Reston, VA.

Degens E. T., Kempe S., and Richey J. E. (eds.) (1991) *Bio-Geochemistry of Major World Rivers*. Wiley, New York.

Drever J. I. (1988) *The Geochemistry of Natural Waters*, 2nd edn. Prentice Hall, Englewood Cliff, NJ, 437pp.

Drever J. I. (1994) The effect of land plants on weathering rates of silicate minerals. *Geochim. Cosmochim. Acta* **58**, 2325–2332.

Drever J. I. and Clow D. W. (1995) Weathering rates in catchments. In *Chemical Weathering Rates of Silicate Minerals*, Reviews in Mineralogy (eds. A. F. White and S. L. Brantley). Mineralogical Society of America, Washington, DC, vol. 31, pp. 463–483.

Drever J. I. and Zobrist J. (1992) Chemical weathering of silicate rocks as a function of elevation in the southern Swiss Alps. *Geochim. Cosmochim. Acta* **56**, 3209–3216.

Flintrop C., Hohlmann B., Jasper T., Korte C., Podlaha O. G., Sheele S., and Veizer J. (1996) Anatomy of pollution: rivers of North Rhine-Westphalia, Germany. *Am. J. Sci.* **296**, 58–98.

Gaillardet J., Dupré B., Allègre C. J., and Négrel P. (1997) Chemical and physical denudation in the Amazon River Basin. *Chem. Geol.* **142**, 141–173.

Gaillardet J., Dupré B., Louvat P., and Allègre C. J. (1999) Global silicate weathering and CO_2 consumption rates deduced from the chemistry of large rivers. *Chem. Geol.* **159**, 3–30.

Garrels R. M. and Mackenzie F. T. (1971) *The Evolution of Sedimentary Rocks*. W. W. Norton, New York.

Gibbs R. J. (1970) Mechanism controlling world water chemistry. *Science* **170**, 1088–1090.

Guyot J. L. (1993) Hydrogéochimie des fleuves de l'Amazonie bolivienne. Etudes et Thèses, ORSTOM, Paris, 261pp.

Hem J. D. (1989) *Study and Interpretation of the Chemical Characteristics of Natural Water*. US Geol. Surv., Water Supply Pap. 2254, USGS, Reston, VA.

Holland H. D. (1978) *The Chemistry of the Atmosphere and Oceans*. Wiley-Interscience, New York.

Huh Y. (1998/1999) The fluvial geochemistry of the rivers of Eastern Siberia: I to III. *Geochim. Cosmochim. Acta* **62**, 1657–1676; **62**, 2053–2075; **63**, 967–987.

ICPR (International Commission for the Protection of the Rhine) (1975) ICPR/IKSR Bundesh f. Gewässerkunde, Koblenz (Germany).

Kimstach V., Meybeck M., and Baroudy E. (eds.) (1998) *A Water Quality Assessment of the Former Soviet Union*. E and FN Spon, London.

Kobayashi J. (1959) Chemical investigation on river waters of southeastern Asiatic countries (report I). The quality of waters of Thailand. Bericht. Ohara Inst. für Landwirtschaft Biologie, vol. 11(2), pp. 167–233.

Kobayashi J. (1960) A chemical study of the average quality and characteristics of river waters of Japan. Bericht. Ohara Inst. für Landwirstchaft Biologie, vol. 11(3), pp. 313–357.

Kump L. R. and Arthur M. A. (1997) Global chemical erosion during the Cenozoic: weatherability balances the budget. In *Tectonic and Climate Change* (ed. W. F. Ruddiman). Plenum, New York, pp. 399–426.

Li Y. H. (1981) Geochemical cycles of elements and human perturbations. *Geochim. Cosmochim. Acta* **45**, 2073–2084.

Likens G. E., Bormann F. H., Pierce R. S., Eaton J. S., and Johnson N. M. (1977) *Biogeochemistry of a Forested Ecosystem*. Springer, Berlin.

Livingstone D.A. (1963) *Chemical Composition of Rivers and Lakes*. Data of Geochemistry. US Geol. Surv. Prof. Pap. 440 G, chap. G, G1–G64.

Ludwig W., Amiotte-Suchet P., Munhoven G., and Probst J. L. (1998) Atmospheric CO_2 consumption by continental erosion: present-day controls and implications for the last glacial maximum. *Global Planet. Change* **16–17**, 107–120.

Meybeck M. (1979) Concentration des eaux fluviales en éléments majeurs et apports en solution aux océans. *Rev. Géol. Dyn. Géogr. Phys.* **21**(3), 215–246.

Meybeck M. (1986) Composition chimique naturelle des ruisseaux non pollués en France. *Sci. Geol. Bull.* **39**, 3–77.

Meybeck M. (1987) Global chemical weathering of surficial rocks estimated from river dissolved loads. *Am. J. Sci.* **287**, 401–428.

Meybeck M. (1993) Riverine transport of atmospheric carbon: sources, global typology and budget. *Water Air Soil Pollut.* **70**, 443–464.

Meybeck M. (1994) Origin and variable composition of present day riverborne material. In *Material Fluxes on the Surface of the Earth. Studies in Geophysics* (ed. Board on Earth Sciences and Resources–National Research Council). National Academic Press, Washington, DC, pp. 61–73.

Meybeck M. (1996) River water quality, global ranges time and space variabilities. *Vehr. Int. Verein. Limnol.* **26**, 81–96.

Meybeck M. (2002) Riverine quality at the Anthropocene: propositions for global space and time analysis, illustrated by the Seine River. *Aquat. Sci.* **64**, 376–393.

Meybeck M. and Ragu A. (1996) River Discharges to the Oceans. An assessment of suspended solids, major ions, and nutrients. Environment Information and Assessment Report. UNEP, Nairobi, 250p.

Meybeck M. and Ragu A. (1997) Presenting Gems Glori, a compendium of world river discharge to the oceans. *Int. Assoc. Hydrol. Sci. Publ.* **243**, 3–14.

Meybeck M., Chapman D., and Helmer R. (eds.) (1989) *Global Fresh Water Quality: A First Assessment.* Basil Blackwell, Oxford, 307pp.

Meybeck M., de Marsily G., and Fustec E. (eds.) (1998) *La Seine en Son Bassin.* Elsevier, Paris.

Miller J. P. (1961) *Solutes in Small Streams Draining Single Rock Types, Sangre de Cristo Range, New Mexico.* US Geol. Surv., Water Supply Pap., 1535-F.

Millot R., Gaillardet J., Dupré B., and Allègre C. J. (2002) The global control of silicate weathering rates and the coupling with physical erosion: new insights from rivers of the Canadian Shield. *Earth Planet. Sci. Lett.* **196**, 83–98.

Rabalais N. N. and Turner R. E. (eds.) (2001) *Coastal Hypoxia.* Coastal Estuar. Studies, **52**, American Geophysical Union.

Reeder S. W., Hitchon B., and Levinston A. A. (1972) Hydrogeochemistry of the surface waters of the Mackenzie river drainage basin, Canada: I. Factors controlling inorganic composition. *Geochim. Cosmochim. Acta* **36**, 825–865.

Stallard R. F. (1995a) Relating chemical and physical erosion. In *Chemical Weathering Rates of Silicate Minerals,* Reviews in Mineralogy (eds. A. F. White and S. L. Brantley). Mineralogical Society of America, Washington, DC, pp. 543–562.

Stallard R. F. (1995b) Tectonic, environmental and human aspects of weathering and erosion: a global review using a steady-state perspective. *Ann. Rev. Earth Planet. Sci.* **23**, 11–39.

Stallard R. F. and Edmond J. M. (1981) Geochemistry of the Amazon: 1. Precipitation chemistry and the marine contribution to the dissolved load at the time of peak discharge. *J. Geophys. Res.* **86**, 9844–9858.

Stallard R. F. and Edmond J. M. (1983) Geochemistry of the Amazon: 2. The influence of geology and weathering environment on the dissolved load at the time of peak discharge. *J. Geophys. Res.* **86**, 9844–9858.

USGS (1909–1965) *The Quality of Surface Waters of the United States.* US Geol. Surv., Water Supply Pap., 236, 237, 239, 339, 363, 364, 839, 970, 1953, Reston, VA.

Viers J., Dupré B., Polvé M., Schott J., Dandurand J. L., and Braun J. J. (1997) Chemical weathering in the drainage basin of a tropical watershed (Nsimi-Zoetele site, Cameroon): comparison between organic poor and organic rich waters. *Chem. Geol.* **140**, 181–206.

Weiler R. R. and Chawla V. K. (1969) Dissolved mineral quality of the Great Lakes waters. *Proc. 12th Conf. Great Lakes Res.: Int. Assoc. Great Lakes Res.*, 801–818.

White A. T. and Blum A. E. (1995) Effects of climate on chemical weathering in watersheds. *Geochim. Cosmochim. Acta* **59**, 1729–1747.

5.09
Trace Elements in River Waters

J. Gaillardet

Institut de Physique du Globe de Paris, France

and

J. Viers and B. Dupré

Laboratoire des Mécanismes et Transferts en Géologie, Toulouse, France

5.09.1 INTRODUCTION

Trace elements are characterized by concentrations lower than 1 mg L^{-1} in natural waters. This means that trace elements are not considered when "total dissolved solids" are calculated in rivers, lakes, or groundwaters, because their combined mass is not significant compared to the sum of Na^+, K^+, Ca^{2+}, Mg^{2+}, H_4SiO_4, HCO_3^-, CO_3^{2-}, SO_4^{2-}, Cl^-, and NO_3^-. Therefore, most of the elements, except about ten of them, occur at trace levels in natural waters. Being trace elements in natural waters does not necessarily qualify them as trace elements in rocks. For example, aluminum, iron, and titanium are major elements in rocks, but they occur as trace elements in waters, due to their low mobility at the Earth's surface. Conversely, trace elements in rocks such as chlorine and carbon are major elements in waters.

The geochemistry of trace elements in river waters, like that of groundwater and seawater, is receiving increasing attention. This growing interest is clearly triggered by the technical advances made in the determination of concentrations at lower levels in water. In particular, the development of inductively coupled plasma mass spectrometry (ICP-MS) has considerably improved our knowledge of trace-element levels in waters since the early 1990s. ICP-MS provides the capability of determining trace elements having isotopes of interest for geochemical dating or tracing, even where their dissolved concentrations are extremely low.

The determination of trace elements in natural waters is motivated by a number of issues. Although rare, trace elements in natural systems can play a major role in hydrosystems. This is particularly evident for toxic elements such as aluminum, whose concentrations are related to the abundance of fish in rivers. Many trace elements have been exploited from natural accumulation sites and used over thousands of years by human activities. Trace elements are therefore highly sensitive indexes of human impact from local to global scale. Pollution impact studies require knowledge of the natural background concentrations and knowledge of pollutant behavior. For example, it is generally accepted that rare earth elements (REEs) in waters behave as good analogues for the actinides, whose natural levels are quite low and rarely measured. Water quality investigations have clearly been a stimulus for measurement of toxic heavy metals in order to understand their behavior in natural systems.

From a more fundamental point of view, it is crucial to understand the behavior of trace elements in geological processes, in particular during chemical weathering and transport by waters. Trace elements are much more fractionated by weathering and transport processes than major elements, and these fractionations give clues for understanding the nature and intensity of the weathering + transport processes. This has not only applications for weathering studies or for the past mobilization and transport of elements to the ocean (potentially recorded in the sediments), but also for the possibility of better utilization of trace

elements in the aqueous environment as an exploration tool.

In this chapter, we have tried to review the recent literature on trace elements in rivers, in particular by incorporating the results derived from recent ICP-MS measurements. We have favored a "field approach" by focusing on studies of natural hydrosystems. The basic questions which we want to address are the following: What are the trace element levels in river waters? What controls their abundance in rivers and fractionation in the weathering + transport system? Are trace elements, like major elements in rivers, essentially controlled by source-rock abundances? What do we know about the chemical speciation of trace elements in water? To what extent do colloids and interaction with solids regulate processes of trace elements in river waters? Can we relate the geochemistry of trace elements in aquatic systems to the periodic table? And finally, are we able to satisfactorily model and predict the behavior of most of the trace elements in hydrosystems?

An impressive literature has dealt with experimental works on aqueous complexation, uptake of trace elements by surface complexation (inorganic and organic), uptake by living organisms (bioaccumulation) that we have not reported here, except when the results of such studies directly explain natural data. As continental waters encompass a greater range of physical and chemical conditions, we focus on river waters and do not discuss trace elements in groundwaters, lakes, and the ocean. In lakes and in the ocean, the great importance of life processes in regulating trace elements is probably the major difference from rivers.

Section 5.09.2 of this chapter reports data. We will review the present-day literature on trace elements in rivers to show that our knowledge is still poor. By comparing with the continental abundances, a global mobility index is calculated for each trace element. The spatial and temporal variability of trace-element concentrations in rivers will be shown to be important. In Section 5.09.3, sources of trace elements in river waters are indicated. We will point out the great diversity of sources and the importance of global anthropogenic contamination for a number of elements. The question of inorganic and organic speciation of trace elements in river water will then be addressed in Section 5.09.4, considering some general relationships between speciation and placement in the periodic table. In Section 5.09.5, we will show that studies on organic-rich rivers have led to an exploration of the "colloidal world" in rivers. Colloids are small particles, passing through the conventional filters used to separate dissolved and suspended loads in rivers. They appear as major carriers of trace elements in rivers and considerably complicate aqueous-speciation calculation. Finally, in Section 5.09.6, the significance of interactions between solutes and solid surfaces in river waters will be reviewed. Regulation by surfaces is of major importance for a great range of elements. Although for both colloids and surface interactions, some progress has been made, we are still far from a unified model that can accurately predict trace-element concentrations in natural water systems. This is mainly due to our poor physical description of natural colloids, surface site complexation, and their interaction with solutes.

5.09.2 NATURAL ABUNDANCES OF TRACE ELEMENTS IN RIVER WATER

Trace-element abundances in the dissolved load of a number of worldwide rivers, classified by continents, are compiled in Table 1, along with the corresponding references. These data have been measured after filtration of the river sample using either 0.2 μm or 0.45 μm filters. We have tried to report, as far as possible, "natural" river systems by discarding data from highly polluted systems. For the Amazon and Orinoco rivers, we have reported data for the dissolved load filtered using filters or membranes with smaller pore size. A major feature of trace-element abundances in river waters is that they are strongly dependent upon filtration characteristics. For simplicity, elements have been ranked in alphabetical order. Most often, data for large rivers are a compendium of data derived from different bibliographical sources and therefore have not been measured on the same river sample. However, in the case of the Amazon basin, two different analyses, corresponding to two different sampling dates are reported for the major tributaries. These comparisons show the high degree of variability of trace-element concentrations in river waters through time. When time series are available, a mean value is given in Table 1. It is apparent from these data that the Americas and Africa are the places where most of our information on natural levels of trace elements in waters has been obtained. By contrast, rivers of Asia have received relatively little attention, except for a number of key elements of geochemical importance such as osmium, strontium, and uranium.

North America. A substantial literature exists on the trace-element levels in North American rivers. We have focused in Table 1 on the largest systems. Smaller and pristine watersheds from northern Canada have also been reported for comparison. The St. Lawrence system is represented by the St. Lawrence itself and by the Ottawa River, which, due to its status as an international standard of river water (SLRS 4),

appears as the most often measured river. The Mistassini River is a black (high humic acid content) river discharging into St. Jean Lake and represents the North American end-member in terms of trace-element levels. Other large river systems include the Mackenzie and Peel rivers, the Fraser, the Columbia, Mississippi, Connecticut and Hudson rivers. Small rivers from the Mackenzie Basin have been reported, because they drain relatively pristine regions of North America with well-characterized geological substratum. The Indin River drains the Slave Province, dominated by old granites. The Beatton River is a typical black river and suspended sediment-rich river draining the interior sedimentary platform. Other rivers drain the western Cordillera, composed of volcanic and volcaniclastic rocks (Upper Yukon and Skeena rivers) or sedimentary rocks (Peel river).

Europe. Situation is far less favorable in Europe. Fewer data are available in the standard scientific literature and most of the rivers are strongly impacted by pollution. Only the rivers from the northern part of Europe (Kola peninsula) reported by Pokrovski and Schott (2002) are probably close to the natural background. The enhancement of dissolved concentrations for a number of elements due to human activities is apparent from the data on the Seine River. Data for REEs in the Rhine River and streams from the Vosges mountains in Alsace and central Europe have been reported for comparison. All these locations are known to have suffered from acid deposition in the very recent past.

Africa. Recently published studies have considerably improved our knowledge of trace-element concentrations in the rivers of central Africa. We have reported in Table 1 data for the Congo–Zaire River (the second largest river in the world in terms of discharge) and its main tributaries (Ubangui, Kasai, and Zaire), data from the upper Niger and Douna (the main tributary of the Niger river) and from smaller rivers in Cameroon. The Mengong River is a small stream that has been shown to have extreme concentrations of certain trace elements. All the rivers reported for Africa drain the humid and forested part of the continent.

South America. Due to its importance in terms of water discharge and to generally high trace-element levels, the Amazon river system has been well documented for a number of elements, including their seasonal variations. In Table 1, we have reported different analyses of the Amazon River and its major tributaries to show their temporal variability at a given location. Results of ultrafiltration experiments for the Amazon and Orinoco rivers demonstrate that the concentration of a number of elements in waters depends on filtration pore size. Finally, data for

Table 1 Database of trace element concentration in the dissolved load (<0.2 μm) of rivers. All concentrations in ppb (μg L^{-1}) except for Ra (fg L^{-1}) and Os (pg L^{-1}), DOC, TSS, TDS are Dissolved Organic Carbon (mg L^{-1}), total suspended solid (mg L^{-1}), and total dissolved solutes (mg L^{-1}). Water discharge and surface area are in m^3 s^{-1} and 10^3 Km^2, respectively.

Element	References	pH	DOC	TSS	TDS	Discharge	Surface area	Ag	Al	As
Africa										
Oubangui	1	6.81		30	36.1	3,500	475		12	
Zaire	1	5.98		31	48.68	17,000	1,660		46	
Kasai	1	6.35		17		11,000	900		51	
Congo at Brazzaville	1, 2, 3	6.4		21		39,100	3,500		76	
Niger	2, 4	7.00	1.5	43	44	907	141		76	
Douna	2, 4									
Nyong	5	5.60	23	5	22	340	29		215	
Sanaga	6	7.43	3.82				133		29	0.17
Nyong	6	5.88	14.1			340	29		159	0.11
Mengong	6	4.62	24						480	0.11
Europe										
Seine at Paris, Fr.	7			40	400	260	44		16	2.71
Garonne River Fr.	8, 9	8.01		128	400	540	55			
Rhine in Alsace, Fr.	10	7.10	1.95						<50	
Vosges Stream, Fr.	10	5.10	2.03						76	
Harz Mountains, Ger.	11								1,080	0.37
Kalix River, Sweden, 1991	12, 45	6.95	3.4			296	24		27	
Kalix River, 1977 May	12					296	24			
Idel river	13	6.85	12.3			2	1		44	0.21
N. America										
St. Lawrence	2, 14, 15, 16, 46	8.00		11	183	10,700	1,020		15	0.91
SLRS 4	17								53	0.70
Ottawa	14								67	0.45
Mistassini, Can.	14	5.50	26	5		195	10		174	0.12
Mackenzie	2, 14, 18, 19	8.10	5.8	300	226	9,000	1,680		18	0.50
Peel, Can.	14	8.10	12.9	250	167	690	71		22	0.35
Indin River, Can.	14	6.70			12	8	2		28	0.14
Beatton, Can.	14	7.20		810	68				91	0.80
Upper Yukon, Can.	2, 14	7.70							25	0.62
Skeena, Can.	14	7.50	4.48	60	53	962	42		33	0.21
Fraser River, Can.	19, 20	7.34	3.39	175	93	3630	238		19	0.52
Columbia River	8, 19	7.85		64	115	5941	670			
Californian Streams	21									
Connecticut	22		3.16		70	540	25			
Hudson River	22, 23, 46	7.00			126	621	35	0.004		
Upper Mississippi	24	7.70		462						

River	References	pH						
Missouri	24, 25	7.70						2,332
Ohio	25	8.20						177
Illinois	25							102
Mississippi at Mouth	3, 15, 19, 22, 25, 46	7.80			2,980	18,400	280	860
S. America								
Amazon, mean value	**3, 15, 26, 27, 28, 29, 46, 48**	**6.89**	**5.05**	**9.4**	**6,100**	**205,000**	**44**	**182**
Amazon <0.2 μm	**19, 30**	**7.10**		**6.2**	**6,100**	**205,000**	**44**	**182**
Amazon <100 kDa	30		2.96	0.8				
Amazon <5 kDa	30		1.11	0.5				
Negro < 0.2 μm	26	4.85		113.9				
Negro < 0.2 μm	30	5.87	7.17	97.0				
Solimoes < 0.2 μm	26	7.10		171.4				
Solimoes < 0.2 μm	30	7.66	2.76	5.8				
Madeira < 0.2 μm	26	6.73		2.6				
Madeira < 0.2 μm	30	7.54	11.1	4.3				
Trompetas < 0.2 μm	26	6.10		39.2				
Trompetas < 0.2 μm	30	6.54	1.85	7.9				
Tapajos < 0.2 μm	26	6.68		15.2				
Tapajos < 0.2 μm	30	7.45	1.48	4.0				
Rio Beni at Riberalta	31				243	8,262		
Mamore	31				599	8,392		
Rio Beni at Rurrenabaque	31				68	2,025		
Orinoco < 0.2 μm	19, 27, 29, 30, 46, 48	6.51	5.42	61.8	1100	35,900	25	132
Orinoco < 10 kDa	27, 30		4.3	5.6				
Caroni at Cuidad Bolivar	19, 27, 29, 30, 32	5.58	3.71	16.1				
Asia								
Ob	33		7.4–9.95		2,990	13,500	126	40
Yenisei	33		4		2,500	19,800	112	23
Lena	2, 19, 34	7.60		0.15	1,200	19,100	112	30
Changjiang	2, 3, 9, 19, 27, 35, 36, 37, 3	7.80		0.83	1,808	29,400	221	520
Huanghe	2, 15, 38, 39, 40, 47	8.30		2.00	752	1300	460	27,000
Xijiang	2, 15, 47	7.70			437	11,500	161	190
Ganges	19, 29, 41, 42, 43, 46, 47, 4	7.70			1,050	15,600	182	1,100
Mekong	2, 15, 47	7.80			795	14,800	263	321
Brahmaputra	9, 19, 41, 42, 46, 47, 48	7.40			580	16,100	101	1,060
Indus	9, 41, 44, 47	7.80			960	2880	302	2,780
Shinano, Jpn.	44, 47	7.11			11	475		
World average				**32**		1,200		
Riverine flux (kt yr^{-1})				**0.62**		23		

Table 1 (continued).

Element	B	Be	Ba	Cd	Ce	Co	Cr	Cs	Cu	Dy	Er	Eu	Fe	Ga
Africa														
Oubangui			17		0.4920	0.077	0.533	0.008		0.043	0.029	0.0160	60	
Zaire			24.8		0.6610	0.075	0.386	0.0026		0.053	0.033	0.0220	202	
Kasai			20		0.4520	0.0580	0.4	0.009		0.038	0.025	0.0140	108	
Congo at Brazzaville	3.1		30		0.6890	0.0594	0.501	0.016		0.006	0.033	0.0170	179	
Niger	3.2				0.1710	0.0400	0.450	0.0100	0.630				105	
Douna														
Nyong			18		1.3200	0.3637			0.952	0.119	0.071	0.0255	241	
Sanaga			27		0.1780	0.0590			2.030	0.015	0.016		31	
Nyong			19		0.8060	0.2530			1.397	0.086	0.055		174	0.0220
Mengong			24		0.8274	0.4307				0.059	0.027		614	0.1087
Europe														
Seine at Paris, Fr.	25.0		32	0.0600	0.0600	0.1800	11.460		3.530				302	
Garonne River Fr.					0.0810						0.004	0.0016		
Rhine in Alsace, Fr.					0.0096					0.002	0.001	0.0003		
Vosges Stream, Fr.					0.0930					0.020	0.012	0.0087		
Harz Mountains, Ger.		0.61	13	0.4200		0.2600	<0.85		0.820					
Kalix River, Sweden, 1991					0.2170								525	
Kalix River, 1977 May														
Idel river	39.0		5	0.0200	0.2430	0.0840	1.07	0.0112	0.456	0.013	0.008	0.0039	666	0.0088
N. America														
St. Lawrence	24.8		23	0.0114	0.0600	0.0632		0.0052	0.936	0.005	0.004	0.0030	111	
SLRS 4	6.0	0.008	13	0.0140	0.3600	0.0480	0.37	0.0090	1.930	0.024	0.013	0.0080	108	0.0119
Ottawa	3.3		15	0.0207	0.5599	0.0746		0.0061	1.144	0.040	0.023	0.0112	112	
Mistassini, Can.	11.8		8	0.0873	1.1771	0.1221		0.0070	1.578	0.036	0.022	0.0120	170	
Mackenzie			56	0.1838	0.0266	0.0682	0.375	0.0066	1.609	0.003		0.0023	119	
Peel, Can.			62	0.0347	0.0310	0.1426	0.294	0.0037	1.043	0.001	0.003	0.0019	152	
Indin River, Can.			3		0.1305	0.0158			0.841				50	
Beatton, Can.			47	0.1206	0.4750	0.1515	0.741	0.0028	2.594	0.072	0.056	0.0290	739	
Upper Yukon, Can.			50	0.0906	0.0311	0.0616	0.241	0.0049	1.306	0.009		0.0010	102	
Skeena, Can.			13	0.0194	0.0637	0.0794		0.0006	1.077				52	0.0083
Fraser River, Can.			15		0.0600	0.0800	2.1		1.040	0.009		0.0030	47	
Columbia River					0.0583							0.0016		
Californian Streams														0.001–0.006
Connecticut					0.0258					0.004	0.003	0.0009	10	
Hudson River					0.0621					0.013	0.008	0.0026		
Upper Mississippi			73						1.850					

Missouri			80						2.010					
Ohio			32						1.741					
Illinois			59						1.984					
Mississippi at Mouth	37.8		62		0.0074				1.60–2.24	0.004	0.004	0.0007		
S. America														
Amazon, mean value	**6.1**	**0.0095**	**21**	**0.1781**	**0.2180**	**0.1766**	**0.717**		**1.463**	**0.033**	**0.018**	**0.0104**		**0.0174**
Amazon <0.2 μm			**28**		**0.0680**					**0.011**	**0.006**	**0.0027**	**43**	**0.0217**
Amazon < 100 kDa			26		0.0200					0.004	0.002	0.0007	23	
Amazon < 5 kDa			17		0.0067					0.000	0.000		13	
Negro < 0.2 μm			*6*		*0.4150*	*0.1241*			*0.399*	*0.023*	*0.015*	*0.0088*		*0.0050*
Negro < 0.2 μm			7		0.5853					0.030	0.018	0.0078	117	
Solimoes < 0.2 μm	*3.8*		*28*		*0.3630*	*0.1643*			*1.542*	*0.044*	*0.028*	*0.0150*	*351*	*0.0390*
Solimoes < 0.2 μm			29		0.0528					0.007	0.004	0.0019	53	
Madeira < 0.2 μm	*3.4*		*18*		*0.1380*	*0.0176*			*0.863*	*0.024*	*0.013*	*0.0083*	*18*	*0.0025*
Madeira < 0.2 μm			32		0.0074					0.001	0.001		26	0.0169
Trompetas < 0.2 μm	*1.5*		*14*		*0.9080*	*0.1274*			*0.269*	*0.044*	*0.028*	*0.0105*	*87*	*0.0059*
Trompetas < 0.2 μm			15		0.1300					0.007	0.005	0.0017	30	
Tapajos < 0.2 μm			*21*		*0.1150*	*0.0195*			*0.227*	*0.012*	*0.008*	*0.0033*		*0.0033*
Tapajos < 0.2 μm			18		0.0277					0.003	0.002	0.0007	11	0.0178
Rio Beni at Riberalta			30	0.0081					1.517					
Mamore			4	0.0091					1.997					
Rio Beni at Rurrenabaque			23	0.0011					0.710					
Orinoco < 0.2 μm		0.009	8		0.5207			0.007–0.013		0.056	0.031	0.0140	142	0.1176
Orinoco < 10 kDa			8		0.1703					0.020	0.012	0.0047	15	0.1143
Caroni at Cuidad Bolivar	2.4	0.0135	7		0.1443			0.006		0.012	0.006	0.0032	16	0.1027
Asia														
Ob				0.0006–0.0008					1.8–2.4				24–36	
Yenisei				0.0012–0.0018					1.39–1.91				14–17.8	
Lena	4.7			0.0089					0.755				24.3	
Changjiang	12.5			0.0033	0.1150				1.66			0.0050	31	
Huanghe	150.0			0.0011–0.0055		0.0059–0.0295			0.96–1.6				1.4–25	
Xijiang	6.0													
Ganges	17.8	0.00056												
Mekong	15.0													
Brahmaputra	20.9													
Indus					0.0024					0.001	0.001	0.0002		
Shinano, Jpn.					0.0834					0.012	0.007	0.003		
World average	***10.2***	***0.0089***	***23***	***0.08***	***0.2620***	***0.148***	***0.7***	***0.011***	***1.48***	***0.03***	***0.02***	***0.0098***	***66***	***0.03***
Riverine flux (kt yr^{-1})	380	*0.33*	*860.2*	*3*	*9.800*	*5.5*	*26*	*0.4*	*55*	*1.1*	*0.75*	*0.37*	*2470*	*1.1*

(continued)

Table 1 (continued).

Element	Gd	Ge	Hf	Ho	La	Li	Lu	Mn	Mo	Nb	Nd	Ni	Os (pg L⁻¹)	P
Africa														
Oubangui	0.0510		0.0042	0.0090	0.249		0.0040				0.277	1.15		
Zaire	0.0630		0.0057	0.0110	0.349		0.0040				0.360	1.02		
Kasai	0.0470		0.0038	0.0080	0.189		0.0030				0.241	0.41		
Congo at Brazzaville	0.0660	0.0066	0.0067	0.0120	0.319		0.0045				0.350	0.934	6.7	
Niger			0.0030		0.091						0.085	0.29	5.3	
Douna								0.50						
Nyong	0.1343			0.0199	0.538		0.0080	29.72			0.690	0.70		
Sanaga	0.0240				0.09			0.44			0.084			
Nyong	0.094	0.0065		0.0165	0.349		0.0105	22.61			0.505	1.18		
Mengong	0.0551	0.0799		0.0116	0.348		0.0053	20.02			0.416	5.04		
Europe														
Seine at Paris, Fr.	0.0088			0.0016	0.030		0.0006	3.76			0.030	5.06	41.8	
Garonne River Fr.	0.0025			0.0004	0.047		0.0004				0.038			
Rhine in Alsace, Fr.	0.0037			0.0041	0.005		0.0014				0.005			
Vosges Stream, Fr.					0.153						0.245			
Harz Mountains, Ger.					0.480	2.00		48.00				0.92		
Kalix River, Sweden, 1991					0.155			9.40						
Kalix River, 1977 May														
Idel river	0.0190	0.0082		0.0027	0.151	0.80	0.0015	22.80	0.112		0.141	0.35		2.67
N. America														
St. Lawrence	0.0059	0.0031	0.0031	0.0013	0.029		0.0006	6.28	1.292	0.0021	0.038	1.33	22.8	
SLRS 4	0.0342	0.0100		0.0047	0.287	0.54	1.9000	3.37	0.210		0.269	0.82		
Ottawa	0.0593	0.0086	0.0034	0.0078	0.411		0.0037	14.86	0.199	0.0045	0.411	0.83		
Mistassini, Can.	0.0618	0.0058	0.0070	0.0071	0.635		0.0025	11.31	0.039	0.0107	0.547	0.47		
Mackenzie	0.0019			0.0005	0.002	4.60		1.28	1.067	0.0012	0.019	1.83	25.5	
Peel, Can.	0.0116			0.0020	0.002		0.0024	4.54	1.078	0.0019	0.004	2.68		
Indin River, Can.					0.099	0.91		1.89			0.091	0.64		1.82
Beaton, Can.	0.1599	0.0049		0.0101	0.090			2.98	0.301	0.0069	0.042	5.14		
Upper Yukon, Can.	0.0136	0.0196	0.1106	0.0010	0.001	0.64	0.0006	2.29	1.055	0.0019	0.007	10.39	24.7	
Skeena, Can.		0.0014			0.051	0.35		5.37	0.418		0.081	0.91		
Fraser River, Can.	0.0110				<0.05	1.05		5.40	1.330		0.044	1.86		
Columbia River	0.0065	0.0138		0.0009	0.030	1.46	0.0007				0.023			
Californian Streams														
Connecticut	0.0047				0.021		0.0008				0.020			
Hudson River	0.0190						0.0007				0.060			
Upper Mississippi								0.41	1.114			1.66		

Missouri								0.44	1.613				1.53
Ohio								0.46	1.258				1.12
Illinois								0.70	2.314				2.92
Mississippi at Mouth	0.0042	0.0219			0.008	10.00	0.0006	0.66–1.82	1.63–2.69		0.011	1.12–1.77	
S. America													
Amazon, mean value	**0.0356**	**0.0048**	**0.0064**		**0.106**	**0.91**	**0.0020**	**50.73**	**0.175**	**0.0020**	**0.136**	**0.74**	**4.6**
Amazon <0.2 μm	**0.0123**	**0.0074**	**0.0021**		**0.032**	**2.46**	**0.0009**	**3.31**		**0.0009**	**0.042**		
Amazon <100 kDa	0.0043	0.0076	0.0007		0.010		0.0003	2.90		0.0003	0.013		
Amazon <5 kDa	0.0007	0.0061	0.0001		0.006			2.00			0.003		14.45
Negro <0.2 μm	0.0350		0.0050		0.151		0.0016	8.24			0.172	0.21	
Negro <0.2 μm	0.0432	0.0046	0.0061		0.208		0.0023	7.35			0.211		
Solimoes <0.2 μm	0.0490		0.0093		0.166		0.0037	14.56			0.226	0.92	
Solimoes <0.2 μm	0.0089		0.0017		0.050	1.02	0.0006	6.54			0.032		
Madeira <0.2 μm	0.0260		0.0053		0.054		0.0014				0.100	0.57	
Madeira <0.2 μm	0.0018	0.0041	0.0003		0.005	1.18	0.0001	3.29			0.005		
Trompetas <0.2 μm	0.0485		0.0093		0.266		0.0037	8.62			0.309	0.12	
Trompetas <0.2 μm	0.0102	0.0049	0.0016		0.044	0.41	0.0008	1.36			0.053		2.57
Tapajos <0.2 μm	0.0114		0.0020		0.228		0.0009	1.34			0.072	0.22	
Tapajos <0.2 μm	0.0037	0.0053	0.0007		0.016		0.0004	0.46			0.018		
Rio Beni at Riberalta								4.13	0.380			0.91	
Mamore								113.52	0.240			1.11	
Rio Beni at Rurrenabaque								2.37	0.218			0.79	
Orinoco <0.2 μm	0.0737		0.0107		0.177	0.32	0.0043	6.82			0.289		11.09
Orinoco <10 kDa	0.0256		0.0040		0.049		0.0018	5.24			0.094		6.1
Caroni at Cuidad Boliver	0.0147		0.0021		0.067	0.16	0.0009	5.57			0.078	3.3	6.61
Asia													
Ob												1.24–1.42	
Yenisei												0.52–0.55	8.2
Lena						1.33						0.38	13.9
Changjiang		0.0122			0.005	3.44	0.0020	1.00			0.070	0.15	42.1
Huanghe								0.55–2.2				0.30–0.59	8.3
Xijiang						3.47							32.0
Ganges													17.2
Mekong													9.9
Brahmaputra						2.61							11.2
Indus					0.003		0.0002				0.003		
Shinano, Jpn.					0.037		0.0016				0.050		
World average	*0.04*	*0.0068*	*0.0059*	*0.0071*	*0.120*	*1.84*	*0.0024*	*34*	*0.420*	*0.0017*	*0.152*	*0.801*	*9.0*
Riverine flux (kt yr⁻¹)	*1.5*	*0.25*	*0.22*	*0.27*	*4.5*	*69*	*0.09*	*1270*	*16*	*0.063*	*5.7*	*30*	*0.33.10-3*

(continued)

Table 1 (continued).

Element	Pb	Pd	Pr	Ra (fg L^{-1})	Re	Rb	Sb	Sc	Se	Sm	Sr	Ta	Tb	Th	Ti
Africa															
Oubangui			0.069			2.7		0.055		0.0600	15.0		0.0070	0.042	
Zaire			0.093			3.9		0.067		0.0820	21.0		0.0100	0.056	
Kasai			0.052			2.7		0.062		0.0470	10.5		0.0060	0.023	
Congo at Brazzaville			0.089			3.1		0.087		0.0620	11.5		0.0097	0.065	
Niger	0.039					3.86					26.4			0.013	
Douna															
Nyong			0.179			4.18				0.1362	9.7		0.0178	0.121	
Sanaga			0.024			6.16					30.3			0.012	0.231
Nyong			0.114			3.68				0.1210	12.4			0.111	0.199
Mengong			0.096			0.73				0.0780	17.9			0.137	5.808
Europe															
Seine at Paris, Fr.	0.220					1.40		1.340			227.0			0.010	
Garonne River Fr.			0.005							0.0082			0.0012		
Rhine in Alsace, Fr.			0.001							0.0012			0.0003		
Vosges Stream, Fr.			0.049							0.0500			0.0046		
Harz Mountains, Ger.	3.800														
Kalix River, Sweden, 1991						5.90	0.190				20.0				
Kalix River, 1977 May															
Idel river	0.119		0.037			0.96	0.027			0.0240	16.8		0.0025	0.022	1.070
N. America															
St. Lawrence	0.233		0.009			1.04	0.205			0.0067	177.2		0.0010	0.004	0.509
SLRS 4	0.084	0.021	0.069	2		1.53	0.270		0.230	0.0574	28.2		4.3000		1.460
Ottawa	0.105		0.104			1.55	0.057			0.0705	50.7		0.0073	0.027	1.854
Mistassini, Can.	0.113		0.149	48		1.14	0.023			0.0783	11.4		0.0085	0.041	2.278
Mackenzie	0.771		0.012			0.66	0.121			0.0055	237.8	0.0009		0.634	0.423
Peel, Can.	1.129		0.012			0.36	0.120			0.0149	154.0	0.0029	0.0025	0.588	0.574
Indin River, Can.		0.001	0.023			1.77	0.005			0.0152	10.5				0.112
Beatton, Can.	0.269		0.132			0.30	0.102			0.1849	62.7	0.1484	0.0289	1.054	1.200
Upper Yukon, Can.	0.818		0.006			0.90	0.150			0.0056	162.2		0.0010	0.988	0.768
Skeena, Can.		0.028	0.015			0.20	0.044			0.0230	78.4				0.372
Fraser River, Can.	0.078		0.011			0.91	0.053	0.141		0.0110	108.0		0.0012		0.680
Columbia River			0.010							0.0435					
Californian Streams															
Connecticut										0.0042					
Hudson River										0.0119					
Upper Mississippi	0.008			4–31		1.24									

	1	2	3	4	5	6	7	8	9	10	11	12	13	14	15
Missouri	0.006					0.93									
Ohio	0.007					0.87									
Illinois	0.035					0.94									
Mississippi at Mouth	0.011–0.016			5–30		1.17				0.0030					0.006
S. America															
Amazon, mean value	**0.064**					**1.49**	**0.061**	**1.540**		**0.0349**	**25.8**		**0.0043**		
Amazon <0.2 μm	**0.031**	**0.009**		**9–31**	**0.00020**	**1.89**	**0.051**	**1.580**		**0.0100**	**51.2**		**0.0017**	**0.006**	
Amazon <100 kDa	0.009					1.79		1.550		0.0041	47.2		0.0007		
Amazon <5 kDa	0.003					1.29					31.3				
Negro < 0.2 μm	*0.170*	*0.047*				*1.13*				*0.0380*	*3.6*		*0.0040*		
Negro < 0.2 μm		0.054				1.73		0.900		0.0390	4.2		0.0056	0.053	
Solimoes < 0.2 μm	*0.151*	*0.052*				*1.59*				*0.0520*	*45.7*		*0.0067*		
Solimoes < 0.2 μm		0.007				1.69		1.770		0.0082	61.5		0.0014	0.010	
Madeira < 0.2 μm	*0.005*	*0.022*				*1.34*				*0.0311*	*19.2*		*0.0048*	*0.001*	
Madeira < 0.2 μm		0.001				1.94		1.510		0.0014	55.5		0.0002	0.002	
Trompetas < 0.2 μm	*0.052*	*0.080*				*2.95*				*0.0596*	*6.7*		*0.0061*	*0.121*	
Trompetas < 0.2 μm		0.013				4.04		1.260		0.0094	9.6		0.0012	0.016	
Tapajos < 0.2 μm	*0.061*	*0.017*				*2.75*				*0.0181*	*9.9*		*0.0015*	*0.002*	
Tapajos < 0.2 μm		0.004				2.08		1.410		0.0040	6.5		0.0005		
Rio Beni at Riberalta						1.01					42.9				
Mamore						1.44					31.4				
Rio Beni at Rurrenabaque					0.00083	0.90					48.3				
Orinoco < 0.2 μm		0.062		12–17		1.50	0.032–0.050	0.560		0.0682	8.0		0.0098	0.073	
Orinoco < 10 kDa		0.020				1.43		0.620		0.0234	7.5		0.0035	0.026	
Caroni at Cuidad Bolivar		0.019				1.13	0.019–0.020	0.530		0.0153	2.9		0.0020	0.017	
Asia															
Ob	0.011–0.017														
Yenisei	0.005–0.006														
Lena	0.019														
Changjiang	0.054						0.22–0.23			0.0150	210				
Huanghe	0.010–4.1			50							1140				
Xijiang											110				
Ganges				45–90											
Mekong					0.00170						90				
Brahmaputra					0.00011						298				
Indus				31						0.0007	59				
Shinano, Jpn.										0.0110	324				
World average	*0.079*	*0.028*	*0.04*	*24*	*0.0004*	*1.63*	*0.07*	*1.2*	*0.07*	*0.036*	*60.0*	*0.0011*	*0.0055*	*0.041*	*0.489*
Riverine flux (kt yr^{-1})	*3*	*1.5*	*1.05*	*0.9·10⁻⁶*	*0.015*	*60.962*	*2.6*	*45*	*2.6*	*1.3*	*2240*	*0.04*	*0.2*	*1.5*	*18*

(continued)

Table 1 (continued).

Element	Tl	Tm	U	V	W	Y	Yb	Zn	Zr
Africa									
Oubangui		0.0040	0.055				0.0240		
Zaire		0.0050	0.071				0.0270		
Kasai		0.0030	0.027				0.0190		
Congo at Brazzaville		0.0035	0.049				0.0290		
Niger			0.020	0.590				0.89	0.120
Douna									
Nyong			0.029	0.645		0.0870	0.0597		0.395
Sanaga		0.0085	0.028			0.4610		1.02	0.038
Nyong		0.0085	0.022			0.2821	0.0530	1.81	0.355
Mengong		0.0051	0.022				0.0311	3.12	0.592
Europe									
Seine at Paris, Fr.			0.820	2.850		0.0500		4.98	
Garonne River Fr.		0.0006	0.750				0.0036		
Rhine in Alsace, Fr.							0.0018		
Vosges Stream, Fr.							0.0120		
Harz Mountains, Ger.			0.060	0.400		1.4000		27.00	
Kalix River, Sweden, 1991			0.090						
Kalix River, 1977 May	0.0400								
Idel river		0.0015	0.038	0.442		0.0920	0.0079	6.30	0.130
N. America									
St. Lawrence	0.0076	0.0006	0.373	0.439		0.0320	0.0029	2.58	0.022
SLRS 4		0.0002	0.050	0.350		0.1460	0.0120	1.24	0.120
Ottawa		0.0035	0.072	0.341		0.2173	0.0201	3.53	0.086
Mistassini, Can.		0.0025	0.022	0.324		0.2033	0.0191	3.79	0.047
Mackenzie		0.0016	0.730	0.253		0.0313	0.0073	0.50	0.054
Peel, Can.		0.0011		0.236		0.0574		0.88	0.038
Indin River, Can.				0.009		0.0533		1.52	0.037
Beatton, Can.		0.0087		0.398		0.8936		1.34	0.710
Upper Yukon, Can.				0.347		0.0283	0.0040	2.29	0.041
Skeena, Can.				0.106		0.1412			0.048
Fraser River, Can.			0.330	0.390		0.0690			
Columbia River							0.0045		
Californian Streams					0.1–180				
Connecticut							0.0047		

Hudson River				0.0091		
Upper Mississippi		1.285	2.055		0.21	
Missouri		1.142	0.638		0.12	
Ohio		0.333	0.581		0.17	
Illinois		1.404	1.770		0.98	
Mississippi at Mouth		0.62–1.3	0.82–1.84	0.0044	0.18–0.35	
S. America						
Amazon, mean value	**0.0033**	**0.052**	**0.703**	**0.0159**	**0.45**	
Amazon <0.2 μm	**0.0009**	**0.055**		**0.0051**	**0.76**	**0.027**
Amazon <100 kDa	0.0003	0.022		0.0016		0.004
Amazon <5 kDa		0.004			0.80	
Negro < 0.2 μm	*0.0024*	*0.019*		*0.0100*	*1.80*	
Negro < 0.2 μm	0.0025	0.034		0.0169	1.21	0.068
Solimoes < 0.2 μm	*0.0045*	*0.040*		*0.0214*	*2.35*	
Solimoes < 0.2 μm	0.0006	0.050		0.0037	3.01	0.008
Madeira < 0.2 μm	*0.0025*	*0.023*		*0.0092*	*0.67*	
Madeira < 0.2 μm	0.0001	0.026		0.0007	0.67	0.001
Trompetas < 0.2 μm	*0.0041*	*0.044*		*0.0264*	*1.15*	
Trompetas < 0.2 μm	0.0006	0.024		0.0043	1.16	0.026
Tapajos < 0.2 μm	*0.0013*	*0.019*		*0.0055*	*1.02*	
Tapajos < 0.2 μm	0.0003	0.015		0.0019	0.75	0.003
Rio Beni at Riberalta		0.033			0.46	
Mamore		0.042			0.27	
Rio Beni at Rurrenabaque		0.060			0.40	
Orinoco < 0.2 μm	0.0043	0.049			1.75	0.105
Orinoco < 10 kDa	0.0018	0.023			2.42	0.029
Caroni at Cuidad Bolivar	0.0009	0.012			1.53	0.070
Asia						
Ob						
Yenisei						
Lena						
Changjiang		1.100		0.0080	0.36	
Huanghe		7.500			0.039–0.078	
Xijiang					0.065–0.32	

(continued)

Table 1 (continued).

Element	Tl	Tm	U	V	W	Y	Yb	Zn	Zr
Ganges			2.000						
Mekong									
Brahmaputra			1.000			0.0009			
Indus			4.940			0.0071			
Shinano, Jpn.									
World average		*0.0033*	*0.372*	*0.71*	*0.1*	*0.0400*	*0.0170*	*0.60*	*0.039*
Riverine flux (kt yr^{-1})		0.12	14	27	3.7	1.5	0.6	23	1.5

(1) Dupré et al. (1996), (2) Levasseur et al. (1999), (3) Froelich et al. (1985), (4) Picouet et al. (2001), (5) Viers et al. (2000), (6) Viers et al. (1997), (7) Roy (1996), (8) Keasler and Loveland (1982), (9) Chabaux et al. (2001), (10) Tricca et al. (1999), (11) Frei et al. (1998), (12) Ingri et al. (2000), (13) Pokrovski and Schot (2002), (14) Gaillardet et al. (2003), (15) Lemarchand et al. (2000), (16) Andrae and Froeclich (1985), (17) Yeghicheyan et al. (2000), (18) Vigier et al. (2001), (19) Huh et al. (1998), (20) Cameron et al. (1995), (21) Johannesson et al. (1999), (22) Sholkovitz (1995), (23) Benoit (1995), (24) Shiller (1997), (25) Shiller and Mao (2000), (26) Gaillardet et al. (1997), (27) Yee et al. (1987), (28) Seyler and Boaventura (2002), (29) Brown et al. (1992a), (30) Deberdt et al. (2002), (31) Elbaz-Poulichet et al. (1999), (32) Edmond et al. (1995), (33) Dai and Martin (1995), (34) Martin et al. (1993), (35) Zhang et al. (1998), (36) Shiller and Boyle (1985), (37) Edmond et al. (1985), (38) Huang et al. (1988), (39) Zhang (1994), (40) Zhang et al. (1993), (41) Sharma et al. (1999), (42) Sarin et al. (1990), (43) Dalai et al. (2001), (44) Goldstein and Jacobsen (1988), (45) Porcelli et al. (1997), (46) Chabaux et al. (2003), (47) Gaillardet et al. (1999a), (48) Colodner et al. (1993).

some Andean tributaries of the Madeira River have been reported. They show remarkably similar levels to those of the Amazon. Trace-element data for rubidium, caesium, barium, and uranium, have been measured for the Guyana shield.

Asia. Large rivers of Asia are clearly the less well documented in terms of trace-element concentrations. This is mainly due to their low abundances of trace elements, probably related to their high pH character. A couple of studies have focused on the riverine input of metals to the Arctic and Pacific oceans. Himalayan rivers have not been documented for REEs (except the Indus river), but have been analyzed for particular elements such as strontium, uranium, osmium, and radium. There is clearly a need for data on trace elements in the rivers of Asia, particularly in the highly turbid peri-Himalayan rivers.

5.09.2.1 Range of Concentrations of Trace Elements in River Waters

Based on the compilation of Table 1, an attempt has been made to compute the mean value of trace-element input to the estuaries and ocean by selecting the largest rivers (Congo, Niger, Sanaga, Seine, Garonne, Kalix, Idel, St. Lawrence, Mackenzie, Peel, Fraser, Columbia, Connecticut, Hudson, Mississippi, Amazon, Orinoco, Ob, Yenisei, Lena, Changjiang, Huanghe, Xijiang, Ganges, Mekong, Brahmaputra, and Indus). This calculation is based on the assumption that our database is representative of most of the rivers and a total water discharge of $3.74 \times 10^4 \, \text{km}^{-3} \, \text{yr}^{-1}$ (E. K. Berner and R. A. Berner, 1996).

This mean value should be considered as a first-order approximation, because for the majority of elements, the number of analyses is small. Because our best information on trace concentrations is from humid tropical areas, the mean values proposed here are dominated by rivers such as the Amazon, Orinoco, and Congo, and may therefore be overestimates for a number of elements (such as REEs). For trace metals, the role of global pollution is an unresolved issue that may also contribute to enhancing the mean value proposed in Table 1 (especially for trace metals). These values compare relatively well with the previous estimates of Martin and Meybeck (1979) for selected trace elements, with the exception of lower concentrations for some metals in the present study.

The results of Table 1 are summarized in Figure 1, in which elements are ranked according to the order of magnitude of their concentration in river water. Trace-element concentrations in river waters span 10 orders of magnitude, from

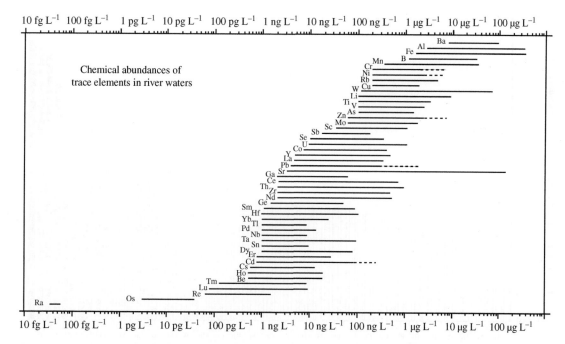

Figure 1 Graphical representation of the order of magnitude of natural trace-element concentrations in the river dissolved load. World average values derived from Table 1.

elements present in extremely low concentrations such as radium (on the order of fg L^{-1}) to elements present in relatively high concentrations such as iron, aluminum, and barium whose concentration range exceeds 10 μg L^{-1}. In between, the vast majority of trace elements have dissolved concentrations between 1 ng L^{-1} and 1 μg L^{-1} (heavy REEs (HREEs), transition metals). The platinum group elements (PGEs) are represented by osmium and show very low concentrations, ~10 pg L^{-1}. Very few data exist for PGEs in river waters.

5.09.2.2 Crustal Concentrations versus Dissolved Concentrations in Rivers

The abundances of trace elements in rivers depends both on their abundances in the continental crust and their mobility during weathering and transport. In order to depict a global "solubility" trend of trace elements, dissolved concentrations (Cw) are normalized to those of the upper continental crust (Cc) (Figure 2). Data from the continental crust are from Li (2000). In this figure, major elements in river waters are also shown and all normalized concentrations are compared to the value for sodium. It is important to note that the Cw/Cc ratio is a global mobility index rather than a solubility index because, as will be shown below, a number of very different processes contribute to the occurrence of trace elements in river dissolved load. In addition, for a

number of rarely measured elements, the concentrations in the upper continental crust may well not be correct. The graph of Figure 2 is therefore a first-order approach.

A rough classification of trace-element mobility in river waters can be drawn. The first group comprises the highly mobile elements, having mobility close to or greater than that of sodium. It consists of chlorine, carbon, sulfur, rhenium, cadmium, boron, selenium, arsenic, antimony, molybdenum, calcium, magnesium, and strontium. The case of palladium is uncertain because few data are available, and its enrichment in Figure 2 could well be a result of an incorrect continental crust concentration. The following group of moderately mobile elements includes uranium, osmium, silicon, lithium, tungsten, potassium, manganese, barium, copper, radium, rubidium, cobalt, and nickel. Their mobility is ~10 times less than that of sodium. The third group of elements contains the REEs, zinc, chromium, yttrium, vanadium, germanium, thorium, lead, caesium, beryllium, gallium, iron, and hafnium. Their mobility is 10–100 times less than that of sodium. We will call them the nonmobile elements. Finally, the last category, the most immobile elements, includes niobium, titanium, zirconium, aluminum, and tantalum, with mobility indexes more than 100 times lower than that of sodium. Depending on weathering, soil and river conditions, it is clear that some elements of these groups can pass to another group (e.g., for highly variable elements such as aluminum, iron, rubidium, lithium, and manganese), but the picture

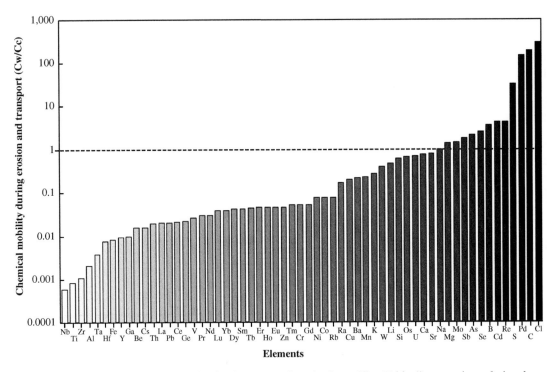

Figure 2 Normalization diagram of dissolved concentrations in rivers (Cw, Table 1) to continental abundances (Cc, Li, 2000). All values are compared to that of Na. This graph shows the increasing mobility of the elements in the weathering + transport processes, from the left to the right.

described here should be considered as an average global trend. At present, the database of trace elements in river waters is too incomplete, but there is probably much information on the processes that control the distribution of trace elements in river waters to be gained by addressing the variability of trace-element concentrations in rivers. Part of this variability corresponds to seasonal variability (see below).

5.09.2.3 Correlations between Elements

Generally, trace-element variations are not independent, and a number of authors have reported good correlations among trace elements or between trace elements and major elements. These correlations allow us to isolate groups of elements that present similar behavior during weathering and transport.

Interrelationships between trace elements and major elements or within the group of trace elements give information not only on their origin, but also about the mechanisms controlling the transport of the elements in rivers.

A clear distinction exists between trace elements whose abundance follows the abundance of major elements or TDS, and those whose concentrations are decoupled from the variations of major ions. This is illustrated by the dendrogram (Figure 3) deduced from a cluster analysis of trace-element

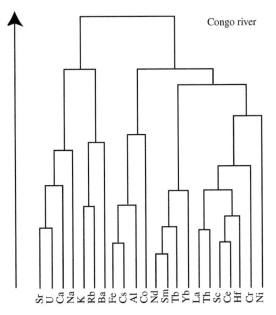

Figure 3 Dendrograms of cluster analysis of the trace-element concentrations in the dissolved load of the Congo rivers (Dupré *et al.*, 1996) showing affinity groupings of elements.

concentration in the river waters of the Congo Basin (from Dupré *et al.*, 1996). In the Congo waters, three categories are revealed by a cluster analysis. In the first set of elements, uranium, rubidium, barium, and strontium are closely related to variations of major elements. By contrast, REE variations are not correlated with those of the previous set but rather co-vary with thorium, scandium, and chromium (elements of the second set). A third set of elements showing good inter-correlation consists of caesium, cobalt, iron, and aluminum. The behavior of the latter elements is decoupled from that of the major elements but close to that of the REE group. Elements from both second and third sets show decreasing concentrations with increasing pH, in contrast to elements of the first set.

On a global scale, alkali and alkaline earth trace elements are strongly correlated to either sodium, calcium, or TDS in rivers (Edmond *et al.*, 1995, 1996). Beryllium is an alkaline-earth element with a peculiar behavior. As shown by Brown *et al.* (1992a,b), it is easily adsorbed by particle surfaces. Like beryllium, caesium has a behavior that contrasts with that of the other alkali elements. In the Upper Beni river (Bolivia) good correlations between molybdenum, zinc, cadmium, rubidium, strontium, barium and major solutes have been reported, indicating that under those conditions, trace metals can also co-vary with major elements (Elbaz-Poulichet *et al.*, 1999). From the largest tributaries of the Amazon basin, Seyler and Boaventura (2001) reported that vanadium, copper, arsenic, barium, and uranium concentrations are strongly positively correlated with major elements and pH. A correlation of vanadium with HCO_3 has also been reported by Shiller and Mao (2000) in Californian streams. Other trace metals, in nonorganic contexts, have been shown to follow major elements, e.g., the correlation between copper and silica concentrations reported by Zhang and Huang (1992) for the Huanghe River.

Selenium concentrations correlate with those of calcium and sulfate in the Orinoco river system (Yee *et al.*, 1987). Similarly, co-variations of tin or antimony with sulfate are documented by Cameron *et al.* (1995) in the Fraser river system. Little information is available on the behavior of the platinum group elements. Osmium and rhenium appear to correlate with the variations of major elements (Levasseur *et al.*, 1999; Dalai *et al.*, 2001). A good correlation between rhenium concentration and the sum of major cations is reported by Dalai *et al.* (2001) for the rivers of Yamuna and Ganga river systems in India. More generally, elements known to be present as oxyanions in oxidized waters also exhibit some coherence with major solute variations. A good correlation between tungsten or boron concentrations and chlorine has been reported by Johannesson *et al.* (1999) in rivers from California and Nevada, suggesting an essentially conservative behavior for these elements in the watershed and a

strong contribution from evaporite weathering. Correlation between boron and major solutes is also reported by Lemarchand *et al.* (2000) at a global scale. In California and Nevada, the correlation between molybdenum or vanadium with chlorine is poor compared to tungsten or boron, suggesting that secondary processes operate for the regulation of these elements. From the 56 rivers analyzed for germanium by Froelich *et al.* (1985), a relatively good correlation between germanium and silicon on a global scale is observed, provided contaminated rivers are disregarded.

The behavior of uranium varies considerably across samples studied. Palmer and Edmond (1993) have reported increasing trends of uranium concentration with river alkalinity in the Orinoco, Amazon and Ganga river basins, showing the importance of limestone and black shale dissolution for the control of uranium concentrations in river waters. The association of uranium and major soluble elements is also reported by Elbaz-Poulichet *et al.* (1999) for the upper Amazonian basins of Bolivia. However, under the organic-rich conditions of Scandinavian rivers (Porcelli *et al.*, 1997) or African rivers (Viers *et al.*, 1997), uranium concentrations can be decoupled from those of major elements, due to the existence of a colloidal fraction of uranium.

The best examples of decoupling between major solutes and trace elements in river waters come from the REEs. In most cases, dissolved REE concentrations are insensitive to major solute concentrations or even increase when solute concentrations are low. Parameters such as dissolved organic carbon (DOC) and pH appear to control the REE concentrations, and those of a number of associated elements, in river waters. Figure 4 is from Deberdt *et al.* (2002) and shows the extensive database of neodymium concentrations in river water (<0.2 μm or 0.45 μm) as a function of pH. Given the very good correlation coefficients between the different REE elements, this correlation indicates that the lowest concentrations of REEs in river waters are found in the rivers having the highest pH. This graph explains why the majority of dissolved REE concentrations published so far are for African or Northern rivers of low pH. The rivers of Asia, being much influenced by carbonate dissolution, have high pH and alkalinity values and therefore present very low REE concentrations.

Correlated with REE concentrations in river waters are elements such as thorium, yttrium, and, to a lesser extent, iron, aluminum, gallium, zirconium, manganese, and zinc (Laxen *et al.*, 1984; Shiller and Boyle, 1985; Dupré *et al.*, 1996; Viers *et al.*, 1997; Pokrovski and Schott, 2002). The close association of these elements in river waters with DOC has been shown by several authors (Perdue *et al.*, 1976; Sholkovitz *et al.*, 1978;

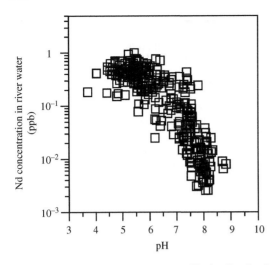

Figure 4 Nd concentrations measured in the dissolved phase (<0.2 μm) of a great variety of rivers draining different regions of the world. This graph is based on the compilation of Deberdt *et al.* (2002) and shows the global control of dissolved REE concentrations by pH.

Sholkovitz, 1978; Elderfield *et al.*, 1990; Viers *et al.*, 1997; Ingri *et al.*, 2000; Tricca *et al.*, 1999; Deberdt *et al.*, 2002).

5.09.2.4 Temporal Variability

The temporal variability of trace-element concentrations has been poorly addressed in the literature. The variability of trace-element concentration may be important even on a diel basis. Brick and Moore (1996) have reported diel variations, for example, in the Upper Clark Fork river, Montana. This river is a high-gradient river rich in metals because of mining. Correlating with pH variations between day and night, dissolved manganese and zinc concentrations increase two- and three-fold at night, while major solutes and water discharge show no evidence of variations.

Tenfold daily to weekly variations of manganese have been described in the Kalix River (Ponter *et al.*, 1992) between winter (minimum concentrations) and snow-melt in May. These variations are attributed to the input of manganese-rich mire waters (due to rapid oxidation–reduction processes) and are shown to be associated with significant variations of dissolved REEs, aluminum, and iron. A sevenfold increase of lanthanum is observed and is associated with the increase of water discharge and DOC concentrations and the decrease of pH occurring at spring flood (Figure 18; Ingri *et al.*, 2000). It is important to note that even the REE patterns are affected, changing from a relatively flat to an LREE-enriched pattern when water discharge increases. In the Kalix River, these variations in the dissolved load are associated with variations in particulate concentrations. A similar

cycling of manganese and iron concentrations has been reported for the St. Lawrence system by Cossa *et al.* (1990).

A similar seasonal variability is observed in the organic-rich rivers of Africa. Viers *et al.* (2000) have classified the trace elements as a function of their variability (standard deviation/mean value) during the period 1994–1997. While major elements show the lowest variability (except potassium), trace elements show high variability, ranging from 80% for rubidium and manganese to 40–50% for DOC, REEs, thorium, aluminum, and iron. The temporal evolution of elements such as thorium or REEs strictly mimics that of DOC and shows the highest concentrations during the high-water stage period. Major elements follow the reverse tendency.

The variability of very large river systems has been addressed for the Mississippi (Shiller and Boyle, 1987; Shiller, 1997, 2002) and the Amazon (Seyler and Boaventura, 2002). In the Mississippi, the largest amplitude of dissolved concentrations is observed for manganese (50 times increase with water discharge increase) and iron (eightfold variation). All other trace elements analyzed show lower concentration variability (from 1.5-fold for barium and rubidium to fourfold for lead and molybdenum). Certain trace elements vary in phase with manganese and iron, being maximum at high-water stages, e.g., zinc and lead, some others show an opposite seasonal behavior, such as vanadium, molybdenum, uranium and, to a lesser extent, copper, nickel, and cadmium. REEs (Shiller, 2002) co-vary with manganese or zinc, the LREEs showing much more variability than the HREEs. These seasonal trends for the lower Mississippi are remarkably similar to those observed for the Kalix River. In the Amazon system, a time series of trace-element concentrations at Obidos (800 km upstream from the mouth of the Amazon) shows (Seyler and Boaventura, 2002) that several patterns of variation exist. As in the Mississippi river, manganese shows the largest variations (10-fold) and exhibits maximum concentrations at high-water stage (but one month after the water peak). Elements such as cadmium (×8) and cobalt (×4) show similar patterns. The opposite trend is observed for antimony, molybdenum, copper, strontium, barium, vanadium, and major elements, with maximum concentrations at low-water stage and less variability. The elements uranium, rubidium, nickel, and chromium do not show significant or systematic variations during the hydrological cycle. So far, variations of REEs, aluminum, and iron have not been documented in the Amazon.

The conclusion of this rapid review of the literature on temporal variations of trace-element abundances in river waters is that the range of variation can be very large. Temporal series are thus necessary not only to compute riverine budgets to the oceans but also to understand the processes controlling trace-element concentrations. Manganese and iron appear to be the most variable elements in rivers, as well as a number of elements whose chemistries seem to be associated with those of manganese and iron. As will be discussed later, manganese and iron variations are consistent with redox changes within the river system. We would stress that, in order to understand controls on trace-element concentrations in watersheds and to refine load budgets of elements to the ocean, future studies will have to take into account temporal variations.

5.09.2.5 Conservative Behavior of Trace Elements in River Systems

The question of seasonal variability of trace elements in river systems raises the question of their conservative or nonconservative behavior during the mixing of tributaries. Major elements and trace elements of high mobility indexes generally have conservative behavior during the mixing of tributaries. This is apparent for the Amazon River (Seyler and Boaventura, 2002). Conversely, if redox cycling and associated exchange with solid phases, related to temperature or biological activity within the river system, control the temporal variations of manganese, iron, and associated elements, it is highly probable that their behavior will not be conservative. In the Mississippi River, seasonal changes in trace-element concentrations are not explained by the hydrological mixing of tributaries, except for barium and uranium (Shiller and Boyle, 1987; Shiller, 1997). The same conclusion is reached from times-series comparison in the Amazon basin by Seyler and Boaventura (2002). The downstream evolution of major and trace elements along the Solimoes and Amazon rivers (from 3,000 km upstream to the river mouth) illustrates conservative and nonconservative behavior in the Amazon system. The patterns of manganese, vanadium, and nickel are shown on Figure 5. While vanadium, like the major elements, shows a downstream decrease of concentration, consistent with the inputs of the dilute lowland rivers, manganese and nickel show increasing and constant concentrations, respectively. In the case of manganese and nickel, a substantial source is necessary that cannot be represented by the minor input of lowland tributaries, but rather by processes releasing these elements into the dissolved load.

5.09.2.6 Transport of Elements

Rivers transport material both in dissolved form and as solid load (suspended matter and

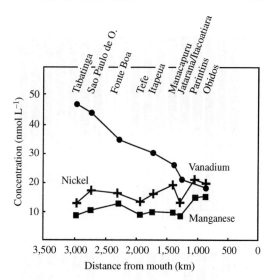

Figure 5 Downstream evolution of vanadium, nickel, and manganese concentrations measured in the dissolved load of the Amazon river showing different possible behaviors of trace elements in river systems. Major solutes and TDS follow the similar pattern to that of vanadium and correspond to a dilution of Andean waters by waters from the Amazonian lowlands (source Seyler and Boaventura, 2002).

bottom sands). The dominant form of transport of trace elements per liter of river water depends on both the mobility of the element in the weathering + transport process and on the amount of solids transported annually by the river. Two contrasted examples of big rivers from the Amazon basin are shown on Figure 6. The Solimoes River is characterized by relatively high concentrations of suspended sediments (230 mg L^{-1}), whereas the Rio Negro is characterized by very low suspended sediment concentrations (less than 10 mg L^{-1}). This discrepancy is related to contrasting weathering regimes in the two basins. In the Solimoes, even the most mobile elements, such as sodium, are transported dominantly in a solid form. This also applies to the highly turbid rivers of Asia. Conversely, rivers from humid tropical regions such as the Rio Negro can show a significant fraction of their trace elements transported in the dissolved load, both because the suspended matter concentration is low (less than 10 mg L^{-1}) and because their dissolved load is generally enriched in the so-called "nonmobile and highly insoluble" elements of Figure 2. In the Rio Negro, half of the REEs, thorium, and aluminum are transported in the dissolved load.

If TSS is total suspended sediment concentration in the river, K_d, the ratio of concentration in the suspended sediments over dissolved concentration, then the proportion D of a given element in the dissolved phase can be

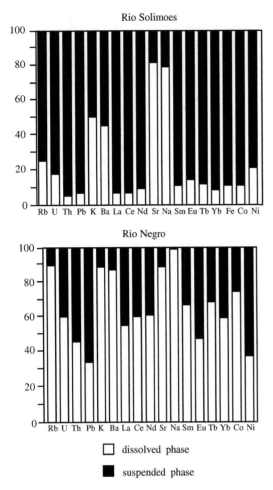

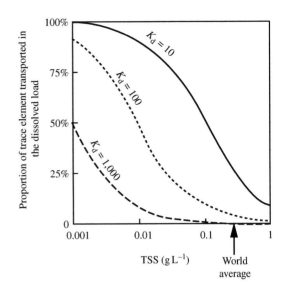

Figure 7 Theoretical proportion of dissolved transport as a function of the concentration of riverine suspended sediments for three values of the ratio of concentration in the suspended load over concentration in the dissolved load. $K_d = 10$, 100, and 1,000 correspond to the most mobile elements (Na, B, Re, Se, As, Sb), the intermediate elements (Cu, Ni, Cr, Ra) and highly immobile elements (Th, Al, Ti, Zr) respectively (see Figure 2).

Figure 6 Proportion of elements transported in dissolved (<0.2 μm) and particulate forms by contrasted rivers of the Amazon Basin: the Solimoes (a turbid river, mostly influenced by the Andes) and the Rio Negro (a typical lowland, black river with very low suspended sediments yields).

expressed as

$$D = \frac{1}{1 + K_d \cdot \text{TSS}}$$

The typical values of K_d can be estimated using Table 1 and average values of concentrations in suspended sediments (Gaillardet *et al.*, 1999b). Except for the most mobile elements, K_d values are not very different from the continental crust normalized concentrations shown on Figure 2. We have plotted in Figure 7 the proportion D as a function of TSS for three types of elements: thorium (immobile element), nickel (intermediate), and boron (mobile element), corresponding to the K_d values of 1,000, 100, and 10, respectively. This graph shows that at the world average value of TSS (350 mg L^{-1}), the dominant form of transport is the solid form. Only the most mobile elements can be significantly transported in a dissolved form.

5.09.3 SOURCES OF TRACE ELEMENTS IN AQUATIC SYSTEMS

Continental crust is the ultimate source of trace elements in hydrologic systems. Trace elements are introduced in the river basin by rock weathering, atmospheric dry and wet deposition and by anthropogenic activities. The sketch diagram of Figure 8 summarizes the natural and anthropogenic sources of elements in aquatic environments (modified after Foster and Charlesworth, 1996).

5.09.3.1 Rock Weathering

The chemical weathering of rocks results in the release of the most soluble elements and, in the case of silicate rocks, leaves a residue, which, conversely, is enriched in the insoluble elements. The normalization of soil materials or river suspended sediments, taken as a natural average of the finest soil materials, allows a first-order examination of the behavior of trace elements during chemical weathering. For example, the suspended sediments of some large unpolluted rivers have been normalized to the upper-crust concentrations (Gaillardet *et al.*, 1999b) in Figure 9. Elements are classified according to their magmatic compatibility. Most of the trace elements analyzed on those patterns have

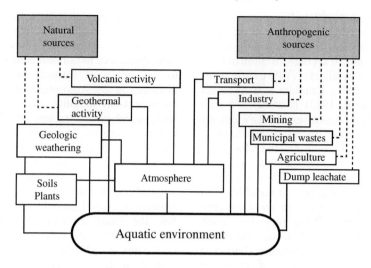

Figure 8 Pathways of trace elements to the aquatic system.

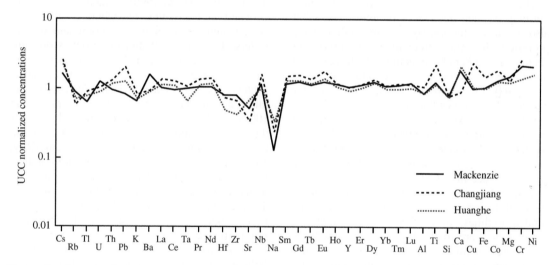

Figure 9 Concentrations measured in the suspended sediments of large river systems normalized to the upper continental crust UCC (Li, 2000). The order of element on the *x*-axis is an order of magmatic compatibily (source Gaillardet *et al.*, 1999b).

continental-like abundances. Only a few elements are depleted. They include soluble elements (rubidium, uranium, barium, and major elements) and elements concentrated in heavy minerals (zirconium, hafnium). The complementary reservoirs are, respectively, the dissolved phase and bottom sands. For elements that are not depleted by chemical weathering and transport processes, their ratio in the particles is expected to be similar to that of the source rock from which they are derived. For example, the Sm/Nd ratio of suspended sediments is a good integrated signature of the rocks that have undergone weathering throughout the drainage basin. The Sm/Nd isotopic system is therefore insensitive to the weathering history of sediment, and it records episodes of crustal formation rather than

weathering + deposition cycles. This property is the basis of the use of neodymium isotopes as tracers of crustal growth.

In the dissolved load, the preferential dissolution of the different types of lithology or mineral can cause large variations in trace-element abundances in rivers. A classical tracer of rock dissolution is strontium. The isotopic ratio of strontium in river waters is, to a first approximation, explained by mixing between different sources. Rain, evaporite dissolution, and carbonate and silicate weathering are the predominant sources of strontium in rivers (see Chapter 5.12). Examples of the use of Ca/Sr, Na/Sr, and Sr isotopic ratios to quantify the proportion of strontium derived from carbonate weathering in large basins can be found in Négrel *et al.* (1993)

and Gaillardet *et al.* (1999a). Levasseur (1999) have used isotopic ratios of osmium in river waters to determine the origin of osmium in the largest rivers. As a global average, osmium in river waters is derived from silicates (14%), carbonates (55%), and shales (31%). Uranium in river basins (Palmer and Edmond, 1993) is associated with carbonate dissolution. A number of trace elements also appear to be good tracers of black shale weathering (Peucker-Ehrenbrink and Hannigan, 2002; Dalai *et al.*, 2001, Pierson-Wickman *et al.* (2000) Chabaux *et al.*, 2001). These rocks are enriched in elements that have a strong affinity for organic matter and are stable in reducing conditions (uranium, metals, PGE); their dissolution can lead to substantial enrichments. The correlation between tungsten and chlorine in California streams reflects the importance of evaporite dissolution to the riverine budget of tungsten (Johannesson *et al.*, 2000).

Trace-element abundance in rivers depends not only on their abundance in source rocks but also on the weathering style. Some authors have proposed that trace-element ratios involving chemically similar elements might be fractionated during the weathering of silicates and could serve as a proxy for weathering intensity. It has, for example, been shown that significant variations in the Ge/Si or Ga/Al ratios occur in river waters and that these variations are related to the degree of weathering (Murnane and Stallard, 1990; Shiller and Frilot, 1995). A weathering-limited regime would produce high Ga/Al and low Ge/Si in the river dissolved load because of slight differences in solubility between elements. In transport-limited regimes, these ratios should approach those of the source rocks. However, as we will see later, the chemistry of the river, especially organic-matter complexation, can significantly obscure the weathering signal. Shiller and Mao (2000) have shown this for vanadium. Although vanadium in rivers is essentially derived from silicate weathering, variations in the V/Si ratios are observed due to minor inputs from sulfide dissolution, increased solubilization of vanadium by dissolved organic matter, and variations of the vanadium content of source rocks. In spite of these complications, research is needed to evaluate the use of trace element in rivers as tracers of specific types of mineral/rock weathering or as indexes of weathering regimes. A vast majority of dissolved trace elements in natural waters are not provided by rock weathering alone; other sources will be discussed below.

5.09.3.2 Atmosphere

Trace elements are transported in the atmosphere in the form of wind-blown soil particles, fine

volcanic products, sea salts, ashes from forest fires, and biogenic aerosols. The atmospheric input of trace elements can be significant, depending on their abundance in rain and aerosol solubility. Due to the ability of aerosols to travel over long distances in the atmosphere, it is difficult, based on present-day sampling, to determine the natural background concentrations of elements in rain water and aerosols. In remote sites of Niger and the Ivory Coast, Freydier *et al.* (1998) have shown that the emission of terrigeneous particles of crustal composition is the main source of trace elements in atmospheric deposition of this region. Where the amount of crustal particles in the atmosphere is low, the contribution of other sources becomes evident (ocean, vegetation, human activities). In both sites, a 50-fold enrichment of zinc with respect to the continental crust is observed in rainwater, possibly derived from the vegetation or from remote industrial sources. In the Amazon basin (Artaxo *et al.*, 1990), biogenic aerosols are enriched in potassium, phosphorus, sulfur, and zinc, which can explain the enrichment of rain waters. The impact of human activities on rainwater chemistry has been clearly demonstrated in large cities and even in regions far from industrial centers (e.g., Berg *et al.*, 1994). In Western Europe, rain waters in Paris have concentrations more than 100 times those in the mean continental crust for the elements cadmium, zinc, lead, copper, bismuth, nickel, chromium, cobalt, and vanadium (Roy, 1996). In those samples, arguments based on lead isotopes have shown that 100% of the excess of metals can be attributed to anthropogenic sources. These elements are introduced into the atmosphere by coal-burning or metal smelting that produces fine particles. The atmosphere is therefore extremely sensitive to anthropogenic contamination. Nriagu (1979) has calculated that global atmospheric emissions increased exponentially over the twentieth century.

5.09.3.3 Other Anthropogenic Contributions

Apart from the atmosphere, there are a number of potential point and non-point sources of contamination in hydrological systems. The industrial revolution, since the beginning of the twentieth century, has caused a drastic increase in the exploitation and processing of metals, resulting in their release into the environment and the release of associated elements with no economic value (e.g., arsenic). In southwestern France, Schäfer and Blanc (2002) have shown that the geochemistry of suspended sediments of the largest rivers can be related to the occurrence of ore deposits in the upper basins of the Massif Central and Pyrenees. The release of metals is also associated with the use of metals and other trace elements in paints, cements, pharmaceutical products, wood treatment, water treatment, plastics and electronic items, and fertilizers. These activities can release trace metals directly into the river catchment (Foster and Charlesworth, 1996; Luck and Ben Othman, 1998; De Caritat *et al.*, 1996). For example, positive gadolinium anomalies have been reported in the dissolved load of European rivers and attributed to the use of gadopentetic acid in magnetic resonance imagining (Bau and Dulski, 1996). In the Seine River (Roy, 1996), lead isotope studies have shown that almost half of the lead transported by the river (adsorbed and dissolved) is of anthropogenic origin. In Mediterranean catchments, it has been shown by Luck and Ben Othman (1998) that lead and other metals brought in by rain waters are stored in soils and remobilized during floods as particles. In the Rhine River, the concentration of metals associated with suspended sediments decreases by a factor of 3 between low water discharge and high water discharge (Foster and Charlesworth, 1996). The same effect is observed in the rivers of southwestern France (Schäfer and Blanc, 2002) and is interpreted as a "dilution" of pollution by natural erosion of uncontaminated soil particles at high water discharges. Over longer timescales, there are a few sites where metal pollution has been monitored in rivers. Records from the river Rhine are available in the Netherlands (Foster and Charlesworth, 1996). A steady (between 2 and 20 times) increase in the concentrations of zinc, lead, chromium, copper, arsenic, nickel, cobalt, cadmium, and mercury is observed over the twentieth century and a slight decline after 1975, clearly showing the sensitivity of aquatic systems to human activities. For many trace metals, anthropogenic contributions from all sources far exceeds natural levels. Nriagu and Pacyna (1988) estimated that the man-induced mobilization of arsenic, cadmium, copper, mercury, molybdenum, nickel, lead, antimony, selenium, vanadium, and zinc far exceeds the natural fluxes. For arsenic, cadmium, lead, selenium, and mercury they report enrichment factors due to human activities of 3, 7.6, 24.1, 2.8, and 11.3, respectively, with respect to the natural levels.

Atmospheric acid deposition influences the mobility of trace elements, depending on the source rocks. Central Europe has been strongly affected by acid deposition during the twentieth century. Frei *et al.* (1998) have shown that the concentrations of trace elements in the Upper Ecker drainage and in the Northern Harz Mountains (Germany) are high and are related to the neutralizing capacity of the source rocks. Acidic waters have the highest trace-element concentrations.

As a conclusion, the origin of trace elements in river waters can be very diverse. Elucidating the sources and quantifying the proportion of elements derived from each of these sources is not an easy task, and it has been addressed by the use of isotopic ratios (strontium, neodymium, lead) and enrichment factors of trace elements with respect to well-characterized reservoirs (upper continental crust, ocean). However, we are far from being able to trace the origin of all elements in natural waters, especially when their biogeochemical cycle is complex and involves biomass or an atmospheric subcycle. In addition, human activities have led to a generalized perturbation of element abundances in the atmosphere, soils, and waters; it is very difficult, if not impossible, to assess the natural levels of trace elements in river waters and particles. It is, for a number of elements such as metals, difficult to estimate their natural rate of transport in rivers. The largest rivers with high water discharge and active erosion are probably the best-suited rivers for assessing the natural background of elements in rivers.

The behavior of trace elements in aquatic systems not only depends on the sources but it is also strongly controlled by the soil and in-stream processes, particularly through aqueous (organic and inorganic) complexation and reactions with solids.

5.09.4 AQUEOUS SPECIATION

The speciation of a given element is the molecular form under which it is transported in hydrosystems. Like some of the major elements, trace elements are transported in surface waters in complexed forms. Aqueous complexes correspond to the association of a cation and an anion or neutral molecule, called the ligand. Ligands can be inorganic and organic (see Chapter 5.10). It is rarely possible to measure the concentration of an individual complex (Pereiro and Carro Díaz, 2002; Ammann, 2002). Most of the techniques used to measure trace-element concentrations give the total concentration, whatever the speciation. The proportion of elements corresponding to the different chemical forms has therefore to be calculated, depending on the total concentration, pH, Eh conditions, the major element chemical composition of the water, and the complexation constants of the assumed complexes. Knowing the concentrations of the individual species of a given element in waters is of crucial importance in order to predict its toxicity and bioavailability. For example, the toxicity of metals depends more on their chemical form in waters rather than their total concentration (Morel and Hering, 1993), in particular, because complexation may enhance or inhibit adsorption on surfaces.

In river waters, inorganic ligands are essentially H_2O, OH^-, HCO_3^-, CO_3^{2-} and, to a lesser extent,

Cl^-, SO_4^{2-}, F^-, and NO_3^-. Organic ligands are small organic weak bases such as oxalate or acetate and low-molecular-weight humic acids containing phenolate and carboxylate groups.

Speciation calculations are performed according to mass balance equations and mass law equations corresponding to the different complexation reactions. As an example, we show in Figure 10 the calculated speciation of thorium in pure water (Langmuir and Herman, 1980). The situation would be more complex in natural waters, depending on the abundance and nature of the inorganic and organic ligands, but no free thorium is predicted and $Th(OH)_4^0$ is still the dominant complex at common river water pH.

Complexation constants of trace elements in aqueous solution are determined using a great variety of techniques (including conductimetry, potentiometry, spectrophotometry, solvent extraction, calorimetry, ion exchange) and are reported in numerous publications (see Chapter 5.02). Critical studies reporting complexation constants can be found in Baes and Mesmer (1976), Martell and Smith (1977), or Martel and Hancock (1996). The values for complexation reactions are still a matter of debate, given the high possible number of complexes between trace elements and natural ligands, especially poorly known organic ligands. A number of computer codes have been developed in order to calculate the speciation of trace elements in waters (see Chapter 5.02). These computer codes basically use the mass action laws of the different complexation reactions, activity coefficients and conservation equations (e.g., WATEQF, Plummer et al., 1984).

A number of authors have investigated the interesting issue of relating the aqueous speciation rules of trace elements to the periodic table (Turner et al., 1981; Langmuir, 1997).

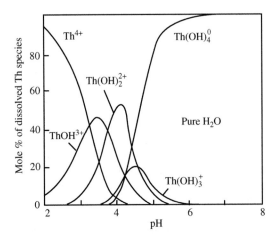

Figure 10 An example of speciation calculation. Distribution of thorium complexes versus pH at 25 °C with $\Sigma Th = 0.01 \ \mu g \ L^{-1}$ in pure water (after Langmuir and Herman, 1980).

The classical "Goldschmidt plot" plots the charge of the cation (in the complex) versus the cationic radius. Three groups of elements can be distinguished: those forming oxyanions in solution (carbon, arsenic, boron, nitrogen,...), elements forming hydroxycations and hydoxyanions (thorium, aluminum, iron, titanium, lead) and elements forming free cations or aquocations (sodium, potassium, silver). More precisely, following the Born theoretical treatment of ion–water interaction (see Turner et al., 1981), the ability of cations to attract electrons can be approximated by the polarizing power, Z^2/r, where Z is the ion charge and r the ionic radius. Figure 11 is a graph of the intensity of hydrolysis

$$M^{n+} + H_2O = (MOH)^{(n-1)+} + H^+$$

as a function of the polarizing parameter and allows a classification of trace elements in solution.

A high polarizing power (>1.3) means that the element easily attracts electrons and is fully hydrolyzed. The speciation of such elements is in the form $MO_p(OH)_q$ and depends on pH, via the dissociation constant. Examples of such elements and their associated acid form are arsenic, boron, chromium, germanium, molybdenum, niobium, tantalum, rhenium, antimony, silicon, selenium, uranium, and vanadium. They exist, for example, in the form of $As(OH)_3$, $B(OH)_3$, $Ge(OH)_4$, $Sn(OH)_4$, CrO_4^{2-}, MoO_4^{2-}, WO_4^{2-}, ReO_4^{2-},..., and behave as strong or weak acids in solution. These elements are known as fully hydrolyzed elements and are belong to groups III—VII in the periodic table.

Elements with intermediate and low Z^2/r are capable of producing free cationic forms in solution. Their speciation is a question of competition among the major ligand species (Cl^-, SO_4^{2-}, OH^-, CO_3^{2-}). These elements are the alkalis, alkaline earths, REEs, and metals. They can be classified according to their tendency to form covalent bonds. The concept of hard and soft ions was introduced by Pearson and is equivalent to the concept of (a)- and (b)-type cations introduced by Ahrland (Langmuir, 1997). The cations are Lewis acids (electron acceptor) and the ligands, Lewis bases (electron donor). "Soft" means that the electrons of the species can easily be deformed whereas "hard" means that the species are rigid. Soft species easily form covalent bonds. Hard species prefer electrostatic bonding. Soft cations form strong covalent bonds with soft ligands, and hard acids form strong ionic bonds with hard ligands (Langmuir, 1997). Complexes between hard (soft) acids and soft (hard) bases are weak and rare. Alkaline earths, La, Ti^{4+}, Y^{3+}, Sc^{3+}, Ga^{3+}, In^{3+}, Sn^{4+}, Cr^{3+}, Mn^{3+}, Fe^{3+}, Co^{3+}, U^{4+}, Th^{4+}, and actinides are classified as hard acids and form strong complexes

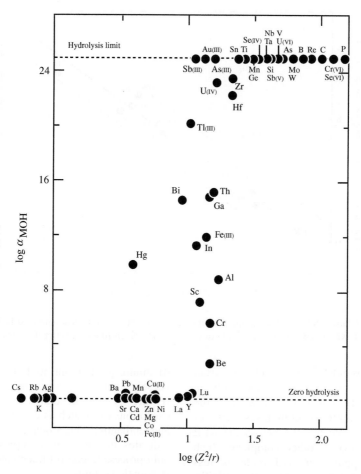

Figure 11 The "Born plot" showing the extent of hydrolysis of a given cation (α_{MOH} quantifies the affinity with OH^-) as a function of the polarizing power (Z^2/r) where r and Z are, respectively, the radius and charge of the cation (after Turner *et al.*, 1981).

with the major ligands found in natural freshwaters: F^-, OH^-, SO_4^-, CO_3^{2-}, or HCO_3^- (hard bases). Complexes with Cl^- (hard–soft ligands) are weaker. Soft acids comprise Cu^+, Ag^+, Au^+, Au^{3+}, Pd^{2+}, Cd^{2+},..., while Zn^{2+}, Fe^{2+}, Co^{2+}, Ni^{2+}, Cu^{2+}, Pb^{2+} are classified as borderline acids due to their intermediate behavior. Turner *et al.* (1981) have reported correlations between the log of complex stability constants for all major freshwater ligands and $Z^2/(r + 0.85)$. Significant linear correlations between the strength of the complex and the electrostatic parameter are observed. Hard ligands (such as F^-, SO_4^{2-}), exhibit a rapid increase of stability constant with polarizing power. Soft ligands, such as Cl^-, show only a slow increase of the affinity constants with $Z^2/(r + 0.85)$. Intermediate ligands (OH^-, CO_3^{2-}) show a clear increase of association constant with the polarizing parameter. For example, Figure 12 represents the case of carbonate complexes. Such relations are remarkable, as they relate the complexation constants of trace elements in solution to relatively simple properties of the periodic table.

The results of cation speciation calculations for freshwaters lead to clear conclusions (Turner *et al.*, 1981). The cations with low polarizing numbers are very weakly complexed in freshwaters and exist dominantly as free ions (lithium, strontium, rubidium, caesium, barium). Conversely, elements of high polarizing power (hafnium, thorium, aluminum, scandium,...) have their speciation dominated by hydrolyzed species and correspond to the elements of the transition zone in the Born graph (Figure 11). A third group of elements corresponds to soft cations (Ag^+, $Au(I)$, $Tl(I)$) in association with soft ligands (essentially Cl^-). The last group of elements consists of the REEs, transition metals, cadmium, which exist both as free ions and carbonate or hydroxyl complexes, depending on the pH. The speciation of REEs in natural waters has been modeled by several authors (Cantrell and Byrne, 1987; Wood, 1990; Millero, 1992; Lee and Byrne, 1993; Johannesson *et al.*, 1996, and references therein, Liu and Byrne, 1997). The dicarbonato ($Ln(CO_3)_2^-$) and carbonato ($LnCO_3^+$) REE carbonate complexes are shown to be

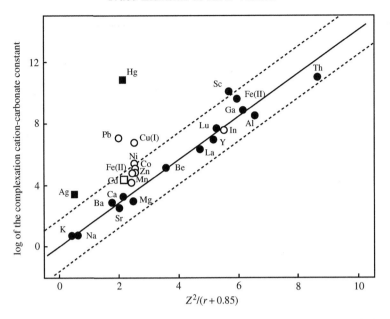

Figure 12 Linear relation between the complexation constant of cations and carbonate in solution as a function of $Z^2/(r + 0.85)$, where 0.85 is an empirical correction of the Born function (after Turner *et al.*, 1981).

the dominant forms of REEs in groundwaters from Nevada and California, having pH values between 7 and 9. Stability constants show that the formation of REE carbonate complexes increases with increasing atomic number across the REE suite. The formation of other complexes such as sulfate, phosphate, or hydroxide is less important but also tends to favor the HREEs. Fractionation of REEs between source rocks and associated groundwaters is not observed in groundwaters from Nevada and California (Johannesson *et al.*, 1996), but HREE enrichments occur in some cases, reflecting the dominance of HREE carbonate complexes, more stable in solution than the LREE carbonate complexes.

Organic speciation has been poorly documented, compared to inorganic speciation. However, as shown by several authors, a good correlation is observed between the first hydrolysis constant and the association of the cation with small organic molecules such as oxalate (Langmuir, 1997). Because humic molecules are large, poorly characterized molecules, they considerably complicate speciation calculations in freshwaters. Most often, these molecules can no longer be considered as dissolved ligands, but rather as "colloids" (see Chapter 5.10).

Finally, it is interesting to note that the relation between the global mobility of trace elements in the aquatic system (Figure 2) and the speciation patterns summarized above is not direct. Although most of the fully hydrolyzed species appear as mobile elements (phosphorus, carbon, rhenium, boron, arsenic, molybdenum), some elements in this family are amongst the least mobile elements

(titanium, germanium, niobium, tantalum). Conversely, the elements with zero hydrolysis (Figure 11) are generally mobile elements but exceptions exist such as caesium, rubidium, lead, and lanthanum. This absence of correlation clearly indicates that aqueous complexation is not the only process to control trace-element abundances in aquatic systems.

5.09.5 THE "COLLOIDAL WORLD"

It is now well known that trace-element concentrations in continental waters depend on the size of the pore filters used to separate the particulate from the dissolved fraction. This is apparent in Table 1, where results from the Amazon and Orinoco are reported using two filtration sizes: the conventional 0.2 μm filtration and filtration with membranes of smaller cutoff size (ultrafiltration). These results suggest the presence in solution of very small (submicrometric) particles that pass through filters during filtration. The view that trace elements can be separated into "particulate" and "dissolved" fractions can thus no longer be held; this has led authors to operationally define a colloidal fraction (0.20 μm or 0.45 μm to 1 nm) and a truly dissolved fraction (<1 nm) (e.g., Buffle and Van Leeuwen, 1992; Stumm, 1993). The existence of a colloidal phase has a major influence on the speciation calculation schemes presented above (based only on aqueous complexation), as the apparent solubility of trace elements will be enhanced by the presence of colloids. The dynamics of colloids also completely change the way reactive solute transport is modeled and are an issue for water quality standards since the toxicity of trace elements depends not only on their abundance and speciation but also on their bioavailability.

Although, strictly speaking, colloids consist of small particles and could be treated as a particular case of reactive surfaces, we summarize below the recent literature on the nature and reactivity of these phases.

5.09.5.1 Nature of the Colloids

The separation and characterization of submicron-sized particles in water is difficult, in particular because of artifacts from sampling and concentration techniques. Lead *et al.* (1997) have presented a critical review of the different techniques for separation and analysis of colloids (filtration, dialysis, centrifugation, but also voltametry, gels (DET/DGT), field-flow fractionation, SPLITT). Ultrafiltration membranes have been developed with nominal cutoff sizes ranging from the thousands of daltons (Da) to hundreds of thousands of daltons, which have been used to separate the colloidal pool into several fractions.

Figure 13 (after Buffle and Van Leeuwen, 1992) shows an example of the distribution of mineral and organic colloids as a function of size from ångstrom to micrometer scales. This diagram shows that there is no clear boundary between dissolved and colloidal substances or between colloids and particulates.

Colloids can be organic or inorganic. Even if they are not separated from the dissolved load by classical filtration, colloids have the physicochemical properties of a solid. Colloids are finely divided amorphous substances or solids with very high specific surface areas and strong adsorption capacities. It is shown by Perret *et al.* (1994) for the Rhine River that the colloids contribute less than 2% of the total particle volume and mass, but represent a dominant proportion of the available surface area for adsorption of pollutants. The abundance of colloids, their fate, through coagulation and sedimentation processes in natural waters therefore control the abundance of a number of elements.

The most common mineral colloids are metallic oxyhydroxides (e.g., mainly iron-, aluminum- and manganese-oxyhydroxides), clays, and siliceous phases. The reaction between trace elements in water and surface hydroxyl groups ($\equiv S-OH$) is, to a first order, analogous to the formation of aqueous complexes. Organic colloids are mainly humic and fulvic acid (humic substances) and derive from an incomplete degradation of soil organic matter (see Chapter 5.10). Humic substances represent 70–90% of DOC in wetland areas (Thurman, 1985) and are usually responsible for the brown color of water. Despite their heterogeneity and complexity, humic substances are characterized by similar functional groups (carboxyls, quinones, phenolic OH groups) and the presence of aliphatic and aromatic components (Stevenson, 1994). Metal binding in natural

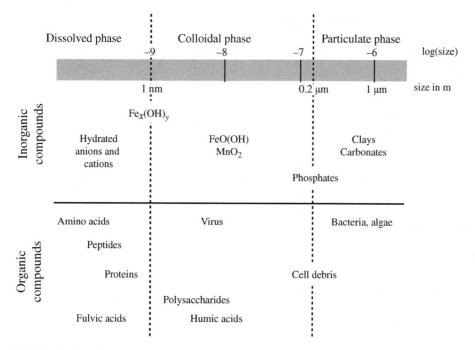

Figure 13 Distribution of mineral and organic colloids as a function of size in aquatic systems (after Buffle and Van Leeuwen, 1992).

environments is, however, mainly related to carboxylic and phenolic groups (Perdue *et al.*, 1984).

Inorganic and organic colloids are most often intimately associated. Association of organic colloids, with clays (Koshinen and Harper, 1990) and with iron oxyhydroxides (Tipping, 1981; Gu *et al.*, 1995; Herrera Ramos and McBride, 1996) are commonly described in natural waters and have been the subject of laboratory experiments. For example, Perret *et al.* (1994) have sampled, fractionated, and characterized submicron particles from the Rhine and showed that these colloids consist of a tight association between organic and inorganic material.

5.09.5.2 Ultrafiltration of Colloids and Speciation of Trace Elements in Organic-rich Rivers

Several authors have recently performed ultrafiltration of natural waters (river, groundwater, estuaries) in order to determine the role of colloids in the transport of major and trace elements. As noted by some authors (Sholkovitz, 1995; Horowitz *et al.*, 1996; Hoffmann *et al.*, 2000), these experiments require caution; mass balance calculations must be performed to check the ultrafiltration procedure, as some truly dissolved elements can be retained inside the filtration system.

The main advantage of ultrafiltration techniques is that they permit a direct determination of the association constants between colloids and trace elements in natural systems. This is illustrated by a series of papers dedicated to organic-rich tropical rivers of the Nyong River Basin (Cameroon, Africa) (Viers *et al.*, 1997, 2000; Dupré *et al.*, 1999). The Nyong basin river waters exhibit low major cation concentrations (i.e., sodium, potassium, calcium, magnesium) but high concentrations of some trace elements (aluminum, iron, thorium, zirconium, REEs), silica, and DOC. Figure 14 shows the behavior of some trace elements (samarium, aluminum, strontium), silicon, and organic carbon during successive filtrations through decreasing pore size membranes (0.2 μm, 300 kDa, 5 kDa, and 1 kDa) of the Awout River (Dupré *et al.*, 1999). Concentrations decrease when samples are filtered through progressively finer pore-size membranes. Three patterns of element concentration as a function of DOC (taken as an index of colloid abundance) are observed. Silica shows no concentration variation as a function of DOC. The second type of behavior is that of yttrium and REEs. The relationship between these elements and DOC defines a straight line going through zero, which means that they are progressively and completely removed from the solution jointly with DOC. A group of elements composed of rubidium, strontium, copper, cobalt, manganese, chromium, vanadium, and major cations presents the same type of relation but a certain amount remains into the solution when all DOC is removed. These linear relationships suggest that the ability of these elements to form complexes

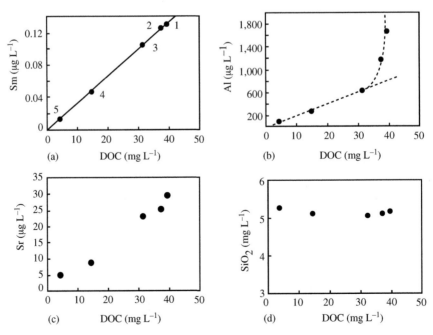

Figure 14 Variation of (a) Sm, (b) Al, (c) Sr, and (d) SiO_2 concentrations as a function of DOC concentrations in the different filtrates of the Awout River. 1: <0.20 μm (FF), 2: <0.20 μm (TF), 3: <300 kDa (TF), 4: <5 kDa (TF), 5: <1 kDa (TF). FF and TF mean frontal filtration and tangential filtration, respectively (source Dupré *et al.*, 1999).

with organic colloids remains constant over the whole molecular size range of the colloid materials examined. For some other elements, plots of (aluminum, iron, gallium, thorium, uranium) concentration versus DOC exhibit a nonlinear relationship. This suggests that, for this set of elements, the binding capacities (i.e., the quantity of element bound to organic material per unit mass of carbon) in the organic colloids are a function of the size of colloids. Several authors, with contradictory conclusions, have discussed this hypothesis. Lakshman *et al.* (1993) separated, by ultrafiltration, three fractions of a fulvic acid extracted from a soil (<500 Da; 500–1,000 Da; 1,000–10,000 Da) and demonstrated that the aluminum binding constant of organic material increases with the molecular weight of the fractions. Conversely, Burba *et al.* (1995), working on humic substance fractions (<1,000 Da; 1,000–5,000 Da; 5,000–10,000 Da), showed that the lower-molecular-weight fractions exhibit the highest complexing capacity for copper. Eyrolle *et al.* (1993, 1996) obtained similar results.

One of the major results of ultrafiltration techniques conducted on the Nyong basin rivers is the decrease of major cation (strontium, rubidium, barium) concentrations with decreasing filter pore size. This result is unexpected, as these cations are known to have a very low affinity for organic matter. Eyrolle *et al.* (1996) observed the same intriguing concentration decrease of cations (sodium, magnesium, potassium, calcium) during ultrafiltration of organic-rich river waters from South America and proposed that these cations were incorporated in plant debris. Dupré *et al.* (1999), alternatively, suggested that there are artifacts associated with the ultrafiltration of organic-rich waters. In the pH range of most natural waters (4–8), the surfaces of organic colloids are negatively charged. During the ultrafiltration process, the coating of the membrane by organic colloids disturbs the charge equilibrium of the solution. As a consequence, retention of free cations occurs at the membrane surface in order to maintain charge balance. This view is supported by the results of isotopic tracing (Dupré *et al.*, 1999). The natural isotopic ratio of strontium ($^{87}Sr/^{86}Sr$) and the isotopic ratios of strontium ($^{86}Sr/^{84}Sr$) and barium ($^{138}Ba/^{135}Ba$) in spiked and filtered samples are constant. This shows that there is one pool of strontium and that strontium, barium and by extension, the other cations are present in an exchangeable form at the surface of colloids. In order to avoid the artifact, that would lead to an incorrect determination of complexation constants, Dupré *et al.* (1999) performed ultrafiltrations at low pH or by adding to the solution a strongly complexing cation such as lanthanum. In both cases, colloids are neutralized and no filtration artifact occurs, allowing Dupré *et al.* (1999) to calculate metal-humate stability constants by comparing the abundances of elements in the different ultrafiltration fractions with a speciation model (BALANCE, Akinfiev, 1976). Stability constants are determined in order to fit the results from the ultrafiltration. The log K values calculated in this way for the metal-humate complexes decrease in the order: Al, Ga, Fe, Th, U, Y, REEs (more than 7) $\gg$ Cr (5.5) $\gg$ Co (3) > Rb, Ba, Sr, Mn, Mg ($\cong$2). As shown in Figure 15, a remarkable relationship is observed between the values calculated for element-humate stability constant and the first hydrolysis constant of the element. The same kind of linear relationship was used by Martell and Hancock (1996), to predict the formation constant for metal ion complexes with unidendate organic ligands. Figure 15 shows, however, that the elements yttrium and REEs appear to lie outside the general trend, having a higher stability constant than metals that have similar first hydrolysis constants. This result is not in agreement with the literature for the complexation of REEs with simple organic acids, which shows that REEs have stability constants comparable to those of copper or zinc (Martell and Hancock, 1996). The filtration experiments performed at low pH and with high lanthanum concentration show that yttrium and the REEs exhibit a very distinct behavior as a function of concentration. This result is in good agreement with a study by Hummel *et al.* (1995) on the binding of radionuclides (americium, curium, and europium) by humic substances. These authors

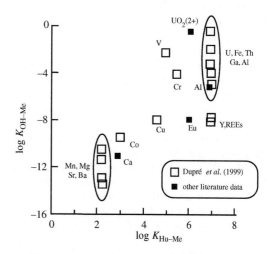

Figure 15 Correlation between metal-humate stability constants (log K_{Hu-Me}) and the first hydrolysis constant (log K_{OH-Me}) of the corresponding metal (Dupré *et al.*, 1999). Squares correspond to humic-metal association constants determined by ultrafiltration of organic-rich waters. K_{Hu-Me} corresponds to the conditional constant of the reaction $Me^{n+} + Hu^- \rightarrow MeHu^{(n-1)+}$.

showed that "complexation at trace-metal concentration on humic acids is rather different than complexation at high metal loading." They explained this discrepancy by the presence of two different sites having different intrinsic constants. In particular, there is a small amount of strong complexing sites, which raises the overall complexation constant at low concentration levels. The same explanation holds for the experiments of Dupré *et al.* (1999), who proposed two types of site for REEs: a first site with a large stability constant ($\log K > 7$) and a second site with a small stability constant, comparable to that of copper ($\log K \cong 4.5$). The first kind of site, present in low concentration, plays a role when there is a low REE concentration, whereas the second site, present in higher concentration, operates at high REE content. The first site has not been characterized, but the second may be related to the carboxylic functional group.

This example illustrates the complexity of colloid–solute interactions in river waters.

5.09.5.3 The Nonorganic Colloidal Pool

In the tropical organic-rich waters of Cameroon, inorganic colloids also explain the behavior of some elements. The relationship between aluminum concentration and DOC, for example, (Figure 14) reflects the combination of two linear relationships with different slopes. The part with the larger slope ($[Al] > 700 \, \mu g \, L^{-1}$) can be assigned to the retention of kaolinitic particles by the 300 kDa membrane. Using X-ray diffraction (XRD), transmission electron microscopy (TEM), Fourier-transform infrared spectroscopy (FTIR), electron paramagnetic resonance spectroscopy (EPR), and visible diffuse reflectance spectroscopy (DRS) techniques performed on the same samples, Olivié-Lauquet *et al.* (1999, 2000) showed that iron and manganese in the colloidal fraction were both in the form of hydroxides and organic complexes. They also documented the presence of euhedral particles of kaolinite in the colored waters ($<0.22 \, \mu m$). This result suggests that only the first part of the curve ($[Al] < 700 \, \mu g \, L^{-1}$) corresponds to aluminum bound to humic substances. Even if organic matter plays a key role in controlling trace-element levels in Cameroon, these results emphasize the difficulty of deciphering the roles of organic and inorganic colloids since metallic oxyhydroxide and organic matter are intimately associated in the solution and in the colloidal phase.

On Earth, organic-rich rivers occur in boreal regions as well as in wet tropical areas. The role of colloids in boreal organic-rich rivers has been addressed by Pokrovski and Schott (2002) and Ingri *et al.* (2000). For rivers of Karelia (northern Russia) Pokrovski and Schott (2002) reported that, conversely to what has been observed in the tropics (Viers *et al.*, 1997), 90% of the organic matter is concentrated in the smallest filtrate (1–10 kDa) in the form of fulvic acids. Pokrovski and Schott (2002) showed the presence of two types of colloids: one type composed of iron oxyhydroxides and another composed of organic matter. The elements aluminum and iron do not show any significant relation with organic matter, which indicates their control by inorganic colloids (aluminosilicates and/or metallic oxyhydroxides). Three groups of elements are deduced by the authors: (i) those not affected by the ultrafiltration procedure and which are present in the form of true ions (i.e., calcium, magnesium, lithium, sodium, potassium, strontium, barium, rubidium, caesium, silicon, boron, arsenic, antimony, molybdenum); (ii) those present in the fraction smaller than 1–10 kDa under the form of inorganic or organic complexes (manganese, cobalt, nickel, copper, zinc, cadmium and for some rivers lead, chromium, yttrium, HREEs, uranium); and (iii) those strongly associated with large iron colloids (phosphorus, aluminum, gallium, REEs, lead, vanadium, chromium, tungsten, titanium, germanium, zirconium, thorium, uranium). No relation between uranium and iron or DOC was reported by Pokrovski and Schott (2002) for the rivers of Karelia, in contrast to the observations of Porcelli *et al.* (1997) for the Kalix river in Sweden, where between 30% and 90% of uranium is associated with organic colloids of >10 kDa molecular weight.

These preliminary investigations on the role of colloids in high-latitude rivers clearly shows differences compared to rivers from the tropics, even if the total dissolved organic content is similar. This observation shows that there is a potential interesting climatic control on the nature and dynamics of colloids.

5.09.5.4 Fractionation of REEs in Rivers

The geochemistry of REEs has proven to be extremely powerful for the study of the genesis of igneous rocks (Henderson, 1984; Taylor and McLennan, 1985). Several attempts have been made over the last decades to generalize their use to surface water geochemistry (Martin *et al.*, 1976; Goldstein and Jacobsen, 1988; Elderfield *et al.*, 1990; Sholkovitz, 1992, 1995; Gaillardet *et al.*, 1997). These papers have clearly shown that REEs are among the less mobile elements in weathering and transport by waters. They explained variations in concentrations of REEs in river waters by the existence of colloids that enhance their apparent solubility. REEs are therefore excellent tracers for the dynamics of the colloidal pool in rivers. We have already mentioned that the concentrations of neodymium measured in the <0.2 μm fraction of rivers decrease with pH. This decrease in pH is associated with a fractionation of REEs. High-pH river waters have the most enriched HREE patterns, close to that of seawater (Gaillardet *et al.*, 1997). The REE patterns of seawater have been modeled (see Sholkovitz, 1995 for references); it has been concluded that fractionation for lanthanum to lutetium is a result of gradual differences in affinity of REEs for adsorption to particles and for complexation with ligands in solution. LREEs are preferentially adsorbed on surfaces while HRREs are preferentially complexed with carbonate in solution. The behavior of REEs in river water is far less constrained, due to the diversity of river chemistries and the complexity of solution–colloid interactions. Qualitatively at least, REE concentrations and fractionation in river waters are the result of pH-dependent reactions in solution and at the interface with colloids. Again, ultrafiltration techniques have proved to be extremely useful.

A recent compilation on REE data in river waters has been made by Deberdt *et al.* (2002). Figure 16 presents the REE concentrations in the smaller-pore-size ultrafiltrates (from 3 kDa to 100 kDa) normalized to those determined in the corresponding solution fraction (<0.2 μm) for different types of rivers. For all these rivers, the concentrations of REEs measured in the solution fraction are strongly to moderately reduced by ultrafiltration. This confirms that REEs are strongly controlled by colloidal matter whatever the pH conditions.

The comparison of REE concentrations for the <0.2 μm fraction with the lower-filter-size fraction shows that there is no unique pattern of colloidal material when rivers of different pH and different environments are compared. Ultrafiltration experiments conducted by Deberdt *et al.* (2002) on rivers from the Amazon and Orinoco basins as well as on Cameroon Rivers show slightly depleted LREE patterns to flat REE patterns when the colloidal fraction is normalized to the bulk solution. The results obtained by Sholkovitz (1995) and Ingri *et al.* (2000) for rivers

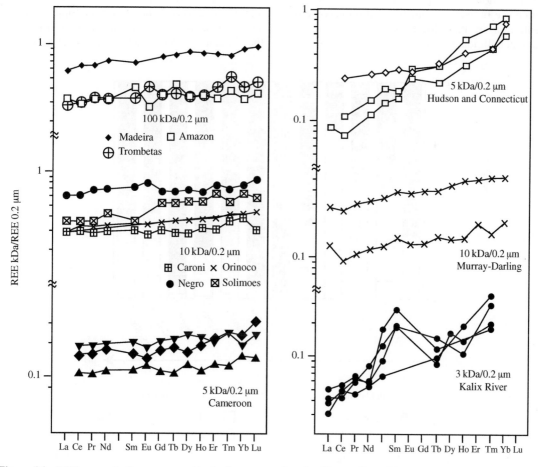

Figure 16 REE concentrations measured in the lower pore size ultrafiltrates (from 3 kDa to 100 kDa) normalized to those found in the solution (<0.2 μm) fraction in a panel of world rivers (source Deberdt *et al.*, 2002).

of boreal environments, the Mississippi, Hudson, and Connecticut rivers, and for the Murray river by Douglas *et al.* (1995) show that the smaller ultrafiltration fractions (e.g., <5 kDa or 3 kDa) are more clearly enriched in HREEs compared with the solution fraction (<0.20 μm). In these rivers, colloids are therefore LREE enriched and the enrichment of HREEs in the truly dissolved fractions is explained by the authors as the direct consequence of aqueous complexation with common inorganic ions such as carbonate, phosphate, fluoride, nitrate, sulfate, and chloride (Sholkovitz, 1995), which tends to enhance their solubility. Therefore, there is no systematic relationship between LREE depletion and pH or DOC, although more work is necessary to establish the precise relationship between filtered and ultrafiltered REE patterns and river chemistry on a global scale.

Using the REE concentration obtained in the <5 kDa fraction and assuming it to be the free REE concentration (REE^{3+}), Deberdt *et al.* (2002) deduced the speciation of REE in the whole <0.20 μm fraction using the complexation constants with the most common inorganic ligands in solution and taking into account organic colloidal complexation (carboxylic and phenolic sites). The difference between calculated and observed concentration in the ultrafiltrates is attributed to mineral colloidal material.

The sum of dissolved REE in solution can be described as follows:

$$[REE]_{colinorg}$$
$$= [REE]_{<0.3\ \mu m} - [REE]_{free}$$
$$\times \left(1 + \sum \frac{K_L a_L^n \gamma_{REE}}{\gamma_{REE(L)_a}} + \sum \frac{K_U a_U^n \gamma_{REE}}{\gamma_{REE(U)_a}} \right)$$

where the subscript *colinorg* refer to inorganic colloids, L and U to solution ligands and humic substances. K are the complexation constants, a and γ are, respectively, the activity and activity coefficients. These calculations demonstrated the general and unexpected predominance of the mineral colloidal fraction in natural river waters (Figure 17). Mineral colloids account for more than 60% of the total neodymium content whatever the pH range (moderately acid to basic) in the rivers studied (Caroni, Orinoco, Amazon, Madeira, Solimoes). Colloidal neodymium can be adsorbed on submicron colloids such as iron oxyhydroxides or clay minerals (e.g., kaolinite), or can occur as submicron colloids such as REE phosphate or carbonate minerals, which contain structural REEs. Note that the variable origin of neodymium in waters seems be supported by the nonhomogeneous neodymium isotopic composition measured in the different filtrates of the Mengong (Viers and Wasserburg, 2002).

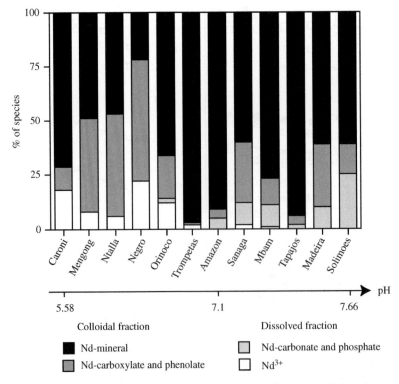

Figure 17 Speciation of Nd in a number of world rivers calculated in the solution fraction (<0.2 μm) by considering inorganic complexes (carbonate and phosphate), mineral and organic colloidal Nd species (dicarboxylic and diphenolic acids) (source Deberdt *et al.*, 2002).

As the physical and chemical properties (e.g., complexation constants) of the hypothetical phases cited above remain poorly documented, it is not possible so far to model the abundances of mineral colloids-related REEs.

This attempt to distinguish between the organic and mineral colloidal pools is highly approximate, but it does show that understanding the abundance of REEs and, by extension, other "insoluble" elements requires a better physical characterization of colloidal phases and association constants with individual elements. Experimental work aimed at determining the association constants of elements with surface sites coupled with field data are necessary. The diversity of colloids, the diversity of associations between elements and colloids, and the existence of filtration artifacts are intrinsic impediments toward a better description and prediction of trace-element concentration in river waters.

5.09.5.5 Colloid Dynamics

As colloids are a major carrier of trace elements in river waters, their behavior (coagulation, adsoprtion, and oxidation) as function of pH, increasing solid load, and ionic strength is crucial to predict and model the behavior of trace elements. Ingri *et al.* (2000) have clearly shown that the strong seasonality of lanthanum abundances in the Kalix River is totally controlled by the dynamics of organic and inorganic colloids (Figure 18). McKnight and Bencala (1989, 1990) and McKnight *et al.* (1989) coupled the use of lithium as a conservative tracer and iron, aluminum, and DOC concentrations in acidic mountain streams in the Colorado Rocky Mountains to show that colloid processes (e.g., precipitation of aluminum at stream confluences, sorption of dissolved organic material by hydrous iron and aluminum oxides in a stream confluence) were controlling element concentrations along the river channel. Finally, the importance of colloid dynamics is revealed in estuaries. Siberian rivers (Ob and Yenisei) have low suspended sediment load and the behavior of metals (cadmium, lead, copper, nickel, iron) in the estuary is dominated by the coagulation/sedimentation processes of riverine colloids. Clearly, the input of rivers to the ocean cannot be predicted without paying attention to the behavior of riverine colloids. For example, Dai and Martin (1995) report that copper (or nickel), although associated with colloids, behaves conservatively in the estuarine mixing zone, whereas organic carbon does not. They interpret this unexpected behavior by suggesting that colloidal organic matter is composed of at least two fractions: a refractory fraction and a labile fraction, which is removed during mixing with seawater. It is possible that the refractory

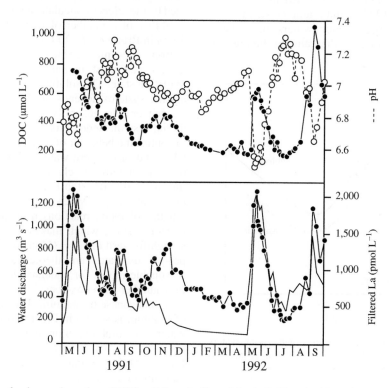

Figure 18 Dissolved organic carbon (DOC; <0.7 μm), filtered La (<0.45 μm), pH, and water discharge in the Kalix River (after Ingri *et al.*, 2000).

fraction may be the degradation products of diatoms, given the good co-variance between silicon and copper in the estuary.

It is interesting to come back to the figure of continental crust-normalized abundances in river waters (Figure 2) in the light of the speciation considerations developed in this section. It seems that the affinity for OH$^-$ and organic-complexation sites decreases from the highly immobile elements (e.g. aluminum, gallium), to the nonmobile elements (REEs, zinc,...), moderately mobile elements and mobile elements. The fact that the elements having the highest affinity for complexation by sites on colloids are the least mobile on a global average indicates that processes other than aqueous complexation control their abundance in aquatic systems.

5.09.6   INTERACTION OF TRACE ELEMENTS WITH SOLID PHASES

5.09.6.1   Equilibrium Solubility of Trace Elements

When a solution equilibrates with a mineral containing a trace element as a major component, the concentration in the solution can be predicted by thermodynamics. In nature, these phases are mostly aluminum, iron, and manganese oxides. A good case study describing solubility of aluminum and iron in the mineral phase is available from the Nsimi-Zoetele watershed in Cameroon (Viers *et al.*, 1997). Aluminum concentrations measured in organic-poor waters have been plotted on a solubility diagram on Figure 19.

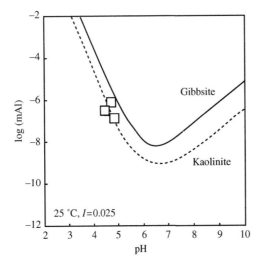

Figure 19 Saturation state of organic-poor waters from the Nsimi-Zoétélé watershed with respect to secondary aluminous phases (i.e., gibbsite, kaolinite). log(mAl) represents the logarithm of the Al concentration measured in the 5 kDa filtrate (source Viers *et al.*, 1997).

Ultrafiltered (5000Da) organic-rich waters similarly plot on the solubility curves of pure well-crystallized kaolinite (observed in the soil profiles). The data suggests that both aluminum and iron concentrations in the waters of the Nsimi watershed are controlled by the solubility of mineral phases produced by rock weathering. Another example is given by Fox (1988), in which dissolved iron concentrations in the Delaware river were determined by dialysis followed by atomic absorption spectroscopy. The concentrations of iron were near saturation with colloidal ferric hydroxides. This is not, however, the case for other rivers such as the Amazon, Zaire, Negro, and Mullica, in which filterable iron concentrations are much higher than values corresponding to equilibrium with ferric hydroxides. The presence of colloids that are not removed by the filtration explains this feature.

For solubility equilibrium to predict the aqueous concentration of a trace element, thermodynamic equilibrium is required and the solid phases must also be identified. In the case of oxyhydroxides and carbonates, it is reasonable to assume a close approach to equilibrium because the characteristic reaction times of dissolution of these minerals are in the range of a few days to a few hundred years (Bruno *et al.*, 2002). For silicate phases the assumption of thermodynamic equilibrium is more problematic due to the low reaction rates compared to the residence time of waters in hydrosystems. Examples of codes and database used by modelers to calculate the speciation and the solubility of a number of trace elements can be found in Bruno *et al.* (2002).

Iron, like manganese, uranium, and other metals, can exist in nature in more than one oxidation state. The solubility of such elements in aquatic systems depends on the concentration of oxygen, or more generally on the redox state of the fluvial system. Redox processes have been shown to control the behavior of a number of elements in lakes and in the pore waters of sediments where pe–pH diagrams are classically used to calculate the domains of stability of the solids that precipitate (e.g., MnO_2, $Fe(OH)_3$, FeS_2). The kinetics of redox reactions in natural systems is generally slow, which makes the use of pe less straightforward than pH (Michard, 1989). Even if pe–pH diagrams are theoretical, they offer a reasonable model for understanding the behavior of a number of redox-sensitive elements such as iron, manganese or uranium. The pe–pH diagram of uranium is shown in Figure 20 for a CO_2-rich water. The solubility of uranium is sensitive to the presence of vanadium and the diagram of Figure 20 is no longer valid for water with vanadium concentrations above 0.1 mg L^{-1}.

Most often, running waters are oxygenated, but anoxic conditions can prevail locally in

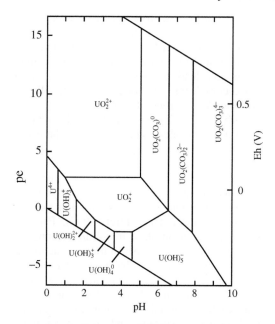

Figure 20 Distribution of dissolved uranium species in the system $U-O-H_2O-CO_2$ at 25 °C, assuming a P_{CO_2} of 0.01 atm (after Drever, 1997).

watersheds (e.g., in lakes, in mires or anoxic soils) and affect the whole river system. Redox reactions can be abiotic or biologically mediated but, as stated by Morel and Hering (1993), "life is by nature a redox process." The ultimate reductant in nature is organic matter made by photosynthetic organisms, while the ultimate oxidant is molecular oxygen (Morel and Hering, 1993). Light and temperature have a clear impact on the rates of redox reactions: examples of the importance of redox reactions in river systems on the regulation of dissolved trace elements and temporal variations are given below.

Nonphotosynthetic iron photoreduction and oxidation have been shown to control the diel cyclicity of dissolved ferrous ion concentrations in a small acidic mountain stream (McKnight *et al.*, 1989). The cycling of manganese and iron concentrations in the Kalix watershed has been especially documented by Ponter *et al.* (1992), Ingri and Widerlund (1994), Ingri *et al.* (2000), and Porcelli *et al.* (1997). The Kalix River is an organic-rich Scandinavian river dominated by the coniferous forest. Mires cover 20% of the drainage area. The concentration of suspended matter is very low with peak concentrations of ~ 10 mg L^{-1}. Tenfold seasonal variations of manganese concentrations in the dissolved load are essentially due to the input of manganese-rich lake waters after breakup of the ice in June. The manganese concentrations are then regulated in the river during summer by precipitation of manganese oxyhydroxide phases in the suspended sediments. This precipitation of

manganese-rich oxides occurs when temperature gets higher than 15 °C and is probably biologically mediated. The precipitation of iron-rich particles is also reported in the Kalix River during summer and this leads to a depletion of iron in solution.

Similar variations of manganese and iron concentrations are observed in larger river systems and are attributed to similar mechanisms. According to Shiller (1997), the strong seasonality of iron and manganese concentrations in the Mississippi traces a redox cyclicity. Within the drainage basin of the Mississippi, as in the Kalix river system, a number of environments are likely to have variable redox conditions through time, including mires and bogs, stagnant streams, bottom waters of stratified lakes and reservoirs, and river-bed sediments.

5.09.6.2 Reactions on Surfaces

Rather than a clear distinction between colloids and suspended material, separated by a 0.2 μm filtration, a continuum of particle radii exists between the smallest and the largest particles in water. Generally, the particle size distribution follows a power-law function in the form

$$N(r) = Ar^{-p}$$

with p close to 4 on the range $1-100$ μm in aquatic systems (Morel and Herring, 1993). The chemical reactivity of these particles is inversely correlated to their size, so that the smallest particles play the major role. The amount of suspended sediments in rivers fluctuates widely from a few milligrams per liter (lowland rivers) up to several grams per liter, for example, for the circum-Himalayan rivers. In rivers, particles encompass a wide range of chemical and mineralogical composition (biological debris, organic substances, oxyhydroxides, clays, rock fragments). These sediments are either produced *in situ* or most generally are derived from the erosion of soils.

Like colloidal material, surfaces have complexing sites for trace elements and the same formalism as that described for colloids can be used. Understanding the partitioning of metals and more generally trace elements between water and solids is crucial for fundamental studies on transport, bioavailability, and fate of trace elements in river systems. For example, the spatial and temporal trends of metal or radionuclide partitioning between dissolved and suspended solids is a major issue for understanding and predicting the pathways of pollutants in the environment. As a consequence, an impressive literature focuses on experimental studies of trace-element adsorption/desorption on synthetic surfaces (mostly hydrous oxides). However, field-based studies aimed at assessing the importance of adsorption mechanisms as a regulating process of trace-element concentrations in natural waters are sparse. We will briefly summarize some of the major ideas that emerged from experimental studies before reviewing the few "natural" studies.

5.09.6.3 Experimental Adsorption Studies

The uptake of trace elements by surfaces is due to the presence of complexing sites similar to those complexing ions in solution (e.g., OH, COOH groups). The difficulty of studying the trace-element–surface interaction is the geometry of the sites at the solid surface. The different processes possibly occurring at the surface and influencing the ion complexation are described in Morel and Herring (1993). In particular, the physical proximity of sites and the long-range nature of electrostatic interactions are major differences between reactions in solution and on solid surfaces. There are several types of solids capable of interacting with solutes: oxides surfaces, carbonate and sulfide surfaces, organic surfaces, and clay mineral surfaces. Most of the literature on the experimental determination of adsorption properties is based on metal oxides (iron, manganese, aluminum, silicon). Oxide surfaces are covered with surface hydroxyl groups represented by (S–OH). Following Schindler and Stumm (1987), the adsorption of metals involves one or two surface hydroxyls:

$$S-OH + M^{z+} = S-OM^{(z-1)+} + H^{+}$$

The adsorption of anions is generally by ligand exchange involving one or two hydroxyls:

$$S-OH + L^{-} = S-L^{+} + OH^{-}$$

These reactions can be treated in the same way as equilibria in solution. Equilibrium adsorption constants can thus be defined, which quantify the affinity of the cation/anion for the surface.

For example, in the case of cation sorption, the adsorption constant will be

$$K_{ads} = [S-OM^{(z-1)+}] \cdot [H^{+}]/[S-OH] \cdot [M^{z+}]$$

where the "constant" K_{ads} takes into account electrostatic interactions at the mineral surface (Drever, 1997). It follows that adsorption of cations and anions is strongly dependent upon pH. Cation adsorption increases with increasing pH. The range of pH at which adsorption starts depends on both the acid–base properties of the surface and the metal adsorption constant. The percentage of adsorption on FeOOH for selected cations as a function of pH is shown in Figure 21(a) (Sigg *et al.*, 2000). Conversely, the adsorption of anions decreases with increasing pH. Very often, the exchange of ligands is

associated with acid–base reactions in solution. Figure 21(b) shows some selected adsorption curves for anions on FeOOH. The association of a ligand in solution, a trace metal in solution and a surface can lead to the formation of ternary surface complexes. In this case, a metal–ligand association is adsorbed on the surface.

Finally, finely divided hydrous oxides of iron, aluminum, manganese, and silicon are the dominant sorbents in nature because they are common in soils and rivers, where they tend to coat other particles. This is the reason why numerous laboratory researchers have been studying the uptake of trace elements by adsorption on hydrous oxides (Dzomback and Morel, 1990). Partition coefficients (concentration in solid/concentration in the solution) for a number of trace elements and a great variety of surfaces have been determined. The comparison of these experimental K_d with natural K_d values should give information on the nature of the material on which trace elements adsorb in natural systems and allow quantitative modeling.

The comparison of experimental adsorption coefficients and measured partition coefficients in natural systems remains difficult because partition coefficients are often defined with respect to bulk solid concentration and not the desorbable element concentration. Except in a few case studies of river systems having very low detrital suspended sediments concentration (e.g., iron in the Kalix River, Ingri *et al.* (2000)), there is no agreement between adsorption and experimental partition coefficients. Another difficulty has been pointed out by Benoit (1995) and Benoit and Rozan (1998) and is known as the "particle concentration effect" (PCE). The PCE is a decline in partition coefficient as suspended particulate matter concentration increases in the river, although thermodynamics predicts no variation of K_d with solute or particle concentration. PCE occurs owing to both the increasing contribution of colloidal material as water discharge and particle discharge increase (Morel and Gschwend, 1987) and the contribution of coarser particles (with lower specific surface area and complexing site density) in periods of high suspended sediment concentrations and high water discharge. The PCE illustrates quite well the inherent difficulties in comparing experimental and field approaches.

5.09.6.4 Adsorption on Hydrous Oxides in River Systems

A series of studies have shown that adsorption onto hydrous oxides is a major mechanism regulating the concentration of trace elements in river waters. Several authors have described

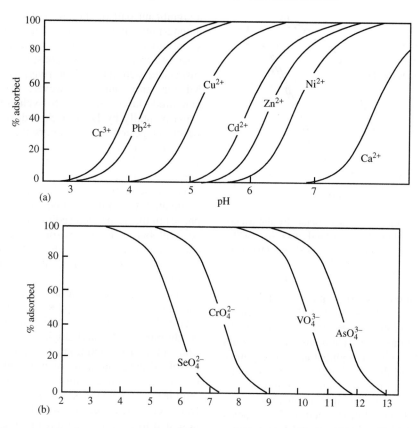

Figure 21 Proportion of cations and anions adsorbed at the surface of hydrous oxides, at high ratio of adsorption sites to adsorbing cations. Adsorption constants are from Dzomback and Morel (1990) (after Sigg *et al.*, 2000).

the formation of freshwater Fe–Mn coatings on gravels in well-oxygenated rivers. The use of the radioactive tracers ^{60}Co, ^{90}Sr, and ^{137}Cs (Cerling and Turner, 1982) clearly demonstrated the rapid and nonreversible (in oxic conditions) adsorption of cobalt and caesium onto river Fe–Mn coatings. Reversible adsorption of strontium was also observed.

A study of iron, cadmium and lead mobility in remote mountain streams of California by Erel *et al.* (1990) showed that the excess of atmospheric pollution-derived lead and cadmium is rapidly removed downstream. The comparison of truly dissolved, colloidal, and surface particle concentrations measured in the stream with the results of a model of equilibrium adsorption indicates that the mechanism of removal in this organic-poor environment is essentially by uptake onto hydrous iron oxides. The experimentally determined partition coefficients (Dzomback and Morel, 1990) explain the behavior of lead; however, they fail to explain the cadmium removal. It is proposed by the authors that cadmium is taken up by surfaces other than hydrous iron oxides.

Another attempt at modeling trace metals (copper, nickel, lead, and zinc) in river water is presented by Mouvet and Bourg (1983) for the Meuse River. In this study, K_d partition coefficients (between adsorbed and dissolved species) are calculated, based on field measurements and compared to model calculations in which the stability constants were estimated from the adsorption of trace metals by sediments of the Meuse River in laboratory experiments. The adsorption curves of copper, zinc, and cadmium by the Meuse River bottom sediments are shown in Figure 22. Under the natural conditions, most of the copper and lead are adsorbed onto the river suspended sediments. Nickel is the least solid-bound element and zinc has an intermediate behavior. The results of the model show a general agreement between predicted and observed partition coefficients (Figure 22). Especially for the periods of high suspended sediment concentrations (TSS > 30 mg L^{-1}), adsorption onto suspended solids explains relatively well the measured river concentrations. When the suspended sediment concentrations are low, the calculated percentages of adsorbed elements are underestimated, suggesting the importance of the association of organic material and suspended material in the river. The poor affinity of nickel in the Meuse River for suspended sediments is due to its dominant aqueous speciation as the $NiCO_3$ complex.

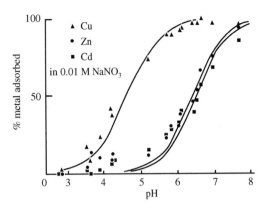

Figure 22 Adsorption of trace metals by the Meuse River bottom sediments: measured data in NaNO₃ 0.01 M and interpretation (curves) in terms of adsorption.

The adsorption of thorium has been documented experimentally by several studies (Langmuir and Herman, 1980, and references therein). As with other cationic trace elements, sorption onto surfaces is maximum at high pH and corresponds, in the absence of dissolved organic ligands, to the scavenging of hydroxy complexes. Interestingly, it has been shown for thorium that the presence of organic ligands in solution, such as fulvic acids and EDTA, inhibits adsorption and favors desorption, indicating that the affinity of thorium for organic sites is greater than for surface sites. However, our current knowledge of adsorption coefficients for the natural materials is not sufficient to allow a quantitative modeling of the role of thorium adsorption on surfaces. Because the stability of manganese and iron oxides is redox sensitive, the behavior of trace elements controlled by adsorption on these surfaces is expected to vary with redox state. Two examples of this are given below.

In the Clark Fork river, Montana, Brick and Moore (1996) showed that the diel cyclicity of dissolved manganese and zinc and acid-soluble particulate aluminum, iron, manganese, copper, and zinc are related to diel changes in biogeochemical processes, such sorption or redox changes, related to pH and dissolved oxygen cycles caused by photosynthesis and respiration in the river. The highest pH values, associated with the lowest dissolved concentrations, are consistent with this scenario. It is also possible that the evapotranspiration by stream-bank vegetation could change the contribution of the hyporheic zone (reduced waters sampled from piezometers beneath the river) as a function of time and induce the observed variations in trace-element concentrations. The uptake of alkali and alkaline-earth elements on suspended iron and manganese oxides in the Kalix River has been documented by Ingri and Widerlund (1994). In this river,

a significant part of sodium, potassium, calcium, magnesium, and also trace elements such as strontium and barium are scavenged by manganese and iron oxides that constitute most of the suspended sediments of the river. The calculated Kd for strontium is, however, two orders of magnitude larger than the Kd values determined experimentally by Dzomback and Morel (1990), which suggested to the authors that DOC might play an additional role in the scavenging of alkali and alkaline-earth cations. The Kalix River is, however, not representative of a "world average" river, as the composition of its suspended load is dominated by minerals that are usually minor phases in the suspended sediments of large rivers. At a global scale, the adsorption of alkali and alkaline-earths does not affect the conservative behavior of alkali and alkaline-earths in river systems.

The strong seasonal variability of trace concentrations in the Mississippi River water has been ascribed to the local dynamic balance of redox conditions (Shiller and Boyle, 1997). The variations in redox conditions not only affects the dissolved concentrations of iron and manganese but also the concentrations of elements sorbed onto iron and manganese oxides such as zinc and lead. Reducing conditions will destabilize manganese and iron oxides and release zinc and lead in waters. Conversely, the reduced forms of molybdenum, vanadium, and uranium are more readily adsorbed on sediment surfaces (Emerson and Huested, 1991); these elements will tend to be scavenged under reducing conditions. If the variations in the redox state of rivers are bacterially mediated, then the seasonal variations of elements such as zinc, lead, molybdenum, vanadium, or uranium in rivers are of biological origin and should be affected by nutrient supply and pollution.

5.09.6.5 The Sorption of REEs: Competition between Aqueous and Surface Complexation

The particle/solution interactions of REEs have attracted the attention of a number of workers trying to model the REE pattern of seawater or groundwaters (e.g., Turner *et al.*, 1981; Erel and Stolper, 1992; Byrne and Kim, 1990). Freshwater systems are more complex and, as of early 2000s, no model taking into account complexation by colloids, surface adsorption and complexation by inorganic ligands has been attempted. The question of the adsorption of REEs onto suspended solids in freshwaters has been addressed by Elderfield *et al.* (1990) and Sholkovitz (1995).

Suspended sediments from the Connecticut and Mississippi rivers were leached with acetic acid or seawater in order to remove the more labile (adsorbed) fraction of REEs on sediments. The so-mobilized REEs have very low concentrations. Although significant differences exist in the leached REE patterns, the desorption of REEs from suspended sediments preferentially releases LREEs. The enrichment of dissolved HREEs in estuarine waters also confirms the preferential scavenging of LREEs on surfaces. This allowed Sholkovitz (1995) to conclude that the composition of REEs in the dissolved form of rivers is mainly controlled by surface reactions. The trend of an overall LREE enrichment in the adsorbed component relative to the dissolved component is in agreement with quantitative thermodynamic models of REE adsorption (e.g., Byrne and Kim, 1990; Erel and Morgan, 1991; Erel and Stolper, 1992) in which the competitive complexation of REEs between the solution and the oxygen-donor groups (e.g., carbonate, hydroxide, phosphate) on particle surfaces is postulated. Erel and Stolper (1992) reported a linear relationship between adsorption constants and the first hydroxide binding constants, which is supported by the data of Dzomback and Morel (1990). A strong cerium anomaly in the Mississippi River dissolved load is attributed by Sholkovitz (1995) to the oxidation of dissolved Ce(III) to particulate Ce(IV) oxides in the suspended sediments of the river. These reactions would produce a cerium anomaly only in rivers of high pH and with abundant surface areas on suspended particles. Bau (1999) has experimentally observed oxidative scavenging of cerium during the sorption of REEs onto iron oxyhydroxides.

5.09.6.6 Importance of Adsorption Processes in Large River Systems

The particles transported by large rivers are a complex mixing of primary minerals, carbonates, clays, oxides and biogenic remains. The assessment of adsorption processes in controlling the levels of trace elements in large rivers has been documented by a couple of studies that will be described below.

Various chemical extraction techniques have been introduced in order to selectively remove metals from the different adsorption or complexation sites of natural sediments (e.g., Tessier *et al.*, 1979; Erel *et al.*, 1990; Leleyter *et al.*, 1999). It is, for example, shown by Leleyter *et al.* (1999) that between 20% and 60% of REE in various suspended river sediments are removed by successive extractions by water, by $Mg(NO_3)_2$ (exchangeable fraction), sodium acetate (acid-soluble fraction), $NH_2OH + HCl$ (manganese oxide dissolution); ammonium oxalate (iron oxide dissolution) and a mixture of $H_2O_2 + HNO_3$ (oxidizable fraction). The complexity of

the extraction procedure and the absence of consensus on the specificity of the reagents for particular mineral phases make the results of sequential extraction procedures difficult to interpret.

The dissolved concentrations of zinc in the Yangtze, Amazon, and Orinoco rivers have been shown to strongly decrease with pH, between 5 and 8.5 (Shiller and Boyle, 1985). This decrease is similar to that obtained in an experiment in which the pH of unfiltered Mississippi River water was adjusted to various pH values, and to isotherms of zinc adsorption on various natural and synthetic metal oxides surfaces (Figure 23). According to Shiller and Boyle (1985), alkaline rivers have less dissolved zinc because they are more suspended sediment-rich that acidic rivers. K_d values between adsorbed (acid-leachable) zinc and dissolved zinc approach 10^5 for the Mississippi, which indicates that the Mississippi transports 10 times more adsorbed zinc than dissolved zinc.

Zhang *et al.* (1993) have examined trace-element data from *in situ* measurements and laboratory experiments. The analysis of both dissolved concentrations and bulk particulate

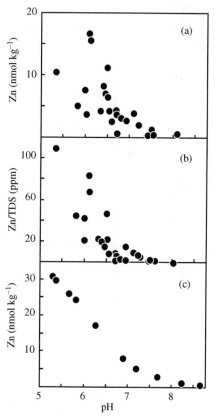

Figure 23 (a) Dissolved Zn concentrations and (b) dissolved Zn concentrations normalized to TDS as a function of pH in large rivers from the Changjiang, Amazon and Orinoco river basins. (c) Zn in pH-adjusted aliquots of Mississippi River (after Shiller and Boyle, 1985).

concentrations allowed them to calculate "partition coefficients." They showed, in the Huanghe River, significant variations in K_d for the different elements. The lowest K_d values are reported for copper and the highest for iron. Parameters such as pH, temperature and total suspended matter concentrations affected trace metal partitioning coefficients in the Huanghe, although the dominant effect was due to pH. As expected, adsorption of metals was favored at high pH. These experiments showed a great affinity of the Huanghe suspended solids for trace metals such as cadmium, copper and lead and showed that the kinetics of adsorption are rapid. Thus, river particulates may have the potential of regulating trace metal inputs to aquatic systems from pollution.

Beryllium is an alkaline-earth elements whose behavior drastically differs from that of the other alkaline-earth elements. Its low mobility in natural waters is attributed to its affinity for surfaces. Laboratory experiments have been performed to examine the partitioning of ^7Be between sediments from natural systems and water (You *et al.*, 1989). The partition coefficient depends strongly on pH in the range 2–7. The curve of K_d as a function of pH (Figure 24) can be explained by a thermodynamic model by taking into account beryllium speciation in freshwater and the

partition coefficient of each species. Data on ^{10}Be and ^9Be measured in the Orinoco system (Brown *et al.*, 1992b) also fit well the model curve and experimental data. Note that the K_d values given here are the partition coefficients between soluble beryllium and exchangeable beryllium (leached from sediments with hydroxylamine). The high K_d value of beryllium clearly shows that beryllium is transported mainly in adsorbed form. The K_d values measured for beryllium between surfaces and dissolved load show that for the sediment yields present in the Orinoco, 90% of beryllium is transported in a suspended form. The low mobility of beryllium in river waters is therefore due to its attraction for surfaces. Similar to beryllium is the case of caesium, an alkali metal whose behavior in freshwaters differs from that of sodium because of its adsorption properties.

5.09.6.7 Anion Adsorption in Aquatic Systems

Boron is present in freshwater as nonionized boric acid and negatively charged borate ion $(B(OH)_4^-)$. The ability of boron to adsorb on surfaces is a well-known characteristic of its geochemical cycle at the surface of the Earth. Spivack *et al.* (1987) determined partition coefficients of boron in the marine sediments off

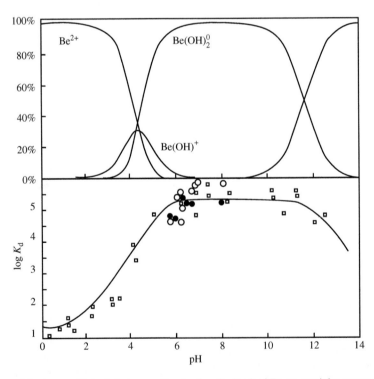

Figure 24 Bulk partitioning coefficients between dissolved and adsorbed Be on particles measured in the Orinoco for ^9Be (filled circles) and ^{10}Be (open circles). Open squares are experimental distribution coefficients determined by You *et al.* (1989) by adsorption of Be onto riverine particles. The line is deduced from a speciation-dependent model of adsorption (after Brown *et al.*, 1992b).

the Mississippi and showed that K_d ($B_{adsorbed}$/$B_{aqueous}$) is close to 1.5. The two isotopes of boron do not behave similarly during adsorption on surfaces: the light isotope is preferentially attached to surfaces. This adsorption of boron onto particles has a significant effect on the boron isotopic composition of seawater, but is probably not significant in freshwaters due to the relatively low amount of dissolved boron in freshwaters.

Among the other trace anions, the concentration of arsenic in natural waters is probably controlled by solid–solution interaction (Smedley and Kinniburgh, 2002). In sediments, the element that most frequently correlates with arsenic is iron and numerous studies have been reported on the sorption of arsenic (as arsenate ion, $HAsO_4^{2-}$) onto iron oxides, manganese oxides or aluminum oxides (see references in Smedley and Kinniburgh (2002)). K_d values up to 10^6 (L kg^{-1}) are found for the experimental adsorption of arsenic onto hydrous ferric oxides. The desorption of arsenic, like the desorption of other anion-forming elements (vanadium, boron, molybdenum, selenium, and uranium), especially from iron oxides, is one of the key processes invoked to explain high levels of arsenic in natural waters and associated toxicity problems. This desorption is favored at pH values above 8 and when iron oxides undergo reductive dissolution.

5.09.6.8 Adsorption and Organic Matter

The extrapolation of experimental studies on trace-metal adsorption to natural waters is difficult: a particular problem is the formation of both ternary surface complexes involving dissolved organic matter and the aqueous complexation of trace elements by dissolved organic matter. The number of studies trying to shed light on the complexity of these interactions by combined studies of organic matter in the suspended phase and the dissolved phase is incredibly low.

Sholkovitz and Copland (1980) used the organic-rich water of the Luce River and a tributary of the Luce River, Scotland, to investigate the relations between pH-related adsorption and organic complexation in solution. They concluded that, under their experimental conditions, the control of organic matter is much more important for determining the levels of dissolved trace elements in the river than adsorption onto the suspended sediments (composed either of clays plus organics or iron oxides plus organics). These results are not in agreement with, and are even opposite to, those inferred from artificial solutions and chemical models (see Sholkovitz and Copland (1980) for references).

Shafer *et al.* (1997) addressed the influence of dissolved organic carbon, suspended particulates and hydrology on the concentration and partitioning of trace metals in two contrasting (one agricultural, one forested) Wisconsin (USA) watersheds. As expected, the forested watershed had higher DOC levels (mean value of 7 mg L^{-1} versus 4 mg L^{-1}), but lower suspended particulate matter (SPM) (mean value of 2 mg L^{-1} versus 18 mg L^{-1}). The high concentrations of DOC and the low concentrations of potential inorganic ligands in the two rivers suggested that the dominant speciation of aluminum, cadmium, copper, lead, and zinc in solution was organic. This study shows the partitioning preferences of selected metals in these two DOC-rich and SPM-poor watersheds. Lead appears as the most strongly adsorbed element with an intermediate affinity for dissolved organic matter. Kd values for lead are uniform in low-DOC and high-SPM waters, but somewhat variable in high-DOC and low clay environments. Copper exhibits the opposite behavior. It shows a stronger affinity for DOC than for surfaces. As a consequence, copper partitioning is rather insensitive to variations in DOC concentration. The intermediate affinity of zinc for SPM and DOC makes zinc partitioning highly sensitive to changes in either SPM or DOC. Accordingly, Kd values are ranked in the order Pb > Zn > Cd > Cu. Note that the trends observed in these two Wisconsin watersheds are probably not representative of larger systems, which generally have higher suspended sediment concentrations.

As shown by Laxen (1985), the adsorption of cadmium, copper, and nickel onto hydrous ferric oxides in the presence of dissolved humic substances is significantly modified. The adsorption of cadmium and nickel is enhanced by the presence of humics. The model favored by the authors is the complexation of metals with adsorbed humics, which is stronger than binding with the dissolved humics. The adsorption of copper is enhanced but more sensitive to the competitive formation of soluble copper–humic complexes. An important implication of these experiments is that metals will be more strongly adsorbed at low pH values in the presence of humics than in their absence. Radiotracer experiments have shown that, in the ocean, interactions of organic-rich colloids and particles (through sorption and coagulation) occur in estuaries. This mechanism is termed "colloidal pumping" and plays a crucial role in the fate of many trace elements in estuarine waters (Wen *et al.*, 1997). The extent of this mechanism in rivers has not been explored.

Finally, one should not end this section without mentioning the uptake of trace elements by living biomass. This process can be of quantitative importance in river systems of low velocity or draining lakes or stagnant water bodies. For example, a study in the St. Lawrence river (Quémerais and Lum, 1997) has shown that an uptake of cadmium by phytoplankton occurs in the summer. In lakes, the uptake of trace elements by living organisms has been demonstrated for a number of trace elements with known biological activity (copper, zinc, iron, molybdenum, manganese), or unknown biological function (cadmium, arsenic) (cf. Sigg *et al.*, 2000).

5.09.6.9 Particle Dynamics

Regulation of concentration of trace elements by surfaces plays a key role in river water systems. This applies to both colloids and larger particles. Over the past decades, the physicochemical understanding of sorption processes has considerably improved. Mass transfer codes (see Chapter 5.02) such as PHREEQE (Parkhurst *et al.*, 1990), PHREEQC (Parkhurst, 1995), SOLMINEQ.88 (Kharaka *et al.*, 1988), MINTEQA2 (Allison *et al.*, 1991), MINTEQ(4.00) (Eary and Jenne, 1992), MINEQL + (Schecher and McAvoy, 1991), and EQ3/6 (Wollery, 1992) perform specialized calculations and predict changes in solution chemistry caused by different processes such as dissolution/precipitation, ion exchange/adsorption, or mixing of water masses. However, so far, these models, which explain the results from the laboratory experiments well, are not able, in general, to predict the pathways of trace elements in the natural system. This is due to the complexity of the natural systems characterized by the diversity of substrates, of surface sites, of ligands in solution, and the abundance of organic material, the presence of colloids that pass through filters and preclude the strict separation between aqueous and solid phase concentrations, and finally the dynamics of colloids and particles within the river system. The transfer of experimental data to natural systems is not straightforward because natural surfaces are far more complex than the experimentally synthesized ones. Studies of the kinetics of adsorption/ desorption of trace elements on natural sediments should be encouraged, even if severe experimental artifacts exist. More data on concentrations of dissolved, adsorbed and colloidal material in natural systems are necessary, particularly on a worldwide scale, in order to improve our knowledge of trace-element behavior in natural waters and to improve the quality of modeling and the accuracy of prediction. Because trace metals are carried on the finer fraction of sediments in river systems (Horovitz, 1991), due to a surface area effect, the sedimentary dynamics and the post-depositional stability of sediments (residence times) in watersheds are important factors to take into account for the transport of trace elements. For example, the preferential deposition of fine-grained material in floodplains, lakes or reservoirs will strongly influence the pathways of metals through catchments. An integrated approach of chemistry, geochemistry, and geomorphology is therefore in order.

5.09.7 CONCLUSION

This review focuses on the concentration of trace elements occurring in river waters, based on the data obtained from natural sources. The determination of concentrations of many trace elements in river waters is facilitated in particular by the technical advances of ICP-MS allowing rapid measurement of a large number of elements. Trace-element measurements are not only motivated by the requirements of pollution and toxicology studies, but also by the need for better understanding and modeling of the behavior of elements during continental weathering and transport, which constitute the two major aspects of the geological cycle.

So far, the global systematics of trace elements in river waters are still poorly understood and limited either to industrial countries, where the natural levels are overwhelmed by anthropogenic inputs, or to regions of the world where rivers are organic-rich, namely, the tropics and the high-latitude areas. Continents such as Asia, Australia and the numerous Pacific peninsulas have been very poorly investigated, except for a few elements of geochemical importance (osmium, uranium, thorium, strontium).

Trace-element concentrations in river waters span over 10 orders of magnitude, similar to the range of crustal abundances. However, the normalization of trace-element levels in river waters to mean continental abundances shows that the abundance of trace elements in river waters depends not only on their continental abundances, but also on their mobility in weathering and transport processes. Normalized abundances of elements in river water vary over five orders of magnitude from highly immobile elements such as titanium, niobium, zirconium, aluminum, and tantalum to highly mobile elements such as selenium, rhenium, boron, arsenic, molybdenum, and sodium. An important feature of trace-element concentrations in river water is a large variability in space and time, even over very short periods of time (days). In addition, the abundance of elements such as REEs is a function of river chemistry as exemplified by the worldwide correlation observed between dissolved neodymium and pH. It is a major characteristic of trace-element concentrations in river waters that they not only depend on their abundance in the source rocks but that they are also strongly dependent on the chemical conditions prevailing in the river. There are a few exceptions to this rule, notably the most mobile elements that correlate fairly well with the major elements.

The mobility of trace elements in river waters results from a complex combination of several factors: their solubility in water, the input to the system of nonweathering sources such as atmospheric and/or anthropogenic sources, the ability of the elements to be complexed by fine (<0.2 μm) colloidal material, and the affinity of the elements for solids (adsorption, co-precipitation, solubility equilibrium). The "chemical" solubility of elements depends on the solubility of the minerals that contain the elements and is approximately related to the atomic and molecular properties of the elements. Elements of small atomic radii tend to form polycharged oxyanions that make them soluble (e.g., boron, arsenic, tungsten, vanadium). Elements of higher atomic radii, and of +1 or +2 valency tend to be soluble as free cations (sodium, potassium, barium, lithium, strontium). Elements such as aluminum, REEs, thorium, and iron fall in between and form poorly soluble oxyanions or hydroxyanions. Nonweathering inputs comprise atmospheric inputs, anthropogenic inputs or plant inputs. Owing to the global present-day contamination of the atmosphere by anthropogenic emissions (e.g., combustion dusts, detrital particles, industrial aerosols, gases etc.), it is difficult to estimate the natural input of trace elements to hydrosystems, but even in the most remote areas, rain can be a significant source of elements transported by the rivers. In the industrialized countries and to a lesser extent, at a global scale, clear evidence of direct contamination of rivers by anthropogenic sources exists, especially for heavy metals and other metals.

Although it controls the mobility of trace elements in water, the aqueous speciation of many elements in river waters is not well known, principally when organic complexes are involved. However, one of the most striking characteristics of the behavior of trace elements for which chemical aqueous species are known, is that their concentrations are usually not predictable by classical speciation calculations involving the major inorganic or organic ligands in solution. This is due to two competitive mechanisms. The first is the existence in rivers, especially in rivers of low pH, of a colloidal phase made of intimately coupled organic and inorganic material, which can pass through the pores of filtration membranes (conventionally 0.2 μm; and therefore analyzed and considered as dissolved material). The more the colloidal material "contaminates" the dissolved phase, the higher the concentrations of chemically insoluble elements will be. Although colloidal material is poorly known, and its affinity for trace elements still debated, the affinity of organic colloid for trace metals increases with the first hydrolysis constant of the metal. The competing mechanism that tends to lower the levels of "dissolved" trace

elements is the affinity of the elements for particles larger than 0.2 μm, which leads to the removal of these elements from the dissolved phase during filtration. In rivers, a great variety of surfaces can exist and uptake mechanisms can be very diverse. Adsorption or co-precipitation on hydrous oxides, organic particles or clays, and uptake by living organisms are possible mechanisms, and they are not easy to distinguish. Although colloids are small particles and should obey the same formalism as surface complexation on larger particles, our understanding and modeling of colloidal and surface uptake in natural hydrosystems is far from being clear and complete. Surface sorption, like aqueous complexation, increases with the first hydrolysis constant. Although, an impressive literature exists on the interaction between solutes and hydrous oxides, the use of the experimentally derived affinity constants adapting to the natural systems is a difficult task. The principal obstacle in this attempt is the complexity of naturally occurring material, the poor physical description we have of colloids, the potential experimental artifacts associated with the isolation of colloids (fractionation, coagulation), and the extremely variable chemistry of rivers from one continent to another. The pH-sensitivity of stability of colloids in the aquatic system, as well as the varying extent of adsorption of ions as a function of pH, can explain the dependence of concentration of the trace elements in river water on pH and other chemical variables.

Much work is clearly needed to gain an understanding of the control on concentrations of trace elements in river waters and aquatic systems. It seems clear that further studies should focus on the physical characterization of the colloid pool and comprehension of its dynamics. Estuarine studies have shown that the behavior of elements in the mixing zones between rivers and ocean water is controlled by the flocculation, coagulation, or degradation of colloidal material. The technological advances in filtration membranes should allow a better size fractionation and isolation of colloids. Nanofiltration and dialysis are other promising techniques that need to be developed and should lead to a better characterization of the "true" dissolved pool and of the colloidal-size material. So far, trace-element concentrations have been measured in a restricted number of environments, which may not be typical. Extended investigations on rivers of high pH, high suspended material yields, and mountainous rivers, or rivers of variable climatic or lithologic settings are necessary to get a better world-scale overview of trace-element levels and controlling parameters. From a more theoretical point of view, the relationships between the periodic table properties and the geochemical behavior of trace element in natural hydrosystems, through organic and inorganic complexation, solid uptake and affinity for life in river systems need to be explored.

Finally, the recent technical progress in the new generation of multicollector ICP–MS instruments that enable the measurement of the isotopic ratios of a number of trace metals, whose isotopes are fractionated by complexation processes, is a new challenge for aquatic sciences, which, no doubt, will improve our knowledge and prediction of trace-element dynamics in hydrosystems.

ACKNOWLEDGMENTS

J. I. Drever and M. Meybeck provided the first impulse for this chapter. We thank O. Pokrovski, J. Schott, J. L. Dandurand, S. Deberdt, R. Millot, P. Seyler, E. Lemarchand, and B. Bourdon for their help and discussions. This chapter benefited from a detailed review by J. I. Drever. This is IPGP contribution no. 1947.

REFERENCES

Allison J. D., Brown D. S., and Novo-Gradac K. J. (1991) *MINTEQA2: A Geochemical Assessment Data Base and Test Cases for Environmental Systems*, vers. 3.0 user's manual. Report EPA/600/3-91/-21. Athens, GA, US EPA.

Ammann A. A. (2002) Speciation of heavy metals in environmental water by ion chromatography coupled to ICP-MS. *Anal. Bioanal. Chem.* **372**, 448–452.

Akinfiev N. N. (1976) "Balance": IBM computer code for calculating mineral aqueous solution-gas equilibria. *Geochemistry* **6**, 882–890 (in Russian).

Artaxo P., Maenhaut W., Storms H., and Van Grieken R. (1990) Aerosol characteristics and sources for the Amazon basin during the wet season. *J. Geophys. Res.* **95**, 16971–16985.

Baes C. F. and Mesmer R. E. (1976) *The Hydrolysis of Cations*. Wiley, New York, 489pp.

Bau M. (1999) Scavenging of dissolved yttrium and rare earths by precipitating iron hydroxide: experimental evidence for Ce oxidation, Y–Ho fractionation and lanthanide tetrad effect. *Geochim. Cosmochim. Acta* **63**, 67–77.

Bau M. and Dulski P. (1996) Anthropogenic origin of positive gadolinium anomalies in river waters. *Earth. Sci. Planet. Lett.* **143**, 345–355.

Benoit G. (1995) Evidence of the particle concentration effect for lead and other metals in fresh waters based on ultrafiltration technique analyses. *Geochim. Cosmochim. Acta* **59**(13), 2677–2687.

Benoit G. and Rozan T. F. (1998) The influence of size distribution on the particle concentration effect and trace metal partitioning in rivers. *Geochim. Cosmochim. Acta* **63**(1), 113–127.

Berg T., Royset O., and Steinnes E. (1994) Trace element in atmospheric precipitation at Norvegian background stations (1989–1990) measured by ICP-MS. *Atmos. Environ.* **28**, 3519–3536.

Berner E. K. and Berner R. A. (1996) Global environment: water, air, and geochemical cycles. Prentice-Hall, Englewood Cliffs, NJ, 376p.

Brick C. M. and Moore J. N. (1996) Diel variation of trace metals in the upper Clark Fork river, Montana. *Environ. Sci. Technol.* **30**, 1953–1960.

Brown E. T., Measures C. I., Edmond J. M., Bourlès D. L., Raibeck G. M., and Yiou F. (1992a) Continental inputs of beryllium to the oceans. *Earth Planet. Sci. Lett.* **114**, 101–111.

Brown E., Edmond J. M., Raibeck G. M., Bourlès D. L., Yiou F., and Measures C. I. (1992b) Beryllium isotope geochemistry in tropical river basins. *Geochim. Cosmochim. Acta* **56**, 1607–1624.

Bruno J., Duro L., and Grivé M. (2002) The applicability and limitations of thermodynamic geochemical models to stimulate trace element behavior in natural waters: lessons from natural analogue studies. *Chem. Geol.* **190**, 371–393.

Buffle J. and Van Leeuwen H. P. (1992) Environmental particles: 1. In *Environmental Analytical and Physical Chemistry Series*. Lewis Publishers, London, 554p.

Burba P., Shkinev V., and Spivakov B. Y. (1995) On-line fractionation and characterisation of aquatic humic substances by means of sequential-stage ultrafiltration. *J. Anal. Chem.* **351**, 74–82.

Byrne R. H. and Kim K. H. (1990) Rare earth element scavenging in seawater. *Geochim. Cosmochim. Acta* **54**, 2645–2656.

Cameron E. M., Hall G. E. M., Veizer J., and Roy Krouse H. (1995) Isotopic and elemental hydrogeochemistry of a major river system: Fraser river, British Columbia, Canada. *Chem. Geol.* **122**, 149–169.

Cantrell K. J. and Byrne R. H. (1987) Rare earth element complexation by carbonate and oxalate ions. *Geochim. Cosmochim. Acta* **51**, 597–605.

Cerling T. E. and Turner R. R. (1982) Formation of freshwater Fe–Mn coatings on gravel and the behavior of ^{60}Co, ^{90}Sr, and ^{137}Cs in a small watershed. *Geochim. Cosmochim. Acta* **46**, 1333–1343.

Chabaux F., Riotte J., Clauer N., and France-Lanord Ch. (2001) Isotopic tracing of the dissolved U fluxes in Himalayan rivers: implications for the U oceanic budget. *Geochim. Cosmochim. Acta* **65**, 3201–3217.

Chabaux F., Riotte J., and Dequincey O. (2003) U–Th–Ra fractionation during weathering and river transport. In *U-series Geochemistry and Applications*, Rev. Mineral. Geochem. (eds. B. Bourdon, S. Turner, G. Henderson, and C. C. Lundstrom). Min. Soc. Am. and Geochem. Soc. (in press).

Colodner D., Sachs J., Ravizza G., Turekian K., Edmond J., and Boyle E. (1993) The geochemical cycle of rhenium: a reconnaissance. *Earth Planet. Sci. Lett.* **117**, 205–221.

Cossa D., Tremblay G. H., and Gobeil C. (1990) Seasonality in iron and manganese concentrations in the St. Lawrence river. *Sci. Tot. Environ.* **97/98**, 185–190.

Dai M. and Martin J.-M. (1995) First data on trace metal level and behaviour in two major Arctic river-estuarine systems (Ob and Yenisey) and in the adjacent Kara Sea, Russia. *Earth Planet. Sci. Lett.* **131**, 127–141.

Dalai T. K., Singh S. K., Triverdi J. R., and Krishnaswami S. (2001) Dissolved rhenium in the Yamuna river system and the Ganga in the Himalaya: role of black shale weathering on the budgets of Re, Os, and U in rivers and CO_2 in the atmosphere. *Geochim. Cosmochim. Acta* **66**(1), 29–43.

Deberdt S., Viers J., and Dupre B. (2002) New insights about the rare earth elements (REE) mobility in river waters. *Bull. Soc. Geol. France* **173**(n°2), 147–160.

De Caritat P., Reimann C., Ayras M., Niskavaara H., Chekushin V. A., and Pavlov V. A. (1996) Stream water geochemistry from selected catchments on the Kola peninsula (NW Russia) and the neighbouring areas of Finland and Norway: 1. Element levels and sources. *Aquat. Geochem.* **2**, 149–168.

Douglas G. B., Gray C. M., Hart B. T., and Beckett R. (1995) A strontium isotopic investigation of the origin of suspended particulate matter (SPM) in the Murray-Darling river system, Australia. *Geochim. Cosmochim. Acta* **59**, 3799–3815.

Drever J. I. (1997) *Geochemistry of Natural Waters*, 3rd edn. Prentice-Hall, Englewood Cliffs, NJ.

Dupré B., Gaillardet J., and Allègre C. J. (1996) Major and trace elements of river-borne material: the Congo Basin. *Geochim. Cosmochim. Acta* **60**, 1301–1321.

Dupré B., Viers J., Dandurand J. L., Polvé M., Bénézeth P., Vervier P., and Braun J. J. (1999) Major and trace elements associated with colloids in organic-rich river waters: ultrafiltration of natural and spiked solutions. *Chem. Geol.* **160**, 63–80.

Dzomback D. A. and Morel F. M. M. (1990) *Surface Complexation Modelling*. Wiley, New York.

Eary L. E. and Jenne E. A. (1992) *Version 4.00 of the MINTEQ Code*. Report PNL-8190/UC-204. Richland, WA, Pacific Northwest Laboratory.

Edmond J. M., Spivack A., Grant B. C., Hu Ming-Hui, Chen Zexiam, Chen Sung, and Zeng Xiushau (1985) Chemical dynamics of the Changjiang estuary. *Cont. Shelf. Res.* **4** (1/2) 17–36.

Edmond J. M., Palmer M. R., Measures C. I., Grant B., and Stallard R. F. (1995) The fluvial geochemistry and denudation rate of the Guyana shield in Venezuela, Columbia and Brazil. *Geochim. Cosmochim. Acta* **59**, 3301–3325.

Edmond J. M., Palmer M. R., Measures C. I., Brown E. T., and Huh Y. (1996) Fluvial geochemistry of the eastern slope of the northeastern Andes and its foredeep in the drainage of the Orinoco in Colombia and Venezuela. *Geochim. Cosmochim. Acta* **60**, 2949–2975.

Elbaz-Poulichet F., Seyler P., Maurice-Bourgoin L., Guyot J. L., and Dupuy C. (1999) Trace element geochemistry in the upper Amazon drainage basin (Bolivia). *Chem. Geol.* **157**, 319–334.

Elderfield H., Upstill-Goddard R., and Sholkovitz E. R. (1990) The rare earth element in rivers, estuaries, and coastal seas and their significance to the composition of ocean waters. *Geochim. Cosmochim. Acta* **54**, 971–991.

Emerson S. R. and Huested S. S. (1991) Ocean anoxia and the concentrations of molybdenum and vanadium in seawater. *Mar. Chem.* **34**, 177–196.

Erel Y. and Morgan J. J. (1991) The effect of surface reactions on the relative abundances of trace metals in deep sea waste. *Geochim. Coschim. Acta* **55**, 1807–1813.

Erel Y. and Stolper E. M. (1992) Modelling of rare-earth element partitioning between particles and solution in aquatic environments. *Geochim. Cosmochim. Acta* **57**, 513–518.

Erel Y., Morgan J. J., and Patterson C. C. (1990) Natural levels of lead and cadmium in remote mountain stream. *Geochim. Cosmochim. Acta* **55**, 707–719.

Eyrolle F., Fevrier D., and Benaim J. U.-Y. (1993) Etude par D. P. A. S. V. de l'aptitude de la matière organique colloïdale a fixer et a transporter les métaux: exemples de bassins versants en zone tropicale. *Environ. Technol.* **14**, 701–717.

Eyrolle F., Benedetti M. F., Benaim J.-Y., and Fevrier D. (1996) The distributions of colloidal and dissolved organic carbon, major elements and trace elements in small tropical catchments. *Geochim. Cosmochim. Acta* **60**, 3643–3656.

Foster I. D. L. and Charlesworth S. M. (1996) Heavy metals in the hydrological cycle: trends and explanation. *Hydrol. Process.* **10**, 227–261.

Fox L. E. (1988) The solubility of colloidal ferric hydroxide and its relevance to iron concentrations in river water. *Geochim. Cosmochim. Acta* **52**, 771–777.

Frei M., Bielert U., and Heinrichs H. (1998) Effects of pH, alkalinity and bedrock chemistry on metal concentrations of springs in an acidified catchment (Ecker dam, Harz Mountains, FRG). *Chem. Geol.* **170**, 221–242.

Freydier R., Dupré B., and Lacaux J. P. (1998) Precipitation chemistry in intertropical Africa. *Atmos. Environ.* **32**, 749–765.

Froelich P. N., Hambrick G. A., Andreae M. O., Mortlock R. A., and Edmond J. M. (1985) The geochemistry of inorganic germanium in natural waters. *J. Geophys. Res.* **90**(C1), 1133–1141.

Gaillardet J., Dupré B., Allègre C. J., and Négrel P. (1997) Chemical and physical denudation in the Amazon river basin. *Chem. Geol.* **142**, 141–173.

Gaillardet J., Dupré B., Louvat P., and Allègre C. J. (1999a) Global silicate weathering of silicates estimated from large river geochemistry. *Chem. Geol. (Spec. Issue Carbon Cycle 7)* **159**, 3–30.

Gaillardet J., Dupré B., and Allegre C. J. (1999b) Geochemistry of large river suspended sediments: silicate weathering or crustal recycling ? *Geochim. Cosmochim. Acta* **63**(23/24), 4037–4051.

Gaillardet J., Millot R., and Dupré B. (2003) Trace elements in the Mackenzie river basin (in preparation).

Goldstein S. J. and Jacobsen S. B. (1988) Rare earth elements in river waters. *Earth Planet. Sci. Lett.* **89**, 35–47.

Gu B., Schmitt J., Chen Z., Liang L., and McCarthy J. F. (1995) Adsorption and desorption of different organic matter fractions on iron oxide. *Geochim. Cosmochim. Acta.* **59**, 219–229.

Henderson P. (1984) General geochemical properties and abundances of the rare earth elements. In *Developments in Geochemistry: 2. Rare Earth Element Geochemistry* (ed. P. Henderson). Elsevier, Amsterdam, pp. 1–32.

Herrera Ramos A. C. and McBride M. B. (1996) Goethite dispersibility in solutions of variable ionic strength and soluble organic matter content. *Clay Clays Min.* **44**, 286–296.

Hoffmann S. R., Shafer M. M., Babiarz C. L., and Armstrong D. E. (2000) A critical evaluation of tangential-flow ultrafiltration for trace metals studies in freshwater systems: 1. Organic carbon. *Environ. Sci. Technol.* **34**, 3420–3427.

Horowitz A. J., Lum K. R., Garbarino J. R., Gwendy E. M. H., Lemieux C., and Demas C. R. (1996) The effect of membrane filtration on dissolved trace element concentrations. *Water Air Soil Pollut.* **90**, 281–294.

Huang W. W., Martin J. M., Seyler, Zhang J., and Zhong X. M. (1988) Distribution and behavior of arsenic in the Huanghe (Yellow river) estuary and Bohai Sea. *Mar. Chem.* **25**, 75–91.

Huh Y., Chan L. H., Zhang L., and Edmond J. M. (1998) Lithium and its isotopes in major world rivers: implications for weathering and the oceanic budget. *Geochim. Cosmochim. Acta* **62**, 2039–2051.

Hummel W., Glaus M., Van Loon L. R. (1995) Binding of radionuclides by humic substances: the "Conservative Roof" approach. In *Proceedings of an NEA Workshop, September 14–16, 1994, Bad Zurzach, Switzerland*. OECD documents, ISBN no. 92-94, 251–262.

Ingri J. and Widerlund A. (1994) Uptake of alkali and alkaline-earth elements on suspended iron and manganese in the Kalix River, northern Sweden. *Geochim. Cosmochim. Acta* **58**, 5433–5442.

Ingri J., Widerlund A., Land M., Gustafsson Ö., Andersson P., and Öhlander B. (2000) Temporal variations in the fractionation of the rare earth elements in a boreal river: the role of colloidal particles. *Chem. Geol.* **166**, 23–45.

Johannesson K., Stetzenbach K., Hodge V. F., and Lyons W. B. (1996) Rare earth element complexation behavior in circumneutral pH groundwaters: assessing the role of carbonate and phosphate ions. *Earth Planet. Sci. Lett.* **139**, 305–319.

Johannesson K. H., Lyons W. B., Graham E. Y., and Welch C. A. (2000) Oxyanion concentrations in eastern Sierra Nevada rivers: 3. Boron, Molybdenum, Vanadium, and Tungsten. *Aquat. Geochem.* **6**, 19–46.

Kharaka Y. K., Gunter W. D., Aggarwal P. K., Perkins E. H., and Debraal J. D. (1988) SOLMINEQ: 88. A computer program for geochemical modelling of water rock interactions. In *US Geological Survey Water Resources Inv. 88-4227, Meno Park, CA: US Geological Survey*.

Keasler K. M. and Loveland W. D. (1982) Rare earth elemental concentrations in some Pacific Northwest rivers. *Earth Planet. Sci. Lett.* **61**, 68–72.

Koshinen W. C. and Harper S. S. (1990) The retention process: mechanisms. In *Pesticides in the Soil Environment: Processes, Impacts, and Modelling*. Book Ser. 2 (ed. H. H. Cheng). Soil Science Society of America, Madison, WI, pp. 51–77.

Lakshman S., Mills R., Patterson H., and Cronan C. (1993) Apparent differences in binding site distributions and aluminum(III) complexation for three molecular weight fractions of a coniferous soil fulvic acid. *Anal. Chim. Acta* **282**, 101–108.

Langmuir D. (1997) *Aqueous Environmental Geochemistry*. Prentice-Hall, New Jersey, 600p.

Langmuir D. and Herman J. S. (1980) The mobility of thorium in natural waters at low temperatures. *Geochim. Cosmochim. Acta* **44**, 1753–1766.

Laxen D. P. H. (1985) Trace metals adsorption/co-precipitation on hydrous ferric oxide under realistic conditions. *Water Res.* **19**, 1229–1236.

Laxen D. P. H., Davison X., and Wook C. (1984) Manganese chemistry in rivers and streams. *Geochim. Cosmochim. Acta* **48**, 2107–2111.

Lead J. R., Davison W., Hamilton-Taylor J., and Buffle J. (1997) Characterizing colloidal material in natural waters. *Aquat. Geochem.* **3**, 213–232.

Lee J. H. and Byrne R. H. (1993) Complexation of trivalent rare earth elements by carbonate ions. *Geochim. Cosmochim. Acta* **57**, 295–302.

Leleyter L., Probst J. L., Depetris P., Haida S., Mortatti J., Rouault R., and Samuel J. (1999) Distribution des terres rares dans les sediments fluviaux: fractionnement entre les phases labiles et résiduelles. *C. R. Acad. Sci. Paris* **329**, 45–52.

Lemarchand D., Gaillardet J., Lewin E., and Allègre C. J. (2000) Boron isotopes river fluxes: limitation for sea-water pH reconstruction over the last 100 Myr. *Nature* **408**, 951.

Levasseur S. (1999) Contribution à l'étude du cycle externe de l'osmium. PhD Thesis, University of Paris 7.

Levasseur S., Birck J. L., and Allègre C. J. (1999) The osmium riverine flux and the oceanic mass balance of osmium. *Earth Planet. Sci. Lett.* **174**, 7–23.

Li Y. (2000) A compendium of geochemistry. In *From Solar Nebula to the Human Brain*. Princeton University Press.

Liu X. and Byrne R. H. (1997) Rare earth and yttrium phosphate solubilities in aqueous solution. *Geochim. Cosmochim. Acta* **61**, 1625–1633.

Luck J. M. and Ben Othman D. (1998) Geochemistry and water dynamics: II. Trace metals ad Pb–Sr isotopes as tracers of water movements and erosion processes. *Chem. Geol.* **150**, 263–282.

Martin J. M. and Meybeck M. (1979) Elemental mass balance of materiel carried by world major rivers. *Mar. Chem.* **7**(2), 173–206.

Martin J. M., Hogdahl O., and Philippot J. C. (1976) Rare earth element supply to the ocean. *J. Geophys. Res.* **81**(18), 3119–3124.

Martin J. M., Guan D. N., Elbaz-Poulichet F., Thomas A. J., and Gordeev V. V. (1993) Preliminary assessment of the distributions of some trace elements (As, Cd, Cu, Fe, Ni, Pb, Zn) in a pristine aquatic environment: the Lean River estuary (Russia). *Mar. Chem.* **43**, 185–199.

Martel A. E. and Hancock R. D. (1996) Metal complexes in Aqueous solutions. Plenum, New York, 253pp.

Martell A. E. and Smith R. M. (1977) Critical stability constants. In *Other Organic Ligands*. Volume 3, Plenum, New York.

McKnight D. M. and Bencala K. E. (1989) Reactive iron transport in an acidic mountain stream in Summit County, Colorado: a hydrologic perspective. *Geochim. Cosmochim. Acta* **53**, 2225–2234.

McKnight D. M. and Bencala K. E. (1990) The chemistry of iron, aluminium, and dissolved organic material in three acidic, metal-enriched, mountains streams, as controlled by watershed and in-stream processes. *Water Resour. Res.* **26**(n°12), 3087–3100.

McKnight D. M., Kimball B. A., and Bencala K. E. (1989) Iron photoreduction and oxydation in an acidic mountain stream. *Science* **240**, 637–640.

Michard G. (1989) *Equilibres chimiques dans les eaux naturelles*. Edition Publisud, Paris.

Millero F. J. (1992) Stability constants for the formation of rare earth inorganic complexes as a function of ionic strength. *Geochim. Cosmochim. Acta* **56**, 3123–3132.

Morel F. M. M. and Gschwend P. M. (1987) The role of colloids in the partitioning of solutes in natural waters. In *Aquatic Surface Chemistry* (ed. W. Stumm). Wiley, New York, pp. 405–422.

Morel F. M. M. and Hering J. G. (1993) *Principles and Applications of Aquatic Chemistry*. Wiley, New York.

Mouvet C. and Bourg A. C. M. (1983) Speciation (including adsorbed species) of copper, lead, nickel, and zinc in the Meuse River: observed results compared to values calculated with a chemical equilibrium computer program. *Water Res.* **6**, 641–649.

Murnane R. J. and Stallard R. J. (1990) Germanium and silicon in rivers of the Orinoco drainage basin. *Nature* **344**, 749–752.

Négrel P., Allègre C. J., Dupré B., and Lewin E. (1993) Erosion sources determined by inversion of major and trace element ratios in river water: the Congo Basin case. *Earth Planet. Sci. Lett.* **120**, 59–76.

Nriagu J. O. (1979) Global inventory of natural and anthropogenic emissions of trace metals to the atmosphere. *Nature* **279**, 409–411.

Nriagu J. O. and Pacyna J. M. (1988) Quantitative assessment of worldwide contamination of air, water, and soils by trace metals. *Nature* **333**, 134–139.

Olivié-Lauquet G., Allard T., Benedetti M., and Muller J. P. (1999) Chemical distribution of trivalent iron in riverine material from a tropical ecosystem: a quantitative EPR study. *Water Res.* **33**, 2276–2734.

Olivié-Lauquet G., Allard T., Bertaux J., and Muller J. P. (2000) Crystal chemistry of suspended matter in a tropical hydrosystem, Nyong basin (Cameroon, Africa). *Chem. Geol.* **170**, 113–131.

Palmer M. R. and Edmond J. M. (1993) Uranium in river water. *Geochim. Cosmochim. Acta* **57**, 4947–4955.

Parkhurst D. L. (1995) *Users Guide to PHREEQC: A Computer Program for Speciation, Reaction-path, Advective-transport, and Inverse Geochemical Calculations*. US Geological Survey Water Resources Inv. Report, 95-4227.

Parkhurst D. L., Thorstenson D. C., and Plummer N. L. (1990) *PHREEQE: A Computer Program for Geochemical Calculations*. Rev. US Geological Survey Water Resources Inv. Report, 80-96.

Perdue E. M., Beck K. C., and Reuter J. H. (1976) Organic complexes of iron and aluminum in natural waters. *Nature* **260**, 418–420.

Perdue E. M., Reuter J. H., and Parrish R. S. (1984) A statistical model of proton binding by humus. *Geochim. Cosmochim. Acta* **48**, 1257–1263.

Pereiro R. and Carro Díaz A. (2002) Speciation of mercury, tin, and lead compounds by gas chromatography with microwave-induced plasma and atomic-emission detection (GC–MIP–AED). *Anal. Bioanal. Chem.* **372**, 74–90.

Perret D., Newman M. E., Negre J. C., Chen Y., and Buffle J. (1994) Submicron particles in the Rhine river: I. Physico-chemical characterization. *Water Res.* **28**, 91–106.

Peucker-Ehrenbrink and Hannigan (2000) Effects of black shale weathering on the mobility of rhenium and platinum group elements. *Geology* **28**, 475–478.

Picouet C., Dupré B., Orange D., and Valladon M. (2001) Major and trace element geochemistry of the upper Niger (Mali): physical and chemical weathering rates and CO_2 consumption. *Chem. Geol.* 93–124.

Pierson-Wickman A. C., Reisberg L., and France-Lanord C. (2000) The Os isotopic composition of Himalayan river bedloads and bedrocks: importance of black shales. *Earth Planet. Sci. Lett.* **176**, 201–216.

Plummer L. E., Jones B. F., and Truesdell A. H. (1984) *WATEQF—A FORTRAN IV Version of WATEQ: A Computer Program for Calculating Chemical Equilibria of Natural Waters.* Rev. US Geological Survey Water Resources Inv. Report 76-13. US Geological Survey, Reston, VA.

Pokrovski O. S. and Schott J. (2002) Iron colloids/organic matter associated transport of major and trace elements in small boreal rivers and their estuaries (NW Russia). *Chem. Geol.* **190**, 141–181.

Ponter C., Ingri J., and Boström K. (1992) Geochemistry of manganese in the Kalic river, northern Sweden. *Geochim. Cosmochim. Acta* **56**, 1485–1494.

Porcelli D., Andersson P. S., Wasserburg G. J., Ingri J., and Baskaran M. (1997) The importance of colloids and mires for the transport of uranium isotopes through the Kalix River watershed and Baltic Sea. *Geochim. Cosmochim. Acta* **61**, 4095–4113.

Quémerais B. and Lum K. R. (1997) Distribution and temporal variation of Cd in the St. Lawrence river basin. *Aquat. Sci.* **59**, 243–259.

Roy S. (1996) Utilisation des isotopes du plomb et du strontium comme traceurs des apports atnthropiques et naturels dans les précipitations et rivières du bassin de Paris. PhD Thesis, Université Paris 7.

Sarin M. M., Krishnaswami S., Somayajulu B. L. K., and Moore W. S. (1990) Chemistry of U, Th, and Ra isotopes in the Ganga–Brahmaputra river system: weathering processes and fluxes to the Bay of Bengal. *Geochim. Cosmochim. Acta* **54**, 1387–1396.

Schäfer J. and Blanc G. (2002) Relationship between ore deposits in river catchments and geochemistry of suspended particulate matter from six rivers in southwest France. *Sci. Tot. Environ.* **298**(1–3), 103–118.

Schecher W. D. and McAvoy D. C. (1991) *MINEQL + : A Chemical Equilibrium Program for Personal Computers,* user's manual ver. 2.1. Edgewater, MD, Environ. Res. Software.

Schindler P. W. and Stumm W. (1987) The surface chemistry of oxides, hydroxides and oxide minerals. In *Aquatic Surface Chemistry* (ed. W. Stumm). Wiley, New York, pp. 83–110.

Seyler P. and Boaventura G. (2001) Trace metals in the mainstem river. In *The Biogeochemistry of the Amazon Basin and its Role in a Changing World* (eds. M. McClain, R. L. Victoria, and J. E. Richey). Oxford University Press, Oxford, pp. 307–327.

Seyler P. and Boaventura G. (2002) Distribution and partition of trace elements in the Amazon basin. In *Hydrological Processes, Special Issue of International symposium on Hydrological and Geochemical Processes in Large Scale River Basins*, Nov. 15–19, 1999, Manaus, Brésil.

Shafer M. M., Overdier J. T., Hurley J. P., Armstrong J., and Webb D. (1997) The influence of dissolved organic carbon, suspended particulates, and hydrology on the concentration, partitioning and variability of trace metals in two contrasting Wisconsin watersheds (USA). *Chem. Geol.* **136**, 71–97.

Sharma M., Wasserburg G. J., Hofmann A. W., and Chakrapani G. J. (1999) Himalayan uplift and osmium isotopes in oceans and rivers. *Geochim. Cosmochim. Acta* **63**, 4005–4012.

Shiller A. M. (1997) Dissolved trace elements in the Mississippi River: seasonal, interannual, and decadal variability. *Geochim. Cosmochim. Acta* **51**(20), 4321–4330.

Shiller A. M. (2002) Seasonality of dissolved rare earth elements in the lower Mississippi River. *Geochem. Geophys. Geosys.* **3**(11), 1068.

Shiller A. M. and Boyle E. (1985) Dissolved zinc in rivers. *Nature* **317**, 49–52.

Shiller A. M. and Boyle E. (1987) Variability of dissolved trace metals in the Mississippi River. *Geochim. Cosmochim. Acta* **51**, 3273–3277.

Shiller A. M. and Boyle E. (1997) Trace elements in the Mississippi River delta outflow region: behavior at high discharge. *Geochim. Cosmochim. Acta* **55**, 3241–3251.

Shiller A. M. and Mao L. (2000) Dissolved vanadium in rivers: effects of silicate weathering. *Chem. Geol.* **165**, 13–22.

Shiller A. M. and Frilot D. M. (1995) The geochemistry of gallium relative to aluminium in Californian streams. *Geochim. Cosmochim. Acta* **60**, 1323–1328.

Sholkovitz E. R. (1978) The flocculation of dissolved Fe, Mn, Al, Cu, Ni, Co, and Cd during estuarine mixing. *Earth Planet. Sci. Lett.* **41**, 77–86.

Sholkovitz E. R. (1992) Chemical evolution of REE: fractionation between colloidal and solution phases of filtered river water. *Earth Planet. Sci. Lett.* **114**, 77–84.

Sholkovitz E. R. (1995) The aquatic chemistry of rare earth elements in rivers and estuaries. *Aquat. Chem.* **1**, 1–34.

Sholkovitz E. R. and Copland D. (1980) The coagulation, solubility and adsorption properties of Fe, Mn, Cu, Ni, Cd, Co, and humic acids in a river water. *Geochim. Cosmochim. Acta* **45**, 181–189.

Sholkovitz E. R., Boyle E. R., and Price N. B. (1978) Removal of dissolved humic acid and iron during estuarine mixing. *Earth Planet. Sci. Lett.* **40**, 130–136.

Sigg L., Behra P., and Stumm W. (2000) *Chimie des Milieux Aquatiques*, Dunod, Parris, 3rd edn.

Smedley P. L. and Kinniburg D. G. (2002) A review of the source, behavior and distribution of arsenic in natural waters. *Appl. Geochem.* **17**, 517–568.

Spivack A. J., Palmer M. R., and Edmond J. M. (1987) The sedimentary cycle of the boron isotopes. *Geochim. Cosmochim. Acta* **51**, 1939–1949.

Stevenson F. J. (1994) *Humus Chemistry: Genesis, Composition, Reactions,* 2nd edn. Wiley, New York.

Stumm W. (1993) Aquatic colloids as chemical reactants: surface structure and reactivity. *Coll. Surf.* **A73**, 1–18.

Taylor S. R. and McLennan S. M. (1985) *The Continental Crust: Its Composition and Evolution.* Blackwell, Oxford, 312p.

Tessier A., Campbell P. G. C., and Bisson M. (1979) Sequential extraction procedure for the speciation of particulate trace metals. *Anal. Chem.* **51**, 844–851.

Thurman E. M. (1985) *Organic Geochemistry of Natural Waters.* Nijhoff and Junk publishers, Dordrecht.

Tipping E. (1981) The adsorption of humic substances by iron oxides. *Geochim. Cosmochim. Acta* **45**, 191–199.

Tricca A., Stille P., Steinmann M., Kiefel B., Samuel J., and Eikenberg J. (1999) Rare earth elements and Sr and Nd isotopic compositions of dissolved and suspended loads from small river systems in the Vosges mountains (France), the river Rhien and groundwater. *Chem. Geol.* **160**, 139–158.

Turner D. R., Whitfield M., and Dickson A. G. (1981) The equilibrium speciation of dissolved components in freshwater and seawater at 25 °C and 1 atm pressure. *Geochim. Cosmochim. Acta* **45**, 855–881.

Viers J. and Wasserburg G. J. (2002) Behavior of Sm and Nd in a lateritic soil profile. *Geochim. Cosmochim. Acta* (in press).

Viers J., Dupré B., Polvé M., Schott J., Dandurand J. L., and Braun J. J. (1997) Chemical weathering in the drainage basin of a tropical watershed (Nsimi-Zoetele site, Cameroon) comparison between organic-poor and organic-rich waters. *Chem. Geol.* **140**, 181–206.

Viers J., Dupré B., Deberdt S., Braun J. J., Angeletti B., Ndam Ngoupayou J., and Michard A. (2000) Major and traces elements abundances, and strontium isotopes in the Nyong basin rivers (Cameroon) constraints on chemical weathering processes and elements transport mechanisms in humid tropical environments. *Chem. Geol.* **169**, 211–241.

Vigier N., Bourdon B., Turner S., and Allègre C. J. (2001) Erosion timescales derived from U-decay series measurements in rivers. *Earth Planet. Sci. Lett.* **193**, 549–563.

Wollery T. J. (1992) EQ3/6, A software package for geochemical modelling of aqueous systems: package overview and installation guide (ver. 7.0). UCRL-MA-110662: Part I. Lawrence Livermore Natl. Lab.

Wood S. A. (1990) The aqueous geochemistry of the rare earth elements and yttrium. *Chem. Geol.* **82**, 159–186.

Wen L. S., Santschi P. H., and Tang D. (1997) Interactions between radioactively labelled colloids and natural particles: evidence for colloidal pumping. *Geochim. Cosmochim. Acta* **61**, 2867–2878.

Yee H. S., Measures C. I., and Edmond J. M. (1987) Selenium in the tributaries of the Orinoco in Venezuela. *Nature* **326**, 686–689.

Yeghicheyan D., Carignan J., Valladon M., Bouhnik Le Coz M., Le Cornec F., Castrec Rouelle M., Robert M., Aquilina L., Aubry E., Churlaud C., Dia A., Deberdt S., Dupré B., Freydier R., Gruau G., Hénin O., de Kersabiec A. M., Macé J., Marin L., Morin N., Petitjean P., and Serrat E. (2001) A compilation of silicon and thirty one trace elements measured in the natural river water reference SLRS-4 (NRC-CNRS). *Geostand. Newslett.* **35**(2–3), 465–474.

You C. F., Lee T., and Li Y. H. (1989) The partition of Be between soil and water. *Chem. Geol.* **77**, 105–118.

Zhang J. (1994) Geochemistry of trace metals from Chinese river/estuary systems: an overview. *Estuar. Coast. Shelf Sci.* **41**, 631–658.

Zhang J. and Huang W. W. (1992) Dissolved trace metals in the Huanghe, the most turbid river large river in the world. *Water Res.* **27**(1), 1–8.

Zhang J., Huang W. W., and Wang J. H. (1993) Trace element chemistry of the Huanghe (Yellow river), China: examination of the data from *in situ* measurements and laboratory approach. *Chem. Geol.* **114**, 83–94.

Zhang C., Wang L., Zhang, S., and Li X. (1998) Geochemistry of rare earth elements in the mainstream of the Yangtze river, China. *Appl. Geochem.* **13**, 451–462.

5.10

Dissolved Organic Matter in Freshwaters

E. M. Perdue and J. D. Ritchie

Georgia Institute of Technology, Atlanta, GA, USA

5.10.1 INTRODUCTION

5.10.1.1 Terminology

Organic matter in freshwaters exists as dissolved molecules, colloids, and particles. It is appropriate to regard these distinctions as dynamic, however, because organic matter can be interconverted readily between these forms by dissolution and precipitation, sorption and desorption, aggregation and disaggregation, etc. Dissolved organic matter (DOM), the subject of this chapter, is defined operationally as the fraction of organic matter in a water sample that passes through a 0.45 μm filter. In the authors' opinion, the scientific literature on organic matter in freshwaters will be better reflected in this review, if data are considered without regard to the manner in which water samples may have been filtered. This more general approach is warranted because:

- many submicron colloids and some micro-organisms can pass through 0.45 μm filters;
- the effective pore size of a 0.45 μm filter is usually unknown, because it is decreased by partial clogging during the filtration of a water sample;
- some important studies have been conducted on unfiltered samples or on samples that were filtered through other types of filters; and
- some important studies have been conducted on samples that were concentrated with ultrafiltration (UF), nanofiltration (NF), or reverse osmosis (RO) membranes.

As methods for fractionation and isolation of organic matter in freshwaters have evolved, and as the intensity of research has waxed and waned in various academic disciplines, a rich and potentially confusing nomenclature has evolved for organic matter in freshwaters. Some of the more commonly encountered descriptors and their associated acronyms, if any, are yellow organic acids (YOAs), aquatic humus, DOM, and natural organic matter (NOM). Regardless of the terminology used in the original literature, the organic matter in freshwaters is referred to as DOM in this review, except when it is necessary to be more specific.

5.10.1.2 Analytical Measurements

Prior to the publication of its 13th edition in 1971, *Standard Methods for the Examination of Water and Wastewater*, which has been published since 1905 by the American Public Health Association and other co-sponsoring agencies, did not contain any analytical method for measurement of organic carbon in water. Since 1971, several analytical methods for measurement of organic carbon in water have been published in *Standard Methods* (e.g., Clesceri *et al.*, 1989).

Table 1 Analytical forms of carbon in natural waters.

Name	Acronym
Total carbon	TC
Inorganic carbon	IC
Total organic carbon	TOC
Dissolved organic carbon	DOC
Nondissolved organic carbon	NDOC
Purgeable organic carbon	POC
Nonpurgeable organic carbon	NPOC

All of these methods are based ultimately on quantification of the CO_2 that is formed by wet oxidation or combustion of the organic matter in a water sample.

Carbon in natural waters is classified as either inorganic or organic carbon, and the organic carbon is further classified on the basis of particle size or volatility. Depending on the particular combination of filtration, acidification, purging, and oxidation steps leading to an analytical measurement, a wide variety of fractions of carbon can be quantified. Many of the common analytical descriptors used in new editions of *Standard Methods* are summarized in Table 1. In the scientific literature, NDOC is most often referred to as particulate organic carbon (also POC), so two very different fractions of organic matter share a common acronym—a source of potential confusion and misinterpretation of the literature. Subsequently in this review, POC will only be used in reference to particulate organic carbon.

The most commonly reported analytical parameters are TOC and DOC. In some cases, TOC is determined indirectly from measured values of TC and IC. In other cases, TOC is determined directly after purging an acidified sample with an inert gas to remove IC. The two methods generally yield essentially the same result; however, if a sample contains a significant amount of purgeable organic carbon, the direct approach will underestimate TOC. When these approaches are applied to samples that have been filtered through a 0.45 μm filter, the resulting analytical parameter is DOC. In this review, no attempt has been made to distinguish between directly and indirectly measured values of TOC or DOC. The median DOC concentrations of a variety of types of freshwaters are given in Table 2.

5.10.1.3 The Major Areas of Research Interest

Since the early 1950s, the level of awareness of organic matter as an active participant in the geochemistry, environmental chemistry, and ecology of freshwaters has grown enormously. The approximately exponential growth of the published literature on organic matter in natural waters is readily demonstrated. When

Table 2 Typical concentrations of organic carbon in freshwaters.[a]

Water type	TOC (μmol L^{-1})
Groundwater	60
River	580
Oligotrophic Lake	180
Eutrophic Lake	1,000
Marsh	1,420
Bog	2,750

[a] After Thurman (1985).

Table 3 Publications on organic matter in freshwaters (1900–2002).[a]

Time period	Number of publications
1900–1959	26
1960–1969	44
1970–1979	278
1980–1989	542
1990–1999	2,221
2000–2002	2,009

[a] Search conducted in November, 2002.

Chemical Abstracts is searched for the following keywords: (i) "yellow organic acids," (ii) "aquatic humus," (iii) "dissolved organic matter," (iv) "natural organic matter," (v) "total organic carbon," and (vi) "dissolved organic carbon," the results in Table 3 are obtained. Because many publications describing organic matter in natural waters may not contain the exact keywords used in this search (e.g., "humic acids (HAs)," "fulvic acids (FAs)," "nonliving organic matter," and "organic colloids" have not been considered here), the estimates given here are undoubtedly too low. The point of this cursory search of the scientific literature is simply to demonstrate the near-exponential growth of the published literature on organic matter in natural waters, which, if continued, will lead to publication of at least 8,000 papers in the first decade of twenty-first century.

The results in Table 3 document the explosive growth in research on organic matter in freshwaters. The great increase in research activity in the 1970s is due to major discoveries that occurred almost simultaneously in several technical and scientific disciplines. Probably the most important factor was the discovery that chlorination of water causes the production of chloroform and other disinfection by-products through reactions between chlorine and humic substances in the raw water (Rook, 1974) and the nearly simultaneous report of those same halogenated compounds in human blood plasma and drinking water (Dowty *et al.*, 1975). At the same time, there was a growing body of evidence that the chemical speciation of metal cations in natural waters is affected significantly by their interactions with organic matter (e.g., Beck *et al.*, 1974; Gardiner, 1974; Reuter and Perdue, 1977). Even more interestingly, laboratory studies revealed clearly that the biological uptake and acute toxicity of trace metals are very strongly affected by their chemical speciation (e.g., Zitko *et al.*, 1973; Pagenkopf *et al.*, 1974; Sunda and Guillard, 1976). It was further recognized in the 1970s that organic matter is an essential component of microbial food webs in natural waters, because it is a growth substrate for bacteria that are then consumed by higher trophic levels (Pomeroy, 1974). During the 1970s, with a growing awareness of human impacts on global climate, Garrels *et al.* (1975) published one of the first comprehensive efforts to balance the global carbon cycle, including the riverine flux of organic carbon from the continents to the oceans.

In the first decade of the twenty-first century, research on all of these issues continues to expand rapidly. Each of these areas of research will be reviewed briefly; however, the largest part of this chapter will be devoted to a synthesis and analysis of the published literature on the chemical properties of organic matter in natural waters. Among the properties to be presented are molecular weight and/or size distributions, elemental composition, acidic functional groups, carbon distribution from ^{13}C nuclear magnetic resonance (NMR) studies, concentrations of amino acids, sugars, and lignin-derived phenols. A goal of this effort is a statistically based description of the average chemical composition (and range of variation from the average) of organic matter in freshwaters.

5.10.2 INVENTORIES AND FLUXES

Prior to the 1970s, TOC and DOC concentrations were not often measured for freshwaters. In the *Handbook of Geochemistry*, Turekian (1969) cites an earlier estimate of 3.6×10^4 km^3 yr^{-1} for the annual discharge of water from the continents to the oceans; however, he does not attempt to estimate the average organic carbon content of those waters. Table 4 contains selected estimates of the global fluxes of organic carbon (TOC, DOC, and/or POC) from the continents to the oceans. These estimates were most often obtained by scaling the average or aggregate behavior of a representative set of rivers or drainage basins to the global scale. The scaling factor is most often either the total annual discharge of water into the oceans or the total global land area being drained by rivers, so these estimates, when provided, are also included in Table 4. To facilitate intercomparisons, all originally published dimensional units have been converted to a common set of units in Table 4

Table 4 Selected estimates of global fluxes of organic carbon to the oceans.

Reference	Land area (10^6 km^2)	Discharge (km^3 yr^{-1})	TOC flux (Tmol yr^{-1})	DOC flux (Tmol yr^{-1})	POC flux (Tmol yr^{-1})
Garrels *et al.* (1975)		36,000	16.5	10.7	5.8
Kempe (1979)		37,700	15.7	10.2	5.5
Schlesinger and Melack (1981)		42,000	30.8		
Schlesinger and Melack (1981)	113.0		34.2		
Meybeck (1981, 1982)	99.9	37,400	31.5	17.9	14.9
Degens *et al.* (1991)	125.9	35,319	27.9		
Ludwig *et al.* (1996)	106.3	38,170	31.5	17.1	14.4
Aitkenhead and McDowell (2000)				30.1	
Aitkenhead and McDowell (2000)				30.2	

and in the following discussion. Global average concentrations of DOC, POC, and TOC in this section are most often calculated from the ratio of annual carbon export to annual water flux, so they are discharge-weighted average values.

5.10.2.1 Estimates of Carbon Fluxes—1970s

Garrels *et al.* (1975) estimated the annual riverine fluxes of DOC and POC to be 10.7 Tmol yr^{-1} and 5.8 Tmol yr^{-1}, respectively; however, the basis of these estimates was not stated in their paper. The corresponding riverine flux of TOC is 16.5 Tmol yr^{-1}. Assuming that these fluxes of organic carbon include the contributions of both river water and groundwater (3.2×10^4 km^3 yr^{-1} and 4,000 km^3 yr^{-1}), the corresponding global average concentrations of DOC, POC, and TOC in freshwaters discharging to the oceans would be 297 μmol L^{-1}, 161 μmol L^{-1}, and 458 μmol L^{-1}, respectively.

Working with updated data for annual discharge of rivers (3.77×10^4 km^3 yr^{-1}) from Baumgartner and Reichel (1975), Kempe (1979) revised the estimates of Garrels *et al.* (1975) slightly to obtain annual riverine fluxes of DOC, POC, and TOC of 10.2 Tmol yr^{-1}, 5.5 Tmol yr^{-1}, and 15.7 Tmol yr^{-1}, respectively. The corresponding global average concentrations of DOC, POC, and TOC are 271 μmol L^{-1}, 146 μmol L^{-1}, and 417 μmol L^{-1}, respectively.

5.10.2.2 Estimates of Carbon Fluxes—1980s

As more complete data sets for TOC and DOC appeared in the literature, more sophisticated estimates of the annual riverine fluxes of DOC and POC from the continents to the oceans were developed. Schlesinger and Melack (1981) used two approaches to estimate the annual riverine flux of organic carbon. First, they noted that, for about a dozen rivers for which data were available at that time, a log–log plot of the total annual organic carbon load versus the total annual

riverflow was highly linear. They applied this relationship to the 50 largest rivers in the world, which together account for 43% of the global total annual river flow (4.2×10^4 km^3 yr^{-1}), and they obtained an estimate of 30.8 Tmol yr^{-1} for the annual flux of TOC, which corresponds to a global average TOC concentration of 733 μmol L^{-1}. Schlesinger and Melack (1981) also noted that the annual riverine flux of organic carbon from terrestrial watersheds varies considerably with the type of ecosystem. Grouping data by 10 types of ecosystems and scaling in proportion to the global area of each type of ecosystem, they obtained a second estimate of 34.2 Tmol yr^{-1} for the annual flux of TOC from the continents to the oceans. This corresponds to a global average TOC concentration of ~814 μmol L^{-1}.

Meybeck (1981, 1982) used a similar approach. He grouped the available data by six climatic zones, scaled in proportion to the global area of each climatic zone, and obtained an estimate of 479 μmol L^{-1} for the global average DOC concentration. The corresponding annual flux of DOC from the continents to the oceans is 17.9 Tmol yr^{-1}. Meybeck also noted that the organic carbon content of particulate matter in rivers decreases with increasing concentration of total suspended solids, which in turn increases with increasing riverine discharge. He grouped data into nine classes according to the total concentration of suspended solids, estimated the average annual riverine flux of POC for each group of rivers, and scaled globally to obtain a final estimate of ~14.9 Tmol yr^{-1}, which corresponds to a global average POC concentration of 398 μmol L^{-1}. Interestingly, Meybeck (1981, 1982) produced an independent estimate of TOC flux and concentration, using the same method that he used for DOC, but applied to a much larger data set. The estimated annual riverine flux of TOC is 31.5 Tmol yr^{-1}, which corresponds to a global average TOC concentration of 842 μmol L^{-1}. These values are only slightly smaller than the corresponding values that can be obtained by summing the DOC and POC data.

5.10.2.3 Estimates of Carbon Fluxes—1990s and 2000s

Degens *et al.* (1991), in a summary of nearly a decade of research under the auspices of the Scientific Committee on Problems of the Environment (SCOPE), estimated the annual flux of TOC from each continent (except Australia) as the total discharge from the continent multiplied by the mean TOC concentration of rivers on that continent for which they present tabulated data. Summing the separate continental contributions, they estimated that the annual flux of TOC from the continents (excluding Australia) to the oceans is 27.9 Tmol yr^{-1}. If this flux is coupled with their estimated global river runoff of 3.53×10^4 km^3 yr^{-1}, the corresponding global average TOC concentration in rivers is 790 µmol L^{-1}. They advocate caution in the interpretation of such estimates and in intercomparisons with other estimates. The authors state that many earlier measurements of TOC concentrations relied on surface samples rather than depth-integrated samples and used a variety of nonstandard methods of analysis. They caution further that estimates of average riverine discharge must be interpreted with care because substantial seasonal and interannual variations in the water volume of rivers are known to occur.

Ludwig *et al.* (1996) developed empirical relationships between published carbon fluxes of a large set of major world rivers and the climatic, biologic, and geomorphologic patterns characterizing the drainage basins of those rivers, as extracted from various ecological databases. They modeled discharge in terms of average temperature, mean annual precipitation total, and basin slope, resulting in an estimated total river discharge of 3.82×10^4 km^3 yr^{-1}. They found that the flux of DOC is mainly a function of discharge, basin slope, and the carbon content of soils in the drainage basin. The estimated annual riverine flux of DOC is 17.1 Tmol yr^{-1}, which corresponds to a global average DOC concentration of 448 µmol L^{-1}. Ludwig *et al.* (1996) found that the carbon content of suspended solids is a decreasing nonlinear function of the total concentration of suspended solids, which is, in turn, a function of discharge, rainfall intensity, and basin slope. The estimated annual riverine flux of POC is 14.4 Tmol yr^{-1}, which corresponds to a global average POC concentration of 377 µmol L^{-1}. Combining the estimates for DOC and POC, the estimated annual riverine flux of TOC is 31.5 Tmol yr^{-1}, which corresponds to a global average TOC concentration of 825 µmol L^{-1}.

Aitkenhead and McDowell (2000) used annual fluxes of DOC from 164 watersheds, which were grouped into 15 biome types, to estimate mean DOC fluxes according to biome type. When soil C/N ratios from existing databases were also used to estimate mean soil C/N ratios according to biome type, they discovered a strong linear relationship ($r^2 = 0.992$, $n = 15$) between the mean annual fluxes of DOC and the mean soil C/N ratios of those biome types. They used this relationship to predict annual riverine fluxes of DOC from three terrestrial ecosystems, given only the soil C/N ratios of those ecosystems, and their predictions underestimated observed fluxes by only 3.2–4.5%. When soil C/N ratios were grouped and averaged according to the six climatic zones used by Meybeck (1981, 1982), their linear model predicted an annual riverine flux of DOC to the oceans of 30.1 Tmol yr^{-1}. When they reorganized the soil C/N ratios according to the 10 types of ecosystems used by Schlesinger and Melack (1981), they estimated an annual riverine flux of DOC to the oceans of 30.2 Tmol yr^{-1}. These estimates are nearly a factor of 2 greater than the highest prior estimated flux of DOC, but they agree quite well with prior estimated fluxes of TOC. In fact, Aitkenhead and McDowell (2000) actually compared their predictions with those of Meybeck (1981, 1982) and Schlesinger and Melack (1981), both of which were fluxes of TOC, not DOC. The authors of this review are uncertain how the estimates of Aitkenhead and McDowell (2000) should be interpreted, but they are nonetheless included in Table 4.

Overall, the earliest estimate of Garrels *et al.* (1975) and the related estimate of Kempe (1979) are much lower than all subsequent estimates. The estimates of Degens *et al.* (1991) for discharge and TOC flux, which do not include the contributions of Australia, are somewhat lower than other recent estimates. Even if the model estimates of Ludwig *et al.* (1996) for Australia are added to the estimates of Degens *et al.* (1991), their estimates of discharge and TOC flux increase to only ~3.59×10^4 km^3 yr^{-1} and 28.6 Tmol yr^{-1}, respectively. Given the fact that their data set was generated during recent focused international efforts using the state-of-the-science methodologies to measure the discharge and TOC concentrations for many of the world's major rivers, the estimates of Degens *et al.* (1991) cannot be dismissed. It thus seems quite likely that the annual riverine flux of TOC from the continents to the oceans is 28–31 Tmol yr^{-1}, which corresponds to a global average TOC concentration of 730–880 µmol L^{-1}. Based on the limited number of global estimates of DOC and POC, the global average DOC/POC ratio is ~1.20.

5.10.3 CHEMICAL AND BIOLOGICAL INTERACTIONS

Starting in the early 1950s, there has been a growing awareness of the multitude of interactions between DOM and other chemical and biological components of freshwaters. This trend is illustrated by the evolving coverage of the subject of metal–organic interactions in the first, second, and third editions of *Aquatic Chemistry* (Stumm and Morgan, 1970, 1981, 1996). These three editions contain 2, 7, and 13 index entries, respectively, for humic or fulvic acid, and the amount of discussion devoted to the interaction of DOM with metal cations increased gradually from the first to the third edition. In the first edition, the authors argued, on the basis of binding strengths of metal cations with simple monomeric ligands, that DOM is very unlikely to play an important role in complexation of trace metal cations. By the third edition, however, the authors concluded that the metal binding properties of monomeric ligands do not provide a reliable representation of the true situation, and they discussed a variety of compositional, structural, and thermodynamic reasons why metal complexation by HAs and FAs cannot be described adequately using simply monomeric ligands as models. They concluded that humic substances may be important in the regulation of some trace metals (e.g., $Hg(II)$, $Cu(II)$, and $Pb(II)$) in natural waters.

In Section 5.10.1.3, the explosive growth of research and publications dealing with DOM in freshwaters in the 1970s was attributed to the nearly simultaneous recognition that DOM is the major source of disinfection by-products during chlorination of water, that complexation with DOM significantly affects the chemical speciation of metal cations in natural waters, that biological effects of trace metal cations are correlated with their chemical speciation, and that DOM is an essential component of microbial food webs in natural waters. We will be brief, drawing heavily upon and directing the reader to more specialized published reviews of each research topic.

5.10.3.1 DOM and the Acid–Base Chemistry of Freshwaters

From the earliest studies of DOM (e.g., Shapiro, 1957), its acidic character has been noted and quantified. Beck *et al.* (1974) reported carboxyl and phenolic contents for nine isolated samples of riverine DOM. They noted that acid titration curves of low-pH waters showed no inflection points, from which they concluded that the acidic fraction of river water organic matter consists of a mixture of substances with gradational differences in dissociation constants, i.e., DOM contains a complex mixture of acidic functional groups. They noted that Gran plots (Gran, 1950) of those titration data were straight lines whose slopes deviated slightly from that which would be expected for simple dilution of the acid titrant, i.e., nonideal slopes of Gran plots are an indication that the acidic functional groups of DOM are being titrated. They also noted that, if Gran alkalinity was attributed entirely to the carbonate system, the corresponding dissolved inorganic carbon (DIC) was much greater than DIC values that were measured directly using a carbon analyzer, indicating that the acidic functional groups of DOM contribute significantly to Gran alkalinities in organic-rich river waters. Finally, they noted that the partially dissociated acidic functional groups of DOM contribute significantly to charge balance in organic-rich river waters. Cronan *et al.* (1978) also postulated that DOM was an important source of acidity in natural waters of low ionic strength, and they noted the apparent "anion deficit" in those waters, and Lee and Brosset (1978) further investigated the slopes of Gran plots, with a goal of extracting information regarding the nature of the acids being titrated.

Many experimental and modeling studies have shown that 50% of the carboxyl groups in DOM are dissociated in the pH range of 3.6–3.9 (e.g., Perdue and Lytle, 1983; Oliver *et al.*, 1983; Ephraim *et al.*, 1986; Tipping and Hurley, 1992; Koopal *et al.*, 1994). These studies and models also indicate that 34–40% of the carboxyl groups remain dissociated at pH 3—the lowest pH to which samples are usually titrated in Gran titrations. The end point of a Gran titration commonly lies in the pH range of 4.5–5.6, depending on experimental conditions (e.g., Cantrell *et al.*, 1990), so a significant portion of the carboxyl groups of DOM will remain deprotonated and thus undetected at a typical end point of an alkalinity titration.

Without any doubt, the greatest impetus since the early 1980s for the study of the acid–base properties of DOM was the large-scale research effort which was initiated in the 1980s to investigate the impact of acidic precipitation on terrestrial and aquatic ecosystems (e.g., Linthurst *et al.*, 1986; Messer *et al.*, 1986; Herlihy *et al.*, 1991; Munson and Gherini, 1993). Long-term decreases in pH and alkalinity in many lakes and rivers were attributed to increased rates of input of natural and/or anthropogenic acids. Detailed chemical analyses were conducted on a large number of low-alkalinity natural waters, nearly always using Gran plots to analyze titration data to determine alkalinity. In general, the effect of DOM was most clearly manifested in the frequently observed excess of cationic charge and in the observation that Gran alkalinities are greater than carbonate alkalinity (calculated from pH and DIC). These observations are consistent with inclusion of some, but not all, of the organic charge from carboxyl groups of DOM in the measured Gran alkalinities. Many papers have been published in which the undetected organic "strong acidity" has been either correlated with the concentration of DOC or estimated using empirical models of the acid–base chemistry of DOM. Cantrell *et al.* (1990) showed that the Gran ANC error (or strong organic acidity) is actually the residual ANC at the end point of an alkalinity titration, and that this quantity ($\sim$4.5 μeq (mg C)$^{-1}$) is slightly smaller than the actual concentration of undetected organic charge ($\sim$5.5 μeq (mg C)$^{-1}$) when Gran titrations are conducted down to pH 3. From the median carboxyl content and percent carbon of DOM (see Sections 5.10.4.5.3 and 5.10.4.4.1, respectively), it follows that $\sim$55% of the carboxyl groups of DOM will be undetected in a Gran titration.

In addition to papers already cited, the papers of Henriksen and Seip (1980), Molværsmyr and Lund (1983), Barnard and Bisogni (1985), Eshleman and Hemond (1985), Driscoll *et al.* (1989), Kahl *et al.* (1989), Sullivan *et al.* (1989), Hemond (1990), Hongve (1990), Grzyb (1995), and Köhler *et al.* (2001) are recommended for further information on the role of DOM in the acid–base chemistry of freshwaters. Further general discussion of the role of organic acids in aquatic ecosystems may be found in Perdue and Gjessing (1990), and Ritchie and Perdue (2003) have recently summarized the acid–base properties of HAs, FAs, and NOM from both aquatic and terrestrial environments. The published literature for the concentrations of carboxyl and phenolic groups in DOM is summarized in Section 5.10.4.5 of this chapter.

5.10.3.2 DOM and UV/Visible Radiation

Some of the molecules of which DOM is comprised contain chromophores and fluorophores, through which DOM interacts with UV and visible radiation. Absorbed photons transfer energy to DOM, causing the molecules in DOM which contain chromophores or fluorophores to be excited to higher electronic energy states. The absorbed energy can be utilized in photochemical reactions of DOM, transferred to other constituents of natural waters, or released to the environment as heat (absorbance) or longer-wavelength radiation (fluorescence). Excellent general reviews of this subject have been written by Blough and Green (1995) and by Miller (1998).

DOM contains many nonidentical chromophores, whose collective attenuation of incident UV and visible light causes the absorbance of DOM to decrease with increasing wavelength. The most common expression of this relationship is the E_4/E_6 ratio, which is simply the ratio of absorbances at 465 nm and 665 nm. Samples containing a larger proportion of chromophores that absorb at longer wavelengths (smaller E_4/E_6 ratios) are believed to have higher average molecular weights and to contain a higher proportion of HAs (e.g., de Haan and de Boer, 1987; de Haan, 1993). Chin *et al.* (1994) found that both aromaticities and weight-average molecular weights of humic substances correlated much more strongly with number-average molar absorptivities at 280 nm than with E_4/E_6 ratios. Both Korshin *et al.* (1996) and Kitis *et al.* (2001a) reported that UV absorbance at 254 nm correlates very well with trihalomethane formation potential (THMFP) (see Section 5.10.3.6). For DOM and related materials, the decrease of absorbance with increasing wavelength of incident light can be closely approximated by a simple exponential function (Zepp and Schlotzhauer, 1981; Bricaud *et al.*, 1981). Blough and Green (1995) summarized many similar observations and linked this exponential relationship to the familiar E_4/E_6 ratio. The papers of Davies-Colley and Vant (1987), Edwards and Cresser (1987), Battin (1998), and Lean (1998) are recommended for additional discussion of the absorbance spectrum of DOM and its correlation with the TOC concentrations of natural waters.

As stated previously, DOM also contains a variety of fluorophores. In general, DOM and related materials absorb UV and visible light over a wavelength range of $\sim$300–500 nm and emit fluorescent light at wavelengths somewhat longer than that of the incident light (e.g., Senesi *et al.*, 1989). Fluorescence spectra are much more highly structured than absorbance spectra, often exhibiting several distinct maxima of fluorescence intensity with respect to both excitation and emission wavelength. The state of the science in this field is the acquisition and interpretation of total luminescence spectra, in which emission spectra are recorded over a wide range of wavelengths as the excitation wavelength is gradually scanned over the desired range (e.g., Goldberg and Weiner, 1989; Coble, 1996; Mobed *et al.*, 1996; McKnight *et al.*, 2001). Because DOM absorbs light at both the excitation and emission wavelengths, raw data must be corrected for inner filtering effects. The work of Mobed *et al.* (1996) is very highly recommended for a rigorous examination of total luminescence spectra of DOM and related materials and for a clear illustration of the necessity for basing interpretations of the solution chemistry of DOM on corrected spectra. Hemmingsen and McGown (1997) have extended this work, using phase-resolved fluorescence intensity to explore the fluorescent lifetimes of the complex mixture of fluorophores in HAs.

Marhaba *et al.* (2000) have measured the total luminescence spectra of the six fractions of DOM that can be obtained using the fractionation procedure of Leenheer and Huffman (1976), which is described in Section 5.10.4.2.1.

The role of DOM in aquatic photochemistry is discussed in general reviews by Brezonik (1994), Morel and Hering (1993), and Stumm and Morgan (1996). DOM may undergo direct photolysis, in which its electronically excited constituents undergo some kind of chemical reaction(s). Alternatively, DOM may act as a photosensitizing agent by transferring energy from its electronically excited constituents to other components of natural waters. When a solution of DOM is exposed to solar radiation, its color fades gradually in a process known as photobleaching (e.g., Grzybowski, 2000; Osburn *et al.*, 2001). During photobleaching of DOM, inorganic carbon (CO_2 and CO) and inorganic nitrogen (NH_4^+) are formed (Morris and Hargreaves, 1997; Zuo and Jones, 1997; Gao and Zepp, 1998). Morris and Hargreaves (1997) also reported that photobleaching can destroy the chromophores in DOM without complete mineralization to inorganic carbon. Several studies (e.g., Miles and Brezonik, 1981; Gao and Zepp, 1998) have demonstrated clearly that iron plays a key role in photobleaching reactions. When photobleaching of DOM takes place, the bioavailability of DOM is increased (see Section 5.10.3.5). This topic has been reviewed by Moran and Zepp (1997). Detailed reviews of the aquatic photochemistry of DOM by Zepp (1988), Cooper *et al.* (1989), and Hoigné *et al.* (1989) are highly recommended for further insight into this aspect of the geochemistry of DOM in freshwaters.

5.10.3.3 DOM and Chemical Speciation of Trace Metal Cations

Even though very few publications dealing with DOM were published prior to 1970 (see Table 3 and related discussion), there was a growing awareness that the chemistry of some trace metal cations was somehow influenced by DOM. Early evidence consisted mainly of correlations between the concentrations of metal ions and DOM. For example, Shapiro (1964, 1966) and Lamar (1968) examined the relationship between organic color and iron in natural surface waters. Beck *et al.* (1974) and Perdue *et al.* (1976) reported that the concentrations of both iron and aluminum were linearly correlated with the concentration of DOM in an organic-rich river, and they presented evidence for competition between the two metals for binding sites in DOM. Benes *et al.* (1976) also investigated the relationship between aluminum and DOM, and Driscoll (1980) and Johnson *et al.* (1981) also

observed linear relationships between the concentrations of aluminum and DOM in freshwaters.

The relationships that were implied by these empirical correlations were substantiated when analytical methods that could distinguish between inorganic and organically complexed forms of trace metals were applied to natural waters. Ion selective electrodes (ISEs) can be used to distinguish between a free metal cation and its complexed forms (e.g., Smith and Manahan, 1973; Gardiner, 1974; Buffle *et al.*, 1977; Turner *et al.*, 1986; Cabaniss and Shuman, 1988a,b); however, the typical detection limit of $\sim 10^{-6}$ M has limited the direct application of this method to natural waters. For this reason, most studies in which ISEs have been used are conducted on reconstituted solutions, not on unmodified natural waters. Recent advances in the fabrication and performance of ISEs (Labuda *et al.*, 1994; Sokalski *et al.*, 1997; Bakker *et al.*, 2001) should lead to more frequent application of this method to natural waters.

Metal cations that have more than one accessible oxidation state in aqueous solution can be analyzed by anodic stripping voltammetry to distinguish between labile and nonlabile forms of the metal cation (Shuman and Woodward, 1973; Chau and Lum-Shue-Chan, 1974; Chau *et al.*, 1974; Xue and Sigg, 1994, 1998). The method is far more sensitive than conventional ISE measurements, but "labile" metal includes both the free metal ion and some portion of the organically complexed metal. When competing ligands are used, it is possible to back-calculate the chemical speciation of the metal cation in the original water (e.g., Xue and Sigg, 1998). Similar information can be obtained using cation-exchange methods, and cation-exchange methods have the added advantage of being applicable to metal cations that cannot be oxidized or reduced in aqueous solution (Driscoll, 1984; Backes and Tipping, 1987; Brown *et al.*, 1999). Cathodic stripping voltammetry can also be used to distinguish between labile and nonlabile forms of a metal (van den Berg, 1984a,b; Xue and Sigg, 1993; Qian *et al.*, 1998). This method has excellent sensitivity, is applicable if either the metal or the competing ligand can be reduced at the electrode, and the chemical speciation of the metal cation in the original water can be back-calculated.

A powerful, but less frequently used, method for studying the complexation of trace metal cations by DOM is lanthanide ion probe spectroscopy (Dobbs *et al.*, 1989). In this method, fluorescence intensities of Eu(III) at two emission wavelengths are used to determine the proportions of complexed and free Eu(III). When a competing metal cation is added to the solution, its complexation by DOM causes the fluorescence intensity of free Eu(III) to increase and that of complexed Eu(III) to decrease, from which the extent of complexation of the competing metal cation can be determined. The method is best suited for laboratory measurements on reconstituted samples.

In several studies (e.g., Larive *et al.*, 1996; Li *et al.*, 1998; Otto *et al.*, 2001a,b), [113]Cd NMR spectrometry has been shown to provide far greater insight into the nature of the ligands to which Cd(II) is bound in Cd–DOM complexes and the rates of exchange reactions between the free and various complexed forms of Cd(II). It has been shown clearly that chemical exchange is quite fast at acidic-to-neutral pH and that it becomes much slower at alkaline pH. It has also been shown that Cd(II) is bound almost exclusively to oxygen donor atoms under conditions of fast exchange and that nitrogen donor atoms become much more obviously involved in cadmium binding at higher pH. As this method develops further, it is anticipated that far more information will be forthcoming concerning the factors that control the distribution and exchange of Cd(II) among the complex mixture of binding sites in DOM. Detection limits are still very poor for NMR-based methods, so this technique can only be used with reconstituted solutions containing relatively high concentrations (10^{-4} M) of Cd(II).

In field studies of metal binding by DOM, experimental conditions are established by the chemical matrix of the natural sample, and the major challenges are (i) measuring the total concentration of a trace metal ion, (ii) distinguishing between free and complexed forms of the metal ion, and (iii) conducting all phases of the study using clean techniques to avoid contamination of the sample (Ahlers *et al.*, 1990; Benoit, 1994). In laboratory studies, metal binding by DOM can be studied over a range of experimental conditions. Both field and laboratory studies often culminate in an effort to summarize the acquired knowledge of metal binding by DOM through the development of chemical speciation models, the best of which consider competitive effects of H^+, other metal ions, and other ligands (e.g., Cabaniss and Shuman, 1988a,b; Susetyo *et al.*, 1991; Tipping and Hurley, 1992; Koopal *et al.*, 1994). These models have reached a substantial level of sophistication, and their predictive capabilities are steadily improving.

Three modern and reasonably successful models of metal–DOM interactions are the competitive Gaussian distribution model (Susetyo *et al.*, 1991), models V and VI (Tipping and Hurley, 1992; Tipping, 1994, 1998), and the family of nonideal competitive adsorption (NICA) models (Koopal *et al.*, 1994; Benedetti *et al.*, 1995; Benedetti *et al.*, 1996; Milne *et al.*, 2001). It is generally understood today that DOM contains many nonidentical binding sites, so the central modeling challenge is to describe quantitatively the relative concentrations and strengths of these binding sites. All three classes of models assume that DOM contains two broad distributions of binding sites, presumably attributable mainly to carboxyl and phenolic hydroxyl groups. All three classes of models use distributions of proton binding sites that are symmetrically distributed around average log K values, and all three classes of models use some kind of range parameter to control the width of the distribution of log K values for proton binding. The three classes of models thus describe the shape and size of a distribution of binding sites with three parameters (total concentration, average log K, range parameter). The models use somewhat different methods to generate the basic distributions of proton and metal binding sites, and they also differ significantly in the treatment of the effect of ionic strength on cation binding.

Since the early 1970s, amazing advances have been seen in analytical methods for measuring and modeling the effect of DOM on the distribution of free and/or complexed forms of many metal ions in natural waters. The body of scientific evidence today indicates that Al(III), Fe(III), Cu(II), Hg(II), Pb(II), and Ag(I) are strongly complexed by DOM in many freshwaters. The degree of complexation of Zn(II), Cd(II), and Ni(II) by DOM is sensitive to competition with major cations such as Ca^{2+} and Mg^{2+}, so these metals are likely to be more completely bound by DOM in relatively soft waters. For most other trace metal cations, complexation by DOM is relatively less important. For further general reading on metal speciation in natural waters, including metal complexation by DOM and related materials, the reader is referred to Warren and Haake (2001) and the edited volumes by Christman and Gjessing (1983), Bernhard *et al.* (1986), Kramer and Allen (1988), Allen *et al.* (1993, 1998), and Taillefert and Rozan (2002), and the book by Buffle (1988).

5.10.3.4 DOM and Biological Activity of Trace Metal Cations

In the early 1970s, Zitko *et al.* (1973) noted that HAs reduced the activity of the free Cu^{2+} ion, as measured by the copper ion-selective electrode, and that this effect was quite well correlated with the observed toxicity of Cu(II) to salmon. Soon thereafter, Pagenkopf *et al.* (1974) also reported on the effect of complexation on the toxicity of copper to fishes. Sunda and Guillard (1976) demonstrated that the activity of the free Cu^{2+} ion in synthetic growth media containing the metal complexing agent TRIS (2-amino-2-hydroxymethyl-1,3-propanediol) was an excellent predictor of both cellular uptake of copper and the toxicity of copper to two species of phytoplankton. Sunda and Lewis (1978) and Sunda et al. (1978) extended this research to natural organic ligands and their effect on the toxicity of copper to phytoplankton and the toxicity of cadmium to grass shrimp. A few years later, Rueter and Morel (1981) modeled the competitive effects of Cu(II) and Zn(II) on the uptake of silicon by diatoms using a chemical equilibrium approach. Petersen (1982) demonstrated that the joint effects of Cu(II) and Zn(II) on the growth rate of a freshwater alga could be described by a chemical equilibrium model in which both uptake and toxicity were expressed as functions of the concentrations of free Cu^{2+} and Zn^{2+} ions. Formal statements of the chemical equilibrium hypothesis for binding of metal cations to biological surfaces followed soon thereafter (Morel, 1983; Pagenkopf, 1983).

The toxicity of a metal cation to biota is generally determined using bioassay experiments, in which progressively higher concentrations of the metal cation are added to the solution where the test organism is being grown until some percentage of the test organisms (e.g., 50%) are killed. In virtually all such studies, even those that consider the chemical speciation of the metal cation, it is assumed implicitly that the added metal cation is the actual toxicant. The chemical equilibrium approach adds a new twist to the interpretation of such experiments—one that is very often overlooked. Whenever any metal cation is added to a solution containing complexing ligands and other competing metal cations, the activities of *all* competing cations are increased to some extent. This raises the possibility that the biological response is actually due to changes in the chemical speciation of some metal cation other than the cation whose toxicity is under investigation (e.g., Morel, 1983). For example, Rueter et al. (1987) showed that the biological response in bioassays of the toxicity of Al(III) to a freshwater alga was actually caused by Cu(II) cations which were being displaced from complexing ligands by the added Al(III). In a related example of the same chemical principle, Harrison and Morel (1983) demonstrated that additions of the non-nutrient metal Cd(II) stimulated growth of a marine diatom through competitive displacement of Fe(III), which was a growth-limiting nutrient in that study.

The growing body of scientific evidence that DOM affects both the chemical speciation of metal cations and their toxicities to aquatic biota has regulatory implications (Allen and Hansen, 1996; Bergman and Dorward-King, 1997). Building on the early models of Morel (1983) and Pagenkopf (1983) and the more recent research and modeling by Playle et al. (1993a,b), the newest and most comprehensive model linking metal speciation, DOM, and toxicity to aquatic biota has emerged (Di Toro et al., 2001; Santore et al., 2001). The so-called biotic ligand model (BLM) merges a chemical model of metal binding to biological surfaces with an existing chemical speciation model that was developed initially to describe metal–DOM complexation in natural waters (Tipping, 1994). A recent comprehensive review and summary of the state of the science in modeling metal speciation as it pertains to toxicity (Gorsuch et al., 2002) is highly recommended for further information on this important topic.

5.10.3.5 Bioavailability of DOM

One of the oldest and most widely applied methods for estimation of the bioavailability of DOM in freshwaters is the measurement of biological oxygen demand (BOD). This method is routinely applied by water treatment facilities to characterize the possible impact of their effluents on the concentration of dissolved oxygen in streams receiving those effluents. BOD is also one of the general water quality parameters that are widely reported in routine surveys of water quality in freshwaters. When combined with measurements of TOC and chemical oxygen demand (COD), BOD measurements provide some insight into the degree to which the DOM in a freshwater can be utilized by heterotrophic organisms in those waters. Because BOD reflects the heterotrophic activity of a population of microorganisms, the results of the analysis may depend on the source and history of that population.

A useful paradigm in the discipline of riverine ecology is the "river continuum concept" (Vannote et al., 1980), in which it is postulated that the chemical diversity of soluble organic compounds will be greatest in first-order streams and will decrease steadily along the continuum to higher-order streams. It is further postulated that downstream communities of heterotrophs are structured to capitalize on the inefficient utilization of DOM in upstream reaches of a riverine ecosystem. These concepts lead naturally to the hypothesis that the bioavailability of DOM should generally decrease from the headwaters of a stream to its ultimate confluence with a larger body of water. This hypothesis has been tested directly on filter-sterilized river water samples (Leff and Meyer, 1991) and on reconstituted samples of isolated, desalted DOM (Sun et al., 1997). In both studies, bioavailability was expressed in terms of the maximum increase of microbial biomass, normalized to the concentration of DOC in a water sample. In both studies, bioavailability was found to decrease from headwaters to downstream sites, and blackwater tributaries contained the least bioavailable DOM.

Sun *et al.* (1997) demonstrated that the bioavailability of a DOM sample was a function of its bulk elemental composition, increasing with increasing H/C and N/C ratios and decreasing with increasing O/C ratio. It was shown further by Sun *et al.* (1997) that downstream trends in bioavailability and elemental composition were both consistent with selective utilization of aliphatic components of DOM. Vallino *et al.* (1996) developed a simple model that relates bioavailability of DOM to the average degree of reduction of the DOM. The degree of reduction, which is calculated from bulk elemental composition, increases with increasing H/C ratio and decreases with increasing O/C and N/C ratios. Hopkinson *et al.* (1998) found that these two approaches, and several others as well, successfully ranked the bioavailabilities of DOM samples from a number of US rivers. Hunt *et al.* (2000) measured bioavailabilities of four isolated humic substances, and they found that the increase in bacterial cell counts was a linear function of the N/C ratios of the humic substances. They also presented evidence that bacterial growth was nitrogen limited in their experiments—a possibility that was excluded in the earlier study by Sun *et al.* (1997), in which the defined culture medium contained ~20 μmol L^{-1} each of NH$_4^+$ and NO$_3^-$.

As described briefly in Section 5.10.3.2, photobleaching of DOM not only forms CO_2 and CO, but also renders the residual DOM more bioavailable (e.g., Moran and Zepp, 1997). The general subject of bioavailability of DOM has been reviewed thoroughly in two recent books on DOM in aquatic environments (Hessen and Tranvik, 1998; Findlay and Sinsabaugh, 2003). Each of these highly recommended books contains several up-to-date chapters devoted to the subject of bioavailability of DOM.

5.10.3.6 DOM and Disinfection By-products

Since the discovery by Rook (1974) that chloroform and other halogenated organic compounds are generated when natural waters are disinfected with chlorine, the reactions of DOM with chlorine and other disinfectants have been studied extensively. Many disinfection by-products have now been identified (e.g., Christman *et al.*, 1983; Kronberg *et al.*, 1991; Becher *et al.*, 1992), their concentrations in municipal water systems have been measured (Krasner *et al.*, 1989), THMFP has been measured for many sources of DOM (Jolley *et al.*, 1983; Collins *et al.*, 1986; Reckhow *et al.*, 1990; Owen *et al.*, 1993; Krasner *et al.*, 1996; Minear and Amy, 1996), and adverse effects on health and reproduction have been investigated (Graves *et al.*, 2001).

The methods used most often to fractionate DOM (see Section 5.10.4.2) have been used to obtain fractions of DOM whose reactivities with chlorine and other disinfectants have been studied to identify those fractions whose removal during pretreatment should be further optimized. Noot *et al.* (1989) reviewed published studies on the Ames mutagenicity of various DOM samples and found that Ames mutagenicity varied with isolation method, fractionation method, and solvents used to extract DOM from source materials. According to Noot *et al.* (1989), hydrophobic acid and hydrophobic neutral fractions were most active to the Ames test. Collins *et al.* (1986), Watt *et al.* (1996), and Li *et al.* (2001) observed that THMFP tends to be greatest for hydrophobic acids and least for hydrophilic fractions of DOM. Singer and Chang (1989), Reckhow *et al.* (1990), and Watt *et al.* (1996) observed that HAs have greater THMFP than do FAs. Stevens *et al.* (1989), Reckhow *et al.* (1990), and Becher *et al.* (1992) observed that the yield of disinfection by-products decreases with increasing pH.

It is now very well established that DOM is the major source of trihalomethanes and other disinfection by-products in disinfected water. In fact, the measurement of THMFP is now a routine monitoring task in the water treatment industry, and suppliers in the US are required to advise consumers of the concentrations of trihalomethanes and other disinfection by-products in drinking water. Efforts to remove DOM from waters before they are chlorinated have driven much of the research that has led to advances in membrane-based methods of isolation of DOM from water (see the discussion of UF, NF, etc., in Section 5.10.4.2.2). Nikolaou and Lekkas (2001) have recently reviewed many aspects of the reactions of DOM with chlorine and other disinfectants. They review the relationships between reactivity of DOM (i.e., formation of disinfection by-products) and the chemical properties of DOM and several types of fractions of DOM. They also discuss the formation and potentially adverse effects of several classes of disinfection by-products. Urbansky and Magnuson (2002) have reviewed the subject of disinfection by-products, including a brief discussion of DOM. Both of these reviews are recommended for further up-to-date details on the role of DOM in the formation of disinfection by-products.

5.10.4 CHEMICAL PROPERTIES

DOM in freshwaters is a complex mixture of organic compounds, each of which has a unique elemental composition, molecular weight, and set of chemical and physical properties.

The observed properties of DOM are reasonably expected to be some kind of weighted average of the properties of the individual compounds of which DOM is comprised. Ideally, DOM could be studied at this level—the level of individual compounds. Practically, the mixture is too complex for this approach, so DOM is studied at the level of elemental composition and at the level of functional groups and structural subunits, the latter of which include the amino acids, sugars, and lignin-derived phenols that can be liberated from DOM by hydrolysis or oxidative degradation.

This section of the review of DOM in freshwaters will examine its isolation by XAD resins and by membranes, the molecular weight distribution of DOM, its elemental composition, its acidic functional groups, its distribution of carbon among structural subunits, and the low-molecular-weight molecules (amino acids, sugars, and lignin-derived phenols) that are liberated from DOM by hydrolysis or oxidative degradation. For each of these subtopics, a statistical summary of published data will be presented.

The statistical summary for each topic will take the form of a multipart figure consisting of three-to-four main sections. The top section of each figure will contain one or more graphs in which the cumulative percentage of observations is plotted on the *y*-axis and the experimental range of the parameter(s) of interest is/are plotted on the *x*-axis. Just below the graph(s) is a table that provides a routine statistical analysis (number of observations, range of values, median, mean, SD, etc.) for the data that are plotted in the upper part of the figure. Below the table of statistics is a highly compressed list of references from which the raw data were obtained. Most of these references are not cited in the text of this review and are not included at the end of the chapter; however, they are provided in each figure so that the analysis presented here can be reconstructed and extended in the future. The optional section of each figure following the condensed list of references will be used for footnotes pertaining to the figure. The authors of this review have attempted to summarize enough of the published data to obtain representative statistics, but it is probable that some important data have been overlooked.

5.10.4.1 Shapiro's Yellow Organic Acids

Before reviewing the developments since the 1950s in our understanding of the chemical properties of organic matter in freshwaters, it is worthwhile to summarize briefly the seminal paper by Shapiro (1957) on the YOAs of lake water. This paper probably provided a more

detailed chemical description of organic matter in freshwaters than all prior research on these materials. The ensuing half century of research has seen the development of a wide array of powerful analytical, spectroscopic, and computational tools, the publication of thousands of papers, and the expenditure of many millions of dollars in this field of research. The paper by Shapiro (1957) therefore serves as a kind of "time capsule" for the experimental and conceptual state of the science at that time and as a benchmark against which the scientific progress up until the early 2000s can be judged.

Using a combination of evaporation, drying, dissolution in organic solvents, etc., Shapiro (1957) isolated ~50% of the organic carbon from lake water in the form of a freeze-dried solid product containing only 1.4% inorganic ash. Working with this material, he conducted a wide variety of experiments, from which he produced the following results and conclusions for the YOAs in lake water:

- the elemental composition is: 54.56% carbon, 5.64% hydrogen, 39.12% oxygen, and 1.23% nitrogen;
- the average molecular weight from isothermal distillation in CH_3OH is 456 Da;
- the average empirical formula is $C_{50}H_{62}O_{27}N$, but a formula of $C_{21}H_{26}O_{11}$ is more consistent with the average molecular weight (assuming that nitrogen is an impurity);
- the equivalent weight is 228 Da (for which the corresponding carboxyl content is 4.39 meq g^{-1});
- if nitrogen is an impurity, then, on average, YOAs are dicarboxylic acids (otherwise, the empirical formula contains approximately five carboxyl groups);
- UV–visible absorbance varies reversibly with pH, increasing as pH increases.
- at constant pH, the absorbances of solutions obey Beer's law up to a concentration of at least 70 mg L^{-1};
- infrared spectra of the YOAs and their methylated derivatives give direct confirmation of the presence of carboxylic acids and suggest that additional OH groups are also present;
- when excited with long-wavelength UV light, the YOAs produce a greenish-yellow fluorescence, and some chromatographic fractions exhibit orange-colored fluorescence;
- X-ray diffraction measurements indicate that the YOAs are amorphous materials;
- a variety of colorimetric tests, reactions with bromine water, and catalytic hydrogenation give contradictory evidence for phenols and/or enols, indicating that aromatic structures are of relatively minor importance, if they are present at all;

- chromatographic separations indicate that YOAs are a potentially complex mixture, whose chromatographic zones probably represent variants of a single compound or small number of related compounds;
- YOAs in lake water are most similar to soil extracts, indicating that soil is their most likely source; however, it is probable that YOAs will be found wherever vegetation decomposes;
- the stimulation of algal growth by additions of $5-50$ mg L^{-1} of YOAs to growth media is attributed to increased bioavailability of iron, which results from stabilization of aqueous solutions of Fe(III) for up to four weeks without any precipitation of Fe(III) hydroxide; and
- oxidation of the YOAs in refluxing acidic $KMnO_4$ for 2 h destroys color and fluorescence, but it does not destroy the infrared absorption of this material, from which it is concluded that at least part of this material may be difficult for organisms to utilize and that caution must be exercised in assessing DOM by wet oxidation methods.

Since the 1950s, isolation and fractionation methods have advanced dramatically, and very sophisticated spectroscopic methods have been used to probe the composition and structure of organic matter in freshwaters. Even so, many of the observations and conclusions of Shapiro (1957) have survived the test of time.

5.10.4.2 Isolation and Fractionation of DOM

Whether studying DOM at the level of elemental composition or at the level of functional groups and structural subunits, it is clearly advantageous to be able to isolate DOM from water and to remove any inorganic solutes that were present in the water. A variety of methods have been used for this purpose, among which the most common are methods in which DOM is extracted from water by adsorption on XAD resins and methods in which water is removed from DOM using UF, NF, or RO.

5.10.4.2.1 Isolation and fractionation by XAD resins

The XAD resins (Rohm and Haas, Philadelphia, PA, USA) are a series of nonionic macroporous copolymers that differ in pore size, surface area, and polarity (Aiken, 1988). Their generally large specific surface areas and more-or-less reversible adsorption of organic solutes from aqueous solution have made them well suited for isolation of selected fractions of DOM from freshwaters.

The two most widely used XAD resins are XAD-2 and XAD-8, which are a styrene divinylbenzene and an acrylic ester, respectively. XAD-8 is the more polar of these two resins. When filtered freshwater is acidified to pH 2 and passed through a column of XAD-2 or XAD-8, virtually all color and a significant fraction of the DOC are adsorbed on the resin, thus concentrating the organic matter and isolating it from more polar organic and inorganic constituents of the water. The adsorbed organic matter can be eluted from the resin at high pH and further purified, concentrated, and freeze-dried, according to the needs of the researcher.

XAD resins are commonly used to isolate humic substances from natural waters. Even though recoveries of DOM using XAD resins are not especially high, much that is known (or thought to be known) about the role of DOM in many environmental processes has been derived from studies on DOM samples that were isolated and/or fractionated using XAD resins. This method was developed initially by Weber and Wilson (1975), who used XAD-2 resin to isolate humic materials from a blackwater pond and river, and Mantoura and Riley (1975), who examined the operational characteristics of the adsorption and desorption of humic materials with XAD-2 resin. Leenheer and Huffman (1976) extended the basic isolation method using XAD resins to develop a more comprehensive fractionation scheme for classification of organic solutes in water. Their scheme employed XAD-2 and XAD-8 resins, cation- and anion-exchange resins, and aqueous HCl and NaOH solutions to separate DOM into hydrophilic and hydrophobic fractions, each of which was further separated into acidic, basic, and neutral fractions. Their scheme was intended only to classify DOM according to the relative proportions of these six fractions (expressed in terms of DOC), not to isolate freeze-dried products for further analysis and characterization. The approach was limited to some extent by irreversible adsorption of portions of the DOM on the adsorbents that were used. Leenheer (1981) extended the original classification scheme to the preparative scale, by which significant quantities of the individual fractions could be isolated for further analysis and research.

Thurman and Malcolm (1981) proposed a large-scale preparative method for isolation of aquatic humic substances, in which XAD-8 resin is used to adsorb humic substances from acidified water samples. The aquatic humic substances are then back-eluted with NaOH, and fractionated into aquatic HAs (insoluble at pH 1) and aquatic FAs (soluble at pH 1). The two fractions are further purified to remove inorganic solutes and freeze-dried separately.

Many authors have used this approach to isolate HAs and FAs from freshwaters. Malcolm and MacCarthy (1992) extended the method of Thurman and Malcolm (1981) by passing the effluent from the XAD-8 column through a column of XAD-4 resin to collect an additional fraction of DOM, which they termed the "XAD-4 acids." They found that only ~50% of the XAD-4 acids could be desorbed with dilute base solutions. Since the mid-1990s, this tandem two-resin procedure is the most commonly employed resin-based method of isolation of DOM from freshwaters.

The statistical summary in Figure 1 is organized in terms of the classification scheme of Leenheer and Huffman (1976), rather than in terms of HAs, FAs, and XAD-4 acids. Data for FAs and HAs are incorporated into the data set for hydrophobic acids, and data for XAD-4 acids are incorporated into the data set for hydrophilic acids. The statistical summary in Figure 1 is presented as percentages of DOC, because original data were generally expressed on that basis. It will be assumed in this discussion that all six fractions of DOM have similar carbon contents. It is evident that hydrophobic bases are rarely reported and, when they are measured,

they account for only ~1% of DOM. Median values for the percentage of hydrophobic neutral compounds, hydrophilic bases, and hydrophilic neutral compounds in DOM from freshwaters are 6%, 5%, and 6%, respectively. Thus, the basic and neutral fractions account collectively for 18% of DOM. The remaining DOM is distributed in a 2:1 ratio between hydrophobic acids (56%) and hydrophilic acids (28%). The publications of Aiken et al. (1979, 1992), Aiken (1985, 1988), Malcolm et al. (1989), Boerschke et al. (1996), and Knulst et al. (1997) are recommended for further details on the use of XAD resins for isolation and fractionation of DOM in freshwaters.

If data for the 31 FAs that are included in Figure 1 are processed separately, the median, mean, and SD for the percent of FAs in DOM are 46%, 46%, and 13%, respectively. If data for the 21 HAs in Figure 1 are processed separately, the corresponding values are 14%, 13%, and 5%, respectively. On average, the ratio of FAs to HAs is ~3:1 in freshwaters. The sum of the median percentages of HAs and FAs in DOM from freshwaters is 60%, which is in good agreement with the overall median percentage (56%) of hydrophobic acids in DOM from freshwaters (see Figure 1).

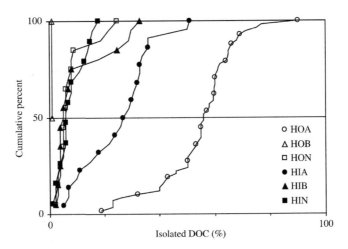

Fraction	Obs.	Isolated DOC (%)			
		Range	Median	Mean	SD
Hydrophobic acids (HOA)	58	19.0–90.0	56.0	54.4	13.8
Hydrophobic bases (HOB)	2	1.0–1.0	1.0	1.0	0.0
Hydrophobic neutrals (HON)	20	2.0–25.0	6.0	7.8	6.5
Hydrophilic acids (HIA)	22	5.0–51.0	28.0	25.8	12.9
Hydrophilic bases (HIB)	20	2.0–33.0	5.0	10.3	10.4
Hydrophilic neutrals (HIN)	19	1.0–18.0	6.0	7.9	5.1

References. *Ann. Chim.*: **74**: 257–267. *Can. J. Chem.*: **74**: 2460–2470. *Environ. Int.*: **18**: 597–607; **18**: 621–629; **24**: 251–255. *Limnol. Oceanogr.*: **31**: 739–754; **45**: 1088–1096. *Org. Geochem.*: **4**: 27–35; **18**: 563–573. *Wat. Res.*: **36**: 2357–2371. *Wat. Sup.*: **11**: 79–90. *In* Averett et al. (1989): 13–19. *In* Aiken et al. (1985): 363–385.

Figure 1 Recoveries of organic carbon from freshwaters by the XAD resin methods.

5.10.4.2.2 Isolation and fractionation by membranes

Both organic and inorganic solutes can be concentrated from natural waters using UF, NF, or RO membranes. In all three methods, a *feed solution* consisting of water and aqueous solutes is placed under pressure and passed across a semipermeable membrane, where the feed solution is separated into a *permeate solution* (relatively lower concentrations of solutes) and a *retentate solution* (relatively higher concentrations of solutes). As the feed solution is processed, the retentate solution is recycled back to the sample reservoir and the permeate solution is discarded. More feed solution is added either continuously or discontinuously to the sample reservoir. As more feed solution is processed, the concentrations of all solutes that are well rejected by the membrane gradually increase in the sample reservoir.

The major practical difference between UF, NF, and RO membranes is their respective abilities to "reject" solutes. Rejection is generally defined as

$$R = 1 - C_P/C_F \qquad (1)$$

where C_P and C_F are the concentrations of solute in the permeate and feed solutions, respectively. The rejection of any given solute follows the basic order of RO > NF > UF. It is also generally true that, for all of these membranes, rejection of solutes follows the basic order of polyvalent ions > monovalent ions > neutral molecules. Furthermore, at least for UF membranes, there is clear evidence that the rejection of humic substances decreases with increasing ionic strength (Kilduff and Weber, 1992).

The fundamental problem that arises if membranes are used to concentrate DOM from natural waters is the co-concentration of inorganic solutes. The ideal membrane would allow all inorganic solutes to pass through the membrane ($R = 0$) while retaining all of the DOM ($R = 1$). Unfortunately, membranes that have relatively low R values for inorganic solutes also have relatively low R values for DOM. The best that can be achieved with a single type of membrane is relatively good removal of inorganic solutes and relatively good recovery of DOM.

The percent recovery of DOM by membrane processes is strongly dependent on both the value of R and on the water concentration factor (the ratio of total volume of feed water to final volume of retentate solution). Letting W be the water concentration factor, it can be shown that the percent recovery of a solute is predicted to be

Percent recovery

$$= (100/W)[1/(1 - R)]\{1 - R\exp[-(1 - R)$$
$$\times (W - 1)]\} \qquad (2)$$

Either trying to concentrate the sample too much or working with a membrane having a low value of R will lead to lower recoveries of DOM by membrane-based methods.

A statistical summary of many of the results that have been published for isolation of DOM using UF and RO membranes is provided in Figure 2. Because UF membranes with a wide variety of nominal molecular weight cutoffs have been used, the plots and tabulated results in Figure 2 reflect the overall retention of DOC above each specified nominal molecular weight cutoff. Above nominal molecular weight cutoffs of 100 kDa, 10 kDa, and 1 kDa, the median recoveries of DOM are 31%, 53%, and 71%, respectively. As the percent retention of DOM increases, so does the likelihood that the isolated material is representative of the whole DOM. The publications of Gjessing (1970), Schindler and Alberts (1974), Buffle et al. (1978), Benner and Hedges (1993), and Hedges et al. (1994) are recommended for further details on the general use of UF membranes to isolate DOM from freshwaters.

If the median results in Figure 2 are taken literally, then 31% of the constituents of DOM have molecular weights that are greater than 100 kDa, 22% have molecular weights in the range of 10–100 kDa, 18% have molecular weights in the range of 1–10 kDa, and 29% have molecular weights that are less than 1 kDa. The minimum possible weight-average molecular weight of DOM (see later discussion) would be ~34 kDa ($0.31 \times 100 + 0.22 \times 10 + 0.18 \times 1 + 0.29 \times 1$). As will become evident in Section 5.10.4.3, this value is almost certainly at least one order of magnitude too high. Several careful studies of the use of UF membranes to concentrate organic solutes have concluded that UF membranes retain organic anions whose molecular weights are much smaller than the nominal molecular weight cutoffs of the UF membranes (Kwak et al., 1977; Aiken, 1984; Kilduff and Weber, 1992). At best, the average molecular weights of fractions of DOM that are isolated using UF membranes will increase generally with increasing nominal molecular weight cutoffs of the UF membranes. It is inappropriate, however, to use nominal molecular weight cutoffs of UF membranes as surrogates for average molecular weights of those DOM fractions.

The median recovery of DOM using RO membranes is ~90%, and the range of results is much narrower than is the case for UF membranes. The high recoveries of DOM that are obtained using RO membranes suggest strongly that the recovered DOM is representative of the DOM in the original source waters. Under typical operating conditions, filtered feed

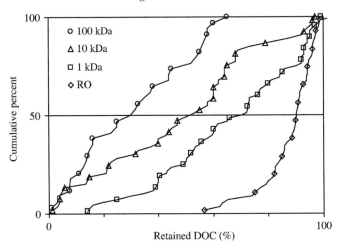

Membrane cutoff	Obs.	Retained DOC (%)			
		Range	Median	Mean	SD
100 kDa	34	1.0–65.0	31.0	31.0	20.0
10 kDa	53	1.3–97.0	53.0	48.4	28.2
1 kDa	68	14.3–99.0	71.3	66.2	24.0
Reverse osmosis	55	56.7–104.0	90.1	87.8	9.8

References. *Arch. Hydrobiol.*: **110** : 589–603. *Environ. Int.*: **25**: 367–371. *Environ. Sci. Technol.*: **4**: 437–438; **18**: 978–981; **25**: 4289–4294. *Hydrologie*: **32**: 190–200. *Limnol. Oceanogr.*: **41**: 41–51; **42**: 714–721; **45**: 1088–1096. *Org. Geochem.*: **4**: 27–35. *Sci. Tot. Environ.*: **15**: 207–216; **113**: 159–177; **229**: 53–64. *Talanta*: **45**: 977–988. *Wat. Res.*: **24**: 911–916; **25**: 1033–1037; **29**: 471–1477; **32**: 3108–3124; **33**: 2363-2373; **35**: 985–996. *Wat. Sci. Technol.*: **40**: 89–93.

Figure 2 Recoveries of organic carbon from freshwaters by ultrafiltration and reverse osmosis.

water can be processed rapidly without chemical pre-treatment, and, if a cation-exchange resin is used continuously to remove polyvalent cations from the concentrated sample, a freshwater sample can often be concentrated up to 30-fold *in the field*. The principal weakness of the RO method is the co-concentration of sulfate anions and dissolved silica, neither of which is easily separated from the concentrated DOM. The publications of Serkiz and Perdue (1990), Clair *et al.* (1991), Sun *et al.* (1995), and Kitis *et al.* (2001b) are recommended for further details on the general use of RO membranes to isolate DOM from freshwaters.

Several studies have coupled RO with fractionation methods for isolation and fractionation of DOM. Crum *et al.* (1996) used RO membranes to concentrate DOM, and then they used a series of UF membranes to fractionate the concentrated DOM into three size fractions. Ma *et al.* (2001) used RO membranes to concentrate DOM, and then they used XAD-8 resin (see earlier discussion) to fractionate the concentrated DOM into HAs, FAs, and hydrophilic acids.

Other studies have compared the DOM isolated by using RO membranes with DOM isolated by

other methods. For example, Gjessing *et al.* (1999) compared the concentration of DOM using RO membranes with evaporative concentration of DOM, in which water is removed under vacuum at a temperature of less than 30 °C. Solid samples that were obtained by these methods were redissolved and compared with original samples from several lakes. Somewhat better recoveries of DOM were obtained using evaporative concentration, but the method is simply not suited for concentrating large volumes of water. They recommended that the international DOM community should use the RO-isolation technique. Maurice *et al.* (2002) compared numerous properties of the DOM in several water samples with corresponding properties of DOM samples that were isolated using RO membranes or using the tandem XAD resin method (Malcolm and MacCarthy, 1992). Numerous differences in the chemical properties of the various samples were noted, with some indications of condensation or coagulation during the RO process and some indications of ester hydrolysis during the XAD process.

NF membranes have been evaluated in the water treatment field both for softening water and for removal of DOM prior to chlorination.

In such applications, the permeate solution is the desired product, and the retentate solution, which contains most of the DOM, is a waste product that is generally discarded. Accordingly, Figure 2 does not contain any results for the recovery of DOM using NF membranes. It is, nonetheless, possible to estimate the likely range of recoveries using NF membranes from published data for the rejection (see Equation (1)) of DOM by those membranes. Amy *et al.* (1990) obtained an average rejection of $85 \pm 6\%$ for removal of TOC using NF membranes to treat a river water and a groundwater that was derived from recharged secondary effluent. Tan and Sudak (1992) tested two RO membranes (brackish water-type) and two NF membranes in a 336 h study and found that the percent rejection of TOC by the two NF membranes (95.0 ± 1.7 and 95.7 ± 1.2) was only slightly less than the results obtained for the two RO membranes (98.5 ± 0.9 and 97.9 ± 1.0). Keck (1994) reported that the percent rejection of TOC by an NF membrane increased from 97% to 99% over a 10-fold range in TOC concentration of the feed water but was not affected by a twofold increase in applied pressure. Siddiqui *et al.* (2000) obtained an average rejection of $90 \pm 3\%$ for removal of TOC using NF membranes to treat four source waters in short-term (1 h) tests. They reported problems with membrane fouling in long-term (30 d) tests. Collectively, these studies reported 35 measurements of percent rejection of DOM by NF membranes. Percent rejection ranged from 76% to 99%, with a median value of 90% and a mean and SD of $90 \pm 5\%$. In comparison, the mean percent rejection of DOM by RO membranes is ~99%. The percent recovery of DOM depends strongly on both the percent rejection of DOM by a membrane and on the degree to which a water sample is concentrated (see Equation (2)). The percent recovery of DOM using NF membranes is thus expected generally to be lower than with RO membranes, and the difference should increase with increasing water concentration factor.

5.10.4.3 Average Molecular Weights

5.10.4.3.1 *Molecular weight distributions and averages*

A complex mixture such as DOM contains many organic compounds, so the "molecular weight" of DOM is actually a distribution of molecular weights. Distributions of molecular weights are often characterized by their first, second, and third moments, namely the number-average (M_n), weight-average (M_w), and z-average

(M_z) molecular weights of the distribution. These three averages are defined as

number-average:

$$M_{\mathrm{n}} = \sum w_i \Big/ \sum n_i = \sum n_i M_i \Big/ \sum n_i \qquad (3)$$

weight-average:

$$M_{\mathrm{w}} = \sum w_i M_i \Big/ \sum w_i = \sum n_i M_i^2 \Big/ \sum n_i M_i \qquad (4)$$

z-average: $\qquad M_z = \sum n_i M_i^3 \Big/ \sum n_i M_i^2 \qquad (5)$

where n_i, M_i, and w_i are the number of moles, molecular weight, and mass of the ith component of a mixture (note that $w_i = n_i M_i$). As a very simple example, consider a mixture containing 1 mol each of compounds with molecular weights of 500 Da, 1,000 Da, and 3,000 Da. For this mixture, M_n, M_w, and M_z are 1,500 Da, 2,278 Da, and 2,744 Da, respectively.

Experimental methods for measurement of average molecular weights may lead naturally to one or more of these averages, and the comparison of average molecular weights that were obtained using different experimental methods is a potential source of confusion. The vast majority of studies have yielded number-average molecular weights (M_n) and/or weight-average molecular weights (M_w) for unfractionated DOM and/or chromatographic fractions of DOM. Excellent background papers by Lansing and Kraemer (1935), Glover (1975), Wershaw and Aiken (1985), Swift (1989a), and Cabaniss *et al.* (2000) are recommended for further general details on the various average molecular weights of complex mixtures such as DOM.

5.10.4.3.2 *Colligative and noncolligative methods*

An impressive number of methods have been used to study the distribution of molecular weights in DOM. These methods may be classified as either colligative or noncolligative methods (see subsequent discussion). In either case, experimental methods can be applied to either the whole sample or chromatographic fractions of the whole sample. The colligative methods most widely applied to DOM are vapor pressure osmometry (VPO) and cryoscopy (CRY). The noncolligative methods most widely applied to DOM are gel permeation chromatography (GPC) and size exclusion chromatography (SEC), UF, flow field–flow fractionation (FFF), UV scanning ultracentrifugation (UV-UCGN), diffusivimetry (DIFF), UV molar absorptivity (MA-UVS), and multi-angle laser light scattering (MALLS). The series of review papers by Swift (1989b), Clapp *et al.* (1989), Aiken and Gillam (1989), Wershaw (1989), De Nobili *et al.* (1989), Duxbury (1989), and Chen and Schnitzer (1989) provides a comprehensive review of many of these methods.

Colligative methods are based on the variation of some property of a *solvent* with its mole fraction in a solution (e.g., melting point, boiling point, vapor pressure, and osmotic pressure). For example, according to Raoult's law, the vapor pressure of a solution containing a nonvolatile solute is directly proportional to the mole fraction of solvent. The nature of the nonvolatile solute is unimportant—only its mole fraction matters. This relationship is the basis of VPO. When colligative methods are applied to DOM, it is necessary to make proper corrections to the raw data for the H^+ that forms from dissociation of acidic functional groups (e.g., Hansen and Schnitzer, 1969; Reuter and Perdue, 1981; Gillam and Riley, 1981) and for any water-soluble inorganic impurities (Na^+, SO_4^{2-}, H_4SiO_4, etc.) that might be present in a sample of DOM. If colligative measurements are made in organic solvents such as tetrahydrofuran, additional corrections may be required for adsorbed water in the "dried" DOM (Aiken and Malcolm, 1987; Marinsky and Reddy, 1990). The failure to make such corrections results in substantial underestimation of M_n values of DOM samples. Only the first moment (M_n) of a molecular weight distribution can be obtained using colligative methods. Even so, because the resulting M_n values are independent of the physical properties of DOM, they serve as a very reliable reference against which M_n values that are obtained from noncolligative methods can be judged.

Noncolligative methods are based directly on the properties (e.g., sizes, shapes, densities, diffusion coefficients) of *solute* molecules. Most noncolligative methods yield a distribution of apparent molecular weights, from which M_n and M_w values can be extracted. When raw data from noncolligative methods are converted into molecular weight distributions, it is generally necessary to make, explicitly or implicitly, one or more simplifying assumptions about the physical properties of the constituents of DOM. For example, in SEC the elution times of standard solutes in the SEC system are used to construct calibration curves, from which the elution times of DOM samples are converted into apparent molecular weight distributions. Perminova *et al.* (1998) have shown that such molecular weight distributions of humic substances are shifted toward much higher molecular weights with decreasing charge density of the standard materials used to calibrate the SEC system.

Several prior studies (Aiken and Malcolm, 1987; Chin and Gschwend, 1991; Chin *et al.*, 1994; Cabaniss *et al.*, 2000; Maurice *et al.*, 2002) have collectively applied VPO in H_2O, VPO in tetrahydrofuran (THF), and SEC in H_2O to FAs from five freshwaters in the US. The three methods yield overall average M_n values of 690 ± 89, 639 ± 126, and $1,078 \pm 288$, respectively, for those five FAs. VPO clearly yields lower M_n values than does SEC. The fact that VPO measurements in H_2O (where corrections for dissociated H^+ were required) and in THF (where no corrections were made) yield very similar M_n values is a strong argument in favor of these lower M_n values. The M_n values from SEC are almost certainly somewhat high. The tendency of SEC-derived molecular weight distributions of DOM to be biased high has been attributed to the widespread use of UV detectors in SEC systems, which overestimate the concentrations of high molecular weight components of DOM and underestimate the concentrations of low-molecular-weight components (Chin *et al.*, 1994). TOC measurements reportedly provide less biased estimates of the relative concentrations of low and high molecular weight components of DOM, allowing more accurate determination of molecular weight distributions (Her *et al.*, 2002).

5.10.4.3.3   Statistical summary of average molecular weights

A statistical summary of many of the results that have been published for M_n and M_w values of FAs, HAs, and NOM in freshwaters is given in Figure 3. In the tabulated statistics, results are organized according to both the type of material and the experimental methodology, which more clearly indicates the degree of variability among methods. The M_n and M_w of NOM that were obtained using MALLS are an order of magnitude greater than values obtained for NOM by all other methods. These values will not be considered further in this analysis. The median M_n and M_w values for FAs are similar to those of NOM and slightly more than one-half of the corresponding values for HAs. The overall median values of M_n for FAs, HAs, and NOM are about 834 Da, 1,700 Da, and 1,100 Da, respectively. The corresponding median values of M_w are about 1,805 Da, 3,310 Da, and 1,590 Da. By comparison, the average molecular weight of 34 kDa that is implied from the median retention of DOM on selected UF membranes (see earlier discussion of Figure 2) is at least one order of magnitude too high. As expected for polydisperse materials, M_w/M_n ratios are greater than 1 for all these materials.

A closer examination of M_n values from colligative and noncolligative methods reveals

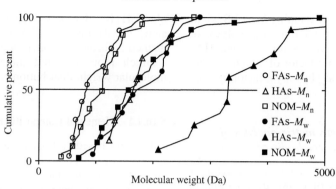

Method[a]	Sample	Type[b]	Obs.	Molecular weight (Da)			
				Range	Median	Mean	SD
SEC/HPSEC	FAs	M_n	11	639–1,790	1,180	1,096	362
FFF	FAs	M_n	7	980–1,666	1,160	1,296	262
CRY/VPO	FAs	M_n	14	540–900	633	678	118
FFF	HAs	M_n	6	1,320–2,374	1,750	1,837	402
VPO	HAs	M_n	1	1,220	1,220	1,220	0
SEC/HPSEC	NOM	M_n	31	400–2700	1,109	1,107	471
FFF	NOM	M_n	7	890–1,760	910	1,133	350
VPO	NOM	M_n	1	614	614	614	0
MALLS	NOM	M_n	2	15,050–16,595	15,823	15,823	1,092
SEC/HPSEC	FAs	M_w	14	980–2,430	1,672	1,740	522
FFF	FAs	M_w	6	1,240–2,800	1,997	1,984	612
UV–UCGN	FAs	M_w	4	950–2,260	1,815	1,710	620
SEC/HPSEC	HAs	M_w	2	2,600–3,320	2,960	2,960	509
FFF	HAs	M_w	6	2,090–4,390	3,293	3,387	808
UV–UCGN	HAs	M_w	4	2,710–6,590	4,005	4,328	1,640
SEC/HPSEC	NOM	M_w	37	784–2,743	1,700	1,684	530
FFF	NOM	M_w	7	1,030–4,900	1,470	2,227	1,512
DIFF	NOM	M_w	9	700–3,400	2,300	2,089	862
MA–UVS	NOM	M_w	4	728–1,330	982	1,005	249
MALLS	NOM	M_w	11	15,000–57,800	22,400	25,564	12,607

References. *Anal. Chim. Acta*: **364**: 203–221. *Chem. Geol.*: **19**: 285–293. *Coll. Surf. A*: **120**: 87–100; **181**: 289–301. *Environ. Int.*: **25**: 275–284. *Environ. Sci. Technol.*: **21**: 289–295; **26**: 1853–1858; **31**: 132–136; **34**: 1103–1109. *Geochim. Cosmochim. Acta*: **45**: 2017–2022; **51**: 2177–2184; **54**: 131–138; **55**: 1309–1317. *J. Membr. Sci.*: **164**: 89–110. *Limnol. Oceanogr.*: **44**: 1316–1322. *Org. Geochem.*: **33**: 269–279. *Sep. Sci. Technol.*: **30**: 1435–1453. *Soil Sci.*: **132**: 191–199. *Wat. Res.*: **33**: 2265–2276; **36**: 925–932; **36**: 2357–2371. *Wat. Sci. Technol.*: **40**: 307–314.

[a] SEC: size exclusion chromatography; HPSEC: high-pressure size exclusion chromatography; FFF: flow field–flow fractionation; CRY: cryoscopy; VPO: vapor pressure osmometry; MALLS: multi-angle laser light scattering; UV-UCGN: ultraviolet scanning ultracentrifugation; DIFF: diffusivimetry; MA-UVS: molar absorptivity-ultraviolet spectrophotometry.
[b] Number-average (M_n) and weight-average (M_w) molecular weights.

Figure 3 Molecular weights of freshwater FAs, HAs, and NOM.

the anticipated bias in noncolligative methods that use UV absorbance as a surrogate for concentration in SEC measurements (see earlier discussion). Even so, the overall median M_n values of FAs from colligative and noncolligative methods vary by less than a factor of 2 (633–1,170 Da). The ranges of variation in overall median M_n values of HAs (1,220–1,750 Da) and NOM

(614–1,100 Da) are even smaller. Considering that the colligative methods are independent of the physical properties of DOM and the noncolligative methods exploit collectively several different physical properties of DOM, the median M_n values in Figure 3 are remarkably consistent.

Only noncolligative methods can provide estimates of M_w values. The wide variety of noncolligative methods, excluding MALLS, yield remarkably consistent median M_w values in Figure 3. Median M_w values for FAs and HAs vary by much less than a factor of 2, and median M_w values for NOM vary by only a factor of 2.3.

5.10.4.3.4 Mass spectrometric analysis of molecular weight distributions

Direct measurements of molecular weights of individual constituents of DOM are possible, in principle, using mass spectrometry. Such measurements have appeared in the scientific literature since the 1980s, and it has generally been observed that the most abundant ions are found at m/z values of 200–400 Da (e.g., Schulten, 1987, 1999; Kramer *et al.*, 2001). These relatively low m/z values may be the result of fractionation of the sample by selective volatilization of low-molecular-weight compounds, fragmentation of molecular ions into smaller ions, and/or the formation of multiply charged ions. Stenson *et al.* (2002) have combined electrospray ionization with Fourier transform ion cyclotron resonance mass spectrometry to measure molecular weight distributions of aquatic FAs and HAs. The authors provided evidence that multiply charged ions were not abundant, so the observed m/z distributions could be interpreted as molecular weight distributions. They found peaks at every nominal mass up to 3,000 Da, and each "peak" consisted of a complex mixture of ions whose masses differed by fractional mass units. Each of these individual high-resolution peaks in a cluster of peaks with the same nominal mass includes an unknown number of isomers. The most abundant ions were found at molecular weights of 500–1,000 Da. Stenson *et al.* (2002) did not calculate M_n or M_w values, but the observed range of maximum peak intensities seems likely to be consistent with M_n and M_w values that have been determined by colligative and noncolligative methods (see Figure 3).

The study by Stenson *et al.* (2002) clearly illustrates the extremely high complexity of the mixtures of compounds known as FAs and HAs. At the same time, however, they discovered and analyzed several repeating patterns that were observed across the entire m/z range. Through the entire observable mass range, they found peaks at mass intervals corresponding to $-CH_2-$ groups, as is observed in a homologous series of molecules. In addition, they observed regular intervals between peaks that correspond to the mass of two hydrogen atoms, as would be found in saturated–unsaturated pairs of molecules (e.g., $-CH_2CH_2-$ and $-CH=CH-$ or $-CH_2OH$ and $-CHO$). Other paired peaks were attributed to net replacement of oxygen atoms by CH_4 units (e.g., replacement of $-C_2H_5$ by $-CHO$) and replacement of nitrogen atoms by CH units (e.g., $-NH_2$ to $-CH_3$). As this method is further exploited in the study of DOM, a much more detailed understanding of its molecular weight distribution can be anticipated.

5.10.4.4 Elemental Composition

Element-based approaches are commonly used to estimate reservoirs and fluxes of organic carbon, nitrogen, and phosphorous in aquatic systems, mainly because the measurements can be conducted directly on water samples without isolation of the DOM. Such approaches provide insight into the processing of carbon, nitrogen, and phosphorous in aquatic ecosystems; however, the real power of element-based approaches is only realized when an elemental analysis that includes hydrogen and oxygen is performed. Such an elemental analysis can be conducted with acceptable accuracy only on dry, low-ash samples (Huffman and Stuber, 1985).

Many aspects of the redox chemistry, acid–base chemistry, and metal complexation chemistry of DOM are manifestations of the H/C and O/C ratios of the DOM. When a complete elemental composition is available, it is even possible to estimate quite well the relative proportions of aliphatic carbon, aromatic carbon, and other forms of carbon in DOM (Perdue, 1984, 1998; Wilson *et al.*, 1987). From a biological and ecological perspective, the structural and compositional information that can be derived from a complete elemental analysis have been shown to relate directly to the bioavailability of DOM to bacterial assemblages in river water (Sun *et al.*, 1997; Vallino *et al.*, 1996; Hunt *et al.*, 2000).

In this section of this review, it will occasionally be useful to compare the elemental composition (and derived parameters) of DOM with those of the major constituents of terrestrial and aquatic biomass, from which DOM is ultimately derived. Lignins, which are found only in terrestrial biomass, have an average empirical formula of $C_{139}H_{148}O_{52}$ (after Brauns, 1960). Lipids, proteins, and sugars, which occur in all living matter, have average empirical formulas of $C_{18}H_{34}O_2$, $C_{106}H_{168}O_{34}N_{28}S_1$, and $C_6H_{10}O_5$, respectively (Hedges *et al.*, 2002).

5.10.4.4.1 Mass percentages of carbon, hydrogen, nitrogen, oxygen, and sulfur

A statistical summary of many of the results that have been published for the elemental composition of DOM is provided in Figure 4. All data from original sources have been

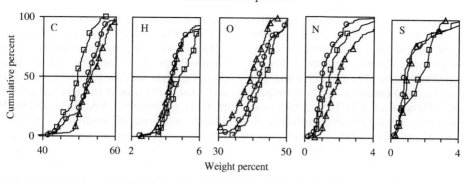

Element	Sample	Obs.	Weight percent[a]			
			Range	Median	Mean	SD
Carbon	FAs	117	41.4–62.7	52.3	52.1	4.2
	HAs	107	38.7–62.7	53.3	53.4	3.9
	NOM	57	42.3–57.2	49.6	49.5	3.3
Hydrogen	FAs	117	2.5–8.1	4.4	4.6	1.0
	HAs	107	2.6–8.2	4.3	4.5	1.0
	NOM	57	3.6–7.9	4.8	5.0	1.0
Oxygen[b]	FAs	117	27.5–52.1	41.9	41.5	4.9
	HAs	107	23.5–47.2	39.1	38.5	4.9
	NOM	57	34.3–52.6	43.5	43.0	4.1
Nitrogen	FAs	117	0.2–9.2	1.0	1.3	1.0
	HAs	107	0.6–9.8	1.9	2.4	1.7
	NOM	57	0.4–5.4	1.4	1.7	1.0
Sulfur	FAs	43	0.2–4.3	0.8	1.2	1.0
	HAs	36	0.3–3.2	0.9	1.2	0.8
	NOM	8	0.5–4.7	1.9	2.0	1.3

References. *Anal. Chim. Acta*: **169**: 87–98; **362**: 299–308; **364**: 203–221; **392**: 333–341; **418**: 205–215. *Ann. Chim.*: **74**: 257–267. *Aus. J. Soil Sci.*: **16**: 41–52. *Can. J. Chem.*: **74**: 2460–2470. *Chem. Geol.*: **12**: 113–126. *Chemosphere*: **34**: 1693–1704; **38**: 2913–2928. *Environ. Sci. Technol.*: **17**: 412–417; **20**: 904–911; **21**:2 43–248. *Environ. Int.*: **18**: 609–620; **20**: 387–390; **25**: 145–159. *Geoderma*: **49**: 241–254. *Geochim. Cosmochim. Acta*: **45**: 2017–2022; **56**: 1753–1757. *Limnol. Oceanogr.*: **2**: 161–179; **31**: 739–754; **42**: 714–721. *Org. Geochem.*: **18**: 563–537; **21**: 443–451. *Talanta*: **55**: 733–742. *Wat. Res.*: **7**:911–916; **9**: 1079–1084; **30**: 1502–1516; **32**: 597–608; **32**: 3108–3124; **33**: 2265–2276; **35**: 985–996; **36**: 2357–2371. *Wat. Supply*: **11**: 79–90. *In* Clapp *et al.* (1996): 41–46; 305–316; 389–398. *In* Frimmel *et al.* (2002): 3–38. *In* Wong *et al.* (1983): 751–772.

[a] Elemental data reported on a dry, ash-free basis. [b] Oxygen calculated as the % mass difference (100–%C–%N–%H) for 74 FAs, 71 HAs acids, and 49 NOM samples.

Figure 4 Elemental compositions of freshwater FAs (○), HAs (△), and NOM (□).

converted, if necessary, to a dry, ash-free basis. The data have been subdivided into separate compilations for FAs, HAs, and NOM. All data sets included results for carbon, hydrogen, nitrogen, and oxygen; however, fewer than one-third of the data sets also included results for sulfur. The statistical results in Figure 4 indicate that NOM typically contains the least carbon and the most hydrogen and oxygen, which is often interpreted as evidence for a greater proportion of carbohydrate-like moieties in NOM. For carbon, hydrogen, and oxygen, median values and mean values are very similar; however, mean values exceed median values significantly for both

nitrogen and sulfur. The results in Figure 4 are generally comparable to those reported by Rice and MacCarthy (1991) in a previously published statistical analysis of the elemental compositions of humic substances, which includes about half as many observations for HAs and FAs from freshwaters. The degree to which the two data sets overlap cannot be ascertained, because references to source data were not provided by Rice and MacCarthy (1991). They also found very little difference between median and mean values for carbon, hydrogen, and oxygen, and reported that mean values exceeded median values significantly for both nitrogen and sulfur.

What is most striking in Figure 4 and the earlier paper by Rice and MacCarthy (1991) is the narrowness of the distributions of elemental composition for the major elements (carbon, hydrogen, and oxygen). The relative SDs (SD expressed as a percentage of the mean value) for carbon, hydrogen, and oxygen in Figure 4 are only 7.5%, 15.9%, and 11.6%, respectively. For the less abundant elements (nitrogen and sulfur), relative SDs are much larger (70.6% and 74.0%, respectively). Carbon, hydrogen, and oxygen are found in all four of the major constituents of biomass; however, both nitrogen and sulfur are ultimately derived from proteins. Although DOM is clearly more complex than an unaltered mixture of the major constituents of biomass (see next two sections), the greater relative SDs for nitrogen and sulfur may ultimately be a consequence of local variations in the proteinaceous content of biomass.

5.10.4.4.2 *Atomic ratios*

The statistical summary of the atomic ratios of H/C, O/C, N/C, and S/C for FAs, HAs, and NOM in Table 5 is based on the source data for Figure 4. The median atomic ratios in Table 5 are slightly lower than those reported by Rice and MacCarthy (1991). As might be anticipated from the mass percentages in Figure 4, NOM has slightly greater H/C and O/C ratios than do FAs and HAs. Overall, though, the similarities are much more striking than the differences.

Another view of the elemental compositions of FAs, HAs, and NOM and the major constituents of biomass is provided in Figure 5, which is known as a van Krevelen plot. The elemental compositions of FAs, HAs, and NOM overlap considerably in Figure 5. This overlap is consistent with the median elemental compositions in

Figure 4 and the median atomic ratios of H/C and O/C in Table 5. The *biomass triangle* defined by lipids, lignins, and sugars delimits the possible compositions of all biomass. Aquatic organisms do not contain lignins, so their compositions are restricted to the smaller triangle defined by lipids, proteins, and sugars. If FAs, HAs, and NOM were simply mixtures of unaltered constituents of biomass, their compositions would have to plot inside or on the edges of the biomass triangle. It is strikingly evident that this is not the case. The locus of points is shifted below (lower H/C) and to the right (higher O/C) of the line that represents a mixture of lignins and sugars. The overall conversion of biomass to materials whose chemical compositions plot outside the biomass triangle can be formally represented by reactions in which biomass reacts with a less-than-stoichiometric amount of O_2 to form some CO_2 and H_2O and more highly oxidized organic matter. FAs, HAs, and NOM are, on average, more highly oxidized than the major constituents of biomass. The YOAs of Shapiro (1957) also plot outside the biomass triangle (H/C = 1.23; O/C = 0.54), but very near the mixing line of lignins and sugars.

5.10.4.4.3 *Average oxidation state of organic carbon*

If hydrogen, oxygen, nitrogen, and sulfur are assumed to have oxidation states of +1, −2, −3, and −2, respectively, it is readily shown that the average oxidation state of carbon in a pure organic compound or a complex mixture such as DOM is

$$Z_C = -(H/C) + 2(O/C) + 3(N/C) + 2(S/C)$$
$$(6)$$

When the average elemental ratios of lignins, lipids, proteins, and sugars are substituted into

Table 5 Atomic ratios of H/C, O/C, N/C, and S/C for FAs, HAs, and NOM from freshwaters.

Ratio	Sample	Obs.	Range	Median	Mean	Std. Dev.
	FAs	117	0.575–1.740	1.031	1.045	0.199
H/C	HAs	107	0.567–1.882	0.978	1.018	0.265
	NOM	57	0.823–1.893	1.170	1.207	0.230
	FAs	117	0.340–0.942	0.593	0.607	0.120
O/C	HAs	107	0.284–0.728	0.538	0.546	0.092
	NOM	57	0.471–0.933	0.640	0.657	0.104
	FAs	117	0.003–0.138	0.016	0.022	0.017
N/C	HAs	107	0.009–0.165	0.032	0.039	0.029
	NOM	57	0.006–0.104	0.025	0.030	0.019
	FAs	43	0.001–0.031	0.005	0.008	0.007
S/C	HAs	36	0.002–0.024	0.006	0.008	0.006
	NOM	8	0.004–0.035	0.014	0.015	0.010

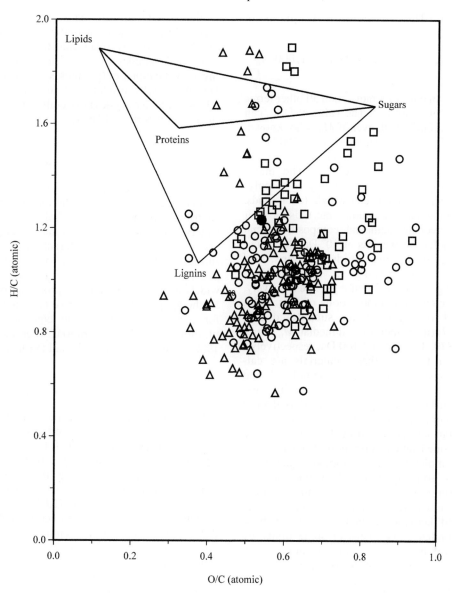

Figure 5   van Krevelen plot of the average atomic H/C and O/C ratios of major components of biomass, FAs, HAs, NOM, and YOAs (O—FAs; △—HAs; □—NOM; ● Shapiro, 1957).

Equation (6), the average oxidation states of carbon in these major constituents of biomass are −0.32, −1.67, −0.13, and 0.0, respectively. In contrast, when the atomic ratios for FAs, HAs, and NOM from the source data for Figure 4 are substituted into Equation (6), the median oxidation states of organic carbon in freshwater FAs, HAs, and NOM are calculated to be +0.29, +0.25, and +0.33, respectively. As expected from the loci of data points in Figure 5, freshwater FAs, HAs, and NOM differ only slightly from each other but are substantially more oxidized than the major constituents of biomass. These data are consistent with the early diagenetic alteration of biomass, in

which the incorporation of oxygen from O_2 exceeds the loss of oxygen in CO_2 and H_2O.

5.10.4.4.4   *Unsaturation and aromaticity*

The degree of unsaturation in an organic compound, which arises from the presence of rings and π-bonds in its structure, can be calculated from its elemental composition, if its molecular weight is known. Perdue (1984) demonstrated that such calculations can be extended rigorously to a complex mixture of organic compounds if its bulk chemical composition and number-average molecular weight (M_n)

are known. This relationship is given in Equation (7):

$$U_{total} = C_{total} + \tfrac{1}{2}N_{total} - \tfrac{1}{2}H_{total} + 1,000/M_n \tag{7}$$

where M_n is in units of g mol^{-1} and all other terms are in units of mmol g^{-1}. The value of U_{total} can be estimated fairly well, even if M_n is not known. For example, as M_n increases from 500 Da to infinity, the last term in Equation (7) only changes from 2 mmol g^{-1} to 0 mmol g^{-1}. The major constituents of biomass exist mainly as biopolymers of rather high molecular weight (even lipids, which include suberins, cutins, etc.) Using their respective average elemental compositions and assumed molecular weights of 10^4 Da, the calculated U_{total} of lignins, lipids, proteins, and sugars are 24.6 mmol g^{-1}, 3.6 mmol g^{-1}, 15.0 mmol g^{-1}, and 6.2 mmol g^{-1}, respectively. When the elemental compositions of FAs, HAs, and NOM in Figure 4 are substituted into Equation (7) (using their overall median M_n values of about 840, 1,700, and 1,100 Da, respectively), the median U_{total} of these materials are calculated to be 23 mmol g^{-1}, 24.7 mmol g^{-1}, and 18.8 mmol g^{-1}, respectively. There is far less variation in U_{total} among these materials than in the major components of biomass.

The total unsaturation from Equation (7) can be apportioned into contributions from alicyclic structural moieties such as furanose and pyranose rings, aromatic structural moieties such as benzene rings, and carbonyl-containing moieties such as carboxylic acids, esters, amides, aldehydes, and ketones. Olefinic moieties such as the unsaturated alkyl chains of some lipids are thought to be of only minor importance. Equation (7) can thus be rewritten as

$$U_{total} = C_{al}(U/C)_{al} + C_{ar}(U/C)_{ar} + C_{C=O}(U/C)_{C=O} \tag{8}$$

where C_{al}, C_{ar}, and $C_{C=O}$ are the concentrations (in mmol g^{-1}) of aliphatic, aromatic, and carbonyl-containing moieties in an organic compound or mixture, and the (U/C) ratios are the average unsaturation-to-carbon ratios in the various structural classes. Aliphatic moieties probably have (U/C) ratios somewhere between 0 and 1/6, the latter ratio being found in pyranose forms of sugars. Aromatic moieties will have (U/C) ratios of 4/6, if all aromatic carbon occurs in benzene rings. All carbonyl-containing moieties will have (U/C) ratios of 1/1.

Upper limits on the aromaticities of FAs, HAs, and NOM can be obtained from Equation (8). Using the median carboxyl contents of FAs, HAs, and NOM (see Figure 6 and related discussion) and assuming that all other unsaturation must occur in aromatic rings, the *maximum* median aromaticities

of these materials are 61%, 65%, and 49%, respectively. Actual aromaticities will be lower if DOM contains any other moieties that contain C=O bonds (amides, esters, aldehydes, ketones) or any alicyclic moieties (pyranoses, furanoses, etc.). This topic will be addressed further in Section 5.10.4.8, which provides an overall summary of the chemical properties of DOM.

5.10.4.5 Acidic Functional Groups

The acid–base properties of DOM are of intrinsic interest because acidic functional groups contribute to the acid–base balance of natural waters, affect complexation and transport of dissolved metals, and interact with mineral surfaces. The concentrations of carboxyl and phenolic functional groups are among the most widely measured and reported properties of DOM. Methodologically, there are two basic approaches for measuring acidic group content—indirect titrations and direct titrations (Perdue *et al.*, 1980; Perdue, 1985; Ritchie and Perdue, 2003).

5.10.4.5.1 Indirect titration methods

Indirect titrations have the advantage of simplicity—titration of a filtered reaction mixture to a fixed pH end point following a 24 h equilibration with a weak or strong base. Total acidity is determined using the strong base Ba(OH)$_2$, and carboxyl content is determined using the weak base Ca(CH$_3$CO$_2$)$_2$. Phenolic contents are calculated in indirect methods as the difference between total acidity and carboxyl content. Dubach *et al.* (1964) and van Dijk (1966) have pointed out, however, that these methods cannot actually distinguish between functional groups on the basis of structure but only on the basis of acidic strength.

Although originally developed for research purposes on soil and coal-derived HAs (Blom *et al.*, 1957; Brooks and Sternhell, 1957; Schnitzer and Gupta, 1965), indirect titration methods were also applied to aquatic samples in the 1970s. For example, Beck *et al.* (1974) analyzed 10 unfractionated DOM samples from a blackwater river in southeastern Georgia (USA), and Weber and Wilson (1975) analyzed three FAs from a darkly colored river and pond in New Hampshire (USA). It was later recognized (Sposito and Holtzclaw, 1979; Perdue *et al.*, 1980) that incomplete removal of a humic substance from a Ca(CH$_3$CO$_2$)$_2$ reaction mixture causes carboxyl content to be overestimated. Davis (1982) found that incomplete removal of a humic substance from a Ba(OH)$_2$ reaction mixture causes total acidity to be underestimated. The critical nature of the filtration step in indirect titrations was further discussed by Perdue (1985).

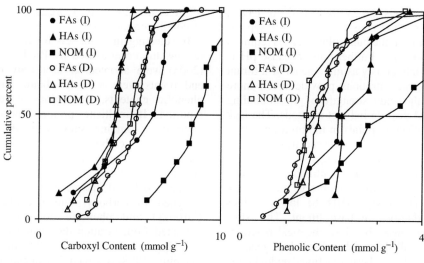

Group (method)	Sample	Obs.	Acidic group content (mmol g⁻¹)[a]			
			Range	Median	Mean	SD
Carboxyl (indirect)	FAs	8	1.9–8.1	6.5	5.7	2.0
	HAs	8	1.1–5.1	4.4	3.9	1.3
	NOM	11	6.0–10.7	8.9	8.7	1.4
Carboxyl (direct)	FAs	73	2.2–8.9	5.4	5.1	1.2
	HAs	22	1.6–5.9	4.2	4.0	1.0
	NOM	11	2.7–10.0	5.2	5.2	2.0
Phenolic (indirect)	FAs	8	1.5–4.3	2.2	2.4	0.9
	HAs	8	2.1–3.7	2.5	2.6	0.6
	NOM	11	1.0–8.3	3.5	3.6	2.0
Phenolic (direct)	FAs	65	0.5–5.1	1.6	1.7	0.8
	HAs	22	1.0–3.0	1.8	1.8	0.5
	NOM	6	1.3–3.6	1.5	1.9	0.9

References. *Anal. Chem.*: **55**: 410–411. *Anal. Chim. Acta*: **257**: 35–39; **392**: 333–341. *Aqua. Geochem.*: **4**: 3–48. *Geochim. Cosmochim. Acta* : **38**: 341–364; **67**: 85–96. *Geoderma*: **49**: 241–254. *Environ. Sci. Technol.*: **20**: 354–366; **20**: 904–911; **20**: 1028–1032; **23**: 356–362; **27**: 846–856; **27**: 1182–1189; **27**: 2015–2022; **32**: 3346–3355; **35**: 2049–2059. *Wat. Res.*: **9**: 1079–1084; **30**: 1502–1516; **35**: 985–996. *In* Aiken *et al.* (1985): 105–145; 181–209. *In* Averett *et al.* (1989): 209–229. *In* Clapp *et al.* (1996): 41–46. *In* Christman and Gjessing (1983): 83–106. *In* Frimmel *et al.* (2002): 3–38. *In* Holcomb (1979). *In* Reuter (1980).

[a] Carboxyl and phenolic contents reported on a dry, ash-free basis.

Figure 6 Carboxyl and phenolic contents of freshwater FAs, HAs, and NOM.

Even in the absence of filtration-related problems, indirect titrations do not provide insight into the strengths of acidic functional groups, because results are measured at only the two pH end points.

5.10.4.5.2 Direct titration methods

Although less commonly employed than indirect titration methods, soil scientists also used direct titrations methods to study the acid–base chemistry and metal interactions of soil organic matter (Pommer and Breger, 1960; Posner, 1966; Gamble, 1970, 1972; van Dijk, 1971; Khalaf *et al.*, 1975;

Sposito *et al.*, 1977; Sposito and Holtzclaw, 1977). Aside from the study of YOAs by Shapiro (1957), direct titrations of freshwater DOM were uncommon until the late 1970s (Wilson and Kinney, 1977; Reuter and Perdue, 1977; Choppin and Kullberg, 1978). Since the early 1980s, direct titrations have been much more often used than indirect titrations to quantify the acidic functional groups of DOM even though there is no standardized protocol (Santos *et al.*, 1999). Experimental conditions for conducting direct titrations (i.e., titer, length of titration, ionic strength) are left to the discretion of the researcher.

Direct titrations are well suited to DOM, because the method does not require the separation of DOM from the bulk solution, as is required in indirect methods. Furthermore, because pH is monitored continuously as increments of a titrant are added, direct titrations provide much more detail about the acidic strengths of proton binding sites in DOM than do indirect methods. The lack of distinct inflection points in the titration curves of DOM is a result of the broad spectrum of acidic strengths of its functional groups.

Because pH is monitored continuously in direct titrations, it is possible to observe that, above pH 6–7, pH slowly drifts to lower values following additions of base titrant. This slow drift may be caused by slow chemical reactions occurring at alkaline pH values that generate additional acidity. These reactions undoubtedly occur during indirect titrations, but they cannot be observed because pH is not monitored continuously through the 24 h equilibration time. The generation of "new" acidity during a titration with base causes hysteresis in reverse titrations, i.e., forward and reverse titration curves are not superimposable. Although there is no consensus regarding the cause of pH drift and hysteresis, several studies suggest that the "new" acidity is produced by alkaline hydrolysis of esters (Bowles *et al.*, 1989; Antweiler, 1991).

5.10.4.5.3 Carboxyl and phenolic contents of DOM

It is not possible to use data from a direct titration to quantify unambiguously the separate contributions of carboxyl and phenolic groups to the total acidity of DOM. Instead, arbitrary pH end points are chosen or mathematical models are used to extract that information. For example, it is often assumed that all carboxyl groups (and no phenolic groups) are titrated by pH 8.0 and that one-half of phenolic groups are titrated between pH 8.0 and pH 10.0 (Thurman, 1985; Bowles *et al.*, 1989). Most mathematical models of proton binding by DOM fall into two general classes—discrete multisite models and continuum models. It is beyond the scope of this review to examine these and other classes of models, but the publications of Eberle and Feuerstein (1979), Eshleman and Hemond (1985), Paxéus and Wedborg (1985), Dzombak *et al.* (1986), Fish *et al.* (1986), Leuenberger and Schindler (1986), Cabaniss and Shuman (1988a,b), and Kramer and Davies (1988) are representative of the use of discrete models, and the publications of Perdue and Lytle (1983), Perdue *et al.* (1984), Tipping and Hurley (1992), Tipping (1998), Koopal *et al.* (1994), Benedetti *et al.* (1995, 1996), and

Milne *et al.* (2001) are representative of the use of continuum models.

In selecting the data sets for inclusion in this review, the criteria used by Ritchie and Perdue (2003) were adopted. If arbitrarily chosen pH end points were used to obtain carboxyl and phenolic contents, the values were used as reported. If multisite proton binding models were used, binding sites with $pKa < 8$ were classified as carboxyl groups, and all other binding sites were classified as phenols. If continuum models were used, the concentrations of the two broad classes of binding sites were assigned to carboxyl and phenolic contents. Carboxyl and phenolic contents that were extracted from titration data using either model V (Tipping and Hurley, 1992) or model VI (Tipping, 1998) are included in this review due to the widespread use of models V and VI with regard to metal complexing by aquatic and terrestrial humic substances. Carboxyl and phenolic contents from indirect titrations were accepted as reported in the literature.

A statistical analysis of the carboxyl and phenolic contents of FAs, HAs, and NOM is given in Figure 6. Separate statistical descriptions are provided for the results of indirect and direct titrations. Perhaps because of the aforementioned difficulties in indirect titrations of DOM, most of the reported carboxyl and phenolic contents were obtained using direct titrations.

Indirect and direct estimates of carboxyl content in HAs were comparable, perhaps because HAs are more completely precipitated in the equilibrated reaction mixture of an indirect titration and can thus be removed by filtration prior to the critical titration step. FAs and NOM tend to remain in solution in the equilibrated reaction mixture of an indirect titration, and thus indirect estimates of their carboxyl contents are greater than direct estimates. Problems with removal of reacted DOM from the equilibrated reaction mixture during a measurement of total acidity should cause total acidity to be underestimated. Alternatively, if alkaline hydrolysis of esters and related reactions generate additional acidity during the 24 h equilibration time at very high pH, then total acidity may be overestimated in indirect methods. If total acidity is overestimated or underestimated, then phenolic content will also be overestimated or underestimated. Judging from the median results in Figure 6, the generation of additional acidity may be quite important in the indirect $Ba(OH)_2$ method, because indirect estimates of phenolic content are 1.4–2.3 times greater than direct estimates for FAs, HAs, and NOM. Direct titrations are usually conducted rapidly, so the likelihood of base-catalyzed side reactions that generate additional acidity is minimized (although still observable).

The authors believe that direct methods provide more consistent estimates of the concentrations of carboxyl and phenolic groups (Ritchie and Perdue, 2003), and subsequent analysis will be only consider the results of direct titrations.

FAs and NOM have very similar median carboxyl contents, which are ~22% greater than the median carboxyl content of HAs. All three materials have very similar phenolic contents.

One interesting aspect of the median results in Figure 6 is that phenolic groups account for only 23%, 30%, and 22% of the total acidities of FAs, HAs, and NOM, respectively. In models V and VI (Tipping and Hurley, 1992; Tipping, 1998), which are widely used to describe the pH dependence of metal complexation equilibria in natural waters, 33% of total acidity is attributed to phenolic groups. It is likely that these models overestimate the contribution of phenolic groups to acid–base and metal complexation chemistry of DOM.

5.10.4.6 Carbon Distribution from ^{13}C NMR Spectrometry

Carbon-13 NMR spectrometry has evolved into one of the most powerful tools for studying functional groups and structural subunits in DOM. The method is applicable to aqueous and nonaqueous solutions of DOM, and to solid samples (if cross-polarization with magic angle spinning (CP-MAS) techniques are used). The chemical shift range of ^{13}C in organic compounds lies in the range of ~0–220 ppm, relative to $Si(CH_3)_4$. The historical development of these methods and their application to humic substances from soils and waters have been described in detail by Wilson (1987) and by Malcolm (1989). More recent applications of both solution-state and solid-state ^{13}C NMR to the same set of samples have been published by Haiber *et al.* (2002) and Lankes and Lüdemann (2002), respectively.

Most studies, as of the early 2000s, have yielded one-dimensional spectra, in which the proportions of the major structural subunits of DOM are estimated by integrating spectra between user-specified limits of chemical shift. Although such results are often endowed with quantitative significance, the popular CP-MAS method reportedly overestimates sp^3-hybridized carbon in alkyl and alkoxy groups and underestimates sp^2-hybridized carbon in aromatic, carboxyl, and carbonyl groups (e.g., Mao *et al.*, 2000, *op. cit.*). The number of structural subunits and the chemical shift range of each structural subunit vary widely from one study to another, depending on the interests of the author and the appearance of the spectra.

5.10.4.6.1 Processing source data

In this review, the chemical shift range of ^{13}C is divided into the following five subunits:

- carbonyl (aldehydes and ketones)—190–220 ppm;
- carboxyl (carboxylic acids, esters, and amides)—160–190 ppm;
- aromatic (unsubstituted and substituted)—110–160 ppm;
- alkoxy (alcohols, hemiacetals, ethers, etc.)—60–110 ppm; and
- alkyl (methyl, methylene, and methine)—0–60 ppm.

Published results in which different chemical shift ranges and/or different numbers of structural subunits were reported have been converted into these five subunits using the following simple approximation method:

(i) the original spectrum is approximated as a sequence of isosceles triangles whose bases correspond to the chemical shift ranges of the structural subunits used in the original work and whose heights are adjusted to match the peak areas of the structural subunits in the original work; and

(ii) vertical lines are drawn at the chemical shift limits of the five structural subunits used in this review, and the approximate spectrum is re-integrated between the new limits.

5.10.4.6.2 Statistical summary of ^{13}C NMR results

A statistical summary of more than 82 re-integrated ^{13}C NMR spectra of FAs, HAs, and NOM is given in Figure 7. The results are striking. Freshwater FAs, HAs, and NOM have remarkably similar distributions of carbon among the five structural subunits. For all three materials, organic carbon is distributed in the following order of abundance: alkyl > alkoxy ≈ aromatic > carboxyl ≫ carbonyl.

Mahieu *et al.* (1999) conducted a similar statistical survey of the ^{13}C NMR spectra of several hundred whole soils and soil size fractions, soil-derived HAs and FAs, and aquatic FAs. They used four structural subunits: carbonyl—160–220 ppm; aromatic—110–160 ppm; O-alkyl—50–110 ppm; and alkyl—0–50 ppm. Their carbonyl subunit includes the carbonyl and carboxyl structural subunits used in this review, and their boundary between alkyl and O-alkyl subunits is 50 ppm, rather than 60 ppm. For the 31 aquatic FAs in that study, the mean percentages of carbonyl, aromatic, O-alkyl, and alkyl carbon were 24%, 21%, 20%, and 35%, respectively. These percentages are very close to the mean values in Figure 7 for FAs in freshwaters (21%, 25%, 23%, and 32%, respectively). The degree to

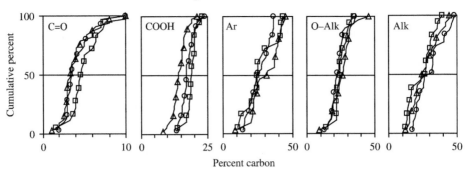

Carbon group	Sample	Obs.	Percent carbon			
			Range	Median	Mean	SD
Carbonyl (C=O)	FAs	31	2.0–10.0	3.6	4.2	1.8
	HAs	33	1.0–10.0	3.4	4.0	1.9
	NOM	18	1.5–12.6	4.7	5.3	2.3
Carboxyl (COOH)	FAs	37	13.4–24.0	17.0	17.4	2.4
	HAs	38	8.20–21.0	14.1	14.4	2.7
	NOM	19	13.7–23.0	19.0	19.0	2.0
Aromatic (Ar)	FAs	31	14.3–36.4	25.2	26.1	7.4
	HAs	32	8.9–44.8	31.5	30.6	9.8
	NOM	18	10.4–43.0	24.7	26.8	10.8
Alkoxy (O-Alk)	FAs	31	13.0–35.4	22.6	22.9	5.4
	HAs	32	9.0–45.3	26.5	25.7	7.2
	NOM	18	14.0–34.3	24.2	24.9	4.6
Alkyl (Alk)	FAs	31	16.9–48.2	31.6	30.2	10.0
	HAs	32	11.9–44.4	24.8	26.5	9.5
	NOM	18	12.0–39.4	26.7	24.1	9.0

References. *Can. J. Chem.*: **74**: 2460–2470; **75**: 14–27. *Environ. Sci. Technol.*: **25**: 1160–1164; **26**: 107–116; **31**: 8–12. *Geochim. Cosmochim. Acta*: **56**: 1753–1757. *Geoderma*: **49**: 241–254. *J. Anal. Appl. Pyr.*: **58**: 349–359. *Org. Geochem.*: **11**: 273–280; **18**: 563–573; **24**: 859–873. *Talanta*: **55**: 733–742. *Wat. Res.*: **32**: 3108–3124; **33**: 2265–2276. *In* Clapp *et al.* (1996): 389–398. *In* Frimmel *et al.* (2002): 96–114 *In* Thorn *et al.* (1992).

Figure 7 ^{13}C NMR-based carbon distributions of freshwater FAs (○), HAs (△), and NOM (□).

which the two data sets overlap cannot be ascertained, because references to source data were not provided by Mahieu *et al.* (1999).

5.10.4.7 Biomolecules in DOM

In Section 5.10.4.4 of this review, it was noted that DOM is ultimately derived from the major constituents of terrestrial and aquatic biomass—lignins, lipids, proteins, and sugars. Lignins are formed only in vascular plants, and they are quite resistant to microbial degradation. As such, they are an excellent chemical tracer for allochthonous inputs of organic matter into natural waters. Lipids, proteins, and sugars are found in all biota, and they are microbially utilized much more rapidly than lignins. Accordingly, lipids, proteins, and sugars can provide some insight into the level of heterotrophic activity in natural waters, but they are probably less useful as chemical tracers

of allochthonous and autochthonous inputs of organic matter into natural waters.

Despite the general complexity of DOM, which has been illustrated and discussed at length in this review, several percent of DOM can be attributed directly to lignins, lipids, proteins, and sugars, which are found in freshwaters both as their chemically bound forms and as their free forms. Mostly, these constituents of DOM exist as chemically bound forms, from which their respective monomers (lignin-derived phenols, simple carboxylic acids, amino acids, and mono-saccharides) can be liberated by vigorous hydro-lytic and/or oxidative degradation.

Thurman (1985) provided a comprehensive assessment of the distribution of identifiable organic compounds in natural waters. Total hydrolyzable amino acids accounted for 2–3% of DOC in rivers and 3–13% of DOC in eutrophic lakes. Total hydrolyzable sugars accounted for 5–10% of DOC in rivers and 8–12% of DOC in lakes. Thurman (1985) also summarized the published results for low-molecular-weight carboxylic acids in natural waters. The volatile and nonvolatile fatty acids together may account for 6% of the DOC in freshwaters. More polar low-molecular-weight carboxylic acids are less abundant, but they may collectively account for another 2% of DOC. Other types of identifiable organic compounds occur at trace levels in freshwaters. The identifiable organic compounds that were reviewed by Thurman (1985) accounted collectively for more than 20% of DOM in rivers and lakes.

Kaplan and Newbold (2003) have provided an updated review of the abundances of low-molecular-weight organic compounds in freshwaters, with emphasis on their metabolism in stream ecosystems. They reported that the DOC in an average river includes ~1% free sugars, 0.05% free amino acids, 3.3% carboxylic acids, 0.02% phenols, and 0.2% hydrocarbons. Collectively, these identifiable compounds account for less than 7% of the DOC in unpolluted freshwaters.

Benner (2003) estimated that total hydrolyzable amino acids and neutral sugars account for 1–3% and 1–2%, respectively, of the DOC in rivers. In a humic lake dominated by allochthonous inputs of organic carbon, amino acids and sugars accounted for slightly higher percentages of DOC (3.2% and 4.5%); however, amino acids and sugars accounted for much higher percentages of DOC (10–20% and 4–24%) in two lakes that are dominated by autochthonous inputs of carbon. Together, these compound classes account for 2–5% of DOC in rivers, 14–44% of DOC in clear lakes, and 8% of DOC in humic lakes.

The preceding reviews provide ranges and average total concentrations of the bound and free forms of amino acids and sugars in freshwaters. Very little information is provided about individual compounds within each compound class. The statistical summaries that are presented in this section describe not only the overall concentrations of the classes of compounds in freshwaters but also the relative proportions of individual compounds in each class of compounds. Moreover, the full distribution of results and summary statistics are presented for each individual compound.

5.10.4.7.1 *Amino acids*

The analysis of amino acids can be conducted on unmodified samples to quantify free amino acids, or total hydrolyzable amino acids can be quantified by conducting the analysis on samples that have been hydrolyzed for 24 h in 6 M HCl at 100 °C. In either case, amino acids must be derivatized prior to quantitative analysis. Derivatization is used to render amino acids less polar and/or more volatile or to attach chromophores or fluorophores that increase analytical sensitivity. Derivatives may be formed before or after the individual amino acids are separated chromatographically, depending on the method of separation (ion-exchange chromatography, gas chromatography, high-performance liquid chromatography, etc.). The hydrolysis and derivatization of amino acids have some undesired consequences. Trypotophan is destroyed during acid hydrolysis, so its concentration is rarely reported in amino acid analyses. Proline does not have a primary amino group in its structure, so its reactions with derivatizing agents may yield products that are not detected or underestimated by spectrophotometric and fluorometric methods of detection. Another problem in quantitative analysis of amino acids is contamination, even in commonly employed reagent chemicals such as concentrated HCl. Another significant source of contamination is human fingerprints, which contain ~10% (w/w) amino acids (Hare and St. John, 1988); very high levels of serine and ornithine may be indicative of this source of contamination (Oro and Skewes, 1965; Hamilton, 1965).

A statistical summary of the relative abundances of 18 amino acids (mol kmol^{-1}) and their overall concentrations in freshwaters (mmol L^{-1}) and in DOM (percent of C) is given in Figure 8. The plots and statistical summaries for the relative abundances of amino acids were generated from the combined data sets for total hydrolyzable amino acids (THAAs) and total free amino acids (TFAAs). It should be noted that neither the median nor the mean relative abundances of tabulated amino acids in Figure 8 sum to 1,000 mol kmol^{-1}. Median values are not generally constrained to do so, and neither are mean values, if statistics for the individual amino acids are based on different numbers of observations. In addition, some amino acid analyses include data for tryptophan and/or several nonessential amino acids (γ-aminobutyric acid, α-aminobutyric acid, diaminopimelic acid, β-alanine, citrulline, and hydroxyproline). These amino acids are usually a very minor fraction of the amino acids in freshwaters, and these results were omitted from Figure 8. Statistical summaries for the total concentration of amino acids in freshwaters and in DOM were obtained on the separate data sets for THAAs and TFAAs.

The individual amino acids in Figure 8 are listed in order of decreasing median abundance in freshwaters. The five most abundant amino acids (glycine, aspartic acid, alanine, glutamic acid, and serine) account for ~60% of the amino acids in freshwaters. Aromatic and basic amino acids account collectively for ~10% of the amino acids in freshwaters, and sulfur-containing amino acids account for ~1.5%.

Dissolved Organic Matter in Freshwaters

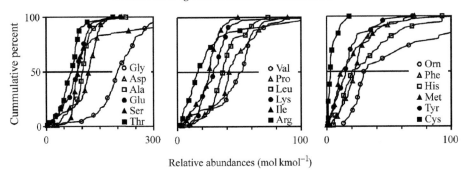

Relative abundances (mol kmol^{-1})

Amino acid	Obs.		Relative abundances (mol kmol^{-1})			
	THAAs[a]	TFAAs[a]	Range	Median	Mean	SD
Glycine (Gly)	69	19	26–450	200	206	64
Aspartic (Asp)	69	19	20–212	117	113	38
Alanine (Ala)	69	18	4–223	102	98	37
Glutamic (Glu)	69	17	14–208	88	84	36
Serine (Ser)	69	19	31–483	88	126	94
Threonine (Thr)	69	19	2–187	73	70	33
Valine (Val)	69	16	2–145	52	49	24
Proline (Pro)	50	2	10–89	42	45	17
Leucine (Leu)	68	19	3–73	37	38	14
Ornithine (Orn)	25	9	13–190	31	55	45
Lysine (Lys)	69	9	9–149	29	31	18
Isoleucine (Ile)	67	19	4–49	26	24	10
Phenylalanine (Phe)	68	19	2–70	23	24	14
Histidine (His)	56	18	5–93	20	28	20
Arginine (Arg)	52	15	2–117	17	26	25
Tyrosine (Tyr)	61	19	0.5–69	15	17	14
Methionine (Met)	60	16	0.2–108	12	21	20
Cysteine (Cys)	13	0	1–18	3	5	5

	Obs.	Concentration			
		Range	Median	Mean	SD
THAAs (μmol L^{-1})	51	0.12–23.2	1.3	4.1	6.2
TFAAs (μmol L^{-1})	21	0.05–1.8	0.3	0.6	0.5
% DOC as THAAs	59	0.42–10.4	1.8	2.2	1.8
% DOC as TFAAs	14	0.02–1.2	0.1	0.3	0.4

References. *Arch. Hydrobiol.*: **115**: 499–521. *Biogeochem.*: **30**: 77–97. *Geochim. Cosmochim. Acta*: **36**: 867–883; **38**: 341–364. *Environ. Sci. Technol.*: **15**: 224–228. *Environ. Technol.*: **15**: 901–916. *Freshwater Biol.*: **16**: 255–268. *Hydrobiol.*: **199**: 201–216. *Limnol. Oceanogr.*: **32**: 97–111; **34**: 531–542; **39**: 743–761; **42**: 39–44; **45**: 775–788. *Microb. Ecol.*: **10**: 301–316. *Org. Geochem.*: **23**: 343–353. *Wat. Res.*: **30**: 1502–1516.

[a] Total hydrolyzed amino acids (THAAs) and total free amino acids (TFAAs).

Figure 8 Distribution and concentrations of amino acids in freshwaters.

Consider the subset of amino acids in Figure 8 whose median relative abundances are greater than 3%. With the exception of serine and ornithine, the amino acids in this subset have very similar mean and median values and relative SDs of 31–49%. For serine and ornithine, mean values greatly exceed median values, and their relative SDs are 75% and 82%, respectively. If the data sets for TFAAs and THAAs are analyzed separately, it is found that the median relative abundance of serine in TFAAs (28%) is 3.7 times greater than in THAAs (7.6%). A similar situation exists for ornithine, whose median relative abundance in TFAAs (10.2%) is 3.5 times greater than in THAAs (2.9%). This unusual 3.5–3.7-fold enrichment of ornithine and serine in the data set for TFAAs is strongly indicative of contamination by human fingerprints during at least some of the measurements of TFAAs (Oro and Skewes, 1965; Hamilton, 1965).

The reported concentrations of THAAs and TFAAs in freshwaters range over three orders of magnitude, with median concentrations of $1.3 \ \mu\text{mol L}^{-1}$ and $0.3 \ \mu\text{mol L}^{-1}$, respectively. The median percentages of DOC in the form of THAAs and TFAAs are 1.8% and 0.1%, respectively. These median concentrations of THAAs and TFAAs in freshwaters and as a percentage of DOC are roughly consistent with the earlier estimates of Thurman (1985), Kaplan and Newbold (2003), and Benner (2003).

5.10.4.7.2 Sugars

The analysis of sugars can be conducted on unmodified samples to quantify free sugars, or total hydrolyzable sugars can be quantified by conducting the analysis on samples that have been subjected to acid hydrolysis. Conditions vary, but typical conditions include pretreatment with 72% H_2SO_4 for 24 h, followed by heating for several hours in $0.5 \ \text{M} \ H_2SO_4$ at 100 °C. The acidic reaction mixture is usually neutralized with $BaCO_3(s)$, which offers the advantages of exact neutralization of acidity without over-titration and quantitative removal of SO_4^{2-} as $BaSO_4(s)$. After removal of the insoluble organic residues and $BaSO_4(s)$, the liberated sugars are quantified.

The solution chemistry of sugars is inherently complex, because each sugar exists in aqueous solution as a dynamic mixture of its open-chain form and anomeric pairs of pyranose and/or furanose rings. This complex chemistry is the result of reversible reactions between carbonyl and hydroxyl groups within a sugar molecule, in which aldohexoses such as glucose form cyclic hemiacetals and ketohexoses such as fructose form cyclic hemiketals. When a sugar is derivatized using a reagent that attacks its carbonyl group, which only exists in the open-chain form of the sugar, only a single derivative is formed. As the open-chain form is consumed in the derivatization reaction, it is replenished by the reversible reactions with cyclic forms until all forms of the sugar have been converted into the derivative of the open-chain form. When a sugar is derivatized using a reagent that attacks its hydroxyl groups, the reversible reactions between the open-chain and cyclic forms are interrupted by derivatization of the anomeric hydroxyl group, and multiple derivatives are formed from a single sugar. This problem can be avoided by reducing sugars to their corresponding alditols, which only exist in open-chain form, and then derivatizing and

quantifying the alditols (e.g., Sweet and Perdue, 1982). This approach is quite good for aldohexoses such as glucose, which are reduced to form unique alditols, but it is inappropriate for analysis of ketohexoses such as fructose, because they are reduced to form diastereomeric mixtures of alditols.

Both sugars and their alditols are derivatized to render them less polar and/or more volatile or to attach chromophores or fluorophores that increase analytical sensitivity. Derivatives are generally formed before the individual sugars are separated chromatographically (e.g., Cowie and Hedges, 1984). Direct analysis of underivatized sugars using high-performance liquid chromatography with pulsed amperometric detection has been used (e.g., Mopper *et al.*, 1992; Tranvik and Jørgensen, 1995; Jahnel *et al.*, 1998; Cheng and Kaplan, 2001).

A statistical analysis of the relative proportions of selected sugars and their concentrations in freshwaters and in DOM is given in Figure 9. Five of the sugars (glucose, galactose, rhamnose, mannose, and fucose) are hexoses, and the other four sugars (xylose, arabinose, ribose, and lyxose) are pentoses. Using the median relative abundances of these sugars, four of the hexoses and two of the pentoses are relatively abundant, with glucose being most abundant of all. These six sugars account for 95% of sugars, leaving only 5% of sugars in the form of the one remaining hexose (fucose) and two remaining pentoses (ribose and lycose). The six most abundant sugars have similar median and mean values and relative SDs that range from 22% to 55%.

The concentration of total hydrolyzable sugars in freshwaters ranges over more than three orders of magnitude, with a median concentration of $\sim2.6 \ \mu\text{mol L}^{-1}$. The median percentage of DOC that occurs as sugars is $\sim3\%$, which agrees well with the estimate of Benner (2003) for rivers and humic lakes, but which is about a factor of 3 lower than the estimates of Thurman (1985) for rivers and lakes.

5.10.4.7.3 Lignin-derived phenols

Lignins have much greater diversity in biomass than do proteins and carbohydrates. They are pseudo-random, three-dimensional polymers, rather than relatively simple linear polymers. The phenolic monomers from which lignin is formed are covalently linked by various combinations of carbon–carbon and carbon–oxygen bonds involving the aromatic rings of the phenols and their three-carbon side chains. The resulting polymers are much more complex than the repeating amide bonds of proteins and the repeating hemiacetal bonds in polysaccharides. These structural complexities are believed to

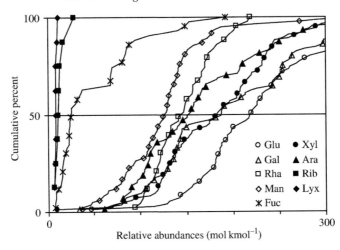

Sugar	Obs.	Relative abundances (mol kmol^{-1})			
		Range	Median	Mean	SD
Glucose (Glu)	71	47–591	220	264	140
Galactose (Gal)	85	35–875	188	204	113
Rhamnose (Rha)	43	94–219	147	147	33
Mannose (Man)	84	37–357	127	126	51
Fucose (Fuc)	43	8–192	26	50	46
Xylose (Xyl)	84	9–400	183	184	67
Arabinose (Ara)	87	61–455	153	167	77
Ribose (Rib)	8	7–28	12	13	7
Lyxose (Lyx)	8	7–11	8	9	1
	Obs.	Concentration			
		Range	Median	Mean	SD
Total sugars (μmol L^{-1})	104	0.08–20.0	2.6	3.1	2.7
% DOC as sugars	95	0.08–35.1	3.0	7.1	8.4

References. *Biogeochem.*: 30: 77–97; 29: 89–105. *Environ. Sci. Technol.*: 16: 692–698. *Geochim. Cosmochim. Acta*: 66: 955–962. *Hydrobiol.*: 113: 179–187. *Limnol. Oceanogr.*: 32: 39–44; 39: 743–761. *Micro.Ecol.*: 31: 41–55. *Org. Geochem.*: 32: 597–611. *Wat. Res.*: 30: 1502–1516. *In* Drozd *et al.* (1997): 635–640.

Figure 9 Distribution and concentrations of sugars in freshwaters.

impart to lignin a relatively high resistance against microbial decomposition, presumably because the sheer variety of bonds that must be cleaved to decompose lignin is perhaps beyond the enzymatic capabilities of any single organism.

Woody and nonwoody forms of gymnosperms and angiosperms use different mixtures of phenolic monomers to synthesize their respective lignins. The lignin signature is sufficiently distinct and well preserved during early diagenesis that the mixture of lignin-derived phenols released from particles in water and sediments by oxidation in alkaline CuO(s) can be used to distinguish between the four major classes of plant precursors (Hedges and Mann, 1979a). Because lignins only occur in terrestrial vascular plants, they are unequivocal indicators of allochthonous inputs of organic matter into natural waters (Hedges and

Parker, 1976). In this context, lignin-derived phenols are used to estimate the flux of terrestrial organic matter to ocean sediments (Hedges and Mann, 1979b) and the degree of diagenetic alteration of lignin-containing moieties in terrestrially derived organic matter (Ertel and Hedges, 1984).

Lignin-derived phenols occur primarily as chemically bound moieties in DOM, from which they are liberated by oxidation in alkaline CuO(s) for 3 h at 170 °C. The liberated phenols are usually converted into trimethylsilyl derivatives, which are then separated using gas chromatography with flame ionization detection (e.g., Hedges and Parker, 1976; Benner and Opsahl, 2001). In some instances, the underivatized phenols have been separated using high-performance liquid chromatography and detected spectrophotometrically (Lobbes *et al.*, 1999, 2000).

The literature on lignin-derived phenols is heavily influenced by the collected works of Hedges and co-workers. They classified lignin-derived phenols according to the patterns of substitution of methoxyl groups on the aromatic rings and the number of carbons in the aliphatic side chain of each phenol. *p*-Hydroxyl phenols include *p*-hydroxybenzoic acid, *p*-hydroxybenz-aldehyde, and *p*-hydroxyacetophenone, all of which have no ring methoxyl groups and a one- or two-carbon side chain. Vanillyl phenols include vanillic acid, vanillin, and acetovanillone, all of which have one ring methoxyl group and a one- or two-carbon side chain. Syringyl phenols include syringic acid, syringealdehyde, and acetosyringone, all of which have two ring methoxyl groups and a one- or two-carbon side chain. Cinnamyl phenols include *p*-coumaric acid and ferulic acid, both of which have less than two ring methoxyl groups and a three-carbon side chain. The concentrations of the individual phenols and the four groups of phenols are usually given in units of mg per 100 mg of organic carbon. *p*-Hydroxyl phenols are not derived uniquely from vascular plants, so they are less often used for diagnostic purposes, even though their concentrations are often reported along with the vanillyl, syringyl, and cinnamyl phenols. Finally, the sum of vanillyl, syringyl, and cinnamyl phenols is often reported in studies of lignin-derived phenols.

The statistical summary of lignin-derived phenols in freshwaters is given in Figure 10. The published literature contains more information for the four groups of phenols than for the individual lignin-derived phenols. Accordingly, individual compounds are not listed in Figure 10. Instead, the common groups of *p*-hydroxyl (P), vanillyl (V), syringyl (S), and cinnamyl (C) phenols are used in the statistical analysis. Vanillyl phenols are approximately twice as abundant as the syringyl and the *p*-hydroxyl phenols. Cinnamyl phenols are significantly lower in abundance than the other three classes. From the plot for concentrations of the V, S, C, and P groups of phenols in Figure 10, it is evident that the distributions are skewed by infrequent but very high abundances of all four groups of phenols. As would be expected, mean values are 50% greater than median values.

The ratios S/V and C/V are widely used to relate lignin-derived phenols to biochemical precursors. Generally, woody plants have very low C/V ratios, and angiosperms generally have much greater S/V ratios than do gymnosperms. Some papers in the literature provide the mass ratios of the four groups of phenols but not the concentrations of individual groups of phenols. These results have been combined with mass ratios that are calculated from more complete data sets to examine the statistical distribution of mass ratios of these phenols. The median S/V and C/V ratios in Figure 10 indicate that the average freshwater contains lignin-derived phenols originate mainly from a mixture of woody angiosperms and woody gymnosperms, which is not at all surprising.

The total concentration of lignin-derived phenols (V + S + C) in freshwaters ranges over three orders of magnitude, with a median concentration of $\sim$10 μg L^{-1}. Collectively, the lignin-derived phenols account for $\sim$0.6% of DOC.

5.10.4.8 Overall Summary of the Chemical Properties of DOM

The statistical summaries in Figures 1–10 were derived from published results for many samples of DOM, and these samples of DOM are reasonably expected to exhibit spatial and temporal variability. Each measured property of DOM is a weighted-average property of the bulk mixture, and both determinate and indeterminate experimental errors are generally associated with its measurement. For any observable property in Figures 1–10, the characteristics of the distribution of observed values (range, median, mean, and SD), therefore, reflect both experimental errors and the natural variability of DOM. It is possible, or perhaps probable, that no compound in a DOM sample actually has a set of chemical properties that match those of the bulk mixture (see the simple example of average molecular weights that follows Equation (5)). Even so, it is instructive to combine and analyze the assorted median results in Figures 1–10 to obtain a more integrated perspective on the chemical properties of DOM.

Resin-based methods of isolation and fractionation of DOM indicate that more than 80% of DOM is distributed in a 2 : 1 ratio of hydrophobic acids (56%) and hydrophilic acids (28%). The hydrophobic acids are further distributed in an $\sim$3 : 1 ratio of FAs (46%) and HAs (14%). Membrane-based methods of isolation and fractionation of DOM indicate that molecular weights of DOM fractions range from <1 kDa to more than 100 kDa; however, these results are almost certainly too high.

A summary of the median values of selected chemical properties of FAs, HAs, and NOM is given in Table 6, which includes both the properties for which complete statistical summaries were presented in Figures 3–10 and other derived or modeled properties that provide additional insight into the similarities and differences among these materials. The derived parameters include the average oxidation state of carbon (Z_C), which is calculated from Equation (6), and total unsaturation (U_{total}), which is calculated from Equation (7).

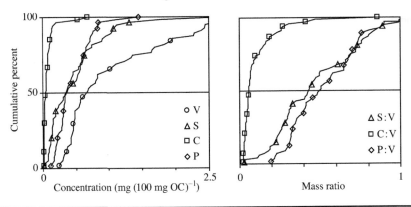

Lignin phenol group	Obs.	Concentration (mg (100 mg OC)$^{-1}$)			
		Range	Median	Mean	SD
Vanillyl[a] (V)	57	0.24–3.18	0.68	1.02	0.78
Syringyl[b] (S)	55	0.02–2.88	0.36	0.50	0.50
Cinnamyl[c] (C)	54	0.01–0.68	0.04	0.07	0.11
p-Hydroxy[d] (P)	57	0.12–1.46	0.36	0.45	0.27
Total lignin phenol[e]	55	0.59–6.66	1.41	2.06	1.47
	Obs.	Mass ratio (Relative to vanillyl content)			
		Range	Median	Mean	SD
Syringyl (S:V)	68	0.03–1.75	0.43	0.50	0.32
Cinnamyl (C:V)	68	0.02–0.86	0.06	0.11	0.13
p–Hydroxy (P:V)	55	0.19–1.22	0.51	0.54	0.23
	Obs.	Concentration			
		Range	Median	Mean	SD
Total lignin[e] (μg L^{-1})	55	0.42–39.4	9.7	10.7	9.8
% DOC as lignin	55	0.24–3.12	0.6	1.0	0.7

References. *Geochim. Cosmochim. Acta*: **64**: 2973–2983. *Limnol. Oceanogr.*: **31**: 739–754. *Org. Geochem.*: **32**: 597–611. *Science*: **223**: 485–487. *In* Sun (1993).

[a] Vanillic acid + vanillin + acetovanillone. [b] Syringic acid + syringaldehyde + acetosyringone. [c] p-Coumaric acid + ferulic acid. [d] p-Hydroxybenzoic acid + p-hydroxybenzaldehyde + p-hydroxyacetophenone. [e] S + C + V

Figure 10 Distribution and concentrations of lignin-derived phenols in freshwaters.

For purposes of retrospective comparison, the corresponding results for YOAs from lake water (Shapiro, 1957) are included in Table 6.

The probabilistic model of Perdue (1984) has been applied to the median compositions of YOAs, FAs, HAs, and NOM in Table 6 to obtain estimates for the most probable distribution of carbon in these materials. This model uses elemental composition, carboxyl content, and M_n to predict the most probable percentages of aliphatic, aromatic, and "excess" carbon of a complex mixture such as DOM. In this model, aliphatic carbon includes all sp^3-hybridized carbon and is thus equivalent to the sum of alkoxy

and alkyl carbon from ^{13}C NMR measurements. Aromatic carbon includes all sp^2-hybridized carbon in aromatic rings and thus corresponds directly to aromatic carbon from ^{13}C NMR measurements. Excess carbon includes carboxylic acids and ketones, whose concentrations are often measured independently, and all other sp^2-hybridized carbon (alkenes, esters, and amides), so excess carbon is equivalent to the sum of carbonyl and carboxyl carbon from ^{13}C NMR measurements.

A cursory examination of the results in Table 6 indicates that YOAs are chemically quite similar to FAs, HAs, and NOM; however,

Table 6 Median Properties of YOAs[a], FAs, HAs, and NOM in freshwaters.

Parameter	Units	YOAs	FAs	HAs	NOM
Carbon	mass %	54.6	52.3	53.3	49.6
Hydrogen	mass %	5.6	4.4	4.3	4.8
Oxygen	mass %	39.1	41.9	39.1	43.5
Nitrogen	mass %	1.2	1.0	1.9	1.4
Sulfur	mass %	ND	0.8	0.9	1.9
M_n^b	g mol^{-1}	456	633	1220	614
M_n^c	g mol^{-1}	ND	1170	1750	1101
M_w^d	g mol^{-1}	ND	1805	3310	1590
Carboxyl[e]	mmol g^{-1}	4.4	5.4	4.2	5.2
Phenolic[e]	mmol g^{-1}	ND	1.6	1.8	1.5
Carbonyl[f]	% of C	ND	4	3	5
Carboxyl[f]	% of C	ND	17	14	19
Aromatic[f]	% of C	ND	25	32	25
Alkoxy[f]	% of C	ND	23	26	24
Alkyl[f]	% of C	ND	32	25	27
Amino acids	% of C	ND	ND	ND	1.8
Sugars	% of C	ND	ND	ND	3.0
Lignin phenols	% of C	ND	ND	ND	0.6
Z_C^g	eq mol^{-1}	−0.10	0.27	0.24	0.25
U_{total}^h	mmol g^{-1}	20	22	25	18
Carbonyl + carboxyl[i]	% of C	22	29	28	25
Aromatic[i]	% of C	23	31	34	24
Alkoxy + alkyl[i]	% of C	55	40	38	51

[a] Shapiro (1957). [b] number-average molecular weight (colligative). [c] number-average molecular weight (non-colligative). [d] weight-average molecular weight (noncolligative). [e] estimated by direct titrations. [f] estimated by ^{13}C NMR. [g] average oxidation state of carbon (Equation (6)). [h] total unsaturation (Equation (7)). [i] probabilistic estimates (Perdue, 1984).

small differences can be discerned. For example, the %C and %H are slightly greater and the %O is slightly smaller in YOAs than in FAs, HAs, and NOM. The carboxyl content of YOAs is close to that of HAs but substantially lower than those of FAs and NOM. The average oxidation state of organic carbon, which is a function of elemental composition, is much lower in YOAs than in FAs, HAs, or NOM. The most probable distribution of carbon, also a function of elemental composition, indicates that YOAs contain a greater percentage of "alkoxy + alkyl" and smaller percentages of "aromatic" and "carbonyl + carboxyl" structural subunits than FAs, HAs, and NOM. All of these observations support the hypothesis that YOAs are somewhat enriched in less polar components of DOM. Given that YOAs were ultimately an ethyl acetate-soluble fraction of DOM, it is not surprising to find that they are less polar than other fractions of DOM. The somewhat lower M_n of YOAs, relative to FAs and NOM, may also be a consequence of the isolation procedure.

Among FAs, HAs, and NOM, similarities are more striking than differences. This result is somewhat surprising, because FAs and HAs are isolated from the hydrophobic acid fraction (~56% of DOM) in the XAD resin-based isolation procedure. NOM is logically expected to consist of a 56 : 44 mixture of these materials and other

presumably more polar constituents of DOM. The median results in Table 6 indicate that there is a tendency for the properties of FAs to lie between those of HAs and NOM, so it is perhaps appropriate to focus on the properties that are most different for HAs and NOM. The elemental abundances of hydrogen and oxygen, the ^{13}C NMR-based abundances of carbonyl and carboxyl subunits, the average oxidation state of carbon, and the probabilistic estimate of the sum of alkyl and alkoxy carbon are smallest for HAs and greatest for NOM. The elemental abundance of carbon, all types of average molecular weights, total unsaturation, and the probabilistic estimate of aromatic carbon are greatest for HAs and smallest for NOM.

In addition to such direct comparisons between samples, it is worthwhile to examine the internal consistency of the median properties of each type of material. Because the distribution of organic carbon among major structural subunits is provided both by ^{13}C NMR spectrometry and by the probabilistic model of Perdue (1984), those results serve as a useful starting point for examining the internal consistency of the results in Table 6. It should be noted that, for half of the samples examined by Perdue (1984), the most probable percentages of aromatic carbon were within 3% of the values measured by ^{13}C NMR spectrometry. Furthermore, when Wilson *et al.* (1987) used the model to predict the percentage of aliphatic carbon in eight hypothetical organic molecules whose structures contained from 16% to 92% aliphatic carbon, they found that the most probable percentages of aliphatic carbon were within 5% of the theoretical values for five of the eight structures. If discrepancies of greater than 5% between ^{13}C NMR-based and probabilistic model-based estimates of the distribution of organic carbon are found in Table 6, then the median results may not be internally consistent.

From the results in Table 6, the differences between ^{13}C NMR-based and probabilistic model-based estimates of "carbonyl + carboxyl", "aromatic", and "alkoxy + alkyl" are −8%, −5%, and +14% for FAs, −11%, −3%, and +13% for HAs, and −2%, +1%, and 0% for NOM. The discrepancies observed for most forms of carbon in FAs and HAs are greater than 5%, and they are consistent with theory and observation for CP-MAS ^{13}C NMR spectrometry, in which the ratio of sp^2-hybridized carbon to sp^3-hybridized carbon is significantly underestimated (Mao *et al.*, 2000). For NOM, the two approaches yield highly consistent estimates of the distribution of organic carbon. It is perhaps significant that most of the ^{13}C NMR spectra for NOM samples in Figure 7 were obtained on aqueous solutions, not on solid samples, i.e., the CP-MAS method was not used.

The systematic underestimation of unsaturated structural subunits in FAs and HAs by CP-MAS ^{13}C NMR methods is also revealed when ^{13}C NMR data in Table 6 are used to estimate U_{total} and %O. Each structural subunit has a characteristic molar ratio of unsaturation-to-carbon (U/C) and oxygen-to-carbon (O/C), and these ratios can be used to convert the carbon distribution into distributions of unsaturation and oxygen, from which U_{total} and %O can be calculated (see the example in Equation (8) for U_{total}). Unsaturation definitely exists in carbonyl, carboxyl, aromatic, and alkoxy subunits, and it might also exist in alkyl subunits in the form of alicyclic structural moieties. Neglecting the latter possibility and assigning U/C ratios of 1/1, 1/1, 4/6, and 1/6 to the first four structural subunits, U_{total} values for FAs, HAs, and NOM are predicted from ^{13}C NMR results in Table 6 to be 18 mmol g^{-1}, 19 mmol g^{-1}, and 18 mmol g^{-1}, respectively. The ^{13}C NMR-based estimates of U_{total} for FAs and HAs are clearly much lower than the tabulated values in Table 6 (which are calculated from elemental composition using Equation (7)). The ^{13}C NMR-based estimate of U_{total} for NOM is the same as the tabulated value in Table 6.

^{13}C NMR data can also be used to predict the oxygen contents of FAs, HAs, and NOM. Carbonyl subunits have O/C ratios of 1/1, carboxyl subunits have O/C ratios of 2/1 for carboxyl groups and esters and O/C ratios of 1/1 for amides, aromatic subunits have O/C ratios of ~1/6 (see following

discussion), alkoxy subunits in sugar-like moieties have O/C ratios of ~5/6, and alkyl subunits do not contain O. Given these assumptions, and assuming that all nitrogen occurs as amides, the %O in FAs, HAs, and NOM are calculated to be 41%, 40%, and 43%, respectively. All of these values are in quite good agreement with the median mass percentages of oxygen for FAs, HAs, and NOM in Table 6. It seems that the underestimation of carbonyl, carboxyl, and aromatic carbon is fortuitously compensated by overestimation of alkoxy carbon, insofar as oxygen content is concerned.

The ^{13}C NMR-based estimates of carboxyl subunits in FAs, HAs, and NOM are 7.4 mmol g^{-1}, 6.2 mmol g^{-1} and 7.9 mmol g^{-1}, respectively (after conversion to these units), and these values are considerably greater than the carboxyl contents that were obtained by direct titrations (5.4 mmol g^{-1}, 4.2 mmol g^{-1}, and 5.2 mmol g^{-1}, respectively. The ^{13}C NMR-based estimates of the carboxyl contents of FAs, HAs, and NOM are actually closer to their total acidities (7.0 mmol g^{-1}, 6.0 mmol g^{-1}, and 6.7 mmol g^{-1}). A very similar trend has been observed by Hatcher *et al.* (1981). The results of Mao et al. (2000) and the preceding analysis of median compositions in Table 6 suggest that the discrepancy between potentiometric and ^{13}C NMR-based estimates of carboxyl content should be even greater. Whether or not that is the case, the difference between potentiometric and ^{13}C NMR-based estimates of carboxyl content is an indication of the presence of esters and/or amides, both of which have peaks in the same chemical shift range as carboxyl groups. The hypothesis that the downward drift of pH during direct titrations and the hysteresis in paired forward-reverse titrations may be caused by alkaline hydrolysis of esters is supported by these findings.

The preceding analysis has established that the ^{13}C NMR-based and probabilistic model-based estimates of the aromatic carbon content of NOM are in excellent agreement. The median NOM sample contains 1.5 mmol g^{-1} of phenolic hydroxyl groups and 10.3 mmol g^{-1} of aromatic carbon. The ratio of phenolic hydroxyl groups to aromatic carbon in NOM is ~1 : 7, so the median aromatic ring contains slightly less than one phenolic hydroxyl group. Because of the rather large discrepancy between ^{13}C NMR-based and probabilistic model-based estimates of the aromatic carbon content of FAs and HAs, predicted ratios of phenolic hydroxyl groups to aromatic carbon will depend on which estimate of aromatic carbon is used. The median aromatic rings of FAs and HAs are predicted to contain 0.72–0.88 and 0.70–0.76 phenolic hydroxyl groups, respectively.

The median concentration of amino acids accounts for 1.8% of the carbon in NOM, from which it can be further calculated that amino acids account for only ~18% of the nitrogen in NOM. The median concentration of sugars accounts for 3.0% of the carbon in NOM, from which it can be further calculated that only ~12% of the alkoxy carbon in NOM is in the form of hydrolyzable sugars. Finally, lignin-derived phenols account for only 0.6% of the carbon in NOM, so much of the aromatic carbon of NOM is in structures other than lignin-derived phenols. Collectively, these biomolecules account for less than 6% of the carbon in NOM. Even if an additional 3% of carbon is attributed to low-molecular-weight organic acids, which have not been considered in this review (Thurman, 1985; Kaplan and Newbold, 2003), 90% of the carbon in NOM must reside in unknown classes of compounds.

5.10.5 SUMMARY AND CONCLUSIONS

DOM in freshwaters has been the focus of research of many scientists in many disciplines for ~50 yr. Great advances were made in the mid-1970s that led to the exponentially growing literature in this field. It is likely that more than 8,000 new papers will be published in this decade.

From several fundamentally independent approaches, it is now estimated that the global flux of organic carbon from continents to the oceans is 28–31 Tmol yr^{-1}. Assuming an annual riverine water flux of $(3.53–3.82) \times 10^4 \, km^3 \, yr^{-1}$, the global average TOC concentration in rivers is 730–880 $\mu mol \, L^{-1}$. On a global scale, the average concentrations of DOC and POC are ~400–480 $\mu mol \, L^{-1}$ and 330–400 $\mu mol \, L^{-1}$, respectively.

DOM is often fractionated into six fractions: hydrophobic acids, bases, and neutral compounds, and hydrophilic acids, bases, and neutral compounds. In the median freshwater, more than 80% of DOC is distributed in a 2 : 1 ratio between hydrophobic acids and hydrophilic acids, and less than 20% of DOC is evenly distributed between hydrophilic bases and the two neutral fractions. Very little DOC is in the hydrophobic base fraction. In the median freshwater, ~60% of DOC is generally distributed in a 3 : 1 ratio between FAs and HAs.

UF and RO offer an alternative means of isolating DOM from freshwaters. With UF membranes having nominal molecular weight cutoffs of 100 kDa, 10 kDa, and 1 kDa, the median recoveries of DOM are 31%, 53%, and 71%. It is possible to show by a simple calculation that the average molecular weight of DOM in freshwaters would be at least 34 kDa, if actual molecular weights of DOM fractions equaled the nominal molecular weight cutoffs of the UF membranes by which those fractions are retained. With RO, the median recovery of DOM is 90%.

Colligative and noncolligative methods have been used to determine average molecular weights and molecular weight distributions of DOM from freshwaters. Using colligative methods, the median number-average molecular weights of FAs, HAs, and NOM are 633 Da, 1,220 Da, and 614 Da, respectively. Using noncolligative methods, the median number-average molecular weights of these three materials are 1,170 Da, 1750 Da, and 1,101 Da. Weight-average molecular weights for these fractions are 1,805 Da, 3,310 Da, and 1,590 Da, respectively. Given that these number-average and weight-average molecular weights were determined by six quite different experimental methods, the results are remarkably consistent. Without any doubt, the value of 34 kDa that can be estimated from retention of DOM by UF membranes is at least an order of magnitude too high.

The median elemental compositions of FAs, HAs, and NOM are rather similar, and so consequently are properties such as the average oxidation state of organic carbon and the total unsaturation of these materials. Most of the elemental compositions that were included in this review lie outside the "biomass triangle" on a van Krevelen plot, indicating that DOM is not simply a residue of biomass. It has clearly gained oxygen and lost hydrogen, as would be expected for partial oxidation of biomass to CO_2 and H_2O.

This partially oxidized material contains 4.2–5.4 mmol g^{-1} of carboxylic acid functional groups, which account in part for its generally acidic character, its charge distribution as a function of pH, its complexation of metal ions, and its interaction with mineral surfaces in the environment. Phenolic hydroxyl groups are present at concentration of ~1.5–1.8 mmol g^{-1} in FAs, HAs, and NOM. The ratio of carboxyl-to-phenolic groups is thus ~3 : 1, which is significantly greater than the ratio of 2 : 1 that is commonly used in some mathematical models of the acidity of these materials.

^{13}C NMR measurements provide insight into the distribution of organic carbon among carbonyl, carboxyl, aromatic, alkoxy, and alkyl structural subunits in FAs, HAs, and NOM. All three materials are remarkably similar, and the most probable distribution for all three materials is: carbonyl (5%), carboxyl (22%), aromatic (30%), alkoxy (20%), and alkyl (23%). The ^{13}C NMR results are somewhat inconsistent with predictions of the distribution of organic carbon that arise from consideration of elemental composition, carboxyl content, number-average molecular weight, etc. The evidence is clear that carbonyl, carboxyl, and probably aromatic structural subunits are systematically under-estimated and alkoxy and alkyl structural subunits are overestimated by ^{13}C NMR measurements.

The major constituents of biomass are present at low levels in DOM. The median freshwater contains ~ 1.3 μmol L^{-1} of total hydrolyzable amino acids, the most abundant of which are glycine (20%), aspartic acid (12%), alanine (10%), glutamic acid (9%), and serine (9%). Collectively, the amino acids account for 1.8% of the organic carbon in DOM and ~ 18% of its nitrogen. Free amino acids also exist in fresh-waters, but the median concentration of 0.3 μmol L^{-1} is a factor of 4–5 lower than the concentration of bound amino acids. Elevated levels of serine and ornithine in the data set for free amino acids strongly suggest that some samples may have been contaminated by human fingerprints.

Total hydrolyzable sugars are present in freshwaters at a median concentration of 2.6 μmol L^{-1}. Glucose (22%), galactose (19%), xylose (18%), arabinose (15%), and rhamnose (15%) are most abundant. Collectively, the total hydrolyzable sugars account for 3.0% of the organic carbon in DOM. Finally, the DOM in freshwaters also contains lignin-derived phe-nols at a median concentration of ~ 10 μg L^{-1}. Vanillyl phenols are the most abundant class of lignin-derived phenols, followed by syringyl and *p*-hydroxy phenols. Cinnamyl phenols are found only at quite low concentrations. Collectively, the lignin-derived phenols account for 0.6% of the organic carbon in DOM. Altogether, the organic carbon in amino acids, sugars, and lignin-derived phenols accounts for less than 6% of the organic carbon found in DOM in freshwaters.

Even with the tremendous effort that has been expended since the mid-1970s, DOM remains something of a mystery. Looking back to the seminal paper by Shapiro (1957), one can but marvel at what small advances have been made by the collective efforts of so many scientists throughout the world. This fact alone testifies to the unrivaled chemical complexity of this material that occurs so ubiquitously in the Earth's soils, sediments, and natural waters. The continued pursuit of its chemical nature will surely engage scientists for many years.

REFERENCES

Ahlers W. W., Reid R. M., Kim P. J., and Hunter K. A. (1990) Contamination-free sample collection and handling proto-cols for trace elements in natural fresh waters. *Austral. J. Mar. Freshwater Res.* **41**, 713–720.

Aiken G. R. (1984) Evaluation of ultrafiltration for determining molecular weight of fulvic acid. *Environ. Sci. Technol.* **18**, 978–981.

Aiken G. R. (1985) Isolation and concentration techniques for aquatic humic substances. In *Humic Substances in Soil, Sediment, and Water: Geochemistry, Isolation, and Charac-terization* (eds. G. R. Aiken, D. M. McKnight, R. L. Wershaw, and P. MacCarthy). Wiley, New York, pp. 363–385.

Aiken G. R. (1988) A critical evaluation of the use of macroporous resins for the isolation of aquatic humic substances. In *Humic Substances and Their Role in the Environment*. Dahlem Workshop Report (eds. F. H. Frimmel and R. F. Christman). Wiley, Chichester, pp. 15–28.

Aiken G. R. and Gillam A. H. (1989) Determination of molecular weights of humic substances by colligative property measurements. In *Humic Substances II: In Search of Structure*. (eds. M. H. B. Hayes, P. MacCarthy, R. L. Malcolm, and R. S. Swift) Wiley, Chichester, pp. 515–544.

Aiken G. R. and Malcolm R. L. (1987) Molecular weight of aquatic fulvic acids by vapor pressure osmometry. *Geochim. Cosmochim. Acta* **51**, 2177–2184.

Aiken G. R., Thurman E. M., Malcolm R. L., and Walton H. F. (1979) Comparison of XAD macroporous resins for the concentration of fulvic acid from aqueous solution. *Anal. Chem.* **51**, 1799–1803.

Aiken G. R., McKnight D. M., Wershaw R. L., and MacCarthy P. (1985) *Humic Substances in Soil, Sediment, and Water. Geochemistry, Isolation, and Characterization*. Wiley, New York, 692pp.

Aiken G. R., McKnight D. M., Thorn K. A., and Thurman E. M. (1992) Isolation of hydrophilic organic acids from water using nonionic macroporous resins. *Org. Geochem.* **18**, 567–573.

Aitkenhead J. A. and McDowell W. H. (2000) Soil C : N ratio as a predictor of annual riverine DOC flux at local and global scales. *Global Biogeochem. Cycles* **14**, 127–138.

Allen H. E. and Hansen D. J. (1996) The importance of trace metal speciation to water quality criteria. *Water Environ. Res.* **68**, 42–54.

Allen H. E., Perdue E. M., and Brown D. S. (1993) *Metals in Groundwater*. Lewis Publishers, Boca Raton, FL, 437pp.

Allen H. E., Garrison A. W., and Luther G. W. (1998) *Metals in Surface Waters*. Ann Arbor Press, Ann Arbor, MI, 262pp.

Amy G., Alleman B. C., and Cluff C. B. (1990) Removal of dissolved organic matter by nanofiltration. *J. Environ. Eng.—ASCE* **116**, 200–205.

Antweiler R. C. (1991) The hydrolysis of Suwannee River fulvic acid. In *Organic Substances and Sediments in Water: Volume 1. Humics and Soils* (ed. R. A. Baker). Lewis Publishers, Chelsea, MI, pp. 163–177.

Averett R. C., Leenheer J. A., McKnight D. M., and Thorn K. A. (1989) *Humic Substances in the Suwannee River, GA: Interaction, Properties, and Proposed Structures*, USGS Open File Report 87-557. United States Government Printing Office, Washington, DC, 377pp.

Backes C. A. and Tipping E. (1987) An evaluation of the use of cation-exchange resin for the determination of organically-complexed Al in natural acid waters. *Int. J. Environ. Analyt. Chem.* **30**, 135–143.

Bakker E. and Pretsch E. (2001) Potentiometry at trace levels. *Trends Anal. Chem.* **20**, 11–19.

Battin T. J. (1998) Dissolved organic matter and its optical properties in a blackwater tributary of the upper Orinoco river. Venezuela. *Org. Geochem.* **28**, 561–569.

Barnard T. E. and Bisogni J. J. (1985) Errors in Gran function analysis of titration data for dilute acidified waters. *Water Res.* **19**, 393–399.

Baumgartner A. and Reichel E. (1975) *The World Water Balance— Mean Annual Global, Continental, and Maritime Precipitation, Evaporation and Run-off*. Elsevier, Amsterdam, 179pp.

Becher G., Ovrum N. M., and Christman R. F. (1992) Novel chlorination by-products of aquatic humic substances. *Sci. Tot. Environ.* **117/118**, 509–520.

Beck K. C., Reuter J. H., and Perdue E. M. (1974) Organic and inorganic geochemistry of some coastal plain rivers of the southeastern United States. *Geochim. Cosmochim. Acta* **38**, 341–364.

Benedetti M. F., Milne C. J., Kinniburgh D. G., Van Riemsdijk W. H., and Koopal L. K. (1995) Metal ion binding to humic substances: application of the Non-ideal Competitive Adsorption model. *Environ. Sci. Technol.* **29**, 446–457.

Benedetti M. F., Van Riemsdijk W. H., and Koopal L. K. (1996) Humic substances considered as a heterogeneous Donnan gel phase. *Environ. Sci. Technol.* **30**, 1804–1813.

Benes P., Gjessing E. T., and Steinner E. (1976) Interactions between humus and trace elements in fresh water. *Water Res.* **10**, 711–716.

Benner R. (2003) Molecular indicators of the bioavailability of dissolved organic matter. In *Aquatic Ecosystems: Interactivity of Dissolved Organic Matter* (eds. S. E. G. Findlay and R. L. Sinsabaugh). Elsevier, San Diego, pp. 97–119.

Benner R. and Hedges J. I. (1993) A test of the accuracy of freshwater DOC measurements by high-temperature catalytic oxidation and UV-promoted persulfate oxidation. *Mar. Chem.* **41**, 161–165.

Benner R. and Opsahl S. (2001) Molecular indicators of the sources and transformations of dissolved organic matter in the Mississippi River plume. *Org. Geochem.* **32**, 597–611.

Benoit G. (1994) Clean techniques measurement of Pb, Ag and Cd in freshwater: a redefinition of metal pollution. *Environ. Sci. Technol.* **28**, 1987–1991.

Bergman H. L. and Dorward-King E. J. (1997) *Reassessment of Metals Criteria for Aquatic Life Protection.* SETAC Press, Pensacola, FL, 114pp.

Bernhard M., Brinckman F. E., and Sadler P. J. (1986) *The Importance of Chemical "Speciation" in Environmental Processes*, Dahlem Workshop Report. Springer, Berlin, 763pp.

Blom L., Edelhausen L., and Van Krevelen D. W. (1957) Chemical structure and properties of coal XVII—Oxygen groups in coal and related products. *Fuel* **36**, 135–153.

Blough N. V. and Green S. A. (1995) Spectroscopic characterization and remote sensing of nonliving organic matter. In *The Role of Nonliving Organic Matter in the Earth's Carbon Cycle*, Dahlem Workshop Report (eds. R. G. Zepp and C. Sonntag). Wiley, Chichester, pp. 23–45.

Boerschke R. C., Gallie E. A., Belzile N., Gedye R. N., and Morris J. R. (1996) Quantitative elemental and structural analysis of dissolved organic carbon fractions from lakes near Sudbury, Ontario. *Can. J. Chem.* **74**, 2460–2470.

Bowles E. C., Antweiler R. C., and MacCarthy P. (1989) Acid–base titrations and hydrolysis of fulvic acid from the Suwannee River. In *Humic Substances in the Suwannee River, GA: Interaction, Properties, and Proposed Structures*, USGS Open File Report 87-557 (eds. R. C. Averett, J. A. Leenheer, D. M. McKnight, and K. A. Thorn), United States Government Printing Office, Washington, DC, pp. 209–229.

Brauns F. R. (1960) *The Chemistry of Lignin*. Academic Press, New York, 804pp.

Brezonik P. L. (1994) *Chemical Kinetics and Process Dynamics in Aquatic Systems*. Lewis Publishers, Boca Raton, FL, 754pp.

Bricaud A., Morel A., and Prieur L. (1981) Absorption by dissolved organic matter of the sea (yellow substance) in the UV and visible domains. *Limnol. Oceanogr.* **16**, 43–53.

Brooks J. D. and Sternhell S. (1957) Chemistry of brown coals. *Austral. J. Appl. Sci.* **8**, 206.

Brown G. K., MacCarthy P., and Leenheer J. A. (1999) Simultaneous determination of Ca, Cu, Ni, Zn, and Cd binding strengths with fulvic acid fractions by Schubert's method. *Anal. Chim. Acta* **402**, 169–181.

Buffle J. (1988) *Complexation Reactions in Aquatic Systems: An Analytical Approach*. Ellis Horwood Ltd., 692pp.

Buffle J., Greter F.-L., and Haerdi W. (1977) Measurement of complexation of humic and fulvic acids in natural waters with lead and copper ion-selective electrodes. *Anal. Chem.* **49**, 216–222.

Buffle J., Deladoey P., and Haerdi W. (1978) The use of ultrafiltration for the separation and fractionation of organic ligands in fresh waters. *Anal. Chim. Acta* **101**, 339–357.

Cabaniss S. E. and Shuman M. S. (1988a) Copper binding by dissolved organic matter: I. Suwannee River fulvic acid equilibria. *Geochim. Cosmochim. Acta* **52**, 185–193.

Cabaniss S. E. and Shuman M. S. (1988b) Copper binding by dissolved organic matter: II. Variation in type and source of organic matter. *Geochim. Cosmochim. Acta* **52**, 195–200.

Cabaniss S. E., Zhou Q., Maurice P. A., Chin Y.-P., and Aiken G. R. (2000) A log-normal distribution model for the molecular weight of aquatic fulvic acids. *Environ. Sci. Technol.* **34**, 1103–1109.

Cantrell K. J., Serkiz S. M., and Perdue E. M. (1990) Evaluation of acid neutralizing capacity data for solutions containing natural organic acids. *Geochim. Cosmochim. Acta* **54**, 1247–1254.

Chau Y. K. and Lum-Shue-Chan K. (1974) Determination of labile and strongly bound metals in lake water. *Water Res.* **8**, 383–388.

Chau Y. K., Gächter R., and Lum-Shue-Chan K. (1974) Determination of the apparent complexing capacity of lake water. *J. Fish. Res. Board Can.* **31**, 1515–1519.

Chen Y. and Schnitzer M. (1989) Sizes and shapes of humic substances by electron microscopy. In *Humic Substances II: In Search of Structure* (eds. M. H. B. Hayes, P. MacCarthy, R. L. Malcolm, and R. S. Swift). Wiley, Chichester, pp. 621–638.

Cheng X. and Kaplan L. A. (2001) Improved analysis of dissolved carbohydrates in stream water with HPLC-PAD. *Anal. Chem.* **73**, 458–461.

Chin Y.-P. and Gschwend P. M. (1991) The abundance, distribution, and configuration of porewater organic colloids in recent sediments. *Geochim. Cosmochim. Acta* **55**, 1309–1317.

Chin Y.-P., Aiken G. R., and O'Loughlin E. (1994) Molecular weight, polydispersity, and spectroscopic properties of aquatic humic substances. *Environ. Sci. Technol.* **26**, 1853–1858.

Choppin G. R. and Kullberg L. (1978) Protonation thermodynamics of humic acid. *J. Inorg. Nucl. Chem.* **40**, 651–654.

Christman R. F. and Gjessing E. T. (1983) *Aquatic and Terrestrial Humic Materials*. Ann Arbor Science, Ann Arbor, MI, 538pp.

Christman R. F., Norwood D. L., Millington D. S., and Johnson J. D. (1983) Identity and yields of major halogenated products of aquatic fulvic acid chlorination. *Environ. Sci. Technol.* **17**, 625–628.

Clair T. A., Kramer J. R., Sydor M., and Eaton D. (1991) Concentration of aquatic dissolved organic matter by reverse osmosis. *Water Res.* **25**, 1033–1037.

Clapp C. E., Emerson W. W., and Olness A. E. (1989) Sizes and shapes of humic substances by viscosity measurements. In *Humic Substances II: In Search of Structure* (eds. M. H. B. Hayes, P. MacCarthy, R. L. Malcolm, and R. S. Swift). Wiley, Chichester, pp. 497–514.

Clapp C. E., Hayes M. H. B., Senesi N., and Griffith S. M. (1996) *Humic Substances and Organic Matter in Soil and Water Environments: Characterization, Transformations, and Interactions. Proceedings of the 7th International Conference of the International Humic Substances Society, University of the West Indies, St. Augustine, Trinidad and Tobago, July 3–9, 1994.* IHSS, St. Paul, MN, 493pp.

Clesceri L. S., Greenberg A. E., and Trusell R. R. (eds.) (1989) *Standard Methods for the Examination of Water and Wastewater*, 17th edn. American Public Health Association, Washington, DC.

Coble P. G. (1996) Characterization of marine and terrestrial DOM in seawater using excitation-emission matrix spectroscopy. *Mar. Chem.* **51**, 325–346.

Collins M. R., Amy G. L., and Steelink C. (1986) Molecular weight distribution, carboxylic acidity, and humic substances content of aquatic organic matter: implications of removal during water treatment. *Environ. Sci. Technol.* **20**, 1028–1032.

Cooper W. J., Zika R. G., Petasne R. G., and Fischer A. M. (1989) Sunlight-induced photochemistry of humic substances in natural waters: major reactive species. In *Aquatic Humic Substances: Influence on Fate and Treatment of Pollutants.* Advances in Chemistry Series 219 (eds. I. H. Suffet and P. MacCarthy). American Chemical Society, Washington, DC, pp. 333–362.

Cowie G. L. and Hedges J. I. (1984) Determination of neutral sugars in plankton, sediments, and wood by capillary gas chromatography of equilibrated isomeric mixtures. *Anal. Chem.* **58**, 497–504.

Cronan C. S., Reiners W. A., Reynolds R. C., and Lang G. E. (1978) Forest floor leaching: contributions from mineral, organic, and carbonic acids in New Hampshire subalpine forests. *Science* **200**, 309–311.

Crum R. H., Murphy E. M., and Keller C. K. (1996) A nonadsorptive method for the isolation and fractionation of natural dissolved organic carbon. *Water Res.* **30**, 1304–1311.

Davies-Colley R. J. and Vant W. N. (1987) Absorption of light by yellow substance in freshwater lakes. *Limnol. Oceanogr.* **32**, 416–425.

Davis J. A. (1982) Adsorption of natural dissolved organic matter at the oxide/water interface. *Geochim. Cosmochim. Acta* **46**, 2381–2393.

Degens E. T., Kempe S., and Richey J. E. (1991) Summary: biogeochemistry of major world rivers. In *SCOPE 42, Biogeochemistry of Major World Rivers* (eds. E. T. Degens, S. Kempe, and J. E. Richey). Wiley, Chichester, pp. 323–347.

de Haan H. (1993) Solar UV-light penetration and photodegradation of humic substances in peaty lake water. *Limnol. Oceanogr.* **38**, 1072–1076.

de Haan H. and de Boer T. (1987) Applicability of light absorbance and fluorescence as measures of concentration and molecular size of dissolved organic carbon in humic Lake Tjeukemeer. *Water Res.* **21**, 731–734.

De Nobili M., Gjessing E., and Sequi P. (1989) Sizes and shapes of humic substances by gel chromatography. In *Humic Substances II: In Search of Structure* (eds. M. H. B. Hayes, P. MacCarthy, R. L. Malcolm, and R. S. Swift). Wiley, Chichester, pp. 561–597.

Di Toro D. M., Allen H. E., Bergman H. L., Meyer J. S., Paquin P. R., and Santore R. C. (2001) Biotic ligand model of the acute toxicity of metals: I. Technical basis. *Environ. Toxicol. Chem.* **20**, 2383–2396.

Dobbs J. C., Susetyo W., Knight F. E., Castles M. A., Carreira L. A., and Azarraga L. V. (1989) Characterization of metal binding sites in fulvic acids by lanthanide ion probe spectroscopy. *Anal. Chem.* **61**, 483–488.

Dowty B., Carlisle D., Laseter J. L., and Storer J. (1975) Halogenated hydrocarbons in New Orleans drinking water and blood plasma. *Science* **187**, 75–77.

Driscoll C. T. (1980) Chemical characterization of some dilute acidified lakes and streams in the Adirondack region of New York state. PhD Thesis, Cornell University.

Driscoll C. T. (1984) A procedure for the fractionation of aqueous aluminum in dilute acidic waters. *Int. J. Environ. Analyt. Chem.* **16**, 267–283.

Driscoll C. T., Fuller R. D., and Schecher W. D. (1989) The role of organic acids in the acidification of surface waters in the eastern US. *Water Air Soil Pollut.* **43**, 21–40.

Drozd J., Gonet S. S., Senesi N., and Weber J. (1997) *The Role of Humic Substances in the Ecosystems and in Environmental Protection. Proceedings of the 8th Meeting of the International Humic Substances Society, Wroclaw, Poland, September 9–14, 1996.* Polish Society of Humic Substances, Wroclaw, Poland, 1002pp.

Dubach P., Mehta N. C., Jakab T., and Martin F. (1964) Chemical investigations on soil humic substances. *Geochim. Cosmochim. Acta* **28**, 1567–1578.

Duxbury J. M. (1989) Studies of the molecular size and charge of humic substances by electrophoresis. In *Humic Substances II: In Search of Structure* (eds. M. H. B. Hayes, P. MacCarthy, R. L. Malcolm, and R. S. Swift). Wiley, Chichester, pp. 593–620.

Dzombak D. A., Fish W., and Morel F. M. M. (1986) Metal–humate interactions: 1. Discrete ligand and continuous distribution models. *Environ. Sci. Technol.* **20**, 669–675.

Eberle S. H. and Feuerstein W. (1979) On the pK spectrum of humic acid from natural waters. *Naturwissenschaften* **66**, 572–573.

Edwards A. C. and Cresser M. S. (1987) Relationships between ultraviolet absorbance and total organic carbon in two upland catchments. *Water Res.* **21**, 49–56.

Ephraim J., Alegret S., Mathuthu A., Bicking M., Malcolm R. L., and Marinsky J. A. (1986) A united physiochemical description of the protonation and metal ion complexation equilibria of natural organic acids (humic and fulvic acids): 2. Influence of polyelectrolyte properties and functional group heterogeneity on the protonation equilibria of fulvic acid. *Environ. Sci. Technol.* **30**, 354–366.

Ertel J. R. and Hedges J. I. (1984) The lignin component of humic substances: distribution among soil and sedimentary humic, fulvic, and base-insoluble fractions. *Geochim. Cosmochim. Acta* **48**, 2065–2074.

Eshleman K. N. and Hemond H. F. (1985) The role of organic acids in the acid–base status of surface waters at Bickford watershed, Massachusetts. *Water Resour. Res.* **21**, 1503–1510.

Findlay S. E. G. and Sinsabaugh R. L. (eds.) (2003) *Aquatic Ecosystems—Interactivity of Dissolved Organic Matter.* Academic Press, San Diego, 512pp.

Fish W., Dzombak D. A., and Morel F. M. M. (1986) Metalhumate interactions: 2. Application and comparison of models. *Environ. Sci. Technol.* **20**, 676–683.

Frimmel F. H., Abbt-Braun G., Heumann K. G., Hock B., Ludemann H.-D., and Spiteller M. (eds.) (2002) *Refractory Organic Substances in the Environment.* Wiley-VCH, Weinheim, 546pp.

Gamble D. S. (1970) Titration curves of fulvic acid: the analytical chemistry of a weak acid polyelectrolyte. *Can. J. Chem.* **48**, 2662–2669.

Gamble D. S. (1972) Potentiometric titration of fulvic acid: equivalence point calculations and acidic functional groups. *Can. J. Chem.* **50**, 2680–2690.

Gao H. and Zepp R. G. (1998) Factors influencing photoreactions of dissolved organic matter in a coastal river of the southeastern United States. *Environ. Sci. Technol.* **32**, 2940–2946.

Gardiner J. (1974) The chemistry of cadmium in natural water: I. A study of cadmium complex formation using the cadmium specific-ion electrode. *Water Res.* **8**, 23–30.

Garrels R. M., Mackenzie F. T., and Hunt C. (1975) *Chemical Cycles and the Global Environment.* Kaufmann, Los Altos, pp. 73–80.

Gillam A. H. and Riley J. P. (1981) Correction of osmometric number-average molecular weights of humic substances for dissociation. *Chem. Geol.* **33**, 355–366.

Gjessing E. T. (1970) Ultrafiltration of aquatic humus. *Environ. Sci. Technol.* **4**, 437–438.

Gjessing E. T., Egeberg P. K., and Hakedal J. (1999) Natural organic matter in drinking water—the "NOM-Typing Project," background and basic characteristics of original water samples and NOM isolates. *Environ. Int.* **25**, 145–159.

Glover C. A. (1975) Absolute colligative property measurements. In *Polymer Molecular Weights* (ed. P. E. Slade, Jr.) Dekker, New York, pp. 79–159.

Goldberg M. C. and Weiner E. R. (1989) Fluorescence measurements of the volume, shape, and fluorophore composition of fulvic acid from the Suwannee River. In *Humic Substances in the Suwannee River, Georgia: Interactions, Properties, and Proposed Structures*, USGS Open File Report 87-557 (eds. R. C. Averett, J. A. Leenheer, D. M. McKnight, and K. A. Thorn). United States Government Printing Office, Washington, DC, pp. 179–204.

Gorsuch J. W., Janssen C. R., Lee C. M., and Reiley M. C. (eds.) (2002) *Special Issue: The Biotic Ligand Model for Metals—Current Research, Future Directions, Regulatory Implications. Comp. Biochem. Physiol.* **133C**, 343pp.

Gran G. (1950) Determination of the equivalent point in potentiometric titrations. *Acta Chem. Scand.* **4**, 559–577.

Graves C. G., Matanowski G. M., and Tardiff R. G. (2001) Weight of evidence for an association between adverse reproductive and developmental effects of exposure to disinfection by-products: a critical review. *Reg. Toxicol. Pharmacol.* **34**, 103–124.

Grzyb K. R. (1995) NOAEM (natural organic anion equilibrium model): a data analysis algorithm for estimating functional properties of dissolved organic matter in aqueous environments: Part 1. Ionic component speciation and metal association. *Org. Geochem.* **23**, 379–390.

Grzybowski W. (2000) Effect of short-term sunlight irradiation on absorbance spectra of chromophoric organic matter dissolved in coastal and riverine water. *Chemosphere* **40**, 1313–1318.

Haiber S., Herzog H., Buddrus J., Burba P., and Lambert J. (2002) Quantification of substructures of refractory organic substances by means of nuclear magnetic resonance. In *Refractory Organic Substances in the Environment* (eds. F. H. Frimmel, G. Abbt-Braun, K. G. Heumann, B. Hock, H.-D. Ludemann, and M. Spiteller). Wiley-VCH, Weinheim, Germany, pp. 115–127.

Hamilton P. B. (1965) Amino acids on hands. *Nature* **205**, 284–285.

Hansen E. H. and Schnitzer M. (1969) Molecular weight measurements of polycarboxylic acids in water by vapor pressure osmometry. *Anal. Chim. Acta* **46**, 247–254.

Hare P. E. and St. John P. A. (1988) Detection limits for amino acids in environmental samples. In *Detection in Analytical Chemistry. Importance, Theory, and Practice* (ed. L. A. Currie). ACS Press, Washington, DC, pp. 275–285.

Harrison G. and Morel F. M. M. (1983) Antagonism between cadmium and iron in the marine diatom *Thalassiosira weissflogii. J. Phycol.* **19**, 495–507.

Hatcher P. G., Schnitzer M., Dennis L. W., and Maciel G. E. (1981) Aromaticity of humic substances in soils. *Soil Sci. Soc. Am. J.* **45**, 1089–1094.

Hedges J. I. and Mann D. C. (1979a) The characterization of plant tissues by their lignin oxidation products. *Geochim. Cosmochim. Acta* **43**, 1803–1807.

Hedges J. I. and Mann D. C. (1979b) The lignin geochemistry of marine sediments from the southern Washington coast. *Geochim. Cosmochim. Acta* **43**, 1809–1818.

Hedges J. I. and Parker P. L. (1976) Land-derived organic matter in surface sediments from the Gulf of Mexico. *Geochim. Cosmochim. Acta* **40**, 1019–1029.

Hedges J. I., Cowie G. L., Richey J. E., Quay P. D., Benner R., Strom M., and Forsberg B. R. (1994) Origins and processing of organic matter in the Amazon River as indicated by carbohydrates and amino acids. *Limnol. Oceanogr.* **39**, 743–761.

Hedges J. I., Baldock J. A., Gelinas Y., Lee C., Peterson M. L., and Wakeham S. G. (2002) The biochemical and elemental compositions of marine plankton: a NMR perspective. *Mar. Chem.* **78**, 47–63.

Hemmingsen S. L. and McGown L. B. (1997) Phase-resolved fluorescence and lifetime characterization of commercial humic substances. *Appl. Spectrosc.* **51**, 921–929.

Hemond H. F. (1990) Acid neutralizing capacity, alkalinity, and acid–base status of natural waters containing organic acids. *Environ. Sci. Technol.* **24**, 1486–1489.

Henriksen A. and Seip H. M. (1980) Strong and weak acids in surface waters of southern Norway and southwestern Scotland. *Water Res.* **14**, 809–813.

Her N., Amy G., Foss D., and Cho J. (2002) Variations of molecular weight estimation by HP-Size Exclusion Chromatography with UVA versus online DOC detection. *Environ. Sci. Technol.* **36**, 3393–3399.

Herlihy A. T., Kaufmann P. K., and Mitch M. E. (1991) Stream chemistry in the Eastern United States: 2. Current sources of acidity in acidic and low acid-neutralizing capacity streams. *Water Resour. Res.* **27**, 629–642.

Hessen D. O. and Tranvik L. J. (eds.) (1998) *Aquatic Humic Substances: Ecology and Biogeochemistry*. Springer, Berlin, 346pp.

Hoigné J., Faust B. C., Haag W. R., Scully F. E., Jr., and Zepp R. G. (1989) Aquatic humic substances as sources and sinks of photochemically produced transient reactants. In *Aquatic Humic Substances: Influence on Fate and Treatment of Pollutants*, Advances in Chemistry Series 219 (eds. I. H. Suffet and P. MacCarthy). American Chemical Society, Washington, DC, pp. 363–381.

Holcomb D. W. (1979) Chemical characterization of swamp peat humic substances. MS Thesis, Georgia Institute of Technology, 63pp.

Hongve D. (1990) Shortcomings of Gran titration procedures for determination of alkalinity and weak acids in humic water. *Water Res.* **24**, 1305–1308.

Hopkinson C. S., Buffam I., Hobbie J., Vallino J., Perdue M., Eversmeyer B., Prahl F., Covert J., Hodson R., Moran M. A., Smith E., Baross J., Crump B., Findlay S., and Foreman K. (1998) Terrestrial Inputs of organic matter to coastal ecosystems: an intercomparison of chemical characteristics and bioavailability. *Biogeochemistry* **43**, 211–234.

Huffman E. W. D. and Stuber H. A. (1985) Analytical methodology for elemental analysis of humic substances. In *Humic Substances in Soil, Sediment, and Water: Geochemistry, Isolation, and Characterization* (eds. G. R. Aiken, D. M. McKnight, R. L. Wershaw, and P. MacCarthy). Wiley, New York, pp. 433–455.

Hunt A. P., Parry J. D., and Hamilton-Taylor J. (2000) Further evidence of elemental composition as an indicator of the bioavailability of humic substances to bacteria. *Limnol. Oceanogr.* **45**, 237–241.

Jahnel J. B., Ilieva P., and Frimmel F. H. (1998) HPAE-PAD—a sensitive method for determination of carbohydrates. *Fres. J. Anal. Chem.* **360**, 827–829.

Johnson N. M., Driscoll C. T., Eaton J. S., Likens G. E., and McDowell W. H. (1981) "Acid rain," dissolved aluminum and chemical weathering at the Hubbard Brook Experimental Forest, New Hampshire. *Geochim. Cosmoshim. Acta* **45**, 1421–1437.

Jolley R. L., Brungs W. A., Cotruvo J. A., Cumming R. B., Mattice J. S., and Jacobs V. A. (1983) *Water Chlorination. Environmental Impact and Health Effects: Vol. 4, Book 1, Chemistry and Water Treatment*. Ann Arbor Science, Ann Arbor, MI. 1491pp.

Kahl J. S., Norton S. A., MacRae R. K., Haines T. A., and Davis R. B. (1989) The influence of organic acidity on the acid–base chemistry of surface waters in Maine, USA. *Water Air Soil Pollut.* **46**, 221–233.

Kaplan L. A. and Newbold J. D. (2003) The role of monomers in stream ecosystem metabolism. In *Aquatic Ecosystems: Interactivity of Dissolved Organic Matter* (eds. S. E. G. Findlay and R. L. Sinsabaugh). Elsevier, San Diego, pp. 97–119.

Keck D. W. (1994) Removal of dissolved natural organic matter from a synthetic groundwater with nanofiltration. MS Thesis, Georgia Institute of Technology, 65pp.

Kempe S. (1979) Carbon in the freshwater cycle. In *SCOPE 13: The Global Carbon Cycle*. (eds. B. Bolin, E. T. Degens, S. Kempe, and P. Ketner). Wiley, Chichester, pp. 317–342.

Khalaf K. Y., MacCarthy P., and Gilbert T. W. (1975) Application of thermometric titrations to the study of soil organic matter: II. Humic acids. *Geoderma* **14**, 331–340.

Kilduff J. E. and Weber W. J. (1992) Transport and separation of organic macromolecules in ultrafiltration processes. *Environ. Sci. Technol.* **26**, 569–577.

Kitis M., Karanfil T., Kilduff J. E., and Wigton A. (2001a) The reactivity of natural organic matter to disinfection by-products formation and its relation to specific ultraviolet absorbance. *Water Sci. Technol.* **43**, 9–16.

Kitis M., Kilduff J. E., and Karanfil T. (2001b) Isolation of dissolved organic matter (DOM) from surface waters using reverse osmosis and its impact on the reactivity of DOM to formation and speciation of disinfection by-products. *Water Res.* **35**, 2225–2234.

Knulst J. C., Boerschke R. C., and Loemo S. (1997) Differences in organic surface microlayers from an artificially acidified and control lake, elucidated by XAD-8/XAD-4 tandem separation and solid state ^{13}C NMR spectroscopy. *Environ. Sci. Technol.* **32**, 8–12.

Köhler S. J., Lofgren S., Wilander A., and Bishop K. (2001) Validating a simple equation to predict and analyze organic anion charge in Swedish low ionic strength surface waters. *Water Air Soil Pollut.* **130**, 799–804.

Koopal L. K., Van Riemsdijk W. H., de Wit J. C. M., and Benedetti M. F. (1994) Analytical isotherm equations for multicomponent adsorption to heterogeneous surfaces. *J. Colloid Interface Sci.* **166**, 51–60.

Korshin G. V., Li C.-W., and Benjamin M. M. (1996) Use of UV spectroscopy to study chlorination of natural organic matter. In *Water Disinfection and Natural Organic Matter: Characterization and Control*, ACS Symposium Series 649 (eds. R. A. Minear and G. L. Amy). American Chemical Society, Washington, DC, pp. 182–195.

Kramer J. R. and Allen H. E. (1988) *Metal Speciation—Theory, Analysis, and Application*. Lewis Publishers, Chelsea, 357pp.

Kramer J. R. and Davies S. S. (1988) Estimation of non-carbonato protolytes for selected lakes in the Eastern Lakes Survey. *Environ. Sci. Technol.* **22**, 182–185.

Kramer R. W., Kujawinski E. B., Zang X., Green-Church K. B., Jones R. B., Freitas M. A., and Hatcher P. G. (2001) Studies of the structure of humic substances by electrospray ionization coupled to a quadrupole-time of flight (QQ-TOF) mass spectrometer. In *Humic Substances: Structures, Models and Functions* (eds. E. A. Ghabbour and G. Davies). The Royal Society of Chemistry, Cambridge, England, pp. 95–107.

Krasner S. W., Croué J.-P., Buffle J., and Perdue E. M. (1996) Three approaches for characterizing NOM. *J. Am. Water Works. Assoc.* **88**, 66–79.

Krasner S. W., McGuire M. J., Jacangelo J. G., Patania N. L., Reagan K. M., and Aieta E. M. (1989) The occurrence of disinfection by-products in US drinking water. *J. Am. Water Works Assoc.* **81**, 41–53.

Kronberg L., Christman R. F., Singh R., and Ball L. M. (1991) Identification of oxidized and reduced forms of the strong bacterial mutagen (Z)-2-chloro-3-(dichloromethyl)-4-oxo-butenoic acid (MX) in extracts of chlorine-treated water. *Environ. Sci. Technol.* **25**, 99–104.

Kwak J. C. T., Nelson R. W. P., and Gamble D. S. (1977) Ultrafiltration of fulvic and humic acid, a comparison of stirred cell and hollow fiber techniques. *Geochim. Cosmochim. Acta* **41**, 993–996.

Labuda J., Vaníčková M., Uhlemann E., and Mickler W. (1994) Applicability of chemically modified electrodes for determination of copper species in natural waters. *Anal. Chim. Acta* **284**, 517–523.

Lamar W. L. (1968) Evaluation of organic color and iron in natural surface waters. *US Geol. Surv. Prof. Pap.* 600D, D24–D29.

Lankes U. and Lüdemann H.-D. (2002) Structural characterization of refractory organic substances by solid-state high-resolution 13C and 15N nuclear magnetic resonance. In *Refractory Organic Substances in the Environment* (eds. F. H. Frimmel, G. Abbt-Braun, K. G. Heumann, B. Hock, H.-D. Ludemann, and M. Spiteller). Wiley-VCH, Weinheim, pp. 98–114.

Lansing W. D. and Kraemer E. O. (1935) Molecular weight analysis of mixtures by sedimentation equilibrium on the Svedberg ultracentrifuge. *J. Am. Chem. Soc.* **57**, 1369–1377.

Larive K. C., Rogers A., Morton M., and Carper W. R. (1996) ^{113}Cd NMR binding studies of Cd-fulvic acid complexes: evidence of fast exchange. *Environ. Sci. Technol.* **30**, 2828–2831.

Lean D. (1998) Attenuation of solar radiation in humic waters. In *Aquatic Humic Substances, Ecological Studies 133* (eds. D. O. Hessen and L. J. Tranvik). Springer, Berlin, pp. 109–124.

Lee Y. H. and Brosset C. (1978) The slope of Gran's plot: a useful function in the examination of precipitation, the water-soluble part of airborne particles, and lake water. *Water Air Soil Pollut.* **10**, 457–469.

Leenheer J. A. (1981) Comprehensive approach to preparative isolation and fractionation of dissolved organic carbon from natural waters and wastewaters. *Environ. Sci. Technol.* **15**, 578–587.

Leenheer J. A. and Huffman E. W. D., Jr. (1976) Classification of organic solutes in water by using macroreticular resins. *J. Res. US Geol. Surv.* **4**, 737–751.

Leff L. G. and Meyer J. L. (1991) Biological availability of dissolved organic carbon along the Ogeechee River. *Limnol. Oceanogr.* **36**, 315–323.

Leuenberger B. and Schindler P. W. (1986) Application of integral pK spectrometry to the titration curve of fulvic acid. *Anal. Chem.* **58**, 1471–1474.

Li C., Benjamin M. M., and Korshin G. V. (2001) The relationship between TOX formation and spectral changes accompanying chlorination of pre-concentrated or fractionated NOM. *Water Res.* **36**, 3265–3272.

Li J., Perdue M., and Gelbaum L. T. (1998) Using cadmium-113 NMR spectrometry to study metal complexation by natural organic matter. *Environ. Sci. Technol.* **32**, 483–487.

Linthurst R. A., Landers D. H., Eilers J. M., Brakke D. F., Overton W. S., Meier E. P., and Crowe R. E. (1986) *Characteristics of Lakes in the Eastern United States: Volume I. Population Descriptions and Physico-chemical Relationships*, EPA/600/4–86/007a, US Environmental Protection Agency, Washington, DC, 136pp.

Lobbes J. M., Fitznar H. P., and Kattner G. (1999) High-performance liquid chromatography of lignin-derived phenols in environmental samples with diode array detection. *Anal. Chem.* **71**, 3008–3012.

Lobbes J. M., Fitznar H. P., and Kattner G. (2000) Biogeochemical characteristics of dissolved and particulate organic matter in Russian rivers entering the Arctic Ocean. *Geochim. Cosmochim. Acta* **64**, 2973–2983.

Ludwig W., Probst J.-L., and Kempe S. (1996) Predicting the oceanic input of organic carbon by continental erosion. *Global Biogeochem. Cycles* **10**, 23–41.

Ma H., Allen H. E., and Yin Y. (2001) Characterization of isolated fractions of dissolved organic matter from natural waters and a wastewater effluent. *Water Res.* **35**, 985–996.

Mahieu N., Powlson D. S., and Randall E. W. (1999) Statistical analysis of published carbon-13 CPMAS NMR spectra of soil organic matter. *Soil Sci. Soc. Am. J.* **63**, 307–319.

Malcolm R. L. (1989) Applications of solid-state ^{13}C NMR spectroscopy to geochemical studies of humic substances. In *Humic Substances II: In Search of Structure* (eds. M. H. B. Hayes, P. MacCarthy, R. L. Malcolm, and R. S. Swift). Wiley, Chichester, pp. 339–372.

Malcolm R. L. and MacCarthy P. (1992) Quantitative evaluation of XAD-8 and XAD-4 resins used in tandem for removing organic solutes from water. *Environ. Int.* **18**, 597–607.

Malcolm R. L., Aiken G. R., Bowles E. C., and Malcolm J. D. (1989) Isolation of fulvic and humic acids from the Suwannee River. In *Humic Substances in the Suwannee River, Georgia: Interactions, Properties, and Proposed Structures*, USGS Open File Report 87-557 (eds. R. C. Averett, J. A. Leenheer, D. M. McKnight, and K. A. Thorn). United States Government Printing Office, Washington, DC, pp. 23–35.

Mantoura R. F. C. and Riley J. P. (1975) The analytical concentration of humic substances from natural waters. *Anal. Chim. Acta* **76**, 97–106.

Mao J.-D., Hu W.-G., Schmidt-Rohr K., Davies G., Ghabbour E. A., and Xing B. (2000) Quantitative characterization of humic substances by solid-state carbon-13 nuclear magnetic resonance. *Soil Sci. Soc. Am. J.* **64**, 873–884.

Marhaba T. F., Van D., and Lippincott R. L. (2000) Rapid identification of dissolved organic matter fractions in water by spectral fluorescent signatures. *Water Res.* **34**, 3543–3550.

Marinsky J. A. and Reddy M. M. (1990) Vapor-pressure osmometric study of the molecular weight and aggregation tendency of a reference-soil fulvic acid. *Anal. Chim. Acta* **232**, 123–130.

Maurice P. A., Pullin M. J., Cabaniss S. E., Zhou Q., Namjesnik-Dejanovic K., and Aiken G. R. (2002) A comparison of surface water natural organic matter in raw filtered water samples, XAD, and reverse osmosis isolates. *Water Res.* **36**, 2357–2371.

McKnight D. M., Boyer E. W., Westerhoff P. K., Doran P. T., Kulbe T., and Andersen D. T. (2001) Spectrofluorometric characterization of dissolved organic matter for indication of precursor organic material and aromaticity. *Limnol. Oceanogr.* **46**, 38–48.

Messer J. J., Ariss C. W., Baker J. R., Drousé S. K., Eshleman K. N., Kaufmann P. R., Linthurst R. A., Omernik J. M., Overton W. S., Sale M. J., Schonbrod R. D., Stambaugh S. M., and Tuschall J. R., Jr. (1986) *National Surface Water Survey: National Stream Survey, Phase I—Pilot Survey*, EPA/600/4-86/026, US Environmental Protection Agency, Washington, DC, 179pp.

Meybeck M. (1981) River transport of organic carbon to the ocean. In *Flux of Organic Carbon by Rivers to the Ocean: CONF 8009140 UC-11*. US Department of Energy, Washington, DC, pp. 219–269.

Meybeck M. (1982) Carbon, nitrogen, and phosphorus transport by world rivers. *Am. J. Sci.* **282**, 401–450.

Miles C. J. and Brezonik P. L. (1981) Oxygen consumption in humic-colored waters by a photochemical ferrous–ferric catalytic cycle. *Environ. Sci. Technol.* **15**, 1089–1095.

Miller W. L. (1998) Effects of UV radiation on aquatic humus: photochemical principles and experimental considerations. In *Aquatic Humic Substances, Ecological Studies 133* (eds. D. O. Hessen and L. J. Tranvik). Springer, Berlin, pp. 125–143.

Milne C. J., Kinniburgh D. G., and Tipping E. (2001) Generic NICA-Donnan model parameters for proton binding by humic substances. *Environ. Sci. Technol.* **35**, 2049–2059.

Minear R. A. and Amy G. L. (1996) *Disinfection By-products in Water Treatment: The Chemistry of Their Formation and Control*. Lewis Publishers, Boca Raton, FL, 394pp.

Mobed J. J., Hemmingsen S. L., Autry J. L., and McGown L. B. (1996) Fluorescence characterization of IHSS humic substances: total luminescence spectra with absorbance correction. *Environ. Sci. Technol.* **30**, 3061–3065.

Molværsmyr K. and Lund W. (1983) Acids and bases in freshwaters. Interpretation of results from Gran plots. *Water Res.* **17**, 303–307.

Mopper K., Schultz C. A., Chevolot L., Germain C., Revuelta R., and Dawson R. (1992) Determination of sugars in unconcentrated seawater and other natural waters by liquid chromatography and pulsed amperometric detection. *Environ. Sci. Technol.* **26**, 133–138.

Moran M. A. and Zepp R. G. (1997) Role of photoreactions in the formation of biologically labile compounds from dissolved organic matter. *Limnol. Oceanogr.* **42**, 1307–1316.

Morel F. M. M. (1983) *Principles of Aquatic Chemistry*. Wiley-Interscience, New York, pp. 300–308.

Morel F. M. M. and Hering J. G. (1993) *Principles and Applications of Aquatic Chemistry*. Wiley-Interscience, New York, 588pp.

Morris D. P. and Hargreaves B. R. (1997) The role of photochemical degradation of dissolved organic carbon in regulating the UV transparency of three lakes on the Pocono Plateau. *Limnol. Oceanogr.* **42**, 239–249.

Munson R. K. and Gherini S. A. (1993) Influence of organic acids on the pH and acid-neutralizing capacity of Adirondack lakes. *Water Resour. Res.* **29**, 891–899.

Nikolaou A. D. and Lekkas T. D. (2001) The role of natural organic matter during formation of chlorinated by-products: a review. *Acta Hydrochim. Hydrobiol.* **29**, 63–77.

Noot D. K., Anderson W. B., Daignault S. A., Williams D. T., and Huck P. M. (1989) Evaluating treatment processes with the Ames mutagenicity assay. *J. Am. Water Works Assoc.* **81**, 87–102.

Oliver B. G., Thurman E. M., and Malcolm R. L. (1983) The contribution of humic substances to the acidity of colored natural waters. *Geochim. Cosmochim. Acta* **47**, 2031–2035.

Oro J. and Skewes H. B. (1965) Free amino acids on human fingers: the question of contamination in microanalysis. *Nature* **207**, 1042–1045.

Osburn C. L., Zagarese H. E., Morris D. P., Hargreaves B. R., and Cravero W. E. (2001) Calculation of spectral weighting functions for the solar photobleaching of chromophoric dissolved organic matter in temperate lakes. *Limnol. Oceanogr.* **46**, 1455–1467.

Otto W. H., Burton S. D., Carper W. R., and Larive C. K. (2001a) Examination of cadmium(II) complexation by the Suwannee River fulvic acid using ^{113}Cd NMR relaxation measurements. *Environ. Sci. Technol.* **35**, 4900–4904.

Otto W. H., Carper W. R., and Larive C. K. (2001b) Measurement of cadmium(II) and calcium(II) complexation by fulvic acids using ^{113}Cd NMR. *Environ. Sci. Technol.* **35**, 1463–1468.

Owen D. M., Amy G. L., and Chowdhury Z. K. (1993) *Characterization of Natural Organic Matter and its Relationship to Treatablility*. American Water Works Association, Denver, CO, 250pp.

Pagenkopf G. K. (1983) Gill surface interaction model for trace-metal toxicity to fishes: role of complexation, pH, and water hardness. *Environ. Sci. Technol.* **17**, 342–347.

Pagenkopf G. K., Russo R. C., and Thurston R. V. (1974) Effect of complexation on toxicity of copper to fishes. *J. Fish. Res. Board Can.* **31**, 462–465.

Paxéus N. and Wedborg M. (1985) Acid–base properties of aquatic fulvic acid. *Anal. Chim. Acta* **169**, 87–98.

Perdue E. M. (1984) Analytical constraints on structural features of humic substances. *Geochim. Cosmochim. Acta.* **48**, 1435–1442.

Perdue E. M. (1985) Acidic functional groups of humic substances. In *Humic Substances in Soil, Sediment, and Water: Geochemistry, Isolation, and Characterization* (eds. G. R. Aiken, D. M. McKnight, R. L. Wershaw, and P. MacCarthy). Wiley, New York, pp. 493–526.

Perdue E. M. (1998) Chemical composition, structure, and metal binding properties. In *Aquatic Humic Substances: Ecology and Biogeochemistry* (eds. D. O. Hessen and L. Tranvik). Springer, Berlin, pp. 41–61.

Perdue E. M. and Gjessing E. T. (1990) *Organic Acids in Aquatic Ecosystems*. Dahlem Workshop Reports. Wiley, Chichester, 345pp.

Perdue E. M. and Lytle C. R. (1983) Distribution model for binding of protons and metal ions by humic substances. *Environ. Sci. Technol.* **17**, 654–660.

Perdue E. M., Beck K. C., and Reuter J. H. (1976) Organic complexes of iron and aluminium in natural waters. *Nature* **260**, 418–420.

Perdue E. M., Reuter J. H., and Ghosal M. (1980) The operational nature of acidic functional group analyses and its impact on mathematical descriptions of acid–base equilibria in humic substances. *Geochim. Cosmochim. Acta* **44**, 1841–1851.

Perdue E. M., Reuter J. H., and Parrish R. S. (1984) A statistical model of proton binding by humus. *Geochim. Cosmochim. Acta* **48**, 1257–1263.

Perminova I. V., Frimmel F. H., Kovalevskii D. V., Abbt-Braun G., Kudryavtse A. V., and Hesse S. (1998) Development of a predictive model for calculation of molecular weight of humic substances. *Water Res.* **32**, 872–881.

Petersen R. (1982) Influence of copper and zinc on the growth of a freshwater alga, *Scenedesmus quadricauda*: the significance of chemical speciation. *Environ. Sci. Technol.* **16**, 443–447.

Playle R. C., Dixon D. G., and Burnison K. (1993a) Copper and cadmium binding to fish gills: modification by dissolved organic carbon and synthetic ligands. *Can. J. Fish. Aquat. Sci.* **50**, 2667–2677.

Playle R. C., Dixon D. G., and Burnison K. (1993b) Copper and cadmium binding to fish gills: estimates of metal-gill stability constants and modeling of metal accumulation. *Can. J. Fish. Aquat. Sci.* **50**, 2678–2687.

Pomeroy L. R. (1974) The ocean's food web, a changing paradigm. *Bioscience* **24**, 499–504.

Pommer A. M. and Breger I. A. (1960) Potentiometric titration and equivalent weight of humic acid. *Geochim. Cosmochim. Acta* **20**, 30–44.

Posner A. M. (1966) The humic acids extracted by various reagents from a soil: Part 1. Yield, inorganic components, and titration curves. *J. Soil Sci.* **17**, 65–78.

Qian J., Xue H. B., Sigg L., and Albrecht A. (1998) Complexation of cobalt by natural ligands in freshwater. *Environ. Sci. Technol.* **32**, 2043–2050.

Reckhow D. A., Singer P. C., and Malcolm R. L. (1990) Chlorination of humic materials: by-product formation and chemical interpretations. *Environ. Sci. Technol.* **24**, 1655–1665.

Reuter J. H. (1980) *Chemical and Spectroscopic Characterization of Humic Substances Derived from River Swamps in the Floodplains of Southeastern US Coastal Streams.* Technical Completion Report USDI/OWRT Project No. B-132-GA, 66pp.

Reuter J. H. and Perdue E. M. (1977) Importance of heavy metal-organic interactions in natural waters. *Geochim. Cosmochim. Acta* **41**, 325–334.

Reuter J. H. and Perdue E. M. (1981) Calculation of molecular weights of humic substances from colligative data: application to aquatic humus and its molecular size fractions. *Geochim. Cosmochim. Acta* **45**, 2017–2022.

Rice J. A. and MacCarthy P. (1991) Statistical evaluation of the elemental composition of humic substances. *Org. Geochem.* **17**, 635–648.

Ritchie J. D. and Perdue E. M. (2003) Proton binding study of standard and reference fulvic acids, humic acids, and natural organic matter. *Geochim. Cosmochim. Acta* **67**, 85–96.

Rook J. J. (1974) Formation of haloforms during chlorination of natural waters. *Water Treat. Exam.* **23**, 234–243.

Rueter J. G. and Morel F. M. M. (1981) The interaction between zinc deficiency and copper toxicity as it affects the silicic acid uptake in *Thalassiosira pseudonana*. *Limnol. Oceanogr.* **26**, 67–73.

Rueter J. G., Jr., O'Reilly K. T., and Petersen R. R. (1987) Indirect aluminum toxicity to the green alga *Scenedesmus*

through increased cupric ion activity. *Environ. Sci. Technol.* **21**, 435–438.

Santore R. C., Di Toro D. M., Paquin P. R., Allen H. E., and Meyer J. S. (2001) Biotic ligand model of the acute toxicity of metals: I. Application to acute copper toxicity in freshwater fish and Daphnia. *Environ. Toxicol. Chem.* **20**, 2397–2402.

Santos E. B. H., Esteves V. I., Rodrigues J. P. C., and Duarte A. C. (1999) Humic substances' proton binding equilibria: assessment of errors and limitations of potentiometric data. *Anal. Chim. Acta* **392**, 333–341.

Schindler J. L. and Alberts J. J. (1974) Analysis of organic–inorganic associations of four Georgia reservoirs. *Arch. Hydrobiol.* **74**, 429–440.

Schlesinger W. H. and Melack J. M. (1981) Transport of organic carbon in the world's rivers. *Tellus* **33**, 172–187.

Schnitzer M. and Gupta U. C. (1965) Determination of acidity in soil organic matter. *Soil Sci. Soc. Am. Proc.* **29**, 274–277.

Schulten H.-R. (1987) Pyrolysis and soft ionization mass spectrometry of aquatic/terrestrial humic substances and soils. *J. Analyt. Appl. Pyrol.* **12**, 149–186.

Schulten H.-R. (1999) Analytical pyrolysis and computational chemistry of aquatic humic substances and dissolved organic matter. *J. Analyt. Appl. Pyrol.* **49**, 385–415.

Senesi N., Miano T. M., Provenzano M. R., and Brunetti G. (1989) Spectroscopic and compositional comparative characterization of I.H.S.S. reference and standard fulvic and humic acids of various origin. *Sci. Tot. Environ.* **81/82**, 143–156.

Serkiz S. M. and Perdue E. M. (1990) Isolation of dissolved organic matter from the Suwannee River using reverse osmosis. *Water Res.* **24**, 911–916.

Shapiro J. (1957) Chemical and biological studies on the yellow organic acids of lake water. *Limnol. Oceanogr.* **2**, 161–179.

Shapiro J. (1964) Effect of yellow organic acids on iron and other metals in water. *J. Am. Water Works Assoc.* **56**, 1062–1082.

Shapiro J. (1966) The relation of humic color to iron in natural waters. *Verh. Int. Ver. Limnol.* **16**, 477–484.

Shuman M. S. and Woodward G. P. (1973) Chemical constants of metal complexes from a complexometric titration followed with anodic stripping voltammetry. *Anal. Chem.* **45**, 2032–2035.

Siddiqui M., Amy G., Ryan J., and Odem W. (2000) Membranes for the control of natural organic matter from surface waters. *Water Res.* **34**, 3355–3370.

Singer P. C. and Chang S. D. (1989) Correlations between trihalomethanes and total organic halides formed during water treatment. *J. Am. Water Works Assoc.* **81**, 61–65.

Smith M. J. and Manahan S. E. (1973) Copper determination in water by standard addition potentiometry. *Anal. Chem.* **45**, 836–839.

Sokalski T., Ceresa A., Zwick T., and Pretsch E. (1997) Large improvement of the lower detection limit of ion-selective polymer membrane electrodes. *J. Am. Chem. Soc.* **119**, 11347–11348.

Sposito G. and Holtzclaw K. M. (1977) Titration studies on the polynuclear, polyacidic nature of fulvic acid extracted from sewage sludge–soil mixtures. *Soil Sci. Soc. Am. J.* **41**, 330–335.

Sposito G. and Holtzclaw K. M. (1979) Copper(II) complexation by fulvic acid extracted from sewage sludge as influenced by nitrate versus perchlorate background ionic media. *Soil Sci. Soc. Am. J.* **43**, 47–51.

Sposito G., Holtzclaw K. M., and Keech D. A. (1977) Proton binding in fulvic acid extracted from sewage sludge–soil mixtures. *Soil Sci. Soc. Am. J.* **41**, 1119–1125.

Stenson A. C., Landing W. M., Marshall A. G., and Cooper W. T. (2002) Ionization and fragmentation of humic substances in electrospray ionization Fourier transform-ion cyclotron resonance mass spectrometry. *Anal. Chem.* **74**, 4397–4409.

Stevens A. A., Moore L. A., and Miltner R. J. (1989) Formation and control of non-trihalomethane disinfection by-products. *J. Am. Water Works Assoc.* **81**, 54–60.

Stumm W. and Morgan J. J. (1970) *Aquatic Chemistry.* 1st edn. Wiley-Interscience, New York, 583pp.

Stumm W. and Morgan J. J. (1981) *Aquatic Chemistry.* 2nd edn. Wiley-Interscience, New York, 780pp.

Stumm W. and Morgan J. J. (1996) *Aquatic Chemistry.* 3rd edn. Wiley-Interscience, New York, 1022pp.

Sullivan T. J., Driscoll C. T., Gherini S. A., Munson R. K., Cook R. B., Charles D. F., and Yatsko C. P. (1989) Influence of aqueous aluminum and organic acids on measurement of acid neutralizing capacity in surface waters. *Nature* **338**, 408–410.

Sun L. (1993) Isolation, characterization, and bioavailability of dissolved organic matter in natural waters. PhD Thesis, Georgia Institute of Technology, 139pp.

Sun L., Perdue E. M., and McCarthy J. F. (1995) Using reverse osmosis to obtain organic matter from surface and ground waters. *Water Res.* **29**, 1471–1477.

Sun L., Perdue E. M., Meyer J. L., and Weis J. (1997) Use of elemental composition to predict bioavailability of dissolved organic matter in a Georgia river. *Limnol. Oceanogr.* **42**, 714–721.

Sunda W. and Guillard R. R. L. (1976) The relationship between cupric ion activity and the toxicity of copper to phytoplankton. *J. Mar. Res.* **34**, 511–529.

Sunda W. G. and Lewis J. A. M. (1978) Effect of complexation by natural organic ligands on the toxicity of copper to a unicellular alga, *Monochrysis lutheri.* *Limnol. Oceanogr.* **23**, 870–876.

Sunda W. G., Engel D. W., and Thoutte R. M. (1978) Effect of chemical speciation on toxicity of cadmium to grass shrimp, *Palaemonetes pugio*: importance of free Cd ion. *Environ. Sci. Technol.* **12**, 409–413.

Susetyo W., Carreira L. A., Azarraga L. V., and Grimm D. M. (1991) Fluorescence techniques for metal–humic interactions. *Fres. J. Anal. Chem.* **339**, 624–635.

Sweet M. S. and Perdue E. M. (1982) Concentration and speciation of dissolved sugars in river water. *Environ. Sci. Technol.* **16**, 692–698.

Swift R. S. (1989a) Molecular weight, size, shape, and charge characteristics of humic substances: some basic considerations. In *Humic Substances II: In Search of Structure* (eds. M. H. B. Hayes, P. MacCarthy, R. L. Malcolm, and R. S. Swift). Wiley, Chichester, pp. 449–465.

Swift R. S. (1989b) Molecular weight, shape, and size of humic substances by ultracentrifugation. In *Humic Substances II: In Search of Structure* (eds. M. H. B. Hayes, P. MacCarthy, R. L. Malcolm, and R. S. Swift). Wiley, Chichester, pp. 467–495.

Taillefert M. and Rozan T. F. (2002) *Environmental Electrochemistry. Analyses of Trace Element Biogeochemistry,* ACS Symposium Series 811. American Chemical Society, Washington, DC, 412pp.

Tan L. and Sudak R. G. (1992) Removing color from a groundwater source. *J. Am. Water Works Assoc.* **84**, 79–87.

Thorn K. A., Folan D. W., and MacCarthy P. (1992) *Characterization of the International Humic Substances Society Standard and Reference Fulvic Acids by Solution State Carbon-13 and Hydrogen-1 Nuclear Magnetic Resonance Spectrometry.* Water Resources Investigations Report 89-4196, U.S.G.S. Denver, 93pp.

Thurman E. M. (1985) *Organic Geochemistry of Natural Waters.* Martinus Nijhoff/Dr. W. Junk Publishers, Lancaster, 497pp.

Thurman E. M. and Malcolm R. L. (1981) Preparative isolation of aquatic humic substances. *Environ. Sci. Technol.* **15**, 463–466.

Tipping E. (1994) WHAM—a chemical equilibrium model and computer code for waters, sediments, and soils incorporating a discrete site/electrostatic model of ion binding by humic substances. *Comput. Geosci.* **20**, 973–1023.

Tipping E. (1998) Humic ion-binding Model VI: an improved description of the interactions of protons and metal ions with humic substances. *Aquat. Geochem.* **4**, 3–48.

Tipping E. and Hurley M. A. (1992) A unifying model of cation binding by humic substances. *Geochim. Cosmochim. Acta* **56**, 3627–3641.

Tranvik L. J. and Jørgensen N. O. G. (1995) Colloidal and dissolved organic matter in lake water: carbohydrate and amino acid composition, and ability to support bacterial growth. *Biogeochem.* **30**, 77–97.

Turekian K. K. (1969) The oceans, streams, and atmosphere. In *Handbook of Geochemistry* (ed. K. H. Wedepohl). Springer, Berlin, vol. 1, pp. 297–323.

Turner D. R., Varney M. S., Whitfield M., Mantoura R. F. C., and Riley J. P. (1986) Electrochemical studies of copper and lead complexation by fulvic acid: I. Potentiometric measurements and critical comparison of metal binding models. *Geochim. Cosmochim. Acta* **50**, 289–297.

Urbansky E. T. and Magnuson M. L. (2002) Analyzing drinking water for disinfection by-products. *Anal. Chem.* **74**, 260A–271A.

Vallino J. J., Hopkinson C. S., and Hobbie J. E. (1996) Modeling bacterial utilization of dissolved organic matter: optimization replaces Monod growth kinetics. *Limnol. Oceanogr.* **41**, 1591–1609.

van den Berg C. M. G. (1984a) Determination of the complexing capacity and conditional stability constants of complexes of copper(II) with natural organic ligands in seawater by cathodic stripping voltammetry of copper-catechol complex ions. *Mar. Chem.* **15**, 1–18.

van den Berg C. M. G. (1984b) Determination of copper in sea water by cathodic stripping voltammetry of complexes with catechol. *Anal. Chim. Acta* **164**, 195–207.

van Dijk H. (1966) Some physico-chemical aspects of the investigation of humus. In *The Use of Isotopes in Soil Organic Matter Studies.* Pergamon, Oxford, pp. 129–141.

van Dijk H. (1971) Cation binding of humic acids. *Geoderma* **5**, 53–67.

Vannote R. L., Minshall G. W., Cummins K. W., Sedell J. R., and Cushing C. E. (1980) The river continuum concept. *Can. J. Fish. Aquat. Sci.* **37**, 130–137.

Warren L. A. and Haake E. A. (2001) Biogeochemical controls on metal behaviour in freshwater environments. *Earth. Sci. Rev.* **54**, 261–320.

Watt B. E., Malcolm R. L., Hayes M. H. B., Clark N. W. E., and Chipman J. K. (1996) Chemistry and potential mutagenicity of humic substances in waters from different watersheds in Britain and Ireland. *Water Res.* **30**, 1502–1516.

Weber J. H. and Wilson S. A. (1975) The isolation and characterization of fulvic acid and humic acid from river water. *Water Res.* **9**, 1079–1084.

Wershaw R. L. (1989) Sizes and shapes of humic substances by scattering techniques. In *Humic Substances II: In Search of Structure* (eds. M. H. B. Hayes, P. MacCarthy, R. L. Malcolm, and R. S. Swift). Wiley, Chichester, pp. 545–559.

Wershaw R. L. and Aiken G. R. (1985) Molecular size and weight measurements of humic substances. In *Humic Substances in Soil, Sediment, and Water. Geochemistry, Isolation, and Characterization* (eds. G. R. Aiken, D. M. McKnight, R. L. Wershaw, and P. MacCarthy). Wiley, New York, pp. 477–492.

Wilson M. A. (1987) *NMR Techniques and Applications in Geochemistry and Soil Chemistry.* Pergamon, Oxford.

Wilson D. E. and Kinney P. (1977) Effects of polymeric charge variations on the proton-metal ion equilibria of humic materials. *Limnol. Oceanogr.* **22**, 281–289.

Wilson M. A., Vassallo A. M., Perdue E. M., and Reuter J. H. (1987) Compositional and solid-state nuclear magnetic resonance study of humic and fulvic acid fractions of soil organic matter. *Anal. Chem.* **59**, 551–558.

Wong C. S., Boyle E., Bruland K. W., Burton J. D., and Goldberg E. D. (1983) *Trace Metals in Seawater.* Plenum, New York, 920pp.

Xue H.-B. and Sigg L. (1993) Free cupric ion concentration and Cu(II) speciation in a eutrophic lake. *Limnol. Oceanogr.* **38**, 1200–1213.

Xue H.-B. and Sigg L. (1994) Zinc speciation in lake waters and its determination by ligand exchange with EDTA and differential pulse anodic stripping voltammetry. *Anal. Chim. Acta* **284**, 505–515.

Xue H.-B. and Sigg L. (1998) Cadmium speciation and complexation by natural organic ligands in fresh water. *Anal. Chim. Acta* **363**, 249–259.

Zepp R. G. (1988) Environmental photoprocesses involving natural organic matter. In *Humic Substances and Their Role in the Environment*, Dahlem Workshop Report (eds. F. H. Frimmel and R. F. Christman). Wiley, Chichester, pp. 193–214.

Zepp R. G. and Schlotzhauer P. F. (1981) Comparison of photochemical behavior of various humic substances in water: III. Spectroscopic properties of humic substances. *Chemosphere* **10**, 479–486.

Zitko P., Carson W. V., and Carson W. G. (1973) Prediction of incipient lethal levels of copper to juvenile Atlantic salmon in the presence of humic acid by cupric electrode. *Bull. Environ. Contam. Toxicol.* **10**, 265–271.

Zuo Y. and Jones R. D. (1997) Photochemistry of natural dissolved organic matter in lake and wetland waters—production of carbon monoxide. *Water Res.* **31**, 850–858.

5.11

Stable Isotope Applications in Hydrologic Studies

C. Kendall and D. H. Doctor

United States Geological Survey, Menlo Park, CA, USA

5.11.1 INTRODUCTION

The topic of stream flow generation has received considerable attention over the last two decades, first in response to concern about "acid rain" and more recently in response to the increasingly serious contamination of surface and shallow groundwaters by anthropogenic contaminants. Many sensitive, low-alkalinity streams in North America and Europe are already acidified (see Chapter 9.10). Still more streams that are not yet chronically acidic may undergo acidic episodes in response to large rainstorms and/or spring snowmelt. These acidic events can seriously damage local ecosystems. Future climate changes may exacerbate the situation by affecting biogeochemical controls on the transport of water, nutrients, and other materials from land to freshwater ecosystems.

New awareness of the potential danger to water supplies posed by the use of agricultural chemicals and urban industrial development has also focused attention on the nature of rainfall-runoff and recharge processes and the mobility of various solutes, especially nitrate and pesticides, in shallow systems. Dumping and spills of other potentially toxic materials are also of concern because these chemicals may eventually reach streams and other public water supplies. A better understanding of hydrologic flow paths and solute sources is required to determine the potential impact of contaminants on water supplies, develop management practices to preserve water quality, and devise remediation plans for sites that are already polluted.

Isotope tracers have been extremely useful in providing new insights into hydrologic processes, because they integrate small-scale variability to give an effective indication of catchment-scale processes. The main purpose of this chapter is to provide an overview of recent research into the use of naturally occurring stable isotopes to track the movement of water and solutes in hydrological systems where the waters are relatively fresh: soils, surface waters, and shallow groundwaters. For more information on shallow-system applications, the reader is referred to Kendall and McDonnell (1998). For information on groundwater systems, see Cook and Herczeg (2000).

5.11.1.1 Environmental Isotopes as Tracers

Environmental isotopes are naturally occurring (or, in some cases, anthropogenically produced) isotopes whose distributions in the hydrosphere can assist in the solution of hydrological and biogeochemical problems. Typical uses of environmental isotopes include the identification of sources of water and solutes, determination of water flow paths, assessment of biologic cycling of nutrients within the ecosystem, and testing flow path, water budget, and geochemical models developed using hydrologic or geochemical data.

Environmental isotopes can be used as tracers of waters and solutes in shallow low-temperature environments because:

(i) Waters that were recharged at different times, were recharged in different locations, or that followed different flow paths are often isotopically distinct; in other words, they have distinctive "fingerprints."

(ii) Unlike most chemical tracers, environmental isotopes are relatively conservative in reactions with the bedrock and soil materials. This is especially true of oxygen and hydrogen isotopes in water; the waters mentioned above retain their distinctive fingerprints until they mix with other waters.

(iii) Solutes in the water that are derived from atmospheric sources are often isotopically distinct from solutes derived from geologic and biologic sources within the catchment.

(iv) Both biological cycling of solutes and water/rock reactions often change isotopic ratios in the solutes in predictable and recognizable directions; these processes often can be reconstructed from the isotopic compositions.

The applications of environmental isotopes as hydrologic tracers in low-temperature, freshwater systems fall into two main categories: (i) tracers of the water itself (water isotope hydrology) and (ii) tracers of the solutes in the water (solute isotope biogeochemistry). These classifications are by no means universal but they are useful conceptually and often eliminate confusion when comparing results using different tracers.

Water isotope hydrology focuses on the isotopes that form water molecules: the oxygen isotopes (oxygen-16, oxygen-17, and oxygen-18) and the hydrogen isotopes (protium, deuterium, and tritium). These isotopes are ideal tracers of water sources and movement because they are constituents of water molecules, not something that is dissolved in the water like other tracers that are commonly used in hydrology (e.g., dissolved species such as chloride).

Oxygen and hydrogen isotopes are generally used to determine the source of the water (e.g., precipitation versus groundwater in streams, recharge of evaporated lake water versus snowmelt water in groundwater). In catchment research, the main use is for determining the contributions of "old" and "new" water to high flow (storm and snowmelt runoff) events in streams. "Old" water is defined as the water that existed in a catchment prior to a particular storm or snowmelt period. Old water includes groundwater, soil water, and surface water. "New" water is either rainfall or snowmelt, and is defined as the water that triggers the particular storm or snowmelt runoff event.

Solute isotope biogeochemistry focuses on isotopes of constituents that are dissolved in the water or are carried in the gas phase. The most commonly studied solute isotopes are the isotopes of carbon, nitrogen, and sulfur. Less commonly investigated stable, nonradiogenic isotopes include lithium, chloride, boron, and iron.

Although the literature contains numerous case studies involving the use of solutes (and sometimes solute isotopes) to trace water sources and flow paths, such applications include an implicit assumption that these solutes are transported conservatively with the water. Unlike the isotopes in the water molecules, the ratios of solute isotopes can be significantly altered by reaction with geological and/or biological materials as the water moves through the system. While the utility of solutes in the evaluation of rainfall-runoff processes has been repeatedly demonstrated, in a strict sense, *solute isotopes only trace solutes*. Solute isotopes also provide information on the reactions that are responsible for their presence in the water and the flow paths implied by their presence.

5.11.1.2 Isotope Fundamentals

5.11.1.2.1 Basic principles

Isotopes are atoms of the same element that have different numbers of neutrons but the same number of protons and electrons. The difference in the number of neutrons between the various isotopes of an element means that the various isotopes have different masses. The superscript number to the left of the element abbreviation indicates the number of protons plus neutrons in the isotope. For example, among the hydrogen isotopes, deuterium (denoted as ^2H or D) has one neutron and one proton. This is

approximately twice the mass of protium (^1H), whereas tritium (^3H) has approximately three times the mass of protium.

The stable isotopes have nuclei that do not decay to other isotopes on geologic timescales, but may themselves be produced by the decay of radioactive isotopes. Radioactive (unstable) isotopes have nuclei that spontaneously decay over time to form other isotopes. For example, ^{14}C, a radioisotope of carbon, is produced in the atmosphere by the interaction of cosmic-ray neutrons with stable ^{14}N. With a half-life of ~5,730 yr, ^{14}C decays back to ^{14}N by emission of a beta particle. The stable ^{14}N produced by radioactive decay is called "radiogenic" nitrogen. This chapter focuses on stable, nonradiogenic isotopes. For a more thorough discussion of the fundamentals of isotope geochemistry, see Clark and Fritz (1997) and Kendall and McDonnell (1998).

5.11.1.2.2 Isotope fractionation

For elements of low atomic numbers, the mass differences between the isotopes of an element are large enough for many physical, chemical, and biological processes or reactions to "fractionate" or change the relative proportions of various isotopes. Two different types of processes—equilibrium isotope effects and kinetic isotope effects—cause isotope fractionation. As a consequence of fractionation processes, waters and solutes often develop unique isotopic compositions (ratios of heavy to light isotopes) that may be indicative of their source or of the processes that formed them.

Equilibrium isotope-exchange reactions involve the redistribution of isotopes of an element among various species or compounds in a closed, well-mixed system at chemical equilibrium. At isotope equilibrium, the forward and backward reaction rates of any particular isotope are identical. This does not mean that the isotopic compositions of two compounds at equilibrium are identical, but only that the ratios of the different isotopes in each compound are constant for a particular temperature. During equilibrium reactions, the heavier isotope generally preferentially accumulates in the species or compound with the higher oxidation state. For example, during sulfide oxidation, the sulfate becomes enriched in ^{34}S relative to sulfide (i.e., has a more positive $\delta^{34}S$ value); consequently, the residual sulfide becomes depleted in ^{34}S.

Chemical, physical, and biological processes can be viewed as either reversible *equilibrium* reactions or irreversible unidirectional *kinetic* reactions. In systems out of chemical and isotopic equilibrium, forward and backward reaction rates are not identical, and isotope reactions may, in fact, be unidirectional if reaction products become physically isolated from the reactants. Such reaction rates are dependent on the ratios of the masses of the isotopes and their vibrational energies, and hence are called *kinetic* isotope fractionations.

The magnitude of a kinetic isotope fractionation depends on the reaction pathway, the reaction rate, and the relative bond energies of the bonds being severed or formed by the reaction. Kinetic fractionations, especially unidirectional ones, are usually larger than the equilibrium fractionation factor for the same reaction in most low-temperature environments. As a rule, bonds between the lighter isotopes are broken more easily than equivalent bonds of heavier isotopes. Hence, the light isotopes react faster and become concentrated in the products, causing the residual reactants to become enriched in the heavy isotopes. In contrast, reversible equilibrium reactions can produce products heavier or lighter than the original reactants.

Biological processes are generally uni-directional and are excellent examples of "kinetic" isotope reactions. Organisms preferentially use the lighter isotopic species because of the lower energy "costs," resulting in significant fractionations between the substrate (which generally becomes isotopically heavier) and the biologically mediated product (which generally becomes isotopically lighter). The magnitude of the fractionation depends on the reaction pathway utilized and the relative energies of the bonds being severed and formed by the reaction. In general, slower reaction steps show greater isotopic fractionation than faster steps because the organism has time to be more selective. Kinetic reactions can result in fractionations very different from, and typically larger than, the equivalent equilibrium reaction.

Many reactions can take place either under purely equilibrium conditions or be affected by an additional kinetic isotope fractionation. For example, although evaporation can take place under purely equilibrium conditions (i.e., at 100% humidity when the air is still), more typically the products become partially isolated from the reactants (e.g., the resultant vapor is blown downwind). Under these conditions, the isotopic compositions of the water and vapor are affected by an additional kinetic isotope fractionation of variable magnitude.

The partitioning of stable isotopes between two substances A and B can be expressed by using the isotopic fractionation factor α (alpha):

$$\alpha = R_p/R_s \qquad (1)$$

where R_p and R_s are the ratios of the heavy to light isotope (e.g., $^2H/^1H$ or $^{18}O/^{16}O$) in the *product* and *substrate* (reactant), respectively.

Values for α tend to be very close to 1. An isotope enrichment factor, ε, can be defined as

$$\varepsilon_{p-s} = (\alpha - 1) \times 1,000 \qquad (2)$$

5.11.1.2.3 Rayleigh fractionation

The Rayleigh equation is an exponential relation that describes the partitioning of isotopes between two reservoirs as one reservoir decreases in size. The equations can be used to describe an isotope fractionation process if: (i) the material is continuously removed from a mixed system containing molecules of two or more isotopic species (e.g., water with ^{18}O and ^{16}O, or sulfate with ^{34}S and ^{32}S); (ii) the fractionation accompanying the removal process at any instant is described by the fractionation factor α; and (iii) α does not change during the process. The general form of the Rayleigh equation is

$$R = R_o f^{(\alpha-1)} \qquad (3)$$

where R_o is the isotope ratio (e.g., $^{18}O/^{16}O$) of the original substrate, and f is the fraction of remaining substrate. A commonly used approximate version of the equation is

$$\delta \cong \delta_o + \varepsilon \ln(f) \qquad (4)$$

where δ_o is the initial isotopic composition. This approximation is valid for δ_o values near 0 and positive ε values less than $\sim +10‰$.

In a strict sense, the term "Rayleigh fractionation" should only be used for chemically *open* systems where the isotopic species removed at every instant are in thermodynamic and isotopic equilibrium with those remaining in the system at the moment of removal. Furthermore, such an "ideal" Rayleigh distillation is one where the reactant reservoir is finite and well mixed, and does not re-react with the product (Clark and Fritz, 1997). However, the term "Rayleigh fractionation" is commonly applied to equilibrium closed systems and kinetic fractionations as well because the situations may be computationally identical. Isotopic fractionations are strongly affected by whether a system is open or closed.

5.11.1.2.4 Terminology

Stable isotopic compositions are normally reported as delta (δ) values in parts per thousand (denoted as ‰ or permil, per mil, per mille, or per mill) enrichments or depletions relative to a standard of known composition. The term "δ" is spelled and pronounced delta, not del. δ values are calculated by

$$\delta \, (\text{in } ‰) = (R_{sample}/R_{standard} - 1) \times 1,000 \qquad (5)$$

where "R" is the ratio of the heavy to light isotope. A positive delta value means that the isotopic ratio of the sample is higher than that of the standard; a negative value means that the isotope ratio of the sample is lower than that of the standard.

Stable oxygen and hydrogen isotopic ratios are normally reported relative to the "standard mean ocean water" (SMOW) standard (Craig, 1961b) or the equivalent Vienna-SMOW (V-SMOW) standard. Carbon, nitrogen, and sulfur stable isotope ratios are reported relative to the Pee Dee Belemnite (PDB) or Vienna-PDB (VPDB), ambient air (AIR), and Canyon Diablo Troilite (CDT) standards, respectively, as defined later. The use of the "V" before SMOW or PDB indicates that the measurements were calibrated on normalized per mil scales (Coplen, 1996).

The δ values for stable isotopes are determined using isotope ratio mass spectrometry, typically either using gas-source stable isotope mass spectrometers (e.g., hydrogen, carbon, nitrogen, oxygen, sulfur, chlorine, and bromine) or solid-source mass spectrometry (e.g., lithium, boron, and iron). The analytical precisions are small relative to the ranges in δ values that occur in natural earth systems. Typical one standard deviation (1σ) analytical precisions for oxygen, carbon, nitrogen, and sulfur isotopes are in the 0.05–0.2‰ range; typical precisions for hydrogen isotopes are generally poorer, from 0.2‰ to 1.0‰, because of the low $^2H:^1H$ ratio of natural materials.

5.11.1.3 Causes of Isotopic Variation

5.11.1.3.1 *Isotope fractionations during phase changes*

The most important phase changes in hydrological systems are associated with water as it condenses, evaporates, and melts. At equilibrium, the isotope fractionation between two coexisting phases is a function of temperature. As the phase change proceeds, the δ values of both the reactant and product change in a regular fashion. For example, as water evaporates, the $\delta^{18}O$ and δ^2H values of the residual water increase, and the δ values of the resulting vapor also increase. The lower the humidity, the greater the kinetic fractionation associated with the phase change. The changes in their δ values can be described using the Rayleigh equations. The Rayleigh equation applies to an open system from which material is removed continuously under condition of a constant fractionation factor.

The isotope enrichment achieved can be very different in closed versus open systems. For example, Figure 1 shows the changes in the $\delta^{18}O$

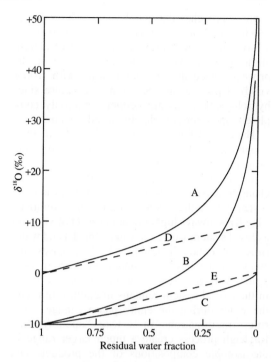

Figure 1 Isotopic change under open- and closed-system Rayleigh conditions for evaporation with a fractionation factor $\alpha = 1.01$ for an initial liquid composition of $\delta^{18}O = 0$. The $\delta^{18}O$ of the remaining water (solid line A), the instantaneous vapor being removed (solid line B), and the accumulated vapor being removed (solid line C) all increase during single-phase, open-system, evaporation under equilibrium conditions. The $\delta^{18}O$ of water (dashed line D) and vapor (dashed line E) in a two-phase closed system also increase during evaporation, but much less than in an open system; for a closed system, the δ values of the instantaneous and cumulative vapor are identical (after Gat and Gonfiantini, 1981).

of water and vapor during both open-system (solid lines) and closed-system (dashed lines) evaporation with a constant fractionation factor $\alpha_{l-v} = 1.010$ (i.e., the newly formed vapor is always 10‰ lower than the residual water). During open-system evaporation where the vapor is continuously removed (i.e., isolated from the water), as evaporation progresses (i.e., $f \rightarrow 0$), the $\delta^{18}O$ of the remaining water (solid line A) becomes higher. The $\delta^{18}O$ of the instantaneously formed vapor (solid line B) describes a curve parallel to that of the remaining water, but lower than it (for all values of f) by the precise amount dictated by the fractionation factor for ambient temperature, in this case by 10‰. For higher temperatures, the α value would be smaller and the curves closer together. The integrated curve, giving the isotopic composition of the accumulated vapor thus removed, is shown as solid line C. Mass-balance considerations require that the isotope content of the total accumulated vapor approaches the initial water $\delta^{18}O$ value as $f \rightarrow 0$. Hence, any process that can be modeled as a Rayleigh fractionation will not exhibit fractionation between the product and source if the process proceeds to completion (with 100% yield). This is a rare occurrence in nature since hydrologic systems are neither commonly completely open nor completely closed systems.

The dashed lines in Figure 1 show the $\delta^{18}O$ of vapor (E) and water (D) during equilibrium evaporation in a closed system (i.e., where the vapor and water are in contact for the entire phase change). Note that the $\delta^{18}O$ of vapor in the open system where the vapor is continuously removed (line B) is always higher than the $\delta^{18}O$ of vapor in a closed system where the vapor (line E) and water (line D) remain in contact. In both cases, the evaporation takes place under equilibrium conditions with $\alpha = 1.010$, but the cumulative vapor in the closed system remains in equilibrium with the water during the entire phase change. As a rule, fractionations in a true "open-system" Rayleigh process create a much larger range in the isotopic compositions of the products and reactants than in closed systems. This is because of the lack of back-reactions in open systems. Natural processes will produce fractionations between these two "ideal" cases.

5.11.1.3.2   *Mixing of waters and/or solutes*

Waters mix conservatively with respect to their isotopic compositions. In other words, the isotopic compositions of mixtures are intermediate between the compositions of the end-members. Despite the terminology (the δ notation and units of ‰) and common negative values, the compositions can be treated just like any other chemical constituent (e.g., chloride content) for making mixing calculations. For example, if two streams with known discharges (Q_1, Q_2) and known $\delta^{18}O$ values ($\delta^{18}O_1, \delta^{18}O_2$) merge and become well mixed, the $\delta^{18}O$ of the combined flow (Q_T) can be calculated from

$$Q_T = Q_1 + Q_2 \qquad (6)$$

$$\delta^{18}O_T Q_T = \delta^{18}O_1 Q_1 + \delta^{18}O_2 Q_2 \qquad (7)$$

Another example: any mixing proportions of two waters with known $\delta^{18}O$ and δ^2H values will fall along a tie line between the compositions of the end-members on a $\delta^{18}O$ versus δ^2H plot.

What is *not* so obvious is that on many types of X–Y plots, mixtures of two end-members will not necessarily plot along straight lines but instead along *hyperbolic curves* (Figure 2(a)). This has been explained very elegantly by Faure (1986), using the example of $^{87}Sr/^{86}Sr$ ratios. The basic principle is that mixtures of two components that have different isotope ratios (e.g., $^{87}Sr/^{86}Sr$ or $^{15}N/^{14}N$) and different concentrations of the element in question (e.g., strontium or nitrogen) form hyperbolas when plotted on diagrams with coordinates of isotope ratios versus concentration. As the difference between the elemental concentrations of two components (end-members) approaches 0, the hyperbolas flatten to lines. The hyperbolas are concave or convex, depending on whether the component with the higher isotope ratio has a higher or lower concentration than the other component. Mixing hyperbolas can be transformed into straight lines by plotting isotope ratios versus the inverse of concentration ($1/C$), as shown in Figure 2(b). One way to avoid curved lines on plots of δ values versus solute concentrations is to plot molality instead of

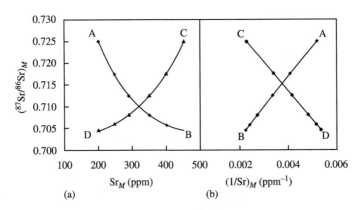

Figure 2 (a) Hyperbolas formed by the mixing of components (waters or minerals) A and B with different Sr concentrations and Sr isotope ratios ($^{87}Sr/^{86}Sr$). If the concentrations of Sr in A and B are identical, the mixing relation would be a straight line; otherwise, the mixing relations are either concave or convex curves, as shown. (b) Plotting the reciprocals of the strontium concentrations transforms the mixing hyperbolas into straight lines. If the curves in (a) were the result of some fractionation process (e.g., radioactive decay) that is an exponential relation, plotting the reciprocals of the Sr concentrations would not produce straight lines (after Faure, 1986).

molarity for the solute. Mixing and fractionation lines on $\delta^{18}O$ versus δ^2H plots are generally straight lines, because the concentrations of oxygen and hydrogen in water are essentially constant except for extremely saline brines.

Although the topic is rarely discussed, the isotope concentration of a sample is not necessarily equal to the isotope activity. For example, the isotope activity coefficient for water-O or water-H can be positive or negative, depending on solute type, molality, and temperature of the solution. The isotopic compositions of waters and solutes can be affected significantly by the concentration and types of salts, because the isotopic compositions of waters in the hydration spheres of salts and in regions farther from the salts are different (see Horita (1989) for a good discussion of this topic). In general, the only times when it is important to consider isotope activities are for low pH, high SO_4, and/or high magnesium brines because the activity and concentration δ values of these waters (δ_a and δ_c) are significantly different. For example, the difference ($\delta^2H_a - \delta^2H_c$) between the activity and concentration δ values for sulfuric acid solutions in mine tailings is $\sim +16\permil$ for $2\ m$ solutions. For normal saline waters (e.g., seawater), the activity coefficients for $\delta^{18}O$ and δ^2H are essentially equal to 1.

Virtually all laboratories report $\delta^{18}O$ activities (not concentrations) for water samples. The δ^2H of waters may be reported in either concentration or activity δ values, depending on the method used for preparing the samples for analysis. Methods that involve quantitative conversion of the H in H_2O to H_2 produce δ_c values. Methods that analyze H_2O by equilibrating it with H_2 (or with CO_2), and then analyzing the equilibrated gas for isotopic composition produce δ_a values. "Equilibrate" in this case means letting the liquid and gas reach isotopic equilibrium at a constant, known temperature. $\delta^{13}C$, $\delta^{15}N$, and $\delta^{34}S$ preparation methods do not involve equilibration and, hence, these are δ_c values. To avoid confusion, laboratories and research papers should always report the method used.

5.11.1.3.3 *Geochemical and biological reactions*

Reactions in shallow systems that frequently occur under equilibrium isotope conditions include (i) solute addition and precipitation (e.g., oxygen exchange between water and carbonate, sulfate, and other oxides, carbonate dissolution and precipitation) and (ii) gas dissolution and exchange (e.g., exchange of hydrogen between water and dissolved H_2S and CH_4, oxygen isotope exchange between water and CO_2). However, most of the reactions occurring in the soil zone, in

shallow groundwater, and in surface waters are biologically mediated, and hence produce non-equilibrium isotope fractionations. These reactions may behave like Rayleigh fractionations in that there may be negligible back-reaction between the reactant and product, regardless of whether the system is open or closed. This arises due to the slow rates of many inorganic isotopic equilibrium exchange reactions at near-surface environmental conditions (i.e., low temperature, low pressure, pH 6–9).

Nonequilibrium isotopic fractionations typically result in larger ranges of isotopic compositions between reactants and products than those under equivalent equilibrium conditions. An example of this process is biologically mediated denitrification (reduction) of nitrate to N_2 in groundwater. The N_2 produced may be lost to the atmosphere, but even if it remains in contact with the residual nitrate, the gas does not re-equilibrate with the nitrate nor is there a biologically mediated back-reaction.

Graphical methods are commonly used for determining whether the data support an interpretation of mixing of two potential sources or fractionation of a single source. Implicit in such efforts is often the idea that mixing will produce a "straight line" connecting the compositions of the two proposed end-members whereas fractionation will produce a "curve." However, as shown in Figure 3(a), both mixing and fractionation (in this case, denitrification) can produce curves (Mariotti *et al.*, 1988), although both relations can look linear for small ranges of concentrations. However, the equations describing mixing and fractionation processes are different and, under

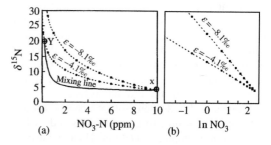

Figure 3 (a) Theoretical evolution of the $\delta^{15}N$ and the nitrate-N concentration during mixing (solid line) of two waters X and Y, and during an isotope fractionating process (e.g., denitrification of water X with an NO_3 concentration of 10 ppm). Denitrification for $\varepsilon = -4.1\permil$ results in a curve (dashed line) that ends at Y. Two different enrichment factors are compared: $\varepsilon = -4.1\permil$ and $\varepsilon = -8.1\permil$. The data points represent successive 0.1 increments of mixing or denitrification progress. (b) Plotting the natural log of the concentrations for a fractionation process yields straight lines, different for different ε values (after Mariotti *et al.*, 1988).

favorable conditions, the process responsible for the curve can be identified. This is because Rayleigh fractionations are exponential relations (Equation (3)), and plotting δ values versus the natural log of concentration (C) will produce a straight line (Figure 3(b)). If an exponential relation is not observed and a straight line is produced on a δ versus $1/C$ plot (as in Figure 2(b)), this supports the contention that the data are produced by simple mixing of two end-members.

5.11.2 TRACING THE HYDROLOGICAL CYCLE

In most low-temperature, near-surface environments, stable hydrogen and oxygen isotopes behave conservatively. This means that as water molecules move through the subsurface, chemical exchange between the water and oxygen and hydrogen in the organic and inorganic materials through which flow occurs will have a negligible effect on the overall isotope ratios of the water (an exception is within certain geothermal systems, especially in carbonates, where large amounts of oxygen exchange between the water and rock can occur). Water isotopes are useful tracers of water flow paths, especially in confined groundwater systems dominated by Darcian flow in which a source of water with a distinctive isotopic composition forms a "plume" in the subsurface. This phenomenon forms the basis of the science of *isotope hydrology*, or the use of stable water isotopes in hydrological studies.

Although tritium also exhibits insignificant reaction with geologic materials, it does change in concentration over time because it is radioactive and decays with a half-life of ~12.4 yr. The main processes that dictate the oxygen and hydrogen isotopic compositions of waters in a catchment are: (i) phase changes that affect the water above or near the ground surface (evaporation, condensation, and melting) and (ii) simple mixing at or below the ground surface.

Oxygen and hydrogen isotopes can be used to determine the source of groundwaters and surface waters. For example, $\delta^{18}O$ and δ^2H can be used to determine the contributions of old and new water to a stream or shallow groundwater during periods of high runoff because the rain or snowmelt (new water) that triggers the runoff is often isotopically different from the water already in the catchment (old water). This section briefly explains why waters from different sources often have different isotopic compositions. For more detailed discussions of these and other environmental isotopes, the reader can consult texts such as Kendall and McDonnell (1998) and Cook and Herczeg (2000).

5.11.2.1 Deuterium and Oxygen-18

5.11.2.1.1 Basic principles

Craig (1961a) observed that $\delta^{18}O$ and δ^2H (or δD) values of precipitation that has not been evaporated are linearly related by

$$\delta^2H = 8\delta^{18}O + 10 \qquad (8)$$

This equation, known as the "global meteoric waterline" (GMWL), is based on precipitation data from locations around the globe. The slope and intercept of the "local meteoric waterline" (LMWL) for rain from a specific catchment or basin can be different from the GMWL. The *deuterium excess* (*d* excess, or *d*) parameter has been defined to describe these different meteoric waterlines (MWLs), such that

$$d = \delta^2H - 8\delta^{18}O \qquad (9)$$

On $\delta^{18}O$ versus δ^2H plots, water that has evaporated from open surfaces (e.g., ponds and lakes), or mixed with evaporated water, plots below the MWL along a trajectory typically with a slope between 2 and 5 (Figure 4). The slope of the evaporation line and the isotopic evolution of the reservoir are strongly dependent upon the humidity under which evaporation occurs, as well as temperature and wind speed (Clark and Fritz, 1997), with lower slopes found in arid regions.

5.11.2.1.2 Precipitation

The two main factors that control the isotopic character of precipitation at a given location are the temperature of condensation of the precipitation and the degree of rainout of the air mass (the ratio of water vapor that has already condensed into precipitation to the initial amount of water vapor in the air mass). The $\delta^{18}O$ and δ^2H of precipitation are also influenced by altitude, distance inland along different storm tracks, environmental conditions at the source of the vapor, latitude, and humidity. Progressive rainout as clouds move across the continent causes successive rainstorms to have increasingly lower δ values (Figure 5). At any point along the storm trajectory (i.e., for some specific fraction f of the total original vapor mass), $\delta^{18}O$ and δ^2H of the residual fraction of vapor in the air mass can be calculated using the Rayleigh equation.

At a given location, the seasonal variations in $\delta^{18}O$ and δ^2H values of precipitation and the weighted average annual $\delta^{18}O$ and δ^2H values of precipitation do not vary greatly from year to year. This happens because the annual range and sequence of climatic conditions (temperatures, vapor source, direction of air mass movement, etc.) generally vary within a predictable

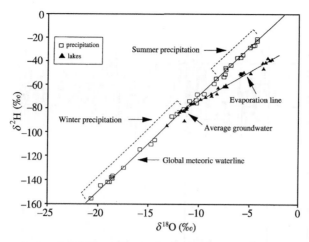

Figure 4 Isotopic compositions (δ^2H versus δ^{18}O) of precipitation and lake samples from northern Wisconsin (USA). The precipitation samples define the LMWL ($\delta D = 8.03\delta^{18}O + 10.95$), which is virtually identical to the GMWL. The lake samples plot along an evaporation line with a slope of 5.2 that intersects the LMWL at the composition of average local groundwater, with additional data (after Kendall *et al.*, 1995a).

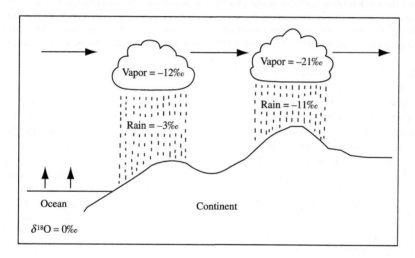

Figure 5 Origin of variations in the δ^{18}O of meteoric waters as moisture rains out of air masses moving across the continents.

range. Rain in the summer months has higher δ values than rain in the winter months due to the difference in mean seasonal air temperatures.

Superimposed on the seasonal cycles in precipitation δ values are storm-to-storm and intrastorm variations in the δ^{18}O and δ^2H values of precipitation (Figure 6). These variations may be as large as the seasonal variations. It is this potential difference in δ values between the relatively uniform old water and variable new water that permits isotope hydrologists to determine the contributions of old and new water to a stream during periods of high runoff (Sklash *et al.*, 1976). See Section 5.11.2.3 for more details about temporal and spatial variability in the δ^{18}O and δ^2H of precipitation.

The only long-term regional network for collection and analysis of precipitation for δ^2H

and δ^{18}O in the USA (and world) was established by a collaboration between the International Atomic Energy Agency (IAEA) and the World Meteorological Organization (WMO). The discovery that changes in the δ^{18}O of precipitation over mid- and high-latitude regions during 1970s to 1990s closely followed changes in surface air temperature (Rozanski *et al.*, 1992) has increased the interest in revitalizing a global network of isotopes in precipitation (GNIP), sites where precipitation will be collected and analyzed for isotopic composition.

5.11.2.1.3 Shallow groundwaters

Shallow groundwater δ^{18}O and δ^2H values reflect the local average precipitation values but are modified to some extent by selective recharge and fractionation processes that may alter the δ^{18}O

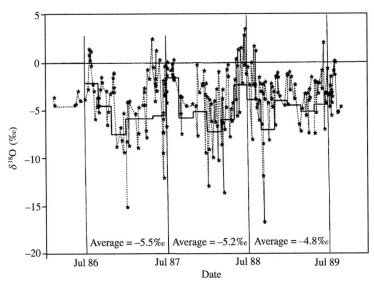

Figure 6 The $\delta^{18}O$ of storms at Panola Mountain, Georgia (USA) shows a 20‰ range in composition. Superimposed on the interstorm variability is the expected seasonal oscillation between heavy $\delta^{18}O$ values in the summer and lighter values in the winter caused by seasonal changes in temperature, storm track, and rain-out. The volume-weighted $\delta^{18}O$ values for two-month intervals are shown by heavy solid lines. The volume-weighted averages for the years 1987, 1988, and 1989 were each calculated from July 1 to June 30.

and δ^2H values of the precipitation before the water reaches the saturated zone (Gat and Tzur, 1967). These processes include (i) evaporation of rain during infiltration, (ii) selective recharge (e.g., only from major storms or during snowmelt), (iii) interception of rain water (Kendall, 1993; DeWalle and Swistock, 1994) and snow (Claassen and Downey, 1995) by the tree canopy, (iv) exchange of infiltrating water with atmospheric vapor (Kennedy *et al.*, 1986), and (v) various post-depositional processes (e.g., differential melting of snowpack or evaporation during infiltration).

The isotopic fractionation during evaporation is mainly a function of ambient humidity. Evaporation under almost 100% humidity conditions is more or less equivalent to evaporation under closed-system conditions (i.e., isotopic equilibrium is possible), and data for waters plot along a slope of 8 (i.e., along the LMWL). Evaporation at 0% humidity describes open-system evaporation. These two contrasting humidities result in the different fractionations (Figure 1). An open-surface water reservoir that is strongly evaporated will show changing $\delta^{18}O$ and δ^2H values along a trajectory with a slope of typically 5, beginning at the initial composition of the water on the LMWL, and extending out below the LMWL (Figure 4).

Once the rain or snowmelt passes into the saturated zone, the δ values of the subsurface water moving along a flow path change only by mixing with waters that have different isotopic compositions. The conventional wisdom regarding the isotopic composition of groundwater has long been that homogenizing effects of recharge and dispersive processes produce groundwater

with isotopic values that approach uniformity in time and space, approximating a damped reflection of the precipitation over a period of years (Brinkmann *et al.*, 1963). Although generally the case for porous media aquifers that are dominated by Darcian flow, this may not always hold true, particularly in shallow groundwater systems where preferential flow paths are common, hydraulic conductivities are variable, and waters may have a large range of residence times (see Section 5.11.2.3.2). Thus, one should always test the assumption of homogeneity by collecting water over reasonable spatial and temporal scales.

5.11.2.1.4 Deep groundwaters and paleorecharge

Deep groundwaters can be very old, often having been recharged thousands of years in the past. Since the $\delta^{18}O$ and δ^2H values of precipitation contributing to aquifer recharge are strongly dependent on the temperature and humidity of the environment, and these δ values generally behave conservatively in the subsurface, the δ values of deep groundwaters reflect the past climatic conditions under which the recharge took place. Thus, water isotopic composition can be a powerful addition to groundwater dating tools for the identification of paleorecharge. These tracers can be especially useful for the identification of groundwater recharged during the last glacial period of the Pleistocene, when average Earth surface temperatures were dramatically lower than during the Holocene.

Confined aquifers with relatively high porosity (such as sandstones layered between clay aquitards) are the best groundwater systems for the preservation of paleorecharge signals in the isotopic composition of the groundwater. Note that water isotopes do not provide an "age" of the groundwater sampled. Instead, they provide a tool by which to distinguish between modern recharge and recharge from some other time in the past, or from some other source.

An independent means of establishing groundwater age is always needed before making conclusions about the history of recharge from stable water isotope data (Fontes, 1981). Estimation of flow velocities and recharge rates via physically based groundwater flow models may lead to theoretical "ages" of groundwater in such settings. However, stable isotopic data should not be used to estimate ages without employing a chemically based means of groundwater dating, such as is afforded through a variety of radiometric dating techniques as well as noble gas methods (see Chapter 5.15).

Several studies and a number of reviews have demonstrated the use of water isotopes for identifying paleorecharge groundwater. These include Fontes (1981), Dutton and Simpkins (1989), Fontes *et al.* (1991, 1993), Deák and Coplen (1996), Coplen *et al.* (2000), and Bajjali and Abu-Jaber (2001). The basic principle underlying the technique is that precipitation formed under temperature conditions that were much cooler than at present will be reflected in old groundwater that has significantly lower δ values. On a plot of $\delta^{18}O$ versus δ^2H, precipitation from cooler climates has more negative δ values than precipitation from warmer climates. In general, locations and times with cooler and more humid conditions have lower d excess values.

Climate changes in arid regions are often readily recognized in paleowaters. Strong variability in humidity causes large shifts in the deuterium excess (or d excess), and corresponding shifts in the vertical position of the LMWL (Clark and Fritz, 1997). The value of the d excess in water vapor increases as the relative humidity under which evaporation takes place decreases. Since air moisture deficit is positively correlated with temperature, d excess should be positively correlated with temperature. High d excess in groundwater indicates that the recharge was derived partly from water evaporated under conditions of low relative humidity.

Comparison of paleowater isotopic data must be made to the modern LMWL; if a long-term record of local precipitation is not available, an approximation can be obtained from the nearest station of GNIP and Isotope Hydrology Information System (ISOHIS), established by the Isotope Hydrology Section of the IAEA, Vienna, Austria.

The combined use of chemical indicators and stable isotopes is especially powerful for identifying paleowaters. Chlorine concentrations and chloride isotopes are often applied to estimate paleorecharge processes, whereas noble gas techniques have been applied to the derivation of paleotemperatures of recharge for very old groundwaters (Stute and Schlosser, 2000: see Chapter 5.15). Saline paleowaters related to evaporitic periods in the geologic past can be recognized by δ values that plot significantly below the MWL because of evaporative enrichment. Secondary brines resulting from the dissolution of salts by modern recharge will plot on or very near to the MWL (Fontes, 1981).

Only in very limited cases may the isotopes of water be useful as proxies for paleotemperatures (e.g., in high-latitude ice cores). The isotopic content of precipitation depends upon many factors, not just temperature. Thus, the lack of a significant difference between the δ values of modern precipitation and groundwaters dated to times corresponding with glacial maxima does not necessarily imply a poor age-dating technique. Humidity, isotopic content of water vapor sources, and precipitation distribution all play a role in the isotopic composition of meteoric recharge. In the absence of independent age-dating methods, altitude effects may be misinterpreted as paleorecharge signals, particularly in cases where rivers having high-altitude recharge areas are responsible for a large portion of groundwater recharge taking place at lower elevations downstream. One cannot expect a simple relationship between the stable isotopic content of meteoric recharge and average temperatures for past periods of different climate. Also, dispersion and molecular diffusion may cause blurring of differences in isotopic composition between Holocene and pre-Holocene groundwaters.

5.11.2.1.5 Surface waters

Figure 7 shows the spatial distribution of $\delta^{18}O$ and δ^2H values of surface waters in the USA (Kendall and Coplen, 2001). These maps were generated by the analysis of more than 4,800 depth- and width-integrated stream samples from 391 sites sampled bimonthly or quarterly for 2.5–3 yr between 1984 and 1987 by the USGS (Coplen and Kendall, 2000). Spatial distributions of $\delta^{18}O$ and δ^2H values are very similar to each other and closely match topographic contours. The ability of this data set to serve as a proxy for the isotopic composition of modern precipitation in the USA is supported by the excellent agreement between the river data set and the compositions of adjacent precipitation monitoring sites, the strong spatial coherence of the

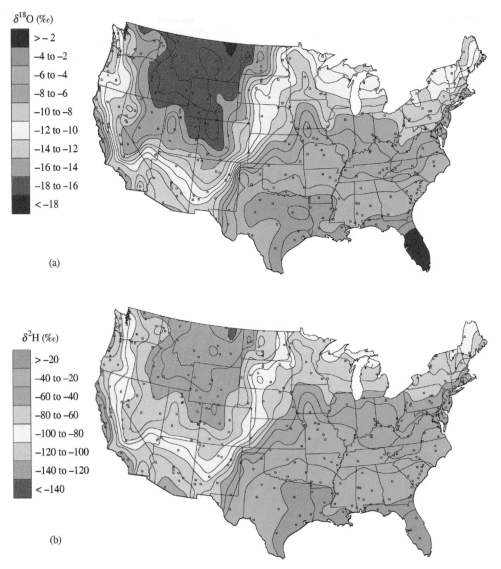

Figure 7 Spatial variation in the: (a) $\delta^{18}O$ and (b) δ^2H of rivers in the USA (source Kendall and Coplen, 2001).

distributions of $\delta^{18}O$ and δ^2H, the good correlations of the isotopic compositions with climatic parameters, and the good agreement between the "national" MWL generated from unweighted analyses of samples from the 48 contiguous states of $\delta^2H = 8.11 \; \delta^{18}O + 8.99 \; (r^2 = 0.98)$ with the unweighted global MWL of the IAEA-GNIP sites of $\delta^2H = 8.17\delta^{18}O + 10.35$ (Kendall and Coplen, 2001).

Although many isotope hydrologists would agree that such maps should be constructed from1 long-term annual averages of precipitation compositions, there are some advantages to stream samples as indicators of precipitation compositions. First, stream samples are relatively easy to obtain by taking advantage of existing long-term river monitoring networks. Second, stream water is a better spatial and temporal integrator of the isotopic composition of

precipitation intercepted by a large drainage basin than recent precipitation collected at a single location in the basin. Third, since groundwater is probably the dominant source of stream flow in most basins, the river provides information on the composition of waters that have infiltrated the soils to recharge the groundwater system. Finally, the remains of biota that lived in the soil, lakes, and streams that are used to reconstruct paleohydrology and paleoclimates record the integrated isotopic signal of recharge water, not the isotopic compositions of the precipitation.

Water in most rivers has two main components: (i) recent precipitation that has reached the river either by surface runoff, channel precipitation, or by rapid flow through shallow subsurface flow paths; and (ii) groundwater. The relative contributions of these sources differ in each watershed or basin, and depend on the physical setting of the drainage basin (e.g., topography, soil type, depth to bedrock, vegetation, fractures, etc.), climatic parameters (e.g., precipitation amount, seasonal variations in precipitation, temperature, potential evapotranspiration, etc.), and human activities (e.g., dams, reservoirs, irrigation usages, clearing for agriculture, channel restructuring, etc.).

The $\delta^{18}O$ and δ^2H of rivers will reflect how the relative amounts of precipitation and groundwater vary with time, and how the isotopic compositions of the sources themselves change over time. Seasonal variations will be larger in streams where recent precipitation is the main source of flow, and smaller in streams where groundwater is the dominant source. As the basin size increases, the isotopic compositions of river waters are also increasingly affected by evaporation (Gat and Tzur, 1967). Local precipitation events are an important component of river water in the headwaters of large basins. For example, the average amount of new water in small, forested watersheds during storms is ~40%, although during storms the percentage can be higher (Genereux and Hooper, 1998). However, in the lower reaches, local additions of precipitation can be of minor importance (Friedman *et al.*, 1964; Salati *et al.*, 1979), except during floods (Criss, 1999). For example, analysis of long-term tritium records for seven large (5,000–75,000 km^2) drainage basins in the USA indicated that ~60% ± 20% of the river water was less than 1 yr old (Michel, 1992).

The dual nature of river water (partly recent precipitation, partly groundwater) can be exploited for studying regional hydrology or climatology. Under favorable circumstances, knowledge of the isotopic compositions of the major water sources can be used to quantify the time-varying contributions of these sources to river water (Sklash *et al.*, 1976; Kendall and McDonnell, 1998). Alternatively, if the isotopic composition of base flow is thought to be a good representation of mean annual precipitation (Fritz, 1981), then the $\delta^{18}O$ and δ^2H of rivers sampled during low flow can integrate the composition of rain over the drainage areas and be useful for assessing regional patterns in precipitation related to climate.

5.11.2.2 Tritium

Tritium (^3H) is a radiogenic and radioactive isotope of hydrogen with a half-life of 12.4 yr. It is an excellent tracer for determining timescales for the mixing and flow of waters because it is considered to be relatively conservative geochemically, and is ideally suited for studying processes that occur on a timescale of less than 100 yr. Tritium content is expressed in tritium units (TU), where 1 TU equals 1 ^3H atom in 10^{18} atoms of hydrogen.

Prior to the advent of atmospheric testing of thermonuclear devices in 1952, the tritium content of precipitation was probably in the range of 2–8 TU (Thatcher, 1962); this background concentration is produced by cosmic ray spallation. While elevated tritium levels have been measured in the atmosphere since 1952, tritium produced by thermonuclear testing ("bomb tritium") has been the dominant source of tritium in precipitation. A peak concentration of several thousand TU was recorded in precipitation in the northern hemisphere in 1963, the year that the atmospheric test ban treaty was signed. After 1963, the tritium levels in precipitation began to decline gradually because of radioactive decay and the cessation of atmospheric testing.

The simplest use of tritium is to check whether detectable concentrations are present in the water. Although pre-bomb atmospheric tritium concentrations are not well known, waters derived exclusively from precipitation before 1953 would have maximum tritium concentrations of ~0.1–0.4 TU by 2003. For waters with higher tritium contents, some fraction of the water must have been derived since 1953; thus, the tritium concentration can be a useful marker for recharge since the advent of nuclear testing.

Distinct "old" and "new" water tritium values are required for storm and snowmelt runoff hydrograph separation. In small (first-order) catchments where the average residence time (the average time it takes for precipitation to enter the ground and travel to the stream) of the old water is on the order of months, the old and new water tritium concentrations will not likely be distinguishable. However, in some larger catchments with longer residence times, old and new waters may be distinctive as a result of the gradual decline in precipitation tritium values since 1963 and the even more gradual decline in groundwater tritium values.

Tritium measurements are frequently used to calculate recharge rates, rates or directions of subsurface flow, and residence times. For these purposes, the seasonal, yearly, and spatial variations in the tritium content of precipitation must be accurately assessed. This is difficult to do because of the limited data available, especially before the 1960s. For a careful discussion of how to calculate the input concentration at a specific location, see Michel (1989) and Plummer *et al.* (1993). Several different approaches (e.g., piston-flow, reservoir, compartment, and advective-dispersive models) to modeling tritium concentrations in groundwater are discussed by Plummer *et al.* (1993). The narrower topic of using environmental isotopes to determine residence time is discussed briefly below.

If the initial concentration of ^3H in the atmosphere is not known, an alternative dating method is to analyze waters for both ^3H and ^3He. Because ^3H decays to ^3He, it is possible to use the tritiogenic ^3He component of ^3He in groundwater as a quantitative tracer of the age of the water since it was separated from the atmosphere.

5.11.2.3 Determination of Runoff Mechanisms

A major uncertainty in hydrologic and chemical modeling of catchments has been the quantification of the contributions of water and solutes from various hydrologic pathways. Isotope hydrograph separation apportions storm and snowmelt hydrographs into contributing components based on the distinctive isotopic signatures carried by the two or more water components (e.g., new precipitation, soil water, and groundwater). The basic principle behind isotope hydrograph separation is that, if the isotopic compositions of the sources of water contributing to stream flow during periods of high runoff are known and are different, then the relative amounts of each source can be determined. Since these studies began in the 1960s, the overwhelming conclusion is that old water is by far the dominant source of runoff in humid, temperate environments (Sklash *et al.*, 1976; Pearce *et al.*, 1986; Bishop, 1991). Buttle (1998) provides a succinct review of watershed hydrology.

5.11.2.3.1 *Isotope hydrograph separation and mixing models*

Until the 1970s, the term "hydrograph separation" meant a graphical technique that had been used for decades in predicting runoff volumes and timing. For example, the graphical separation technique introduced by Hewlett and Hibbert (1967) is commonly applied to storm hydrographs (i.e., stream flow versus time graphs) from forested catchments to quantify "quick flow" and "delayed flow" contributions. According to this technique, the amounts of quick flow and delayed flow in stream water can be determined simply by considering the shape and timing of the discharge hydrograph.

Isotope hydrograph separation apportions storm and snowmelt hydrographs into contributing components based on the distinctive isotopic signatures carried by the old and new water components. Hence, the method allows the calculation of the relative contributions of new precipitation and older groundwater to stream flow. A major limitation of the tracer method is that we cannot directly determine how the water reaches the stream (i.e., geographic source of the

water) nor the actual runoff generation mechanism from knowledge of the temporal sources of water. In a very real sense, the isotope hydrograph separation technique is still a "black box" method that provides little direct information about what is actually going on in the subsurface. However, in combination with other types of information (e.g., estimates of saturated areas or chemical solute compositions of water along specific flow paths), runoff mechanisms can sometimes be inferred.

Hydrograph separations are typically performed with only one isotope, both to save on the cost of analyses and because of the high correlation coefficient between $\delta^{18}O$ and δ^2H ($r^2 > 0.95$; Gat, 1980). However, analysis of both isotopes will often prove very beneficial, and is highly recommended.

Two-component mixing models. Isotope hydrograph separation normally involves a two-component mixing model for the stream. The model assumes that water in the stream at any time during storm or snowmelt runoff is a mixture of two components: new water and old water.

During base-flow conditions (the low-flow conditions that occur between periods of storm and snowmelt runoff), the water in a stream is dominated by old water. The chemical and isotopic character of stream water at a given location during base flow represents an integration of the old-water discharged from upstream. During storm and snowmelt runoff events, however, new water is added to the stream. If the old and new water components are chemically or isotopically different, the stream water becomes changed by the addition of the new water. The extent of this change is a function of the relative contributions by the old and new water components.

The contributions of old and new water in the stream at any time can be calculated by solving the mass-balance equations for the water and tracer fluxes in the stream, provided that the stream, old water, and new water tracer concentrations are known:

$$Q_s = Q_o + Q_n \qquad (10)$$

$$C_s Q_s = C_o Q_o + C_n Q_n \qquad (11)$$

where Q is discharge, C refers to tracer concentration, and the subscripts "s," "o," and "n" indicate the stream, the old water components, and the new water components, respectively. If stream samples are taken at a stream gauging station, the actual volumetric contributions of old and new water can be determined. If no discharge measurements are available, old and new water contributions can be expressed as percentages of total discharge.

Although the simple two-component mixing model approach to stream hydrograph separation does not directly identify the actual runoff generation mechanisms, the model can sometimes allow the hydrologist to evaluate the importance of a given conversion process in a catchment. For example, if a rapid conversion mechanism (partial-area overland flow, saturation overland flow, or perhaps subsurface flow through macropores) is the dominant conversion process contributing to a storm runoff hydrograph, the isotopic content of the stream will generally reflect mostly new water in the stream. Conversely, if a slow conversion mechanism (Darcian subsurface flow) is dominant in producing the storm runoff, the isotopic content of the stream should indicate mostly old water in the stream. Evaluation of the flow processes can be enhanced by applying the mixing model to water collected from subsurface runoff collection pits, overland flow, and macropores.

Sklash and Farvolden (1982) listed five assumptions that must hold for reliable hydrograph separations using environmental isotopes:

(i) the old water component for an event can be characterized by a single isotopic value or variations in its isotopic content can be documented;

(ii) the new water component can be characterized by a single isotopic value or variations in its isotopic content can be documented;

(iii) the isotopic content of the old water component is significantly different from that of the new water component;

(iv) vadose water contributions to the stream are negligible during the event or they must be accounted for (use an additional tracer if isotopically different from groundwater); and

(v) surface water storage (channel storage, ponds, swamps, etc.) contributions to the stream are negligible during the runoff event.

The utility of the mixing equations for a given high runoff period is a function mainly of the magnitude of $(\delta_o - \delta_n)$ relative to the analytical error of the isotopic measurements, and the extent to which the aforementioned assumptions are indeed valid (Pearce *et al.*, 1986; Genereux, 1998). Clearly the relative amounts of new versus old water are affected by many environmental parameters such as: size of the catchment, soil thickness, ratio of rainfall rate to infiltration rate, steepness of the watershed slopes, vegetation, antecedent moisture conditions, permeability of

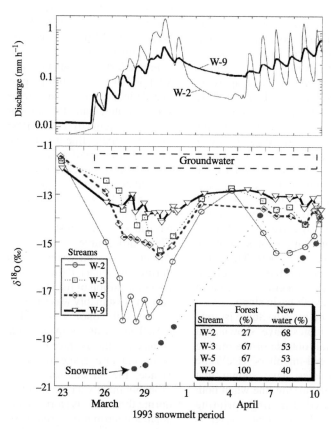

Stream	Forest (%)	New water (%)
W-2	27	68
W-3	67	53
W-5	67	53
W-9	100	40

Figure 8 Early spring samples from four streams at Sleepers River Watershed, Vermont, have $\delta^{18}O$ values intermediate between the compositions of snowmelt collected in pan lysimeters and groundwater. Diurnal fluctuations in discharge correlate with diurnal changes in $\delta^{18}O$, especially at W-2. W-2, a 59 ha agricultural basin, shows much greater contributions from snowmelt than the other three catchments. The three mixed agricultural/forested nested catchments—W-9 (47-ha), W-3 (837-ha), and W-5 (11,125)—show increasing contributions from new snowmelt as scale increases (after Shanley *et al.*, 2001).

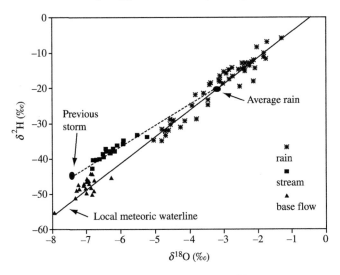

Figure 9 Storm flow samples from the Mattole River, California, have $\delta^{18}O$ and δD values that do not plot along the mixing line (solid line) between the two supposed sources: new rain and pre-storm base flow. Therefore, there must be another source of water, possibly soil water derived from a previous storm, contributing significant amounts of water to storm flow. The stream samples plot along a mixing line (dashed line) between the composition of average new rain and the composition of previous rain (after Kennedy *et al.*, 1986).

the soil, amount of macropores, and storage capacity of the catchment.

Figure 8 shows an example of a two-component hydrograph separation for four catchments at Sleepers River Watershed, Vermont USA (Shanley *et al.*, 2001). During the snowmelt period, groundwater appeared to have an approximately constant $\delta^{18}O$ value of $-11.7 \pm 0.3‰$, whereas snowmelt collected in snow lysimeters ranged from $-20‰$ to $-14‰$. The stream samples had $\delta^{18}O$ values intermediate between the snowmelt and groundwater. Two-component isotope separations showed that the meltwater inputs to the stream ranged from 41% to 74%, and generally increased with catchment size (41–11,125 ha). Another multicatchment isotope hydrograph study showed no correlation of basin size and the percent contribution of subsurface flow to stream flow (Sueker *et al.*, 2000). This study found that subsurface flow positively correlated with the amount of surficial material in the basin and negatively correlated with basin slope.

Three-component mixing models. Some isotope hydrograph separation studies have shown that the stream $\delta^{18}O$ values fall outside of the mixing line defined by the two "end-members" (DeWalle *et al.*, 1988). Other studies have shown that stream δ values fall off the mixing line when both $\delta^{18}O$ and $\delta^{2}H$ are used for fingerprinting (Kennedy *et al.*, 1986). If the stream water is not collinear with the two suspected end-members, either the end-members are inaccurate or there must be another component (probably soil water) that plots above or below the mixing line for the other components (Kennedy *et al.*, 1986). In this case, three-component models using

silica or some other chemical tracer can be used to estimate the relative contributions of the sources to stream water (DeWalle *et al.*, 1988; Wels *et al.*, 1991; Hinton *et al.*, 1994; Genereux and Hooper, 1998).

Figure 9 shows that stream water samples at the Mattole River in California plot above the mixing line defined by the two supposed sources of the stream water, namely, pre-storm base flow and new rain (Kennedy *et al.*, 1986). As often is the case, the mixing line is collinear with the LMWL. Thus, no mixture of waters from these two sources of water can produce the stream water. This means that either (i) there is an additional source of water (above the line) which contributes significantly to stream flow or (ii) some process in the catchment has caused waters to shift in isotopic composition. This third source of water is often soil water. Although no soil-water samples were collected during the January 1972 storm at the Mattole River, the composition of the last big storm to hit the area (which presumably was a major source of recharge to the soil zone) was analyzed for $\delta^{18}O$ and $\delta^{2}H$. The stream samples plot along a mixing line connecting the average composition of the earlier storm and the average composition of January rain. Hence, the stream samples on Figure 9 were probably mainly a mixture of soil water recharged during the previous storm and new rainwater, with smaller contributions from base flow.

If only $\delta^{18}O$ *or* $\delta^{2}H$ had been analyzed, the stream water would have plotted intermediate between the two end-members (consider the projections of these water compositions onto the $\delta^{18}O$ and $\delta^{2}H$ scales) and a two-component isotope hydrograph separation would have seemed feasible. Only when both isotopes were analyzed did it become apparent that there was a mass-balance problem. Hence, we recommend analysis of both oxygen and hydrogen isotopes on a subset of stream water and end-member samples to ensure that an end-member or process is not overlooked.

Sometimes it is obvious that the two-component model is inadequate: rain combined with glacial meltwater (Behrens *et al.*, 1979), for example. For rain-on-snow events where snow cores are used to assess the new-water composition, a two-component model is also inadequate (Wallis *et al.*, 1979). However, if the catchment is snow-covered, overland flow is minimal, the surface area of the stream is small, and several snowmelt lysimeters are used to catch sequential samples of new water derived from infiltration of both rain and melting snow, a two-component model may be sufficient. It is difficult to judge *a priori* when the isotopic difference between soil water and groundwater is sufficient to require a three-component model (Kennedy *et al.*, 1986; DeWalle *et al.*, 1988; Swistock *et al.*, 1989; Hinton *et al.*, 1994).

DeWalle *et al.* (1988) used a three-component model for the special case where the discharge of one component was known. Wels *et al.* (1991) used a three-component model for the case where the isotopic compositions of new overland and new subsurface flow were identical, and the silica contents of new and old subsurface flow were identical. Ogunkoya and Jenkins (1993) used a model where one component was time varying. Hinton *et al.* (1994) listed five assumptions required for three-component models using two tracers.

5.11.2.3.2 Temporal and spatial variability in end-members

The universal applicability of the simple two-component mixing model has been frequently challenged (Kennedy *et al.*, 1986; Rodhe, 1987; Sklash, 1990; McDonnell *et al.*, 1991; Bishop, 1991; Kendall and McDonnell, 1993). These challenges address the five assumptions governing the use of the two-component mixing model given in Sklash and Farvolden (1982) and listed above.

First, the new water component is rarely constant in isotopic composition during events. Temporal variations in rain isotopic composition have been found to have an appreciable effect on hydrograph separations (McDonnell *et al.*, 1990). Second, investigators have concluded that in some catchments, soil water is an important contributor to storm runoff and is isotopically and/or chemically distinct from groundwater (Kennedy *et al.*, 1986; DeWalle *et al.*, 1988; Hooper *et al.*, 1990); in these cases, a three-component mixing model may be useful. In some catchments, the shallow soil and groundwaters are so heterogeneous in

isotopic composition that isotope hydrograph separations are ill-advised (McDonnell *et al.*, 1991; Ogunkoya and Jenkins, 1991).

None of these challenges to the simple two-component mixing model approach has caused any significant change in the basic conclusion of the vast majority of the isotope and chemical hydrograph studies to date, namely, that *most storm flow in humid-temperate environments is old water* (Bishop, 1991). There is, however, a need to address the potential impact of natural isotopic variability on the use of isotopes as tracers of water sources. Most of these concerns can be addressed by applying more sophisticated two- and three-component mixing models that allow for variable isotopic signatures (e.g., Harris *et al.*, 1995).

The "new" water component. The isotopic composition of rain and snow can vary both spatially and temporally during storms. In general, rain becomes progressively more depleted in ^{18}O and ^2H during an event because these isotopes preferentially rain out early in the storm. Successive frontal and convective storms may have more complex isotopic variations. Intrastorm rainfall compositions have been observed to vary by as much as 90‰ in δ^2H at the Maimai watershed in New Zealand (McDonnell *et al.*, 1990), 16‰ in $\delta^{18}O$ in Pennsylvania, USA (Pionke and DeWalle, 1992) and 15‰ in $\delta^{18}O$ in Georgia, USA (Kendall, 1993).

Snowmelt also varies in isotopic composition during the melt period and is commonly different from that of melted snow cores (Hooper and Shoemaker, 1986; Stichler, 1987; Taylor *et al.*, 2001). Part of the changes in snowmelt isotopic composition can be attributed to distinct isotopic layers in the snow, rain events on snow, and isotopic fractionation during melting. In general, the snowmelt has low $\delta^{18}O$ values early in the season but the $\delta^{18}O$ values become progressively higher as the pack melts (Shanley *et al.*, 2001; Taylor *et al.*, 2001).

The variations in "new" water isotopic content are assessed by collecting sequential samples of rain or through fall during rainstorms, and samples of meltwater from snow lysimeters during snowmelt. The method of integrating this information into the separation equations depends on the magnitude and rate of temporal variations in the "new" water isotopic content and the size of the catchment. If a varying "new" water value is used, and the catchment is large enough that travel times are significant, one must account for the travel time to the stream and to the monitoring point.

In forested areas, rain intercepted by the tree canopy can have a different isotopic composition than rain in open areas (Saxena, 1986; Kendall, 1993; DeWalle and Swistock, 1994). On average, through fall during individual storms at a site in

Georgia (USA) had $\delta^{18}O$ values 0.5‰ higher and δ^2H values 3.0‰ higher than rain. Site-specific differences in canopy species, density, and microclimate resulted in an average $\delta^{18}O$ range of 0.5‰ among 32 collectors over a 0.04 km^2 area for the same storm, with a maximum range of 1.2‰ (Kendall, 1993). Other sites appear to have little spatial variability (Swistock *et al.*, 1989). Because of this potential for spatial variability, putting out several large collectors and combining the through fall for isotopic analysis is advisable (DeWalle *et al.*, 1988).

Different rainfall weighting methods can substantially affect estimates of new/old waters in storm runoff in basins with large contributions of new water (McDonnell *et al.*, 1990). Use of sequential rain values is probably the best choice in very responsive catchments or in catchments with high proportions of overland flow. When rain intensities are low and soils drain slowly, current rain may not infiltrate very rapidly and thus use of the cumulative approach (i.e., running average) is probably more realistic (Kendall *et al.*, 2001a).

In large watersheds and watersheds with significant elevation differences, both the "old" water and "new" water δ values may show large spatial variability; generally, the higher elevations will have lower δ values. Such large or steep watersheds would require many sampling stations to adequately assess the variability in old and new waters. Spatial variations in the $\delta^{18}O$ of up to 5‰ were observed for rain storms in a 2,500 km^2, 35-station network in Alabama, USA (Kendall and McDonnell, 1993). Multiple collectors are especially critical for assessing temporal and spatial variability in both the timing and composition of snowmelt. Consequently, simple isotope hydrograph separation techniques are best suited for small first- and second-order catchments.

Preferential storage of different time fractions of rain in the shallow soil zone can be important because of the large amounts of rain potentially going into storage. For a storm that dropped 12 cm of rain over a 1 day period at Hydrohill, a 500 m^2 artificial catchment in China, 45% of the total rain went into storage (Kendall *et al.*, 2001a), with almost 100% of the first 2 cm of rain going into storage before runoff began.

The "old" groundwater component. Groundwater in catchments can frequently be heterogeneous in isotopic composition due to variable residence times. For example, groundwater at Hydrohill catchment showed a 5‰ range in $\delta^{18}O$ after a storm (Kendall *et al.*, 2001a). Groundwaters at the Sleepers River watershed in Vermont (USA) showed a 4‰ isotopic stratification in most piezometer nests during snowmelt (McGlynn *et al.*, 1999). Consequently, even if a large number of wells are sampled to characterize the potential variability in the groundwater composition, base

flow is probably a better integrator of the water that actually discharges into the stream than any average of groundwater compositions. Hence, it may be safer to use pre-storm base flow, or interpolated intrastorm base flow (Hooper and Shoemaker, 1986), or ephemeral springs as the old-water end-member. If groundwater is used for the end-member, researchers need to check the effects of several possible choices for composition on their separations.

One major assumption in sampling base flow to represent the "old" water component is that the source of groundwater flow to the stream during storms and snowmelt is the same as the source during base flow conditions. It is possible that ephemeral springs remote from the stream or deeper groundwater flow systems may contribute differently during events than between events, and if their isotopic signatures differ from that of base flow, the assumed "old" water isotopic value may be incorrect. Although the occurrence of such situations could be tested by hydrometric monitoring and isotopic analyses of these features, their quantification could be difficult. Recent tracer studies showing the wide range of groundwater residence times in small catchments illustrate the problems with simple source assumptions (McDonnell *et al.*, 1999; Kirchner *et al.*, 2001; McGlynn *et al.*, 2003).

Soil water. The role of soil water in storm flow generation in streams has been intensively studied using the isotope tracing technique (DeWalle *et al.*, 1988; Buttle and Sami, 1990). Most isotope studies have lumped soil-water and groundwater contributions together, which may mask the importance of either component alone. However, Kennedy *et al.* (1986) observed that as much as 50% of the stream water in the Mattole catchment was probably derived from rain from a previous storm stored in the soil zone (Figure 9). If there is a statistically significant difference between the isotopic compositions of soil and groundwater, then they need to be considered separate components (McDonnell *et al.*, 1991).

The isotopic composition of soil water may be modified by a number of processes including direct evaporation, exchange with the atmosphere, and mixing with infiltrating water (Gat and Tzur, 1967). Although transpiration removes large amounts of soil water, it is believed to be nonfractionating (Zimmerman *et al.*, 1967). Wenner *et al.* (1991) found that the variable composition of the percolating rain became largely homogenized at depths as shallow as 30 cm due to mixing with water already present in the soil.

Intrastorm variability of soil water of up to 1.5‰ in $\delta^{18}O$ was observed during several storms at Panolan Mountain Georgia, and a 15‰ range in δ^2H was seen during a single storm at Maimai, New Zealand (Kendall and McDonnell, 1993).

Analysis of over 1,000 water samples at Maimai (McDonnell *et al.*, 1991) showed a systematic trend in soil-water composition in both downslope and downprofile directions. Multivariate cluster analysis also revealed three distinct soil water groupings with respect to soil depth and catchment position, indicating that the soil water reservoir is poorly mixed on the timescale of storms.

If the isotopic variability of rain, through fall, meltwater, soil water, and groundwater is significant at the catchment scale (i.e., if hill-slope waters that are variable in composition actually reach the stream during the storm event) or if transit times are long and/or variable, then simple two- and three-component, constant composition, mixing models may not provide realistic interpretations of the system hydrology. One approach is to develop models with variable end-members (Harris *et al.*, 1995). Another possible solution is to include alternative, independent, chemical or isotopic methods for determining the relative amounts of water flowing along different subsurface flow paths. Recent comparisons of isotopic and chemical approaches include Rice and Hornberger (1998) and Kendall *et al.* (2001a).

Additional isotopic tracers include radon-222 (Genereux and Hemond, 1990; Genereux *et al.*, 1993); beryllium-7 and sulfur-35 (Cooper *et al.*, 1991); carbon-13 (Kendall *et al.*, 1992); strontium-87 (Bullen and Kendall, 1998); and nitrogen-15 (Burns and Kendall, 2002). In particular, carbon isotopes, possibly combined with strontium isotopes, appear to be useful for distinguishing between deep and shallow flow paths and, combined with oxygen isotopes, for distinguishing among water sources. Finally, a number of mathematical models can be employed that incorporate the variability of input tracer concentrations, and which attempt to reproduce observed output concentrations (Kirchner *et al.*, 2001; McGlynn *et al.*, 2003). Some of these techniques will be discussed in the following section.

5.11.2.4 Estimation of Mean Residence Time

Arguably the most important quantity that can be estimated in hydrologic studies is the age of the water that has been sampled. Water age has utility for determining hydrological processes in catchments (as discussed previously), as well as direct implications for management of larger groundwater resources via determination of aquifer sustainability and migration rates of contaminants. Thus, the estimation of mean residence time (MRT) of groundwater in a catchment or in an aquifer is an important goal. The discussion above demonstrated that the stable isotopes of the water molecule can be used to determine—qualitatively—the relative ages of water masses moving through a watershed using the mean input

concentration of the tracer (i.e., mean isotopic composition of precipitation). While this method is imperfect, the signals observed in isotope data collected in time series are clear and informative. These signals can be used in a more quantitative manner to calibrate flow and transport models. While it is beyond the scope of this chapter to thoroughly review modeling techniques in hydrology, it is worth noting some of the recent advances in the combined use of geochemical data and hydrometric data for improving and calibrating predictive models. The topic of groundwater residence time is discussed in detail in Chapter 5.15.

One approach taken to model flow in catchments by using isotopic tracers is the *lumped-parameter* modeling approach. The approach is based upon the use of a mathematical convolution integral for the prediction of tracer output at sampling points that are assumed to integrate flow processes occurring over some (undefined) spatial and (defined) temporal scale. The convolution is performed between a defined tracer input function observed or estimated from tracer data in time series (such as the isotopic composition of precipitation) and a chosen model function representative of the manner in which transit times are distributed in the system according to a steady-state equation of the general form

$$C_o(t) = \int_0^\infty C_i(t - \tau)g(\tau)d\tau \qquad (12)$$

where $C_o(t)$ and $C_i(t)$ are output and input tracer concentrations, τ is the residence time, and $g(\tau)$ is the system response or weighting function for the tracer (Unnikrishna *et al.*, 1995; Maloszewski and Zuber, 1996). Commonly used forms of the weighting function ($g(\tau)$) include the piston flow model (PFM), the exponential model (EM), the dispersion model (DM), and the combined exponential-piston flow model (EPM). Details of these weighting function models are provided in Maloszewski and Zuber (1996).

In the lumped-parameter approach, there is no notion *a priori* about the geometry of the flow paths in the system. This contrasts with the deterministic or distributed modeling approach, which strives to apply governing equations of flow to an estimated geometric flow field. In the latter approach, it is assumed that the properties controlling flow (such as hydraulic conductivity, transmissivity, porosity, etc.) can be adequately estimated at all points in the catchment model, regardless of scale or depth. In addition, deterministic models are founded in Darcian flow, which in fact may contribute in small degree to measured tracer responses in surface water and shallow groundwater systems. The lumped-parameter approach, however, makes no assumption about the geometry of the system. Instead, the factors that give rise to observed (measured) tracer output responses to observed (measured) tracer inputs are lumped together within a small number of model parameters.

These model parameters may or may not have any direct physical meaning, depending largely upon individual interpretation. The parameters do, however, represent a real means by which to characterize responses, predict flows, and facilitate comparisons among systems. These models are sometimes called "black box" models, reflecting the fact that nothing is assumed known about the flow system, and the entire system is treated as being homogeneous with respect to the distribution of flow lines. Some who espouse their use have softened the moniker to "gray-box" models, in order to reflect the notion that significant knowledge about the system being studied is required in order to (i) choose an appropriate model formulation, (ii) make meaningful interpretations of model output, and (iii) to apply the models wisely. In any case, it is clear that, where tracer inputs and outputs from a hydrologic system are measurable on the timescale of their major variability, lumped-parameter modeling has been the most widely applied approach. Recent reviews and applications of lumped-parameter modeling for MRT estimation in catchments have been conducted by Maloszewski and Zuber (1996), Unnikrishna *et al.* (1995), McGuire *et al.* (2002), and McGlynn *et al.* (2003).

Using the lumped-parameter modeling approach with $\delta^{18}O$ data, McGuire *et al.* (2002) calculated MRTs of 9.5 and 4.8 months for two watersheds in the central Appalachian Mountains of Pennsylvania (USA). Soil-water residence time at a depth of 100 cm was estimated to be ~2 months, indicating that the shallow groundwater was rapidly mobilized in response to precipitation events. In the steeper Maimai catchment of New Zealand, Stewart and McDonnell (1991) reported MRTs calculated using δ^2H data ranging from 12 d to greater than 100 d. Shallow suction lysimeters showed the most rapid responses (12–15 d). MRTs increased with increasing soil depth, such that lysimeters at 40 cm depth showed MRTs of 50–70 d, and lysimeters at 800 cm depth showed residence times of ~100 d or more.

5.11.3 CARBON, NITROGEN, AND SULFUR SOLUTE ISOTOPES

This section presents a discussion of the fundamentals of three major solute isotope systems: carbon, nitrogen, and sulfur. Recent comprehensive reviews of the geochemistry of several other solute isotope tracers in hydrologic systems include those by Faure (1986),

Kendall and McDonnell (1998), and Cook and Herczeg (2000). As discussed above, water isotopes often provide useful information about residence times and relative contributions from different water sources; these data can then be used to make hypotheses about water flow paths. Solute isotopes can provide an alternative, independent isotopic method for determining the relative amounts of water flowing along various subsurface flow paths. However, the least ambiguous use of solute isotopes is for tracing the relative contributions of potential solute sources to groundwater and surface water.

5.11.3.1 Carbon

Over the last several decades, the decline in alkalinity in many streams in Europe and in northeastern USA as a result of acid deposition has been a subject of much concern (Likens *et al.*, 1979). The concentration of bicarbonate, the major anion buffering the water chemistry of surface waters and the main component of dissolved inorganic carbon (DIC) in most stream waters, is a measure of the "reactivity" of the watersheds and reflects the neutralization of carbonic and other acids by reactions with silicate and carbonate minerals encountered by the acidic waters during their residence in watersheds (Garrels and Mackenzie, 1971). Under favorable conditions, carbon isotopes of DIC can be valuable tools by which to understand the biogeochemical reactions controlling carbonate alkalinity in groundwater and watersheds (Mills, 1988; Kendall *et al.*, 1992; see Chapter 5.14).

The $\delta^{13}C$ of DIC can also be a useful tracer of the seasonal and discharge-related contributions of different hydrologic flow paths to surface stream flow (Kendall *et al.*, 1992) and to shallow groundwater discharge from springs in carbonate and fractured rock terrains (Deines *et al.*, 1974; Staniaszek and Halas, 1986; Doctor, 2002). In many carbonate-poor catchments, waters along shallow flow paths in the soil zone have characteristically low $\delta^{13}C_{DIC}$ values reflecting the dominance of soil CO_2 contributions of DIC; waters along deeper flow paths within less weathered materials have intermediate $\delta^{13}C_{DIC}$ values characteristic of carbonic-acid weathering of carbonates (Bullen and Kendall, 1998). In carbonate terrains, the extent of chemical evolution of the $\delta^{13}C$ of carbon species in groundwater can be used to distinguish between groundwaters traveling along different flow paths (Deines *et al.*, 1974).

The carbon isotopic composition of dissolved organic carbon (DOC) has also been successfully applied to studies of hydrologic and biogeochemical processes in watersheds. Humic and fulvic acids are the primary components of DOC in natural waters, and thus contribute to their overall acidity. The carbon isotopic composition of DOC reflects its source, and can provide insight into processes that control the loading of DOC to streams and groundwater. In addition, DOC is a primary nutrient and hydrologic processes that affect the fluxes of DOC are important for a better understanding of the carbon cycle.

5.11.3.1.1 Carbon isotope fundamentals

Carbon has two stable naturally occurring isotopes: ^{12}C and ^{13}C, and ratios of these isotopes are reported in ‰ relative to the standard Vienna Pee Dee Belemnite (V-PDB) scale. Carbonate rocks typically have $\delta^{13}C$ values of $0 \pm 5‰$. There is a bimodal distribution in the $\delta^{13}C$ values of terrestrial plant organic matter resulting from differences in the photosynthetic reaction utilized by the plant, with $\delta^{13}C$ values for C_3 and C_4 plants averaging $\sim -25‰$ and $-12‰$, respectively (Deines, 1980). Soil organic matter has a $\delta^{13}C$ comparable to that of the source plant material, and changes in C_3 and C_4 vegetation types will result in a corresponding change in the $\delta^{13}C$ value of the soil organic matter. Therefore, recent changes in the relative proportion of C_3 and C_4 plants can be detected by measuring the carbon isotopic composition of the current plant community and soil organic matter. The $\delta^{13}C$ values of DIC in catchment waters are generally in the range of $-5‰$ to $-25‰$. Values more negative than $-30‰$ usually indicate the presence of oxidized methane. The $\delta^{18}O$ of DIC is not a useful tracer of alkalinity sources or processes because it exchanges rapidly with oxygen in water.

The primary reactions that produce DIC are: (i) weathering of carbonate minerals by acidic rain or other strong acids; (ii) weathering of silicate minerals by carbonic acid produced by the dissolution of biogenic soil CO_2 by infiltrating rain water; and (iii) weathering of carbonate minerals by carbonic acid. The first and second reactions produce DIC identical in $\delta^{13}C$ to the composition of either the reacting carbonate or carbonic acid, respectively, and the third reaction produces DIC with a $\delta^{13}C$ value exactly intermediate between the compositions of the carbonate and the carbonic acid. Consequently, without further information, DIC produced solely by the third reaction is identical to DIC produced in equal amounts from the first and second reactions.

If the $\delta^{13}C$ values of the reacting carbon-bearing species are known and the $\delta^{13}C$ of the stream DIC determined, in theory we can calculate the relative contributions of these two sources of carbon to the production of stream DIC and carbonate alkalinity, assuming that: (i) there are no other sources or sinks for carbon, and (ii) calcite dissolution occurs under closed-system

conditions (Kendall, 1993). With additional chemical or isotopic information, the $\delta^{13}C$ values can be used to estimate proportions of DIC derived from the three reactions listed above. Other processes that may complicate the interpretation of stream water $\delta^{13}C$ values include CO_2 degassing, carbonate precipitation and dissolution, exchange with atmospheric CO_2, carbon uptake and release by aquatic organisms, oxidation of DOC, methanogenesis, and methane oxidation. Correlation of variations in $\delta^{13}C$ with major ion chemistry (especially HCO_3^-), P_{CO_2}, redox conditions, and with other isotope tracers such as $\delta^{34}S$, $^{87}Sr/^{86}Sr$, and ^{14}C may provide evidence for such processes (Clark and Fritz, 1997; Bullen and Kendall, 1998).

5.11.3.1.2 $\delta^{13}C$ of soil CO_2 and DIC

The biosphere, and particularly a number of soil processes, have a tremendous influence on the $\delta^{13}C$ of DIC in hydrologic systems. Soil CO_2 is comprised mainly of a mixture of atmospherically derived ($\delta^{13}C = -7‰$ to $-8‰$) and microbially respired CO_2. Respiration is a type of biologic oxidation of organic matter, and as such produces CO_2 with approximately the same $\delta^{13}C$ value as the organic matter (Park and Epstein, 1961); hence, areas dominated by C_3 plants should have soil CO_2 with $\delta^{13}C$ values around $-25‰$ to $-30‰$.

Several workers have demonstrated that diffusion of CO_2 gas through the soil causes a fractionation in the CO_2 that exits from the soil surface (Dörr and Münnich, 1980; Cerling et al., 1991; Davidson, 1995; Amundson et al., 1998; see Chapter 5.01). These workers draw an important distinction between soil CO_2 (present in the soil) and soil respired CO_2 (gas that has passed across the soil surface and into the atmosphere). Simultaneous measurements of shallow soil CO_2 (i.e., <0.5 m) and soil respired CO_2 show that the $\delta^{13}C$ of soil respired CO_2 is often $\sim 4‰$ lower than soil CO_2 (Cerling et al., 1991; Davidson, 1995). This magnitude of fractionation is similar to what would be expected by diffusion processes in air (4.4‰), given the theoretical diffusion coefficients of $^{12}CO_2$ and $^{13}CO_2$ (Cerling et al., 1991; Davidson, 1995).

While diffusion of CO_2 across the soil-atmosphere boundary may play a dominant role in influencing the $\delta^{13}C$ of shallow soil CO_2, it is generally recognized that the $\delta^{13}C$ values of soil CO_2 depend upon several additional factors, including P_{CO_2} of the soil zone and of the atmosphere, depth of soil CO_2 sampling, and differences in $\delta^{13}C$ of soil-respired CO_2 from the decomposition of organic matter, the latter of which is occasionally observed to have progressively higher $\delta^{13}C$ values with increasing depth (Davidson, 1995; Amoitte-Suchet *et al.*, 1999). Seasonal changes in the P_{CO_2} of soils are likely to be a controlling factor on the isotopic composition of soil CO_2, and therefore ought to be measured with the appropriate frequency (Rightmire, 1978). Given that diffusion processes and mixing with atmospheric CO_2 are most likely to influence the $\delta^{13}C$ of soil CO_2 in the upper 0.5 m of soils, sampling soil CO_2 at a depth of greater than 0.5 m is recommended for estimating the $\delta^{13}C$ of DIC in soil waters when direct soil water sampling is not feasible.

The $\delta^{13}C$ of soil CO_2 can also be affected by fermentation of organic matter, followed by methanogenesis, which produces methane with $\delta^{13}C$ values ranging from $-52‰$ to $-80‰$ (Stevens and Rust, 1982). As fermentation progresses, the $\delta^{13}C$ of the residual CO_2 or DIC by-products becomes progressively higher (Carothers and Kharaka, 1980); $\delta^{13}C$ values higher than $+10‰$ are not uncommon. Thus, very positive $\delta^{13}C$ of DIC can be a useful indicator of methanogenesis in natural waters. Oxidation of methane, alternatively, produces CO_2 with approximately the same composition as the original methane, with $\delta^{13}C$ values generally much lower than $-30‰$.

The evolution of the isotopic compositions of carbon-bearing substances in uncontaminated systems where carbon is derived from carbonate minerals and soil CO_2 is bounded between two limiting cases: (i) open systems, where carbonate reacts with water in contact with a gas phase having a constant P_{CO_2}, and (ii) closed systems, where the water is isolated from the CO_2 reservoir before carbonate dissolution (Deines *et al.*, 1974; Clark and Fritz, 1997). Both of the extremes assume water residence times long enough for significant isotope exchange between the gas and the aqueous phase to take place.

If DIC in soil water or groundwater exchanges isotopically with other carbon-bearing species, then the isotopic signatures may be blurred and conservative mixing of two distinctive end-member compositions cannot be assumed. The predominant inorganic carbon species at typical soil pH values of ~5 is carbonic acid. The equilibrium isotope fractionation between CO_2 and carbonic acid at 25 °C is 1‰ (Deines *et al.*, 1974). Assuming isotopic equilibrium among all DIC-bearing species (H_2CO_3, HCO_3^-, CO_3^{2-}), the total DIC resulting from the dissolution of calcite ($\delta^{13}C = 0‰$) by carbonic acid ($\delta^{13}C = -22‰$) has a $\delta^{13}C = -11‰$. If this dissolution occurs under open-system conditions in the vadose zone, the DIC would exchange with the soil CO_2 reservoir ($-21‰$) and reach a $\delta^{13}C$ value of ~$-22‰$, thus eliminating any carbon isotopic evidence that half the DIC was derived from dissolution of calcite. The addition of strontium isotopic measurements can sometimes shed light on the proportion of carbonate-derived DIC. If the $\delta^{87}Sr$ of calcite were distinctive relative to the $\delta^{87}Sr$ of other catchment minerals, dissolution of the calcite may have left its "signature" in the $\delta^{87}Sr$ of the water (Bullen and Kendall, 1998; see Chapter 5.12).

The carbon in subsurface waters that flow into streams is not in chemical and isotopic equilibrium with the atmosphere. For example, CO_2 concentrations in the soil zone are often as high as 5%. Therefore, because the atmospheric concentration is ~0.03%, CO_2 is rapidly lost as soil water and shallow groundwater seeps into a stream. Laboratory experiments performed by Mook (1968) indicate that the $\delta^{13}C$ of DIC rapidly increases by ~0.5‰ during degassing. Furthermore, isotopic exchange between DIC and atmospheric CO_2 is inevitable. In streams with pH values from 5 to 6 and temperatures of 20 °C, the equilibrium $\delta^{13}C$ of stream DIC should be around $-8‰$. Hence, if the residence time of water in the stream is long enough, the $\delta^{13}C$ of DIC will gradually approach $-8‰$. However, equilibrium isotopic exchange between stream DIC and atmospheric CO_2 does not appear to be achieved in the first- and second-order streams of forested catchments due to short stream-water residence times.

Additional in-stream processes can affect the $\delta^{13}C$ of DIC (Figure 10). Assimilation of DIC by aquatic organisms through photosynthesis produces organic material with a $\delta^{13}C$ ~ 30‰ lower than the carbon utilized (Rau, 1979; Mook and Tan, 1991), resulting in an increase in the $\delta^{13}C$ of the remaining DIC. Dissolution of carbonate minerals in-stream will also tend to increase the $\delta^{13}C$ of DIC. In contrast, precipitation of calcite will cause a decrease in the $\delta^{13}C$ of the remaining DIC, due to the equilibrium fractionation between calcite and DIC of ~2‰.

If any of these soil-zone or in-stream processes is a significant source or sink of carbon, it may complicate the interpretation of stream $\delta^{13}C$ values (Kendall, 1993). However, correlation of variations in $\delta^{13}C$ with changes in hydrology, chemistry, or other isotopes such as strontium may provide evidence as to whether such processes are significant. For example, lack of any systematic increase in $\delta^{13}C$ downstream, particularly in the summer when flow is slow, would argue against significant exchange of stream DIC with atmospheric CO_2. Similarly, low pH of stream water would rule out precipitation of calcite as a means to decrease $\delta^{13}C$ of stream DIC. Finally, a strong positive correlation between $\delta^{13}C$ and DIC concentration of stream water, and a typically negative correlation between these parameters and $\delta^{87}Sr$ together, strongly suggest calcite dissolution (Bullen and Kendall, 1998).

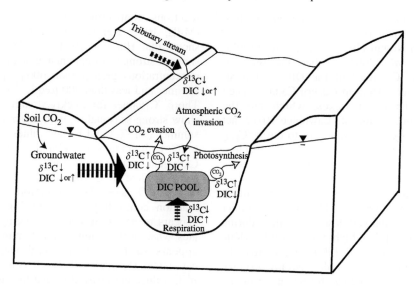

Figure 10 A schematic showing how various in-stream processes can affect the $\delta^{13}C$ of DIC (after Atekwana and Krishnamurthy, 1998).

5.11.3.1.3 Tracing sources of carbonate alkalinity in rivers

Several studies in the last decade have focused on the origin and cycling of carbon in large river systems using stable isotope techniques (e.g., Taylor and Fox, 1996; Yang *et al.*, 1996; Atekwana and Krishnamurthy, 1998; Aucour *et al.*, 1999; Karim and Veizer, 2000; Kendall *et al.*, 2001b; Hélie *et al.*, 2002). These studies have shown that carbon isotopes are useful indicators of biogeochemical reactions taking place within rivers along their courses of flow, especially concerning the relative amounts of aquatic photosynthesis and respiration.

The processes of primary concern that affect the $\delta^{13}C$ signatures of DIC in large rivers are: (i) CO_2 exchange with the atmosphere, (ii) dissolution/precipitation of carbonate minerals, and (iii) photosynthesis and respiration *in situ*. In particular, mixing between tributary waters of different DIC concentrations and $\delta^{13}C$ values has a great effect on the overall DIC composition of a large river. The studies to date generally suggest that upstream reaches and tributaries are responsible for the primary pool of DIC supplied to the main stem of a large river, with *in situ* recycling of that DIC pool controlling $\delta^{13}C$ compositions further downstream. Smaller tributaries tend to exhibit lower $\delta^{13}C$ of DIC, while further downstream in the main river stem positive shifts in the $\delta^{13}C$ of DIC occur due to photosynthetic uptake of CO_2 and equilibration with the atmosphere. The effects of these processes can vary seasonally. For example, riverine DIC can decrease while pool $\delta^{13}C_{DIC}$ increases during the summer because of photosynthesis, whereas DIC concentrations can increase and $\delta^{13}C_{DIC}$ decrease during the late

fall as photosynthesis declines and in-stream decay and respiration increases (Kendall, 1993; Atekwana and Krishnamurthy, 1998; Bullen and Kendall, 1991).

Several attempts have been made to model these processes. Taylor and Fox (1996) studied the Waimakariri River in New Zealand. They modeled equilibrium of DIC with atmospheric CO_2 using a chemical mass-balance model that accounts for the kinetics of CO_2 equilibration between the aqueous and gas phase. Their results show that the measured $\delta^{13}C$ values of the river water cannot be explained solely by equilibrium with atmospheric CO_2, and that an addition of biogenic CO_2 is necessary to account for the measurements. While their model represents a step forward in kinetic considerations of atmospheric CO_2 exchange in river systems, their model does not account for DIC uptake by phytoplankton via photosynthesis, which would cause biological recycling and increases in $\delta^{13}C$ of the residual riverine DIC pool.

Aucour *et al.* (1999) applied a diffusion-based model to account for their observations of the $\delta^{13}C$ value of DIC in the Rhône River, France. Their model predicts that 15–60 min are required to establish equilibrium between dissolved and atmospheric CO_2 in the Rhône. Using $\delta^{13}C = -12.5‰$ for the starting riverine DIC and $\delta^{13}C = -8‰$ for atmospheric CO_2, they predict that it would take between 0.2 d and 2.0 d to attain the observed average $\delta^{13}C$ value of $-5‰$ in the downstream river DIC, depending upon the mean depth of the river. In addition, Aucour *et al.* (1999) observed an inverse trend between $\delta^{13}C$ of DIC and the concentration of DIC in the Rhône River. They attribute this to conservative mixing between the main stem Rhône and its tributaries.

Weiler and Nriagu (1973), Yang *et al.* (1996), and Hélie *et al.* (2002) studied the seasonal changes in $\delta^{13}C$ values of the DIC in the St. Lawrence River system in eastern Canada. They found that the Great Lakes, which are the upstream source of the St. Lawrence River, act as a strong biogeochemical buffer to the $\delta^{13}C_{DIC}$ value of the water in the river's upstream reaches. However, the main stem river downstream shows pronounced seasonal variability in the $\delta^{13}C$ of DIC, with high $\delta^{13}C$ values near isotopic equilibrium with atmospheric CO_2 in the summer during low flow, and very low $\delta^{13}C$ values during the winter and spring high flow periods (Hélie *et al.*, 2002). Hélie *et al.* (2002) report that the $\delta^{13}C$ of DIC in the Great Lakes outflow average $-0.5‰$ to $-1.0‰$, whereas further downstream two large tributaries have a strong influence on the $\delta^{13}C$ of DIC. These rivers—the Ottawa River and the Mascouche River—exhibit much lower $\delta^{13}C$ values than the main stem St. Lawrence, on average $-8.8‰$ and $-12.1‰$, respectively. Hélie *et al.* (2002) conclude that during low water levels in the summer, the Great Lakes provide nearly 80% of the river flow, and the $\delta^{13}C$ of DIC reflects that of atmospheric equilibrium with water in the Great Lakes. During higher flows in the other seasons, the $\delta^{13}C$ of the downstream St. Lawrence River is derived from the influx of the tributaries, which provide a pool of DIC with low $\delta^{13}C$ to the river. In addition, a striking increase in the P_{CO_2} (in excess of atmospheric equilibrium) of the river water was observed during the winter months, indicating a greater proportion of groundwater discharge contribution to the river. In contrast, very low P_{CO_2} values (near atmospheric equilibrium) were observed during the summer, when photosynthetic uptake of CO_2 is at its peak.

5.11.3.1.4 *Tracing sources of carbonate alkalinity in catchments*

As discussed previously, the headwater tributaries of large river systems in temperate climates have been shown to contain carbonate alkalinity that has a lower $\delta^{13}C$ than in the main stem portion of the river downstream. The most likely explanation for this observation is that soil water and/or groundwater with DIC with a low $\delta^{13}C$ derived from terrestrial biogenic CO_2 is being supplied to the low-order streams. Mixing between tributaries and the main stem of large rivers plays a dominant role in controlling the $\delta^{13}C$ value of DIC; biological recycling of carbon in-stream and, to a lesser extent, equilibration with atmospheric CO_2 are also important.

Under favorable conditions, carbon isotopes can be used to understand the biogeochemical

reactions controlling alkalinity in watersheds (Mills, 1988; Kendall *et al.*, 1992). Figure 11 shows that Hunting Creek (in Maryland, USA) shows a strong seasonal change in alkalinity, with concentrations greater than $600~\mu eq~L^{-1}$ in the summer and less than $400~\mu eq~L^{-1}$ in the winter. The $\delta^{13}C$ values for weekly stream samples also show strong seasonality, with lower values in the summer $(-13‰)$ and higher values in the winter $(-5‰)$, inversely correlated with alkalinity. Carbonates found in fractures in the altered basalt bedrock have $\delta^{13}C$ values of $\sim -5‰$ and soil-derived CO_2 is $\sim -21‰$. Therefore, the dominant reaction controlling stream DIC in the winter appears to be strong acid weathering of carbonates, whereas the dominant reaction in the summer appears to be carbonic acid weathering of carbonates (Kendall, 1993). Of course, with just $\delta^{13}C$ values the possibility that the summer $\delta^{13}C$ values instead reflect approximately equal contributions of DIC produced by strong acid weathering of calcite $(-5‰)$ and carbonic acid weathering of silicates $(-21‰)$ cannot be ruled out. However, the presence of carbonate in the bedrock strongly influences weathering reactions, and the stream chemistry during the summer suggests that silicate weathering has only a minor effect.

Although variations in stream $\delta^{18}O$ during storm events can easily be interpreted in terms of relative contributions of "new" and "old" waters in the two catchments, the $\delta^{18}O$ data do not reveal how the water is delivered to the stream or how these flow paths change during the event. Because $\delta^{13}C$ of DIC has been shown to be sensitive to differences in reactions occurring along shallow versus deep flow paths (Kendall *et al.*, 1992), samples were analyzed for both $\delta^{13}C$ and $\delta^{18}O$. During storm events, the $\delta^{18}O$ shows small gradual changes in water sources in response to changes in rain intensity, in contrast to rapid oscillations in $\delta^{13}C$. For example, during a single three-day storm event in March 1987, $\delta^{13}C$ in both streams oscillated rapidly over a 3‰ range, whereas $\delta^{18}O$ varied only gradually and less than 1‰ (Figure 12). The oscillations in $\delta^{13}C$ apparently reflect more complex changes in flow paths, as well as mixing among waters derived from various flow paths, than is apparent based on $\delta^{18}O$ alone (Kendall *et al.*, 1992).

5.11.3.1.5 *Carbon-14*

The radioactive isotope of carbon, ^{14}C, is continuously being produced by reaction of cosmic ray neutrons with ^{14}N in the atmosphere and decays with a half-life of 5,730 yr. ^{14}C concentrations are usually reported as specific activities (disintegrations per minute per gram of carbon relative to a standard) or as percentages of

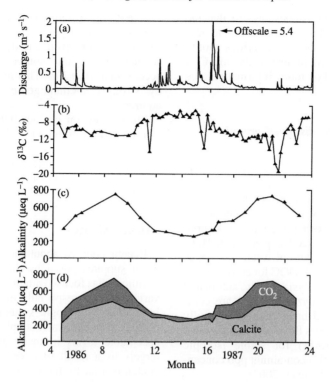

Figure 11 Determining seasonal changes in the sources of alkalinity by using the $\delta^{13}C$ of stream DIC: (a) discharge at Hunting Creek in the Catoctin Mountains, Maryland (USA), 1986–1987; (b) $\delta^{13}C$ of stream DIC collected weekly; (c) alkalinity; and (d) estimation of the relative contributions of carbon from CO_2 and calcite to stream alkalinity using $\delta^{13}C$ of calcite = $-5\permil$ and $\delta^{13}C$ of CO_2 = $-21\permil$; shaded areas show the relative proportions of carbon sources (after Kendall, 1993).

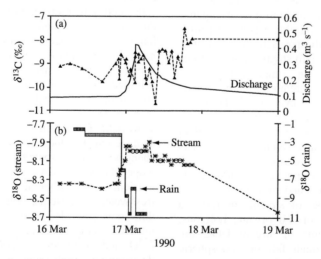

Figure 12 Rain and stream $\delta^{18}O$ and stream $\delta^{13}C$ for a storm March 16–20, 1990 at Hunting Creek, Maryland, USA: (a) $\delta^{13}C$ values and discharge and (b) $\delta^{18}O$ values of rain and stream water (note difference in the scales). Although the gradual shifts in stream $\delta^{18}O$ suggest equally gradual changes in water sources during the storm, the $\delta^{13}C$ values suggest rapidly fluctuating contributions of old waters of different $\delta^{13}C$ values in response to small changes in rainfall intensity.

modern ^{14}C concentrations. Under favorable conditions, ^{14}C can be used to date carbon-bearing materials. However, because the "age" of a mixture of waters of different residence times is not very meaningful, the least ambiguous hydrologic use of ^{14}C is as a tracer of carbon sources. The ^{14}C of DIC can be a valuable check on conclusions derived using $\delta^{13}C$ values.

For example, if the $\delta^{13}C$ values suggest that the dominant source of carbon is from carbonic acid weathering of silicates, the ^{14}C activities should be high reflecting the young age of the soil CO_2 (Schiff *et al.*, 1990). For more information, see Chapter 5.15.

5.11.3.1.6 Sources of dissolved organic carbon

The use of carbon isotopes to study DOC is becoming more prevalent due to technological advances in mass spectrometry. DOC generally occurs in natural waters in low concentrations, typically ranging between 0.5 ppm and 10 ppm carbon (Thurman, 1985; see Chapter 5.10). Thus, several liters to tens of liters of water were once necessary to extract enough DOC for conventional dual gas-inlet isotopic analysis. Today, automated total organic carbon analyzers (TOCs) are commercially available, and have been successfully interfaced with continuous flow isotope ratio mass spectrometers (CF-IRMS) for stable isotopic measurements of samples containing ppb concentrations of DOC (e.g., St-Jean, 2003).

^{14}C combined with $\delta^{13}C$ has been used to study the origin, transport, and fate of DOC in streams and shallow groundwater in forested catchments (Schiff *et al.*, 1990; Wassenaar *et al.*, 1991; Aravena *et al.*, 1992; Schiff *et al.*, 1997; Palmer *et al.*, 2001). Two dominant pools tend to contribute to DOC in surface waters: (i) a younger, more labile pool derived from recent organic matter in soils; and (ii) an older, more refractory pool held within groundwater (Schiff *et al.*, 1997). DOC that is mobilized into streams during storm events is generally young compared to that of groundwater feeding the stream during base flow, indicating that extensive cycling of the labile DOC takes place in catchment soils. Schiff *et al.* (1990) report DOC turnover times of less than 40 yr in streams, lakes, and wetlands of Canada as determined from ^{14}C measurements. Palmer *et al.* (2001) report similarly young ages of DOC (post-1955) in soil pore waters and stream waters.

In general, the DOC of streams in forested catchments is formed in soil organic horizons, while minimal amounts of stream DOC are accounted for by through fall or atmospheric deposition (Schiff *et al.*, 1990). Riparian flow paths can account for the greatest proportion of DOC exported to headwater streams (Hinton *et al.*, 1998). The $\delta^{13}C$ of stream DOC can also serve to distinguish between terrigenous-derived DOC and that derived *in situ* in streams in cases where decomposition of phytoplankton is a major DOC source (Wang *et al.*, 1998).

Aravena and Wassenaar (1993) found that while most high molecular weight DOC in shallow unconfined groundwaters has its source in recent organic carbon present in the upper soil, deeper confined aquifers may obtain a significant amount of DOC from the degradation of buried sedimentary organic sources (peats or carbonaceous shales) *in situ*. This may be accompanied by strongly methanogenic conditions in the aquifer.

Much work on characterizing changes in the isotopic composition of DOC and DIC within aquatic systems lies ahead. Carbon isotopes promise to shed much light on the primary biogeochemical processes involved in the carbon cycle.

5.11.3.2 Nitrogen

Recent concern about the potential danger to water supplies posed by the use of agricultural chemicals has focused attention on the mobility of various solutes, especially nitrate and pesticides, in shallow hydrologic systems. Nitrate concentrations in public water supplies have risen above acceptable levels in many areas of the world, largely as a result of overuse of fertilizers and contamination by human and animal waste. The World Health Organization and the United States Environmental Protection Agency have set a limit of $10 \, \text{mg L}^{-1}$ nitrate (as N) for drinking water because high-nitrate water poses a health risk, especially for children, who can contract methemoglobinemia (blue-baby disease). High concentrations of nitrate in rivers and lakes can cause eutrophication, often followed by fish-kills due to oxygen depletion. Increased atmospheric loads of anthropogenic nitric and sulfuric acids have caused many sensitive, low-alkalinity streams in North America and Europe to become acidified. Still more streams that are not yet chronically acidic could undergo acidic episodes in response to large rain storms and/or spring snowmelt. These acidic "events" can seriously damage sensitive local ecosystems. Future climate changes may exacerbate the situation by affecting biogeochemical controls on the transport of water, nutrients, and other materials from land to freshwater ecosystems. Nitrogen isotope ($\delta^{15}N$) data can provide needed information about nitrogen sources and sinks. Furthermore, nitrate $\delta^{18}O$ and $\delta^{17}O$ are promising new tools for determining nitrate sources and reactions, and complement conventional uses of $\delta^{15}N$.

5.11.3.2.1 Nitrogen isotope fundamentals

There are two stable isotopes of N: ^{14}N and ^{15}N. Since the average abundance of ^{15}N in air is a very constant 0.366% (Junk and Svec, 1958), air (AIR) is used as the standard for reporting $\delta^{15}N$ values. Most terrestrial materials have $\delta^{15}N$ values between $-20‰$ and $+30‰$. The dominant source of nitrogen in most natural ecosystems is the atmosphere ($\delta^{15}N = 0‰$). Plants fixing N_2 from the atmosphere have $\delta^{15}N$ values of $\sim 0‰$ to $+2‰$, close to the $\delta^{15}N$ value of atmospheric N_2. Most plants have $\delta^{15}N$ values in the range of $-5‰$ to $+2‰$ (Fry, 1991). Other sources of nitrogen include fertilizers produced from atmospheric nitrogen with $\delta^{15}N$ values of $0 \pm 3‰$, and animal manure with nitrate $\delta^{15}N$ values generally in the range of $+10‰$ to $+25‰$; rock sources of nitrogen are generally negligible.

Biologically mediated reactions (e.g., assimilation, nitrification, and denitrification) strongly control nitrogen dynamics in the soil, as briefly described below. These reactions almost always result in ^{15}N enrichment of the substrate and ^{15}N depletion of the product.

5.11.3.2.2   Oxygen isotopes of nitrate

The evaluation of sources and cycling of nitrate in the environment can be aided by analysis of the oxygen isotopic composition of nitrate. Oxygen has three stable isotopes: ^{16}O, ^{17}O, and ^{18}O. The $\delta^{18}O$ and $\delta^{17}O$ values of nitrate are reported in per mil (‰) relative to the standard SMOW or VSMOW. There is almost an 80‰ range in $\delta^{18}O$ values, corresponding to a 30‰ "normal" range in $\delta^{15}N$ values (Figure 13). Most of the spread in $\delta^{18}O$ values is caused by precipitation samples. However, there is also considerable variability in $\delta^{18}O$ values produced by nitrification of ammonium and organic matter, and in the $\delta^{18}O$ values of nitrate present in soils, streams, and groundwaters that contain mixtures of atmospheric and microbial nitrate. Quantifying mixtures of sources of nitrate and identifying transformations of inorganic nitrogen in the subsurface can

be difficult with only one isotope tracer ($\delta^{15}N$). $\delta^{18}O$ and $\delta^{17}O$ of NO_3 provide two additional tracers for deciphering the nitrogen cycle.

5.11.3.2.3   Tracing atmospheric sources of nitrate

Using $\delta^{15}N$. Complex chemical reactions in the atmosphere result in a large range of $\delta^{15}N$ values of nitrogen-bearing gases and solutes depending on the compound involved, the season, meteorological conditions, ratio of NH_4^+ to NO_3^- in the precipitation, types of anthropogenic inputs, proximity to pollution sources, distance from ocean, etc. (Hübner, 1986). Natural atmospheric sources of nitrogen-bearing gases (e.g., N_2O, HNO_3, NH_3, etc.) include volatilization of ammonia from soils and animal waste (with fractionations as large as $-40‰$), nitrification and denitrification in soils and surface waters, and production in thunderstorms from atmospheric N_2. Anthropogenic sources include chemical processing and combustion of fossil fuels in automobiles and power plants. The $\delta^{15}N$ values of atmospheric NO_3^- and NH_4^+ are usually in the range of $-15‰$ to $+15‰$ (Figure 13). Extremely low $\delta^{15}N$ values for NO_3^- can be expected near chemical plants because of sorption of NO_x gases (with high $\delta^{15}N$ values) in exhaust scrubbers (Hübner, 1986). In general, the NO_3^- in rain appears to have a higher $\delta^{15}N$ value than the coexisting NH_4^+. Although precipitation often contains subequal quantities of ammonium and nitrate, most of the atmospheric nitrogen that reaches the soil surface is in the form of nitrate, because ammonium is preferentially retained by the canopy relative to atmospheric nitrate (Garten and Hanson, 1990). Soil nitrate is preferentially assimilated by tree roots relative to soil ammonium (Nadelhoffer *et al.*, 1988).

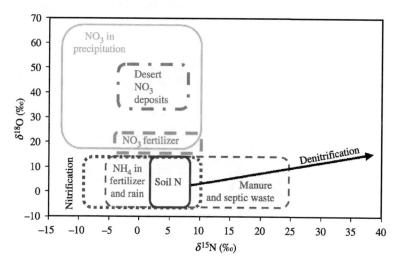

Figure 13 Typical ranges of nitrate $\delta^{18}O$ and $\delta^{15}N$ values for major nitrate reservoirs (after Kendall, 1998).

Considerable attention has been given to nitrogen oxides (and sulfur oxides) in the atmosphere because of their contributions to acid rain. Several studies suggest that different isotopic techniques can be used to differentiate among different types of atmospherically derived nitrate and ammonium. For example, Heaton (1990) found that NO_x derived from vehicle exhaust had $\delta^{15}N$ values of $-2‰$ to $-11‰$, whereas power plant exhaust was $+6‰$ to $+13‰$. The differences in $\delta^{15}N$ values were not caused by differences in the $\delta^{15}N$ of the fuels combusted; instead, they reflect differences in the nature of the combustion and NO_x reaction processes.

Using $\delta^{18}O$. There are limited data on the $\delta^{18}O$ of nitrate in atmospheric deposition, with little known about possible spatial or temporal variability, or their causes. The bimodal distribution of nitrate $\delta^{18}O$ values in precipitation in North America presents moderate evidence of at least two sources and/or processes affecting the compositions. The lower mode is centered around values of $+22‰$ to $+28‰$, and the higher mode has values centering around $+56‰$ to $+64‰$ (Kendall, 1998). Several studies in the USA (e.g., Kendall, 1998; Williard, 1999; Pardo *et al.*, in press) have observed seasonal and spatial differences in the $\delta^{18}O$ of nitrate in precipitation (much higher in winter than summer). These differences may be partially related to variations in the contributions of nitrate from different sources (low δ values partially from vehicle exhaust, and high δ values partially from power plant exhaust). Or the dominant cause may be seasonal changes in atmospheric oxidative processes (Michalski *et al.*, in review). If atmospheric nitrate from power plant plumes and dispersed vehicle sources generally forms via different oxidative pathways or different extents of penetration into the atmosphere, it could account for the apparent correlation of $\delta^{18}O$ values and types of anthropogenic sources.

Further evidence of the correlation of power plant emissions and high nitrate $\delta^{18}O$ values is the observation that all reported nitrate $\delta^{18}O$ values of precipitation in Bavaria, which has high concentrations of nitrate in precipitation, many acid-rain-damaged forests, and is downwind of the highly industrialized parts of Central Europe, are $>50‰$; whereas samples from Muensterland, farther from the pollution sources in Central Europe, have considerably lower $\delta^{18}O$ values (Kendall, 1998). A recent study of nitrate isotopes in Greenland glaciers has verified that pre-industrial $\delta^{18}O$ values were in the range of $+20‰$ to $+30‰$ (Savarino *et al.*, 2002), suggesting that relatively recent changes in atmospheric chemistry related to anthropogenic emissions of reactive NO_x are responsible for the high ozone $\delta^{18}O$ values that apparently cause the high $\delta^{18}O$ of nitrate.

Using $\delta^{17}O$. In all terrestrial materials, there is a constant relation between the $\delta^{18}O$ and $\delta^{17}O$ values of any given substance, because isotope fractionations are mass dependent. On a three-isotope plot with $\delta^{17}O$ on the x-axis and $\delta^{18}O$ on the y-axis (Figure 14), mass *dependent* fractionations (MDF) for nitrate result in values defined by the relation: $\delta^{17}O = 0.52\delta^{18}O$, whereas mass *independent* fractionations (MIF) cause values that deviate from this relation. For nitrate, the mass-independent relation is described by $\Delta^{17}O = \delta^{17}O - 0.52\delta^{18}O$ (Michalski *et al.*, 2002). The term "Δ" is pronounced "cap delta," which is short for "capital delta." Hence, MDF results in $\Delta^{17}O = 0$, whereas an MIF results in $\Delta^{17}O \neq 0$. $\delta^{17}O$ and $\Delta^{17}O$ values reflect $^{17}O/^{16}O$ ratios, and are reported in per mil relative to SMOW.

Recent investigations show that ozone and many oxides derived from high atmospheric processes have "excess ^{17}O" (beyond the $\delta^{17}O$ expected from the $\delta^{18}O$ value) derived from a mass-independent fractionation. Further work shows that all atmospherically derived nitrate is

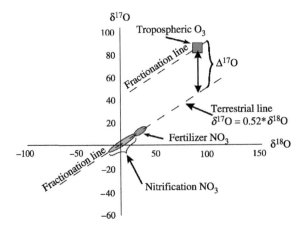

Figure 14 Schematic of relationship between $\delta^{18}O$ and $\delta^{17}O$ values (after Michalski *et al.*, 2002).

labeled by characteristic nonterrestrial $\delta^{17}O$ signatures (Michalski *et al.*, 2002; Michalski *et al.*, in review). One especially noteworthy feature of $\Delta^{17}O$ values is that they are not affected by any terrestrial fractionating processes such as denitrification and assimilation. Seasonal variations in atmospheric $\Delta^{17}O$ (from $+20‰$ to $+30‰$) observed in southern California were explained by a shift from nitric acid production by the $OH + NO_2$ reaction, which is predominant in the spring and summer, to N_2O_5 hydrolysis reactions that dominate in the winter (Michalski *et al.*, in review). This shift is driven largely by temperature variability and NO_x concentrations (Michalski *et al.*, in review), both of which should vary in plume (power plant) and dispersed (automobile) conditions. Ice-core samples from a glacier in Greenland showed higher $\Delta^{17}O$ values during the 1880s due to large biomass burning events in North America (Savarino *et al.*, 2002). The collection of preliminary findings suggests that atmospheric source characterization may be possible when $\delta^{17}O$ is considered in combination with $\delta^{18}O$ and $\delta^{15}N$ measurements, if nitrogen sources from different combustion processes (i.e., with different oxidation chemistries and perhaps degrees of NO_x penetration and recycling in the stratosphere) have different $\delta^{17}O$ values.

5.11.3.2.4 Fertilizer and animal waste sources of nitrogen

Synthetic fertilizers have $\delta^{15}N$ values that are uniformly low reflecting their atmospheric source (Figure 13), generally in the range of $-4‰$ to $+4‰$; however, some fertilizer samples have shown a total range of $-8‰$ to $+7‰$ (see compilations by Hübner, 1986; Macko and Ostrom, 1994). Organic fertilizers (which include the so-called "green" fertilizers such as cover crops and plant composts, and liquid and solid animal waste) generally have higher $\delta^{15}N$ values and a much wider range of compositions (generally $+2$ to $+30‰$) than inorganic fertilizers because of their more diverse origins. Note that the $\delta^{15}N$ of nitrate in fertilized soils may not be the same as the fertilizer. Amberger and Schmidt (1987) determined that nitrate fertilizers have distinctive $\delta^{18}O$ and $\delta^{15}N$ values. All three oxygens in fertilizer nitrate are derived from atmospheric O_2 ($+22$ to $+24‰$), and hence the $\delta^{18}O$ of the nitrate is in this range.

It has often been observed that consumers (microbes to invertebrates) are $2-3‰$ enriched in ^{15}N relative to their diet. The increases in $\delta^{15}N$ in animal tissue and solid waste relative to diet are due mainly to the excretion of low $\delta^{15}N$ organics in urine or its equivalent (Wolterink *et al.*, 1979). Animal waste products may be further enriched in ^{15}N because of volatilization of ^{15}N-depleted ammonia, and subsequent oxidation of much of the residual waste material may result in nitrate with a high $\delta^{15}N$ (Figure 13). By this process, animal waste with a typical $\delta^{15}N$ value of about $+5‰$ is converted to nitrate with $\delta^{15}N$ values generally in the range of $+10‰$ to $+20‰$ (Kreitler, 1975, 1979); human and other animal waste become isotopically indistinguishable under most circumstances (an exception is Fogg *et al.*, 1998).

5.11.3.2.5 Soil sources of nitrogen

The $\delta^{15}N$ of total soil nitrogen is affected by many factors including soil depth, vegetation, climate, particle size, cultural history, etc.; however, two factors, drainage and influence of litter, have a consistent and major influence on the $\delta^{15}N$ values (Shearer and Kohl, 1988). Soils on lower slopes and near saline seeps have higher $\delta^{15}N$ values than well-drained soils (Karamanos *et al.*, 1981), perhaps because the greater denitrification in more boggy areas results in high-$\delta^{15}N$ residual nitrate. Surface soils beneath bushes and trees often have lower $\delta^{15}N$ values than those in open areas, presumably as the result of litter deposition (Shearer and Kohl, 1988). Fractionations during litter decomposition in forests result in surface soils with lower $\delta^{15}N$ values than deeper soils (Nadelhoffer and Fry, 1988). Gormly and Spalding (1979) attributed the inverse correlation of nitrate-$\delta^{15}N$ and nitrate concentration beneath agricultural fields to increasing denitrification with depth.

5.11.3.2.6 Processes affecting nitrogen isotopic compositions

Irreversible (unidirectional) kinetic fractionation effects involving metabolic nitrogen transformations are generally more important than equilibrium fractionation effects in low-temperature environments. Many biological processes consist of a number of steps (e.g., nitrification: organic-N $\rightarrow NH_4^+ \rightarrow NO_2^- \rightarrow NO_3^-$). Each step has the potential for fractionation, and the overall fractionation for the reaction is highly dependent on environmental conditions including the number and type of intermediate steps, sizes of reservoirs (pools) of various compounds involved in the reactions (e.g., O_2, NH_4^+), soil pH, species of the organism, etc. Hence, estimation of fractionations in natural systems is very complex.

A useful "rule of thumb" is that most of the fractionation is caused by the so-called "rate-determining step"—which is the slowest step. This step is commonly one involving a large pool of substrate where the amount of material actually used is small compared to the size of the reservoir. In contrast, a step that is not rate determining generally involves a small pool of some compound that is rapidly converted from reactant to product; because the compound is converted to product as soon as it appears, there is no fractionation at this step. The isotopic compositions of reactant and product during a multistep reaction where the net fractionation is controlled by a single rate-determining step can be successfully modeled with Rayleigh equations. For more details, see recent reviews by Kendall (1998), Kendall and Aravena (2000), and Böhlke (2002).

N-fixation refers to processes that convert unreactive atmospheric N_2 into other forms of nitrogen. Although the term is usually used to mean fixation by bacteria, it has also been used to include fixation by lightning and, more importantly, by human activities (energy production, fertilizer production, and crop cultivation) that produce reactive nitrogen (NO_x, NH_y, and organic nitrogen). Fixation of atmospheric N_2 commonly produces organic materials with $\delta^{15}N$ values slightly less than 0‰. A compilation by Fogel and Cifuentes (1993) indicates measured fractionations ranging from $-3‰$ to $+1‰$. Because these values are generally lower than the values for organic materials produced by other mechanisms, low $\delta^{15}N$ values in organic matter are often cited as evidence for N_2 fixation.

Assimilation generally refers to the incorporation of nitrogen-bearing compounds into organisms. Assimilation, like other biological reactions, discriminates between isotopes and generally favors the incorporation of ^{14}N over ^{15}N. Measured values for apparent fractionations caused by assimilation by microorganisms in soils show a range of $-1.6‰$ to $+1‰$, with an average of $-0.52‰$ (compilation by Hübner, 1986). Fractionations by vascular plants show a range of $-2.2‰$ to $+0.5‰$ and an average of $-0.25‰$, relative to soil organic matter (Mariotti et al., 1980). Nitrogen uptake by plants in soils causes only a small fractionation and, hence, only slightly alters the isotopic composition of the residual fertilizer or soil organic matter. The $\delta^{15}N$ of algae in rivers in the USA is generally $\sim 4‰$ lower than that of the associated nitrate (Kendall, unpublished data).

Mineralization is usually defined as the production of ammonium from soil organic matter. This is sometimes called *ammonification*, which is a less confusing term. Mineralization usually causes only a small fractionation ($\sim 1‰$) between soil organic matter and soil ammonium. In general, the $\delta^{15}N$ of soil ammonium is usually within a few per mil of the composition of total organic nitrogen in the soil.

Nitrification is a multistep oxidation process mediated by several different autotrophic organisms for the purpose of deriving metabolic energy; the reactions produce acidity. In general, the extent of fractionation is dependent on the size of the substrate pool (reservoir). In nitrogen-limited systems, the fractionations are minimal. The total fractionation associated with nitrification depends on which step is rate determining. Because the oxidation of nitrite to nitrate is generally quantitative (rapid) in natural systems, this is generally not the rate-determining step, and most of the nitrogen fractionation is probably caused by the slow oxidation of ammonium by *Nitrosomonas*. In soils, overall N-nitrification fractionations have been estimated to range between 12‰ and 29‰ (Shearer and Kohl, 1986).

Volatilization is the term commonly used for the loss of ammonia gas from surficial soils to the atmosphere; the ammonia gas produced has a lower $\delta^{15}N$ value than the residual ammonium in the soil. Volatilization in farmlands results from applications of urea and manure to fields, and occurs within piles of manure; the resulting organic matter may have $\delta^{15}N$ values $>20‰$ because of ammonia losses. Ammonium in precipitation derived from volatilazation from waste lagoons can have a distinctive low $\delta^{15}N$ value.

Denitrification is a multistep process with various nitrogen oxides (e.g., N_2O, NO) as intermediate compounds resulting from the chemical or biologically mediated reduction of nitrate to N_2. Depending on the redox conditions, organisms will utilize different oxidized materials as electron acceptors in the general order: O_2, NO_3^-, SO_4^{2-}. Although microbial denitrification does not occur in the presence of significant amounts of oxygen, it can occur in anaerobic pockets within an otherwise oxygenated sediment or water body (Koba et al., 1997).

Denitrification causes the $\delta^{15}N$ of the residual nitrate to increase exponentially as nitrate concentrations decrease, and causes the acidity of the system to decrease. For example, denitrification of fertilizer nitrate that originally had a distinctive $\delta^{15}N$ value of $+0‰$ can yield residual nitrate with much higher $\delta^{15}N$ values (e.g., $+15‰$ to $+30‰$) that are within the range of compositions expected for nitrate from a manure or septic-tank source. Measured enrichment factors (apparent fractionations) associated with denitrification range from $-40‰$ to $-5‰$; hence, the $\delta^{15}N$ of the N_2 is lower than that of the nitrate by about these values. The N_2 produced by denitrification results in *excess N_2 contents* in groundwater; the $\delta^{15}N$ of this N_2 can provide useful information about sources and processes (Vogel et al., 1981; Böhlke and Denver, 1995).

5.11.3.2.7   Processes affecting the $\delta^{18}O$ of nitrate

Nitrification. When organic nitrogen or ammonium is microbially oxidized to nitrate (i.e., nitrified), the three oxygens are derived from two sources: H_2O and dissolved O_2. If the oxygen is incorporated with no fractionation, the $\delta^{18}O$ of the resulting nitrate is intermediate between the compositions of the two oxygen sources. Andersson and Hooper (1983) showed that oxidation of NH_4^+ to NO_2^- incorporated one atom from water and one atom from dissolved oxygen. Work by Aleem *et al.* (1965) indicates that the additional oxygen atom incorporated during the oxidation of NO_2^- to NO_3^- originates entirely from H_2O. This combined work suggests that the oxygen isotopic composition of NO_3^- formed during autotrophic nitrification ($NH_4^+ \rightarrow NO_3^-$) would have the composition of

$$\delta^{18}O_{NO_3} = \tfrac{2}{3}\delta^{18}O_{H_2O} + \tfrac{1}{3}\delta^{18}O_{O_2} \quad (13)$$

For waters with $\delta^{18}O$ values in the normal range of $-25‰$ to $+4‰$, the $\delta^{18}O$ of soil nitrate formed from *in situ* nitrification of ammonium should be in the range of $-10‰$ to $+10‰$, respectively (Figure 13). For highly evaporated water ($+20‰$), the $\delta^{18}O$ of nitrate could be as high as about $+21‰$ (Böhlke *et al.*, 1997). The oxygen isotopic composition of dissolved O_2 reflects the effects of three primary processes: (i) diffusion of atmospheric oxygen ($\sim+23.5‰$) in the subsurface, (ii) photosynthesis—resulting in the addition of O_2 with a low $\delta^{18}O$ similar to that of water, and (iii) respiration by microbes—resulting in isotopic fractionation and higher $\delta^{18}O$ values for the residual O_2.

The above model makes four critical assumptions: (i) the proportions of oxygen from water and O_2 are the same in soils as observed in laboratory cultures; (ii) there are no fractionations resulting from the incorporation of oxygen from water or O_2 during nitrification; (iii) the $\delta^{18}O$ of water used by the microbes is identical to that of the bulk soil water; and (iv) the $\delta^{18}O$ of the O_2 used by the microbes is identical to that of atmospheric O_2.

Many studies have used measurement of $\delta^{18}O$ of NO_3^- in freshwater systems for assessing sources and cycling (see below). Many of them find that the $\delta^{18}O$ of microbial NO_3^- appears to be a few per mil higher than expected for the equation and assumptions above (Kendall, 1998). A variety of explanations have been offered for these high nitrate-$\delta^{18}O$ values: (i) nitrification in contact with soil waters with higher-than-expected $\delta^{18}O$ values because of evaporation (Böhlke *et al.*, 1997) or seasonal changes in rain $\delta^{18}O$ (Wassenaar, 1995); (ii) changes in the proportion of oxygen from H_2O and O_2 sources (Aravena *et al.*, 1993); (iii) nitrification using O_2 that has a high $\delta^{18}O$ due to heterotrophic respiration (Kendall, 1998); and (iv) nitrification that occurs simultaneously via both heterotrophic and autotrophic pathways (Mayer *et al.*, 2001). At this time, the mechanisms responsible for

generating the isotopic composition of NO_3^- in nature are poorly understood.

Denitrification. Denitrification is the process that poses most difficulties for simple applications of nitrate isotopes. Hence, for successful applications of nitrate isotopes for tracing sources, it is critical to (i) determine if denitrification has occurred, and, if so (ii) determine what was the initial isotopic composition of the nitrate (which is a necessary prerequisite for later attempts to define sources). There are several geochemical methods for identifying denitrification in groundwater, and distinguishing it from mixing:

(i) Geochemical evidence of a reducing environment (e.g., low dissolved O_2, high H_2S, etc.).

(ii) Hyperbolic versus exponential relationships between $\delta^{15}N$ and NO_3^- (i.e., mixing is a hyperbolic function, whereas denitrification is exponential). Hence, if mixing of two sources is responsible for the curvilinear relationship between $\delta^{15}N$ and NO_3^-, plotting $\delta^{15}N$ versus $1/NO_3^-$ will result in a straight line (Figure 2). In contrast, if denitrification (or assimilation) is responsible for the relationship, plotting $\delta^{15}N$ versus ln NO_3^- will produce a straight line (Figure 3). Analysis of dissolved N_2 (produced by the denitrification of NO_3^-) for $\delta^{15}N$ to show that there are systematic increases in the $\delta^{15}N$ of N_2 with decreases in NO_3^- and increases in the $\delta^{15}N$ of NO_3^- (e.g., a "Rayleigh equation" relationship like Figure 1).

(iii) Analysis of the NO_3^- for $\delta^{18}O$ as well as $\delta^{15}N$ to see if there is a systematic increase in $\delta^{18}O$ with increase in $\delta^{15}N$ and decrease in NO_3^-, consistent with denitrification. On $\delta^{18}O$ versus $\delta^{15}N$ plots, denitrification produces lines with slopes of ~0.5 (Figure 13).

(iv) No change in $\Delta^{17}O$ during a process that has $\delta^{18}O$ and/or $\delta^{15}N$ trends characteristic of denitrification (if material has $\Delta^{17}O > 0$).

5.11.3.2.8 Small catchment studies

The main application of nitrate $\delta^{18}O$ has been for the determination of the relative contributions of atmospheric and soil-derived sources of nitrate to shallow groundwater and small streams. This problem is intractable using $\delta^{15}N$ alone because of overlapping compositions of soil and atmospherically derived nitrate, whereas these nitrate sources have very distinctive nitrate $\delta^{18}O$ values (Figure 13). Durka *et al.* (1994) analyzed nitrate in precipitation and spring samples from several German forests for both isotopes, and found that the $\delta^{18}O$ of nitrate in springs was correlated with the general health of the forest, with more healthy or limed forests showing lower $\delta^{18}O$ values closer to the composition of microbially produced nitrate, and severely damaged forests showing higher $\delta^{18}O$ values values indicative of major contributions of atmospheric nitrate to the system.

A similar application is the determination of the source of nitrate in early spring runoff. During early spring melt, many small catchments experience episodic acidification because of large pulses of nitrate and hydrogen ions being flushed into the streams. There has been some controversy over the source of this nitrate. A number of studies have found that much of the nitrate in early runoff from small catchments is microbial rather than atmospheric (Kendall *et al.*, 1995b; Burns and Kendall, 2002; Campbell *et al.*, 2002; Sickman *et al.*, 2003; Pardo *et al.*, in press). For example, during the 1994 snowmelt season in Loch Vale watershed in Colorado (USA), almost all the stream samples had nitrate $\delta^{18}O$ and $\delta^{15}N$ values within the range of pre-melt and soil waters, and significantly different from the composition of almost all the snow samples. Therefore, the nitrate eluted from the snowpack appeared to go into storage, and most of the nitrate in stream flow during the period of potential acidification was apparently derived from pre-melt sources (Kendall *et al.*, 1995b). During subsequent years, half or more of the nitrate in the stream was microbial in origin, and probably originated from shallow groundwater in talus deposits (Campbell *et al.*, 2002).

In contrast to these forested catchment studies, atmospheric nitrate appears to be a major contributor to stream flow in urban catchments. A pilot study by Ging *et al.* (1996) to determine the dominant sources of nitrate in storm runoff in suburban watersheds in Austin, Texas (USA) found that during base flow (when chloride was high), nitrate had high $\delta^{15}N$ and low $\delta^{18}O$ values; in contrast, during storms (when chloride was low), nitrate had low $\delta^{15}N$ and high $\delta^{18}O$ values. The strong correspondence of $\delta^{15}N$ and $\delta^{18}O$ values during changing flow conditions, and the positive correlation of the percentage of impervious land-cover and the $\delta^{18}O$ of nitrate, suggests that the stream composition can be explained by varying proportions of two end-member compositions (Figure 15), one dominated by atmospheric nitrate (and perhaps fertilizer) that is the major source of water during storms and the other a well-mixed combination of sewage and other nitrate sources that contributes to base flow (Silva *et al.*, 2002). A recent study of nitrate from large rivers in the Mississippi Basin has shown that atmospheric nitrate is also a significant source of nitrate to large undeveloped and urban watersheds (Battaglin *et al.*, 2001; Chang *et al.*, 2002).

5.11.3.2.9 Large river studies

Two recent studies in the USA evaluated whether the combination of nitrate $\delta^{18}O$ and $\delta^{15}N$ would allow discrimination of watershed sources of nitrogen and provide evidence for denitrification. A pilot study in the Mississippi Basin (Battaglin *et al.*, 2001; Chang *et al.*, 2002) showed that large watersheds with different land uses (crops, animals, urban, and undeveloped) had overlapping moderately distinguishable differences in nutrient isotopic compositions (e.g., nitrate $\delta^{18}O/\delta^{15}N$ and particulate organic matter (POM) $\delta^{15}N/\delta^{13}C$). A study in large rivers in the northeastern part of the USA found strong positive correlations between $\delta^{15}N$ and the amounts of waste nitrogen (Mayer *et al.*, 2002)

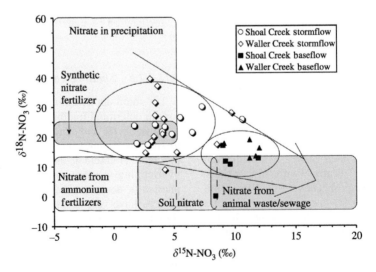

Figure 15 $\delta^{15}N$ versus $\delta^{18}O$ values of nitrate during storm flow and base flow conditions from Waller and Shoal Creeks, Austin, Texas, superimposed on common fields of nitrate from various sources. Ellipses indicate two standard deviations from average values. Arrows point in the direction of the base flow (subsurface) nitrate end-member (source Silva *et al.*, 2002).

Prior nitrogen mass-balance studies in the Mississippi Basin and in the northeastern USA rivers suggested appreciable losses of nitrogen via denitrification, especially in the headwaters. However, neither study found isotopic evidence for denitrification, perhaps because of continuous mixing with new nitrate, the small extent of denitrification, or the low fractionations resulting from diffusion-controlled (i.e., benthic) denitrification (Sebilo *et al.*, 2003). A detailed study of denitrification in various stream orders in the Seine River system (France) during summer low flow conditions indicates that extent of denitrification determined by examination of the shifts in the $\delta^{15}N$ of residual nitrate provides only a minimum estimate of denitrification (Sebilo *et al.*, 2003).

A useful adjunct to tracing nitrogen sources and sinks in aquatic systems with nitrate isotopes is the analysis of POM for $\delta^{15}N$, $\delta^{13}C$, and $\delta^{34}S$. In many river systems, most of the POM is derived from *in situ* production of algae. Even if an appreciable percent of the POM is terrestrial detritus, the C:N value of the POM and the $\delta^{15}N$ and $\delta^{13}C$ can, under favorable conditions, be used to estimate the percent of POM that is algae, and its isotopic composition (Kendall *et al.*, 2001b). The $\delta^{15}N$ and $\delta^{13}C$ (and $\delta^{34}S$) of the POM reflect the isotopic compositions of dissolved inorganic nitrogen, carbon, and sulfur in the water column. These compositions reflect the sources of nitrogen, carbon, and sulfur to the system, and the biogeochemical processes (e.g., photosynthesis, respiration, denitrification, and sulfate reduction) that alter the isotopic compositions of the dissolved species (Figure 16). Hence, the changes in the isotopic composition can be used to evaluate a variety of in-stream processes that might affect the interpretation of nitrate $\delta^{15}N$ and $\delta^{18}O$ (Kendall *et al.*, 2001b). The $\delta^{15}N$ of algae may even serve as an integrator for the $\delta^{15}N$ of nitrate; the $\delta^{15}N$ of algae appears to be ~4‰ lower than that of the associated nitrate (Battaglin *et al.*, 2001; Kendall, unpublished data).

5.11.3.2.10 *Subsurface waters*

Applications of $\delta^{15}N$ to trace relative contributions of fertilizer and animal waste to groundwater (Kreitler, 1975; Kreitler and Jones, 1975; Kreitler *et al.*, 1978; Gormly and Spalding, 1979) are complicated by a number of biogeochemical reactions, especially ammonia volatilization, nitrification, and denitrification. These processes can modify the $\delta^{15}N$ values of nitrogen sources prior to mixing and the resultant mixtures, causing estimations of the relative contributions of the sources of nitrate to be inaccurate. The combined use of $\delta^{18}O$ and $\delta^{15}N$ should allow better resolution of these issues. Analysis of the $\delta^{15}N$ of both nitrate and N_2 (e.g., Böhlke and Denver, 1995) provides an effective means for investigating denitrification. There have been only a few studies thus far using $\delta^{15}N$ and $\delta^{18}O$ to study nitrate sources and cycling mechanisms in groundwater. However, since it is likely that more studies in the future will utilize a multi-isotope or multitracer approach, the discussion below will concentrate on dual isotope studies.

The first dual isotope investigation of groundwater, in municipal wells downgradient from heavily fertilized agricultural areas near Hanover (Germany), successfully determined that the decreases in nitrate away from the fields was caused by microbial denitrification, not mixing with more dilute waters from nearby forests (Böttcher *et al.*, 1990). They found that low concentrations of nitrate were associated with high $\delta^{18}O$ and $\delta^{15}N$ values, and vice versa, and

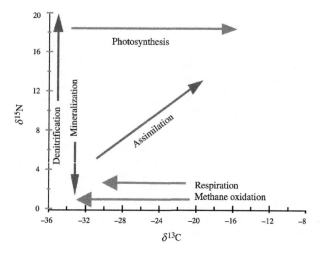

Figure 16 Schematic of in-stream biogeochemical processes that affect the $\delta^{15}N$ and $\delta^{13}C$ of algae. Arrows indicate the general effect of each process on the isotopic composition of the resulting algae.

that changes in $\delta^{18}O$ and $\delta^{15}N$ values along the flow path were linearly related, with a slope of ~0.5. The linear relation between the isotope values and the logarithm of the fraction of residual nitrate (i.e., the data fit a Rayleigh equation) indicated that denitrification with constant enrichment factors for oxygen (−8.0‰) and nitrogen (−15.9‰) was responsible for the increases in $\delta^{18}O$ and $\delta^{15}N$.

There have been several subsequent dual isotope fertilizer studies. The high concentrations of NO_3^- in shallow groundwater in the Abbotsford aquifer, British Columbia (Canada), were attributed to nitrification of poultry manure, with lesser amounts of ammonium fertilizers (Wassenaar, 1995). A study of denitrification in a riparian zone showed a higher slope (~0.7) for the relative fractionation of $\delta^{18}O$ to $\delta^{15}N$ (Mengis *et al.*, 1999). Nitrate from 10 major karst springs in Illinois during four different seasons was found to be mainly derived from nitrogen fertilizer (Panno *et al.*, 2001).

The applicability of $\delta^{18}O$ and $\delta^{15}N$ of nitrate and other tracers to delineate contaminant plumes derived from domestic septic systems was evaluated by Aravena *et al.* (1993), in a study within an unconfined aquifer beneath an agricultural area in Ontario (Canada). They found that $\delta^{15}N$ of nitrate, $\delta^{18}O$ of water, and water chemistry (especially sodium) were effective for differentiating between the plume and native groundwater. The differences in $\delta^{15}N$ values reflect differences in dominant nitrogen sources (human waste versus mixed fertilizer/manure). There was good delineation of the plume with water-$\delta^{18}O$ values, reflecting differences in residence time of water within the shallow unconfined aquifer, and the deeper confined aquifer that supplies water to the house and the plume. There was no significant difference between the $\delta^{18}O$ of nitrate in the plume and in local groundwater; the $\delta^{18}O$ values suggest that nitrification of ammonium, from either human waste or agricultural sources, is the source of the nitrate. Another study of a septic plume determined that use of a multi-isotope approach (using $\delta^{13}C$ of DIC, and $\delta^{18}O$ and $\delta^{34}S$ of sulfate in addition to nitrate $\delta^{18}O$ and $\delta^{15}N$) provided valuable insight into the details of the processes affecting nitrate attenuation in groundwater (Aravena and Robertson, 1998).

5.11.3.3 Sulfur

Several important reviews of the use of stable sulfur isotopes as hydrologic and biogeochemical tracers have been produced since the mid-1980s. Krouse and Grinenko (1991) provide a very comprehensive evaluation of stable sulfur isotopes as tracers of natural and anthropogenic sulfur.

Krouse and Mayer (2000) present an in-depth review and several case studies of the use of stable sulfur and oxygen isotopes of sulfate as hydrologic tracers in groundwater. In addition, the use of sulfur isotopes for investigations of hydrological processes in catchments has been reviewed by Mitchell *et al.* (1998). Thus, what is presented here is just a brief overview of the application of sulfur isotopes in hydrologic studies.

Much of the work using sulfur isotopes in surface hydrology to date has been driven by the need to understand the effects of atmospheric deposition on sulfur cycling in the natural environment, particularly in forest ecosystems. This is in response to increased sulfur loadings to terrestrial ecosystems from anthropogenic sulfur emissions, as sulfur is a dominant component of "acid rain." In addition to studying ecosystem responses to sulfur deposition from atmospheric sources, there is often a need to identify sulfur sources and transformations along flow pathways in groundwater. Bacterially mediated sulfate reduction is one of the most widespread biogeochemical processes occurring in groundwater systems. Elevated sulfate concentrations in groundwater and production of H_2S gas in aquatic systems can be both a nuisance and a health hazard, yet are ubiquitous. The stable isotopes of sulfur and of oxygen in sulfate are useful tracers in such studies (Clark and Fritz, 1997; Krouse and Mayer, 2000). A more applied use of stable sulfur isotopes in terms of groundwater risk management is for identification of arsenic contamination sources and biological remediation efficacy of organic pollutants.

5.11.3.3.1 Sulfur isotope fundamentals

Sulfur has four stable isotopes: ^{32}S (95.02%), ^{33}S (0.75%), ^{34}S (4.21%), and ^{36}S (0.02%) (MacNamara and Thode, 1950). Stable isotope compositions are reported as $\delta^{34}S$, ratios of $^{34}S/^{32}S$ in per mil relative to the standard CDT. The general terrestrial range is +50‰ to −50‰, with rare values much higher or lower.

Geologic sources of sulfur include primary and secondary sulfide minerals as well as chemically precipitated sulfate minerals. In sedimentary rocks, sulfur is often concentrated. Examples include pyritic shales, evaporites, and limestones containing pyrite and gypsum in vugs and lining fractures. In many cases, these rocks can be both the source of organic matter as well as sulfate for the biogeochemical reduction of sulfate to occur.

There is a large range in $\delta^{34}S$ of sedimentary sulfide minerals, from −30‰ to +5‰ (Migdisov *et al.*, 1983; Strauss, 1997). Crystalline rocks of magmatic origin show a more narrow range, from 0‰ to +5‰. Volcanic rocks are similar;

Sulfate minerals

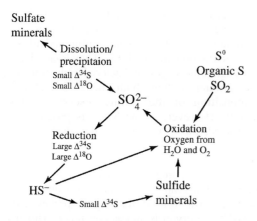

Figure 17 Schematic showing the relative fractionations in $\delta^{18}O$ and $\delta^{34}S$ caused by different reactions. $\Delta\delta^{34}S$ and $\Delta\delta^{18}O$ denote the fractionations ($\delta-\delta$ values) caused by the reactions (after Krouse and Mayer, 2000).

however, some have values up to $+20‰$, indicating recycling of oceanic sulfate at subduction zones.

Figure 17 shows schematically the major transformations of sulfur in aquatic systems and their related isotopic fractionations. Processes that do not significantly fractionate sulfur isotopes include: (i) weathering of sulfide and sulfate minerals, (ii) adsorption–desorption interactions with organic matter, and (iii) isotopic exchange between SO_4^{2-} and HS^- or H_2S in low-temperature environments. Precipitation of sulfate minerals is accompanied by only slight isotopic fractionation, and the precipitates are generally enriched in the heavier isotopes (^{34}S and ^{18}O) relative to the residual sulfate in solution (Holser and Kaplan, 1966; Szaran *et al.*, 1998).

The primary process by which sulfur is fractionated in hydrologic environments is via biologically mediated sulfur transformations. The most important of these is the dissimilatory sulfate reduction (DSR) reaction

$$2H^+ + SO_4^{2-} + 4H_2 \rightarrow H_2S + 4H_2O \quad (14)$$

which is facilitated by anaerobic bacteria of the *Desulfovibrio* and *Desulfotomaculum* genera (Mitchell *et al.*, 1998). These sulfate-reducing bacteria utilize dissolved sulfate as an electron acceptor during the oxidation of organic matter. Bacterial reduction of SO_4^{2-} is the primary source of the variability of SO_4^{2-} isotopic compositions observed in natural aquatic systems. During sulfate reduction, the bacteria produce H_2S gas that has a $\delta^{34}S$ value $\sim 25‰$ lower than the sulfate source (Clark and Fritz, 1997). Consequently, the residual pool of SO_4^{2-} becomes progressively enriched in ^{34}S. Thus, long-term anaerobic conditions in groundwater ought to be reflected

in the $\delta^{34}S$ values of SO_4^{2-} that are significantly higher than that of the sulfate source.

As this process is accompanied by oxidation of organic matter, there is a concomitant shift in the $\delta^{13}C$ of DIC toward that of the organic source due to the production of CO_2. The result is that these simultaneous biogeochemical processes—sulfate reduction and oxidation of organic matter—can be useful in assessing the biological activity within groundwater bodies (Clark and Fritz, 1997). Several studies have demonstrated the utility of sulfur isotopes in the assessment of biological remediation of organic pollutants in groundwater (Bottrell *et al.*, 1995; Spence *et al.*, 2001; Schroth *et al.*, 2001). Using sulfur, carbon, and oxygen isotopes in concert may also facilitate the tracing of groundwater flow paths as well as mixing processes in hydrogeologic systems, thanks to the wide separation in $\delta^{34}S$ and $\delta^{18}O$ values of SO_4^{2-} and in the $\delta^{13}C$ of DIC. However, it is important to note that under conditions of low concentrations of reactive organic matter, reoxidation of mineral sulfides can lead to constant recycling of the dissolved sulfur pool. In such cases, increases in $\delta^{34}S$ do not occur in conjunction with reaction progress since the sulfide concentration in solution is held approximately constant (Spence *et al.*, 2001; Fry, 1986).

Aside from bacterially mediated reduction, the isotopic composition of sulfate is controlled by (i) the isotopic composition of sulfate sources, (ii) isotope exchange reactions, and (iii) kinetic isotope effects during transformations. Exchange of sulfur isotopes between SO_4^{2-} and HS^- or H_2S is not thought to be of significance in most groundwater systems; however, microbial activity may enhance this exchange. Under favorable conditions, $\delta^{34}S$ can also be used to determine the reaction mechanisms responsible for sulfide oxidation (Kaplan and Rittenberg, 1964; Fry *et al.*, 1986, 1988) and sulfate reduction (Goldhaber and Kaplan, 1974; Krouse, 1980).

5.11.3.3.2 *Oxygen isotopes of sulfate*

The evaluation of sources and cycling of sulfate in the environment can be aided by analysis of the oxygen isotopic composition of sulfate. Oxygen has three stable isotopes: ^{16}O, ^{17}O, and ^{18}O. The $\delta^{18}O$ and $\delta^{17}O$ values of sulfate are reported in per mil relative to the standard SMOW or VSMOW. The $\delta^{18}O$ of sulfate shows considerable variations in nature, and recent studies indicate that the systematics of cause and effect remain somewhat obscure and controversial (Van Stempvoort and Krouse, 1994; Taylor and Wheeler, 1994). Reviews of applications of $\delta^{18}O$ of sulfate include Holt and Kumar (1991) and Pearson and Rightmire (1980). When in isotopic equilibrium at 0 °C, aqueous sulfate has a $\delta^{18}O$ value $\sim 30‰$

higher than water (Mizutani and Rafter, 1969). However, the rate of oxygen isotope exchange between sulfate and water is very slow at low temperatures and normal pH levels. Even in acidic rain of pH 4, the "half-life" of exchange is ~1,000 yr (Lloyd, 1968). Thus, aqueous sulfate is generally out of isotopic equilibrium with the host water, although there have been exceptions (e.g., Nriagu *et al.*, 1991; Berner *et al.*, 2002). Oxygen isotope exchange between SO_4^{2-} and H_2O is significant only under geothermal conditions, and in such cases may be useful as a geothermometer (Krouse and Mayer, 2000).

Depending upon the reaction responsible for sulfate formation, between 12.5% and 100% of the oxygen in sulfate is derived from the oxygen in the environmental water; the remaining oxygen comes from O_2 (Taylor *et al.*, 1984). Holt *et al.* (1981) studied mechanisms of aqueous SO_2 oxidation and the corresponding $\delta^{18}O$ of the SO_4^{2-} produced. They concluded that three of the four oxygen atoms that give rise to the $\delta^{18}O_{SO_4}$ are derived from the oxygen in the water, and the rest is derived from dissolved O_2 gas; atmospheric O_2 is +23.8‰ (Horibe *et al.*, 1973).

Despite the complexities of the mechanisms that establish the $\delta^{18}O$ of aqueous sulfate, it is clear that this measurement is a useful additional parameter by which sources of SO_4^{2-} and sulfur cycling can be evaluated in aquatic systems. Use of a dual isotope approach to tracing sources of sulfur (i.e., measurement of $\delta^{18}O$ and $\delta^{34}S$ of sulfate) will often provide better separation of potential sources of sulfur and, under favorable conditions, provide information on the processes responsible for sulfur cycling in the ecosystem. Some examples of studies that have examined the sources of SO_4^{2-} to groundwater using a dual isotope approach include Yang *et al.* (1997), Dogramaci *et al.* (2001), and Berner *et al.* (2002).

A study by Berner *et al.* (2002) suggests that as the bacterial reduction of sulfate progresses, there is variable sensitivity in the sulfur and oxygen isotopic fractionation in SO_4^{2-}. They found that, in the beginning stages of bacterial reduction in a groundwater system, both the $\delta^{34}S$ and $\delta^{18}O$ composition of the sulfate reservoir change rapidly. However, as the SO_4^{2-} is progressively consumed, the $\delta^{18}O$ value of the residual SO_4^{2-} approaches a constant value, while the $\delta^{34}S$ continues to increase (Figure 18). Hence, $\delta^{18}O$ of sulfate is a more sensitive indicator at the initial stages of bacterial SO_4^{2-} reduction, whereas $\delta^{34}S$ is more sensitive for describing the process when the SO_4^{2-} reservoir is almost consumed. This is in accordance with the findings of Fritz *et al.* (1989), who observed similar behavior and attributed it to the involvement of sulfite during the biologically mediated reaction and oxygen-isotope exchange with the water.

5.11.3.3.3 *Tracing atmospheric deposition*

Using $\delta^{34}S$ and $\delta^{18}O$. Atmospheric sulfur $\delta^{34}S$ values are typically in the range of −5‰ to +25‰ (Krouse and Mayer, 2000). Seawater sulfate has a $\delta^{34}S$ value of +21‰. Atmospheric sulfate derived from marine sources will show $\delta^{34}S$ values between +15‰ and +21‰, while emissions of atmospheric sulfur from biogenic reduction (e.g., H_2S) have lower values.

Atmospheric deposition has been repeatedly shown to be the major source of sulfur deposition in many forest ecosystems, especially in regions significantly impacted by acid deposition (Mitchell *et al.*, 1992b). Sulfur inputs to forests derived from mineral weathering generally play a minor role; however, this is dependent upon the local geology. Regions underlain by sulfide-rich

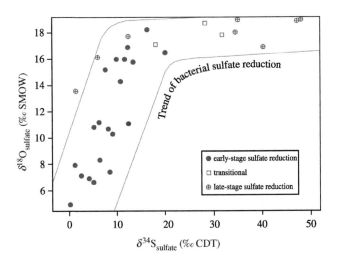

Figure 18 During sulfate reduction, both $\delta^{18}O$ and $\delta^{34}S$ of sulfate change rapidly during the early stages of the reaction, but there is little change in the $\delta^{18}O$ during the late stages (after Berner *et al.*, 2002).

shales can have mineral inputs exceeding those derived from the atmosphere, while weathering of ore minerals and evaporites also contributes sulfur. Nriagu and Coker (1978) report that the $\delta^{34}S$ of precipitation in central Canada varies seasonally from about +2‰ to +9‰, with the low values caused mainly by biological sulfur, whereas the high values reflect sulfur from fossil fuels.

The spatial distribution of $\delta^{34}S$ and the relative contributions from marine versus continental (including anthropogenic combustion) sources in Newfoundland (Canada) have been monitored by analyzing the $\delta^{34}S$ of rainfall and of lichens that obtain all their sulfur from the atmosphere (Wadleigh et al., 1996; Wadleigh and Blake, 1999). Wadleigh et al. (1996) report rainwater sulfate with $\delta^{34}S$ and $\delta^{18}O$ of ~+4‰ and +11‰, respectively, for a continental end-member representative of long-range sulfate transport. The marine end-member has $\delta^{34}S$ of +21‰ and $\delta^{18}O$ of +9.5‰, representing sulfate in precipitation derived from sea spray. The study of epiphytic lichens (Wadleigh and Blake, 1999) yielded a "bulls-eye" $\delta^{34}S$ contour plot showing low values in the interior of the island that are probably related to anthropogenic point sources, and progressively higher (more marine) values towards the coasts. These studies suggest that the study area is influenced by both marine (high $\delta^{34}S$ values) and continental sources (lower $\delta^{34}S$ values), with the possibility of anthropogenic influence from fossil-fuel powered plants.

Using $\delta^{17}O$. Atmospheric sulfate (aerosol and rainfall) has recently been found to have a mass independent isotopic composition, with excess ^{17}O over what would have been expected based on the $\delta^{18}O$ of sulfate (Lee et al., 2001). On a three-isotope plot with $\delta^{17}O$ on the x-axis and $\delta^{18}O$ on the y-axis, mass-dependent fractionation (MDF) result in sulfate values defined by the relation: $\delta^{17}O = 0.52\delta^{18}O$, whereas MIF causes values that deviate from this relation. For sulfate, the mass-independent relation is described by $\Delta^{17}O = \delta^{17}O - 0.52\delta^{18}O$ (Lee et al., 2001). Hence, MDF results in $\Delta^{17}O = 0$, whereas an MIF results in $\Delta^{17}O \neq 0$. $\delta^{17}O$ and $\Delta^{17}O$ values reflect $^{17}O/^{16}O$ ratios, and are reported in per mil relative to SMOW.

The causes of the ^{17}O anomalies of sulfate and other oxygen-bearing materials in the atmosphere and in extraterrestrial materials are the topic of active research. Aqueous-phase oxidation of S(IV) causes an MIF in sulfate only if the oxidant has an MIF oxygen isotopic composition (Lee et al., 2001). Hence, the anomalous $\delta^{17}O$ values of atmospheric sulfate appear to be caused by interactions with either O_3 or H_2O_2, since thus far they are the only known oxidants in the atmosphere with mass-independent oxygen isotopic compositions (Johnson and Thiemens, 1997).

The discovery of the anomalous oxygen isotopic compositions of atmospheric sulfate provides a new means for identifying sulfate of atmospheric origin. Rainwater and aerosols from southern California were found to have $\Delta^{17}O$ values in the range of 0‰ to +1.5‰ (Lee et al., 2001). The average $\Delta^{17}O$ of snow sulfate in the Rocky Mountains (Colorado, USA) was +1.3‰. Sulfate in ice cores, massive sulfate deposits, and Dry Valley soils from various locations also have MIF (Bao et al., 2000; Lee et al., 2001). There appears to be seasonality in the $\Delta^{17}O$ of sulfate in precipitation, with higher values in the winter and lower values in the summer, probably due to seasonal changes in climatic effects that favor aqueous phase S(IV) oxidation in winter relative to summer (Lee and Thiemens, 2001).

Sulfate $\delta^{17}O$ can be used to identify the relative contributions of atmospheric sulfate versus terrestrial biological or geologic sources of sulfate to streams. In the Rocky Mountains, stream-water sulfate had $\Delta^{17}O$ values of +0.2‰ to +0.9‰ (Johnson et al., 2001). Isotope mass-balance calculations suggest that 20–40% of the sulfate was atmospheric in origin, considerably lower than expected (Johnson et al., 2001). Combined studies using $\delta^{34}S$, $\delta^{18}O$, $\delta^{17}O$, and ^{35}S in alpine watersheds in the Rocky Mountains show that sulfur and oxygen cycling processes appear to be decoupled (Kester et al., 2002), with atmospheric sulfur and oxygen having different residence times in the catchment.

Using sulfur-35. Sulfur-35 (^{35}S) is a naturally produced radioactive tracer (half-life = 87 d) that can be used to trace the movement of atmospherically derived sulfate in the environment. It is formed in the atmosphere from cosmic ray spallation of ^{40}Ar (Peters, 1959), and deposits on the Earth's surface in precipitation or as dryfall. It can be used both to trace the timescales for movement of atmospheric sulfate through the hydrosphere and, in ideal cases, to trace the movement of young (<1 yr) water. It is an especially useful tracer in regions away from the ocean where sulfate concentrations are relatively low.

The isotope was originally used by Cooper et al. (1991) to trace the movement of young sulfate in an arctic watershed. They found that sulfur deposited as precipitation is strongly adsorbed within the watershed, and that most sulfur released to stream flow is derived from longer-term storage in soils, vegetation, or geologic materials. Further work (Michel et al., 2000) showed that ^{35}S could be an especially effective isotope in tracing movement of young sulfate through an alpine watershed at Loch Vale watershed in the Rocky Mountains (Colorado, USA). They found that ~50% of the atmospheric sulfate deposited in that watershed was retained less than one year. This work is now being expanded to include the use of the ^{17}O

anomaly to try to obtain a better estimate of the real age of atmospheric sulfate leaving the watershed (Kester *et al.*, 2002). The isotope also appears to be an effective tracer for sulfate cycling in small alpine lakes (Michel *et al.*, 2002).

Sulfur-35 was also used to determine the source of the sulfate in stream water in a highly polluted watershed in the Czech Republic (Novak *et al.*, 2003). The sulfate input to this watershed has decreased over the past decade because of new controls on sulfur emissions, but the watershed continues to export sulfate far in excess of the atmospheric loading at the present time. Measurement of ^{35}S in runoff indicates that none of the recently deposited sulfate is exiting in the system. Apparently the atmospheric sulfate interacts with the soil layers and consequently takes more than a year to be removed from the watershed (Novak *et al.*, 2003).

5.11.3.3.4 *Sulfur in catchment surface waters*

Sulfur occurs naturally in both organic and inorganic forms. In most forested catchments, the sulfur content of the catchment is largely in the organic form stored within pools of organic matter (Mitchell *et al.*, 1998). Carbon-bonded sulfur in organic compounds is the dominant organic form, while sulfate is the primary form of inorganic sulfur in surface water systems. Organic sulfur is generally not labile; thus, fluxes of sulfur in catchments are often related solely to movement and transformation of inorganic sulfur, including adsorption–desorption interactions with organic matter (Mitchell *et al.*, 1992a, 1994). Catchments that contain a significant geologic component of sulfur-bearing minerals will likely have sulfur budgets that are dominated by weathering inputs instead of atmospheric inputs.

Because sulfur isotope ratios are strongly fractionated by biogeochemical processes, there has been concern over whether δ^{34}S could be used effectively to separate sources of sulfur in shallow hydrological systems. Sulfur appears to be affected by isotope fractionation processes in some catchments (Fuller *et al.*, 1986; Hesslein *et al.*, 1988; Andersson *et al.*, 1992), but not in others (Caron *et al.*, 1986; Stam *et al.*, 1992). Andersson *et al.* (1992) explain the decreases in sulfate concentrations and increases in δ^{34}S of stream flow during the summer by sulfate reduction in stagnant pools. Stam *et al.* (1992) suggest that the extent of fractionation might be a function of water residence time in the catchment, with steep catchments showing less fractionation. They note that increases in δ^{34}S of stream sulfate during the winter may be a result of micropore flow during the snow-covered period, instead of the more typical macropore-flow characteristic of storms.

Intensive investigations of the sulfur dynamics of forest ecosystems in the last decade can be attributed to the dominant role of sulfur as a component of acidic deposition. Studies in forested catchments include Fuller *et al.* (1986), Mitchell *et al.* (1989), Stam *et al.* (1992), and Andersson *et al.* (1992). Sulfur with a distinctive isotopic composition has been used to identify pollution sources (Krouse *et al.*, 1984), and has been added as a tracer (Legge and Krouse, 1992; Mayer *et al.*, 1992, 1993). Differences in the natural abundances can also be used in systems where there is sufficient variation in the δ^{34}S of ecosystem components. Rocky Mountain lakes (USA), thought to be dominated by atmospheric sources of sulfate, have different δ^{34}S values than lakes believed to be dominated by watershed sources of sulfate (Turk *et al.*, 1993).

5.11.3.3.5 *Sulfur in groundwater*

Sulfur in groundwater is primarily in the oxidized sulfate species, SO_4^{2-}. Sulfate in groundwater can have several sources. These include: (i) dissolution of evaporite sulfate minerals such as gypsum and anhydrite, (ii) oxidation of sulfide minerals, (iii) atmospheric deposition, and (iv) mineralization of organic matter.

Production of sulfate in shallow groundwater via the oxidation of sulfide minerals yields δ^{34}S values of the sulfate that are very similar to those of the source sulfides, because there is little isotope fractionation during near-surface, low-temperature sulfide mineral oxidation (Toran and Harris, 1989). Although the sulfur isotopic composition of sulfide minerals can vary widely in geologic materials, the isotopic compositions of sulfur sources in a regional study area often fall within a narrow range (Krouse and Mayer, 2000). Thus, sulfur isotopes can be useful for identifying sources of sulfur species in groundwater as well as biogeochemical transformations along flow paths.

For example, δ^{34}S and δ^{18}O of SO_4^{2-} was used to assess mixing between a vertically stacked aquifer system in contact with a salt dome located in northern Germany (Berner *et al.*, 2002). Two major sulfate pools were identified based upon their isotopic compositions: (i) SO_4^{2-} from the dissolution of evaporite minerals, and (ii) SO_4^{2-} derived from atmospheric deposition and from the oxidation of pyrite. Using both the δ^{34}S and δ^{18}O of the SO_4^{2-}, zones of variable groundwater mixing and significant amounts of bacterial sulfate reduction were identified. The SO_4^{2-} derived from the dissolution of the rock salt in the highly saline deep brine showed nearly constant δ^{34}S values between +9.6‰ and 11.9‰ and δ^{18}O between +9.5‰ and 12.1‰, consistent with the Permian evaporite deposits. Sulfate in near-surface unconfined groundwater derived from riverine recharge showed values more typical of meteoric sulfur mixed with oxidized sulfide minerals ($\delta^{34}S = +5.2‰$; $\delta^{18}O = +8.2‰$).

One complicating factor is the possibility of sulfur recycling through subsequent oxidation–reduction of biologically reduced SO_4^{2-}. H_2S gas produced by the biological reduction of SO_4^{2-} may become reoxidized, especially in near-surface aquifers or in mixing zones with more oxidizing waters. The oxidation of this H_2S will produce a secondary SO_4^{2-} pool with a much lower $\delta^{34}S$ than the original source sulfate. This can result in a depth gradient in which higher $\delta^{34}S$ values of sulfate are found at depth (partially reducing environment) while lower $\delta^{34}S$ values of sulfate are found in the more oxidizing environment closer to the surface or within mixing zones.

Some recent studies have shown that the identification of arsenic-bearing sulfides and their contribution to arsenic loadings in groundwater and surface water used for drinking supply can also be assessed using $\delta^{34}S$ (Sidle, 2002; Schreiber et al., 2000).

5.11.4 USE OF A MULTI-ISOTOPE APPROACH FOR THE DETERMINATION OF FLOW PATHS AND REACTION PATHS

Flow paths are the individual runoff pathways contributing to surface flow in a catchment. These result from runoff mechanisms that include, but are not limited to, saturation-excess overland flow, Hortonian overland flow, near-stream groundwater ridging, hill-slope subsurface flow through the soil matrix or macropores, and shallow organic-layer flow. Knowledge of hydrologic flow paths in catchments is critical to the preservation of public water supplies and the understanding of the transport of point and non-point-source pollutants (Peters, 1994).

Stable isotopes such as ^{18}O and ^2H are an improved alternative to traditional nonconservative chemical tracers, because waters are often uniquely labeled by their isotopic compositions (Sklash and Farvolden, 1979; McDonnell and Kendall, 1992), often allowing the separation of waters from different sources (e.g., new rain versus old pre-storm water). However, studies have shown that flow paths cannot be identified to a high degree of certainty using $\delta^{18}O$ or δ^2H data and simple hydrograph separation techniques, because waters within the same flow path can be derived from several different sources (Ogunkoya and Jenkins, 1991). Thus, a number of plausible runoff mechanisms can be consistent with the isotope data. The need to incorporate flow path dynamics is recognized as

a key ingredient in producing reliable chemical models (Robson et al., 1992).

Reactive solute isotopes such as ^{13}C, ^{34}S, and ^{87}Sr can provide valuable information about flow paths (not water sources) and reaction paths useful for geochemical and hydrologic modeling precisely, because they can reflect the reaction characteristics of and taking place along specific flow paths (Bullen and Kendall, 1998). These reactive solute isotopes can serve as additional thermodynamic constraints in geochemical computer models such as NETPATH (Plummer et al., 1991) for eliminating possible geochemical reaction paths (Plummer et al., 1983; see Chapter 5.14).

As an example, water isotopes were used to determine the actual flow paths by which water discharging from Crystal Lake (Wisconsin, USA) mixed with regional groundwater and local rain recharge while flowing across a narrow isthmus separating the lake from Big Muskellunge Lake (Krabbenhoft et al., 1992, 1993; Bullen et al., 1996). Solute isotopes such as ^{87}Sr and ^{13}C, along with chemical data, were then used to identify the geochemical reactions taking place along these flow paths.

The waters flowing along mineralogically distinctive horizons are sometimes distinctively labeled by their chemical composition and by the isotopic compositions of solute isotopes such as ^{13}C, ^{87}Sr, ^{34}S, ^{15}N, etc. For example, waters flowing through the soil zone often have $\delta^{13}C$ values that are depleted in ^{13}C relative to deeper groundwaters because of biogenic production of carbonic acid in organic soils (Kendall, 1993). These same shallow waters can also have distinctive lead and strontium isotopic compositions (Bullen and Kendall, 1998).

A multi-isotope approach can be particularly useful in tracing the reactions specific to a flow path (Krabbenhoft et al., 1992, 1993). The solute isotope tracers can be an especially powerful tool because they usually are affected by a smaller number of processes than chemical constituents, making interpretation of changes in isotopic composition less ambiguous than the simultaneous changes in solute concentrations (Kendall, 1993). In particular, the strengths of one isotope may compensate for the weakness of another. For example, one of the concerns with using $\delta^{13}C$ to trace carbon sources in catchments is that possible isotopic exchange of subsurface DIC with soil or atmospheric CO_2 could allow the $\delta^{13}C$ of DIC to be "reset" so that the characteristic "geologic" signature of the calcite-derived carbon is lost (Kendall et al., 1992; Kendall, 1993). Since strontium isotopes are not affected by this kind of exchange, the same samples can be analyzed for strontium isotopes; if the strontium signature of calcite is distinctive, the correlation of changes in

^{87}Sr/^{86}Sr and changes in δ^{13}C of subsurface waters could provide evidence that calcite dissolution was occurring and that exchange was not a problem (Bullen and Kendall, 1998).

If the chemical composition of waters along a specific flow path is related to topography, mineralogy, initial water composition, and antecedent moisture conditions, then within a uniform soil layer, the degree of geochemical evolution is some function of residence time. In this case, it may be possible to link geochemical evolution of waters along specific flow paths to hydrologic models such as TOPMODEL (Beven and Kirkby, 1979) by the testable assumption that residence time—and hence degree of rock/water reaction—is primarily a function of topography. Alternatively, because TOPMODEL is used to predict the different contributions from the different soil horizons along the stream channel as the topographic form of the hill-slope changes (Robson et al., 1992), these predictions can be tested with appropriate isotope and chemical data. Future advances in watershed modeling, both conceptually and mathematically, will require the combined use of chemical and isotopic tracers for calibration and validation.

5.11.5 SUMMARY AND CONCLUSIONS

The dominant use of isotopes in shallow hydrologic systems in the last few decades has been to trace sources of waters and solutes. Generally, such data were evaluated with simple mixing models to determine how much water or solute was derived from either of two (sometimes three) constant-composition sources. The world does not seem this simple anymore. With the expansion of the field of isotope hydrology in the last decade, made possible by the development and increased availability of automated sample preparation and analysis systems for mass spectrometers, we have documented considerable heterogeneity in the isotopic compositions of rain, soil water, groundwater, and solute sources. In addition, hydrologists who utilize geochemical tracers recognize that the degree of variability observed is highly dependent upon the sampling frequency, and more effort is being placed on event-based studies with high-frequency sampling. We are still grappling with how to deal with this heterogeneity in our hydrologic and geochemical models. A major challenge is to use the variability as signal, not noise; the isotopes and chemistry are providing very detailed information about sources and reactions in shallow systems, and the challenge now is to develop appropriate models to use the data. In the past, much reliance was placed upon the stable isotopes of water (δ^{18}O, δ^2H) to reveal all of the information about the hydrologic processes taking place in a catchment. Today, we acknowledge that the best approach is to combine as many tracers as possible, including solutes and solute isotopes. This integration of chemical and isotopic data with complex hydrologic and geochemical models constitutes an important frontier of hydrologic research.

ACKNOWLEDGMENTS

The authors would like to thank Thomas D. Bullen, Michael G. Sklash, and Eric A. Caldwell for their contributions to chapters that were extensively mined to produce an early draft of this chapter: Kendall et al. (1995a), Bullen and Kendall (1998), and Kendall and Caldwell (1998).

REFERENCES

Aleem M. I. H., Hoch G. E., and Varner J. E. (1965) Water as the source of oxidant and reductant in bacterial chemosynthesis. *Biochemistry* **54**, 869–873.

Amberger A. and Schmidt H. L. (1987) Naturliche isotopengehalte von nitrat als indikatoren fur dessen herkunft. *Geochim. Cosmochim. Acta* **51**, 2699–2705.

Amoitte-Suchet P., Aubert D., Probst J. L., Gauthier-Lafaye F., Probst A., Andreux F., and Viville D. (1999) δ^{13}C pattern of dissolved inorganic carbon in a small granitic catchment: the Strengbach case study (Vosges mountains, France). *Chem. Geol.* **159**, 129–145.

Amundson R., Stern L., Baisden T., and Wang Y. (1998) The isotopic composition of soil and soil-respired CO_2. *Geoderma* **82**, 83–114.

Andersson K. K. and Hooper A. B. (1983) O_2 and H_2O are each the source of one O in NO_2 produced from NH_3 by Nitrosomonas: ^{15}N–NMR evidence. *FEBS Lett.* **164**(2), 236–240.

Andersson P., Torssander P., and Ingri J. (1992) Sulphur isotope ratios in sulphate and oxygen isotopes in water from a small watershed in central Sweden. *Hydrobiologia* **235/236**, 205–217.

Aravena R. and Robertson W. D. (1998) Use of multiple isotope tracers to evaluate denitrification in groundwater: case study of nitrate from a large-flux septic system plume. *Ground Water* **31**, 180–186.

Aravena R. and Wassenaar L. I. (1993) Dissolved organic carbon and methane in a regional confined aquifer, southern Ontario, Canada: carbon isotope evidence for associated subsurface sources. *Appl. Geochem.* **8**, 483–493.

Aravena R., Schiff S. L., Trumbore S. E., Dillon P. J., and Elgood R. (1992) Evaluating dissolved inorganic carbon cycling in a forested lake watershed using carbon isotopes. *Radiocarbon* **34**(3), 636–645.

Aravena R., Evans M. L., and Cherry J. A. (1993) Stable isotopes of oxygen and nitrogen in source identification of nitrate from septic systems. *Ground Water* **31**, 180–186.

Atekwana E. A. and Krishnamurthy R. V. (1998) Seasonal variations of dissolved inorganic carbon and δ^{13}C of surface waters: application of a modified gas evolution technique. *J. Hydrol.* **205**, 26–278.

Aucour A., Sheppard S. M. F., Guyomar O., and Wattelet J. (1999) Use of ^{13}C to trace origin and cycling of inorganic carbon in the Rhône River system. *Chem. Geol.* **159**, 87–105.

Bajjali W. and Abu-Jaber N. (2001) Climatological signals of the paleogroundwater in Jordan. *J. Hydrol.* **243**, 133–147.

Bao H., Campbell D. A., Bockheim J. G., and Thiemens M. H. (2000) Origins of sulphate in Antarctic dry-valley soils as deduced from anomalous ^{17}O compositions. *Nature* **407**, 499–502.

Battaglin W. A., Kendall C., Chang C. C. Y., Silva S. R., and Campbell D. H. (2001) *Chemical and Isotopic Composition of Organic and Inorganic Samples from the Mississippi River and its Tributaries, 1997–98.* USGS Water Resources Investigation Report **01-4095**, 57p.

Behrens H., Moser H., Oerter H., Rauert W., Stichler W., Ambach W., and Kirchlechner P. (1979) Models for the runoff from a glaciated catchment area using measurements of environmental isotope contents. In *Isotope Hydrology 1978* (ed. IAEA). IAEA, Vienna, pp. 829–846.

Berner Z. A., Stüben D., Leosson M. A., and Klinge H. (2002) S- and O-isotopic character of dissolved sulphate in the cover rock aquifers of a Zechstein salt dome. *Appl. Geochem.* **17**, 1515–1528.

Beven K. J. and Kirkby M. J. (1979) A physically based, variable contributing model of basin hydrology. *Hydrol. Sci. Bull.* **10**, 43–69.

Bishop K. H. (1991) Episodic increases in stream acidity, catchment flow pathways and hydrograph separation. PhD Thesis, University Cambridge.

Böhlke J. K. (2002) Groundwater recharge and agricultural contamination. *Hydrol. J.* **10**, 153–179.

Böhlke J. K. and Denver J. M. (1995) Combined use of groundwater dating, chemical, and isotopic analyses to resolve the history and fate of nitrate contamination in two agricultural watersheds, Atlantic coastal plain, Maryland. *Water Resour. Res.* **31**, 2319–2339.

Böhlke J. K., Eriksen G. E., and Revesz K. (1997) Stable isotope evidence for an atmospheric origin of desert nitrate deposits in northern Chile and southern California USA. *Isotope Geosci.* **136**, 135–152.

Böttcher J., Strebel O., Voerkelius S., and Schmidt H. L. (1990) Using isotope fractionation of nitrate–nitrogen and nitrate–oxygen for evaluation of microbial denitrification in a sandy aquifer. *J. Hydrol.* **114**, 413–424.

Bottrell S. H., Hayes P. J., Bannon M., and Williams G. M. (1995) Bacterial sulfate reduction and pyrite formation in a polluted sand aquifer. *Geomicrobiol. J.* **13**(2), 75–90.

Brinkmann R., Eichler R., Ehhalt D., and Munnich K. O. (1963) Über den deuterium-gehalt von niederschlags-und grundwasser. *Naturwissenschaften* **19**, 611–612.

Bullen T. D. and Kendall C. (1991) $^{87}Sr/^{86}Sr$ and ^{13}C as tracers of interstream and intrastorm variations in water flowpaths, Catoctim Mountain, MD. *Trans. Am. Geophys. Union* **72**, 218.

Bullen T. D. and Kendall C. (1998) Tracing of weathering reactions and water flowpaths: a multi-isotope approach. In *Isotope Tracers in Catchment Hydrology* (eds. C. Kendall and J. J. McDonnell). Elsevier, Amsterdam, pp. 611–646.

Bullen T. D., Krabbenhoft D. P., and Kendall C. (1996) Kinetic and mineralogic controls on the evolution of groundwater chemistry and $^{87}Sr/^{86}Sr$ in a sandy silicate aquifer, northern Wisconsin. *Geochim. Cosmochim. Acta* **60**, 1807–1821.

Burns D. A. and Kendall C. (2002) Analysis of $\delta^{15}N$ and $\delta^{18}O$ to differentiate NO_3-sources in runoff at two watersheds in the Catskill Mountains of New York. *Water Resour. Res.* **38** 14-1–14-11.

Buttle J. M. (1998) Fundamentals of small catchment hydrology. In *Isotope Tracers in Catchment Hydrology* (eds. C. Kendall and J. J. McDonnell). Elsevier, Amsterdam, pp. 1–49.

Buttle J. M. and Sami K. (1990) Recharge processes during snowmelt: an isotopic and hydrometric investigation. *Hydrol. Process.* **4**, 343–360.

Campbell D. H., Kendall C., Chang C. C. Y., Silva S. R., and Tonnessen K. A. (2002) Pathways for nitrate release from an alpine watershed: determination using $\delta^{15}N$ and $\delta^{18}O$. *Water Resour. Res.* **38** 9-1–9-11.

Caron F. A., Tessier A., Kramer J. R., Schwarcz H. P., and Rees C. E. (1986) Sulfur and oxygen isotopes of sulfate in precipitation and lakewater, Quebec, Canada. *Appl. Geochem.* **1**, 601–606.

Carothers W. W. and Kharaka Y. K. (1980) Stable carbon isotopes of HCO_3^- in oil-field waters- implications

for the origin of CO_2. *Geochim. Cosmochim. Acta* **44**, 323–332.

Cerling T. E., Solomon D. K., Quade J., and Bowman J. R. (1991) On the isotopic composition of carbon in soil carbon dioxide. *Geochim. Cosmochim. Acta* **55**, 3404–3405.

Chang C. C. Y., Kendall C., Silva S. R., Battaglin W. A., and Campbell D. H. (2002) Nitrate stable isotopes: tools for determining nitrate sources and patterns among sites with different land uses in the Mississippi Basin. *Can. J. Fish. Aquat. Sci.* **59** 1874–1885.

Claassen H. C. and Downey J. S. (1995) A model for deuterium and oxygen-18 isotope changes during evergreen interception of snowfall. *Water Resour. Res.* **31**(3), 601–618.

Clark I. D. and Fritz P. (1997) *Environmental Isotopes in Hydrogeology.* CRC Press, New York.

Cook P. and Herczeg A. L. (eds.) (2000) *Environmental Tracers in Subsurface Hydrology*, Kluwer Academic, Norwell, MA.

Cooper L. W., Olsen C. R., Solomon D. K., Larsen I. L., Cook R. B., and Grebmeier J. M. (1991) Stable isotopes of oxygen and natural and fallout radionuclides used for tracing runoff during snowmelt in an arctic watershed. *Water Resour. Res.* **27**, 2171–2179.

Coplen T. B. (1996) New guidelines for reporting stable hydrogen, carbon and oxygen isotope-ratio data. *Geochim. Cosmochim. Acta* **60**, 3359–3360.

Coplen T. B. and Kendall C. (2000) *Stable Hydrogen and Oxygen Isotope Ratios for Selected Sites of the US Geological Survey's Nasqan and Benchmark Surface-water Networks.* US Geol. Surv., Open-file Report. 00-160, 424.

Coplen T. B., Herczeg A. L., and Barnes C. (2000) Isotope engineering—using stable isotopes of the water molecule to solve practical problems. In *Environmental Tracers in Subsurface Hydrology* (eds. P. Cook and A. Herzceg). Kluwer Academic, Norwell, MA, pp. 79–110.

Craig H. (1961a) Isotopic variations in meteoric waters. *Science* **133**, 1702–1703.

Craig H. (1961b) Standard for reporting concentrations of deuterium and oxygen-18 in natural waters. *Science* **133**, 1833.

Criss R. E. (1999) *Principles of Stable Isotope Distribution.* Oxford University Press, New York, pp. 123–128.

Davidson G. R. (1995) The stable isotopic composition and measurement of carbon in soil CO_2. *Geochim. Cosmochim. Acta* **59**, 2485–2489.

Deák J. and Coplen T. B. (1996) Identification of Pleistocene and Holocene groundwaters in Hungary using oxygen and hydrogen isotopic ratios. In *Isotopes in Water Resources Management*, IAEA, Vienna, vol. 1, 438p.

Deines P. (1980) The isotopic composition of reduced organic carbon. In *Handbook of Environmental Isotope Geochemistry* (eds. P. Fritz and J. Ch. Fontes). Elsevier, Amsterdam, vol. 1, pp. 329–406.

Deines P., Langmuir D., and Harmon R. S. (1974) Stable carbon isotope ratios and the existence of a gas phase in the evolution of carbonate ground waters. *Geochim. Cosmochim. Acta* **38**, 1147–1164.

DeWalle D. R. and Swistock B. E. (1994) Differences in oxygen-18 content of throughfall and rainfall in hardwood and coniferous forests. *Hydrol. Process.* **8**, 75–82.

DeWalle D. R., Swistock B. R., and Sharpe W. E. (1988) Three-component tracer model for stormflow on a small Appalachian forested catchment. *J. Hydrol.* **104**, 301–310.

Doctor D. H. (2002) The hydrogeology of the classical karst (Kras) aquifer of southwestern Slovenia. PhD Dissertation, University of Minnesota.

Dogramaci S. S., Herczeg A. L., Schiff S. L., and Bone Y. (2001) Controls on $\delta^{34}S$ and $\delta^{18}O$ of dissolved SO_4 in aquifers of the Murray Basin, Australia and their use as indicators of flow processes. *Appl. Geochem.* **16**, 475–488.

Dörr H. and Münnich K. O. (1980) Carbon-14 and carbon-13 in soil CO_2. *Radiocarbon* **22**, 909–918.

Durka W., Schulze E. D., Gebauer G., and Voerkelius S. (1994) Effects of forest decline on uptake and leaching of deposited nitrate determined from ^{15}N and ^{18}O measurements. *Nature* 372, 765–767.

Dutton A. R. and Simpkins W. W. (1989) Isotopic evidence for paleohydrologic evolution of ground-water flowpaths, southern Great Plains, United States. *Geology* 17(7), 653–656.

Faure G. (1986) *Principles of Isotope Geology*, 2nd edn. Wiley, New York, 589p.

Fogel M. L. and Cifuentes L. A. (1993) Isotope fractionation during primary production. In *Organic Geochemistry* (eds. M. H. Engel and S. A. Macko). Plenum, New York, pp. 73–98.

Fogg G. E., Rolston D. E., Decker D. L., Louie D. T., and Grismer M. E. (1998) Spatial variation in nitrogen isotope values beneath nitrate contamination sources. *Ground Water* 36, 418–426.

Fontes J.-C. (1981) Palaeowaters. In *Stable Isotope Hydrology—Deuterium and Oxygen-18 in the Water Cycle*. Technical Reports Series no. 2101. IAEA, Vienna, pp. 273–302.

Fontes J.-C., Andrews J. N., Edmunds W. M., Guerre A., and Travi Y. (1991) Paleorecharge by the Niger River (Mali) deduced from groundwater geochemistry. *Water Resour. Res.* 27(2), 199–214.

Fontes J.-C., Gasse F., and Andrews J. N. (1993) Climatic conditions of Holocene groundwater recharge in the Sahel zone of Africa. In *Isotope Techniques in the Study of Past and Current Environmental Changes in the Hydrosphere and the Atmosphere*. IAEA, Vienna, pp. 271–292.

Friedman I., Redfield A. C., Schoen B., and Harris J. (1964) The variation of the deuterium content of natural waters in the hydrologic cycle. *Rev. Geophys.* 2, 1–124.

Fritz P. (1981) River waters. In *Stable Isotope Hydrology: Deuterium and Oxygen-18 in the Water Cycle*. IAEA, Vienna, pp. 177–201.

Fritz P., Basharmal G. M., Drimmie R. J., Ibsen J., and Qureshi R. M. (1989) Oxygen isotope exchange between sulphate and water during bacterial reduction of sulphate. *Chem. Geol.* 79, 99–105.

Fry B. (1986) Stable sulphur isotopic distributions and sulphate reduction in lake-sediments of the Adirondack Mountains, New York. *Biogeochemistry* 2(4), 329–343.

Fry B. (1991) Stable isotope diagrams of freshwater foodwebs. *Ecology* 72, 2293–2297.

Fry B., Cox J., Gest H., and Hayes J. M. (1986) Discrimination between ^{34}S and ^{32}S during bacterial metabolism of inorganic sulfur compounds. *J. Bacteriol.* 165, 328–330.

Fry B., Ruf W., Gest H., and Hayes J. M. (1988) Sulfur isotope effects associated with oxidation of sulfide by O_2 in aqueous solution. *Chem. Geol.* 73, 205–210.

Fuller R. D., Mitchell M. J., Krouse H. R., Syskowski B. J., and Driscoll C. T. (1986) Stable sulfur isotope ratios as a tool for interpreting ecosystem sulfur dynamics. *Water Air Soil Pollut.* 28, 163–171.

Garrels R. M. and Mackenzie F. T. (1971) *Evolution of Sedimentary Rocks*. W. W. Norton.

Garten C. T., Jr. and Hanson P. J. (1990) Foliar retention of ^{15}N-nitrate and ^{15}N-ammonium by red maple (Acer rubrum) and white oak (Quercus alba) leaves from simulated rain. *Environ. Exp. Bot.* 30, 33–342.

Gat J. R. (1980) The isotopes of hydrogen and oxygen in precipitation. In *Handbook of Environmental Isotope Geochemistry* (eds. P. Fritz and J. Ch. Fontes). Elsevier, Amsterdam, pp. 21–47.

Gat J. R. and Gonfiantini R. (eds.) (1981) *Stable Isotope Hydrology—Deuterium and Oxygen-18 in the Water Cycle*. Technical Reports Series #210. IAEA, Vienna, 337pp.

Gat J. R. and Tzur Y. (1967) Modification of the isotopic composition of rainwater by processes which occur before groundwater recharge. *Isotope Hydrol.: Proc. Symp.* IAEA, Vienna, pp. 49–60.

Genereux D. P. (1998) Quantifying uncertainty in tracer-based hydrograph separations. *Water Resour. Res.* 34, 915–920.

Genereux D. P. and Hemond H. F. (1990) Naturally occurring radon 222 as a tracer for stream flow generation: steady state methodology and field example. *Water Resour. Res.* 26, 3065–3075.

Genereux D. P. and Hooper R. P. (1998) Oxygen and hydrogen isotopes in rainfall-runoff studies. In *Isotope Tracers in Catchment Hydrology* (eds. C. Kendall and J. J. McDonnell). Elsevier, Amsterdam, pp. 319–346.

Genereux D. P., Hemond H. F., and Mulholland P. J. (1993) Use of radon-222 and calcium as tracers in a three-end-member mixing model for stream flow generation on the west fork of Walker branch watershed. *J. Hydrol.* 142, 167–211.

Ging P. B., Lee R. W., and Silva S. R. (1996) Water chemistry of Shoal Creek and Waller Creek, Austin Texas, and potential sources of nitrate. U.S. Geological Survey Water Resources Investigations, Rep. #96-4167.

Goldhaber M. B. and Kaplan I. R. (1974) The sulphur cycle. In *The Sea* (ed. E. D. Goldberg). Wiley, vol. 5, pp. 569–655.

Gormly J. R. and Spalding R. F. (1979) Sources and concentrations of nitrate-nitrogen in ground water of the Central Platte region, Nebraska. *Ground Water* 17, 291–301.

Harris D. M., McDonnell J. J., and Rodhe A. (1995) Hydrograph separation using continuous open-system isotope mixing. *Water Resour. Res.* 31, 157–171.

Heaton T. H. E. (1990) ^{15}N/^{14}N ratios of NO_x from vehicle engines and coal-fired power stations. *Tellus* 42B, 304–307.

Hélie J.-F., Hillaipre-Marcel C., and Rondeau B. (2002) Seasonal changes in the sources and fluxes of dissolved inorganic carbon through the St. Lawrence River—isotopic and chemical constraint. *Chem. Geol.* 186, 117–138.

Hesslein R. H., Capel M. J., and Fox D. E. (1988) Stable isotopes in sulfate in the inputs and outputs of a Canadian watershed. *Biogeochemistry* 5, 263–273.

Hewlett J. D. and Hibbert A. R. (1967) Factors affecting the response of small watersheds to precipitation in humid areas. *Proc. 1st Int. Symp. Forest Hydrol.* 275–290.

Hinton M. J., Schiff S. L., and English M. C. (1994) Examining the contributions of glacial till water to storm runoff using two-and three-component hydrograph separations. *Water Resour. Res.* 30, 983–993.

Hinton M. J., Schiff S. L., and English M. C. (1998) Sources and flowpaths of dissolved organic carbon in two forested watersheds of the Precambrian Shield. *Biogeochemistry* 41, 175–197.

Holser W. T. and Kaplan I. R. (1966) Isotope geochemistry of sedimentary sulfates. *Chem. Geol.* 1, 93–135.

Holt B. D. and Kumar R. (1991) Oxygen isotope fractionation for understanding the sulphur cycle. In *Stable Isotopes: Natural and Anthropogenic Sulphur in the Environment*, SCOPE 43 (Scientific Committee on Problems of the Environment) (eds. H. R. Krouse and V. A. Grinenko). Wiley, pp. 27–41.

Holt B. D., Cunningham P. T., and Kumar R. (1981) Oxygen isotopy of atmospheric sulfates. *Environ. Sci. Tech.* 15, 804–808.

Hooper R. P. and Shoemaker C. A. (1986) A comparison of chemical and isotopic hydrograph separation. *Water Resour. Res.* 22, 1444–1454.

Hooper R. P., Christophersen N., and Peters N. E. (1990) Modelling streamwater chemistry as a mixture of soilwater end-members—an application to the Panolan Mountain catchment, Georgia, USA. *J. Hydrol.* 116, 321–343.

Horibe Y., Shigehara K., and Takakuwa Y. (1973) Isotope separation factors of carbon dioxide-water system and isotopic composition of atmospheric oxygen. *J. Geophys. Res.* 78, 2625–2629.

Horita J. (1989) Analytical aspects of stable isotopes in brines. *Chem. Geol.* 79, 107–112.

Hübner H. (1986) Isotope effects of nitrogen in the soil and biosphere. In *Handbook of Environmental Isotope Geochemistry, 2b, The Terrestrial Environment* (eds. P. Fritz and J. C. Fontes). Elsevier, Amsterdam, pp. 361–425.

Johnson J. C. and Thiemens M. (1997) The isotopic composition of tropospheric ozone in three environments. *J. Geophys. Res.* **102**, 25395–25404.

Johnson C. A., Mast M. A., and Kester C. L. (2001) Use of $^{17}O/^{16}O$ to trace atmospherically deposited sulfate in surface waters: a case study in alpine watersheds in the Rocky Mountains. *Geophys. Res. Lett.* **28**, 4483–4486.

Junk G. and Svec H. (1958) The absolute abundance of the nitrogen isotopes in the atmosphere and compressed gas from various sources. *Geochim. Cosmochim. Acta* **14**, 234–243.

Kaplan I. R. and Rittenberg S. C. (1964) Microbiological fractionation of sulphur isotopes. *J. Gen. Microbiol.* **34**, 195–212.

Karamanos E. E., Voroney R. P., and Rennie D. A. (1981) Variation in natural ^{15}N abundance of central Saskatchewan soils. *Soil. Sci. Soc. Am. J.* **45**, 826–828.

Karim A. and Veizer J. (2000) Weathering processes in the Indus River Basin: implications from riverine carbon, sulfur, oxygen and strontium isotopes. *Chem. Geol.* **170**, 153–177.

Kendall C. (1993) *Impact of Isotopic Heterogeneity in Shallow Systems on Stormflow Generation*. PhD Dissertation, University of Maryland, College Park.

Kendall C. (1998) Tracing nitrogen sources and cycling in catchments. In *Isotope Tracers in Catchment Hydrology* (eds. C. Kendall and J. J. McDonnell). Elsevier, Amsterdam, pp. 519–576.

Kendall C. and Aravena R. (2000) Nitrate isotopes in groundwater systems. In *Environmental Tracers in Subsurface Hydrology* (eds. P. Cook and A. Herzceg). Kluwer Academic, Norwell, MA, pp. 261–298.

Kendall C. and Caldwell E. A. (1998) Fundamentals of isotope geochemistry. In *Isotope Tracers in Catchment Hydrology* (eds. C. Kendall and J. J. McDonnell). Elsevier, Amsterdam, pp. 51–86.

Kendall C. and Coplen T. B. (2001) Distribution of oxygen-18 and deuterium in river waters across the United States. *Hydrol. Process.* **15**, 1363–1393.

Kendall C. and McDonnell J. J. (1993) Effect of intrastorm heterogeneities of rainfall, soil water and groundwater on runoff modelling. In *Tracers in Hydrology*, Int. Assoc. Hydrol. Sci. Publ. #215, July 11–23, 1993 (eds. N. E. Peters *et al.*) Yokohama, Japan, pp. 41–49.

Kendall C. and McDonnell J. J. (eds.) (1998) *Isotope Tracers in Catchment Hydrology*. Elsevier, 839pp. http://www.rcamnl.wr.usgs.gov/isoig/isopubs/itchinfo.html.

Kendall C., Mast M. A., and Rice K. C. (1992) Tracing watershed weathering reactions with $\delta^{13}C$. In *Water–Rock Interaction*, Proc. 7th Int. Symp., Park City, Utah (eds. Y. K. Kharaka and A. S. Maest). Balkema, Rotterdam, pp. 569–572.

Kendall C., Sklash M. G., and Bullen T. D. (1995a) Isotope tracers of water and solute sources in catchments. In *Solute Modelling in Catchment Systems* (ed. S. Trudgill). Wiley, Chichester, UK, pp. 261–303.

Kendall C., Campbell D. H., Burns D. A., Shanley J. B., Silva S. R., and Chang C. C. Y. (1995b) Tracing sources of nitrate in snowmelt runoff using the oxygen and nitrogen isotopic compositions of nitrate. In *Biogeochemistry of Seasonally Snow-covered Catchments*, Proc., July 3–14, 1995 (eds. K. Tonnessen, M. Williams, and M. Tranter). Int. Assoc. Hydrol. Sci., Boulder Co., pp. 339–347.

Kendall C., McDonnell J. J., and Gu W. (2001a) A look inside 'black box' hydrograph separation models: a study at the Hydrohill catchment. *Hydrol. Process.* **15**, 1877–1902.

Kendall C., Silva S. R., and Kelly V. J. (2001b) Carbon and nitrogen isotopic compositions of particulate organic matter in four large river systems across the United States. *Hydrol. Process.* **15**, 1301–1346.

Kennedy V. C., Kendall C., Zellweger G. W., Wyermann T. A., and Avanzino R. A. (1986) Determination of the components of stormflow using water chemistry and environ-mental isotopes, Mattole River Basin, California. *J. Hydrol.* **84**, 107–140.

Kester C. L., Johnson C. A., Mast M. A., Clow D. W., and Michel R. L. (2002) Tracing atmospheric sulfate through a subalpine ecosystem using ^{17}O and ^{35}S. *EOS, Trans., AGU.* **83**, no. 47.

Kirchner J. W., Feng X., and Neal C. (2001) Catchment-scale advection and dispersion as a mechanism for fractal scaling in stream tracer concentrations. *J. Hydrol.* **254**, 82–101.

Koba K., Tokuchi N., Wada E., Nakajima T., and Iwatsubo G. (1997) Intermittent denitrification: the application of a 15N natural abundance method to a forested ecosystem. *Geochim. Cosmochim. Acta* **61**, 5043–5050.

Krabbenhoft D. P., Bullen T. D., and Kendall C. (1992) Isotopic indicators of groundwater flow paths in a northern Wisconsin aquifer. *AGU Trans.* **73**, 191.

Krabbenhoft D. P., Bullen T. D., and Kendall C. (1993) Use of multiple isotope tracers as monitors of groundwater-lake interaction. *AGU Trans.* **73**, 191.

Kreitler C. W. (1975) Determining the source of nitrate in groundwater by nitrogen isotope studies: Austin, Texas, University of Texas, Austin, Bureau Econ. Geol. Rep. Inv. #83.

Kreitler C. W. (1979) Nitrogen-isotope ratio studies of soils and groundwater nitrate from alluvial fan aquifers in Texas. *J. Hydrol.* **42**, 147–170.

Kreitler C. W. and Jones D. C. (1975) Natural soil nitrate: the cause of the nitrate contamination of groundwater in Runnels County, Texas. *Ground Water* **13**, 53–61.

Kreitler C. W., Ragone S. E., and Katz B. G. (1978) $^{15}N/^{14}N$ ratios of ground-water nitrate, Long Island, NY. *Ground Water* **16**, 404–409.

Krouse H. R. (1980) Sulfur isotopes in our environment. In *Handbook of Environmental Isotope Geochemistry* (eds. P. Fritz and J. Ch. Fontes). Elsevier, Amsterdam, pp. 435–471.

Krouse H. R. and Grinenko V. A. (eds.) (1991) *Stable Isotopes: Natural and Anthropogenic Sulphur in the Environment*, SCOPE 43 (Scientific Committee on Problems of the Environment). Wiley, UK.

Krouse H. R. and Mayer B. (2000) Sulphur and oxygen isotopes in sulphate. In *Environmental Tracers in Subsurface Hydrology* (eds. P. Cook and A. Herzceg). Kluwer Academic, Norwell, MA, pp. 195–231.

Krouse H. R., Legge A., and Brown H. M. (1984) Sulphur gas emissions in the boreal forest: the West Whitecourt case study V: stable sulfur isotopes. *Water Air Soil Pollut.* **22**, 321–347.

Lee C. C.-W. and Thiemens M. H. (2001) Use of $^{17}O/^{16}O$ to trace atmospherically deposited sulfate in surface waters: a case study in alpine watersheds in the Rocky Mountains. *Geophys. Res. Lett.* **28**, 4483–4486.

Lee C. C.-W., Savarino J., and Thiemens M. H. (2001) Mass independent oxygen isotopic composition of atmospheric sulfate: origin and implications for the present and past atmosphere of Earth and Mars. *Geophys. Res. Lett* **28**, 1783–1786.

Legge A. H. and Krouse H. R. (1992) An assessment of the environmental fate of industrial sulphur in a temperate pine forest ecosystem (Paper 1U22B.01). In *Critical Issues in the Global Environment*, vol. 5. Ninth World Clean Air Congress Towards Year 2000.

Likens G. E., Wright R. F., Galloway J. N., and Butler T. J. (1979) Acid rain. *Sci. Am.* **241**, 43–50.

Lloyd R. M. (1968) Oxygen isotope behavior in the sulfate-water system. *J. Geophys. Res.* **73**, 6099–6110.

Macko S. A. and Ostrom N. E. (1994) Pollution studies using nitrogen isotopes. In *Stable Isotopes in Ecology and Environmental Science* (eds. K. Lajtha and R. M. Michener). Blackwell, Oxford, pp. 45–62.

MacNamara J. and Thode H. G. (1950) Comparison of the isotopic composition of terrestrial and meteoritic sulfur. *Phys. Rev.* **78**, 307–308.

Maloszewski P. and Zuber A. (1996) Lumped parameter models for interpretation of environmental tracer data. In *Manual on Mathematical Models in Isotope Hydrologeology* (IAEA TECHDOC 910). IAEA, Vienna, pp. 9–58.

Mariotti A., Landreau A., and Simon B. (1988) ^{15}N isotope biogeochemistry and natural denitrification process in groundwater: application to the chalk aquifer of northern France. *Geochim. Cosmochim. Acta* 52, 1869–1878.

Mariotti A., Pierre D., Vedy J. C., Bruckert S., and Guillemot J. (1980) The abundance of natural nitrogen 15 in the organic matter of soils along an altitudinal gradient (Chablais, Haute Savoie, France). *Catena* 7, 293–300.

Mayer B., Fritz P., and Krouse H. R. (1992) Sulphur isotope discrimination during sulphur transformations in aerated forest soils. In *Workshop Proceedings on Sulphur Transformations in Soil Ecosystems* (eds. M. J. Hendry and H. R. Krouse). National Hydrology Research Symposium No. 11, Saskatoon, Sakatchewan, November 5–7, 1992, pp. 161–172.

Mayer B., Krouse H. R., Fritz P., Prietzel J., and Rehfuess K. E. (1993) Evaluation of biogeochemical sulfur transformations in forest soils by chemical and isotope data. In *Tracers in Hydrology*, IAHS Publ. No. 215, pp. 65–72.

Mayer B., Bollwerk S. M., Mansfeldt T., Hütter B., and Vezier J. (2001) The oxygen isotopic composition of nitrate generated by nitrification in acid forest floors. *Geochim. Cosmochim. Acta* 65(16), 2743–2756.

Mayer B., Boyer E., Goodale C., Jawoarski N., van Bremen N., Howarth R., Seitzinger S., Billen G., Lajtha K., Nadelhoffer K., Van Dam D., Hetling L., Nosil M., Paustian K., and Alexander R. (2002) Sources of nitrate in rivers draining 16 major watersheds in the northeastern US: isotopic constraints. *Biochemistry* 57/58, 171–192.

McDonnell J. J. and Kendall C. (1992) Isotope tracers in hydrology—report to the hydrology section. *EOS, Trans., AGU* 73, 260–261.

McDonnell J. J., Bonell M., Stewart M. K., and Pearce A. J. (1990) Deuterium variations in storm rainfall: implications for stream hydrograph separations. *Water Resour. Res.* 26, 455–458.

McDonnell J. J., Stewart M. K., and Owens I. F. (1991) Effect of catchment-scale subsurface mixing on stream isotopic response. *Water Resour. Res.* 27, 3065–3073.

McDonnell J. J., Rowe L., and Stewart M. (1999) A combined tracer-hydrometric approach to assessing the effects of catchment scale on water flowpaths, source and age. *Int. Assoc. Hydrol. Sci. Publ.* 258, 265–274.

McGlynn B., McDonnell J. J., Shanley J., and Kendall C. (1999) Riparian zone flowpath dynamics. *J. Hydrol.* 222, 75–92.

McGlynn B., McDonnell J., Stewart M., and Seibert J. (2003) On the relationships between catchment scale and streamwater mean residence time. *Hydrol. Process.* 17, 175–181.

McGuire K. J., DeWalle D. R., and Gburek W. J. (2002) Evaluation of mean residence time in subsurface waters using oxygen-18 fluctuations during drought conditions in the Mid-Appalachians. *J. Hydrol.* 261, 132–149.

Mengis M., Schiff S. L., Harris M., English M. C., Aravena R., Elgood R. J., and MacLean A. (1999) Multiple geochemical and isotopic approaches for assessing ground water NO_3-elimination in a riparian zone. *Ground Water* 37, 448–457.

Michalski G., Scott Z., Kaibling M., Thiemens M. H. First measurements and modeling of $\Delta^{17}O$ in atmospheric nitrate: application for evaluation of nitrogen deposition and relevance to paleoclimate studies. *GRL* (in review).

Michalski G., Savarino J., Bohlke J. K., and Thiemans M. (2002) Determination of the total oxygen isotopic composition of nitrate and the calibration of a Del ^{17}O nitrate reference material. *Anal. Chem.* 74 4989–4993.

Michel R. L. (1989) Tritium deposition over the continental United States, 1953–1983. In *Atmospheric Deposition*, International Association of Hydrological Sciences, Oxford, UK, pp. 109–115.

Michel R. L. (1992) Residence times in river basins as determined by analysis of long-term tritium records. *J. Hydrol.* 130, 367–378.

Michel R. L., Campbell D. H., Clow D. W., and Turk J. T. (2000) Timescales for migration of atmospherically derived sulfate through an alpine/subalpine watershed, Loch Vale, Colorado. *Water Resour. Res.* 36, 27–36.

Michel R. L., Turk J. T., Campbell D. H., and Mast M. A. (2002) Use of natural 35S to trace sulphate cycling in small lakes, Flattops wilderness area, Colorado, USA. *Water Air Soil Pollut. Focus* 2/2, 5–18.

Migdisov A. A., Ronov A. B., and Grinenko V. A. (1983) The sulphur cycle in the lithosphere. In *The Global Biogeochemical Sulphur Cycle*, SCOPE 19 (eds. M. V. Ivanov and J. R. Freney). Wiley, pp. 25–95.

Mills A. L. (1988) Variations in the delta C-13 of stream bicarbonate: implications for sources of alkalinity. MS Thesis, George Washington University.

Mitchell M. J., Driscoll C. T., Fuller R. D., David M. B., and Likens G. E. (1989) Effect of whole-tree harvesting on the sulfur dynamics of a forest soil. *Soil Sci. Soc. Am.* 53, 933–940.

Mitchell M. J., Burke M. K., and Shepard J. P. (1992a) Seasonal and spatial patterns of S, Ca, and N dynamics of a northern hardwood forest ecosystem. *Biogeochemistry* 17, 165–189.

Mitchell M. J., David M. B., and Harrison R. B. (1992b) Sulfur dynamics in forest ecosystems. In *Sulfur Cycling on the Continents*, SCOPE 48 (eds. R. W. Howarth, J. W. B. Stewart, and M. V. Ivanov). Wiley, pp. 215–254, Chap. 9.

Mitchell M. J., David M. B., Fernandez I. J., Fuller R. D., Nadelhoffer K., Rustad L. E., and Stam A. C. (1994) Response of buried mineral soil-bags to three years of experimental acidification of a forest ecosystem. *Soil Sci. Soc. Am. J.* 58, 556–563.

Mitchell M. J., Krouse H. R., Mayer B., Stam A. C., and Zhang Y. (1998) Use of stable isotopes in evaluating sulfur biogeochemistry of forest ecosystems. In *Isotope Tracers in Catchment Hydrology* (eds. C. Kendall and J. J. McDonnell). Elsevier, Amsterdam, pp. 489–518.

Mizutani Y. and Rafter T. A. (1969) Oxygen isotopic composition of sulphates: Part 4. Bacterial fractionation of oxygen isotopes in the reduction of sulphate and in the oxidation of sulphur. *NZ J. Sci.* 12, 60–67.

Mook W. G. (1968) Geochemistry of the stable carbon and oxygen isotopes of natural waters in the Netherlands. PhD Thesis, University of Groningen, Netherlands.

Mook W. G. and Tan F. C. (1991) Stable carbon isotopes in rivers and estuaries. In *Biogeochemistry of Major World Rivers* (eds. E. T. Degens, S. Kempe, and J. E. Richey). Wiley, pp. 245–264.

Nadelhoffer K. J. and Fry B. (1988) Controls on natural nitrogen-15 and carbon-13 abundances in forest soil organic matter. *Soil Sci. Soc. Am. J.* 52, 1633–1640.

Nadelhoffer K. J., Aber J. D., and Melillo J. M. (1988) Seasonal patterns of ammonium and nitrate uptake in nine temperate forest ecosystems. *Plant Soil* 80, 321–335.

Novak M., Michel R. L., Prechova E., and Stepanovan M. (2003) The missing flux in a 35S budget of a small catchment. *Water Air Soil Pollut. Focus* (in press).

Nriagu J. O. and Coker R. D. (1978) Isotopic composition of sulphur in precipitation within the Great Lakes Basin. *Tellus* 30, 365–375.

Nriagu J. O., Rees C. E., Mekhtiyeva V. L., Yu A., Lein P., Fritz R. J., Drimmie R. G., Pankina R. W., Robinson R. W., and Krouse H. R. (1991) Hydrosphere. In *Stable Isotopes—Natural and Anthropogenic Sulphur in the Environment*, Chap. 6, SCOPE 43 (Scientific Committee on Problems of the Environment) (eds. H. R. Krouse and V. A. Grinenko). Wiley, UK, pp. 177–265.

Ogunkoya O. O. and Jenkins A. (1991) Analysis of runoff pathways and flow distributions using deuterium and stream chemistry. *Hydrol. Process.* 5, 271–282.

Ogunkoya O. O. and Jenkins A. (1993) Analysis of storm hydrograph and flow pathways using a three-component hydrograph separation model. *J. Hydrol.* **142**, 71–88.

Palmer S. M., Hope D., Billett M. F., Dawson J. J. C., and Bryant C. L. (2001) Sources of organic and inorganic carbon in a headwater stream: evidence from carbon isotope studies. *Biogeochemistry* **52**, 321–338.

Panno S. V., Hackley K. C., Hwang H. H., and Kelly W. R. (2001) Determination of the sources of nitrate contamination in karst springs using isotopic and chemical indicators. *Chem. Geol.* **179**, 113–128.

Pardo L. H., Kendall C., Pett-Ridge J., and Chang C. C. Y. Evaluating the source of streamwater nitrate using ^{15}N and ^{18}O in nitrate in two watersheds in New Hampshire, *Hydrol. Processes* (in press).

Park R. and Epstein S. (1961) Metabolic fractionation of ^{13}C and ^{12}C in plants. *Plant Physiol.* **36**, 133–138.

Pearce A. J., Stewart M. K., and Sklash M. G. (1986) Storm runoff generation in humid headwater catchments: 1. Where does the water come from? *Water Resour. Res.* **22**, 1263–1272.

Pearson F. J. and Rightmire C. T. (1980) Sulfur and oxygen isotopes in aqueous sulfur compounds. In *Handbook of Environmental Isotope Geochemistry* (eds. P. Fritz and J.-C. Fontes). Elsevier, Amsterdam, pp. 179–226.

Peters B. (1959) Cosmic-ray produced radioactive isotopes as tracers for studying large-scale atmospheric circulation. *J. Atmos. Terr. Phys.* **13**, 351–370.

Peters N. E. (1994) Hydrologic studies. In *Biogeochemistry of Small Catchments: A Tool for Environmental Research*, SCOPE Report 51 (eds. B. Moldan and J. Cerny). Wiley, UK, chap. 9, pp. 207–228.

Pionke H. B. and DeWalle D. R. (1992) Intra- and inter-storm ^{18}O trends for selected rainstorms in Pennsylvania. *J. Hydrol.* **138**, 131–143.

Plummer L. N., Parkhurst D. L., and Thorstenson D. C. (1983) Development of reaction models for ground-water systems. *Geochim. Cosmochim. Acta* **47**, 665–686.

Plummer L. N., Prestemon E. C., and Parkhurst D. L. (1991) An Interactive Code (NETPATH) for Modelling Net Geochemical Reactions along a Flow Path. *USGS Water-resources Inves. Report* 91-4078, 227pp.

Plummer L. N., Michel R. L., Thurman E. M., and Glynn P. D. (1993) Environmental tracers for age dating young ground water. In *Regional Ground-water Quality* (ed. W. M. Alley). V. N. Reinhold, New York, pp. 255–294.

Rau G. (1979) Carbon-13 depletion in a subalpine lake: carbon flow implications. *Science* **201**, 901–902.

Rice K. C. and Hornberger G. M. (1998) Comparison of hydrochemical traccers to estimate source contributions to peak flow in a small, forested, headwater catchment. *Water Resour. Res.* **34**, 1755–1766.

Rightmire C. T. (1978) Seasonal variations in pCO$_2$ and ^{13}C of soil atmosphere. *Water Resour. Res.* **14**, 691–692.

Robson A., Beven K. J., and Neal C. (1992) Towards identifying sources of subsurface flow: a comparison of components identified by a physically based runoff model and those determined by chemical mixing techniques. *Hydrol. Process.* **6**, 199–214.

Rodhe A. (1987) The origin of stream water traced by oxygen-18. PhD Thesis, Uppsala University, Department, Phys. Geogr., Div. Hydrol., Sweden. Rep Ser. A., No. 41.

Rozanski K., Araguas-Araguas L., and Gonfiantini R. (1992) Relation between long-term trends of oxygen-18 isotope composition of precipitation and climate. *Science* **258**, 981–985.

Salati E., Dall'Olio A., Matsui E., and Gat J. R. (1979) Recycling of water in the Amazon basin: an isotopic study. *Water Resour. Res.* **515**(5), 1250–1258.

Savarino J., Alexander B., Michalski G. M., and Thiemens M. H. (2002) Investigation of the oxygen isotopic composition of nitrate trapped in the Site A Greenland ice core.

Presented at the AGU Spring Meeting, Washington, DC. *EOS, Trans., AGU.*

Saxena R. K. (1986) Estimation of canopy reservoir capacity and oxygen-18 fractionation in throughfall in a pine forest. *Nordic Hydrol.* **17**, 251–260.

Schiff S. L., Aravena R., Trumbore S. E., and Dillon P. J. (1990) Dissolved organic carbon cycling in forested watersheds: a carbon isotope approach. *Water Resour. Res.* **26**, 2949–2957.

Schiff S. L., Aravena R., Trumbore S. E., Hinton M. J., Elgood R., and Dillon P. J. (1997) Export of DOC from forested catchments on the Precambrian shield of central Ontario: clues from ^{13}C and ^{14}C. *Biogeochemistry* **36**, 43–65.

Schreiber M. E., Simo J. A., and Freiberg P. G. (2000) Stratigraphic and geochemical controls on naturally occurring arsenic in groundwater, eastern Wisconsin, USA. *Hydrogeol. J.* **8**, 161–176.

Schroth M. H., Kleikemper J., Bollinger C., Bernasconi S. M., and Zeyer J. (2001) *In situ* assessment of microbial sulfate reduction in a petroleum-contaminated aquifer using push-pull tests and stable sulfur isotope analyses. *J. Contamin. Hydrol.* **51**, 179–195.

Sebilo M., Billen G., Grably M., and Mariotti A. (2003) Isotopic composition of nitrate-nitrogen as a marker of riparian and benthic denitrification at the scale of the whole Seine River system. *Biogeochemistry* **63**, 35–51.

Shanley J. B., Kendall C., Smith T. M., Wolock D. M., and McDonnell J. J. (2001) Controls on old and new water contributions to stream flow at some nested catchments in Vermont, USA. *Hydrol. Process.* **16**, 589–609.

Shearer G. and Kohl D. (1986) N$_2$ fixation in field settings, estimations based on natural 15N abundance. *Austral. J. Plant Physiol.* **13**, 699–757.

Shearer G. and Kohl D. H. (1988) ^{15}N method of estimating N$_2$ fixation. In *Stable Isotopes in Ecological Research* (eds. P. W. Rundel, J. R. Ehleringer, and K. A. Nagy). Springer, New York, pp. 342–374.

Sickman J. Q., Leydecker A., Chang C. C. Y., Kendall C., Melack J. M., Lucero D. M. and Schimel J. (2003) Mechansims uderlying export of N from high-elevation catchments during seasonal transitions. *Biogeochemistry* **64**, 1–24.

Sidle W. C. (2002) $^{18}OSO_4$ and $^{18}OH_2O$ as prospective indicators of elevated arsenic in the Goose River ground-watershed, Maine. *Environ. Geol.* **42**, 350–359.

Silva S. R., Ging P. B., Lee R. W., Ebbert J. C., Tesoriero A. J., and Inkpen E. L. (2002) Forensic applications of nitrogen and oxygen isotopes of nitrate in an urban environment. *Environ. Foren.* **3**, 125–130.

Sklash M. G. (1990) Environmental isotope studies of storm and snowmelt runoff generation. In *Process Studies in Hillslope Hydrology* (eds. M. G. Anderson and T. P. Burt). Wiley, UK, pp. 401–435.

Sklash M. G. and Farvolden R. N. (1979) The role of groundwater in storm runoff. *J. Hydrol.* **43**, 45–65.

Sklash M. G. and Farvolden R. N. (1982) The use of environmental isotopes in the study of high-runoff episodes in streams. In *Isotope Studies of Hydrologic Processes* (eds. E. C. Perry, Jr. and C. W. Montgomery). Northern Illinois University Press, DeKalb, Illinois, pp. 65–73.

Sklash M. G., Farvolden R. N., and Fritz P. (1976) A conceptual model of watershed response to rainfall, developed through the use of oxygen-18 as a natural tracer. *Can. J. Earth Sci.* **13**, 271–283.

Spence M. J., Bottrell S. H., Thornton S. F., and Lerner D. N. (2001) Isotopic modelling of the significance of bacterial sulphate reduction for phenol attenuation in a contaminated aquifer. *J. Contamin. Hydrol.* **53**, 285–304.

St-Jean G. (2003) Automated quantitative and isotopic (^{13}C) analysis of dissolved inorganic carbon and dissolved organic carbon in continuous-flow using a total organic carbon analyzer. *Rapid Commun. Mass Spectrom.* **17**, 419–428.

Stam A. C., Mitchell M. J., Krouse H. R., and Kahl J. S. (1992) Stable sulfur isotopes of sulfate in precipitation and stream solutions in a northern hardwood watershed. *Water Resour. Res.* **28**, 231–236.

Staniaszek P. and Halas S. (1986) Mixing effects of carbonate dissolving waters on chemical and $^{13}C/^{12}C$ compositions. *Nordic Hydrol.* **17**, 93–114.

Stevens C. M. and Rust F. E. (1982) The carbon isotopic composition of atmospheric methane. *J. Geophys. Res.* **87**, 4879–4882.

Stewart M. K. and McDonnell J. J. (1991) Modeling base flow soil water residence times from deuterium concentration. *Water Resour. Res.* **27**(10), 2681–2693.

Stichler W. (1987) Snowcover and snowmelt process studies by means of environmental isotopes. In *Seasonal Snowcovers: Physics, Chemistry, Hydrology* (eds. H. G. Jones and W. J. Orville-Thomas). Reidel, pp. 673–726.

Strauss H. (1997) The isotopic composition of sedimentary sulfur through time. *Palaeogeogr. Paleaoclimatol. Paleaoecol.* **132**, 97–118.

Stute M. and Schlosser P. (2000) Atmospheric noble gases. In *Environmental Tracers in Subsurface Hydrology* (eds. P. Cook and A. Herzceg). Kluwer Academic, Dordrecht.

Sueker J. K., Ryan J. N., Kendall C., and Jarrett R. D. (2000) Determination of hydrological pathways during snowmelt for alpine/subalpine basins, Rocky Mountain National Park, Colorado. *Water Resour. Res.* **36**, 63–75.

Swistock B. R., DeWalle D. R., and Sharpe W. E. (1989) Sources of acidic stormflow in an Appalachian headwater stream. *Water Resour. Res.* **25**, 2139–2147.

Szaran J., Niezgoda H., and Halas S. (1998) New determination of oxygen and sulphur isotope fractionation between gypsum and dissolved sulphate. *RMZ-mater. Geoenviron.* **45**, 180–182.

Taylor B. E. and Wheeler M. C. (1994) Sulfur- and oxygen-isotope geochemistry of acid mine drainage in the western US: Field and experimental studies revisited. In *Environmental Geochemistry of Sulfide Oxidation*, ACS Symposium Series 550 (eds. C. N. Alpers and D. W. Blowes). American Chemical Society, Washington, DC, pp. 481–514.

Taylor B. E., Wheeler M. C., and Nordstrom D. K. (1984) Oxygen and sulfur compositions of sulphate in acid mine drainage: evidence for oxidation mechanisms. *Nature* **308**, 538–541.

Taylor C. B. and Fox V. J. (1996) An isotopic study of dissolved inorganic carbon in the catchment of the Waimakariri River and deep ground water of the North Canterbury Plains, New Zealand. *J. Hydrol.* **186**, 161–190.

Taylor S., Feng X., Kirchner J. W., Osterhuber R., Klaue B., and Renshaw C. E. (2001) Isotopic evolution of a seasonal snowpack and its melt. *Water Resour. Res.* **37**, 759–769.

Thatcher L. L. (1962) *The Distribution of Tritium Fallout in Precipitation Over North America*. International Association of Hydrological Sciences (IAHS), Publication No. 7, Louvain, Belgium, pp. 48–58.

Thurman E. M. (1985) *Organic Geochemistry of Natural Waters*. Martinus Nijhof, Dr W Junk Publishers, Dordrecht.

Toran L. and Harris R. F. (1989) Interpretation of sulfur and oxygen isotopes in biological and abiological sulfide oxidation. *Geochim. Cosmochim. Acta* **53**(9), 2341–2348.

Turk J. T., Campbell D. H., and Spahr N. E. (1993) Use of chemistry and stable sulfur isotopes to determine sources of trends in sulfate of Colorado lakes. *Water Air Soil Pollut.* **67**, 415–431.

Unnikrishna P. V., McDonnell J. J., and Stewart M. L. (1995) Soil water isotopic residence time modelling. In *Solute Modelling in Catchment Systems* (ed. S. T. Trudgill). Wiley, New York, pp. 237–260.

Van Stempvoort D. R. and Krouse H. R. (1994) Controls of ^{18}O in sulfate: a review of experimental data and application to specific environments. In *Environmental Geochemistry of Sulfide Oxidation*, ACS Symp. Ser. 550 (eds. C. N. Alpers and D. W. Blowes). American Chemical Society, Washington, DC, pp. 446–480.

Vogel J. C., Talma A. S., and Heaton T. H. E. (1981) Gaseous nitrogen as evidence for denitrification in groundwater. *J. Hydrol.* **50**, 191–200.

Wadleigh M. A. and Blake D. M. (1999) Tracing sources of atmospheric sulphur using epiphytic lichens. *Environ. Pollut.* **106**, 265–271.

Wadleigh M. A., Schwarcz H. P., and Kramer J. R. (1996) Isotopic evidence for the origin of sulphate in coastal rain. *Tellus* **48B**, 44–59.

Wallis P. M., Hynes H. B. N., and Fritz P. (1979) Sources, transportation, and utilization of dissolved organic matter in groundwater and streams. Sci. Ser. 100, Inland Waters Directorate, Water Quality Branch, Ottawa, Ont.

Wang Y., Huntington T. G., Osher L. J., Wassenaar L. I., Trumbore S. E., Amundson R. G., Harden J. G., McKnight D. M., Schiff S. L., Aiken G. R., Lyons W. B., Aravena R. O., and Baron J. S. (1998) Carbon cycling in terrestrial environments. In *Isotope Tracers in Catchment Hydrology* (eds. C. Kendall and J. J. McDonnell). Elsevier, Amsterdam, pp. 577–610.

Wassenaar L. (1995) Evaluation of the origin and fate of nitrate in the Abbotsford Aquifer using the isotopes of ^{15}N and ^{18}O in NO_3^-. *Appl. Geochem.* **10**, 391–405.

Wassenaar L. I., Aravena R., Fritz P., and Barker J. F. (1991) Controls on the transport and carbon isotopic composition of dissolved organic carbon in a shallow groundwater system, Central Ontario, Canada. *Chem. Geol.* **87**, 39–57.

Weiler R. R. and Nriagu J. O. (1973) Isotopic composition of dissolved inorganic carbon in the Great Lakes. *J. Fish. Res. Board Can.* **35**, 422–430.

Wels C., Cornett R. J., and Lazerte B. D. (1991) Hydrograph separation: a comparison of geochemical and isotopic tracers. *J. Hydrol.* **122**, 253–274.

Wenner D. B., Ketcham P. D., and Dowd J. F. (1991) Stable isotopic composition of waters in a small piedmont watershed. In *Geochemical Society Special Publication No. 3* (eds. H. P. Taylor, Jr., J. R. O'Neil, and I. R. Kaplan), pp. 195–203.

Williard K. W. J. (1999) Factors affecting stream nitrogen concentrations from mid-appalachian forested watersheds. PhD Dissertation, Penn. State University.

Wolterink J. J., Williamson H. J., Jones D. C., Grimshaw T. W., and Holland W. F (1979) Identifying sources of subsurface nitrate pollution with stable nitrogen isotopes. U.S. Environmental Protection Agency, EPA-600/4-79-050, 150p.

Yang C., Telmer K., and Veizer J. (1996) Chemical dynamics of the St-Lawrence riverine system: δD_{H_2O}, $\delta^{18}O_{H_2O}$, $\delta^{13}C_{DIC}$, δ^{34}sulfate, and dissolved $^{87}Sr/^{86}Sr$. *Geochim. Cosmochim. Acta* **60**(5), 851–866.

Yang W., Spencer R. J., and Krouse H. R. (1997) Stable isotope composition of waters and sulfate species therein, Death Valley, California, USA: implications for inflow and sulfate sources, and arid basin climate. *Earth Planet. Sci. Lett.* **147**, 69–82.

Zimmerman U., Ehhalt D., and Munnich K. O. (1967) Soil-water movement and evapotranspiration: changes in the isotopic composition of water. In *Isotopes in Hydrology*. IAEA, Vienna, pp. 567–584.

5.12
Radiogenic Isotopes in Weathering and Hydrology

J. D. Blum

The University of Michigan, Ann Arbor, MI, USA

and

Y. Erel

The Hebrew University, Jerusalem, Israel

5.12.1 INTRODUCTION AND OVERVIEW OF RELEVANT ISOTOPE SYSTEMS

There are a small group of elements that display variations in their isotopic composition, resulting from radioactive decay within minerals over geological timescales. These isotopic variations provide natural fingerprints of rock–water interactions and have been widely utilized in studies of weathering and hydrology. The isotopic systems that have been applied in such studies are dictated by the limited number of radioactive parent–daughter nuclide pairs with half-lives and isotopic abundances that result in measurable differences in daughter isotope ratios among common rocks and minerals. Prior to their application to studies of weathering and hydrology, each of these isotopic systems was utilized in geochronology and petrology. As in the case of their original introduction into geochronology and petrology, isotopic systems with the highest concentrations of daughter isotopes in common rocks and minerals and systems with the largest observed isotopic variations were introduced first and have made the largest impact on our understanding of weathering and hydrologic processes. Although radiogenic isotopes have helped elucidate many important aspects of weathering and hydrology, it is important to note that in almost every case that will be discussed in this chapter, our fundamental understanding of these topics came from studies of variations in the concentrations of major cations and anions. This chapter is a "tools chapter" and thus it will highlight applications of radiogenic isotopes that have added additional insight into a wide spectrum of research areas that are summarized in almost all of the other chapters of this volume.

The first applications of radiogenic isotopes to weathering processes were based on studies that sought to understand the effects of chemical weathering on the geochronology of whole-rock samples and geochronologically important minerals (Goldich and Gast, 1966; Dasch, 1969; Blaxland, 1974; Clauer, 1979, 1981; Clauer *et al.*, 1982); as well as on the observation that radiogenic isotopes are sometimes preferentially released compared to nonradiogenic isotopes of the same element during acid leaching of rocks (Hart and Tilton, 1966; Silver *et al.*, 1984; Erel *et al.*, 1991). A major finding of these investigations was that weathering often results in anomalously young Rb–Sr isochron ages, and discordant Pb–Pb ages. Rubidium is generally retained relative to strontium in whole-rock samples, and in some cases radiogenic strontium and lead are lost preferentially to common strontium and lead from weathered minerals.

The most widely utilized of these isotopic systems is Rb–Sr, followed by U–Pb. The K–Ar system is not directly applicable to most studies of rock–water interaction, because argon is a noble gas, and upon release during mineral weathering mixes with atmospheric argon, limiting its usefulness as a tracer in most weathering applications. Argon and other noble gas isotopes have, however, found important applications in hydrology (see Chapter 5.15). Three other isotopic systems commonly used in geochronology and petrology include Sm–Nd, Lu–Hf, and Re–Os. These parent and daughter elements are in very low abundance and concentrated in trace mineral phases. Sm–Nd, Lu–Hf, and Re–Os have been used in a few weathering studies but have not been utilized extensively in investigations of weathering and hydrology.

The decay of ^{87}Rb to ^{87}Sr has a half-life of 48.8 Gyr, and this radioactive decay results in natural variability in the ^{87}Sr/^{86}Sr ratio in rubidium-bearing minerals (e.g., Blum, 1995). The trace elements rubidium and strontium are geochemically similar to the major elements potassium and calcium, respectively. Therefore, minerals with high K/Ca ratios develop high ^{87}Sr/^{86}Sr ratios over geologic timescales. Once released into the hydrosphere, strontium retains its isotopic composition without significant fractionation by geochemical or biological processes, and is therefore a good tracer for sources and cycling of calcium. The decay of ^{235}U to ^{207}Pb, ^{238}U to ^{206}Pb, and ^{232}Th to ^{208}Pb have half-lives of 0.704 Gyr, 4.47 Gyr, and 14.0 Gyr, respectively, and result in variations in the ^{207}Pb/^{204}Pb, ^{206}Pb/^{204}Pb, and ^{208}Pb/^{204}Pb ratios (e.g., Blum, 1995). Uranium-234 has a half-life of 0.25 Myr and the ratio ^{234}U/^{238}U approaches a constant secular equilibrium value in rocks and minerals if undisturbed for ~1 Myr. Differences in this ratio are often observed in solutions following rock–water interaction and have been used in studies of weathering and hydrology. Uranium and thorium tend to be highly concentrated in the trace accessory minerals such as zircon, monazite, apatite, and sphene, which therefore develop high ^{206}Pb/^{204}Pb, ^{207}Pb/^{204}Pb, and ^{208}Pb/^{204}Pb ratios. Once released into the hydrosphere, lead retains its isotopic composition without significant geochemical or biological fractionation and tends to generally follow the chemistry of iron in soils and aqueous systems (Erel and Morgan, 1992). The use of the U–Th disequilibrium series as a dating tool falls outside the scope of this chapter and is reviewed in Chapters 6.14 and 6.17 as well as Chapter 3.15. The decay of ^{147}Sm to ^{143}Nd, ^{176}Lu to ^{176}Hf, and ^{187}Re to ^{187}Os have half-lives of 106 Gyr, 35.7 Gyr, and 42.3 Gyr, respectively, and result in natural variability in the ^{144}Nd/^{143}Nd, ^{176}Hf/^{177}Hf, and ^{187}Os/^{188}Os ratios (e.g., Blum, 1995). Neodymium is a rare earth element (REE), hafnium is a transition metal with chemical similarities to zirconium, and osmium is a platinum group element. The geochemical behaviors of these elements in the hydrosphere are largely determined by these chemical affinities.

5.12.2 ANALYTICAL METHODS, MIXING EQUATIONS, AND PREVIOUS REVIEWS

The analytical methods and mixing equations used in studies of radiogenic isotopes in weathering and hydrology followed directly from research applications in geochronology, petrology, and cosmochemistry that utilize the same isotopic systems. Analytical methods are thoroughly reviewed in several textbooks on isotope geology, including Faure (1986), Geyh and Schleicher (1990), and Dickin (1995). Mixing equations are reviewed in Faure (1986), Dickin (1995), and Albarède (1995). Several recent books and review articles have summarized some of the analytical methods and mixing equations as they apply specifically to studies of weathering and hydrology (Stille and Shields, 1997; Kendall and Caldwell, 1998; Bullen and Kendall, 1998; Kraemer and Genereux, 1998; Capo *et al.*, 1998; Nimz, 1998; Bierman *et al.*, 1998; Stewart *et al.*, 1998). The reader is referred to these previous publications for analytical methods and mixing equations relevant to radiogenic isotope studies. This chapter reviews applications of the methods to studies of weathering and hydrology.

5.12.3 OVERVIEW OF APPLICATIONS AND ORGANIZATION OF CHAPTER

The most significant scientific contributions of radiogenic isotopes to our understanding of weathering and hydrology can be organized into the three general subject areas: (i) identification of mineral dissolution reactions; (ii) differentiation between atmospheric- and weathering-derived cations in ecosystems; and (iii) tracing of hydrologic flow paths and subsurface mixing. Radiogenic isotopes provide a useful tracer of which mineral(s) are dissolving in mixed-mineral weathering experiments, soil-weathering profiles, and in surface and groundwaters. Radiogenic isotopes have been

particularly useful at differentiating between atmospheric- and weathering-derived inputs of base cations and toxic metals to soils and in the study of their cycling in ecosystems. Radiogenic isotopes have also found extensive applications in hydrology, allowing the isotopic characterization of various water sources, flow paths, and mixing.

Applications of radiogenic isotopes specifically to weathering and hydrology amounted to only a few dozen studies prior to 1990. During the 1990s and early 2000s, radiogenic isotopes were applied to a wide range of investigations and the number of publications expanded to well over two hundred at the time of this writing. In this chapter we will provide an overview of the breadth of applications of radiogenic isotopes, as well as a comprehensive review of the peer-reviewed literature on this topic; abstracts and conference proceedings are generally excluded from inclusion in this chapter. We have highlighted a few studies in each section and reproduced key figures from the literature that provide particularly good examples of how radiogenic isotopes have been used to elucidate geochemical and hydrologic processes.

5.12.4 IDENTIFICATION OF SPECIFIC MINERAL WEATHERING REACTIONS

Mineral dissolution is the process by which minerals react with near-surface water and re-equilibrate thermodynamically to the conditions present near the Earth's surface (see Chapters 5.01–5.07, and 5.18). Weathering moderates the concentration of CO_2 in the atmosphere, releases nutrients, provides cation-exchange capacity in soils, and largely determines the inorganic element geochemistry of surface and groundwaters (see Chapters 5.08, 5.09, 5.16, and 5.17). Weathering rates have been determined following three general methodologies: (i) laboratory dissolution experiments (see Chapter 5.03), (ii) soil profile element depletion studies (see Chapters 5.01, 5.04, and 5.05), and (iii) mass balances on small lithologically simple catchments (see Chapters 5.04 and 5.05). Each of these three types of study has an extensive and mature literature; radiogenic isotopes have been applied to such studies relatively recently, and in each case have elucidated important aspects of weathering processes that were not previously well understood.

5.12.4.1 Laboratory Dissolution Experiments

We begin our discussion of the use of radiogenic isotopes in weathering studies with laboratory experiments, because they provide the simplest and most well-constrained application. Laboratory dissolution experiments using major elements are discussed in detail in Chapter 5.03. Several research groups have integrated radiogenic isotope techniques into these studies, yielding important new information on: (i) the importance of accessory mineral inclusions in what were previously considered "pure" mineral starting materials; (ii) the differential rate of release of cations from contrasting mineralogic sites within single minerals; and (iii) the weathering rate of individual minerals in multi-mineral experiments.

5.12.4.1.1   Strontium isotopes

As far as we are aware, the first laboratory experiments that utilized radiogenic isotopes in the study of mineral dissolution were performed by Zuddas *et al.* (1995) and Seimbille *et al.* (1998). Most mineral dissolution experiments are performed at 25 °C, whereas these experiments were performed at 180 °C and thus could be considered "hydrothermal" experiments. In these experiments powdered granite was reacted with an artificial "hydrothermal fluid" containing major element concentrations near equilibrium with the granitic mineral assemblage. Combining strontium isotopic and major element mixing equations, Seimbille *et al.* (1998) demonstrated that plagioclase dissolved 3–4 times faster than biotite and 10–20 times faster than K-feldspar under these conditions. White *et al.* (1999) performed similar experiments using powdered granite in a flow-through column with deionized water at 5 °C, 17 °C, and 35 °C. Strontium isotope ratios were used to demonstrate that at the lower temperatures strontium release to solution from reacting minerals was not congruent with respect to major cation concentrations. It was proposed that this was due to either the extreme sensitivity of strontium isotope ratios to the dissolution of trace calcium-bearing phases with low $^{87}Sr/^{86}Sr$, the preferential retention of strontium relative to potassium in biotite, or the preferential leaching of strontium with low $^{87}Sr/^{86}Sr$ from plagioclase or K-feldspar. This study suggests that $^{87}Sr/^{86}Sr$ ratios derived from granitoid weathering may be temperature dependent, with lower ratios in warmer climates (White *et al.*, 1999; Figure 1).

The first single-mineral dissolution experiments to utilize radiogenic isotopes investigated the dissolution of the feldspars bytownite, microcline, and albite in flow-through cells at a pH 3 and at 25 °C (Brantley *et al.*, 1998). Solutions from major element experiments (Stillings and Brantley, 1995) were reanalyzed for strontium and rubidium concentrations and $^{87}Sr/^{86}Sr$ ratios with the goal of

comparing strontium release with major element patterns. Strontium release early in the experiments was found to be neither stoichiometric nor isotopically identical to the bulk mineral and was attributed to the dissolution of mineral inclusions in the feldspars and/or to leaching of cations from mineral defect sites. Once the experiments had reached steady-state dissolution, the $^{87}Sr/^{86}Sr$ ratio became equal to that of the bulk material. This study demonstrated the importance of minute trace phases in controlling the $^{87}Sr/^{86}Sr$ released by weathering of freshly exposed minerals.

Taylor *et al.* (2000a,b) used flow-through column reactors to study overall dissolution kinetics and strontium release from biotite, phlogopite, and labradorite at 25 °C. Strontium isotopes elucidated several processes including: (i) the importance of trace calcite inclusions in biotite and phlogopite during the early stages of weathering (Figure 2); (ii) the more rapid release of interlayer cations from biotite and phlogopite compared to octahedral and tetrahedral cations; and (iii) that labradorite dissolution is congruent, with stoichiometric strontium release unaffected by solution saturation state or preferential release from mineral defect sites. A common theme of all these experimental studies is that trace inclusions of calcite and other nonsilicate phases are common in silicate mineral specimens, and strontium isotopes are very sensitive to their presence. In the early stages of weathering, the $^{87}Sr/^{86}Sr$ of weathering solutions may be significantly affected by reactive trace phases and differ from bulk mineral values. However, once a steady state is reached, the dissolution of calcite inclusions is limited by the rate at which the inclusions are exposed to solution by dissolution of the host silicate mineral, and their influence is greatly diminished.

5.12.4.1.2   Lead isotopes

The behavior of lead isotopes during the dissolution of silicate minerals has not been extensively studied in laboratory experiments. To the best of our knowledge, there has been only one attempt to study systematically the release of lead isotopes during the dissolution of silicates. Harlavan and Erel (2002) studied the release of lead and REEs during the dissolution of a crushed granite subjected to sequential leaching with dilute acids, and the results were compared with lead and REE data from soils developed on the same bedrock (Harlavan *et al.*, 1998; Harlavan and Erel, 2002). During the early stages of granitoid dissolution, lead and REEs were preferentially released from some of the accessory

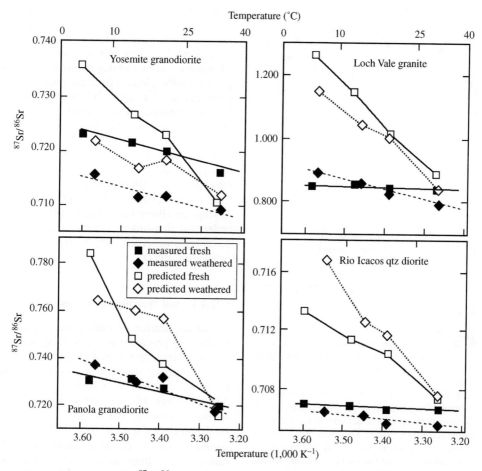

Figure 1 Temperature versus the $^{87}Sr/^{86}Sr$ of effluent from fresh and weathered samples of crushed granitoids (from the four locations noted) during weathering experiments. Predicted values are those based on Na and K concentrations and the assumption of stoichiometric dissolution of plagioclase and biotite. The deviation between predicted and measured $^{87}Sr/^{86}Sr$ values at low temperature demonstrates that $^{87}Sr/^{86}Sr$ cannot be used in this simple way to accurately identify the relative rate of plagioclase to biotite weathering at low temperatures. This is due to either preferential retention of radiogenic Sr, or enhanced dissolution of nonradiogenic Sr from trace Ca-bearing phases such as calcite included in plagioclase or K-feldspar. The data suggest that $^{87}Sr/^{86}Sr$ ratios derived from granitoid weathering are expected to be temperature dependent, with lower ratios in warmer climates (source White *et al.*, 1999).

phases (i.e., allanite, sphene, and apatite), leading to higher $^{206}Pb/^{207}Pb$ and $^{208}Pb/^{207}Pb$ ratios and different REE patterns in solution than in the rock itself. Three stages of rock dissolution were identified (Figure 3). Among the accessory phases, allanite dissolution dominated the first stage and $^{208}Pb/^{207}Pb$ ratios in solution increased and approached the values of allanite. The isotopic ratios of lead and the REE patterns indicated that in the second stage, dissolution of apatite and sphene became more significant. In the third stage, the isotopic ratios of lead and the REE patterns reflected the depletion of accessory phases and the increasing dominance of feldspar dissolution. It was also argued in this study that biotite dissolution was significantly more rapid than hornblende dissolution under the experimental conditions of acid leaching.

5.12.4.1.3 *Uranium isotopes*

The release of uranium isotopes during the dissolution of rocks under laboratory conditions has been studied by several research groups in an attempt to understand the processes that release uranium from host minerals to the hydrologic cycle (Rosholt *et al.*, 1963; Chalov and Merkulova, 1966; Szalay and Samsoni, 1969; Kigoshi, 1971; Kovalev and Malyasova, 1971; Fleischer and Raabe, 1978; Fleischer, 1980, 1982; Moreira-Nordemann, 1980; Zielinski *et al.*, 1981; Michel, 1984; Lathan and Schwarcz, 1987; Eyal and Olander, 1990; Bonotto and Andrews, 1998; Bonotto *et al.*, 2001). These studies were ultimately motivated by the goal of improving geochemical exploration for uranium and better understanding the long-term risks associated with

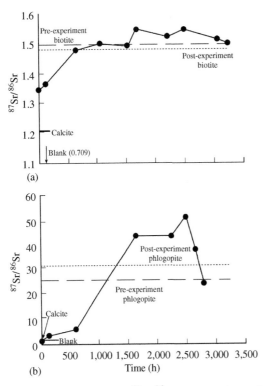

(a)

(b)

Figure 2 Time versus $^{87}Sr/^{86}Sr$ of experimental effluents during (a) biotite and (b) phlogopite dissolution experiments along with bulk mineral, trace calcite, and experimental blank values. The changing $^{87}Sr/^{86}Sr$ through time can be explained by an initial release of Sr from calcite inclusions (source Taylor *et al.*, 2000b).

radioactive waste disposal. The experiments involved leaching of ground rocks with various reagents in an effort to mimic natural weathering conditions. The most important observation of the laboratory studies was that in many cases ^{234}U was preferentially released from the rock into solution compared to ^{235}U and ^{238}U. Several mechanisms have been proposed to account for this observed uranium fractionation. They include α-decay recoil damage to the crystal lattice, resulting in paths for rapid diffusion (Kigoshi, 1971), an enhanced rate of ^{234}U auto-oxidation to U(VI) (Rosholt *et al.*, 1963), and preferential release of ^{234}U from interstitial oxides and cryptocrystalline aggregates (Brown and Silver, 1955). Both zero-order and first-order uranium removal processes have been observed and discussed (Lathan and Schwarcz, 1987).

5.12.4.2 Soil Weathering Studies

The application of radiogenic isotopes in soil weathering profiles is a considerably more mature field than the application to experimental studies. Soil weathering studies using major elements are discussed in detail in Chapters 5.01, 5.04, and 5.05. Studies of radiogenic isotope behavior during weathering began with a comparison of fresh and altered igneous rocks by Dasch (1969) in which it was found that whole-rock $^{87}Sr/^{86}Sr$ ratios were either unaffected or increased in weathered rock as compared to fresh rock.

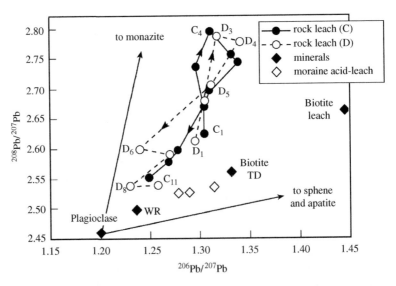

Figure 3 Acid leaches of ground granitoid samples from the Sierra Nevada, California, show three stages of accessory phase dissolution. The initial leach (C_1, D_1) is a mixture of major and accessory phases. The successive leaches show first an increasing influence of monazite and/or allanite dissolution (C_1-C_4, D_1-D_3), followed by an increase in the importance of sphene and apatite (C_4-C_6, D_3-D_4), and finally the depletion of accessory phases and the evolution of the leach composition to a value closer to that of plagioclase (C_6-C_{11}, D_4-D_9). Also plotted are the average $^{206}Pb/^{207}Pb$ and $^{208}Pb/^{207}Pb$ values of whole rock (WR) and mineral separates from the literature (TD = total digest), and the acid leach fractions of soil samples and biotite from the Sierra Nevada, California (source Harlavan and Erel, 2002).

The first measurements of radiogenic isotopes in soils used strontium in order to determine the relative importance of atmospheric dust versus rock weathering as inputs of strontium (and by inference calcium) to forested ecosystems (Graustein and Armstrong, 1983; Gosz *et al.*, 1983). The primary focus of these studies was on forest ecosystem processes rather than mineral weathering *per se*; therefore, a more detailed description of these and other similar studies that followed will be deferred to a later section of this chapter on differentiating atmospheric- from weathering-derived cations in ecosystems.

5.12.4.2.1 Strontium isotopes

Radiogenic isotope studies with a primary focus on mineral weathering in soil profiles began with the use of granitic soil chronosequences developed on glacial moraines and alluvial terraces to investigate changes in mineral weathering patterns through time during soil development. Blum and Erel (1995, 1997) measured the strontium isotopic composition of the ion-exchange complex and bulk soils in six soil profiles developed on parent material of the same composition but ranging in age from 0.40 kyr to 300 kyr in the Wind River Mountains, Wyoming. The $^{87}Sr/^{86}Sr$ of soil digests did not vary appreciably with age, but $^{87}Sr/^{86}Sr$ of the cation-exchange pool dropped systematically with age from 0.795 to 0.711, indicating a strong sensitivity to the early and rapid depletion of highly radiogenic strontium from biotite in the soil parent material (Figure 4). Biotite was estimated to weather eight times faster than plagioclase in the young soils, but five times slower than plagioclase in the oldest soils. An implication of this work was that global glaciations could significantly elevate the average riverine $^{87}Sr/^{86}Sr$ ratio, providing a mechanism linking global glaciations with increases in the marine $^{87}Sr/^{86}Sr$ record (Blum, 1997).

Bullen *et al.* (1997) carried out a strontium isotope study of a granitic soil chronosequence developed on alluvial parent materials 0.2–3,000 kyr in age in the Sierra Nevada Mountains, California. A significant difference between this study and study of Blum and Erel (1995, 1997) was that the glacial till parent materials studied by Blum and Erel (1995, 1997) represented freshly ground mineral fragments, whereas the alluvial parent materials studied by Bullen *et al.* (1997) were sands that had been sorted, winnowed, and partially weathered during transport and deposition by alluvial processes. Bullen *et al.* (1997) found that biotite in the soil parent materials had lost most of its radiogenic strontium during transport. Therefore, while Blum and Erel (1995, 1997) found that the dominant control on the $^{87}Sr/^{86}Sr$

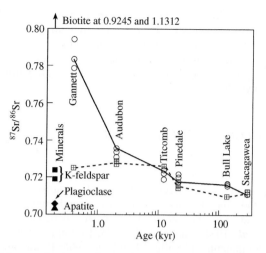

Figure 4 Age of moraines in the Wind River Mountains, Wyoming, on which soils of a chronosequence have formed (with the names of the glacial advances labeled) versus the $^{87}Sr/^{86}Sr$ of the B-horizon soil exchangeable fraction (circles) and C-horizon total soil digests (open squares). Also plotted are analyses of mineral separates from the granitoid bedrock in the area (filled symbols). The elevated $^{87}Sr/^{86}Sr$ in the younger soils was attributed to the release of radiogenic Sr as biotite is altered to form hydrobiotite and vermiculite (source Blum and Erel, 1997).

released by weathering was the weathering of biotite to hydrobiotite and vermiculite, Bullen *et al.* (1997) found that this transformation was already complete in the soil parent materials; therefore, the $^{87}Sr/^{86}Sr$ was controlled dominantly by weathering of K-feldspar and plagioclase. Radiogenic $^{87}Sr/^{86}Sr$ values in some soils were attributed to incongruent leaching of strontium from K-feldspar.

Innocent *et al.* (1997) used strontium isotopes to investigate weathering and ion-exchange processes in tropical laterites developed on basalt in Brazil. This study showed that most of the rock-derived strontium was released in the earliest stages of weathering in the tropical weathering environment, leading to the control of strontium isotope ratios by radiogenic rain waters or groundwater originating on other geological substrates. Yang *et al.* (2000) used the $^{87}Sr/^{86}Sr$ ratio of calcite in Chinese loess paleosols as a proxy for chemical weathering intensity associated with the monsoonal climate. They found a good correlation with magnetic susceptibility and the marine $\delta^{18}O$ record over a 5–150 kyr timescale. In another study, Jacobson *et al.* (2002a) used strontium isotopes along with major elements to study weathering of a Himalayan glacial moraine chronosequence developed on glacial till that was mostly silicate but contained ~1% carbonate. The study used strontium isotopes to demonstrate that carbonate minerals, even in low abundance, can control the dissolved flux of strontium and calcium emanating from stable landforms comprised of mixed silicate and carbonate for tens of thousands of years after exposure of landforms to the weathering environment.

5.12.4.2.2 *Lead isotopes*

Lead isotopes have been extensively used to trace the behavior and transport of lead in soils, to distinguish between natural and anthropogenic lead, and to trace the sources of anthropogenic lead (e.g., Ault *et al.*, 1970; Gast, 1970; Gulson *et al.*, 1981; Bacon *et al.*, 1995; Erel *et al.*, 1997; Steinmann and Stille, 1997; Erel, 1998; Brannvall *et al.*, 2001; Teutsch *et al.*, 2001; Emmanuel and Erel, 2002). Hart and Tilton (1966) were the first to report a systematic difference between the isotopic composition of lead in bulk rocks compared to that released from the same rocks by weathering. The behavior of lead isotopes during rock and mineral weathering was further studied by Erel *et al.* (1994) and Harlavan *et al.* (1998), using the same granitic glacial moraine soil chronosequence in Wyoming studied by Blum and Erel (1995, 1997). The major findings of these studies were that the isotopic composition of lead released by rock weathering changed

systematically with soil age and with the degree of weathering, reflecting preferential initial release of lead from a radiogenic mineral weathering pool (probably composed of easily weathered accessory phases, interstitial oxides, and cryptocrystalline aggregates (Brown and Silver, 1955)). With the progression of chemical weathering intensity, it was observed that a higher proportion of the lead released from the rock originated from plagioclase. Hence, these findings provided a complementary view of the weathering process in soils compared to the view provided by strontium isotope studies (Blum and Erel, 1997).

5.12.4.2.3 *Uranium isotopes*

Variation of the $^{234}U/^{238}U$ activity ratio in soils as a reflection of weathering processes has been studied for more than 30 years (e.g., Rosholt *et al.*, 1963; Rosholt and Bartel, 1969; Stuckless *et al.*, 1977). This topic in general, and the use of $^{234}U/^{238}U$ versus $^{230}Th/^{238}U$ plots as a means of characterizing the course of weathering in soils, is thoroughly summarized by Osmond and Ivanovich (1992; Figure 5). Moreira-Nordemann (1980) used variations in uranium isotope ratios ($^{234}U/^{238}U$) as a proxy for regional weathering rates. In that study measurements of $^{234}U/^{238}U$ values in river water, weathered rock

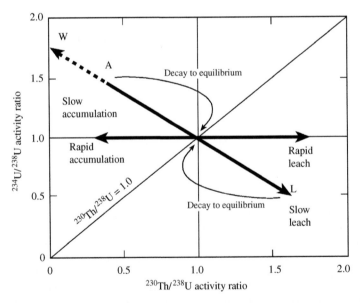

Figure 5 A schematic plot of the $^{234}U/^{238}U$ activity ratio versus the $^{230}Th/^{238}U$ activity ratio in rock and soil samples. On this diagram leaching and precipitation of U results in straight line vectors, while decay toward equilibrium results in curved lines. The point (1, 1) represents equilibrium for both activity ratios. The light diagonal line represents equilibrium between ^{234}U and ^{230}Th; below and to the right the region of U leaching, above and to the left the region of U precipitation (slow accum. and rapid accum.). On this diagram, preferential leaching (rapid and slow) of ^{234}U (because of recoil processes) produces a leached solid with an activity ratio at L and a resulting water with a U activity ratio at W (connected by heavy diagonal line). If the mobilized U is reprecipitated, it will accumulate in the vicinity of A. Rapid U leaching without preferential mobilization of ^{234}U is shown as a heavy horizontal line (source Osmond and Ivanovich, 1992).

(or soil C-horizon), and bedrock were combined to provide an estimate for the weathering rate of bedrock in a study catchment. Several papers have been published on the fate of uranium in soils (e.g., Lowson *et al.*, 1986; Frindik and Vollmer, 1999; Romero *et al.*, 1999; von Gunten *et al.*, 1999). For example, von Gunten *et al.* (1999) showed that values of $^{234}U/^{238}U$ change with the stability of soil minerals in a weathered lateritic uranium ore body. The least stable, amorphous iron and manganese oxyhydroxides had the lowest $^{234}U/^{238}U$ values due to preferentially release of ^{234}U to solution. The more stable crystalline oxide minerals goethite and hematite had $^{234}U/^{238}U$ activity ratios around 1, and the most stable Al-silicate minerals had values well above 1. The primary motivation for most soil studies has been to understand the behavior of uranium and other radioactive nuclides near ore deposits and radioactive waste disposal sites.

5.12.4.2.4 *Neodymium isotopes*

Relatively few studies have dealt with changes in neodymium isotopes during the progression of chemical weathering, although some have considered the link between weathering and the marine neodymium isotope record (Vance and Burton, 1999). Ohlander *et al.* (2000) found that in the process of weathering, neodymium is preferentially released from minerals with a lower Sm/Nd ratio than bulk soil, leading to lower $^{143}Nd/^{144}Nd$ values in the initial stages of weathering. Aubert *et al.* (2001) combined the study of strontium and neodymium isotopes and REEs in

soils, sediments, and waters from the Strengbach Catchment, France, to investigate granitic soil weathering processes (Figure 6). Strontium and neodymium isotopes indicated that: (i) stream suspended load originated from the finest soil size fraction (<20 μm); (ii) stream suspended load contained ~3% strontium and neodymium from the trace mineral apatite; and (iii) a large fraction of the strontium and neodymium in the dissolved load originated from leaching of apatite. Other studies have used neodymium isotopes to identify the source material for pedogenic carbonate (e.g., Borg and Banner, 1996), and in conjunction with other geochemical measurements to trace the behavior of REEs in polluted soils (e.g., Steinmann and Stille, 1997). Some recent studies have used strontium and neodymium isotopes to investigate the relative proportions of marine aerosol, mineral weathering, and eolian dust inputs to soil cation-exchange pools. These studies will be discussed in a later section of this chapter on differentiating atmospheric from weathering-derived cations in ecosystems.

5.12.4.2.5 *Osmium and hafnium isotopes*

Peucker-Ehrenbrink and Blum (1998) conducted a study of soil weathering processes utilizing the Re–Os isotopic system on the same granitic soil chronosequence in Wyoming used for strontium (Blum and Erel, 1997) and lead (Erel *et al.*, 1994; Harlavan *et al.*, 1998) isotopic studies. Peucker-Ehrenbrink and Blum (1998) pointed out that preferential weathering of biotite and rapid oxidation of magnetite in the soil

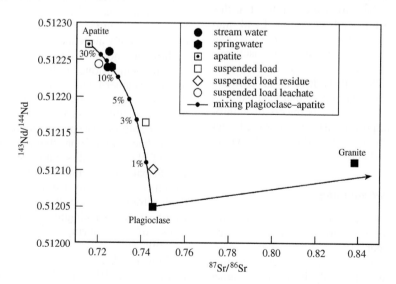

Figure 6 Sr and Nd isotopic compositions of granitoid bedrock, minerals, waters, and sediments from the Strengbach Catchment, France. A mixing calculation is shown with the percentage apatite in apatite–plagioclase mixtures. Sediments are dominated by plagioclase, whereas the waters and sediment leachate derive 10–30% of their Nd and Sr from the trace mineral apatite (source Aubert *et al.*, 2001).

environment largely controls the isotopic composition of osmium and the Re/Os ratio released during granitoid weathering. They suggested that high-latitude Precambrian shields may be important source areas of radiogenic osmium in seawater, and proposed that weathering of glacial tills exposed after deglaciation of Precambrian rocks surrounding the North Atlantic might result in elevated $^{187}Os/^{186}Os$ in rivers draining these areas. van de Flierdt *et al.* (2002) compared the hafnium isotopic composition of various rocks types and minerals with that of seawater, and suggested that during periods of intense mechanical weathering due to glaciations, zircons are weathered more efficiently, resulting in the release of highly nonradiogenic hafnium to the oceans.

5.12.4.3 Springs and Small Streams

In earlier sections we discussed the use of radiogenic isotopes to study mineral weathering in both laboratory experiments and in soils. In many of these studies the isotopic composition of the element of interest released by weathering was estimated in soils by measuring the composition of the labile pool in the soil (the acid leach or cation-exchange fractions). Another means of estimating the isotopic composition released by weathering is by measuring the isotopic composition of that element dissolved in springs and small streams draining a catchment underlain by a single bedrock lithology. In the absence of significant atmospheric inputs, the dissolved element is often expected to be in equilibrium with the soil labile pool. Among the various isotopic systems the most widely utilized in this way are strontium and uranium. Lead and neodymium isotopes have been used in several studies in conjunction with strontium isotopes, and a limited number of studies have been conducted utilizing exclusively lead or neodymium isotopes. It is important to note that lead isotopes have been used extensively as tracers of anthropogenic lead transport in surface and groundwater reservoirs, but these studies do not fall within the scope of this chapter.

5.12.4.3.1 *Strontium isotopes*

Two of the most significant early investigations of strontium isotopes in streams were by Fisher and Stueber (1976) and Wadleigh *et al.* (1985), who established that waters draining crystalline bedrock generally have much higher $^{87}Sr/^{86}Sr$ and lower strontium concentration than waters draining areas dominated by carbonate bedrock. They also demonstrated that even a small amount of carbonate in a drainage basin strongly influences the strontium isotopic composition of stream waters. Faure (1986) provides an excellent review of research on strontium in surface waters up until the mid-1980s.

Strontium isotopes were first applied to studies of deep saline groundwater brines in the late 1980s; a discussion of these studies can be found in the later section of this chapter on hydrology. Blum *et al.* (1993) used stream water $^{87}Sr/^{86}Sr$ to infer relative weathering rates of silicate minerals in small paired granitic catchments in the Sierra Nevada Mountains, California, that were either recently (12 kyr ago) glaciated or nonglaciated. The sensitivity of the dissolved $^{87}Sr/^{86}Sr$ ratio to biotite weathering was used to determine that glaciation had the effect of accelerating biotite weathering (i.e., the transformation to hydrobiotite and vermiculite) compared to plagioclase weathering in nonglaciated versus glaciated catchments. Another study with the primary focus on the interpretation of stream water $^{87}Sr/^{86}Sr$ to determine mineral weathering rates was carried out in two Scottish catchments, one underlain by andesite and the other by schist/granulite (Bain and Bacon, 1994). These authors investigated the $^{87}Sr/^{86}Sr$ ratios of total digests of soils and rock as well as stream- and rainwater, and concluded that plagioclase was weathering preferentially from the andesite compared to K-feldspar and chlorite, whereas the relative rate of dissolution of minerals in the schist/granulite could not be determined due to insufficient information on mineral compositions.

A study by Bullen *et al.* (1996) integrated experimental dissolution experiments, groundwater geochemical measurements, and a groundwater flow-path analysis to investigate mineral weathering in the subsurface near several Wisconsin lakes. Combining strontium isotopes and elemental analysis of waters and aquifer minerals along a shallow groundwater flow-path, the authors were able to demonstrate a dramatic change in the relative weathering rates of silicate minerals along the flow path. Shallower and more dilute waters had compositions indicative of dominantly plagioclase dissolution. Deeper and higher ionic strength waters had compositions that indicated a suppression of plagioclase dissolution and dominance by K-feldspar and biotite weathering. This application of strontium isotopes is an excellent demonstration of the power of the strontium isotope tracer to distinguish changes in the relative rates of weathering of complex mineral assemblages.

Clow *et al.* (1997), Blum *et al.* (1998), and Horton *et al.* (1999) used strontium isotopes and cation chemistry of stream waters to determine the proportion of dissolved ions originating from silicate versus carbonate dissolution reactions in catchments that were predominantly silicate, but also contained small (<1%) amounts of carbonate in the bedrock. In each case it was found that trace amounts of disseminated carbonate in crystalline silicate rocks could have a dominant effect on the strontium isotopic compositions and calcium fluxes from small catchments. The use of strontium isotopes in these small catchment studies have helped to solve the puzzle of why stream waters draining silicate bedrock catchments often display Ca/Na and Ca/Sr ratios far in excess of what would be released by weathering of the abundant silicate minerals in the soil parent materials (Figure 7). Thus, as in the case of experimental weathering studies and soil profile studies, we find that a major contribution of strontium isotopes as tracers of weathering is the identification of inputs from highly reactive trace minerals.

Jacobson *et al.* (2002b) combined strontium isotope and major element studies of 40 small streams in the Southern Alps of New Zealand to examine the climatic and tectonic controls on the relative weathering rates of carbonate and silicate minerals and documented a strong coupling between physical and chemical denudation. Peters *et al.* (2003) combined an experimental dissolution study with a stream catchment study by applying powdered wollastonite (CaSiO$_4$) to a stream channel in the Hubbard Brook Experimental Forest, New Hampshire. The interpretation of the stream water chemistry response to this application required the ability to separate calcium fluxes from the normal stream background, from calcium derived by wollastonite dissolution. A large contrast in the strontium isotopic composition between the stream before application and the strontium released from wollastonite allowed

Peters *et al.* (2003) to quantify the amount of calcium in the stream flux that was derived from each of these two sources, even though the dissolution of calcium and silicon from the wollastonite was not stoichiometric.

5.12.4.3.2 Lead isotopes

The isotopic composition of lead released to stream water by rock weathering was studied by Erel *et al.* (1990, 1991). During spring snowmelt a pristine mountain stream in the Sierra Nevada Mountains, California, was observed to contain mostly anthropogenic lead, but in the fall during base flow most of the lead in the stream water was from bedrock weathering. Erel *et al.* (1991) found that the isotopic value of the released lead was more radiogenic than the value of lead in the rock, and was similar to the value of lead released from the rock by a weak acid leach. Since the studied watershed was glaciated ~12 kyr ago exposing fresh rock surfaces, it was suggested that radiogenic lead was preferentially released from the granitic bedrock during the early stages of rock weathering. The preferential release of radiogenic lead was attributed to one or more of the following mechanisms: (i) rapid dissolution of some of the radiogenic accessory phases (e.g., apatite, monazite; Silver *et al.* (1984)); (ii) α-recoil damage to the uranium-enriched minerals (Kigoshi, 1971); or (iii) the presence of radiogenic lead-enriched interstitial oxides and cryptocrystalline aggregates (Brown and Silver, 1955). Luck and Ben Othman (1998) studied both lead and strontium isotopes in dissolved and particulate

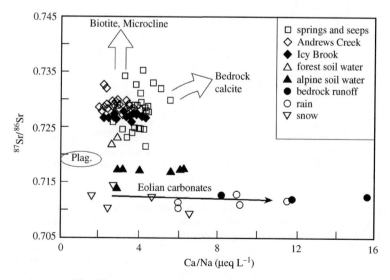

Figure 7 Ca/Na ratio versus ^{87}Sr/^{86}Sr ratio for waters and minerals from the granitoid Loch Vale, Colorado watershed. High Ca/Na ratios are indicative of a significant contribution from calcite dissolution to the dissolved cation chemistry of the waters. Sr isotopes allow differentiation of inputs from bedrock calcite versus eolian carbonate dust derived from sedimentary rocks in surrounding areas (source Clow *et al.*, 1997).

matter of a small Mediterranean karstic watershed in France during flooding and nonflooding periods. They were able to use these isotopic systems to estimate changes in physical and chemical weathering within the watershed in response to the water flux.

5.12.4.3.3 Uranium isotopes

Uranium isotopes have been used widely in surface water systems to trace water sources and pathways as well as rock weathering processes (Moore, 1967; Moreira-Nordemann, 1980; Scott, 1982; Sarin *et al.*, 1990; Tuzova and Novikov, 1991; Osmond and Ivanovich, 1992; Plater *et al.*, 1992; Palmer and Edmond, 1993; Pande *et al.*, 1994; Zhao *et al.*, 1994; Lienert *et al.*, 1994; Chabaux *et al.*, 1998; Riotte and Chabaux, 1999; Hakam *et al.*, 2001). Some of these studies have addressed the effect of rock weathering and rock type (especially carbonates versus silicates) on the $^{234}U/^{238}U$ ratio (e.g., Moreira-Nordemann, 1980; Toulhoat and Beaucaire, 1993; Pande *et al.*, 1994), and others have emphasized the links between the $^{234}U/^{238}U$ ratio and the hydrology of the studied drainages (Lienert *et al.*, 1994; Chabaux *et al.*, 1998; Riotte and Chabaux, 1999; Hakam *et al.*, 2001). For example, Riotte and Chabaux (1999) attributed variations in the $^{234}U/^{238}U$ ratio to changes in

the water flow-paths in the Strengbach catchment in France, and pointed out the complexity of the interpretation of the $^{234}U/^{238}U$ ratio as a tracer of rock weathering processes. Nonetheless, they effectively combined the use of $^{87}Sr/^{86}Sr$ and $^{234}U/^{238}U$ ratios to trace the rock types that control the composition of the stream waters as they flow over various compositions of bedrock (Figure 8).

5.12.4.3.4 Neodymium isotopes

Tricca *et al.* (1999) combined the analysis of neodymium and strontium isotopes in major and small rivers to investigate the origin of REEs in river waters from small catchments in the Rhine Valley. They observed that in the Rhine River the isotopic composition of neodymium was similar to the average continental crustal value, but in smaller, less evolved streams, the isotopic composition of neodymium reflected the breakdown of accessory minerals, such as apatite, enriched with middle REEs. Andersson *et al.* (2001) studied the effect of weathering on the isotopic composition of neodymium in the Kalix River in northern Sweden. They reported that the isotopic composition of neodymium had a narrow range and that, in general, the isotopic composition of neodymium in river water was lower than that of the local bedrock, probably reflecting preferential weathering of low Sm/Nd minerals.

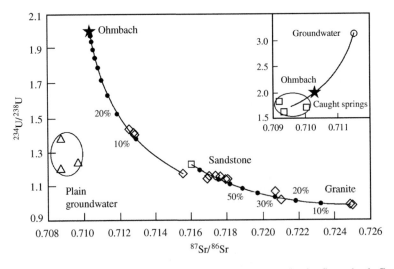

Figure 8 Evolution of the U and Sr isotopic composition of stream water in the Strengbach Catchment, France (diamonds) with mixing curves between granite and sandstone bedrock end-members. Also shown is a mixing curve between Strengbach stream water (where it flows from granite and sandstone bedrock into carbonate bedrock) and a carbonate end-member. The carbonate end-member is represented by the Ohmbach stream (star), which is located in the nearby carbonate-bedrock Ombach Catchment. The mixing line proportions presented here are based on both U and Sr data (concentrations and isotopic ratios). Inset: The Ohmbach-like component can be explained as a mixture of waters from surface springs (Caught Springs—squares) and the deep groundwater sampled in the Ohmbach catchment (circles). Plain groundwater refers to groundwater samples collected from the Rhine plain; their contrasting composition is used as evidence that Rhine groundwaters do not feed the Strengbach stream within the study area (source Riotte and Chabaux, 1999).

These observations are similar to those made by Ohlander *et al.* (2000) in a study of neodymium isotopes in soils developed on tills in northern Sweden.

5.12.4.3.5 *Osmium isotopes*

Pegram *et al.* (1994) studied the isotopic composition of osmium released from river sediments and pointed out that the $^{187}Os/^{188}Os$ of the dissolved river flux to the ocean is more radiogenic than the $^{187}Os/^{188}Os$ of the currently eroding continental crust, implying a significant osmium contribution from marine and terrestrial peridotites. Sharma and Wasserburg (1997) and Sharma *et al.* (1999) studied the behavior of osmium and its isotopes in major rivers draining the Himalayas. Although the isotopic composition of osmium in major rivers is beyond the scope of this chapter, we do note that these authors came to the conclusion that the Himalayas do not provide an unusually high flux of osmium or highly radiogenic osmium to the oceans. Chesley *et al.* (2000) studied the osmium and strontium isotopic record of Himalayan paleorivers and observed that $^{187}Os/^{188}Os$ and $^{87}Sr/^{86}Sr$ ratios covary, implying a common source. They provided evidence that large and rapid changes in the riverine fluxes and isotopic ratios of Himalayan rivers must have taken place during the last 5 Myr.

5.12.4.4 Proxies for Global Weathering Rates

Among the isotopic systems discussed in this chapter, strontium is the only one that has a long enough residence time in the ocean to unequivocally record changes in global weathering rates. For example, lead isotopes in marine manganese nodules, studied extensively since the early 1990s, have been mostly used to trace changes in ocean circulation rather than weathering rates. The isotopic composition of strontium in seawater is controlled largely by a balance between riverine inputs to the ocean (which result from continental weathering), marine hydrothermal inputs from the seafloor, and diagenetic fluxes from marine sediments (e.g., Brass, 1976; Palmer and Elderfield, 1985). Because marine carbonate rocks faithfully record the history of marine $^{87}Sr/^{86}Sr$ through geological time, and because the ~2 Myr residence time of strontium in the ocean results in a single global marine $^{87}Sr/^{86}Sr$ value, strontium is a potentially powerful proxy for changes in global weathering processes. In some studies, global weathering rates have been estimated from the marine strontium isotope record by assuming that hydrothermal and diagenetic strontium fluxes and $^{87}Sr/^{86}Sr$ ratios are constant through time,

and that the $^{87}Sr/^{86}Sr$ ratio of continental weathering has remained equal to the weighted average value of present-day riverine fluxes. It has further been assumed in some studies that the $^{87}Sr/^{86}Sr$ ratio of continental carbonate weathering remains equal to the average value for marine carbonates exposed on the continents, and that the average silicate weathering value could be calculated by difference after removing the carbonate contribution from major river fluxes. These assumptions allowed the development of models using variations in the marine $^{87}Sr/^{86}Sr$ ratio to estimate changes in global weathering rates due to tectonic uplift and global glaciations. Some workers suggested that changes in marine $^{87}Sr/^{86}Sr$ are caused not only by changes in silicate weathering rates, but also by changes in the $^{87}Sr/^{86}Sr$ of the average global riverine $^{87}Sr/^{86}Sr$ due to changes in the proportion of various silicate rocks and minerals weathering at any given time. Mountain building episodes, development of flood basalt provinces, sea-level change, and removal of soils by continental glaciations have all been proposed as mechanisms that might influence the proportion and age of silicate rocks exposed. Hodell (1994) provides an excellent review of the extensive literature on this topic through the early 1990s.

Several assumptions inherent in these earlier models have been challenged during the 1990s, as additional information has become available from studies of the systematics of strontium isotope release during weathering (summarized in this chapter). Recent contributions have pointed out some new complexities in the interpretation of the marine $^{87}Sr/^{86}Sr$ ratio as a silicate weathering proxy, including observations that: (i) strontium released during the early stages of weathering from fresh mineral surfaces can have a considerably different $^{87}Sr/^{86}Sr$ than bulk rocks or minerals (Blum and Erel, 1995; Brantley *et al.*, 1998; Taylor and Lasaga, 1999; White *et al.*, 1999; Taylor *et al.*, 2000a,b); (ii) trace amounts of carbonate or apatite weathering can dominate riverine strontium fluxes even in areas of silicate bedrock (Clow *et al.*, 1997; Blum *et al.*, 1998, 2002; Harris *et al.*, 1998; Horton *et al.*, 1999; Aubert *et al.*, 2001; Jacobson *et al.*, 2002a,b); (iii) carbonates may, in some cases, yield high fluxes of strontium with very high $^{87}Sr/^{86}Sr$, particularly in the Himalayan Mountains (Palmer and Edmond, 1992; Quade *et al.*, 1997; Harris *et al.*, 1998; Galy *et al.*, 1999; English *et al.*, 2000; Karim and Veizer, 2000; Jacobson *et al.*, 2002a); and (iv) hot springs (Evans *et al.*, 2001) and groundwater fluxes (Basu *et al.*, 2001) may be a significant part of the continental strontium fluxes to the oceans. It has become clear that the simple assumptions made in earlier work must be modified in order to derive meaningful estimates of weathering from the marine strontium isotope record. This presents an enormous challenge that has not been achieved as yet, but continued research will likely lead to a more quantitative understanding of the relationship between the marine strontium isotope record, global weathering rates, and other global geochemical cycles.

5.12.5 DIFFERENTIATING ATMOSPHERIC- FROM WEATHERING-DERIVED CATIONS IN ECOSYSTEMS

5.12.5.1 Soil Carbonates

Strontium isotope studies primarily focusing on the origin of soil carbonates began with a study by Quade *et al.* (1995), which demonstrated that calcium and strontium in soil carbonates within ~100 km of the southern coast of Australia were derived dominantly from marine dust and sea-spray with very little contribution from the soil substrate. This work was followed by similar studies of soil carbonate formed near the shoreline in Hawaii (Capo *et al.*, 2000; Whipkey *et al.*, 2000) and Morocco (Hamidi *et al.*, 1999), in the inland setting of Central Spain (Chiquet *et al.*, 1999), and in the southwestern USA (Capo and Chadwick, 1999; Naiman *et al.*, 2000). Each of these studies used strontium isotopes to demonstrate the overwhelming importance of atmospheric dust inputs (and the recycling of soil carbonate) to soil carbonate formation. Quade *et al.* (1997) and Chesley *et al.* (2000) studied the strontium isotopic composition of paleosol carbonate from ancestral Himalayan river deposits to reconstruct a record of changing riverine $^{87}Sr/^{86}Sr$ over the past 20 Myr. A dramatic increase in the riverine $^{87}Sr/^{86}Sr$ during the Late Miocene was attributed to exhumation of high $^{87}Sr/^{86}Sr$ meta-limestones in the Himalayas, and it was suggested that this may have influenced the marine $^{87}Sr/^{86}Sr$ record.

5.12.5.2 Base Cation Nutrients

5.12.5.2.1 Strontium isotopes

Quantifying the sources and rates of input of base cation nutrients (calcium, magnesium, potassium, and sodium) to forest ecosystems is an important goal in forest biogeochemistry, particularly when seeking to understand the recovery from environmental disturbances such as acid rain and forest clear-cutting. The earliest study to use isotopes as an indicator of atmospheric inputs to soils was by Dymond *et al.* (1974), who used strontium isotope measurements of micas in Hawaiian soils to determine that a significant proportion of the potassium input to Hawaiian soils was from deposition of dust transported

across the Pacific Ocean from Asia. Straughan *et al.* (1981) demonstrated that when fly ash produced by coal combustion is deposited onto soil, the strontium released from the fly ash is readily available to plants and can dominate the plant $^{87}Sr/^{86}Sr$ ratio with as little as 0.2% fly ash by weight in the soil.

Two nearly simultaneous investigations of atmospheric inputs to forested ecosystems of the Sangre de Cristo Mountains, New Mexico, were the first to fully integrate strontium isotopes into forest biogeochemical studies (Gosz *et al.*, 1983; Graustein and Armstrong, 1983). These studies took advantage of a large contrast in $^{87}Sr/^{86}Sr$ between the Precambrian granitic bedrock of the Sangre de Cristo mountains and the Phanerozoic sedimentary and volcanic rocks of the surrounding areas, which are the major source of atmospheric dust to the ecosystems. Strontium isotope and concentration measurements of a wide range of ecosystem components allowed quantification of the proportion of atmospheric-derived (>75%) and weathering-derived (<25%) strontium to forest vegetation (Gosz *et al.*, 1983; Graustein and Armstrong, 1983). Both research groups later expanded upon their earlier pioneering strontium isotope forest biogeochemistry studies of New Mexico watersheds (Graustein, 1989; Gosz and Moore, 1989). Åberg and Jacks (1985), Åberg *et al.* (1989), and Jacks *et al.* (1989) used a similar approach to study weathering and atmospheric inputs of calcium in several granitic stream catchments in Sweden. Differences in the $^{87}Sr/^{86}Sr$ of atmospheric and weathering inputs allowed the calculation of a calcium output budget, which suggested that excess calcium in streams was caused by loss from the cation-exchange pool due to soil acidification. A major source of uncertainty in all of these early studies was the $^{87}Sr/^{86}Sr$ released by weathering of ancient granite composed of minerals with highly contrasting $^{87}Sr/^{86}Sr$ ratios.

Miller *et al.* (1993) applied the strontium isotope methodology described above to investigate the response of the Whiteface Mountain, New York, forest ecosystem to atmospheric deposition of acidity and other pollutants. The soil parent material was anorthosite, which is comprised mostly of plagioclase and which has low Rb/Sr and $^{87}Sr/^{86}Sr$, minimizing uncertainties related to variation in $^{87}Sr/^{86}Sr$ among minerals. In this study it was determined that 50–60% of the strontium in the vegetation and organic soil pools was of atmospheric origin—presumably largely from coal-combustion fly-ash and dust related to cement production and farming. An ecosystem box-model was used to estimate that the annual stream export of strontium was ~70% from mineral weathering inputs and ~30% from soil cation-exchange reactions. Bailey *et al.* (1996)

used a similar methodology at the Cone Pond Watershed, New Hampshire, to investigate inputs and transport of base cations in another ecosystem perturbed by acid deposition. A large variability in the strontium isotopic composition of the various minerals in the bedrock, and spatial variability of soil parent material, complicated assessment of the $^{87}Sr/^{86}Sr$ ratio released to solution by weathering reactions. The authors were unable to make definitive estimates of cation-exchange depletion of calcium, but explored the range of possibilities using a variety of assumptions for the $^{87}Sr/^{86}Sr$ ratio of the weathering end-member. Using their "most likely" estimate of the weathering end-member, Bailey *et al.* (1996) suggested that the cation-exchange pool was being depleted significantly and was in danger of being exhausted.

Probst *et al.* (2000) measured strontium isotopes and major element concentrations of the various hydrochemical reservoirs in the Strengbach Catchment, France. Based on chemical budgets and a rock–water interaction model used to estimate the weathering end-member composition, they concluded that ~50% of the dissolved strontium in stream water was atmospherically derived. Aubert *et al.* (2002b)

incorporated neodymium isotope and REE concentration analyses into the Stengbach Catchment study and were able to show that atmospheric contributions of strontium and neodymium in throughfall and soil solution ranged from 20% to 70%. Blum *et al.* (2000, 2002) integrated strontium isotopes and Ca/Sr ratios into a study of the forest biogeochemistry of a small catchment in the Hubbard Brook Experimental Forest, New Hampshire. A four-step digestion procedure on soil parent material was used to determine the composition of soil calcium reservoirs for each soil horizon. The mineral apatite was found to be depleted from the Oa-, E-, Bh-, and Bs1-horizons but to be the dominant source of weatherable calcium in the Bs2- and C-horizons (Blum *et al.*, 2002). Moreover, apatite-derived calcium was inferred to be utilized to a larger extent by ectomycorrhizal tree species, suggesting that mycorrhizae may accelerate the weathering of apatite. Mass balance estimates suggest that of the calcium leaving the catchment in stream water, ~30% was atmospheric, ~35% was from silicate weathering, and ~35% was from apatite (Figure 9).

Kennedy *et al.* (1998) and Chadwick *et al.* (1999) applied strontium isotopes to trace the

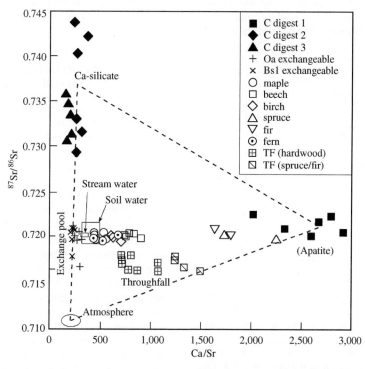

Figure 9 Ca/Sr versus $^{87}Sr/^{86}Sr$ for waters, C-horizon soil digests, soil-exchangeable fraction, and vegetation from the granitoid W-1 watershed at Hubbard Brook, New Hampshire. C-horizon digests 1, 2, and 3 were done in sequence and are progressively more aggressive acid digestions. Foliage (with species shown in key), soil-exchange pool, and water compositions each represent mixing between three sources: apatite leached from the soil parent material, silicate mineral weathering of the soil parent material, and atmospheric deposition. TF is throughfall, which represents a mixture of Ca and Sr leached from foliage and deposited from the atmosphere. Spruce and fir trees appear to access to a larger extent Ca and Sr from the apatite soil pool (source Blum *et al.*, 2002).

atmospheric input of calcium to the strongly marine-influenced ecosystems of the Hawaiian Islands. Using a soil chronosequence developed on basaltic lava flows varying in age from 0.3 kyr to 4,100 kyr, they found that the exchangeable soil pool (and tree foliage) was dominated by weathering-derived strontium at sites <10 kyr in age, where weathering rates far exceeded atmospheric inputs. As weathering rates declined with age, the system became dominated by marine inputs at all sites >10 kyr in age (Figure 10). Vitousek *et al.* (1999) expanded upon this work to study strontium isotopes in the leaves of trees from 34 Hawaiian forests (with varying precipitation rates) developed on young soils. Weathering was found to supply most of the strontium in most of the sites, but atmospheric sources supplied 30–50% of the strontium in the wettest sites and those closest to the ocean. Stewart *et al.* (2001) similarly used strontium isotopes to investigate the relative proportions of weathering and atmospherically derived strontium in 150 kyr soils along a precipitation gradient in Hawaii. They also found a transition between weathering-dominated and rainfall-dominated sources of plant-available strontium with increasing precipitation.

Kennedy *et al.* (2002) applied strontium isotopes to differentiate between weathering and atmospherically derived sources of strontium and calcium in a maritime forest in Chile, unaffected by atmospheric pollutants. In this study area they found that the exchangeable soil pool and foliage of the dominant tree species were mostly atmospherically derived strontium (>80%). These workers also applied an artificially enriched ^{84}Sr tracer to the forest floor to study plant uptake and leaching losses of strontium. They observed strong retention of the tracer in the upper soil horizons, yet no significant uptake into foliage nearly two years after treatment.

5.12.5.2.2 Lead and neodymium isotopes

Brimhall *et al.* (1991) used lead isotopes in zircons within a bauxite profile from Western Australia to differentiate between zircons derived from the underlying bedrock and zircons of eolian origin. Borg and Banner (1996) applied both neodymium and strontium isotopes to constrain the sources of soil developed on carbonate bedrock. Using these isotopes and Sm/Nd ratios, they were able to delineate the importance of atmospheric versus bedrock contributions in controlling the composition of the soil. Kurtz *et al.* (2001) used neodymium and strontium isotopes to determine the amount of Asian dust in a Hawaiian soil chronosequence. They found that the basaltic bedrock isotope signatures in soils had, in many cases, been completely overprinted by dust additions, demonstrating the profound effect of Asian dust on soil nutrient supplies.

5.12.6 TRACING HYDROLOGIC FLOW PATHS AND SUBSURFACE MIXING

The general geochemistry of groundwater is discussed in detail in Chapter 5.15. Here we highlight the uses of radiogenic isotopes that are dissolved in groundwater as a tool in hydrologic investigations. None of the radiogenic isotopes discussed in this chapter are truly conservative (i.e., nonreactive) tracers of groundwater. Instead, the radiogenic isotope compositions are controlled to varying degrees by mineral dissolution and precipitation along water flow-paths. In some cases the isotopic composition changes very little along flow paths and can be used as a simple tracer of flow path and mixing of distinct waters. In other instances, dissolution and isotopic exchange during flow can be used as an asset, yielding useful hydrologic information. Strontium has been the most frequently used radiogenic isotope tracer in hydrology, because it is very soluble in groundwater and has a large range in isotopic composition in freshwaters. Uranium is also quite soluble and numerous studies have been published in which uranium isotopes and U–Th disequilibrium products were utilized to follow subsurface flow of water and mixing of different subsurface waters (e.g., Osmond and Cowart, 1976, 1992),

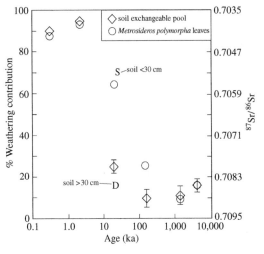

Figure 10 The ^{87}Sr/^{86}Sr ratio of the soil-exchangeable pool and foliage for a chronosequence of soils developed on Hawaiian basaltic lava flows. Also plotted is the result of a calculation of the percent weathering contribution needed to explain the ^{87}Sr/^{86}Sr ratio—assuming that all Sr is derived from basaltic weathering and marine atmospheric deposition. Sr isotopes clearly demonstrate the transition of the forest ecosystems from weathering dominated Sr to atmospherically dominated Sr as soils mature and easily weathered Sr is removed from soils (source Kennedy *et al.*, 1998).

sometimes in conjunction with strontium isotopes (Cao *et al.*, 1999). The use of U–Th disequilibrium series in tracing ground waters falls outside the scope of the current chapter, but an excellent review of this topic is provided by Osmond and Cowart (1992).

Lead, neodymium, and osmium have not been used widely as tracers in hydrology, because they have very low solubilities and are highly reactive with aquifer materials. Lead isotopes have been used in groundwater mostly to distinguish between natural and anthropogenic lead sources (e.g., Whitehead *et al.*, 1997), where the natural end-member was assumed to equal the isotopic value of lead released from the bulk rock. Although this assumption is probably incorrect for many groundwater systems (Harlavan *et al.*, 1998), the large excess of anthropogenic lead in most cases overwhelms small uncertainties in the isotopic composition of lead released from the bedrock. Combining analyses of several isotopic tracers has proved to be an effective approach, and several research groups have successfully combined neodymium, lead, and/or uranium isotopes with strontium isotopes in order to trace the migration of groundwaters (e.g., Doe *et al.*, 1966; Stuckless *et al.*, 1991; Toulhoat and Beaucaire, 1993; Mariner and Young, 1996; Barbieri and Voltaggio, 1998; Tricca *et al.*, 1999; Roback *et al.*, 2001).

Hydrologic applications of radiogenic isotopes can be broadly grouped into: (i) regional-scale studies of groundwater flow-paths and mixing over hundreds of kilometers; (ii) local-scale studies of groundwater flow-paths and mixing over tens of kilometers; and (iii) studies of the hydrology of springs and small streams in closed catchments. In addition to studies of radiogenic isotopes dissolved in waters, some studies have taken advantage of secondary minerals deposited in the subsurface to infer changes in water chemistry and hydrology over geologic time. Many of the earliest investigations of radiogenic isotopes in groundwater focused on the origin of oil field brines and hydrothermal fluids. These are mentioned only briefly here, because the emphasis of this chapter is on freshwater hydrology. They are discussed in Chapters 5.16 and 5.17. Doe *et al.* (1966) were among the first investigators to utilize strontium and lead isotopes in brines and potential source rocks to constrain the source of metals in hydrothermal fluids. Later studies by Stettler (1977), Stettler and Allègre (1979), and Elderfield and Greaves (1981) modeled the relative contributions of mantle and crustal sources to hydrothermal brines as a result of fluid–rock interaction. Early studies that utilized strontium isotopes in regional aquifers to determine the sources of salts in brines include Chaudhuri (1978), Sass and Starinsky (1979), Starinsky *et al.* (1983a,b),

Stueber *et al.* (1987, 1984, 1993), Banner *et al.* (1989) and Smalley *et al.* (1998). McNutt *et al.* (1990) and Franklyn *et al.* (1991) utilized strontium isotope ratios in deep brines from mines and drill holes in the crystalline Canadian Shield to establish that dissolution of plagioclase was the dominant water–rock interaction controlling the strontium isotopic composition. They utilized differences in $^{87}Sr/^{86}Sr$ between pockets of brine to study mixing of different brines as well as mixing with meteoric waters. Similarly, Toulhoat and Beaucaire (1993) used lead and uranium isotopes in groundwater to trace the release of uranium and lead by the dissolution of minerals in uranium ore deposits.

5.12.6.1 Regional Scale Groundwater Studies

5.12.6.1.1 Strontium isotopes

We begin our detailed discussion of the literature with the first study that utilized strontium isotopes to investigate hydrologic flow paths and mixing in a freshwater aquifer. In a study of the Australian Artesian Basin, Collerson *et al.* (1988) used strontium isotopes to characterize various water masses and investigated their subsurface flow and mixing over a ~ 500 km length scale. Strontium isotopes were subsequently applied to investigations of the sources of salts and flow paths of groundwaters in South Australia (Ullman and Collerson, 1994; Lyons *et al.*, 1995). Musgrove and Banner (1993) combined the use of major ions, as well as oxygen, hydrogen, and strontium isotopes in groundwaters from three adjacent regional flow regimes in the mid-continental US to delineate large-scale fluid flow and mixing processes. $^{87}Sr/^{86}Sr$ ratios were useful in delineating between end-member waters, but there was some evidence for modification of $^{87}Sr/^{86}Sr$ by interaction of groundwater with silicate minerals in the aquifers, which may have obscured some of the mixing relationships between end-member waters.

Strontium isotopes have been applied to several studies of groundwater evolution in limestone aquifers. $^{87}Sr/^{86}Sr$ and Ca/Sr ratios have provided insight into calcite dissolution and recrystallization along flow paths, the transformation of aragonite to calcite, and the fluxes of strontium from soils into carbonate aquifers (Banner *et al.*, 1994, 1996; Banner, 1995). Dissolution and recrystallization of calcite along flow paths can compromise ^{14}C dating of groundwater by diluting ^{14}C with aquifer carbon. Bishop *et al.* (1994), in a study of the Lincolnshire Limestone aquifer, England, introduced a methodology, whereby changes in $^{87}Sr/^{86}Sr$ and $\delta^{13}C$ with flow down the hydraulic gradient could be used to estimate the amount of carbon dilution and allow more accurate correction of ^{14}C ages of waters. Johnson and DePaolo (1996) presented a method for inversion of $^{87}Sr/^{86}Sr$ and ^{14}C data to obtain carbonate dissolution rates and dissolution-corrected water velocities and applied this model to the Lincolnshire Limestone aquifer using the data of Bishop *et al.* (1994).

Naftz *et al.* (1997) used strontium isotope ratios in groundwaters from southeastern Utah to determine if re-injected oil-field brines or water from a deeper aquifer was the source of salinity to a shallower aquifer that is an important source of drinking water. Distinct $^{87}Sr/^{86}Sr$ ratios in the various waters allowed the authors to verify that the oil-field brine was not the source of salinity, whereas the deeper aquifer was a plausible source. Armstrong *et al.* (1998) used $^{87}Sr/^{86}Sr$ and Ca/Sr ratios to study groundwater mixing and rock–water interaction in the Milk River Aquifer, Alberta, Canada. This study elucidated the importance of the variable reactivity of different sedimentary rocks along differing groundwater flow paths. The $^{87}Sr/^{86}Sr$ of recharging groundwater was modified by the local lithology, causing distinct geochemical patterns along varying flow paths. This could be verified by calculating pore-water fluid evolution paths for the various aquifer materials. Woods *et al.* (2000) used strontium isotopes and major elements to study the chemical evolution and movement of groundwater along flow paths in a coastal limestone aquifer in North Carolina. Variations in $^{87}Sr/^{86}Sr$ were explained largely by mixing of groundwater infiltrating from a surficial aquifer. Johnson *et al.* (2000) used strontium isotopes in groundwater in the Snake River Plain, Idaho, to identify groundwater "slow-flow zones" and "fast-flow zones" in a largely basaltic fracture flow aquifer (Figure 11).

5.12.6.1.2 Uranium isotopes

Osmond and Cowart (1992) and Nimz (1998) recently reviewed the literature on the use of uranium isotopes in groundwater studies. These reviews cite early studies, such as Spiridonov *et al.* (1969), Cherdyntsev (1971), and Osmond *et al.* (1974), as well as more recent papers. As in the case of uranium isotope studies in the soil environment, it has been widely observed that the $^{234}U/^{238}U$ ratio in groundwaters deviates from the secular equilibrium value (Osmond and Cowart, 1992). The highest $^{234}U/^{238}U$ ratios have been recorded in slow-moving groundwaters because of the preferential mobilization of ^{234}U in systems with overall low uranium concentrations (Cowart and Osmond, 1980). In fast moving groundwater systems such as karstic carbonate bedrock, the $^{234}U/^{238}U$ activity ratios are usually much lower (Osmond *et al.*, 1974; Osmond and Cowart, 1976), although some high ratios have been observed (e.g., Kronfeld *et al.*, 1975, 1994).

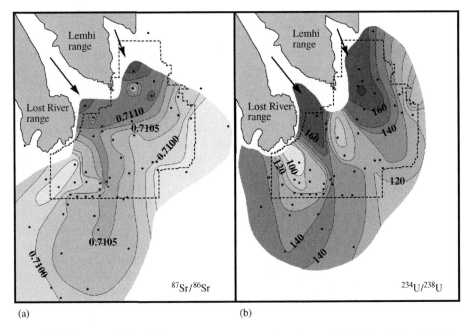

(a)   (b)

Figure 11 (a) $^{87}Sr/^{86}Sr$ and (b) $^{234}U/^{238}U$ ratios of groundwater in and around the Idaho National Engineering and Environmental Laboratory on the Snake River Plain. High $^{87}Sr/^{86}Sr$ and $^{234}U/^{238}U$ waters enter the Snake River system (arrows) and react with lower $^{87}Sr/^{86}Sr$ and $^{234}U/^{238}U$ basalt in the study area. Lower $^{87}Sr/^{86}Sr$ and $^{234}U/^{238}U$ groundwaters occur in "slow-flow" zones due to prolonged interaction with basalts whereas "fast-flow" zones have higher $^{87}Sr/^{86}Sr$ and $^{234}U/^{238}U$ (sources Johnson *et al.*, 2000; Roback *et al.*, 2001).

Andrews and Kay (1982, 1983) measured uranium concentrations and $^{234}U/^{238}U$ activity ratios in several aquifers. They observed that changes in the redox conditions of the aquifer strongly affect uranium concentrations in the groundwater and that these changes are sometimes accompanied by changes in the $^{234}U/^{238}U$ activity ratios. They proposed that in certain cases the isotopic composition of uranium could be explained by the sealing of uranium-bearing surfaces by rock cementation.

Several studies have utilized uranium isotopes to classify aquifers, to trace the movement of groundwater, to investigate the mixing of groundwaters with surface water, and to identify the penetration of seawater into aquifers (e.g., Kraemer, 1981; Lienert *et al.*, 1994). Other studies have investigated patterns of rock–water interactions, the dating of groundwater, and the secondary accumulation of uranium (e.g., Banner *et al.*, 1990; Chalov, 1989; Cowart and Osmond, 1977; Hussain, 1995). Many papers have also dealt with hydrothermal systems and with brines where the isotopic composition of uranium was affected by the long residence time of the water in the subsurface environment. For a comprehensive list of these studies, see Osmond and Cowart (1992) and Nimz (1998).

Roback *et al.* (2001) analyzed the same samples from the strontium isotope study of Johnson *et al.* (2000) in the Snake River Plain, described in the previous section, for $^{234}U/^{238}U$ activity ratios. This was done in order to trace the lateral distribution and possible mixing of groundwaters in the eastern Snake River aquifer. Uranium isotopes in conjunction with strontium isotopes further clarified the groundwater "slow-flow zones" and "fast-flow zones" identified by Johnson *et al.* (2000). These companion studies in the Snake River Plain nicely demonstrate the advantages of radiogenic isotope ratios, which are unaffected by solute losses due to mineral precipitation, ion exchange, adsorption, or evaporation, and thus provide a clearer picture of groundwater flow patterns than can be obtained from the study of solute concentrations alone (Figure 11).

5.12.6.2 Local-scale Groundwater Studies

5.12.6.2.1 *Strontium isotopes*

Several studies have used radiogenic isotopes to study groundwater questions on relatively short (or local) length scales (<2 km). Stuckless *et al.* (1991) combined the use of strontium and uranium isotope measurements, in studies of groundwaters and secondary calcite deposits in fault zones at Yucca Mountain, Nevada. They tested whether veins formed by upwelling of deep-seated waters

or infiltration of surface waters and found that vein deposits were isotopically distinct from any of the groundwaters in the area suggesting that they were formed by infiltrating rather than upwelling waters. A later study of $^{87}Sr/^{86}Sr$ in vein fillings from drill holes in the area came to a similar conclusion as the earlier study (Marshall *et al.*, 1992). However, a subsequent modeling study suggested that a conclusive indication of the paleowater table elevation is not possible because of the effects of water–rock interaction (Johnson and DePaolo, 1994).

Katz and Bullen (1996) used strontium along with hydrogen, oxygen, and carbon isotopes to study the interaction between groundwater, lake water, and aquifer minerals around a seepage lake in mantled karst terrain of northern Florida. They were able to effectively identify mixing of lake water leakage with groundwater and to study processes of mineral–water interaction. Johnson and DePaolo (1997a) used strontium isotopes to investigate groundwater flow-paths and velocities beneath the Lawrence Berkeley National Laboratory, California, by analyzing well waters and aquifer materials. Variations in $^{87}Sr/^{86}Sr$ in the subsurface were interpreted as a cation-exchange front coupled with the dissolution of aquifer minerals. The data were inverted using a one-dimensional transport–dissolution–exchange model (Johnson and DePaolo, 1997b), yielding average flow velocities consistent with ^{14}C measurements.

Peterman and Wallin (1999) used strontium isotope ratios in groundwater to study a fracture flow system in granitic bedrock on an island in southeastern Sweden. Strontium isotope ratios were used to help constrain three distinct end-member waters that include precipitation infiltration, old saline water in low-conductivity zones, and Baltic Sea water. Systematic trends in groundwater composition were attributed to mixing of these components. Siegel *et al.* (2000) explored the use of strontium and lead isotopes to identify sources of water beneath a large municipal landfill on Staten Island, New York. Lead isotope ratios varied widely and were not useful in distinguishing the origin of waters in the groundwater system. Strontium isotope ratios were very sensitive to seawater inputs and were used to argue that many deep wells were not contaminated by landfill leachates (Siegel *et al.*, 2000). Bohlke and Horan (2000) used strontium isotopes to trace inputs of fertilizer to groundwater systems in small agricultural watersheds in Maryland. This study pointed out the potential for fertilizer application to interfere with the natural geochemistry of strontium and raised the possibility that strontium can be useful for tracing non-point-sources in catchments.

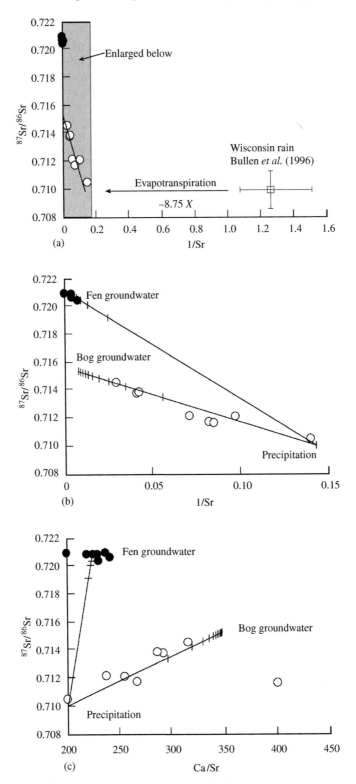

Figure 12 $^{87}Sr/^{86}Sr$ versus (a, b) 1/Sr and (c) Ca/Sr for groundwater samples and precipitation from Lost River in the Glacial Lake Agassiz Peatlands, Minnesota. Mixing trends (with 10% increments) are shown for mixing of evaporated precipitation with distinct groundwater end-members for the adjacent bog and fen sites (source Hogan *et al.*, 2000).

Hogan *et al.* (2000) used $^{87}Sr/^{86}Sr$ and Ca/Sr ratios to study groundwater discharge and precipitation recharge in bog-fen systems of the Glacial Lake Agassiz Peatland in northern Minnesota. Precipitation recharge pore waters had a very distinct composition similar to precipitation, whereas groundwater discharge had higher $^{87}Sr/^{86}Sr$ and Ca/Sr ratios reflecting deep flow through Precambrian crystalline bedrock. A clear chemical differentiation between these end-member waters allowed elucidation of mixing relationships and flow paths (Hogan *et al.*, 2000; Figure 12).

5.12.6.2.2 *Uranium isotopes*

Uranium isotopes have also been used to study local groundwater problems, in particular those related to the mobility of uranium in groundwater near radioactive waste disposal sites and the role of colloids in controlling the subsurface movement of uranium and its decay products (Ivanovich *et al.*, 1988; Short *et al.*, 1988; Suksi *et al.*, 2001; Toulhoat *et al.*, 1996; Hussain, 1995; Gomes and Cabral, 1981). Paces *et al.* (2002) used extensive measurements of $^{234}U/^{238}U$ in order to determine patterns of recharge and groundwater flow in the vicinity of Yucca Mountain, Nevada. Based on $^{234}U/^{238}U$ data they suggested that the aquifer beneath Yucca Mountain is dominated by local recharge and is not affected by the regional groundwater flow. This and the studies of Stuckless *et al.* (1991), Marshall *et al.* (1992), and Johnson and DePaolo (1994) mentioned above are important applications of radiogenic isotopes to contaminant hydrology, as they shed light on the public and scientific controversy surrounding the appropriateness of the development of a high-level nuclear waste repository at Yucca Mountain.

5.12.6.3 Springs and Small Streams

5.12.6.3.1 *Strontium isotopes*

Neumann and Dreiss (1995) and Barbieri and Morotti (2003) used strontium isotopes in spring-waters from the Mono Basin, California, and Monte Vulture, Italy, respectively, to elucidate the subsurface mixing of waters that had traveled through different lithologies. Three relatively recent studies have utilized strontium isotopes in studies of first-order streams and the shallow groundwater feeding into them in order to gain insight into water sources and flow paths during varying hydrologic conditions. Ben Othman *et al.* (1997) studied the strontium isotope, major element, and trace element compositions of a small Mediterranean catchment during a series of rain events over a four-day period. Strontium isotopes proved very effective at differentiating water originating as runoff, from deeper Jurassic carbonate aquifers, and from shallower Miocene marl aquifers. Land *et al.* (2000) used $^{87}Sr/^{86}Sr$, Ba/Sr, and Ca/Sr ratios to distinguish between soil water, shallow groundwater, and deep groundwater and to model the contributions from these sources in samples collected weekly over a one-year period. $^{87}Sr/^{86}Sr$ ratios of these end-member waters were used to test the validity of a stream water source model. Aubert *et al.* (2002a) measured $^{87}Sr/^{86}Sr$ in stream water samples collected weekly, and explored concentration–discharge relationships for $^{87}Sr/^{86}Sr$. Higher $^{87}Sr/^{86}Sr$ ratios were observed during high flow periods, suggesting that flow paths change within the catchment soils depending on moisture conditions. This study demonstrated that changing hydrologic conditions must be carefully considered when sampling stream water for $^{87}Sr/^{86}Sr$ in studies of weathering fluxes. Hogan and Blum (2003) used $^{87}Sr/^{86}Sr$, Ba/Sr, and $\delta^{18}O$ to study water flow-paths during storm events sampled hourly in a small granitic catchment. Hydrograph separations with $\delta^{18}O$ were used to quantify the amount of "new water," while $^{87}Sr/^{86}Sr$ was used to separate "old water" into separate soil water and groundwater components, each with distinct $^{87}Sr/^{86}Sr$ values. In contrast, Bain *et al.* (1998) found no consistent change in $^{87}Sr/^{86}Sr$ with discharge during a storm event in a small andesitic catchment.

5.12.6.3.2 *Uranium isotopes*

Only a few studies have been performed using $^{234}U/^{238}U$ ratios as tracers in springs and small streams, and these have mainly used uranium isotopes to trace the interaction of groundwater with surface water (e.g., Lienert *et al.*, 1994). Lienert *et al.* (1994) attempted to link changes in uranium behavior to decreases in the anthropogenic inputs of phosphorous, which in turn affects biological activity and redox conditions within waters. As discussed in an earlier section of this chapter, the $^{234}U/^{238}U$ ratio has also been used to trace changes in the water flow-paths within the Strengbach watershed (Riotte and Chabaux, 1999). In another study, Kronfeld and Rosenthal (1981) used uranium isotopes to trace the movement of water within the Bet-Shean-Harod Valley, Israel. Finally, Barbieri and Voltaggio (1998) combined the use of uranium and strontium isotopes to study the hydrology of the Sangemini (Italy) mineral-water springs, and based on the combined use of these isotopic systems they were able to put constraints on the extent of the aquifer and the springwater flow paths.

5.12.7 SUMMARY

Radiogenic isotopes have proven to be an important and powerful tool in investigations of many aspects of weathering and hydrology. The general absence of isotope fractionation of heavy radiogenic isotopes in nature gives these tracers many advantages over major and trace element ratios. Well over 200 articles have been written in this topical area, and it is now evident that this methodology provides important scientific insights, and will increasingly become a routine tool in studies of weathering and hydrology. Strontium isotopes have unquestionably become the most commonly used radiogenic isotope tracer because of the large variability in isotopic composition and the interest in tracing sources and cycling of the analog element calcium. The most notable applications are: (i) use in identifying the dissolution of trace calcium-bearing phases in experiments, soils, and along groundwater flow-paths; (ii) use in differentiating atmospheric from weathering sources of calcium to ecosystems; and (iii) use in differentiating distinct subsurface waters that have interacted with contrasting aquifer materials. The mechanisms controlling differences in strontium isotope ratios in the environment and the hydrogeochemical behavior of strontium have become reasonably well understood and as a result, strontium will be increasingly used in a routine manner in weathering and hydrologic studies.

Lead isotopes have found limited applications in the study of the weathering of uranium and thorium-rich accessory phases in laboratory experiments and in soils. The preferential release of ^{234}U compared to ^{238}U has been more widely used as a tracer of weathering reactions, as a tracer of the geochemical behavior of uranium, and as a tracer of groundwater sources and mixing. Neodymium has proven useful in a few studies as a tracer of the weathering release of trace phases such as apatite, as well as of inputs of atmospheric dust to soils. Osmium has been used in only one weathering study, as of early 2000s, and was useful in inferring rates of dissolution of magnetite in crystalline rocks. In general, the use of neodymium, hafnium, and osmium isotopes in weathering and hydrology is in its infancy, and much additional research will be needed to gain a thorough understanding of the behavior of these systems and to ascertain their usefulness in more routine investigations. We expect that major breakthroughs in the use of radiogenic isotopes in weathering and hydrology will increasingly rely on the combined use of several isotopic systems together, which yield contrasting information and insight into any given scientific application.

REFERENCES

Åberg G. and Jacks G. (1985) Estimation of the weathering rate by $^{87}Sr/^{86}Sr$ ratios. *Geol. Foren. Stock. Forh.* **107**, 289–290.

Åberg G., Jacks G., and Hamilton P. J. (1989) Weathering rates and Sr-87/Sr-86 ratios—an isotopic approach. *J. Hydrol.* **109**, 65–78.

Albarède F. (1995) Introduction to geochemical modeling. Cambridge University Press, Cambridge, 543pp.

Andersson P. S., Dahlqvist R., Ingri J., and Gustafsson O. (2001) The isotopic composition of Nd in a boreal river: a reflection of selective weathering and colloidal transport. *Geochim. Cosmochim. Acta* **65**, 521–527.

Andrews J. N. and Kay R. L. F. (1982) $^{234}U/^{238}U$ activity ratios of dissolved uranium in groundwaters from a Jurassic limestone aquifer in England. *Earth Planet. Sci. Lett.* **57**, 139–151.

Andrews J. N. and Kay R. L. F. (1983) The U content and $^{234}U/^{238}U$ activity ratios of dissolved uranium in groundwaters from some Triassic sandstones in England. *Isot. Geosci.* **1**, 101–117.

Armstrong S. C., Sturchio N. C., and Hendry M. J. (1998) Strontium isotopic evidence on the chemical evolution of pore waters in the Milk River Aquifer, Alberta, Canada. *Appl. Geochem.* **13**, 463–475.

Aubert D., Stille P., and Probst A. (2001) REE fractionation during granite weathering and removal by waters and suspended loads: Sr and Nd isotopic evidence. *Geochim. Cosmochim. Acta* **65**, 387–406.

Aubert D., Probst A., Stille P., and Viville D. (2002a) Evidence of hydrological control of Sr behavior in stream water (Strengbach catchment, Vosges mountains, France). *Appl. Geochem.* **17**, 285–300.

Aubert D., Stille P., Probst A., Gauthier-Lafaye F., Pourcelot L., and Del Nero M. (2002b) Characterization and migration of atmospheric REE in soils and surface waters. *Geochim. Cosmochim. Acta* **66**, 3339–3350.

Ault W. U., Senchal R. G., and Erlebach W. E. (1970) Isotopic composition as a natural tracer of lead in the environment. *Environ. Sci. Technol.* **4**, 305–313.

Bacon J. R., Berrow M. L., and Shand C. A. (1995) The use of isotopic composition in field studies of lead in upland Scottish soils (UK). *Chem. Geol.* **124**, 125–134.

Bailey S. W., Hornbeck J. W., Driscoll C. T., and Gaudett H. E. (1996) Calcium inputs and transport in a base-poor forest ecosystem as interpreted by Sr isotopes. *Water Resour. Res.* **32**, 707–719.

Bain D. C. and Bacon J. R. (1994) Strontium isotopes as indicators of mineral weathering in catchments. *Catena* **22**, 201–214.

Bain D. C., Midwood A. J., and Miller J. D. (1998) Strontium isotope ratios in streams and the effect of flow rate in relation to weathering in catchments. *Catena* **32**, 143–151.

Banner J. L. (1995) Application of the trace element and isotope geochemistry of strontium to studies of carbonate diagenesis. *Sedimentology* **42**, 805–824.

Banner J. L., Wasserburg G. J., Dobson P. F., Carpenter A. B., and Moore C. H. (1989) Isotopic and trace element constraints on the origin and evolution of saline groundwaters from central Missouri. *Geochim. Cosmochim. Acta* **52**, 383–398.

Banner J. L., Wasserburg G. J., Chen J. H., and Moore C. H. (1990) $^{234}U-^{238}U-^{230}Th-^{232}Th$ systematics in saline groundwaters from central Missouri. *Earth Planet. Sci. Lett.* **101**, 296–312.

Banner J. L., Musgrove M., and Capo R. C. (1994) Tracing ground-water evolution in a limestone aquifer using Sr isotopes: effects of multiple sources of dissolved ions and mineral-solution reactions. *Geology* **22**, 687–690.

Banner J. L., Musgrove M., Asmerom Y., Edwards R. L., and Hoff J. A. (1996) High-resolution temporal record of Holocene ground-water chemistry: tracing links between climate and hydrology. *Geology* **24**, 1049–1053.

Barbieri M. and Morotti M. (2003) Hydrogeochemistry and strontium isotopes of spring and mineral waters from Monte Vulture volcano, Italy. *Appl. Geochem.* **18**, 117–125.

Barbieri M. and Voltaggio M. (1998) Applications of Sr isotopes and U-series radionuclides to the hydrogeology of Sangemini area (Terni, central Italy). *Mineral. Petrograph. Acta* **41**, 119–126.

Basu A. R., Jacobsen S. B., Poreda R. J., Dowling C. B., and Aggarwal P. K. (2001) Large groundwater strontium flux to the oceans from the Bengal Basin and the marine strontium isotope record. *Science* **293**, 1470–1473.

Ben Othman D., Luck J. M., and Tournoud M. G. (1997) Geochemistry and water dynamics: application to short time-scale flood phenomena in a small Mediterranean catchment: I. Alkalis, alkali-earths, and Sr isotopes. *Chem. Geol.* **140**, 9–28.

Bierman P. R., Albrecht A., Bothner M. H., Brown E. T., Bullen T. D., Gray L. B., and Turpin L. (1998) Erosion, weathering, and sedimentation. In *Isotope Tracers in Catchment Hydrology* (eds. C. Kendall and J. J. McDonnell). Elsevier, Amsterdam, pp. 647–678.

Bishop P. K., Smalley P. C., Emery D., and Dickson J. A. D. (1994) Strontium isotopes as indicators of the dissolving phase in a carbonate aquifer: implications for ^{14}C dating of groundwater. *J. Hydrol.* **154**, 301–321.

Blaxland A. B. (1974) Geochemistry and geochronology of chemical weathering, Butler Hill Granite, Missouri. *Geochim. Cosmochim. Acta* **38**, 843–852.

Blum J. D. (1995) Isotope decay data. In *Global Earth Physics, A Handbook of Physical Constants* (ed. T. J. Ahrens). American Geophysical Union, pp. 271–282.

Blum J. D. (1997) The effect of late Cenozoic glaciation and tectonic uplift on silicate weathering rates and the marine ^{87}Sr/^{86}Sr record. In *Tectonic Uplift and Climate* (ed. W. Ruddiman). Plenum, New York, pp. 259–288.

Blum J. D. and Erel Y. (1995) A silicate weathering mechanism linking increases in marine ^{87}Sr/^{86}Sr with global glaciation. *Nature* **373**, 415–418.

Blum J. D. and Erel Y. (1997) Rb–Sr isotope systematics of a granitic soil chronosequence: the importance of biotite weathering. *Geochim. Cosmochim. Acta* **61**, 3193–3204.

Blum J. D., Erel Y., and Brown K. (1993) ^{87}Sr/^{86}Sr ratios of Sierra Nevada stream waters: implications for relative mineral weathering rates. *Geochim. Cosmochim. Acta* **58**, 5019–5025.

Blum J. D., Gazis C. A., Jacobson A. D., and Chamberlain C. P. (1998) Carbonate versus silicate weathering in the Raikhot watershed within the high Himalayan crystalline series. *Geology* **26**, 411–414.

Blum J. D., Taliaferro E. H., Weisse M. T., and Holmes R. T. (2000) Changes in Sr/Ca, Ba/Ca and ^{87}Sr/^{86}Sr ratios between trophic levels in two forest ecosystems in the northeastern USA. *Biogeochemistry* **49**, 87–101.

Blum J. D., Klaue A., Nezat C. A., Driscoll C. T., Johnson C. E., Siccama T. G., Eagar C., Fahey T. J., and Likens G. E. (2002) Mycorrhizal weathering of apatite as an important calcium source in base-poor forest ecosystems. *Nature* **417**, 729–731.

Bohlke J. K. and Horan M. (2000) Strontium isotope geochemistry of groundwaters and streams affected by agriculture, Locust Grove, MD. *Appl. Geochem.* **15**, 599–609.

Bonotto D. M. and Andrews J. N. (1998) The laboratory evaluation of the transfer of ^{234}U and ^{238}U to the waters interacting with carbonates and implications to the interpretation of field data. *Min. Mag.* **62A**, 187–188.

Bonotto D. M., Andrews J. N., and Darbyshire D. P. F. (2001) A laboratory study of the transfer of ^{234}U and ^{238}U during water-rock interactions in the Carnmenellis granite

(Cornwall, England) and implications for the interpretation of field data. *Appl. Radiat. Isotopes* **54**, 977–994.

Borg L. E. and Banner J. L. (1996) Neodymium and strontium isotopic constraints on soil sources in Barbados, West Indies. *Geochim. Cosmochim. Acta* **60**, 4193–4206.

Brannvall M. L., Kurkkio H., Bindler R., Emteryd O., and Renberg I. (2001) The role of pollution versus natural geological sources for lead enrichment in recent lake sediments and surface forest soils. *Environ. Geol.* **40**, 1057–1065.

Brantley S. L., Chesley J. T., and Stillings L. L. (1998) Isotopic ratios and release rates of strontium measured from weathering feldspars. *Geochim. Cosmochim. Acta* **62**, 1493–1500.

Brass G. W. (1976) The variation of marine ^{87}Sr/^{86}Sr ratio during phanerozoic time: interpretation using a flux model. *Geochim. Cosmochim. Acta* **40**, 721–730.

Brimhall G. H., Chadwick O. A., Lewis C. J., Compston W., Williams I. S., Danti K. J., Dietrich W. E., Power E. M., Hendricks D., and Bratt J. (1991) Deformational mass transport and invasive processes in soil evolution. *Science* **255**, 695–702.

Brown H. and Silver L. T. (1955) The possibility of obtaining long-term supplies of uranium, thorium, and other substances of uranium from igneous rocks. In *Proceedings of Conference on Peaceful Use of Atomic Energy, Geneva*, IAEA, pp. 91–95.

Bullen T. D., Krabbenhoft D. P., and Kendall C. (1996) Kinetic and mineralogic controls on the evolution of groundwater chemistry and ^{87}Sr/^{86}Sr in a sandy silicate aquifer, northern Wisconsin, USA. *Geochim. Cosmochim. Acta* **60**, 1807–1821.

Bullen T. D. and Kendall C. (1998) Tracing of weathering reactions and water flowpaths: a multi-isotope approach. In *Isotope Tracers in Catchment Hydrology* (eds. C. Kendall and J. J. McDonnell). Elsevier, Amsterdam, pp. 611–646.

Bullen T., White A., Blum A. E., Harden J., and Schultz M. (1997) Chemical weathering of a soil chronosequence on granitoid alluvium: II. Mineralogic and isotopic constraints on the behavior of strontium. *Geochim. Cosmochim. Acta* **61**, 291–306.

Cao H., Cowart J. B., and Osmond J. K. (1999) Uranium and strontium isotopic geochemistry of karst waters, Leon Sinks geological area, Leon County, Florida. *Cave Karst Sci.* **26**, 101–106.

Capo R. C. and Chadwick O. A. (1999) Sources of strontium and calcium in desert soil and calcrete. *Earth Planet. Sci. Lett.* **170**, 61–72.

Capo R. C., Stewart B. W., and Chadwick O. A. (1998) Strontium isotopes as tracers of ecosystem processes: theory and methods. *Geoderma* **82**, 173–195.

Capo R. C., Whipkey C. E., and Chadwick O. A. (2000) Pedogenic origin of dolomite in a basaltic weathering profile, Kohala Peninsula, Hawaii. *Geology* **28**, 271–274.

Chabaux F., Riotte J., Benedetti M., Boulegue J., Gerard M., and Ildefonse P. (1998) Uranium isotopes in surface waters from the Mount Cameroon: tracing water sources or basalt weathering? *Min. Mag.* **62A**, 296–297.

Chadwick O. A., Derry L. A., Vitousek P. M., Huebert B. J., and Hedin L. O. (1999) Changing sources of nutrients during four million years of ecosystem development. *Nature* **397**, 491–497.

Chalov P. I. and Merkulova K. I. (1966) Comparative rate of oxidation of ^{234}U and ^{238}U atoms in certain minerals. *Dokl. Akad. Nauk.* **167**, 146–148.

Chalov P. I., Merkulova K. I., and Mamyrov U. I. (1989) Fractionation of even isotopes of uranium in the process of its separation upon various sorbents from some ground water springs. *Geokhimiya* **27**, 588–592.

Chaudhuri S. (1978) Strontium isotopic composition of several oil field brines from Kansas and Colorado. *Geochim. Cosmochim. Acta* **42**, 329–331.

Cherdyntsev V. V. (1971) *Uranium-234*. Israel Program for Scientific Translations, Jerusalem, 234pp.

Chesley J. T., Quade J., and Ruiz J. (2000) The Os and Sr isotopic record of Himalayan paleorivers: Himalayan tectonics and influence on ocean chemistry. *Earth Planet. Sci. Lett.* **179**, 115–124.

Chiquet A., Michard A., Nahon D., and Hamelin B. (1999) Atmospheric input vs. *in situ* weathering in the genesis of calcretes: an Sr isotope study at Galvez (central Spain). *Geochim. Cosmochim. Acta* **63**, 311–323.

Clauer N. (1979) Relationship between the isotopic composition of strontium in newly formed continental clay minerals and their source material. *Chem. Geol.* **31**, 325–334.

Clauer N. (1981) Strontium and argon isotopes in naturally weathered biotite, muscovite, and feldspars. *Chem. Geol.* **31**, 325–334.

Clauer N., O'Neil J. R., and Bonnot-Courtois C. (1982) The effect of natural weathering on the chemical and isotopic compositions of biotites. *Geochim. Cosmochim. Acta* **46**, 1755–1762.

Clow D. W., Mast M. A., Bullen T. D., and Turk J. T. (1997) Strontium 87/strontium 86 as a tracer of mineral weathering reactions and calcium sources in an alpine/subalpine watershed, Loch Vale, Colorado. *Water Resour. Res.* **33**, 1335–1351.

Collerson K. D., Ullman W. J., and Torgersen T. (1988) Ground waters with unradiogenic $^{87}Sr/^{86}Sr$ ratios in the Great Artesian Basin, Australia. *Geology* **16**, 59–63.

Cowart J. B. and Osmond J. K. (1977) Uranium isotopes in groundwater: their use in prospecting for sandstone-type uranium deposits. *J. Geochem. Explor.* **8**, 365–379.

Cowart J. B. and Osmond J. K. (1980) *Uranium Isotopes in Groundwater as a Prospecting Technique.* US Dept. Energy Report, GJBX 119, 112 pp.

Dasch E. J. (1969) Strontium isotopes in weathering profiles, deep-sea sediments, and sedimentary rocks. *Geochim. Cosmochim. Acta* **33**, 1521–1552.

Dickin A. P. (1995) *Radiogenic Isotope Geology.* Cambridge University Press, Cambridge, 452 pp.

Doe B. R., Hedge C. E., and White D. E. (1966) Preliminary investigation of the source of lead and strontium in deep geothermal brines underlying the Salton Sea geothermal area. *Econ. Geol.* **61**, 462–483.

Dymond J., Biscaye P. E., and Rex R. W. (1974) Eolian origin of mica in Hawaiian soils. *Geol. Soc. Am. Bull.* **85**, 37–40.

Elderfield H. and Greaves M. J. (1981) Strontium isotope geochemistry of Icelandic geothermal system and implications for sea water chemistry. *Geochim. Cosmochim. Acta* **45**, 2201–2212.

Emmanuel S. and Erel Y. (2002) Implications from concentrations and isotopic data for Pb partitioning processes in soils. *Geochim. Cosmochim. Acta* **66**, 2517–2527.

English N. B., Quade J., DeCelles P. G., and Garzione C. N. (2000) Geologic control of Sr and major element chemistry in Himalayan rivers, Nepal. *Geochim. Cosmochim. Acta* **64**, 2549–2566.

Erel Y. (1998) Mechanisms and velocities of anthropogenic Pb migration in Mediterranean soils. *Environ. Res.* **78**, 112–117.

Erel Y. and Morgan J. J. (1992) The relationships between rock-derived Pb and Fe in natural waters. *Geochim. Cosmochim. Acta* **56**, 4157–4167.

Erel Y., Patterson C. C., Scott M. J., and Morgan J. J. (1990) Transport of industrial lead in snow through soil to stream water and groundwater. *Chem. Geol.* **85**, 383–392.

Erel Y., Morgan J. J., and Patterson C. C. (1991) Transport of natural lead and cadmium in a remote mountain stream. *Geochim. Cosmochim. Acta* **55**, 707–721.

Erel Y., Harlavan Y., and Blum J. D. (1994) Lead isotope systematics of granitiod weathering. *Geochim. Cosmochim. Acta* **58**, 5299–5306.

Erel Y., Veron A., and Halicz L. (1997) Tracing the transport of anthropogenic lead in the atmosphere and in soils using isotopic ratios. *Geochim. Cosmochim. Acta* **61**, 4495–4505.

Evans M. J., Derry L. A., Anderson S. P., and France-Lanord C. (2001) Hydrothermal source of radiogenic Sr to Himalayan rivers. *Geology* **29**, 803–806.

Eyal Y. and Olander D. R. (1990) Leaching of uranium and thorium from monazite: I. Initial leaching. *Geochim. Cosmochim. Acta* **54**, 1867–1877.

Faure G. (1986) *Principles of Isotopic Geology.* Wiley, New York, 589pp.

Fisher R. S. and Stueber A. M. (1976) Strontium isotopes in selected streams within the Susquehanna River Basin. *Water Resour. Res.* **12**, 1061–1068.

Fleischer R. L. (1980) Isotopic disequilibria of uranium: alpha recoil damage and preferential solution effects. *Science* **207**, 979–981.

Fleischer R. L. (1982) Nature of alpha-recoil damage: evidence from preferential solution effects. *Nuclear Tracks* **6**, 35–42.

Fleischer R. L. and Raabe O. G. (1978) Recoiling alpha-emitting nuclei—mechanisms for uranium series disequilibrium. *Geochim. Cosmochim. Acta* **42**, 973–978.

Franklyn M. T., McNutt R. H., Camineni D. C., Gascoyne M., and Frape S. K. (1991) Groundwater $^{87}Sr/^{86}Sr$ values in the Eye-Dashwa Lakes pluton, Canada: evidence for plagioclase-water reaction. *Chem. Geol.* **86**, 111–122.

Frindik O. and Vollmer S. (1999) Particle-size dependent distribution of thorium and uranium isotopes in soil. *J. Radioanalyt. Nuclear Chem.* **241**, 291–296.

Galy A., France-Lanord C., and Derry L. A. (1999) The strontium isotopic budget of Himalayan rivers in Nepal and Bangladesh. *Geochim. Cosmochim. Acta* **63**, 1905–1925.

Gast P. W. (1970) Isotopic composition as a natural tracer of lead in the environment. *Environ. Sci. Technol.* **4**, 313–314.

Geyh M. A. and Schleicher H. (1990) *Absolute Age Determination.* Springer, Berlin, 503pp.

Goldich S. S. and Gast P. W. (1966) Effects of weathering on the Rb–Sr and K–Ar ages of biotite from the Morton gneiss, Minnesota. *Earth Planet. Sci. Lett.* **1**, 372–375.

Gomes F. V. M. and Cabral F. C. F. (1981) Utilization of natural uranium isotopes for the study of ground water in Bambui limestone aquifer, Bahia. *Revista Brasileira de Geociencias* **11**, 179–184.

Gosz J. R. and Moore D. I. (1989) Strontium isotope studies of atmospheric inputs to forested watersheds in New Mexico. *Biogeochemistry* **8**, 115–134.

Gosz J. R., Brookins D. G., and Moore D. I. (1983) Using strontium isotope ratios to estimate inputs to ecosystems. *Bioscience* **33**, 23–30.

Graustein W. C. (1989) $^{87}Sr/^{86}Sr$ ratios measure the sources and flow of strontium in terrestrial ecosystems. In *Stable Isotopes in Ecological Research* (eds. P. W. Rundel, J. R. Ehleringer, and K. A. Nagy). Springer, New York, pp. 491–512.

Graustein W. C. and Armstrong R. L. (1983) The use of $^{87}Sr/^{86}Sr$ ratios to measure atmospheric transport into forested watersheds. *Science* **219**, 289–292.

Gulson B. L., Tiller K. G., Mizon K. J., and Merry R. H. (1981) Use of lead isotopes in soils to identify the source of lead contamination near Adelaide, South Australia. *Environ. Sci. Technol.* **15**, 691–696.

Hakam O. K., Choukri A., Reyss J. L., and Lferde M. (2001) Determination and comparison of uranium and radium isotopes activities and activity ratios in samples from some natural water sources in Morocco. *J. Environ. Radioact.* **57**, 175–189.

Hamidi E. M., Nahon D., McKenzie J. A., Michard A., Colin F., and Kamel S. (1999) Marine Sr (Ca) input in Quaternary volcanic rock weathering profiles from the Mediterranean coast of Morocco; Sr isotopic approach. *Terra Nova* **11**, 157–161.

Harlavan Y. and Erel Y. (2002) The release of Pb and REE from granitoids by the dissolution of accessory phases. *Geochim. Cosmochim. Acta* **66**, 837–848.

Harlavan Y., Erel Y., and Blum J. D. (1998) Systematic changes in lead isotopic composition with soil age in glacial granitic terrains. *Geochim. Cosmochim. Acta* **62**, 33–46.

Harris N., Bickle M., Chapman H., Fairchild I., and Bunbury J. (1998) The significance of Himalayan rivers for silicate weathering rates: evidence from the Bhote Kosi tributary. *Chem. Geol.* **144**, 205–220.

Hart S. R. and Tilton G. R. (1966) The isotope geochemistry of strontium and lead in Lake Superior sediments and water. In *The Earth Beneath the Continents*. American Geophysical Union, Washington, DC, vol. 10.

Hodell D. A. (1994) Editorial: progress and paradox in strontium isotope stratigraphy. *Paleoceanography* **9**, 395–398.

Hogan J. F. and Blum J. D. (2003) Tracing hydrologic flowpaths in a small watershed using variations in $^{87}Sr/^{86}Sr$, [Ca]/[Sr], [Ba]/[Sr] and $\delta^{18}O$. *Water Resour. Res.* (in press).

Hogan J. F., Blum J. D., Siegel D. I., and Glaser P. H. (2000) $^{87}Sr/^{86}Sr$ as a tracer of groundwater discharge and precipitation recharge in the Glacial Lake Agassiz Peatlands, northern Minnesota. *Water Resour. Res.* **36**, 3701–3710.

Horton T. W., Chamberlain C. P., Fantle M., and Blum J. D. (1999) Chemical weathering and lithologic controls of water chemistry in a high-elevation river system: Clark's fork of the Yellowstone River, Wyoming and Montana. *Water Resour. Res.* **35**, 1643–1656.

Hussain N. (1995) Supply rates of natural U–Th series radionuclides from aquifer solids into groundwater. *Geophys. Res. Lett.* **22**, 1521–1524.

Innocent C., Michard A., Malengreau N., Loubet M., Noack Y., Benedetti M., and Hamelin B. (1997) Sr isotopic evidence for ion-exchange buffering in tropical laterites from the Parana, Brazil. *Chem. Geol.* **136**, 219–232.

Ivanovich M., Duerden P., Payne T., Nightingale T., Longworth G., Wilkins M.A., Hasler S. E., Edgehill R. B., Cockayne D. J., and Davey B. G. (1988) Natural analogue study of the distribution of uranium series radionuclides between colloid and solute phases in hydrological systems. DOE report AERE-R. 12975/DOE/RW/88.076.

Jacks G., Aberg G., and Hamilton P. J. (1989) Calcium budgets for catchments as interpreted by strontium isotopes. *Nordic Hydrol.* **20**, 85–96.

Jacobson A. D., Blum J. D., Chamberlain C. P., Poage M. A., and Sloan V. F. (2002a) The Ca/Sr and Sr isotope systematics of a Himalayan glacial chronosequence: carbonate versus silicate weathering rates as a function of landscape surface age. *Geochim. Cosmochim. Acta* **66**, 13–27.

Jacobson A. D., Blum J. D., Chamberlain C. P., Craw D., and Koons P. O. (2002b) Climatic versus tectonic controls on weathering in the New Zealand Southern Alps. *Geochim. Cosmochim. Acta* **66**, 3417–3429.

Johnson T. M. and DePaolo D. J. (1994) Interpretation of isotopic data in groundwater systems: model development and application to Sr isotope data from Yucca Mountain. *Water Resour. Res.* **30**, 1571–1587.

Johnson T. M. and DePaolo D. J. (1996) Reaction-transport models for radiocarbon in groundwater: the effects of longitudinal dispersion and the use of Sr isotope ratios to correct for water-rock interaction. *Water Resour. Res.* **32**, 2203–2212.

Johnson T. M. and DePaolo D. J. (1997a) Rapid exchange effects on isotope ratios in groundwater systems: 1. Development of a transport–dissolution–exchange model. *Water Resour. Res.* **33**, 187–195.

Johnson T. M. and DePaolo D. J. (1997b) Rapid exchange effects on isotope ratios in groundwater systems: 2. Flow investigation using Sr isotope ratios. *Water Resour. Res.* **33**, 197–209.

Johnson T. M., Roback R. C., McLing T. L., Bullen T. D., DePaolo D. J., Doughty C., Hunt R. J., Smith R. W., Cecil L. D., and Murrell M. T. (2000) Groundwater "fast paths" in the Snake River Plain aquifer: radiogenic isotope ratios as natural groundwater tracers. *Geology* **28**, 871–874.

Karim A. and Veizer J. (2000) Weathering processes in the Indus River Basin: implications from riverine carbon, sulfur, oxygen, and strontium isotopes. *Chem. Geol.* **170**, 153–177.

Katz B. G. and Bullen T. D. (1996) The combined use of $^{87}Sr/^{86}Sr$ and carbon and water isotopes to study the hydrochemical interaction between groundwater and lakewater in mantled karst. *Geochim. Cosmochim. Acta* **60**, 5075–5087.

Kendall C. and Caldwell E. A. (1998) Fundamentals of isotope geochemistry. In *Isotope Tracers in Catchment Hydrology* (eds. C. Kendall and J. J. McDonnell). Elsevier, Amsterdam, pp. 51–86.

Kennedy M. J., Chadwick O. A., Vitousek P. M., Derry L. A., and Hendricks D. M. (1998) Changing sources of base cations during ecosystem development, Hawaiian Islands. *Geology* **26**, 1015–1018.

Kennedy M. J., Hedin L. O., and Derry L. A. (2002) Decoupling of unpolluted temperate forests from rock nutrient sources revealed by natural $^{87}Sr/^{86}Sr$ and ^{84}Sr tracer addition. *Proc. Natl. Acad. Sci.* **99**, 9639–9644.

Kigoshi K. (1971) Alpha-recoil ^{234}Th: dissolution into water and the $^{234}U/^{238}U$ disequilbrium in nature. *Science* **173**, 47–48.

Kovalev V. P. and Malyasova Z. V. (1971) The content of mobile uranium in extrusive and intrusive rocks of the eastern margins of the South Minusinsk Basin. *Geokhimiya* **7**, 855–865.

Kraemer T. F. (1981) ^{234}U and ^{238}U concentration in brine from geopressured aquifers of the northern Gulf of Mexico Basin. *Earth Planet. Sci. Lett.* **56**, 210–216.

Kraemer T. F. and Genereux D. O. (1998) Applications of uranium- and thorium-series radionuclides in catchment hydrology studies. In *Isotope Tracers in Catchment Hydrology* (eds. C. Kendall and J. J. McDonnell). Elsevier, Amsterdam, pp. 679–722.

Kronfeld J. and Rosenthal E. (1981) Uranium isotope as a natural tracer of waters of the Bet-Shean-Harod Valley, Israel. *Israel J. Hydrol.* **22**, 77–88.

Kronfeld J., Gradsztan E., Muller H. W., Radin J., Yaniv A., and Zach R. (1975) Excess ^{234}U: an aging effect in confined water. *Earth Planet. Sci. Lett.* **27**, 189–196.

Kronfeld J., Vogel J. C., and Talma A. S. (1994) A new explanation for extreme $^{234}U/^{238}U$ disequilibria in a dolomitic aquifer. *Earth Planet. Sci. Lett.* **123**, 81–93.

Kurtz A. C., Derry L. A., and Chadwick O. A. (2001) Accretion of Asian dust to Hawaiian soils: isotopic, elemental and mineral mass balance. *Geochim. Cosmochim. Acta* **65**, 1971–1983.

Land M., Ingri J., Andersson P. S., and Ohlander B. (2000) Ba/Sr, Ca/Sr and $^{87}Sr/^{86}Sr$ ratios in soil water and groundwater: implications for relative contributions to stream water discharge. *Appl. Geochem.* **15**, 311–325.

Lathan A. G. and Schwarcz H. P. (1987) On the possibility of determining rates of removal of uranium from crystalline igneous rocks using U-series disequilibria: 1. A U-leach model, and its applicability to whole rock data. *Appl. Geochem.* **2**, 55–65.

Lienert C., Short S. A., and von Gunten H. R. (1994) Uranium infiltration from a river to shallow groundwater. *Geochim. Cosmochim. Acta.* **58**, 5455–5463.

Lowson R. T., Short S. A., Davey B. G., and Gray D. J. (1986) $^{234}U/^{238}U$ and $^{230}Th/^{234}U$ activity ratios in mineral phases in a lateritic weathered zone. *Geochim. Cosmochim. Acta* **50**, 1697–1702.

Luck J. and Ben Othman D. (1998) Geochemistry and water dynamics: II. Trace metals and Pb–Sr isotopes as tracers of water movements and erosion processes. *Chem. Geol.* **150**, 263–282.

Lyons W. B., Tyler S. W., Gaudette H. E., and Long D. T. (1995) The use of strontium isotopes in determining groundwater mixing and brine fingering in a playa spring zone, Lake Tyrrell, Australia. *J. Hydrol.* **167**, 225–239.

Mariner R. H. and Young H. W. (1996) Lead and strontium isotopes indicate deep thermal-aquifer in Twin Falls, Idaho, area. *Feder. Geotherm. Res. Prog. Update* **4**, 135–140.

Marshall D. D., Whelan J. F., Peterman Z. E., Futa K., Mahan S. A., and Struckless J. S. (1992) Isotopic studies of fracture coatings at Yucca Mountain, Nevada, USA. In *Proceedings 7th Water–Rock Interaction Symposium Park City, UT* (eds. Y. K. Kharaka and A. S. Maest). A. A. Balkema, Roterdam, pp. 737–740.

McNutt R. H., Frape S. K., Fritz P., Jones M. G., and MacDonald I. M. (1990) The ^{87}Sr/^{86}Sr values of Canadian shield brines and fracture minerals with applications to groundwater mixing, fracture history, and geochronology. *Geochim. Cosmochim. Acta* **54**, 205–215.

Michel J. (1984) Redistribution of uranium and thorium series isotopes during isovolumetric weathering of granite. *Geochim. Cosmochim. Acta* **48**, 1249–1255.

Miller E. K., Blum J. D., and Friedland A. J. (1993) Determination of soil exchangeable-cation loss and weathering rates using Sr isotopes. *Nature* **362**, 438–441.

Moore W. S. (1967) Amazon and Mississippi river concentration of uranium, thorium, and radium isotopes. *Earth Planet. Sci. Lett.* **2**, 231–234.

Moreira-Nordemann L. M. (1980) Use of ^{234}U/^{238}U disequilibrium in measuring chemical weathering rate of rocks. *Geochim. Cosmochim. Acta* **44**, 103–108.

Musgrove M. and Banner J. L. (1993) Regional groundwater mixing and the origin of saline fluids—mid-continent, United States. *Science* **259**, 1877–1882.

Naftz D. L., Peterman Z. E., and Spangler L. E. (1997) Using delta ^{87}Sr values to identify sources of salinity to a freshwater aquifer, Greater Aneth Oil Field, Utah USA. *Chem. Geol.* **141**, 195–209.

Naiman Z., Quade J., and Patchett P. J. (2000) Isotopic evidence for eolian recycling of pedogenic carbonate and variations in carbonate dust sources throughout the southwest United States. *Geochim. Cosmochim. Acta* **64**, 3099–3109.

Neumann K. and Dreiss S. (1995) Strontium 87/strontium 86 ratios as tracers in groundwater and surface waters in Mono Basin, California. *Water Resour. Res.* **31**, 3183–3193.

Nimz G. J. (1998) Lithogenic and cosmogenic tracers in catchment hydrology. In *Isotope Tracers in Catchment Hydrology* (eds. C. Kendall and J. J. McDonnell). Elsevier, Amsterdam, pp. 247–290.

Ohlander B., Ingri J., Land M., and Schoberg H. (2000) Change of Sm–Nd isotope composition during weathering of till. *Geochim. Cosmochim. Acta* **64**, 813–820.

Osmond J. K. and Cowart J. B. (1976) The theory and uses of natural uranium isotopic variations in hydrology. *Atom. Energy Rev.* **14**, 621–679.

Osmond J. K. and Cowart J. B. (1992) Ground water. In *Uranium Series Disequilibrium: Application to Earth, Marine, and Environmental Sciences* (eds. M. Ivanovich and R. S. Harmon). Oxford University Press, Oxford, pp. 290–333.

Osmond J. K. and Ivanovich M. (1992) Uranium-series mobilization and surface hydrology. In *Uranium Series Disequilbrium: Application to the Earth, Marine, and Environmental Sciences* (eds. M. Ivanovich and R. S. Harmon). Oxford University Press, Oxford, pp. 259–288.

Osmond J. K., Kaufman M. I., and Cowart J. B. (1974) Mixing volume calculations, sources and aging trends of Floridan aquifer water by uranium isotopic methods. *Geochim. Cosmochim. Acta* **38**, 1083–1100.

Paces J. B., Ludwig K. R., Peterman Z. E., and Neymark L. A. (2002) ^{234}U/^{238}U evidence for local recharge and patterns of ground-water flow in the vicinity of Yucca Mountain, Nevada, USA. *Appl. Geochem.* **17**, 751–779.

Palmer M. R. and Edmond J. M. (1992) Controls over the strontium isotope composition of river water. *Geochim. Cosmochim. Acta* **56**, 2099–2111.

Palmer M. R. and Edmond J. M. (1993) Uranium in river water. *Geochim. Cosmochim. Acta* **56**, 4947–4955.

Palmer M. R. and Elderfield H. (1985) Sr isotope composition of sea water over the past 75 Myr. *Nature* **314**, 526–528.

Pande K., Sarin M. M., Trivedi J. R., Krishnaswami S., and Sharma K. K. (1994) The Indus River system (India–Pakistan): major-ion chemistry, uranium and strontium isotopes. *Chem. Geol.* **116**, 245–259.

Pegram W. J., Esser B. K., Krishnaswami S., and Turekian K. K. (1994) The isotopic composition of leachable osmium from river sediments. *Earth Planet. Sci. Lett.* **128**, 591–599.

Peterman Z. E. and Wallin B. (1999) Synopsis of strontium isotope variations in groundwater at Aspo, southern Sweden. *Appl. Geochem.* **14**, 939–951.

Peters S. C., Blum J. D., Driscoll C. T., and Likens G. E. (2003) Dissolution of wollastonite during the experimental manipulation of a forested catchment. *Biogeochemistry* (in press).

Peucker-Ehrenbrink B. and Blum J. D. (1998) Re–Os isotope systematics and weathering of Precambrian crustal rocks: implications for the marine osmium isotope record. *Geochim. Cosmochim. Acta* **62**, 3193–3203.

Plater A. J., Ivanovich M., and Dugdale R. E. (1992) Uranium series disequilibrium in river sediments and waters: the significance of anomalous activity ratios. *Appl. Geochem.* **7**, 101–110.

Probst A., El Gh'mari A., Aubert D., Fritz B., and McNutt R. (2000) Strontium as a tracer of weathering processes in a silicate catchment polluted by acid atmospheric inputs, Strengbach, France. *Chem. Geol.* **170**, 203–219.

Quade J., Chivas A. R., and McCulloch M. T. (1995) Strontium and carbon isotope tracers and the origins of soil carbonate in South Australia and Victoria. In *Arid-Zone Paleoenvironments*, Palaeogeogr, Palaeoclimatol. Palaeoecol. 113 (ed. A. R. Chivas), pp. 103–117.

Quade J., Roe L., DeCelles P. G., and Ojha T. P. (1997) The late Neogene ^{87}Sr/^{86}Sr record of lowland Himalayan rivers. *Science* **276**, 1828–1831.

Riotte J. and Chabaux F. (1999) (^{234}U/^{238}U) activity ratios in freshwaters as tracers of hydrological processes: the Strengbach watershed (Vosges, France). *Geochim. Cosmochim. Acta* **63**, 1263–1275.

Roback R. C., Johnson T. M., McLing T. L., Murrell M. T., Luo S. D., and Ku T. L. (2001) Groundwater flow patterns and chemical evolution in Snake River Plain aquifer in the vicinity of the INEEL: constraints from ^{234}U/^{238}U and ^{87}Sr/^{86}Sr isotope ratios. *Geol. Soc. Am. Bull.* **113**, 1133–1141.

Romero G. E. T., Ordonez R. E., Esteller A. M. V., and Reyes G. L. R. (1999) Uranium behaviour through the unsaturated zone in soil. In *Environmental Radiochemical Analysis*, Special Publication, 234 (ed. G. W. Newton). Royal Society of Chemistry, pp. 143–151.

Rosholt J. N. and Bartel A. J. (1969) Uranium, thorium, and lead systematics in Granite Mountains, Wyoming. *Earth Planet. Sci. Lett.* **7**, 141–147.

Rosholt J. N., Shields W. R., and Garner E. L. (1963) Isotope fractionation of uranium in sandstone. *Science* **139**, 224–226.

Sarin M., Krishnaswami S., Somayajulu B. L. K., and Moore W. S. (1990) Chemistry of U, Th, and Ra isotopes in the Ganga–Brahamputra river system: weathering processes and fluxes to the bay of Bengal. *Geochim. Cosmochim. Acta* **54**, 1387–1396.

Sass E. and Starinsky A. (1979) Behaviour of strontium in subsurface calcium chloride brines: southern Israel and Dead Sea rift valley. *Geochim. Cosmochim. Acta* **43**, 885–895.

Scott M. R. (1982) The chemistry of U- and Th-series nuclides in rivers. In *Uranium Series Disequilbrium: Application to Earth, Marine, and Environmental Sciences* (eds. M. Ivanovich and R. S. Harmon). Oxford University Press, Oxford, pp. 181–202.

Seimbille F., Zuddas P., and Michard G. (1998) Granite-hydrothermal interaction: a simultaneous estimation of the mineral dissolution rate based on the isotopic doping technique. *Earth Planet. Sci. Lett.* **157**, 183–191.

Sharma M. and Wasserburg G. J. (1997) Osmium in the rivers. *Geochim. Cosmochim. Acta* **61**, 5411–5416.

Sharma M., Wasserburg G. J., Hofmann A. W., and Chakrapani G. J. (1999) Himalayan uplift and osmium isotopes in oceans and rivers. *Geochim. Cosmochim. Acta* **63**, 4005–4012.

Short S. A., Lowson R. T., and Ellis J. (1988) $^{234}U/^{238}U$ and $^{230}Th/^{234}U$ activity ratios in the colloidal phases of aquifers in lateritic weathered zones. *Geochim. Cosmochim. Acta* **52**, 2555–2563.

Siegel D. I., Bickford M. E., and Orrell S. E. (2000) The use of strontium and lead isotopes to identify sources of water beneath the Fresh Kills Landfill, Staten Island, New York, USA. *Appl. Geochem.* **15**, 493–500.

Silver L. T., Woodhead J. A., Williams I. S., and Chappell B. W. (1984) *Uranium in Granites from the Southwestern United States: Actinide Parent–Daughter Systems, Sites and Mobilization.* Department of Energy, Report DE-AC13-76GI01664. 380pp.

Smalley P. C., Raheim A., Dickson J. A. D., and Emery D. (1998) $^{87}Sr/^{86}Sr$ in waters from the Lincolnshire Limestone aquifer, England, and the potential of natural strontium isotopes as a tracer for a secondary recovery seawater injection process in oilfields. *Appl. Geochem.* **3**, 591–600.

Spiridonov A. I., Sultankhodzhayev A. N., Surganova N. A., and Tyminskiy V. G. (1969) Some results of the study of uranium isotopes ($^{234}U/^{238}U$) in ground water of the artesian basin in the Tashkent area. *Uzbekiston Geologiya Zhurnali* **4**, 82–84.

Starinsky A., Bielski M., Ecker A., and Steinitz G. (1983a) Tracing the origin of salts in groundwater by Sr isotopic composition (the crystalline complex of the southern Sinai, Egypt). *Chem. Geol.* **41**, 257–267.

Starinsky A., Bielski M., Lazar B., Steinitz G., and Raab M. (1983b) Strontium isotope evidence on the history of oilfield brines, Mediterranean Coastal Plain, Israel. *Geochim. Cosmochim. Acta* **47**, 687–695.

Steinmann M. and Stille P. (1997) Rare earth element behavior and Pb, Sr, Nd isotope systematics in a heavy metal contaminated soil. *Appl. Geochem.* **12**, 607–623.

Stettler A. (1977) $^{87}Rb/87Sr$ systematics of a geothermal water–rock association in the Massif Central, France. *Earth Planet. Sci. Lett.* **34**, 432–438.

Stettler A. and Allègre C. J. (1979) $^{87}Rb–^{87}Sr$ constraints on the genesis and evolution of the Cantal contitntal volcanic system (France). *Earth Planet. Sci. Lett.* **44**, 269–278.

Stewart B. W., Capo R. C., and Chadwick O. A. (1998) Quantitative strontium isotope models for weathering, pedogenesis and biogeochemical cycling. *Geoderma* **82**, 173–195.

Stewart B. W., Capo R. C., and Chadwick O. A. (2001) Effects of rainfall on weathering rate, base cation provenance, and Sr isotope composition of Hawaiian soils. *Geochim. Cosmochim. Acta* **65**, 1087–1099.

Stille P. and Shields G. (1997) *Radiogenic Isotope Geochemistry of Sedimentary and Aquatic Systems.* Springer, Berlin, 217pp.

Stillings L. L. and Brantley S. L. (1995) Feldspar dissolution at 25°C and pH 3: reaction stoichiometry and the effect of cations. *Geochim. Cosmochim. Acta* **59**, 1483–1496.

Straughan I. R., Elseewi A. A., Kaplan I. R., Hurst R. W., and Davis T. E. (1981) Fly ash-derived strontium as an index to monitor deposition from coal-fired power plants. *Science* **212**, 1267–1269.

Stuckless J. S., Bunker C. M., Bush C. A., Doering W. P., and Scott J. H. (1977) Geochemical and petrological studies of uraniferous granite from Granite Mountains, Wyoming. *US Geol. Surv. J. Res.* **5**, 61–81.

Stuckless J. S., Peterman Z. E., and Muhs D. R. (1991) U and Sr isotopes in ground water and calcite, Yucca Mountain, Nevada: evidence against upwelling water. *Science* **254**, 551–554.

Steuber A. M., Pushkar P., and Hetherington E. A. (1984) A strontium isotope study of Smackover brines and associated solids, southern Arkansas. *Geochim. Cosmochim. Acta* **48**, 1637–1649.

Stueber A. M., Pushkar P., and Hetherington E. A. (1987) A strontium isotopic study of formation waters from the Illinois Basin, USA. *Appl. Geochem.* **2**, 477–494.

Stueber A. M., Walter L. M., Huston T. J., and Pushkar P. (1993) Formation waters from Mississippian–Pennsylvanian reservoirs, Illinois Basin, USA: chemical and isotopic constraints on evolution and migration. *Geochim. Cosmochim. Acta* **57**, 763–784.

Suksi J., Rasilainen K., Casanova J., Ruskeeniemi T., Blomqvist R., and Smellie J. (2001) U-series disequilibria in a groundwater flow route as an indicator of uranium migration processes. *J. Contamin. Hydrol.* **47**, 187–196.

Szalay S. and Samsoni Z. (1969) Investigation of the leaching of the uranium from crushed magmatic rocks. *Geochemistry* **14**, 613–623.

Taylor A. S. and Lasaga A. C. (1999) The role of basalt weathering in the Sr isotope budget of the oceans. *Chem. Geol.* **161**, 199–214.

Taylor A. S., Blum J. D., and Lassaga A. C. (2000a) The dependence of labradorite dissolution and Sr isotope release rates on solution saturation state. *Geochim. Cosmochim. Acta* **64**, 2389–2400.

Taylor A. S., Blum J. D., Lasaga A. C., and MacInnis I. N. (2000b) Kinetics of dissolution and Sr release during biotite and phlogopite weathering. *Geochim. Cosmochim. Acta* **64**, 1191–1208.

Teutsch N., Erel Y., Halicz L., and Banin A. (2001) The distribution of natural and anthropogenic lead in Mediterranean soils. *Geochim. Cosmochim. Acta* **65**, 2853–2864.

Toulhoat P. and Beaucaire C. (1993) Geochemistry of water crossing the Cigar Lake uranium deposit (Saskatchewan, Canada), and use of uranium and lead isotopes as ore guides. *Can. J. Earth Sci.* **30**, 754–763.

Toulhoat P., Gallien J. P., Louvat D., and Moulin V. (1996) Preliminary studies of groundwater flow and migration of uranium isotopes around the Oklo natural reactors (Gabon). *J. Contamin. Hydrol.* **21**, 3–17.

Tricca A., Stille P., Steinmann M., Kiefel B., and Samuel J. (1999) Rare earth elements and Sr and Nd isotopic compositions of dissolved and suspended loads from small river systems in the Vosges Mountains (France), the river Rhine and groundwater. *Chem. Geol.* **160**, 139–158.

Tuzova T. V. and Novikov V. N. (1991) Uranium isotope-related features of streamflow formation for Pyandzh River. *Water Resour. Res.* **18**, 59–65.

Ullman W. J. and Collerson K. D. (1994) The Sr-isotope record of late-Quaternary hydrologic changes around Lake Frome, South Australia. *Austral. J. Earth Sci.* **41**, 37–45.

Vance D. and Burton K. (1999) Neodymium isotopes in planktonic foraminifera: a record of the response of continental weathering and ocean circulation rates to climate change. *Earth Planet. Sci. Lett.* **173**, 365–379.

van de Flierdt T., Frank M., Lee D., and Halliday A. N. (2002) Glacial weathering and the hafnium isotope composition of seawater. *Earth Planet. Sci. Lett.* **198**, 167–175.

Vitousek P. M., Kennedy M. J., Derry L. A., and Chadwick O. A. (1999) Weathering versus atmospheric sources of strontium in ecosystems on young volcanic soils. *Oecologia* **121**, 255–259.

von Gunten H. R., Roessler E., Lowson R. T., Reid P. D., and Short S. A. (1999) Distribution of uranium and thorium series radionuclides in mineral phases of a weathered lateritic transect of a uranium ore body. *Chem. Geol.* **160**, 225–240.

Wadleigh M. A., Veizer J., and Brooks C. (1985) Strontium and its isotopes in Canadian rivers: fluxes and global implications. *Geochim. Cosmochim. Acta* **49**, 1727–1736.

Whipkey C. E., Capo R. C., Chadwick O. A., and Stewart B. W. (2000) The importance of sea spray to the cation budget of a coastal Hawaiian soil: a strontium isotope approach. *Chem. Geol.* **168**, 37–48.

White A. F., Blum A. E., Bullen T. D., Vivit D. V., Schulz M., and Fitzpatrick J. (1999) The effect of temperature on experimental and natural chemical weathering rates of granitoid rocks. *Geochim. Cosmochim. Acta* **63**, 3277–3291.

Whitehead K., Ramsey M. H., Maskall J., Thornton I., and Bacon J. R. (1997) Determination of the extent of anthropogenic Pb migration through fractured sandstone using Pb isotope tracing. *Appl. Geochem.* **12**, 75–81.

Woods T. L., Fullagar P. D., Spruill R. K., and Sutton-Lynn C. (2000) Strontium isotopes and major elements as tracers of ground water evolution: example from the upper Castle Hayne Aquifer of North Carolina. *Ground Water* **38**, 762–771.

Yang J., Chen J., An Z., Shields G., Tao X., Zhu H., Ji J., and Chen Y. (2000) Variations in $^{87}Sr/^{86}Sr$ ratios of calcites in Chinese loess: a proxy for chemical weathering associated with the East Asian summer monsoon. *Palaeogeogr. Palaeoclimatol. Palaeoecol.* **157**, 151–159.

Zhao S., Liu M., Qiao G., Wu A., and Li C. (1994) A study and application of uranium isotopes in underground and surface water from Liangcheng area, southern Shandong. *Acta Petrologica Sinica* **10**, 202–210.

Zielinski R. A., Peterman Z. E., Stuckless J. S., Rosholt J. N., and Nkomo I. T. (1981) The chemical and isotopical record of rock–water interaction in the Sherman Granite, Wyoming and Colorado. *Contrib. Mineral. Petrol.* **78**, 209–219.

Zuddas P., Seimbille F., and Michard G. (1995) Granite-fluid interaction at near-equilibrium conditions: experimental and theoretical constraints from Sr contents and isotopic ratios. *Chem. Geol.* **121**, 145–154.

5.13

Geochemistry of Saline Lakes

B. F. Jones and D. M. Deocampo

US Geological Survey, Reston, VA, USA

5.13.1 INTRODUCTION

Saline lakes are important environmental features, with significant geochemical impacts on ecology, water resources, and economic activity around the world. Ancient sediments within saline lake basins also offer important sedimentary archives of past climates and tectonics, because such deposits can record significant hydrologic variations.

Saline lake geochemistry is the product of a complex system involving meteoric precipitation, weathering, groundwater, evaporation, precipitation–dissolution reactions, and biotic activity. Its study is therefore an inherently interdisciplinary effort, and substantial reviews of the subject have emphasized relevant aspects of the hydrology (Rosen, 1994), mineralogy (Spencer, 2000), sedimentology (Hardie *et al.*, 1978; Smoot and Lowenstein, 1991), and evolutionary pathways taken by progressively evaporated waters (Eugster and Hardie, 1978).

The purpose of this paper is to summarize the geochemistry of saline lake basins throughout the world, from dilute inflow to evaporated brine. Following the general approach of Eugster and Hardie (1978), we will review the theoretical background of the evolution of closed basin waters, and then selected field examples that are representative of the major water types. In this work, we have assigned an increased importance to the effect of magnesium salts on brine evolution pathways, and have incorporated this into new models of brine evolution and evaporite precipitation. We follow Chapter 5.16 in referring to waters with total dissolved solids greater than 3.5×10^4 mg L^{-1} as true "brines," and those with from 1 mg L^{-1} to 10^4 mg L^{-1} as "brackish."

5.13.2 REGIONAL OCCURRENCE

Saline lakes are found in all the continents (Figure 1). Hydrological closure that can lead to

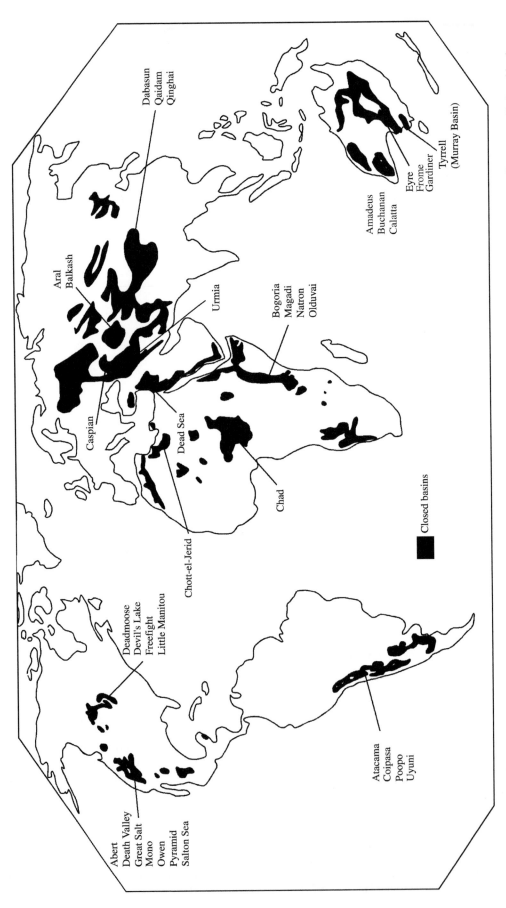

Figure 1 Global distribution of major hydrologically closed watersheds. Important saline lakes discussed in the text are approximately located according to overall closed basin areas. Saline lake areas of Antarctica (described in Section 5.13.6) are not shown (after Cooke and Warren, 1973; Smoot and Lowenstein, 1991).

Table 1 Salt content, volume, and elevation of the world's largest saline lakes.

Lake name	Salt mass ($\times 10^{15}$ g)	Volume (km³)	Elevation (masl)
Caspian Sea	1,016	78,200	-28
Dead Sea	56	188	-393
Aral Sea	10.7	1,020	53
Lake Urmia	10.4	45	1275
Lake Issyk-kul	10.0	1,730	1,608
Kara Bogaz Bay	7.0	20	-28
Great Salt Lake	5.4	19	1,240
Lake Van	4.6	206	1,646
Lake Eyre	2.3	23	-10
Lake Turkana	NA	251	360
Lake Balkhash	NA	118	341
Lake Chad	NA	44	285

salt buildup may occur on the flanks of continental uplifts, within extensional basins, in intermontane basins, and in glaciated terrain (Table 1). Most of the world's saline lake water is found in the Caspian Sea and in the other lakes of western Asia, located in foreland basins associated with Cenozoic orogeny. Old, deep rift basins such as the East African Rift also host large saline lakes, such as Lake Turkana. All other saline lakes are much smaller in volume than these examples, but they provide important evidence of water–mineral interactions in their respective geologic, hydrologic, and climatic settings.

Inasmuch as evaporation in excess of precipitation plays such an important role in the development of saline lakes, it is not at all surprising that these features are to be found in the semi-arid or arid regions of the world. However, in the driest parts of the true deserts of the globe, such as the Arabian or Atacama deserts, there is likely to be insufficient precipitation even for the development of any surface drainage system, much less a highly ephemeral water body (Cooke and Warren, 1973). Under severe drought conditions, the most likely location of a saline lake, or even a playa, is at the periphery of such areas, where inflow originates outside the low gradient margins, and evaporative moisture loss becomes more intense toward the central part. The largest arid to semi-arid regions illustrate this situation. The steppes of central Asia not only host the Caspian Sea and the Aral Sea, but also Lake Urmia on the southwest, and Lake Balkash to the east (Kelts and Shahrabi, 1986; Micklin, 1988; Verzilin and Utsal, 1991; Kosarev and Yablonskaya, 1994). In China, the Quinghai Lake is just east of the Gobi desert (Spencer *et al.*, 1990; Zheng *et al.*, 1993; Zheng, 1997; Duan and Hu, 2001). On the margins of the Sahara are Lake Chad to the south, and the Chott-el-Jerid and related playas of Tunisia and Algeria to the north (Gac *et al.*, 1977; Bryant *et al.*, 1994).

On the eastern edge of the Kalahari desert in southern Africa are the alkali marshes and ponds of the Okavango delta (McCarthy *et al.*, 1998). In the central part of the South American Altiplano, Lake Poopo is the present terminus of outflow from Lake Titicaca to the north, but the giant salars of Coipasa and Uyuni further south are the remnants of a much larger lake system that occupied the entire region in the Pleistocene (Rettig *et al.*, 1980; Risacher and Fritz, 1991a,b). Similarly, in North America, the Great Salt and Sevier Lakes in the east, Pyramid, Mono, and Walker lakes in the west, Abert and Harney Lakes in the north, and the Salton Sea in the south all are near the margins of the Great Basin in the western United States, and all were part of significant paleolakes in glacial times (Spencer *et al.*, 1985a,b). The large, historically flooded playa of Lake Eyre (as well as "Lakes" Frome, Torrens, and Gardiner) to the southeast (Bowler, 1986), are in the Australian outback, whereas the greatest concentration of very shallow, acid saline lakes of the deeply weathered crystalline "Yilgarn Block" terrain occurs near the western margin (Mann, 1983; MacArthur *et al.*, 1991).

Several of the world's largest saline lakes are found at elevations exceeding 1 km, such as Lake Van (Turkey) and the Great Salt Lake (USA) reflecting the tendency for these systems to form in tectonically active regions. Even those large saline lakes nearer to sea level, or below, such as the Caspian and Dead Seas, are associated with active tectonism. As we will discuss below, the tectonic setting of saline lakes is an important control on their geology and geochemistry.

5.13.3 PHYSICAL CONDITIONS AFFECTING GEOCHEMISTRY

5.13.3.1 Hydrology

Three basic conditions must be met for a saline lake to form and persist (Eugster and Hardie, 1978). First, outflow must be absent or severely restricted to ensure hydrological closure. Second, evaporation must exceed or approximate inflow. Finally, inflow must be sufficient to sustain a permanent body of water at or very close to the surface. Probably the most favorable locations for saline lakes are in the rain-shadows of mountain ranges or highland areas providing the catchment for precipitation, whereas the adjoining arid basin floor is exposed to substantial evaporation. In other areas of much lower relief, shallow basins can be the focus of local discharge and evaporation from regionally extensive groundwater systems.

Under any circumstances, to maintain a permanent saline body of water, surface or groundwater inflow must closely approximate, but not exceed, the rate of evaporation (Figure 2). If maintained

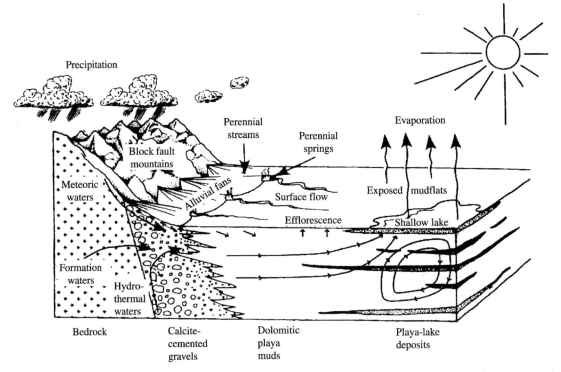

Figure 2 Major hydrologic features of hydrologically closed basins (after Eugster and Hardie, 1975) (reproduced with permission of the Geological Society of America from *Geol. Soc. Am. Bull.* **1975**, *86*, 319–334). Flow lines have been added here beneath the playa lake to indicate the possibility of salinity-driven density circulation, and the interaction between fresh meteoric groundwaters and recirculating evolved brines (sources Duffy and Al-Hassan, 1988; Rosen, 1994).

primarily by surface inflow with typical seasonal variation, the lake level will fluctuate significantly and the hydrochemistry will be strongly affected by peripheral areas, such as in marshes, mudflats, and deltas. If a saline lake is fed principally by groundwater, its chemistry will be more affected by internal processes. Saline lakes, therefore, represent the simple, completely closed end-member of the playa types discussed by Rosen (1994), where the balance between inflow and evaporation is maintained over time, allowing dissolved solids to accumulate in the water column. Because their only outlet is evaporation, changes in the hydrologic budget for a basin, such as precipitation, runoff, leakage, or evaporation, will be balanced by a change in the surface area of the lake (Almendinger, 1990). To accommodate a change in surface area, closed lakes can fluctuate drastically in their level, thereby leaving sedimentological, biological, and geochemical evidence of their paleohydrology. Such hydrologic responses produce particularly dramatic changes in lake level where topographic gradients are steep.

The sources of water and solutes for saline lake brines are mainly direct precipitation, associated surface flow, or groundwater (Figures 2 and 3). Groundwater can be derived from the local or regional meteoric system, interstitial water of sediments, or deep basinal or hydrothermal

fluids (Rosen, 1994). Interstitial fluids in playa settings are usually more concentrated than surface waters, and may have chemistries dramatically different from those of inflow sources (Jones *et al.*, 1969). Hydrothermal discharge, though small in volume, can be appreciable in its contribution of solutes (Spencer *et al.*, 1990; Renaut and Tiercelin, 1994). Areal variations in precipitation, runoff, river discharge, and groundwater input can influence the chemical evolution of saline lake waters because of exposure to different lithologies. Although lithologic composition is very important, the hydraulic conductivity, aquifer residence times and dissolution rates, and leakages in the lake system can also be very significant determinants of the hydrochemistry of closed basins (Donovan and Rose, 1994). Despite topographic closure, the ratio of subsurface water outflow to inflow, or "leakage ratio," can have a substantial effect on evaporite precipitation and solute fractionation by delaying the salinity development (Wood and Sanford, 1990, 1995; Sanford and Wood, 1991). In a leaky system, the rate of solute loss can be faster than solute contribution by weathering, so that the water chemistry is dominated by precipitation and/or aerosols. Bowler (1986) has pointed out that the frequency of wetting and drying modifies the rate of evaporative loss

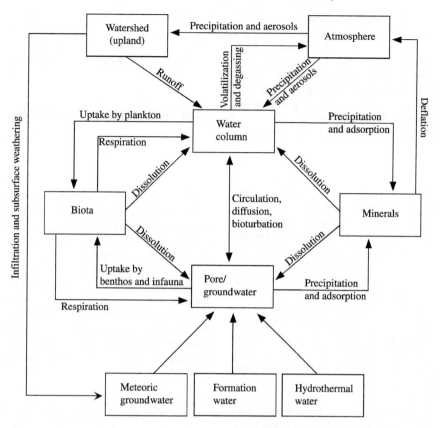

Figure 3 Schematic flow diagram for major fluxes of solutes in a hydrologically closed basin.

of interstitial waters from an exposed basin surface, which, in turn, affects the solute evolution of interstitial and local groundwaters (Figure 3). In coastal lakes variably connected to the sea, such as in the Bahamas (Teeter, 1995), Tasmania (Buckney and Tyler, 1976), or the west coast of Antarctica, the source, and the proportion, of solutes is uniform, but differences in physical properties or conditions (e.g., temperature) can produce a wide range of salinities and solute fractionation.

Hydrologically speaking, closed basin lakes and their surface or groundwater inflow are dominated by the reaction of natural waters with the lithologies of the surrounding drainage basin. Hydrologic setting and processes affect secondary modifications, such as the simple mixing of two or more chemically different inflow waters, and pore-water reactions and residence times (Jones *et al.*, 1969; Lerman and Jones, 1973). Strong seasonality, meromictic stratification, and multiple inflow sources from variable host lithologies all enhance the importance of mixing processes within the lake environment.

5.13.3.2 Climate

Climate plays a critical role in the water balance of all lakes, especially closed-basin lakes.

The amount of inflow to a closed basin, including precipitation, must be closely balanced by evaporative loss in the basin in order to produce elevated salinities. Either excess or insufficiency in precipitation compared to evaporation will prompt the hydrologic response discussed above (Almendinger, 1990).

In addition to the average precipitation inflow into closed basins, seasonality has important impacts on the chemistry of lakes. Highly seasonal or monsoonal systems experience dramatic hydrologic fluctuations, and these can have important effects on the chemistry of saline lakes. For example, short episodes of monsoonal flooding in saline and alkaline basins may provide the majority of a year's input of dissolved Ca^{2+} into the central basin, because during the rest of the year Ca^{2+} can be removed by carbonate and sulfate precipitation in marginal lake environments. The effects of seasonality may be attenuated by a large groundwater flux, depending on local hydrologic conditions (Torgersen *et al.*, 1986).

Ambient temperature is also an important climatic control on lake systems, not only with regard to effects on the local biota, but also, as Last (1999) has pointed out, on changes in mineral thermodynamics that may produce different mineralogies in concentrated sulfo-carbonate waters. Temperature also plays an important role both in simple evaporation from open water, and in the more complicated evapotranspiration from vegetated waters (Linacre *et al.*, 1970; Gilman, 1994). In Antarctic saline lakes, salinity-driven freezing point suppression can allow brines to reach temperatures below $-10\,°C$ (Doran *et al.*, 2003), and low temperatures may also play a role in the freeze-out of minerals in other settings.

The sensitivity of closed basin lake systems to climate makes them important indicators of climate change, and important sedimentary archives of past climates (Street-Perrott and Roberts, 1983; Johnson, 1996). Over time, changes in the amount and seasonality of precipitation can affect not only the overall salinity of lakes, but also the relative proportions of meteoric, ground, and hydrothermal waters (Renaut and Tiercelin, 1994). Such hydrochemical variations can be reflected in the evaporite mineralogy of closed basin lakes, and can provide important indications of past climate (Li *et al.*, 1997). Bowler (1986) has presented a valuable summary (and diagram) of five successive stages in basin development that represent a time sequence of increasing aridity in the Pleistocene. This sequence, with its mineralogic and morphometric implications, is based on a number of Australian examples, especially Lake Frome. This work has clearly demonstrated that a knowledge of the interaction of groundwater with surface water and the subsequent diagenesis of sediments is of critical importance in deciphering paleoenvironments (Torgersen *et al.*, 1986).

5.13.3.3 Geology

The lithology of rocks and sediments available for weathering and groundwater residence times are fundamental controls on the initial chemistry of dilute inflow into closed basins. As discussed below, the chemical pathways taken by an evolving brine and the evaporite mineralogies that are produced are in large part determined by their initial chemistry. Understanding the geology of a closed basin is, therefore, a necessary first step in attempts to understand the chemistry of its waters. Because the geology of a watershed is greatly affected by tectonics, certain patterns are evident in the distribution of brine types among the closed basins of the world. For example, the calc-alkaline volcanic activity that has dominated East African geology since the Miocene has produced widespread sodium carbonate waters (Jones *et al.*, 1977). In contrast, the Northern Plains of North America are underlain by Paleozoic to Mesozoic evaporitic sedimentary rocks and Quaternary till, which contribute strongly sulfatic waters (Last, 1999). In tectonically complex regions such as the North American

Great Basin (Jones, 1966) or the Volga–Caspian Basin (Clauer *et al.*, 1998), multiple sources of chemically distinct waters are generally associated with corresponding geologic settings.

5.13.4 HYDROCHEMICAL VARIATION

A major component of groundwater in the inflow to a basin is required to avoid the vagaries of climate and seasonality (Bowler, 1986; Rosen, 1994). The most concentrated saline lake waters result from a closely maintained balance between inflow and evaporation. Where dependent on sparse atmospheric precipitation and runoff, the salinity of a saline lake can vary by more than three orders of magnitude, and such conditions most commonly produce "dry lakes" or playas.

The intensive (non-mass-dependent) parameters of saline lake chemistry can vary nearly as much as the salinity. The pH is usually close to neutral for saline waters containing mostly chloride and sulfate salts. Waters associated with Ca–Mg carbonates are mildly alkaline (pH = 7.5–8.5); higher pH (up to 12) can be associated with alkali carbonate solutes and relatively high silica content. Acid saline lake waters with pH as low as 2–3 can be obtained in the absence of carbonate or with low silicate buffer capacity (Hines *et al.*, 1992). Saline waters near the extremes of pH typically also have low P_{CO_2}, as much as three orders of magnitude below atmospheric values. However, P_{CO_2} levels can be two to three orders of magnitude higher than atmospheric values in environments where microbial respiration exceeds degassing to the atmosphere. In these settings, such as lake-marginal wetlands and sublacustrine pore waters, P_{CO_2} has a major effect on pH, thereby affecting all mineral stabilities (Cerling, 1996; Deocampo and Ashley, 1999; Deocampo, 2001).

Except for the lower solubility of O_2, redox conditions in saline lakes are less directly related to increased salinity. With a decrease in heterotrophs, primary productivity reaches very high levels, leading to considerable decay and low redox conditions in lacustrine sediment. Physical conditions can produce reduced circulation and, aided by salinity contrast, meromixis (stratification), which is normally accompanied by low redox conditions in bottom waters. Oxidation–reduction reactions are particularly important in controlling the distribution of minor metals in closed basin lake systems. The chemical stratification that is so common in large saline lakes is critical in maintaining redox boundaries, across which reductive dissolution of metal oxides and anaerobic degradation of organic matter takes place. These conditions then influence solute speciation involving both dissolved inorganic and organic ligands, and the reactions with the solid phases controlling metal solubility. Although much attention has been given to such effects in marine systems, which could be applied to closed lakes with similar chemistries, little information is available for sulfo-carbonate saline lakes.

In one detailed study, Domagalski *et al.* (1990) examined reactivity, mobility, organic complexation, mineral transformation, and precipitation in three saline lakes of the western Great Basin, USA. Two alkaline ($Na-CO_3-Cl-SO_4$) lake systems — Walker Lake, NV and Mono Lake, CA — and one non-alkaline ($Na-Mg-Cl-SO_4$) system — Great Salt Lake — were studied (Figure 4). As in the ocean, saline lake metals initially associated with iron–manganese oxides and organic matter react to lowered redox potential by forming sulfide solids and humic acid complexes. In Walker Lake, where no stratification currently exists, and where dissolved oxygen persists throughout the water column, metals are held near the sediment–water interface by iron–manganese oxides, which dissolve only slowly in deeper anoxic sediments. In contrast, in stratified, near-neutral pH Great Salt Lake, sediment particles are entrained in dense bottom brines, where maximum organic decay and H_2S production occur. Pyrite formation occurs near the sediment–water interface, and renders most metals insoluble. The normally stratified, more alkaline Mono Lake, with higher sulfide levels in the bottom waters, gives rise to metal assimilation by metastable iron monosulfide near the sediment–water interface, but the high pH hinders pyrite formation and permits further metal reactivity in the pore fluids.

Ultimately, solute concentrations are controlled by the lithology of the source rocks, evaporation, solute losses to minerals (Table 2), and the biota. As in all natural waters, the composition of saline lakes is dominated by less than 10 major solutes, especially Na^+, K^+, Ca^{2+}, Mg^{2+}, Cl^-, SO_4^{2-}, dissolved inorganic carbon as $HCO_3^- + CO_3^{2-}$, and rarely, SiO_2. Of these, Na^+ is the most common cation, and is usually almost the only cation in solution. Compositional trends can be described almost entirely in terms of the anions in saline waters of the western Great Basin of the United States, which show about as much chemical variability as any region of the world (Figure 5; Hutchinson, 1957; Jones, 1966). Eugster and Hardie (1978) illustrated the hydrochemical variation of closed basin waters using solute data plotted on contiguous triangular plots of the major cations, summing sodium + potassium, and anions, summing carbonate species. These plots not only reveal the predominance of sodium, but also indicate the ubiquity of all the major anions. Eugster and Hardie (1978) used subdivisions of the triangular plots to classify

the analyses of saline waters according to the relative abundance of the principal cations and anions, largely in groups of three or four. As discussed below, Bodine and Jones (1986) found that it was more useful to classify saline waters according to normative analyses of their principal anion–cation associations based on relative mineral solubility.

In addition to the chemical evolution of saline lake waters brought about by evaporative concentration and mineral precipitation, a number of other processes can fractionate solutes in these systems. One of the most important processes involves wetting and drying cycles and the precipitation–dissolution of efflorescent crusts (Jones and Vandenburgh, 1966; Drever and Smith, 1978). With capillary pumping and the evaporation of near-surface fluids under desert conditions, all solutes are precipitated together. Deflation may then remove some mineral components, especially sodium carbonates, which are usually porous and easily eroded (Bowler, 1986; Connell and Dreiss, 1995). Such deflated dust may also be deposited in aqueous environments, such as marshes adjacent to mudflats, producing locally elevated sodium and alkalinity. Under subsequent wet conditions, fractional dissolution returns only the most soluble constituents to solution, leaving behind the alkaline-earth carbonates and silica, typically as coatings or interstitial fillings (Smith and Drever, 1976).

Sulfate reduction is another common process affecting the chemical evolution of saline lake waters. Suitable conditions are readily set up by the high productivity and decay of salt tolerant organisms, and are commonly enhanced by the stratification developed in saline lake brines (Domagalski *et al.*, 1990). Sulfate reduction, bidirectional diffusion, and the nature of related diagenetic carbonate reactions in pore fluids of anaerobic sediments have been examined in the Devils Lake system by Komor (1992, 1994). Komor described bidirectional sulfate diffusion to a near-surface zone of sulfate reduction, not only downward from the water column but also upward from deeper layers containing appreciably higher pore fluid sulfate concentrations, which are a remnant of low stands 25 years earlier. In addition to sulfate reduction in stratified lake bottom-waters and below the sediment–water interface, reduction can also occur in marshes and mudflats. In productive marshlands sulfate may be reduced before substantial evaporation of the waters. This affects the evaporation path early in the evolution of the brines. The process may be disrupted by oxygenation of the waters by large mammals such as hippopotami (Wolanski and Gereta, 1999; Deocampo, 2002).

Hydrogen and oxygen isotopes fractionate during evaporation. This affects the isotopic

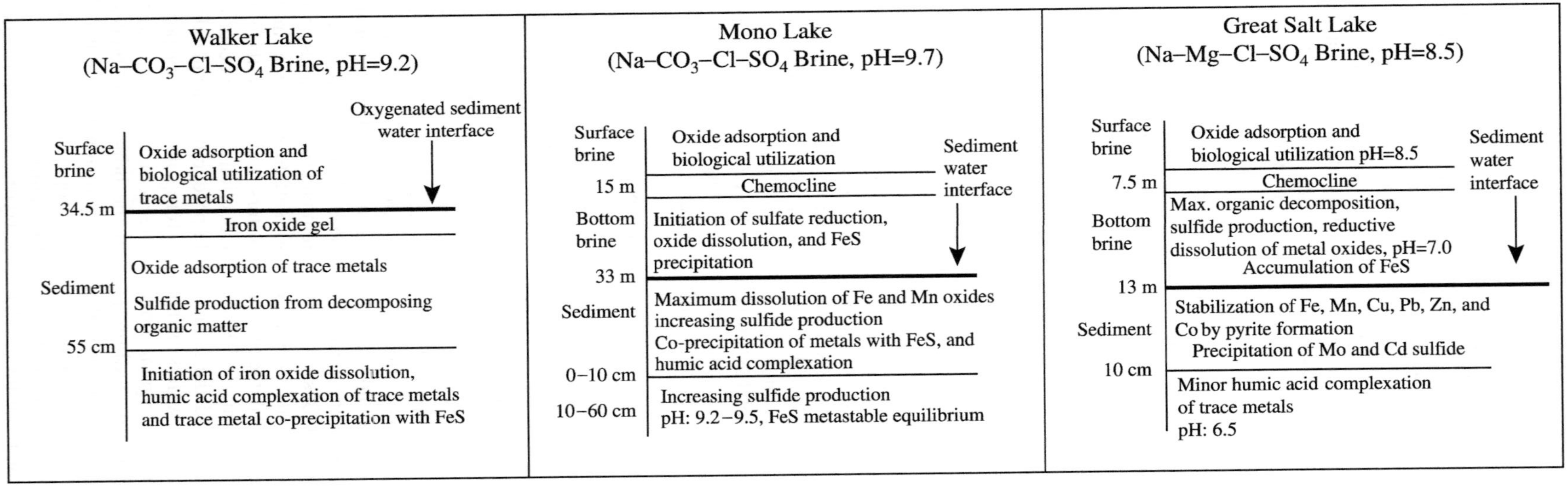

Walker Lake
(Na–CO$_3$–Cl–SO$_4$ Brine, pH=9.2)

Oxygenated sediment water interface →

Surface brine	Oxide adsorption and biological utilization of trace metals
34.5 m	Iron oxide gel
	Oxide adsorption of trace metals
Sediment	Sulfide production from decomposing organic matter
55 cm	Initiation of iron oxide dissolution, humic acid complexation of trace metals and trace metal co-precipitation with FeS

Mono Lake
(Na–CO$_3$–Cl–SO$_4$ Brine, pH=9.7)

Sediment water interface →

Surface brine	Oxide adsorption and biological utilization
15 m	Chemocline
	Initiation of sulfate reduction, oxide dissolution, and FeS precipitation
Bottom brine	Maximum dissolution of Fe and Mn oxides increasing sulfide production Co-precipitation of metals with FeS, and humic acid complexation
33 m	
Sediment	Increasing sulfide production pH: 9.2–9.5, FeS metastable equilibrium
0–10 cm	
10–60 cm	

Great Salt Lake
(Na–Mg–Cl–SO$_4$ Brine, pH=8.5)

Sediment water interface →

Surface brine	Oxide adsorption and biological utilization pH=8.5
7.5 m	Chemocline
Bottom brine	Max. organic decomposition, sulfide production, reductive dissolution of metal oxides, pH=7.0 Accumulation of FeS
13 m	Stabilization of Fe, Mn, Cu, Pb, Zn, and Co by pyrite formation Precipitation of Mo and Cd sulfide
Sediment	Minor humic acid complexation of trace metals pH: 6.5
10 cm	

Figure 4 Summary of oxidation–reduction conditions in three saline lakes of the Great Basin, USA (source Domagalski *et al.*, 1990) (reproduced by The Geochemical Society from *Fluid–Mineral Interactions: A Tribute to H. P. Eugster* **1990**, 2, 315–354).

Table 2 Names and chemical formulas for major evaporite minerals from saline lakes.

Name	Chemical formulas
Carbonates	
Calcite	$CaCO_3$
Aragonite	$CaCO_3$
Dolomite	$CaMg(CO_3)_2$
Huntite	$CaMg_3(CO_3)_4$
Magnesite	$MgCO_3$
Hydromagnesite	$4(MgCO_3)\cdot Mg(OH)_2\cdot 4H_2O$
Gaylussite	$CaCO_3\cdot Na_2CO_3\cdot 5H_2O$
Trona	$Na_2CO_3\cdot NaHCO_3\cdot 2H_2O$
Thermonatrite	$Na_2CO_3\cdot H_2O$
Nahcolite	$NaHCO_3$
Natron	$Na_2CO_3\cdot 10H_2O$
Sulfates	
Gypsum	$CaSO_4\cdot 2H_2O$
Anhydrite	$CaSO_4$
Mirabilite	$Na_2SO_4\cdot 10H_2O$
Thenardite	Na_2SO_4
Epsomite	$MgSO_4\cdot 7H_2O$
Hexahydrite	$MgSO_4\cdot 6H_2O$
Starkeyite	$MgSO_4\cdot 4H_2O$
Kieserite	$MgSO_4\cdot H_2O$
Bloedite	$Na_2Mg(SO_4)2\cdot 4H_2O$
Glauberite	$Na_2SO_4\cdot CaSO_4$
Kainite	$KCl\cdot MgSO_4\cdot 3H_2O$
Schoenite	$K_2SO_4\cdot MgSO_4\cdot 6H_2O$
Leonite	$K_2SO_4\cdot MgSO_4\cdot 4H_2O$
Polyhalite	$K_2SO_4\cdot MgSO_4\cdot 2CaSO_4\cdot 2H_2O$
Burkeite	$Na_6(SO_4)_2CO_3$
Chlorides	
Halite	$NaCl$
Hydrohalite	$NaCl\cdot 2H_2O$
Sylvite	KCl
Antarcticite	$CaCl_2\cdot 6H_2O$
Bischofite	$MgCl_2\cdot 6H_2O$
Tachyhydrite	$CaMg_2Cl_6\cdot 12H_2O$
Carnallite	$KCl\cdot MgCl_2\cdot 6H_2O$

may be reconstructed after applying appropriate mineralization fractionation functions. Complex diagenetic histories involving different fluid chemistries can sometimes be reconstructed by discriminating between the different isotopic signatures of co-occurring saline lake precipitates, such as carbonates, authigenic clays, and chert, as shown by Hay and Kyser (2001).

The stable carbon isotopic composition of water and associated solids is less directly affected by evaporative concentration. In general, factors affecting $\delta^{13}C$ in other lacustrine environments apply in saline lakes as well, with negative values associated with the oxidation of organic detritus, and positive values produced by carbonate dissolution and atmospheric gas exchange. The highest positive values are produced by methanogenesis (Wigley *et al.*, 1978). The exceptions to this enrichment are the carbonate-rich brines, in which the dissolved inorganic carbon reservoir is so large that its isotopic signature is little affected by peripheral processes. In such settings, large carbon reservoir effects are usually encountered, complicating efforts to use radiocarbon to date waters, carbonate, or organic matter within the basin.

The stable isotopic composition of sulfur ($\delta^{34}S$) can be strongly affected by the reduction of sulfate, as commonly occurs in anoxic hypolimnetic bottom waters. The process of reduction involves $\delta^{34}S$ fractionation. Since it is biologically mediated, fractionation coefficients are highly variable due to biological complications (Kaplan and Rittenberg, 1964). During the reduction of sulfate, isotopically light sulfide is produced, leaving residual pore waters that are enriched in ^{34}S (Anderson and Arthur, 1983; Kaplan, 1983). Under certain conditions the values of $\delta^{34}S$ can allow one to distinguish sulfur sources, i.e., seawater from volcanic thermal waters or atmospheric pollutants (Kellogg *et al.*, 1972; Krouse, 1980), and can provide some indication of the microbial mediation of reductive processes in pore waters (Komor, 1994).

The isotopes of strontium ($^{87}Sr/^{86}Sr$ ratios) can be used as groundwater flow tracers, because groundwaters tend to assume the isotopic composition of the aquifer (Collerson *et al.*, 1988; McNutt *et al.*, 1990). In closed lake basins, this can be a valuable indicator of the sources of the groundwater flux to the lake (Neumann and Dreiss, 1995).

signatures of closed-basin lakes profoundly (see Chapter 5.11). During the course of evaporation, the lighter isotopes are favored in the vapor and in molecular diffusion through the air boundary layer (Craig and Gordon, 1965; Merlivat and Coantic, 1975; Gat, 1995). As a result, lake waters become progressively enriched in the heavier isotopes. In the Dead Sea, $\delta^{18}O$ becomes enriched by over 5‰ relative to the Jordan River inflow, whereas δD can increase by nearly 10‰ (Gat, 1995). Similar isotopic enrichment is found in East African lakes (Casanova and Hillaire-Marcel, 1992). The steady-state isotopic signature is fundamentally controlled by inflow composition and hydrology. Over time, changes in the isotopic signature of closed basin lakes may therefore be related to changes in the water balance of a basin (Gat, 1995). If these signatures are recorded in ancient chemical or biochemical precipitates, either carbonates or silicates, the paleochemistry of a basin

5.13.5 BRINE EVOLUTION

The chemical evolution of saline lake waters begins with the acquisition of solutes in dilute inflow, primarily through atmospheric input and chemical weathering reactions. The solute composition of these waters is dictated by

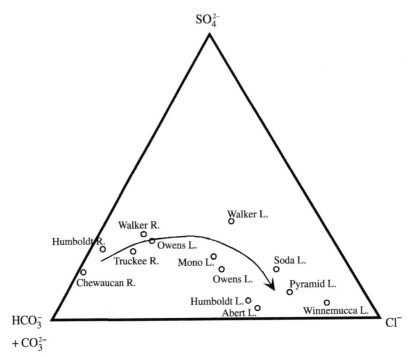

Figure 5 Anion compositional trends in the Great Basin, USA. The arrow indicates general trend of increasing compositional evolution of waters (Hutchinson (1957); reproduced from *A Treatise on Limnology*, **1957**).

the principal lithologies of the drainage basin and their relative distribution. The ultimate pathway of solute evolution of any saline lake water can be associated with a few fundamental rock types, their mode of reaction with dilute natural waters, and the resulting relation of major cations and anions. The characterization of water reactions (Figure 6) shows the effects of different primary mineral solution mechanisms. Sulfide oxidation reactions are placed separately to emphasize the acid reversal of alkaline hydrolysis reactions. This can have the important effect of reversing bicarbonate/alkaline-earth ratios, as well as the addition of sulfate to the solutions. In fact, as explained below, it is the initial $HCO_3/(Ca + Mg)$ ratios that are most important in the early stages of the evolution of saline waters.

In the course of evaporative evolution, it is usually valuable to compare individual solute concentrations to a conservative solute. In general, this is chloride in less concentrated brines, and bromide in waters supersaturated with respect to halite (Eugster and Jones, 1979). Solute behavior plotted this way takes one of the several diagnostic routes as shown in Figure 7. In the later stages of chemical evolution, the ratios with respect to the alkalies and sulfo-chloride become important, and it is advantageous to consider associations within the entire major solute matrix. A useful tool for this purpose is normative analysis (Jones and Bodine, 1987). The computer program SNORM (Bodine and Jones, 1986) calculates the

equilibrium salt assemblage expected for any water taken to dryness at 25 °C and atmospheric P_{CO_2}. Thus, regardless of their initial composition, saline waters can be compared in terms of the total salt assemblage to be expected on complete evaporation, including the presence or absence of characteristic double salts. Examples of the so-called "simple salts" are given in Table 4, where multiphase mineral assemblages are simplified to cation–anion associations. Such analyses permit the evaluation of the effects of processes other than evaporative concentration on the chemical evolution of saline water by recognizing the deviation from the projected assemblage.

By the time surface or groundwater inflow reaches the periphery of the closed basin floor, rock–water reactions are superceded by other processes as the principal controls on water composition. Although evaporative concentration normally plays the dominant role leading up to mineral formation, other factors include mixing, degassing (particularly of CO_2), and temperature changes. However, our discussion will first treat the solute evolution expected to accompany sequential mineral saturation and precipitation. The sequence is most readily illustrated with a flow chart similar to (but simplified from) that used by Eugster and Hardie (1978), Figure 8.

As is to be expected from relative solubility, the first minerals to form from concentrating waters are the alkaline-earth carbonates. The equilibrium precipitation of calcite also illustrates

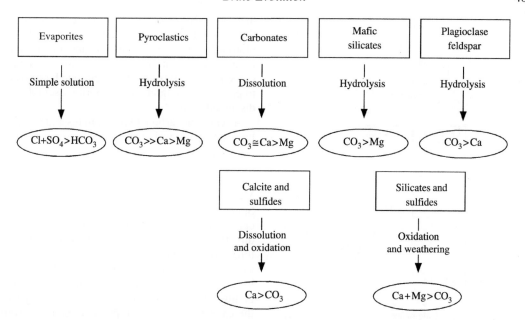

Figure 6 Schematic representation of major fluid types produced by weathering of different rock types, with relative solubility decreasing towards the right. Solutes (Ca, Mg, SO$_4$, CO$_3$) refer to total aqueous species in equivalents, and CO$_3$ refers to all aqueous CO$_2$ species. Sulfide weathering is represented on a second level to demonstrate the possibility of competing weathering processes that can reverse the (Ca + Mg)/CO$_3$ ratio, an important early determinant of brine evolutionary pathways.

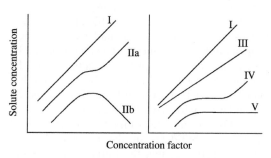

Figure 7 Schematic logarithmic plot of solute concentration versus the degree of evaporative concentration. Type I solutes are conservative, and show no evidence for loss to mineral precipitation, such as typically found with Na$^+$. Solute types IIa and IIb are examples of paired ions, one a cation and the other an anion, that precipitate together as a mineral. The slopes of both ions decrease while the mineral precipitates, and once the less abundant ion is depleted (IIb), the slope of the more abundant ion resumes its increase. Type III solutes are removed gradually from solution over the whole range of evaporative concentration, perhaps by multiple processes, such as kinetically-inhibited mineral precipitation, adsorption, or degassing. Type IV solutes are removed only during the middle stages of evaporative concentration, which may occur with surface adsorption of K$^+$, or reduction of SO$_4^{2-}$. Uncharged solutes, such as SiO$_2$, that remain constant at the solubility limit of their corresponding solid are represented by the Type V curve (Eugster and Jones (1979); reproduced by permission from *Am. J. Sci.*, **1979**, *297*, 609–631).

a major factor in the solute evolution of saline lake waters: the "chemical divide" (Hardie and Eugster, 1970). For the precipitation of pure calcite, it is clear that Ca and CO$_3$ must contribute to the precipitate and be lost from solution in equal proportion. At the same time, the solubility product ([Ca] × [CO$_3$]) must be constant, so that as the mineral precipitates, unequal proportions in solution will dictate the dominant constituent on further concentration, i.e., increasing Ca will decrease the amount of CO$_3$, or vice versa. Thus, the early precipitation of calcite determines whether the remaining solution becomes carbonate rich or carbonate poor. A complete derivation and detailed description of the "chemical divide" concept is given in Drever (1997).

Probably the simplest quantitative means of applying the "chemical divide" concept of Hardie and Eugster (1970) to the prediction of solute evolution pathways is through the use of the "Spencer Triangle" (Jones and Bodine, 1987; Smoot and Lowenstein, 1991; Spencer, 2000). This technique takes advantage of the fact that in a triangular diagram of the system Ca–SO$_4$–(HCO$_3$ + CO$_3$), a line connecting a point representing a particular solute distribution and a particular precipitate composition will define, in terms of the three-component system, the direction of the chemical evolution of the water (Figure 9). The join between CaCO$_3$ and the SO$_4$ apex represents the calcite divide, and the join

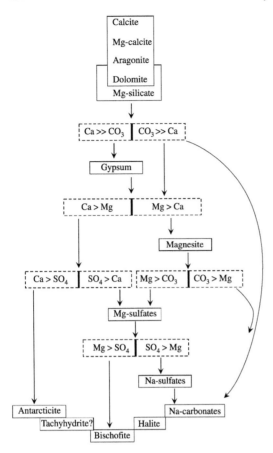

Figure 8 Flowchart of idealized evaporative evolution of a closed-basin brine, with relatively greater evaporative concentration toward the bottom. Solutes (Ca, Mg, SO_4, CO_3) refer to total charged aqueous species in equivalents, and CO_3 refers to all aqueous CO_2 species. Solute conditions are indicated by dashed boxes, and mineral precipitates are indicated by solid boxes (after Eugster and Hardie, 1978) (reproduced by permission of the Geological Society of America from *Geol. Soc. Am. Bull.* **1975**, *86*, 319–334).

from $CaCO_3$ to $CaSO_4$ represents the gypsum divide. Waters will move away from these divides as the respective minerals precipitate. Sodium and chloride are considered to be ubiquitous. Each triangular field is labeled according to the principal solutes left in the final fluid. Representative starting solution compositions are also plotted in each field as examples of characteristic solute assemblages. These include the average composition of the World River (Livingstone, 1963), seawater at the point of initial calcite saturation (Li *et al.*, 1997), and the Malad hot springs inflow to the Great Salt Lake system. On calcite precipitation, solution compositions plotting in the lower triangle will evolve away from the mineral composition point to become alkali sulfo-carbonate brines. Water chemistries plotting in the central triangular field, however, will evolve away from calcite composition toward

the Ca–SO_4 boundary, precipitate gypsum, and then shift toward the sulfate corner. Solutions plotting in the upper part of the diagram are progressively enriched in calcium as calcium carbonate and gypsum are precipitated, ultimately to produce an antarcticite evaporite assemblage as discussed below.

As calcium carbonate precipitates and solutions become progressively more concentrated, it becomes important to consider the solute evolution of magnesium. We use the Spencer Triangle approach to do this, replacing the Ca-vertex with Mg (Figure 10). In a similar fashion as the Ca Spencer Triangle, precipitation of a mineral will cause the migration of solute compositions directly away from that mineral's endpoint in the three-component space. In this manner, most surface waters, which begin somewhere in the lower right area of the triangle, will evolve along a pathway toward the upper left, moving away from the calcite or Mg-calcite point. In contrast to the Ca-triangle, however, pathways due to $CaCO_3$ precipitation are not constrained within individual fields, because calcium is independent of the system. The four arrows on the Mg-triangle (Figure 10) represent four conditions of varying Ca/Mg ratios, that produce Ca–Mg carbonates and magnesium carbonates at different times. Low-calcium waters will produce magnesium-rich carbonates earlier than high-calcium waters, and the timing of this is difficult to predict, because it is subject to poorly understood kinetics. In general, however, the more calcite that can be produced, the farther toward the upper left of the diagram the water will evolve. Once calcium has been exhausted, if CO_3 still remains, magnesite precipitation will drive the fluid away from the Mg–CO_3 axis, and determine the molar ratios of Mg/SO_4 or SO_4/CO_3 in the residual fluid. It follows, therefore, that the initial relative abundance of calcium and magnesium, and the timing of calcium depletion relative to magnesium (subject to kinetics) are critical factors in determining if the subsequent brine will be sulfate or carbonate rich. This process of decreased Ca/Mg ratios due to calcium carbonate precipitation is seen clearly in Lake Balkash, Kazakhstan (Verzilin and Utsal, 1991; Petr, 1992). Here the Ca/Mg of the lake waters decreases steadily with distance from the fluvial input of the Ili River, even as evaporative concentration dramatically increases the absolute magnesium content of the solution (Figure 11). Muller *et al.* (1972) examined the formation and diagenesis of inorganic Ca–Mg carbonates from 25 lakes in Europe, the Middle East, and Africa. Salinity ranged up to more than 400 g L^{-1} in the Turkish lakes (Lake Tuz and vicinity). These have been described in more detail by Irion (1973). The precipitation of high-Mg calcite was associated with solutions

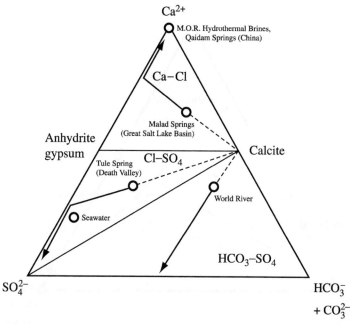

Figure 9 The "Spencer Triangle", ternary phase diagrams in the system $Ca^{2+}-SO_4^{2-}-(HCO_3^- + CO_3^{2-})$, with three sample compositions indicated. The pathways taken by each sample represent evolution of the water composition upon precipitation of calcite, with subsequent precipitation of gypsum for the Tule and Malad Springs samples (sources Jones and Bodine, 1987; Spencer *et al.*, 1990).

with an Mg/Ca ratio between 2 and 7. At a ratio greater than 12, aragonite was precipitated and remained stable. At very high Mg/Ca ratios and high magnesium concentrations, hydrous magnesium carbonate was found. Diagenetic carbonate in sediment is restricted to lakes where the Mg/Ca ratio leads to the precipitation of Mg-calcite. An Mg/Ca ratio >7 in the pore fluid leads to the formation of dolomite, and a ratio >40 converts dolomite to huntite or magnesite. Muller and Wagner (1978) have found up to 20 mol.% $MgCO_3$ in calcite in the relatively dilute (but historically brackish) Lake Balaton in Hungary. Baltres and Medesan (1978) have reported up to 28 mol.% Mg in biogenic calcite from Lake Techirghiol in Romania, which was formerly connected to the Black Sea.

The interplay between Ca–Mg carbonates and aqueous Ca/Mg ratios is especially well illustrated in a sediment core from the Great Salt Lake during the relatively rapid decrease in the level of Lake Bonneville levels at the end of the last glacial period (Spencer *et al.*, 1984). In this core interval the proportion of carbonate mineral increases steadily, accompanied by an increase in the magnesium content of calcite to ~11 mol.%, at which point aragonite becomes the principal carbonate phase. Thereafter, dolomite also appears in the sediment column, although it is not clear how much of this material is detrital in origin. In the later core intervals that represent a Holocene saline lake, levels fluctuated but were not greatly different from historical levels. The

magnesium and the silica content of the clay fraction increases progressively, apparently the result of kerolitic (hydrous talc) interstratification in detrital smectite (Jones and Spencer, 1999). The flow diagram (Figure 8) reflects the apparent overlap in the formation of magnesium carbonate and silicate at the near-neutral pH conditions of the Great Salt Lake. Obviously, the formational precedence of these two phases will be affected by the relative abundance of aqueous silica and the alkalinity of the system, not the concentration of magnesium alone. For example, in the highly alkaline waters of East Africa, Mg-clays seem to precipitate early, sometimes even from relatively dilute waters (Stoessell and Hay, 1978; Deocampo *et al.*, 2000), although these may subsequently dissolve in CO_2-rich, sublacustrine pore waters (Cerling, 1996).

Von Damm and Edmond (1984) utilized the lakes of the Ethiopian and northern Kenya rift zones to examine "reverse weathering" (the formation of authigenic clay minerals), because here evaporative concentration had not proceeded to the extent that salt precipitation interfered with a mass balance approach. They found that ~60% of an alkalinity deficit could be accounted for by processes other than carbonate precipitation, and concluded that solute magnesium was lost as rapidly to clay as solute calcium was to carbonate. This situation, particularly in volcanic terrain, was also initially recognized at saline Lake Abert, Oregon, by Jones and VanDenburgh (1966).

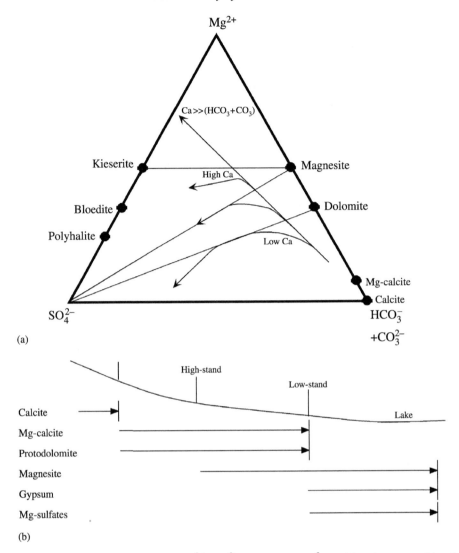

Figure 10 (a) Ternary diagram in the system $Mg^{2+}-SO_4^{2-}-(HCO_3^- + CO_3^{2-})$, with pathways resulting from initial precipitation of calcite, then precipitation of dolomite and magnesite. The path taken and the ultimate fate of the brine is dependent on how rapidly Ca^{2+} is depleted, relative to Mg^{2+}. Kieserite ($MgSO_4 \cdot H_2O$) is used here to represent all the hydration states of $MgSO_4$. (b) Schematic distribution of calcium- and magnesium-bearing minerals in the Basque Lakes. modified from Nesbitt (1974). Precipitates farther along the evolutionary pathways (e.g. gypsum, Mg-sulfates) are found toward the center of the lake basin (after Nesbitt, 1974).

Jones (1986) and Banfield *et al.* (1991) have suggested that the principal mode of formation of interstratified magnesium smectites in saline lakes maintained primarily by sediment-bearing surface water inflow is by the topotactic growth (as on a template) of magnesium silicate (kerolite, or hydrous talc) on ultrafine clay detritus. The purity of the resulting magnesium silicate depends on the relative amount of detrital versus precipitated clay, as well as on the composition of inflow. The spring-fed pond sediments of the Amargosa Desert, Nevada, contain purer kerolitic smectite than most of the clays of the Great Salt Lake (including glacial Lake Bonneville; Jones and Spencer, 1999), or the clay fractions of Lake Chad examined by Gac *et al.* (1977).

Similarly pure beds of kerolite and sepiolite are found in the modern and Pleistocene groundwater wetlands of Amboseli, Kenya (Stoessell and Hay, 1978; Hay and Stoessell, 1984; Hay *et al.*, 1995). The concentration of dissolved silica is also important; at higher ratios of SiO_2 to Mg, chain-structure clay (sepiolite-palygorskite) can precipitate directly from solution, as is also the case at Amargosa.

After the Ca–Mg carbonates, and again reflecting relative solubility, the next major "chemical divide" is created by the next most soluble phase, gypsum. At this stage, the carbonate and silica contents of the evaporating solution can be sufficient to drive the calcium and magnesium to very low concentrations, and preclude the

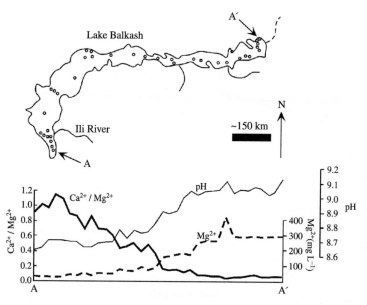

Figure 11 Changes in water chemistry in Lake Balkash as a function of distance from fluvial input of the Ili River. Precipitation of calcite causes Ca/Mg to decrease as waters become more evaporatively concentrated (after Versilin and Utsal, 1991).

precipitation of additional alkaline-earth carbonate minerals; any further precipitates involve only alkali metal salts. This is the usual situation in volcanic terrains dominated by silicate hydrolysis, such as the African rift valleys and the northwest portion of the Great Basin (US). However, sulfate exceeds carbonate even after precipitation of gypsum, sufficient carbonate and magnesium may remain to form magnesite. Thereafter, the proportion of magnesium and carbonate will determine whether or not a magnesium sulfate mineral forms, or signal a direct concentration progression to the precipitation of an alkali salt. These pathways are characteristic of the northern Great Plains of North America (Last, 1999). If the calcium concentration exceeds magnesium and sulfate (the Ca/Mg and Ca/SO$_4$ divides) following gypsum precipitation, the eventual result is an "antarcticite" (CaCl$_2 \cdot$6H$_2$O) brine, which is rare with surface salines, but not uncommon in deep sedimentary basin brines from the dissolution of bitterns or from their chemical evolution (Carpenter, 1978). Antarcticite was named for saline lakes in the "dry" valleys of Antarctica, but it is also found in Bristol playa in the Mojave Desert of California (Rosen, 1989, 1991), a basin that contains at least 300 m of interbedded clastics and evaporites (Handford, 1982).

The last major pathway illustrated in the flow chart of Figure 8 involves saline waters dominated by sulfate and magnesium, which, in addition to chloride and sodium, are the principal constituents left after concentration has exceeded saturation conditions for Ca–Mg carbonates and gypsum. After the precipitation of magnesium sulfates,

which can include double salts containing sodium (such as bloedite) or potassium (such as polyhalite), the proportions of Mg and SO$_4$ (Mg/SO$_4$ divide) dictate whether the final fluid is characterized by MgCl$_2$ (bischofite) or simply sodium salts, especially halite. Along this pathway, temperature controls whether magnesium or sodium sulfate is the earliest mineral in the precipitation sequence. Although the thermodynamic effects of the temperature shift are subtle, they can have important effects on evaporite mineral assemblages (Figure 12). Between about 30 °C and 45 °C, the magnesium salts are less soluble (Last, 1999), and thus may be the earlier precipitates, as found in central Spain (Ordonez and Garcia del Cura, 1994). In contrast, the hydrated sodium sulfate (mirabilite) can be expected to precede in the saline pond deposits of the Canadian prairies (Last, 1999). Details of the final solute evolution and precipitate sequence in these types of lake systems are difficult to follow because of the relatively small size and local complexity.

The solute and salt sequence for the final stages of seawater evaporation has been determined computationally (Eugster *et al.*, 1980), and in commercial solar salt ponds (Hermann *et al.*, 1973). The only equivalent information on a saline lake system of considerable size has been provided for the Great Salt Lake (GSL) area, also from evaporation ponds (Jones *et al.*, 1997) and by direct computation. Kohler (2002), utilized the computer model of Moller *et al.* (1997), allowing for the dominance of halite in the GSL system, and worked out a precipitation sequence essentially of Ca-sulfate to Mg-sulfate to MgCl$_2$,

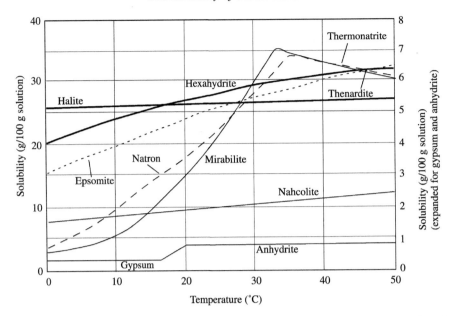

Figure 12 Temperature effects on the solubility of several important evaporite mineral phases, from Last (1999). Note that above ~28 °C, mirabilite is more soluble than epsomite (source Last, 1999) (reproduced from the Geological Survey of Canada from *Geol. Surv. Can. Bull.* **1999**, *534*, 23–55).

complete with the appropriate intermediate alkali double-salts: glauberite with calcium and sodium; bloedite with magnesium and sodium; and polyhalite, kainite, and carnallite with magnesium, and potassium (Figure 13). The former three were not identified in the evaporation ponds, and were probably back-reacted, as predicted in the model. For a nearly closed basin with some leakage or periodic overflow (such as in the GSL–Bonneville system), differences may occur in the relative proportion of the mineral precipitates, but the sequence will remain the same. That is, the openness of the basin will determine how far along the solute evaporation path the brine will evolve (Bowler, 1986; Sanford and Wood, 1991).

5.13.6 EXAMPLES OF MAJOR SALINE LAKE SYSTEMS

It is important to recognize that the models of brine evolution discussed above are idealized, and that the actual chemistry in the field is usually complicated by peculiarities in hydrodynamics, biotic activity, mineral kinetics, and diagenesis. Nevertheless, we find general agreement between the model simulations based on thermodynamics and simple stoichiometry, and examples of major saline lake systems.

In terms of volume, the large saline lakes of central and western Asia far exceed all others on Earth (Table 1). The Caspian Sea itself is the world's largest inland water body. The northern Caspian, in the area of its major inflow, the Volga, is less than 50 m deep. The central part of the sea

is up to 790 m deep; in the southern basin, separated by a ridge that is an extension of the Caucasus Mountains, depths over 950 m are reached. Most of the eastern side of the sea has a higher salinity because of increased evaporation and decreased runoff. There is a fairly continuous sinking of the denser surface water, which keeps the bottom water aerated and prevents chemical stratification (Dickey, 1968). The pH of the entire sea is alkaline. The pH of the surface water ranges from 8.3 to 8.6, the bottom-water pH is relatively constant at 7.7. The sediment-laden river input from the tectonically active Caucasus on the western side has high silica content.

Because the Caspian Sea was connected with the Black Sea basin prior to the last glaciation, its waters have retained compositional similarities to the original seawater (Tables 3 and 4), and hence have been affected appreciably by sulfatic river inflow (Bruyevich, 1938; Blinov, 1962; Kosarev and Yablonskaya, 1994). Continental waters, mostly draining the Volga watershed and constituting over 75% of the inflow to the Caspian Sea, deliver calcium sulfate- and bicarbonate-rich waters to the basin. With evaporative concentration, the rapid depletion of carbonate by oölite and calcareous mud precipitation (Kosarev and Yablonskaya, 1994) promotes widespread gypsum precipitation in the shallows of the Kara Bogaz Gulf (Dickey, 1968; Buineviya *et al.*, 1978), which has no surface inflow except what is driven into it by wind and intense evaporation. This lack of inflow also allows Mg^{2+} and Na^+ to become enriched, eventually leading to the precipitation of magnesium and sodium sulfates and even halite (Klenova, 1948), under more

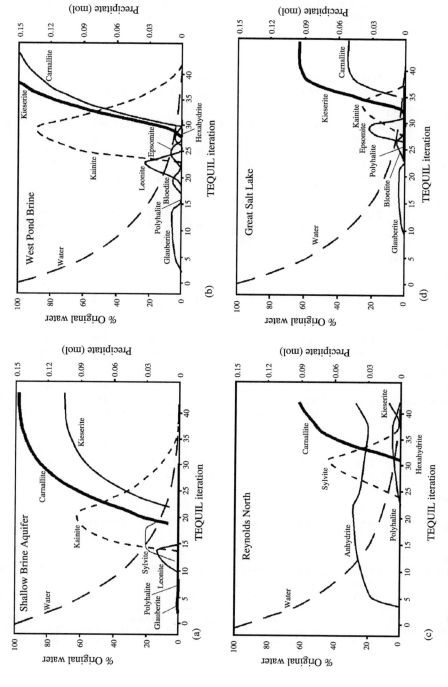

Figure 13 TEQUIL model (Moller *et al.*, 1997) plots illustrating quantities of salts precipitated (in moles) as a function of fixed increment of solution evaporated for four bodies of brine in the Bonneville desert (initial solution chemistry and model output after Kohler, 2002). Note that relative abundances reflect the concentration of the starting fluid, which is appreciably greater for the West Pond and Shallow Brine Aquifer solutions than for the Great Salt Lake or Reynolds North crustal pore fluids. Of most significance is the ratio of sulfate and chloride salts. In this regard, note the similarities between the Great Salt Lake and West Pond brines. In contrast, the greater association of sulfate with calcium in the Reynolds North fluids leads to higher proportion of chloride to sulfate salts in the Shallow Brine Aquifer.

Table 3 Major ion concentrations discussed in text and data source for SNORM model results shown in Table 2. All results in mg L^{-1}.

Locality	Na	K	Ca	Mg	SO$_4$	Cl	HCO$_3$ + CO$_3$	$\sum$
Western Asia[a]								
Aral Sea	2,450	54	460	603	3,380	3,840	187	10,970
Lake Balkhash (delta)	Na + K = 267		57	74	350	228	300	1,280
Lake Balkhash (center)	Na + K = 1,360		14	290	1,690	1,110	790	5,250
Caspian Sea	2,280	70	270	540	2,100	3,860	200	9,320
Caspian Sea tributaries	38.4	2.0	87	29	260	37	110	563
Caspian Sea hot springs	1,250	68	605	90	1,480	2,320	0	5,810
Qaidam Basin, China[b]								
Nyer Co	28,000	20,440	920	76,080	16,900	272,970	0	415,310
Qinghai	46,800	16,540	610	15,320	56,550	88,110	4,760	186,570
Great Salt Lake Basin, USA[c]								
Salt Lake	85,700	4,550	319	8,050	17,400	147,000	327	263,350
Bear River	105	9.2	70	34	44	148	324	730
Sevier River	760	15	140	170	1,000	900	1,100	4,090
Malad Springs	15,000	790	830	230	480	26,000	480	43,800
West Pond	48,000	2,800	2,800	4,300	11,000	100,000	n.a.	168,900
Lake Bogoria Basin, Kenya[d]								
Lake Bogoria	24,600	435	12	<1	330	5,680	30,000	61,060
Sandai River	80	4.5	27	6.5	4.7	40	240	400
Loburu hot springs	1,480	28	1.0	<1	55	230	2,750	4,540
Northern Great Plains, USA and Canada[e]								
Big Coulee Riv.	46	10	48	19	170	14	120	430
Souris River	51	13	42	26	150	14	126	420
Sheyenne River	37	7.5	47	18	97	12	110	330
Des Lacs River	21	9.0	34	16	110	12	90	290
Little Manitou Lake	480	20	22	703	741	455	18	2,440
Freefight Lake	1,310	64	11	500	960	260	201	3,310
Devils Lake	800	100	64	200	1,560	390	260	3,370
Death Valley, USA[f]								
Amargosa River	970	47	32	18	756	935	245	3,000
Mormon Point Springs	27,420	1,660	3,540	1,020	2,580	53,450	62	89,730
DV-40	76,400	1,010	820	243	4,100	117,000	50	199,620
Australia[g]								
Yilgarn	71	1.6	0.6	7.7	15	131	0	230
Lake Tyrrell	77,160	660	514	7,100	15,450	131,000	105	231,990
Lake Eyre	115,160	368	642	7,670	18,030	189,000	192	331,060
Dead Sea Basin[h]								
Jordan River	224	32	129	95	97	762	1,530	2,870
Zohar Springs	20,500	2,580	11,400	22,500	1,550	119,900	130	178,560
Dead Sea	39,330	6,500	17,750	40,450	760	212,600	290	317,680
Other Great Basin, USA[i]								
Bristol Dry Lake	65,090	2,310	104,060	4,590	96	210,270	30	386,450
Mono Lake	3,000	153	9.4	121	2,170	2,080	2,700	10,230
Salton Sea	11,000	180	1,000	1,300	10,000	17,000	200	40,680

[a] Blinov (1962), Kosarev and Yablonskaya (1994), and Clauer *et al.* (1998). [b] Spencer *et al.* (1990) and Zheng (1997). [c] Spencer *et al.* (1985a), Kohler (2002), and USGS (2002). [d] Renaut and Tiercelin (1994). [e] Komor (1994), Last (1994a), and USGS (2002). [f] Li *et al.* (1997). [g] Johnson (1980), Mann (1983), and Herczeg and Lyons (1991). [h] Neev and Emery (1967), Abed *et al.* (1990), and Vengosh and Rosenthal (1994). [i] Rosen (1991) and USGS, (2002).

evaporated conditions, such as in the Aral Sea prior to its drying out (Rubanov, 1984). Clauer *et al.* (1998) have also shown that thermal discharge in the region, such as near Baku, contributes calcium- and chlorine-rich waters to the basin that carry isotopic evidence of its origin as a deep brine.

Prior to its severe desiccation in recent years, the Aral Sea was the fourth largest body of inland water in the world (Table 1). It is located in the desert to the east of the Caspian Sea and with which it was connected prior to the last glaciation by overflow. The region has a moisture deficit of

Table 4 SNORM model output, in terms of "simple salt" cation–anion associations in mole %.

Locality	Na₂Cl₂	K₂Cl₂	CaCl₂	MgCl₂	Na₂SO₄	K₂SO₄	CaSO₄	MgSO₄	Na₂CO₃	CaCO₃	MgCO₃
Western Asia[a]											
Aral Sea	59.0	0.19	0.0	0.4	0.0	0.58	12.7	25.4	0.0	0.0	1.7
Lake Balkhash (delta)	33.7	0.0	0.0	0.0	22.8	0.0	13.8	2.1	0.0	0.0	27.6
Lake Balkhash (center)	38.4	0.0	0.0	0.0	32.3	0.0	0.8	10.0	0.0	0.0	18.4
Caspian Sea	62.4	1.1	0.0	5.7	0.0	0.0	8.5	20.3	0.0	0.0	2.0
Caspian Sea tributaries	12.5	0.0	0.0	0.0	7.3	0.6	51.4	6.5	0.0	0.0	21.6
Caspian Sea hot springs	58.0	1.9	0.2	7.9	0.0	0.0	32.0	0.0	0.0	0.0	0.0
Qaidam Basin, China[b]											
Nyer Co	15.0	6.5	0.0	74.0	0.0	0.0	0.6	3.8	0.0	0.0	0.0
Qinghai	54.3	10.9	0.0	2.7	0.0	0.4	0.8	30.9	0.0	0.0	0.0
Great Salt Lake Basin, USA[c]											
Salt Lake	83.7	2.4	0.0	5.9	0.0	0.0	0.3	7.7	0.0	0.0	0.1
Bear River	44.3	0.0	0.0	0.0	0.1	2.4	6.1	0.0	0.0	23.3	23.8
Sevier River	49.5	0.0	0.0	0.0	11.2	0.7	12.9	8.1	0.0	0.0	17.6
Malad Springs	89.0	2.8	3.3	2.6	0.0	0.0	1.3	0.0	0.0	1.0	0.0
West Pond	87.6	3.4	0.9	6.2	0.0	0.0	2.0	0.0	0.0	0.0	0.0
Lake Bogoria Basin, Kenya[d]											
Lake Bogoria	14.4	0.5	0.0	0.0	0.2	0.5	0.0	0.0	84.0	0.1	0.0
Sandai River	22.0	0.7	0.0	0.0	0.5	1.4	0.0	0.0	41.0	24.6	9.8
Loburu Hot	11.8	0.0	0.0	0.0	1.1	1.1	0.0	0.0	85.8	0.1	0.1
Northern Great Plains, USA and Canada[e]											
Big Coulee Riv.	3.2	0.0	0.0	0.0	9.8	2.4	0.0	0.0	0.1	52.5	32.1
Souris River	8.8	0.0	0.0	0.0	28.0	3.5	0.0	0.0	6.5	22.1	31.2
Sheyenne River	5.1	0.0	0.0	0.0	25.9	0.0	0.0	0.0	24.0	18.8	26.2
Des Lacs River	5.8	0.0	0.0	0.0	9.3	17.3	0.2	0.0	0.0	42.1	25.3
Little Manitou Lake	22.9	0.0	0.0	0.0	1.7	1.0	2.3	71.2	0.0	0.0	0.9
Freefight Lake	12.5	0.0	0.0	0.0	29.8	3.5	0.5	50.5	0.0	0.0	3.2
Devils Lake	20.7	0.0	0.0	0.0	40.3	4.5	5.7	19.1	0.0	0.0	9.7
Death Valley, USA[f]											
Amargosa River	57.2	0.0	0.0	0.0	31.5	2.6	0.0	0.0	2.1	3.4	3.2
Mormon Point Springs	79.8	2.8	8.3	5.6	0.0	0.0	3.4	0.0	0.0	0.1	0.0
DV-40	97.4	0.0	0.0	0.0	0.0	0.7	1.2	0.6	0.0	0.0	0.0
Australia[g]											
Yilgarn	78.9	0.7	0.0	12.6	0.0	0.0	3.5	4.3	0.0	0.0	0.0
Lake Tyrrell	82.8	0.7	0.0	8.0	0.0	0.0	0.3	8.2	0.0	0.0	0.0
Lake Eyre	87.6	0.7	0.0	5.0	0.0	0.0	0.6	6.0	0.0	0.0	0.1
Dead Sea Basin[h]											
Jordan River	39.3	4.7	0.0	1.5	0.0	0.0	4.2	0.0	0.0	21.8	29.8
Zohar Springs	26.4	2.8	16.8	49.7	0.0	0.0	1.2	0.0	0.0	0.03	0.03
Dead Sea	28.1	2.7	14.2	54.6	0.0	0.0	0.3	0.0	0.0	0.04	0.04
Other Great Basin, USA[i]											
Bristol Dry Lake	51.7	1.7	44.7	1.8	0.0	0.0	0.1	0.0	0.0	0.0	0.0
Mono Lake	39.4	0.0	0.0	0.0	12.8	3.1	0.0	0.0	44.4	0.02	0.3
Salton Sea	69.4	0.0	0.0	0.0	5.4	0.7	7.8	16.2	0.0	0.0	0.5
Seawater	77.4	1.7	0.0	11.2	0.0	0.0	3.4	5.9	0.0	0.0	0.4

Source: SNORM program from Bodine and Jones (1986).
[a] Blinov, 1962; Kosarev and Yablonskaya, 1994; Clauer *et al.*, 1998. [b] Spencer *et al.*, 1990; Zheng, 1997. [c] Spencer *et al.*, 1985a; Kohler, 2002; USGS, 2002). [d] Renaut and Tiercelin, 1994. [e] Komor, 1994; Last, 1994a; USGS, 2002. [f] Li *et al.*, 1997. [g] Johnson, 1980; Mann, 1983; Herczeg and Lyons, 1991. [h] Neev and Emery, 1967; Abed *et al.*, 1990; Vengosh and Rosenthal, 1994. [i] Rosen, 1991; USGS, 2002.

at least 0.2 m. The sea is (or has been) fed by two large rivers, the Amu Darya and the Syr Darya, which rise in the Pamir mountains hundreds of kilometers to the south. In recent times the river flow has been substantially reduced by irrigation and evaporation before reaching the sea. Considerable calcium carbonate, and subsequently gypsum, has been precipitated from the small amount of inflow that has reached the sea. Between 1960 and 1987, the level of the sea dropped by nearly 30 m, its area decreased by 40%, volume diminished by 66%, average depth dropped to 9 m, and its average salinity rose to 27 g L^{-1} (Micklin, 1988). The solute composition of the residual Aral Sea waters is now dominated by sodium and magnesium sulfate, with a large amount of alkali carbonate, and a distinctly lower proportion of NaCl than in the Caspian Sea. In historical times, much was made of the clarity and color of the Aral Sea water, which was attributed to flocculation associated with a high CaSO$_4$ content. The marls of the central part of the sea are extremely fine grained, sulfide bearing, and very high (>50%) in CaCO$_3$ (Dickey, 1968).

Some of the lakes in the Qaidam Basin of western China produce potash salts such as sylvite and carnallite, but lack magnesium sulfates (Spencer *et al.*, 1990; Zheng *et al.*, 1993; Zheng, 1997; Duan and Hu, 2001). As Spencer *et al.* (1990) showed, these highly soluble potassium minerals can be produced in terrestrial environments where hydrothermal Ca-chloride waters mix with terrestrial Na-bicarbonate waters. In such a mixture, sulfate becomes depleted due to gypsum precipitation, allowing Na–K chloride saturation to be reached. Although the volume of Ca-chloride waters contributed to potash salt-producing lakes may be minimal, the solute loads of these brines profoundly affect the evolution of the waters, providing calcium for removal of CO$_3$ and SO$_4$ by the precipitation of carbonates and gypsum.

Probably the world's most famous saline lake is the Dead Sea, occupying a tectonic rift valley significantly below sea level (−415 m; Tables 3 and 4). The origin of its solute composition has been the subject of considerable attention and has been summarized by Starinsky (1974) and Vengosh *et al.* (1991). They used boron isotopes and boron concentrations, plus chlorine and lithium isotope ratios, to suggest that the brines of the Dead Sea and on-shore hydrothermal springs are products of the interaction of evaporated seawater with marine (largely carbonate) sediments. Subsequently, Yechieli *et al.* (1996), using ^{14}C and ^{36}Cl analyses, proposed that saline groundwaters in the Dead Sea area are the result of the infiltration of brines with a significant rainwater component from a precursor lake, for example the fossil Lake Lisan, which covered the area in the past to a much higher elevation than the present Dead Sea, and went through several evaporation stages. This lake contained a significant rainwater component, besides the original salts from the ancient sea. Abu-Jaber (1998), utilizing PHRQPITZ equilibrium modeling (Plummer *et al.*, 1988), indicated that a simple mixture of present saline groundwaters and 50 times evaporated Zohar hot spring water could account for the Dead Sea salts. He suggested that the hot spring waters have a deeply

circulated "meteoric-continental" origin, rather than from seawater, but Hardie (1990, 1991) has proposed that hydrothermal and/or oilfield brine-type diagenetic reactions operating on marine fluids (formation of dolomite, gypsum, albite, and/or chlorite) probably produced the initial solutes of the Dead Sea system.

Another unusual area supporting lakes containing near-neutral, high chloride brines is the "dry" valleys area of Antarctica (Matsubaya *et al.*, 1979). Unlike smaller lakes of the western coastal areas connected, or nearly connected, with the sea, and of related solute composition, Lake Bonney in the Taylor Valley has been flooded with freshwater. This produced a pronounced stratification modified only by diffusional mixing. The low-temperature concentration of seawater takes place by evaporation or by freeze-drying (which precipitates mirabilite at 4 times seawater salinity, and hydrohalite at 8 times). Lakes Vanda and Don Juan Pond in the Wright Valley are strongly stratified in both salinity and isotope ratios like Lake Bonney, but they are characterized by a high concentration of calcium relative to sodium and magnesium, apparently due to inputs of low-temperature hydrothermal solutions that interacted at depth with volcanic or crystalline rocks (Lyons and Mayewski, 1993).

The Salton Sea of southern California could be considered the largest man-made saline lake in the world. Although the present water body was created by overflow from the Colorado River in the course of railroad construction in 1905, the playa basin was undoubtedly occupied by a shallow desert lake several times in the geologically recent past. In the formation of the present body of water, Arnal (1961) has estimated that 45% of the total salt tonnage was derived from the pre-existing salina. Arnal (1961) also calculated the association of solutes in a similar way to the normative analysis of SNORM, and concluded that the major ion composition was similar to seawater (70–79% NaCl; 8.2–14.4% Mg-salt), except that the Mg-salt was predominantly a sulfate, rather than a chloride (Tables 3 and 4). Concomitantly, there is twice as much CaSO$_4$, and half the amount of K-salt as in seawater. Insoluble residue determinations indicate that up to 40% of the bottom sediment is composed of calcium carbonate and some gypsum. The increase in the organic carbon content and nutrients in the modern lake reflect its progressive eutrophication (Schroeder *et al.*, 2002).

Major saline systems at the borders of the Sahara include the large, ephemeral playas of Tunisia and Algeria, and the large, areally fluctuating, Lake Chad. Bryant *et al.* (1994) have described the principal control of the geochemistry of the Chott el Jerid in Tunisia by runoff from Cretaceous through Quaternary sediments dominated by marine evaporites. Eugster and Hardie (1978) have summarized the large number of evaporitic processes involved at Lake Chad from the main lake evaporation of major river inflow to fractional crystallization of alkali salts in interdunal depressions that comprise much of the northeast shore of the lake.

There are many closed-basin lakes in the South American altiplano (Risacher and Fritz, 1991a), where dilute inflow is strongly affected by aqueous $Na-HCO_3$ produced by the weathering of volcanics (Rettig *et al.*, 1980). Despite the initially high alkalinity in the inflow, lake brines are commonly dominated by chloride and sulfate which is probably a product of the oxidation of native sulfur derived from fumeroles (Risacher and Fritz, 1991a). Other important processes have been documented in these lakes, such as the spectacular development of magnesium-rich smectite produced by the reaction of diatoms with alkaline waters (Badaut and Risacher, 1983). These lakes have also produced records of Quaternary paleo-environments based on evaporite mineral assemblages largely dominated by halite and gypsum (Risacher and Fritz, 1991b; Bobst *et al.*, 2001), although other fresher periods are recorded by carbonates (Valero-Garces *et al.*, 1999, 2001).

Several other closed-basin regions are useful as examples of particular brine types. These include the Great Salt Lake (chloride), the lakes of East Africa (carbonate), the northern Great Plains of North America (sulfate), Death Valley (mixed), and the Australian "outback" (acid-chloride). Representative major ion compositions and SNORM model results for these examples are presented in Tables 3 and 4, respectively.

5.13.6.1 Chloride—Great Salt Lake

The Great Salt Lake, Utah, USA, the largest saline lake in North America, is probably the most studied and most well-documented body of saline water in the world today, with the exception of the ocean itself. It is discussed here as representative of a closed-basin system whose chemistry reflects solute inputs from the weathering of rocks ranging in age from Precambrian and Paleozoic in the high mountains on the east to Tertiary and Quaternary sediments in the desert basins to the west. It is the remnant of a much larger lacustrine body (Lake Bonneville), that occupied the basin during the Pleistocene, just as large pluvial lakes occupied other parts of the Great Basin and other parts of the globe (e.g., Lakes Titicaca, Eyre, Chad, etc.). Spencer *et al.* (1985a) have examined the geochemistry of the Great Salt Lake in detail for the historic period 1850–1982, and Spencer *et al.* (1985b) have developed a model for its hydrochemical evolution over the last 30,000 years.

Gwynn (2002) has summarized the most recent information.

The water balance of Great Salt Lake depends predominantly on the inflow from three major rivers draining the ranges to the east, basin evaporation, and precipitation directly on the lake (Arnow and Stephens, 1990). Major solute inputs can be attributed to calcium-bicarbonate-type river waters mixing with sodium chloride type springs, which are in part hydrothermal and part peripheral recycling agents for NaCl (Figure 14). Spencer *et al.* (1985a) have noted that prior to 1930, the lake concentration inversely tracked lake volume, which reflected climatic variation in the drainage. However, since that time salt precipitation, primarily halite and mirabilite, and dissolution have periodically modified lake brine chemistry and have led to density stratification and the formation of brine pockets of different composition because of fractional crystallization and resolution. Construction of a railway causeway has restricted circulation, nearly isolating the northern from the southern part of the lake, which receives over 95% of the inflow. This has led to halite precipitation in the north (Gwynn, 2002). Widespread halite precipitation has also occurred prior to 1959, especially in the southern area of the lake, associated with the most severe droughts (Spencer *et al.*, 1985a). Spencer *et al.* (1985a) have also described the presence of a subaqueous ridge, which probably separated the lake into two basins at very low lake stands in the distant past. These conditions emphasize brine differentiation, mixing, and fractional precipitation of salts as major factors in solute evolution, especially in relatively shallow systems. The work at Great Salt Lake has also highlighted the role of pore fluids as sources and sinks of solutes in the lake, depending on the concentration gradient. In the Great Salt Lake system, diagenetic reactions in pore fluids produce the occasional formation and dissolution of gypsum, which is not usually included in the salt precipitation sequence, because calcium levels are very low after carbonate precipitation. Alkalinity and low calcium concentrations are nearly constant at aragonite saturation levels. However, in sediments with a sufficient rate of decay of organic matter and with high sulfate pore fluid, local dissolution of carbonate responding to elevated P_{CO_2} can supply enough calcium necessary for gypsum saturation. As CO_2 is lost, the reaction is reversed, leaving only disrupted strata to mark the change. In addition to carbonate, significant amounts of dissolved magnesium and potassium are also lost to clays through diagenetic pore fluid reactions (Spencer *et al.*, 1984, 1985a; Jones and Spencer, 1999).

Spencer *et al.* (1985b) used sedimentologic and biostratigraphic evidence obtained from gravity cores (Spencer *et al.*, 1984) to develop

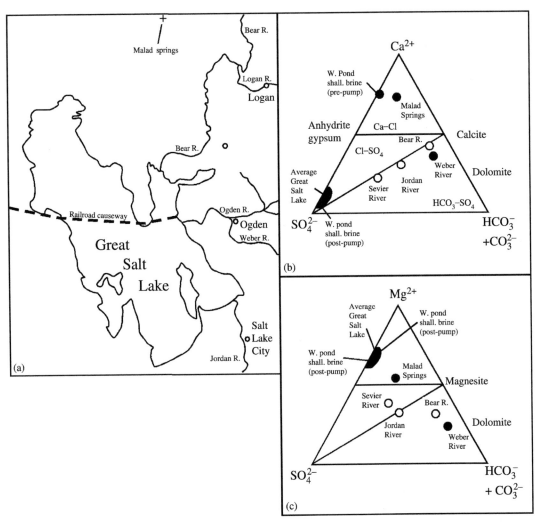

Figure 14 (a) Location of the major tributaries to the Great Salt Lake. (b) Ternary phase diagram ("Spencer Triangle") in the system $Ca-SO_4-(HCO_3 + CO_3)$, showing the relationship between dilute river inflow, the brines of Great Salt Lake and the Bonneville Flats (West Pond pre- and post-pumping), and the Malad Springs. Although the river samples are not geographically linked, the arrangement of points along a trajectory directly away from calcite, toward the sulfate corner, suggests that the precipitation of this mineral affects the chemical evolution of the waters. (c) Ternary diagram in the system $Mg-SO_4-(HCO_3 + CO_3)$, with points again suggestive of calcite or dolomite precipitation (sources Spencer *et al.*, 1985a; Kohler, 2002 (reproduced by permission of J. F. Kohler and Utah Dept. of Natural Resources from *Great Salt Lake, An Overview* **2002**, 487–498); and USGS, 2002).

a geochemical model for Great Salt Lake going back to ~30 ka. This period includes the presence of a greatly expanded freshwater lake period (Lake Bonneville) peaking ~19 ka and bracketed by fluctuations around saline conditions not dissimilar to those of the present day. During freshwater periods high discharge inflow was dominated by calcium-bicarbonate-type river waters, whereas saline stages were associated with a greater proportion of input from low discharge, NaCl-rich springs. An evolution in lake composition to NaCl domination is illustrated by halite-free mirabilite deposits during a relatively rapid lake level decrease ~8–10 ka. In contrast, historic droughts have yielded mainly halite. Pre-drawdown carbonate was found to be

primarily calcite, whose magnesium content rises to 11% during the period of lake level decline, and then shifts to aragonite with a small amount of dolomite. Significant amounts of solutes were lost to sediments by pore fluid diffusion. Accounting for this permitted a reasonable balance to be made between solute input over time and the quantities now found in solution and in sediments. Excess amounts of calcium, carbonate, and silica can be attributed to detrital input.

According to Kelts and Shahrabi (1986), Lake Urmia in northwestern Iran, one of the larger saline lakes in the world, resembles Great Salt lake in morphology, water chemistry, and sediments. The carbonate precipitates are largely aragonite as fecal pellets, thin crusts, or oölites.

Some dolomite occurs in peripheral shallows and flats, also various types of gypsum that are also found in lakes of the American southwest (Bradbury, 1971).

5.13.6.2 Carbonate—East Africa

Hydrologically closed basins are common in East Africa. They are typically associated with Cenozoic rift-related depressions. Basement metamorphic rocks of granitic composition are shallow across the region, except in areas that have been buried by Cenozoic volcanic rocks and tephra. Rift basin lakes range in size from large, deep lakes such as Lake Turkana (Yuretich and Cerling, 1983; Cerling, 1996), to shallow, smaller lakes such as Lake Magadi (Jones *et al.*, 1977), controlled by a combination of tectonic subsidence and basin segmentation due to the distribution of volcanism (Grove, 1986). Regional hydrology is controlled primarily by rift-related structures. In some areas there is evidence for regional-scale flow, such as toward the hydrologic terminus of Lake Magadi (Jones *et al.*, 1977).

The dilute inflow to the East African basins acquires most of its alkalinity by the rapid hydrolysis of volcanic glass and lavas, producing high initial Na^+, SiO_2, and HCO_3^- concentrations (Jones *et al.*, 1977). Waters in the region are therefore nearly exclusively of the $Na–CO_3$ or $Na–CO_3–Cl$ type. Most of the other solutes are lost to carbonate or silicate precipitation (Jones *et al.*, 1977; Beadle, 1981; Renaut and Tiercelin, 1994). Sulfate in East African waters is often removed from solution during evaporative concentration, probably due to reduction, especially in lake-marginal wetlands (Deocampo and Ashley, 1999).

Areally close compositional contrast is seen in the recent rift area of Djibouti. On the western border, the terminal Lake Abhe is fed by the sodium-carbonate-type Awash River from the Ethiopian volcanic highlands, whereas the Asal Lake near an arm of the Gulf of Aden is fed by seawater through recent fault zones and hydrothermal inputs in addition to local runoff, and precipitates halite and gypsum (Fontes *et al.*, 1979).

Carbonate precipitates of variable calcium and magnesium content are widespread throughout the region. Abundant diatoms and phytoliths seem to buffer amorphous silica, which maintains levels near or slightly below amorphous silica solubility (Deocampo and Ashley, 1999). Evaporite mineralogies are dominated by carbonate phases such as trona and gaylussite, with alkali silicates such as Na-silicates and zeolites precipitated from the most evolved brines (Hay, 1976; Renaut and Tiercelin, 1994; Hay and Kyser, 2001). The origin of "Magadi"-type chert (Hay, 1968) has been the subject of recent discussion, and appears to result either from leaching of alkali silicates or from biomineralization in siliceous waters (Behr, 2002). Unlike thermal springs of other regions or oil field brines, hot springs in the East African rift tend to lack a $CaCl_2$ SNORM signature, indicating that these waters instead represent recycled surface brines (Deocampo and Ashley, 1999). The simple salt assemblages precipitated from these waters are dominated by the products of weathering of volcaniclastic material (Table 4).

5.13.6.3 Sulfate—Northern Great Plains, North America

A large area of internal drainage is found across the Canadian provinces of Alberta, Saskatchewan, and Manitoba, extending south into the American states of Montana and North Dakota (Figure 15). This area is underlain predominantly by Paleozoic carbonates and evaporites, and Mesozoic to Cenozoic siliciclastic rocks (Last, 1999). These sedimentary rocks are overlain by unconsolidated sediments, mostly Quaternary glacial and fluviolacustrine deposits (Klassen, 1989). Landforms associated with the last deglaciation control the hydrology of the region. Ice-marginal incision features provide many of the basins for modern lakes (Kehew and Teller, 1994).

The sedimentary deposits that underlie most of the region produce dilute inflow with a sulfo-carbonate chemistry. Last (1999) has indicated that 95% of the lakes of the Great Plains are of the sulfo-carbonate type. Early precipitation of carbonates and subsequently gypsum produces evolved brines nearly devoid of Ca^{2+}, but still enriched in Mg^{2+} (Last, 1992). Carbonate precipitates are widespread, with higher magnesium content, including lacustrine dolomite, reflecting waters with elevated Mg/Ca ratios and higher overall salinity (Last, 1990; Vance *et al.*, 1997). As these carbonates precipitate, waters evolve along the calcium carbonate and Ca–Mg carbonate trajectories, represented schematically by the Big Coulee, Souris, and Sheyenne Rivers (Figure 15). Ultimately, the waters produced by this evolution are Na–Mg sulfate brines, and this is reflected in mineral assemblages dominated by bloedite, epsomite, and mirabilite (Last, 1994a,b, 1999).

As Last (1999) has pointed out, the lack of integrated drainage patterns over large areas makes precise definition of watersheds and drainage divides difficult, but groundwater is clearly a critical factor in the regional geolimnology. LaBaugh (1988) has indicated that interaction with local groundwater flow systems affects changes in concentration as much as the difference between average annual evaporation and precipitation on a regional scale. Donovan and Rose's (1994) detailed study of evaporative groundwater-fed lakes in

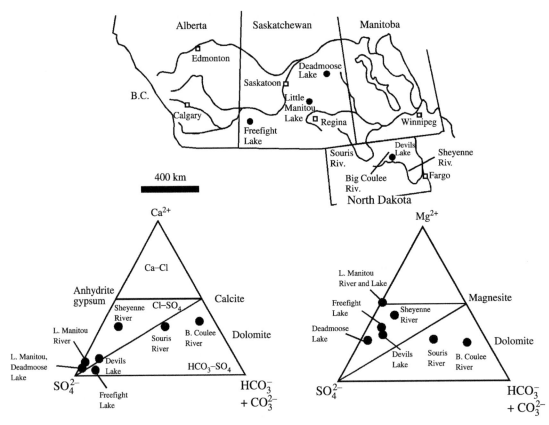

Figure 15 Location of the major lakes and rivers of the northern Great Plains, with ternary-phase diagrams in the systems $Ca^{2+}-SO_4^{2-}-(HCO_3^- + CO_3^{2-})$, and $Mg^{2+}-SO_4^{2-}-(HCO_3^- + CO_3^{2-})$. Again, although the river samples are not geographically linked, they outline a pathway suggestive of the effect of the precipitation of Ca- and Mg-carbonates (sources Last, 1994a; Komor, 1994; and USGS, 2002).

the glaciated Great Plains region near the junction of Montana, North Dakota, and Saskatchewan demonstrated that hydrochemical variability depended on the path of groundwater inflow and on subsequent surface mixing and evaporation. Differing depths of groundwater circulation promoted variations in aquifer solute source. Lakes in shallow surface depressions obtained water from local shallow groundwater, whereas lakes in deep or broad topographic lows received additional input from greater depth. In the area of study, the relative dominance of sulfate or carbonate (and/or sodium and magnesium) was found to be indicative of source. The compositional contrast between groundwater input and pre-existing surface brines, in addition to solute fractionation accompanying evaporation and mineral precipitation, was also important. Specific surface hydrologic effects can be seen in wide seasonal variations in lake-water composition, and the common development of chemical stratification in the water column (discussed in detail by Last, 1999). The path of relatively conservative solutes (Na and SO_4) follows the one predicted by evaporation of typical shallow and/or intermediate groundwaters, from which carbonates have been precipitated as aragonite or

magnesium calcite under near-atmospheric CO_2 pressure (Donovan and Rose, 1994). There are many examples of this type of evaporation in the Canadian Great Plains (Last, 1999). The Devils Lake drainage, North Dakota, provides a significant example of surface evaporation driving a compositional trend from magnesium calcite to dolomite to bedded mirabilite in East Stump Lake at the end of the chain (Callender, 1968).

Several small saline lakes that are fed almost exclusively by groundwaters occur on the semiarid Interior Plateau of central British Columbia (Nesbitt, 1974, 1990; Renaut, 1994) between the Coast Mountains and the Columbia Rocky Mountain ranges, and in the semi-arid plains of central Spain (Carenas *et al.*, 1982; Pueyo and De la Pena, 1991). The most saline of these lakes follow a solute development and mineral precipitation pattern similar to many of the saline lakes of the northern Great Plains, evolving through Na-sulfate to Mg-sulfate dominant brines. Other, small, Mg-sulfate or sodium sulfo-carbonate saline lakes are found in northern (Anderson, 1958a,b) or central Washington State (Bennett, 1962; Edmundson and Anderson, 1965).

An example of the hydrochemical importance of lithologic variation in solute source is the difference between the Canadian Great Plains lakes and the Sand Hill Lake region of western Nebraska. Though solute concentration is very dependent on local groundwater flow paths in the same way, the solute composition in the Sand Hill Lake region reflects the mineralogy of the feldspathic sands, and the lakes' position in low points of an extensive stabilized dunefield. In fully closed ponds, this leads to some of the highest potassium concentrations anywhere in the world (Bradley and Rainwater, 1956; Gosselin *et al.*, 1994).

5.13.6.4 Mixed Anions—Death Valley

At present Death Valley, California (Figure 16), does not contain a perennial lake, but it is an excellent example of a closed-basin system in which several major water types occur (Li *et al.*, 1997). Upland weathering of a wide range of bedrock lithologies produces $Na-HCO_3$ and $Na-SO_4-HCO_3-Cl$ waters typical of meteoric input from the Amargosa River, springs, and groundwaters in the north and central parts of the Valley, including the big springs in the Funeral Mountains discharging from the intermontane Paleozoic aquifer underlying much of southern Nevada (Winograd and Thordarson, 1975). In contrast, springs and wells to the south of the central Death Valley salt pan are $Na-Ca-Cl$ type waters, probably produced diagenetically by water–rock interaction and hydrothermal activity at depth (Hardie, 1990), and probably associated with a 15 km deep magma chamber. These waters yield a residual $CaCl_2$ normative signature (Table 4), and are similar to the $CaCl_2$ brines of the Qinghai Basin of northern China (Spencer *et al.*, 1990; Zheng, 1997). The $Na-Cl-SO_4$ waters typical of Death Valley evolved brines are thought to be produced by the evaporative evolution of a mixture of these two end-members, which are represented by the Mormon Point

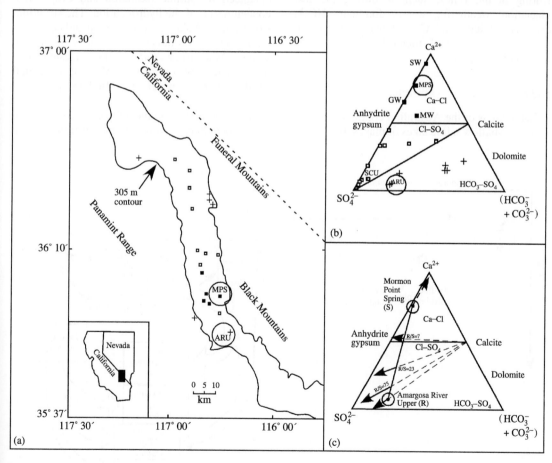

Figure 16 (a) Location of Death Valley, USA: (+) dilute $Na-HCO_3-SO_4$ waters; (□) $Cl-SO_4$ waters typical of Death Valley brines; and (■) $Ca-Cl$ spring waters, associated with a magma chamber at depth. (b) and (c) Ternary-phase diagrams in the system $Ca^{2+}-SO_4^{2-}-(HCO_3^-+CO_3^{2-})$ showing how various brine compositions can be acquired by mixing end-member $Ca-Cl$ and $Na-HCO_3-SO_4$ waters, with subsequent calcite precipitation (source Li *et al.*, 1997) (reproduced by permission of the Geological Society of America from *Geol. Soc. Am. Bull.* **1997**, *109*, 1361–1371).

418   *Geochemistry of Saline Lakes*

Spring and the Upper Amargosa River samples. This evolution involves the precipitation first of carbonate, and then sulfate species to draw down calcium and magnesium, producing the pathways that result in SO_4–Cl brines. Li *et al.* (1997) used the joins between points on the triangular diagram to demonstrate that mixing ratios of Amargosa River to Mormon Point Spring waters would range from 7% to 75%. The somewhat zonal distribution of water types with respect to the fan toes, saltpan, and estimated position of subsurface magma is apparent over kilometer scales (Figure 16).

Evaporite assemblages produced by these waters are dominated by carbonates and sulfates, and analyses of cores show that there have been cyclical changes over the past 100 ka due to variations in the ratio of Ca–Cl (diagenetic) to Na–HCO_3 (meteoric) waters (Li *et al.*, 1997). During dry times, such as the present, meteoric input is relatively low, allowing calcium to exceed HCO_3 and eventually precipitate with sulfate as gypsum and glauberite rather than as a carbonate. During humid times, such as during the last glaciation, weathering waters dominate, HCO_3 exceeds calcium, and Ca-sulfates cannot form; hence, mirabilite and thenardite become the dominant precipitates.

5.13.6.5 Acid Lakes—Australia

Large areas of the Australian continent contain scattered shallow saline lakes or playas, the largest of which is Lake Eyre in the province of South Australia (Jankowski and Jacobson, 1989). Eyre is fed by seepage at the southern end of a large regional discharge zone in central Australia. Most of the time it is a highly ephemeral feature, containing significant amounts of water only about once every 10 years. The geochemistry is dominated by Na–Cl and Ca–SO_4. These ions comprise such a large percentage of the solutes (Johnson, 1980) that they suggest selective resolution of marine evaporites. Indeed, many of the closed basins of Australia contain acid Na–Mg–SO_4–Cl saline waters with strong normative affinities to seawater (Table 4). The most extreme cases are found in the Yilgarn block of Western Australia (Mann, 1983; Salama, 1994); here some major solute analyses of the lake waters produce salt norms virtually identical to the marine norm (Table 4). At the same time, pH measurements can reach values below 3, probably due to a thick silcrete regolith that provides none of the buffer capacity of underlying crystalline bedrock, perhaps enhanced by sulfide oxidation and evaporative concentration of the hydrogen ion. The strontium isotope mass balance in the region indicates that marine aerosols could account for 97% of the total solutes (MacArthur *et al.*, 1991), consistent with the earlier observations of compositional similarities to seawater (e.g., Bodine and Jones, 1986).

An important example of solute composition and geochemical evolution in Australia is the

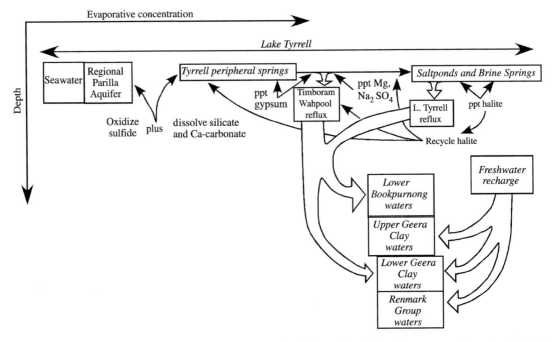

Figure 17 Schematic stratigraphic, hydrochemical, and hydrodynamic relationships suggested by calculated salt norms for the Lake Tyrrell system and pore waters for the central Murray Basin Piangil West 2 borehole. For piezometric map of the area, see Macumber (1983).

Lake Tyrrell system (Figure 17) of the central Murray Basin (Long *et al.*, 1992a,b; Jones *et al.*, 1994). Most of the saline surface and associated shallow groundwaters in the central Murray Basin have a distinctly marine character despite a limited history of marine incursions. Some of the Tyrrell waters show the NaCl increase and SO_4 salt decrease expected with evaporative concentration and gypsum precipitation in an ephemeral saline lake or playa. Normative analyses of the subsurface waters as well are compatible with the dilution of variably fractionated marine bitterns somewhat depleted in sodium salts, similar to evolved brines of Lake Tyrrell. The basin geochemistry is most readily explained by the input of marine aerosols, which have been leached into shallow groundwaters basin-wide. Such an explanation can readily account for the meteoric isotope composition of the saline waters in the basin. At the same time the groundwater hydrophysics, stable isotopes, and salt norms can help in determining the elements of a complex system such as Lake Tyrrell and its satellites (Figure 17).

Following the general playa lake hydrology described by Jacobsen and Jankowski (1988) for the Amadeus Basin of central Australia, Arakel and Hongjun (1994) examined the evaporite precipitation sequence in Calatta Lake, part of a chain of playa lakes to the east of Amadeus. Seasonal efflorescent crusts are characterized by laterally intergrading halite and Na_2SO_4 salts (mirabilite/thenardite). The detailed sequence, which can include schoenite, sylvite, and epsomite in local pools, is a consequence of groundwater discharge pattern and prevailing climatic conditions, particularly temperature, so that it differs significantly from simple evaporation. The system demonstrates the effectiveness of brine recycling, saline mineral diagenesis, and dissolution–precipitation in a desert climate (Herczeg and Lyons, 1991). Similar to the environments described by Last (1999) in the Canadian prairies, Arakel and Hongjun (1994) have shown that the precipitation of temperature-dependent sulfate minerals can proceed independently of halite crystallization from ponded brines. Mineralogic zonation in playa efflorescent crusts is therefore not entirely dependent on evaporative concentration. An exception to the similarity of the composition of Australian saline lake waters to evaporated seawater is Lake Buchanan, a closed basin lake, which lies in the Great Dividing Range of central Queensland in northeast Australia (Chivas *et al.*, 1986). Here the solute Ca/Mg ratio is about unity, the Br/Cl ratio is anomalously low, and the major solute composition reflects the clay-rich sandstones of the drainage area. Thus, marine aerosols and recycling are not the source of the salts in all Australian saline lake systems.

5.13.7 SUMMARY

Saline lakes are important environmental features in continental basins worldwide. They have major implications for water resources, economic activity, and paleoclimate studies. Water and solute sources for saline lake brines are direct precipitation, surface flow, local or regional meteoric groundwaters, interstitial water of sediments, or deep basinal fluids. Solute concentrations are controlled by the weathering of source rocks, evaporation, and solute losses to minerals and biota. Aquifer residence times, rock dissolution rates, leakage ratios, and dissolution–precipitation fractionation are also important factors. With evaporation, inflow waters generally reach supersaturation first with respect to alkaline-earth carbonates, the first "chemical divide." Subsequent chemical divides occur for gypsum, magnesite, and sulfates of magnesium and sodium, as shown in our revised brine evolution flowchart. Brine evolution along different flow paths is therefore controlled by the molar ratio of solutes, eventually producing diagnostic mineral assemblages.

The "Spencer Triangle" is a simple representation of the effect of chemical divides on brine evolution, and a valuable tool for assessing solute behavior in the system $Ca–SO_4–HCO_3$. Such plots show that the initial solution chemistry is a good predictor of the resulting mineral assemblage. A similar triangle in which the calcium vertex is replaced with magnesium shows that the initial Ca/Mg ratio, and the timing of calcium depletion relative to magnesium are critical factors in determining whether the subsequent brine will be dominated by sulfate or carbonate. These two triangles together demonstrate the principal major ion behaviors responsible for solution chemistries in saline lakes worldwide. In addition, normative calculations on the major ion concentrations of representative saline lake waters can provide specific measures of their hydrochemical evolution.

ACKNOWLEDGMENTS

We are especially grateful to the volume editor, J. I. Drever, and to B. Kimball, USGS, Salt Lake City, for manuscript review. We also wish to thank the following for helpful comments: Profs. C. J. Bowser, U. Wisc. (ret.), G. M. Ashley, Rutgers U., T. K. Lowenstein, SUNY Binghamton, Drs. E. Callender (ret.) and W. Wood, USGS, Reston, and Dr. Y. Yechieli of the Hydrologic

Service of Israel. We also appreciate the support of Jim Kohler and Bill White, Bureau of Land Management, Salt Lake City, and the assistance of Marge Shapira in manuscript preparation. This work was supported in part by National Science Foundation grant INT-0202612.

REFERENCES

Abed A., Al-Sbaeay I., and Khbeis S. (1990) The Dead Sea: a recent survey of its waters. *Proc. 3rd Jordanian Geol. Conf.*, vol. 3, pp. 446–472.

Abu-Jaber N. S. (1998) A new look at the chemical and hydrological evolution of the Dead Sea. *Geochim. Cosmochim. Acta* **62**, 1471–1479.

Almendinger J. E. (1990) Groundwater control of closed-basin lake levels under steady-state conditions. *J. Hydrol.* **112**, 293–318.

Anderson G. C. (1958a) Seasonal characteristics of two saline lakes in Washington. *Limnol. Oceanogr.* **3**, 51–68.

Anderson G. C. (1958b) Some limnological features of a shallow saline meromictic lake. *Limnol. Oceanogr.* **3**, 259–270.

Anderson T. F. and Arthur M. A. (1983) Stable isotopes of oxygen and carbon and their applications to sedimentologic and paleoenvironmental problems. *Stable Isotopes in Sedimentary Geology.* SEPM Short Course 10, (eds. M. A. Arthur, T. F. Anderson, I. R. Kaplan, J. Veizer, and L. S. Land). Society for Sedimentary Research (SEPM), Tulsa.

Arakel A. V. and Hongjun T. (1994) Seasonal evaporite sedimentation in desert playa lakes of the Karinga Creek drainage system, central Australia. In *Sedimentology and Geochemistry of Modern and Ancient Saline Lakes* (eds. R. W. Renaut and W. M. Last). Society for Sedimentary Geology (SEPM), Tulsa, pp. 91–100.

Arnal R. E. (1961) Limnology, sedimentation, and microorganisms of the Salton Sea, California. *Geol. Soc. Am. Bull.* **72**, 427–478.

Arnow T. and Stephens D. (1990) Hydrologic characteristics of the Great Salt Lake, Utah: 1847–1986. *US Geol. Surv. Water-Supply Pap.* **2332**, 32pp.

Badaut D. and Risacher F. (1983) Authigenic smectite on diatom frustules in Bolivian saline lakes. *Geochim. Cosmochim. Acta* **47**, 363–375.

Baltres A. and Medesan A. (1978) High-magnesium calcite in fecal pellets of *Artemia salina* from Techirghiol Lake. *Sedim. Geol.* **20**, 281–290.

Banfield J. F., Jones B. F., and Veblen D. R. (1991) An AEM–TEM study of weathering and diagenesis, Abert Lake, Oregon: II. Diagenetic modification of the sedimentary assemblage. *Geochim. Cosmochim. Acta* **55**, 2795–2810.

Beadle L. C. (1981) *The Inland Waters of Tropical Africa.* Longman, London, 475 pp.

Behr H.-J. (2002) Magadiite and magadi chert: a critical analysis of the silica sediments in the Lake Magadi Basin, Kenya. In *Sedimentation in Continental Rifts* (eds. R. W. Renaut and G. M. Ashley). Society for Sedimentary Research (SEPM), Tulsa, pp. 257–273.

Bennett W. A. G. (1962) Saline lake deposits in Washington. *State Washington Div. Min. Geol. Bull.* **49**, 129 pp.

Blinov L. K. (1962) The physico-chemical properties of Caspian waters and their comparable characteristics. *Trudi Gos Ocean. Inst.* **68**, 7–28. (in Russian).

Bobst A. L., Lowenstein T. K., Jordan T. E., Godfrey L. V., Ku T. L., and Luo S. (2001) A 106 ka paleoclimate record from drill core of the Salar de Atacama, northern Chile. *Palaeogeogr. Palaeoclimatol. Palaeoecol.* **173**, 21–42.

Bodine M. W., Jr. and Jones B. F. (1986) The salt norm: a quantitative chemical–mineralogical characterization of natural waters. *US Geol. Surv. Water Res. Inv. Rep.* 86-4086.

Bowler J. M. (1986) Spatial variability and hydrologic evolution of Australian lake basins: analogue for Pleistocene hydrologic change and evaporite formation. *Palaeogeogr. Palaeoclimatol. Palaeoecol.* **54**, 21–41.

Bradbury J. P. (1971) Limnology of Zuni Salt Lake, New Mexico. *Geol. Soc. Am. Bull.* **82**, 379–398.

Bradley E. and Rainwater F. H. (1956) Geology and groundwater resources of the Upper Niobrara River basin, Nebraska and Wyoming. *US Geol. Surv. Water-Supply Pap.* 1368, USGS, Washington, DC.

Bruyevich S. V. (1938) Hydrochemical features of the Caspian Sea. *Prioda* **4**, 16–27. (in Russian).

Bryant R. G., Drake N. A., Millington A. C., and Sellwood B. W. (1994) The chemical evolution of the brines of Chott El Djerid, southern Tunisia, after an exceptional rainfall event in January 1990. In *Sedimentology and Geochemistry of Modern and Ancient Saline Lakes* (eds. R. W. Renaut and W. M. Last). Society for Sedimentary Geology (SEPM), Tulsa, pp. 3–12.

Buckney R. T. and Tyler P. A. (1976) Chemistry of salt lakes and other waters of the sub-humid regions of Tasmania. *Austral. J. Mar. Freshwater Res.* **27**, 359–366.

Buineviya N. A., Lepeshkob I. N., and Luchkov B. P. (1978) Formation of saline deposits of the type found in Kora-Bogaz Bay. *Kompleksnoe ispolzovanie mineralnogo cirya* **11**, 69–74. (in Russian).

Callender E. (1968) The post-glacial sedimentology of Devils Lake, North Dakota. PhD Dissertation, University of North Dakota.

Carenas B., Marfil R., and de la Pena J. A. (1982) Modes of formation and diagnostic features of Recent gypsum in a continental environment, La Mancha (Spain). *Est. Geol.* **38**, 345–359.

Carpenter A. B. (1978) Origin and chemical evolution of brines in sedimentary basins. *Okl. Geol. Surv. Circular* **79**, 60–77.

Casanova J. and Hillaire-Marcel C. (1992) Chronology and paleohydrology of Late Quaternary high lake levels in the Manyara Basin (Tanzania) from Isotopic Data (^{18}O, ^{13}C, ^{14}C, Th/U) on fossil stromatolites. *Quat. Res.* **38**, 205–226.

Cerling T. E. (1996) Pore water chemistry of an alkaline lake: Lake Turkana. In *The Limnology, Climatology, and Paleoclimatology of the East African Lakes* (eds. T. C. Johnson and E. O. Odada). Gordon and Breach, Amsterdam, pp. 225–240.

Chivas A. R., DeDeckker P., Nind M., Thiriet D., and Watson G. (1986) The Pleistocene paleoenvironmental record of Lake Buchanan: an atypical Australian playa. *Palaeogeogr. Palaeoclimatol. Palaeoecol.* **54**, 131–152.

Clauer N., Zuppi G. M., Blanc G., Toulkeridis T., and Gasse F. (1998) Compositions chimiques et isotopiques d'eaux de la mer Caspienne et de tributaires de la region de Makachkala (Russie): premieres donnees sur le fonctionnement d'un systeme endoreique particulier. *Sci. de ter. plan.* **327**, 17–24.

Collerson K. D., Ullman W. J., and Torgersen T. (1988) Ground waters with unradiogenic ^{87}Sr/^{86}Sr ratios in the Great Artesian Basin, Australia. *Geology* **16**, 59–63.

Connell T. L. and Dreiss S. J. (1995) Chemical evolution of shallow groundwater along the northeast shore of Mono Lake, California. *Water Resour. Res.* **31**, 3171–3182.

Cooke R. U. and Warren A. (1973) *Geomorphology in Deserts.* University of California Press, Berkeley.

Craig H. and Gordon L. I. (1965) Deuterium and oxygen-18 variations in the ocean and marine atmosphere. In *Stable Isotopes in Oceanographic Studies and Paleotemperatures.* (ed. E. Tongion). Spoleto, pp. 1–130.

Deocampo D. M. (2001) Groundwater wetlands in East Africa: geochemistry, sedimentology, and Plio-Pleistocene deposits. PhD Dissertation, Rutgers University.

Deocampo D. M. (2002) Sedimentary structures generated by *Hippopotamus amphibius* in a lake-margin wetland, Ngorongoro Crater, Tanzania. *Palaios* **17**, 212–217.

Deocampo D. M. and Ashley G. M. (1999) Siliceous islands in a carbonate sea: Modern and Pleistocene spring-fed wetlands in Ngorongoro Crater and Olduvai Gorge, Tanzania. *J. Sedim. Res.* **69**, 974–999.

Deocampo D. M., Hay R. L., Ashley G. M., Kyser T. K., and Liutkus C. M. (2000) Lacustrine clay diagenesis in northern Tanzania, with paleoenvironmental application at Olduvai Gorge. *Geol. Soc. Am. Abstr. Prog.* **32**, A-366.

Dickey (1968) Contemporary non-marine sedimentation in Soviet central Asia. *AAPG Bull.* **52**, 2396–2421.

Domagalski J. L., Eugster H. P., and Jones B. F. (1990) Trace metal geochemistry of Walker, Mono, and Great Salt Lakes. In *Fluid–Mineral Interactions: A Tribute to H. P. Eugster.* Special Publication 2 (eds. R. J. Spencer and I. M. Chou). Geochemical Society, San Antonio, pp. 315–354.

Donovan J. J. and Rose A. W. (1994) Geochemical evolution of lacustrine brines from variable-scale groundwater circulation. *J. Hydrol.* **154**, 35–62.

Doran P. T., Fritsen C. H., McKay C. P., Priscu J. C., and Adams E. E. (2003) Formation and character of an ancient 19-m ice cover and underlying trapped brine in an "ice-sealed" east Antarctic lake. *Proc. Natl. Acad. Sci.* **100**, 26–31.

Drever J. I. (1997) *The Geochemistry of Natural Waters.* Prentice Hall, Englewood Cliffs.

Drever J. I. and Smith C. L. (1978) Cyclic wetting and drying of the soil zone as an influence on the chemistry of ground water in arid terrains. *Am. J. Sci.* **278**, 1448–1454.

Duan Z. and Hu W. (2001) The accumulation of potash in a continental basin: the example of the Qarhan Saline Lake, Qaidam Basin, West China. *Euro. J. Mineral.* **13**, 1223–1233.

Duffy C. J. and Al-Hassan S. (1988) Groundwater circulation in a closed desert basin: topographic scaling and climatic forcing. *Water Resour. Res.* **24**, 1675–1688.

Edmundson W. T. and Anderson G. C. (1965) Some features of saline lakes in central Washington. *Limnol. Oceanogr.* **10**, 87–96.

Eugster H. P. and Hardie L. A. (1975) Sedimentation in an ancient playa-lake complex: the Wilkins Peak Member of the Green River Formation of Wyoming. *Geol. Soc. Am. Bull.* **86**, 319–334.

Eugster H. P. and Hardie L. A. (1978) Saline lakes. In *Lakes: Chemistry, Geology, Physics* (ed. A. Lerman). Springer, New York, pp. 237–293.

Eugster H. P. and Jones B. F. (1979) Behavior of major solutes during closed-basin brine evolution. *Am. J. Sci.* **279**, 609–631.

Eugster H. P., Harvie C. E., and Weare J. H. (1980) Mineral equilibria in the six-component sea water system, Na–K–Mg–SO_4–Cl–H_2O, at 25 °C. *Geochim. Cosmochim. Acta* **44**, 1335–1348.

Fontes J. C., Florkowski T., Pouchan P., and Zuppi G. M. (1979) *Isotopes in Lake Studies.* IAEA, Viemma.

Gac J. Y., Droubi A., Fritz B., and Tardy Y. (1977) Geochemical behavior of silica and magnesium during evaporation of waters in Chad. *Chem. Geol.* **19**, 215–228.

Gat J. R. (1995) Stable isotopes of fresh and saline lakes. In *Physics and Chemistry of Lakes* (eds. A. Lerman and J. Gat). Springer, New York, pp. 139–166.

Gilman K. (1994) *Hydrology and Wetland Conservation.* Wiley, New York.

Gosselin D. C., Sibray S., and Ayers J. (1994) Geochemistry of K-rich alkaline lakes, Western Sandhills, Nebraska, USA. *Geochim. Cosmochim. Acta* **58**, 1403–1418.

Grove A. T. (1986) Geomorphology of the African rift system. *Sedimentation in the African Rifts.* Special Publication 25, (eds. L. E. Frostick, R. W. Renaut, I. Reid and J. J. Tiercelin). Geological Society, London, pp. 9–18.

Gwynn J. W. (2002) Great Salt Lake, Utah: chemical and physical variations of the brine and effects of the SPRR causeway, 1966–1996. In *Great Salt Lake, an Overview of Change* (ed. J. W. Gwynn). Utah Department of Natural Resources, pp. 87–106.

Handford C. R. (1982) Sedimentology and evaporite genesis in a Holocene continental-sabkha playa basin—Bristol Dry Lake, California. *Sedimentology* **29**, 239–253.

Hardie L. A. (1990) The roles of rifting and hydrothermal $CaCl_2$ brines in the origin of potash evaporites: a hypothesis. *Am. J. Sci.* **284**, 193–240.

Hardie L. A. (1991) On the significance of evaporites. *Ann. Rev. Earth Planet. Sci.* **19**, 131–168.

Hardie L. A. and Eugster H. P. (1970) The evolution of closed-basin brines. *Spec. Publ. Min. Soc. Am.* **3**, 273–290.

Hardie L. A., Smoot J. P., and Eugster H. P. (1978) Saline lakes and their deposits: a sedimentological approach. In *Modern and Ancient Lake Sediments* (eds. A. Matter and M. E. Tucker). International Association of Sedimentologists, Oxford, pp. 7–41.

Hay R. L. (1968) Chert and its sodium-silicate precursors in sodium-carbonate lakes in East Africa. *Contrib. Mineral. Petrol.* **17**, 255–274.

Hay R. L. (1976) *Geology of the Olduvai Gorge.* University of California Press, Berkeley.

Hay R. L. and Kyser T. K. (2001) Chemical sedimentology and paleoenvironmental history of Lake Olduvai, a Pliocene lake in northern Tanzania. *Geol. Soc. Am. Bull* **113**, 1505–1521.

Hay R. L. and Stoessell R. K. (1984) Sepiolite in the Amboseli Basin of Kenya: a new interpretation. In *Palygorskite-Sepiolite, Occurrences, Genesis, and Uses.* Developments in Sedimentology 37 (eds. A. Singer and E. Galan). Elsevier, Amsterdam, pp. 125–136.

Hay R. L., Hughes R. E., Kyser T. K., Glass H. D., and Liu J. (1995) Magnesium-rich clays of the Meerschaum mines in the Amboseli Basin, Tanzania and Kenya. *Clays Clay Min.* **43**, 455–466.

Herczeg A. L. and Lyons W. B. (1991) A chemical model for the evolution of Australian sodium chloride lake brines. *Palaeogeogr. Palaeoclimatol. Palaeoecol.* **84**, 43–53.

Hermann A. G., Knake D., Schneider J., and Peters H. (1973) Geochemistry of modern seawater and brines from salt pans: main components and bromide distribution. *Contrib. Mineral. Petrol.* **40**, 1–24.

Hines M. E., Lyons W. B., Lent R. M., and Long D. T. (1992) Sedimentary biogeochemistry of an acidic, saline ground-water discharge zone in Lake Tyrrell, Victoria, Australia. In *The Geochemistry of Acid Groundwater Systems.* Chem. Geol. (eds. W. E. B. Lyons *et al.*) **96**, 53–65.

Hutchinson G. E. (1957) *A Treatise on Limnology: Volume 1.* Wiley, New York.

Irion G. (1973) Die anatolischen Salzseen, ihr Chemismus und die Entstehung ihrer chemischen Sedimente. *Arch. Hydrobiol.* **71**, 517–557.

Jacobsen G. and Jankowski J. (1988) Evolutionary model for Lake Amadeus playa brines, Central Australia. In *Fluvial Sedimentology. Mem. Can. Soc. Pet. Geol.* (ed. A. D. Miall) CSPG, Calgary, **5**, 543–576.

Jankowski J. and Jacobson G. (1989) Hydrochemical evolution of regional groundwaters to playa brines in central Australia. *J. Hydrol.* **108**, 123–173.

Johnson M. (1980) The origin of Australias salt lakes. *Dept. Min. Res. Dev., Geol. Surv. N.S. Wales* **19**, 221–266.

Johnson T. C. (1996) Sedimentary processes and signals of past climatic change in the large lakes of the East African Rift Valley. In *The Limnology, Climatology, and Paleoclimatology of the East African Lakes* (eds. T. C. Johnson and E. O. Odada). Gordon and Breach, Amsterdam.

Jones B. F. (1966) Geochemical evolution of closed basin water in the western Great Basin. In *2nd Symposium on Salt* (ed. J. L. Ran). Northern Ohio Geological Society, pp. 181–200.

Jones B. F. (1986) Clay mineral diagenesis in lacustrine sediments. In *Studies in Diagenesis.* (ed. F. A. Mumpton). *US Geol. Surv. Bull.* vol. 1578, pp. 291–300.

Jones B. F. and Bodine M. W., Jr. (1987) Normative salt characterization of natural waters. In *Saline Water and Gases in Crystalline Rocks.* Geol. Ass. Can. Spec. Pap. 33 (eds. P. Fritz and S. K. Frape) 5–18.

Jones B. F., Carmody R. and Frape S. K. (1997) Variations in principal solutes and stable istopes of Cl and S on evaporation of brines form the Great Salt Lake, Utah. *Geological Society of America*, Abstracts with programs, vol. 29, no. 6, p. 261.

Jones B. F. and Spencer R. J. (1999) Clay mineral diagenesis at Great Salt Lake, Utah, USA. *5th International Symposium on the Geochemistry of the Earth's Surface, Reykjavik, Iceland*. Balkema, Rotterdam, pp. 293–297.

Jones B. F. and Vandenburgh A. S. (1966) Geochemical influences on the chemical character of closed basins. *IAHS Symp. Garda, Hydrol. Lakes Reservoirs* **70**, 435–446.

Jones B. F., VanDenburgh A. S., Truesdell A. H., and Rettig S. L. (1969) Interstitial brines in playa sediments. *Chem. Geol.* **4**, 253–262.

Jones B. F., Eugster H. P., and Rettig S. L. (1977) Hydrochemistry of the Lake Magadi basin, Kenya. *Geochim. Cosmichim. Acta* **41**, 53–72.

Jones B. F., Hanor J. S., and Evans W. R. (1994) Sources of dissolved salts in the central Murray Basin, Australia. *Chem. Geol.* **111**, 135–154.

Kaplan I. R. (1983) Stable isotopes of sulfur, nitrogen and deuterium in Recent marine environments. In *Stable Isotopes in Sedimentary Geology*, SEPM Short Course. Society for Sedimentary Research (SEPM), Tulsa, pp. 2-1–2-108.

Kaplan I. R. and Rittenberg S. C. (1964) Microbiological fractionation of sulphur isotopes. *J. Gen. Microbiol.* **34**, 195–212.

Kehew A. E. and Teller J. T. (1994) History of late glacial runoff along the southwestern margin of the Laurentide ice sheet. *Quat. Sci. Rev.* **13**, 859–877.

Kellogg W. W., Cadle R. D., Allen E. R., Lazrus A. L., and Martell E. A. (1972) The sulfur cycle. *Science* **175**, 587–596.

Kelts K. and Shahrabi M. (1986) Holocene sedimentology of hypersaline Lake Urmia, Northwestern Iran. *Paleogeogr. Paleoclimatol. Paleoecol.* **54**, 105–130.

Klassen R. W. (1989) Quaternary geology of the southern Canadian Interior Plains. In *Quaternary Geology of Canada and Greenland* (ed. R. J. Fulton). Geological Survey of Canada, Ottawa, pp. 138–173.

Klenova M. V. (1948) *Geologiya moray [Geology of the sea]*. Gosudarstvennoe Uchebno-Pedagogicheskoe Izdatel'stvo (in Russian).

Kohler J. F. (2002) Effects of the West Desert Pumping Project on the near-surface brines in a portion of the Great Salt Lake Desert, Tooele, and Box Elder Counties, Utah. In *Great Salt Lake, an Overview of Change* (ed. J. W. Gwynn). Utah Department of Natural Resources, Salt Lake City, pp. 487–498.

Komor S. C. (1992) Bidirectional sulfate diffusion in saline-lake sediments: evidence from Devils Lake, northeast North Dakota. *Geology* **20**, 319–322.

Komor S. C. (1994) Bottom-sediment chemistry in Devils Lake, northeast North Dakota. In *Sedimentology and Geochemistry of Modern and Ancient Saline Lakes* (eds. R. W. Renaut and W. M. Last). Society for Sedimentary Geology (SEPM), Tulsa, pp. 21–32.

Kosarev A. N. and Yablonskaya E. A. (1994) *The Caspian Sea*. SPB Academic Publishing, The Hague.

Krouse H. R. (1980) Sulphur isotopes in our environment. In *Handbook of Environmental Isotope Geochemistry* (eds. P. Fritz and J. Ch. Fontes). Elsevier, New York, pp. 435–472.

LaBaugh J. W. (1988) Relation of hydrologic setting to chemical characteristics of selected lakes and wetlands within a climate gradient in the North-Central United States. *Verh. Int. Ver. Limnol.* **23**, 131–137.

Last W. M. (1990) Lacustrine dolomite—an overview of modern, Holocene, and Pleistocene occurrences. *Earth Sci. Rev.* **27**, 221–263.

Last W. M. (1992) Chemical composition of saline and subsaline lakes of the northern Great Plains, western Canada. *Int. J. Salt Lake Res.* **1**, 47–76.

Last W. M. (1994a) Deep-water evaporite mineral formation in lakes of western Canada. In *Sedimentology and Geochemistry of Modern and Ancient Saline Lakes* (eds. R. W. Renaut and W. M. Last). Society for Sedimentary Geology (SEPM), Tulsa, pp. 51–60.

Last W. M. (1994b) Paleohydrology of playas in the northern Great Plains: perspectives from Palliser's triangle. In *Paleoclimate and Basin Evolution of Playa Systems*. Special Paper 289 (ed. M. R. Rosen). Geological Society of America, Boulder, pp. 69–80.

Last W. M. (1999) Geolimnology of the Great Plains of western Canada. In *Holocene Climate and Environmental Change in the Palliser Triangle: A Geoscientific Context for Evaluating the Impacts of Climate Change on the Southern Canadian Prairies* (eds. D. S. Lemmen and R. E. Vance). *Geol. Surv. Can. Bull.* Geological Survey of Canada, Ottawa, vol. 534, pp. 23–55.

Lerman A. and Jones B. F. (1973) Transient and steady-state salt transport between sediments and brine in closed lakes. *Limnol. Oceanogr.* **18**, 72–85.

Li J., Lowenstein T. K., and Blackburn I. R. (1997) Responses of evaporite mineralogy to inflow water sources and climate during the past 100 k.y. in Death Valley. California. *Geol. Soc. Am. Bull.* **109**, 1361–1371.

Linacre E. T., Hicks B. B., Sainty G. R., and Grauze G. (1970) The evaporation from a swamp. *Agri. Meteorol.* **7**, 375–386.

Livingstone D. A. (1963) Chemical composition of rivers and lakes. In *Data of Geochemistry*. (ed. M. Fleischer). *US Geol. Surv. Prof. Pap. 440-G* 6th edn. USGS, Washington, DC.

Long D. T., Fegan N. E., Lyons W. B., Hines M. E., Macumber P. G., and Giblin A. M. (1992a) Geochemistry of acid brines: Lake Tyrrell, Victoria, Australia. In *The Geochemistry of Acid Groundwater Systems* (eds. W. E. B. Lyons et al.). Chem. Geol. **96**, 33–52.

Long D. T., Fegan N. E., McKee J. D., Lyons W. B., Hines M. E., Macumber P. G., et al. (1992b) Formation of alunite, jarosite and hydrous iron oxides in a hypersaline system: Lake Tyrrell, Victoria, Australia. In *The Geochemistry of Acid Groundwater Systems* (eds. W. E. B. Lyons et al.). Chem. Geol. **96**, 183–202.

Lyons W. B. and Mayewski P. A. (1993) The geochemical evolution of terrestrial waters in the Antarctic: the role of rock-water interactions. Physical and Biogeochemical Processes in Antarctic Lakes. *Antarct. Res. Ser.* **59**, 135–143.

MacArthur J. M., Turner J. V., Lyons W. B., Osborn A. O., and Thirlwall M. F. (1991) Hydrochemistry on the Yilgarn Block, Western Australia: ferrolysis and mineralisation in acidic brines. *Geochim. Cosmochim. Acta* **55**, 1273–1288.

Macumber P. (1983) Interaction of groundwater and surface systems in north Victoria. PhD Dissertation, Melbourne University.

Mann A. W. (1983) Hydrochemistry and weathering on the Yilgarn Block, Western Australia—ferrolysis and heavy metals in continental brines. *Geochim. Cosmochim. Acta* **47**, 181–190.

Matsubaya O., Sakai H., Torii T., Burton H., and Kerry K. (1979) Antarctic saline lakes-stable isotope ratios, chemical compositions and evolution. *Geochim. Cosmochim. Acta* **43**, 7–25.

McCarthy T. S., Bloem A., and Larkin P. A. (1998) Observations on the hydrology and geohydrology of the Okavango Delta. *S. Afr. J. Geol.* **101**, 101–117.

McNutt R. H., Frape S. K., Fritz P., Jones M. G., and MacDonald I. M. (1990) The $^{87}Sr/^{86}Sr$ values of Canadian Shield brines and fracture minerals with applications to groundwater mixing, fracture history and geochronology. *Geochim. Cosmochim. Acta* **54**, 205–215.

Merlivat L. and Coantic M. (1975) Study of mass transfer at the air-water interface by an isotopic method. *J. Geophys. Res.* **80**, 3455–3464.

Micklin P. P. (1988) Desiccation of the Aral Sea: a water management disaster in the Soviet Union. *Science* **241**, 1170–1176.

Moller N., Weare J. H., Duan Z., and Greenberg J. P. (1997) Chemical models for optimizing geothermal energy production. US Department of Energy Technical Site (www.doe.gov), Research Summaries—Reservoir Technology.

Muller G. and Wagner F. (1978) Holocene carbonate evolution in Lake Balaton (Hungary): a response to climate and impact of man. *Spec. Publ. Int. Ass. Sedim.* **2**, 57–81.

Müller G., Irion G., and Förstner U. (1972) Formation and diagenesis of inorganic Ca–Mg carbonates in the lacustrine environment. *Naturwis.* **59**, 158–164.

Neev D. and Emery K. (1967) The Dead Sea: depositional processes and environments of evaporites. *Bulletin 41*, Geological Survey of Israel, Jerusalem, 147. pp.

Neumann K. and Dreiss S. J. (1995) Strontium 87/strontium 86 ratios as tracers in groundwater and surface waters in Mono Basin California. *Water Resour. Res.* **31**, 3183–3194.

Nesbitt H. W. (1974) The study of some mineral-aqueous solution interactions. PhD Dissertation, Johns Hopkins University.

Nesbitt H. W. (1990) Groundwater evolution, authigenic carbonates and sulphates of the Basque Lake No. 2 basin, Canada. In *Fluid–Mineral Interactions: A Tribute to H. P. Eugster* (eds. R. J. Spencer and I.-M. Chou). Geochemical Society, San Antonio, pp. 355–371.

Ordonez S. and Garcia del Cura M. A. (1994) Deposition and diagenesis of sodium–calcium sulfate salts in the Tertiary saline lakes of the Madrid Basin, Spain. In *Sedimentology and Geochemistry of Modern and Ancient Saline Lakes* (eds. R. W. Renaut and W. M. Last). Society for Sedimentary Geology (SEPM), Tulsa, pp. 229–238.

Petr T. (1992) Lake Balkhash, Kazakhstan. *Int. J. Salt Lake Res.* **1**, 21–46.

Plummer L. N., Parkhurst D. L., Fleming G., and Dunkle S. (1988) PHRQPITZ, a computer program incorporating Pitzer's equations for calculation of geochemical reactions in brines. *US Geol. Surv. Water Res. Inv. Rep.* WRI88-4153.

Pueyo J. J. and De la Pena J. A. (1991) Los Lagos salinos Espanoles: Sedimentologia, hidroquimica y diagenesis. In *Genesis de formaciones evaporiticas: Modelos Andinos e Ibericos* (ed. J. J. Pueyo). Universitat de Barcelona Estudi-General, Barcelona, pp. 163–192.

Renaut R. W. (1994) Carbonate and evaporite sedimentation at Clinton Lake, British Columbia, Canada. In *Paleoclimate and Basin Evolution of Playa Systems*. Special Paper 289 (ed. M. R. Rosen). Geological Society of America, Boulder, pp. 49–68.

Renaut R. W. and Tiercelin J. J. (1994) Lake Bogoria, Kenya Rift Valley—a sedimentological overview. In *Sedimentology and Geochemistry of Modern and Ancient Saline Lakes* (eds. R. W. Renaut and W. M. Last). SEPM (Society for Sedimentary Geology), Tulsa, pp. 101–123.

Rettig S. L., Jones B. F., and Risacher F. (1980) Geochemical evolution of brines in the salar of Uyuni, Bolivia. *Chem. Geol.* **30**, 57–79.

Risacher F. and Fritz B. (1991a) Geochemistry of Bolivian salars, Lipez, southern Altiplano: origin of solutes and brine evolution. *Geochim. Cosmochim. Acta* **55**, 687–705.

Risacher F. and Fritz B. (1991b) Quaternary geochemical evolution of the salars of Uyuni and Coipasa, Central Altiplano, Bolivia. *Chem. Geol.* **90**, 211–231.

Rogers D. B. and Dreiss S. J. (1995) Saline groundwater in Mono Basin, California: 1. Distribution. *Water Resour. Res.* **31**, 3151–3170.

Rosen M. R. (1989) Sedimentologic, geochemical, and hydrologic evolution of an intracontinental, closed-basin playa (Bristol Dry Lake, California): a model for playa development and its implications for paleoclimate. PhD Dissertation, University of Texas at Austin.

Rosen M. R. (1991) Sedimentologic and geochemical constraints on the evolution of Bristol Dry Lake Basin, California, USA. *Palaeogeogr. Palaeoclimatol. Palaeoecol.* **84**, 229–257.

Rosen M. R. (1994) The importance of groundwater in playas: a review of playa classifications and the sedimentology and hydrology of playas. In *Paleoclimate and Basin Evolution of Playa Systems*. Special Paper 289 (ed. M. R. Rosen). Geological Society of America, Boulder, pp. 1–18.

Rubanov I. V. (1984) Sulfate-bearing sediments of the Aral Sea, their structure and composition. *Litol. Polez. Iskop* **1**, 117–125. (in Russian).

Salama R. B. (1994) The evolution of saline lakes in the relict drainage of the Yilgarn River, Western Australia. In *Sedimentology and Geochemistry of Modern and Ancient Saline Lakes* (eds. R. W. Renaut and W. M. Last). SEPM (Society for Sedimentary Geology), Tulsa, pp. 189–202.

Sanford W. E. and Wood W. W. (1991) Brine evolution and mineral deposition in hydrological open evaporite basins. *Am. J. Sci.* **291**, 687–710.

Schroeder R. A., Orem W. H., and Kharaka Y. K. (2002) Chemical evolution of the Salton Sea, California: nutrient and selenium dynamics. *Hydrobiologia* **473**, 1–23.

Smith C. L. and Drever J. I. (1976) Controls on the chemistry of springs at Teels Marsh, Mineral Country, Nevada. *Geochim. Cosmochim. Acta* **40**, 1081–1093.

Smoot J. and Lowenstein T. (1991) Depositional environments of non-marine evaporites. In *Evaporites, Petroleum, and Mineral Resources: Developments in Sedimentology* (ed. J. Melvin). Elsevier, New York, vol. 50, pp. 189–384.

Spencer R. J. (2000) Sulfate Minerals in Evaporite Deposits. In *Sulfate Minerals, Reviews in Mineralogy and Geochemistry* (eds. C. N. Alpers, J. L. Jambor, and D. K. Nordstrom). Mineralogical Society of America and Geochemical Society, Washington, DC, vol. 40, pp. 173–192.

Spencer R. J., Baedecker M. J., Eugster H. P., Forester R. M., Goldhaber M. B., Jones B. F., Kelts K., McKenzie J., Madsen B. B., Rettig S. L., Rubin M., and Bowser C. J. (1984) Great Salt Lake, and precursors, Utah: the last 30,000 years. *Contrib. Mineral. Petrol.* **86**, 321–334.

Spencer R. J., Eugster H. P., Jones B. F., and Rettig S. L. (1985a) Geochemistry of Great Salt Lake, Utah: I. Hydrochemistry since 1850. *Geochim. Cosmochim. Acta* **49**, 727–737.

Spencer R. J., Eugster H. P., and Jones B. F. (1985b) Geochemistry of Great Salt Lake, Utah: II. Pleistocene-Holocene Evolution. *Geochim. Cosmochim. Acta* **49**, 739–747.

Spencer R. J., Lowenstein T. K., Casas E., and Pengxi Z. (1990) Origin of potash salts and brines in the Qaidam Basin, China. In *Fluid–Mineral Interactions: A Tribute to H. P. Eugster* (eds. R. J. Spencer and I.-M. Chou). Geochemical Society, San Antonio, pp. 395–408.

Starinsky A. (1974) Relation between Ca-chloride brines and sediment rocks in Israel. PhD Dissertation, Hebrew University (in Hebrew with English summary).

Stoessell R. K. and Hay R. L. (1978) The geochemical origin of sepiolite and kerolite at Amboseli, Kenya. *Contrib. Mineral. Petrol.* **65**, 255–267.

Street-Perrott F. A. and Roberts N. (1983) Fluctuations in closed-basin lakes as an indicator of past atmospheric circulation patterns. In *Variations in the Global Water Budget* (eds. F. A. Street-Perrott, M. Berman, and R. Ratchliffe). Reidel, Dordrecht, pp. 331–346.

Teeter J. W. (1995) Holocene saline lake history, San Salvador Island, Bahamas. In *Terrestrial and Shallow Marine Geology of the Bahamas and Bermuda*. GSA Special Paper 300 (eds. H. A. Curran and B. White). Geological Society of America, Boulder, pp. 117–124.

Torgersen T., DeDekker P., Chivas A. R., and Bowler J. M. (1986) Salt lakes: a discussion of processes influencing paleoenvironmental interpretation and recommendations for future study. *Palaeogeogr. Palaeoclimatol. Palaeoecol.* **54**, 7–19.

USGS (2002) National Water Information System (http://waterdata.usgs.gov/nwis/qwdata) (unpublished).

Valero-Garces B. L., Grosjean M., Kelts K., Schreier H., and Messerli B. (1999) Holocene lacustrine deposition in the Atacama Altiplano: facies models, climate and tectonic forcing. *Palaeogeogr. Palaeoclimatol. Palaeoecol.* **151**, 101–125.

Valero-Garces B. L., Arenas C., and Delgado-Huertas A. (2001) Depositional environments of quaternary lacustrine travertines and stromatolites from high-altitude Andean lakes, northwestern Argentina. *Can. J. Earth Sci.* **38**, 1263–1283.

Vance R. E., Last W. M., and Smith A. J. (1997) Hydrologic and climatic implications of a multidisciplinary study of late Holocene sediment from Kenosee Lake, southeastern Saskatchewan, Canada. *J. Paleolimnol.* **18**, 365–393.

Vengosh A. and Rosenthal E. (1994) Saline groundwater in Israel: its bearing on the water crisis in the country. *J. Hydrol.* **156**, 389–430.

Vengosh A., Starinsky A., Kolodny Y., and Chivas A. R. (1991) Boron isotope geochemistry as a tracer for the evolution of brines and associated hot springs from the Dead Sea, Israel. *Geochim. Cosmochim. Acta* **55**, 1689–1695.

Verzilin N. N. and Utsal K. R. (1991) Mineral composition of sediments (Lake Balkash). In *Lake History of the Sevan, Issyk-kul, Balkash, Zaisan, and Aral Sea Basins* (eds. D. V. Sevastyanov *et al.*). NAUKA (in Russian), Moscow.

Von Damm K. L. and Edmond J. M. (1984) Reverse weathering in the closed-basin lakes of the Ethiopian rift. *Am. J. Sci.* **284**, 835–862.

Wigley T. M. L., Plummer L. N., and Pearson F. J. (1978) Mass transfer and carbon isotope evolution in natural water systems. *Geochim. Cosmochim. Acta* **42**, 1117–1139.

Winograd I. J. and Thordarson W. (1975) Hydrogeologic and hydrochemical framework, south-central Great Basin, Nevada-California, with special reference to the Nevada Test Site. *US Geol. Surv. Prof. Pap.*, 712-C.

Wolanski E. and Gereta E. (1999) Oxygen cycle in a hippo pool, Serengeti National Park, Tanzania. *Afr. J. Ecol.* **37**, 419–423.

Wood W. W. and Sanford W. E. (1990) Ground-water control of evaporite deposition. *Econ. Geol.* **85**, 1226–1235.

Wood W. W. and Sanford W. E. (1995) Eolian transport, saline lake basins, and groundwater solutes. *Water Resour. Res.* **31**, 3121–3129.

Yechieli Y., Ronen D., and Kaufman A. (1996) The source and age of groundwater brines in the Dead Sea area, as deduced from ^{36}Cl and ^{14}C. *Geochim. Cosmochim. Acta* **60**, 1909–1916.

Yuretich R. F. and Cerling T. E. (1983) Hydrogeochemistry of Lake Turkana, Kenya: mass balance and mineral reactions in an alkaline lake. *Geochim. Cosmochim. Acta* **47**, 1099–1109.

Zheng M. (1997) *An Introduction to Saline Lakes on the Qinghai-Tibet Plateau.* Kluwer Academic, London.

Zheng M., Tang J., Liu J., and Zhang F. (1993) Chinese saline lakes. *Hydrobiologia* **267**, 23.

5.14
Geochemistry of Groundwater

F. H. Chapelle

US Geological Survey, Columbia, SC, USA

5.14.1 INTRODUCTION

Groundwater geochemistry is concerned with documenting the chemical composition of groundwater produced from different aquifer systems, and with understanding the processes that control this composition. This chapter provides a brief overview of the development of the science of groundwater geochemistry, and gives a series of examples of how aquifer lithology and hydrology affect groundwater composition. This, in turn, provides an introduction to the principles of mass balance, equilibrium chemistry, and microbiology that have proved useful in understanding the composition of groundwater in a variety of geologic settings. While the natural processes that affect groundwater geochemistry are complex, observational and quantitative methodologies have been developed that help to elucidate them. This chapter summarizes the more widely used methodologies in groundwater geochemistry, and shows how they have been applied to solve various hydrologic problems in both pristine and contaminated aquifer systems.

5.14.1.1 Historical Overview

Humans have used springs and wells as sources of water for thousands of years. Very early in human history, it was noted that the taste and chemical quality of spring and well waters differ from place to place (Back *et al.*, 1995). Some springs, for example, produced water that was cool, crisp, and excellent for drinking. Other springs produced hot, salty, or sulfurous waters that were undrinkable. Because springwaters were so important as sources of water, people paid careful attention to these chemical differences. Many cultures developed elaborate mythologies to explain why springwaters differed so much in their chemical quality. It was common, for example, to rationalize these differences in moral terms. Good-quality drinking water was associated with good deities, and poor-quality waters were associated with evil deities (Chapelle, 2000).

It was also noted that water from certain springs or wells appeared to have medicinal qualities (Burke, 1853). It was entirely logical to assume that these medicinal effects reflected the different composition of springwaters. One early motivation for studying groundwater geochemistry was to learn what dissolved constituents were present in springwaters, and to understand how these constituents imparted medicinal effects to the waters. Steel (1838), for instance, identified iodine in the springwaters of Saratoga Springs, NY, and associated this constituent with a variety of beneficial medicinal properties. It is not certain, however, whether Steel recognized that iodine in the springwaters could cure or prevent simple goiters, a medicinal use of iodine that later became widely known.

Similar practical concerns stimulated the earliest quantitative chemical analyses of groundwater. For example, Rogers (1917) noticed that oil-field brines associated with petroleum hydrocarbons lacked dissolved sulfate, whereas oil-field brines without associated petroleum hydrocarbons contained abundant sulfate. Rogers (1917) therefore raised the possibility that water chemistry could be useful in prospecting for oil. Other studies in the early twentieth century noted that water chemistry did not vary randomly. Rather, there were systematic changes in chemistry as water flowed along regional hydrologic gradients. For example, Renick (1924) noticed that shallow groundwater in the recharge areas of the Fort Union Formation of Montana were "hard," i.e., contained relatively high concentrations of calcium and magnesium. Downgradient, however, these waters became "soft," and contained relatively high concentrations of sodium. Renick (1924) attributed these changes to "base exchange" (i.e., cation exchange) processes that removed calcium and magnesium from solution and replaced it with sodium. Correlating water-chemistry changes with directions of groundwater flow, and attempting to identify specific chemical reactions occurring in aquifers, was an approach followed by several subsequent investigations (Cedarstrom, 1946; Chebotarev, 1955). This approach was formalized by Back (1966), who coined the term "hydrochemical facies" to refer to water chemistry changes commonly observed in the direction of groundwater flow.

The observation that water chemistry tended to change systematically and predictably suggested that the chemical and or biological processes involved in these changes could be identified. During the next thirty years, a number of approaches to studying these processes, and learning how to quantify their effects dominated groundwater geochemistry. These approaches assumed that waters entering aquifers are initially relatively unmineralized rainwater or snowmelt. Since these waters did contain dissolved carbon dioxide, the resulting carbonic acid attacks minerals present in aquifers and dissolves them. Thus, the chemistry of groundwaters could be explained in terms of initially nonequilibrium conditions moving toward chemical equilibrium by reacting with aquifer minerals. This equilibrium approach was formalized by Garrels and Christ (1965), and provided a quantitative framework within which the composition of groundwater could be understood. Application of the equilibrium approach soon showed that, once the minerals being dissolved were identified, it was possible to apply mass-balance constraints to calculate the amount of each mineral that was dissolved to produce groundwater of a particular composition (Garrels and Mackenzie, 1967). This mass-balance approach was soon expanded to include balances of electrons (accounting for redox reactions) and isotopes, and was formalized within a rigorous mathematical framework (Helgeson et al., 1970; Plummer, 1977; Plummer and Back, 1980; Parkhurst et al., 1982; Plummer et al., 1983). This approach combined the concepts of chemical equilibrium and material balance, and came to be known as "geochemical modeling" (Plummer, 1984; see Chapter 5.02).

Applications of this quantitative approach soon demonstrated that there were many more mineral phases present in aquifers, including mineral phases with variable chemical compositions, than could be included in any one geochemical model. It was possible to construct a virtually limitless number of geochemical models that explained the observed water chemistry of even the simplest aquifer. The nonunique nature of these geochemical models is a reflection of the mineralogic complexity of aquifers and reflects the fact that aquifers are never at true thermodynamic equilibrium. In spite of this, much insight into the processes that control groundwater chemistry has been gained by applying the geochemical modeling approach.

5.14.1.2 Lithology, Hydrology, and Groundwater Geochemistry

In the early 1960s, Feth et al. (1964) published a compendium of analyses of springwaters in the

Sierra Nevada Mountains. This study suggested that the chemistry of the springwaters reflected the progressive alteration of plagioclase feldspar to kaolinite, which in turn produced the solutes present in groundwater. This hypothesis could readily be tested using the concepts of equilibrium chemistry and mass balance. In testing this hypothesis, Garrels and Mackenzie (1967) showed that the observed concentrations of sodium, silica, and the water pH were similar to those calculated by assuming that plagioclase was progressively dissolved by carbon dioxide-bearing waters. It was relatively easy to start with a given springwater and subtract quantities of calcium, sodium, and silica appropriate to the stoichiometry of feldspars and/or other reactive minerals. In this manner, it was possible to estimate the amount of minerals that had dissolved in order to generate the observed water composition.

The mass-balance approach (see Chapter 5.04) was important because it provides a rational explanation for the chemical variability of groundwater. A general statement of the principle of conservation of mass for the chemical composition of groundwater flowing along a flow path is

$$\text{initial water} + \text{dissolving minerals}$$
$$- \text{precipitating minerals} = \text{final water} \quad (1)$$

The mass-balance approach clearly shows that groundwater composition is determined by the kinds and amounts of minerals that dissolve and/or precipitate in an aquifer.

5.14.1.3 Mineral Dissolution and Precipitation

The presence or absence of particular minerals in an aquifer places obvious constraints on which minerals may dissolve or precipitate (Equation (1)). A more subtle question has to do with mineral solubility. If a mineral is present but is insoluble, it cannot be the source of dissolved solutes. Mineral solubility constraints, in turn, can be addressed using the principles of equilibrium chemistry.

The *saturation index* (SI) of a particular mineral in the presence of water is defined as

$$SI = \log IAP/K \quad (2)$$

where IAP is the ion activity product and K is the equilibrium constant for a particular mineral dissolution/precipitation reaction. Take, for example, the dissolution of calcite:

$$CaCO_3 = Ca^{2+} + CO_3^{2-} \quad (3)$$

The IAP of this reaction is given by

$$IAP = a_{Ca^{2+}} \, a_{CO_3^{2-}} \quad (4)$$

and the equilibrium constant is given by

$$a_{Ca^{2+}} \, a_{CO_3^{2-}} / a_{CaCO_3} = K \quad (5)$$

At equilibrium, and assuming the solid phase ($CaCO_3$) has an activity of 1 (unit activity), it is clear that

$$IAP = K \quad (6)$$

so that

$$IAP/K = 1 \quad (7)$$

or

$$SI = \log IAP/K = 0 \quad (8)$$

An SI index of zero indicates that a mineral is in thermodynamic equilibrium with the solution. If the SI is less than zero, the solution is undersaturated with respect to the mineral and that mineral may dissolve. Conversely, an SI greater than zero indicates the solution is oversaturated with respect to the mineral and that the mineral may precipitate. These considerations are useful in constraining the kinds of reactions that are possible in groundwater systems.

By taking into account aquifer lithology, direction of groundwater flow, and mineral solubility constraints, a great deal can be learned about groundwater geochemistry. These considerations can be taken into account in a variety of ways, ranging from entirely qualitative to highly constrained, rigorously mathematical approaches. The approach taken generally depends on the goals of the individual investigator. Regardless of the approach taken, however, the geochemistry of groundwater reflects the lithologic and hydrologic characteristics of different aquifers. The following sections, which describe the groundwater geochemistry of different aquifer systems in the United States, are designed to illustrate how lithology and hydrology affect the composition of groundwater, and how qualitative and quantitative approaches can be used to understand the groundwater geochemistry of different hydrologic systems.

5.14.2 GROUNDWATER GEOCHEMISTRY IN SOME COMMON GEOLOGIC SETTINGS

5.14.2.1 A Basalt Lithology: The Snake River Aquifer of Idaho

One of the principal difficulties in understanding groundwater geochemistry is due to the chemical heterogeneity of most aquifer systems. The rocks and sediments that form most aquifers were often deposited over long periods of time and under a variety of geologic conditions. Thus, the chemical composition of the rock or sediment matrix of aquifers is seldom constant. There are, however, examples of aquifer systems that are relatively homogeneous in mineral content and composition. The basalt aquifers of the Snake River Plain in Idaho are one of these relatively homogeneous systems. As such, it provides an excellent example of the effects of lithology on groundwater chemistry.

The Snake River Plain marks the track of the Yellowstone Hotspot, a deep mantle plume that has been in place and active since the Miocene ($\sim$18 Myr). As the North American Plate moved westward over the plume, at a rate 2.5–8.0 cm yr^{-1}, the surface manifestations of the associated volcanism moved eastward from what is now Nevada, along the Snake River Plain, and is now located underneath Yellowstone National Park in Wyoming (Armstrong, 1971; Embree *et al.*, 1982; Pierce and Morgan, 1992). The later stages of volcanism have been characterized by flood basalts, which have filled the valley created by the collapse of underlying magma chambers (Pierce and Morgan, 1992). As a result, the Snake River Plain is now underlain by several hundred meters of basalt flows, which are themselves underlain by rhyolitic volcanic rocks.

The Snake River Plain aquifer system, which has formed in the near-surface basalts, is largely unconfined. Groundwater moves from northeast to southwest until it discharges to the Snake River, mostly from a series of springs located between Twin Falls and King Hill (Lindholm *et al.*, 1983). The aquifer is recharged by atmospheric precipitation that falls on the topographically high areas to the east, north, and south of the Snake River Valley. This recharge pattern drives groundwater flow from east to west (Figure 1). The highly fractured and rubbly nature of the basalts reflects their rapid cooling at land surface. Consequently, the aquifer exhibits very high transmissivities of $\sim$10^5 m^2 d^{-1}. This high transmissivity, combined with the relatively large amounts of recharge that enter the aquifer, leads to rates of groundwater flow that average 3 m d^{-1}. As a consequence of these rapid groundwater flow rates, the average residence time of water in the aquifer is relatively low, 200–250 yr.

The groundwater geochemistry of the Snake River Plain aquifer system has been described by Wood and Low (1988). It is an excellent example of how a knowledge of aquifer mineralogy can

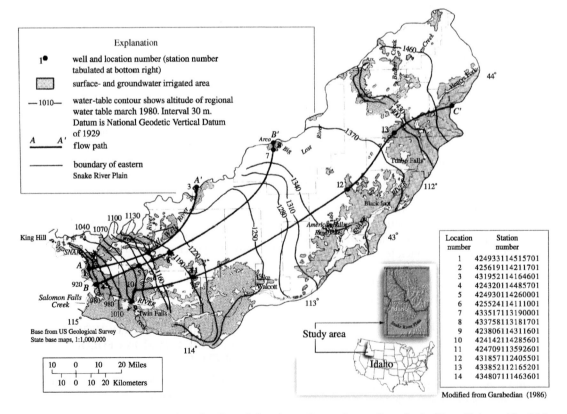

Figure 1 Map showing the location of wells and directions of groundwater flow, Snake River Plain aquifer, Idaho (source Wood and Low, 1988).

Table 1 Representative analyses of groundwater from the Snake River Plain, Idaho.

Station no. (from Figure 2)	Distance along flow path (km)	pH (units)	Dissolved oxygen (mg L⁻¹)	Calcium (mg L⁻¹)	Magnesium (mg L⁻¹)	Sodium (mg L⁻¹)	Potassium (mg L⁻¹)	Bicarbonate (mg L⁻¹)	Chloride (mg L⁻¹)	Sulfate (mg L⁻¹)	Silica (mg L⁻¹)
14	0	8.0	7.8	36	13	20	3.2	180	13	14	45
13	30	7.5	9.4	69	18	15	3.3	280	16	46	25
12	65	7.5	9.6	47	9.3	4.8	1.0	180	1.5	17	16
11	160	7.8	8.7	44	18	30	4.3	190	42	50	33
10	190	7.5	8.7	62	28	37	5.6	280	55	72	39

Source: Wood and Low (1988).

explain the observed composition of groundwaters. Some representative analyses of Snake River Plain groundwater are shown in Table 1. The water is characterized by moderate concentrations of calcium (36–69 mg L⁻¹), magnesium (9–28 mg L⁻¹), sodium (5–37 mg L⁻¹), bicarbonate (180–280 mg L⁻¹), chloride (1.5–55 mg L⁻¹), and sulfate (14–72 mg L⁻¹). The pH of the water is relatively high (7.5–8.0), and it contains relatively high concentrations of dissolved oxygen (7.8–9.6 mg L⁻¹), and relatively high concentrations of dissolved silica (16–45 mg L⁻¹).

Wood and Low (1988) began their investigation of Snake River Plain groundwater geochemistry by describing in detail the mineralogic composition of the basalt aquifer. Using bulk chemical analyses, mineralogic analyses, and scanning electron micrograph (SEM) images, they showed that the most reactive minerals present in the basalt were olivine, plagioclase feldspar, pyroxenes, and pyrite. The SEM images clearly showed that these minerals were actively dissolving in the aquifer. The SEM images also showed that secondary calcite, montmorillonite, and silica were being precipitated. It is a simple matter to estimate qualitatively the amounts of minerals that dissolved and precipitated in order to produce the observed groundwater geochemistry. Wood and Low (1988) normalized the amount of dissolution and precipitation to the annual water discharge from the aquifer, and reported the total amount of mineral dissolution/precipitation occurring throughout the aquifer in units of 10^9 mol yr⁻¹. In addition, they normalized the relative amounts of olivine and pyroxene dissolved based on petrographic observations. Specifically, they assumed that 76% of dissolved magnesium comes from olivine, and 24% of magnesium comes from pyroxene.

The procedure used by Wood and Low (1988) illustrates several important features of the mineralogic control on groundwater geochemistry in this aquifer system. First, while olivine comprises a relatively small percentage of the aquifer material, its dissolution has quite a large effect on the water chemistry, producing much of the observed magnesium and bicarbonate. Second, while calcite was not an important component of the original basalt, calcite precipitation is an important control on concentrations of calcium and bicarbonate, and on the pH of groundwater.

Equilibrium calculations based on mineral saturation indices also show how the concentration of solutes that are present in trace amounts are influenced by mineral dissolution/precipitation reactions. For example, the concentration of barium in groundwater appears to be buffered by the saturation index of barite ($BaSO_4$) (Figure 2).

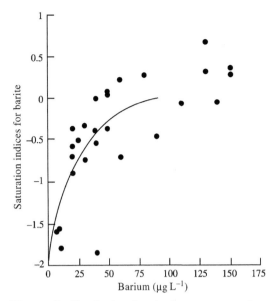

Figure 2 Graph showing barium concentrations plotted versus the saturation index of the mineral barite calculated for Snake River Plain groundwater (source Wood and Low, 1988).

The Snake River Plain aquifer is also an example of an aquifer in which groundwater chemistry is not strongly affected by directions and rates of groundwater flow. There are no well-defined changes in solute concentrations along the aquifer flow path (Table 1). As we shall see in subsequent examples, this insensitivity to directions of groundwater flow is unusual, and is due to the relatively rapid nature of ongoing dissolution and precipitation reactions in the Snake River Plain aquifer.

5.14.2.2 A Stratified Drift–Arkosic Sandstone Lithology: The Bedrock Aquifer of New England

The basaltic aquifers of the Snake River Plain are unusual in that the mineralogy of the aquifer matrix is (relatively) uniform. The basalts are composed of minerals (i.e., olivine, pyroxene, and plagioclase) that are relatively unstable in the presence of water at low temperature ($\sim15\,^\circ\text{C}$), and their rates of reaction with carbon dioxide-bearing waters are relatively rapid. In addition, the rate and direction of groundwater flow in the Snake River Plain aquifer can be determined with fair confidence. This combination of attributes makes it relatively easy to apply a simple mass-balance approach to document the geochemical processes that determine the groundwater geochemistry. But most aquifers are lithologically heterogeneous, are composed of minerals that react with water at slower rates, and are

characterized by uncertain directions and rates of groundwater flow. It is worth asking, therefore, whether the simple mass-balance approach is applicable to more complex hydrologic systems. One such highly complex hydrologic system, characterized by glacial sediments overlying arkosic bedrock, underlies part of New England (Rogers, 1987). As such, it provides an opportunity to apply the mass-balance approach to a more complex hydrologic system.

The study area described by Rogers (1987) is in the central lowlands of Connecticut. The area is underlain by lithified sedimentary rocks of the Newark Supergroup (Mesozoic) that are locally referred to as "bedrock." The bedrock is mantled by "stratified drift" and till deposits of Pleistocene age. The glacial deposits are primarily medium- to coarse-grained sands interbedded with gravels and silts. In places, these glacial sands and gravels are productive aquifers which are locally tapped for groundwater. The bedrock underlying the glacial deposits is typically faulted and fractured, and is characterized by partings along bedding planes. Groundwater movement in bedrock is primarily along these secondary fractures and bedding planes, and individual wells produce ~0–750 gal ($1\,\text{gal} = 3.785 \times 10^{-3}\,\text{m}^3$) of water per minute. Because this hydrologic system is highly complex and heterogeneous, it is difficult to delineate individual flow paths other than to say that flow is generally from areas of higher elevation to areas of lower elevation.

The mineralogy of the bedrock–glacial drift aquifers is similarly complex. Rocks of the Newark Supergroup consist primarily of quartz (44%), potassium feldspar (20%), plagioclase feldspar (10%), and clays and micas (18%). The bedrock is cemented with hematite (3.5%) and in places by calcite (2.5%). The glacial drift deposits are often derived from the bedrock and, other than having a higher proportion of quartz, have a similar mineralogy as the bedrock.

In spite of this mineralogic and hydrologic complexity, the groundwater geochemistry exhibits some uniformity. The mean concentration of the major ions present in stratified drift and bedrock aquifers is shown in Table 2. These data show that calcium, sodium, bicarbonate, and silica are the most abundant solutes, with lower concentrations of magnesium, sulfate, and chloride. The pH of the water varies from 6.4 to 8.1, and tends to be lower in groundwater containing low concentrations of dissolved solids (Table 2). SI calculated from the analytical data for chalcedony range from -0.3 to $+0.4$, and the SI for calcite range from -3.0 to $+2.0$ (Figure 3). Interestingly, the saturation indices for calcite are more tightly grouped around 0.0 (equilibrium with calcite) in bedrock groundwater than glacial drift groundwater. Because of the generally

Table 2 Mean concentrations of dissolved constituents in the stratified drift and bedrock aquifers of Connecticut.

Sample group number	Calcium (mmol L^{-1})	Magnesium (mmol L^{-1})	Sodium (mmol L^{-1})	Potassium (mmol L^{-1})	Chloride (mmol L^{-1})	Sulfate (mmol L^{-1})	Bicarbonate (mmol L^{-1})	SiO$_2$ (mmol L^{-1})	pH (units)
Water from stratified drift aquifers									
G1	0.085	0.062	0.11	0.010	0.073	0.024	0.21	0.20	6.4
G2	0.24	0.063	0.13	0.013	0.093	0.17	0.40	0.16	6.9
G3	0.61	0.14	0.24	0.022	0.23	0.16	0.96	0.25	7.5
G4	0.81	0.084	0.24	0.016	0.16	0.33	1.20	0.23	8.1
Water from bedrock aquifers									
B1	0.36	0.064	0.16	0.012	0.13	0.16	0.55	0.19	7.4
B2	0.41	0.068	0.11	0.020	0.086	0.081	1.00	0.24	7.9
B3	0.66	0.12	0.29	0.013	0.14	0.090	1.50	0.27	8.1

After Rogers (1987).

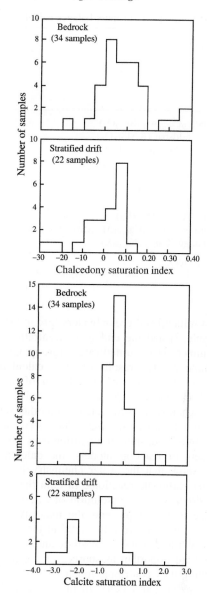

Figure 3 Graph showing saturation indices for chalcedony and calcite in the stratified drift/bedrock aquifers of Connecticut (source Rogers, 1987).

longer residence time of groundwater in bedrock aquifers, this suggests that concentrations of calcium and bicarbonate are regulated by equilibrium with calcite, and suggests that the solubility of chalcedony regulates the concentration of dissolved silica.

Dissolution of feldspars is a logical source of dissolved silica, calcium, sodium, and potassium in groundwater. Similarly, the reaction of carbon dioxide-charged water with silicate minerals is a logical source of bicarbonate. Rogers (1987) examined these and other hypotheses using a mass-balance approach. In these calculations, chloride and sulfate were not considered, and the beginning concentrations were considered to be

Table 3 Summary of mass-balance calculations for the stratified drift and bedrock aquifers of Connecticut.

Mineral phase	The reaction model is identified by a number				
	Stratified drift (mmol kg^{-1} H$_2$O)		Bedrock (mmol kg^{-1} H$_2$O)		
	1	2	3	4	5
Albite (Na$_{0.9}$Ca$_{0.1}$Al$_{1.1}$Si$_{2.9}$O$_8$)	0.048	0.044	0.133	0.124	
Potassium feldspar (KAlSi$_3$O$_8$)	0.006	0.006	0.001	0.001	0.001
Augite (Ca$_{0.7}$Mg$_{0.9}$Fe$_{0.3}$Al$_{0.1}$Si$_{1.9}$O$_6$)	0.004		0.062		0.062
Hornblende (Na$_{0.5}$Ca$_2$Mg$_{3.5}$Fe$_{0.5}$Al$_{1.8}$Si$_7$O$_{22}$)(OH$_2$)		0.006		0.016	
Calcite (CaCO$_3$)	0.703	0.708	0.277	0.284	0.321
Carbon dioxide (CO$_2$)	0.287	0.282	0.693	0.686	0.649
Chalcedony (SiO$_2$)	−0.112	−0.094			
Kaolinite (Al$_2$Si$_2$O$_5$(OH)$_4$)	−0.031	−0.033	0.161	0.116	0.026
Ca-montmorillonite (Ca$_{0.17}$Al$_{2.33}$Si$_{3.67}$O$_{10}$)(OH$_2$)			−0.025	−0.171	−0.025
Ion exchange (Ca$_{0-1.0}$Na$_{2.0}$X)					0.060[a]

After Rogers (1987).

[a] Indicates sodium entering and calcium being removed from solution.
Positive values mean dissolution and negative ones mean precipitation.

the less mineralized groundwaters (group G1 and B1, Table 2) and the ending concentrations the more mineralized groundwaters (group G4 and B4, Table 2). The results of these calculations (Table 3) show that the dissolution of aquifer minerals can indeed explain the observed changes in water chemistry. However, several other sets of reactions can explain the water chemistry equally well (Table 3). In spite of the differences between these sets of reaction models, the calculations clearly show that the water chemistry is dominated by carbonate reactions. In both the glacial and bedrock waters, dissolution of calcite in the presence of CO$_2$ represents the majority of the total mass transfer from the solid to the aqueous phases. While dissolution of albite and potassium feldspar are important as sources of sodium and potassium, they contribute much less to the total mass transfer than calcite. In fact, if dissolution of albite is not included in the mass balance (model 5, Table 3), it is still possible to balance the observed water chemistry. One of Rogers' (1987) major conclusions from this mass-balance exercise was that the rate of calcite dissolution is a factor of 16 faster than the rates of the associated silicate minerals. These findings are similar to those of Wood and Low (1988) in the Snake River Plain aquifer, which showed that the dissolution/precipitation of calcite was one of the most important controls on groundwater geochemistry.

The study of Rogers (1987) indicates that, in spite of the mineralogic and hydrologic complexity of this aquifer system, mass-balance calculations can be useful in understanding the important mineral dissolution/precipitation reactions controlling groundwater geochemistry. Thus, the utility of the mass-balance approach does not necessarily depend on the simplicity of aquifer hydrology and mineralogy. The case

studies of the Snake River Plain basalt aquifers and the bedrock/stratified drift aquifers, however, show that mass-balance models are not unique. It can be argued, in fact, that the nonunique nature of mass-balance models injects more uncertainty into understanding groundwater geochemistry than lithologic and hydrologic uncertainties. This nonuniqueness of mass-balance models is an important characteristic of methods for studying groundwater geochemistry. This is discussed in detail by Bricker *et al.* (see Chapter 5.04), and will be examined further in the following example.

5.14.2.3 A Carbonate Lithology: The Floridan Aquifer of Florida

Aquifers that consist of carbonate rocks, which include limestones (CaCO$_3$) and dolomites (CaMg(CO$_3$)$_2$), are among the most productive sources of groundwater in the world. The Floridan aquifer, which underlies all of Florida, most of Georgia, and part of South Carolina, yields more than 3 billion gallons per day of freshwater. Because of its economic importance, and because the chemical quality of groundwater produced from the Floridan aquifer differs noticeably from place to place, it has been carefully studied by groundwater geochemists (Hanshaw *et al.*, 1965; Back and Hanshaw, 1970; Plummer, 1977; Plummer and Back, 1980; Plummer *et al.*, 1983; Sprinkle, 1989). These studies, in addition to elucidating the processes controlling the chemical quality of Floridan aquifer water, were also important in the development of methods for studying groundwater geochemistry.

The Floridan aquifer system consists of the Upper and Lower aquifers, which are separated by a less permeable confining unit (Miller, 1986). The Upper Floridan, which consists of the Tampa, Suwannee, and Ocala limestones as well as the Avon Park Formation, is more widely used for water supply than the Lower Floridan, and is the focus of this brief overview. The pre-pumping potentiometric surface of the Upper Floridan aquifer is shown in Figure 4. It indicates that the aquifer is recharged in the topographically high areas of central Florida, and that groundwater flows radially away from the recharge area. A cross-section from Polk City, which is located in the potentiometric high in the center of the Florida peninsula, directly south through Ft. Meade, Wauchula, and Arcadia is shown in Figure 5 (Plummer, 1977). This cross-section shows that in

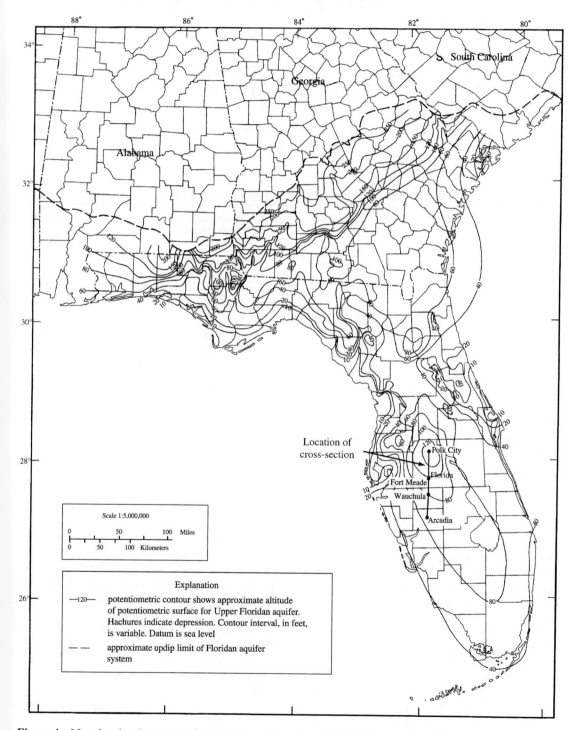

Figure 4 Map showing the prepumping potentiometric surface of the Floridan aquifer and locations of wells shown in Figure 5 (after Sprinkle, 1989).

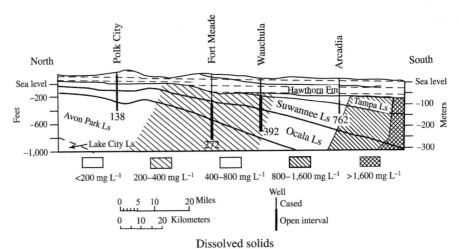

Figure 5   Cross-section showing location of wells and concentrations of dissolved solids in the Floridan aquifer (source Plummer, 1977).

Table 4   Water chemistry of the Upper Floridan aquifer.

Constituent	Initial water (Polk City) (mg L^{-1})	Intermediate water (Ft. Meade) (mg L^{-1})	Final water (Wauchula) (mg L^{-1})
SiO$_2$	12.0	16.0	18.0
Ca^{2+}	34.0	58.0	66.0
Mg^{2+}	5.6	17.0	29.0
Na$^+$	3.2	6.1	8.3
K$^+$	0.5	0.7	2.0
HCO$_3^-$	124.0	163.0	168.0
SO$_4^{2-}$	2.4	71.0	155.0
Cl$^-$	4.5	9.0	10.0
F$^-$	0.1	0.4	0.7
NO$_3^-$	0.1	0.1	0.0
H$_2$S	0.0		1.2
pH (units)	8.0	7.75	7.69
T (°C)	23.8	26.6	25.4
δ^{13}C (per mil)	−14.3	−10.8	−8.5
^{14}C (% modern)	34.3	17.3	4.4
δ^{34}S (per mil)	−14.0		+24.9
δ^{34}S$_{(H_2S)}$ (per mil)			−32.9

Source: Plummer *et al.* (1983).

the aquifer recharge area near Polk City, the groundwater is characterized by relatively low concentrations of dissolved solids (<200 mg L^{-1}). As the water moves south along the regional hydrologic gradient, the concentration of dissolved solids progressively increases. In this freshwater portion of the aquifer this increase in dissolved solids reflects the dissolution of minerals as the water moves downgradient. Analyses of groundwater from Polk City, Ft. Meade, and Wauchula are given in Table 4 (Plummer *et al.*, 1983). These show that calcium, magnesium, and bicarbonate are the principal dissolved solids. As groundwater nears coastal

areas, sodium and chloride concentrations increase due to the mixing of freshwater with seawater that is present in coastal and offshore portions of the Upper Floridan aquifer (Figure 5).

The lithology of the Floridan aquifer system consists predominantly of calcite and dolomite. However, the aquifer also contains nodular gypsum (CaSO$_4$·2H$_2$O), anhydrite (CaSO$_4$), pyrite (FeS$_2$), ferric hydroxide (FeOOH), and traces of lignitic organic matter. Plummer (1977) used a mass-balance approach to estimate the amount of calcite, dolomite, gypsum, and carbon dioxide dissolution needed to explain the evolution of groundwater as it flowed downgradient. This mass balance defined changes in solute concentrations and the dissolution of minerals as

$$\sum_{p=1}^{P} \alpha_p b_{p,k} = \Delta m_{\text{tot},k}, \quad k = 1,...,J \quad (9)$$

which states that the molar change in any element k, $\Delta m_{\text{tot},k}$, along a flow path is equal to the sum of all the sources and sinks for element k. Furthermore, this equation states that this relationship holds for 1 through J elements. The sources and sinks may include mineral dissolution, precipitation, microbial degradation, gas transfer and so forth, of P phases (minerals, gases) along a flow path where α_p is the number of moles reacting and b is the stoichiometric coefficient for the element in the pth mineral phase. Values for $\Delta m_{\text{tot},k}$ are derived from groundwater chemistry data.

If it is assumed that calcite, dolomite, gypsum, and carbon dioxide are the phases to be considered, mass balance can be described by four linear equations of the form given by Equation (9). Simultaneous solution of these four equations for water chemistry changes between unmineralized rainwater to the water composition of Polk City yields the mass balance:

$$\sum_{p=1}^{P} \alpha_p = 0.482\text{CaCO}_3 + 0.230\text{CaMg(CO}_3)_2$$
$$+ 0.025\text{CaSO}_4\cdot2\text{H}_2\text{O} + 1.095\text{CO}_2$$
$$(10)$$

Similar solutions of these mass-balance equations can be used to explain water chemistry changes from Polk City and Ft. Meade, and from Ft. Meade to Wauchula (Plummer, 1977), as well as other parts of the Floridan aquifer (Sprinkle, 1989).

One reason why mass-balance models of groundwater geochemistry are useful, as

illustrated by this example (Plummer, 1977), is that they can explain regional changes in water chemistry (Table 4). In this case, the net dissolution of carbonate minerals and gypsum increases concentrations of total dissolved solids as water flows along the hydrologic gradient (Figure 5). An inherent problem with the mass-balance approach, however, is that aquifers are much more lithologically complex then can be fully taken into account. The mass balance given above for the Floridan aquifer, for example, considers just four mineral phases and four dissolved constituents. Clearly, this is a simplification of the natural system. The question is, how much information is being obscured by making such simplifications?

This important question was addressed for the Floridan aquifer, using the same hydrologic data (Table 4) in a subsequent study by Plummer *et al.* (1983). In this study, the lithologic complexities of the aquifer were addressed by considering more mineral phases (pyrite, ferric hydroxide, organic carbon, and methane), by considering variations in the composition of calcite (which may contain up to 5 mol.% magnesium), dolomite (which may contain different proportions of Ca^{2+}, Mg^{2+}, and Fe^{2+}), and by considering the balance of sulfur and carbon isotopes. Because the possible number of minerals (termed "plausible phases") exceeded the number of chemical constituents in groundwater, it was necessary to construct multiple sets of mass-balance models. As the Floridan is a confined aquifer in this part of the hydrologic system, it was initially assumed that the system was closed to sources of carbon dioxide (models 1–6; Table 5). However, while each of the mass-balance models constructed was capable of explaining the observed major ion water chemistry (Table 5), they were not consistent with the observed isotopic ratios of carbon and sulfur.

Table 5 Results of mass-balance calculations for the Floridan aquifer assuming a system closed to CO_2.

Plausible phases	Reaction model number					
	1	*2*	*3*	*4*	*5*	*6*
CaCO_3	−2.45	−19.79				
$\text{CaMg(CO}_3)_2$	0.96					
$\text{CaSO}_4\cdot2\text{H}_2\text{O}$	2.29	2.29	2.24	2.06	−7.60	6.13
CO_2						
CH_2O	1.32	1.32	1.21	1.03		
FeOOH	0.33	0.33	0.25	0.37	−4.66	4.44
FeS_2	−0.33	−0.33	−0.31		4.61	
$\text{Ca}_{0.95}\text{Mg}_{0.05}\text{CO}_3$		19.26				
$\text{Ca}_{0.98}\text{Mg}_{0.02}\text{CO}_3$			−2.68	−2.49	7.61	−6.74
$\text{Ca}_{1.05}\text{Mg}_{0.90}\text{Fe}_{0.05}\,(\text{CO}_3)_2$			1.13	1.13	0.90	1.22
FeS				−0.43		−4.50
CH_4					−8.62	5.09

Source: Plummer *et al.* (1983).
Values of α_p in mmol kg^{-1} H$_2$O. Positive values mean dissolution and negative ones mean precipitation.

Table 6 Results of mass-balance calculations for the Floridan aquifer assuming a system open to CO_2.

Plausible phases	Reaction model number					
	7	8	9	10	11	12
$CaCO_3$	−1.84	−19.18				
$CaMg(CO_3)_2$	0.96					
$CaSO_4 \cdot 2H_2O$	1.68	1.68	1.68	1.68	1.68	1.68
CO_2	0.53	0.53	0.49	0.47	0.57	0.56
CH_2O	0.17	0.17	0.16	0.18		
$FeOOH$	0.03	0.03	−0.03	−0.002	−0.03	−0.002
FeS_2	−0.03	−0.03	−0.03		−0.03	
$Ca_{0.95}Mg_{0.05}CO_3$		19.26				
$Ca_{0.98}Mg_{0.02}CO_3$			−2.09	−2.09	−2.09	−2.09
$Ca_{1.05}Mg_{0.90}Fe_{0.05}(CO_3)_2$			1.12	1.12	1.12	1.12
FeS				−0.05		−0.05
CH_4					0.08	0.09

Source: Plummer *et al.* (1983).
Values of α_p in mmol kg^{-1} H_2O. Positive values mean dissolution and negative ones mean precipitation.

This led the investigators to reconsider the assumption that the system was closed to carbon dioxide. Again, several possible mass-balance models were tested assuming that the system was partially open to carbon dioxide (Table 6). Of these possible mass-balance models, one (model 7) was consistent with most aspects of the water chemistry, including the net precipitation of calcite and pyrite, the net dissolution of gypsum, dolomite, and ferric hydroxide, the oxidation of organic matter, and the dissolution of carbon dioxide. It was also consistent with the observed $\delta^{13}C$ of dissolved inorganic carbon (DIC). This, in turn, suggested that hitherto neglected processes—microbial oxidation of organic matter coupled to the reduction of ferric hydroxides and sulfate—were important geochemical processes in the Floridan aquifer.

This exercise (Plummer *et al.*, 1983) illustrates the power of the mass-balance approach to understanding groundwater geochemistry. By carefully considering the lithologic complexities of the aquifer and by systematically noting the inconsistencies of the resulting mass and isotopic balances, it was possible to learn something about the system (the importance of microbial carbon dioxide production) that was not known previously. The hypothesis testing approach to understanding groundwater geochemistry, as laid out by Plummer *et al.* (1983) for the Floridan aquifer, was a major contribution to modern methodology for studying groundwater geochemistry.

5.14.2.4 A Coastal Plain Lithology: The Black Creek Aquifer of South Carolina

The hypothesis-testing approach for using mass-balance models to understand groundwater geochemistry, introduced by Plummer *et al.* (1983),

has been widely applied to a number of aquifer systems, sometimes with surprising and unexpected results. An example of how this approach can lead to unexpected and counter-intuitive results was given by McMahon and Chapelle (1991b) in a study of an aquifer in the coastal plain of South Carolina.

The Black Creek aquifer crops out in the upper Coastal Plain in a narrow band that runs parallel to the Fall Line (Figure 6). The aquifer consists of unconsolidated clastic sediments and carbonate shell material of upper Cretaceous age (Campanian–Maastrichtian) that dip eastward away from the outcrop area. The aquifer is recharged directly by rainfall in the outcrop area, and by vertical leakage from overlying confining beds downgradient of the outcrop area. Groundwater flow proceeds in a southeasterly direction (Figure 6), and water chemistry changes significantly in the direction of flow (Table 7). Near the outcrop area, groundwater is undersaturated with respect to calcite, but as water flows downgradient, the water asymptotically approaches saturation with respect to calcite (Figure 7). Furthermore, concentrations of DIC increase from less than 1 mM to more than 10 mM (Figure 8). This large increase in DIC is accompanied by a striking 1 : 1 increase in the concentration of sodium (Figure 8), and the $\delta^{13}C$ of inorganic carbon increases from less than −25‰ near the outcrop area to approximately −5‰ downgradient (Figure 9).

The origin of sodium bicarbonate waters, as illustrated by the Black Creek aquifer, is a classic problem in groundwater geochemistry. Numerous early investigators, beginning with Renick (1924), followed by the studies of Cedarstrom (1946) and Foster (1950) have considered the origin of sodium bicarbonate groundwater. These studies used simple mass balance, stoichiometric relationships, and experimental studies to explain

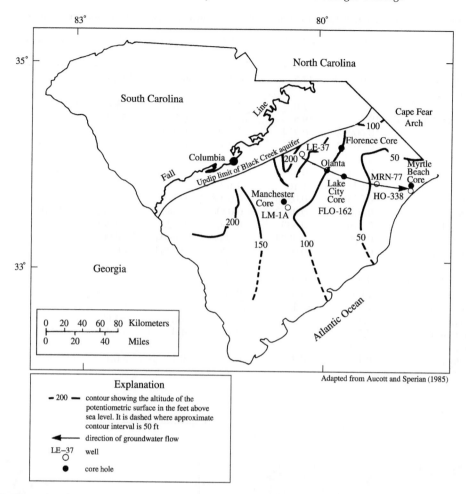

Figure 6 Map showing the location of wells and directions of groundwater flow, Black Creek aquifer, South Carolina (source McMahon and Chapelle, 1991b).

observed water chemistry. The stoichiometry of calcite dissolution combined with cation exchange of calcium for sodium yields a 1 : 1 molar ratio of sodium and DIC (if bicarbonate is the predominant inorganic carbon species, as occurs when pH values are between 6.4 and 10.3)

$$CaCO_3 + CO_2 + H_2O + Na_2 \cdot Ex$$
$$\rightarrow 2Na^+ + 2HCO_3^- + Ca \cdot Ex \quad (11)$$

where Ex denotes exchange sites associated with clays in the aquifer matrix. This overall process was thought to explain the observed 1 : 1 correspondence of sodium and bicarbonate concentrations in Black Creek aquifer water (Foster, 1950). However, because the $\delta^{13}C$ of shell material, which is the source of dissolving calcite, is $\sim 0\%o$, and because the $\delta^{13}C$ of organic matter that produces carbon dioxide is approximately $-25\%o$, application of isotope balance indicates that the $\delta^{13}C$ of DIC in groundwater should approach $-12.5\%o$. But, as Figure 9 shows, the $\delta^{13}C$ of DIC approaches $-5\%o$. Clearly, there is an

inconsistency between the mass balance implied by Equation (11) and the required isotope balance.

McMahon and Chapelle (1991b) approached this inconsistency by using the hypothesis testing approach of Plummer *et al.* (1983). Earlier studies (McMahon and Chapelle, 1991a) had shown that the pore waters of confining beds overlying and underlying the Black Creek aquifer contain relatively high ($\sim 100 \mu M$) concentrations of dissolved organic acid anions, primarily acetate and formate, whereas aquifer water contains relatively low concentrations of organic acid anions ($\sim 1 \mu M$). In addition, confining bed pore waters contain higher concentrations of sulfate ($\sim 100 \mu M$) than aquifer water ($\sim 5 \mu M$). This, in turn, suggested that diffusion of dissolved organic carbon (DOC) and sulfate into the aquifer drove ongoing sulfate reduction at clay–sand contacts (McMahon *et al.*, 1992). Could it be that diffusion-driven exchange of organic carbon from confining beds (where most of the organic matter in this system resides) could partially explain the

Table 7 Chemical composition of Black Creek aquifer groundwater and confining bed pore waters.

	LE-37	OLANTA	LM-IA	FLO-162	HO-338	Confining bed 38 m	Confining bed 67 m
Distance down-gradient (km)	10.0	42.5	50.0	69.0	135.0		
pH (units)	4.5	7.6	8.4	8.6	8.4		
Temperature (°C)	19.5	19.5	18.0	20.2	23.0		
Ca (mmol L^{-1})	0.00998	0.274	0.324	0.0170	0.0599	0.157	0.177
Mg (mmol L^{-1})	0.00823	0.132	0.0905	0.00400	0.0370	0.0580	0.0410
Na (mmol L^{-1})	0.0478	0.135	0.826	1.48	12.6	2.26	2.83
K (mmol L^{-1})	0.0281	0.384	0.187	0.0480	0.128	0.169	0.307
DIC (mmol L^{-1})	0.675	1.16	0.170	1.24	10.4	2.57	2.70
Cl (mmol L^{-1})	0.0536	0.0536	0.0508	0.116	2.48	0.0880	0.105
SO$_4$ (mmol L^{-1})	0.0729	0.0802	0.0708	0.0820	0.00208	0.101	0.381
SiO$_2$ (mmol L^{-1})	0.216	0.649	0.350	0.328	0.266		
Al (mmol L^{-1})	0.0200	<0.0004	<0.0004	<0.0004	0.00185		
Fe (mmol L^{-1})	0.0134	0.00394	0.0005	0.0004	0.0004		
Sr (mmol L^{-1})			0.0025				
H$_2$S (as S) (mg L^{-1})	<0.03	<0.03	<0.03	<0.03	<0.03		
CH$_4$(μmol L^{-1})	0.05	0.05	0.08				
δ^{13}C of DIC (per mil)	−27.8	−18.1	−12.8				
Aragonite$_{si}$	−7.7	−1.1	−0.1	−1.6	−1.6		
Low-magnesium calcite$_{si}$	−7.5	−0.9	0.1	−1.4	−1.4		
Chalcedony$_{si}$	−0.1	0.4	0.1	0.1	0.1		

Source: McMahon and Chapelle (1991b).

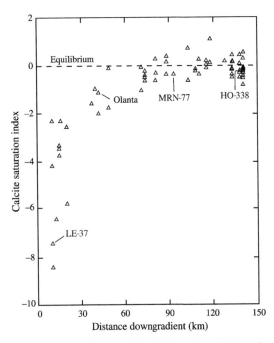

Figure 7 Calcite saturation indices plotted along the direction of groundwater flow in the Black Creek aquifer (source McMahon and Chapelle, 1991b).

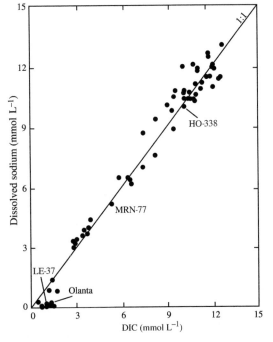

Figure 8 Observed 1 : 1 correspondence of DIC and sodium in the Black Creek aquifer of South Carolina (source McMahon and Chapelle, 1991b).

evolution of aquifer water chemistry in this system?

The locations of representative wells along the flow path are shown in Figure 7, and analyses of groundwater associated with these wells are shown in Table 7. In addition to well-water chemistry, Table 7 shows the composition of pore water in confining beds overlying the Black Creek aquifer at the Lake City Core Hole (Figure 7). These analytical data were used to

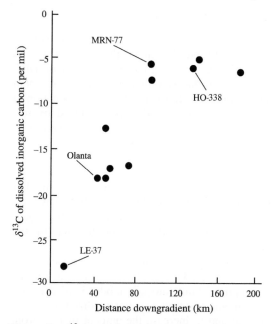

Figure 9 $\delta^{13}C$ composition of DIC in Black Creek aquifer water (source McMahon and Chapelle, 1991b).

Table 8 Calculated mass transfer for hypothesis 1.

Mineral phase	Mass transfer	
	Olanta to MRN-77	MRN-77 to HO-338
$CaCO_3$	2.297	2.531
NaX/CaX^a	2.536	2.550
NaX/MgX^b	0.0061	0.152
$Illite^c$	−0.472	−0.186
$Glauconite^d$	0.0020	0.0458
SiO_2	1.211	0.136
$Pyrite^e$	−0.0080	−0.0914
CO_2	1.796	2.165
CH_2O^f	0.0363	0.427
Seawater	0.0	0.0043

Source: McMahon and Chapelle (1991b).
[a] Ca-for-Na cation exchange. [b] Mg-for-Na cation exchange.
[c] $K_{0.6}Mg_{0.25}Al_{2.5}Si_{3.5}O_{10}$. [d] $K_2Fe_2Al_6(Si_4O_{10})3(OH)_{12}$. [e] FeS_2.
[f] Organic carbon, oxidation state +1.
Positive values mean dissolution and negative ones mean precipitation.

construct material-balance models designed to test a series of hypotheses. These are:

Hypothesis 1. Mass balance given by the stoichiometry of Equation (11) can explain the major ion and carbon isotope composition of Black Creek aquifer water.

Hypothesis 2. Diffusion of DOC and sulfate from confining bed pore waters provides sources of electron donor (organic carbon) and electron acceptor (sulfate). Carbon dioxide produced by this reaction drives shell material dissolution/calcite cement precipitation which can explain the major ion and carbon isotope composition of Black Creek aquifer water.

Hypothesis 3. Diffusion of DIC together with DOC, sulfate, and cations from confining bed pore waters to the Black Creek aquifer provides sources of electron donor (organic carbon) and electron acceptor (sulfate) for microbial metabolism and additional inorganic carbon to drive low-magnesium calcite precipitation. The combination of magnesium-calcite dissolution from shell material driven by microbially produced carbon dioxide, and the precipitation of more thermo-dynamically stable low-magnesium calcite cement in the aquifer, can explain major ion and carbon isotope composition of Black Creek aquifer water.

Each of these hypotheses is testable using mass-balance and isotopic-balance considerations.

5.14.2.4.1 Hypothesis 1

The mass balance implied by hypothesis 1 (Table 8) shows several inconsistencies with

what is known about this hydrologic system. First, this model requires a source of carbon dioxide (CO_2) independent of organic matter (CH_2O) oxidation to achieve mass balance. This requirement stems from the lack of electron acceptors (ferric iron, sulfate) needed to drive oxidation of organic matter, and the lack of significant methane production (in which DIC acts as an electron acceptor). Since carbon dioxide in low temperature groundwater systems is produced only by microbial metabolism, it would seem logical that there must be a source of electron acceptors to support this metabolism. In addition, isotope balance using the mass-balance coefficients calculated from hypothesis 1 predicts $\delta^{13}C$ of DIC values at wells MRN-77 and HO-338 of −12.9‰ and −12.5‰, respectively. However, the measured $\delta^{13}C$ values at these wells are −5.6‰ and −6.1‰, respectively. Thus, based on both mass-balance and isotopic-balance considerations, hypothesis 1 can be ruled out.

5.14.2.4.2 Hypothesis 2

Another possible explanation for the observed water chemistry is that confining beds provide sources of both organic carbon (electron donor) and sulfate (electron acceptor), which drives shell material dissolution in the Black Creek aquifer. The relatively large pool of solid organic carbon present in confining beds adjacent to the Black Creek aquifer is subject to fermentation, which in turn produces relatively high concentrations of organic acids in confining bed pore waters (McMahon and Chapelle, 1991b). Confining bed pore waters contain higher concentrations of sulfate than adjacent aquifer waters (Table 7). Oxidation of organic acids as they diffuse to the aquifer, coupled to reduction of sulfate as it diffuses to the aquifer, would produce carbon dioxide. This carbon dioxide, in turn, could drive dissolution/precipitation of carbonate shell material in the aquifer. Because carbon-12 tends to be used preferentially in carbonate dissolution/precipitation reactions, a process called Rayleigh distillation (Plummer *et al.*, 1983; see Chapter 5.11) causes the $\delta^{13}C$ of DIC to become progressively heavier. These coupled processes, therefore, might account for the observed mass and carbon isotope balance in this system.

The mass balance implied by hypothesis 2 is shown in Table 9. Again, however, while this model accounts for mass balance, predicted $\delta^{13}C$ of DIC values are not consistent with observed values. Isotope balance at wells MRN-77 and HO-338 using the mass balance of Table 9 predicts $\delta^{13}C$ of DIC values of $-25.6‰$ and $-24.5‰$, which are at variance with the observed values of $-5.6‰$ and $-6.1‰$, respectively. The relatively light predicted values reflect the light isotopic signature of organic matter in the confining beds (McMahon and Chapelle, 1991a), and the fact that the model of hypothesis 2 draws much of the inorganic carbon from oxidation of this organic matter. Rayleigh distillation accompanying shell material dissolution and calcite cement precipitation could account for the observed carbon isotope values (Plummer *et al.*, 1983), but it would require over 50 mM of shell material to dissolve per liter of water, and with an equal amount of calcite cement to precipitate in the downgradient part of the aquifer. This, in turn, would cause complete cementation of the Black Creek aquifer. While secondary calcite cement is present in the downgradient portion of the aquifer, it cements less than 10% of the aquifer by volume. Thus, hypothesis 2 can be ruled out by mass-balance and isotope-balance considerations.

5.14.2.4.3 Hypothesis 3

Analysis of pore waters associated with confining beds (Table 7) shows that the water also contains relatively high concentrations of DIC. This suggests that as solid organic matter present in confining beds is fermented with the production of organic acids, carbon dioxide produced by fermentation reacts with shell material in confining beds, producing relatively high DIC concentrations. Unlike hypothesis 2, the stoichiometry of dissolution suggests that half of the confining bed DIC would come from shell material and half from organic matter. Diffusion of sulfate, DIC, and organic acids to the Black Creek aquifer would significantly alter the isotope balance of the system.

The mass balance implied by hypothesis 3 is shown in Table 10. The net result of the assumptions built into this model is to decrease the amount of organic matter oxidized to carbon dioxide and to increase the amount of DIC from shell material dissolution. This, in turn, decreases the amount of shell material dissolution/calcite cement precipitation needed to achieve isotope balance. Between Olanta and MRN-77, the amount of dissolution/precipitation needed for isotope balance is 2.0 mmol $CaCO_3$ kg^{-1} of H_2O, and 25 mmol $CaCO_3$ from MRN-77 to HO-338. This, in turn, implies that 1–13 vol.% of the aquifer would be cemented by calcite, which is roughly in line with observed calcite cementation.

The overall carbon balance in the Black Creek aquifer implied by hypothesis 3 is shown in Table 11. According to this balance, shell material has contributed 83–87% of the carbon to DIC in the aquifer, compared to 13–17% from organic carbon. Of the total carbon added from shell material and organic carbon, ~74% was subsequently removed as calcite cement in the

Table 9 Calculated mass transfer for hypothesis 2.

Mineral phase	Mass transfer	
	Olanta to MRN-77	MRN-77 to HO-338
$CaCO_3$	−7.196	−10.102
NaX/CaX[a]	−6.957	−10.084
NaX/MgX[b]	−1.0019	−1.170
Illite[c]	−4.504	−5.471
Glauconite[d]	1.212	1.631
SiO_2	0.808	0.393
Pyrite[e]	−2.427	−3.262
Pore water 1	11.326	15.225
Pore water 2	20.076	26.316
Seawater	0.0	0.0043

Source: McMahon and Chapelle (1991b).
Footnote symbols 'a' to 'e' are same as in Table 8.
Positive values mean dissolution and negative ones mean precipitation.

Table 10 Calculated mass transfer for hypothesis 3.

Mineral phase	Mass transfer	
	Olanta to MRN-77	MRN-77 to HO-338
$CaCO_3$	0.526	0.0193
NaX/CaX[a]	0.951	0.282
NaX/MgX[b]	−0.105	0.0062
Illite[c]	−1.138	−1.060
Glauconite[d]	0.0692	0.134
SiO_2	2.737	2.137
Pyrite[e]	−0.142	−0.268
Pore water 1	0.662	1.247
Pore water 2	1.116	1.462
Seawater	0.0	0.0043

Source: McMahon and Chapelle (1991b).
Footnote symbols 'a' to 'e' are same as in Table 8.
Positive values mean dissolution and negative ones mean precipitation.

Table 11 Sources and sinks of organic and inorganic carbon to Black Creek aquifer groundwater.

	Olanta to MRN-77		MRN-77 to HO-338	
	mmol kg^{-1} H$_2$O	% of total	mmol kg^{-1} H$_2$O	% of total
Carbon source				
Shell material aquifer	12.526	78	14.0193	73
Shell material confining bed	1.471	9	1.928	10
Organic carbon aquifer	0.0	0	0.0	0
Organic carbon confining bed	2.133	13	3.175	17
Seawater	0.0	0	0.00998	<1
Carbon sink				
Calcite cement	12.0	74	14.0	73

Source: McMahon and Chapelle (1991b).

aquifer. This relatively large component of carbonate dissolution–precipitation is reflected in the relatively heavy δ^{13}C of DIC observed in the downgradient portion of the aquifer. Confining beds contributed 22–27% of the total carbon input, compared to 73–78% from the aquifer.

5.14.2.4.4 Evaluation of hypotheses

It is clear from this hypothesis testing approach that an infinite number of mass-balance scenarios can be constructed to explain the observed water-chemistry changes in the Black Creek aquifer. This nonuniqueness/uncertainty is a general characteristic, and is an unavoidable characteristic of material-balance calculations in systems that are not at full thermodynamic equilibrium. However, much can still be learned by carefully considering the implications of each material balance. Each of the hypotheses considered above produced testable predictions. Hypothesis 1 could not account for observed carbon isotope values, and did not predict the presence of calcite cement in the aquifer. Similarly, for hypothesis 2 to account for isotopic balance, it would require almost complete cementation of the aquifer by secondary calcite cement. Of the three hypotheses considered, hypothesis 3 gave the most plausible explanation of carbon isotope balance and observed calcite cementation of the aquifer. This does not, however, mean that hypothesis 3 is "correct." It simply provides an explanation for the observed geochemical features of the Black Creek aquifer, which includes aquifer water and confining bed pore-water composition, aquifer and confining bed mineralogy, and the observed presence and abundance of secondary calcite cements.

The chief utility of this hypothesis testing is that it indicates that microbial processes occurring in the aquifer (sulfate reduction) as well as microbial processes in confining beds (organic matter fermentation) have an important impact on the geochemistry of groundwater in adjacent aquifers.

These nonintuitive results could not have been arrived at without undertaking a quantitative material balance.

5.14.3 REDUCTION/OXIDATION PROCESSES

Reduction–oxidation (redox) processes affect the chemical composition of groundwater in all aquifer systems. In particular, redox processes affect the mobility of organic chemicals and metals in both pristine and contaminated systems. Thus, methods for characterizing redox processes are an important part of groundwater geochemistry. The purpose of this section is to review equilibrium and kinetic frameworks for documenting the spatial and temporal distribution of redox processes in groundwater systems.

5.14.3.1 The Equilibrium Approach

The traditional approach for characterizing redox processes in groundwater is based on conventions and methods developed in classical physical chemistry (Sillén, 1952). Back and Barnes (1965) used platinum electrode measurements to determine the Eh of groundwater samples. This approach was systematized by Stumm and Morgan (1981), who suggested that the theoretical activity of electrons in aqueous solution (p_e), could be used by direct analogy to hydrogen ion activity (pH) as a "master variable" to describe redox processes. In this treatment, the p_e of a water sample is a linear function of Eh ($p_e = 16.9$Eh at 25 °C).

The definition of Eh, and thus p_e, is given by the Nernst equation, in which the Eh of a solution is related to concentrations of aqueous redox couples at chemical equilibrium and the voltage of a standard hydrogen electrode (E^0). For example, when concentrations of aqueous Fe^{3+} and Fe^{2+} are at equilibrium, Eh is defined as

$$\text{Eh} = E^0 + \frac{2.303RT}{nF}\log\frac{a_{\text{Fe}^{3+}}}{a_{\text{Fe}^{2+}}} \qquad (12)$$

Equation (12) illustrates an important point. Eh *is uniquely defined only when a system is at thermodynamic equilibrium* (Drever, 1982, p. 257). If the activities of Fe^{3+} and Fe^{2+} ions are not at equilibrium, an infinite number of apparent Eh values can be calculated or measured with a platinum electrode, but will not be "Eh" as defined by Equation (12). Similarly, if the Eh defined by the Fe^{3+}/Fe^{2+} pair is not the same as that defined by the SO_4^{2-}/H_2S pair or the O_2/H_2O pair, it is impossible to define the Eh of the solution as a whole.

In the 1960s and 1970s, when groundwater systems were thought to be largely sterile environments devoid of microbial life, assuming equilibrium or near-equilibrium conditions seemed to be quite reasonable. However, in the early 1980s it became clear that groundwater systems contained active, respiring, reproducing microorganisms (Wilson *et al.*, 1983; Chapelle *et al.*, 1987). Furthermore, it gradually became clear that many of the important redox processes occurring in groundwater systems were catalyzed by microorganisms (Baedecker *et al.*, 1988; Lovley *et al.*, 1989; Chapelle and Lovley, 1992). This realization coincided with growing evidence that measuring the Eh of groundwaters was problematic. In particular, it was shown that Eh measurements with platinum electrodes were not consistent with Eh calculated from the Nernst equation, and that different redox couples gave widely different Eh values (Lindberg and Runnells, 1984). There are several reasons for these problems. These include:

(i) The aqueous activity of free electrons in water are so low ($\sim 10^{-55}$ M) that they are essentially zero (Thorstenson, 1984). Thus, while electron activity is a thermodynamically definable quantity, it is not measurable in the same way that hydrogen ion activity (pH) is.

(ii) The pH electrode responds to aqueous concentrations of hydrogen ions. In contrast, the Eh electrode responds to *electron transfers* between solutes (Thorstenson, 1984). A platinum Eh electrode, therefore, readily responds to concentrations of Fe^{2+} and Fe^{3+} because they react rapidly and reversibly with platinum. Because solutes such as oxygen, carbon dioxide, and methane react sluggishly on a platinum surface, the Eh electrode is relatively insensitive to the O_2/H_2O and CO_2/CH_4 redox couples.

(iii) Groundwater usually contains multiple redox couples such as O_2/H_2O, Fe^{3+}/Fe^{2+}, SO_4/H_2S, and CO_2/CH_4 that are not in mutual equilibrium.

(iv) Microorganisms cannot actively respire and reproduce unless there is available free energy to drive their metabolism. That is, microorganisms *require that their immediate environment* not *be at thermodynamic equilibrium*. Thus, using Eh to describe redox processes driven by microbial processes violates the underlying equilibrium assumption of Eh.

In light of these difficulties, it is not surprising that Eh measurements in groundwater systems are so often problematic.

5.14.3.2 The Kinetic Approach

Equilibrium considerations are not the only way to describe redox processes in groundwater systems. The metabolism of microorganisms is based on the cycling of electrons from electron donors (often organic carbon) to electron acceptors such as molecular oxygen, nitrate, ferric iron, sulfate, carbon dioxide, or other mineral electron acceptors. This flow of electrons is capable of doing work. Microorganisms capture this electrical energy, convert it to chemical energy, and use it to support their life functions. If it is assumed that redox processes in groundwater systems are driven predominantly by microbial metabolism, it becomes possible to describe these processes by the cycling of electron donors, electron acceptors, and intermediate products of microbial metabolism. Because this is an inherently nonequilibrium, kinetic description, it has been termed the "kinetic approach" (Lovley *et al.*, 1994).

A kinetic description of redox processes in groundwater systems includes two components. These are documenting (i) the source of electrons (electron donor) that supports microbial metabolism, and (ii) the final sink for electrons (electron acceptors) that supports microbial metabolism. In many groundwater systems, identifying electron donors is not a difficult problem since particulate or DOC is the most common source of electrons for subsurface microorganisms. Another common source of electrons is the aerobic oxidation of sulfide minerals, which facilitates the growth of sulfide-oxidizing microorganisms such as *Thiobacillus* sp. Alternatively, the aerobic oxidation of ferrous iron can facilitate the growth of iron-oxidizing bacteria such as *Gallionella* sp. A more difficult problem is determining the terminal electron accepting processes (TEAPs) that occur in a system. This problem is made even more difficult by the inherent heterogeneity of groundwater systems. This heterogeneity causes both spatial (Chapelle and Lovley, 1992) and temporal (Vroblesky and Chapelle, 1994) variations in TEAPs.

5.14.3.3 Identifying TEAPs

Studies in aquatic sediment microbiology have clearly demonstrated that microbially mediated redox processes tend to become segregated into discrete zones. When this happens, the observed sequence of redox zones follows a predicable pattern. At the sediment–water interface, oxic metabolism predominates. This oxic zone is underlain by zones dominated by nitrate reduction, manganese reduction, and ferric iron reduction (Froelich *et al.*, 1979). In organic rich marine sediments, a sulfate-reducing zone overlies a zone dominated by methanogenesis (Martens and Berner, 1977).

The mechanisms causing the observed segregation of redox zones remained unclear for a long time. However, studies with pure cultures of methanogens and sulfate reducers (Lovley and Klug, 1983), followed by studies with aquatic sediments (Lovley and Klug, 1986; Lovley and Goodwin, 1988), showed that redox zonation resulted from the ecology of aquatic sediments. In anoxic sediments, organic matter oxidation is carried out by food chains in which fermentative microorganisms partially oxidize organic matter with the production of fermentation products such as acetate and hydrogen. These fermentation products are then consumed by terminal electron-accepting microorganisms such as Fe(III) reducers or sulfate reducers. Because Fe(III) reduction produces more energy per mole of acetate or hydrogen oxidation, Fe(III) reducers are able to lower environmental concentrations of these fermentation products below levels required by less efficient sulfate reducers. Thus, when Fe(III) is available, Fe(III) reducers can outcompete sulfate reducers for available hydrogen, and sequester the majority (although generally not all) of the available electron flow. Similarly, when sulfate is available, sulfate reducers can outcompete methanogens for available hydrogen, and sequester the majority of electron flow. Thus, by considering concentrations of potential electron acceptors, and by considering concentrations of dissolved hydrogen (Chapelle *et al.*, 1995), it is often possible to deduce the distribution of TEAPs in groundwater systems (Figure 10).

5.14.3.4 Comparison of Equilibrium and Kinetic Approaches

A comparison of the equilibrium (Eh) and kinetic (TEAPs) approaches to describe redox processes in a petroleum hydrocarbon-contaminated aquifer was given by Chapelle *et al.* (1996). In this study, Eh measurements were made with a platinum electrode, and the results plotted on a standard Eh–pH diagram (Sillén, 1952). The results of this analysis are shown in Figure 11. Based on this analysis, it can be concluded that Fe(III) reduction is the predominant redox process, as none of the measured Eh values are sufficiently negative to indicate sulfate reduction

or methanogenesis. In contrast, measured hydrogen concentrations (Figure 12) indicate the presence of methanogenesis near the water table surrounded by zones dominated by sulfate reduction and Fe(III) reduction. By considering other redox-sensitive solutes (Figure 10), including the depletion of potential electron acceptors such as oxygen (<0.01 mg L^{-1}), sulfate (<1.0 mg L^{-1}), and nitrate (<0.02 mg L^{-1}) inside the plume, as well as the generation of sulfide (~1.0 mg L^{-1}), methane ($5-20$ mg L^{-1}), and Fe^{2+} ($3-20$ mg L^{-1}), it is clear that methanogenesis and sulfate reduction are occurring (Figure 12) in addition to Fe(III)-reduction indicated in Figure 11. In some groundwater systems, therefore, the kinetic approach appears to give a more complete evaluation of ambient redox processes than the more traditional equilibrium approach.

5.14.4 CONTAMINANT GEOCHEMISTRY

The principles and methods used to understand groundwater geochemistry were largely developed by considering pristine aquifer systems (Back, 1966; Plummer *et al.*, 1983). In the 1970s, however, it became clear that decades of unregulated disposal of industrial wastes had contaminated shallow groundwater underlying thousands of sites throughout the world (Volume 9 of this Treatise). Many scientists involved with assessing and remediating this environmental contamination had been trained in the use of the mass-balance and equilibrium methods used in groundwater geochemistry. It was natural, therefore, to apply the mass-balance approach to problems of environmental contamination. This section gives a brief overview of the application of these principles to petroleum hydrocarbon and chlorinated solvent contamination of groundwater systems.

5.14.4.1 Petroleum Hydrocarbon Contamination

During the 1980s, enormous efforts were made in order to assess and monitor petroleum hydrocarbon contamination of groundwater in the United States and Europe. From this mass of information came several unanticipated and surprising results. It was widely observed, for example, that plumes of petroleum-hydrocarbon contaminated groundwater stopped expanding over time and assumed a dynamic steady-state configuration. Perhaps the best-documented example of this behavior was a crude oil spill in northern Minnesota near the town of Bemidji (Baedecker *et al.*, 1988; Cozzarelli *et al.*, 1999). In 1979 an oil pipeline ruptured and spilled 1,670 m^3 of crude oil on the land surface. During the

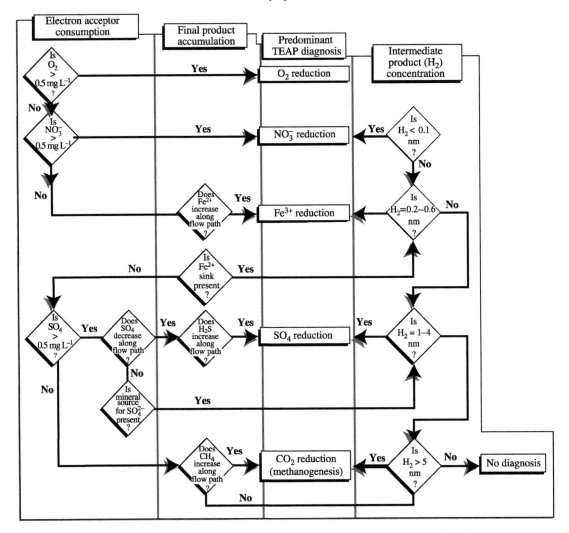

Figure 10 Flowchart for deducing operative TEAPs in groundwater systems (source Chapelle *et al.*, 1995).

following year, oil migrated downward and formed a lens floating on the water table. The site was instrumented with observation wells and monitored throughout the 1980s and 1990s. A plume of dissolved hydrocarbons, principally composed of soluble benzene, toluene, ethylbenzene, and xylene (BTEX) compounds, developed downgradient of the oil lens. However, by 1985, the BTEX plume appeared to stop spreading, extending only ~150 m downgradient of the oil lens. This result, was unanticipated by many scientists. Its explanation, however, came directly from mass-balance considerations (Baedecker *et al.*, 1993). The dynamic steady state of the plume reflected a balance between the rate at which soluble hydrocarbons were leached into the groundwater, and the rate at which biodegradation processes consumed the hydrocarbons (Lovley *et al.*, 1989; Baedecker *et al.*, 1993).

Similar observations were being made in other hydrologic systems. In a controlled experimental release to a shallow sandy aquifer, Barker *et al.* (1987) showed that naturally occurring biodegradation processes effectively removed benzene, toluene, and all three xylene isomers from solution in about one year. Barker *et al.* (1987) used the term *natural attenuation* to describe the combined dilution, dispersion, sorption, and biodegradation processes that caused contaminant concentrations to decrease. Similarly, Chiang *et al.* (1989) showed significant losses of benzene over a three-year period in a contaminated aquifer underlying a gas plant, and attributed this loss to natural attenuation processes.

On a much larger scale, a study of 7,167 municipal supply wells in California showed that, while leaking gasoline was by far the most common contaminant being released into groundwater systems, benzene was found in only 10 wells (Hadley and Armstrong, 1991). This study concluded that biodegradation processes were actively consuming BTEX compounds, and that

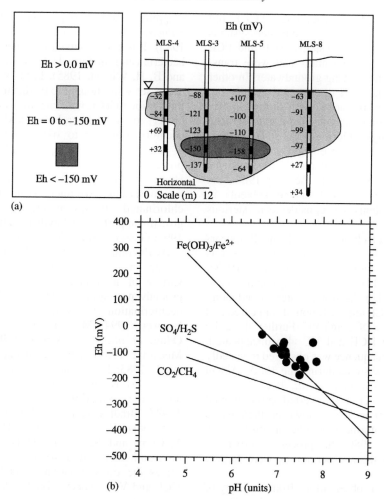

Figure 11 Measured Eh values in a cross-section of a contaminant plume, and the Eh values plotted on an Eh–pH diagram (source Chapelle *et al.*, 1996).

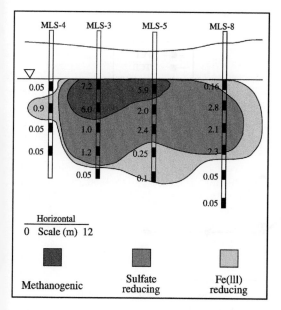

Figure 12 Concentrations of hydrogen in a cross-section of a contaminant plume, and the associated diagnosis of TEAPs (source Chapelle *et al.*, 1996).

these processes were protecting California groundwater supplies from widespread petroleum hydrocarbon contamination. These and other studies showed that natural biodegradation processes are far more important in affecting the migration of soluble petroleum hydrocarbons than had been foreseen in the early 1980s, and that they are a major factor in the remediation of environmental contaminants (see Chapter 9.12; US EPA, 1999).

5.14.4.2 Chlorinated Solvent Contamination

Mass-balance considerations, in particular the observed consumption of contaminants, were useful in showing the importance of biodegradation processes for limiting the mobility of petroleum hydrocarbons in groundwater systems. The mass-balance approach also contributed to our understanding of the environmental fate of chlorinated solvents in groundwater systems. In the 1980s, the observed behavior of chlorinated ethenes such as trichloroethene (TCE) and perchloroethene (PCE) in groundwater systems were puzzling. In some systems, TCE and PCE acted like conservative solutes and were transported readily with flowing groundwater. In other systems, TCE and PCE disappeared rapidly but were apparently replaced with more lightly chlorinated ethenes such as *cis*-dichloroethene (DCE) and vinyl chloride (VC). To further complicate matters, these lightly chlorinated ethenes apparently accumulated in some systems and disappeared in others. In the early 1980s, the environmental fate of chlorinated solvents in groundwater systems was considered mysterious and unpredictable.

An example of chlorinated solvent disappearance, from a site in Kings Bay, Georgia was described by Chapelle and Bradley (1998). Initially, PCE and TCE were the only contaminants present (Figure 13). As groundwater moved along the flow path, PCE and TCE both disappeared and were replaced by DCE and VC. Further along the flow path, both DCE and VC also disappeared. This same basic sequence was observed repeatedly in groundwater systems during the 1980s (Vogel *et al.*, 1987; Barrio-Lage *et al.*, 1987). Clearly, transformation processes converted one compound into another. Just as clearly, however, there was a net loss of all chlorinated ethenes along the flow path. In the late 1980s, the process or processes contributing to these observations were largely unknown.

Soon thereafter, observations from a variety of sources began to explain this behavior. It was shown experimentally that pure cultures of methanogenic microorganisms were capable of stripping chlorine groups from PCE and TCE by *reductive dechlorination* (Vogel and McCarty, 1985). It was also shown that methane-oxidizing bacteria can transform TCE as well (J. T. Wilson and B. H. Wilson, 1985). Initially, both of these processes were thought to be due to the accidental interaction of TCE with enzyme systems designed either to reduce carbon dioxide (reductive dechlorination) or to oxidize methane (Vogel *et al.*, 1987; Barrio-Lage *et al.*, 1987). Both processes were initially termed *co-metabolic*. However, further research has shown a class of previously unknown microorganisms that can use chlorinated ethenes as electron acceptors in growth-supporting metabolism (DiStefano *et al.*, 1991; Gossett and Zinder, 1996).

By the mid-1990s, it was clear that chlorinated ethenes in groundwater systems are subject to a variety of microbial degradation processes in groundwater systems. These include reductive dechlorination (Barrio-Lage *et al.*, 1987, 1990; Bouwer, 1994; McCarty and Semprini, 1994; Odum *et al.*, 1995; Vogel, 1994; Vogel and McCarty, 1985), aerobic oxidation (Hartmans *et al.*, 1985; Davis and Carpenter, 1990; Phelps *et al.*, 1991; Bradley and Chapelle, 1996, 1998a,b), anaerobic oxidation (Bradley and Chapelle, 1998b; Bradley *et al.*, 1998), and aerobic co-metabolism (J. T. Wilson and B. H. Wilson, 1985; McCarty and Semprini, 1994). In many field environments, initial reductive dechlorination drives the transformation of PCE and TCE to DCE and VC, respectively. The latter in turn are transformed into carbon dioxide, chloride and water by the combined effects of methane-oxidizing co-metabolism and anaerobic oxidation.

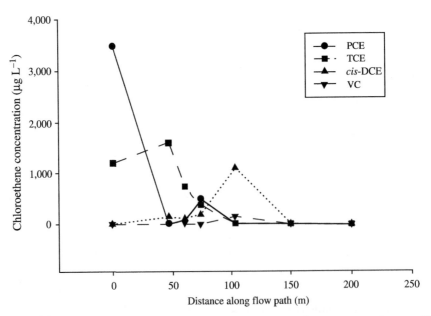

Figure 13 Observed transformation of chlorinated ethenes in a groundwater system (source Chapelle and Bradley, 1998).

Furthermore, because these biodegradation processes are all reduction/oxidation processes, the efficiency of biodegradation is very sensitive to ambient redox conditions (Chapelle, 1996).

A simple use of mass-balance considerations that shows this net transformation of chlorinated ethenes is the observed accumulation of chloride in contaminant plumes. For example, in a study of a large plume of TCE-contaminated groundwater at Dover Air Force Base, Delaware, Witt *et al.* (2002) showed that the concentration of chloride increased as TCE concentrations decreased, suggesting the net transformation of the chlorine in TCE to chloride. This chloride "tracer" of chlorinated ethene biodegradation is of use for mass-balance calculations to demonstrate biodegradation processes in the field. These field observations formed the basis of experimental studies under controlled laboratory conditions that documented the many and varied microbial processes that transform and destroy chlorinated ethenes in groundwater systems.

5.14.5 SUMMARY AND CONCLUSIONS

This brief overview shows that while the scope of groundwater geochemistry has changed significantly over time, the themes of practicality and the mass-balance approach have remained constant. The major questions of groundwater geochemistry have come directly from practical, human problems. Whether it was understanding the apparent medicinal properties of well and springwaters (Steel, 1838), helping to prospect for petroleum hydrocarbons (Rogers, 1917), explaining the distribution of "hard" and "soft" groundwater (Renick, 1924), investigating the origin of high concentrations of sodium and bicarbonate (Cedarstrom, 1946; Foster, 1950), showing how chemical changes of groundwater were related to directions of groundwater flow (Back, 1966), understanding the composition of springwaters (Garrels and Mackenzie, 1967), or documenting biodegradation of environmental contaminants (Baedecker *et al.*, 1993), all involved practical questions. The guiding principle used to address these questions has always been the notion of mass balance, that groundwater chemistry reflects the difference of what is put into solution minus what is taken out. Rigorous application of the mass-balance approach, and constraining material balances for thermodynamic consistency (Plummer *et al.*, 1983) has led to many unexpected and surprising discoveries. One of these is the realization that nonequilibrium, microbially mediated processes have an important effect on groundwater chemistry (Chapelle, 2001). In recent years, the mass-balance approach has helped identify microbial and nonmicrobial transformations that serve to detoxify environmental contaminants (US EPA, 1999), and has served to identify numerous remediation technologies for dealing with contamination problems. There is every reason to expect that the twin principles of practicality and mass balance will continue to be major themes in the development of groundwater geochemistry.

REFERENCES

Armstrong R. L. (1971) K–Ar chronology of Snake River Plain, Idaho (abs): Geological Society of America Abstracts with Programs, p. 366.

Back W. (1966) Hydrochemical facies and groundwater flow patterns in northern part of Atlantic Coastal Plain. *US Geol. Surv. Prof. Pap.* **498-A**, 42pp.

Back W. and Barnes I. (1965) Relation of electrochemical potentials and iron content to groundwater flow patterns. *US Geol. Surv. Prof. Pap.* **498-C**, 16pp.

Back W. and Hanshaw B. B. (1970) Comparison of chemical hydrogeology of the carbonate peninsulas of Florida and Yucatan. *J. Hydrol.* **10**, 330–368.

Back W., Landa E. R., and Meeks L. (1995) Bottled water, spas, and early years of water chemistry. *Ground Water* **33**, 605–613.

Baedecker M. J., Siegel D. I., Bennett P. C., and Cozzarelli I. M. (1988) The fate and effects of crude oil in a shallow aquifer: 1. The distribution of chemical species and geochemical facies. In *US Geological Survey Toxic Substances Hydrology Program*, Proceedings of the Technical Meeting, Phoenix, Arizona, September 26–30, 1988 (eds. G. E., Mallard and S. E. Ragone). US Geological Survey Water-Resources Investigations Report 88-4220, Reston, VA, pp. 13–20.

Baedecker M. J., Cozzarelli I. M., Eganhouse R. P., Siegel D. I., and Bennett P. C. (1993) Crude oil in a shallow sand and gravel aquifer: III. Biogeochemical reactions and mass balance modeling in anoxic ground water. *Appl. Geochem.* **8**, 569–586.

Barker J. F., Patrick G. C., and Major D. (1987) Natural attenuation of aromatic hydrocarbons in a shallow sand aquifer. *Ground Water Monitoring Rev.* **7**, 64–71.

Barrio-Lage G. A., Parsons F. Z., Nassar R. S., and Lorenzo P. A. (1987) Biotransformation of trichloroethene in a variety of subsurface materials. *Environ. Toxicol. Chem.* **6**, 571–578.

Barrio-Lage G. A., Parsons F. Z., Barbitz R. M., Lorenzo P. L., and Archer H. E. (1990) Enhanced anaerobic biodegradation of vinyl chloride in ground water. *Environ. Toxicol. Chem.* **9**, 403–415.

Bouwer E. J. (1994) Bioremediation of chlorinated solvents using alternate electron acceptors. In *Handbook of Bioremediation* (eds. R. D. Norris, R. E. Hinchee, R. Brown, P. L. McCarty, J. T. Wilson, M.. Reinhard, E. J. Bouwer, R. C. Bordon, T. M. Vogel, J. M. Thomas, and C. H. Ward). Lewis Publishers, Boca Raton, pp. 149–175.

Bradley P. M. and Chapelle F. H. (1996) Anaerobic mineralization of vinyl chloride in Fe(III)-reducing aquifer sediments. *Environ. Sci. Technol.* **30**, 2084–2086.

Bradley P. M. and Chapelle F. H. (1998a) Effect of contaminant concentration on aerobic microbial mineralization of DCE and VC in stream-bed sediments. *Environ. Sci. Technol.* **32**, 553–557.

Bradley P. M. and Chapelle F. H. (1998b) Microbial mineralization of VC and DCE under different terminal electron accepting conditions. *Anaerobe* **4**, 81–87.

Bradley P. M., Chapelle F. H., and Wilson J. T. (1998) Field and laboratory evidence for intrinsic biodegradation of vinyl chloride contamination in a Fe(III)-reducing aquifer. *J. Contamin. Hydrol.* **31**, 111–127.

Burke W. (1853) *The Virginia Mineral Springs with Remarks on Their Use, The Diseases to Which They are Applicable, and in Which They are Contra-indicated.* Ritchies and Dunnavant, Richmond, VA, 376pp.

Cedarstrom D. J. (1946) Genesis of groundwaters in the coastal plain of Virginia. *Econ. Geol.* **41**(3), 218–245.

Chapelle F. H. (1996) Identifying redox conditions that favor the natural attenuation of chlorinated ethenes in contaminated groundwater systems. In *Symposium on Natural Attenuation of Chlorinated Organics in Ground Water*, EPA/540/R-96/509, pp. 17–20.

Chapelle F. H. (2000) *The Hidden Sea: Ground Water, Springs, and Wells.* The National Ground Water Association, Westerville, OH, 232pp.

Chapelle F. H. (2001) *Groundwater Microbiology and Geochemistry*, 2nd edn. Wiley, New York, 477pp.

Chapelle F. H. and Bradley P. M. (1998) Selecting remediation goals by assessing the natural attenuation capacity of groundwater systems. *Bioremed. J.* **2**, 227–238.

Chapelle F. H. and Lovley D. R. (1992) Competitive exclusion of sulfate-reduction by Fe(III)-reducing bacteria: a mechanism for producing discrete zones of high-iron ground water. *Ground Water* **30**, 29–36.

Chapelle F. H., Zelibor J. L., Grimes D. J., and Knobel L. L. (1987) Bacteria in deep coastal plain sediments of Maryland: a possible source of CO_2 to ground water. *Water Resour. Res.* **23**(8), 1625–1632.

Chapelle F. H., McMahon P. B., Dubrovsky N. M., Fujii R. F., Oaksford E. T., and Vroblesky D. A. (1995) Deducing the distribution of terminal electron-accepting processes in hydrologically diverse groundwater systems. *Water Resour. Res.* **31**, 359–371.

Chapelle F. H., Haack S. K., Adriaens P., Henry M. A., and Bradley P. M. (1996) Comparison of Eh and H_2 measurements for delineating redox processes in a contaminated aquifer. *Environ. Sci. Technol* **30**, 3565–3569.

Chebotarev I. I. (1955) Metamorphism of natural waters in the crust of weathering. *Geochim. Cosmochim. Acta* **8**(Part 1), 22–48; **8**(Part 2), 137–170; **8**(Part 3), 198–212.

Chiang C. Y., Salanitro J. P., Chai E. Y., Colthart J. D., and Klein C. L. (1989) Aerobic biodegradation of benzene, toluene, and xylene in a sandy aquifer: data analysis and computer modeling. *Ground Water* **27**, 823–834.

Cozzarelli I. M., Herman J. S., Baedecker M. J., and Fischer J. M. (1999) Geochemical heterogeneity of a gasoline-contaminated aquifer. *J. Contamin. Hydrol.* **40**, 261–284.

Davis J. W. and Carpenter C. L. (1990) Aerobic biodegradation of vinyl chloride in groundwater samples. *Appl. Environ. Microbiol.* **56**, 3870–3880.

DiStefano T. D., Gossett J. M., and Zinder S. H. (1991) Reductive dechlorination of high concentrations of tetrachloroethene to ethene by an anaerobic enrichment culture in the absence of methanogenesis. *Appl. Environ. Microbiol.* **57**, 2287–2292.

Drever J. I. (1982) *The Geochemistry of Natural Waters.* Prentice Hall, Englewood Cliffs, NJ, 388pp.

Embree G. F., McBroome L. A., and Soherty D. J. (1982) Preliminary stratigraphic framework of the Pliocene and Miocene rhyolite, Eastern Snake River Plain. *Idaho Bureau Mines Geol. Bull.* **26**, 333–346.

Feth J. H., Robertson C. E., and Polzer W. L. (1964) Sources of mineral constituents in water from granitic rocks in Sierra Nevada, California and Nevada. *US Geol. Surv. Water-Supply Pap.* **1535-I**, 70p.

Froelich P. N., Klinkhammer G. P., Bender M. L., Luedtke N. A., Heath G. R., Cullen D., and Dauphin P. (1979) Early oxidation of organic matter in pelagic sediments of the eastern equatorial Atlantic: suboxic diagenesis. *Geochim. Cosmochim. Acta* **43**, 1075–1090.

Foster M. D. (1950) The origin of high sodium bicarbonate waters in the Atlantic and Gulf Coast Plains. *Geochim. Cosmochim. Acta* **1**(1), 33–48.

Garrels R. M. and Christ C. L. (1965) *Solutions, Minerals, and Equilibria.* Freeman-Cooper, San Francisco, CA, 450pp.

Garrels R. M. and Mackenzie F. T. (1967) Origin of the chemical compositions of some springs and lakes. In *Equilibrium Concepts in Natural Water Chemistry*, Advances in Chemistry Series 67 (ed. R. F. Gould). American Chemical Society, Washington, DC, pp. 222–242.

Gossett J. M. and Zinder S. H. (1996) Microbiological aspects relevant to natural attenuation of chlorinated ethenes. In *Symposium on Natural Attenuation of Chlorinated Organics in Ground Water*, Denver, Co, EPA/540/R-96/509, pp. 10–13.

Hadley P. W. and Armstrong R. (1991) "Where's the benzene?"—Examining California groundwater quality surveys. *Ground Water* **29**, 35–40.

Hanshaw B. B., Back W., and Rubin M. (1965) Radiocarbon determinations for estimating groundwater flow velocities in central Florida. *Science* **148**, 494–495.

Hartmans S., deBont J. A. M., Tramper J., and Luyben K. C. A. M. (1985) Bacterial degradation of vinyl chloride. *Biotechnol. Lett.* **7**, 383–388.

Helgeson H. C., Brown T. H., Nigrini A., and Jones T. A. (1970) Calculation of mass transfer in geochemical processes involving aqueous solutions. *Geochim. Cosmochim. Acta* **34**, 569–592.

Lindberg R. D. and Runnells D. D. (1984) Groundwater redox reactions: an analysis of equilibrium state applied to Eh measurements and geochemical modeling. *Science* **225**, 925–927.

Lindholm G. F., Garabedian S. P., Newton G. D., and Whitehead R. L. (1983) *Configuration of the Water Table, March 1980 in the Snake River Plain Regional Aquifer System, Idaho and Eastern Oregon.* US Geological Survey Open-file Report 82-1022, scale 1:1,000,000.

Lovley D. R. and Goodwin S. (1988) Hydrogen concentrations as an indicator of the predominant terminal electron-accepting reactions in aquatic sediments. *Geochim. Cosmochim. Acta* **52**, 2993–3003.

Lovley D. R. and Klug M. J. (1983) Sulfate reducers can outcompete methanogens at freshwater sulfate concentrations. *Appl. Environ. Microbiol.* **45**, 187–192.

Lovley D. R. and Klug M. J. (1986) Model for the distribution of methane production and sulfate reduction in freshwater sediments. *Geochim. Cosmochim. Acta* **50**, 11–18.

Lovley D. R., Baedecker M. J., Lonergan D. J., Cozzarelli I. M., Phillips E. J. P., and Siegel D. I. (1989) Oxidation of aromatic contaminants coupled to microbial iron reduction. *Nature* **339**, 297–299.

Lovley D. R., Chapelle F. H., and Woodward J. C. (1994) Use of dissolved H_2 concentrations to determine distribution of microbially catalyzed redox reactions in anoxic groundwater. *Environ. Sci. Technol.* **28**, 1205–1210.

Martens C. S. and Berner R. A. (1977) Interstitial water chemistry of anoxic Long Island Sound sediments: I. Dissolved gases. *Limnol. Oceanogr.* **22**, 10–25.

McCarty P. L. and Semprini L. (1994) Groundwater treatment for chlorinated solvents. In *Handbook of Bioremediation* (eds. R. D. Norris, R. E. Hinchee, R. Brown, P. L. McCarty, J. T. Wilson, M.. Reinhard, E. J. Bouwer, R. C. Bordon, T. M. Vogel, J. M. Thomas, and C. H. Ward). Lewis Publishers, Boca Raton, pp. 87–116.

McMahon P. B. and Chapelle F. H. (1991a) Microbial production of organic acids in aquitard sediments and its role in aquifer geochemistry. *Nature* **349**, 233–235.

McMahon P. B. and Chapelle F. H. (1991b) Geochemistry of dissolved inorganic carbon in a Coastal Plain aquifer: 2. Modeling carbon sources, sinks, and $\delta^{13}C$ evolution. *J. Hydrol.* **127**, 109–135.

McMahon P. B., Chapelle F. H., Falls W. F., and Bradley P. M. (1992) Role of microbial processes in linking sandstone diagenesis with organic-rick clays. *J. Sedimen. Petrol.* **62**, 1–10.

Miller J. A. (1986) Hydrogeologic framework of the Floridan aquifer system in Florida and in parts of Georgia, South Carolina, and Alabama. *US Geol. Surv. Prof. Pap. 1403-B*, 91pp.

Odum J. M., Tabinowski J., Lee M. D., and Fathepure B. Z. (1995) Anaerobic biodegradation of chlorinated solvents: comparative laboratory study of aquifer microcosms. In *Handbook of Bioremediation* (ed. R. D. Norris, R. E. Hinchee, R. Brown, P. L. McCarty, J. T. Wilson, M. Reinhard, E. J. Bouwer, R. C. Bordon, T. M. Vogel, J. M. Thomas, and C. H. Ward). Lewis Publishers, Boca Raton, pp. 17–24.

Parkhurst D. L., Plummer L. N., and Thorstenson D. C. (1982) BALANCE—a computer program for calculation of chemical mass balance. *US Geol. Surv. Water Res. Investigations Report 82-14.*

Phelps T. J., Malachowsky K., Schram R. M., and White D. C. (1991) Aerobic mineralization of vinyl chloride by a bacterium of the order *Actinomycetales*. *Appl. Environ. Microbiol.* **57**, 1252–1254.

Pierce K. L. and Morgan L. A. (1992) The track of the Yellowstone hot spot: volcanism, faulting, and uplift. *Geol. Soc. Am. Memoir 179.*

Plummer L. N. (1977) Defining reactions and mass transfer in part of the Floridan aquifer. *Water Resour. Res.* **13**, 801–812.

Plummer L. N. (1984) Geochemical modeling: a comparison of forward and reverse methods. In *First Canadian/American Conference on Hydrogeology. Practical Applications of Ground Water Geochemistry* (eds. B. Hitchon and E. I. Walick). National Water Well Association, Dublin, OH, pp. 149–177.

Plummer L. N. and Back W. (1980) The mass balance approach: application to interpreting the chemical evolution of hydrologic systems. *Am. J. Sci.* **280**, 130–142.

Plummer L. N., Parkhurst D. L., and Thorstenson D. C. (1983) Development of reaction models for groundwater systems. *Geochim. Cosmochim. Acta* **4**, 665–686.

Renick B. C. (1924) Base exchange in ground water by silicates as illustrated in Montana. *US Geol. Surv. Water Supply Pap.* **520-D**, 53–72.

Rogers G. S. (1917) Chemical relations of the oil-field waters in San Joaquin Valley, California. *US Geol. Surv. Bull.* **653**, 93–99.

Rogers R. J. (1987) Geochemical evolution of groundwater in stratified-drift and arkosic bedrock aquifers in north central Connecticut. *Water Resour. Res.* **23**, 1531–1545.

Sillén L. G. (1952) Redox diagrams. *J. Chem. Educat.* **29**, 600–608.

Sprinkle C. L. (1989) Geochemistry of the Floridan aquifer system in Florida and in parts of Georgia, South Carolina, and Alabama. *US Geol. Surv. Prof. Pap.* **1403-I**, 105pp.

Steel J. H. (1838) *An Analysis of the Mineral Waters of Saratoga and Ballston*. G. M. Davison, Saratoga Springs, NY, 203pp.

Stumm W. and Morgan J. J. (1981) *Aquatic Chemistry*. Wiley, New York, 780pp.

Thorstenson D. C. (1984) The concept of electron activity and its relation to redox potentials in aqueous geochemical systems. *US Geol. Surv. Open File Report* 84-072.

US Environmental Protection Agency (US EPA) (1999) Final OSWER Monitored Natural Attenuation Policy (Oswer Directive 9200.4-17P). *United States Environmental Protection Agency, Office of Solid Waste and Emergency Response.*

Vogel T. M. (1994) Natural bioremediation of chlorinated solvents. In *Handbook of Bioremediation* (eds. R. D. Norris, R. E. Hinchee, R. Brown, P. L. McCarty, J. T. Wilson, M. Reinhard, E. J. Bouwer, R. C. Bordon, T. M. Vogel, J. M. Thomas, and C. H. Ward). Lewis Publishers, Boca Raton, pp. 201–225.

Vogel T. M. and McCarty P. L. (1985) Biotransformation of tetrachloroethylene to trichloroethylene, dichloroethylene, vinyl chloride, and carbon dioxide under methanogenic conditions. *Appl. Environ. Microbiol.* **49**, 1080–1083.

Vogel T. M., Criddle C. S., and McCarty P. L. (1987) Transformation of halogenated aliphatic compounds. *Environ. Sci. Technol.* **21**, 722–736.

Vroblesky D. A. and Chapelle F. H. (1994) Temporal and spatial changes of terminal electron-accepting processes in a petroleum hydrocarbon-contaminated aquifer and the significance for contaminant biodegradation. *Water Resour. Res.* **30**, 1561–1570.

Wilson J. T. and Wilson B. H. (1985) Biotransformation of trichloroethylene in soil. *Appl. Environ. Microbiol.* **49**, 242–243.

Wilson J. T., McNabb J. F., Balkwill D. L., and Ghiorse W. C. (1983) Enumeration and characterization of bacteria indigenous to a shallow water-table aquifer. *Ground Water* **21**, 134–142.

Witt M. E., Klecka G. M., Lutz E. J., Ei T. A., Grosso N. R., and Chapelle F. H. (2002) Natural attenuation of chlorinated solvents at Area 6 Dover Air Force Base: groundwater biogeochemistry. *J. Contamin. Hydrol.* **57**, 61–80.

Wood W. W. and Low W. H. (1988) Solute geochemistry of the Snake River Plain Regional Aquifer System, Idaho and Eastern Oregon. *US Geol. Surv. Prof. Pap.* **1408-D**, 79pp.

5.15
Groundwater Dating and Residence-time Measurements

F. M. Phillips

New Mexico Tech, Socorro, NM, USA

and

M. C. Castro

University of Michigan, Ann Arbor, USA

5.15.1 INTRODUCTION

Water is the key to the geochemistry of the upper crust and the surface of the Earth. Water interacts extensively with minerals through hydrolysis, dissolution, precipitation, interface solute layers, etc., and is necessary for life, which directly and indirectly affects the chemistry of Earth systems. The local characterization of aqueous geochemical reactions is a natural and straightforward extension of classical chemistry. However, neither the composition nor the evolution of the upper layers of the Earth can be explained by local-scale geochemistry. It is the open-system nature of geochemistry in this environment that provides most of its explanatory power, as well as its intellectual challenge. The fundamental aspects of near-surface geochemistry such as weathering reactions, precipitation of sedimentary minerals, diagenesis, and metamorphic evolution all depend on transport of mass in and out of these systems. In most cases, moving water is the transport agent.

Geochemical constituents are transported by the circulation of the Earth's three major near-surface dynamic systems: the oceans, the atmosphere, and continental water. This chapter deals with the third of these, specifically with subsurface water. Surface water is an important medium for transporting geochemical constituents on a global scale, and is relatively easily accessed and quantified. In contrast, subsurface water is not easy to access, and

fluxes through subsurface systems are very difficult to measure. To some extent, subsurface fluxes can be estimated using methods based on the physics of water flow (e.g., numerical modeling of flow systems), but the very large variability of permeability imparts a high uncertainty to the results. Fortunately, measurement of certain geochemical constituents in water can define residence times and fluxes in subsurface systems. This chapter will explore the geochemical means that are available to help quantify subsurface fluxes, and will evaluate the implications of their application to a variety of specific systems.

5.15.2 NATURE OF GROUNDWATER FLOW SYSTEMS

5.15.2.1 Driving Forces

Water moves from areas of higher total fluid energy to areas of lower energy. Kinetic energy, mechanical energy (e.g., compression), gravitational potential, interfacial potential, and chemical potential can drive fluid flow. In general, velocities in the subsurface are low enough, so that kinetic energy can be neglected. Water, therefore, moves from areas of higher to ones of lower fluid potential (Hubbert, 1940). In systems that are only partially saturated with water (e.g., the vadose zone and oil and gas reservoirs), the combination of gravitational potential and interfacial potential (capillary action) serves to drive flow. In relatively shallow (less than a few kilometers) saturated groundwater systems, gravitational potential is usually dominant. In deeper settings, additional forces—such as compression due to compaction or tectonic movement, pressure generated by chemical reactions, or chemical potentials developed across semipermeable shales—may also be important (Neuzil, 1995, 2000; see Chapter 5.16).

5.15.2.2 Topographic Control on Flow

The fundamentals of the major subsurface water cycle are straightforward. Water precipitates out of the atmosphere onto areas of relatively high topography. Some of this water infiltrates into the subsurface. Because it has greater gravitational potential energy than water at lower elevations, it flows downward, eventually re-emerging at the surface either at the lowest elevation within the region, or at higher elevations where low-permeability barriers force it to the surface.

Topographically driven groundwater flow is generally more important for geochemical processes than flow produced by the alternative mechanisms mentioned above. This is because the circulation of meteoric water, driven ultimately by solar energy, is a process that provides a continuous source of subsurface water that can carry large fluxes of solutes. Although other driving forces, such as sediment compaction or tectonic compression, may, for periods of time, provide water velocities comparable to topographically driven flow at similar depths, the total water flux is limited to the pore volume of the rock undergoing compression.

One frequently underappreciated aspect of topographically driven flow is the pervasive extent of water circulation through the shallow crust. Fairly subdued topography can drive appreciable water flow to depths of several kilometers. Under favorable circumstances, more significant relief can produce circulation to depth below 5 km (Person and Baumgartner, 1995; Person and Garven, 1992). Permeability generally decreases with depth (Manning and Ingebritsen, 1999), and this is one factor that tends to limit groundwater fluxes at considerable depth. Another fairly common limitation is provided by brines at depth; topographically driven freshwater is generally unable to displace brines that are below the discharge point of a groundwater basin (see Chapter 5.17).

5.15.2.3 Hydraulic Conductivity and Its Variability

The flow of groundwater and other subsurface fluids is described by Darcy's law (Darcy, 1856). Hydraulic conductivity, the constant in Darcy's law, has an extremely wide range in natural materials, extending over ~16 orders of magnitude (Freeze and Cherry, 1979; Manning and Ingebritsen, 1999). Although, in any particular subsurface setting, the range will probably be much less, variations of five or more orders of magnitude are common. To compound the difficulties of quantifying the behavior of fluids in such heterogeneous systems, access to the subsurface is generally very limited and actual measurements of hydraulic conductivity are generally sparse. Thus, although the physics of groundwater flow is well understood and amenable to quantitative analysis, the great variability of properties and the lack of constraints on their values render such analyses, subject to great uncertainty. It is in this perspective that the utility of geochemical tracers for water residence time can be appreciated. A sequence of residence-time measurements along a flow path essentially integrates the velocity history of the groundwater and thus provides a very valuable constraint on the permeability structure.

The structure of the heterogeneity in hydraulic conductivity is typically complex. The fundamentals of the geological controls on conductivity have been recognized for many years. For example, Meinzer (1923) codified long-standing ideas of high- and low-conductivity units termed "aquifers" and "aquitards." However, continued research has shown that these are considerable oversimplifications of a typically very complex subsurface permeability structure. An alternative view is to characterize the permeability structure as essentially fractal (Neuman, 1990, 1995); in other words, as the spatial scale of measurement increases, the variability of hydraulic conductivity, and the size of units of differing conductivity, also increases in some proportional fashion, without any apparent upper limit. Further refinement of these ideas would conceptualize hydraulic conductivity as being characterized by nested scales of variability, such that the increase in variability might be large for certain ranges of spatial scale and small for others (Gelhar, 1993). For example, consider an aquifer formed by deposition in the valley of a meandering river, dominated by sand channel deposits, but also containing many fine-grained overbank deposits. At a scale smaller than that of a single channel or overbank unit, the permeability might be relatively uniform, but as the sampling volume increases to incorporate bounding units of contrasting lithology, the variability would increase dramatically. However, further increase of the volume to incorporate many similar units over the thickness of the entire aquifer would result in little additional variability.

This complex and hierarchical variation of hydraulic conductivity results in complex subsurface flow paths. It is not possible to incorporate a complete and reliable description of actual permeability structures in numerical models. Although geochemical tracers cannot provide a map of the permeability variations, they can provide important clues to the nature of the variations, and they can help to discriminate between competing hypotheses.

5.15.2.4 Scales of Flow Systems

Both permeability distributions and topographic variations can be considered as some form of hierarchical fractal process (Boufadel *et al.*, 2000). The interaction of hierarchical variability in both the main driving force and the main control on flow results in a nested, hierarchical system of flow regimes (Freeze and Witherspoon, 1966, 1967; Tóth, 1963). These are commonly classified as vadose-zone scale, local scale, aquifer scale, and regional scale, going from smallest to largest (Figure 1).

5.15.2.4.1 *Vadose-zone scale*

The vadose zone forms an essential, though frequently overlooked, component of the subsurface hydrologic system (Stephens, 1996). Fluxes

are generally vertical (usually downward, but upward in discharge areas and in many desert regions, and also seasonally upward in many recharge areas). The scales of the flow systems range from a few centimeters to as much as several hundred meters in deserts and mountainous karstic regions. Understanding downward fluxes is of considerable importance because of the need to quantify groundwater recharge and to assess the potential for aquifer contamination. However, this quantification is difficult because of the nonlinear nature of the laws governing unsaturated flow and the great degree of seasonal variability in fluxes.

5.15.2.4.2 *Local scale*

Local flow systems dominate groundwater circulation at shallow depths (down to as much as 100 m) in humid regions (i.e., having significant diffuse areal recharge), unless the topography is very flat (Tóth, 1963). Such systems vary from ~100 m to several kilometers in horizontal scale. It is the discharge from such systems that supports surface-water flow in streams and rivers under all but storm conditions. The primary control on

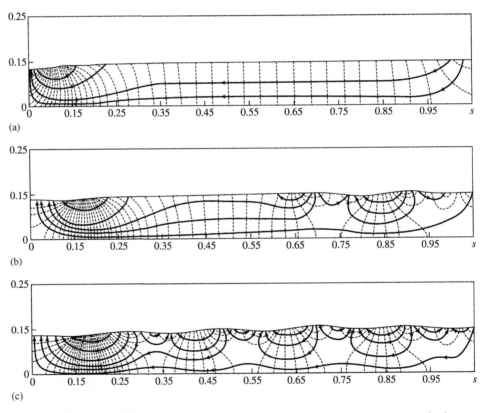

Figure 1 Illustration of the development of increasingly complex flow systems as topography becomes more complex. Contours of hydraulic head are indicated by dashed lines and groundwater flow lines by solid arrows. Scale is arbitrary, but might correspond to 100 km in the horizontal direction. In (a), smooth topography produces a regional-scale flow system. In (b) and (c) increasing local topography creates a mixture of intermediate and local-scale flow systems superimposed on the regional one (Freeze and Witherspoon, 1967) (reproduced by permission of American Geophysical Union from *Water Resour. Res.* **1967**, *3*, 623–634).

groundwater flow in such systems is generally considered to be topographic, with recharge at higher elevations (ridges or hills) and flow to immediately adjacent streams or rivers (Buttle, 1998). The water table generally represents a subdued replica of the topography. Such systems provide a large fraction of the groundwater supply for drinking and other uses, and are particularly susceptible to contamination.

5.15.2.4.3 Aquifer scale

Many hydrogeologic investigations are carried out at the scale of the aquifer, a map-scale geological unit capable of yielding significant amounts of water. Aquifer scale may vary from a few kilometers to hundreds of kilometers. Although aquifers may often span the horizontal scale of the relevant flow system, they generally do not encompass the full vertical extent of the system. Although aquifer-scale studies often do not provide a complete representation of the flow system, they are very important because they focus on the units that provide most of the water supply, and for which the most data are available.

5.15.2.4.4 Regional scale

Many areas exhibit trends of elevation at the continental, or subcontinental scale (hundreds to thousands of kilometers). Unless pervasive permeability barriers are present, the topographic gradients will drive deep circulation systems over these continental scales (Figure 1) (Freeze, and Witherspoon, 1966; Person and Baumgartner, 1995). Although both the depth and the quality of water in such systems limit their human consumption, they are critical for understanding the origins of diagenesis, many ore deposits, and petroleum reservoirs (Ingebritsen and Sanford, 1998).

5.15.2.5 Sources of Solutes at Various Scales

The chemical (and isotopic) composition of subsurface water shows a fair degree of predictability based on the hydrogeological environment in which it is found (Drever, 1997). This predictability arises from distinctive geochemical reactions and sources of solutes that characterize each of the flow-system scales described above. Although these are treated in more detail elsewhere in the *Treatise on Geochemistry*, they are summarized below because they are an integral part of the origin and interpretation of geochemical indicators of residence time and flow path.

5.15.2.5.1 Meteoric (recharge)

Meteoric (atmospheric origin) waters that enter the subsurface environment are not devoid of solutes, although they are generally quite dilute (E. K. Berner and R. A. Berner, 1996). They contain ionic solutes, dissolved gases, and isotopic signatures that are derived from atmospheric sources. The most important ionic solutes are generally of marine origin. These especially include Na^+, Cl^-, and SO_4^{2-}. They also contain N_2, O_2, and CO_2 from equilibration with the atmosphere, producing an oxidizing and weakly acidic solution. In areas having minimal soil–water or rock–water interaction after infiltration of precipitation, these characteristics of infiltrating precipitation produce dilute Na^+ or Ca^{2+}–HCO_3^- subsurface waters. These constituents, particularly Cl^-, which tends to act conservatively in groundwater, can under some circumstances be used to infer residence times.

In addition to these "ordinary" constituents, infiltrating meteoric waters may contain more exotic solutes that are particularly useful for tracing of water flow paths and residence times. Prominent among these are radionuclides produced by the action of cosmic rays on the atmosphere. The most commonly employed are ^3H, ^{14}C, and ^{36}Cl, but many others are described in this chapter. Another category of useful tracers is that produced by human activity, whose atmospheric concentration histories are generally known. Those frequently employed for subsurface tracing include ^3H, ^{14}C, and ^{36}Cl produced by atmospheric nuclear-weapons testing, ^{85}Kr and ^{129}I released from nuclear reactors, and chlorofluorocarbons and SF_6 released by industrial and commercial users. Certain other "trace solutes" can be used to estimate residence times by virtue of temporal fluctuations in input concentrations that can be related to known fluctuations in environmental conditions. The trace molecular species of water $^1H_2\,^{18}O$ and $^2H\,^1H\,^{16}O$ are introduced into the atmosphere by oceanic and continental evapotranspiration and are precipitated along with the more common $^1H_2\,^{16}O$. The proportion of these rare isotopic species is governed by a variety of environmental conditions, primarily by temperature (see Chapter 5.11). Similarly, the concentrations of the atmospheric noble gases are temperature dependent. If fluctuations of temperature in the recharge area are known, then subsurface transport times can be inferred if corresponding spatial fluctuations can be observed in the subsurface water. All of these tracers share the characteristic that can then be used to estimate the time since entry into subsurface water systems.

5.15.2.5.2 *Weathering*

The chemical composition of most shallow groundwater reflects that of atmospheric precipitation, modified by additional solutes derived from weathering. Weathering reactions generally result from the thermodynamic instability of minerals formed in the deep subsurface (or the oceans) in the low-temperature, low-pressure, dilute, oxidizing, and acidic conditions typical of vadose zones and shallow groundwater. The solutes added to infiltrating precipitation through these reactions vary greatly, depending on the mineralogy of the solid phase and the environmental conditions (Drever, 1997). Owing to the slow kinetics of many weathering reactions (particularly those involving aluminosilicate reactions), solute concentrations can continue to increase and evolve over time (Rademacher *et al.*, 2001). These reactions lead to water at the vadose, local, and often aquifer scales that is more basic, less oxidizing, and more concentrated than precipitation, but still generally containing less than $\sim 1,000$ mg L^{-1}. Solutes introduced from these reactions are generally not used as residence-time tracers, but they can complicate the use of atmospheric tracers that also have weathering sources (e.g., ^{14}C and ^{32}Si).

5.15.2.5.3 *Diagenetic*

As the scale of flow systems increases from the local to the regional, groundwater moves to depths where pressures are much higher, temperatures higher, and redox conditions generally more reducing. Under these conditions, waters that have achieved near-equilibrium under vadose or shallow aquifer environments begin to react again with the solid phases present. In some cases these reactions simply take the form of recrystallization to higher $P-T$ phases, but in others solutes are dissolved at one point along a flow path and precipitated at another point, producing cementation or mineralization of the rocks through which they are flowing. Increases of temperature, particularly, tend to produce total solute concentrations that are higher than in groundwaters in shallower flow systems. Anion concentration tends to evolve from HCO_3^- dominance to Cl^- dominance along regional flow paths (Chebotarev, 1955); cation evolution is less predictable. One particularly important aspect of deep rock–water interaction is that highly soluble evaporite minerals (principally halite) are uncommon near the surface of the Earth, because they have generally already been dissolved by circulating meteoric water. Such minerals are more common in the deep subsurface where circulation rates are slow, and thus

regional-scale groundwaters frequently acquire very high dissolved solids through interaction with evaporites. Certain distinctive solutes introduced from the matrix along regional-scale flow paths can be used for estimating residence times. These include certain isotopes of He, Ne, and Ar, and ^{129}I.

5.15.2.5.4 *Connate*

In addition to evaporites, groundwaters on regional flow paths frequently encounter connate waters, i.e., residual solutions retained in the subsurface from the time that formations in the system were deposited in the ocean. These waters often show Na^+-Cl^- dominance (see Chapter 5.16). Such waters were previously thought to have been largely swept out of sedimentary basins by circulation of meteoric water (Clayton *et al.*, 1966), but this analysis ignored the high density of these brines, which makes them very resistant to displacement by flowing dilute groundwater (Knauth and Beeunas, 1986). Connate waters (or other saline formation waters) influence the composition of topographically driven groundwater mainly by upward diffusion of solutes. Because of their generally great age and immobility, they are not amenable to most groundwater dating methods, with the possible exception of ^{129}I (Moran *et al.*, 1995).

5.15.2.5.5 *"Basement waters"*

Highly saline water of distinctive isotopic composition is often found in environments of such depth and low permeability that flow rates must be extremely low or zero. These waters are often characterized by $Ca^{2+}-Cl^-$ compositions and stable isotope composition "above" the meteoric waterline (see Chapter 5.17). They apparently result from water–rock equilibration over very long periods of time. Geochemically, they have little influence on waters in active circulation systems, but mobile isotopes of the noble gases diffusing upward from this environment can be a powerful tool for understanding the flow systems into which they move. The noble-gas isotopes can also provide clues to the histories of these nearly static waters.

5.15.3 SOLUTE TRANSPORT IN SUBSURFACE WATER

The objective of applying geochemical tracers to subsurface water systems is to understand their dynamics. The tracers themselves are solutes. In order to correctly understand the distribution of the tracers, and to apply them to analyze the system dynamics, a basic understanding of subsurface transport processes is necessary. Transport of subsurface solutes can be ascribed to three processes: advection, diffusion, and dispersion. This section gives a brief introduction to these transport processes. More extensive treatments can be found in Bredehoeft and Pinder (1973), Domenico and Schwartz (1998), and Phillips (1991).

5.15.3.1 Fundamental Transport Processes

5.15.3.1.1 Advection

Advection is mechanical transport of solutes along with the bulk flux of the water. It is driven by the gradient in the total mechanical energy of the solution, just as the water flux is driven. For most circumstances, this means the gradient in gravitation potential energy. Advective fluxes are simply the product of the water flux from Darcy's law with the solute concentration:

$$\bar{J}_a = C\bar{\bar{K}}\nabla h = C\bar{q}$$

where $\bar{J}_a$ is the advective solute flux, C the volume concentration, $\bar{\bar{K}}$ the hydraulic conductivity tensor, ∇h the hydraulic head gradient, and $\bar{q}$ the specific discharge. Advection transports all solutes at the same rate.

5.15.3.1.2 Diffusion

Diffusion differs fundamentally from advection in that it is entropy driven, following the principle that solutes within a system will redistribute themselves so as to maximize the entropy of the system. Diffusive transport is, therefore, governed by concentration gradients:

$$\bar{J}_d = D'_d \nabla C$$

where $\bar{J}_d$ is the diffusive flux, D'_d the effective diffusion coefficient (accounting for the effects of porosity and tortuosity), and ∇C the concentration gradient. Since the concentration distributions of various solutes in a single system need not be identical, different solutes generally have different diffusive fluxes.

The relative importance of advection and diffusion is of fundamental significance for understanding solute transport in various settings. This can be assessed using the Peclet number

$$Pe = \frac{q_s d_m}{D'_d}$$

where q_s is the average linear velocity of the water, d_m the mean grain diameter, and D'_d the effective diffusion coefficient. For values of Pe less than ~ 1, diffusion dominates, but above this value advection dominates (Perkins and Johnston, 1963). Application of this equation to the various flow systems described above indicates that for typical local or aquifer-scale systems, advection dominates decisively over diffusion. However, at the very slow flow velocities, characteristic of the deeper portions of regional flow systems, diffusion may become dominant. Diffusion may also dominate in shallower flow systems where permeabilities are very low, such as clay-rich tills or shales, or in unusual circumstances where hydraulic gradients are negligible. In general, diffusion is an effective mechanism for redistributing solutes over short distances under most circumstances, but is a significant transport mechanism at aquifer or regional scales only when the time available for transport is also long.

5.15.3.1.3 Dispersion

There is an additional mechanism of transport in flowing water termed dispersion. The dispersive flux is given by

$$\bar{J}_{disp} = \bar{\bar{D}}\nabla C$$

where $\bar{\bar{D}}$ is the dispersion coefficient tensor and is given by

$$\bar{\bar{D}} = \bar{\bar{\alpha}}q_s + D'_d$$

where $\bar{\bar{\alpha}}$ is the dispersivity tensor. Although the equation above appears to treat dispersion and diffusion as separate and additive processes, they are, in fact, linked. An increased diffusion coefficient may appear to increase dispersion, but will actually decrease it so long as the linear velocity term in the equation is not very small. Dispersion is a mixing process that is actually driven by differential advection. Small-scale differences in velocity along different flow paths produce increasing variation in the concentration as a function of position. However, diffusion tends to equalize concentrations between high- and low-velocity portions of the flow path and thus limits the amount of differential transport due to mechanical dispersion. The additive diffusion coefficient term on the right-hand side simply ensures that concentration-gradient-driven transport goes to the diffusive flux as the velocity goes to zero.

Dispersivity is treated as a tensor, because the magnitude of velocity variations is greatest in the direction of flow and least transverse to the direction (usually vertically transverse, if the flow is horizontal). This means that, for equal concentration gradients, the magnitude of the dispersive flux will also be greatest in the direction of flow (longitudinal dispersion) and smallest perpendicular to it (transverse dispersion).

5.15.3.2 Advection–Dispersion Equation

In order to obtain a comprehensive description of the transport of solutes in subsurface water, it is necessary to consider all three transport mechanisms. This can be accomplished by linking the three transport flux equations above through the continuity equation, yielding the advection–dispersion equation for steady-state flow:

$$\overline{\overline{D}}\nabla^2 C - \overline{q_s}\cdot\nabla C + \sum_i R_i = \frac{\partial C}{\partial t}$$

where R_i are chemical or nuclear reactions affecting the constituent modeled. The advection–dispersion equation can be solved analytically for simple boundary conditions, and such solutions have been used frequently in interpreting the distribution of age tracers (Sudicky and Frind, 1981; Zuber, 1986). However, numerical models provide a more flexible and realistic means of implementing the advection–dispersion equation for groundwater systems, which rarely correspond to the simplifications required for analytical solutions.

5.15.3.3 Interaction between Hydrogeological Heterogeneity and Transport

Section 5.15.2.3 above emphasized the high degree of heterogeneity of permeability that is characteristic of most geological materials, and the role of hierarchical scales of organization of the permeability in determining average flow paths. These characteristics also play a critical role in the transport of solutes in the subsurface. The role they play depends on the scale of the transport.

5.15.3.3.1 Small-scale transport and effective dispersion

Dispersion has often been treated as a constant parameter, with values of the dispersivity on the order of centimeters (Perkins and Johnston, 1963). For typical shallow groundwater flow velocities, this places dispersive transport on the same order as diffusion. However, extensive field tracer tests have demonstrated that dispersion is much more effective than these values would indicate and, furthermore, that effective dispersivity tends to increase with the scale of the tracer test (Gelhar, 1986; Schwartz, 1977; Sudicky and Frind, 1981). This behavior derives from the fundamentally quasifractal nature of hydraulic heterogeneity

described above. Small-scale variations in pore structure tend to spread solutes over limited distances. This spreading is then amplified as the solute-carrying groundwater encounters progressively larger-scale heterogeneities with increased flow distance. The end result of these processes is that at typical shallow groundwater velocities, longitudinal dispersion is typically one or more orders of magnitude more effective in transporting solutes than is diffusion. Furthermore, the effectiveness of the process increases as the transport scale increases. The amount of mixing and consequent spreading of concentration fronts is highly dependent on the local hydrogeology, which controls the permeability structure. Methodologies for moving from an understanding of the geological environment to the ability to predict the permeability structure, and from that ability to the prediction of dispersive behavior, have not been achieved (Weissmann *et al.*, 1999).

5.15.3.3.2 Large-scale transport and mixing

As the scale is increased from the aquifer to the regional, solute transport becomes even more difficult to predict. At smaller scales a somewhat homogeneous spatial distribution of high- and low-permeability facies can often be assumed. This leads to more uniform relative contributions of advection and diffusion to total dispersion. However, as the spatial scale increases, mixing processes extend over units that may have greatly differing distributions of high- and low-permeability facies. For example, transport processes in a sandstone aquifer may be dominated by advection, but in the adjoining shale aquitards, diffusion may dominate. The boundaries between units of differing transport characteristics are not necessarily abrupt or of simple geometry.

Several approaches have been taken to dealing with solute transport in general, and transport of geochemical dating tracers in particular, through these large-scale heterogeneous systems. The earliest approach was to assume such drastic contrasts that low-permeability units could be treated as impermeable and impervious to diffusion (Hanshaw and Back, 1974; Pearson and White, 1967). This is equivalent to treating the permeable units as a sealed tube. This conceptual model is clearly an oversimplification, inasmuch as we know that cross-formational water fluxes are common, and that diffusion coefficients of low-permeability units are not drastically smaller than those of higher permeability units. The second of these shortcomings was acknowledged in models that treated diffusion of tracers into adjacent aquitards as a boundary condition for transport equations (Sanford, 1997; Sudicky and Frind, 1981). Although an improvement, this conceptualization is also limited by the assumption that transport in low-permeability units is diffusion dominated (Phillips *et al.*, 1990). In many cases advection may still dominate, and this carries important implications for the spatial distribution of radionuclide tracers or transient tracer pulses. For example, diffusion will selectively move $H^{14}CO_3^-$ out of aquifers and into aquitards, along the concentration gradient set up by ^{14}C decay within the aquitards, but will not induce any net transport of stable HCO_3^-, thus mimicking the effects of radioactive decay on the $^{14}C/C$ ratio. Advection, in contrast, will transport both species at the same rate.

Any truly comprehensive and accurate mathematical description of large-scale solute transport must incorporate both the effects of encountering progressively larger scales of hydraulic heterogeneity on differential advective transport and the progressive increase in the importance of diffusion as larger volumes of low-permeability rock are contacted. It must account for the three-dimensional interaction between the hydraulic head field and the spatial configuration of heterogeneities, and the effect of this interaction on the fundamental advective and diffusive transport mechanisms. The spatial distributions of permeability in any such model must have a rigorous basis in hydrogeology. This is a formidable undertaking, and we are far from accomplishing it. However, environmental geochemical tracers provide one of the few realistic approaches to obtaining data on solute transport at these scales, and thus represent a critical component to the solution of this fundamental hydrological/geochemical problem.

5.15.3.4 Groundwater Dating and the Concept of "Groundwater Age"

Suppose one holds an igneous rock specimen in which all of the mineral grains crystallized at very nearly the same time, for which the crystallization time was short compared to the period since solidification, and which has not experienced any reheating or alteration since initial crystallization. One could then measure some time-dependent property of the specimen (e.g., the amount of ^{40}Ar accumulated from the decay of ^{40}K) and infer a reasonably well-defined "crystallization age" for the rock. Unfortunately, an analogous set of conditions is quite unlikely to hold rigorously for any sample bottle full of groundwater. As previously emphasized by Davis and Bentley (1982), the dynamic nature of groundwater systems renders such simplistic scenarios unlikely. This review has emphasized that, due to the

complex interaction of irregular topographic forcing with extremely heterogeneous distributions of permeability, groundwater flow paths are complex and difficult to predict, often resulting in mixing on various scales. When large-scale advective mixing is combined with differential transport of different species by diffusion and dispersion, acting over a range of scales, the expectation of simple closed-system behavior for parcels of groundwater is clearly naïve.

Geochemical species that can be used to gain information on groundwater residence times are much better thought of as groundwater "age tracers" than they are as "dating methods." In some cases (e.g., wells sampling local-scale flow systems in relatively homogeneous materials), closed-system behavior may be approximated, and "dates" can be interpreted to reflect residence time in a straightforward fashion. In general, however, concentrations of age tracers should be viewed as data upon which inferences regarding the residence-time structure of the three-dimensional groundwater flow system can be based. Application of several tracers will generally give much better constrained inferences than those using only a single tracer (Castro *et al.*, 2000; Mazor, 1976, 1990). Numerical transport models offer by far the most flexible and comprehensive approach to interpreting the tracer data (Castro and Goblet, 2003; Dinçer and Davis, 1984; Park *et al.*, 2002).

Given the complex flow paths and mixing of waters recharged at different times, even the basic meaning of "groundwater age" can be ambiguous. Recently, several authors (Bethke and Johnson, 2002a,b; Goode, 1996) have addressed this problem by treating groundwater age as a parameter that accumulates in all water in a system by a linear buildup per unit mass per unit time, termed "age mass." The "age" of the groundwater can then be rigorously defined by the accumulated age mass, equivalent to the average of the residence time of every water molecule in the sample. Age mass is a completely conservative quantity whose total does not vary with time in basins that have been at hydraulic steady state for long periods. In such a system, the total age mass is simply "the total water mass" × "the average residence time" (total mass divided by rate of recharge) × "the age mass production rate" (1 kg age mass $(kg\ water)^{-1}\ yr^{-1}$).

More traditional approaches to treating the age of groundwater systems have tended to over-emphasize the advective-dominated portions of the system. The age mass approach forces consideration of both advective and diffusive transport through the entire system. This is nicely illustrated by the observation of Bethke and Johnson (2002b) that the accumulation of age mass in the water of an aquifer that is sandwiched between two aquitards is, counter intuitively, independent of the exchange rate with the water in the aquitards. The reason is that if the exchange rate is high, longer residence-time water is rapidly moved into the aquifer, but the average age of the water entering is not great (due to the rapid exchange). Alternatively, if the exchange rate is low, only small amounts of aquitard water enter, but the average age of the aquitard water is great. The average age of the water exiting the aquifer depends only on the total mass of water in the system and the total water flux through the system, not on the proportion of aquifer to aquitard, or the transport characteristics of the two lithologies.

This approach to conceptualizing residence-time distributions in groundwater systems has important practical implications for interpreting age tracer measurements. For example, the pattern of radiogenic helium concentration in the aquifer system described above would strongly resemble that of the theoretical age mass distribution. Specifically, the concentration of helium would increase much more rapidly with distance along the aquifer than a simple calculation based on the linear flow rate through the aquifer would indicate. It would increase more rapidly because helium would diffuse out of the aquitards into the aquifer. If this additional source of helium is not accounted for, the velocity in the aquifer inferred from the helium distribution would be much slower than the actual velocity. Similarly, a radionuclide tracer (such as ^{14}C or ^{36}Cl) would diffuse into the aquitard, along with the water, also producing an incorrectly low apparent linear velocity in the aquifer.

Although a merely conceptual application of the age mass approach can provide important insights into a simple system such as the one described above, its distribution in an actual groundwater system is a complex function of advective and diffusive controls, like any other solute. Thus, for actual applications the most promising approach is to use numerical models to calibrate the age mass distribution using results from a number of age tracers. Regardless of whether the goal of a study is to define groundwater velocity distributions, or to understand subsurface transport processes, the approach must be based on an understanding of the dynamics of the entire groundwater system.

5.15.4 SUMMARY OF GROUNDWATER AGE TRACERS

5.15.4.1 Introduction

Age tracers for groundwater and the vadose zone can be divided into three basic classes:

(i) radionuclides of atmospheric origin that can be used with the radiometric decay equation to "date" the time since recharge; (ii) stable constituents originating with recharge that have known patterns of age with time and can hence be used to infer residence time; and (iii) constituents that are added to groundwater in the subsurface at an approximately known rate, whose accumulation can thus be used to gain information on the residence time of the water.

5.15.4.2 Radionuclides for Age Tracing of Subsurface Water

5.15.4.2.1 Argon-37

Among the unstable isotopes that have been used as natural tracers in groundwater, ^{37}Ar has the shortest half-life (35 d; Lal and Peters, 1967; Loosli and Oeschger, 1969). Argon-37 is produced by cosmic rays in the atmosphere. Since its half-life is extremely short, most groundwaters have completely lost the atmospheric ^{37}Ar signal that was initially present in the recharge water. Measured ^{37}Ar in groundwater thus reflects the subsurface production rate of ^{37}Ar and the transfer efficiency from rocks to groundwater. The dominant source of subsurface ^{37}Ar is through the following neutron reaction with ^{40}Ca: ^{40}Ca(n,α)^{37}Ar. The neutron flux originates either directly from spontaneous fission of uranium or from (α,n) reactions within the matrix rock. At present, ^{37}Ar activities can be measured down to 100 μBq per liter (STP) of argon by gas proportional counting together with ^{39}Ar (see below). For comparison, atmospheric ^{37}Ar activity is 50 μBq per liter (STP) of argon (Loosli and Oeschger, 1969); activities as high as 0.2 Bq per liter (STP) of argon have been measured in groundwater in the Stripa granite, in Sweden (Loosli *et al.*, 1989). When used as a natural tracer of groundwater, ^{37}Ar is used in combination with a number of other tracers (e.g., ^{39}Ar, ^{85}Kr, see below) in order to provide more reliable estimates of groundwater residence times (Loosli *et al.*, 1989, 2000).

5.15.4.2.2 Sulfur-35

Sulfur-35 is formed from the spallation of argon in the atmosphere. It has a half-life of 87 d (Lal and Peters, 1967). After oxidation to sulfate, it is deposited on the land surface. Because of its relatively long residence time in the upper atmosphere and its short half-life, specific activities of ^{35}S are low at the time of deposition. Due to the anionic form of ^{35}SO$_4^{2-}$ it is relatively conservative in soil water and groundwater and can be used to estimate residence times in shallow, rapidly circulating systems. It has seen limited application in hydrology because of its low specific activity in groundwater, requiring large sample volumes, and the concomitant requirement for low background sulfate sample waters (Michel, 2000).

Most applications have mainly been to low-sulfate environments with rapid turnover, especially to studying the behavior of sulfate during snowmelt (Michel *et al.*, 2000). It has proved useful in estimating storage time of groundwater contributing to stream flow, showing that even in apparently low-storage alpine systems, a preponderant fraction of the flow can be derived from groundwater with residence times of several years or more (Suecker *et al.*, 1999).

5.15.4.2.3 Krypton-85

Anthropogenic ^{85}Kr (half-life 10.76 yr) was first released to the atmosphere in considerable amounts in the 1950s and 1960s by atmospheric nuclear-weapons testing. Today, it is continuously produced through fission of uranium and plutonium and released to the atmosphere, mainly during fuel rod reprocessing in nuclear power plants. As shown in Figure 2(d), this artificially produced ^{85}Kr has completely overwhelmed natural cosmogenic production of ^{85}Kr, and has created a steady increase in the activity of this isotope in the atmosphere, particularly in the northern hemisphere (Weiss *et al.*, 1992). The southern hemisphere, without significant sources of this isotope, exhibits an average ^{85}Kr activity 20% lower than the northern hemisphere. Such variations need to be considered when interpreting ^{85}Kr activities in groundwater. Because ^{85}Kr activities in groundwater are very low (1 L of water in equilibrium with the atmosphere has a ^{85}Kr activity of only 8.6 mBq), analysis of this isotope is relatively difficult (Fairbank *et al.*, 1998; Rozanski and Florkowski, 1979; Smethie and Mathieu, 1986; Thonnard *et al.*, 1997). Generally, ^{85}Kr is employed as a natural tracer in combination with other isotopes and, particularly, with ^{3}H, which has a similar half-life. Both tracers are used to estimate the age of rather young ($\leq$40 yr) waters. When dispersion is small in a particular groundwater system, the age of groundwater can be estimated directly from ^{85}Kr activities (e.g., Ekwurzel *et al.*, 1994; Smethie *et al.*, 1992).

5.15.4.2.4 Tritium

Tritium is produced in the atmosphere by cosmic-ray spallation of nitrogen (Lal and Peters, 1967). Tritium decays to ^{3}He with a half-life of 4,500 $\pm$ 8 d (equivalent to 12.32 $\pm$ 0.02 yr) (Lucas and Unterweger, 2000). Atmospherically produced tritium reacts rapidly to form tritiated water: ^{3}HHO. Tritiated water is precipitated out of the atmosphere together with ordinary water. The tritium concentration in natural water is commonly expressed in Tritium Units (TU), where 1 TU is equal to one molecule of ^{3}HHO per 10^{18} molecules of H_2O. This is equivalent to a specific activity of 0.1181 Bq (kg water)$^{-1}$.

Tritium is commonly measured by two methods: β-counting or ^{3}He ingrowth. Natural levels of tritium are only marginally measurable by direct β-counting methods, but fortunately tritium is highly enriched over ordinary H in the H_2 produced by the electrolysis of water. β-Detection, either by gas proportional counting or liquid scintillation counting, can then easily measure natural tritium (Taylor, 1981). A newer approach, with generally higher sensitivity, is to degass the water, then allow ^{3}He to accumulate for up to a year, and measure the ^{3}He using mass spectrometry (Clarke *et al.*, 1976). The excellent properties of ^{3}HHO as a tracer of water and the relative ease of measurement have encouraged a very wide spectrum of applications of tritium to waters with a residence time of less than 50 yr.

The average global production of tritium is $\sim$2,500 atom m^{-2} s^{-1} (Solomon and Cook, 2000). The deposition rate of the tritium varies with latitude, but it is also mixed with the bulk of precipitation originating from the ocean (which has a very low tritium content), and thus the average tritium content of precipitation tends to vary inversely with annual precipitation. Natural tritium in precipitation varies from $\sim$1 TU in oceanic high-precipitation regions to as high as 10 TU in arid inland areas.

Since the 1950s natural tritium deposition has been swamped by the pulse of tritium released by atmospheric nuclear-weapons testing, followed by other anthropogenic releases (Gat, 1980). Significant tritium was released by the fission and thermonuclear devices tested in the mid-to-late 1950s, but it was the stratospheric thermonuclear-weapons testing of the early 1960s that produced the peak pulse, as large as 5,000 TU in the mid-latitude northern hemisphere (Figure 2(a)). Because most of the testing was in the northern hemisphere, southern hemisphere fallout was much less. This pulse has decayed in a quasi-exponential fashion since that time and is now beginning to approach natural levels. It has provided an invaluable transient tracer for shallow hydrological systems.

Application of tritium to hydrologic problems was first proposed by Libby (Bergmann and Libby, 1957; Libby, 1953). Early applications (Allison and Holmes, 1973; Carlston *et al.*, 1960) focused on tracing the bomb-tritium pulse through

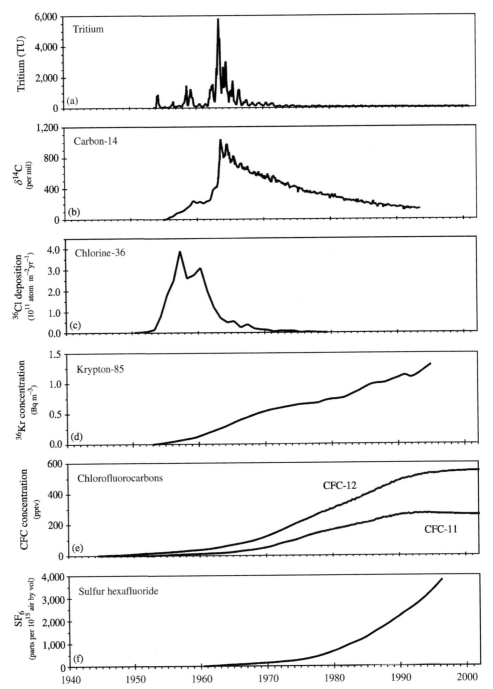

Figure 2 Global environmental histories of hydrologic-cycle tracers that have been strongly affected by human activities: (a) tritium (data from measurements at Ottawa, Canada, by the International Atomic Energy Agency; http:// isohis.iaea.org/); (b) carbon-14 (data for 1955–1962 from Kalin (2000), data for 1963–1991 from Nydal *et al.* (1996)) http://cdiac.esd.ornl.gov/epubs/ndp/ndp057/ndp057.htm); (c) chlorine-36 (data from Synal *et al.* (1990); (d) krypton-85 (data from Loosli *et al.*, 2000); (e) chlorofluorocarbons (data for 1944–1978 from Plummer and Busenberg (2000); data for 1978–2003 from Cunnold *et al.* (1997); http://cdiac.esd.ornl.gov/ftp/ale_gage_Agage/); and (f) sulfur hexafluoride (data from Plummer and Busenberg, 2000).

aquifers. This was soon extended to estimating recharge by measuring tritium profiles through the vadose zone (Schmalz and Polzer, 1969; Vogel *et al.*, 1974). Measurement of tritium profiles through the English Chalk vadose zone proved an important key to understanding matrix diffusion of solutes in fractured media (Foster, 1975). During the 1980s the bomb-tritium pulse in groundwater began to lose its definition. The combined effects of decay and dispersion tended to reduce bomb-peak levels to values similar to that of precipitation at that time. Fortunately, the limitations imposed by this loss of the tracer pulse were countered by the introduction of the ^{3}H/^{3}He method (Tolstikhin and Kamensky, 1968). By measuring first the tritiogenic ^{3}He and then the ^{3}H content by ^{3}He ingrowth, both the initial ^{3}H content and the time of decay can be calculated. The power of this approach for reconstructing residence times has been elegantly demonstrated by Schlosser *et al.* (1988) and Solomon *et al.* (1992, 1993). The ^{3}H/^{3}He method has since become a standard tool in the investigation of the dynamics of shallow aquifers (Beyerle *et al.*, 1999; Ekwurzel *et al.*, 1994; Puckett *et al.*, 2002). However, ^{3}He is also produced by fission of uranium and nucleogenic processes in the subsurface, and corrections for these sources must sometimes be employed (Andrews and Kay, 1982a,b).

5.15.4.2.5 Silicon-32

Silicon-32 is produced in the atmosphere by cosmic-ray spallation of ^{40}Ar (Lal *et al.*, 1960). It decays to ^{32}P with a half-life of 140 ± 6 yr (Morgenstern *et al.*, 1996), although there is some uncertainty regarding the exact value (Nijampurkar *et al.*, 1998). It is oxidized and incorporated into precipitation as silicic acid. The global deposition rate is ~2 atom m^{-2} s^{-1} (Kharkar *et al.*, 1966; Morgenstern *et al.*, 1996; Nijampurkar *et al.*, 1998). This produces activities in natural water in the range of 2–20 mBq m^{-3} (Morgenstern, 2000). Silicon-32 does not behave in a conservative fashion after it begins to move through the subsurface. It reacts with silicate minerals, presumably by exchange, and ^{32}Si concentrations are markedly reduced during transport through the vadose zone (Fröhlich *et al.*, 1987; Morgenstern *et al.*, 1995).

Silicon-32 is most sensitively measured using liquid scintillation counting. Silicon must be extracted from ~1 m^3 of water. The daughter ^{32}P is allowed to grow into equilibrium over several months, then is milked with stable phosphorus for scintillation counting (Morgenstern, 2000). Very low counting backgrounds are necessary.

Silicon-32 has many attractive aspects for age tracing of groundwater. Its half-life makes it very useful for filling the gap between ^{3}H (12.3 yr) and ^{14}C (5,730 yr). It has a very small subsurface production (Florkowski *et al.*, 1988). However, the disadvantage of unpredictable loss to the solid phase and analytical difficulties have limited the number of its applications (Lal *et al.*, 1970; Nijampurkar *et al.*, 1966). Morgenstern (2000) has proposed that most of the nonconservative

behavior is produced in the vadose zone and that ^{32}Si may be useful for residence-time determinations in aquifers if the initial activity in the recharge area can be determined. However, until this can be demonstrated, the applications of the isotope will remain limited.

5.15.4.2.6 Argon-39

In contrast to ^{37}Ar, ^{39}Ar in groundwater has its origin in the atmosphere where it is produced by cosmic rays, and enters the water cycle as a fraction of the dissolved argon during groundwater recharge. The activity of ^{39}Ar in groundwater is extremely small, a fact that complicates its analysis (Loosli *et al.*, 2000). Modern atmospheric ^{39}Ar activity is 16.7 mBq m^{-3} of air (Loosli, 1983). Two thousand liters of water in equilibrium with the atmosphere at 20 °C have an ^{39}Ar activity of only 1.2 mBq. When subsurface production of ^{39}Ar (resulting from the reaction ^{39}K(n,p)^{39}Ar) can be neglected, this isotope with a half-life of 269 yr has been successfully used in combination with other tracers, in particular with ^{14}C to estimate groundwater ages in sedimentary basins (e.g., Andrews *et al.*, 1984; Purtschert, 1997). However, this procedure can be complicated in crystalline terrains with high uranium and thorium contents and hence a non-negligible subsurface production of ^{39}Ar (Andrews *et al.*, 1989; Lehmann *et al.*, 1993).

5.15.4.2.7 Carbon-14

Carbon-14 is produced in the atmosphere by a low-energy cosmic-ray neutron reaction with nitrogen. It decays back to ^{14}N with a half-life of 5,730 yr. The production rate is ~2 × 10^4 atom m^{-2} s^{-1}, the highest of all cosmogenic radionuclides. The rate is so high, because nitrogen is the most abundant element in the atmosphere and also has a very large thermal neutron absorption cross-section. Radiocarbon activity in the atmosphere is the result of complicated exchanges between terrestrial reservoirs (primarily vegetation), the ocean, and the atmosphere. The combination of varying cosmic-ray fluxes (due to both varying solar and terrestrial magnetic field modulations) and varying exchange between reservoirs has caused the atmospheric activity of radiocarbon to fluctuate by approximately a factor of 2 over the past 3 × 10^4 yr (Bard, 1998). The current specific activity is 0.23 Bq (g C)$^{-1}$, rendering radiocarbon easily measurable by gas proportional or liquid scintillation counting, or by accelerator mass spectrometry (AMS). Radiocarbon measurements are usually reported as "percent modern carbon," indicating the sample specific activity as a percentage of the 0.23 Bq $(g C)^{-1}$ modern atmospheric specific activity. In addition to the natural cosmogenic production of ^{14}C, the isotope was also released in large amounts by atmospheric nuclear-weapons testing in the 1950s and 1960s (Figure 2(b)).

Natural radiocarbon was first detected by Libby in the mid-1940s (Arnold and Libby, 1949; Libby, 1946), but the first applications to subsurface hydrology were not attempted for another decade (Hanshaw et al., 1965; Münnich, 1957; Pearson, 1966). These early investigators discovered that radiocarbon shows clear and systematic decreases with flow distance that can be attributed to radiodecay, but also exhibits the effects of carbonate mineral dissolution and precipitation reactions. Quantification of residence time is not possible without correction for additions of nonatmospheric carbon. Numerous approaches to this problem were attempted, including simple empirical measurements (Vogel, 1970), simplified chemical equilibrium mass balance (Tamers, 1975), mass balance using $\delta^{13}C$ as an analogue for ^{14}C (Ingerson and Pearson, 1964), and combined chemical/$\delta^{13}C$ mass balance (Fontes and Garnier, 1979; Mook, 1980).

The correction methods cited above were mainly intended to account for carbonate reactions in the vadose zone during recharge, although in practice they have often been applied to reactions along the groundwater flow path as well. It is now the general consensus that the preferred approach to treating the continuing reactions of dissolved inorganic carbon during flow is geochemical mass transfer/equilibrium models (Kalin, 2000; Zhu and Murphy, 2000). The most commonly employed model is NETPATH (see Chapters 5.02 and 5.14; Plummer et al., 1991). This uses a backward-calculated solute mass balance, constrained by simple equilibrium considerations, to calculate mass transfers of solutes from phase to phase between two sample points.

The vast majority of groundwater radiocarbon investigations have extracted the dissolved inorganic carbon from the water for measurement. However, dissolved organic carbon (DOC) has been sampled in a limited number of studies (Murphy et al., 1989; Purdy et al., 1992; Tullborg and Gustafsson, 1999). In principle, sampling DOC could avoid the complex geochemistry of inorganic carbon. In reality, DOC consists of a large number of organic species from various sources (with different original ^{14}C activities), of varying chemical stability, and of varying reactivity with the solid phase. Routine use of DOC for groundwater age tracing may one day show advantages over inorganic carbon, but this will require considerable work separating and identifying the appropriate component of the DOC for this application.

Radiocarbon is an indispensable tool in the age tracing of groundwater. Both its ease of use and its half-life make it the method of choice for many investigations. However, results must always be approached with caution. The reliability of results depends strongly on the complexity of the geochemistry. In some cases groundwater may show little evidence of chemical evolution after recharge and measured radiocarbon activities can be accepted at face value. In other cases, the isotopic composition of carbon may be strongly altered by numerous surface reactions that are difficult to quantify, and interpretations may be speculative, at best. Interpretations must be evaluated on a case-by-case basis.

5.15.4.2.8 Krypton-81

With a half-life of 2.29×10^5 yr, ^{81}Kr, still in its "infancy" as a natural tracer, has had the potential to be an excellent tool for dating old groundwater ever since it was first detected in the atmosphere (Loosli and Oeschger, 1969). In addition to its constant atmospheric concentration, anthropogenic and subsurface production of ^{81}Kr have been shown to be negligible (e.g., Collon et al., 2000; Lehmann et al., 1993). Nearly all ^{81}Kr in groundwater results from the interaction of cosmic rays with nuclei in the Earth's atmosphere, in particular, through neutron capture by ^{80}Kr and spallation of heavier krypton isotopes. However, its activity (0.11 nBq in water in equilibrium with the atmosphere) is lower than that of any other tracer that has been used. The analytical challenge to measure this isotope is, therefore, extreme. The first measurements of ^{81}Kr in groundwater were reported in the Milk River aquifer in Canada (Lehmann et al., 1991; Thonnard et al., 1997), but an elaborate multistep enrichment procedure made it difficult to quantify the results, though an estimated groundwater age of 1.4×10^5 yr was obtained. A new measurement technique based on accelerator mass spectrometry (AMS) using positive krypton ions coupled to a cyclotron was reported by Collon et al. (2000). In this study, ^{81}Kr was measured in groundwater in the Great Artesian Basin (GAB) in Australia and, for the first time, definite determinations of water residence times were made based on the atmospheric $^{81}Kr/Kr$ ratio. Krypton dissolved in surface water in contact with the atmosphere has an atmospheric $^{81}Kr/Kr$ ratio of $(5.20 \pm 0.4) \times 10^{-13}$. Observed reductions of isotope ratios in groundwater were interpreted as being due to radioactive decay since recharge (Collon et al., 2000; Lehmann et al., 2002b).

5.15.4.2.9 Chlorine-36

Chlorine-36 is produced in the atmosphere by cosmic-ray spallation of ^{40}Ar (Lal and Peters, 1967). It decays to ^{36}Ar with a half-life of $301,000 \pm 4,000$ yr (Bentley *et al.*, 1986a). The globally averaged production rate is between 20 atom m^{-2} s^{-1} and 30 atom m^{-2} s^{-1} (Phillips, 2000). Meteoric ^{36}Cl then mixes with stable oceanic chlorine in the atmosphere, diluting the ^{36}Cl greatly near coastlines, but less inland. The resultant specific activities of chlorine in atmospheric deposition range from less than 20 μBq (g Cl)$^{-1}$ to more than 1,000 μBq (g Cl)$^{-1}$. These specific activities are at the lower limits of the sensitivity of any form of decay counting, and require large masses of chlorine that may be difficult to obtain from dilute waters. Routine application of ^{36}Cl to age tracing of groundwater is possible only because of the high sensitivity and much smaller sample size permitted by AMS (Elmore *et al.*, 1979). Because an isotopic ratio is measured by mass spectrometry, the atomic ratio, atoms ^{36}Cl per 10^{15} atoms Cl, is normally used in ^{36}Cl studies instead of specific activity.

Application of ^{36}Cl to earth-science problems was attempted as early as the mid-1950s (Davis and Schaeffer, 1955), but little use was made of the method until the advent of AMS in 1979. Since the analytical barrier was overcome, there have been regular applications to long-residence-time aquifers around the world (mainly large sedimentary basins). These include systems in Australia (Bentley *et al.*, 1986b; Cresswell *et al.*, 1999; Torgersen *et al.*, 1991), North America (Davis *et al.*, 2003; Nolte *et al.*, 1991; Phillips *et al.*, 1986; Purdy *et al.*, 1996), Europe (Pearson *et al.*, 1991; Zuber *et al.*, 2000), Asia (Balderer and Synal, 1996; Cresswell *et al.*, 2001), and Africa (Kaufman *et al.*, 1990).

The major complications in using ^{36}Cl to estimate the residence time of groundwater are introduction of subsurface chloride and subsurface production of ^{36}Cl. High chloride concentrations arising from connate waters, evaporite dissolution, or possibly rock–water interaction are common in sedimentary basins and deep crystalline rocks. These may either diffuse upward or be advected into aquifers by cross-formational flow. In general, this subsurface-source chloride is not ^{36}Cl free, because ^{36}Cl is produced at low levels (compared to meteoric chloride) in the subsurface by absorption by ^{35}Cl of thermal neutrons produced by uranium and thorium series decay and uranium fission (Lehmann *et al.*, 1993; Phillips, 2000). The effects of radiodecay and of mixing with low ^{36}Cl ratio subsurface chloride must be separated in order to evaluate residence times correctly. Park *et al.* (2002) have recently evaluated in detail the effects of various mixing scenarios. Although simple mixing models may often provide useful approximations of residence time (Bentley *et al.*, 1986a; Phillips, 2000), the best approach, as for all age tracing methods, is comprehensive simulation of flow and transport for the entire groundwater system (Bethke and Johnson, 2002b; Castro and Goblet, 2003).

In addition to the natural cosmogenic production of ^{36}Cl in the atmosphere, a very large pulse of ^{36}Cl was also introduced by the testing of thermonuclear weapons between 1954 and 1958 (Bentley *et al.*, 1982; Phillips, 2000; Zerle *et al.*, 1997). This pulse had much less of a tail than the bomb radiocarbon and tritium pulses, because the atmospheric residence time of chloride is shorter (Figure 2(c)). Chlorine-36 can be applied to groundwater studies in a fashion analogous to tritium, but has proved most useful in tracing water movement through the vadose zone, because, unlike tritium and radiocarbon, it is not volatile and thus does not disperse in the vapor phase (Cook *et al.*, 1994; Phillips *et al.*, 1988; Scanlon, 1992).

5.15.4.2.10 Iodine-129

Iodine-129 has major sources in both surface and subsurface environments. It is produced in the atmosphere by cosmic-ray spallation of xenon and in the subsurface by the spontaneous fission of uranium. Iodine-129 of subsurface origin is released to the surface environment through volcanic emissions, groundwater discharge, and other fluxes. Due to its very long half-life, 15.7 Ma, these sources are well mixed in the oceans and surface environment, producing a specific activity of ~5 μBq (g I)$^{-1}$ (equivalent to a ^{129}I/I ratio of ~10^{-12}). Due to its low activity, the preferred detection method of ^{129}I is AMS (Elmore *et al.*, 1980).

The very long half-life of ^{129}I raises the possibility of dating groundwater on timescales greater than a million years (Fabryka-Martin *et al.*, 1985). For a few subsurface environments that approximate closed systems, including brines incorporated in salt domes (Fabryka-Martin *et al.*, 1985), and connate brine reservoirs (Fehn *et al.*, 1992; Moran *et al.*, 1995), and iodine in oil (Liu *et al.*, 1997), this has proved successful. However, due to the complex subsurface geochemistry of iodine, subsurface fissionogenic production which can be difficult to quantify, and the release of iodine from subsurface organic material, determination of the residence time of water in dynamic groundwater systems has generally not proved practicable (Fabryka-Martin, 2000; Fabryka-Martin *et al.*, 1991).

Iodine-129 has numerous anthropogenic sources. It was globally distributed by the atmospheric nuclear-weapons testing of the 1950s and 1960s, and has continued to be released in large amounts by nuclear technology, especially nuclear fuel reprocessing (Wagner *et al.*, 1996). As a result, environmental levels of the ^{129}I/I ratio have increased to values in the range 10^{-10}–10^{-7}. This anthropogenic pulse can be used in a fashion similar to bomb tritium. It has proved to be especially useful in identifying releases from nuclear reprocessing plants and other nuclear facilities (Moran *et al.*, 2002; Oktay *et al.*, 2000; Rao and Fehn, 1999).

5.15.4.3 Stable, Transient Tracers

5.15.4.3.1 Chlorofluorocarbons and sulfur hexafluoride

Several industrial chemicals have been created that are highly volatile, resistant to degradation, act conservatively in groundwater, are unusual or absent in the natural environment, and can be detected at very low concentrations. If these compounds show a relatively regular increase in atmospheric concentration, then their concentration in shallow groundwater systems can be correlated with the time of recharge of the water, and residence times can be determined.

One such class of compounds is the chlorofluorocarbons (CFCs), halogenated alkanes that have been used in refrigeration and other industrial applications since the 1930s (Plummer and Busenberg, 2000). The atmospheric lifetimes of the common CFC compounds range from 50 yr to 100 yr. Their concentrations increased in a quasi-exponential fashion from the 1950s through the late 1980s (Figure 2(e)). Unfortunately, CFCs catalyze reactions in the stratosphere that deplete atmospheric ozone. This resulted in international agreements to limit global CFC production drastically. In response the rate of atmospheric CFC increase has leveled off and has started to decrease. This complex concentration history has resulted in a degree of ambiguity in residence-time estimates for water recharged since ~1990.

The advantages of CFC dating are a lack of sensitivity to dispersion and mixing, due to the gradual and monotonic atmospheric concentration history (at least through the 1990s), and the virtual year-to-year dating sensitivity. The major complications of CFC dating are the time lag involved in the diffusion of the atmospheric signal through vadose zones thicker than ~10 m, minor-to-moderate sorption of some CFC species, and microbial degradation of some CFCs under strongly reducing conditions (Plummer and Busenberg, 2000).

Application of CFCs for groundwater studies was first explored in the early 1970s (Thompson, 1976; Thompson *et al.*, 1974), but was not widely employed until important proof-of-concept papers were published in the early 1990s (Böhlke *et al.*, 2002; Busenberg and Plummer, 1991; Dunkle *et al.*, 1993; Ekwurzel *et al.*, 1994). Since that time the method has been widely used in shallow groundwater studies. It has found particular application in studies tracing the source and fate of agricultural contaminants (Böhlke *et al.*, 2002) and surface–groundwater interaction (Beyerle *et al.*, 1999).

Sulfur hexafluoride (SF$_6$) is a nearly inert gas than has been widely used as a gas-phase electrical insulator. It has a very long atmospheric lifetime, and is detectable to very low levels by gas chromatography using the electron capture detector. These characteristics render it useful for estimating groundwater residence times in a fashion similar to CFCs, for the period from ~1970 to the early 2000s (Busenberg and Plummer, 2000). The major advantage over CFCs is that the atmospheric concentration of SF$_6$ is continuing to increase monotonically (Figure 2(f)). One potential disadvantage is that relatively high levels of apparently natural background SF$_6$ have been detected in areas of volcanic and igneous rock (Busenberg and Plummer, 2000).

5.15.4.3.2 Atmospheric noble gases and stable isotopes

The solubility of ordinary atmospheric noble gases (neon, argon, krypton, and xenon) in water is temperature dependent (Benson, 1973). The stable isotope composition of precipitation (δ^2H, δ^{18}O) also depends on temperature. If variations in these constituents can be related to a known history of temperature variation, then groundwater residence times can be estimated (Stute and Schlosser, 1993).

This approach to residence-time estimation has been most commonly applied on two very different timescales. The first is in utilizing the annual temperature cycle. The second is the global glacial–interglacial transition at ~15 ka. The approach of taking samples along a groundwater flow path and matching it to the seasonal recharge temperature signal generally works best when applied to aquifer recharge from a surface-water body (Beyerle *et al.*, 1999; Sugisaki, 1961), because under conditions of general vertical recharge the vadose zone damps much of the annual signal, and vertical mixing during water-table fluctuations smoothes the annual signal.

The noble-gas concentration method has been applied to investigating the temperature history on the glacial–interglacial timescale around the world (Stute and Schlosser, 1993). Although it is most commonly used in conjunction with independent dating of the groundwater, the magnitude and timing of the noble-gas temperature signal are well enough established, so that, when other dating means are not available or successful, it can be used to define the position of water recharged during the climate transition (Blavoux *et al.*, 1993; Clark *et al.*, 1997; Zuber *et al.*, 2000).

5.15.4.3.3 Nonatmospheric noble gases

The following sections comprise a brief description of noble-gas isotopes (^{3}He, ^{4}He, ^{21}Ne, and ^{40}Ar), where measured concentrations in groundwaters were found to be in excess of solubility equilibrium with the atmosphere (air-saturated water (ASW)). Such excesses allow the use of these isotopes as natural tracers of groundwater flow (as opposed to those such as ^{20}Ne, ^{22}Ne, ^{36}Ar, and ^{38}Ar that exhibit an atmospheric contribution only).

(i) Helium isotopes

Helium-3. The concentration of ^{3}He in groundwater frequently exceeds that of ASW values (cf. Section 5.15.4.3.2). These excesses can be up to several orders of magnitude greater than the ASW concentration in old groundwater (e.g., Castro *et al.*, 1998a,b). They can derive from a number of ^{3}He sources. Typically, with the exception of recharge areas and superficial aquifers with fast-flowing (young) waters, the main source of ^{3}He over time is nucleogenic. Nucleogenic production of ^{3}He is the result of a series of reactions (Andrews and Kay, 1982a,b; Lal, 1987; Mamyrin and Tolstikhin, 1984; Morrison and Pine, 1955): (a) spontaneous and neutron-induced fission of uranium isotopes and reaction of α-particles with the nuclei of light elements give rise to a subsurface neutron flux; (b) some of these neutrons reach epithermal energies and react with nuclei of the light isotope of lithium (^{6}Li) to produce α-particles and tritium:

$$^6\text{Li}(3p,3n,3e^-)+n\rightarrow\alpha(2p,2n)+{}^3\text{H}(1p,2n,1e^-)$$

(c) ^{3}H decays (half-life of 12.32 yr) by the emission of β^- (decay of a neutron into a proton and electron) to the stable isotope ^{3}He:

$$^3\text{H}\xrightarrow{\beta^-}{}^3\text{He}\ (2p,1n,2e^-)$$

Reactions between the light isotopes of lithium and epithermal neutrons yield almost all terrestrial nucleogenic ^{3}He, including that found in excess of ASW in most groundwater reservoirs. Nucleogenic ^{3}He can also be produced through ^{7}Li reactions ($^7\text{Li}(\alpha,{}^3\text{H})^8\text{Be}$; $^3\text{H}(\beta^-){}^3\text{He}$; $^7\text{Li}(\gamma,{}^3\text{H})^4\text{He}$, $^3\text{H}(\beta^-){}^3\text{He}$), but the amount produced is negligible compared to that produced through ^{6}Li

(Gerling *et al.*, 1971; Kunz and Schintlmeister, 1965; Mamyrin and Tolstikhin, 1984). Excesses of nucleogenic ^{3}He in groundwater can result from *in situ* production (i.e., produced in the reservoir rock of the aquifer itself) and/or can be produced in deeper layers or deeper crust (external source). In the latter case, ^{3}He must be transported to the upper aquifers either through advection, dispersion, and/or diffusion. The concentration of ^{3}He in groundwater will typically increase with distance from the recharge area and/or over time. In tectonically active areas a mantle component of ^{3}He may also be present (Oxburgh *et al.*, 1986), reaching groundwater reservoirs either directly through igneous intrusions or through water transport processes such as advection, dispersion, and diffusion.

In addition to the nucleogenic and mantle component, tritiogenic ^{3}He resulting from β-decay of natural (produced mainly in the upper atmosphere through the bombardment of nitrogen by the flux of neutrons in cosmic radiation) and bomb ^{3}H (as a result of thermonuclear testing in the 1950s and 1960s) (see Section 5.15.4.2.4) can be present. This bomb-tritiogenic ^{3}He component, when significant, is present in very young (tens of years), fast-flowing waters.

Helium-4. This isotope is also commonly found in excess to ASW concentrations in groundwater; excesses are typically higher in older than in younger groundwaters. Unlike ^{3}He excesses, ^{4}He result directly from radioactive decay of ^{235}U, ^{238}U, and ^{232}Th. The radioactive decay of uranium and thorium series elements yields α-particles (2p, 2n) that rapidly acquire two electrons, thus turning into stable ^{4}He (2p, 2n, 2e$^-$). Because production of ^{4}He is neither dependent on the nuclei of light elements nor on neutron flux, the radiogenic ^{4}He production rate in the crust tends to be greater and more homogeneous than the nucleogenic production of ^{3}He (Martel *et al.*, 1990). As a result, the observed excesses of radiogenic ^{4}He in groundwater tend to be more uniform than those of nucleogenic ^{3}He. Assuming a homogeneous element distribution in the rocks, the radiogenic production rate of ^{4}He is given by

$$P(^4\text{He}) = 5.39 \times 10^{-18}[\text{U}] + 1.28$$
$$\times 10^{-18}[\text{Th}]\ \text{mol}\ \text{g}_{\text{rock}}^{-1}\ \text{yr}^{-1}$$

where [U] and [Th] are the concentrations of U and Th in the rock in ppm (Steiger and Jäger, 1977). Radiogenic ^{4}He can be produced *in situ* or have an origin external to the aquifer.

^{3}He/^{4}He ratio. The measurement of the ^{3}He/^{4}He ratio (R) in groundwaters is particularly important, because it reflects the relative importance of crustal (radiogenic/nucleogenic) and mantle helium sources. Typical crustal rock ^{3}He/^{4}He production ratios have been estimated

to be 1×10^{-8} (Mamyrin and Tolstikhin, 1984). Other authors (e.g., Kennedy *et al.*, 1984) consider a wider range of typical crustal production ratios, with $0.01 < R/R_a < 0.1$ (atmospheric ${}^3\mathrm{He}/{}^4\mathrm{He}$ ratio $R_a = 1.384 \times 10^{-6}$; Clarke *et al.*, 1976). Much higher ${}^3\mathrm{He}/{}^4\mathrm{He}$ values have been observed in the mantle; the most common is 1.2×10^{-5} (Craig and Lupton, 1981). However, at specific sites such as Hawaii, mantle ${}^3\mathrm{He}/{}^4\mathrm{He}$ ratios can be up to a factor of 5 higher than this common mid-ocean ridge basalt (MORB) value (Allègre *et al.*, 1983; Craig and Lupton, 1976; Kyser and Rison, 1982; Rison and Craig, 1983).

In order to use helium isotopes as natural tracers of groundwater flow and, in particular, in order to estimate groundwater ages, it is necessary to separate different nonatmospheric helium components for both isotopes. The major components are ${}^3\mathrm{He}$ and ${}^4\mathrm{He}$ from atmospheric equilibrium and from incorporation of "excess air" (Heaton and Vogel, 1981; Weyhenmeyer *et al.*, 2000), ${}^3\mathrm{He}$ and ${}^4\mathrm{He}$ of crustal origin (i.e., nucleogenic and radiogenic), ${}^3\mathrm{He}$ of mantle origin, and tritiogenic ${}^3\mathrm{He}$. Methods for separating these components have been described by Stute *et al.* (1992b), Weise (1986), and Weise and Moser (1987).

(ii) Neon-21

Neon-21 excesses relative to ASW concentrations have also been observed in old groundwaters. However, the excesses are not comparable to those observed for helium; they are only in the order of a few to a few tens of percent. This excess is attributed to the nucleogenic production of ${}^{21}\mathrm{Ne}$ in crustal rocks. The reaction ${}^{18}\mathrm{O}(\alpha, \mathrm{n}){}^{21}\mathrm{Ne}$ accounts for 97% of the total nucleogenic ${}^{21}\mathrm{Ne}$. Magnesium accounts for the remaining production of nucleogenic ${}^{21}\mathrm{Ne}$ via the reaction ${}^{24}\mathrm{Mg}(\mathrm{n}, \alpha){}^{21}\mathrm{Ne}$ (Wetherill, 1954). Although only one study has reported the presence of nucleogenic ${}^{21}\mathrm{Ne}$ in groundwaters (Castro *et al.*, 1998b), it is possible that, as more measurements of ${}^{21}\mathrm{Ne}$ in old deep groundwaters become available, the presence of nucleogenic ${}^{21}\mathrm{Ne}$ will prove to be more common in groundwater reservoirs than is known at present. Numerous studies have reported considerable excesses of nucleogenic ${}^{21}\mathrm{Ne}$ in oil and gas fields (Ballentine and O'Nions, 1991; Ballentine *et al.*, 1996; Hiyagon and Kennedy, 1992; Kennedy *et al.*, 1985). There is a general consensus that noble gases found in these systems are transported to oil and gas reservoirs by groundwater. If so, similar ${}^{21}\mathrm{Ne}$ excesses should be expected in neighboring groundwaters in amounts close to those in gas and oil fields.

The presence of nucleogenic ${}^{21}\mathrm{Ne}$ is generally identified through comparison with the atmospheric ${}^{21}\mathrm{Ne}/{}^{22}\mathrm{Ne}$ ratio of 0.0290. A typical crustal production ${}^{21}\mathrm{Ne}/{}^{22}\mathrm{Ne}$ ratio is 0.47 ± 0.02 (Shukolyukov *et al.*, 1973; Kennedy *et al.*, 1990). The total present-day production rate of ${}^{21}\mathrm{Ne}$ has been calculated for typical crustal rock by Bottomley *et al.* (1984) to be

$$P({}^{21}\mathrm{Ne}) = (3.572 \times 10^{-25})[\mathrm{U}] + 1.767 \times 10^{-25}[\mathrm{Th}]$$

where [U] and [Th] are the concentrations of U and Th in the rock in ppm. The values for isotopic composition and decay constants are taken from Steiger and Jäger (1977).

A ${}^{21}\mathrm{Ne}$ mantle component has not been identified in groundwaters. Its presence was reported for the first time in 1991 in hydrocarbon gas reservoirs within the Vienna Basin (Ballentine and O'Nions, 1991).

(iii) Argon-40

Excess ${}^{40}\mathrm{Ar}$, up to 35% of ASW, has also been found in deep, old groundwaters in sedimentary basins (Castro *et al.*, 1998b; Torgersen *et al.*, 1989). Radiogenic production of ${}^{40}\mathrm{Ar}$ through radioactive decay of ${}^{40}\mathrm{K}$ by electron capture is responsible for these excesses. Eleven percent of ${}^{40}\mathrm{K}$ decay produces ${}^{40}\mathrm{Ar}$; the remainder produces ${}^{40}\mathrm{Ca}$ by the emission of β^--particles (Faure, 1986). The presence of a radiogenic ${}^{40}\mathrm{Ar}$ component is easily identified and quantified by comparison with the ${}^{40}\mathrm{Ar}/{}^{36}\mathrm{Ar}$ atmospheric ratio (295.5; Ozima and Podosek, 1983). The highest measured value of this ratio due to the presence of radiogenic ${}^{40}\mathrm{Ar}$ in groundwaters is 471.5. At present, no resolvable mantle component has been identified.

Assuming a homogeneous element distribution in the rocks, the radiogenic production rate of ${}^{40}\mathrm{Ar}$ is

$$P({}^{40}\mathrm{Ar}) = 1.73 \times 10^{-22}[\mathrm{K}] \ \mathrm{mol} \ \mathrm{g}_{\mathrm{rock}}^{-1} \ \mathrm{yr}^{-1}$$

where [K] is the concentration of K in rocks in ppm, using current values of isotopic composition and decay constants from Steiger and Jäger (1977). In a manner similar to the helium and neon isotopes, ${}^{40}\mathrm{Ar}$ in groundwater reservoirs can result both from release of ${}^{40}\mathrm{Ar}$ from minerals forming the aquifer and from production and release deeper in the crust followed by upward transport.

5.15.5 LESSONS FROM APPLYING GEOCHEMICAL AGE TRACERS TO SUBSURFACE FLOW AND TRANSPORT

5.15.5.1 Introduction

The most common and traditional tracers used in groundwater studies in the last few decades (e.g., ${}^3\mathrm{H}$, ${}^{14}\mathrm{C}$, and ${}^{36}\mathrm{Cl}$), as well as some relatively new and promising tracers for use in quantitative groundwater dating (e.g., ^{32}Si, ^{81}Kr, and ^{4}He), have been discussed above. The practical applications and information to be gained from these environmental tracers are multiple and differ, depending on the tracers to be used and on the hydrogeological problem to be treated. Typically and depending on the scale of the study area (e.g., regional versus local) as well as on the hydraulic properties of the formation(s) (e.g., high versus low to very low hydraulic conductivity), certain tracers will be better suited than others. Radioisotopes with short half-lives (e.g., ^{85}Kr and ^{3}H) are suitable for dating young/fast-flowing waters ($\leq$40 yr). These radioisotopes are, therefore, ideal for the study of aquifers with high hydraulic conductivities or, alternatively, to the study of recharge areas in regional aquifer systems (Schlosser *et al.*, 1989; Solomon *et al.*, 1993). Tracers with intermediate half-lives (e.g., ^{14}C and ^{36}Cl) are typically used to date submodern ($>$1,000 yr) and old groundwaters up to 10^6 yr. They are suited to the study of aquifers of intermediate hydraulic conductivity, i.e., study at the aquifer scale and certain regional systems (Jacobson *et al.*, 1998; Pearson and White, 1967; Phillips *et al.*, 1986). In contrast, stable tracers such as ^{4}He have no age limitations and can be used within all age ranges; as a result, they are particularly suited to the study of regional groundwater systems where very old waters may be present, as well as the study of groundwater flow dynamics in aquitards where water can be almost immobile (Andrews *et al.*, 1985; Andrews and Lee, 1979; Bethke *et al.*, 1999; Castro *et al.*, 1998a, 2000; Torgersen and Ivey, 1985). Silicon-32 potentially covers an age from a few tens of years to 1,000 yr, a range that is problematic for most tracers.

Two main approaches to derive information from such tracers in a groundwater system are possible: (i) independent and direct use of one particular tracer by analyzing variations in its concentration in space and/or its evolution over time and (ii) a more indirect use through incorporation of those tracers into analytical or numerical transport models. Although both approaches allow estimation of velocity fields, groundwater ages, recharge rates, and identification of mixing of different water bodies, the second approach will generally provide more complete information for a particular groundwater flow systems, as it usually allows for a better representation of the real system (see below). In all cases, the use of a "multitracer" approach (i.e., the use of several tracers simultaneously) provides more definitive answers compared to those based on only one single tracer.

In the following sections, we will briefly describe both approaches, their limitations and pitfalls, as well as information essential to their proper use. These sections will be followed by a discussion of several case studies that represent groundwater flow at different scales, ranging from regional to local, as well as within the vadose zone.

5.15.5.2 Approaches

5.15.5.2.1 *Direct groundwater age estimation*

This method is commonly used with radioisotope tracers that undergo radioactive decay. The tracer concentration will decrease over time following an exponential decay law:

$$\frac{dN}{dt} = -\lambda N$$

where N is the number of atoms of the radioactive tracer, t the time elapsed since radioactive decay started at some initial time t_0, and λ is the decay constant of the tracer. Integrating this equation yields

$$N(t) = N_0\, e^{-\lambda t}$$

where N_0 is the initial number of radioactive tracer atoms at time $t = 0$. This equation can be rearranged and expressed in terms of time elapsed since decay started:

$$t = -\frac{1}{\lambda} \ln \frac{N_t}{N_0}$$

If radioactive decay is the dominant process causing reduction/change in tracer concentration, time t represents the "groundwater age" (Cook and Böhlke, 2000). It corresponds to the time that has elapsed since the water became isolated from the Earth's atmosphere, i.e., the time elapsed since recharge took place (Busenberg and Plummer, 1992; Davis and Bentley, 1982). One must be cautious in interpreting such groundwater age estimates. Such ages may represent an "apparent" rather than the "real" groundwater age (Park *et al.*, 2002). A variety of processes may create these discrepancies. For example, ^{14}C-determined groundwater ages may be influenced by the introduction of inorganic "dead" carbon by the dissolution of calcite or dolomite, resulting in apparent groundwater that are too old. Numerous geochemical models have been developed to account for ^{14}C mass transfer and its impact on groundwater ^{14}C concentration (e.g., Fontes and Garnier, 1979; Mook, 1976; Pearson and White, 1967; Plummer *et al.*, 1991; Tamers, 1967). Assumptions inherent in such geochemical models introduce new sources of uncertainty. Some authors, therefore, prefer to the use the term "model age" (Hinkle, 1996). Groundwater mixing, diffusion, or dispersion may also give rise to erroneous age estimates. One way to identify the effects of groundwater mixing when using ^{14}C ages is to use age constraints from the concentration of CFCs or tritium in tandem with the ^{14}C data.

Complications in the estimation of groundwater ages are not unique to ^{14}C. For example, diffusion can create major errors in estimated ages when applying the $^{3}H/^{3}He$ method in areas where the groundwater velocity is less than ~ 0.01 m yr^{-1}. In such cases ^{3}He will diffuse upwards, and little information regarding travel times will be preserved in the ^{3}He concentrations (Schlosser *et al.*, 1989). Similar complications are inherent in the use of other tracers as well (see, e.g., Phillips, 2000; Sudicky and Frind, 1981; Varni and Carrera, 1998).

Some conservative tracers, such as ^{4}He, have also been used for the direct estimation of groundwater ages (Andrews *et al.*, 1982; Bottomley *et al.*, 1984; Marine, 1979; Torgersen, 1980). The general concept of using ^{4}He as a dating tool is simple, and opposite to that of radioactive tracers. As ^{4}He is produced, minerals release it into groundwater. The longer the groundwater in contact with these minerals, the greater the ^{4}He accumulation in groundwater. If ^{4}He release rates can be estimated (either through direct measurement or through comparison of $^{3}He/^{4}He$ ratios in waters and host rocks; e.g., Castro *et al.*, 2000; Solomon *et al.*, 1996); it should be possible to estimate groundwater residence times as the latter should be proportional to ^{4}He concentrations. However, these methods have systematically led to apparent inconsistencies in water ages calculated using other tracers. In particular, ^{4}He methods typically yield water ages that are much older than ^{14}C ages, sometimes by up to several orders of magnitude (e.g., Andrews and Kay, 1982a; Andrews *et al.*, 1982; Bottomley *et al.*, 1984). Numerous authors have observed that the ^{4}He excesses in groundwater are generally larger than can be supported by the steady-state release of ^{4}He from the reservoir rocks to groundwater. Such ^{4}He excesses have been explained by the presence of an upward flux originating mainly in the deep crust (Bethke *et al.*, 1999; Castro *et al.*, 1998a,b, 2000; Marty *et al.*, 1993; Stute *et al.*, 1992b; Torgersen and Clarke, 1985; Torgersen and Ivey, 1985). Not considering the contribution of ^{4}He from external sources is probably the reason for many of the discrepancies between ages estimated from ^{4}He data and those estimated from other tracers (Castro *et al.*, 1998b). The presence of such an external flux invalidates the use of the ^{4}He method, except in areas where *in situ* production is the only source of the observed excesses, or where its contribution can be determined precisely. Although important information can be gained from such direct age estimation methods, it

is important to view these groundwater ages with a critical eye. Accounting for multiple sources of ^{4}He within a complex hydrological system requires the incorporation of data for the concentration of this tracer into a transport model. Helium-4 can be used as a powerful quantitative dating tool for groundwater (Castro *et al.*, 1998a).

5.15.5.2.2 Modeling techniques: analytical versus numerical

As mentioned above, direct groundwater age estimation methods may not be well suited for all study areas or with all groundwater tracers. Often, groundwater age estimates are more easily and correctly achieved through incorporation data for natural tracers in analytical and/or numerical transport models. This is particularly true when multiple sources contribute to particular tracers, or when an accurate hydrogeological representation is needed for complex groundwater systems.

Traditionally, the use of analytical models in environmental tracer studies has been far more widespread than numerical models. There are a number of reasons for this. Analytical models are easier to use and manipulate than numerical ones; they require less hydrodynamic information and/or field data; and the time required to build a transport model is much less than for a numerical model. Multiple examples of such models can be found in the literature for various tracers: ^{3}H, ^{4}He, ^{14}C, and ^{36}Cl (e.g., Castro *et al.*, 2000; Nolte *et al.*, 1991; Schlosser *et al.*, 1989; Solomon *et al.*, 1996; Stute *et al.*, 1992b; Torgersen and Ivey, 1985). Generally, these models are either applied to a single aquifer in porous or fractured media or to one particular area within the aquifer such as recharge or discharge areas.

The reliability of transport model results depends on how well the model approximates the field situation. They are limited in their ability to incorporate complexities such as heterogeneity in hydraulic properties and three-dimensional flow paths. In order to deal with more realistic situations, or when simulating transport of a tracer at the regional scale (e.g., the scale of an entire sedimentary basin), it is necessary to solve the mathematical model approximately using numerical techniques.

Many different kinds of transport models have been developed, and have been extensively described in the literature (Bethke *et al.*, 1993; Bredehoeft and Pinder, 1973; Goblet, 1999; Konikow and Bredehoeft, 1978; Prickett *et al.*, 1981; Simunek *et al.*, 1999). Although the incorporation of environmental tracers in numerical transport models is not new (see, e.g., Pearson *et al.*, 1983), the effort has been made only recently to simulate the transport of these tracers, in particular stable and conservative tracers such as ^{3}He, ^{4}He, ^{21}Ne, and ^{40}Ar, at a regional scale (e.g., Bethke *et al.*, 1999; Castro *et al.*, 1998a; Zhao *et al.*, 1998). In an attempt to reduce uncertainties present in results obtained independently through groundwater flow models, transport models simulating the distribution of ^{14}C and ^{4}He have been coupled with the latter to obtain recharge rates and groundwater residence times (Castro and Goblet, 2003; Zhu, 2000).

Caution should also be used when estimating hydraulic properties and groundwater ages through transport models. Although analytical and numerical models (in particular) allow for a much better representation of the geometry and the heterogeneity of hydraulic properties of groundwater systems, a number of assumptions and simplifications still have to be made when imposing boundary conditions and hydraulic parameters on the flow and transport domain. These assumptions give rise to sources of uncertainty, resulting from (i) the inability to properly and fully describe the internal properties and boundary conditions of a groundwater system and (ii) the lack of sufficient information for all of the parameters to be included in such models (see, e.g., Konikow and Bredehoeft, 1992; Maloszewski and Zuber, 1993). In this sense, estimated groundwater ages will also carry an error/uncertainty that is difficult to assess. Furthermore, numerical models with a large number of adjustable parameters inherently produce nonunique results. Confidence in model results can be increased through the use/simulation of multiple tracers within the same groundwater flow system and the use of coupled water flow and transport models.

5.15.5.2.3 *Identification of sources and sinks of a particular tracer*

Independent of the method used, proper identification of all sources and sinks of a particular tracer is essential for estimating groundwater residence times. If all important sources (e.g., internal production and external sources) and sinks (e.g., radioactive decay) are not taken into account, erroneous water ages will be obtained. For example, exclusion of an external source of ^{4}He will lead to great discrepancies between ^{4}He and ^{14}C "ages." Commonly, most tracers have multiple sources and sinks (e.g., Loosli *et al.*, 2000; Phillips, 2000; Solomon, 2000; Solomon and Cook, 2000). Some sources have a dominant impact on tracer concentrations; others have only a minor or negligible impact. Moreover, when considering an entire aquifer, one source may be dominant in one part of the system and negligible in others. For example, *in situ* production of ^{3}He and ^{4}He is of

major importance near recharge areas, and of lesser importance in basin centers and in discharge areas (Castro *et al.*, 2000). Heterogeneities of hydraulic properties within a formation may also affect the relative impact of a particular source or sink in different parts of the same system; it is, therefore, necessary to account fully for such variations. The use of numerical methods facilitates this task, as variations may be imposed through boundary conditions of the transport model (Bethke *et al.*, 1999; Castro *et al.*, 1998a).

5.15.5.2.4 *Defining boundary conditions (modeling approach)*

The imposition of proper boundary conditions is another essential component of groundwater age estimation when using a modeling approach. The task is to account for the effects of areas outside the region of interest on the system being modeled (de Marsily, 1986; Domenico and Schwartz, 1998). Three main types of boundary conditions can be considered in a transport model: (i) imposed concentrations within a particular region—such a boundary condition is independent of the groundwater flow regime; (ii) imposed mass flux entering or exiting the system; and (iii) imposed sink or source terms for a tracer within the modeled domain. When conducting simulations in transient state, the imposition of an initial condition is required. Inappropriate model boundary conditions can lead to important differences between actual and simulated flow regimes and thus can lead to erroneous age estimate in real hydrological systems.

5.15.6 TRACERS AT THE REGIONAL SCALE

5.15.6.1 Introduction

This section deals with environmental tracers applied to regional systems, i.e., to the study of groundwater flow regimes in entire sedimentary systems. We will focus on studies employing numerical modeling that have been conducted in two classical sedimentary basin-scale systems: the Paris Basin in France (Castro *et al.*, 1998a,b) and the Great Artesian Basin (GAB) in Australia (Bethke *et al.*, 1999). These multi-layered aquifer systems are ideal to illustrate the behavior of conservative tracers such as noble gases and their relation to depth and recharge distance. The Paris Basin study illustrates how a conservative tracer such as helium can be used as a quantitative dating tool through the use of coupled transport and groundwater flow models, and how a multitracer approach (e.g., ^{3}He, ^{4}He, ^{21}Ne, and ^{40}Ar) can be used to identify dominant transport processes such as advection, dispersion, and/or diffusion. Both studies illustrate the impact of external ^{4}He fluxes and hydraulic conductivities on ^{4}He concentrations. In addition, Bethke *et al.* (1999) have demonstrated the impact of diffusion coefficients on aquitard concentrations. Applications of the direct dating method using ^{81}Kr, ^{36}Cl, and ^{129}I in the GAB are also presented (e.g., Bentley *et al.*, 1986b; Lehmann *et al.*, 2002). These are an important complement to numerical modeling studies.

5.15.6.2 Examples of Applications

5.15.6.2.1 *Noble-gas isotopes—Paris Basin*

The Paris Basin, situated in northeastern and central France, is an intracratonic depression with a diameter of 600 km and maximum depth of 3,200 m. The system consists of seven major aquifers (from top to bottom: Ypresian, Albian, Neocomian, Portlandian, Lusitanian, Dogger, and Trias) separated by low to very low hydraulic conductivity aquitards (cf. Figures 3(a) and (b)). Aquifers are recharged at outcrops to the east and southeast, and gravitational flow is toward the northwest. Concentrations and isotopic compositions of helium, neon, and argon measured in five of these aquifers show excess ^{3}He (^{3}He)$_{exc}$, ^{4}He (^{4}He)$_{exc}$, and ^{40}Ar, as well as vertical concentration gradients of these isotopes throughout the basin (Castro *et al.*, 1998a,b). Water of the Dogger and the Trias formations also has a ^{21}Ne excess above ASW values (Figure 4). The dominant source of these excesses is radiogenic/nucleogenic production in the deep crust underlying the basin. This external flux is

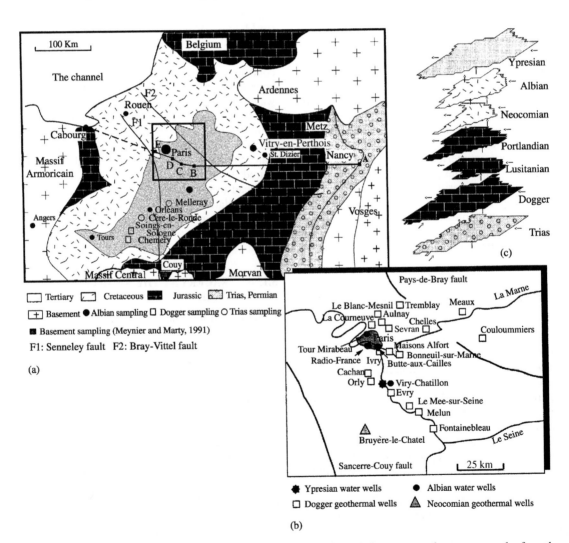

Figure 3 Simplified diagram of the Paris Basin: (a) locations of the sampled areas; central square—samples from the Paris region (Meynier and Marty, 1990); (b) central square enlarged; and (c) simplified diagram of the main aquifers in the Paris Basin—arrows show exchanges of water (entering and leaving) between the aquifers (Castro *et al.*, 1998b) (reproduced by permission of American Geophysical Union from *Water Resour. Res.* **1998**, *34*, 2443–2467).

responsible for the vertical concentration gradients. *In situ* production in the basin appears to be a minor source (13% at most) of these isotopes. They are transported vertically throughout the entire basin by advection, dispersion, and diffusion. The spatial and vertical variability throughout the basin of the $^4He/^{40}Ar$ (0.69–70) and the $^{21}Ne/^{40}Ar$ ((8–23) $\times 10^{-7}$) ratios, as compared to the crustal production ratios of 4 ± 3 and 0.96×10^{-7}, respectively, allow for identification of the dominant transport processes of each isotope through aquitards. An interesting feature of the $^4He/^{40}Ar$ ratio was observed in the Dogger, where it exhibits a great spatial variability. Values close to the crustal production ratio occur in water sampled near major faults (where transport by advection is dominant), but much higher values are observed away from faulted areas (Figure 5). In the Trias, which is the deepest aquifer in direct contact with the bedrock, all $^4He/^{40}Ar$ ratios are similar to the crustal production ratio,

suggesting that ^4He movement in aquitards is more independent of advection than ^{40}Ar and thus is transported mostly by diffusion, whereas ^{40}Ar is primarily transported by advection (see discussion by Castro *et al.*, 1998a). Neon-21 reflects an intermediate situation. The behavior of this isotope reflects differences in diffusion coefficients in water and differences in vertical concentration gradients. Similar $^4He/^{40}Ar$ deviations have also been observed in the GAB (Torgersen *et al.*, 1989). Numerical simulations of ^4He transport in the Paris Basin allowed the calculation of the advective, dispersive, and diffusive flux of ^4He throughout the aquitards (cf. Figure 6). They confirmed that diffusion is the main transport mechanism of ^4He through these formations, and that these fluxes decrease toward the surface. This decrease is directly related to a decrease in concentration gradient that is related to dilution by recharge water carrying atmospheric ^4He. This dilution effect, as well as

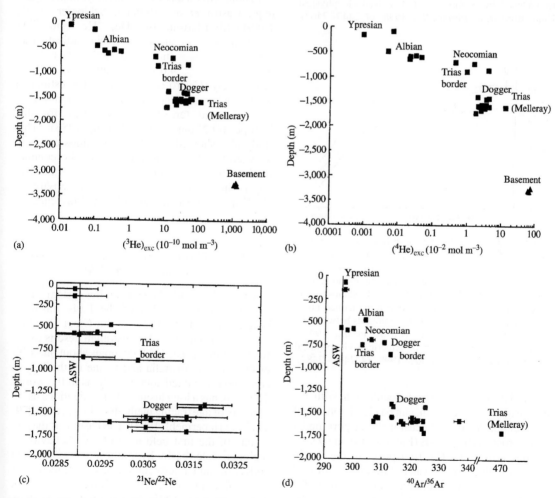

Figure 4 Evolution (versus depth) of: (a) $(^3He)_{exc}$ concentrations in the water; (b) $(^4He)_{exc}$; (c) $(^{21}Ne/^{22}Ne)$ ratio; and (d) $(^{40}Ar/^{36}Ar)$ ratio. The different aquifers are shown. All data are from Castro *et al.* (1998a) (solid squares) except those from the bedrock in figures 6(a) and (b) (Couy) (after Meynier and Marty, 1990), represented by solid triangles. The ASW value is given for $(^{21}Ne/^{22}Ne)$ and $(^{40}Ar/^{36}Ar)$ (Castro *et al.*, 1998b) (reproduced by permission of American Geophysical Union from *Water Resour. Res.* **1998**, *34*, 2443–2467).

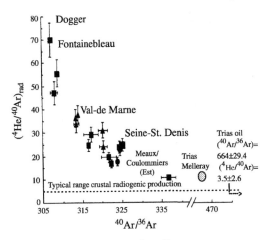

Figure 5 Evolution of $(^4He/^{40}Ar)_{rad}$ ratio versus $^{40}Ae/^{36}Ar$ in Dogger samples; the value of the Trias sample at Melleray is also given, together with the typical range of radiogenic crustal production values; the position of the mean value measured in oil (Pinti and Marty, 1995) is also shown (Castro *et al.*, 1998b) (reproduced by permission of American Geophysical Union from *Water Resour. Res.* **1998**, *34*, 2443–2467).

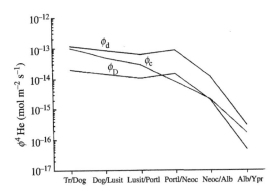

Figure 6 Calculated values of diffusive ϕ_d, advective ϕ_C, and dispersive ϕ_D fluxes of ^4He in each aquitard. Values are estimated on the same vertical line at the center of the basin using ^4He concentrations and vertical velocities obtained by fitting the model. Names on the *x*-axis indicate the aquitards between the named units. For example, Tr/Dog refers to the aquitards between the Trias and Dogger aquifers (Castro *et al.*, 1998a) (reproduced by permission of American Geophysical Union from *Water Resour. Res.* **1998**, *34*, 2467–2484).

the variation of the ^4He concentration with recharge distance and depth, can be clearly observed in the calibrated two-dimensional transport model along cross-section A–E (Figures 3(a) and 7).

Simulation of ^4He transport in the basin coupled with a groundwater flow model allowed the estimation of hydraulic conductivities for the different aquifers and, consequently, the estimation of groundwater residence times. Average turnover times for different aquifers are highly variable, ranging from 8,700 yr for the shallowest aquifer (Ypresian) to 30 Myr for the deepest

(Trias). These calculated turnover times were derived from modeling ^4He concentrations in the Dogger aquifer ($\sim 10^5$ yr for a corresponding thickness of 20 m). They agree with pure groundwater flow model results (Wei *et al.*, 1990). However, these ^4He groundwater ages are much shorter than those estimated from ^4He data by Marty *et al.* (1993). This discrepancy is due to the neglect of diffusive transport of ^4He by Marty *et al.* (1993). When vertical transport by diffusion is important, the amount of the tracer that moves upward into each aquifer is large compared to transport by advection only. Diffusive transport increases tracer concentrations, and thus leads to "apparent older" residence times if it is assumed that the tracer concentration is only proportional to groundwater age. Diffusion is particularly important in the deeper aquifers within a sedimentary basin. In shallow aquifers the opposite can be true. Significant amounts of ^3He and ^4He can leave the system to the atmosphere by diffusion. This leads to "apparent younger ages" (e.g., Castro *et al.*, 1998a,b; Schlosser *et al.*, 1989). Simulations of ^3He, ^4He, and ^{40}Ar concentrations in the Paris Basin indicate that their crustal fluxes (in units of mol m^{-2} yr^{-1}) are of 4.33×10^{-13}, 4.0×10^{-6}, and 2.52×10^{-7}, respectively (Castro *et al.*, 1998b). The estimated basin ^4He flux is of the same order of magnitude as the estimated terrestrial crustal flux for this isotope (O'Nions and Oxburgh, 1983). This study also showed how $^4He/^{40}Ar$ ratios can help to determine low to very low hydraulic conductivities. Based on the value of this ratio, an upper limit for the hydraulic conductivity of 10^{-11} m s^{-1} was determined for the Lias aquitard.

5.15.6.2.2 *Great Artesian Basin—noble-gas isotopes, ^{36}Cl, and ^{81}Kr*

The GAB, a multi-layered aquifer system with a structure similar to that of the Paris Basin, is located in Australia between the Great Dividing Range to the east and the Australian deserts to the west (Figure 8). The GAB occupies about one-fifth of the area of Australia and is one of the largest contiguous confined aquifer systems in the world. The basin has a long history of investigations using a variety of age tracing methods (Airey *et al.*, 1979; Calf and Habermehl, 1983). The basin was the site of one of the first field tests of ^{36}Cl behavior in regional-scale basin hydraulics, and ^{36}Cl data collection and interpretation has continued ever since (Bentley *et al.*, 1986b; Love *et al.*, 2000; Torgersen *et al.*, 1991). In general, the data show the expected spatial variation: a relatively smooth, quasi-exponential decrease in ^{36}Cl concentration and $^{36}Cl/Cl$ ratio with flow distance (Figure 9). The distribution of ages calculated in these

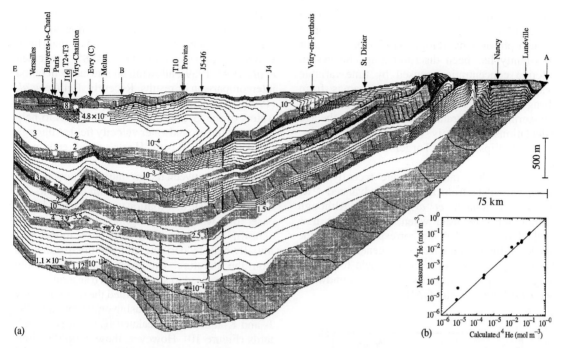

(a)

(b)

Figure 7 (a) Distribution of ^{4}He concentration contours in mol m^{-3}. Contours for values of 10^{-1}, 10^{-2}, 10^{-3}, 10^{-4}, and 10^{-5} mol m^{-3} are marked. Except for the curve for 1.1×10^{-1} mol m^{-3} in the Trias, all other contours express constant concentration variations of 1 unit inside each order of magnitude. The measured value of each sample is given together with some contours whose values are close to the measured ones. In these cases, only the molar number is given; (b) measured ^{4}He values plotted as a function of calculated values (Castro *et al.*, 1998a) (reproduced by permission of American Geophysical Union from *Water Resour. Res.* **1998**, *34*, 2467–2484).

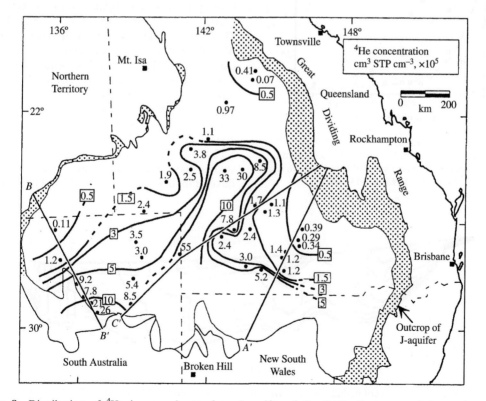

Figure 8 Distribution of ^{4}He in groundwater from J-aquifer of the GAB. Contours and data points show ^{4}He concentrations (10^{-5} cm^3 STP cm^{-3}) measured by Torgersen *et al.* (1992); contour lines are interpretive. Straight lines show flow paths A–A$'$, B–B$'$, and C–C$'$ (Bethke *et al.*, 1999) (reproduced by permission of American Geophysical Union from *J. Geophys. Res.* **1999**, *104*, 12999–13010).

studies appears to be in general agreement with the understanding of flow in the basin developed through physical hydrology. The main source of uncertainty has been discriminating changes in chloride concentration caused by time-variable evapotranspiration in the recharge areas from changes caused by diffusion of chloride out of low-permeability strata, since these affect the age calculations differently (Andrews and Fontes, 1993; Phillips, 2000; Torgersen and Phillips, 1993).

The ^{36}Cl flow rate data were recently compared with data from the newly advanced ^{81}Kr method (Lehmann *et al.*, 2002b). Although caution must be used when evaluating a new method such as ^{81}Kr, the correspondence of the residence-time estimates was generally good. Further, application of ^{81}Kr may help to resolve long-standing uncertainties in the groundwater age distribution of the GAB.

Parallel to the collection of age tracing data using cosmogenic radionuclides, radiogenic subsurface-produced noble gases have also been sampled (Torgersen and Clarke, 1985; Torgersen *et al.*, 1992, 1989), allowing an unusually detailed intercomparison of results. Recently, Bethke *et al.* (1999) have simulated ^4He transport in groundwaters of the GAB, using the ^{36}Cl results to help constrain the groundwater velocity field. Although (unlike the Paris Basin) ^4He was not directly used to estimate groundwater residence times, the relationship between the groundwater flow regime in the GAB, transport of ^4He within the basin, and the resulting ^4He distribution was examined. In this case, the water flow regime was considered "known," and the groundwater flow model was initially built by deriving all hydraulic parameters from the literature. The groundwater flow model results were used to simulate ^4He transport.

Bethke *et al.* (1999) simplified the hydrogeology of the GAB somewhat by using only four units, the J- and K-aquifers intercalated between two aquitards (Figure 10). However, these simplifications do not affect the response of ^4He to variations in the external fluxes or hydraulic conductivities imposed on aquitards, or on assumed diffusion coefficients in these formations.

The ^4He concentrations in the J-aquifer in simulations by Bethke *et al.* (1999) were obtained from previous investigations (Torgersen and Clarke, 1985; Torgersen *et al.*, 1992, 1989). These show ^4He excesses with respect to the ASW that are larger in the center of the basin than at the margins (Figure 8). For water less than 5×10^4 yr old, excesses were attributed to *in situ* production; for older waters ($>10^5$ yr), the dominant ^4He source was attributed to crustal degassing. Torgersen and Clarke (1985) estimated the crustal ^4He degassing flux in the GAB to be 5.12×10^{-18} mol cm^{-2} s^{-1}. A number of previous studies have reported average ^4He continental fluxes, ranging from 0.5×10^{-18} mol cm^{-2} s^{-1} to

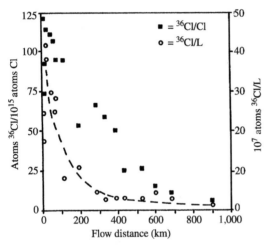

Figure 9 Variation of the ^{36}Cl concentration ratio with distance from the eastern margin of the GAB (Phillips, 1993) (reproduced by permission of Elsevier from *Appl. Geochem.* **1993**, *8*, 643–647).

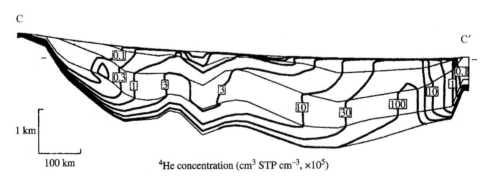

Figure 10 Calculated model of helium transport for cross-section through GAB along flow line C–C′ in the central subbasin, from the Great Dividing Range to near Lake Eyre. Vertical exaggeration is ~100 times. Stratigraphic units, bottom to top, are J-aquifer, confining layers, K-aquifer, and overlying beds. Note logarithmic spacing of ^4He contours (Bethke *et al.*, 1999) (reproduced by permission of American Geophysical Union from *J. Geophys. Res.* **1999**, *104*, 12999–13010).

13×10^{-18} mol cm^{-2} s^{-1}. The latter is a high estimate of the whole-crustal flux (see Torgersen *et al.*, 1989). Radiogenic ^{40}Ar excesses (compared to the ASW) were also observed, with ^{40}Ar/^{36}Ar ratios up to 325.91 (Torgersen *et al.*, 1989). As in the Paris Basin, high deviations of the ^{4}He/^{40}Ar ratio (up to 55) compared to the crustal production ratio also occur. In this case, differential release of ^{4}He and ^{40}Ar from rock to water, and a spatially and temporally variable release of the two nuclides due to tectonic stresses and igneous intrusions were offered as explanations for these deviations. Transport by diffusion was not considered by Torgersen *et al.* (1989).

Hydrologic residence times estimated by Bethke *et al.* (1999) range from 1.5×10^5 yr to 2×10^6 yr, in the central portion of the basin (cross-section C–C') based on the previous ^{36}Cl studies, and also on physical hydrology. Here, groundwater flows to the west and southwest, and ^{4}He concentrations, calculated assuming a crustal flux to the J-aquifer of 3×10^{-18} mol cm^{-2} s^{-1}, increase to the southeast (Figure 10). The most notable outcome of their simulations was that a set of two-dimensional models, in which hydraulic conductivity was estimated based on ^{36}Cl data, produced a reasonable match to the general trends and irregularities in the ^{4}He distribution. These results support the combined application of multiple tracers and numerical transport modeling as a key to understanding the fluxes of water and geochemically important solutes in the geological environment. Results of sensitivity analyses show the ^{4}He concentrations at the top of the J-aquifer that were predicted for differing fluxes of ^{4}He into the base of the aquifer system (Figure 11). It is clear that the ^{4}He concentration at the top of the J-aquifer increases with higher crustal fluxes. By contrast, an increase in the hydraulic conductivity within aquifers leads to a decrease in ^{4}He concentrations, owing to dilution by recharge water carrying atmospheric ^{4}He. An increase in diffusion coefficients within aquitards will have an opposite effect on upper aquifers, as more ^{4}He will be delivered not only through advection but also through diffusion. Bethke *et al.* (1999) concluded that a crustal flux of 3×10^{-18} mol cm^{-2} s^{-1} best explains the ^{4}He concentrations in the GAB. This value is about half of the production rate for the whole crust.

5.15.6.2.3 Implications at the regional scale

Simultaneous combination of a number of natural tracer concentrations (e.g., He, Ne, Ar, and ^{36}Cl), isotopic ratios, and modeling of groundwater flow and mass transport within entire sedimentary systems leads to a tremendous improvement in knowledge of tracer behavior and in the dynamics of these waters as a whole. This understanding would not be possible without such a multitracer modeling approach in a system where extreme contrasts in hydraulic conductivity are present. For example, initial studies of noble gases in the Paris Basin concentrated on analyzing the concentration of gases in one aquifer (Marty *et al.*, 1988). This approach did not indicate the external contribution of the noble gases (helium in particular). Noble gases were thought to be solely the result of *in situ* production. Groundwater ages were estimated based on *in situ* production rates, and the aquifer was thought to be a "closed," nearly static system. Subsequent sampling at multiple levels across the Paris Basin (at different depths and recharge distances) and isotopic analysis revealed that the groundwater system is very dynamic and interactive from a geochemical and hydrodynamic point of view. This dynamism is evident both at the scale of aquifers and aquitards within the system, and at the larger scale of the Earth's continental crust. Indeed, the horizontal and vertical distribution of ^{4}He/^{40}Ar and ^{21}Ne/^{40}Ar within the deepest aquifers emphasize the importance of active, advective "crustal-scale" transport (Castro *et al.*, 1998b). Concentrations and isotopic ratios also reveal combined upward movement via advection, dispersion, and diffusion of gases through successive formations. The horizontal and vertical variations of ^{4}He/^{40}Ar and ^{21}Ne/^{40}Ar highlight a dominant upward diffusive transport of helium isotopes through low and very low permeability formations. This illustrates how such ratios provide insight into the ranges of hydraulic conductivities within such formations, information that is unobtainable from independent hydrodynamic studies. For example, strongly fractionated ^{4}He/^{40}Ar ratios indicate the

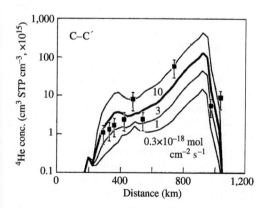

Figure 11 Variation in ^{4}He concentration at top of J-aquifer along flow line C–C' as a function of basal ^{4}He flux. Bold line corresponds to Figure 10. Measured values are from Torgersen and Clarke (1985) and Torgersen *et al.* (1992) (Bethke *et al.*, 1999) (reproduced by permission of American Geophysical Union from *J. Geophys. Res.* **1999**, *104*, 12999–13010).

presence of underlying aquitards with hydraulic conductivities $\sim 10^{-11}$ m s^{-1} or less. Combined analyses of isotope ratios (actually, absence of fractionated isotope ratios) can identify areas where advective transport dominates over diffusion, as in areas where underlying formations exhibit relatively large vertical hydraulic conductivities or where they are close to important faults. In contrast, strongly fractionated isotope ratios indicate that diffusion is the dominant transport process.

At the origin of these regional studies, there is the question of the "age" of groundwaters and our quest to "accurately determine groundwater ages." Knowledge of the age of a groundwater implies:

- understanding of how a particular groundwater system functions;
- all the hydrodynamic parameters in a particular system, with relative accuracy (e.g., distribution of hydraulic head, hydraulic conductivity, and porosity);
- recharge (infiltration) rates;
- understanding of how our groundwater system relates to the "outside world" through the imposition of boundary conditions represented through numerical modeling; and
- understanding and quantification of water flow and mass interactions taking place between subsystems (aquifers and aquitards) within our major groundwater system of interest.

In one simple phrase, to be able to truly determine the age of a single sample of groundwater implies that we are able to understand precisely and fully the functioning of the system from both physical and geochemical points of view. Such a goal has not been accomplished. The combined regional multitracer/modeling approach represents one new important step in this (iterative) process. Perhaps more than at any other scale, regional studies exemplify the interdependence of geochemical tracers and processes driving the dynamics of groundwater, in that they integrate the long-term response of quasistatic formations (aquitards) with that of much more dynamic formations (aquifers).

5.15.7 TRACERS AT THE AQUIFER SCALE

5.15.7.1 Introduction

There are many examples in the literature where a variety of tracers such as ^{37}Ar, ^{39}Ar, ^{36}Cl, ^{14}C, ^{3}He, and ^{4}He have been used to date groundwater at the aquifer scale. Notable examples include the Ojo Alamo aquifer in the San Juan Basin, New Mexico (Phillips *et al.*, 1989; Stute *et al.*, 1995), the Stampriet and Uitenhage aquifers in Southern Africa (Heaton *et al.*, 1983; Vogel *et al.*, 1982),

the Carrizo aquifer in Texas (Andrews and Pearson, 1984; Castro and Goblet, 2003; Castro *et al.*, 2000; Pearson and White, 1967), and the Milk River aquifer in Ontario, Canada (Andrews *et al.*, 1991; Fröhlich *et al.*, 1991; Hendry and Schwartz, 1988; Nolte *et al.*, 1991; Phillips *et al.*, 1986; Schwartz and Muehlenbachs, 1979). Here, we illustrate the use of natural tracers in the Carrizo aquifer, where the measured tracers include ^{14}C, ^{3}He, ^{4}He, atmospheric noble gases, and ^{36}Cl. Confined aquifers such as the Carrizo are ideal to observe, analyze, and discuss groundwater mixing with surrounding formations. In addition, while these two aquifers have been the object of extensive study for several decades, our understanding of tracer behavior in these systems is still an ongoing process (Heaton *et al.*, 1983).

5.15.7.2 Carrizo Aquifer

The Carrizo aquifer, a major groundwater system, is part of a thick regressive sequence of fluvial, deltaic, and marine terrigenous units in the Rio Grande Embayment area of South Texas along the northwestern margin of the Gulf Coast Basin (Figure 12(a)). This aquifer has been particularly well studied for its content of environmental tracers in Atascosa and McMullen Counties south of San Antonio (Andrews and Pearson, 1984; Castro and Goblet, 2003; Castro *et al.*, 2000; Cowart, 1975; Cowart and Osmond, 1974; Pearson and White, 1967). It is overlain by the Recklaw formation that is composed of shales with thin interbedded sands, and is underlain by the Wilcox Group mainly composed of clay, shale, and lenticular beds of sand (Figure 12(b)). Rainfall recharges the aquifer outcrop at the northern part of Atascosa County, and groundwater flow is to the southeast. Discharge occurs by upward cross-formational leakage driven by high fluid pressure, as well as by fault-related permeability pathways that are mainly present in the south of McMullen County.

The Carrizo was one of the earliest aquifers for which ^{14}C was employed as a tracer (Pearson and White, 1967). The results of the ^{14}C measurements showed a consistent increase of age along flow paths. The early ^{14}C results were checked about 30 years later using measurements of atmospheric noble-gas concentrations (Stute *et al.*, 1992a). These indicated that water recharged prior to a radiocarbon age of $\sim 1.2 \times 10^4$ yr was colder by $\sim 5\,^{\circ}$C than the modern mean annual temperature. Although this age is $\sim 30\%$ younger than indicated by independent chronology for the last glacial maximum, it generally confirms the radiocarbon results. Further confirmation comes from measurement of ^{36}Cl in the same area (Bentley *et al.*, 1986a). The water is too young to show much ^{36}Cl decay.

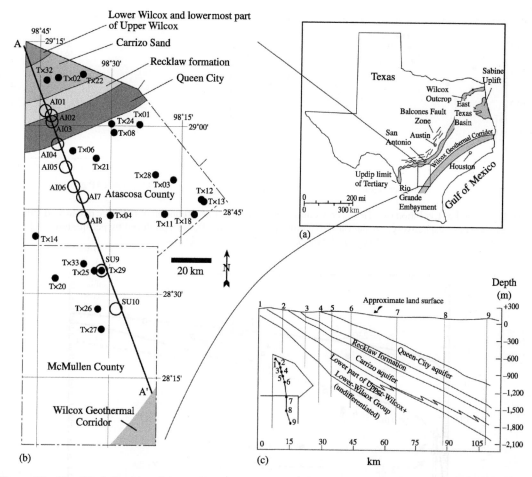

Figure 12 Simplified diagram of the studied area and formations in southwestern Texas. (a) Geographical and tectonic context of the area of investigation within Texas (after Bebout *et al.*, 1978; Hamlin, 1988). (b) Detailed area of investigation: latitude and longitude are indicated, as well as the location of cross-section A–A' in which simulations of groundwater flow and mass transport were carried out; the locations of all ⁴He sampled sites in Atascosa and McMullen counties (cf. Castro *et al.*, 2000) are also indicated as well as that of the wells for which hydraulic head measurements are used for calibration of the groundwater flow model, closed and open circles, respectively. (c) General schematic representation of cross-section in the area for the four studied formations: the Lower-Wilcox and lower part of the Upper-Wilcox formations undifferentiated, the Carrizo aquifer, the Recklaw formation, and the Queen-City aquifer (Castro and Goblet, 2003) (reproduced by permission of American Geophysical Union from *Water Resour. Res.* **2003**, *39*, 1172).

The ³⁶Cl/Cl ratio in the region documenting the noble-gas paleotemperature decrease actually shows an increase in the ³⁶Cl/Cl. Bentley *et al.* (1986a) explain this by comparison with the curve of distance from the coastline, which varied as sea level decreased during the glacial period. Chloride concentration in precipitation decreases as with increasing distance to the coast. The good correspondence further supports the evidence for distribution of flow velocities in the Carrizo.

This well-supported understanding of flow velocities should aid in the interpretation of the pattern of helium-isotope concentrations. Helium-3 and helium-4 concentrations are in excess by up to one and two orders of magnitude compared to the ASW, respectively (Castro *et al.*, 2000). Concentrations increase with recharge

distance and ¹⁴C ages (e.g., Figure 13). Separation of helium components shows that ~99% of the excess helium—(³He)$_{exc}$ and (⁴He)$_{exc}$—is of terrigenic origin due to the addition of nucleogenic ³He and radiogenic ⁴He at a ³He/⁴He production ratio of 7×10^{-8}. From this, and through analyses of ³He/⁴He ratios in groundwater and in reservoir rocks, it is apparent that most of the helium has an origin external to the aquifer, and that *in situ* production makes a smaller and variable contribution. Simulation and calibration of ⁴He transport using a simple analytical model (Castro *et al.*, 2000) illustrates the *in situ* contribution variations with recharge distance and with depth within the aquifer (Figure 14). Modeling results (dashed line, Figure 14) show that *in situ* production can account for ⁴He concentrations in the top of the aquifer if

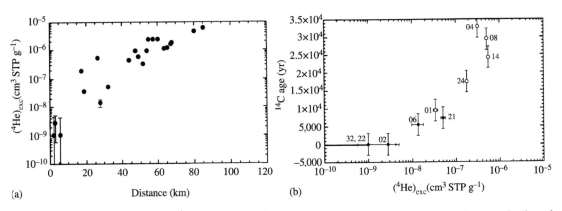

(a)   (b)

Figure 13 (a and b) Evolution of $(^4He)_{exc}$ concentrations in the water as a function of recharge distance, depth, and calculated ^{14}C ages, respectively, for the Carrizo aquifer (Castro *et al.*, 2000) (reproduced by permission of Elsevier from *Appl. Geochem.* **2000**, *15*, 1137–1167).

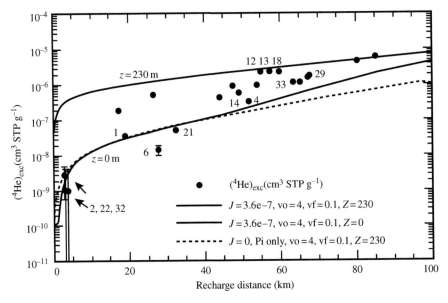

Figure 14 Calculated $(^4He)_{exc}$ concentration curves for the bottom ($z = 230$ m) and for the top of the aquifer ($z = 0$ m) (upper and lower plain line, respectively) as a function of the recharge distance for the Carrizo aquifer. The measured $(^4He)_{exc}$ concentration values are also shown as well as the ^4He concentration curve resulting from *in situ* production only (dashed line) (Castro *et al.*, 2000) (reproduced by permission of Elsevier from *Appl. Geochem.* **2000**, *15*, 1137–1167).

the distance from the recharge area is ≤40 km. Similar to the Paris Basin (Section 5.15.6.2.1), this contribution decreases toward the discharge area (~27%). At the bottom of the aquifer the *in situ* contribution represents no more than ~15% of total ^4He.

The analytical model was calibrated with an external flux to the bottom of the aquifer of 3.6×10^{-7} cm^3 STP cm^{-2} yr^{-1}. It is important to note that this flux represents only part of the local integrated crustal production as a consequence of downstream ^4He dilution in the deeper aquifer. Thus, this flux is smaller than the average crustal flux calculated for the GAB by Torgersen and Clarke (1985) (3.62×10^{-6} cm^3 STP cm^{-2} yr^{-1}) or that calculated for the Paris Basin by

Castro *et al.* (1998a) (9×10^{-6} cm^3 STP cm^{-2} yr^{-1}). Using an average velocity calculated from previous studies of 1.6 m yr^{-1}, tests of sensitivity to the ^4He external flux were carried out (e.g., Figures 15(a)–(e)). Some aspects of these tests are worth noting: differences between bottom and top concentrations decrease with decreasing flux. As the flux decreases, *in situ* production becomes more important, ^4He accumulation becomes independent of vertical and horizontal position in the aquifer, and varies mainly with time. For the smallest flux of 3.59×10^{-9} cm^3 STP cm^{-2} yr^{-1}, ^4He concentrations are similar in the top and bottom of the aquifer as *in situ* production becomes almost entirely responsible for the ^4He concentrations,

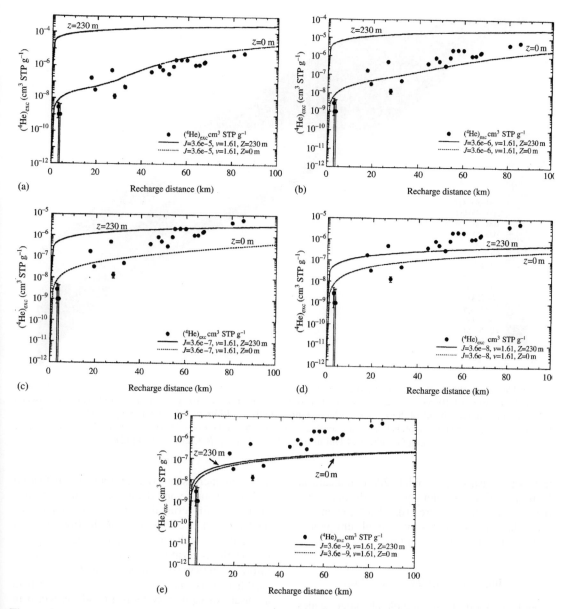

Figure 15 Model results for sensitivity tests of the ^{4}He flux value for the Carrizo aquifer—(a–e): calculated $(^4\text{He})_{exc}$ curves for five different orders of magnitude of the flux value, between 3.6×10^{-5} cm^3 STP cm^2 yr^{-1} and 3.6×10^{-9} cm^3 STP cm^2 yr^{-1} and a constant velocity value of 1.61 m yr^{-1}. The calculated concentration curves for the bottom ($z = 230$ m) and for the top ($z = 0$ m) of the aquifer are represented. The measured $(^4\text{He})_{exc}$ concentration values in the aquifer are also shown (Castro *et al.*, 2000) (reproduced by permission of Elsevier from *Appl. Geochem.* **2000**, *15*, 1137–1167).

and the external flux is negligible. This illustrates the interplay between internal versus external sources of ^{4}He. In the study by Castro *et al.* (2000), it was only possible to calibrate ^{4}He concentrations by imposing an exponential decrease in the groundwater velocities between recharge and discharge areas, with initial and final velocities of 4.0 m yr^{-1} and 0.1 m yr^{-1}, respectively.

The analytical modeling of Castro *et al.* (2000) is typical of an aquifer-scale approach. Comparison of a more recent simulation of ^{4}He transport using a finite element model with a more accurate representation of boundary conditions (i.e., a step toward a comprehensive aquifer-system or even basin-scale approach) has shown representation by a simple analytical model to be inadequate for this system (Castro and Goblet, 2003). Although the finite element model has confirmed an exponential decrease in Carrizo velocities, the external flux value was one order of magnitude higher in the analytical model than that obtained through numerical modeling. Advective groundwater ages estimated from numerical simulation of ^{4}He

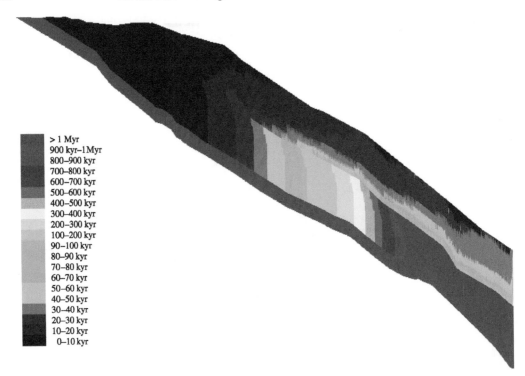

> 1 Myr
900 kyr–1 Myr
800–900 kyr
700–800 kyr
600–700 kyr
500–600 kyr
400–500 kyr
300–400 kyr
200–300 kyr
100–200 kyr
90–100 kyr
80–90 kyr
70–80 kyr
60–70 kyr
50–60 kyr
40–50 kyr
30–40 kyr
20–30 kyr
10–20 kyr
0–10 kyr

Figure 16 Distribution of calculated advective water ages (kyr) in the system. Water age contours correspond to constant variations of 10^4 yr between 0 Myr and 0.1 Myr and variations of 10^5 yr for time periods varying between 0.1 Myr and 1 Myr, each one of these intervals being represented by a different color, from the youngest (dark blue) to the oldest (red), which corresponds to ages higher than 1 Myr (Castro and Goblet, 2003) (reproduced by permission of American Geophysical Union from *Water Resour. Res.* **2003**, *39*, 1172).

transport (Castro and Goblet, 2003) indicate an increase in Carrizo water ages with distance from the recharge zone (Figure 16). Clearly, even with multiple tracers, issues of model uniqueness continue to affect interpretations.

5.15.7.3 Implications at the Aquifer Scale

Perhaps more than studies at other scales, modeling of groundwater flow at the aquifer scale has revealed the utility of combining atmospheric-source radioisotope or stable tracers (e.g., ^{14}C, atmospheric noble gases) with subsurface-sourced stable tracers. In comparison to the regional scale, detailed knowledge of the permeability structure is often needed. Independent information can often be obtained from hydraulic tests or from semiquantitative interpretation of aquifer properties based on hydrogeological descriptions. Concentrations of natural tracers reveal not only the history of the tracer itself (e.g., its sources), but also the patterns of water dynamics within a particular aquifer (e.g., variation in water velocity is the horizontal direction). Some generalizations regarding patterns that are very commonly observed from the numerous studies of noble gases in aquifers in sedimentary systems can now be drawn. These

are: (i) an increase in concentration with increased recharge distance and depth and (ii) the presence of two main crustal sources (radiogenic or nucleogenic), i.e., *in situ* production and an external flux with variable contributions depending on the depositional context. This is general information that can now be considered to be well established, as it has been observed consistently in a variety of aquifer systems around the world in diverse sedimentary settings, as discussed above. If the external flux and *in situ* production of noble gases in general, and ^4He in particular, in a specific aquifer system are relatively constant, then the pattern of variation in concentration along one particular aquifer, such as a rapid versus a slow increase in concentrations with increased recharge distance, will yield information on the spatial variation of hydraulic parameters, and those of hydraulic conductivity in particular.

At the aquifer scale, the most important contribution of age tracers is probably reduction in the nonuniqueness of numerical models. Lack of uniqueness stems, among other things, from inadequate knowledge of the distribution of hydraulic properties within groundwater systems and from poor constraints on boundary conditions (Konikow and Bredehoeft, 1992; Maloszewski and Zuber, 1993). Commonly, groundwater flow models are calibrated on measured hydraulic head and/or hydraulic conductivity values. However, a good match does not prove the validity of the model, because the nonuniqueness of model solutions means that a good comparison can be achieved with an inadequate or erroneous model. For example, in the case of the Carrizo aquifer, Castro and Goblet (2003) show that variations of hydraulic conductivity up to two orders of magnitude in the Carrizo aquifer and in the overlying confining layer lead to similar calculated hydraulic heads. No clear-cut arguments are available to validate one groundwater flow scenario over a different one. This is due, in part, to lack of information regarding recharge (infiltration) rates in the study area. In contrast, when tested with a ^{4}He transport conceptual model, all groundwater flow calibrated scenarios except one failed to reproduce a coherent ^{4}He transport behavior in the system. This highlights the importance of contributions of a tracer such as ^{4}He for determining which model most closely replicates natural conditions. Without the information provided by this tracer, it would be very difficult to have a precise idea of the groundwater flow conditions of this system, and, as a result, on the velocity distribution of those waters as well as water age distribution.

5.15.8 TRACERS AT THE LOCAL SCALE

5.15.8.1 Introduction

There are numerous examples in the literature in which environmental tracers have been applied at a local scale, both in porous media and in fractured crystalline rocks. Among these are studies in Liedern/Bocholt, Germany (Schlosser *et al.*, 1988, 1989), the Stripa Granite, Sweden (Andrews *et al.*, 1989; Loosli *et al.*, 1989), in Sturgeon Falls, Ontario, Canada (Milton *et al.*, 1997; Robertson and Cherry, 1989; Solomon *et al.*, 1993), the Delmarva Peninsula (Dunkle *et al.*, 1993; Ekwurzel *et al.*, 1994; Plummer, 1993), and in the Borden aquifer, Ontario (Smethie *et al.*, 1992; Solomon *et al.*, 1992). Most of these studies were conducted in relatively surficial aquifers, and many focused on the application of the tritium/^{3}He and CFC methods, which are well suited for young, fast-flowing groundwater. Tracers such as ^{37}Ar, ^{39}Ar, ^{85}Kr, ^{36}Cl, and ^{4}He were also investigated. Helium-3 confinement, dispersion, diffusion, sorption, and microbial degradation are some of the processes that have been investigated by means of various tracers. As an example, we focus on studies in the Delmarva Peninsula (eastern seaboard of the US). These studies illustrate in a clear and elegant manner how a multitracer approach using

^{3}H/^{3}He, CFCs (CFC-11 and CFC-12), and ^{85}Kr can give insight into the dynamics of groundwater flow systems and the major transport processes that influence the concentration of different tracers. They also illustrate how the use of a multitracer approach using direct water dating methods at a local scale increases confidence in estimated ages.

5.15.8.2 ^{3}H/^{3}He, CFC-11, CFC-12, ^{85}Kr—Delmarva Peninsula

The Atlantic Coastal Plain of the Delmarva Peninsula in the East Coast of the US is composed of Jurassic to Holocene sediments deposited during a series of transgressions and regressions (Figure 17). It is covered by a highly permeable surficial Pleistocene aquifer that is directly recharged by local precipitation. Thirty-three wells, some of which tap this formation, were used to sample ^{3}H/^{3}He (52 samples), CFCs (282 samples), and ^{85}Kr (four samples) (Dunkle *et al.*, 1993; Ekwurzel *et al.*, 1994; Plummer *et al.*, 1993; Reilly *et al.*, 1994).

In wells sampled at several levels, ^{3}H and (^{3}He)$_{tri}$ concentrations, as well as apparent ^{3}H/^{3}He, CFC-11, and CFC-12 ages, increase significantly with depth below the water table (Figure 18). Moreover, most of the apparent CFC, ^{3}H/^{3}He, and ^{85}Kr ages agree with each other to within three years over the entire peninsula (e.g., Figures 19(a) and (b)), indicating that these tracers behave fairly conservatively in this sandy unconfined aquifer. Waters located in discharge areas such as those in well KE Be 170, where a mix of a wide range of water ages intersects the well screen, present large differences between CFC and ^{3}H/^{3}He ages. This illustrates how ages based on different tracers can identify specific recharge versus discharge areas within a groundwater system.

Simulations using different dispersion values and an average vertical velocity of $1~m~yr^{-1}$ in a one-dimensional model (Ekwurzel *et al.*, 1994) showed that ^{3}H/^{3}He ages are more affected by dispersion than CFC ages (Figure 20). This is because the abrupt concentration front presented by the peak-shaped ^{3}H source function is subject to greater dispersion than the relatively gradual, monotonic CFC concentration increase (Figure 2). Such dispersion would result in an apparent increase in ^{3}H/^{3}He ages in postbomb peak waters, and in reduced apparent ages in prebomb peak waters. However, comparison of (^{3}H + ^{3}He) concentrations with the ^{3}H input function shows little smoothing, indicating that dispersion has been negligible (Dunkle *et al.*, 1993; Ekwurzel *et al.*, 1994). This inference is supported by detailed numerical transport modeling

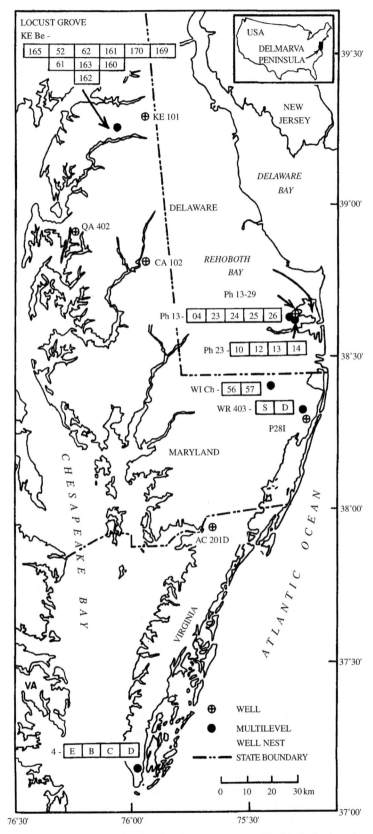

Figure 17 Map of the Delmarva Peninsula, on the east coast of the US. Wells having the same series number (e.g., KE Be) are listed as the series number followed by a dash and well numbers listed in order of increasing depth within the boxes (Ekwurzel *et al.*, 1994) (reproduced by permission of American Geophysical Union from *Water Resour. Res.* **1994**, *30*, 1693–1708).

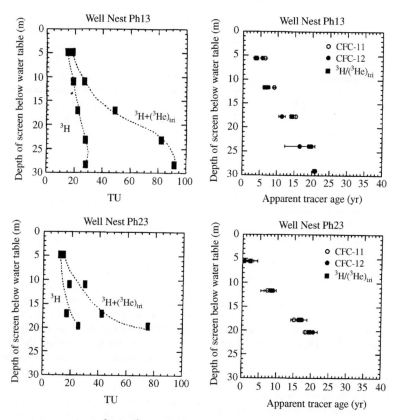

Figure 18   Depth profiles of tritium, $^3H + (^3He)_{tri}$ (tritium units), and apparent tracer ages from well nests Ph13 and Ph23 (Ekwurzel *et al.*, 1994) (reproduced by permission of American Geophysical Union from *Water Resour. Res.* **1994**, *30*, 1693–1708).

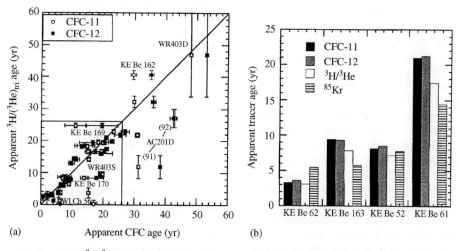

Figure 19   (a) Apparent $^3H/^3He$ age compared toapparent CFC-11 and CFC-12 ages. Diagonal line is a 1:1 reference line. The vertical and horizontal reference lines are 26 yr age lines representing the bomb peak and a 2 yr lag for flow through the unsaturated zone. (b) Apparent CFC-11, CFC-12, $^3H/^3He$, and ^{85}Kr ages for Locust Grove wells sampled in November 1991 (Ekwurzel *et al.*, 1994) (reproduced by permission of American Geophysical Union from *Water Resour. Res.* **1994**, *30*, 1693–1708).

(Reilly *et al.*, 1994). The high correlation observed between ages calculated using the three age tracers also supports this inference. Recharge rates calculated from these different tracers are also in good agreement (Ekwurzel *et al.*, 1994).

5.15.8.3   Implications at the Local Scale

The studies surveyed above, and those at other sites, provide a strong contrast to aquifer and regional scale results. While the results of the

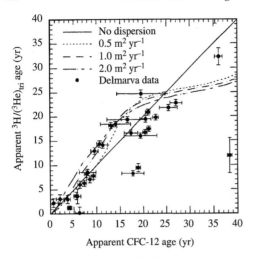

Figure 20 Comparison of $^3H/^3He$ and CFC-12 ages from the Delmarva Peninsula, including effects of dispersion. Average linear velocity of $1\ m\ yr^{-1}$ is plotted with different dispersion coefficients and with a $1:1$ reference line (Ekwurzel *et al.*, 1994) (reproduced by permission of American Geophysical Union from *Water Resour. Res.* **1994**, *30*, 1693–1708).

larger scale studies typically exhibit many enigmatic features and major issues of nonuniqueness, most of the local-scale flow systems appear to allow straightforward interpretations. Most of the studies conducted at sites investigated previously or, by other approaches (Engesgaard *et al.*, 1996; Reilly *et al.*, 1994; Shapiro *et al.*, 1999; Solomon *et al.*, 1992), hold few surprises.

Several aspects of the results from local flow systems are noteworthy. One is the ease and reliability of estimating recharge rates by measurement of vertical profiles of $^3H/^3He$ or CFCs in recharge areas (Böhlke and Denver, 1995; Ekwurzel *et al.*, 1994; Solomon *et al.*, 1993). Another is the apparently relatively minor level of problems created by degassing of volatile tracers from the water table. Sampling near discharge areas (principally streams), however, is somewhat hazardous because of the mixing of deep, upward flow with shallow flow at the point of discharge.

What is most surprising is the relative homogeneity and the dominance of advective flow in most of the investigated systems. Where multiple tracers have been used, it is possible to estimate dispersivity independently. The longitudinal dispersivity values obtained for studies at the 10^2–10^3 m flow scale have ranged from 0.1 m to 0 m (Ekwurzel *et al.*, 1994; Engesgaard *et al.*, 1996; Reilly *et al.*, 1994; Solomon *et al.*, 1993; Szabo *et al.*, 1996). These are significantly smaller than many extrapolations would suggest (Gelhar, 1993; Neuman, 1995). Whether these implications can be widely extended, or whether they result from application of the tracing methods to a restricted and hydrologically unusually homogeneous set of

study areas is uncertain. Although age tracing techniques for local systems have clearly demonstrated their validity, have increased the general understanding of local flow systems, and have aided in the reconstruction and fate of contaminants introduced into these systems, they have not yet reached their potential for elucidating fundamental transport mechanisms in a variety of geological environments. There is much to be learned by future studies for this purpose.

5.15.9 TRACERS IN VADOSE ZONES

Vadose zones are of great importance for both hydrological and geochemical reasons. Hydrologically, they represent the portion of the physical system where precipitation partitions the essential elements of the hydrological cycle: evapotranspirative return to the atmosphere, runoff, and deep infiltration (i.e., aquifer recharge). Quantification of the fluxes below the root zone is thus critical for both closing the water balance at the watershed scale and for analyzing groundwater resources. Geochemically, the vadose zone constitutes the primary zone of interaction between earth materials and precipitation. Knowledge of water fluxes through vadose zones is thus a key to understanding the global solute balance, as well as interactions of climate, geochemistry, and tectonics on the continental scale.

Determination of water fluxes under unsaturated conditions by physical monitoring is often very difficult due to the effects of seasonal fluctuations, problems of installing and maintaining instrumentation, and the nonlinear nature of the variation of water potential and hydraulic conductivity with water content. More direct means of residence-time estimation by tracing solutes thus provide an important alternative to physical monitoring (Allison *et al.*, 1994; Scanlon *et al.*, 1997). The majority of residence-time tracers that rely on radiodecay for their time constant are volatile; therefore, most vadose-zone dating studies employ some form of signal tracing rather than radioactive decay dating.

Conceptually, vadose zones can be divided into two end-members: humid and arid. In humid climates (or in irrigated areas), vadose zones are generally thin (0–10 m), dynamic, and are characterized by great variability in flow paths. Residence times vary from hours to a few years. In contrast, arid-region vadose zones are usually thick (10–1,000 m), with very small water fluxes, and a high degree of uniformity in flow paths. We will give examples from very humid, subhumid (irrigated), and arid vadose-zone environments.

Bonell *et al.* (1998) described the results of detailed monitoring of soil water and runoff in tropical northeast Queensland, Australia.

Figure 21 shows the δ^2H variation of soil water as a function of depth during a three-month period in 1991. Each panel represents a nest of soil water samplers at different positions on a hillslope. Some sites (NC2, NC4, and SC3) show a large range in δ^2H throughout the depth profile. These are in response to heavy precipitation events having differing isotopic compositions. Others (SC1 and SC2) show much variability with time in the upper soil horizons but little at depth, indicating an absence of continuous rapid flow paths to greater than 1 m. Figure 22 shows

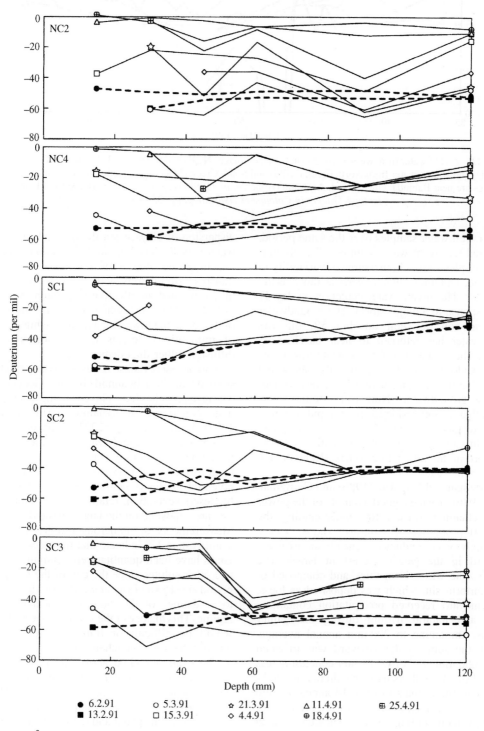

Figure 21 δ^2H values of soil water as a function of depth in vadose-zone profiles from small catchments near Babinda, Queensland, Australia. Numbers in symbol key refer to sampling dates during 1991. The climate is tropical and mean annual precipitation amounts to 4,000 mm (source Bonell *et al.*, 1998).

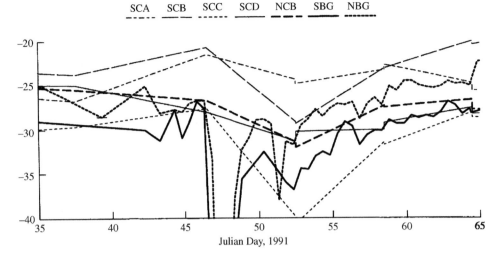

Figure 22 δ^2H of water from wells in the North Creek drainage (NCB) and South Creek drainage (SCA, SCB, SCC, and SCD) and stream water from North Creek (NBG) and South Creek (SBG) as a function of time during 1991. The drainages are near Babinda, Queensland, Australia (Bonell *et al.* (1998) (reproduced by permission of Elsevier from *Isotope Tracers in Catchment Hydrology*, pp. 334–347).

δ^2H for well water samples over the same period, compared to stream water samples. The ground-water samples show a high degree of variability in both space and time, but not nearly as much as the soil water. The stream-water depletion on day 47 is in response to a major storm. The groundwater response can be seen on day 52, but the groundwater has returned to near baseline by day 60. These data, therefore, indicate flow times for the vadose zone of ~5 d, but also show that new recharge is mixing with a large reservoir of shallow groundwater. Numerous studies have shown that a short residence time, rapid response and flow rates, and large temporal and spatial variability in fluxes are characteristic of tracer transport in humid vadose zones (Kendall and McDonnell, 1998).

Gvirtzman and Magaritz (1986) and Gvirtzman *et al.* (1986) provide a good example of the power of geochemical tracers for understanding the transport properties of semi-arid vadose zones. The study was conducted beneath an irrigated vineyard in the Negev Desert in Israel. The vadose zone was ~11 m thick and composed of very uniform silt, which was deposited as loess. The vineyard received precipitation containing bomb and natural tritium during the winter months, amounting to 200 mm annual average. During the summer the vineyard was irrigated with ~650 mm yr^{-1} of well water containing less than 1 TU. The authors used the alternating high- and low-tritium inputs to trace 14 annual tritium cycles down to a depth of 8.75 m (Figure 23), for a mean velocity of 0.62 m yr^{-1}. They determined that the spreading of the annual pulses was largely due to exchange by diffusion with immobile regions within the loess that were created by

dispersion of clays when the original saline pore water was displaced by dilute irrigation water. By matching the shapes of the annual pulses in transport simulations with those from the data, they estimated an effective dispersion coefficient of 5×10^{-11} m^2 s^{-1}, which, due to the uniformity of the loess, the low velocity, and the high volume of immobile water, is mainly attributable to diffusion rather than dispersion.

Arid or desert vadose zones have been much less studied than those in humid climates. Geochemical dating approaches have played an important role in understanding their dynamics (Allison *et al.*, 1994). Walvoord *et al.* (2002a,b) have employed a combination of liquid–vapor flow modeling and modeling of chloride concentration profiles to demonstrate that pervasive upward total fluid potential gradients below the root zone, caused by desert vegetation, retain all water and solutes that are deposited on the land surface (Figure 24). Measurements of numerous profiles within the southwestern US indicate that most desert vadose zones have been quantitatively retaining chloride and other solutes since the end of the last glacial period at ~15 ka (Tyler *et al.*, 1996). This amounts to a "dating" of the duration of a hydrological regime. This history of the vadose-zone hydraulics would be difficult or impossible to ascertain without the use of geochemical tracers.

The examples cited above illustrate the great variability in soil–water movement and resultant geochemical processes in vadose zones. Humid vadose zones are highly dynamic, with large water fluxes and great spatial and temporal variability of those fluxes. Although variability is large at small time and space scales, mixing processes tend to rapidly average out these variations and their

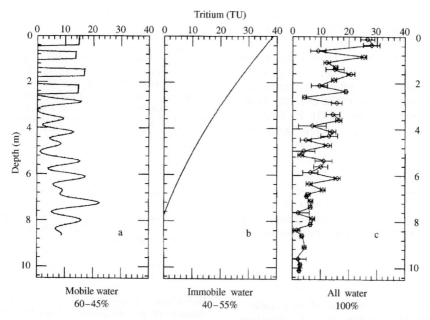

Figure 23 Comparison of reconstructed irrigation water tritium input (left panel), immobile vadose-zone water tritium (center panel), and measured tritium (right panel), all as a function of depth. Study site is an irrigated vineyard in the Negev Desert, Israel (source Gvirtzman and Magaritz, 1986).

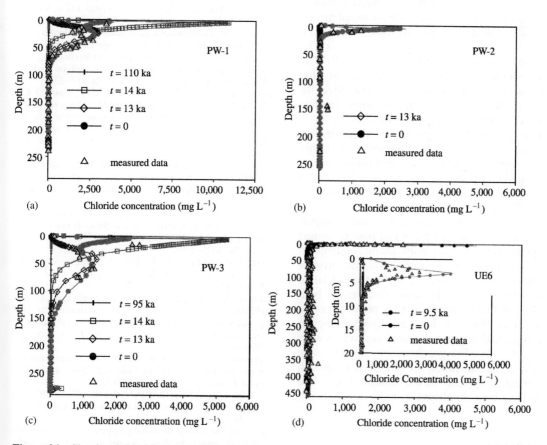

Figure 24 Simulated chloride concentrations as a function of depth, compared to measured data, for deep boreholes at the Nevada Test Site, USA. The simulations indicate that the upper 10–100 m of these vadose zones have been retaining all chloride deposited from the atmosphere onto the land surface for time periods from 9.5 ka to 110 ka (Walvoord *et al.*, 2002a) (Walvoord, 2002) (reproduced by permission of American Geophysical Union from *Water Resour. Res.* **2002**, *38*, 1291).

effects on tracer concentrations. It is clear that these characteristics are very favorable for weathering reactions, and that these are environments of high rates of chemical denudation. Less attention has been devoted to how the small-scale heterogeneity in flow paths and fluxes affects the chemical outcome of weathering averaged over the landscape (Bullen and Kendall, 1998). In contrast, desert landscapes exhibit a remarkable uniformity of hydrologic regime, both in space and time (except in the vicinity of water courses or areas where runoff is focused). Such vadose zones can be regarded as surfaces of "negative weathering"— there is no export of solutes and rather than being depleted of mobile elements with time; they accumulate them from atmospheric deposition.

5.15.10 CONCLUSIONS

During most of the second half of the twentieth century, the emphasis in the study of subsurface water age tracers was on the development of new techniques and methodologies. This phase of the field has slowed greatly. Most tracers not in common use have at least been explored, and most are no longer used, because of major sampling and analytical difficulties. The only tracer that may be poised for a breakthrough in the near future is ^{81}Kr (Collon *et al.*, 2000; Lehmann *et al.*, 2002b). The inert geochemistry and the half-life of this radionuclide render it sufficiently attractive for regional and aquifer scale investigations in spite of the difficulties in sampling and analysis.

The new frontier in groundwater age tracers lies not in technique development, but rather in interpretation within the hydrogeochemical system. Until the 1990s, interpretation has tended to focus on geochemical issues, while minimizing issues of transport and mixing. With the increasing use of three-dimensional flow and transport computer programs, attention can now be turned to quantifying the dynamics of the systems in which the tracers are measured. This initiative at the boundaries of geochemistry and quantitative hydrogeology will yield major advances for both fields. For hydrogeology, it may prove a key to elucidating the link between hydrogeological structures and transport mechanisms over a wide range of scales. For geochemistry, it will finally enable a quantification of the fluxes through geochemical systems in the shallow crust. The tools have been developed; now let us use them!

REFERENCES

Airey P. L., Calf G. E., Campbell B. L., Hartley P. E., Roman D., and Habermehl M. A. (1979) Aspects of the isotope hydrology of the Great Artesian Basin, Australia. In *Isotope Hydrology 1978*. IAEA, Vienna, vol. 1, pp. 205–219.

Allègre C. J., Staudacher T., Sarda P., and Kurz M. D. (1983) Constraints on evolution of Earth's mantle from rare gas systematics. *Nature* **303**, 762–766.

Allison G. B. and Holmes J. W. (1973) The environmental tritium concentrations of underground water and its hydrological interpretation. *J. Hydrol.* **19**, 131–143.

Allison G. B., Gee G. W., and Tyler S. W. (1994) Vadose-zone techniques for estimating groundwater recharge in arid and semiarid regions. *Soil Sci. Soc. Am. J.* **58**, 6–14.

Andrews J. N. and Fontes J.-C. (1993) Comment on "Chlorine 36 dating of very old groundwater: 3. Further results on the Great Artesian Basin, Australia" by T. Torgerson *et al. Water Resour. Res.* **29**, 1871–1874.

Andrews J. N. and Kay R. L. F. (1982a) ^{234}U/^{238}U activity ratios of dissolved uranium in groundwater from a Jurassic limestone aquifer in England. *Earth Planet. Sci. Lett.* **57**, 139–151.

Andrews J. N. and Kay R. L. F. (1982b) Natural production of tritium in permeable rocks. *Nature* **298**, 361–363.

Andrews J. N. and Lee D. J. (1979) Inert gases in groundwater from the Bunter Sandstone of England as indicators of age and paleoclimatic trends. *J. Hydrol.* **41**, 233–252.

Andrews J. N., Giles I. S., Kay R. L. F., Lee D. J., Osmond J. K., Cowart J. B., Fritz P., Barker J. F., and Gale J. (1982) Radioelements, radiogenic helium, and age relationships for groundwaters from the granites at Stripa, Sweden. *Geochim. Cosmochim. Acta* **46**, 1533–1543.

Andrews J. N., Balderer W., Bath A. H., Clausen H. B., Evans G. V., Florkowski T., Goldbrunner J. E., Ivanovich M., Loosli H. H., and Zojer H. (1984) Environmental isotope studies in two aquifer systems: a comparison of ground water dating methods. In *Isotope Hydrology 1983*. IAEA, Vienna, pp. 535–576.

Andrews J. N., Goldbrunner J. E., Darling W. G., Hooker P. J., Wilson G. B., Youngman M. J., Eichinger L., Rauert W., and Stichler W. (1985) A radiochemical, hydrochemical, and dissolved gas study of groundwaters in the Molasse Basin of Upper Austria. *Earth Planet. Sci. Lett.* **73**, 317–332.

Andrews J. N., Davis S. N., Fabryka-Martin J., Fontes J.-C., Lehmann B. E., Loosli H. H., Michelot L., Moser H., Smith B., and Wolf M. (1989) The *in-situ* production of radioisotopes in rock matrices with particular reference to the Stripa granite. *Geochim. Cosmochim. Acta* **53**, 1803–1815.

Andrews J. N., Florkowski T., Lehmann B. E., and Loosli H. H. (1991) Underground production of radionuclides in the Milk River aquifer, Alberta, Canada. *Appl. Geochem.* **6**, 425–434.

Andrews R. W. and Pearson F. J., Jr. (1984) Transport of ^{14}C and uranium in the Carrizo aquifer of South Texas, a natural analog of radionuclide migration. *Mater. Res. Soc. Symp. Proc.* **26**, 1085–1091.

Arnold J. R. and Libby W. F. (1949) Age determinations by radiocarbon content: checks with samples of known age. *Science* **110**, 678–680.

Balderer W. and Synal H.-A. (1996) Application of the chlorine-36 method for the characterisation of the groundwater circulation in tectonic active areas: examples from north western Anatolia/Turkey. *Terra Nova* **8**, 324–333.

Ballentine C. J. and O'Nions R. K. (1991) The nature of mantle neon contributions to Vienna Basin hydrocarbon reservoirs. *Earth Planet. Sci. Lett.* **113**, 553–567.

Ballentine C. J., O'Nions R. K., and Coleman M. L. (1996) A Magnus opus: helium, neon, and argon isotopes in a North Sea oilfield. *Geochim. Cosmochim. Acta* **60**, 831–850.

Bard E. (1998) Geochemical and geophysical implications of the radiocarbon calibration. *Geochim. Cosmochim. Acta* **62**, 2025–2039.

Bebout D. G., Gavenda V. J., and Gregory A. R. (1978) Geothermal resources, Wilcox Group, Texas Gulf Coast. Bur. Econ. Geol., University of Texas at Austin, 82pp.

Benson B. B. (1973) Noble gas concentration ratios as paleotemperature indicators. *Geochim. Cosmochim. Acta* **37**, 1391–1396.

Bentley H. W., Phillips F. M., Davis S. N., Gifford S., Elmore E., Tubbs L. E., and Gove H. E. (1982) Thermonuclear ^{36}Cl pulse in natural water. *Nature* **300**, 737–740.

Bentley H. W., Phillips F. M., and Davis S. N. (1986a) Chlorine-36 in the terrestrial environment. In *Handbook of Environmental Isotope Geochemistry* (eds. P. Fritz and J.-C. Fontes). Elsevier, Amsterdam, vol. 2B, pp. 427–480.

Bentley H. W., Phillips F. M., Davis S. N., Airey P. L., Calf G. E., Elmore D., Habermehl M. A., and Torgersen T. (1986b) Chlorine-36 dating of very old ground water: I. The Great Artesian Basin, Australia. *Water Resour. Res.* **22**, 1991–2002.

Bergmann F. and Libby W. F. (1957) Continental water balance, groundwater inventory and storage times, surface ocean mixing rates, and worldwide water circulation patterns from cosmic ray and bomb tritium. *Geochim. Cosmochim. Acta* **12**, 277–296.

Berner E. K. and Berner R. A. (1996) *Global Environment: Water, Air, and Geochemical Cycles*. Prentice Hall, Upper Saddle River, NJ.

Bethke C. M. and Johnson T. C. (2002a) Ground water age. *Ground Water* **40**, 337–339.

Bethke C. M. and Johnson T. C. (2002b) Paradox of groundwater age: correction. *Geology* **30**, 385–388.

Bethke C. M., Lee M. K., Quinodoz H. A. M., and Kreiling W. N. (1993) *Basin Modeling with Basin2: A Guide to Basin2, B2 Plot, B2 Video, B2 View*. University of Illinois at Urbana-Champaign, Urbana.

Bethke C. M., Zhao X., and Torgersen T. (1999) Groundwater flow and the ^{4}He distribution in the Great Artesian Basin of Australia. *J. Geophys. Res.* **104**, 12999–13010.

Beyerle U., Aeschbach-Hertig W., Hofer M., Imboden D. M., Baur H., and Kipfer R. (1999) Infiltration of river water to a shallow aquifer investigated with ^{3}H/^{3}He, noble gases and CFCs. *J. Hydrol.* **220**, 169–185.

Blavoux B., Dray M., Fehri A., Olive P., Groening M., Sonntag C., Hauquin J. P., Pelissier G., and Pouchan P. (1993) Palaeoclimate and hydrodynamic approach to the Aquitaine Basin deep aquifer (France) by means of environmental isotopes and noble gases. In *Isotope Techniques in the Study of Past and Current Environmental Changes in the Hydrosphere and the Atmosphere*. IAEA, Vienna, pp. 293–305.

Böhlke J. K. and Denver J. M. (1995) Combined use of groundwater dating, chemical, and isotopic analyses to resolve the history and fate of nitrate contamination in two agricultural watersheds, Atlantic coastal plain, Maryland. *Water Resour. Res.* **31**, 2319–2340.

Böhlke J. K., Wanty R., Tuttle M., Delin G., and Landon M. (2002) Denitrification in the recharge area and discharge area of a transient agricultural nitrate plume in a glacial sand outwash aquifer, Minnesota. *Water Resour. Res.* **38**(7), doi: 10.1029/2001WR000663.

Bonell M., Barnes C. J., Grant C. R., Howard A., and Burns J. (1998) High rainfall, response-dominated catchments: a comparative study of experiments in tropical Northeast Queensland with temperate New Zealand. In *Isotope Tracers in Catchment Hydrology* (eds. C. Kendall and J. J. McDonnell). Elsevier, Amsterdam, pp. 334–347.

Bottomley D. J., Ross J. D., and Clarke W. B. (1984) Helium and neon isotope geochemistry of some groundwaters from the Canadian Precambrian Shield. *Geochim. Cosmochim. Acta* **48**, 1973–1985.

Boufadel M. C., Lu S., Molz F. J., and Lavallee D. (2000) Multifractal scaling of the intrinsic permeability. *Water Resour. Res.* **36**, 3211–3222.

Bredehoeft J. D. and Pinder G. F. (1973) Mass transport in flowing groundwater. *Water Resour. Res.* **9**, 194–210.

Bullen T. D. and Kendall C. (1998) Tracing of weathering reactions and water flowpaths: a multi-isotope approach. In *Isotope Tracers in Catchment Hydrology* (eds. C. Kendall and J. J. McDonnell). Elsevier, Amsterdam, pp. 611–646.

Busenberg E. and Plummer L. N. (1991) Chlorofluorocarbons (CCl$_3$F and CCl$_2$F$_2$): use as an age dating tool and hydrologic tracer in shallow ground-water systems. *US Geological Survey, Water Resources Investigations Report 91-4034*.

Busenberg E. and Plummer L. N. (1992) Use of chlorofluorocarbons (CCl$_3$F and CCl$_2$F$_2$) as hydrologic tracers and age-dating tools: the alluvium and terrace system of central Oklahoma. *Water Resour. Res.* **28**, 2257–2284.

Busenberg E. and Plummer L. N. (2000) Dating young groundwater with sulfur hexafluoride: natural and anthropogenic sources of sulfur hexafluoride. *Water Resour. Res.* **36**, 3011–3030.

Buttle J. M. (1998) Fundamentals of small catchment hydrology. In *Isotope Tracers in Catchment Hydrology* (eds. C. Kendall and J. J. McDonnell). Elsevier, Amsterdam, pp. 1–49.

Calf G. E. and Habermehl M. A. (1983) Isotope hydrology and hydrochemistry of the Great Artesian Basin, Australia. In *Isotope Hydrology 1983*. IAEA, Vienna, pp. 397–414.

Carlston C. W., Thatcher L. L., and Rhodehamel E. C. (1960) Trtium as a hydrologic tool, the Wharton tract study. *Int. Assoc. Sci. Hydrol. Publ.* **52**, 503–512.

Castro M. C. and Goblet P. (2003) Calibration of regional groundwater flow models: working toward a better understanding of site-specific systems. *Water Resour. Res.* **39** (6), 1172, doi: 10.1029/2002WR001653.

Castro M. C., Goblet P., Ledoux E., Violette S., and de Marsily G. (1998a) Noble gases as natural tracers of water circulation in the Paris Basin: 2. Calibration of a groundwater flow using noble gas isotope data. *Water Resour. Res.* **34**, 2467–2484.

Castro M. C., Jambon A., de Marsily G., and Schlosser P. (1998b) Noble gases as natural tracers of water circulation in the Paris Basin: 1. Measurements and discussion of their origin and mechanisms of vertical transport in the basin. *Water Resour. Res.* **34**, 2443–2467.

Castro M. C., Stute M., and Schlosser P. (2000) Comparison of ^{4}He ages and ^{14}C ages in simple aquifer systems: implications for groundwater flow and chronologies. *Appl. Geochem.* **15**, 1137–1167.

Chebotarev I. I. (1955) Metamorphism of natural water in the crust of weathering. *Geochim. Cosmochim. Acta* **8**, 22–48, 137–170.

Clark J. F., Stute M., Schlosser P., Drenkard S., and Bonani G. (1997) A tracer study of the Floridan aquifer in southeastern Georgia: implications for groundwater flow and paleoclimate. *Water Resour. Res.* **33**, 281–290.

Clarke W. B., Jenkins W. J., and Top Z. (1976) Determination of tritium by mass spectrometric measurements of ^{3}He. *Int. J. Appl. Radiat. Isotopes* **27**, 515–522.

Clayton R. N., Friedman I., Graf P. L., Mayeda T. K., Meents W. F., and Shimp N. F. (1966) The origin of saline formation waters: I. Isotopic composition. *J. Geophys. Res.* **71**, 3869–3882.

Collon P., Kutschera W., Loosli H. H., Lehmann B. E., Purtschert R., Love A. H., Sampson L., Anthony D., Cole D., Davids B., Morrissey D. J., Sherrill B. M., Steiner M., Pardo R. C., and Paul M. (2000) ^{81}Kr in the Great Artesian Basin, Australia: a new method for dating very old groundwater. *Earth Planet. Sci. Lett.* **182**, 103–113.

Cook P. G. and Böhlke J. H. (2000) Determining time scales for groundwater flow and solute transport. In *Environmental Tracers in Subsurface Hydrology* (eds. P. G. Cook and A. L. Herczeg). Kluwer Academic, Amsterdam, pp. 1–30.

Cook P. G., Jolly I. D., Leaney F. W., Walker G. R., Allan G. L., Fifield L. K., and Allison G. B. (1994) Unsaturated zone tritium and chlorine-36 profiles from southern Australia: their use as tracers of soil water movement. *Water Resour. Res.* **30**, 1709–1719.

Cowart J. B. (1975) Uranium isotope disequilibrium in natural waters, recent studies. *Florida Scientist* **38**, 22.

Cowart J. B. and Osmond J. K. (1974) ^{234}U and ^{238}U in the Carrizo sandstone aquifer of south Texas. In *Isotope Techniques in Groundwater Hydrology*. IAEA, Vienna, vol. II, pp. 149–313.

Craig H. and Lupton J. E. (1976) Primordial neon, helium and hydrogen in oceanic basalts. *Earth Planet. Sci. Lett.* **31**, 369–385.

Craig H. and Lupton J. E. (1981) Helium-3 and mantle volatiles in the ocean and oceanic basalts. In *The Sea* (ed. C. Emiliani). Wiley, New York, pp. 391–428.

Cresswell R. G., Jacobson G., Wischusen J., and Fifield L. K. (1999) Ancient groundwaters in the Amadeus Basin, Central Australia: evidence from the radio-isotope ^{36}Cl. *J. Hydrol.* **223**, 212–220.

Cresswell R. G., Bauld J., Jacobson G., Khadka M. S., Jha M. G., Shrestha M. P., and Regmi S. (2001) A first estimate of ground water ages for the deep aquifer of the Kathmandu Basin, Nepal, using the radioisotope chlorine-36. *Ground Water* **39**, 449–457.

Cunnold D., Weiss R., Prinn R., Hartley D., Simmonds P., Fraser P., Miller B., Alyea F., and Porter L. (1997) GAGE/AGAGE measurements indicating reductions in global emissions of CCl$_3$F and CCl$_2$F$_2$ in 1992–1994. *J. Geophys. Res.* **102**, 1259–1269.

Darcy H. G. (1856) *Les fontaines publiques de la Ville de Dijon*. Victor Dalmont, Paris.

Davis R. J. and Schaeffer O. A. (1955) Chlorine-36 in nature. *Ann. NY Acad. Sci.* **62**, 105–122.

Davis S. N. and Bentley H. W. (1982) Dating groundwater—a short review. In *Nuclear and Chemical Dating Techniques—Interpreting the Environmental Record*, ACS Symposium Series No. 176 (ed. L. A. Currie). American Chemical Society, Washington, DC, pp. 187–222.

Davis S. N., Zreda M., Moysey S., and Cecil L. D. (2003) Chlorine-36 in groundwater of the United States: empirical data. *Hydrogeol. J.* **11**, 217–227.

de Marsily G. (1986) *Quantitative Hydrogeology*. Academic Press, London.

Dinçer T. and Davis G. H. (1984) Application of environmental isotope tracers to modeling in hydrology. *J. Hydrol.* **68**, 95–113.

Domenico P. A. and Schwartz F. W. (1998) *Physical and Chemical Hydrogeology*. Wiley, New York.

Drever J. I. (1997) *The Geochemistry of Natural Waters: Surface and Groundwater Environments*. Prentice Hall, Upper Saddle River, NJ.

Dunkle S. A., Plummer L. N., Busenberg E., Phillips P. J., Denver J. M., Hamilton P. A., Michel R. L., and Coplen T. B. (1993) Chlorofluorocarbons (CCl$_3$F and CCl$_2$F$_2$) as dating tools and hydrologic tracers in shallow groundwater of the Delmarva Peninsula, Atlantic Coastal Plain, United States. *Water Resour. Res.* **29**, 3837–3861.

Ekwurzel B., Schlosser P., Smethie W. M., Jr., Plummer L. N., Busenberg E., Michel R. L., Weppperning R., and Stute M. (1994) Dating of shallow groundwater: comparison of the transient tracers ^{3}H/^{3}He, chlorofluorocarbons, and ^{85}Kr. *Water Resour. Res.* **30**, 1693–1708.

Elmore D., Fulton B. R., Clover M. R., Marsden J. R., Gove H. E., Naylor H., Purser K. H., Kilius L. R., Beukens R. P., and Litherland A. E. (1979) Analysis of ^{36}Cl in environmental water samples using an electrostatic accelerator. *Nature* **277**, 22–25.

Elmore D., Gove H. E., Ferraro R., Kilius L. R., Lee H. W., Chang K. H., Beukens R. P., Litherland A. E., Russo C. J., Purser K. H., Murrell M. T., and Finkel R. C. (1980) Determination of iodine-129 using tandem accelerator mass spectrometry. *Nature* **286**, 138–140.

Engesgaard P., Jensen K. H., Molson J., Frind E. O., and Olsen H. (1996) Large-scale dispersion in a sandy aquifer: simulation of subsurface transport of environmental tritium. *Water Resour. Res.* **32**, 3253–3266.

Fabryka-Martin J. (2000) Iodine-129 as a groundwater tracer. In *Environmental Tracers in Subsurface Hydrology* (eds. P. G. Cook and A. L. Herczeg). Kluwer Academic, Boston, pp. 504–510.

Fabryka-Martin J., Bentley H., Elmore D., and Airey P. L. (1985) Natural iodine-129 as an environmental tracer. *Geochim. Cosmochim. Acta* **49**, 337–348.

Fabryka-Martin J., Whittemore D. O., Davis S. N., Kubik P. W., and Sharma P. (1991) Geochemistry of halogens in the Milk River aquifer, Alberta, Canada. *Appl. Geochem.* **6**, 447–464.

Fairbank W. M., Jr., Hansen C. S., LaBelle R. d., Pan X.-J., Zhang Y., Chamberlin E. P., Nogar N. S., Miller C. M., Fearey B. L., and Oona H. (1998) Photon burst mass spectrometry for the measurement of ^{85}Kr at ambient levels. *Proc. Soc. Photo-Optical Instr. Eng.* **3270**, 174–180.

Faure G. (1986) *Principles of Isotope Geology*. Wiley, New York.

Fehn U., Peters E. K., Tullai-Fitzpatrick S., Kubik P. W., Sharma P., Teng R. T. D., Gove H. E., and Elmore D. (1992) ^{129}I and ^{36}Cl concentrations in waters of the eastern Clear Lake area, California: residence times and source ages of hydrothermal fluids. *Geochim. Cosmochim. Acta* **56**, 2069–2079.

Florkowski T., Morawska L., and Rozanski K. (1988) Natural production of radionuclides in geological formations. *Nucl. Geophys.* **2**, 1–14.

Fontes J.-C. and Garnier J. M. (1979) Determination of the initial ^{14}C activity of the total dissolved carbon—a review of the existing models and a new approach. *Water Resour. Res.* **15**, 399–413.

Foster S. S. D. (1975) The chalk groundwater tritium anomaly—a possible explanation. *J. Hydrol.* **25**, 159–165.

Freeze R. A. and Cherry J. A. (1979) *Groundwater*. Prentice Hall, Englewood Cliffs, NJ.

Freeze R. A. and Witherspoon P. A. (1966) Theoretical analysis of regional groundwater flow: I. Analytical and numerical solutions to the mathematical model. *Water Resour. Res.* **2**, 623–634.

Freeze R. A. and Witherspoon P. A. (1967) Theoretical analysis of regional groundwater flow: II. Effect of water table configuration and subsurface permeability variations. *Water Resour. Res.* **3**, 623–634.

Fröhlich K., Franke T., Gellermann G., Herbert D., and Jordan H. (1987) Silicon-32 in different aquifer types and implications for groundwater dating. In *Proceedings International Symposium on Isotopic Techniques in Water Resources Development*. IAEA, Vienna, pp. 149–163.

Fröhlich K., Ivanovich M., Hendry M. J., Andrews J. N., Davis S. N., Drimmie R. J., Fabryka-Martin J., Florkowski T., Fritz P., Lehmann B., Loosli H. H., and Nolte E. (1991) Application of isotopic methods to dating of very old groundwaters: Milk River aquifer, Alberta, Canada. *Appl. Geochem.* **6**, 465–472.

Gat J. R. (1980) The isotopes of hydrogen and oxygen in precipitation. In *Handbook of Environmental Isotope Geochemistry* (eds. P. Fritz and J. C. Fontes). Elsevier, Amsterdam, pp. 21–34.

Gelhar L. W. (1986) Stochastic subsurface hydrology from theory to applications. *Water Resour. Res.* **22**(suppl.), 135S–145S.

Gelhar L. W. (1993) *Stochastic Subsurface Hydrology*. Prentice Hall, Old Tappan, NJ.

Gerling E. K., Mamyrin B. A., Tolstikhin I. N., and Yaklovleva R. Z. (1971) Isotope composition of helium rocks (in Russian). *Geokhimiya* **5**, 608–617.

Goblet P. (1999) Programme METIS: simulation d'ecoulement et de Transport Miscible in Milieu Poreaux et Fracturé—Notice de conception—Mise à jour 1^{3r}/11/99. *CIG/LHM/RD/99/38*.

Goode D. J. (1996) Direct simulation of groundwater age. *Water Resour. Res.* **32**, 289–296.

Gvirtzman H. and Magaritz M. (1986) Investigation of water movement in the unsaturated zone under an irrigated area using environmental tritium. *Water Resour. Res.* **22**, 635–642.

Gvirtzman H., Ronen D., and Magaritz M. (1986) Anion exclusion during transport through the unsaturated zone. *J. Hydrol.* **87**, 267–283.

Hamlin H. S. (1988) Depositional and groundwater flow systems of the Carrizo-Upper Wilcox, South Texas. *Texas Bureau of Economic Geology, Report of Investigations 175*, Austin.

Hanshaw B. B. and Back W. (1974) Determination of regional hydraulic conductivity through use of ^{14}C dating of groundwater. In *Memoirs de l'Association Internationale des Hydrogeologues*. Montpellier, France, vol. 10, pp. 195–196.

Hanshaw B. B., Back W., and Rubin M. (1965) Radiocarbon determinations for estimating groundwater flow velocities in central Florida. *Science* **148**, 494–495.

Heaton T. H. E. and Vogel J. C. (1981) "Excess air" in groundwater. *J. Hydrol.* **50**, 201–216.

Heaton T. H. E., Talma A. S., and Vogel J. C. (1983) Origin and history of nitrate in confined groundwater in the Western Kalahari. *J. Hydrol.* **62**, 243–262.

Hendry M. J. and Schwartz F. W. (1988) An alternative view on the origin of chemical and isotopic patterns in groundwater from the Milk River aquifer, Canada. *Water Resour. Res.* **24**, 1747–1766.

Hinkle S. R. (1996) Age of groundwater in basalt aquifers near Spring Creek National Fish Hatchery, Skkamania County, Washington. *US Geological Survey, Water-Resources Investigations Report 95-4272*.

Hiyagon G. and Kennedy B. M. (1992) Noble gases in CH_4-rich gas fields, Alberta, Canada. *Geochim. Cosmochim. Acta* **56**, 1569–1588.

Hubbert M. K. (1940) The theory of groundwater motion. *J. Geol.* **48**, 785–944.

Ingebritsen S. E. and Sanford W. E. (1998) *Groundwater in Geologic Processes*. Cambridge University Press, Cambridge.

Ingerson C. W. and Pearson F. J. (1964) Estimation of age and rate of motion of ground-water by the ^{14}C method. In *Recent Researches in the Field of Hydrosphere, Atmosphere, and Nuclear Geochemistry*, Sugawara Festival Volume. Maruzen Company, Tokyo, Japan, pp. 263–283.

Jacobson G., Cresswell R., Wischusen J., and Fifield K. (1998) Arid-zone groundwater recharge and paleorecharge: insights from the radioisotope chlorine-36. *AGSO Res. Lett.* **29**, 1–3.

Kalin R. M. (2000) Radiocarbon dating of groundwater systems. In *Environmental Tracers in Subsurface Hydrology* (eds. P. Cook and A. L. Herczeg). Kluwer Academic, Dordrecht, pp. 111–144.

Kaufman A., Magaritz M., Paul M., Hillaire-Marcel C., Hollus G., Boaretto E., and Taieb M. (1990) The ^{36}Cl ages of the brines in the Magadi-Natron basin, East Africa. *Geochim. Cosmochim. Acta* **54**, 2827–2834.

Kendall C. and McDonnell J. J. (1998) *Isotope Tracers in Catchment Hydrology*. Elsevier, Amsterdam.

Kennedy B. M., Reynolds J. H., and Smith S. P. (1984) Helium isotopes: Lower Geyser Basin, Yellowstone National Park. *EOS* **65**, 304.

Kennedy B. M., Lynch M. A., Reynolds J. H., and Smith S. P. (1985) Intensive sampling of noble gases in fluids at Yellowstone: I. Early overview of the data, regional patterns. *Geochim. Cosmochim. Acta* **49**, 1251–1261.

Kennedy B. M., Hiyagon G., and Reynolds J. H. (1990) Crustal neon: a striking uniformity. *Earth Planet. Sci. Lett.* **98**, 277–286.

Kharkar D. P., Nijampurkar V. N., and Lal D. (1966) The global fallout of Si^{32} produced by cosmic rays. *Geochim. Cosmochim. Acta* **30**, 621–631.

Knauth L. P. and Beeunas M. A. (1986) Isotope geochemistry of fluid inclusions in Permian halite with implications for the isotopic history of ocean water and the origin of saline formation waters. *Geochim. Cosmochim. Acta* **50**, 419–433.

Konikow L. F. and Bredehoeft J. D. (1978) Computer model of two-dimensional solute transport and dispersion in ground water. *US Geological Survey, Techniques of Water Resources Investigations, Book 7, Chapter C2*.

Konikow L. F. and Bredehoeft J. D. (1992) Groundwater models cannot be validated. *Adv. Water Resour.* **15**, 75–83.

Kunz W. and Schintlmeister I. (1965) *Tabellen der atome-kerne*. Akademie-Verlag, Berlin.

Kyser T. K. and Rison W. (1982) Systematics of rare gases in basic lavas and ultramafic xenoliths. *J. Geophys. Res.* **87**, 5611–5630.

Lal D. (1987) Production of ^3He in terrestrial rocks. *Chem. Geol. (Isotope Geosc. Sect.)* **66**, 89–98.

Lal D. and Peters B. (1967) Cosmic ray produced radioactivity on the Earth. In *Handbuch der Physik* (ed. K. Sitte). Springer, Berlin, vol. 46/2, pp. 551–612.

Lal D., Goldberg E. D., and Koide M. (1960) Cosmic-ray-produced ^{32}Si in nature. *Science* **313**, 332–337.

Lal D., Nijampurkar V. N., and Rama S. (1970) Silicon-32 hydrology. In *Isotope Hydrology 1970*. IAEA, Vienna, pp. 847–868.

Lehmann B. E., Loosli H. H., Rauber D., Thonnard N., and Willis R. D. (1991) ^{81}Kr and ^{85}Kr in groundwater, Milk River aquifer, Alberta, Canada. *Appl. Geochem.* **6**, 419–423.

Lehmann B. E., Davis S. N., and Fabryka-Martin J. T. (1993) Atmospheric and subsurface sources of stable and radio-active nuclides used for groundwater dating. *Water Resour. Res.* **29**, 2027–2040.

Lehmann B. E., Love A. H., Purtschert R., Collon P., Loosli H. H., Kutschera W., Beyerle U., Aeschbach-Hertig W., Kipfer R., Frape S., Herczeg A. L., Moran J. E., Tolstikhin I., and Gröning M. (2003) A comparison of groundwater dating with ^{81}Kr, ^{36}Cl, and ^4He in 4 wells of the Great Artesian Basin, Australia. *Earth Planet. Sci. Lett.* **211**, 237–250.

Lehmann B. E., Purtschert R., Loosli H. H., Love A. H., Collon P., Kutschera W., Beyerle U., Aeschbach-Hertig W., Kipfer R., and Frape S. (2002b) Kr-81 calibration of Cl-36 and He-4 evolution in the western Great Artesian Basin, Australia. *Geochim. Cosmochim. Acta* **66**(15A(suppl. 1)), A445.

Libby W. F. (1946) Atmospheric helium three and radiocarbon from cosmic radiation. *Phys. Rev.* **69**, 671–673.

Libby W. F. (1953) The potential usefulness of natural tritium. *Proc. Natl. Acad. Sci.* **39**, 245–247.

Liu X., Fehn U., and Teng R. T. D. (1997) Oil formation and fluid convection in Railroad Valley, NV: a study using cosmogenic isotopes to determine the onset of hydrocarbon migration. *Nucl. Instr. Meth. Phys. Res.* **B123**, 356–360.

Loosli H. H. (1983) A dating method with ^{39}Ar. *Earth Planet. Sci. Lett.* **63**, 51–62.

Loosli H. H. and Oeschger H. (1969) ^{37}Ar and ^{81}Kr in the atmosphere. *Earth Planet. Sci. Lett.* **7**, 67–71.

Loosli H. H., Lehmann B. E., and Balderer W. (1989) Argon-39, argon-37 and krypton 85 isotopes in Stripa groundwaters. *Geochim. Cosmochim. Acta* **53**, 1825–1829.

Loosli H. H., Lehmann B. E., and Smethie W. M., Jr. (2000) Noble gas radioisotopes: ^{37}Ar, ^{85}Kr, ^{39}Ar, ^{81}Kr. In *Environmental Tracers in Subsurface Hydrology* (eds. P. Cook and A. L. Herczeg). Kluwer Academic, Dordrecht, pp. 379–396.

Love A. J., Herczeg A. L., Sampson L., Cresswell R. G., and Fifield L. K. (2000) Sources of chloride and implications for ^{36}Cl dating of old groundwater, southwestern Great Artesian Basin, Australia. *Water Resour. Res.* **36**, 1561–1574.

Lucas L. L. and Unterweger M. P. (2000) Comprehensive review and critical evaluation of the half-life of tritium. *J. Res. Natl. Inst. Stand. Technol.* **105**, 541–549.

Maloszewski P. and Zuber A. (1993) Principles and practice of calibration and validation of mathematical models for the interpretation of environmental tracer data in aquifers. *Adv. Water Resour.* **16**, 173–190.

Mamyrin B. A. and Tolstikhin L. N. (1984) *Helium Isotopes in Nature*. Elsevier, Amsterdam.

Manning C. E. and Ingebritsen S. E. (1999) Permeability of the continental crust: implications of geothermal data and metamorphic systems. *Rev. Geophys.* **37**, 127–150.

Marine W. I. (1979) The use of naturally occurring helium to estimate groundwater velocities for studies of geologic storage of radioactive waste. *Water Resour. Res.* **15**, 1130–1136.

Martel D. J., O'Nions R. K., Hilton D. R., and Oxburgh E. R. (1990) The role of element distribution in production and release of radiogenic helium: the Carnmellis Granite, southwest England. *Chem. Geol.* **88**, 207–221.

Marty B., Criaud A., and Fouillac C. (1988) Low enthalpy geothermal fluids from the Paris sedimentary basin. Characteristics and origin of gases. *Geothermics* **17**, 619–633.

Marty B., Torgersen T., Meynier V., O'Nions R. K., and de Marsily G. (1993) Helium isotope fluxes and groundwater ages in the Dogger aquifer, Paris Basin. *Water Resour. Res.* **29**, 1025–1035.

Mazor E. (1976) Multitracing and multisampling in hydrological studies. In *Interpretation of Environmental Isotope and Hydrochemical Data in Groundwater Hydrology*. IAEA, Vienna, pp. 7–36.

Mazor E. (1990) *Applied Chemical and Isotopic Groundwater Hydrology*. Halsted Press, New York.

Meinzer O. E. (1923) Outline of ground-water hydrology, with definitions. *US Geol. Surv., Water-Supply Pap.* 494.

Meynier V. and Marty B. (1990) Helium isotopes in fluids from the GPF3 scientific borehole at Couy, Cher County, France: implications for deep circulations in the Paris sedimentary basin (abstract), 4th AGU Chapman Conference, Snowbird Utah.

Michel R. A. (2000) Sulphur-35. In *Environmental Tracers in Subsurface Hydrology* (eds. P. G. Cook and A. L. Herczeg). Kluwer Academic, Dordrecht, pp. 502–504.

Michel R. L., Campbell D., Clow D., and Turk J. T. (2000) Timescales for migration of atmospherically derived sulphate through an alpine/subalpine watershed, Loch Vale, Colorado. *Water Resour. Res.* **36**, 27–36.

Milton G. C. D., Milton G. M., Andrews H. R., Chant L. A., Cornett R. J. J., Davies W. G., Greiner B. F., Imanori Y., Koslowsky V. T., Kotzer T., Kramer S. J., and McKay J. W. (1997) A new interpretation of the distribution of bomb-produced chlorine-36 in the environment, with special reference to the Laurentian Great Lakes. *Nucl. Instr. Meth. Phys. Res.* **B123**, 382–386.

Mook W. G. (1976) The dissolution-exchange model for dating groundwater with carbon-14. In *Interpretation of Environmental Isotope and Hydrochemical Data in Groundwater Hydrology*, IAEA, Vienna, pp. 213–225.

Mook W. G. (1980) Carbon-14 in hydrogeological studies. In *Handbook of Environmental Isotope Geochemistry* (eds. P. Fritz and J. C. Fontes). Elsevier, Amsterdam, vol. 1, pp. 49–74.

Moran J. E., Fehn U., and Hanor J. S. (1995) Determination of source ages and migration patterns of brines from the US Gulf Coast basin using ^{129}I. *Geochim. Cosmochim. Acta* **59**, 5055–5069.

Moran J. E., Oktay S. D., and Santschi P. H. (2002) Sources of iodine and iodine 129 in rivers. *Water Resour. Res.* **38**(8), doi: 10.1029/2001WR000622.

Morgenstern U. (2000) Silicon-32. In *Environmental Tracers in Subsurface Hydrology* (eds. P. G. Cook and A. L. Herczeg). Kluwer Academic, Boston, pp. 499–502.

Morgenstern U., Gellermann R., Hebert D., Borner I., Stolz W., Vaikmae R., Rajamae R., and Putnik H. (1995) ^{32}Si in limestone aquifers. *Chem. Geol.* **120**, 127–134.

Morgenstern U., Taylor C. B., Parrat Y., Gaggeler H. W., and Eichler B. (1996) ^{32}Si in precipitation: evolution of temporal and spatial variation and as a dating tool for glacial ice. *Earth Planet. Sci. Lett.* **144**, 289–296.

Morrison P. and Pine J. (1955) Radiogenic origin of helium isotopes in rocks. *Ann. NY Acad. Sci.* **62**, 69–92.

Münnich K. O. (1957) Messungen des ^{14}C-Gehaltes vom hartem Grundwasser. *Naturwisschaften* **44**, 32–33.

Murphy E. M., Davis S. N., Long A., Donahue D., and Jull A. J. T. (1989) ^{14}C in fractions of dissolved organic carbon in ground water. *Nature* **337**, 153–155.

Neuman S. P. (1990) Universal scaling of hydraulic conductivities and dispensivities in geologic media. *Water Resour. Res.* **26**, 1749–1758.

Neuman S. P. (1995) On advective transport in fractal permeability and velocity fields. *Water Resour. Res.* **31**, 1455–1460.

Neuzil C. E. (1995) Abnormal pressures as hydrodynamic phenomena. *Am. J. Sci.* **295**, 742–786.

Neuzil C. E. (2000) Osmotic generation of "anomalous" fluid pressures in geological environments. *Nature* **403**, 182–184.

Nijampurkar V. N., Amin B. S., Kharkar D. P., and Lal D. (1966) "Dating" ground waters of ages younger than 1,000–1,500 years using natural silicon-32. *Nature* **210**, 478–480.

Nijampurkar V. N., Rao D. K., Oldfield F., and Renberg I. (1998) The half-life of ^{32}Si: a new estimate based on varved lake sediments. *Earth Planet. Sci. Lett.* **163**, 191–196.

Nolte E., Krauthan P., Korschinek G., Maloszewski P., Fritz P., and Wolf M. (1991) Measurements and interpretations of ^{36}Cl in groundwater, Milk River aquifer, Alberta, Canada. *Appl. Geochem.* **6**, 435–445.

Nydal R., Lövseth K., Zumbrunn V., and Boden T. A. (1996) Carbon-14 measurements in atmospheric CO_2 from northern and southern hemisphere sites, 1962–1991. *Oak Ridge National Laboratory, US Department of Energy, NDP-057*, Oak Ridge, TN.

Oktay S. D., Santschi P. H., Moran J. E., and Sharma P. (2000) The ^{129}I bomb pulse recorded in Mississippi River Delta sediments: results from isotopes of I, Pu, Cs, Pb, and C. *Geochim. Cosmochim. Acta* **64**, 989–996.

O'Nions R. K. and Oxburgh E. R. (1983) Heat and helium in the Earth. *Nature* **306**, 429–431.

Oxburgh E. R., O'Nions R. K., and Hill R. I. (1986) Helium isotopes in sedimentary basins. *Nature* **324**, 632–635.

Ozima M. and Podosek F. A. (1983) *Noble Gas Geochemistry*. Cambridge University Press, Cambridge, 369pp.

Park J., Bethke C. M., Torgersen T., and Johnson T. M. (2002) Transport modeling applied to the interpretation of groundwater ^{36}Cl age. *Water Resour. Res.* **38**(5), doi: 10.1029/2001WR000399.

Pearson F. J. (1966) Ground-water ages and flow rates by the C^{14} method. PhD, University of Texas.

Pearson F. J., Jr., Balderer W., Loosli H. H., Lehmann B. E., Matter A., Peters T., Schmassmann H., and Gautschi A. (1991) *Applied Isotope Hydrogeology: A Case Study in Northern Switzerland*. Elsevier, New York.

Pearson F. J. J. and White D. E. (1967) Carbon-14 ages and flow rates of water in the Carrizo Sand, Atascosa County, Texas. *Water Resour. Res.* **3**, 251–261.

Pearson F. J. J., Noronha C. J., and Andrews R. W. (1983) Mathematical modeling of the distribution of natural ^{14}C, ^{234}U, and ^{238}U in a regional ground-water system. *Radiocarbon* **25**, 291–300.

Perkins T. K. and Johnston O. C. (1963) A review of diffusion and dispersion in porous media. *Soc. Petrol. Eng. J.* **3**, 70–83.

Person M. A. and Baumgartner L. (1995) New evidence for long-distance fluid migration within the Earth's crust. *Rev. Geophys.* (Supplement, July 1995, US National Report to the International Union of Geodesy and Geophysics 1991–1994), 1083–1092.

Person M. A. and Garven G. (1992) Hydrologic constraints on petroleum generation within continental rift basins: theory and application to the Rhine Graben. *Am. Assoc. Petrol. Geologists Bull.* **76**, 468–488.

Phillips F. M. (1993) Comment on "Reinterpretation of ^{36}Cl data: physical processes, hydraulic interconnections and age estimates in groundwater systems" by E Mazor. *Appl. Geochem.* **8**, 643–647.

Phillips F. M. (2000) Chlorine-36. In *Environmental Tracers in Subsurface Hydrology* (eds. P. Cook and A. L. Herczeg). Kluwer Academic, Dordrecht, pp. 299–348.

Phillips F. M., Bentley H. W., Davis S. N., Elmore D., and Swannick G. B. (1986) Chlorine-36 dating of very old ground water: II. Milk River aquifer, Alberta. *Water Resour. Res.* **22**, 2003–2016.

Phillips F. M., Mattick J. L., Duval T. A., Elmore D., and Kubik P. W. (1988) Chlorine-36 and tritium from nuclear weapons fallout as tracers for long-term liquid and vapor movement in desert soils. *Water Resour. Res.* **24**, 1877–1891.

Phillips F. M., Tansey M. D., Peeters L. A., Cheng S., and Long A. (1989) An isotopic investigation of groundwater in the central San Juan Basin, New Mexico: carbon-14 dating as a basis for numerical flow modeling. *Water Resour. Res.* **25**, 2259–2273.

Phillips F. M., Knowlton R. G., and Bentley H. W. (1990) Comment on Hendry and Schwartz (1988). *Water Resour. Res.* **26**, 1693–1698.

Phillips O. M. (1991) *Flow and Reactions in Permeable Rocks.* Cambridge University Press, Cambridge.

Pinti D. L. and Marty B. (1995) Noble gases in crude oil from the Paris Basin, France: implications for the orgin of fluids and constraints on oil–water–gas interactions. *Geochem. Cosmochim. Acta* **59**, 3389–3404.

Plummer L. N. (1993) Stable isotope enrichment in paleo-waters of the Southeast Atlantic Coastal Plain, United States. *Science* **262**, 2016–2020.

Plummer L. N. and Busenberg E. (2000) Chlorofluorocarbons. In *Environmental Tracers in Subsurface Hydrology* (eds. P. Cook and A. L. Herczeg). Kluwer Academic, Dordrecht, pp. 441–478.

Plummer L. N., Prestemon E. C., and Parkhurst D. L. (1991) An interactive code (NETPATH) for modeling *net* geochemical reactions along a flow path. *US Geological Survey, Water-Resources Investigations Report 91-4078.*

Plummer L. N., Dunkle S. A., and Busenberg E. (1993) Data on chlorofluorocarbons (CCl$_3$F and CCl$_2$F$_2$) as dating tools and hydrologic tracers in shallow groundwater of the Delmarva Peninsula. *US Geological Survey, Open File Report 93-484.*

Prickett T. A., Naymik T. C., and Lonnquist C. G. (1981) A random-walk solute transport model for selected ground-water quality evaluations. *Illinois State Water Survey, Bulletin 65*, Champaign, Illinois.

Puckett L. J., Cowdery T. J., McMahon P. B., Tornes L. H., and Stoner J. D. (2002) Using chemical, hydrologic, and age dating analysis to delineate redox processes and flow paths in the riparian zone of a glacial outwash aquifer-stream system. *Water Resour. Res.* **38**(8), doi: 10.1029/2001WR000396.

Purdy C. B., Burr G. S., Rubin M., Helz G. R., and Mignerey A. C. (1992) Dissolved organic and inorganic ^{14}C concentrations and ages for coastal plain aquifers in southern Maryland. *Radiocarbon* **34**, 654–663.

Purdy C. B., Helz G. R., Mignerey A. C., Kubik P. W., Elmore D., Sharma P., and Hemmick T. (1996) Aquia aquifer dissolved Cl$^-$ and ^{36}Cl/Cl: implications for flow velocities. *Water Resour. Res.* **32**, 1163–1172.

Purtschert R. (1997) Multitracer-Studien in der Hydrologie: Anwendugen im Glattal, am Wellenberg und in Vals. PhD Dissertation, University of Bern.

Rademacher L. K., Clark J. F., Hudson G. B., Erman D. C., and Erman N. A. (2001) Chemical evolution of shallow groundwater as recorded by springs, Sagehen basin: Nevada County, California. *Chem. Geol.* **179**, 37–51.

Rao U. and Fehn U. (1999) Sources and reservoirs of anthropogenic iodine-129 in western New York. *Geochim. Cosmochim. Acta* **63**, 1927–1938.

Reilly T. E., Plummer L. N., Phillips P. J., and Busenberg E. (1994) The use of simulation and multiple environmental tracers to quantify groundwater flow in a shallow aquifer. *Water Resour. Res.* **30**, 412–434.

Rison W. and Craig H. (1983) Helium isotopes and mantle volatiles in Loihi Seamount and Hawaiian Island basalts and xenoliths. *Earth Planet. Sci. Lett.* **66**, 407–426.

Robertson W. D. and Cherry J. A. (1989) Tritium as an indicator of recharge and dispersion in a groundwater system in central Ontario. *Water Resour. Res.* **25**, 1097–1109.

Rozanski K. and Florkowski T. (1979) Krypton-85 dating of groundwater. In *Isotope Hydrology 1979.* IAEA, Vienna, vol. II, pp. 949–961.

Sanford W. E. (1997) Correcting for diffusion in carbon-14 dating of ground water. *Ground Water* **35**, 357–361.

Scanlon B. R. (1992) Evaluation of liquid and vapor water flow in desert soils based on chlorine-36 and tritium tracers and nonisothermal flow simulations. *Water Resour. Res.* **28**, 285–298.

Scanlon B. R., Tyler S. W., and Wierenga P. J. (1997) Hydrologic issues in arid, unsaturated systems and implications for contaminant transport. *Rev. Geophys.* **35**, 461–490.

Schlosser P., Stute M., Dörr H., Sonntag C., and Münnich K. O. (1988) Tritium/^{3}He dating of shallow groundwater. *Earth Planet. Sci. Lett.* **89**, 353–362.

Schlosser P., Stute M., Sonntag C., and Münnich K. O. (1989) Tritiogenic ^{3}He in shallow groundwater. *Earth Planet. Sci. Lett.* **94**, 245–256.

Schmalz B. L. and Polzer W. L. (1969) Tritiated water distribution in unsaturated soils. *Soil Sci.* **108**, 43–47.

Schwartz F. W. (1977) Macroscopic dispersion in porous media: the controlling factors. *Water Resour. Res.* **13**, 743–752.

Schwartz F. W. and Muehlenbachs K. (1979) Isotope and ion geochemistry of groundwaters in the Milk River aquifer, Alberta. *Water Resour. Res.* **15**, 259–268.

Shapiro S. D., LeBlanc D. R., Schlosser P., and Ludin A. (1999) Characterizing a sewage plume using the ^{3}H–^{3}He technique. *Ground Water* **37**, 861–878.

Shukolyukov Y. A., Sharif-Zade V. B., and Ashkinadze G. S. (1973) Neon isotopes in natural gases. *Geochem. Int.*, **10**(2), 346–354.

Simunek J., Sejna M., and van Genuchten M. T. (1999) The HYDRUS-2D software package for simulating the two-dimensional movement of water, heat and multiple solutes in variably-saturated media—version 2.0. *US Department of Agriculture—US Salinity Laboratory*, Riverside, California, 256pp.

Smethie W. M., Jr. and Mathieu G. (1986) Measurement of krypton-85 in the ocean. *Mar. Chem.* **18**, 17–33.

Smethie W. M., Solomon D. K., Schiff S. L., and Mathieu G. G. (1992) Tracing groundwater flow in the Borden aquifer using krypton-85. *J. Hydrol.* **130**, 279–297.

Solomon D. K. (2000) ^{4}He in groundwater. In *Environmental Tracers in Subsurface Hydrology* (eds. P. Cook and A. L. Herczeg). Kluwer Academic, Dordrecht, pp. 425–440.

Solomon D. K. and Cook P. G. (2000) ^{3}H and ^{3}He. In *Environmental Tracers in Subsurface Hydrology* (eds. P. Cook and A. L. Herczeg). Kluwer Academic, Dordrecht, pp. 397–424.

Solomon D. K., Poreda R. J., Schiff S. L., and Cherry J. A. (1992) Tritium and helium-3 as groundwater age tracers in the Borden aquifer. *Water Resour. Res.* **28**, 741–756.

Solomon D. K., Schiff S. L., Poreda R. J., and Clarke W. B. (1993) A validation of the ^{3}H/^{3}He method for determining groundwater recharge. *Water Resour. Res.* **29**, 2951–2962.

Solomon D. K., Hunt A., and Poreda R. J. (1996) Source of radiogenic helium 4 in shallow aquifers: implications for dating young groundwater. *Water Resour. Res.* **32**, 1805–1813.

Steiger R. H. and Jäger E. (1977) Subcommission on geochronology: convention on the use of decay constant in gas and cosmochronology. *Earth Planet. Sci. Lett.* **36**, 359–362.

Stephens D. B. (1996) *Vadose Zone Hydrology*. Lewis Publishers, Boca Raton, FL.

Stute M. and Schlosser P. (1993) Principles and applications of the noble gas paleothermometer. In *Climate Change in Continental Isotopic Records*, Geophysical Monograph 78 (eds. P. K. Swart; K. C. Lohmann, J. A. McKenzie, and S. Savin). American Geophysical Union, Washington, DC, pp. 89–100.

Stute M., Schlosser P., Clark J. F., and Broecker W. S. (1992a) Paleotemperatures in the southwestern United States derived from noble gases in ground water. *Science* **256**, 1000–1002.

Stute M., Sonntag C., Deak J., and Schlosser P. (1992b) Helium in deep circulating groundwater in the Great Hungarian Plain: flow dynamics and crustal and mantle helium fluxes. *Geochim. Cosmochim. Acta* **56**, 2051–2067.

Stute M., Clark J. F., Schlosser P., Broecker W. S., and Bonani G. (1995) A 30,000 yr continental paleotemperature record derived from noble gases dissolved in groundwater from the San Juan Basin, New Mexico. *Quat. Res.* **43**, 209–220.

Sudicky E. A. and Frind E. O. (1981) Carbon-14 dating of groundwater in confined aquifers: implications of aquitard diffusion. *Water Resour. Res.* **17**, 1060–1064.

Suecker J. K., Turk J. T., and Michel R. L. (1999) Use of cosmogenic sulfur-35 for comparing ages of water from three alpine–subalpine basins in the Colorado Front Range. *Geomorphology* **27**, 61–74.

Sugisaki R. (1961) Measurement of effective flow velocity by means of dissolved gasses. *Am. J. Sci.* **259**, 144–153.

Synal H.-A., Beer J., Bonani G., Suter M., and Wölfli W. (1990) Atmospheric transport of bomb-produced ^{36}Cl. *Nucl. Instr. Meth. Phys. Res.* **B52**, 483–488.

Szabo Z., Rice D. E., Plummer L. N., Busenberg E., Drenkard S., and Schlosser P. (1996) Age dating of shallow groundwater with chlorofluorocarbons, tritium/helium 3, and flow path analysis, southern New Jersey coastal plain. *Water Resour. Res.* **32**, 1023–1038.

Tamers M. A. (1967) Radiocarbon ages of groundwater in an arid zone unconfined aquifer. In *Isotope Techniques in the Hydrologic Cycle*, Monograph No. 11 (ed. G. E. Stout). American Geophysical Union, Washington, DC, pp. 143–152.

Tamers M. A. (1975) Validity of radiocarbon dates on ground water. *Geophys. Surv.* **2**, 217–239.

Taylor C. B. (1981) Present status and trands in electrolytic enrichment of low-level tritium in water. In *Methods of Low-level Counting and Spectrometry*. IAEA, Vienna, pp. 303–323.

Thompson G. M. (1976) Trichloromethane: a new hydrologic tool for tracing and dating ground water. PhD, University of Indiana.

Thompson G. M., Hayes J. M., and Davis S. N. (1974) Fluorocarbon tracers in hydrology. *Geophys. Res. Lett.* **1**, 177–180.

Thonnard N., McKay L. D., Cumbie D. H., and Joyner C. P. (1997) Status of laser-based krypton-85 analysis development for dating of groundwater. *Abstr. Prog., Geol. Soc. Am. Ann. Meet.* **29**(6), 78.

Tolstikhin I. N. and Kamensky I. L. (1968) Determination of groundwater age by the T–^{3}He method. *Geochem. Int.* **6**, 810–811.

Torgersen T. (1980) Controls on pore-fluid concentration of helium-4 and radon-222 and calculation of helium-4/radon-222 ages. *J. Geochem. Explor.* **13**, 57–75.

Torgersen T. and Clarke W. B. (1985) Helium accumulation in groundwater: I. An evaluation of sources and the continental flux of crustal ^{4}He in the Great Artesian Basin, Australia. *Geochim. Cosmochim. Acta* **49**, 1211–1218.

Torgersen T. and Ivey G. N. (1985) Helium accumulation in groundwater: II. A model for the accumulation of the crustal

^{4}He degassing flux. *Geochim. Cosmochim. Acta* **49**, 2445–2452.

Torgersen T. and Phillips F. M. (1993) Reply to "Comment on 'Chlorine 36 dating of very old groundwater: 3. Further results on the Great Artesian Basin, Australia' by T. Torgerson *et al.*" by J. N. Andrews and J.-C. Fontes. *Water Resour. Res.* **29**, 1875–1877.

Torgersen T., Kennedy B. M., Hiyagun H., Chiou K. Y., Reynolds J. H., and Clark W. B. (1989) Argon accumulation and the crustal degassing flux of ^{40}Ar in the Great Artesian Basin, Australia. *Earth Planet. Sci. Lett.* **92**, 43–56.

Torgersen T., Habermehl M. A., Phillips F. M., Elmore D., Kubik P., Jones B. G., Hemmick T., and Gove H. E. (1991) Chlorine-36 dating of very old groundwater: III. Further studies in the Great Artesian Basin, Australia. *Water Resour. Res.* **27**, 3201–3214.

Torgersen T., Habermehl M. A., and Clarke W. B. (1992) Crustal helium fluxes and heat flow in the Great Artesian Basin, Australia. *Chem. Geol.* **102**, 139–152.

Tóth J. (1963) A theoretical analysis of groundwater flow in small drainage basins. *J. Geophys. Res.* **67**, 4375–4387.

Tullborg E.-L. and Gustafsson E. (1999) ^{14}C in bicarbonate and dissolved organics—a useful tracer. *Appl. Geochem.* **14**, 927–938.

Tyler S. W., Chapman J. B., Conrad S. H., Hammermeister D. P., Blout D. O., Miller J. J., Sully M. J., and Ginanni J. M. (1996) Soil–water flux in the southern Great Basin, United States: temporal and spatial variations over the last 120,000 years. *Water Resour. Res.* **32**, 1481–1499.

Varni M. and Carrera J. (1998) Simulation of groundwater age distributions. *Water Resour. Res.* **34**, 3271–3282.

Vogel J. C. (1970) Carbon-14 dating of groundwater. In *Isotope Hydrology 1970*. IAEA, Vienna, pp. 225–239.

Vogel J. C., Thilo L., and Van Dijken M. (1974) Determination of groundwater recharge with tritium. *J. Hydrol.* **23**, 131–140.

Vogel J. C., Talma A. S., and Heaton T. H. E. (1982) The age and isotopic composition of groundwater in the Stampriet Artesian Basin, SWA. *National Physical Research Laboratory*, CSIR, Pretoria, South Africa, 49pp.

Wagner M. J. M., Dittrich-Hannen B., Synal H.-A., Suter M., and Schotterer U. (1996) Increase of ^{129}I in the environment. *Nucl. Instr. Meth. Phys. Res.* **B123**, 367–370.

Walvoord M., Phillips F. M., Tyler S. W., and Hartsough P. C. (2002a) Deep arid system hydrodynamics: Part 2. Application to paleohydrologic reconstruction using vadose-zone profiles from the northern Mojave Desert. *Water Resour. Res.* **38**(12), 1291, doi: 10.1029/2001WR000925.

Walvoord M., Plummer M. A., Phillips F. M., and Wolfsberg A. V. (2002b) Deep arid system hydrodynamics: Part 1. Equilibrium states and response times in thick desert vadose zones. *Water Resour. Res.* **38**(12), 1308, doi: 10.0129/2001WR000824.

Walvoord M. A. (2002) A unifying conceptual model to describe water, vapor, and solute transport in deep arid vadose zones. PhD, New Mexico Institute of Mining and Technology.

Wei H. F., Ledoux E., and de Marsily G. (1990) Regional modeling of groundwater flow and salt and environmental tracer transport in deep aquifers in the Paris Basin. *J. Hydrol.* **120**, 341–358.

Weise S. M. (1986) Heliumisotopen-Gehalte im Grumdwasser, Messung und Interpretation. PhD Dissertation, University of Munchen.

Weise S. M. and Moser H. (1987) Groundwater dating with helium isotopes. In *Techniques in Water Resources Development*. IAEA, Vienna, pp. 105–126.

Weiss W. H., Sartorius H., and Stockburger H. (1992) The global distribution of atmospheric krypton-85: a data base for the verification of transport and mixing models.

In *Isotopes of Noble Gases as Tracers in Environmental Studies.* IAEA, Vienna, pp. 105–126.

Weissmann G. S., Carle S. F., and Fogg G. E. (1999) Three-dimensional hydrofacies modeling based on soil surveys and transition probability geostatistics. *Water Resour. Res.* **35**, 1761–1770.

Wetherill G. W. (1954) Variations in the isotopic abundances of neon and argon extracted from radioactive materials. *Phys. Rev.* **96**, 679–683.

Weyhenmeyer C. E., Burns S. J., Waber H. N., Aesbach-Hertig W., Kipfer R., Loosli H. H., and Matter A. (2000) Cool glacial temperatures and changes in moisture source recorded in Oman groundwaters. *Science* **287**, 842–845.

Zerle L., Faestermann T., Knie K., Korschinek G., and Nolte E. (1997) The [41]Ca bomb pulse and atmospheric transport of radionuclides. *J. Geophys. Res.* **102**, 19517–19527.

Zhao X., Fritzel T. L. B., Quinodoz H. A. M., Bethke C. M., and Torgersen T. (1998) Controls on the distribution and isotopic composition of helium in deep ground-water flows. *Geology* **26**, 291–294.

Zhu C. (2000) Estimate of recharge from radiocarbon dating of groundwater and numerical flow and transport modeling. *Water Resour. Res.* **36**, 2607–2620.

Zhu C. and Murphy W. M. (2000) On radiocarbon dating of ground water. *Ground Water* **38**, 802–804.

Zuber A. (1986) On the interpretation of tracer data in variable flow systems. *J. Hydrol.* **86**, 45–57.

Zuber A., Weise S. M., Osenbrück K., Pajnowska H., and Grabczak J. (2000) Age and recharge pattern of water in the Oligocene of the Mazovian basin (Poland) as indicated by environmental tracers. *J. Hydrol.* **233**, 174–188.

5.16
Deep Fluids in the Continents: I. Sedimentary Basins

Y. K. Kharaka

US Geological Survey, Menlo Park, CA, USA

and

J. S. Hanor

Louisiana State University, Baton Rouge, LA, USA

NOMENCLATURE

A number of descriptive terms, including *oil-field brine*, *basinal brine*, *basinal water*, and *formation water*, have been used in the literature to describe deep aqueous fluids in sedimentary basins. No satisfactory overall classification system exists, due to the fact that these waters can be assessed by several different criteria. These include: the salinity of the water, the concentration and origin of various dissolved constituents, and the origin of the H_2O, which is commonly different from that of the solutes. The following terminology has been extracted mainly from Hanor (1987) and from Kharaka and Thordsen (1992). The interested reader should also consult White *et al.* (1963) and Sheppard (1986).

Salinity. Synonymous with *total dissolved solids* (TDS), generally reported in milligrams per liter (mg L^{-1}) as determined either (i) directly by summing measured dissolved constituents or by weighing solid residues after evaporation, or (ii) indirectly from electrical conductivity or spontaneous potential response.

Chlorinity. The dissolved chloride concentration, generally reported in mg L^{-1}.

Formation water. Water present in the pores and fractures of rocks immediately before drilling (Case, 1955). This term is used extensively in the petroleum industry, but has no genetic or age significance.

Brine. Water of salinity higher than that of average seawater, i.e., more than 3.5×10^4 mg L^{-1} TDS. The majority of oil-field waters are brines according to this definition, whereas only a small fraction could be classified as brines based on the definitions of Davis (1964) and Carpenter *et al.*

(1974), which place the lower salinity limit of brines at 1×10^5 mg L^{-1}.

Saline water. Water of salinity $(1-3.5) \times 10^4$ mg L^{-1}.

Brackish water. Water of salinity $(0.1-1) \times 10^4$ mg L^{-1}.

Freshwater. Water of salinity less than 1,000 mg L^{-1}.

Na–Ca–Cl-type water. In this classification scheme, the cations followed by the anions are listed in order of decreasing concentrations. The concentration, commonly in mg L^{-1}, of any ion listed must be $\geq 5\%$ of the concentration of TDS. This is equivalent to $\geq 10\%$ of the total of cations or anions (Kharaka and Thordsen, 1992).

Meteoric water. Water derived from rain, snow, streams, and other bodies of surface water that percolates in rocks and displaces interstitial water that may have been connate, meteoric, or of any other origin. Meteoric water in sedimentary basins is generally recharged at higher elevations along the margins of the basin. The time of last contact with the atmosphere is intentionally omitted from this definition, but may be specified to further define meteoric water. Thus, "Recent meteoric water," "Pleistocene meteoric water," or "Tertiary meteoric water," would indicate the time of last contact with the atmosphere (Kharaka and Carothers, 1986).

Connate water. The word *connate* (Latin for "born-with") was introduced by Lane (1908) to describe what he presumed to be seawater of unaltered chemical composition trapped in the pore spaces of a Proterozoic pillow basalt since the time of extrusion onto the seafloor. The term has since taken on a variety of meanings. While some authors prefer to use connate in its original sense (e.g., Hanor, 1987), others have used it to refer to waters that have been modified chemically and isotopically, but have been out of contact with the atmosphere since their deposition, although they need not be present in the rocks with which they were deposited (e.g., White *et al.*, 1963; Kharaka and Thordsen, 1992). Connate water may be specified as marine connate, if it was deposited with marine sediments.

5.16.1 INTRODUCTION

Pore water with salinities commonly ranging from 5,000 mg L^{-1} to 3×10^5 mg L^{-1} TDS comprises ~20% by volume of most sedimentary basins (e.g., Hanor, 1987; Kharaka and Thordsen, 1992). This water, which is generally sampled while drilling for petroleum or is co-produced with oil and gas, has *in situ* temperatures of ~20 °C to >150 °C and fluid pressures of ~100 bar to >1,000 bar. The chemical and isotopic compositions of this water provide important information on the geochemical, hydrologic, thermal, and tectonic evolution of the Earth's crust. Deep basinal water is an important crustal reservoir of mobile elements, such as the halogens, and fluid and solute fluxes between this water and surface continental water and the oceans are an integral part of the hydrologic and exogenic cycles. Water in sedimentary basins also acts as an intermediate reservoir for volatiles degassing from the lower crust and mantle and as such can be used to study deep-seated processes.

The geochemistry of basinal waters provides insight into a number of important processes that occur within sedimentary basins, especially the (i) generation, transport, accumulation, and production of petroleum; (ii) chemical aspects of mineral diagenesis, including dissolution, precipitation, and the alteration of sediment porosity and permeability; (iii) transport and precipitation of copper, uranium, and especially lead and zinc in sediment-hosted Mississippi-Valley-type ore deposits; (iv) tectonic deformation; (v) transport of thermal energy for geothermal and geopressured–geothermal systems; and (vi) interaction, movement, and ultimate fate of large quantities of liquid hazardous wastes injected into the subsurface (Hanor *et al.*, 1988; Kharaka and Thordsen, 1992; Tuncay *et al.*, 2000).

Interest in the geochemistry of formation waters has risen for two reasons. First, depleted petroleum fields are being investigated as possible repositories for the sequestration of large amounts of liquid CO_2 isolated from power plants (Herzog and Drake, 1998; White *et al.*, 2003). The success of such operations will depend largely on understanding water–mineral–CO_2 interactions (e.g., Gunter *et al.*, 2000). Second, petroleum production, drilling operations, and improperly sealed abandoned wells have caused major contamination of surface and groundwaters and soils in energy-producing states in USA, and probably throughout the world (Richter and Kreitler, 1993; Kharaka *et al.*, 1995). Contamination results mainly from the improper disposal of saline water produced with oil and gas (as of early 2000s, 20–30 billion barrels per year) and from hydrocarbons, and produced water releases caused by equipment failures, vandalism, and accidents. Prior to the institution of federal regulations in the 1970s, produced waters were often discharged into streams, creeks, and unlined evaporation ponds. Because these waters are highly saline and may contain toxic metals, organic and inorganic components, and naturally occurring radioactive material (NORM), including ^{226}Ra and ^{228}Ra, they have caused salt scars and water pollution (Stephenson, 1992; Otton *et al.*, 1997; Kharaka *et al.*, 1999a).

The history of thought on the origin of saline subsurface water dates back to ancient times (White *et al.*, 1963; Hanor, 1983, 1987). The first comprehensive chemical analyses of basinal water only appeared in the late 1800s (e.g., Hunt, 1879). With the rapid development of the oil and gas industry in the early twentieth century, a large database for the composition of formation waters co-produced with hydrocarbons became available. This led to further development of hypotheses regarding the origin of basinal waters (Warren and Smalley, 1994; Gas Research Institute, 1995; Breit *et al.*, 2001). Since the early 1970s there has been a significant expansion in our knowledge and understanding of the properties, interactions, and origin of water in sedimentary basins.

This has come about as a result of (i) improved analytical methodologies that require only a small sample volume for the determination of multi-elements at low concentrations (μg L^{-1} or lower); (ii) increased availability and utilization of data for a variety of stable and radioactive isotopes (Fritz and Fontes, 1980, 1986; Faure, 1986; Clark and Fritz, 1997; Cook and Herczeg, 2000; Chapter 5.15); (iii) major improvements in the chemical thermodynamic data and procedures for applying them to brines and minerals (Johnson *et al.*, 1992; Shock, 1995; Helgeson *et al.*, 1998); and (iv) development and application of detailed geochemical, hydrological, and solute transport codes (Kharaka *et al.*, 1988; Bethke, 1994; Wolery, 1992; Hanor, 2001; Birkle *et al.*, 2002). We now realize that these fluids are much more mobile and that their interactions with rocks are much more complex than previously realized. Also, the discovery of high concentrations (up to 1×10^4 mg L^{-1}) of reactive organic species in these waters has led to a new field of organic–inorganic interactions and has developed bridges between the fields of aqueous fluids, organic matter, and petroleum (Willey et al., 1975; Crossey et al., 1986; Hanor and Workman, 1986; Kharaka et al., 2000).

In this chapter we review what is known about the geochemistry of water in sedimentary basins in the continental and transitional continental oceanic crust. The emphasis is on water below the zone of shallow meteoric groundwater circulation, and on the main processes that are responsible for the modification of the chemical and isotopic composition of these waters including (i) mixing; (ii) dissolution of evaporites, especially halite; (iii) reflux and incorporation of bitterns, the residual water remaining after the precipitation of evaporites; (iv) dissolution and precipitation of minerals other than evaporites; (v) interaction with rocks, principally clays, siltstone, and shale that behave as geological membranes; (vi) activity of bacteria that can survive in sedimentary rocks at temperatures up to $\sim 80\,°C$ (Carothers and Kharaka, 1978); (vii) interactions with organics, including petroleum and solid organic matter; and (viii) diffusion, especially in and near salt domes.

5.16.2 FIELD AND LABORATORY METHODS

Much of the detailed information that has been generated on the composition of deep waters in sedimentary basins has come from the analysis of aqueous fluids co-produced with crude oil and natural gas. A total of ~ 3.5 million oil and gas wells have been drilled in USA alone since 1859; fewer than one million wells are, as of early 2000s, producers (Kharaka and Thordsen, 1992; Breit et al., 2001).

Most sampling takes place at the wellhead rather than down-hole. The fluids are therefore subjected to major reductions in temperature and pressure, to gas loss and to exposure to oxidizing conditions during sampling. The special methods that must be used in sample collection, preservation, and field and laboratory determinations of chemical components and isotopes in formation waters are detailed in Lico et al. (1982) and Kharaka et al. (1985).

Wells selected for sampling must meet the following criteria: they (i) have not been affected by water and CO_2 flooding or chemical treatment, including acidification; (ii) have a single and narrow perforation zone; (iii) produce large amounts of water relative to oil; (iv) produce $> 0.16\,m^3$ of water per $\sim 3 \times 10^4\,m^3$ of gas (10 barrels per million ft^3); and (v) have ports for sampling before the fluid enters a separator. In cases where the objective of the study is to determine disposal options for produced water, sampling from water disposal tanks may be

appropriate to determine the physical and chemical properties of the mixture.

The fluids from petroleum wells are collected in prewashed and prerinsed 8 L or 20 L carboys with a bottom spigot. Water and oil require from five minutes to several hours to separate, depending on the temperature, proportion of water, and the composition of oil and water. Immediately after separation of water from oil, the water is passed through glass wool to remove solids and oil droplets before the samples are collected in separate 125 mL flint-glass bottles with polyseal caps for the field determination of conductance, pH, Eh, alkalinity, and H_2S, and for laboratory determination of the carbon isotopes (Lico et al., 1982).

Filtration and preservation of water samples immediately after collection is important to prevent loss of constituents through precipitation and sorption. Filtration through a $0.45\,\mu m$ filter, using either compressed nitrogen or compressed air as the pressure source, is adequate for determination of the major cations and all of the anions. Filtration through a $0.1\,\mu m$ filter, however, is required for aluminum, mercury, and other trace metals, because colloidal oxyhydroxides of iron and manganese and clay particles can pass through larger pores; these particles would then dissolve upon acidification, increasing the concentration of these trace metals (Kennedy et al., 1974; Kharaka et al., 1987). Filtration and field chemical determinations are better performed in a mobile laboratory equipped with pH meters, a spectrophotometer, and filtration, titration, and other field equipment. Because of the presence of oil, the measurement of Eh is difficult for oil-field waters, even using flow-through cells (Kharaka et al., 1987).

Samples collected for heavy and trace metals (iron, manganese, lead, zinc, and mercury) analyses require additional care to minimize contact of the samples with air during collection and filtration. This is required to prevent oxidation of metals (e.g., Fe^{2+}) and their precipitation as oxyhydroxides leading to co-precipitation and adsorption of other metals. Contact with air is minimized by (i) flushing the air in the carboy with nitrogen or argon or even natural gas; (ii) inserting the tubing from the wellhead as far down in the carboy as possible through a hole drilled through the cap; (iii) filling the carboy completely with the fluids; (iv) plugging the hole in the cap with a rubber stopper after the carboy is filled; (v) minimizing the length of the tygon tubing connecting the filtration unit to the carboy and filling it with formation water prior to filtration; (vi) discarding the first 250 mL of the filtered sample; and (vii) using the next liter of filtered water to rinse the collection bottles (Kharaka et al., 1987).

Samples for the analysis of dissolved organic compounds are filtered through a 0.45 μm Teflon- or silver-filter and stored in amber bottles fitted with Teflon inserts in the caps. Stainless steel filtration units and copper or metal tubing are used for collection and filtration of these samples. Mercuric chloride (40 mg L^{-1} mercury) is added as a bactericide and the filtered samples are stored at ~4 °C until analysis.

New methodologies for the laboratory analysis of cations and metals include the use of inductively coupled plasma emission spectrometry (ICP/ES) or the combination of ICP with mass spectrometry (ICP/MS) (e.g., Ivahnenko et al., 2001). The advantages of plasma techniques include: (i) a wide and linear dynamic concentration range; (ii) multi-element capability; and (iii) relatively free of matrix interferences. The use of ion chromatography (IC), gas chromatography (GC), and GC/MS has increased for the analysis of anions and dissolved organics (Barth, 1987; Kharaka and Thordsen, 1992; Ivahnenko et al., 2001).

Chemical data from drill-stem and wire-line tests are always suspect, because of likely contamination with drilling fluids and of mixing with water from different production zones. Chemical analyses of water from carbonate reservoirs should be carefully examined for signs of contamination. These reservoirs are often stimulated by acid injection, and the contaminating effects of the acid are noticeable for months after treatment. Properly evaluated, chemical data from producing wells may provide concentration values for a limited number of major cations and anions (Hitchon, 1996; Breit et al., 2001). However, the concentrations of many of the dissolved constituents needed for evaluating mineral diagenesis, including field pH, dissolved silica, aluminum, and inorganic alkalinity, are not generally available (Kharaka and Thordsen, 1992).

5.16.2.1 Water from Gas Wells

Chemical analyses from gas wells, especially those from higher-temperature reservoirs, may not represent the true chemical composition of formation waters from the production zone because of dilution by condensed water vapor produced with natural gas. Water vapor condenses because of the drop in temperature and pressure as the gases expand on entering the well. This problem is not generally recognized and is probably responsible for many of the reports of fresh or brackish water in petroleum reservoirs (Kharaka et al., 1985).

5.16.2.2 Information from Wire-line Logs

The salinity of formation waters is often calculated using electrical-resistivity and spontaneous-potential (SP) logs, and the values obtained are reasonable, except in geopressured zones with high shale content (Hearst and Nelson, 1985; Rider, 1996). It is often possible to determine vertical variations in salinity over a distance of several kilometers from a single log.

5.16.3 CHEMICAL COMPOSITION OF SUBSURFACE WATERS

5.16.3.1 Water Salinity

The salinity of pore waters in sedimentary basins varies by approximately five orders of magnitude from a few milligrams per liter in shallow meteoric flow regimes to over 4×10^5 mg L^{-1} in evaporite-rich basins such as the Michigan Basin, USA, and the Williston Basin, USA–Canada. The most saline formation water reported in the literature is a 6.43×10^5 mg L^{-1} Ca–Na–Cl-type brine from the Salina formation of the Michigan Basin (Case, 1945). However, not all sedimentary basins contain brines. Well-known examples include the evaporite-free Central Valley, California, USA (Kharaka and Berry, 1974; Fisher and Boles, 1990), and the Pattani (Lundegard and Trevena, 1990) and Mahakam (Bazin et al., 1997a) basins, Indonesia, where salinities are in general that of seawater or lower.

Salinities in sedimentary basins generally increase with depth, but the rate of increase is highly variable (Table 1 and Figure 1). The variations can be large in waters from different areas of the same basin (Table 2) and even in waters from the same petroleum field (Figure 2). Salinities in some basins, such as the Central Valley, California, and the southern Louisiana and southeastern Texas Gulf Coast, show salinity reversals with depth. Nearly constant water salinities $((1.9–2.4) \times 10^4$ mg L$^{-1})$ have been documented over a depth interval of 700–2,800 m in widely varying lithologies in the North Slope, Alaska (Kharaka and Carothers, 1988).

5.16.3.1.1 Controls on salinity

Spatial variations in salinity put important constraints on the interpretation of the origin of basinal brines and on the quantification of diffusion, advection, and dispersion, which are responsible for subsurface solute transport. For example, lateral salinity plumes have been mapped around a number of shallow Gulf Coast salt domes (e.g., Bennett and Hanor, 1987), providing direct evidence for the dissolution of

Table 1 Chemical composition (mg L^{-1}) of formation waters from Sacramento and San Joaquin Valleys, California; North Slope, Alaska; and the central Mississippi Salt Dome Basin, Mississippi.

	Sacramento		San Joaquin			North Slope		Central Mississippi	
	Grimes	Malton-Black Butte	San Emidio Nose	Wheeler Ridge	Kettleman North Dome	Barrow	Prudhoe Bay	Reedy Creek	West Nancy
Sample #	81-NSV-15	81-NSV-1	74-SEN-3	75-WR-5	912-1	78-AX-52	78-AX-54	84-MS-11	84-MS-1
Well name	GOU4#2	19-1	21-15	21-28	323-21	S. Barrow 5	Arco 13	W.M. Geiger	W.L. West
Prod. zone[a]	Forbes	Forbes	Reef Ridge	Tejon	Lower McAdams	Barrow Sandstone	Sadlerochit Group	Rodessa	Smackover
Depth[b] (m)	2,074	1,524	3,337	2,691	3,520	728	2,820	3,486	4,428
Temp. (°C)	65	58	149	117	141	16	94	102	118
TDS[c]	18,600	21,400	10,900	44,300	10,000	22,100	21,900	320,000	275,000
Li	0.32	0.35	1.95	1.95	3.05	2.1	4.0	35	74
Na	6,830	7,510	4,000	7,450	3,760	7,980	7,600	61,700	54,800
K	35.5	28.4	620	135	92.4	3.0	86	990	6,500
Mg	72	148	7.0	27	3.4	67	20	3,050	3,350
Ca	182	331	67	5,550	30.7	119	182	46,600	33,900
Sr	14.3	18.8	8.0	187	4.4	16.1	20.2	1,920	1,670
Ba	6.4	4.6	4.2	12	3.98	175	3.8	60	48
Fe	0.58	54	0.36	2.8	0.31	5.5	63	465	0.47
NH$_3$	34	30	73	32	8.9	19	17	34	119
F			3.0	2.0	0.3	1.6	1.5	1.5	11.5
Cl	11,000	12,700	3,460	21,450	4,680	11,800	10,600	198,000	170,000
Br	44	74	57	80	45	62	54	2,020	2,080
I	30	66	14	46	27	28	19	17	80
HCO$_3$[d]	359	417	2,870	2,210	1,190	1,710	2,930	206	197
SO$_4$	<0.5	0.9	38	50	0.5	n.d.	69	64	161
H$_2$S	0.07	<0.1	0.02	0.11	3.02	<0.1	<0.1	<0.02	57.4
SiO$_2$	31	18	109	46	128	11	62	28	34
B			92	600	43	42	158	59	342
pH	7.6	7.6	7.7	6.9	7.4	7.2	6.5	5.08	5.48

Source: Kharaka and Thordsen (1992).

[a] Production zones are those used by oil companies.   [b] Depth is depth below ground level of midpoint of perforation.   [c] TDS is calculated total dissolved solids.   [d] HCO$_3$ is field-titrated alkalinity and includes organic and inorganic species.

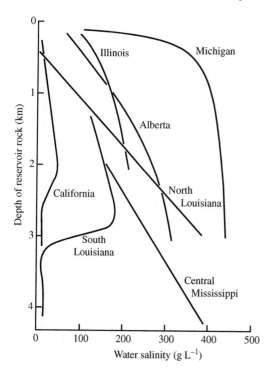

Figure 1 Salinity distribution with depth of the reservoir rocks from several basins in North America. Note the different trends and the reversal of salinity in California and south Louisiana (source Kharaka and Thordsen, 1992).

halite as the source of salinity in these areas. Salinities often increase with depth in basins where there is deep bedded salt and/or deep brines derived from the subaerial evaporation of seawater (Posey and Kyle, 1988; Kharaka and Thordsen, 1992). Diffusive transport (Manheim and Bischoff, 1969; Kharaka, 1986) and dispersive mixing of halite-saturated waters and bittern brines with ambient formation waters and the near-surface recharge of low-TDS meteoric waters into basins (e.g., McIntosh *et al.*, 2002) produce formation waters with a wide range in salinity. These field observations have led to studies involving numerical modeling to investigate the mechanisms and rates of solute transport driven by salt dissolution (e.g., Kharaka, 1986; Ranganathan and Hanor, 1987).

There is also field evidence that halite-derived brines can be transported over long distances in sedimentary basins. For example, the chemical compositions of waters from the Houston–Galveston area, Texas, and several other areas in the northern Gulf of Mexico basin indicate dissolution of halite (Kharaka *et al.*, 1985; Macpherson, 1992). However, in a number of these areas, there are no known salt domes within 50 km of the sampled sites. Large-scale fluid advection is probably the main mechanism for the

transport of dissolved species there, because large fluid potential differences are present in the formations, and numerous faults can act as fluid conduits. Recent advances in quantitative basin modeling (e.g., Ortoleva *et al.*, 1995; Garven, 1995; Person *et al.*, 1996; Wilson *et al.*, 1999) have shown that fluid flow and solute transport can take place on the scale of hundreds of kilometers.

5.16.3.2 Major Cations

Sodium is the dominant cation in oil-field waters. It generally constitutes 70% to >90% of the total cations by mass (Tables 1–3). Calcium is generally the second most abundant cation. Its concentration can rise, especially in Na–Ca–Cl-type waters, to values of up to $\sim 5 \times 10^4$ mg L^{-1} (Table 1). This increase of its concentration with salinity can, however, be different for different parts of the same basin or different basins. The concentrations and proportions of magnesium are generally much lower than those in ocean water and decrease with increasing subsurface temperatures. The concentrations and proportions of strontium, barium, and iron are generally higher than those in ocean water and increase with increasing calcium concentration and chlorinity. The ratios of lithium, potassium, rubidium, and caesium to sodium generally increase with increasing subsurface temperatures; but again, there is a great deal of scatter in the data and their proportions vary from basin to basin (Kharaka and Thordsen, 1992; Hanor, 2001).

Plots of cation concentrations versus chloride differ in slope between the monovalent and divalent cations (Hanor, 1996a, 2001). Both sodium (Figure 3) and potassium show an approximately 1:1 slope on log–log plots, but the divalent cations, magnesium, calcium (Figure 4), and strontium, show approximately 2:1 slopes. This difference in the rate of increase of sodium and calcium with salinity gives rise to the observed progression from Na–Cl to Na–Ca–Cl to Ca–Na–Cl waters with increasing salinity in basinal waters (Hanor, 1987; Wilson and Long, 1993; Davisson and Criss, 1996).

It is generally agreed that most of the chloride in basinal brines has been derived from some combination of the subsurface dissolution of evaporites (e.g., Kharaka *et al.*, 1985; Land, 1997) and the entrapment and/or infiltration of evaporated seawater (e.g., Carpenter, 1978; Kharaka *et al.*, 1987; Moldovanyi and Walter, 1992). Dissolution of halite produces waters dominated by sodium chloride. Evaporation of seawater produces waters having the general trends shown for ion–Br (Figure 5), Na–Cl (Figure 3) and Ca–Cl (Figure 4), but most formation waters have neither the cation (nor anion) composition of an

Table 2 Chemical composition (mg L^{-1}) and production data of formation waters in the geopressured zones from coastal Texas and Louisiana.

	Lafayette, LA		Houston-Galveston, TX		Corpus Christi, TX		McAllen-Pharr, TX	
	Weeks Island	Tigre Lagoon	Chocolate Bayou	Halls Bayou	Portland	East Midway	Pharr	La Blanca
Sample #	77-GG-19	77-GG-55	76-GG-7	76-GG-24	76-GG-63	77-GG-73	77-GG-73	77-GG-117
Well name	St.Un. A#9	Edna Delcambre #1	Angle #3	Houston "FF"#1	Portland A-3	Taylor E-2	Kelly A-1	La Blanca #12 7150
Production Zone [a]	S-Sand	Sand #3	Upper Weiting	Schenck	Morris	Lower Frio	Marks	Sand
Depth [b] (m)	4,275	3,928	3,444	4,161	3,514	3,662	3,018	2,903
Temp. [c] (°C)	117	114	118	150	123	128	127	148
Pressure [d]	43.1	75.8	52.4	80.0	58.0	62.2	52.4	56.6
Fluid production [e]								
Oil/condensate	21.9	0	0.5	3.8	4.8	2.7	0	0.3
Water	56.0	633	6.7	57.9	7.5	0.2	7.1	51.0
Gas	6.1	7.9	5.1	70.8	25.1	4.9	3.2	17.1
TDS [f]	235,700	112,200	73,300	58,100	17,800	36,000	36,600	7,500
Na	78,000	40,000	26,500	20,500	6,500	13,250	9,420	2,680
Li	16	7.1	9.9	15	3.6	4.2	7.5	1.2
K	1,065	265	400	180	68	72	240	46
Rb	3.4	0.8	0.4				0.8	0.1
Cs	11.8	3.5		0.9	0.3	0.5	2.9	3.3
Mg	1,140	270	220	170	15	48	18	150
Ca	10,250	1,860	2,000	800	89	330	4,225	9.6
Sr	920	320	365	170	7.0	23	256	1.5
Ba	185	8.2	290	59	1.4	13	27	<0.1
Fe	84	0.4	10.2	22	2.3	1.6	4.1	
Cl	143,000	67,900	42,700	34,500	9,270	21,000	22,000	3,950
F	0.8	0.8	0.8	32	1.5	7.3	3.9	5.7
Br	419	63	52	11	19	45	78	15
I	18	26	16	91	25	45	22	16
B	44	57	35	13	62	35	105	117
NH$_3$	100	69	29	1.4	5.8	13.5	21.5	4.2
H$_2$S	0.4	0.5	1.2		<0.1	0.04	<0.1	<0.1
HCO$_3$ [g]	450	1,050	455	409	1,600	1,180	114	400
SO$_4$	6.4	220	2.7	16	110	42	7.0	57
SiO$_2$	48	57	87	110	93	132	90	88
pH	6.2	6.3	5.9	6.8	6.8	6.4	6.8	7.3

Source: Kharaka and Thordsen (1992).

[a] Production zones are names used by oil companies. [b] Depth is depth below ground level of midpoint of perforation. [c] Temp. is measured subsurface temperature. [d] Pressure is original bottom-hole pressure in MPa (1 psi = 6.9 kPa). [e] Fluid production is in m^3 d^{-1}. [f] TDS is calculated total dissolved solids. [g] HCO$_3$ is the field titrated alkalinity and includes organic and inorganic species.

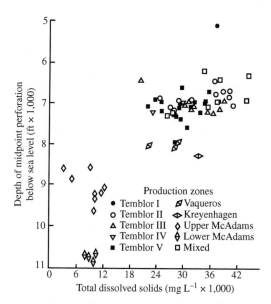

Figure 2 Distribution of salinity of formation waters from Kettleman North Dome oil field, California. Note the much lower salinity of water in the deeper McAdams Formation compared to Temblor Formation (source Kharaka and Berry, 1974).

NaCl solution or of evaporated modern seawater (Stueber and Walter, 1991; Hanor, 2001). During the diagenetic evolution of formation water, there is up to an order of magnitude gain in calcium and strontium (not shown) and up to an order of magnitude loss in magnesium and potassium relative to evaporated modern seawater (Figure 6). As brines derived from the dissolution of halite begin to evolve diagenetically, dissolved sodium is lost and dissolved potassium, magnesium, calcium, and strontium are gained. The observed approximately 1:1 and 2:1 variations in the concentrations of cations versus chloride can be accounted for largely by rock buffering (Hanor, 2001).

Unlike magnesium, calcium, and strontium, there is no significant trend between dissolved barium and chlorinity or salinity. There is, however, a general inverse correlation between barium and SO_4^{2-} (Figure 7), which is consistent with the hypothesis that equilibrium with respect to barite ($BaSO_4$) may be the factor controlling barium concentrations (Kharaka and Berry, 1974; Hanor, 2001).

5.16.3.2.1 Dissolved aluminum

Dissolved aluminum concentrations in subsurface waters are generally less than 0.5 mg L^{-1}; reported higher values are probably due to improper sample treatment in the field (Kharaka et al., 1985, 1987). Determination of dissolved monomeric aluminum requires field filtration through a 0.1 μm or smaller-size filter to prevent contamination with fine clay particles, followed by field solvent extraction as detailed in Lico et al. (1982). There are insufficient data of high quality at present to establish the systematics of dissolved aluminum in most deep sedimentary basins. An alternative approach has been to calculate values for dissolved aluminum using geochemical modeling, assuming fluid equilibrium with respect to muscovite, microcline, albite, or other aluminosilicate minerals that are known to be present in the reservoir rocks. For example, Palandri and Reed (2001) have calculated total dissolved aluminum values from 1×10^{-4} mg L^{-1} to 1×10^{-2} mg L^{-1} for waters from a number of sedimentary basins based on the assumption of equilibrium. Similar calculations by Bazin et al. (1997a,b) for the North Sea and the Mahakam Basin, Indonesia, yield total aluminum concentrations on the order of 3×10^{-3} mg L^{-1}.

The solubility of aluminum in natural waters is commonly treated in terms of $Al_{(aq)}^{3+}$, the aluminum fluoride ($AlF_{n(aq)}^{3-n}$) and hydroxide complexes ($Al(OH)_{n(aq)}^{3-n}$). However, thermodynamic calculations by Kharaka et al. (2000) show that organic acid anions, especially dicarboxylic acid anions, form strong complexes with aluminum. Thermodynamic calculations by Tagirov and Schott (2001) show that in neutral to acidic fluids that contain dissolved fluoride in excess of 1 mg L^{-1} the Al–F complexes $AlF_{n(aq)}^{3-n}$ should dominate aluminum speciation at temperatures below 100 °C and that hydroxy–fluoride complexes of aluminum, especially $Al(OH)_2F_{(aq)}^0$ and $AlOHF_{2(aq)}^0$, should be dominant at temperatures of 100–400 °C. Calculations by these authors show that fluoride concentrations of 2 mg L^{-1}, which are common in basinal waters (Kharaka and Thordsen, 1992; Worden et al., 1999), could be enough to increase total aluminum concentrations as much as two orders of magnitude. Such an increase would adjust calculated aluminum concentrations in the above studies to values on the order of 0.01–1.0 mg L^{-1}. These values are similar to the range of aluminum values of 0.04–0.37 mg L^{-1} determined by direct chemical analysis by Kharaka et al. (1987) in the brines from Mississippi Salt Dome Basin, USA.

5.16.3.2.2 Water pH

There is a general decrease in reported pH values with increasing chlorinity, and formation waters become progressively more acidic. It should be noted that the reported pH values were made in the field or in the laboratory on fluids at Earth surface conditions and are uncorrected for *in situ* pressure–temperature (*P–T*) conditions and the possible loss of volatiles, especially CO_2. The true *in situ* pH values of deep subsurface

Table 3 Chemical composition (mg L^{-1}) of formation waters from High Island Field, offshore Texas.

Sample[a] #	83-TX-1	83-TX-2	83-TX-3	83-TX-5	83-TX-6	83-TX-7	83-TX-8	83-TX-9	83-TX-10	83-TX-11	83-TX-12
Well name	A-11A	A-11B	A-10A	A-9A	C-8A	C-14A	B-14A	B-2A	A-14B	A-12A	A-45-1
Depth[b] (m)	2045	1876	2184	2140	1849	1838	1789	1741	2216	2052	3719
Temp.[c] (°C)	70	55	66	67	53	52	50	49	77	64	98
TDS[d]	74,000	80,000	61,000	159,000	44,000	39,000	138,000	92,000	63,000	88,000	91,000
Li	1.79	1.97	1.74	5.03	0.63	0.58	4.89	2.21	1.79	2.25	5.50
Na	26,300	28,400	19,800	53,700	15,400	13,300	47,700	31,800	22,300	30,400	30,900
K	160	172	139	244	110	103	250	167	143	183	271
Rb	0.15	0.16	0.12	0.24	0.08	0.10	0.19	0.14	0.17	0.16	0.39
Cs	0.09	0.07	0.07	0.29	0.03	0.04	0.24	0.15	0.07	0.14	0.21
Mg	788	869	287	1,270	499	349	1,040	1,010	363	873	482
Ca	1,320	1,370	1,030	3,810	679	850	2,840	2,250	1,130	2,010	2,690
Sr	97	119	59	346	36	27	238	147	66	129	146
Ba	77	102	75	316	40	52	256	171	90	113	200
Mn	0.28	0.19	0.16	0.60	0.09	0.23	0.25	0.43	0.29	0.43	2.63
Fe	7.4	5.8	1.7	17.2	4.3	3.4	11.3	4.4	13.9	5.0	6.5
NH$_3$	140	152	97	206	73	83	182	170	112	182	121
B	23	25	31	33	16	16	40	18	37	24	42
Al (μg L^{-1})				<1	15				<1	<1	3
H$_2$S-titration	0.25	0.68	0.85	0.08	0.00	0.51	0.68	0.51	0.85	0.93	0.76
H$_2$S-electrode	0.04	0.02	0.06	<0.01	0.05	0.04	0.08	0.09	<0.01	0.08	<0.01
Organic (C)	54	60	162	78	73	82		36	209	47	933
Inorganic (C)	89	94	71	34	84	76		64	44	75	38
F	2.3	2.9	1.9	0.53	2.0	5.9	0.81	0.50	13.1	0.81	0.51
Cl	44,400	48,400	33,400	99,000	26,200	23,000	84,600	56,200	37,700	53,800	53,300
Br	143	88.9	102	177	102	287	90.9	108	185	113	118
I	4.0	4.7	4.5	3.1	5.1	4.3	4.6	4.4	5.6	4.3	6.2
SO$_4$	12.4	9.9	16.0	7.4	9.9	9.9	0.5	0.4	14.8	15.3	11.5
SiO$_2$	40.1	38.8	45.4	30.0	40.8	60.0	31.6	32.9	162	41.2	65.8
Alk[e] (HCO$_3$)	494	546	568	308	480	443	293	350	460	402	2,140
pH	7.51	7.39	7.34	7.25	7.59	7.54	7.09	7.11	7.60	6.97	6.83

Source: Kharaka and Thordsen (1992).

[a] Samples from Sigh Island 573 field (Mobil Oil) except sample 83-TX-12, which is from High Island 571 field (Columbia Gas). [b] Depth is that of midpoint of perforation below sea level (depth of water column is ~ 75m).
[c] Temp. is the measured subsurface temperature. [d] TDS = mg L^{-1} total dissolved solids. [e] Alk is field-titrated alkalinity as HCO$_3$, and comprises organic as well as inorganic species.

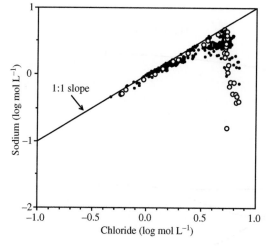

Figure 3 Concentrations of Na in basinal brines as a function of Cl (solid dots). Open circles represent Na–Cl variations in evaporated seawater (source Hanor, 2001).

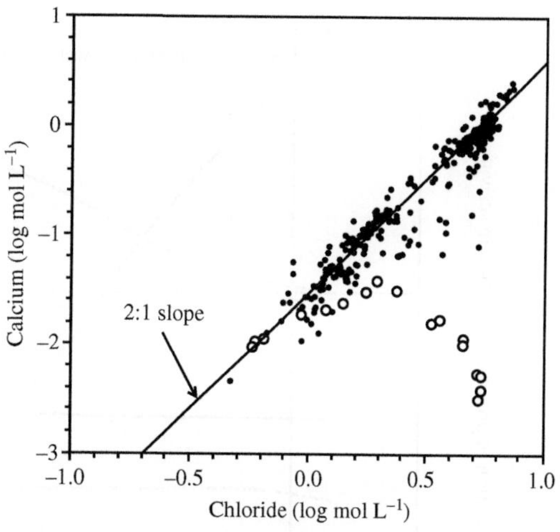

Figure 4 Concentrations of Ca in basinal brines as a function of Cl (solid dots). Open circles represent Ca–Cl variations in evaporated seawater (source Hanor, 2001).

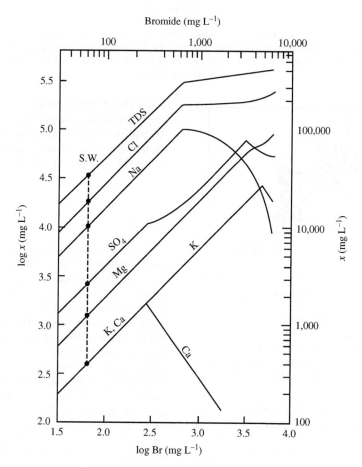

Figure 5 Concentration trends of salinity and several cations and anions in evaporating seawater (after Carpenter, 1978; Kharaka and Thordsen, 1992).

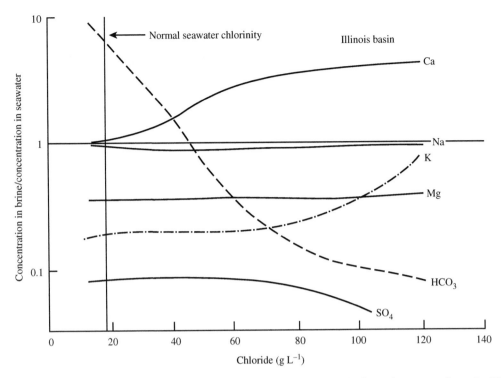

Figure 6 Generalized relative variations in the composition of major ions in formation waters from the Illinois basin. The normalized ratios indicate that relative to seawater, the formation waters are always depleted in SO_4^{2-}, Mg, and K, but that they become progressively enriched in Ca and K and depleted in HCO3 with increasing chlorinity (after Hanor, 1987).

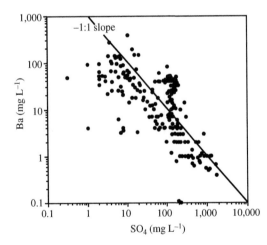

Figure 7 Variation in dissolved Ba with SO_4^{2-} for basinal waters worldwide. Inverse relation suggests buffering by barite (source Hanor, 2001).

fluids are largely unknown, but can be computed from field measurements of pH that have been corrected by titration with the lost CO_2 (Kharaka et al., 1985) or on the basis of assumed equilibrium relations with calcite under subsurface conditions (Merino, 1975, 1979; Palandri and Reed, 2001). The loss of acid volatiles, such as CO_2, generally increases pH by 1–2 pH units (Kharaka et al., 1985). Many of the values for the

in situ pH of highly saline waters of the Smackover formation, Arkansas, calculated on the assumption of fluid equilibrium with respect to calcite, are substantially higher than field pH values (Moldovanyi and Walter, 1992).

5.16.3.2.3 Dissolved silica

Silica concentrations in formation waters range from ~10 mg L^{-1} to ~200 mg L^{-1} as SiO_2 and are controlled primarily by the solubilities of quartz, chalcedony, or cristobalite, which are dependent on subsurface temperature and, to a lesser extent, on pressure (Kharaka and Mariner, 1989). Many authors have noted that although quartz is a nearly ubiquitous phase in sedimentary basins, many basinal waters are not in thermodynamic equilibrium with quartz (Land and Macpherson, 1992b). The slow precipitation kinetics of quartz at low to moderate basinal temperatures results in fluid equilibration with metastable polymorphs of silica such as Opal-A and Opal-CT (Bjørlykke and Egeberg, 1993) or with ultramicrocrystalline (<0.02 mm) quartz (Azaroual et al., 1997). Pore waters can become greatly supersaturated with respect to quartz (Hutcheon, 2000). According to Kharaka and Mariner (1989) and Bjørlykke et al. (1995) quartz equilibrium is achieved when temperatures exceed ~70 °C. In the Mahakam Basin, Indonesia, quartz equilibrium temperatures are on the order of 100–120 °C (Bazin *et al.*, 1997a).

5.16.3.2.4 Boron

The boron cation, B^{3+}, exists in aqueous solution as undissociated boric acid $B(OH)_3$ and as the borate ion $B(OH)_4^-$. Formation waters with pH values lower than 9 favor $B(OH)_3$ as the predominant species, because the pK values for reaction (1)

$$B(OH)_4^- = B(OH)_3 \text{ (aq)} + (OH)^- \qquad (1)$$
$$\underset{\text{borate}}{} \qquad \underset{\text{boric acid}}{}$$

are 4.77 at 25 °C, and 3.42 at 100 °C (Kharaka *et al.*, 1988). Bassett (1977) presented evidence for the existence of polynuclear boron species when boron concentrations are high. Organic–boron complexes may also exist, and organically bound boron has the potential to be used to trace hydrocarbon migration paths in subsurface waters (Mackin, 1987).

Boron is leached from rocks and organic matter, especially at high temperatures. Removal mechanisms for boron include adsorption on clay mineral surfaces at low temperatures (<120 °C) (You *et al.*, 1996) and the incorporation of boron in exchange for tetrahedral silicon during higher-temperature silicate diagenesis (Spivack *et al.*, 1987). Clay minerals play a key role in the boron budget. The clay mineral group illite/smectite (Harder, 1970) contains an order of magnitude more boron than quartz, carbonate, and feldspar. Some types of organic matter contain several hundred ppm boron. Because metasedimentary graphite contains little boron, it is probably released together with hydrogen and oxygen during the thermal degradation of organic compounds (Williams *et al.*, 2001a).

The reported boron concentrations in Gulf Coast (USA) brines range from a few to ~700 mg L^{-1} (Kharaka *et al.*, 1987; Land and Macpherson, 1992a). Dissolved boron shows no correlation with chloride concentration but does show some increase with depth and temperature (Kharaka *et al.*, 1985). The B/Br ratios are highly elevated relative to the seawater evaporation trend for boron and bromide, reflecting derivation of almost all of the boron from rock and/or organic sources.

5.16.3.3 Water–Rock Reactions Controlling Cation Concentrations

Dissolution of halite, as noted earlier, is probably the most important mechanism responsible for the increased sodium (and chloride)

concentration in the very high salinity (>1 × 10^5 mg L^{-1} dissolved solids) brines present in many sedimentary basins where evaporites are, or were present. The concentrations of cations, especially multivalent species, are determined by the origin of the water and by the many chemical, physical, and biological processes that modify the original composition of the water. These processes generally act together to increase or decrease the concentrations of the solutes. The ultimate control on the concentration of a given solute is usually the solubility of the least soluble mineral of the solute. For example, the concentration of calcium in water from a given reservoir may increase because of membrane filtration, albitization of plagioclase feldspar, and/or dolmitization of limestone. Eventually, the water will attain saturation with respect to calcite, the usual ultimate control on the calcium (and carbonate) concentrations.

Congruent and incongruent dissolution and precipitation reactions, other than for halite, which probably control the major cation compositions of formation waters include dolomitization of limestone, resulting in a major increase of calcium and a major decrease of magnesium, as in reaction (2):

$$2CaCO_{3(s)} + Mg^{2+} = CaMg(CO_3)_{2(s)} + Ca^{2+} \qquad (2)$$
$$\underset{\text{calcite}}{} \qquad \underset{\text{dolomite}}{}$$

Albitization of plagioclase feldspar, as in reaction (3) below, also increases calcium concentrations, but lowers the concentration of sodium:

$$Na_{0.7}Ca_{0.3}Al_{1.3}Si_{2.7}O_{8(s)} + 0.6Na^+ + 1.2SiO_{2(s)}$$
$$\underset{\text{andesine}}{} \qquad \underset{\text{quartz}}{}$$
$$= 1.3NaAlSi_3O_8 + 0.3Ca^{2+} \qquad (3)$$
$$\underset{\text{albite}}{}$$

The concentrations of potassium in the samples obtained from the Norphlet Formation in the central Mississippi Salt Dome Basin are those expected from the Louann Salt bittern; potassium values in other samples obtained from reservoirs of Jurassic age are lower by a factor of ~2 (Kharaka *et al.*, 1987). The decrease in the dissolved potassium in these samples is attributed to the formation of authigenic illite and potassium feldspar (Carpenter *et al.*, 1974; Kharaka *et al.*, 1987).

The generally lower magnesium concentrations in formation water, in comparison to that of evaporated seawater, could result from diagenetic formation of chlorite, dolomite, and ankerite (Hower *et al.*, 1976; Boles, 1978). Formation of ankerite becomes important at subsurface temperatures higher than ~120 °C (Boles, 1978).

The concentrations of alkali metals, in the absence of evaporites, are strongly affected by temperature-dependent reactions with clays (transformation of smectite to mixed layer illite/smectite, then with increasing temperature to illite) and feldspar (e.g., Kharaka and Thordsen, 1992). The concentrations of potassium and sodium may be higher or lower than in seawater; the concentrations of lithium and rubidium are generally higher.

The transformation of smectite to mixed layer smectite–illite, and ultimately to illite, with increasing temperature is an extremely important reaction in many sedimentary basins, including the northern Gulf of Mexico Basin (Hower *et al.*, 1976; Boles and Franks, 1979; Kharaka and Thordsen, 1992). The water and solutes released and consumed by this transformation are major factors in the hydrogeochemistry of these basins, because of the enormous quantities of clays involved. Several reactions conserving aluminum or maintaining a constant volume have been proposed for this transformation (Hower *et al.*, 1976; Boles and Franks, 1979). The reaction proposed below (Equation (4)) conserves aluminum and magnesium, and is probably a closer approximation based on the composition of formation waters in these systems:

$$
\begin{aligned}
&10.8H^+ + 3.81K^+ \\
&\quad + 1.69KNaCa_2Mg_4Fe_4Al_{14}Si_{38}O_{100}(OH)_{20} \cdot 10H_2O \\
&\qquad \text{smectite} \\
&= K_{5.5}Mg_2Fe_{1.5}Al_{22}Si_{35}O_{100}(OH)_{20} \\
&\qquad \text{illite} \\
&\quad + 1.59Mg_3Fe_2AlSi_3O_{10}\,(OH)_8 \\
&\qquad \text{chlorite} \\
&\quad + 24.4SiO_{2(s)} + 22.8H_2O \\
&\qquad \text{quartz} \\
&\quad + 1.69Na^+ + 3.38Ca^{2+} + 2.06Fe^{3+} \qquad (4)
\end{aligned}
$$

The Fe^{3+} in reaction (4) will be reduced to Fe^{2+} by organics and some may precipitate as pyrite or ankerite. At any rate, the overall reaction consumes large amounts of potassium and hydrogen and adds important amounts of calcium, sodium, and some Fe^{2+} to the formation water.

5.16.3.3.1 Cation geothermometry

The Li/Na, K/Na, and Rb/Na ratios generally increase with increasing depth (increasing temperatures). The proportions of alkali metals alone or combined with those of alkaline-earth metals (magnesium and calcium in particular), and the concentrations of SiO_2 are so strongly dependent on subsurface temperatures that they have been combined into several chemical geothermometers (Table 4) that can be used to estimate these temperatures (Fournier *et al.*, 1974; Kharaka and Mariner, 1989; Pang and Reed, 1998). The most useful "chemical markers" for increasing subsurface temperatures are the concentrations of silica, boron, and ammonia, and the Li/Mg, Li/Na, and K/Na ratios.

5.16.3.4 Major Anions

At salinities of less than $\sim 1 \times 10^4$ mg L^{-1} and relatively shallow depths, the anionic composition of subsurface water is highly variable and can be dominated by sulfate, bicarbonate, chloride, or even acetate (Hem, 1985; Drever, 1997). Shallow groundwater generally is dominated by sulfate, which is replaced by bicarbonate as the dominant species in deeper meteoric groundwater. Acetate may comprise a large portion of total anions, especially in the $Na-Cl-CH_3COO$-type waters that are present mainly in Cenozoic reservoir rocks at temperatures of 80–120 °C. In these waters, acetate and other organic acid anions (see Section 5.16.3.5) can reach concentrations of up to 1×10^4 mg L^{-1} and contribute up to 99% of the measured alkalinities (Willey *et al.*, 1975; Kharaka *et al.*, 2000).

Chloride is by far the dominant anion in nearly all formation waters having salinities in excess of 3×10^4 mg L^{-1} (Tables 1 and 2). Explaining the origin of saline waters in sedimentary basins is, to some degree, the problem of explaining the origin of the dissolved chloride. Chloride and bromide are fairly closely coupled in their subsurface geochemistry, but the other dissolved halogens, fluoride and iodide, have distinctly different systematics in basinal waters. Sulfate, bicarbonate, and the organic acids provide valuable information on the effects of reactions involving organic matter on formation water chemistry. Other anionic species and weak acids, such as borate and boric acid, may provide information on the degree of water–rock interaction.

5.16.3.4.1 Chloride and bromide

The principal sources of dissolved chloride in the more saline fluids of sedimentary basins include dissolved chloride buried at the time of sediment deposition, chloride derived by refluxing of subaerially evaporated surface brines, chloride derived from subsurface mineral dissolution, principally halite, and marine aerosols. The Cl–Br systematics of sedimentary brines provide useful constraints on interpreting the origin of chloride in these waters (Carpenter, 1978; Kharaka *et al.*, 1987; Kesler *et al.*, 1996).

Although bromide and chloride are both monovalent anions of similar ionic radii ($Br^- = 1.96$ Å, $Cl^- = 1.81$ Å), Cl^- is strongly preferentially partitioned over Br^- into sodium, potassium, and magnesium halogen salts during precipitation (Hanor, 1987; Siemann and Schramm, 2000). During the initial evaporation of seawater, both bromide and chloride increase in concentration in the residual hypersaline waters, and the Br/Cl ratio of these waters does

Table 4 Equations for the general chemical geothermometers and their applicability for use in formation waters from sedimentary basins.

Geothermometer	Equation[a]	Recommendations
Quartz	$t = \dfrac{1,309}{0.41 - \log(k \cdot pf)} - 273$	70–250 °C
	$k = \dfrac{a_{H_2SiO_4}}{a_{H_2O}^2}$; $pf = (1 - 7.862 \times 10^{-5} e^{(3.61 \times 10^{-3} \cdot t)} p)$	
Chalcedony	$t = \dfrac{1,032}{-0.09 - \log(k \cdot pf)} - 273$	30–70 °C
Mg–Li	$t = \dfrac{2,200}{\log(\sqrt{Mg/Li}) + 5.47} - 273$	0–350 °C
Na–K	$t = \dfrac{1,180}{\log(Na/K) + 1.31} - 273$	Do not use in oil-field waters
Na–K–Ca	$t = \dfrac{1,647}{\log(Na/K) + \beta[\log(\sqrt{Ca/Na}) + 2.06] + 2.47} - 273$	Do not use in oil-field waters
	$\beta = 4/3$ for $t < 100$; $= 1/3$ for $t > 100$	
Mg-Corrected,	Same as Na–K–Ca (above) with Mg-corrections	
Na–K–Ca	$t = t_{Na-K-Ca} - \Delta t_{Mg}$	0–350 °C
	For $0.5 < R < 5$	
	$\Delta t_{Mg} = 1.03 + 59.971 \log R + 145.05(\log R)^2$ $-36,711(\log R)^2/T - 1.67 \times 10^7 \log(R/T^2)$;	
	For $5 < R < 50$	
	$\Delta t_{Mg} = 10.66 - 47,415 R + 325.87(\log R)^2$ $-1.032 \times 10^5(\log R)^2/T - 1.968 \times 10^7(\log R)^2/T^2)$ $+1.605 \times 10^7(\log R)^3/T^2$;	
	No correction should be attempted if $R > 50$.	
	$R = \dfrac{Mg}{Mg + 0.61\,Ca + 0.31\,K} \times 100$	
Na/Li	$t = \dfrac{1590}{\log(Na/Li) + 0.779} - 273$	0–350 °C

Source: Kharaka *et al.* (1988).
[a] Concentrations in mg L^{-1}; t is temperature in °C; T in K; p is pressure in bars; a is activity of the subscripted species; Na–K–Ca equation and Mg corrections from Fournier and Potter (1979) (For details see Kharaka and Mariner, 1989).

not vary (Figure 8). When halite saturation is reached, chloride is preferentially precipitated as a constituent of halite. Because only a small fraction of the bromide is incorporated in the halite lattice as Na (Cl, Br), the Br/Cl ratio of the residual brine increases with progressive evaporation (Figure 8). As saturation with respect to K–Mg–Cl salts is reached, the slope of the Br–Cl curve begins to flatten out, because these minerals discriminate against bromide somewhat less than halite. The upper limit to bromide concentrations produced by evaporating seawater is $\sim 0.6 \times 10^4$ mg L^{-1}; the upper limit for chloride is $\sim 2.5 \times 10^5$ mg L^{-1}.

Brines formed by subaerial evaporation of seawater should, in theory, have elevated Br/Cl ratios. Brines formed by the dissolution of halite should have low Br/TDS (Rittenhouse, 1967) and Br/Cl ratios (Carpenter, 1978; Kharaka *et al.*, 1987). Brines representing these end-members and mixtures of these and/or meteoric and/or connate marine waters have been identified in

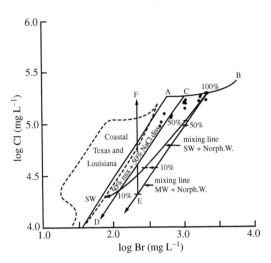

Figure 8 Distribution of Cl and Br in formation waters from the central Mississippi Salt Dome Basin relative to the evaporation line for seawater (SW, A–B) and mixing lines between Norphlet water (Norph. W) and meteoric (MW) and sea (SW) waters. Line E–F gives the trend when the mixture of meteoric and Norphlet waters dissolves halite with 70 ppm Br. Line C–D gives the trend where 50% of the Cl concentration in the mixture of meteoric and Norphlet waters is from dissolution of halite. Note that the samples from Coastal Texas and Louisiana (dashed field) plot in a different field (source Kharaka and Thordsen, 1992).

sedimentary basins on the basis of their Br/Cl and Br/TDS ratios (Kharaka *et al.*, 1987; Hanor, 1987; Worden, 1996).

The high Br/Cl ratios of waters in the Smack-over Formation, USA, for example, support the hypothesis that bromide-rich, subaerially produced brines are important saline end-members in this system (Kharaka *et al.*, 1987; Moldovanyi and Walter, 1992). The central Mississippi Salt Dome Basin provides an excellent example of a system where bittern (residual) water is an important component of the formation water (Table 3). The very high salinities of the brines (up to 3.5×10^5 mg L^{-1} dissolved solids) and their major ion concentrations are directly or indirectly related to their origin as bitterns in the Louann Salt. This conclusion is based on the relation between the "chemical markers" bromide and chloride, sodium and total cations, as well as on the isotopic composition of the water (Carpenter *et al.*, 1974; Stoessell and Carpenter, 1986; Kharaka *et al.*, 1987). The bromide and chloride concentrations are much higher than those expected from dissolution of halite and the values (Figure 8) plot on or below the seawater evaporation line. Kharaka *et al.* (1987) indicate that the samples that plot on or below the seawater evaporation line result from mixing of bitterns with meteoric waters (Figure 8).

The low Br/Cl values of waters in South Louisiana, USA, in contrast, indicate that their high salinity is derived from the dissolution of halite-dominated salt domes (Kharaka *et al.*, 1978, 1985; Hanor, 1987). Other examples include brines in the Paradox Basin (Hanshaw and Hill, 1969) and fluids in Devonian sediments of the Alberta Basin, Canada (Hitchon *et al.*, 1971). The waters of the Norwegian Shelf have an intermediate Br/Cl signature. This supports the conclusion of Egeberg and Aagaard (1989) that they contain at least some contribution from subaerially produced brines in addition to brines generated by the dissolution of halite.

There are several other processes that modify the Cl–Br systematics in formation waters. These include the incongruent dissolution of halite, the incongruent dissolution of chloride salts other than halite, differential rates of molecular diffusion, and the introduction of bromide from organic compounds (Land and Prezbindowski, 1981). Br/Cl ratios in excess of those normally associated with subaerial evaporation may result from the incongruent dissolution of Na–K–Mg–Cl mineral assemblages during progressive burial (Hanor, 1987; Land *et al.*, 1995).

5.16.3.4.2 Iodide

The concentration of dissolved iodide typically ranges from <0.01 mg L^{-1} to >100 mg L^{-1} in basinal waters (Collins, 1975; Worden *et al.*, 1999). An exceptional value of 1,560 mg L^{-1} has been reported for a brine sample from Mississippian limestone of the Anadarko Basin, Oklahoma, where iodide is extracted from the brine commercially (Johnson and Gerber, 1998). There is no correlation between the iodide and chloride concentrations, and the occurrence of iodide appears to be unrelated to either evaporative concentration or to salt dissolution. The probable source of iodide in basinal waters is organic matter. Iodide is an essential trace element in the biological cycle, and it is estimated that ~70% of crustal iodide resides in organic matter in marine sediments (Muramatsu *et al.*, 2001). Iodide is released during the progressive thermal degradation of organic material, and is preferentially partitioned into the aqueous phase as I$^-$ (Collins, 1975; Kharaka and Thordsen, 1992; Muramatsu *et al.*, 2001).

5.16.3.4.3 Fluoride

Fluoride exists in formation waters primarily as fluoride, F$^-$, and cation–fluoride complexes, such as NaF^0, CaF$^+$, and MgF$^+$ (Richardson and Holland, 1979). Concentrations of fluoride in formation waters vary from <1 mg L^{-1}

to >30 mg L^{-1} (Worden *et al.*, 1999). There appears to be a threshold value, $\sim 1 \times 10^5$ mg L^{-1}, for chloride below which fluoride concentrations are typically below 5 mg L^{-1} and above which they are in the range of 10–20 mg L^{-1}. Occasionally they are even higher. Sources of fluoride and controls on the concentrations of fluoride have not been extensively studied. Biogenic fluorapatite, bentonite, and smectitic shales are potential sources (Hitchon, 1995).

The control of fluoride concentration by fluorite (CaF_2) saturation is likely in some waters. Fluorite solubility has been shown to be a complex function of temperature, salinity, and major ion chemistry (Holland and Malinin, 1979; Richardson and Holland, 1979). Hitchon (1995) found that lower salinity waters of the Alberta Basin are generally undersaturated with respect to fluorite, and that there is a gradual increase in fluoride to CaF_2 saturation as temperature and salinity increase.

5.16.3.4.4 Inorganic carbon species

The alkalinity of most formation waters, operationally defined by titration of a given volume of water with H_2SO_4 to an inflection pH, is contributed predominantly by bicarbonate and organic acid anions (Figure 9). The inflection pH for inorganic alkalinity is close to 4.5. For organic acid anions it is at a pH value of 2–3 (Willey *et al.*, 1975; Carothers and Kharaka, 1978). Total inorganic (organic anions are discussed in more detail below) alkalinity, comprised mainly of HCO_3^- and CO_3^{2-} species, is generally less than a few hundred milligrams per liter in waters having salinities in excess of 3×10^4 mg L^{-1}. Alkalinity generally decreases with increasing salinity. There are two main reasons for this

decrease. Both are related to the solubility of carbonate minerals, primarily calcite. First, in a calcium carbonate-buffered system, carbonate alkalinity should decrease with the marked increase in dissolved calcium that occurs with increasing salinity. Second, the increase in H^+ (lower pH) with increasing salinity shifts dissolved carbonate and bicarbonate toward carbonic acid (Equation (5)):

$$HCO_3^- + H^+ = H_2CO_3^0 \qquad (5)$$

5.16.3.4.5 Sulfate

Sulfur can exist in aqueous solution in at least five oxidation states, but data on sulfur species in basinal fluids are limited primarily to sulfate, S(VI), and sulfide, S(−II). Sulfate (SO_4^{2-}) will be discussed here, and hydrogen sulfide (H_2S) and bisulfide (HS^-) will be discussed in Section 5.16.7.

The concentration of SO_4^{2-} in formation waters rarely exceeds 1,000 mg L^{-1} even though it is present in high concentration in seawater (2,700 mg L^{-1}) and in even higher concentrations in residual brines formed by seawater evaporation (Figure 5). Unlike major cations and alkalinity, there is no significant correlation between the concentration of SO_4^{2-} and chloride or salinity, but the solubility of anhydrite decreases rapidly with increasing temperature and provides the ultimate control on SO_4^{2-} concentrations (Kharaka and Thordsen, 1992). The wide variations in sulfate concentrations that exist on the basinal and even the oil-field scale may also reflect rate-controlled processes involving: (i) release of sulfate by the dissolution of sulfate minerals, such as gypsum and anhydrite (Land *et al.*, 1995; Hitchon, 1996) and the oxidation of pyrite

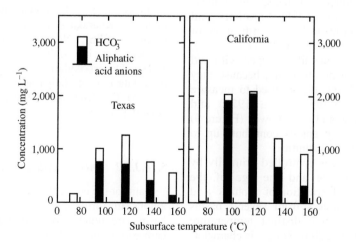

Figure 9 Average concentrations of aliphatic acid anions and bicarbonate alkalinity (as HCO_3^-) in oil-field waters from Texas and California. Note that organic acid anions contribute most of the total alkalinity at temperatures of 80–140 °C (after Carothers and Kharaka, 1978).

(Dworkin and Land, 1996); (ii) dispersive fluid mixing; (iii) precipitation as barite ($BaSO_4$); (iv) removal by bacterial sulfate reduction (BSR) at shallow and deeper zones, particularly in the presence of hydrocarbons (Gavrieli *et al.*, 1995); and (v) removal by thermochemical sulfate reduction (TSR) that becomes important at temperatures >100 °C (Machel, 2001).

5.16.3.5 Reactive Organic Species

Since the beginning of the last quarter of twentieth century there has been considerable interest in the origin and interactions of dissolved reactive organic species in subsurface waters (Willey *et al.*, 1975; MacGowan and Surdam, 1990; Drever and Vance, 1994; Bennett and Larter, 1997; Kharaka *et al.*, 2000; Franks *et al.*, 2001). Geochemical interest in organic species stems from their important role in mineral diagenesis (Surdam *et al.*, 1989; Seewald, 2001). They act as proton donors for a variety of pH-dependent reactions, as pH and Eh buffering agents, and they form complexes with metals such as aluminum, iron, lead, and zinc. They can be used as proximity indicators in petroleum exploration (Kartsev, 1974; Carothers and Kharaka, 1978) and they may serve as possible precursors for natural gas (Kharaka *et al.*, 1983; Drummond and Palmer, 1986).

5.16.3.5.1 Monocarboxylic acid anions

The concentrations of dissolved organic species in oil-field waters (20–200 °C) are much higher than in groundwater. Acetate concentrations may reach values as high as 1×10^4 mg L^{-1} (Kharaka *et al.*, 1986, 2000; MacGowan and Surdam, 1990). Acetate, propionate, butyrate, and valerate were identified as the dominant organic species in these waters (Willey *et al.*, 1975; Carothers and Kharaka, 1978). Prior to the identification of these aliphatic acid anions by Willey *et al.* (1975), these organics were generally grouped with and recorded as bicarbonate alkalinity, because they are titrated with H_2SO_4 that is used to measure field alkalinities. Willey *et al.* (1975) and Carothers and Kharaka (1978) showed (Figure 9) that these organic acid anions contribute up to 99% of the measured alkalinities.

Their concentration is controlled primarily by subsurface temperature and the age of the reservoir rocks. The distribution of aliphatic acid anions in oil-field waters from several basins shows (Figure 10) three distinct temperature zones (Kharaka *et al.*, 2000). Zone 1 is characterized by concentrations of acid anions <500 mg L^{-1} and temperatures <80 °C. The concentration of acetate in this zone is generally

low, and propionate generally predominates. Bacterial degradation is believed to be responsible for the low concentration of organic species in Zone 1 (Carothers and Kharaka, 1978). The highest concentrations of aliphatic acid anions are present in the youngest (Tertiary age) and shallowest reservoir rocks of Zone 2 (temperatures 80–120 °C). Their concentration generally decreases with increasing subsurface temperatures (Figure 10) and with the age of the Zone 2 reservoir rocks. Acetate constitutes more than 90% of acid anions; propionate comprises ~5% of these anions (Carothers and Kharaka, 1978; Lundegard and Kharaka, 1994). The boundary of Zone 3, where no measurable acid anions are present, is placed at ~220 °C. This temperature is obtained by extrapolating the points of Zone 2 (Kharaka *et al.*, 1986).

The decrease in the concentrations of acid anions with increasing temperature (Figure 10) and with the age of the reservoir rocks (Carothers and Kharaka, 1978; Kharaka *et al.*, 2000) indicates that thermal decarboxylation is responsible for the conversion of acid anions to CO_2 and hydrocarbon gases. Further evidence for the importance of thermal decarboxylation in the destruction of acid anions is obtained from the $\delta^{13}C$ values of dissolved bicarbonate, CO_2 and CH_4 in natural gas, and diagenetic calcite and ankerite from Gulf Coast and California basins (Boles, 1978; Carothers and Kharaka, 1980; Lundegard and Land, 1986; Lundegard and Kharaka, 1994). These $\delta^{13}C$ values indicate that the carbon in CO_2 gas, dissolved carbonate, and carbonate cements was derived largely from an organic source.

Experimental studies show that decarboxylation rates for acetic acid are extremely sensitive to temperature and the types of available catalytic surfaces. Rate constants for acetic acid decarboxylation at 100 °C differ by more than 14 orders of magnitude between experiments conducted in stainless steel and in catalytically less active titanium (Kharaka *et al.*, 1983; Drummond and Palmer, 1986). Naturally occurring mineral surfaces provide rather weak catalysts (Bell, 1991). Decarboxylation rates calculated from field data indicate half-life values of 10–60 Myr at 100 °C (Kharaka, 1986; Lundegard and Kharaka, 1994).

5.16.3.5.2 Dicarboxylic acid anions

Data for the concentrations of dicarboxylic acid anions in formation waters from sedimentary basins are much more limited than for monocarboxylic anions. Some of the reported values are controversial (Hanor *et al.*, 1993; Kharaka *et al.*, 1993a). The total reported range is 0–2,640 mg L^{-1} (for discussion and references

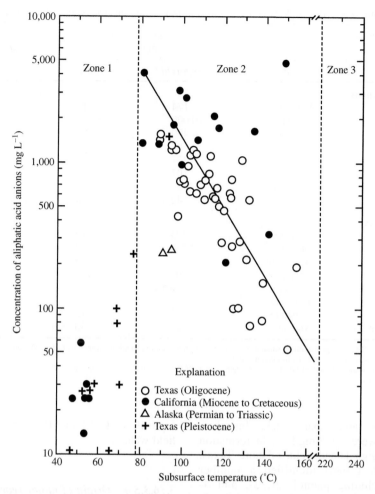

Figure 10 Concentrations of aliphatic acid anions (C_2–C_5) in formation waters from three sedimentary basins. Note that the highest concentrations are at ~80 °C and that they thereafter decrease with increasing temperatures (source Kharaka *et al.*, 1988).

see Kharaka *et al.*, 2000). The highest concentrations of dicarboxylic acid anions are those reported by MacGowan and Surdam (1988, 1990) for water samples from ~40 petroleum wells located mainly in the San Joaquin and Santa Maria basins, California, and in northern Gulf of Mexico basins. They reported values of up to 494 mg L^{-1} oxalate, 2,540 mg L^{-1} malonate, and 66 mg L^{-1} maleate from wells in the North Coles Levee field, San Joaquin Valley Basin, California. Several wells in the North Coles Levee and the nearby Paloma field were resampled by several other investigators (MacGowan and Surdam, 1990; Fisher and Boles, 1990; Kharaka *et al.*, 1993b). These authors found much lower and more typical concentrations of dicarboxylic acid anions (total of ~200 mg L^{-1}). Their concentration is probably limited by a rapid rate of thermal decomposition (MacGowan and Surdam, 1988; Crossey, 1991) and by the low solubility of calcium oxalate and calcium malonate (Kharaka *et al.*, 1986; Harrison and Thyne, 1992).

It is clear from the above discussion that there are large variations and some uncertainty in the reported maximum concentrations of mono- and dicarboxylic acid anions in formation waters. The use of these maximum values leads to erroneous results and conclusions. Maximum reported values together with more reasonable and likely maximum values are listed in Table 5. Only measured concentrations of organic and inorganic species from petroleum wells should be used in rigorous geochemical simulations (Kharaka *et al.*, 1987). If field data are not available, then more reasonable conclusions are obtained by using the likely maximum values of Table 5.

5.16.3.5.3 Other reactive organic species

Data for the concentration of organic species other than the mono- and dicarboxylic acid anions are few. Degens *et al.* (1964) and Rapp (1976) identified several amino acids, including serine,

Table 5   Maximum reported and likely concentrations of mono- and dicarboxylic acid anions in oil-field waters.

Acid anion		Concentration (mg L^{-1})		References[a] (maximum reported)
IUPAC	*Common*	*Reported*	*Likely*	
Monocarboxylic anions				
Methanoate	Formate	174	10	1
Ethanoate	Acetate	10,000	5,000	2
Propanoate	Propionate	4,400	2,000	1
Butanoate	Butyrate	682	500	3
Pentanoate	Valerate	371	200	3
Hexanoate	Caprolate	107	100	4
Heptanoate	Enanthate	99	100	1
Octanoate	Caprylate	42	100	1
Dicarboxylic anions				
Ethanedioate	Oxalate	494	10	1
Propandioate	Malonate	2,540	100	1
Butandioate	Succinate	63	100	4
Pentandioate	Glutarate	95	100	5
Hexandioate	Adipate	0.5	10	4
Heptanedioate	Pimelate	0.6	10	4
Octanedioate	Suberate	5.0	10	4
cis-Butenedioate	Maleic	26	50	1

Source: Kharaka *et al.* (2000). For geochemical simulations, listed likely maximum values from Kharaka *et al.* (2000) are probably more representative of high concentrations in most formation waters.
[a] References: 1 = MacGowan and Surdam (1988); 2 = Surdam *et al.* (1984); 3 = MacGowan and Surdam (1990); 4 = Kharaka *et al.* (1985); 5 = Kharaka *et al.* (2000).

glycine, alanine, and aspartic acid, but their concentrations were <0.3 mg L^{-1}. In formation waters of the High Island field, offshore Texas, Kharaka *et al.* (1986) identified a number of species, including phenol, 2-, 3-, and 4-methylphenol, 2-ethylphenol, 3-, 4-, and 3-, 5-dimethylphenol, cyclohexanone, and 1- and 4-dimethylbenzene. Fisher and Boles (1990) identified various polar aliphatics (fatty acids to C$_9$ with various methyl and ethyl substituents), cyclics (phenols and benzoic acids), and heterocyclics (quinolines). They were able to quantify, at the ppm or sub-ppm level, the concentrations of phenol, methyl-substituted phenols, and benzoic acid. Lundegard and Kharaka (1994) reported data that waters from oil and gas wells in the Sacramento Valley, California, contained the following organic species: phenols (up to 20 mg L^{-1}), 4-methylphenol (up to 2 mg L^{-1}), benzoic acid (up to 5 mg L^{-1}), 4-methylbenzoic acid (up to 4 mg L^{-1}), 2-hydroxybenzoic acid (up to 0.2 mg L^{-1}), 3-hydroxybenzoic acid (up to 1.2 mg L^{-1}), 4-hydroxybenzoic acid (up to 0.2 mg L^{-1}), and citric acid (up to 4 mg L^{-1}). Additional dissolved organic species, including organosulfur compounds, will probably be discovered as analytical procedures improve (Kharaka *et al.*, 1999, 2000).

Significant concentrations of nonpolar, but toxic dissolved organic compounds, including benzene, toluene (up to 60 mg L^{-1} for BTEX), phenols (20 mg L^{-1}), and polyaromatic hydrocarbons (upto 10 mg L^{-1} for PAHs), may be present in oil-field waters.

5.16.3.5.4   Origin of major reactive species

Hydrous pyrolysis experiments of crude oils generated relatively large concentrations of mono- and dicarboxylic acid anions with relative abundances generally similar to those observed in sedimentary basin waters. But analysis of all of the pertinent data indicates that the major part of the organic acid anions in formation waters is probably generated by the thermal alteration of kerogen in source rocks (Kharaka *et al.*, 1993b; Lewan and Fisher, 1994). This conclusion is based on several observations:

(i) the oxygen content of oil (0–1 wt.%) is ~20 times lower than that of kerogen (Tissot and Welte, 1984);

(ii) the yields of organic acid anions per unit weight are approximately two orders of magnitude lower in experiments with oil than in kerogen experiments (Lundegard and Senftle, 1987; Barth *et al.*, 1989; Kharaka *et al.*, 1993b);

(iii) oil is much less abundant than kerogen in sedimentary basins; and

(iv) high concentrations of organic acid anions have been reported from gas fields (e.g., the Sacramento Valley Basin, California), where liquid petroleum has probably never existed (Lundegard and Kharaka, 1994).

5.16.4 ISOTOPIC COMPOSITION OF WATER

The oxygen and hydrogen isotopes of H_2O have become the most useful tools in the study of the origin and evolution of subsurface waters (for detailed reviews, see Shepard, 1986; Kharaka and Thordsen, 1992). Prior to the use of isotopes, it was generally assumed that most of the formation waters in marine sedimentary rocks were connate marine in origin (White *et al.*, 1963). Clayton *et al.* (1966) were the first to use the isotopic composition of H_2O to show that waters in several sedimentary basins are predominantly of local meteoric origin. The connate water was lost by compaction and flushing. The extensive use of isotopes of water, solutes, and associated minerals coupled with studies of the regional geology and paleohydrology have shown that subsurface waters generally have a complicated history and that they are commonly mixtures of waters of different origins (Graf *et al.*, 1966; Connolly *et al.*, 1990; Kharaka and Thordsen, 1992; Birkle *et al.*, 2002).

The distribution and controls on the isotopic composition of present-day precipitation and surface waters, especially in mountainous terrains, are complex (e.g., Kharaka *et al.*, 2002); this topic is covered in detail in Chapter 5.11. This isotopic composition together with data for paleoclimates and regional paleogeography can be used to deduce the isotopic composition of old surface waters, including ocean water. An understanding of these parameters is needed to interpret the origin of deep basin brines (Kharaka and Thordsen, 1992). Reactions between water and minerals, dissolved species, associated gases, and other liquids with which they come into contact can modify the isotopic composition of water, especially the value of $\delta^{18}O$. In addition to mixing of waters of different isotopic composition, the following are the main processes that modify the isotopic composition of formation waters in sedimentary basins: (i) isotopic exchange between water and minerals; (ii) evaporation and condensation; (iii) fractionation caused by the membrane properties of rocks; and (iv) isotopic exchange between water and other fluids, especially petroleum.

5.16.4.1 Formation Waters Derived from Holocene Meteoric Water

The isotopic compositions of waters in some basins indicate that formation waters are related principally to recent local meteoric waters (Clayton *et al.*, 1966; Hitchon and Friedman, 1969; Kharaka and Thordsen, 1992). The δD values of these waters (Figure 11) show a

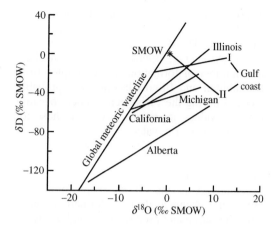

Figure 11 Isotopic compositions of oil-field waters from several basins in North America. Original lines from Clayton *et al.* (1966), Hitchon and Friedman, (1969), and Kharaka *et al.* (1973, 1979). Note that except for Gulf Coast II, the isotope lines intersect the Global Meteoric Line (Craig, 1961) at points with isotope values of present-day meteoric water (Kharaka and Thordsen, 1992).

"hydrogen isotope shift" in addition to the generally observed "oxygen isotope shift." The origins of these shifts vary, but the lines fitted to the δD and $\delta^{18}O$ plots (Figure 11) in each case intersect the global meteoric waterline (Craig, 1961) at values that approximate those of present-day meteoric water in the area of recharge (Clayton *et al.*, 1966; Kharaka and Thordsen, 1992).

Oxygen and hydrogen isotopes of water are sometimes not sufficient by themselves to indicate the origin of water. Kharaka *et al.* (1973) showed that plots of δD values and TDS indicated a connate marine origin for the waters from the McAdams Formation (Eocene) of California and a meteoric origin for those from the overlying formations; the isotope data alone (Figure 11) could indicate a meteoric origin for all the waters as they plot on the same trend. Also, a detailed study (Connolly *et al.*, 1990) of δD and $\delta^{18}O$ values of water and minerals, combined with chloride and strontium concentrations and $^{87}Sr/^{86}Sr$ values, suggests that the waters in the Alberta Basin, Canada, are not all meteoric (Figure 11), but that they define three hydrochemical groups. The upper group is dominated by modern meteoric water; the other two groups are a mixture of old (Laramide age) meteoric water and original bittern or residual brine.

5.16.4.2 Formation Waters from "Old" Meteoric Water

Stable isotopes of water in addition to age determinations based on ^{14}C have shown that the waters in many petroleum fields are "old" meteoric water, i.e., older than Holocene and probably Pleistocene in age (Bath *et al.*, 1978; Kharaka and Thordsen, 1992; Clark and Fritz, 1997). Clayton *et al.* (1966) were the first to show that a number of formation-water samples from the Michigan and Illinois basins were probably recharged during the Pleistocene Epoch, because their $\delta^{18}O$ values are much lower than those for present-day meteoric water.

Kharaka *et al.* (1979) and Kharaka and Carothers (1988) were probably the first to present evidence for meteoric water older than Pleistocene in sedimentary basins. These authors presented isotopic and chemical data for formation waters from exploration and producing oil and gas wells from the North Slope of Alaska. The water samples were obtained from reservoir rocks between 700 m and 2,800 m in depth and in age from Triassic to Mississippian. The waters from all formations, however, are remarkably similar in TDS ($(1.9-2.4) \times 10^4$ mg L^{-1}) and in the concentration of the major cations and anions. The least-squares line through the δD and $\delta^{18}O$ values for these waters (Figure 12) intersects the meteoric water line at δD and $\delta^{18}O$ values of $-65‰$ and $-7‰$, respectively. This line does not pass through the values for standard mean ocean

water (SMOW) or for the present-day meteoric water of the region.

The conclusion drawn by Kharaka and colleagues was that these formation waters were meteoric in origin, but that the recharge occurred when the North Slope had an entirely different climate. The relationship between mean annual temperature and the isotopic composition of meteoric water (Dansgaard, 1964) suggests that the mean annual temperature at the Brooks Range, the most likely recharge area, was $15-20\,°C$ higher at the time of recharge than today. Paleoclimatic indicators show that annual temperatures in northern Alaska were as high as this during the Miocene, as well as throughout most of the Early Cenozoic and the Mesozoic (Wolfe, 1980; Bryant, 1997).

5.16.4.3 Formation Waters of Connate Marine Origin

In several sedimentary basins, geological and hydrodynamic considerations indicate that formation waters were originally connate marine in origin. The Gulf of Mexico is a prime example of such a basin. The extensive drilling for petroleum, especially in the northern half of this basin, has documented a very thick sequence (up to 1.5×10^4 m) of Cenozoic terrigenous shales, siltstones, and sandstones; there are no major unconformities. Fine-grained sediments contain the largest percentage of connate water at the time of deposition (initial porosity up to 80%); most of this water is squeezed out into the interbedded sandstones during compaction (Kharaka and Thordsen, 1992).

Abnormally high fluid pressures (geopressured zones) are another important characteristic of these basins. In the northern Gulf of Mexico basin, geopressured zones with hydraulic pressure gradients higher than hydrostatic (10.5 kPa m^{-1} (0.465 psi ft^{-1})) are encountered at depths that range from ~2,000 m in south Texas to more than 4,000 m in parts of coastal Louisiana (Kharaka *et al.*, 1985). Calculated fluid potentials indicate hydraulic heads that are higher than any recharge areas in the basin, indicating water flow upward and updip. The geologic and paleogeographic history of the northern Gulf of Mexico basin suggest that the flow has always been updip and toward the recharge areas to the north and north-west, indicating that much of the water is connate water squeezed from the marine shales and siltstones of the basin (Kharaka and Carothers, 1986; Land and Fisher, 1987).

The isotopic composition of formation waters from many fields in the northern Gulf of Mexico Basin (Figure 13) fall on a general trend that passes through SMOW and away from the

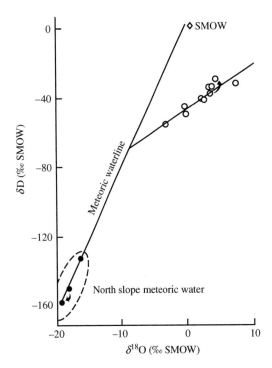

Figure 12 Isotopic composition of formation waters from the North Slope, Alaska. The solid line is the least-squares line drawn through the data. Also shown are values for SMOW, meteoric water in the area, and the Global Meteoric Water Line. Note that the line through the data does not pass through SMOW or the local meteoric water. Arrows indicate corrected values as described in Kharaka and Carothers (1988).

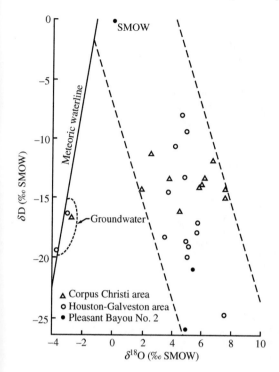

Figure 13 Isotopic composition of formation waters from northern Gulf of Mexico Basin. Note that the trend shows decreased δD values with increasing $\delta^{18}O$ values and that the trend goes through SMOW and away from the local groundwater (Kharaka *et al.*, 1979).

meteoric water of the area. The isotopic data show that the formation waters in the geopressured and normally pressured zones are neither meteoric in origin nor the result of mixing of meteoric with ocean water. Kharaka and Carothers (1986) used mass balance equations to show that isotopic interaction between connate marine water and clay minerals having very light original δD value ($-70\%o$) is responsible for this unusual isotopic trend.

The concentration and isotopic composition of noble gases in natural gas produced from wells in the Gulf of Mexico Basin also support the conclusion that the formation water is marine connate in origin (Mazor and Kharaka, 1981). Finally, age determinations using $^{129}I/I$ ratios for 60 formation water samples from several oil fields in Texas and Louisiana gave ages of 53–55 Ma, pointing to the Wilcox Group shale as the main source of iodide. Moran *et al.* (1995) point out that even older (Mesozoic age) shale could not be ruled out, because of uncertainties in estimating the composition of the fissiogenic component.

5.16.4.4 Bittern Connate Water in Evaporites

Evaporated seawater trapped in precipitated salt and associated sediments and adjacent rocks

(bittern water) is probably an important component of formation waters in several sedimentary Basins (Carpenter *et al.*, 1974; Kharaka *et al.*, 1987; Moldovanyi and Walter, 1992). The central Mississippi Salt Dome Basin is such a basin. The isotopic composition of its formation waters (Figure 14) belongs to three main groups. Samples from rocks of Jurassic age (Smackover and Norphlet Formations) have δD values ($-1\%o$ to $-3\%o$), i.e., approximately equal to that of SMOW; they have $\delta^{18}O$ values ($5.1–7.3\%o$) that are highly enriched in ^{18}O. Samples obtained from rocks of Cretaceous age have δD values ($-9\%o$ to $-13\%o$) that are depleted in deuterium relative to both SMOW and the samples from rocks of Jurassic age. Finally, samples obtained from shallow groundwaters in the area have δD and $\delta^{18}O$ values that plot close to the Global Meteoric Water Line (GMWL). The δ^aD values (Figure 14) give the D/H measurements in terms of the activities of deuterium in solutions calculated using a correction factor from Sofer and Gat (1972).

The isotopic composition of the waters supports the conclusion, based on plots of bromide versus chloride and combinations of other chemical constituents (Carpenter *et al.*, 1974; Stoessell and Carpenter, 1986; Kharaka *et al.*, 1987) that the formation waters in rocks of Jurassic age are evaporated seawater. The δD and $\delta^{18}O$ values of evaporating seawater initially increase with increasing evaporation, but at higher degrees of evaporation the trend reverses and (Figure 14) describes a loop on the δ-diagram (Holser, 1979). Knauth and Beeunas (1986) have shown that line B in Figure 14 is the most likely trajectory for evaporating seawater in the Gulf Coast. This line passes very close to the point giving the δ-values for the water in the Norphlet Formation, which is that expected from evaporating seawater at the point of halite precipitation. The δ-values of the samples from the Smackover Formation were initially close to those from the Norphlet Formation; the $1–2\%o$ shifts in $\delta^{18}O$ values are probably related to isotopic exchange with the enclosing carbonate minerals (Heydari and Moore, 1989).

5.16.4.5 Brines from Mixing of Different Waters

Formation waters in sedimentary basins are highly mobile; this leads to mixing of waters of different age or of different origin and age (Kharaka and Carothers, 1986; Worden *et al.*, 1999; Ziegler and Coleman, 2001). Hitchon and Friedman (1969) made a very detailed isotopic and chemical study of the surface and formation waters of the Western Canada sedimentary

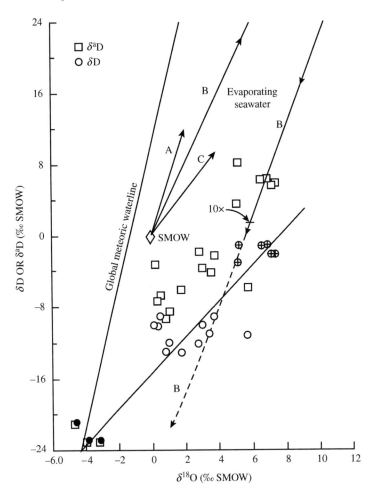

Figure 14 Isotopic composition of shallow groundwater (solid circles) and formation waters from Jurassic (open circles with +) and Cretaceous (open circles) rocks from the central Mississippi Salt Dome Basin. $\delta^a D$ values (squares) were computed from δD values and a correction factor $\Delta\delta$ from Sofer and Gat (1972). Lines originating from SMOW are the range of trajectories for seawater undergoing evaporation, with line B being from Holser's (1979) estimate of evaporating seawater in the Gulf Coast. The least-squares line for the δD and $\delta^{18}O$ values is also indicated (after Kharaka *et al.*, 1987).

basin. Using mass balance calculations for deuterium and TDS, they concluded that the observed distribution of deuterium in the formation waters could best be explained by mixing of diagenetically modified seawater with ~2.9 times its volume of fresh meteoric water. They attributed the ^{18}O enrichment of formation waters to extensive exchange with carbonate rocks (see also Connolly *et al.*, 1990).

Other examples of deep-basin waters related to simple mixing of meteoric water with marine connate waters have been documented in the Dnepr–Donets basin, Ukraine (Vetshteyn *et al.*, 1981) and the Sacramento Valley, California (Berry, 1973; Kharaka *et al.*, 1985). Knauth (1988) used water isotopes and chemical data for waters in the Palo Duro Basin, Texas, to indicate extensive mixing between bittern brines and two

pulses of meteoric water with different isotopic compositions. Mixing between waters of the same origin but affected by different processes is probably common in sedimentary basins. Kharaka *et al.* (1985) used δD and $\delta^{18}O$ values and the chemical composition of formation waters from High Island field, offshore Texas, to show (Figure 15) that the formation waters in Pleistocene reservoir rocks involved mixing of two marine connate waters. One end-member is an essentially unmodified marine connate water of Pleistocene age (samples 6 and 7 in Figure 15); the other (sample 5) is a highly chemically and isotopically modified marine connate water possibly of Miocene age. Approximately the same mixing proportions were obtained using stable isotope values (Figure 15) and concentrations of sodium, chloride, calcium, and lithium.

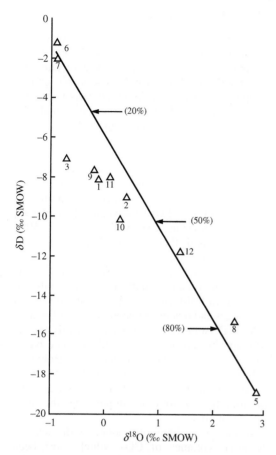

Figure 15 Isotopic composition of formation waters from High Island field, offshore Texas. Percentages represent mixing proportions of the deep brine with Pleistocene connate (sea)water (source Kharaka *et al.*, 1985).

5.16.5 ISOTOPIC COMPOSITION OF SOLUTES

Significant advances in isotope analytical techniques over the last several decades have greatly expanded our knowledge of the isotopic composition of natural waters (Bullen *et al.*, 2001). Highly precise data are now available for isotopic compositions of hydrogen, oxygen, carbon, and sulfur, and a large body of data is available for the strontium, boron, and noble gas isotopes. More recently, data for isotope systematics of lithium, chloride, bromide, and iodide in basinal waters have been accumulating. Applications of isotope geochemistry in studies of sedimentary basins have included identifying the sources of solutes and of H_2O, quantifying the degree of rock–water exchange, tracing fluid flow paths, determining paleotemperatures, and calculating the age and residence time of basinal fluids. The systematics of many isotopic systems, such as those of strontium, have been well worked out for formation waters. Some

systems, such as that of bromide, are in their infancy, and others, such as the stable isotope geochemistry of chloride, are beset by unresolved questions regarding their interpretation.

This section covers the stable isotopes of boron, lithium, carbon, sulfur, chloride, bromide, and strontium and the radiogenic isotopes of chloride and iodide (see Chapter 5.15). The isotopes of noble gases are described in Section 5.16.7. Recent reviews of other isotopic systems, which have been used in subsurface hydrogeology, are given in the volume edited by Cook and Herczeg (2000).

5.16.5.1 Boron Isotopes

Boron has stable isotopes with atomic mass of 10 and 11. These have natural abundances of ~19.82% and ~80.18%, respectively (Palmer and Swihart, 1996; Aggarwal *et al.*, 2000). The analyses of $^{11}B/^{10}B$ isotopic ratios are reported in the standard del notation: $\delta^{11}B‰$.

Natural waters range in $\delta^{11}B$ from $-16‰$ to $+60‰$ relative to the SRM NBS 951 standard. This exceptionally wide range reflects the large variability in the isotopic composition of boron sources and the large fractionation factors attendant to the partitioning of boron between aqueous and solid phases (Barth, 1998). The major factor in the isotopic fractionation of boron is the preference of ^{10}B for tetrahedral coordination with oxygen, either in the borate ion, $B(OH)_4^-$ (aq), or in silicate mineral surface and lattice sites, and the preference of ^{11}B for trigonal coordination with oxygen as in dissolved boric acid, $B(OH)_3$(aq). Under diagenetic conditions, boron is present largely in trigonal coordination in fluids but in tetrahedral coordination in silicates. Fractionation is less mineral specific than coordination specific (Williams *et al.*, 2001b). Lighter isotopes are generally concentrated in the more volatile phase because of their higher vibrational frequency, but boron is an important exception because of the importance of coordination state in the fractionation of its isotopes.

The value of $\delta^{11}B$ in seawater is 40‰, in marine carbonates it is 10–30‰, and in continental rocks and siliciclastic sediments −15‰ to 5‰ (Palmer and Swihart, 1996; Aggarwal *et al.*, 2000). Boron isotopic values for Gulf Coast waters range from +10‰ to +50‰ (Land and Macpherson, 1992a). There is a general decrease in $\delta^{11}B$ and an increase in boron with increasing depth and temperature, reflecting the release of light boron during silicate mineral diagenesis. During deep burial diagenesis some boron is reincorporated into crystal lattices. The $\delta^{11}B$ of diagenetic mineral phases should therefore reflect the $\delta^{11}B$ of the ambient fluid. Since boron substitution for silicon involves the breaking of Si—O bonds, there should be a coincident change in the $\delta^{18}O$ of mineral phases. However, boron fractionation is so large that the value of $\delta^{11}B$ may be more sensitive than $\delta^{18}O$ to fluid rock exchange under diagenetic conditions. Williams *et al.* (2001b) have developed a boron-fractionation equation based on this concept that can possibly be used to determine paleotemperatures.

5.16.5.2 Lithium Isotopes

Lithium has two stable isotopes, ^6Li and ^7Li, which have abundances of 7.5% and 92.5%, respectively. Lithium is a soluble alkali element. Because its ionic radius is small (0.78 Å), it behaves more like magnesium (0.72 Å) than the alkalis. Li^+ tends to substitute for Al^{3+}, Fe^{2+}, and especially for Mg^{2+}. Because of their large relative mass difference, lithium isotopes have the potential to exhibit sizable fractionation, as has been demonstrated by high-precision isotopic analysis.

Lithium isotopic compositions generally are reported in terms of δ^6Li, an unusual formulation because the abundance of the heavier isotope is in the denominator (Chan *et al.*, 2002); δ^6Li is defined by

$$\delta^6Li(\text{‰}) = [(^6Li/^7Li)_{\text{sample}}/(^6Li/^7Li)_{\text{std}} - 1] \times 1,000$$

Lithium in seawater, with an average δ^6Li value of -32‰, is therefore isotopically heavy. Marine sediments, with δ^6Li values of $+1\text{‰}$ to -15‰, are much lighter. Other reported ranges in δ^6Li values for geological materials include those of mid-ocean-ridge basalts, -8‰ to -21‰; hemipelagic clays, -9‰ to -15‰; and continental crustal rocks, -8‰ to -21‰ (Huh *et al.*, 1998). All lithium isotopes are referenced to the NBSL-SVEC Li_2CO_3 standard (Clark and Fritz, 1997).

Chan *et al.* (2002) studied the isotopic composition of lithium in oil-field brines in the Heletz–Kokhav oil field in the southern coastal plain of Israel. The formation waters have chloride concentrations of $(1.8–4.7) \times 10^4$ mg L^{-1} and are thought to represent seawater evaporated into the gypsum stability field and subsequently altered by the dissolution of halite. Lithium concentrations range from 0.97 mg L^{-1} to 2.3 mg L^{-1} and increase with increasing chloride. The δ^6Li values range from -19‰ to -30‰ and are thus lighter than seawater (-32‰), reflecting a rock-dominated source for lithium rather than an evaporated seawater source.

5.16.5.3 Carbon Isotopes

The origin of inorganic and organic carbon in sedimentary basins and the systematics of its isotopes have received a great deal of attention (e.g., Clark and Fritz, 1997). The stable carbon isotopic composition of dissolved carbon species, such as HCO_3^-, $CO_2(aq)$, $CH_4(aq)$, and the organic acids, is normally reported as $\delta^{13}C\text{‰}$ PDB. There are two major reservoirs of carbon in sedimentary basins, marine carbonate having $\delta^{13}C$ values of $\pm4\text{‰}$ and organic carbon with values between -10‰ and -35‰. Bacterial and thermogenic reduction of organic matter to methane produces two isotopically distinct CH_4 reservoirs: biogenic CH_4 with $\delta^{13}C$ of -50‰ to -90‰, and thermogenic CH_4 with $\delta^{13}C$ between -20‰ and -50‰.

Carothers and Kharaka (1980) reported $\delta^{13}C$ values of inorganic carbon dissolved in oil-field waters from California and Texas, and discussed the sources and reactions that yield $\delta^{13}C$ of -20‰ to 28‰. Dissolution of carbonate minerals and the oxidation of reduced carbon both produce bicarbonate as a by-product. Emery and Robinson (1993) reported a range from -60‰ to 10‰, depending on the source of the carbon and the fractionation factors accompanying the production of HCO_3^-.

An increasing body of evidence indicates that the large volumes of CO_2, which have been observed in basins such as the Pannonian Basin, Hungary, the Cooper-Eromanga Basin, Australia, and the South Viking Graben, North Sea, may have deep-seated sources (Wycherley *et al.*, 1999). The isotopic composition of CO_2 released by mantle, deep crustal, and shallow high-temperature processes is estimated to be intermediate between values for inorganic marine carbonate and organic carbon. Wycherley *et al.* (1999) cite the following values: magmatic origin/mantle degassing, -4‰ to -7‰; regional metamorphism, 0‰ to -10‰; contact metamorphism of carbonates, -2‰ to -12‰; and contact metamorphism of coals -10‰ to -20‰. It should be noted, however, that due to the number of sources for carbon of varying $\delta^{13}C$ value and especially due to the ease of isotopic exchange and re-equilibration as well as mixing, carbon isotopes are not highly diagnostic and cannot be used alone to identify carbon sources (Kharaka *et al.*, 1999b). Carbon isotopes in shallow groundwater are discussed by Kendall and Doctor (see Chapter 5.11).

5.16.5.4 Sulfur Isotopes

The isotopic composition of sulfur in geologic materials is reported in the usual del notation as $\delta^{34}S\text{‰}$ (VCDT). The isotopic geochemistry of sulfur is complex, because it exists in several redox states, each with a wide variety of fractionation mechanisms (Seal *et al.*, 2000). The three principal redox states of concern here are sulfur as sulfate (VI), sulfide (−II), and elemental sulfur. Below ~200 °C, the nonbiologic rate of isotopic exchange between dissolved sulfate and sulfide is slow and isotopic equilibrium between the species is rare (Ohmoto and Lasaga, 1982). Kinetic effects therefore dominate the isotopic systematics of sulfur in sedimentary basins. The principal kinetic effects are associated with the reduction of sulfate to elemental sulfur or sulfide S(−II). This can be achieved both microbially with large fractionation effects and thermochemically with minimal fractionation. Fractionation is usually minor during the oxidation of sulfide to sulfate (Seal *et al.*, 2000).

The principal sources of sulfate in formation waters are dissolved marine sulfate, sulfate derived from the dissolution of evaporites, and sulfate formed by the oxidation of sulfides. Sulfate is destroyed by reduction to hydrogen sulfide. The value of $\delta^{34}S$ in gypsum is only ~1.6‰ heavier than sulfate in solutions from which it precipitates. The isotopic composition of sulfur in gypsum in Phanerozoic deposits precipitated from seawater during evaporation thus tracks the secular changes in the isotopic composition of sulfur in seawater (~10‰ to 30‰).

The lighter isotopes of sulfur are preferentially partitioned into sulfide during microbial SO_4^{2-} reduction. Hydrogen sulfide is much lighter than the precursor sulfate, and the residual sulfate is heavier. Sulfate reduction during early diagenesis therefore drives the $\delta^{34}S$ of residual pore-water sulfate toward higher values. Thermochemical reduction of sulfate at higher basinal temperatures, however, typically produces sulfide similar in isotopic composition to the parent sulfate (Machel, 2001).

Dworkin and Land (1996) found that the $\delta^{34}S$ values of sulfate in Frio (Oligocene) formation waters were modified by the addition of light sulfur derived from the oxidation of pyrite. The $\delta^{18}O$ values of dissolved sulfate in these waters fall within the range found in Mesozoic and Cenozoic seawater and are interpreted by Dworkin and Land to reflect sulfate derived from deeper Jurassic evaporites. This conclusion, however, must be qualified because, at temperatures >100 °C, oxygen isotopes equilibrate between oxygen in water and in sulfate (e.g., Kharaka and Mariner, 1989). Gavrieli *et al.* (1995) document two stages in the evolution of the isotopic composition of sulfate-sulfur in oil fields in southwestern Israel. Early shallow bacterial reduction increased the $\delta^{34}S$ of residual dissolved Miocene sulfate from 20‰ to 26‰. In the presence of crude oil, additional sulfate was reduced and increased the $\delta^{34}S$ values to a maximum of 54‰.

5.16.5.5 Chlorine Isotopes

Chlorine has two stable isotopes, ^{35}Cl and ^{37}Cl, and one radioactive isotope, ^{36}Cl (half-life = 0.301 Ma). The stable isotopic composition of chloride in geologic materials is reported in the conventional del notation as $\delta^{37}Cl$. Seawater, which is used as the isotopic standard, has a $\delta^{37}Cl$ of 0‰. Most natural waters have $\delta^{37}Cl$ values between −1‰ and +1‰. However, values of −8‰ have been measured in marine pore waters. Minerals in which chloride substitutes for OH at high temperatures have $\delta^{37}Cl$ values as high as 7‰ (Banks *et al.*, 2000).

During halite precipitation from evaporated seawater, ^{37}Cl is preferentially partitioned into the solid phase. $\delta^{37}Cl$ of the residual brines and of subsequently precipitated halite become progressively lighter. However, during the last stages of evaporation preferential incorporation of ^{35}Cl into potassium and magnesium salts reverses the fractionation trend, and the residual brine and the precipitated salts become isotopically heavier (Eastoe *et al.*, 1999; Banks *et al.*, 2000). The $\delta^{37}Cl$ of halite precipitating from evaporated seawater with a $\delta^{37}Cl$ value of 0‰ is ~0.29‰. Eastoe *et al.* (1999) found a minimum $\delta^{37}Cl$ value of −0.9‰ in brines produced during the laboratory evaporation of seawater at the beginning of the potash facies.

The mechanisms that fractionate chloride isotopes during diagenesis are still not well established (Eastoe *et al.*, 1999). Eggenkamp (1998) found a range from −0.27‰ to −4.96‰ in the $\delta^{37}Cl$ of formation waters from North Sea oil fields. In some fields $\delta^{37}Cl$ decreased with decreasing chloride concentration. Water from oil reservoirs had a much smaller range in $\delta^{37}Cl$, from 0‰ to −1.5‰, over a wide range of chloride concentrations. Bedded salt in the Gulf of Mexico Basin has $\delta^{37}Cl$ values of −0.5‰ to +0.3‰ (Eastoe *et al.*, 2001), values consistent with a $\delta^{37}Cl$ value of 0.0‰ for Jurassic seawater. Eastoe *et al.* (2001) suggest that the slightly heavier values for diapiric salt (0.0–0.5‰) are the result of the incongruent dissolution of halite, which presumably releases lighter chloride preferentially. The $\delta^{37}Cl$ values of formation waters from the Gulf of Mexico Basin range from −1.9‰ to +0.7‰. Waters having $\delta^{37}Cl$ values <0.6‰ are found in siliciclastic strata of Eocene to Miocene age, but not in Plio–Pleistocene sediments or in Mesozoic carbonates, which contain waters of higher $\delta^{37}Cl$ composition. Eastoe *et al.* (2001) invoke differences in the rate of diffusion of ^{35}Cl and ^{37}Cl as a possible fractionation mechanism.

5.16.5.6 Bromine Isotopes

Bromine isotope geochemistry is in its beginning stages. Bromine has two stable isotopes, ^{79}Br and ^{81}Br, having relative mass abundances of 50.686% and 49.314%, respectively (Eggenkamp and Coleman, 2000). Variations in isotopic composition are reported as $\delta^{81}Br$ (SMOB), where SMOB is standard mean oceanic bromide. Several processes have the potential for causing significant fractionation of the bromide isotopes. These include concentration of bromide during the evaporation of seawater, oxidation of bromide to Br_2, and the natural production of organobromide compounds. Eggenkamp and Coleman (2000) found a range of 0.08–1.27‰ in $\delta^{81}Br$ in formation waters of the Norwegian North Sea. There is a negative correlation between $\delta^{81}Br$ and $\delta^{37}Cl$, reflecting differences in fractionation mechanisms of the two isotopic systems.

5.16.5.7 Strontium Isotopes

The strontium isotopic composition of formations waters has shown great utility as a means of identifying sources of strontium in formation waters, the degree of water–rock exchange, and the degree of mixing along regional fluid flow paths (e.g., Armstrong et al., 1998). The strontium isotopic composition of geological materials is expressed as the ratio of $^{87}Sr/^{86}Sr$, which can be measured with great analytical precision. Radiogenic ^{87}Sr is produced by the decay of ^{87}Rb (half-life = 4.88×10^{10} yr). Rocks having high initial concentrations of rubidium, such as granites, are characterized by high $^{87}Sr/^{86}Sr$ ratios. Rocks derived from materials having low rubidium concentrations, such as mantle-derived rocks, have correspondingly low $^{87}Sr/^{86}Sr$ ratios.

Since the Precambrian, the $^{87}Sr/^{86}Sr$ of seawater has fluctuated between ~0.7070 and ~0.7092 as the result of variations in the relative rates of input of ^{87}Sr-enriched strontium from continental weathering and ^{87}Sr-depleted strontium from mantle sources. Fluids in sedimentary basins containing Paleozoic strata typically have $^{87}Sr/^{86}Sr$ ratios in excess of seawater values that are contemporaneous or coeval with the depositional age of the current host sediment. This is well illustrated by the data of Connolly et al. (1990) for the Alberta Basin, Canada. The enrichment is due to the release of strontium attending the alteration of silicates. Due to the significant increase of $^{87}Sr/^{86}Sr$ in seawater since the Jurassic, some formation waters in Cenozoic sedimentary basins actually have $^{87}Sr/^{86}Sr$ ratios lower than those of contemporaneous seawater due to the addition of strontium dissolved from older and deeper sedimentary sources

(e.g., McManus and Hanor, 1993). Precipitates derived from such sources, such as barite in Holocene seafloor vents in the Gulf of Mexico, have lower $^{87}Sr/^{86}Sr$ values less than present-day seawater (Fu, 1998).

5.16.5.8 Radioactive Isotopes

A number of radioactive isotopes produced primarily by cosmic ray interactions in the upper atmosphere, especially ^{14}C (Clark and Fritz, 1997; Mazor, 1997), ^{36}Cl (Andrews et al., 1994; Phillips, 2000), and ^{129}I (Moran et al., 1995; Fabryka-Martin, 2000), as well as dissolved ^{4}He (Torgersen and Clarke, 1985; Solomon, 2000), have been used, in conjunction with data for the stable isotopes and calculated flow rates, for determining the age of natural waters, including fluids in sedimentary basins (Bethke et al., 1999, 2000). The 5.73 ka half-life of ^{14}C at 5.73 ka is so short that it is useful only for dating basinal meteoric water younger than ~40 ka. ^{36}Cl ($t_{1/2} = 0.301$ Ma) is useful for dating water of up to ~2 Ma in age. These isotopic systems are reviewed by Phillips and Castro (see Chapter 5.15).

The ratio of ^{129}I ($t_{1/2} = 15.7$ Ma) to total iodide can be used to estimate ages up to ~80 Ma. The determination of even greater ages is theoretically possible by using ^{4}He generated from the decay of uranium and thorium in rocks (Froehlich et al., 1991). The ages obtained, however, carry large uncertainties, because the radiogenic isotopes can have several sources (e.g., Fabryka-Martin, 2000) and are often subject to fractionation due to isotopic exchange and partitioning. In addition, the origin and age of H_2O in formation waters is generally different from that of the radiogenic and other isotopes used for age determinations (Froehlich et al., 1991; Kharaka and Thordsen, 1992; Clark and Fritz, 1997).

The ratio of ^{129}I to total I, at times in combination with $^{36}Cl/Cl$ ratios, has been successfully used to estimate the residence time of subsurface waters, to trace the migration of brines, and to identify hydrocarbon sources (e.g., Fabryka-Martin, 2000). However, this method can also be used to illustrate the difficulties of dating "very old" formation water. These include: (i) the ratios of $^{129}I/I$ generally are between $1,500 \times 10^{-15}$ and 20×10^{-15}, requiring the use of accelerator mass spectrometry for their measurement; (ii) there are major uncertainties in the correct value of the initial $^{129}I/I$ ratios; (iii) errors in the estimation of the rate of subsurface release of ^{129}I by spontaneous fission of ^{238}U; and (iv) additional (diagenetic) release of ^{127}I from organic-rich sediments (Fabryka-Martin, 2000; Moran et al., 1995). For several of these reasons, the $^{129}I/I$ and $^{36}Cl/Cl$ ratios could not be used to estimate the residence and travel time of water in the Milk River aquifer, Alberta, Canada (Fabryka-Martin, 2000).

The age of brine inclusions in Louisiana salt domes was estimated to be ~8 Ma by ^{129}I/I ratios that assumed no significant *in situ* production of ^{129}I, because of the low uranium concentration in salt (Fabryka-Martin *et al.*, 1985). Moran *et al.* (1995) found that the comparison between measured ratios and the decay curve for hydrospheric ^{129}I results in minimum source ages much older than present host formation ages, indicating migration of brine from older and deeper sources. Corrections for the fissiogenic component of ^{129}I gave Eocene ages (53–55 Ma) for brines residing in Miocene reservoirs. However, Mesozoic sources could not be ruled out, because of uncertainties in estimating the magnitude of the fissiogenic component. Some of the measured ^{129}I/I ratios showed that the brines had resided in formations with locally high uranium values (Moran *et al.*, 1995).

5.16.6 BASINAL BRINES AS ORE-FORMING FLUIDS

Sedimentary formation waters have long been invoked as ore-forming fluids in a number of distinctly different geologic settings. Although ore deposit classification schemes vary, the following have been genetically associated with basinal fluids: (i) Mississippi-Valley-type lead, zinc, copper, barium, and fluoride deposits; (ii) shale-hosted lead, zinc, and barium deposits; (iii) rift-basin and redbed copper deposits; (iv) sandstone-hosted uranium deposits; and (v) carbonate-hosted celestine ($SrSO_4$) deposits.

Much of the information available on the nature of the fluids that were involved in the genesis of these deposits comes from physical and chemical measurements made on fluid inclusions in ore and gangue minerals. For example, chemical analyses of inclusions indicate that many of these fluids were Na–Ca–Cl-brines having salinities in the range of $(1–3) \times 10^5$ mg L^{-1} (Hanor, 1979; Sverjensky, 1986; Giordano and Kharaka, 1994). Mass balance considerations (Barnes, 1979) indicate that concentrations of base and ferrous metals in ore-forming solutions must be at least 1–10 mg L^{-1} to form typical hydrothermal ore deposits. The discussion below emphasizes the mechanisms of solubilization of metals in formation waters. For a comprehensive treatment of the migration of ore-forming fluids and mechanisms of ore precipitation from single or mixed fluids, see Sverjensky (1986), Kharaka *et al.* (1987), and Giordano (2000).

5.16.6.1 Metal-rich Brines

The concentration of heavy metals in oil-field waters, with the exception of iron and manganese, is generally low (Kharaka and Thordsen, 1992; Hitchon *et al.*, 2001). In the case of lead, zinc, and copper, and with the exception of fewer than half a dozen localities worldwide, the concentrations are < 100 μg L^{-1}. The concentrations of rare metals (e.g., mercury, gold, and silver) are generally one to several orders of magnitude lower (e.g., Giordano, 2000).

The central Mississippi Salt Dome Basin is a metal-rich brine locality that has been studied extensively (Carpenter *et al.*, 1974; Kharaka *et al.*, 1987). The brines (Table 4) are Na–Ca–Cl-type waters of extremely high salinity (up to $\sim 3.5 \times 10^5$ mg L^{-1} TDS), but low concentrations of aliphatic acid anions. The metal concentrations in many samples are very high, reaching values ≥ 100 mg L^{-1} for lead, 250 mg L^{-1} for zinc, 500 mg L^{-1} for iron, and 200 mg L^{-1} for manganese (Table 6). The samples with high metal contents have extremely low concentrations (<0.02 mg L^{-1}) of H_2S. Samples that have high concentrations of H_2S have low metal contents that are typical of oil-field waters (Kharaka *et al.*, 1987). Exceptionally rich sources of metals such as redbeds may be needed to provide sufficient amounts of metals for the formation of ore deposits. Redbeds are attractive for this reason and because they contain so little reduced sulfur, and that leached metals will tend to remain in solution.

5.16.6.2 Solubilization of Heavy Metals

A major problem in explaining metal transport in basinal brines is the very low solubility of base metal sulfides, particularly at temperatures less than ~150 °C. For example, the calculated activity product $(a_{Zn^{2+}})(a_{H_2S^0})$ for solutions in equilibrium with sphalerite, $ZnS + 2H^+ = Zn^{2+} + H_2S^0$, using SUPCRT92 (Johnson *et al.*, 1992), is only $10^{-15.1}$ at 100 °C, 500 bar, and at a neutral pH of 6.04. The activity product for lead and H_2S for a solution in equilibrium with galena under these conditions is only $10^{-17.7}$. Thus, aqueous complexes of base metals are required to account for the minimum concentrations of the ore-forming metals. Experimental work and thermodynamic calculations have focused primarily on metal–chloride, metal–bisulfide, and metal–organic complexes as possible solubilizing agents (Sverjensky, 1986; Hanor, 1996b; Giordano, 2000). There is still some disagreement on the relative importance of these complexing agents, although most authors today favor chloride complexing, as discussed below.

Table 6 Selected metal concentrations (mg L^{-1}; *μg L^{-1}) of formation waters from central Mississippi Salt Dome Basin.

Sample ID	Fe	Mn	Pb	Zn	Al*	Cd	Cu*
84-MS-1	137	57.5	8.39	49.6		0.49	<20
84-MS-2	97.4	38.2	0.07	1.22	59	0.05	<20
84-MS-3	61.9	10.6	0.04	0.53	133	0.02	<20
84-MS-4	346	63.9	53.2	222	267	0.83	<20
84-MS-5	407	70.2	60.5	243	42	0.81	61
84-MS-6	284	21.0	26.8	95.1	67	0.63	21
84-MS-7	261	83.5	34.6	172	79	0.86	<20
84-KS-8	194	69.3	22.8	107	132	0.67	34
84-MS-9	0.54	15.5*	<0.5*	12.0*		<0.2*	<0.2
84-MS-10	84.9	44.8	0.08	0.31		0.02	<20
84-MS-11	465	212	70.2	243	367	0.99	21
84-MS-12	65.3	16.4	0.17	0.28		0.03	<20
84-MS-14	0.75	2.78	0.16	0.20		<0.02	<20
84-MS-15	223	53.2	2.28	4.10		0.02	<20
84-MS-16	0.53	9.98	0.02	0.07	142	<0.02	<20
84-MS-18	0.15	3.2*	<0.5*	13*		0.08*	0.61
84-MS-19	0.47	1.64	0.04	0.06		0.03	<20
84-MS-20	0.07	1.46	0.03	0.16		0.05	<20
Field blank*	10.6	1.05	1.04	12.6		0.08	<0.20

Source: Kharaka *et al.* (1987).

5.16.6.3 Bisulfide Complexing

Considerable attention has been paid to complexing of metals by reduced sulfur species. Some of the complexes of zinc and lead which have been considered include: $Zn(HS)_3^-$, $Zn(HS)_4^{2-}$, $Pb(HS)_3^-$, $Pb(HS)_2^0$, and $PbS(H_2S)_2^0$ (Barnes, 1979; Kharaka *et al.*, 1987). Giordano and Barnes (1981) concluded, on the basis of experimental studies, that ore-forming solutions at temperatures less than 200 °C with total dissolved sulfur contents of less than ~1 m (ca. 3.2×10^4 mg L^{-1}) cannot transport significant quantities of lead as bisulfide complexes. Extensive metal complexing by the bisulfide complexes requires much higher pH values than those found in saline formation waters (Kharaka *et al.*, 2000).

5.16.6.4 Organic Complexing

As discussed in Section 5.16.3.5.3, there has been considerable interest in recent years in the possible role of organic ligands as complexing agents of metals in basinal waters (e.g., Kharaka *et al.*, 1987, 2000). Aliphatic acid anions, such as acetate, which are generally the most abundant of the reactive organic species, have received the most attention (Giordano and Kharaka, 1994). There is, however, an inverse correlation between metal content and organic acid concentrations in basinal waters (Hanor, 1994). Concentrations of lead and/or zinc well above 1 mg L^{-1} have been reported only in low-acetate waters. At acetate concentrations greater than ~50 mg L^{-1}, reported metal concentrations are low. Dicarboxylic acid anions form stronger metal–organic complexes, but field data and geochemical modeling indicate that the occurrence of high metal concentrations is not directly related to high concentrations of dissolved organic species (Hanor, 1996b; Kharaka *et al.*, 2000).

5.16.6.5 Chloride Complexing

The solubility of PbS and ZnS in waters containing dissolved chloride is enhanced by the formation of metal–chloro complexes: $MeCl^+$, $MeCl_2^0$, $MeCl_3^-$, and $MeCl_4^{2-}$. Most fluids that have concentrations of dissolved lead and dissolved zinc exceeding 1 mg L^{-1} also have chloride concentrations in excess of ~1×10^5 mg L^{-1} (TDS = 1.7×10^5 mg L^{-1}). A few waters from the Gulf Coast with a chloride concentration between 6×10^4 mg L^{-1} (TDS = 1×10^5 mg L^{-1}) and 1×10^5 mg L^{-1} have metal concentrations in excess of 1 mg L^{-1}. Thermodynamic calculations using Pitzer equations of state show that several properties of typical basinal fluids combine to increase the solubility of PbS and ZnS, at a fixed temperature and activity of dissolved H_2S, by as much as 15 orders of magnitude through an order of magnitude increase in salinity and chlorinity (Hanor, 1996b). These include:

(i) the significant decrease in pH with increasing salinity;

(ii) the onset of the predominance of the tetrachloro complexes ($MeCl_4^{2-}$), whose activities increase by a factor of 10^4–1 with increasing chloride concentration; and

(iii) the strongly nonideal behavior of Cl^-, which results in activity coefficient terms that are significantly greater than unity in very saline waters.

5.16.6.6 Geochemical Modeling of Ore Fluids

Geochemical modeling is a valuable tool for assessing the importance of ligands, including chlorides, bisulfides, and organic acid anions in the transport of significant quantities of metals in ore solutions (Giordano and Kharaka, 1994; Kharaka *et al.*, 2000). The geochemical code used in such modeling, must: (i) have thermo-chemical data for the dominant inorganic and organic species present in formation waters and for the minerals of interest; (ii) include temperature and pressure corrections for the thermochemical data; and (iii) be able to treat high-salinity solutions using Pitzer or comparable ion-interaction formulations.

In geochemical modeling, there have been two approaches used to study the importance of organic ligands, bisulfides, and chlorides in the transport and deposition of metals. In the first approach, simulations are carried out using variable concentrations of known organic and other ligands that are added to model compositions for the ore fluids, especially fluids for Mississippi-Valley-type and redbed-related base metal deposits (Giordano and Kharaka, 1994; Hanor, 2000). In the second approach, Kharaka *et al.* (1987) used the measured chemical composition of metal-rich and metal-poor brines from the central Mississippi Salt Dome Basin to compute the concentrations of sodium, calcium, magnesium, aluminum, iron, lead, and zinc complexed with acetate and other measured organic ligands, as well as with chlorides and bisulfides under subsurface conditions.

Results from both approaches indicate that: (i) high concentrations (~ 100 mg L^{-1}) of lead and zinc can be present in oil-field brines only if the concentration of total H_2S is at the μg L^{-1} level; in such fluids, significant quantities of dissolved lead and zinc can be transported as carboxylate complexes; (ii) lead and zinc in Mississippi-Valley-type ore fluids appear to be transported dominantly as chloride complexes; and (iii) the brines are close to equilibrium with galena and sphalerite under the likely subsurface temperatures, pressures, and pH (Kharaka *et al.*, 1987; Sicree and Barnes, 1996; Giordano, 2000).

5.16.6.7 Fluoride

Fluid inclusion studies have shown that basinal fluids responsible for depositing fluorite ore deposits typically have salinities of 20–26 wt.% NaCl equivalent and high calcium concentrations (Munoz *et al.*, 1991). As noted earlier, concentrations of fluoride typically exceed 10 mg L^{-1} in formation waters having chloride concentrations >100 g L^{-1}; below this threshold, fluoride concentrations are typically <5 mg L^{-1}. This threshold may be related to complexing of fluoride with calcium and magnesium at elevated salinities, which greatly increases the solubility of fluorite (Richardson and Holland, 1979; Munoz *et al.*, 1991; Hitchon, 1995).

5.16.6.8 Barium and Strontium

Although complete solid solution exists, at least metastably, between barite ($BaSO_4$) and celestine ($SrSO_4$), most (Ba, Sr)SO_4 ore deposits consist of one or the other end-member. This fractionation probably reflects the buffering of strontium in basinal brines by silicate–carbonate mineral assemblages. The concentration of strontium in these brines increases roughly exponentially with increasing salinity. Barium concentrations, however, are controlled by saturation with respect to barite (Figure 7).

Basinal fluids with the highest Sr/Ba ratios, and therefore most likely to form celestine rather than barite, have high salinities and moderately high levels of sulfate, which suppress the concentrations of barium. The precursors to such waters can be produced in coastal marine settings by the evaporation of seawater. As these brines reflux through underlying sediments, they can leach substantial amounts of strontium. If these diagenetically altered fluids are then discharged back up into overlying gypsum or anhydrite beds, celestine is apt to precipitate. Many barite deposits have formed where reduced, high barium waters have mixed with oxidized waters high in sulfate (Hanor, 2000).

5.16.7 DISSOLVED GASES

Gases have measurable, but variable, solubilities in aqueous solutions. The aqueous concentration of a volatile species in equilibrium with a pure gas phase at a fixed temperature increases with increasing gas pressure, and decreases with increasing salinity due to the salting out effect (Figure 16). Changes in solubility with temperature are more variable. At a fixed pressure, a volatile species such as CO_2 may display retrograde solubility at low temperatures, but an increase in solubility at higher temperatures (Figure 16). In general, however, the solubility of volatile species increases down a typical basinal depth gradient, in response to

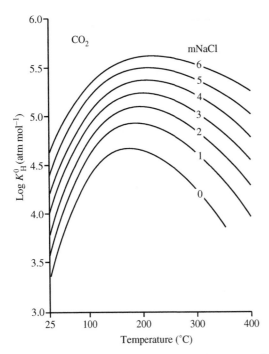

Figure 16 Solubility of CO_2, reported in terms of Henry's law constant K_H^0 (see text) as a function of temperature and salinity of water (after Drummond, 1981).

increasing temperature and pressure. While many volatile species exist in sedimentary basin waters, we will examine only the systematics of dissolved methane, carbon dioxide, hydrogen sulfide, and the noble gases.

5.16.7.1 Methane

Extensive field testing was conducted during the oil crisis of the 1970s to assess the concentration of dissolved CH_4 and of other geopressured–geothermal resources in the northern Gulf of Mexico and in other basins in USA. The tests showed that saline pore fluids to depths exceeding 6 km are saturated with respect to CH_4 (Wallace *et al.*, 1979; Kharaka and Berry, 1980). The high content of dissolved CH_4 in basinal brines is due to the thermocatalytic destruction of kerogen and liquid hydrocarbons, especially in the deeper sections (high T and P) of the basins (Tissot and Welte, 1984).

Experimental work on the solubility of CH_4 (e.g., Haas, 1978; Price, 1979) permits estimation of the solubility of CH_4 over much of the range of pressures, temperatures, and salinities characteristic of sedimentary basins (Kharaka *et al.*, 1988; Spycher and Reed, 1988). High concentrations of CH_4 have been found in the two-phase fluid inclusions in sedimentary ore deposits and in diagenetic mineral overgrowths, and corrections

for the presence of CH_4 have to be made to inclusion homogenization temperatures to derive accurate estimates of filling temperature and pressures (Hanor, 1980).

5.16.7.2 Carbon Dioxide

Dissolved CO_2 in formation waters originates primarily from kerogen and other organic sources, from calcite and other carbonate minerals, and/or from mantle sources (see Section 5.16.5.3). The solubility of CO_2 in water (Figure 16), like that of CH_4, shows a strong dependence on temperature and salinity (Drummond, 1981). CO_2 is much more reactive than CH_4, and plays a more important role in water–CO_2–mineral interactions in the subsurface, especially in the carbonate and pH-sensitive interactions (Kharaka *et al.*, 1985). The concentration of dissolved CO_2 in formation waters can be represented by the fugacity of CO_2 (f_{CO_2}) in a real or fictive gas phase in equilibrium with that of water. The partial pressure and fugacity of CO_2 can be calculated after measuring the CO_2 concentration in a coexisting gas phase at known P and T (Smith and Ehrenberg, 1989; Spycher and Reed, 1988) or by calculating f_{CO_2} assuming equilibrium with respect to calcite (Hutcheon, 2000) (Equation (6)):

$$Ca^{2+} + H_2O + CO_2 = CaCO_3 + 2H^+ \qquad (6)$$

Several authors have noted an increase in f_{CO_2} with increasing temperature in some sedimentary basins, including the Texas Gulf Coast, USA, the Patani Basin, Thailand, and the North Sea (Smith and Ehrenberg, 1989; Lundegard and Trevena, 1990; Hutcheon, 2000). The CO_2 fugacities increase systematically from ~ 0.1 bar at 50 °C to 100 bar at 175 °C. The phase boundary for the reaction of kaolinite with dolomite to form chlorite, calcite, and CO_2 parallels the CO_2-temperature trend observed in the field. This is indicative of CO_2-buffering, but the fugacities predicted are one to two orders of magnitude too low (Hutcheon *et al.*, 1993; Hutcheon, 2000).

5.16.7.3 Hydrogen Sulfide

Reservoirs of natural gas generally contain only small amounts of H_2S, but in some cases contain 2–30% by volume $H_2S(g)$ (Hitchon *et al.*, 2001; Worden and Smalley, 2001). Sources of this H_2S include: (i) degradation of sulfur-rich kerogen, (ii) breakdown of sulfur-rich crude oil, (iii) BSR, and (iv) TSR (Worden and Smalley, 2001; Machel, 2001). Bacterial sulfate reduction is generally limited to temperatures below ~ 80 °C, leaving TSR as the only mechanism that contributes high concentrations of H_2S to natural gas (Worden and Smalley, 2001). Sources of sulfate include aqueous sulfate and sulfate derived from the dissolution of anhydrite.

In clastic sedimentary sequences, where iron-bearing mineral phases are abundant, most of the sulfide can be precipitated as iron sulfide, usually pyrite, and concentrations of H_2S in formation waters in such sequences are often below the limit of detection (Kharaka *et al.*, 1987). A pyrite–siderite or a pyrite–ankerite buffer may be controlling H_2S partial pressures in clastic hydrocarbon reservoirs in the Norwegian Shelf and Gulf Coast (Aagaard *et al.*, 2001). Predicted and measured H_2S partial pressures increase over several orders of magnitude with depth in these basins. In iron-poor carbonate sediments H_2S partial pressures are largely unbuffered, and concentrations of dissolved hydrogen sulfide can approach $1,000$ mg L^{-1} (Wade *et al.*, 1989). Hitchon *et al.* (2001) found a linear relation between the concentration of HS^- in formation waters and the volume percent H_2S in associated natural gas in carbonate reservoirs of the Alberta Basin, Canada.

5.16.7.4 Noble Gases

The noble gas geochemistry of natural waters, including formation waters in sedimentary basins, has been used to determine paleotemperatures in the recharge areas, to evaluate water washing of hydrocarbons, and to identify mantle-derived volatiles (Pinti and Marty, 2000). The dissolved noble gases, helium, neon, argon, krypton, and xenon in sedimentary waters, have four principal sources: the atmosphere, *in situ* radiogenic production, the deep crust, and the mantle. These sources have characteristic chemical and isotopic compositions (Ozima and Podosek, 1983; Kennedy *et al.*, 1997).

The partitioning of atmospheric noble gases (ANGs) between air and surface water follows Henry's law, and noble gas solubility is usually described in terms of the Henry's law coefficient, which varies with temperature and salinity: the concentration of the various noble gases in groundwaters derived from surface waters depends on: (i) the mean annual temperature of recharge, (ii) mean atmospheric pressure, and (iii) the salinity of the recharge water. After correcting for salinity and the elevation of the recharge area, ANG abundances can be used to define the paleotemperature of recharge areas (Mazor and Bosch, 1987).

Dissolved radiogenic noble gases, which have been used in basinal studies, include ^3He, ^4He, $^{20-22}Ne$, ^{40}Ar, and $^{131-136}Xe$ (Ballentine *et al.*, 1996; Pinti and Marty, 2000). The isotope ^4He is produced by alpha decay of ^{235}U, ^{238}U, and ^{232}Th; ^3He is produced by decay of tritium generated from the neutron activation of ^6Li; and ^{40}Ar is produced by electron capture decay of ^{40}K. Of the xenon isotopes, ^{136}Xe is produced by the spontaneous fission of ^{238}U, and ^{129}Xe is produced by beta decay of ^{129}I. Isotopes of neon and other isotopes of argon and xenon are produced by a variety of other nuclear reactions.

Much of the ^{40}Ar produced by decay of ^{40}K is retained within the mineral lattice at temperatures below $250-300\,°C$. In contrast, the closure temperature of ^4He, produced by the decay of ^{235}U, ^{238}U, and ^{232}Th, is usually below $100\,°C$. One would expect, therefore, that noble gases in water from the deeper regions of the crust would have higher $^4He/^{40}Ar$ ratios than noble gases in shallow parts of sedimentary basins.

Basinal fluids often contain significant amounts of mantle-derived noble gas isotopes, although these tend to be masked by high background levels of ANG (O'Nions and Oxburgh, 1988). The most prominent mantle signal is excess ^3He. R_a, the ($^3He/^4He$) ratio of air, is 1.38×10^{-6}. The $^3He/^4He$ ratio of crustal origin is typically <0.03 times R_a. In contrast, mantle-derived $^3He/^4He$ values relative to air are ≥ 8 (Kennedy *et al.*, 1997; Kharaka *et al.*, 1999b). The $^3He/^4He$ ratio in some basinal waters associated with oil fields can be enriched by an order of magnitude relative to the expected values from *in situ* radiogenic production; $^3He/^4He$ and $^{20}Ne/^{22}Ne$ ratios in basinal fluids in excess of their atmospheric values have been taken as evidence for the presence of mantle-derived helium and neon (Ballentine and O'Nions, 1991). Fluids in extensional basins, such as the Pannonian and Paris basins, or in continental rifts have clear mantle-derived noble gas signatures in the form of excess ^3He, ^{129}Xe, and ^{21}Ne (Ballentine and O'Nions, 1991). Basins in which there is active sediment loading, such as the Po Basin, Italy, do not have a clearly defined mantle gas signature (Elliot *et al.*, 1993).

Noble gases exist in formation waters as uncharged, nonpolar species. They preferentially partition from aqueous solutions into nonpolar solvents such as crude oil and natural gas (Kharaka and Specht, 1988). The degree of partitioning depends on such factors as temperature, gas atomic radius, and the salinity of the aqueous phase. Distinct variations in $^{20}Ne/^{36}Ar$, $^{84}Kr/^{36}Ar$, and $^{130}Xe/^{36}Ar$ are produced by fractionation in multiphase fluid systems. These variations have been used as a tool in oil exploration and reservoir evaluation. Pinti and Marty (2000) give detailed examples of these applications in the Pannonian Basin, Hungary and in the Paris Basin, France.

5.16.8 THE INFLUENCE OF GEOLOGICAL MEMBRANES

The ability of soils, clays, and shale (the geological membranes) to serve as semipermeable membranes has been conclusively demonstrated by experimental data (McKelvey and Milne, 1962; Kharaka and Berry, 1973; Fritz and Marine, 1983; Demir, 1988; Whitworth and Fritz, 1994) and field evidence (Bredehoeft *et al.*, 1963; Hanshaw and Hill, 1969; Berry, 1973; Kharaka and Berry, 1974). The chemical composition of water in sedimentary basins can be significantly affected by interaction with geological membranes.

(i) Compacted clays and shale serve as semipermeable membranes that retard or restrict the flow of dissolved chemical species with respect to water. Subsurface water that has flowed through a geological membrane (effluent water) is lower in TDS and has a chemical composition different from that of the original solution (input water) or to the solution remaining in the aquifer on the input side of the membrane (hyperfiltrated water).

(ii) Subsurface waters squeezed from massive shale and siltstones are present in large areas in many sedimentary basins such as the Gulf Coast (Kharaka *et al.*, 1980) and the Central Valley, California (Berry, 1973; Kharaka *et al.*, 1985); these also exhibit increasing "membrane effluent" characteristics with increasing depth (increased compaction). The lowest salinities of the waters in these two basins are in the range $\sim(0.5-1) \times 10^4$ mg L^{-1}. These values are about a quarter of the salinities of formation waters at comparable depths in these basins that were not affected by this process. Laboratory experiments (Kryukov *et al.*, 1962; Kharaka, unpublished data, 1976) have shown that water squeezed from uncompacted clays and shale has the same salinity and chemical composition as the initial solution. As compaction pressure is increased, the salinity of squeezed water decreases and shows other membrane filtration characteristics.

(iii) Clay minerals have cation exchange capacities that range from ~5 (kaolinite) milliequivalents per 100 g to 150 (smectite) milliequivalents per 100 g. Exchange reactions are relatively fast and can modify the composition of subsurface waters.

(iv) Clay transformations, especially the conversion of smectites to illites, are important reactions in many sedimentary basins at temperatures higher than $\sim80\,°C$. The exchange capacity of illite is about half that of smectite. This transformation can therefore result in the uptake of potassium on clays and the release of quantitatively important amounts of adsorbed species to subsurface waters.

Kharaka (1986) report detailed studies of "membrane effluent" effects in the Central Valley,

California, in the northern Gulf of Mexico Basin, and in the North Slope, Alaska. In each basin, the compositions of effluent and hyperfiltrated waters were compared. The "membrane effluent" characteristics that were observed include the following "chemical markers": lower TDS and Ca/Na and Br/Cl ratios, and higher Li/Na, NH$_3$/Na, B/Cl, HCO$_3$/Cl, and F/Cl ratios. These "chemical markers" are similar to those predicted from laboratory studies when these are extrapolated to the temperature, pressure, and hydraulic pressure gradients in sedimentary basins (Kharaka and Smalley, 1976; Haydon and Graf, 1986; Demir, 1988).

Nevertheless, the importance of membrane filtration in modifying the chemical composition of subsurface waters is controversial (Hanor, 1987). Manheim and Horn (1968) discussed the difficulties of producing brines by shale membrane filtration, concluding that the pressure requirements for significant membrane filtration were not encountered in geologic environments. Fluid pressure or hydraulic potentials are needed to force water through shale. However, membrane filtration will lead to an increase in the concentration of solution on the input side of shale relative to the output side. Flow and membrane filtration will cease when the osmotic head thus generated equals the original hydraulic potential. The US Gulf Coast has sometimes been mentioned as an area where production of brine by membrane filtration should be occurring because high fluid-pressure gradients exist across thick regional shale sequences (Graf, 1982; Kharaka, 1986). However, in many regions of the Gulf Coast formation waters of higher salinity are found on the effluent side of these overpressured shales. Processes such as salt dome dissolution rather than membrane effects are the dominant controls on salinity in this area (Hanor, 1994).

Manheim and Horn (1968) are correct in stating that water must pass through shale to be concentrated and diluted by membrane filtration. Flow through shale in sedimentary basins is indeed prevalent (Bredehoeft *et al.*, 1963). Flow is particularly important in two common field situations. The first is encountered in geopressured systems, such as in the US Gulf Coast, in which fluid potentials much higher than hydrostatic values (thousands of meters) are encountered that force water through shale before the development of equal and counteracting osmotic potentials (Kharaka *et al.*, 1980; Graf, 1982). The second situation is encountered in recharging sedimentary basins. Kharaka and Berry (1974) showed that fresh meteoric water recharged at Reef Ridge in the San Joaquin Valley, California, USA, percolates through the Temblor Formation. A portion of this water passes through the overlying McClure Shale. This increases the salinity of water remaining in the Temblor Formation. The salinity of water as it reached the Kettleman North Dome oil field is up to 4×10^4 mg L^{-1}. Of course, the chemical composition of water at Kettleman North Dome is also modified by the other water–rock interactions discussed by Kharaka and Berry (1974) and Merino (1975) and in this chapter.

5.16.9 SUMMARY AND CONCLUSIONS

Formation waters with salinities ranging from ~5,000 mg L^{-1} to >3.5 × 10^5 mg L^{-1} dissolved solids play a major role in the physical and chemical processes that occur in sedimentary basins. Detailed inorganic and organic chemical analyses, together with measurements of the oxygen and hydrogen isotopes in water and of a diverse suite of other stable and radioactive isotopes, have been reported for formation waters from many sedimentary basins worldwide. Application of these data, together with information on data for the local and regional geology and on the recent and paleohydrology, have shown that these fluids are much more mobile, and their origins and interactions with rocks and sediments much more complex than previously realized.

The formation waters in sedimentary basins are dominantly of local meteoric or marine connate origin. However, bittern (residual) water, geologically old meteoric water, and especially waters of mixed origin are important components in many sedimentary basins. The original waters have evolved to Na–Cl, Na–Cl–CH$_3$COO, or Na–Ca–Cl-type waters by a combination of several major processes: (i) dissolution of halite; (ii) incorporation of bitterns; (iii) dissolution and precipitation of minerals other than halite; (iv) interaction of the waters with organic matter present in sedimentary rocks; (v) interaction with rocks, principally clays, siltstones, and shale, that behave as geological membranes; and (vi) diffusion, which appears to be more important than previously thought. The important processes responsible for the chemical evolution of water and mineral diagenesis in each basin can be identified using "chemical markers" and isotopes.

The discovery, in these waters, of high concentrations (up to 1×10^4 mg L^{-1}) of mono- and dicarboxylic acid anions, phenols, and other reactive organic species has led to numerous field and laboratory studies to determine their distribution and importance in inorganic and organic interactions. The observed concentrations are minimum values, because the organics are degraded by bacteria and by thermal decarboxylation reactions. Decarboxylation rates estimated from field data for acetate, the dominant species, give half-life values of 20–60 Myr at 100 °C.

The concentrations of lead, zinc, copper, mercury, and several other metals (excluding iron and manganese) are almost always very low (<1 mg L^{-1}) in oil-field waters, because they are limited by the very low solubility of their respective sulfide minerals. Exceptionally rich sources of metals such as redbeds are probably present in the few localities worldwide, where dissolved metal concentrations are high. Metals leached from these beds precipitate along with available reduced sulfur; dissolved metal concentrations increase to high values only when the concentration of H$_2$S is extremely low (<0.01 mg L^{-1}).

Water–rock interaction in sedimentary basins is and will continue to be a subject of intensive research. The following are areas that are receiving particular attention because of their scientific or economic importance, and/or the arrival of new equipment and techniques:

(i) A plethora of stable and radioactive isotope data is being used to study mineral diagenesis and the origin and age of water and the sources and sinks of solutes in these basins. Determination of accurate fractionation factors and rates of isotopic exchange between waters and minerals under sedimentary conditions are needed for a better interpretation of the isotope data.

(ii) Major advances have been made since the early 1990s documenting the nature, distribution, and organic and inorganic interactions of reactive organic species. Additional investigation is needed, especially in determining the stability of organic–inorganic complexes, rates of decarboxylation under field conditions, and the importance of dissolved organic sulfur and nitrogen compounds.

(iii) Geochemical, hydrologic, and solute transport codes are being applied to sedimentary basins. To obtain reliable information from this endeavor, accurate, detailed data on aqueous and solid phases must be available and more accurate thermodynamic data are needed for clay minerals, minerals of highly variable composition, and brines. Data are also needed for the kinetics of dissolution/precipitation in the field.

(iv) Chemical and physical processes that control the geochemistry of aluminum, especially its transport in formation water.

(v) Evaluation of reaction kinetics by integrating field and reservoir-scale spatial variations in fluid compositions with quantitative fluid flow models and sediment mineralogy.

(vi) Chemical and isotopic evolution of aqueous wastes, including produced water injected into the subsurface. What can be deduced about the fate of these wastes from what is known about the composition and interactions of formation water?

(vii) Chemical and physical processes that decrease formation water salinity.

(viii) Mechanisms by which basinal brines are expelled back up into near surface environments. Can it be done without tectonics? What impact do major brine expulsion events have on the exogenic geochemical cycle?

ACKNOWLEDGMENTS

We would like to thank James I. Drever, Robert Mariner, and James Palandri for reviewing this manuscript and suggesting significant improvements. We would also like to thank Evangelos Kakouros and James Thordsen for helping with figures, tables, and text formatting. Much of Hanor's research on fluids in sedimentary basins has been supported by the National Science Foundation, most recently EAR-985459.

REFERENCES

Aagaard P., Jahren J. S., and Ehrenberg S. N. (2001) H$_2$S-controlling reactions in clastic hydrocarbon reservoirs from the Norwegian Shelf and US Gulf Coast. In *Proceedings of the 10th International Symposium on the Water Rock Interaction* (ed. R. Cidu). A. A. Balkema Publishers, Rotterdam, pp. 129–132.

Aggarwal J. K., Palmer M. R., Bullen T. D., Ragnarsdottir K. V., and Arnorsson S. (2000) The boron isotope systematics of Icelandic geothermal waters: 1. Meteoric water charged systems. *Geochim. Cosmochim. Acta* **64**, 579–585.

Andrews J. N., Edmunds W. M., Smedley P. L., Fontes J. C., Fifield L. K., and Allan G. L. (1994) Chlorine-36 in groundwater as a palaeoclimatic indicator: the East Midlands Triassic sandstone aquifer (UK). *Earth Planet. Sci. Lett.* **122**, 159–171.

Armstrong S. C., Sturchio N. C., and Hendry M. J. (1998) Strontium isotopic evidence on the chemical evolution of pore waters in the Milk River aquifer, Alberta, Canada. *Appl. Geochem.* **13**, 463–475.

Azaroual M., Fouillac C., and Matray J. M. (1997) Solubility of silica polymorphs in electrolyte solutions: II. Activity of aqueous silica and solid silica polymorphs in deep solutions from the sedimentary Paris Basin. *Chem. Geol.* **140**, 167–179.

Ballentine C. J. and O'Nions R. K. (1991) Rare gas constraints on crustal degassing and fluid transport in extensional basins. *Terra Abstr.* **3**, 493–494.

Ballentine C. J., O'Nions R. K., and Coleman M. L. (1996) A Magnus opus: helium, neon, and argon isotopes in a North Sea oilfield. *Geochim. Cosmochim. Acta* **60**, 831–849.

Banks D. A., Gleeson S. A., and Green R. (2000) Determination of the origin of salinity in granite-related fluids; evidence from chloride isotopes in fluid inclusions. *J. Geochem. Explor.* **69–70**, 309–312.

Barnes H. L. (1979) Solubilities of ore minerals. In *Geochemistry of Hydrothermal Ore Deposits*, 2nd edn. (ed. H. L. Barnes). Wiley, New York, pp. 404–460.

Barth S. (1998) Application of boron isotopes for tracing sources of anthropogenic contamination in groundwater. *Water Res.* **32**, 685–690.

Barth T. (1987) Quantitative determination of volatile carboxylic acids in formation waters by isotachophoresis. *Anal. Chem.* **59**, 2232–2237.

Barth T., Borgund A. E., and Hopland A. L. (1989) Generation of organic compounds by hydrous pyrolysis of Kimmeridge oil shale—bulk results and activation energy calculations. *Org. Geochem.* **14**, 69–76.

Bassett R. L. (1977) The geochemistry of boron in thermal waters. PhD Thesis, Stanford University.

Bath A., Edmunds W. M., and Andrews J. N. (1978) Paleoclimatic trends deduced from the hydrochemistry of Triassic sandstone aquifer, United Kingdom. *Isotope Hydrol.* **2**, 545–568. IAEA, Vienna.

Bazin B., Brosse É., and Sommer F. (1997a) Chemistry of oil-field brines in relation to diagenesis of reservoirs: 1. Use of mineral stability fields to reconstruct *in situ* water composition, Example of the Mahakam basin. *Mar. Petrol. Geol.* **14**, 481–495.

Bazin B., Brosse É., and Sommer F. (1997b) Chemistry of oil-field brines in relation to diagenesis of reservoirs: 2. Reconstruction of paleo-water composition for modeling illite diagenesis in the Greater Alwyn area (North Sea). *Mar. Petrol. Geol.* **14**, 497–511.

Bell I. L. (1991) Acetate decomposition in hydrothermal solutions. PhD Thesis, The Pennsylvania State University.

Bennett B. and Larter S. R. (1997) Partition behavior of alkylphenol in crude oil/brine systems under subsurface conditions. *Geochim. Cosmochim. Acta* **61**, 4393–4402.

Bennett S. C. and Hanor J. S. (1987) Dynamics of subsurface salt dissolution at the Welsh Dome, Louisiana Gulf Coast. In *Dynamical Geology of Salt and Related Structures* (eds. I. Lerche and J. J. O'Brien). Academic Press, New York, pp. 653–677.

Berry F. A. F. (1973) High fluid-potentials in the California coast ranges and their tectonic significance. *Am. Assoc. Petrol. Geol. Bull.* **57**, 1219–1249.

Bethke C. M. (1994) *The Geochemist's Workbench: A User's Guide to Rxn, Act2, Tact and Gtplot*. Urbana-Champaign, University of Illinois Hydrogeology Program.

Bethke C. M., Zhao X., and Torgersen T. J. (1999) Groundwater flow and the ^{4}He distribution in the Great Artesian Basin of Australia. *J. Geophys. Res.* **B104**, 12999–13011.

Bethke C. M., Torgersen T., and Park J. (2000) The "age" of very old groundwater: insights from reactive transport models. *J. Geochem. Explor.* **69–70**, 1–4.

Birkle P., Rosillo Aragon J. J., Portugal E., and Fong Aguilar J. L. (2002) Evolution and origin of deep reservoir water at the Activo Luna oil field, Gulf of Mexico, Mexico. *Am. Assoc. Petrol. Geol. Bull.* **86**, 457–484.

Bjørlykke K. and Egeberg P. K. (1993) Quartz cementation in sedimentary basins. *Am. Assoc. Petrol. Geol. Bull.* **77**, 1538–1548.

Bjørlykke K., Aagaard P., Egeberg P. K., and Simmons S. P. (1995) Geochemical constraints from formation water analyses from the North Sea and the Gulf Coast basins on quartz, feldspar, and illite precipitation in reservoir rocks. *Geol. Soc. Spec. Pub.* **86**, 33–50.

Boles J. R. (1978) Active ankerite cementation in the subsurface Eocene of southwest Texas. *Contrib. Mineral. Petrol.* **68**, 13–22.

Boles J. R. and Franks S. G. (1979) Clay diagenesis in Wilcox sandstones of southwest Texas: implications of smectite diagenesis on sandstone cementation. *J. Sedim. Petrol.* **49**, 55–70.

Bredehoeft J. D., Blyth C. R., White W. A., and Maxey G. G. (1963) Possible mechanism for concentration of brines in subsurface formations. *Am. Assoc. Petrol. Geol. Bull.* **47**, 257–269.

Breit G. N., Kharaka Y. K., and Rice C. A. (2001) National database on the chemical composition of formation waters from petroleum wells. *Proceedings of the 7th IPEC Meeting: Environmental Issues and Solutions in Petroleum*

Exploration, Production, and Refining, November 2000 (ed. K. L. Sublette), Albuquerque, NM, Tulsa, OK.

Bryant E. A. (1997) *Climate Processes and Change*. Cambridge University Press, Cambridge.

Bullen T. D., White A. F., Childs C. W., and Horita J. (2001) Reducing ambiguity in isotope studies using a multi-tracer approach. In *Proceedings of the 10th International Symposium on the Water Rock Interaction* (ed. R. Cidu). A. A. Balkema Publishers, Rotterdam, pp. 19–28.

Carothers W. W. and Kharaka Y. K. (1978) Aliphatic acid anions in oil-field waters—implications for origin of natural gas. *Am. Assoc. Petrol. Geol. Bull.* **62**, 2441–2453.

Carothers W. W. and Kharaka Y. K. (1980) Stable carbon isotopes of HCO_3^- in oil-field waters—implications for the origin of CO_2. *Geochim. Cosmochim. Acta* **44**, 323–333.

Carpenter A. B. (1978) Origin and chemical evolution of brines in sedimentary basins. *Okl. Geol. Surv. Circular* **79**, 60–77.

Carpenter A. B., Trout M. L., and Pickett E. E. (1974) Preliminary report on the origin and chemical evolution of lead- and zinc-rich brines in central Mississippi. *Econ. Geol.* **52**, 1191–1206.

Case L. C. (1945) Exceptional Silurian brine near Bay City, Michigan. *Am. Assoc. Petrol. Geol. Bull.* **29**, 567–570.

Case L. C. (1955) Origin and current usage of the term, "connate water." *Am. Assoc. Petrol. Geol. Bull.* **39**, 1879–1882.

Chan L.-H., Starinsky A., and Katz A. (2002) The behavior of lithium and its isotopes in oilfield brines: evidence from the Heletz-Kokhav Field, Israel. *Geochim. Cosmochim. Acta* **66**, 615–623.

Clark I. and Fritz P. (1997) *Environmental Isotopes in Hydrogeology*. CRC Press, Boca Raton, FL.

Clayton R. N., Friedman I., Graf D. L., Mayeda T. K., Meets W. F., and Shimp N. F. (1966) The origin of saline formation waters: I. Isotopic composition. *J. Geophys. Res.* **71**, 3869–3882.

Collins A. G. (1975) *Geochemistry of Oil-field Waters*. Elsevier, New York.

Connolly C. A., Walter L. M., Baadsgaard H., and Longstaff F. (1990) Origin and evolution of formation fluids, Alberta Basin, western Canada sedimentary basin: II. Isotope systematics and fluid mixing. *Appl. Geochem.* **5**, 375–395.

Cook P. G. and Herczeg A. L. (eds.) (2000) *Environmental Tracers in Hydrology*. Kluwer Academic, Boston.

Craig H. (1961) Isotopic variations in meteoric waters. *Science* **133**, 1702–1703.

Crossey L. J. (1991) Thermal degradation of aqueous oxalate species. *Geochim. Cosmochim. Acta* **55**, 1515–1527.

Crossey L. J., Surdam R. C., and Lahann R. (1986) Application of organic/inorganic diagenesis to porosity prediction. In *Relationship of Organic Matter and Mineral Diagenesis*, Special Publication 38 (ed. D. L. Gautier). Society of Economic Paleontologists and Mineralogists, Tulsa, OK, pp. 147–156.

Dansgaard W. (1964) Stable isotopes in precipitation. *Tellus* **16**, 436–468.

Davis S. N. (1964) The Chemistry of Saline Waters by R. A. Krieger. Discussion. *Ground Water* **2**, 51.

Davisson M. K. and Criss R. E. (1996) Na–Ca–Cl relations in basinal fluids. *Geochim. Cosmochim. Acta* **60**, 2743–2752.

Degens E. T., Hunt I. M., Reuter J. H., and Reed W. E. (1964) Data on the distribution of amino acids and oxygen isotopes in petroleum brine waters of various geologic ages. *Sedimentology* **3**, 199–225.

Demir I. (1988) Studies of smectite membrane behavior: electrokinetic, osmotic, and isotopic fractionation processes at elevated pressures. *Geochim. Cosmochim. Acta* **52**, 727–737.

Drever J. I. (1997) *The Geochemistry of Natural Waters: Surface and Groundwater Environments*. Prentice-Hall, Upper Saddle River, NJ.

Drever J. I. and Vance G. F. (1994) Role of soil organic acids in mineral weathering processes. In *Organic Acids in Geological Processes* (eds. E. D. Pittman and M. D. Lewan). Springer, Berlin, pp. 138–161.

Drummond S. E. (1981) Boiling and mixing of hydrothermal fluids: chemical effects on mineral precipitation. PhD Thesis, Pennsylvania State University.

Drummond S. E. and Palmer D. A. (1986) Thermal decarboxylation of acetate: II. Boundary conditions for the role of acetate in the primary migration of natural gas. *Geochim. Cosmochim. Acta* **50**, 813–824.

Dworkin S. I. and Land L. S. (1996) The origin of aqueous sulfate in Frio pore fluids and its implication for the origin of oil-field brines. *Appl. Geochem.* **11**, 403–408.

Eastoe C. J., Long A., and Knauth L. P. (1999) Stable chloride isotopes in the Palo Duro Basin, Texas: evidence for preservation of Permian evaporite brines. *Geochim. Cosmochim. Acta* **63**, 1375–1382.

Eastoe C. J., Long A., Land L. S., and Kyle J. R. (2001) Stable chloride isotopes in halite and brine from the Gulf Coast Basin: brine genesis and evolution. *Chem. Geol.* **176**, 343–360.

Egeberg P. K. and Aagaard P. (1989) Origin and evolution of formation waters from oil fields on the Norwegian shelf. *Appl. Geochem.* **4**, 131–142.

Eggenkamp H. G. M. (1998) The stable isotope geochemistry of the halogens Cl and Br: a review of 15 years development. *Min. Mag.* **62A** (Part 1), 411–412.

Eggenkamp H. G. M. and Coleman M. L. (2000) Rediscovery of classical methods and their application to the measurement of stable bromide isotopes in natural samples. *Chem. Geol.* **167**, 393–402.

Elliot T., Ballentine C. J., O'Nions R. K., and Ricchiuto T. (1993) Carbon, helium, neon, and argon isotopes in a Po Basin (northern Italy) natural gas field. *Chem. Geol.* **106**, 429–440.

Emery D. and Robinson A. (1993) *Inorganic Geochemistry: Applications to Petroleum Geology*. Blackwell Scientific, London.

Fabryka-Martin J. (2000) Iodine-129 as a groundwater tracer. In *Environmental Tracers in Subsurface Hydrogeology* (eds. P. Cook and A. L. Herczeg). Kluwer Academic, Amsterdam, pp. 504–510.

Fabryka-Martin J., Bentley H. W., Elmore D., and Airey P. L. (1985) Natural iodide-129 as an environmental tracer. *Geochim. Cosmochim. Acta* **49**, 337–347.

Faure G. (1986) *Principles of Isotope Geology*, 2nd edn. Wiley, New York.

Fisher J. B. and Boles J. R. (1990) Water–rock interaction in Tertiary sandstones, San Joaquin Basin, California, USA: diagenetic controls on water composition. *Chem. Geol.* **82**, 83–101.

Fournier R. O. and Potter R. W. (1979) Magnesium correction to the Na–K–Ca chemical geothermometer. *Geochim. Cosmochim. Acta* **43**, 1543–1550.

Fournier R. O., White D. E., and Truesdell A. H. (1974) Geochemical indicators of subsurface temperature: I. Basic assumptions. *US Geol. Surv. J. Res.* **2**, 259–262.

Franks S. G., Dias R. F., Freeman K. H., Boles J. R., Holba A. G., Fincannon A. L., and Jordan E. D. (2001) Carbon isotopic composition of organic acids in oilfield waters, San Joaquin Basin, California. *Geochim. Cosmochim. Acta* **65**, 1301–1310.

Fritz P. and Fontes J. Ch. (eds.) (1980) *Handbook of Environmental Isotope Geochemistry*. The Terrestrial Environment, A. Elsevier, Amsterdam, vol. 1.

Fritz P. and Fontes J. Ch. (eds.) (1986) In *Handbook of Environmental Isotope Geochemistry*. The Terrestrial Environment, B. Elsevier, Amsterdam, vol. 2.

Fritz S. J. and Marine I. W. (1983) Experimental support for a predictive osmotic model of clay membranes. *Geochim. Cosmochim. Acta* **47**, 1515–1522.

Froehlich K., Ivanovich M., Hendry M. J., Andrews J. N., Davis S. N., Drimmie R. J., Fabryka-Martin J., Florkowski T., Fritz P., Lehmann B. E., Loosli H. H., and Nolte E. (1991) Application of isotopic methods to dating of very old groundwaters: Milk River aquifer, Alberta, Canada. *Appl. Geochem.* **6**, 465–472.

Fu B. (1998) A study of pore fluids and barite deposits from hydrocarbon seeps: deepwater Gulf of Mexico. PhD Thesis, Louisiana State University.

Garven G. (1995) Continental-scale fluid flow and geologic processes. *Ann. Rev. Earth Planet. Sci.* **24**, 89–117.

Gas Research Institute (1995) *Atlas of Gas-related Produced Water for 1990.* GRI Report 95/0016.

Gavrieli I., Starinsky A., Spiro B., Aizenshat Z., and Nielsen H. (1995) Mechanisms of sulfate removal from subsurface chloride brines: Heletz-Kokhav oilfields, Israel. *Geochim. Cosmochim. Acta* **59**, 3525–3533.

Giordano T. H. (2000) Organic matter as transport agent in ore-forming systems. *Rev. Econ. Geol.* **9**, 133–156.

Giordano T. H. and Barnes H. L. (1981) Lead transport in Mississippi Valley-type ore solutions. *Econ. Geol.* **76**, 2200–2211.

Giordano T. H. and Kharaka Y. K. (1994) Organic ligand distribution and speciation in sedimentary basin brines, diagenetic fluids, and related ore solutions. In *Geofluids: Origin, Migration, and Evolution of Fluids in Sedimentary Basins*, Special Publication No. 78 (ed. J. Palmer). Geological Society of London, London, pp. 175–202.

Graf D. L. (1982) Chemical osmosis and the origin of subsurface brines. *Geochim. Cosmochim. Acta* **46**, 1431–1448.

Graf D. L., Meents W. F., and Shimp N. F. (1966) Chemical composition and origin of saline formation waters in the Illinois and Michigan basins. *Geol. Soc. Am. Spec. Pap.* 87.

Gunter W. D., Perkins E. H., and Hutcheon I. (2000) Aquifer disposal of acid gases: modelling of water–rock reactions for trapping of acid wastes. *Appl. Geochem.* **15**, 1085–1095.

Haas J. A. (1978) *An empirical equation with tables of smoothed solubilities of methane in water and aqueous sodium chloride solutions up to 25 weight percent, 360 °C, and 138 MPa.* US Geological Survey Open-File Report 78–1004.

Hanor J. S. (1979) Sedimentary genesis of hydrothermal fluids. In *Geochemistry of Hydrothermal Ore Deposits* (ed. H. L. Barnes). Wiley, New York, pp. 137–168.

Hanor J. S. (1980) Dissolved methane in sedimentary brines: potential effect on the PVT properties of fluid inclusions. *Econ. Geol.* **75**, 603–609.

Hanor J. S. (1983) Fifty years of development of thought on the origin and evolution of subsurface sedimentary brines. In *Evolution and the Earth Sciences Advances in the Past Half-century* (ed. S. J. Boardman). Kendall/Hunt, Dubuque, pp. 99–111.

Hanor J. S. (1987) *Origin and Migration of Subsurface Sedimentary Brines, Short Course 21.* (SEPM) Society for Sedimentary Geology, Tulsa, OK.

Hanor J. S. (1994) Origin of saline fluids in sedimentary basins. In *Geofluids: Origin and Migration of Fluids in Sedimentary Basins*, Special Publication No. 78 (ed. J. Patnell). Geological Society of London, pp. 151–174.

Hanor J. S. (1996a) Variations in chloride as a driving force in siliciclastic diagenesis. In *Siliciclastic Diagenesis and Fluid Flow: Concepts and Applications*, Special Publication No. 55 (eds. L. J. Crossey, R. Loucks, and M. W. Totten). (SEPM) Society for Sedimentary Geology, Tulsa, OK, pp. 3–12.

Hanor J. S. (1996b) Controls on the solubilization of lead and zinc in basinal brines. In *Carbonate-hosted Lead–Zinc Deposits*, Economic Geology Special Publication 4 (ed. D. F. Sangster). Soc. Econ. Geologists IMC, Littleton, CO, pp. 483–500.

Hanor J. S. (2000) Barite–celestine geochemistry and environments of formation. In *Sulfate Minerals,*

Rev. Mineral. Geochem. **40** (eds. C. Alpers, J. Jambor, and D. K. Nordstrom). pp. 193–275.

Hanor J. S. (2001) Reactive transport involving rock-buffered fluids of varying salinity. *Geochim. Cosmochim. Acta* **65**, 3721–3732.

Hanor J. S. and Workman A. L. (1986) Distribution of dissolved fatty acids in some Louisiana oil-field brines. *Appl. Geochem.* **1**, 37–46.

Hanor J. S., Kharaka Y. K., and Land L. S. (1988) Geochemistry of waters in deep sedimentary basins. *Geology* **16**, 560–561.

Hanor J. S., Land L. S., and Macpherson L. G. (1993) Carboxylic acid anions in formation waters, San Joaquin Basin and Louisiana Gulf Coast, USA—Implications for clastic diagenesis. Critical comment. *Appl. Geochem.* **8**, 305–307.

Hanshaw B. B. and Hill G. A. (1969) Geochemistry and hydrodynamics of the Paradox Basin Region, Utah, Colorado and New Mexico. *Chem. Geol.* **4**, 264–294.

Harder H. (1970) Boron content of sediments as a tool in facies analysis. *Sediment. Geol.* **4**, 153–175.

Harrison W. and Thyne G. D. (1992) Predictions of diagenetic reactions in the presence of organic acids. *Geochim. Cosmochim. Acta* **56**, 565–586.

Haydon P. R. and Graf D. L. (1986) Studies of smectite membrane behavior: Temperature dependence, 20–180 °C. *Geochim. Cosmochim. Acta* **50**, 115–122.

Hearst J. R. and Nelson P. H. (1985) *Well Logging for Physical Properties.* McGraw-Hill, New York.

Helgeson H. C., Owens C. E., Knox A. M., and Richard L. (1998) Calculation of the standard molal thermodynamic properties of crystalline, liquid, and gas organic molecules at high temperatures and pressures. *Geochim. Cosmochim. Acta* **62**, 985–1081.

Hem J. D. (1985) *Study and Interpretation of the Chemical Characteristics of Natural Water.* US Geological Survey Water-supply Paper 2254.

Herzog H. J. and Drake E. M. (1998) CO_2 capture, reuse, and sequestration technologies for mitigating global climate change. In *Proceedings of the 23rd International Technology Conference on Coal Utilization and Fuel Systems.* Clearwater, FL, pp. 615–626.

Heydari E. and Moore C. H. (1989) Burial diagenesis and thermochemical sulfate reduction, Smackover Formation, southeastern Mississippi Salt Basin. *Geology* **17**, 1080–1084.

Hitchon B. (1995) Fluorine in formation waters, Alberta Basin, Canada. *Appl. Geochem.* **10**, 357–367.

Hitchon B. (1996) Rapid evaluation of the hydrochemistry of a sedimentary basin using only 'standard' formation water analysis: example from the Canadian portion of the Williston Basin. *Appl. Geochem.* **11**, 789–795.

Hitchon B. and Friedman I. (1969) Geochemistry and origin of formation waters in the western Canada sedimentary basin: I. Stable isotopes of hydrogen and oxygen. *Geochim. Cosmochim. Acta* **33**, 1321–1349.

Hitchon B., Billings G. K., and Klovan J. E. (1971) Geochemistry and origin of formation waters in the western Canada sedimentary basin: III. Factors controlling chemical composition. *Geochim. Cosmochim. Acta* **35**, 567–598.

Hitchon B., Perkins E. H., and Gunter W. D. (2001) Recovery of trace metals in formation waters using acid gases from natural gas. *Appl. Geochem.* **16**, 1481–1497.

Holland H. D. and Malinin S. D. (1979) The solubility and occurrence of non-ore minerals. In *Geochemistry of Hydrothermal Ore Deposits* (ed. H. L. Barnes). Wiley, New York, pp. 461–508.

Holser W. T. (1979) Trace elements and isotopes in evaporites. In *Reviews in Mineralogy, Marine Minerals* (ed. R. G. Burns). Mineral Society of America, Washington, DC, pp. 295–346.

Hower J., Eslinger E. V., Hower M. E., and Perry E. A. (1976) Mechanism of burial metamorphism of argillaceous sediments: 1. Mineralogical and chemical evidence. *Geol. Soc. Am. Bull.* **87**, 725–737.

Huh Y., Chan L.-H., Zhang L., and Edmond J. M. (1998) Lithium and its isotopes in major world rivers: implications for weathering and the oceanic budget. *Geochim. Cosmochim. Acta* **62**, 2039–2051.

Hunt T. S. (1879) *Chemical and Geological Essays*. Osgood and Company, Boston.

Hutcheon I. (2000) Principles of diagenesis and what drives mineral change: chemical diagenesis. In *Fluids and Basin Evolution*, Short Course Handbook 28 (ed. K. Kyser). Mineralogical Association of Canada, Toronto, pp. 93–114.

Hutcheon I., Shevalier M., and Abercrombie H. J. (1993) pH buffering by metastable mineral-fluid equilibria and evolution of carbon dioxide fugacity during burial diagenesis. *Geochim. Cosmochim. Acta* **57**, 543–562.

Ivahnenko T., Szabo Z., and Gibs J. (2001) Changes in sample collection and analytical techniques and effects on retrospective comparability of low-level concentrations of trace elements in ground water. *Water Res.* **35**, 3611–3624.

Johnson K. S. and Gerber W. R. (1998) Iodide geology and extraction in northwestern Oklahoma. (Proceedings of the 34th Forum on the Geology of Industrial Minerals). *Okl. Geol. Surv. Circular* **102**, 73–79.

Johnson J. W., Oelkers E. H., and Helgeson H. C. (1992) SUPCTR92: a software package for calculating the standard molal thermodynamic properties of minerals, gases, aqueous species, and reactions from 1 bar to 5,000 bar and 0 °C to 1,000 °C. *Comp. Geosci.* **7**, 899–947.

Kartsev V. V. (1974) *Hydrogeology of Oil and Gas Deposits*. National Technical Information Service Report TM3-58022, Moscow.

Kennedy V. C., Zellweger G. W., and Jones B. F. (1974) Filter pore size effects on the analysis of Al, Fe, Mn, and Ti in water. *Water Resour. Res.* **10**, 785–789.

Kennedy B. M., Kharaka Y. K., Evans W. C., Ellwood A., DePaolo D. J., Thordsen J. J., Ambats G., and Mariner R. H. (1997) Mantle fluids in the San Andreas fault system, California. *Science* **278**, 1278–1281.

Kesler S. E., Martini A. M., Appold M. S., Walter L. M., Huston T. J., and Furman F. C. (1996) Na–Cl–Br systematics of fluid inclusions from Mississippi Valley-type deposits, Appalachian Basin: constraints on solute origin and migration paths. *Geochim. Cosmochim. Acta* **60**, 225–233.

Kharaka Y. K. (1986) Origin and evolution of water and solutes in sedimentary basins. In *Proceedings of 3rd Canadian/American Conference on Hydrogeology. Hydrology of Sedimentary Basins: Application to Exploration and Exploitation*. National Water Well Association, Edmonton, Alberta, pp. 173–195.

Kharaka Y. K. and Berry F. A. F. (1973) Simultaneous flow of water and solutes through geological membranes: I. Experimental investigation. *Geochim. Cosmochim. Acta* **37**, 2577–2603.

Kharaka Y. K. and Berry F. A. F. (1974) The influence of geological membranes on the geochemistry of subsurface waters from Miocene sediments at Kettleman North Dome, California. *Water Resour. Res.* **10**, 313–327.

Kharaka Y. K. and Berry F. A. F. (1980) Geochemistry of geopressured geothermal waters from the northern Gulf of Mexico and California basins. *Proceedings of the 3rd International Symposium on Water–Rock Interaction*, 95–96.

Kharaka Y. K. and Carothers W. W. (1986) Oxygen and hydrogen isotope geochemistry of deep basin brines, Chapter 2. In *Handbook of Environmental Isotope Geochemistry* (eds. P. Fritz and J. Ch. Fontes). Elsevier, Amsterdam, vol. II, pp. 305–360.

Kharaka Y. K. and Carothers W. W. (1988) *Geochemistry of Oil-field Waters from the North Slope, Alaska*. US Geological Survey Professional Paper 1399, pp. 551–561.

Kharaka Y. K. and Mariner R. H. (1989) Chemical geothermometers and their application to formation waters from sedimentary basins. In *Thermal History of Sedimentary Basins* (eds. N. D. Naeser and T. H. McCulloh). Springer, New York, pp. 99–117.

Kharaka Y. K. and Smalley W. C. (1976) Flow of water and solutes through compacted clays. *Am. Assoc. Petrol. Geol. Bull.* **60**, 973–980.

Kharaka Y. K. and Specht D. J. (1988) The solubility of noble gases in crude oil at 25–100 °C. *Appl. Geochem.* **3**, 137–144.

Kharaka Y. K. and Thordsen J. J. (1992) Stable isotope geochemistry and origin of water in sedimentary basins. In *Isotope Signatures and Sedimentary Records* (eds. N. Clauer and S. Chaudhuri). Springer, Berlin, pp. 411–466.

Kharaka Y. K., Berry A. F. A., and Friedman I. (1973) Isotopic composition of oil-field brines from Kettleman North Dome oil field, California, and their geologic implications. *Geochim. Cosmochim. Acta* **37**, 1899–1908.

Kharaka Y. K., Brown P. M., and Carothers W. W. (1978) Chemistry of waters in the geopressured zone from coastal Louisiana–implications for geothermal development. *Geotherm. Res. Council Trans.* **2**, 371–374.

Kharaka Y. K., Lico M. S., Wright V. A., and Carothers W. W. (1979) Geochemistry of formation waters from Pleasant Bayou No. 2 well and adjacent areas in coastal Texas. In *Proceedings of the 4th Geopressured–Geothermal Energy Conference*, Austin, TX, pp. 168–193.

Kharaka Y. K., Lico M. S., and Carothers W. W. (1980) Predicted corrosion and scale-formation properties of geopressured–geothermal waters from the northern Gulf of Mexico Basin. *J. Petrol. Technol.* **32**, 319–324.

Kharaka Y. K., Carothers W. W., and Rosenbauer R. J. (1983) Thermal decarboxylation of acetic acid: implications for origin of natural gas. *Geochim. Cosmochim. Acta* **47**, 397–402.

Kharaka Y. K., Hull R. W., and Carothers W. W. (1985) Water–rock interactions in sedimentary basins. In *Relationship of Organic Matter and Mineral Diagenesis*. Short Course 17 (eds. D. L. Gautier, Y. K. Kharaka, and R. C. Surdam). Society of Economic Paleontologists and Mineralogists, Tulsa, OK, pp. 79–176.

Kharaka Y. K., Law L. M., Carothers W. W., and Goerlitz D. F. (1986) Role of organic species dissolved in formation waters in mineral diagenesis. In *Relationship of Organic Matter and Mineral Diagenesis* (ed. D. Gautier). Society of Economic Paleontologists and Mineralogists, Tulsa, OK, Special Volume 38, pp. 111–122.

Kharaka Y. K., Maest A. S., Carothers W. W., Law L. M., Lamothe P. J., and Fries T. L. (1987) Geochemistry of metal-rich brines from central Mississippi Salt Dome Basin, USA. *Appl. Geochem.* **2**, 543–561.

Kharaka Y. K., Gunter W. D., Aggarwal P. K., Perkins E. H., and DeBraal J. D. (1988) SOLMINEQ.88: A Computer Program for Geochemical Modeling of Water–Rock Interactions. *US Geological Survey Water Resources Investigations Report*, 88–4227.

Kharaka Y. K., Ambats G., and Thordsen J. J. (1993a) Distribution and significance of dicarboxylic acid anions in oil-field waters. *Chem. Geol.* **107**, 499–501.

Kharaka Y. K., Lundegard P. D., Ambats G., Evans W. C., and Bischoff J. L. (1993b) Generation of aliphatic acid anions and carbon dioxide by hydrous pyrolysis of crude oils. *Appl. Geochem.* **8**, 317–324.

Kharaka Y. K., Thordsen J. J., and Ambats G. (1995) Environmental degradation associated with exploration for and production of energy sources in USA. In *Proceedings of the 8th International Symposium on the Water Rock Interaction* (eds. Y. K. Kharaka and O. V. Chudaev). A. A. Balkema Publishers, Rotterdam, pp. 25–30.

Kharaka Y. K., Leong L. Y. C., Doran G., and Breit G. N. (1999a) Can produced water be reclaimed?: Experience with the Placerita oil field, California. Environmental

Issues in Petroleum Exploration, Production, and Refining. *Proceedings of the 5th IPEC Meeting*, October, 1998 (ed. K. L. Sublette), CD-ROM format.

Kharaka Y. K., Thordsen J. J., Evans W. C., and Kennedy B. M. (1999b) Geochemistry and hydromechanical interactions of fluids associated with the San Andreas fault system, California. In *Faults and Subsurface Fluid Flow in the Shallow Crust*, AGU Geophys. Monograph Series 113 (eds. W. C. Haneberg, L. B. Goodwin, P. S. Mozley, and J. C. Moore), pp. 129–148.

Kharaka Y. K., Lundegard P. D., and Giordano T. H. (2000) Distribution and origin of organic ligands in subsurface waters from sedimentary basins. *Rev. Econ. Geol.* **9**, 119–132.

Kharaka Y. K., Thordsen J. J., and White L. D. (2002) *Isotope and Chemical Compositions of Meteoric and Thermal Waters and Snow from the Greater Yellowstone National Park Region*. US Geological Survey Open-file Rep.02-194.

Knauth L. P. (1988) Origin and mixing history of brines, Palo Duro Basin Texas, USA. *Appl. Geochem.* **3**, 455–479.

Knauth L. P. and Beeunas M. A. (1986) Isotope geochemistry of fluid inclusions in Permian halite with implications for the isotopic history of ocean water and the origin of saline formation waters. *Geochim. Cosmochim. Acta* **50**, 419–433.

Kryukov P. A., Zhuchkova A. A., and Rengarten E. F. (1962) Change in the composition of solutions pressed from clays and ion exchange resins. *Akademiia Nauk SSSR Earth Sci. Sect.* **144**, 167–169.

Land L. S. (1997) Mass transfer during burial diagenesis in the Gulf of Mexico sedimentary basin: an overview. *Special Publication—SEPM (Society for Sedimentary Geology)* **57**, 29–39.

Land L. S. and Fisher R. S. (1987) Wilcox sandstone diagenesis, Texas Gulf Coast: a regional isotopic comparison with the Frio Formation. In *Diagenesis of Sedimentary Sequences*, Geological Society of America Special Publication 36 (ed. J. D. Marshall). Geological Society of America, Boulder, CO, pp. 219–235.

Land L. S. and Macpherson G. L. (1992a) Origin of saline formation waters, Cenozoic section, Gulf of Mexico sedimentary basin. *Am. Assoc. Petrol. Geol. Bull.* **76**, 1344–1362.

Land L. S. and Macpherson G. L. (1992b) Geothermometry from brine analyses: lessons from the Gulf Coast, USA. *Appl. Geochem.* **7**, 333–340.

Land L. S. and Prezbindowski D. R. (1981) The origin and evolution of saline formation water, Lower Cretaceous carbonates, south-central Texas, USA. *J. Hydrol.* **54**, 51–74.

Land L. S., Eustice R. A., Mack L. E., and Horita J. (1995) Reactivity of evaporites during burial: an example from the Jurassic of Alabama. *Geochim. Cosmochim. Acta* **59**, 3765–3778.

Lane A. C. (1908) Mine waters and their field assay. *Geol. Soc. Am. Bull.*, 501–512.

Lewan M. D. and Fisher J. B. (1994) Organic acids from petroleum source rocks. In *Organic Acids in Geological Processes* (eds. E. D. Pittman and M. D. Lewan). Springer, Berlin, pp. 70–114.

Lico M. S., Kharaka Y. K., Carothers W. W., and Wright V. A. (1982) *Methods for Collection and Analysis of Geopressured Geothermal and Oil-field Waters*. US Geological Survey Water-supply Paper, Report: W 2194.

Lundegard P. D. and Kharaka Y. K. (1994) Distribution of organic acids in subsurface waters. In *Organic Acids in Geological Processes* (eds. E. D. Pittman and M. D. Lewan). Springer, Berlin, pp. 40–69.

Lundegard P. D. and Land L. S. (1986) Carbon dioxide and organic acids: their origin and role in diagenesis, the Texas Gulf Coast Tertiary. In *Relationship of Organic Matter and Mineral Diagenesis*, Society of Economic Paleontologists

and Mineralogists Special Publication 38 (ed. D. L. Gautier). Geological Society of America, Tulsa, OK, pp. 129–146.

Lundegard P. D. and Senftle J. T. (1987) Hydrous pyrolysis: a tool for the study of organic acid synthesis. *Appl. Geochem.* **2**, 605–612.

Lundegard P. D. and Trevena A. S. (1990) Sandstone diagenesis in the Pattani Basin (Gulf of Thailand): history of water–rock interaction and comparison with the Gulf of Mexico. *Appl. Geochem.* **5**, 669–685.

MacGowan D. B. and Surdam R. C. (1988) Difunctional carboxylic acid anions in oilfield waters. *Org. Geochem.* **12**, 245–259.

MacGowan D. B. and Surdam R. C. (1990) Importance of organic–inorganic interactions during progressive clastic diagenesis. In *Chemical Modeling of Aqueous Systems II*, American Chemical Society Symposium Series 416 (eds. D. C. Melchior and R. L. Bassett). Washington, DC, pp. 494–507.

Machel H. G. (2001) Bacterial and thermochemical sulfate reduction in diagenetic settings: old and new insights. *Sediment. Geol.* **140**, 143–175.

Mackin J. E. (1987) Boron and silica behavior in salt-marsh sediments: implications for paleo-boron distributions and the early diagenesis of silica. *Am. J. Sci.* **287**, 197–241.

Macpherson G. L. (1992) Regional variation in formation water chemistry: major and minor elements, Frio Formation fluids, Texas. *Am. Assoc. Petrol. Geol. Bull.* **76**, 740–757.

Manheim F. T. and Bischoff J. L. (1969) Geochemistry of pore water from the Shell Oil Company drill holes on the continental slope of the northern Gulf of Mexico. *Chem. Geol.* **4**, 63–82.

Manheim F. T. and Horn M. K. (1968) Composition of deeper subsurface waters along the Atlantic Continental Margin. *Southeast. Geol.* **9**, 215–231.

Mazor E. (1997) *Chemical and Isotopic Groundwater Hydrology: The Applied Approach*. Marcel Dekker, New York.

Mazor E. and Bosch A. (1987) Noble gases in formation fluids from deep sedimentary basins: a review. *Appl. Geochem.* **2**, 621–627.

Mazor E. and Kharaka Y. K. (1981) Atmospheric and radiogenic noble gases in geopressured–geothermal fluids: northern Gulf of Mexico basin. In *Proceeding of the 5th Conference Geopressured–Geothermal Energy*. Louisiana State University, Baton Rouge, LA, pp. 197–200.

McIntosh J. C., Walter L. M., and Martini A. M. (2002) Pleistocene recharge to midcontinent basins: effects on salinity structure and microbial gas generation. *Geochim. Cosmochim. Acta* **66**, 1681–1700.

McKelvey J. G. and Milne I. H. (1962) The flow of salt through compacted clay. *Clays and Clay Mineralogy* **9**, 248–259.

McManus K. M. and Hanor J. S. (1993) Diagenetic evidence for massive evaporite dissolution, fluid flow, and mass transfer in the Louisiana Gulf Coast. *Geology* **21**, 727–730.

Merino E. (1975) Diagenesis in Tertiary sandstones from Kettleman North Dome, California: I. Diagenetic mineralogy. *J. Sedim. Petrol.* **5**, 320–336.

Merino E. (1979) Internal consistency of a water analysis and uncertainty of the calculated distribution of aqueous species at 25 °C. *Geochim. Cosmochim. Acta* **43**, 1533–1542.

Moldovanyi E. P. and Walter L. M. (1992) Regional trends in water chemistry, Smackover Formation, Southwest Arkansas: geochemical and physical controls. *Am. Assoc. Petrol. Geol. Bull.* **76**, 864–894.

Moran J. E., Fehn U., and Hanor J. S. (1995) Determination of source ages and migration patterns of brines from the US Gulf Coast using ^{129}I. *Geochim. Cosmochim. Acta* **59**, 5055–5069.

Munoz M., Boyce A. J., Courjault-Rade P., Fallick A. E., and Tollon F. (1991) Continental basinal origin of ore veins from southwestern Massif Central fluorite veins (Albigeois, France): evidence for fluid inclusion and stable isotope analysis. *Appl. Geochem.* **14**, 447–458.

Muramatsu Y., Fehn U., and Yoshida S. (2001) Recycling of iodide in fore-arc areas: evidence from the iodide brines in Chiba, Japan. *Earth Planet. Sci. Lett.* **192**, 583–593.

Ohmoto H. and Lasaga A. C. (1982) Kinetics of reactions between aqueous sulfates and sulfides in hydrothermal systems. *Geochim. Cosmochim. Acta* **46**, 1727–1745.

O'Nions R. K. and Oxburgh E. R. (1988) Helium, volatile fluxes, and the development of continental crust. *Earth Planet. Sci. Lett.* **90**, 331–347.

Ortoleva P., Al-Shaieb Z., and Puckette J. (1995) Genesis and dynamics of basin compartments and seals. *Am. J. Sci.* **295**, 345–427.

Otton J. K., Asher-Bolinder S., Owen D. E., and Hall L. (1997) *Effects of Produced Waters at Oilfield Production Sites on the Osage Indian Reservation, Northeastern Oklahoma.* US Geological Survey Open-File Report 97–28.

Ozima M. and Podosek F. A. (1983) *Noble Gas Geochemistry.* Cambridge University Press, Cambridge.

Palandri J. L. and Reed M. H. (2001) Reconstruction of *in situ* composition of sedimentary formation waters. *Geochim. Cosmochim. Acta* **65**, 1741–1767.

Palmer M. R. and Swihart G. H. (1996) Boron mineralogy, petrology, and geochemistry. *Rev. Mineral.* **33**, 709–744.

Pang Z. and Reed M. (1998) Theoretical chemical thermometry on geothermal waters: problems and methods. *Geochim. Cosmochim. Acta* **62**, 1083–1091.

Person M., Raffensperger J. P., Ge S., and Garven G. (1996) Basin-scale hydrogeologic modeling. *Rev. Geophys.* **34**, 61–87.

Phillips F. M. (2000) Chlorine-36. In *Environmental Tracers in Subsurface Hydrology* (eds. P. Cook and A. L. Herczeg). Kluwer Academic, Amsterdam, pp. 299–348.

Pinti D. L. and Marty B. (2000) Noble gases in oil and gas fields: origins and processes. In *Fluids and Basin Evolution* (ed. K. Kyser). Mineralogical Association of Canada. Short Course Handbook, Toronto, vol. 28, pp. 160–196.

Posey H. H. and Kyle J. R. (1988) Fluid-rock interactions in the salt dome environment: an introduction and review. *Chem. Geol.* **74**, 1–24.

Price L. C. (1979) Aqueous solubility of methane at elevated pressures and temperatures. *Am. Assoc. Petrol. Geol. Bull.* **63**, 1527–1533.

Ranganathan V. and Hanor J. S. (1987) A numerical model for the formation of saline waters due to diffusion of dissolved NaCl in subsiding sedimentary basins with evaporites. *J. Hydrol.* **92**, 97–120.

Rapp J. B. (1976) Amino acids and gases in some springs and an oil field in California. *US Geol. Surv. J. Res.* **4**, 227–232.

Richardson C. K. and Holland H. D. (1979) Fluorite deposition in hydrothermal systems. *Geochim. Cosmochim. Acta* **43**, 1327–1336.

Richter B. C. and Kreitler C. W. (1993) *Geochemical Techniques for Identifying Sources of Ground-water Salinization.* CRC Press, Boca Raton, FL.

Rider M. (1996) *The Geological Interpretation of Well Logs*, 2nd edn. Gulf Publishing, Houston.

Rittenhouse G. (1967) Bromine in oil-field waters and its use in determining possibilities of origin of these waters. *Am. Assoc. Petrol. Geol. Bull.* **51**, 2430–2440.

Seal R. R., Alpers C. N., and Rye R. O. (2000) Stable isotope systematics of sulfate minerals. *Rev. Mineral. Geochem.* **40**, 541–602.

Seewald J. S. (2001) Model for the origin of carboxylic acids in basinal brines. *Geochim. Cosmochim. Acta* **65**, 3779–3789.

Sheppard S. M. F. (1986) Characterization and isotopic variations in natural waters. In *Stable Isotopes in High-temperature Geological Processes*. Rev. Mineral. (eds. J. W. Valley, H. P. Taylor, Jr., and J. R. O'Neil), vol. 16, pp. 165–183.

Shock E. L. (1995) Organic acids in hydrothermal solutions: standard molal thermodynamic properties of carboxylic acids and estimates of dissociation constants at high temperatures and pressures. *Am. J. Sci.* **295**, 496–580.

Sicree A. A. and Barnes H. L. (1996) Upper Mississippi Valley district ore fluid model: the role of organic complexes. *Ore Geol. Rev.* **11**, 105–131.

Siemann M. G. and Schramm M. (2000) Thermodynamic modeling of the Br partition between aqueous solutions and halite. *Geochim. Cosmochim. Acta* **64**, 1681–1693.

Smith J. T. and Ehrenberg S. N. (1989) Correlation of carbon dioxide abundance with temperature in clastic hydrocarbon reservoirs: relationship to inorganic chemical equilibrium. *Mar. Petrol. Geol.* **6**, 129–135.

Sofer Z. and Gat J. (1972) Activities and concentrations of ^{18}O in concentrated aqueous salt solutions: analytical and geophysical implications. *Earth Planet. Sci. Lett.* **15**, 232–238.

Solomon D. K. (2000) ^{4}He in groundwater. In *Environmental Tracers in Subsurface Hydrology* (eds. P. Cook and A. L. Herczeg). Kluwer Academic, Amsterdam, pp. 425–440.

Spivack A. J., Palmer M. R., and Edmond J. M. (1987) The sedimentary cycle of the boron isotopes. *Geochim. Cosmochim. Acta* **51**, 1939–1949.

Spycher N. F. and Reed M. H. (1988) Fugacity coefficients of H_2, CO_2, CH_4, H_2O, and of $H_2O-CO_2-CH_4$ mixtures: a virial equation treatment for moderate pressures and temperatures applicable to calculations of hydrothermal boiling. *Geochim. Cosmochim. Acta* **52**, 739–749.

Stephenson M. T. (1992) A survey of produced water studies. *Environ. Sci. Res.* **46**, 1–13.

Stoessell R. K. and Carpenter A. B. (1986) Stoichiometric saturation tests of $NaCl_{1-x}Br_x$ and $KCl_{1-x}Br_x$. *Geochim. Cosmochim. Acta* **50**, 1465–1474.

Stueber A. M. and Walter L. M. (1991) Origin and chemical evolution of formation waters from Silurian–Devonian strata in the Illinois Basin, USA. *Geochim. Cosmochim. Acta* **55**, 309–325.

Surdam R. C., Boese S. W., and Crossey L. J. (1984) The chemistry of secondary porosity. In *Clastic Diagenesis* (eds. D. A. McDonald and R. C. Surdam) *Am. Assoc. Petrol. Geol. Memoir*, **37**, pp. 127–150.

Surdam R. C., Crossey L. J., Hagen E. S., and Heasler H. P. (1989) Organic–inorganic interactions and sandstone diagenesis. *Am. Assoc. Petrol. Geol. Bull.* **73**, 1–23.

Sverjensky D. A. (1986) Genesis of Mississippi Valley-type lead zinc deposits. *Ann. Rev. Earth. Planet. Sci.* **14**, 177–199.

Tagirov B. and Schott J. (2001) Aluminum speciation in crustal fluids revisited. *Geochim. Cosmochim. Acta* **65**, 3965–3992.

Tissot B. P. and Welte D. H. (1984) *Petroleum Formation and Occurrence*. Springer, Berlin.

Torgersen T. and Clarke W. B. (1985) Helium accumulation in groundwater: I. An evaluation of sources and the continental flux of crustal ^{4}He in the Great Artesian Basin, Australia. *Geochim. Cosmochim. Acta* **49**, 1211–1218.

Tuncay K., Park A., and Ortoleva P. (2000) Sedimentary basin deformation: an incremental stress approach. *Tectonophysics* **323**, 77–104.

Vetshteyn V. V., Gavish V. K., and Gutsalo L. K. (1981) Hydrogen and oxygen isotope composition of waters in deep-seated fault zones. *Int. Geol. Rev.* **23**, 302–310.

Wade T. L., Kennicutt M. C., and Brooks J. M. (1989) Gulf of Mexico hydrocarbon seep communities: Part III. Aromatic hydrocarbon concentrations in organisms, sediments, and water. *Mar. Environ. Res.* **27**, 19–30.

Wallace R. H., Kraemer T. F., Taylor R. E., and Wesselman J. B. (1979) *Assessment of Geopressured–Geothermal Resources in the Northern Gulf of Mexico Basin.* US Geological Survey Circular, Report C 0790, pp. 132–155.

Warren E. A. and Smalley P. C. (eds.) (1994) *North Sea Formation Waters Atlas.* Geological Society Memoir No. 15. Geological Society, London.

White D. E., Hem J. D., and Waring G. A. (1963) Chemical composition of subsurface waters. In *Data of Geochemistry*. US Geological Survey Professional Paper 440F.

White C. M., Strazisar B. R., Granite E. J., Hoffman J. S., and Pennline H. W. (2003) 2003 Critical review: seperation and capture of CO_2 from large stationary sources and sequestration in geological formations—coalbeds and deep saline aquifers. *J. Air and Waste Management Assoc.* **53**, 645–715.

Whitworth T. M. and Fritz S. J. (1994) Electrolyte-induced solute permeability effects in compacted smectite membranes. *Appl. Geochem.* **9**, 533–546.

Willey L. M., Kharaka Y. K., Presser T. S., Rapp J. B., and Barnes I. (1975) Short-chain aliphatic acids in oil-field waters of Kettleman Dome Oil Field, California. *Geochim. Cosmochim. Acta* **39**, 1707–1710.

Williams L. B., Hervig R. L., Wieser M. E., and Hutcheon I. (2001a) The influence of organic matter on the boron isotope geochemistry of the Gulf Coast sedimentary basin, USA. *Chem. Geol.* **174**, 445–461.

Williams L. B., Wieser M. E., Fennell J., Hutcheon I., and Hervig R. L. (2001b) Application of boron isotopes to the understanding of fluid–rock interactions in a hydrothermally stimulated oil reservoir in the Alberta Basin, Canada. *Geofluids* **1**, 229–240.

Wilson A. M., Garven G., and Boles J. R. (1999) Paleohydrogeology of the San Joaquin Basin, California. *Bull. Geol. Soc. Am.* **111**, 432–449.

Wilson T. P. and Long D. T. (1993) Geochemistry and isotope chemistry of Ca–Na–Cl brines in Silurian strata, Michigan Basin, USA. *Appl. Geochem.* **8**, 81–100.

Wolery T. J. (1992) *EQ3/6, a Software Package for Geochemical Modeling of Aqueous Systems*. Lawrence Livermore National Laboratory Livermore, CA.

Wolfe J. A. (1980) Tertiary climates and floristic relationships at high latitudes in the northern hemisphere. *Paleogeog. Paleoclimatol. Paleoecol.* **30**, 313–323.

Worden R. H. (1996) Controls on halogen concentrations in sedimentary formation waters. *Min. Mag.* **60**, 259–279.

Worden R. H. and Smalley P. C. (2001) H_2S in North Sea oil fields: importance of thermochemical sulfate reduction in clastic reservoirs. In *Proceedings of the 10th International Symposium on the Water Rock Interaction* (ed. R. Cidu). A. A. Balkema Publishers, Rotterdam, pp. 659–662.

Worden R. H., Coleman M. L., and Matray J. M. (1999) Basin scale evolution of formation waters: a diagenetic and formation water study of the Triassic Chaunoy Formation, Paris Basin. *Geochim. Cosmochim. Acta* **63**, 2513–2528.

Wycherley H., Fleet A., and Shaw H. (1999) Observations on the origins of large volumes of carbon dioxide accumulations in sedimentary basins. *Mar. Petrol. Geol.* **16**, 489–494.

You C. F., Spivack A. J., Gieskes J. M., Martin J. B., and Davisson M. L. (1996) Boron contents and isotopic compositions in pore waters: a new approach to determine temperature-induced artifacts; geochemical implications. *Mar. Geol.* **129**, 351–361.

Ziegler K. and Coleman M. L. (2001) Palaeohydrodynamics of fluids in the Brent Group (Oseberg Field, Norwegian North Sea) from chemical and isotopic compositions of formation waters. *Appl. Geochem.* **16**, 609–632.

5.17
Deep Fluids in the Continents: II. Crystalline Rocks

S. K. Frape and A. Blyth

University of Waterloo, ON, Canada

R. Blomqvist

Geological Survey of Finland, Espoo, Finland

R. H. McNutt

University of Toronto, ON, Canada

and

M. Gascoyne

Gascoyne GeoProjects Inc., Pinawa, MB, Canada

5.17.1 INTRODUCTION

The phrase "crystalline rock" has become a commonly used term to describe igneous and metamorphic rocks. Groundwaters (fluids) found in crystalline rocks can have a highly variable composition similar to fluids from sedimentary environments see (Chapter 5.16). The chemistry of shallow groundwaters is very dilute, composed of a mixture of atmospherically recharged precipitation and gases reacting with the existing minerals to produce dilute chemical dissolved loads (Garrels, 1967; Paces, 1972; see Chapters 5.14 and 5.04). The isotopic signature of shallow groundwaters reflects mixtures of yearly precipitation and therefore attests to a young age and recent meteoric origin. The dissolved load tends to be dominated by calcium and sodium balanced by bicarbonate (Jacks, 1973).

Deeper groundwaters in crystalline rocks tend to be different in both their chemical and isotopic signatures. The chemistry of these fluids can be combinations of the cations of calcium and sodium (occasionally enriched with magnesium) and the chloride and sulfate anions (Jacks, 1978). Similar to sedimentary environments, dissolved loads greater than $100 \, g \, L^{-1}$ are common. The stable isotopic signatures (^{18}O and ^{2}H) of the most concentrated fluids are very different from those found in sedimentary formation fluids, and age-dating techniques have suggested that deep groundwaters are geologically old (Fritz and Frape, 1982).

The detailed characteristics of deep groundwaters in crystalline rocks were relatively unknown until the late 1970s, but the mining industry certainly knew of these highly mineralized, gas filled, often explosive fluids. Reports from the late nineteenth and early twentieth century mention very corrosive fluids found in the silver and copper mines of the Lake Superior region of Canada (Geological Survey of Canada, 1878; Lane, 1914). In the latter part of the twentieth century a new driving force for

exploration of fluids within crystalline rocks came into existence as countries around the world turned to deep rock burial as the solution to radioactive waste disposal (Gascoyne *et al.*, 1995). This led hydrogeologists, geochemists, and mineralogists to examine the fluids and rock environments at proposed respository depths of 500–1,000 m in far more detail at locations worldwide (Chapman and McKinley, 1987).

5.17.1.1 The Crystalline Rock Environment

The environments in which these highly mineralized, gaseous fluids reside are very complicated and often exhibit geologically diverse settings. It is a common phenomenon that heterogeneity of rock types and hydrogeological properties occur at almost every research site studied to date.

5.17.1.1.1 *The rocks*

The crystalline rock environment is composed of a multitude of igneous and metamorphic rock types. The igneous rock compositions and many of the physical properties are controlled by their crystallization history. Ultramafic rocks are composed of mineral assemblages that crystallize at higher temperatures than felsic rock assemblages. Table 1 is a very brief summary of the chemical and mineralogical characteristics of some rock types typically in contact with fluids found in crystalline environments (data taken from Hyndman, 1985). Mafic and ultramafic rocks tend to be composed of mineral phases rich in iron–magnesium–calcium silicates, whereas felsic rocks tend to contain more sodium–potassium–calcium minerals. Fluid/rock interaction is more correctly mineral/fluid interaction, so mineral chemistry must be examined to better understand the geochemical evolution of groundwaters in deep rock systems.

Table 1 Chemistry and mineralogy of representative rock types from crystalline rock environments (analyses normalized to 100%, including MnO, P$_2$O$_5$, water free; oxides are in wt.%).

Type of rock Parameter no. of analyses	Ultramafic (peridotite) n = 287	Mafic (gabbro) n = 1,451	Intermediate (diorite) n = 872	Felsic (granite) n = 2,485
SiO$_2$	45.31	51.06	58.58	72.04
TiO$_2$	0.52	1.17	0.96	0.30
Al$_2$O$_3$	4.38	15.91	16.98	14.42
Fe$_2$O$_3$	4.28	3.10	2.55	1.22
FeO	7.48	7.76	5.13	1.68
MnO	0.26	0.12	0.12	0.05
MgO	31.19	7.68	3.73	0.71
CaO	5.50	9.88	6.66	1.82
Na$_2$O	0.55	2.48	3.60	3.69
K$_2$O	0.30	0.96	1.81	4.12
P$_2$O$_5$	0.11	0.24	0.29	0.12
Minerals	Ol, Py	Py, Pl, Hb, Bi	Pl, Hb, Bi	Pl, Kspar, Bi, Qtz

After Hyndman (1985).
Ol = olivine, Py = pyroxene, Hb = hornblende, Bi = biotite, Pl = plagioclase feldspar, Kspar = potassium feldspar, Qtz = quartz.

5.17.1.1.2 Structural and geomechanical characteristics

The porosity and permeability of crystalline rock masses are influenced by a large variety of processes. Initially, intrusion and cooling of magmatic masses can lead to tectonic and residual stresses. These stresses create a variety of fracture pathways, especially if the rocks are stressed in a brittle fashion (break rather than flow) (Mazurek, 2000). In addition, the protracted erosion of most continental crust has brought many deep-seated crystalline rocks of all types to the present-day Earth surface. These igneous intrusions may be very large composite masses, known as batholiths, or individual intrusions such as "plutons." Plutons are usually thought of as funnel shaped or cylindrical (Wynne-Edwards, 1957; Verhoogen *et al.*, 1970). In most cases, as uplift occurs, further stresses are released, usually in a brittle fashion, as these crystalline rocks are brought into the near-surface environment. The result is the formation of additional porosity and hydraulic pathways within the rock mass. However, secondary hydrothermal and metamorphic reactions can act to further alter the rock mass causing fracture systems to be sealed with secondary clay and zeolite minerals from the alteration of plagioclase and biotite (Mazurek, 2000).

5.17.1.1.3 Hydrogeology

The movement of groundwaters within crystalline environments and the numerous hydrogeologic parameters associated with these environments should have multiple controls. As discussed above, fracture generation and fracture sealing by secondary mineral phases could be a multi-phased process in the life of a crystalline rock mass. Permeability measurements range from 10^{-5} m s^{-1} near surface to 10^{-13} m s^{-1} at depths of 1 km (Gascoyne *et al.*, 1987). In some cases higher conductivities in fracture zones are reported at shallow depths at research sites on the Fennoscandian Shield (Vieno and Nordman, 1999).

Generally, the degree of "openness" of the rock mass is a balance of structural constraints listed above and the extent of fracture-filling mineralogy (Blyth *et al.*, 2000). Parameters such as fracture interconnectivity, hydraulic gradients, density and viscosity of fluids and solubility constraints will profoundly impact the movement of groundwaters in these environments (Mazurek, 2000). Many of the more saline fluids are believed to be very old and stagnant, residing in rock porosity that is poorly interconnected and effectively sealed from shallower flow systems (Frape and Fritz, 1987; Blomqvist, 1990). It has been noted in the mining industry that drilling into these systems can be very dangerous. Several authors report incidences of hundreds of meters of heavy drilling rods being pushed out of drill holes in minutes due to pressure release in pockets of saline fluids. Usually, these events are followed by combustion of gases associated with the fluids. Needless to say, scientific drilling, blasting, and sampling efforts often create extensive hydrogeological disruptions and changes to hydraulic gradients in these systems that change the natural hydraulic and geochemical properties (Frape and Fritz, 1982).

5.17.2 FIELD SAMPLING METHODS

Many techniques ranging from hydrostatic or lithostatic drainage of boreholes in underground openings to very sophisticated pumps and pressurized samplers have been used to obtain fluid samples from these environments (Gascoyne, 2002). Every method has its advantages and disadvantages, some of which will be discussed below.

Perhaps the simplest method, and one frequently used, is to sample from boreholes within mined openings. These openings can be made in experimental research laboratories specifically constructed for characterization of groundwater and chemical conditions or in working gold, sulfide or diamond mines, where exploration boreholes are available for sampling. Boreholes are usually packered shut in the case of research facilities, or in active mines the boreholes can be plugged by taps which allow access to pressurized water samples. Holes may be open and flowing due to hydrostatic or overpressured-gas-induced conditions. Gas-saturated (CH_4, H_2, N_2) boreholes that have been sealed with taps represent an extremely dangerous sampling environment due to potential degassing and the explosive nature of the gases.

The next level of sampling involves placing various devices into the borehole, either in a mined opening or from surface sites. These devices can simply be grab samplers where the device is lowered to the desired level and induced to close and capture a sample. A more sophisticated version is described by Sherwood Lollar *et al.* (1994), where the sampler works on internal timers and valves and is capable of capturing pressurized water and gas samples in narrow exploration boreholes to depths of 1,000 m. Another novel grab sampler was used by Nurmi and Kukkonen (1986) and has produced excellent reconnaissance level samples in several Fennoscandian studies. Their method uses a continuous string of 50 m plastic tubes connected by valves that allow flow in one direction. The tubing is slowly lowered down an open surface borehole to sample the entire water column. Columns of water up to 1,200 m have been sampled using this method with some excellent profiling results showing very different water masses and their relationship to changing rock lithologies in the same borehole. Examples of this kind of profile have been presented by Lahermo and Lampen (1987), Smalley *et al.* (1988), Nurmi *et al.* (1988), and Ruskeeniemi *et al.* (1996), and will be shown later in Figure 6.

The most common groundwater and gas sampling techniques in crystalline rock feature some form of downhole pumping device that lifts the fluid sample to the borehole collar, where standard flow cells and sophisticated sampling techniques are used to process the samples. A wide variety of devices and techniques are available and good reviews can be found in Bottomley *et al.* (1984), Almèn *et al.* (1986),

Axelsen *et al.* (1986), Wikberg (1987), and Lodemann and Wöhrl (1994).

Unfortunately, sampling devices seldom work well in boreholes below 1,000 m depth. In the case of the Kola Deep Hole in northwestern Russia, researchers had to use high-purity water with the drill muds to mix and capture fluids and gases down to depths exceeding 13 km. Due to the high salinities and gas contents in the host rocks, useful chemical data could be extracted from the drilling muds that returned to the surface (Borevsky *et al.*, 1984). The approach in the KTB, German deep research hole, was to case much of the hole, isolate fracture systems at the 4,000 m level, remove the drill water ultimately allowing formation fluids to be sampled, after an extensive pumping period (Lodemann *et al.*, 1998).

In summary groundwater samples have been collected in crystalline environments by a variety of methods. As shown in the next section, a great majority of the samples taken from deep boreholes are saline waters and brines. In open-hole systems the high salinity makes the chemistry useful for the interpretation of fluid characteristics in crystalline rocks. It should be noted that the underground research sites, e.g., Underground Research Laboratory (URL) of Canada, Äspö (Sweden) and many others, use sophisticated packer and pumping systems similar to those described by Ross and Gascoyne (1995) and Black and Chapman (1981) to retrieve pH- and redox-dependent data from discrete structural zones. In most cases where different methods are applied to sampling, especially for more saline fluids, a good agreement has been obtained for the major element and isotopic parameters reported in this chapter.

5.17.3 CHEMISTRY AND ISOTOPIC COMPOSITION OF GROUNDWATERS FROM CRYSTALLINE ENVIRONMENTS

The data in Table 2 are from groundwaters found in a variety of crystalline rock environments on several different continents. The ages and size of the rock complexes vary. The oldest are billions of years, the youngest tens of millions. Some are individual plutons or batholiths; others are massive complexes composed of many rock types of variable age. To attempt to subdivide the data based on crystalline rock type, size of rock complex or rock age would result in so many subdivisions that any discussion of origin and evolution would be unnecessarily cumbersome.

Fluids from crystalline and sedimentary environments are often classified by chemical type (dominant cations and anions), e.g., Ca–Na–Cl water. Often these waters are further divided by

Table 2 Geochemistry of representative groundwater samples from shield sites.

Location	Source	Sample ID	Depth (m)	Water type	TDS ($mg\,L^{-1}$)	Ca ($mg\,L^{-1}$)	Na ($mg\,L^{-1}$)	Mg ($mg\,L^{-1}$)	K ($mg\,L^{-1}$)	Sr ($mg\,L^{-1}$)	Cl ($mg\,L^{-1}$)	Br ($mg\,L^{-1}$)	SO_4 ($mg\,L^{-1}$)	HCO_3 ($mg\,L^{-1}$)	$\delta^{18}O$ (‰)	δ^2H (‰)	^3H (TU)	$\delta^{37}Cl$ (‰)
Finland																		
Outokumpu	1	OKU551/90/1	427	Na–Cl	1,400	196	334	1	3.7	2.1	847	4		9	−16.7	−117		−0.32
	2	OKU741	610	Ca–Cl	12,870	3,380	1,260	10	9.7	20.9	8,030	49	17	74	−14.4	−102		0.01
	2	OKU741	1,004	Ca–Na–Cl	28,120	5,730	2,940	998	34.7	40.5	18,100	120	20	93	−12.9	−80		0.00
Pori	2	PORI-1	111	Na–Cl	2,690	272	669	12	1.0	5.8	1,510	6	74	139	−20.6	−126		0.04
	2	PORI-1	201	Na–Cl	4,880	653	1,100	24	1.6	18.7	2,870	13	167	32	−13.4	−103		−0.19
	3	PO-1	600	Ca–Cl	120,400	36,000	9,500	82	5.0	531.0	73,660	550	19	7	−7.8	−41	1.4	
	3	POI/87	470	Ca–Na–Cl	132,600	36,650	9,420	81	4.8	760.0	85,000	628	32	9	−8.1	−54		0.80
Ylivieska	2	YLI313/92	329	Na–Cl	460	54	95	41	3.9	1.4	129	1	4	129	−13.0	−93		0.15
	2	YLI313/90	511	Na–Ca–Mg–Cl	53,480	5,450	10,100	5,260	57.6	168.0	32,143	244	<1	54	−13.4	−90		0.64
	2	YLI313/88	539	Na–Ca–Mg–Cl	53,660	5,878	10,730	3,366	65.0	186.9	33,000	349	<1	70	−13.1	−46		0.49
	2	YLI313/92	521	Na–Ca–Mg–Cl	65,700	6,420	12,000	4,390	53.3	205.0	42,300	450		53	−13.5	−27		0.14
	3	YLI313/90	587	Na–Mg–Ca–Cl	82,870	8,550	16,000	5,230	96.3	276.0	52,030	625	<1	51	−13.5	−21	4.2	0.33
Enonkoski	2	ENONKOSKI 336	492	Na–Ca–Cl	11,720	1,220	2,616	720	23.0	75.4	6,895	78	<10	33	−14.5	−98		0.73
	2	La-375	419	Na–Ca–Mg–Cl	27,600	3,450	5,560	1,470	45.0	206.0	16,690	184	10	13	−13.2	−81	<6	
Juuka/Miihkali	2	MIIHK116	753	Ca–Na–Cl	34,280	8,050	5,190	1	42.0	33.7	20,500	177	202	78	−10.4	−13		0.00
	2	MIIHK116	490	Ca–Na–Cl	40,680	8,650	5,130	18	48.7	28.8	26,300	187	187	69	−10.4	−28		0.40
	3	Ju/Mi-114	920	Ca–Na–Cl	70,200	16,200	10,800	522	16.0	116.0	42,000	280	190	83	−9.4	−16	<6	
	2	MIIHK116	946	Na–Ca–Cl	134,140	16,700	38,600	228	520.0	156.0	78,700	507	514	189	−11.0	−7		1.06
	3	Ju/Mi-116	1,020	Na–Ca–Cl	166,200	17,000	48,000	230	600.0	188.0	99,500	490	45	146	−11.8	−4	4.3	
Noormarkku	2	R-43	575	Ca–Na–Cl	46,700	10,880	5,670	52	27.0	77.0	29,800	137	2	21	−10.9	−71	3.4	
Mäntsälä	2	MHA1/92	100	Ca–HCO3	100	13	10	4	3.0	0.1	4	0	24	65	−12.1	−88		
	2	MHA2/91	207	Ca–HCO3	200	27	11	7	2.5	0.2	3	0	19	125	−11.5	−84		
	2	Mha-2	455	Ca–Mg–Na–SO4	190	24	12	7	2.5	0.2	2		30	38				
	2	MHA1/92	850	Na–Ca–Cl	4,900	835	1,010	1	11.0	9.4	2,880	17	45	29	−11.2	−75		
	3	Mha-2	855	Na–Ca–Cl	50,300	12,700	6,870	17	18.0	95.0	30,300	253		18	−12.1	−84	13.0	0.24
Olkiluoto	3	OL-KR1	750	Ca–Na–Cl	34,900	6,560	6,880	52	23.0	47.0	21,200	157			−9.0	−70	<6	
Tipasjarvi	2	TIP115/87	523	Ca–Na–Cl	12,510	3,612	893	2	14.0	14.8	7,700	51	57	75	−13.3	−94	<6	
Ruukki	2	PV11/87	298	Na–Ca–Cl	1,530	150	380	22	6.9	1.7	820	6	5	134	−12.2	−93		0.37
Kolari	2	KOL162	525	Na–SO4	2,490	660	58	3	9.0	3.2	135	0	1,600	24	−15.3	−109		0.52
Palmottu	2	NP 357/90/3	242	Na–Cl–SO4	800	37	229	8	3.2	0.3	251		196	62	−13.7	−101		0.17
	2	R348(P)	200	Na–SO4	1,190	25	430	5	2.0	0.3	71	0	580	76	−16.6	−119	0.2	0.65
	2	NP 348,P200	165	Na–SO4	1,460	27	462	5	3.0	0.3	66		838	60	−17.0	−127		0.80
	2	R385	403	Na–Cl	1,620	89	514	15	4.6	1.1	901	8	57	32	−15.6	−116	1.3	0.42

(continued)

Table 2 (continued).

Location	Source	Sample ID	Depth (m)	Water type	TDS (mg L⁻¹)	Ca (mg L⁻¹)	Na (mg L⁻¹)	Mg (mg L⁻¹)	K (mg L⁻¹)	Sr (mg L⁻¹)	Cl (mg L⁻¹)	Br (mg L⁻¹)	SO₄ (mg L⁻¹)	HCO₃ (mg L⁻¹)	δ¹⁸O (‰)	δ²H (‰)	³H (TU)	δ³⁷Cl (‰)
Sweden																		
Aspo	2	KASO3-1566	129	Na–Ca–Cl	2,110	160	583	20	2.2	3.2	1,240	5	33	62	−13.0	−100	0.5	0.17
	2	KLX02-2934	1,155	Ca–Na–Cl	25,210	5,250	3,730	5	10.5	83.8	15,130	125	860	11	−11.4	−83	2.5	0.30
	2	KLX02-2931	1,345	Ca–Na–Cl	50,090	11,200	6,210	3	17.9	191.0	31,230	196	1,024	9	−9.7	−62	0.8	0.72
	2	KLX02-3038	1,385	Ca–Na–Cl	61,240	14,800	7,420	1	32.6	253.0	36,970	509	1,205	42	−9.3	−55	0.8	0.41
Stripa	2,4	V2(17-38)	436	Na–Ca–Cl–HCO₃	2,016	18	56	1	0.4		84		3	54	−12.1	−89		−0.23
	2,4	V2-5(4419)	801	Na–Ca–Cl	300	34	47	0	0.3		190	2	4	16	−13.1	−96	1.0	−0.23
	2,4	V2(69-4)	822	Na–Ca–Cl	790	109	181	0	0.5		440		46	9				0.38
	2,4	V2-4(850514)	814	Na–Ca–Cl	1,170	175	250	0	0.9		640	6	85	6	−13.2	−97	0.3	−0.02
	2,4	V1(81WA203)	815	Na–Ca–Cl	1,203	172	277	0	1.2		630	7	102	9	−12.9	−96	2.3	0.45
Canada																		
East Bull Lake	5	EBL-2	460	Na–Ca–Cl	2,060	228	570	2	1.1	1.5	1,237	3	11		−8.7	−65	48.0	
	5	EBL-4	429	Na–Ca–Cl	4,330	455	1,160	2	3.4	6.2	2,580	11	108	26	−11.9	−80	9.8	
Keweena	6	#6		Ca–Cl	164,400	47,500	13,200	7	70.0	223.0	102,171	829	368		−7.6	−17		
Centennial URL	2	URL16-4-1	85	Na–HCO₃	340	35	52	9	3.5	0.4	18	0	37	184	−11.2	−93	20.5	
	2	M10-1-7	50	Na–HCO₃	350	26	69	3	2.1	0.2	36	0	25	186	−14.5	−105	9.5	
	2	M1A-3-7	265	Na–Cl	700	30	170	2	2.1	0.4	207	1	128	158	−16.6	−120	<0.8	
	2	WN3-90	90	Na–HCO₃	730	25	160	11	3.3	0.5	119	0	82	328	−16.2	−122	<6	
	2	URL15-1-4	125	Na–Cl	750	30	200	4	5.4	0.3	191	1	93	222	−13.6	−105	<6.0	
	2	M11-3-4	290	Na–Cl	1,840	145	440	10	2.8	1.5	590	2	380	276	−18.5	−141	8.8	
	2	URL12-10-19	390	Na–Cl	2,410	156	735	3	2.6	2.3	1,246	3	217	52	−17.5	−128	<0.8	
	2	WN10-3-4	245	Na–Ca–Cl–SO₄	3,210	289	770	27	5.6	4.1	1,350	6	700	71	−17.9	−132	<0.8	
	2	URL12-13-21	605	Ca–Na–Cl–SO₄	4,900	1,070	530	25	10.2	7.7	2,454	12	776	32	−15.4	−114	14.2	
	2	URL14-8	280	Ca–Na–Cl–SO₄	6,660	710	1,800	8	4.9	7.8	3,389	18	732	15	−16.6	−121	<0.8	
	2	M5A-IN8	340	Ca–Na–Cl	7,590	1,480	1,200	12	5.7	10.7	3,980	26	876	32	−17.4	−126	<6	
	2	M14-4-4	370	Ca–Na–Cl–SO₄	11,120	2,600	1,540	18	5.6	16.3	5,800	35	1,138	19	−17.4	−124	<0.8	
	2	WN4-13-20	650	Na–Ca–Cl–SO₄	20,110	2,687	4,890	26	10.5	39.8	11,091	33	1,393	15	−16.0	−117	<0.8	
	2	WN11-17-15	1,000	Na–Ca–Cl–SO₄	31,840	4,930	6,800	27	19.1	61.8	18,944	62	1,105	20	−14.1	−109	3.0	
Matagami	7	R46	1,300	Ca–Cl	81,090	22,600	4,850	1,500	62.5	547.0	50,828	628	5	72				
	7	R36	1,800	Ca–Cl	227,200	61,300	14,800	3,440	338.0	1,460.0	143,581	1,785	405	101				
Norita	7	UN 249		Ca–Na–Cl	12,090	2,640	1,050	525	16.5	107.0	7,600	118	5	38				
	7	4E-85 #1		Ca–Na–Cl	139,100	29,100	13,300	4,200	335.0	960.0	90,000	1,134	5	76				
Sudbury	7	N3651	1,600	Ca–Cl	240,700	65,000	16,880	12	122.0	1,390.0	156,000	1,090	138		−11.3	−33	25.0	
	7,8	N3646A	1,600	Ca–Cl	249,100	63,800	18,900	78	430.0	1,580.0	162,700	1,250	223	58	−10.9	−39	<0.8	
	8	N3640a	1,500	Ca–Cl	250,360	63,800	18,500	24	371.0		166,200	1,200	265		−10.9	−44	14.0	−0.51
Thompson	7,8	T3-2000	610	Ca–Na–Cl	20,340	4,540	2,740	160	32.1	79.7	12,600	111	1	55	−17.9	−127	14.0	
	7,8	2200-1	671	Ca–Na–Cl	20,490	4,840	2,930	233	92.6	98.3	13,700	115	444	28	−15.0	−115	58.0	
	7,8	4000-3	1,220	Ca–Na–Cl	65,440	15,900	6,800	549	45.7	277.0	41,400	405	2	10	−16.0	−102	<0.8	

7,8	Yellowknife	4000-6	1,220	Ca–Na–Cl	101,400	26,800	8,670	637	59.9	558.0	63,800	793	2	15	−15.6	−92	<0.8	
7,8		4000-5	1,220	Ca–Na–Cl	182,600	46,300	17,000	1,960	126.0	910.0	115,000	1,110	107	9	−14.4	−79	17.0	
7,8		4000-4	1,500	Ca–Na–Cl	324,500	64,000	45,000	5,100	199.0	1080.0	207,000	1,760	284	19	−11.7	−53	4.0	
7,8		4500-2	1,372	Ca–Na–Cl	58,340	15,700	9,420	250	110.0	314.0	31,900	324	198	69	−19.9	−145	47.0	
7,8		4500-3	1,372	Ca–Na–Cl	63,140	12,900	10,900	268	111.0	671.0	37,100	358	746	28	−23.0	−163	3.0	
8		YK3464	1,372	Ca–Na–Cl	130,290	26,400	18,800	406	109.0	628.0	82,700	756	428	14	−19.3	−126	<0.8	−0.03
8		YK2042	1,372	Ca–Na–Cl	172,900	39,300	21,700	820	164.0	910.0	109,000	935	54		−18.4	−105	23.0	
8		4500-1c	1,372	Ca–Na–Cl	213,940	49,400	27,900	712	393.0	1190.0	132,800	1,180	8	16	−15.7	−84	1.0	0.00
7		4500-6C	1,372	Ca–Na–Cl	237,100	57,300	32,600	920	495.0	1640.0	142,000	1,520	1	2	−14.4	−71	3.0	
	Russia																	
9	Mir	N 82 [537]	600	Ca–Cl	242,820	46,417	4,025	9,853	4,432.0	882.0	173,338	4,712	38		−2.0	−39		−0.32
9	Udachnaya	Borehole 330	504	Ca–Cl–HCO₃	344,500	70,923	19,870	11,693	12,095.0	1127.0	139,435	3,094			−5.0	−71		−0.20
9	Udachnaya	Borehole 310	834	Ca–Cl	364,020	94,723	5,066	15,306	17,678.0	1600.0	212,280	5,637	38		−8.2	−73		−0.27
9	Mir	Borehole 28	500	Ca–Cl	369,540	86,041	11,294	13,870	7,574.0	1791.0	158,918	3,816	100		−5.2	−52		−0.15
9	Trelyahskaya	Borehole 746	2,777	Ca–Cl	372,530	108,747	10,406	4,005	1,195.0	2279.0	187,668	4,725	183		−6.2	−31		−0.67
	West Europe																	
10	Leuggern	Le	1,643	Na–HCO₃–SO₄	1,050	295	0	8.8			125		263	273				
10	Berghaupten	Berg	392	Na–Cl–HCO₃	1,860	568	4	14.8			447		276	452				
11	Bad Säckingen	GW9-1/86	82	Na–Cl	3,500	1,041	30	81.7			1,761	5	115	525	−9.2	−67		
10	Ohlsbach	Ohlsbach	59	Na–Cl	10,120	2,891	34	252.6	20.5		5,357	18	385	500				
10	Schramberg	Schramb	505	Na–Cl–SO₄	10,680	3,205	57	63.0			3,318		3,206	381				
10	Buehl	Bu	2,535	Na–Ca–Cl–SO₄	201,350	63,900	1,900	503.0	485.0		120,500	726	1,525					
	UK																	
12	Wheal Jane	WJ2	280	Na–Ca–Cl	7,550	995	1,620	23	100.0	16.8	4,460	17	189	24	−5.8	−34	7.0	
12	Wheal Jane	WJ3	400	Na–Ca–Cl	9,930	1,300	2,090	13	132.0	23.0	6,090	19	114	54	−5.6	−37	3.0	
12	South Crofty Mine	SC1	580	Na–Ca–Cl	15,250	1,840	3,520	55	153.0	30.0	9,280	35	129	60	−5.3	−28	6.1	

1 Ivanovich et al. (1992).   2 Author's unpublished data.   3 Blomqvist (1999).   4 Sie and Frape (2002).   5 Bottomley et al. (1999).   6 Kelly et al. (1986).   7 Frape and Fritz (1987).   8 Pearson (1987).
9 Shoukar-Stash et al. (2002).   10 Bucher and Stober (2000).   11 Barth (2000).   12 Edmunds et al. (1987).

their total dissolved solids (TDSs) or chemical load. In this chapter the classification of Davis (1964) is used to describe (i) freshwater, $<1,000$ mg L^{-1} TDS, (ii) brackish water, $1,000-10^4$ mg L^{-1} TDS, (iii) saline water, 10^4-10^5 mg L^{-1} TDS, and (iv) brine $>10^5$ mg L^{-1} TDS. For convenience, many authors often refer to more concentrated fluids in a g L^{-1} format, where, for example, seawater would have a TDS of 35 g L^{-1}. Only a few studies report density of the more saline fluids. Some publications list measured density data (e.g., Frape *et al.*, 1984) and the reader can use this information to calculate a TDS versus density plot that will allow conversion of the data from Table 2. As a semi-quantitative guide the density for the divisions in the Davis (1964) classification at 20 °C would be: fresh, 1.000; brackish, 1.000–1.007; saline, 1.007–1.075; and brine, >1.075.

In the next section, the data from Table 2 are combined with a much larger database of chemical and isotopic analyses taken from numerous published documents (e.g., Blomqvist, 1999) and data previously unpublished by two of the authors (Frape and Blomqvist). The following figures and text are designed as a brief summary of the geochemical and isotopic distribution and trends found in groundwaters from a number of crystalline rock environments.

5.17.3.1 Trends in Major Ions and the Stable Isotopes of Water

5.17.3.1.1 Bicarbonate and depth

Bicarbonate is usually thought to enter the groundwater system as a result of the uptake of CO_2 either from soil zone gases or direct atmospheric inputs (Langmuir, 1971). Additional sources can come from carbonate dissolution. In Figure 1(a) it is apparent that the highest concentrations of bicarbonate are generally in shallow groundwaters. A comparison with chloride versus depth (Figure 1(c)) shows that chloride has the opposite trend. The extremely concentrated bicarbonate fluids from Western European locations are mineral waters from the Black Forest of Germany and nearby areas in France and Switzerland (Balderer *et al.*, 1987; Stober, 1996). These waters contain considerable amounts of CO_2 gas (up to 3,000 mg kg^{-1}) and consequently have a low pH (<6) (Stober and Bucher, 1999; Bucher and Stober, 2000). Although the origin of the CO_2 is not definitively known, it is suspected to come from deep hydrothermal sources in the region (Bucher and Stober, 2000).

In general, bicarbonate concentrations in crystalline groundwaters decrease with depth

(Figure 1(a)). Once in a closed system, it is proposed that a variety of reactions with silicates act to remove HCO_3^- ions usually as precipitated calcite, which is a common fracture-filling mineral in these environments (Gascoyne *et al.*, 1987; Nordstrom *et al.*, 1989a; Blomqvist *et al.*, 1993). At the Äspö underground research site in Sweden, the creation of the mined opening caused surface waters to enter fracture zones and penetrate deeper into the rock. This caused a number of reactions that increased the bicarbonate content (Tullborg and Gustafsson, 1999). The authors used carbon-14 ages as a tracer to identify young and old sources of carbon in the system. They conclude that the increase in bicarbonate resulted from the mixing of three sources: (i) younger atmospherically derived CO_2 (dominantly soil CO_2) pulled to deeper levels in the rock; (ii) oxidation of older organic matter by anaerobic respiration; and (iii) the dissolution of older fracture-filling calcite in the shallow recharge zone. It is probable that similar processes control the system of carbonate equilibria in most crystalline rock environments where active recharge and circulation of freshwaters to depth is occurring.

In deeper systems dominated by calcium-rich saline fluids, it has been shown that both solubility constraints and silicate reactions act to further remove bicarbonate ions as precipitated calcium and magnesium carbonates, often adjusting pH to levels greater than 9 (Barnes and O'Neil, 1971; Fritz *et al.*, 1987a; Clauer *et al.*, 1989). For example, during closed-system dissolution of magnesium olivine (forsterite), a major component of many ultramafic rocks, as the silicate water reaction proceeds water breaks down, H^+ ions are consumed, carbonates precipitate, and hydroxyl ions force the pH to rise (Barnes and O'Neil, 1971; Drever, 1988).

5.17.3.1.2 Sulfate and depth

Sulfate in crystalline groundwaters is a very common constituent, usually increasing in concentration with depth (Figure 1(b)). The sources of sulfate in these environments are numerous (Fritz *et al.*, 1994). Marine incursions have been documented in Finland due to glacial crustal downloading that allowed postglacial flooding of coastal rocks (Nurmi *et al.*, 1988). Based on sulfur-isotopic studies, sedimentary basinal brines are believed to have migrated into some areas at the edge of the Canadian Shield (Gascoyne *et al.*, 1989; Bottomley *et al.*, 1994; Clark *et al.*, 2000). In Figure 1(b), the Canadian samples with sulfate concentration greater than 1,000 mg L^{-1} are from the Lac du Bonnet granite in Manitoba on the edge of the Canadian Shield. Gascoyne *et al.* (1987)

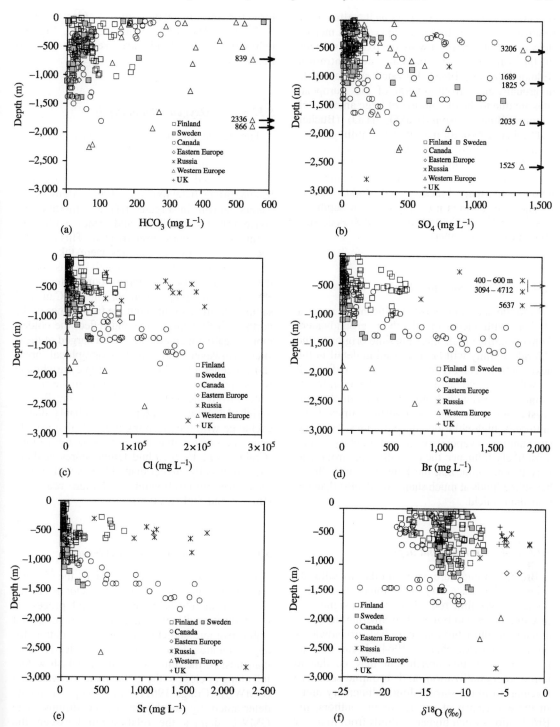

Figure 1 Trends in fluid concentration with depth for crystalline environments in several geographical areas: (a) bicarbonate and depth; (b) sulfate and depth; (c) chloride and depth; (d) bromide and depth; (e) strontium and depth; and (f) oxygen-18 and depth.

present a strong argument that the sulfate in some of these groundwaters resulted from the intrusion of western sedimentary basin brines. The work of Michelot and Fontes (1987) suggests that elevated sulfate levels in the groundwaters at some Swedish sites may have been derived from the

dissolution and penetration of pre-existing evaporites into the shield rocks.

Other sources of sulfate in crystalline environments have been documented to be magmatic or hydrothermal (Hattori and Cameron, 1986; Hattori, 1989). Such sulfates have positive $\delta^{34}S$

signatures (+5‰ to +25‰). As well, sulfate may result from the penetration of oxygenated meteoric waters that then react with and oxidize sulfides in the subsurface to increase concentration (Edmunds *et al.*, 1988). Elevated sulfate levels in groundwaters from several of the European sites are due to the dissolution of large amounts of fracture-filling gypsum ($CaSO_4 \cdot 2H_2O$) (Bucher and Stober, 2000). Similar gypsum infillings have been found in crystalline rocks at Canadian Shield sites (Kamineni, 1983; Mungall *et al.*, 1987). The loss of sulfate by bacterial reduction is commonly reported in many brackish and slightly saline groundwaters, which normally occur at depths of less than 500 m (Edmunds *et al.*, 1988; Pedersen, 1993; Fritz *et al.*, 1994).

5.17.3.1.3 Chloride and depth

Chloride shows the most significant increase of all ions with increasing depth in groundwaters from crystalline rock sites. Possible origins of the chloride salinity will be discussed in detail in the next section. As Figure 1(c) shows, individual crystalline shield areas seem to have different salinity patterns with depth. The Siberian platform crystalline rocks and kimberlite pipes have uniformly high chloride concentrations. Here the surrounding evaporite-rich sedimentary rocks most likely influence the crystalline environment. It is also apparent that the Fennoscandian Shield has saline fluids at much shallower depths than the Canadian Shield.

5.17.3.1.4 Bromide and depth

Bromide trends are very similar to chloride (Figure 1(d)). It is a very minor ion in freshwaters found in crystalline rocks but becomes significant in saline fluids and brines. The Siberian fluids are among the most bromide-enriched groundwaters on the planet (Kapchenko, 1964). It can only be postulated that this enrichment may be due to halite recrystallization and bromine-selective leaching from the surrounding sedimentary units. In the experience of several of the authors, the Br/Cl ratio is often higher in fluids from mafic or ultramafic rock environments and fluids from most crystalline rock environments have a much higher ratio than found in present-day seawater.

5.17.3.1.5 Strontium and depth

Strontium, like bromine, is a minor element in shallow fresh groundwaters (Figure 1(e)) but becomes quite concentrated in saline waters and brines. The $^{87}Sr/^{86}Sr$ ratio is a very useful isotopic

tool for tracing water masses (see Section 5.17.5.1.4). The major cation (Ca, Na, Mg, K) behaviors with depth are very similar to Figure 1(e).

5.17.3.1.6 Oxygen-18 and depth

The range of oxygen-18 values found in groundwaters from crystalline rocks is very similar to that reported for thermal waters (Truesdell and Hulston, 1980) or sedimentary basinal brines (Clayton *et al.*, 1966). Many of the waters shown in Figure 1(f) are mixtures of concentrated saline fluids and brines, with infiltrating local meteoric groundwaters. This is partly because many of the samples were taken from deep mine areas where mining activity has opened the rock and caused artificial groundwater gradients. For example, a number of the Canadian samples from 1,400 m depth are from mines in the northern Yellowknife area and reflect the colder, more negative meteoric waters recharged at these sites. However, individual geographical areas show a general trend towards isotopic enrichment with depth and salinity (Figure 1(f)). In the case of waters from thermal sites in Europe and Russia, enriched isotopic signatures are probably due to exchange with rock forming minerals that usually contain a greater abundance of the heavy isotope (Craig *et al.*, 1956). Fluid temperatures at these sites are usually greater than 100 °C and therefore ideal for mineral/water exchange reactions (Bucher and Stober, 2000).

5.17.3.1.7 Oxygen-18 and deuterium

In most low salinity groundwaters, the ^{18}O and ^{2}H contents are conservative and seldom altered by chemical, physical, or geological processes. Groundwaters closely reflect the average annual stable isotopic signature of the precipitation for any given area. Figures 2(a) and (b) clearly show that most of the dilute fresh and brackish waters fall along or close to the global meteoric waterline (GMWL) (Craig, 1961) with isotopic contents determined by local climatic conditions. The GMWL defines the relationship between the isotopic ratios of the two elements of water ($^{18}O/^{16}O$; $^{2}H/^{1}H$) for surface and near-surface groundwaters around the world. It provides a reference for comparison of local waters and groundwaters and helps identify potential physical and chemical processes if a water sample's isotopic signature deviates from the vicinity of the meteoric line. Modern ocean water essentially has a 0‰ value (VSMOW) for both isotopes, but slight variations do occur depending on salinity and location in the ocean (Clark and Fritz, 1997).

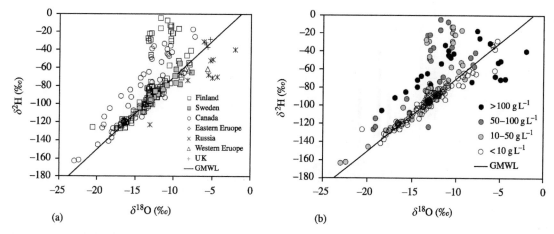

Figure 2 Stable isotopic composition for groundwaters from crystalline rock environments: (a) coded for geographical location and (b) coded for TDSs.

The ocean is thought to have been lighter in ^{18}O in the Archean and Proterozoic and, as a result of exchange with isotopically heavy ocean basaltic rocks, has evolved to its present composition (Veizer *et al.*, 1989, 1992). Surface waters generally plot to the right of the meteoric water line due to evaporative enrichment (Gonfiantini, 1986). Depending on humidity and salinity, evaporative losses tend to drive surface waters and even brines to the right of the GMWL. If these evaporated waters then recharge into a crystalline environment, they often carry this signature with them (Gonfiantini, 1965). Sedimentary basin and hydrothermal fluids, which can be as concentrated as crystalline brines, also plot to the right of the meteoric waterline due to processes such as mineral water exchange in thermal fluids or recharge of surface evaporative fluids (Kharaka and Carothers, 1986; see Chapter 5.16). Some of the Russian sites that plot to the right of the meteoric waterline are mostly likely impacted by local sedimentary fluids.

The most noticeable isotopic difference between saline waters from crystalline rocks and sedimentary formation waters is their position above the meteoric waterline. This is postulated to be due to mineral hydration reactions in a very water-depleted environment (Fritz and Frape, 1982). Several recent studies have suggested that hydration reactions in low water to rock environments can occur and result in increasing salinity. The incorporation of OH^- into primary silicate such as amphiboles and phyllosilicates (where OH^- crystal lattice sites are part of the mineral structure) is suggested as one mechanism for controlling solute concentration (Kullerud, 2000). The formation of secondary OH^- containing mineral phases such as zeolites and clays can also continue to consume water molecules and concentrate the residual fluids both chemically

and isotopically (Bucher and Stober, 2000). Other authors suggest that natural radioactivity in the crystalline rocks can lead to the breakdown of water and the radiolytic enrichment of salts at depth (Vovk, 1987). However, the exact mechanisms that created the isotopic signatures and that impact these waters is still an ongoing debate among hydrogeochemists. According to the present authors, the most deuterium-enriched saline fluids shown in Figure 2(a) are from mafic and ultramafic rocks. Many of the samples, especially from Canada, were taken from active mined openings where the mining activity has created groundwater mixing conditions that introduce fresh local meteoric waters to depth. This appears as dilution and mixing lines from the most concentrated fluids to the local meteoric water intercept on the GMWLs shown in Figures 2(a) and (b) (Fritz and Frape, 1982).

5.17.3.2 Strontium Isotopes

The strontium isotopic distribution for saline waters and brines found in a number of crystalline rock sites is reported as a histogram (Figure 3(a)). The data included in this figure are representative of values and isotopic distributions to be expected from groundwaters found in crystalline environments. Reviews of strontium isotopes in crystalline rocks and waters can be found in McNutt *et al.* (1984, 1990), McNutt (1987, 2000), Smalley *et al.* (1988), and Négrel *et al.* (2001). Strontium isotopes become a valuable tracer in highly saline waters, as strontium concentration can be quite high in these fluids (Table 2 and Figure 1(e)). The role of water/rock interaction and the origin and evolution of strontium isotopic signatures will be addressed in the next section. In most studies it has

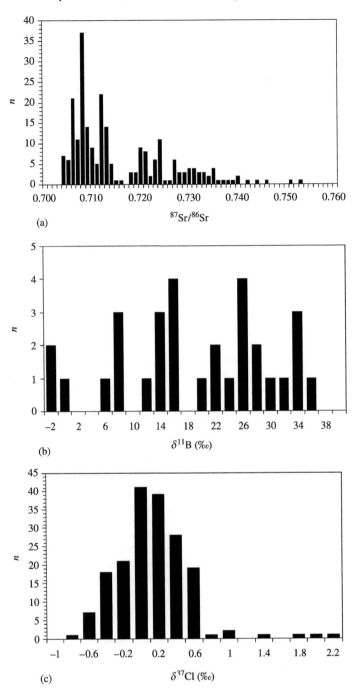

Figure 3 Range of values for various isotopes measured in fluids from crystalline rock environments: (a) strontium 87/86, (b) $\delta^{11}B$, and (c) $\delta^{37}Cl$.

been noted that strontium usually has a very strong correlation with calcium, and the two elements generally are controlled by the same mineral systematics (McNutt *et al.*, 1990).

5.17.3.3 Boron Isotopes

The range of boron isotopic values in fluids from crystalline environments is quite large

and generally, for all data reported, these values are below the seawater value of $+40\permil$ (Figure 3(b)). However, the data set is small due to the difficult analytical procedure to acquire results (Barth, 2000). Published data from various granites worldwide are listed by Barth (2000). Most of the data from earlier studies fall in the range from $-3.2\permil$ to $+6.8\permil$ from Precambrian sites in South America ($+1.6\permil$; Spivack and Edmond, 1987), Western Australia

(−3.2‰ to +6.8‰; Vengosh *et al.*, 1991), and China (−2.4‰ to +3.7‰; Vengosh *et al.*, 1995). Barth (2000) reports for freshwaters values of −3.5‰ to −0.6‰ and for more saline waters values of +6.4‰ to +17.6‰ in crystalline rocks of northern Switzerland and southwestern Germany. The progressive enrichment of $\delta^{11}B$ with salinity and depth is also reported for fresh to saline fluids in granitic rocks from the Vienne research site in France (Casanova *et al.*, 1998, 2001). Similar to Bottomley *et al.* (1994), the authors of the above studies attempt to distinguish seawater/evaporite-derived boron from a crystalline rock source. Studies of several Icelandic geothermal systems showed that in some cases in active high-temperature systems the $\delta^{11}B$ values were less than 0‰ and B was derived from the host basalts (Aggarwal *et al.*, 2000). Lower-temperature Icelandic systems contained B systematics consistent with mixtures of seawater (~+40‰) and leaching of the basaltic rock. Yet other systems showed seawater interaction with low-temperature altered basalts and a shift in the isotopic ratio due to B uptake into calcites, which acts to fractionate the boron isotopes (Aggarwal *et al.*, 2000).

5.17.3.4 Stable Chlorine Isotopes (37/35)

The distribution of the stable isotopes of chlorine (37/35) in crystalline fluids is presented as a final histogram (Figure 3(c)). Seawater is considered to have a uniform isotopic signature worldwide and is arbitrarily assigned a 0‰ value (Kaufmann, 1984). Figure 3(c) shows that the chlorine isotopic signature for groundwaters from crystalline environments has a relatively narrow range of −1.0‰ to +2.0‰. Fluids found in crystalline environments do not have the signature of ocean water (0‰) (Kaufmann *et al.*, 1987) and based on a limited data set, groundwaters from the Fennoscandian Shield show an enrichment over similar fluids from the Canadian Shield (see Section 5.17.5.3.1) (Frape *et al.*, 1995; Sie and Frape, 2002). For comparison, the senior author's unpublished data sets for sedimentary and other environments show the range of chlorine isotopes to be −2‰ to +6‰ for Earth fluids and rocks and about −6‰ to +8‰ for man-made organic compounds. Several published studies on chlorine isotopic signatures suggest that rocks can have a wide range of values. Magenheim *et al.* (1994) found values between 0‰ and +7.5‰ for seafloor basalts (MORB). Other studies have shown hydrothermally altered rocks and minerals and a variety of other mineral phases have values between −1‰ and +2‰ (Eastoe and Guilbert, 1992; Eggenkamp, 1994; Eggenkamp and Schuling, 1995; Boudreau *et al.*, 1997). Rock data from

minerals in the crystalline rocks of Norway (Markl *et al.*, 1997) fall in the range (−1.12‰ to 0.79‰), which overlap much of the fluid data from Figure 3(c).

Finally, the study of Banks *et al.* (2000) presents data from fluid inclusions at two crystalline sites. The $\delta^{37}Cl$ values in mineral inclusions of the Capitan pluton of New Mexico, USA were near 0‰. The authors concluded that an evaporitic source was responsible for the chlorine isotopic signature. Their other site was a batholith in southwestern England and the $\delta^{37}Cl$ signature was approximately +1.8‰. They felt this was more representative of deeper magmatic sources similar to the MORB data reported by Magenheim *et al.* (1994). The data are also similar to water and rock data from similar young granites in Finland (Frape *et al.*, 1998).

5.17.4 GASES FROM CRYSTALLINE ENVIRONMENTS

In the shallow meteorically recharged portion of crystalline environments, atmospheric oxygen, nitrogen, and carbon dioxide are common dissolved constituents in groundwaters. However, the deep saline fluids host a very different suite of gases. The majority of saline fluids reported here contain volumetrically significant amounts of hydrocarbon gas (Fritz *et al.*, 1987b; Pekdeger and Balderer, 1987; Sherwood *et al.*, 1988; Nurmi *et al.*, 1988; Sherwood Lollar *et al.*, 1989, 1993a,b). Most Canadian and Finnish sites studied by Sherwood Lollar *et al.* (1993b) contained large quantities of dissolved methane, C_2, C_3, and C_4 gas, nitrogen and local enrichments of hydrogen.

In a few cases, the isotopic signatures of the hydrocarbon gases were found to be biogenic in origin (Sherwood Lollar *et al.*, 1993a). However, as shown in Figure 4, most of the gases in at least two crystalline shields (Canada and Finland) appear to fall outside of the biogenic isotopic fields (Schoell, 1988). Although Sherwood Lollar *et al.* (1993b) could not totally rule out some biogenic gas mixtures, they felt that the presence of altered ultramafic rock, serpentinization, the presence of abundant volumes of hydrogen gas and isotopic values approaching those reported for mantle-derived gases all favored an abiogenic origin for methane formation.

In some cases, researchers have been able to confirm that microbial-generated methane does exist at some shallow crystalline sites (Sherwood Lollar *et al.*, 1993a). The existence of unique populations of microbes in these environments has been well documented (Pedersen, 2000). Pedersen (2000) points out that foreign microbes introduced into saline crystalline systems do not survive for extended time periods and has reported several

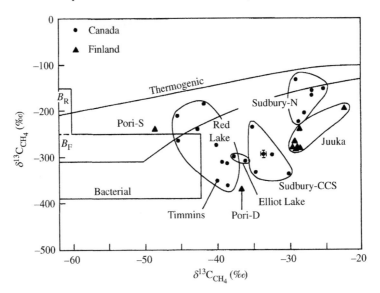

Figure 4 The $\delta^{13}C$ and δ^2H isotopic values for methane gas found dissolved in saline groundwaters from several sites on the Canadian and Fennoscandian Shields (samples for each area are circled). These are compared to the bacterial and thermogenic ranges of Schoell (1988). B_R and B_F are the ranges as outlined by the boxes of bacterial methane production from CO_2 reduction and acetate fermentation, respectively (Lollar *et al.* 1993b) (reproduced by permission of Elsevier from *Geochim. Cosmochim. Acta*, **1993**, *57*, 5087–5097).

new species of microbial life from some of the research sites adding a very new dimension to the potential origin of at least some of the gases from these environments. Also, Tullborg (2000) presents interesting evidence of fossil bacterial casts imbedded in fracture calcites. These calcites have very negative carbon-13 isotopic signatures. Such signatures in calcites are often interpreted as carbon derived from methanogenic processes (Fritz *et al.*, 1989).

As a further note, these dissolved gases bring the added dimension of extreme gas pressures to the saline fluids. Reports from miners of drilling accidents and many exciting sampling scenarios have been common practice for several of the authors of this chapter as a result of very high gas pressures. Similar comments are made throughout reports of the superdeep Kola borehole in Russia. Researchers regularly reported methane and hydrogen gas under high pressures and of hydrogen "boiling" out of the return drilling fluids (Borevsky *et al.*, 1984).

5.17.5 THE ORIGIN AND EVOLUTION OF FLUIDS IN CRYSTALLINE ENVIRONMENTS

The arguments surrounding the origin of highly concentrated fluids found in crystalline rocks can be divided into two schools, the internalists and the externalists. The internalists—the water/rock researchers—favor creating the salinity within

Table 3 Possible origins for highly saline fluids and brines in crystalline rocks.

(A) Water–rock interaction
   (i) dissolution of mineral phases and alteration
   (ii) leakage of fluid inclusions
   (iii) magmatic fluids

(B) Allocthonous sources
   (i) seawater—concentrated seawater
   (ii) evaporative concentration and emplacement of fluid
   (iii) sedimentary formation fluids
   (iv) dissolution of sedimentary evaporite deposits

(C) Hydrothermally driven systems could fit for both sources above

the rock mass or deriving it from the rock. The externalists derive the salinity from allocthonous sources and move highly saline fluids into the rock environment by a number of plausible methods. Table 3 provides a brief summary of some of the possible sources of crystalline rock salinity.

What both schools of thought do agree on is that most of the fresh and brackish groundwaters found in shallow environments are young meteoric waters. These waters usually contain tritium, have local meteoric ^{18}O–^2H isotopic signatures, and are found in highly permeable sections of rock, which are often active flow systems (Frape *et al.*, 1984; Blomqvist *et al.*, 1998; Clark *et al.*, 2000).

5.17.5.1 The Origin of Salinity: The Influence of the Rock

5.17.5.1.1 Halogen-bearing minerals and inclusions

Fluorine, chlorine, and, to a lesser extent, bromine have been studied in many types of crystalline rocks (Kamineni, 1987). Correns (1956) described the major modes of halogen occurrence in rocks as incorporation in mineral lattices, in fluid inclusions and dissolved in mineral glasses. The solubility of halogens in magmas controls their ultimate fate. In many systems where magmatic differentiation occurs, fluorine partitions into the melt displacing chlorine which then becomes available to other mineral and rock phases (Fuge, 1977). The chlorine will most likely end up with alkali metals in alkaline rocks, forming minerals such as sodalite and other similar mineral phases (Fuge, 1977; Carmichael *et al.*, 1974).

Kamineni (1987) presents considerable data for chlorine concentration in a number of mineral types. The data are most likely a combination of chlorine from lattice as well as fluid inclusion sources, although later results in his paper were obtained primarily on lattice incorporated chlorine. The findings are summarized in Table 4. Additionally, data from Kamineni *et al.* (1992) for bromine and chlorine in mafic and ultramafic rocks are included (Figure 5). More recently, it has been suggested that amphiboles such as hornblende [(Ca, Na, K) (Fe$^{2+,3+}$, Mg Al)$_3$ (Si, Al, O$_{22}$, OH)], or phyllosilicates such as biotite (K(Fe, Mg)$_3$ (Si, Al)$_4$ O$_{10}$(OH)$_2$) crystallizing in highly saline solutions can incorporate considerable chlorine into the OH$^-$ lattice site (up to 6 wt.‰ in amphiboles and 7 wt.% in biotites) (Kullerud, 2000). Preferential incorporation of OH$^-$, relative to halogens, in OH$^-$ bearing minerals might also be a means to remove water from the fluid phase, thus creating higher salinity in subsurface fluids (Kullerud, 2000; Bucher and Stober, 2000).

The other major source of halogens in crystalline rocks are the fluid inclusions trapped in numerous mineral phases during crystallization. Fluid inclusions can be trapped within mineral grains or between minerals as grain boundary salts. Most are only a few microns in size but can also be very large (Roedder, 1984). Garrels (1967) first suggested that salinity in igneous rocks might be related to the leakage of fluid inclusions. In general, inclusions would be considered to be of the age of the rock (possibly billions of years in some cases), and therefore the opportunity for post-depositional thermal and mechanical disturbances to fracture the rock and release inclusions is highly likely. Inclusions can have salinities ranging from 0 wt.% to 70 wt.% and compositions which are usually variations of sodium and

Table 4 Summary of the range of chlorine concentration in some minerals.

Mineral	n	Cl (wt.%)
Biotite	219	0.002–0.66
Hornblende	84	0.01–3.6
Other amphiboles	12	0.003–7.24
Apatite	69	0.01–6.8
Sodalite	16	5.4–8.01
Eudialyte	31	0.7–2.2
Scapolite	55	0.1–3.4

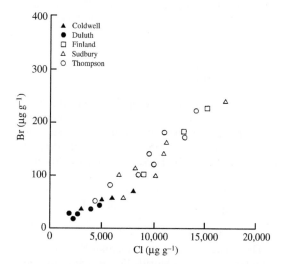

Figure 5 Chloride versus bromide concentration in a suite of mafic and ultramafic rocks from several crystalline field sites (Kamineni *et al.*, 1992) (reproduced by permission of Balkema from *Proc. 7th Int. Symp. Water–Rock Interaction*, **1992**, pp. 801–804).

calcium with chlorine. The study of Nordstrom *et al.* (1989b) on the Stripa granite in Sweden showed that fluid inclusions made up approximately 1% of the porosity. Their leaching experiments and fluid leakage calculations suggested that the slow migration of inclusion salts into the primary rock permeability could account for a large part of the dissolved load. The most concentrated Stripa groundwaters are fairly dilute (<2,000 mg L^{-1}, TDS) compared to many other crystalline sites, and therefore the origin of salinity from leaking inclusions is a possibility at this site (Nordstrom *et al.*, 1989b,c). Subsequently, studies using crush and leach techniques and comparing stable chlorine isotopic signatures between rock and fluid in open fractures have shown that at the Stripa, Sweden site and several other research sites, there is a component of rock matrix/inclusion chloride in the fracture porosity (Sie and Frape, 2002; Blyth and Frape, 2002). Several other studies show the complex nature of the geochemistry found in crystalline rock sites. Irwin and Reynolds (1995) describe multiple stages of fluid trapping in the Stripa granite as evidenced by geochemical analyses of fluid inclusions. They believed that at least some portion of the halogens are produced in an iodine- and bromine-rich rock external to the granite. An external source for part of the Stripa salinity is a very common theme found in most studies published from the site (Fontes *et al.*, 1989). The research of Savoye *et al.* (1998) at a number of crystalline rock sites in Europe provided strong evidence, based on mass balance calculations and inclusion chemistry, that only a very small percentage of salinity at their research sites could be derived from inclusion leakage.

5.17.5.1.2   The role of water–rock interaction

Many authors have proposed that the rock environment, through a series of chemical reactions, will extensively modify fluids in contact with it (Chebotarev, 1955; Collins, 1975; Edmunds *et al.*, 1984, 1987). Early research in this area suggested that long-term evolution of subsurface saline fluids favored the formation of calcium chloride brines that are very old. However, the origin of these brines could not be determined (Krotova, 1958; Rittenhouse, 1967).

The earliest studies of Canadian Shield brines found that calcium chloride brines, with salinities up to 340 g L^{-1}, were ubiquitous at more than 20 sites below depths of 1,000 m (Fritz and Frape, 1982). As these sites were active working mines, the rock mass was disturbed, so there was usually a component of freshwater circulating through the rock mass to depth. This water tended to mix with older brackish to brine solutions, but also seemed

to assimilate major ion chemistry such as Ca, Mg, and Na, and reflect the rock chemistry that it flowed through (Frape *et al.*, 1984). Where freshwaters charged with oxygen and carbon dioxide flowed through mafic rocks, their chemistry became dominated by Ca > Mg > Na and HCO_3 as the dominant anion. In other areas where felsic-granitic rocks occurred the chemistry of the fresh and brackish waters was Na > Ca > Mg > K with HCO_3. The composition of the saline waters and brines also seemed to reflect an influence of rock chemistry. For example, brines from several sites dominated by mafic and ultramafic rocks often have Mg concentrations significantly higher than brines hosted in other rock types (Frape *et al.*, 1984).

In more recent studies from Finland, some interesting relationships between rock type and fluid chemistry were found. Figure 6 shows the geology, water chemistry, and the stable chlorine isotopic signature of water and rock found in the 1,100 m deep Juuka/Miikhali 116 exploration borehole in eastern Finland. This is an open borehole which was sampled using the tube sampler method of Nurmi and Kukkonen (1986). The borehole has been sampled numerous times (Blomqvist, 1990) and was also used in 1987–1988 as part of the gas sampling program of Sherwood Lollar *et al.* (1993b). In several sampling trips to this site the chemical distribution and concentration in the open borehole maintains the stratification seen in Figure 6.

These are low-porosity/low-permeability rocks. The site has a Ca–Na–Cl saline water associated with the mica gneisses and calc silicate rocks (felsic) residing above a more dense Na–Ca–Cl brine with elevated magnesium concentrations. The stratigraphically lower, magnesium-rich brine

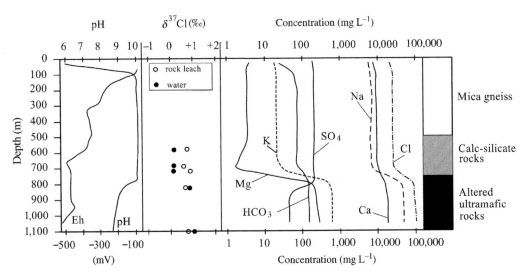

Figure 6   Geology, major element water chemistry, Eh and pH, and stable chlorine isotope analyses of rock and water as functions of depth in drill hole Ju/Mi-116 eastern Finland.

is associated with altered ultramafic rocks. The chlorine isotope signature of the rocks and fluids from the ultramafics is almost identical, suggesting a strong case for long-term equilibrium. The deviation in the chlorine isotopic signature between fluid and rock chlorine isotopic signature for the calc-silicate rocks may be due to the influence of internal or external processes such as seawater intrusion that will be discussed later (Section 5.17.5.3.1).

5.17.5.1.3 Mineral–fluid reactions

Numerous secondary mineral–fluid reactions have been shown to occur in crystalline rock environments (Bucher and Stober, 2000). As mentioned previously, reactions in ultramafic and mafic rocks can have considerable control on pH, carbonate precipitation, and most likely the accumulation of excess H_2 gas (Barnes and O'Neil, 1971; Sherwood Lollar et al., 1993b). A typical reaction (Equation (1)) (not balanced) that would involve olivine dissolution (Drever, 1988) and in closed systems would also consume water and concentrate solutes is as follows:

$$2(Mg, Fe)_2 SiO_4 + 3H_2O = Mg_3 Si_2O_5 (OH)_4 + (Mg, Fe)(OH)_2$$

$$\text{olivine} \qquad\qquad \text{serpentine} \qquad\qquad \text{brucite}$$

$$(1)$$

At low temperature the ferrous-iron-rich brucite $(Mg,Fe)(OH)_2$ can subsequently disproportionate into a magnesium end-member brucite and magnetite, which becomes the stable iron oxide in contact with water at low temperature. The formation of magnetite under certain conditions can result in the release of hydrogen gas. At elevated temperatures in hydrothermal and metamorphic reactions (Equations (2) and (3)), magnesium or potassium can be removed from solution forming the minerals chlorite or sericite:

$$2NaAlSi_3O_8 + 10(Mg, Fe)^{2+} + 2Al^{3+} + 20H_2O$$

albite

$$\rightarrow 2(Mg, Fe)_5 Al_2 Si_3O_{10} (OH)_8 + 2Na^+ + 24H^+ \quad (2)$$

chlorite

$$2Al(Mg, Fe)_5 AlSi_3O_{10}(OH)_8 + 5Al^{3+} + 3Si(OH)_4 + 3K^+ + 2H^+$$

chlorite

$$\rightarrow 3KAl_2AlSi_3O_{10}(OH)_2 + 10(Fe, Mg)^{2+} + 12H_2O \quad (3)$$

sericite

Likewise, the conversion of Ca-plagioclase (with release of $Ca^{2+} + Al^{3+}$) to Na-plagioclase (Equation (4)) is a likely reaction in crystalline environments:

$$2Na^+ + CaAl_2Si_2O_8 + 4SiO_2 \rightarrow 2NaAlSi_3O_8 + Ca^{2+}$$

anorthite quartz albite

$$(4)$$

Alkali feldspars are among the most abundant minerals in the Earth's crust and, due to microtextural strains that occur at temperatures of crystallization, these minerals are very susceptible to exsolution of sodium and potassium and reordering of their elemental composition (Parsons and Lee, 2000). Feldspars have significant concentration of strontium and, because of their high content in crystalline rock, exert important controls on the strontium isotopic signature of the fluids and rocks (McNutt, 2000). In many cases the alkali feldspars (sodium and potassium) have been subjected to dissolution and reprecipitation (Parsons and Lee, 2000) and fluid/feldspar equilibrium is very common (Giggenbach, 1988).

At lower temperatures, as discussed in Chapter 5.16, congruent and incongruent weathering reactions and neoformation of clays and zeolite minerals, which are common secondary mineral phases in crystalline rocks, are postulated to alter fluid and isotopic chemistry (Fritz and Frape, 1982) and increase salinity during hydration reactions (Bucher and Stober, 2000). As an example, reaction (5) represents the formation of zeolites. These well-documented fracture-filling minerals have been found by several authors at Fennoscandian sites, and at research sites in Europe (Bucher and Stober, 2000):

$$Na_4 Ca Al_6Si_{14}O_{40} + 8H_2O \rightarrow 2SiO_2 + CaAl_2Si_4O_{12} \cdot 4 H_2O$$

"plagioclase" quartz laumontite

feldspar

$$+ 4NaAlSi_2O_6 \cdot H_2O \quad (5)$$

analcime

5.17.5.1.4 Strontium isotopes and water–rock interaction

The radiogenic Rb–Sr isotopic system has become a useful parameter in studies of groundwater systems, especially in assessing the origin(s) of the solutes. McNutt (2000) presents a detailed review of strontium isotopic behavior in water with emphasis on water–rock interaction; see also Chapter 5.12.

The isotopic composition of strontium varies with time due to the radioactive β-decay of ^{87}Rb to ^{87}Sr with the very long half-life ($T_{1/2}$) of 48.8 Gyr. The relationship can be expressed as

$$(^{87}Sr/^{86}Sr)_{today}$$

$$= (^{87}Sr/^{86}Sr)_0 + (^{87}Rb/^{86}Sr)_{today}(e^{\lambda t} - 1) \quad (6)$$

The practice is to normalize ^{87}Sr and ^{87}Rb to ^{86}Sr an invariant, stable isotope of Sr (Sr has four isotopes—^{88}Sr, ^{87}Sr, ^{86}Sr, and ^{84}Sr—and only ^{87}Sr varies with time, as the other three are stable). The other notations are:

- today = experimentally measured Sr value of rock or water;

- 0 = Sr isotopic composition of the system at the beginning, i.e., crystallization of a mineral/rock;
- λ = decay constant of $^{87}Rb = 1.42 \times 10^{-11}$ yr^{-1} $(T_{1/2} = \ln 2/\lambda)$;
- t = time from $t = 0$ to the present, i.e., the age of the system.

Figure 7 shows the relationship graphically (taken from McNutt et al., 1990), where $^{87}Sr/^{86}Sr$ is plotted against time. It illustrates some important points.

(i) At the time of the minerals' formation as the rock crystallized, all mineral phases have the same $^{87}Sr/^{86}Sr$ value.

(ii) With the passing of time, each mineral increases its $^{87}Sr/^{86}Sr$ ratio in proportion to its Rb/Sr ratio (more specifically to its $^{87}Rb/^{86}Sr$ ratio; see equation above).

(iii) The rock value is the weighted mean of the mineral values.

(iv) A wide range of present-day values can exist in a rock such as a granite. This is not the case for a rock with all constituent minerals having very low Rb/Sr values (e.g., mafic rocks dominated by Ca-plagioclase, pyroxene, etc.).

When assessing a measured $^{87}Sr/^{86}Sr$ value in a water/ brine sample, a number of variables come into play, including the following.

(i) The mineralogical composition of the host rock. For example, in feldspar-bearing rocks, these minerals control the mass balance for strontium. Plagioclase is low in rubidium but that is not so in the case of alkali feldspars

(microcline, orthoclase, sanidine). Therefore, the latter group has higher $^{87}Sr/^{86}Sr$ values for a given age. In carbonates, strontium is found in calcite/ aragonite or dolomite. Some minerals, particularly the micas, have low total strontium contents but relatively high contents of rubidium and hence high $^{87}Sr/^{86}Sr$ ratios.

(ii) The dynamics of the water system, i.e., the length of time the water is in contact with a given host rock, and/or the possibility of the mixing of water masses from different sources. For initial conditions of dilute water, in a closed system, the ratio in the water will eventually equal the weighted sum of all the mineral phases that have dissolved during the time period in question. In all open systems, the ratio will reflect the more reactive minerals.

(iii) The kinetics of mineral alteration, dissolution, and precipitation, and its relation to the dynamics of the system. These processes do not measurably fractionate the various isotopes of strontium (but see (v) below) so that an alteration product should reflect the isotopic composition of the primary mineral phase. Figure 8 from Franklyn et al. (1991) shows this for the pair plagioclase–epidote. Similarly, while the precipitation of a secondary minerals, such as calcite, aragonite, gypsum, etc., will concentrate strontium and therefore affect the chemical composition of the fluid, they will retain the isotopic signature of the water. Knowledge of the relative rates of mineral dissolution is crucial to any assessment of measured strontium isotopic values in water. Lasaga (1984) showed that there is an enormous range in dissolution rate of rock-forming silicates, spanning five orders of magnitude. Within the minerals of interest here, the sequence is anorthite > nepheline > diopside > albite > microcline > muscovoite > kaolinite. As the feldspars are the main sinks for Sr in crystalline rocks, an understanding of their dissolution behavior is important. An excellent review is provided by Blum and Stillings (1995), wherein they clearly demonstrate how complicated the processes are and that dissolution behavior of plagioclase and alkali feldspars is dependent on a number of physical and chemical factors, including the pH of the solution (slowest at pH 7, more rapid in both acid and basic solutions). Initial-stage release of Sr from feldspars is clearly nonstoichiometric. The role of temperature variation and differential strontium release from feldspars and biotites in granitic rocks is discussed by White et al. (1999). Experiments by Brantley et al. (1998) demonstrate that initial Sr release from feldspars (bytownite > microcline > albite) is different from the long-term steady-state conditions (bytownite > albite > microcline), and therefore the observed $^{87}Sr/^{86}Sr$ value in solution is

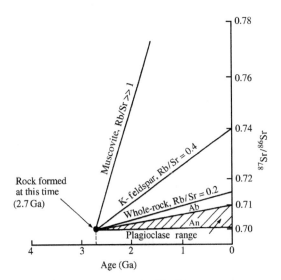

Figure 7 The growth of the $^{87}Sr/^{86}Sr$ ratio over time for the major mineral phases of a granitic rock. Note the much smaller growth rate for feldspars compared to mica. The feldspars contain the bulk of the strontium in such a rock and control the isotopic composition of the saline waters (McNutt et al., 1990) (reproduced by permission of Elsevier from *Geochim. Cosmochim. Acta*, **1990**, *54*, 202–215).

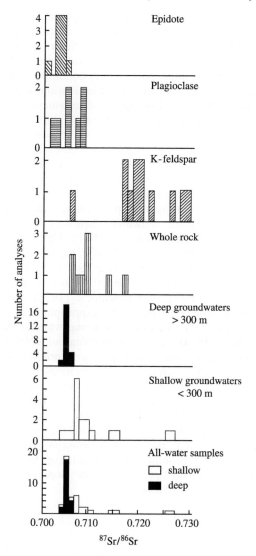

Figure 8 The $^{87}Sr/^{86}Sr$ data for groundwaters, whole-rock samples, and mineral separates from the Eye-Dashwa Lakes pluton, Ontario. Note the agreement in $^{87}Sr/^{86}Sr$ between the deep groundwater and plagioclase, and between plagioclase and its alteration product epidote (Franklyn *et al.*, 1991) (reproduced by permission of Elsevier from *Chem. Geol.* **1991**, *86*, 111–122).

a function of what mineral is dissolving and at what relative rate for a given condition. The initial more rapid dissolution of microcline will lead to higher $^{87}Sr/^{86}Sr$ ratios, as is commonly the case for dilute Ca-HCO$_3$ waters. A similar point was made by Fritz *et al.* (1992), based on thermodynamic–kinetic model calculations involving the weathering of granite.

(iv) The concentration and chemical composition of the water mass before it came into contact with its present host rock. For any water "far from equilibrium," dissolution of minerals will occur and the weighted mean of the dissolving phases will be reflected in the waters' isotopic

composition. However, for a brine (as defined in this chapter), particularly one with a high TDS, dissolution may not be a significant factor and indeed equilibrium considerations may dictate that Sr move from brine to rock and affect the rock's isotopic signature, either by ion exchange with the existing minerals or by precipitation of a secondary mineral in the rock.

(v) The separation of ^{87}Sr from the other Sr isotopes by geological/geochemical processes. It has long been held that one reason certain Rb–Sr minerals, particularly the micas and alkali feldspars, give ages less than the true age of the rock is because ^{87}Sr resides in a K(Rb) lattice site and during processes of alteration, metamorphism, etc., is preferentially lost from that site in comparison to an Rb ion, or a "normal" Sr ion located in a Ca site. This is important for the weathering of granitic rocks, as the experimental data show that very dilute (early stage) waters are more radiogenic than more concentrated (later-stage) waters. Thus, it is important to understand how alkali feldspars and micas dissolve in water, particularly in the early stages. The literature is replete with studies on alkali feldspar dissolution.

(vi) The water–rock ratio. All of the above variables are affected by the water–rock ratio. In a rock-dominated system (intergranular fluids) the water will more readily reflect the rock isotopic value and give a source signature, especially in a closed system. In a water-dominated system (fracture zone) the reverse is true, as the water will reflect the isotopic composition of the more soluble minerals in the fracture system, especially in an open system.

As pointed out above, there are two main schools of thought concerning the origin of brines: water–rock interaction (the internalists) or an allochthonous source (the externalists). The allochthonous sources proposed are modern seawater, fossil seawater (connate), or the dissolution of rocks of marine origin (e.g., carbonates, evaporites) by groundwater which will then carry a seawater isotopic signature, reflecting the age of the rock and the seawater isotopic signature of that time.

The $^{87}Sr/^{86}Sr$ ratios of the world's oceans have varied from 0.7065 to 0.7092 during the Phanerozoic (Burke *et al.*, 1982). If a brine has a signature within that range, then it is permissible to consider a marine isotopic signature. However, one cannot *a priori* assume a marine origin for the strontium, because plagioclase, especially the calcium-rich varieties, frequently have $^{87}Sr/^{86}Sr$ values in the 0.7065–0.7092 range (as can other trace, Ca–Sr minerals such as apatite and fluorite). Therefore, additional mineralogical/geochemical evidence must be brought to bear to argue the marine case. If, however, the isotopic values fall outside of the Phanerozoic seawater values, one must consider the water–rock interaction, either as the only or partial process to explain the values.

Both approaches have validity, indeed the hybrid process may well be common, for example, where an original brine with a seawater signature has been modified by subsequent water–rock interaction, or where two water systems have mixed, one marine in origin and one the result of water–rock interaction (e.g., see Négrel *et al.*, 2001). In conclusion, the measurement of the $^{87}Sr/^{86}Sr$ ratio is a very useful indicator in brine studies, but the values must be evaluated with proper consideration for all the variables that affect water chemistry.

5.17.5.2 The Origin of Salinity: Fluids from External Sources

5.17.5.2.1 *Sources and concentration mechanisms for fluids from external sources*

Potential external sources of concentrated fluids can be found on and adjoining every crystalline rock mass on the planet. Seawater and the derivatives of seawater such as evaporite deposits and sedimentary basin brines are the primary candidates for the external sources of salinity. Dilute seawaters from the Yoldia and Litorina stages of the Baltic Sea ($<10^4$ yr) are recorded as entering crystalline rocks along coastal sections of the Fennoscandian Shield after the last glacial retreat (Nurmi *et al.*, 1988; Lahermo and Lampen, 1987; Smellie and Wikberg, 1991). Such events in coastal areas under the right conditions, such as isostatic rebound, may be very common. With time, and uplift and freshwater infiltration, some or all of these shallow emplaced seawaters may eventually be "flushed" from the rock mass.

The increased concentration of seawater by evaporation in closed or semiconfined basins has been the topic of numerous studies since Usiglio (1849) first studied the evaporation of Mediterranean seawater. Since that time, a number of well-referenced papers have examined the evolution of seawater chemistry during evaporation (Rittenhouse, 1967; Carpenter, 1978; McCaffrey *et al.*, 1987). Kharaka and Hanor (see Chapter 5.16) have reproduced Carpenter's evaporation diagram. Figure 9 is a modified version of Carpenter's (1978) plot showing where the onset of the precipitation of specific mineral phases occurs during evaporation (Matray, 1984). Seawater evaporation creates saline waters and brines that are enriched in sodium and chloride at concentrations below $\sim$350 g L^{-1}; above this concentration, with the precipitation of halite (NaCl), bittern (potash) salts usually of Ca–Mg–Cl–SO$_4$ composition dominate the fluid chemistry. With the precipitation of NaCl, the residual solution rapidly increases in bromine content relative to the chloride ion and thus changes the slope of the line (Figure 9) (Warren, 1999).

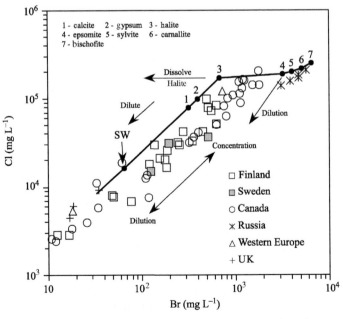

Figure 9 Concentration trends of chloride versus bromide during the evaporation of seawater showing the initial precipitation point of evaporite mineral phases and possible mixing scenarios (after Matray, 1984). Also plotted are the groundwater data found in Table 2 for selected crystalline rock areas.

However, not all highly saline fluids and evaporite deposits are derived from seawater (Eugster, 1970; Eugster and Jones, 1979; see Chapter 5.13).

Nonmarine saline lakes and evaporite deposits are found within continental land masses throughout the world (Hardie and Eugster, 1970; Eugster and Hardie, 1978). One of the most thoroughly studied of these is Great Salt Lake in Utah, where the changing composition and mineral precipitates have been studied since 1850 (Spencer *et al.*, 1985). For nonmarine continental evaporites, the source of the input waters is very important in defining the ingredients ultimately available for evaporative concentration and a number of different common saline water types ranging from $Na-CO_3-Cl$ to $Ca-Mg-Na-Cl$ are possible (Eugster and Hardie, 1978). The interesting theme here is that the local continental rock types and the dominant weathering reactions have a great influence on what is available to solution.

Both marine and nonmarine brines can also be formed at the Earth's surface level by freezing processes under cold climate conditions acting on seawater or isolated continental surface waters (Torii *et al.*, 1979; Matsubaya *et al.*, 1979; Burton, 1981). A number of studies have determined the geochemical evolution of seawater under freezing conditions (Lewis and Thompson, 1950; Nelson and Thompson, 1954; Thompson and Nelson, 1956; Herut *et al.*, 1990). The Nelson and Thompson (1954) study documents changes in ion chemistry with progressive freezing to temperatures of $-30\,°C$. Unlike evaporation, freezing does not produce Ca-carbonate or Ca-sulfate phases in the early stages of concentration (Herut *et al.*, 1990). Continued freezing of seawater causes the precipitation of mirabilite ($Na_2SO_4\cdot10H_2O$) at temperatures of approximately $-8.2\,°C$. When freezing is extended beyond the eutectic point for NaCl ($-20.8\,°C$), hydrohalite ($NaCl\cdot2H_2O$) is precipitated (Nelson and Thompson, 1954). Likewise, nonmarine brines result in mirabilite mineral precipitates (Spencer *et al.*, 1985). Probably the most extreme concentrations for surface waters under frigid conditions are the Don Juan brine pools of Antarctica (Torii and Ossaka, 1965). At temperatures regularly below $-50\,°C$ these pools precipitate the mineral antarcticite ($CaCl_2\cdot6H_2O$) and the solution chemistry would be very similar to some of the most concentrated crystalline rock or sedimentary formation waters described in this chapter. One major control on the mineral phases precipitated, and therefore the residual solution chemistry, appears to be the level to which the yearly temperature may be suppressed under a cold climate scenario such as a glaciation. After an evaporite has been formed at the surface, wind currents and deflation processes can transport salts to locations where they are available to combine with meteoric precipitation and infiltrate into the subsurface. Even the prolonged transport of sea spray can cause salinization of groundwaters in areas where evaporation exceeds precipitation (Starinsky *et al.*, 1983).

5.17.5.2.2 *Emplacement of fluids into crystalline rocks*

In order to facilitate emplacement of concentrated continental surface solutions, seawater or sedimentary formation fluids, there must be a hydraulic gradient and pathways for the solution to penetrate the rock. The pathways exist as large fracture systems that, at various times in a particular rock's history, may be open to the penetration of fluids. Numerous studies of fracture-filling minerals have identified a variety of mineral phases that formed at temperatures less than 200 °C (Bottomley, 1987; Clauer *et al.*, 1989; Tullborg, 1989; Bottomley and Veizer, 1992; Clark and Lauriol, 1992; Blyth *et al.*, 2000). In all cases the authors refer to multiple calcite-filling events in certain fracture systems indicating a repeated brittle failure and mineralogical filling history.

The mechanism for emplacement of surface concentrated brines has been suggested to be density driven when the dense concentrated solutions sink into the crystalline environment (Kelly *et al.*, 1986; Spencer, 1987; Guha and Kanwar, 1987). Such dense solutions may flow laterally from source through deep fracture systems (Möller *et al.*, 1997; Lodemann *et al.*, 1998). Fluids are thought to sink until their descent is balanced by buoyancy created by geothermal gradients, which may even return the fluid towards the surface (Spencer, 1987). Often the solutions can be sealed and trapped, undergo metamorphism and, upon cooling, are influenced by retrograde reactions that alter some chemical components. Additionally, fluid chemistry could be modified by subsurface processes such as membrane filtration (Bredehoeft *et al.*, 1963). These fluids are often released, as erosion and uplift carry the rocks to near surface positions (Kelly *et al.*, 1986; Guha and Kanwar, 1987).

Continental-scale hydrologic forces can control the flow of basinal brines and are another major potential external source of saline fluids that may enter crystalline rock environments. Studies by Bottomley *et al.* (1999) suggest that hydraulic gradients in northwestern Alberta are such that brines are forced from Devonian strata into the underlying Canadian Shield. Similarly, it appears that western Canadian sedimentary basin brines have entered the Canadian Shield in some parts of the Lac du Bonnet batholith, Manitoba (Gascoyne *et al.*, 1987).

5.17.5.2.3   *Major ions and stable isotopes*

Authors who favor the origin of crystalline brines from externally derived fluids point to the position of these fluids on typical ratio plots such as chloride versus bromide (Figure 9). The seawater evaporation line with the points of initial mineral precipitation is shown in Figure 9 (after Rittenhouse, 1967; Matray, 1984). Most of the data from Table 2 are shown in this plot to give the reader a sense of where the existing crystalline groundwater chemistry for many geographical areas falls. The highly concentrated sedimentary formation fluids and seawater concentrated beyond halite precipitation ($>340 \, \text{g L}^{-1}$) plot approximately in the same location in these figures as the most concentrated brines found at depths greater than 1 km in crystalline areas. Some authors point out the uniformity of bromine and chlorine in the most concentrated fluids from many different rock types and suggest that the control of the chemistry is more likely a uniform process like evaporation or, in some cases, freezing out (Bottomley *et al.*, 1994, 1999; Herut *et al.*, 1990). Many models have post-emplacement modification of the brine signatures by the rock for some of the more reactive ions (Kloppmann *et al.*, 2002; Négrel *et al.*, 2001).

As discussed earlier, concentrated brines from crystalline rocks have stable isotopic signatures ($\delta^{18}\text{O}$ and $\delta^2\text{H}$) that occur above the global meteoric waterline, in contrast to sedimentary formation fluids and evaporated seawater. Different scenarios account for the change in the isotopic signature post-emplacement. The brines may have been emplaced at elevated temperatures and, with progressive cooling in a very low water–rock environment, sufficient light isotopes were added to the brine from retrograde reactions of clay minerals that the brines are shifted above the meteroic waterline (Kelly *et al.*, 1986). Alternatively, sedimentary or evaporated fluids were emplaced at times in the past when seawater signatures were much different (Veizer *et al.*, 1989) and therefore a shift across the present-day mctcoric waterline would be expected (Bottomley *et al.*, 1994).

Oxygen isotopes in fracture minerals can be used to determine the origin of past fluids within a rock mass (Blyth *et al.*, 2000). $\delta^{18}\text{O}$ geothermometry on fracture minerals can be used to determine the temperature at which the minerals formed. In order to do this, an estimate of the oxygen isotopic signature of the parent fluid is required. Alternatively, having an independent measure of the temperature at which the mineral formed will allow the investigator to determine the oxygen isotopic signature of the parent fluid. The independent measure of temperature can be accomplished by measuring homogenization temperatures of fluid inclusion in the fracture mineral. The calculated oxygen isotopic signature of the parent fluid can be used to differentiate the source of the water (Taylor, 1987). At the Olkiluoto research site in Finland, Blyth *et al.* (2000) found evidence that basinal brines may have entered the crystalline rock mass during one generation of calcite fracture mineral formation. This brine was thought to be related to Precambrian Jotnian arkosic sandstones located adjacent to the site.

5.17.5.2.4   *Evidence for external sources from B, Sr, Li isotopes*

Boron isotopes can be used as an indication of the source of salinity in crystalline rocks (Bottomley *et al.*, 1984; Barth, 2000; Casanova *et al.*, 2001). In the Canadian Shield, Bottomley *et al.* (1994) sampled 19 operating mines and found that, while boron concentrations were low, relative to seawater or Alberta Basin brines, the $\delta^{11}\text{B}$ values of most of the samples were greater than those for the rock signature and fell on a mixing line between the rock signature and that of seawater (39‰ NBS SRM-951). It is not known why boron concentrations are so low in the brines, relative to seawater: there must be a mechanism, as yet undiscovered, for boron removal. The boron isotopic ratios were generally higher with increasing chlorinity, supporting the conclusions from chloride versus bromide plots (discussed earlier) that most of the chloride in brines is of marine origin.

Similarly, Barth (2000) and Casanova *et al.* (2001) have used boron isotopes, in conjunction with oxygen and deuterium isotopes and other conservative ions, to distinguish between autochthonous and allochthonous fluids in Hercynian granitoids in France, Switzerland, and Germany. $\delta^{11}\text{B}$ signatures of fresh groundwaters were found to be derived from leaching of local rock while the brackish and saline groundwaters were a complex mixture, at least partially derived from overlying evaporites or paleoseawater (Figure 10).

Strontium isotopes have also been used to identify allochthonous sources for saline waters in crystalline rock. $^{87}\text{Sr}/^{86}\text{Sr}$ ratios for deep, saline waters of the Vienne granites (France) show a value that is consistent with Jurassic seawater and not consistent with values expected from equilibration with the rock. A mixing model between Jurassic seawater and crustal end-members can explain the origin of these deep saline groundwaters (Casanova *et al.*, 2001; Négrel *et al.*, 2001).

Lithium isotopic signatures ($\delta^6\text{Li}$) have been used in conjunction with lithium, bromine, strontium, calcium, and chlorine concentrations and ratio plots (discussed earlier), to assess the origin of salinity in deep, saline waters in

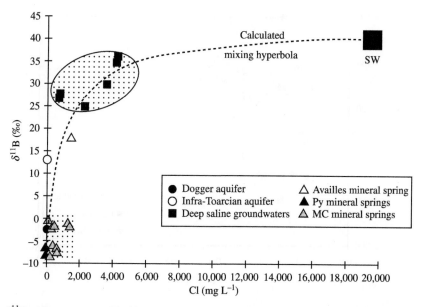

Figure 10 $\delta^{11}B$ values versus chloride content in the Vienne, France groundwaters compared to those of mineral springs from the granitic basement of the Limousin (Availles), the Pyrenees (Py), and the Massif Central (MC) of France. Note the inferred mixing trend between seawater, granitic basement, and sedimentary formations (Casanova *et al.*, 2001) (reproduced by permission of Blackwell Science from *Geofluids*, **2001**, *1*, 91–102).

shield environments. At the Miramar Con gold mine in NWT, Canada, Bottomley *et al.* (1999) found that δ^6Li values of the brines were very close to that of seawater ($-32.3 \pm 0.5\%$ NBS L-SVEC) and very unlike the local rock signatures (-5.4% and -15% in fresh and altered basalt, respectively). Slight depletion of δ^6Li, relative to seawater, may be due to fractionation in secondary clay minerals. The lithium isotopic value was independent of the lithium concentration, indicating a dominance of the seawater source.

5.17.5.3 Examples of Multiple Sources of Salinity

5.17.5.3.1 *Stable chlorine isotope data from the Canadian and Fennoscandian shields*

A plot of chlorine isotopic composition versus depth shows a few additional aspects of the data from two shield environments (Figure 11(a)). Canadian saline waters generally occur at depths below 500 m. Saline or brackish waters are not commonly found in shallow Canadian Shield systems (Frape and Fritz, 1987). Both the Canadian and the Fennoscandian data show a considerable variation in isotopic signature for any particular depth. If the data points were coded for specific sites, there would still be no discernible trends with depth. Finally, some Fennoscandian samples appear to approach the signature of the Baltic Sea, similar to that for the

localities plotted in Figure 11(b), in agreement with studies mentioned earlier that present evidence for paleo-Baltic incursions into coastal areas of Finland and Sweden (Nurmi *et al.*, 1988; Smellie *et al.*, 1995).

The concept that the rocks at a given site can also have a primary, magmatic chloride signature must be considered. Minerals such as amphiboles, micas, and apatite have been shown to contain considerable concentrations of chlorine found both in structural sites (OH^- substitution) and as liquid inclusions, and that the isotopic signature of the mineral bound chlorine can be quite heavy (up to $+4.0\%$) (Magenheim *et al.*, 1994; Eggenkamp, 1994). Work on continental versus ocean rocks by Frape suggests that mafic rocks may have a heavier chlorine signature than felsic rocks. Also, theoretical calculations on fractionation due to chlorine volatilization during volcanic activity suggest that considerable isotopic shifts of both volatiles/condensates and residual fluids may be possible (Kaufmann, 1989).

Figure 11(b) shows the $\delta^{37}Cl$ versus chlorine concentration for the Fennoscandian data plotted by site and coded for oxygen-18 isotopic signature. The most concentrated samples, which are also isotopically enriched in $\delta^{37}Cl$, are found along the line labeled 1 and are generally from ultramafic (serpentinites) and mafic rocks such as the deep parts of Jukka/Miikhali 116 borehole described previously. Sites closer to present-day seawater and less enriched (e.g., Oku) generally have more felsic rocks. Generally, the initial studies by several of the authors indicate that

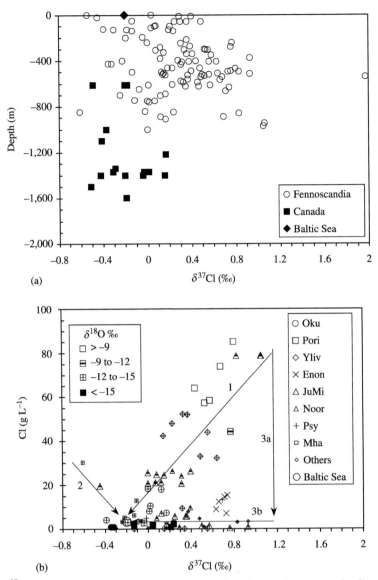

Figure 11 The δ^{37}Cl isotopic signature: (a) versus depth for selected groundwater samples from the Canadian and Fennoscandian and (b) groundwater samples coded for δ^{18}O signature for a number of research sites in Finland (Frape *et al.*, 1996) (reproduced by permission of International Atomic Energy Agency from *Symposium on Isotopes in Water Resources Management*, **1996**, pp. 19–30).

fluids found in mafic rocks may have more enriched stable chlorine isotope signatures than fluids from felsic rocks. The most concentrated, deeper samples at each site do appear to plot in distinct and separate fields on this diagram. As shown in Figure 6, at some sites the host rocks also contain chlorine with enriched isotopic signatures.

In Figure 11(b) many very dilute, shallow samples have depleted oxygen isotopic signatures, possibly representing cold climate or glacially recharged fluids. More complicated mixing scenarios would occur when concentrated brines, such as those found at the Jukka/Miikhali 116 site, are mixed through infiltration of surface water with very dilute, glacial melt waters. The glacial

waters, having little or no chlorine, would not change the δ^{37}Cl signature significantly and the mixed waters would follow a line similar to that labeled 3a in Figure 11(b). With continued ice melting (and before isostatic rebound), the paleo-Baltic Sea covered the site and mixing and/or diffusion of the two δ^{37}Cl end-members could cause a trend similar to line 3b in Figure 11(b).

However, it should be noted that the oxygen isotopic signature of the present Baltic Sea water is approximately −8.5‰. This is very different from the samples with "Baltic like" δ^{37}Cl shown in the figures. One explanation is that the paleo-Baltic δ^{18}O signature would most likely have been much more negative during periods immediately following glacial melting. The source of Baltic salt and thus the $\delta^{37}Cl$ signature is most likely related to the extensive Paleozoic salt deposits in northern Europe and therefore would vary only by the degree of halite ($\sim 0.15‰$) versus Ca–Mg salts ($\sim 0.45‰$) (Eggenkamp, 1994) dissolved during or after a glacial event. Similar glacial melt-water intrusion and mixing scenarios have been described in detail for at least one site in northern Canada (Clark *et al.*, 2000). These results would suggest that there is a potential to alter the chemical and/or isotopic geochemical signature of groundwaters at any particular crystalline site during periods of intense crustal activity such as postglacial uplift.

5.17.5.3.2 *Interpreting Ca/Na versus Br/Cl trends for data from crystalline environments*

Figure 12(a) is a weight ratio plot of the most common cations and the most conservative anions found in saline fluids and brines from crystalline environments. From earlier figures it is apparent that a wide range of values exist for these ions. At first glance, the plot suggests that most fluids from a variety of environments and depths appear to have a limited range of values. However, the extreme values need to be discussed. The Russian samples are brines from crystalline rocks of the Siberian Platform. They have the most enriched bromide values of any brines from crystalline environments. The area has undergone numerous permafrost events and evaporite intrusion and reworking is suspected as the means of concentrating the bromine and probably the calcium in these fluids. The Finnish and Canadian samples with extreme Br/Cl ratios are shallow freshwaters with salinities less than 100 mg L^{-1}. These waters have local modern meteoric isotopic signatures (^{18}O–^2H) and modern-day levels of tritium. It is speculated that both bromine and chlorine are atmospheric in origin and may be of sea-salt or anthropogenic origin. The very enriched Ca/Na samples are from the Lac du Bonnet granite in Manitoba and are believed to be intruded western Canadian sedimentary basin brines (Gascoyne *et al.*, 1987).

Figures 12(b) and (c) expand and divide the data. Although many sites have overlapping data, it is apparent that a large variation of geochemical signature occur in these environments. In many cases (e.g., Ylivieska in Finland; Laxemar in Sweden; or Matagami/Norita in Canada), the samples with the highest Ca/Na or Br/Cl ratios are the deepest and most concentrated at the site. When the other less concentrated or shallower samples are considered, it suggests mixing trends between shallower dilute fluids and the deeper resident brine. Many of the samples in Finland and Sweden group near the Baltic Sea signature, (Figure 12(b)) indicating a probable Baltic seawater component or mixtures of glacial melt-water and seawater. The grouping of Canadian and European dilute NaCl saline waters around the seawater signature (Figure 12(c)) is believed by some authors to indicate a component of paleo-seawater or evaporitic fluid in these sites. The plots do show the wide and diverse nature of the fluids found in these environments, and that almost any combination of ions or isotopes plotted against each other will show the individuality of the geochemical signature for the most concentrated fluids at each site. Usually, these plots also indicate that potential mixing with other end-member fluids has occurred, although it is often much more difficult to determine when such mixing may have happened.

5.17.6 EXAMPLES FROM RESEARCH SITES FOUND IN CRYSTALLINE ENVIRONMENTS

A variety of sites in crystalline rock environments have been studied in detail to gather additional information concerning the evolution of hydrogeological and geochemical properties of the Earth's crust. Five of these studies are briefly described in this chapter. The URL, Canada and the Äspö "Hard-Rock Laboratory," Sweden are research facilities associated with scientific efforts of these countries to understand the geological and hydrogeological environment in crystalline rocks surrounding a potential deep radioactive waste storage facility. The Palmottu site in Finland is a uranium deposit that has been studied by a European/Canadian consortium of scientists as a natural analogue of a radioactive waste repository in granitic rocks. The KTB German deep drill holes and the Kola superdeep well in Russia are two of the deepest drilling sites on the planet. It is hoped that the following brief descriptions will help the reader understand the complex and varied nature of fluids in the crystalline rocks of these sites.

5.17.6.1 Groundwater Composition and Saline Fluids in the Lac du Bonnet Granite Batholith, Manitoba, Canada

The Lac du Bonnet batholith is a 2.6-Gyr-old granite on the western edge of the Canadian Shield, ~ 100 km east of Winnipeg, Manitoba. It has been studied in detail because it is the location of Canada's URL and is a research facility of the Canadian program for disposal of high-level nuclear waste.

Over 100 boreholes have been drilled up to depths of 1 km into the granite to study variations

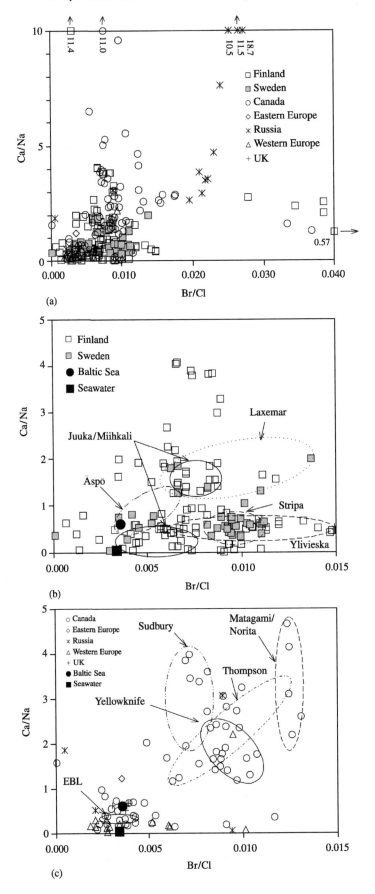

(a)

(b)

(c)

in the geology, rock mechanics, hydrogeology, and geochemistry of the rock. Groundwaters in fractures, faults, and the rock matrix have been sampled using a variety of techniques (Gascoyne, 2002; Ross and Gascoyne, 1995) and have shown distinct zonation with depth.

Shallow groundwaters are generally dilute (TDS < 0.3 g L^{-1}) Ca–Na–H–CO$_3$ waters and occasionally show indications of mixing with Ca–Mg–HCO$_3$–SO$_4$ groundwater from clay-rich overburden sediments that frequently cover the granite. Deeper groundwaters are either dilute Na–Ca–HCO$_3$ waters or contain significant concentrations of Cl and SO$_4$. With greater depth, Na–Ca–Cl–SO$_4$ waters of increasing salinity are found. The most saline fracture groundwaters in the area have been encountered at ~1,000 m depth, and have a salinity of up to 51 g L^{-1}.

The large range in salinity (two orders of magnitude) between groundwaters near the surface and at a depth of 1,000 m is mainly due to increasing concentrations of sodium, calcium, and chlorine with depth, and the differences between each fracture zone are an indication of the limited interconnectivity of these zones (Figure 13). In some locations, shallow groundwaters have a strong chloride signature and indicate discharge of the deeper saline groundwaters at the surface. These areas have been found to have localized discharge of helium gas, a further, more sensitive, indicator of groundwater discharge.

In contrast, ions such as bicarbonate show a gradual decrease, from a surface value of ~250 mg L^{-1} to $<$10 mg L^{-1}, as salinity increases. Sulfate increases to a maximum level of ~1,000 mg L^{-1} and tends to decrease in the more saline fluids. These characteristics indicate solubility control by a sparingly soluble mineral phase (calcite and gypsum, respectively).

The pH of the groundwaters also varies with depth. For depths of 0–200 m, a large range of pH (6.5–9.3) has been determined, the lower values reflecting the influence of acid recharge waters that are rich in dissolved organics and CO$_2$ from the soil zone. Below 200 m, most groundwaters lie in the pH range 7.5–8.8. The redox potentials of the groundwaters (measured as Eh using an electrode sensor) show a general trend of decreasing Eh with depth (from +500 mV at the surface to < -100 mV at depth). Redox is controlled by two main processes: oxidation of

dissolved organics in the shallower zones and the Fe(II)/Fe(III) redox couple in minerals such as pyrite (Fe-sulfide), biotite (Fe-silicate), and magnetite (Fe-oxide), at depth.

Most shallow and dilute groundwaters (i.e., low chlorine concentration) in recharge areas contain ^{3}H to the level of at least 10 TU. Modern ^{3}H content of precipitation is ~15 TU. This indicates that these groundwaters have had a relatively short residence time in the flow system and most were recharged within the last ~50 years or, at least, contain some portion of water that recharged in the last 50 years. Several groundwaters, in the depth range 50–250 m, contain significantly more than 20 TU and may indicate the presence of groundwaters that recharged during atmospheric atom bomb testing in the period 1953–1963.

The relationship between δ^2H and δ^{18}O for local precipitation, recorded at Gimli, Manitoba, for the period 1976–1979 (δ^2H = $8\delta^{18}$O + 7.47) is close to that of global meteoric precipitation. All groundwaters lie close to this line, indicating that they are meteoric in origin. Dilute groundwaters have the isotopic composition of modern precipitation (δ^{18}O = -13‰ to -14‰). This range can be taken to represent the composition of groundwaters recharged during warm-climate periods such as the present. In contrast, brackish groundwaters (in the Cl range of 200 mg L^{-1} to ~8,000 mg L^{-1}) are up to 7‰ lower in δ^{18}O values, suggesting that they were recharged to the bedrock during a colder climate, either glacial or postglacial possibly 10^4 yr ago. This groundwater type is found in the major permeable fault zones in the area, at a depth range of 200–600 m.

At greater salinities the groundwaters have δ^2H and δ^{18}O values comparable to those of modern groundwaters. It is inferred, therefore, that these saline waters were recharged during a warm-climate event, similar to the present and, because they underlie the cold-climate brackish waters, they are likely older.

Groundwaters sampled from sub-vertical fractures near the surface and in fault zones as deep as ~500 m have δ^{13}C values, which range between -21‰ to -11‰ and ^{14}C$_{DIC}$ values between 6 pm C and 80 pm C. Except for a decrease of ^{14}C content with increasing depth, few trends in these data can be seen. Recharging Na–Ca–HCO$_3$ waters have δ^{13}C values of between -18‰ and -11 ‰, and Na–HCO$_3$ and more saline Na–Ca–Cl groundwaters in deeper fracture

Figure 12 The ratio plot Ca/Na versus Br/Cl in fluids found in crystalline environments from a number of geographical areas. (a) All data. Note that the extreme Br/Cl ratios beyond 0.015 belong to shallow groundwaters with TDS $<$100 mg L^{-1} and Siberian deep brines with TDS $>$ 100 g L^{-1}. (b) Sites from Sweden and Finland separated and in some cases named. Note that many sites have specific locations on this plot, although some overlap does occur. (c) Other locations such as Canada and Europe plotted at the same scale. Note the separation of the more concentrated fluids at each site.

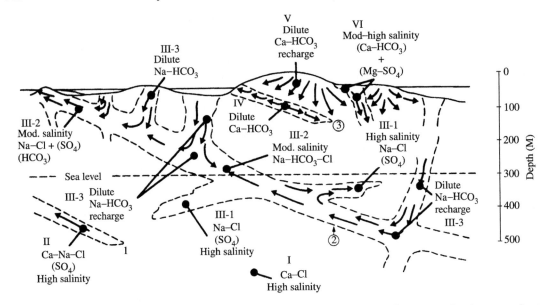

Figure 13 Schematic representation of groundwater compositions at the URL lease area showing groundwater flow paths and geochemical patterns in fracture zones (Gascoyne *et al.*, 1987) (reproduced by permission of Geological Association of Canada from *Saline Water and Gases in Crystalline Rocks*, **1987**, pp. 53–68).

zones have a range of $-21‰$ to $-14‰$, with the large majority of them between approximately $-17‰$ and $-13‰$.

In contrast, ^{14}C shows a more pronounced variation throughout the flow regime: ^{14}C levels in recharging Na–Ca–HCO$_3$ waters vary from 80 pm C in the near surface to $\sim$10 pm C at greater depths. Na–Ca–HCO$_3$ and Na–Ca–Cl groundwaters discharging along the fracture zones have ^{14}C levels between 5 pm C and 54 pm C. These data indicate that ^{14}C is a useful parameter for distinguishing between the shallow, rapidly circulating groundwater and deeper water in fault zones where circulation is slower and more restricted, similar to the findings reported by Tullborg and Gustafsson (1999). Unfortunately, because of the problems of contamination by modern ^{14}C sources (e.g., the atmosphere, residual high-^{14}C drillwater) and because of the influence of water–rock interactions, ^{14}C is of little use in determining the residence times of the deeper, saline groundwaters.

Recently, it has been possible to collect pore fluids from the unfractured granite rock matrix using upward-sloping boreholes at the 420 m level of the URL. These waters are highly saline ($\sim$90 g L^{-1}) CaCl$_2$ fluids that have a unique isotopic signature. Both δ^2H and $\delta^{18}O$ values are about $-15‰$, a composition that places them well to the left of the meteoric water line and also above and to the left of the most saline Canadian Shield mine waters.

The high ^2H content of the pore fluids suggests that they could be derived from undiluted basinal brines whose δ^2H is close to $0‰$. The exchange of ^{18}O with the rock matrix

will shift the fluids to the left, from a composition typical of sedimentary brines, to their current position beyond the meteoric water-line. The pore fluids from the sparsely fractured gray granite are clearly unique in the Lac du Bonnet batholith and differ in a number of aspects from the saline groundwaters found in fractures. Their chemical and isotopic character consistently indicates a prolonged residence time for these fluids in the rock matrix.

To account for the origin of the fracture-hosted groundwaters, some geochemical evidence (high Na/Ca ratios, low Br/Cl ratios and marine-type $\delta^{34}S$ values of the dissolved SO$_4$) indicates that basinal Na–Cl brines may have directly entered the granite through the fracture zones and slowly evolved towards calcium-dominated groundwaters over long periods of geologic time. It is possible, therefore, that the pore fluids are actually basinal brines, derived from these adjacent fracture fluids, that have evolved geochemically and isotopically to their current composition, over periods as long as 10^9 yr. These pore fluids have the same $^{87}Sr/^{86}Sr$ as the whole-rock values which is very rare in water–rock systems. This implies isotopic and possibly chemical equilibrium between fluid and rock (McLaughlin, 1997). To attain such equilibrium does suggest a long residence time for the fluids.

5.17.6.2 Groundwater Composition at the Palmottu Site in Finland

The Palmottu U–Th mineralization is hosted by Proterozoic (1.8 Ga) crystalline bedrock, mainly mica gneiss with granite and granite pegmatite veins (Räisänen, 1986). The mineralized part forms a vertical structure that extends to depths of at least 400 m (Figure 14). Accordingly, it covers both near-surface oxidative conditions, and low-Eh reduced conditions. Ancient hydrothermal alteration of the uraninite deposit led to substantial remobilization of uranium, resulting in the formation of U(IV) silicate (coffinite). Over the last one million years, the site has experienced multiple glaciation events during which a proportion of the uranium has been mobilized under oxidizing conditions, leading to precipitation of U(VI) silicates (uranophane) at shallow depths. Volumetrically significant amounts of

uranium, mainly as U(IV), are also found in association with fracture coatings such as impure calcites that contain clays and iron oxy-hydroxides (Ruskeeniemi *et al.*, 2002).

During the last million years the area has been subjected to several glaciations, the last of which (Weichselian Ice Age) ended 10^4 yr ago (Donner, 1995). It is estimated that the maximum thickness of the continental ice cover in the central part of the Fennoscandian Shield was ~3 km some 2×10^4 yr ago. The weight of the ice depressed the underlying crust by up to 800 m at the center of the glaciated area (Niskanen, 1943; Mörner, 1979). At Palmottu the depression was less, ~300–400 m, but here the isostatic rebound

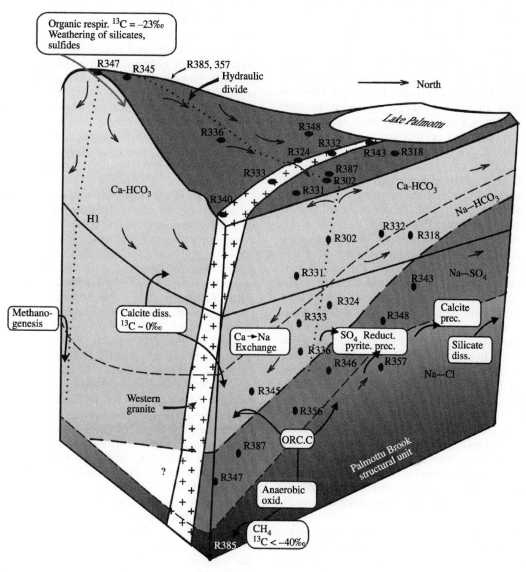

Figure 14 The conceptual hydrogeological model of the Palmottu, Finland research site. The arrows indicate the measured flow directions, the distribution of groundwater types is shown, and some of the measured and inferred geochemical processes are also indicated (Blomqvist *et al.*, 2000) (reproduced by permission of European Commission from *The Palmottu Natural Analogue Project, Phase II: Transport of Radionuclides in a Natural Flow System at Palmottu*, **2000**).

after the deglaciation is still going on at a rate of 4–5 mm yr^{-1} (Ekman, 1996). The retreat of the continental ice cover temporarily stopped for some hundreds of years, caused by slightly cooler climatic conditions. These temporal stops can now be observed as huge accumulations of unconsolidated sediments that were deposited in front of the ice cover, composed of partly washed till, gravel and sand.

Due to the long and complex geological history, the bedrock is cut by numerous generations of fractures. However, most of them are sealed by precipitation of secondary minerals, mainly calcite and various clay minerals. Open, water-conducting fractures are found mainly in the upper part of the bedrock down to 150 m (Figure 14). Subhorizontal fractures, dipping towards the SW or NE, together with variously oriented steeply dipping fractures, form an inter-connected network that allows for a dynamic flow system in the upper part of the bedrock. This is a common feature of the Fennoscandian Shield where stress-release following deglaciation is thought to be a contributory factor to their re-activation. Deeper in the bedrock, open fracturing is rare and, consequently, the hydraulic conductivity is generally very low (Blomqvist et al., 1998).

A site-specific groundwater flow model was constructed based on an integration of structural, hydrogeological, and hydrogeochemical data (Blomqvist et al., 1998, 2000; Pitkänen et al., 1999). The upper 100 m of the bedrock is a zone characterized by dilute groundwater of HCO_3^- type (the upper flow system; Figure 14) with calcium dominant near the surface giving way to sodium in the lower part of the zone. High tritium contents indicate a dynamic flow system, which penetrates to a depth of at least 200 m in the southern part of the site (the dynamic deep flow system). Below the HCO_3 type zones the ground-water is brackish, of $Na-SO_4$ or $Na-Cl$ type, and both the hydrochemistry and isotopic results suggest long residence times, up to 10^4 yr, and stagnant flow conditions (the stagnant flow system). High uranium concentrations (generally 100–500 ppb) are associated with the oxidative HCO_3 groundwaters in the vicinity of the uranium mineralization down to depths of 130 m. At greater depths, clearly reducing conditions (−300 mV) prevail, with low uranium concentrations (below 10 ppb).

A sizeable $\delta^{18}O$ depleted water component is present in the deep $Na-Cl$ and $Na-SO_4$ type groundwaters (additionally in some of the deep $Na-HCO_3$ types). The presence of this water suggests that the system has been open to the incursion of glacial waters to at least 400 m. Irrespective to the origin of these groundwaters, the fact that they are still present at these depths

underlines the stability of the hydrogeochemical system over long periods of geological time under low permeability conditions, at least since the last glaciation $\sim 10^4$ yr ago. At greater depths, beyond the present sampling interval, conditions would be expected to be even more stable (Blomqvist et al., 2000).

5.17.6.3 Groundwater Composition and Hydrogeochemical Aspects of the Äspö Site, Sweden

The Äspö site is located on the southeastern coast of Sweden, ~400 km south of Stockholm. The Äspö Hard Rock Laboratory (HRL) project was initiated in 1986 by the Swedish Nuclear Fuel and Waste Management Company (SKB), with early research consisting of site characterization from surface boreholes. Construction of an access tunnel began in 1990, to obtain detailed geological, hydrogeological, and chemical data as well as providing information for the safety and performance assessment of a nuclear waste repository (Stanfors et al., 1999). Figure 15 is a conceptual composite of the hydrogeological and hydrogeochemical conditions that most likely prevailed at the site before the research activity and opening of the rock mass.

The rocks of the Äspö site are predominantly granitoids of the Trans-Scandinavian Igneous Belt. They were emplaced at a depth of ~15 km during several pulses of magmatism between 1.85 Gyr and 1.65 Gyr ago, but evidence of several post-emplacement uplift and subsidence events has been shown by fission track analysis of apatite (Tullborg et al., 1996). Marine conditions prevailed at the site from the Cambrian to Early Silurian, which gave way to continental conditions by the end of the Silurian. Thick sedimentary cover prevailed from the Late Paleozoic to Early Mesozoic. In the Cretaceous period, a marine transgression prevailed in southern Sweden with sea level up to 300 m higher than present (Hallam, 1984). Late Tertiary and Quaternary glaciation and interglacial stages caused changes in the hydraulic and chemical conditions at the site, which are manifested by brittle fracturing and fracture infilling (Wallin and Peterman, 1999).

The hydrogeology of the Äspö site is controlled by flow in fractures (Figure 15). The hydraulic conductivity of the rock mass is $10^{-8}-10^{-10}$ m s^{-1}, while the hydraulic conductivity of the fractures is $10^{-4}-10^{-6}$ m s^{-1} (Rhén et al., 1997). The irregular water table is locally controlled by sparsely distributed, poorly inter-connected fractures. At depth, the hydraulic pressure head distribution becomes more regular due to control by fracture zones and larger

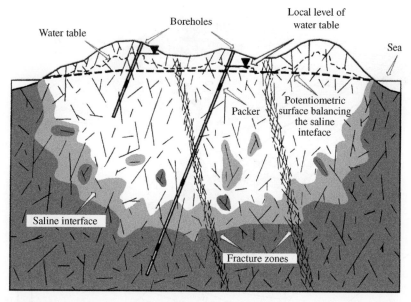

Figure 15 A conceptual diagram of the hydraulic and fracture regime in existence around the Äspö, Sweden research site (Stanfors *et al.*, 1999) (reproduced by permission of Elsevier from *Appl. Geochem.*, **1999**, *14*, 819–834).

fractures that are more conductive. At depths below ~600 m, the flow rates are very small and a stagnant, saline water is present (Stanfors *et al.*, 1999; Louvat *et al.*, 1999). These saline waters are shown in Figure 15 and Table 2.

Geochemical sampling and modeling investigations (Smellie *et al.*, 1995; Laaksoharju *et al.*, 1999) found that the undisturbed groundwater conditions at the Äspö site consisted of: (i) a predominantly fresh, meteoric water in the upper 250 m zone; (ii) a brackish to saline water consisting of both present and ancient Baltic Sea water and glacial melt water at depths of 250–600 m; and (iii) a stagnant, saline water containing brines and possibly ancient glacial water below a depth of 600 m. Fluid inclusion studies by Blyth (2001) and Gehör and Lindblom (2002) showed that grain boundary inclusions and fluid inclusions had Na–Cl and Ca–Cl type fluids up to 21 wt.% equivalent NaCl.

Laaksoharju *et al.* (1999) also noted that construction of the access ramp caused large-scale disruption of the hydrogeological and geochemical conditions at the site. Groundwater inflow to the tunnel caused drawdown to the water table and mixing of groundwater chemistry types. Chlorine and tritium isotopes were used by Blyth and Frape (2002) to show that boreholes initially had an enriched (positive) $\delta^{37}Cl$ value similar to the rock mass and no tritium. After tunnel construction and several underground experiments, the chlorine signature became progressively lighter and tritium values indicated modern water. It was concluded that sequential mixing between brines, shallow groundwaters, and Baltic

Sea water had occurred during and since tunnel construction.

5.17.6.4 The KTB (Continental Deep Drilling Project of Germany)

The KTB deep drilling project resulted in two of the deepest research boreholes drilled in the Earth's crust (Lodemann *et al.*, 1998; Möller *et al.*, 1997). The boreholes were 4,000 m and 9,101 m deep and are located on the western margin of the Bohemian Massif (Weber, 1992). The area contains a variety of rock types that have been tectonically emplaced and thermally altered. Several granitic instrusions are in close proximity to the site and large fracture-thrust belts trend through the area (Figure 16). The site contains both altered sedimentary and crystalline rocks.

During drilling, highly saline fluids were encountered in several zones below 2 km to depths of 9 km. The drilling mud used in the project limited the usefulness of some of the geochemical data, but more reliable data were obtained by Lodemann *et al.* (1998) from a long-term pumping test in the 4,000 m deep pilot hole where a Ca–Na–Cl saline fluid (68 g L^{-1}) and N_2 and CH_4 gases were recovered. The fluids and fracture mineral history from the site indicate a permeable system that does not seem to be in contact with surface groundwater systems. U–Th results and ^{36}Cl content suggested that some of the isotopic and geochemical signatures could not be derived from the KTB rocks, and it was suggested that the nearby Falkenberg granite (2 km, NE)

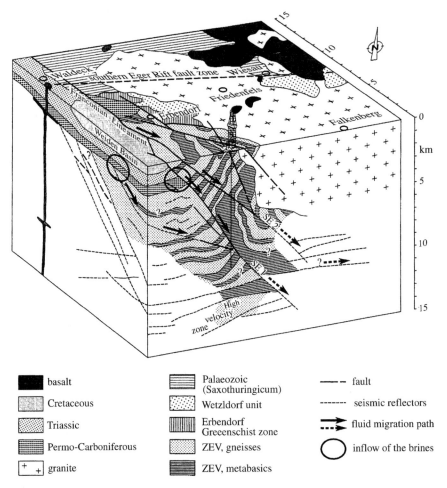

basalt		Palaeozoic (Saxothuringicum)	—— — fault
Cretaceous		Wetzldorf unit	------- seismic reflectors
Triassic		Erbendorf Greeenschist zone	▬▬► fluid migration path
Permo-Carboniferous		ZEV, gneisses	◯ inflow of the brines
granite		ZEV, metabasics	

Figure 16 Block diagram showing the geological structure of the KTB site and the importance of the Franconian linement for fluid migration (Möller *et al.*, 1997) (reproduced by permission of American Geophysical Union from *Am. Geophys. Res.*, **1997**, *102*, 18233–18254).

may be compatible as the source of some of the fluids at this site (Lodemann *et al.*, 1998). Exchange and hydrolysis reactions, particularly with plagioclase feldspars, are invoked to control the chemical evolution to a Ca–Na–Cl saline fluid. Lodemann *et al.* (1998) also show that the strontium isotopic signature of these saline fluids is similar to the range of hydrothermally altered plagioclase feldspars and the alteration mineral epidote from the deepest part of the 4,000 m borehole. The plagioclase–epidote isotopic control is similar to that found by Franklyn *et al.* (1991) from a Canadian Shield site (Figure 8). Interestingly, the $\delta^{18}O$ and $\delta^{2}H$ values of the saline fluids do not shift above the meteoric waterline as found elsewhere. Lodemann *et al.* (1998) believe that the elevated temperatures in the fluids may have caused additional re-equilibration with heavier ^{18}O in mineral phases and altered the original brine signatures. Finally, if the fluids have moved out of the granitic rocks to mix and geochemically evolve with altered basement and sedimentary brines over millions of years.

The scenario adds an interesting aspect to proposed models of the origin and evolution of fluids in deep crystalline environments.

5.17.6.5 The Hydrogeology and Geochemistry of the Kola Superdeep Well

Most of the information presented in this chapter is from sampling sites located at depths of less than 2 km. The Kola Superdeep Well was drilled into the northeastern portion of the Fennoscandian (Baltic) Shield of Russia and is over 13 km deep. The hole was drilled to examine the geology of the Precambrian continental crust in this region (Kozlovsky, 1984). As part of the study, the research team analyzed the fluids, gases, and organic matter extracted from the circulating drill fluids (Borevsky *et al.*, 1984; Karus *et al.*, 1984). As the methodology involved extraction of subsurface fluids and gases from the drilling mud, used to control the pressure encountered during drilling, there is a certain semiquantitative nature to the geochemical determinations.

The generalized hydrogeological section of the SG-3 Kola well and many other drill holes in the area are shown in Figure 17. The well is drilled into metamorphosed sedimentary and volcanic rocks to a depth of 6,835 m. Below this depth the rocks are strongly metamorphosed Archean basement rocks. Porosity is never more than 3% in open fractured zones and less than 1% for most of the geological section. Dilute groundwaters of meteoric origin occur in a highly fractured zone that normally extends to ~800 m depth. Deeper in the well, at locations shown in Figure 17, highly fractured sections of rock occur that contain very chemically concentrated fluids and gases. These permeable zones gave hydrodynamic (hydraulic) testing values of 10^{-7} m d^{-1} and were shown to be very conductive; this is the first time

high-permeability zones at great depths in the crust were documented. Below 9 km the number and thickness of fracture zones decreased, although testing indicated that permeability of some fracture zones was still significant enough to produce concentrated fluids and volatile gases in the return drill mud.

The fluid chemistry in the upper parts of the borehole (<1 km) was Ca–SO$_4$ dominant with TDS usually less that 2 g L^{-1}. From 1 km to 3.5 km the fluids are usually less than 5 g L^{-1} TDS and Na–Ca–Cl–HCO$_3$ in composition. These fluids are slightly reducing ($E_H = -20$ mV) and alkaline (pH = 9.4).

In the deeper zones very concentrated Ca–Na–Cl brines (200–300 g L^{-1}) are common. Locally, magnesium and minor elements such as bromine

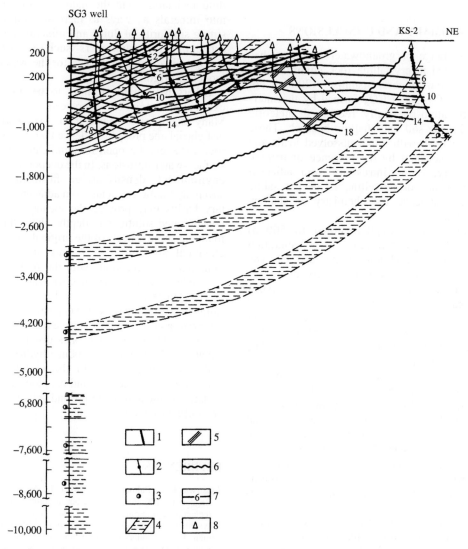

Figure 17 Generalized hydrogeological section of the SG-3 well testing ground. 1—Zones of negative temperature anomalies; 2—zones of intensive fissuring; 3—highly mineralized waters; 4—reservoir rocks of fracture-vein and fracture types; 5—zone of intensive micromovements; 6—boundary between III and IV volcanogenic–sedimentary sheets; 7—temperature, °C; 8—wells (Borevsky *et al.*, 1984) (reproduced from *The Superdeep Well of the Kola Peninsula*, **1984**, pp. 271–287).

and strontium can achieve high concentrations similar to brines elsewhere (Table 2). In the upper parts of the drill hole, gases are usually atmospheric, changing to methane, nitrogen, and hydrogen between 1 km and 4.5 km depth. Below 4.5 km the gases are reported to be dominantly hydrogen, helium, and some carbon dioxide (Borevsky *et al.*, 1984). These researchers describe a very dynamic system to great depths and contend that dehydration and metamorphic reactions would act to continually concentrate the reworked, residual solutions. The limited stable isotope data reported for oxygen-18 and deuterium analyses of the deep fluids are to the right of the global meteoric waterline, possibly due to isotopic exchange with mineral phases under temperature conditions above 100 °C.

5.17.7 SUMMARY AND CONCLUSIONS

A wide range of groundwater chemistry has been recorded in crystalline rock environments. Shallow groundwaters (usually <200 m) are dominantly $Ca-Na-HCO_3$ formed by the interaction of atmospherically recharged meteoric water with the soil and shallow bedrock. These waters are fresh with dilute dissolved loads and young, as indicated by the presence of tritium. Occasionally, saline intrusions from adjacent seawater bodies or upwelling of deeper saline fluids can influence the chemistry of shallow groundwaters.

At intermediate depths (down to 500 m) groundwaters rapidly increase in concentration primarily by the addition of SO_4 and Cl. The concentration of bicarbonate ions decreases because of the precipitation of mineral phases such as calcite. Local variations in chemistry and anions may be due to a variety of rock–water interactions or local processes that result in $Na-SO_4$, $Na-HCO_3$, and $Mg-SO_4$ type waters. The pH begins to rise in this zone and oxygen-consuming reactions and redox mineral controls tend to lower the Eh. The brackish and saline waters found at these intermediate depths have longer residence times. Deep saline waters and brines occur in most locations below depths of 500 m. These fluids are $Ca-Na-Cl$ or $Na-Ca-Cl$ in composition and can have total dissolved loads up to 350 g L^{-1}. Minor elements such as bromide and strontium can here be thousands of milligrams per liter.

Shallow freshwaters and brackish waters have stable isotopic signatures that are interpreted as being meteoric in origin. However, the concentrated brines have $\delta^{18}O-\delta^2H$ signatures that are significantly above the global meteoric waterline, unlike fluids from sedimentary environments. This unique feature of crystalline fluids appears to be a result of mineral–water hydration reactions, although retrograde equilibrium or the changing isotopic composition of the ocean with time are alternative explanations.

Gases associated with the fluids can be atmospheric near surface but change rapidly with depth. Brackish waters often have a component of bacterial hydrogen sulfide from sulfate reduction and bacterial methane. Deep saline waters and brines are dominated by abiogenic methane, other hydrocarbon gases, nitrogen, helium, and locally hydrogen. Hydrogen may be a by-product of mineral–water reactions and appears to become the dominant gas in systems below 3 km.

The sources of the saline fluids are argued by some to be water–rock interaction based on the presence of abundant chloride in mineral phases, fluid inclusions, and the incorporation of water into minerals as ways to increase salinity. The strong similarity of fluids and mineral chemistry, especially for parameters such as strontium isotopes, is cited as evidence for the water–rock model of origin.

Others believe the salinity to be dominantly external in origin and emplaced into the rock. Many examples of seawater incursions after the last glacial age provide evidence that, at times, the rock mass can be open to significant depths for surface-water intrusions. In the deeper more saline environment, density-driven flow/diffusion or strong hydraulic gradients from adjacent sedimentary basins may provide the means to emplace concentrated solutions to depth. These fluids can maintain some aspects of the original isotopic and chemical signatures over long time periods, although some reactions with the rock mass are postulated to modify certain chemical parameters.

The debate on the origin and evolution of the fluids in crystalline rocks is very much an ongoing research area. A number of experimental and research sites such as the ones discussed in this chapter will continue to produce hydrogeochemical information well into the middle of this century as researchers around the globe attempt to understand the hydrogeology and geochemistry of crystalline rock environments.

REFERENCES

Aggarwal J. K., Palmer M. R., Bullen T., Arnórsson S., and Ragnarsdóttir K. V. (2000) The boron isotope systematics of Icelandic geothermal waters: 1. Meteoric water charged systems. *Geochim. Cosmochim. Acta* **64**, 579–585.

Almèn K.-E., Andersson O., Fridh B., Johansson B.-E., Sehlsted M., Hansson K., Olsson O., Nilsson G., and Wikberg P. (1986) *Site Investigation—Equipment for Geological, Geophysical, Hydrogeological, and Hydrochemical Characterization*. Technical Report 86-16, Swedish Nuclear Fuel and Waste Management Company (SKB), Stockholm, Sweden.

Axelsen K., Wikberg P., Andersson L., Nederfeldt K.-G., Lund J., SjöströmT., and Andersson O. (1986) *Equipment for Deep Ground Water Characterization: Calibration and Test Run in Fjällveden*. Status Report AR 84-27, Swedish Nuclear Fuel and Waste Management Company (SKB), Stockholm, Sweden.

Balderer W., Fontes J.-Ch., Michelot J.-L., and Elmore D. (1987) Isotopic investigations of the water–rock system in the deep crystalline rock of northern Switzerland. In *Saline Water and Gases in Crystalline Rocks*. Special Paper 33 (eds. P. Fritz and S. K. Frape). Geological Association of Canada, Memorial University, Newfoundland, pp. 175–195.

Banks D. A., Green R., Cliff R. A., and Yardley B. W. D. (2000) Chlorine isotopes in fluid inclusions: determination of the origins of salinity in magmatic fluids. *Geochim. Cosmochim. Acta* 64, 1785–1789.

Barnes I. and O'Neil J. R. (1971) The relationship between fluids in some fresh alpine-type ultramafics and possible modern sepentinization, western United States. *Geol. Soc. Am. Bull.* 80, 1947–1960.

Barth S. R. (2000) Geochemical and boron, oxygen, and hydrogen isotopic constraints on the origin of salinity in groundwaters from the crystalline basement of the Alpine Foreland. *Appl. Geochem.* 12, 937–952.

Black J. H. and Chapman N. A. (1981) In search of nuclear burial grounds. *New Sci.* 91, 402–404.

Blomqvist R. (1990) Deep groundwaters in the crystalline basement of Finland, with implications for nuclear waste disposal studies. *Geologiska Föreningens i Stockholm Förhandlingar* 112, 369–374.

Blomqvist R. (1999) *Hydrogeochemistry of Deep Groundwaters in the Central Part of the Fennoscandian Shield*. Report YST-101, -41. Geological Survey of Finland. Nuclear Waste Disposal Research.

Blomqvist R., Vuorela P., Nissinen P., Ruskeeniemi T., Frape S. K., and Ivanovich M. (1993) Crustal rebound-related groundwater flow and calcite formation in the crystalline bedrock of the Fennoscandian shield: new observations from Finland. *Extended Abstract for the OECD Meeting*, Workshop on paleohydrogeological methods and their applications for radioactive waste disposal, 5p.

Blomqvist R., Kaija J., Lampinen P., Paananen M., Ruskeeniemi T., Korkealaakso J., Pitkänen P., Ludvigson J.-E., Smellie J., Koskinen L., Floría E., Turrero M. J., Galarza G., Jakobsson K., Laaksoharju M., Casanova J., Grundfelt B., and Hernan P. (1998) *The Palmottu Natural Analogue Project. Phase I: Hydrogeological Evaluation of the Site*. Final report. EUR 18202 EN, 98p. European Commission. Nuclear Science and Technology.

Blomqvist R., Ruskeeniemi T., Kaija J., Ahonen L., Paananen M., Smellie J., Grundfelt B., Pedersen K., Bruno J., Perez del Villar L., Cera E., Rasilainen K., Pitkanen P., Suksi J., Casanova J., Read D., and Frape S. K. (2000) *The Palmottu Natural Analogue Project. Phase II: Transport of Radionuclides in a Natural Flow System at Palmottu*. Final Report EUR 19611 EN, 174p. European Commission. Nuclear Science and Technology.

Blum A. E. and Stillings L. L. (1995) Fieldspar dissolution kinetics. In *Chemical Weathering Rates of Silicate Minerals* (eds. A. F. White and S. L. Brantley). Mineralogical Society of America, Washington, DC. *Riv. Mineralogy*, 31, 291–351.

Blyth A. (2001) *Äspö Hard Rock Laboratory, Matrix Fluid Experiment—Fluid Inclusions Investigation of Quartz*. International Technical Document 01-06, Swedish Nuclear Fuel and Waste Management Company (SKB), Stockholm, Sweden, 20p.

Blyth A. and Frape S. K. (2002) *Evolution of Äspö Groundwaters with Time: Additional Information from Tritium and $\delta^{37}Cl$*. Technical Document 02-18, Swedish Nuclear Fuel and Waste Management Company, Stockholm, Sweden, 17p.

Blyth A., Frape S. K., Blomqvist R., and Nissinen P. (2000) Assessing the past thermal and chemical history of fluids in crystalline rock by combining fluid inclusion and isotopic investigations of fracture calcite. *Appl. Geochem.* 15, 1437–1471.

Borevsky L. V., Vartanyan G. S., and Kulikov T. B. (1984) Hydrological essay. In *The Superdeep Well of the Kola Peninsula* (ed. Ye. A. Kozlovsky). Springer, New York, pp. 271–287, chap. 1.9.

Bottomley D. J. (1987) The isotope geochemistry of fracture calcites from the Chalk River area, Ontario, Canada. *Appl. Geochem.* 2, 81–91.

Bottomley D. J. and Veizer J. (1992) The nature of groundwater in fractured rock: evidence from the isotopic and chemical evolution of recrystallized fracture calcites from the Canadian Precambrian Shield. *Geochim. Cosmochim. Acta* 56, 369–388.

Bottomley D. J., Ross J. D., and Graham B. W. (1984) A borehole methodology for hydrogeochemical investigations in fractured rock. *Water Resour. Res.* 20, 1277–1300.

Bottomley D. J., Gregoire D. C., and Raven K. G. (1994) Saline groundwaters and brines in the Canadian shield: geochemical and isotopic evidence for a residual evaporite brine component. *Geochim. Cosmochim. Acta* 58, 1483–1498.

Bottomley D. J., Katz A., Chan L. H., Starinsky A., Douglas M., Clark I. D., and Raven K. G. (1999) The origin and evolution of Canadian shield brines: evaporite or freezing of seawater? New lithium isotope and geochemical evidence from the Slave craton. *Chem. Geol.* 155, 295–320.

Boudreau A. E., Stewart M. A., and Spivack A. J. (1997) Stable Cl isotopes and origin of high-Cl magmas of the stillwater complex, Montana. *Geology* 25, 791–794.

Brantley S. L., Chesley J. T., and Stillings L. L. (1998) Isotopic ratios and release of rates of strontium measured from weathering feldspars. *Geochim. Cosmochim. Acta* 62, 1493–1500.

Bredehoeft J. D., Blyth C. R., White W. A., and Maxey G. G. (1963) Possible mechanism for concentration of brines in subsurface formations. *Bull. Am. Assoc. Petrol. Geol.* 47, 257–269.

Bucher K. and Stober I. (2000) The composition of groundwater in the continental crystalline crust. In *Hydrogeology of Crystalline Rocks* (eds. I. Stober and K. Bucher). Kluwer Academic, The Netherlands, pp. 141–176.

Burke W. H., Denison R. E., Hetherington E. A., Koepnick R. B., Nelson H. F., and Otto J. B. (1982) Variation of seawater $^{87}Sr/^{86}Sr$ throughout Phanerozoic time. *Geology* 10, 516–519.

Burton H. R. (1981) Chemistry, physics and evolution of Antarctic saline lakes. *Hydrobiologia* 82, 339–362.

Carmichael I. S. E., Turner F. J., and Verhoogen J. (1974) *Igneous Petrology*. McGraw Hill, 739 pp.

Carpenter A. B. (1978) Origin and chemical evolution of brines in sedimentary basins. *Okl. Geol. Surv. Circular* 79, 60–77.

Casanova J., Négrel P., Kaija J., and Blomqvist R. (1998) Constraints added by the strontium and boron isotopes on the geochemical characterization of the Palmottu hydrosystem (Finland). *Min. Mag.* 62A, 278–279.

Casanova J., Négrel P., Kloppmann W., and Aranyossy J. F. (2001) Origin of deep saline groundwaters in the Vienne granitic rocks (France): constraints inferred from boron and strontium isotopes. *Geofluids* 1, 91–102.

Chapman N. A. and McKinley I. G. (1987) *The Geological Disposal of Nuclear Waste*. Wiley, New York, 280 pp.

Chebotarev I. I. (1955) Metamorphism of natural waters in the crust of weathering: 3. *Geochim. Cosmochim. Acta* 8, 198–212.

Clark I. D. and Fritz P. (1997) *Environmental Isotopes in Hydrology*. Lewis Publisher, New York, 328 pp.

Clark I. D. and Lauriol B. (1992) Kinetic enrichment of stable isotopes in cryogenic calcites. *Chem. Geol.* 102, 217–228.

Clark I. D., Douglas M., Raven K. G., and Bottomley D. J. (2000) Recharge and preservation of Laurentide glacial melt water in the Canadian Shield. *Ground Water* 38, 735–742.

Clauer N., Frape S. K., and Fritz B. (1989) Calcite veins of the Stripa granite (Sweden) as records of the origin of the groundwaters and their interactions with the granitic body. *Geochim. Cosmochim. Acta* **53**, 1777–1781.

Clayton R. N., Friedman I., Graf D. L., Mayeda T. K., Meents W. F., and Shimp N. F. (1966) The origin of saline formation waters: 1. Isotopic composition. *J. Geophys. Res.* **71**, 3869–3882.

Collins A. G. (1975) *Geochemistry of Oilfield Waters*. Elsevier, Netherlands, 485pp.

Correns C. W. (1956) The geochemistry of halogens. In *Physics and Chemistry of the Earth* (eds. L. H. Ahrens, K. Rankama, and S. K. Runcorn). Pergamon, Oxford, pp. 181–233.

Craig H. (1961) Isotopic variations in meteoric waters. *Science* **133**, 1702–1703.

Craig H., Boato G., and White D. E. (1956) Isotopic geochemistry of thermal waters. US National Academy of Science, National Resource Council, vol. 400, pp. 29–38.

Davis S. N. (1964) The chemistry of saline waters: discussion. *Ground Water* **2**, 51.

Donner J. (1995) *The Quaternary History of Scandinavia. World and Regional Geology 7*. Cambridge University Press, Cambridge, 200 pp.

Drever J. I. (1988) *The Geochemistry of Natural Waters*. Prentice Hall, NJ, 437 pp.

Eastoe C. J. and Guilbert J. M. (1992) Stable chlorine isotopes in hydrothermal processes. *Geochim. Cosmochim. Acta* **56**, 4247–4255.

Edmunds W. M., Andrews J. N., Burgess W. G., Kay R. L. F., and Lee D. J. (1984) The evolution of saline and thermal groundwaters in the Carnmenellis granite. *Min. Mag.* **48**, 407–424.

Edmunds W. M., Kay R. L. F., Miles D. L., and Cook J. M. (1987) The origin of saline groundwaters in the Carnmenellis granite, Cornwall (UK): further evidence from minor and trace elements. In *Saline Water and Gases in Crystalline Rocks*. Special Paper 33 (eds. P. Fritz and S. K. Frape). Geological Association of Canada, Memorial University, Newfoundland, pp. 127–143.

Edmunds W. M., Andrews J. N., Bromley A. V., Kay R. L. F., Milodowski A., Savage D., and Thomas L. J. (1988) Granite–water interaction in relation to hot dry rock geothermal development. Investigation of the Geothermal Potential of the UK. British Geological Survey, 116 pp.

Eggenkamp H. G. M. (1994) The geochemistry of chlorine isotopes. PhD Geologica Ultraiectina, Mededelingen van de Faculteit Aardwetenschappen, Rijksuniversiteit Utrecht, 116 pp.

Eggenkamp H. G. M. and Schuling R. D. (1995) $\delta^{37}Cl$ variations in selected minerals: a possible tool for exploration. *J. Geochem. Explor.* **55**, 249–255.

Ekman M. (1996) A consistent map of the postglacial uplift of Fennoscandia. *Terra Nova* **8**, 158–165.

Eugster H. P. (1970) Chemistry and origin of the brines of Lake Magadi, Kenya. *Min. Soc. Am. Spec. Paper* **3**, 213–235.

Eugster H. P. and Hardie L. A. (1978) Saline lakes. In *Lakes— Chemisry, Geology, Physics* (ed. A. Lerman). Springer, New York, pp. 237–293.

Eugster H. P. and Jones B. F. (1979) Behavior of major solutes during closed-basin brine evolution. *Am. J. Sci.* **279**, 609–631.

Fontes J. Ch., Fritz P., Louvat D., and Michelot J.-L. (1989) Aqueous sulphates from the Stripa groundwater system. *Geochim. Cosmochim. Acta* **53**, 1783–1789.

Franklyn M. T., McNutt R. H., Kamineni D. C., Gascoyne M., and Frape S. K. (1991) Groundwater $^{87}Sr/^{86}Sr$ values in the Eye-Dashwa Lakes Pluton, Canada: evidence of Plagioclase-water reactions. *Chem. Geol.* **86**, 111–122.

Frape S. K. and Fritz P. (1982) The chemistry and isotopic composition of saline groundwaters from the Sudbury Basin, Ontario. *Can. J. Earth Sci.* **19**, 645–661.

Frape S. K. and Fritz P. (1987) Geochemical trends for groundwater from the Canadian shield. In *Saline Water and Gases in Crystalline Rocks*, Paper 33 (eds. P. Fritz and S. K. Frape). Geological Association of Canada, Memorial University, Newfoundland, pp. 19–38.

Frape S. K., Fritz P., and McNutt R. H. (1984) Water–rock interaction and chemistry of groundwaters from the Canadian shield. *Geochim. Cosmochim. Acta* **48**, 1617–1627.

Frape S. K., Bryant G., Blomqvist R., and Ruskeeniemi T. (1996) Evidence from stable chlorine isotopes for multiple sources of chloride in groundwaters from crystalline shield environments. In *Symposium on Isotopes in Water Resources Management*. United Nations Educational, Scientific and Cultural Organization, Vienna, pp. 19–30.

Frape S. K., Bryant G., Durance P., Ropchan J. C., Doupe J., Blomqvist R., Nissinen P., and Kaija J. (1998) The source of stable chlorine isotopic signatures in groundwaters from crystalline shield rocks. In *Proc. 9th Int. Symp. Water–Rock Interaction* (eds. G. B. Arehart and J. R. Hulston). Taupo, New Zealand; Balkema, Rotterdam, The Netherlands, pp. 223–226.

Fritz B., Clauer N., and Kam M. (1987a) Strontium isotopic data and geochemical calculations as indicators for the origin of saline waters in crystalline rocks. In *Saline Water and Gases in Crystalline Rocks*, Special Paper 33 (eds. P. Fritz and S. K. Frape). Geological Association of Canada, Memorial University, Newfoundland, pp. 121–126.

Fritz B., Richard L., and McNutt R. H. (1992) Geochemical modeling of Sr isotope signatures in the interaction between granitic rocks and natural solutions. In *Proc. 7th Int. Symp. Water–Rock Interaction* (eds. Y. K. Kharaka and A. S. Maest). Balkema, Rotterdam, The Netherlands, pp. 927–930.

Fritz P. and Frape S. K. (1982) Saline groundwaters in the Canadian shield—a first overview. *Chem. Geol.* **36**, 179–190.

Fritz P., Frape S. K., and Miles M. (1987b) Methane in the crystalline rocks of the Canadian shield. In *Saline Water and Gases in Crystalline Rocks*, Special Paper 33 (eds. P. Fritz and S. K. Frape). Geological Association of Canada, Memorial University, Newfoundland, pp. 211–223.

Fritz P., Fontes J. Ch., Frape S. K., Louvat D., Michelot J.-L., and Balderer W. (1989) The isotope geochemistry of carbon in groundwater at Stripa. *Geochim. Cosmochim. Acta* **53**, 1765–1775.

Fritz P., Frape S. K., Drimmie R. J., Appleyard E. C., and Hattori K. (1994) Sulfate in brines in the crystalline rocks of the Canadian shield. *Geochim. Cosmochim. Acta* **58**, 57–65.

Fuge R. (1977) On the behaviour of fluorine and chlorine during magmatic differentiation. *Contrib. Mineral. Petrol.* **61**, 245–249.

Garrels R. M. (1967) Genesis of some ground waters from igneous rocks. *Researches in Geochemistry, Volume 2* (ed. P. H. Abelson). Wiley, New York, pp. 405–420.

Gascoyne M. (2002) *Methods of Sampling and Analysis of Dissolved Gases in Deep Groundwaters*. Working Report, Posiva Oy, Finland.

Gascoyne M., Davison C. C., Ross J. D., and Pearson R. (1987) Saline groundwaters and brines in plutons in the Canadian shield. In *Saline Water and Gases in Crystalline Rocks*. Special Paper 33 (eds. P. Fritz and S. K. Frape). Geological Association of Canada, Memorial Universiy, Newfoundland, pp. 53–68.

Gascoyne M., Purdy A., Fritz P., Ross J. D., Frape S. K., Drimmie R. J., and Betcher R. N. (1989) Evidence for penetration of sedimentary basin brines into an Archean granite of the Canadian shield. In *Proc. 6th Int. Symp. Water–Rock Interaction* (ed. D. L. Miles). Malvern, UK; Balkema, Rotterdam, The Netherlands, pp. 243–245.

Gascoyne M., Stroes-Gascoyne S., and Sargent F. P. (1995) Geochemical influences on the design, construction and operation of a nuclear water vault. *Appl. Geochem.* **10**, 657–671.

Gehör S. and Lindblom S. (2002) Textural, microthermometry and laser ablation ICP-MS investigations of fluid inclusions in drillcore KF0051A01 (5.03–10.95 m). Technical Document 02-13, Swedish Nuclear Fuel and Waste Management Company (SKB), 55 pp.

Geological Survey of Canada (1878) Annual Report, 1877/78: Volume III, Part 2. Geological Survey of Canada.

Giggenbach W. F. (1988) Geothermal solute equilibria. derivation of Na–K–Mg–Ca geoindicators. *Geochim. Cosmochim. Acta* 52, 2749–2765.

Gonfiantini R. (1965) Effetti isotopici nell'evaporazione di acque salate. *Atti Soc. Toscana Sci. Nat. Pisa, A* 72, 550–569.

Gonfiantini R. (1986) Environmental isotopes in lake studies. In *Handbook of Environmental Isotope Geochemistry: Vol. 2. The Terrestrial Environment B* (eds. P. Fritz and J.-C. Fontes). Elsevier, Amsterdam, The Netherlands, pp. 113–168.

Guha J. and Kanwar R. (1987) Vug brines-fluid inclusions: a key to the understanding of secondary gold enrichment processes and the evolution of deep brines in the Canadian shield. In *Saline Water and Gases in Crystalline Rocks*, Special Paper 33 (eds. P. Fritz and S. K. Frape). Geological Association of Canada, Memorial University, Newfoundland, pp. 95–101.

Hallam A. (1984) Pre-quaternary sea-level changes. *Ann. Rev. Earth Planet. Sci.* 12, 205–243.

Hardie L. A. and Eugster H. P. (1970) The evolution of closed-basin brines. Special Paper 3. Mineralogical Society of America, Washington, DC, pp. 273–290.

Hattori K. (1989) Barite–celestite intergrowths in Archean plutons: the product of oxidizing hydrothermal activities related to alkaline intrusions. *Am. Mineral.* 74, 1270–1277.

Hattori K. and Cameron E. M. (1986) Archean magmatic sulfate. *Nature* 319, 5–47.

Herut B., Starinski A., Katz A., and Bein A. (1990) The role of seawater freezing in the formation of subsurface brines. *Geochim. Cosmochim. Acta* 54, 13–21.

Hyndman D. W. (1985) *Petrology of Igneous and Metamorphic Rocks*. McGraw-Hill, New York, 786 pp.

Irwin J. J. and Reynolds J. H. (1995) Multiple stages of fluid trapping in the Stripa granite indicated by laser microprobe analysis of Cl, Be, I, K, U, and nucleogenic Ar, Kr, and Xe in fluid inclusions. *Geochim. Cosmochim. Acta* 59, 355–369.

Ivanovich M., Blomqvist R., and Frape S. K. (1992) Rock/water interaction study in deep crystalline rocks using isotopic and uranium series radionuclide techniques. *Radiohimica Acta* 58/59, 401–408.

Jacks G. (1973) Chemistry of some groundwaters in igneous rocks. *Nordic Hydrol.* 4, 236.

Jacks G. (1978) *Ground Water Chemistry at Depth in Granites and Gneisses*. Technical Report 88, Swedish Nuclear Fuel and Waste Management Company (SKBF/KBS), 28p.

Kamineni D. C. (1983) Sulfur-isotope geochemistry of fracture filling gypsum in an Archean granite near Atikokan, Ontario, Canada. *Chem. Geol.* 39, 263–272.

Kamineni D. C. (1987) Halogen bearing minerals in plutonic rocks: a possible source of chlorine in saline groundwater in the Canadian shield. In *Saline Water and Gases in Crystalline Rocks*, Special Paper 33 (eds. P. Fritz and S. K. Frape). Geological Association of Canada, Memorial University, Newfoundland, pp. 69–79.

Kamineni D. C., Gascoyne M., Melnyk T. W., Frape S. K., and Blomqvist R. (1992) Cl and Br in mafic and ultramafic rocks: significance for the origin of salinity in groundwater. In *Proc. 7th, Int. Symp. Water–Rock Interaction* (eds. Y. K. Kharaka and A. S. Maest). Park City, UT; Balkema, Rotterdam, The Netherlands, 801–804.

Kapchenko L. N. (1964) On the genesis of the deep-seated brines of the Siberian platform. *Geochem. Int.* 4, 1107–1114.

Karus E. W., Kuznetsov O. L., Kuznetsov Yu. P., Kuznetsova L. V., Smirnov Yu. P., Osadtchyi A. P., and Vinogradov E. A. (1984) In *The Superdeep Well of the Kola Peninsula* (ed. Ye. A. Kozlovsky). Springer, Berlin, pp. 339–350.

Kaufmann R. (1984) Chlorine in groundwater: stable isotope distribution. PhD Thesis, University of Arizona, Tucson, Arizona, 132p (unpublished).

Kaufmann R., Frape S. K., Fritz P., and Bentley H. (1987) Chlorine stable isotope composition of Canadian shield brines. In *Saline Water and Gases in Crystalline Rocks*, Special Paper 33 (eds. P. Fritz and S. K. Frape). Geological Association of Canada, Memorial University, Newfoundland, pp. 89–93.

Kaufmann R. (1989) Equilibrium exchange models for chlorine stable isotope fractionation in high temperature environments. In *Proc. 6th Int. Symp. Water–Rock Interaction* (ed. D. L. Miles). Malvern, UK; Balkema, Rotterdam, The Netherlands, pp. 365–368.

Kelly W. C., Rye R. O., and Livnat A. (1986) Saline minewaters of the Keweenaw Peninsula, northern Michigan: their nature, origin, and relation to similar deep waters in Precambrian crystalline rocks of the Canadian shield. *Am. J. Sci.* 286, 281–308.

Kharaka Y. K. and Carothers W. W. (1986) In *Handbook of Environmental Isotope Geochemistry: Vol. 2. The Terrestrial Environment B* (eds. P. Fritz and J. Ch. Fontes). Elsevier, Amsterdam, The Netherlands, pp. 305–360.

Kloppmann W., Girard J.-P., and Négrel P. (2002) Exotic stable isotope compositions of saline waters and brines from the crystalline basement. *Chem. Geol.* 184, 49–70.

Kozlovsky Ye. A. (1984) *The Superdeep Well of the Kola Peninsula*. Springer, Berlin, 558pp.

Krotova V. A. (1958) Conditions of formation of calcium chloride waters in Siberia. *Petrol. Geol.* 2, 545–552.

Kullerud K. (2000) Occurrence and origin of Cl-rich amphibole and biotite in the earth's crust—implications for fluid composition and evolution. In *Hydrogeology of Crystalline Rocks* (eds. I. Stober and K. Bucher). Kluwer Academic, The Netherlands, pp. 205–226.

Laaksoharju M., Tullborg E. L., Wikberg P., Wallin B., and Smellie J. (1999) Hydrogeochemical conditions and evolution at the Äspö HRL, Sweden. *Appl. Geochem.* 14, 835–859.

Lahermo P. W. and Lampen P. H. (1987) Brackish and saline groundwaters in Finland. In *Saline Water and Gases in Crystalline Rocks*, Special Paper 33 (eds. P. Fritz and S. K. Frape). Geological Association of Canada, Memorial University, Newfoundland, pp. 103–109.

Lane A. C. (1914) Mine water composition, an index to the course of ore-bearing currents. *Econ. Geol.* 9, 239–263.

Langmuir D. (1971) The geochemistry of some carbonate ground waters in central Pennsylvania. *Geochim. Cosmochim. Acta* 35, 1023–1045.

Lasaga A. C. (1984) Chemical kinetics of water–rock interactions. *J. Geophys. Res. B* 89, 4009–4025.

Lewis G. J. and Thompson T. G. (1950) The effect of freezing on the sulfate/chlorinity ratio of sea water. *J. Mar. Res.* 9, 211–217.

Lodemann M. and Wöhrl Th. (1994) Technical performance of the pumping test 1991 in the KTB pilot-borehole. *Sci. Drill.* 4, 133–137.

Lodemann M., Fritz P., Wolf M., Ivanovich M., Hansen B., and Nolte E. (1998) On the origin of saline fluids in the KTB (Continental Deep Drilling Project of Germany). *Appl. Geochem.* 13, 653–671.

Louvat D., Michelot J.-L., and Aranyossy J. F. (1999) Origin and residence time of salinity in the Äspö groundwater system. *Appl. Geochem.* 14, 917–925.

Magenheim A. J., Spivack A. J., Volpe C., and Ransom B. (1994) Precise determination of stable chlorine isotope ratios in low-concentration natural samples. *Geochim. Cosmochim. Acta* 58, 3117–3121.

Markl G., Musashi M., and Bucher K. (1997) Chlorine stable isotope composition of granulites from Lofoten Norway: implications for the Cl isotopic composition and for the source of Cl enrichment in the lower crust. *Earth Planet. Sci. Lett.* 150, 95–102.

Matray J.-M. (1984) Hydrochimie et geochime isotopiques des eaux de reservoir pétrolier du trias et du dogger dans le bassin de Paris. DSc Thesis, Universite de Paris-sud-Centre d'Orsay, 55p (unpublished).

Matsubaya O., Sakai H., Torii T., Burton H. R., and Kerry K. (1979) Antarctic saline lakes stable isotope ratios, chemical composition, and evolution. *Geochim. Cosmochim. Acta* **43**, 7–25.

Mazurek M. (2000) Geological and hydraulic properties of water-conducting features in crystalline rock. In *Hydrogeology of Crystalline Rock* (eds. I. Stober and K. Bucher). Kluwer Academic, The Netherlands, pp. 3–26.

McCaffrey M. A., Lazar B., and Holland H. D. (1987) The evaporation path of sea water and the coprecipitation of Br^- and K^+ with halite. *J. Sedim. Petrol.* **57**, 928–937.

McLaughlin R. M. (1997) Boron and strontium isotope study of fluids situated in fractured and unfractured rock of the Lac du Bonnet Batholith, Eastern Manitoba. PhD thesis, McMaster University, 142p (unpublished).

McNutt R. H. (1987) $^{87}Sr/^{86}Sr$ ratios as indicators of water/rock interactions: applications to brines found in Precambrian age rocks from Canada. In *Saline Water and Gases in Crystalline Rocks*, Special Paper 33 (eds. P. Fritz and S. K. Frape). Geological Association of Canada, Memorial University, Newfoundland, pp. 81–88.

McNutt R. H. (2000) Strontium isotopes. In *Environmental Tracers in Subsurface Hydrology.* (eds. P. Cook and A. L. Herczeg). Kluwer Academic, Boston, pp. 234–260.

McNutt R. H., Frape S. K., and Fritz P. (1984) Strontium isotopic composition of some brines from the Precambrian shield of Canada. *Isotope Geosci.* **2**, 205–215.

McNutt R. H., Frape S. K., Fritz P., Jones M. G., and MacDonald I. M. (1990) The $^{87}Sr/^{86}Sr$ values of Canadian shield brines and fracture minerals with applications to groundwater mixing, fracture history, and geochronology. *Geochim. Cosmochim. Acta* **54**, 202–215.

Michelot J.-L. and Fontes J.-C. (1987) Two case studies on the origin of aqueous sulphate. In *Studies on Sulphur Isotope Variations in Nature, Advisory Group Meeting*, International Atomic Energy Agency (IAEA), Vienna, 65–76.

Möller P., Weise S. M., Althaus E., Back W., Behr H. J., Borchardt R., Bräuer K., Drescher J., Erzinger J., Faber E., Hansen B. T., Horn E. E., Huenges E., Kämpf H., Kessels W., Kirsten T., Landwehr D., Lodemann M., Machon L., Pekdeger A., Pielow H.-U., Reutel C., Simon K., Walther J., Weinlich F. H., and Zimmer M. (1997) Paleofluids and recent fluids in the upper continental crust: results from the German continental deep drilling program (KTB). *J. Geophys. Res.* **102**, 18233–18254.

Mörner N.-A. (1979) The Fennoscandian uplift and Late Cenozoic geodynamics; Geological evidence. *GeoJournal* **3**, 287–318.

Mungall J. E., Frape S. K., and Gibson I. L. (1987) Rare-earth abundances in host granitic rocks and fracture-filling gypsum associated with saline groundwaters from a deep borehole, Atikokan, Ontario. *Can. Mineral.* **25**, 539–543.

Négrel P., Casanova J., and Aranyossy J. F. (2001) Strontium isotope systematics used to decipher the origin of groundwaters sampled from granitoids: the Vienne Case (France). *Chem. Geol.* **177**, 287–308.

Nelson K. H. and Thompson T. G. (1954) Deposition of salts from sea water by frigid concentration. *J. Mar. Res.* **13**, 166–182.

Niskanen E. (1943) On the deformation of the Earth's crust under the weight of a glacial ice-load and related phenomena. *Ann. Acad. Sci. Fennicae* **AIII**, 7.

Nordstrom D. K., Ball J. W., Donahoe R. J., and Whittemore D. (1989a) Groundwater chemistry and water–rock interactions at Stripa. *Geochim. Cosmochim. Acta* **53**, 1727–1740.

Nordstrom D. K., Lindblom S., Donahoe R. J., and Barton C. C. (1989b) Fluid inclusions in the Stripa granite and their possible influence on the groundwater chemistry. *Geochim. Cosmochim. Acta* **53**, 1741–1755.

Nordstrom D. K., Olsson T., Carlsson L., and Fritz P. (1989c) Introduction to the hydrogeochemical investigations within the international Stripa project. *Geochim. Cosmochim. Acta* **53**, 1717–1726.

Nurmi P. A. and Kukkonen I. T. (1986) A new technique for sampling water and gas from deep drill holes. *Can. J. Earth Sci.* **23**, 1450–1454.

Nurmi P. A., Kukkonen I. T., and Lahermo P. W. (1988) Geochemistry and origin of saline groundwaters in the Fennoscandian Shield. *Appl. Geochem.* **3**, 185–203.

Pačes T. (1972) Chemical characteristics and equilibrium in natural water—felsic rock—carbon dioxide system. *Geochim. Cosmochim. Acta* **36**, 217–240.

Parsons I. and Lee M. R. (2000) Alkali feldspars as microtextural markers of fluid flow. In *Hydrogeology of Crystalline Rock* (eds. I. Stober and K. Bucher). Kluwer Academic, The Netherlands, pp. 27–52.

Pearson F. J., Jr. (1987) Models of mineral controls on the composition of saline groundwaters of the Canadian, shield. In *Saline Water and Gases in Crystalline Rocks*, Special Paper 33 (eds. P. Fritz and S. K. Frape). Geological Association of Canada, Memorial University, Newfoundland, pp. 39–51.

Pedersen K. (1993) The deep subterranean biosphere. *Earth Sci. Rev.* **34**, 243–260.

Pedersen K. (2000) The hydrogen driven intra-terrestrial biosphere and its influence on the hydrochemical conditions in crystalline bedrock aquifers. In *Hydrogeology of Crystalline Rocks* (eds. I. Stober and K. Bucher). Kluwer Academic, The Netherlands, pp. 249–260.

Pekdeger A. and Balderer W. (1987) The occurrence of saline groundwaters and gases in the crystalline rocks of northern Switzerland. In *Saline Water and Gases in Crystalline Rocks*, Special Paper 33 (eds. P. Fritz and S. K. Frape). Geological Association of Canada, Memorial University, Newfoundland, pp. 157–174.

Pitkänen P., Kaija J., Blomqvist R., Smellie J. A. T., Frape S., Laaksoharju M., Negrel P., Casanova J., and Karhu J. (1999) Hydrogeochemical interpretation of groundwater at Palmottu. Proceedings of the 8th EC-NAWG Workshop, Strasbourg, 23–25 March,1999.

Räisänen E. (1986) Uraniferous granitic veins in the Svecofennian schist belt in Nummi Pusula, Southern Finland. International Atomic Energy Agency (IAEA), Technical Committee Meeting (TC-571) on Uranium Deposits in Magmatic and Metamorphic Rocks, Salamanca, 12p.

Rhén I., Gustafson G., Stanfors R., and Wikberg P. (1997) *Äspö HRL—Geoscientific Evaluation 1997/5. Models Based on Site Characterization 1986–1995*. Technical Report 97-06, Swedish Nuclear Fuel and Waste Management Company (SKB), 428p.

Rittenhouse G. (1967) Bromine in oil-field waters and its use in determining possibilities of origin of these waters. *Am. Assoc. Petrol. Geol. Bull.* **51**, 2430–2440.

Roedder E. (1984) *Fluid Inclusions*. Mineralogical Society of America, Washington, DC. *Riv. Mineralogy* **12**, 644pp.

Ross J. D. and Gascoyne M. (1995) *Methods of Sampling and Analysis of Groundwaters in the Canadian Nuclear Fuel Waste Management Program*. Technical Report 588COG-93-36, Atomic Energy of Canada Limited, Pinawa, Manitoba.

Ruskeeniemi T., Blomqvist R., Lindberg A., Ahonen L., and Frape S. K. (1996) *Hydrogeochemistry of Deep Groundwaters of Mafic and Ultramafic Rocks in Finland*. POSIVA-96-21, Posiva Oy, 123p.

Ruskeeniemi T., Lindberg A., Perez del Villar L., Blomqvist R., Suksi J., Blyth A., and Cera E. (2002) Uranium mineralogy with implications for mobilization of uranium at Palmottu. In *Eighth EC Natural Analogue Working Group Meeting: Proceedings of an International Workshop, Strasbourg, France, March 23–25, 1999* (eds. von H. Maravic and W. R. Alexander). Office for Official

Publications of the European Communities, Luxembourg, pp. 143–154.

Savoye S., Aranyossy J. F., Beaucaire C., Cathelineau M., Louvat D., and Michelot J.-L. (1998) Fluid inclusions in granites and their relationships with present-day groundwater chemistry. *Euro. J. Mineral.* **10**, 1215–1226.

Schoell M. (1988) Multiple origins of methane in the Earth. *Chem. Geology* **71**, 1–10.

Sherwood B., Fritz P., Frape S. K., Macko S. A., Weise S. M., and Welhan J. A. (1988) Methane occurrences in the Canadian shield. *Chem. Geol.* **71**, 223–236.

Sherwood Lollar B., Frape S. K., Drimmie R. J., Fritz P., Weise S. M., Macko S. A., Welhan J. A., Blomqvist R., and Lahermo P. W. (1989) Deep gases and brines of the Candian and Fennoscandian shields—a testing ground for the theory of abiotic methane generation. In *Proc. 6th Int. Symp. Water–Rock Interaction* (ed. D. L. Miles). Malvern, UK; Balkema, Rotterdam, Netherlands, pp. 617–620.

Sherwood Lollar B., Frape S. K., Fritz P., Macko S. A., Welhan J. A., Blomqvist R., and Lahermo P. W. (1993a) Evidence for bacterially generated hydrocarbon gas in Canadian shield and Fennoscandian shield rocks. *Geochim. Cosmochim. Acta* **57**, 5073–5085.

Sherwood Lollar B., Frape S. K., Weise S. M., Fritz P., Macko S. A., and Welhan J. A. (1993b) Abiogenic methanogenesis in crystalline rocks. *Geochim. Cosmochim. Acta* **57**, 5087–5097.

Sherwood Lollar B., Frape S. K., and Weise S. M. (1994) New sampling devices for environmental characterization of groundwater and dissolved gas chemistry (CH_4, N_2, He). *Environ. Sci. Technol.* **28**, 2427.

Shouakar-Stash O., Alexeev S. V., Frape S. K., Alexeeva L. P., and Pinneker E. V. (2002) Geochemistry and stable isotopic signatures of deep groundwaters and brine from the permafrost zone of the Siberian platform, Russia. Geological Society of America, GSA Annual Meeting, Denver, 497p.

Sie P. M. J. and Frape S. K. (2002) Evaluation of the groundwaters from the Stripa mine using stable chlorine isotopes. *Chem. Geol.* **182**, 565–582.

Smalley P. C., Blomqvist R., and Reim A. (1988) Sr isotopic evidence for discrete saline components in stratified ground waters from crystalline bedrock, Outokumpu, Finland. *Geology* **16**, 354–357.

Smellie J. A. T. and Wikberg P. (1991) Hydrogeochemical investigations an Finnsjon, Sweden. *J. Hydrol.* **126**, 129–158.

Smellie J. A. T., Laaksoharju M., and Wikberg P. (1995) Äspö, SE Sweden: a natural groundwater flow model derived from hydrogeochemical observations. *J. Hydrol.* **172**, 147–169.

Spencer R. J. (1987) Origins of Ca–Cl brines in Devonian formations, western Canada sedimentary basin. *Appl. Geochem.* **2**, 373–384.

Spencer R. J., Eugster H. P., Jones B. F., and Rettig S. L. (1985) Geochemistry of Great Salt Lake, Utah: I. Hydrochemistry since 1850. *Geochim. Cosmochim. Acta* **49**, 727–738.

Spivack A. J. and Edmond J. M. (1987) Boron isotope exchange between seawater and the oceanic crust. *Geochim. Cosmochim. Acta* **51**, 1033–1043.

Stanfors R., Rhén I., Tullborg E. L., and Wikberg P. (1999) Overview of geological and hydrogeological conditions of the Äspö hard rock laboratory site. *Appl. Geochem.* **14**, 819–834.

Starinsky A., Bielski M., Ecker A., and Dteinitz G. (1983) Tracing the origin of salts in groundwater by Sr isotopic composition (the crystalline complex of the southern Sinai, Egypt). *Isotope Geosci.* **1**, 257–267.

Stober I. (1996) Hydrogeological investigations in crystalline rocks of the Black Forest, Germany. *Terra Nova* **8**, 255–258.

Stober I. and Bucher K. (1999) Deep groundwater in the crystalline basement of the Black Forest region. *Appl. Geochem.* **14**, 237–254.

Taylor B. E. (1987) Stable isotope geochemistry of ore-forming fluids. In *Short Course in Stable Isotope Geochemistry of*

Low Temperature Fluids (ed. T. K. Kyser). Mineralogical Association of Canada, Toronto, Ontario, pp. 337–418.

Thompson T. G. and Nelson K. H. (1956) Concentration of brines and deposition of salts from water under frigid conditions. *Am. J. Sci.* **254**, 227–238.

Torii T. and Ossaka J. (1965) Antarcticite: a new mineral, calcium chloride hexahydrate, discovered in Antarctica. *Science* **149**, 975–977.

Torii T., Yamagata N., Ossaka J., and Murata S. (1979) *A view of the formation of saline waters in the dry valleys*, Special Issue **13** (ed. T. Nagata). *Mem. Natl. Inst. Polar Res. (Japan)*, pp. 22–33.

Truesdell A. H. and Hulston J. R. (1980) Isotopic evidence of environments of geothermal systems. In *Handbook of Environmental Isotope Geochemistry, Vol. 1, The Terrestrial Environment A* (eds. P. Fritz and J.-C. Fontes). Elsevier, The Netherlands, pp. 179–226.

Tullborg E. L. (1989) $\delta^{18}O$ and $\delta^{13}C$ in fracture calcite used for interpretation of recent meteoric water circulation. In *Proc. 6th Int. Symp. Water–Rock Interaction* (ed. D. L. Miles). Malvern, UK; Balkema, Rotterdam, The Netherlands, pp. 695–698.

Tullborg E. L. (2000) Ancient microbial activity in cyrstalline bedrock—results from stable isotope analyses of fracture calcites. In *Hydrogeology of Crystalline Rocks* (eds. I. Stober and K. Bucher). Kluwer Academic, The Netherlands, pp. 261–269.

Tullborg E. L. and Gustafsson E. (1999) ^{14}C in bicarbonate and dissolved organics—a useful tracer? *Appl. Geochem.* **14**, 927–938.

Tullborg E. L., Larson S. A., and Stiberg J.-P. (1996) *Subsidence and Uplift of the Present Land Surface in the Southeastern Part of the Fennoscandian Shield*. Report 112, Stockholm, Geologiska Foreningens i Stockholm Forhandlingar, pp. 215–225.

Usiglio M. J. (1849) Ètudes sur la composition de l'eau de la Méditerranée et sur l'exploitation des sels qu'elle contient. *Ann. Chim. Phys.* **27**, 172–191.

Veizer J., Hoefs J., Lowe D. R., and Thurston P. C. (1989) Geochemistry of Precambrian carbonates: 2. Archean greenstone belts and Archean seawater. *Geochim. Cosmochim. Acta* **53**, 859–871.

Veizer J., Plumb K. A., Clayton R. N., Hinton R. W., and Grotzinger J. P. (1992) Geochemistry of Precambrian carbonates: V. Late Paleoprotozoic (1.8 ± 0.2 Ga) seawater. *Geochim. Cosmochim. Acta* **56**, 2487–2501.

Vengosh A., Chivas A. R., McCulloch M. T., Starinsky A., and Kolodny Y. (1991) Boron isotope geochemistry of Australian salt lakes. *Geochim. Cosmochim. Acta* **55**, 2591–2606.

Vengosh A., Chivas A. R., Starinsky A., Kolodny Y., Baozhen Z., and Pengxi Z. (1995) Chemical and boron isotope compositions of non-marine brines from the Qaidam Basin, Qinghai, China. *Chem. Geol.* **120**, 135–154.

Verhoogen J., Turner F. J., Weiss L. E., Wahrhaftig C., and Fyfe W. S. (1970) *The Earth: an Introduction to Physical Geology*. Holt, Rinehart and Winston, New York, 748pp.

Vieno T. and Nordman H. (1999) *Safety Assessment of Spent Fuel Disposal in Hastholmen, Kivetty, Olkiluoto, and Romuvaara–Tila-99*. Report 99-07, Posiva Oy, Helsinki, 253p.

Vovk I. F. (1987) Radiolytic salt enrichment and brines in the crystalline basement of the East European Platform. In *Saline Water and Gases in Crystalline Rocks*, Special Paper 33 (eds. P. Fritz and S. K. Frape). Geological Association of Canada, Memorial University, Newfoundland, pp. 197–210.

Wallin B. and Peterman Z. E. (1999) Calcite fracture fillings as indicators of paleohydrology at Laxemar at the Äspö Hard Rock Laboratory, southern Sweden. *Appl. Geochem.* **14**, 953–962.

Warren J. (1999) *Evaporites—Their Evolution and Economics*. Blackwell, London, 438pp.

Weber K. (1992) *Die tektonische Position der KTB-Lokation.* KTB Report 92-2, pp. 455–460.

White A., Blum A., Bullen T., Vivit D., Schulz M., and Fitzpatrick J. (1999) The effect of temperature on experimental and natural chemical weathering rates of granitoid rocks. *Geochim. Cosmochim. Acta* **63**, 3277–3297.

Wikberg P. (1987) The chemistry of deep groundwaters in crystalline rocks. PhD Thesis, The Royal Institute of Technology (KTH), Stockholm, Sweden.

Wynne-Edwards H. R. (1957) The structure of the Westport concordant pluton in the Grenville, Ontario. *J. Geol.* **65**, 649.

5.18
Soils and Global Change in the Carbon Cycle over Geological Time

G. J. Retallack

University of Oregon, Eugene, OR, USA

5.18.1 INTRODUCTION

Soils play an important role in the carbon cycle as the nutrition of photosynthesized biomass. Nitrogen fixed by microbes from air is a limiting nutrient for ecosystems within the first flush of ecological succession of new ground, and sulfur can limit some components of wetland ecosystems. But over the long term, the limiting soil nutrient is phosphorus extracted by weathering from minerals such as apatite (Vitousek *et al.*, 1997a; Chadwick *et al.*, 1999). Life has an especially voracious appetite for common alkali (Na^+ and K^+) and alkaline earth (Ca^{2+} and Mg^{2+}) cations, supplied by hydrolytic weathering, which is in turn amplified by biological acidification (Schwartzmann and Volk, 1991; see Chapter 5.06). These mineral nutrients fuel photosynthetic fixation and chemical reduction of atmospheric CO_2 into plants and plantlike microbes, which are at the base of the food chain. Plants and photosynthetic microbes are consumed and oxidized by animals, fungi, and other respiring microbes, which release CO_2, methane, and water vapor to the air. These greenhouse gases absorb solar radiation more effectively than atmospheric oxygen and nitrogen, and are important regulators of planetary temperature and albedo (Kasting, 1992). Variations in solar insolation (Kasting, 1992), mountainous topography (Raymo and Ruddiman, 1992), and ocean currents (Ramstein *et al.*, 1997) also play a role in climate, but this review focuses on the carbon cycle. The carbon cycle is discussed in detail in Volume 8 of this Treatise.

The greenhouse model for global paleoclimate has proven remarkably robust (Retallack, 2002), despite new challenges (Veizer *et al.*, 2000). The balance of producers and consumers is one of a number of controls on atmospheric greenhouse gas balance, because CO_2 is added to the air from fumaroles, volcanic eruptions, and other forms of mantle degassing (Holland, 1984). Carbon dioxide is also consumed by burial as carbonate and organic matter within limestones and other sedimentary rocks; organic matter burial is an important long-term control on CO_2 levels in the atmosphere (Berner and Kothavala, 2001). The magnitudes of carbon pools and fluxes involved provide a perspective on the importance of soils compared with other carbon reservoirs (Figure 1).

Before industrialization, there was only 600 Gt $(1 Gt = 10^{15} g)$ of carbon in CO_2 and methane in the atmosphere, which is about the same amount as in all terrestrial biomass, but less than half of the reservoir of soil organic carbon. The ocean contained only ~3 Gt of biomass carbon. The deep ocean and sediments comprised the largest reservoir of bicarbonate and organic matter, but that carbon has been kept out of circulation from the atmosphere for geologically significant periods of time (Schidlowski and Aharon, 1992). Humans have tapped underground reservoirs of fossil fuels, and our other perturbations of the carbon cycle have also been significant (Vitousek *et al.*, 1997b; see Chapter 8.10).

Atmospheric increase of carbon in CO_2 to 750 Gt C by deforestation and fossil fuel burning has driven ongoing global warming, but is not quite balanced by changes in the other carbon reservoirs leading to search for a "missing sink" of some 1.8 ± 1.3 Gt C, probably in terrestrial organisms, soils, and sediments of the northern hemisphere (Keeling *et al.*, 1982; Siegenthaler

and Sarmiento, 1993; Stallard, 1998). Soil organic matter is a big, rapidly cycling reservoir, likely to include much of this missing sink.

During the geological past, the sizes of, and fluxes between, these reservoirs have varied enormously as the world has alternated between greenhouse times of high carbon content of the atmosphere, and icehouse times of low carbon content of the atmosphere. Oscillations in the atmospheric content of greenhouse gases can be measured, estimated, or modeled on all timescales from annual to eonal (Figure 2). The actively cycling surficial carbon reservoirs are biomass, surface oceans, air, and soils, so it is no surprise that the fossil record of life on Earth shows strong linkage to global climate change (Berner, 1997; Algeo and Scheckler, 1998; Retallack, 2000a). There is an additional line of evidence for past climatic and atmospheric history in the form of fossil soils, or paleosols, now known to be abundant throughout the geological record (Retallack, 1997a, 2001a). This chapter addresses evidence from fossil soils for global climate change in the past, and attempts to assess the role of soils in carbon cycle fluctuations through the long history of our planet.

5.18.2   APPROACHES TO THE STUDY OF PALEOSOLS

Many approaches to the study of paleosols are unlike those of soil science, and more like soil geochemistry prior to the earlier part of the twentieth century (Thaer, 1857; Marbut, 1935). Such measures of soil fertility as cation exchange capacity and base saturation that are used for characterizing surface soils (Buol *et al.*, 1997) are

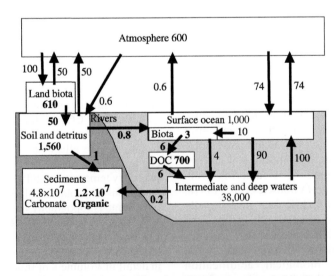

Figure 1   Pools and fluxes of reduced carbon (bold) and oxidized carbon (regular) in Gt in the pre-industrial carbon cycle (sources Schidlowski and Aharon, 1992; Siegenthaler and Sarmiento, 1993; Stallard, 1998).

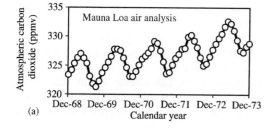

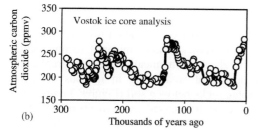

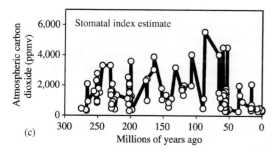

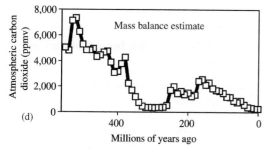

Figure 2 Variation in atmospheric CO_2 composition on a variety of timescales ranging from annual to eonal (a) Keeling *et al.*, 1982; reproduced from *Carbon Dioxide Review* **1982**, 377–385 (b) Petit *et al.*, 1999; reproduced by permission of Nature Publishing Group from *Nature* **1999**, *399*, 429–436 (c) Retallack, 2001d; reproduced by *J. Geol.* **2001**, *109*, 407–426 (d) Berner and Kothavala, 2001; reproduced by permission of American Journal of Science from *Am. J. Sci.* **2001**, *301*, 182–204.

inappropriate for the study of paleosols because of profound modification of the cation exchange complex during burial and lithification of paleosols (Retallack, 1991). Many paleosols are now lithified and amenable to study using petrographic thin sections, X-ray diffraction, electron microprobe, and bulk chemical analysis (Holland, 1984; Ohmoto, 1996; Retallack, 1997a).

5.18.2.1 Molecular Weathering Ratios

Soil formation (see Chapter 5.01) is not only a biological and physical alteration of rocks, but a slow chemical transformation following a few kinds of reactions that seldom reach chemical equilibrium. In many soils, the most important of these reactions is hydrolysis: the incongruent dissolution of minerals such as feldspars to yield clays and alkali and alkaline earth cations in solution. A useful proxy for the progress of this reaction in soils and paleosols is the molar ratio of alumina (representing clay) to the sum of lime, magnesia, soda, and potash (representing major cationic nutrients lost into soil solution). A large database of North American soils (Sheldon *et al.*, 2002) has shown that this ratio is usually less than 2 for fertile soils (Alfisols and Mollisols of Soil Survey Staff, 1999), but more than 2 in less fertile soils (Ultisols). In soils that have been deeply weathered in humid tropical regions for geologically significant periods of time (Oxisols of Soil Survey Staff, 1999), the molar ratio of alumina to bases can reach 100 or more, indicating that the slow progress of hydrolysis has almost gone to completion.

Application of this approach to a Precambrian (1,000 Ma) paleosol from Scotland (Figure 3) showed the expected decrease of hydrolytic weathering down from the surface, and an overall degree of hydrolytic alteration that is modest compared with deeply weathered modern soils (Figure 4). Effects of hydrolysis of this Precambrian paleosol can also be seen in petrographic thin sections and electron microprobe analyses, which document conversion of feldspar into clay (Retallack and Mindszenty, 1994).

Other molar weathering ratios can be devised to reflect leaching (Ba/Sr), oxidation (FeO/Fe_2O_3), calcification ($CaO + MgO/Al_2O_3$), and salinization (Na_2O/K_2O). Two of these ratios reflect differential solubility of chemically comparable elements, but calcification ratio quantifies the accumulation of pedogenic calcite and dolomite, and the ratio of iron of different valence gives reactant and product of iron oxidation reactions. In the Precambrian paleosol illustrated (Figure 4), these molar ratios indicate that the profile was oxidized and well drained, but little leached, calcified or salinized.

Advantages of using molar weathering ratios are their simplicity and precision, free of assumptions concerning parent material composition and changes in volume during weathering and burial compaction. Smooth depth functions of molar weathering ratios (Figure 4) are characteristic of soils and paleosols, whereas parent material heterogeneity is revealed by erratic swings in weathering ratios. Whole-rock chemical analyses are commonly used to calculate molar weathering

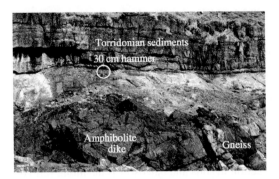

Figure 3 Sheigra paleosol (bleached and reduced zone ~1 m thick to right) under Torridonian (1,000 Ma) alluvial fan deposits, and Staca paleosol at same unfoncormity but on amphibolite (left-hand side) near the hamlet of Sheigra, northwest Scotland (photo courtesy of G. E. Williams; geological age revised by Williams and Schmidt, 1997).

ratios, and thus conflate weathered parts of the soil with unweathered interiors of soil grains. This problem can be circumvented by calculating molar weathering ratios from electron microprobe spot analyses of weathered and unweathered grains within paleosol samples, which can illustrate reaction paths (Bestland and Retallack, 1993).

5.18.2.2 Strain and Mass Transfer Analysis

A full accounting of volume and chemical changes during weathering and burial can be made by assuming that one component of a paleosol has remained stable from the parent material. This method requires measurement of bulk density and identification of a parent material. Alumina, titania, and zirconium are commonly used as stable constituents, with zirconium favored because its mobility can be checked by microscopic examination of pitting of grains of zircon, which is the main soil mineral containing zirconium (Brimhall *et al.*, 1991). The assumption of geochemical stability allows one to calculate volume losses or gains (i.e., strain) of samples from a parent composition and material losses or gains (i.e., mass transfer) of individual chemical elements from a soil or paleosol (see Chapter 5.01). This formulation of strain is especially useful for paleosols, because some component of strain is due to burial compaction, which can be expressed visually (Figure 5).

The Precambrian paleosol illustrated as an example of this approach shows moderate weathering and volume loss with weathering and burial compaction. Most elements were lost from the profile, except potassium, and in one (but not another adjacent) paleosol, iron (Figure 5). This represents a thorough geochemical accounting of

changes relative to zirconium during soil development and burial of this paleosol, but is not at variance with the simpler molar weathering ratio approach, which includes a partial normalization to alumina.

Limitations on calculating strain and mass transfer come mainly from the identification and characterization of the parent material of soils and paleosols. The actual materials from which they weathered no longer exist (Jenny, 1941). The nature of parent materials can be reconstructed by studying the rock or sediment lower within soil or paleosol profiles. Parent material reconstruction can be checked chemically and petrographically for degree of weathering in igneous or metamorphic rocks below a soil (Figures 3 and 4), but is not so easily assessed in sediments or colluvium below a soil. It is difficult to rule out soil formation from a thin sedimentary or colluvial cap to an igneous rock, although large influxes of new material from wind, floods, or landslide will be revealed by positive strain values.

Kinetic modeling approaches (Merino *et al.*, 1993) can be applied to isovolumetric weathering if conservation of volume is supported by textural evidence (Delvigne, 1998). Computer-aided thermodynamic modeling of ancient weathering has also proven useful, especially for Precambrian paleosols (Schmitt, 1999).

5.18.2.3 Analyses of Stable Isotopes of Carbon and Oxygen

Three isotopes of carbon are commonly assayed by mass spectrometer: the common isotope ^{12}C, the rare isotope ^{13}C, and the radiogenic isotope ^{14}C. Radiocarbon is formed in the atmosphere, is incorporated within plants and animals, and is then fossilized as a constituent of carbonates and organic carbon. The progressive radioactive decay of radiocarbon is used for isotopic dating, but unfortunately its abundance decreases to undetectable amounts after ~10^5 yr. In contrast, the stable isotopes ^{12}C and ^{13}C are found in rocks and paleosols of all geological ages. Their relative abundance is commonly reported on a scale of per mil that reflects their ratios normalized to a standard, a fossil belemnite from the Peedee Formation of North Carolina (PDB), or the mean value of modern ocean water (SMOW). These carbon isotopic values ($\delta^{13}C$) are affected by a variety of physical, chemical, and biological processes. The key photosynthetic enzyme of plants, Rubisco, selects the light isotope (^{12}C) preferentially to the heavy isotope (^{13}C), so that plant organic matter is isotopically much lighter ($\delta^{13}C$ more negative) than the atmospheric or oceanic CO_2 from which it was derived.

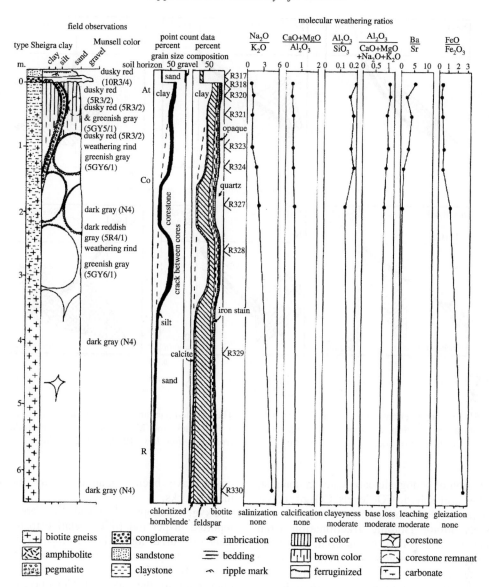

Figure 4 Columnar section (measured in field), petrographic composition (from point counting), and molecular weathering ratios (from major element chemical analyses) of the Sheigra paleosol (Retallack and Mindszenty, 1994) (reproduced by permission of Society for Sedimentary Geology from *J. Sedim. Res.*, **1994**, *A64*, 264–281).

Some plants employ a photosynthetic pathway creating at first a three-carbon phosphoglyceric acid (C_3 or Calvin–Benson photosynthesis). These plants fractionate isotopes more intensely, and so have more negative $\delta^{13}C$ values ($-33‰$ to $-22‰$ PDB) than plants which use a photosynthetic pathway creating at first a four-carbon malic and aspartic acid (C_4 or Hatch–Slack photosynthesis: $-16‰$ to $-9‰$ PDB). Crassulacean acid metabolism (CAM) is yet another photosynthetic pathway, which creates organic matter of intermediate isotopic composition ($-35‰$ to $-11‰$ PDB). Methanogenic microbes are even more extreme in their fractionation of the light isotope ($\delta^{13}C$ down to $-110‰$ and typically $-60‰$ PDB;

Jahren *et al.*, 2001). Today most C_4 plants are tropical grasses, and most CAM plants are submerged aquatic plants and desert succulents. Most other kinds of plants use the C_3 photosynthetic pathway. There is the potential to recognize these various metabolic pathways from the isotopic composition of organic carbon in paleosols and in fossil plants, and in the fossils of animals which ate the plants (Cerling *et al.*, 1997; MacFadden *et al.*, 1999; Krull and Retallack, 2000).

The isotopic composition of carbon in carbonate in paleosols can also be used as a CO_2 paleobarometer (Cerling, 1991). Under high atmospheric CO_2 levels isotopically heavy CO_2 intrudes into soil pores, and can be fixed there by the

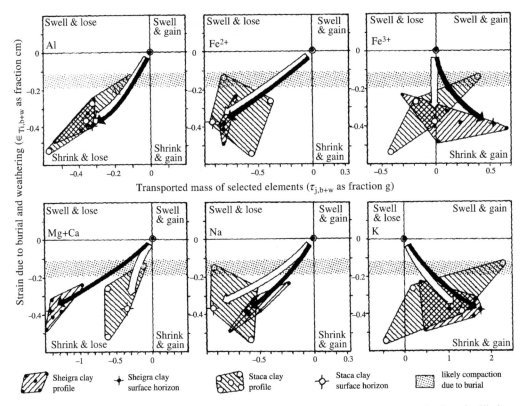

Figure 5 Mass transfer and strain of the Sheigra paleosol. The stippled bars are a range of values for likely strain due to burial compaction, as opposed to pedogenic strain (Retallack and Mindszenty, 1994) (reproduced by permission of Society for Sedimentary Geology from *J. Sedim. Res.*, **1994**, *A64*, 264–281).

precipitation of pedogenic carbonate. In contrast, under low atmospheric CO_2 levels, CO_2 of soil air is isotopically light, because it is respired with relatively minor isotopic fractionation from isotopically light soil plant material, and may, therefore, generate isotopically light pedogenic carbonate. It is also prudent to measure the isotopic composition of organic carbon in the same paleosol, as a guide to the isotopic composition of CO_2 in the ancient atmosphere, because this can vary substantially (Mora *et al.*, 1996; Jahren *et al.*, 2001). A Phanerozoic atmospheric CO_2 curve constructed from a compilation of such data (Ekart *et al.*, 1999) is consistent with independent evidence of CO_2 levels from the stomatal index of fossil leaves during all but a few episodes of catastrophic methane-clathrate dissociation (Retallack, 2001b, 2002).

The attenuation of atmospheric isotopic values within paleosol profiles can also be used to estimate former soil respiration (Yapp and Poths, 1994), sometimes with surprising results, such as the near-modern soil respiration rates inferred from the dramatic attenuation of isotopic values ($\delta^{13}C$) in a Late Ordovician paleosol (Figure 6). In this case, carbonate occluded within pedogenic goethite was analyzed, rather than pedogenic carbonate itself, because this

might have been contaminated by overlying marine rocks.

Oxygen isotopes, ^{16}O and ^{18}O, are usually reported in per mil ($\delta^{18}O$) relative to the same standards used for carbon isotopes (PDB and SMOW). Oxygen isotopes are also fractionated differently by C_3 and C_4 plants because they contribute to the mass of CO_2 taken in for photosynthesis (Farquhar *et al.*, 1993). Oxygen isotopic values are also determined by the composition of water in soil, coming in as rain, and later flowing out as groundwater through buried paleosols (Amundson *et al.*, 1998). Temperature, degree of evaporation, and salinity strongly affect the isotopic composition of oxygen in surface water, and can potentially be inferred from the isotopic composition of oxygen in paleosol carbonates (Mora *et al.*, 1998), paleosol clays (Bird and Chivas, 1993), and fossils in paleosols (Jahren and Sternberg, 2002).

5.18.3　RECORD OF PAST SOIL AND GLOBAL CHANGE

Paleosols have long been recognized in the geological record (Hutton, 1795; Webster, 1826; Buckland, 1837), but their great abundance in terrestrial sedimentary sequences was not appreciated until the 1970s (Allen, 1974; Retallack, 1976). Many variegated red beds, such as the Oligocene Big Badlands of South Dakota, are volumetrically dominated by paleosols (Retallack, 1983). Almost all coal seams are paleosols (Histosols), and these are not the only paleosols in thick coal measure sequences (Retallack, 1994a). Thousands of paleosols of all geological ages have been described since the early 1980s, and there is now the prospect of using them to interpret long-term patterns of environmental and biotic change.

5.18.3.1 Origins of Soil

Soils, like love and home, are difficult to define precisely. If one follows some soil scientists in defining soil as a medium of plant growth (Buol *et al.*, 1997), then the formation of soils began either at the Silurian advent of vascular land plants (Gray, 1993), or at the Cambrian advent of nonvascular land plants (Strother, 2000), or at the Late Precambrian advent of eukaryotic soil algae or algal phycobionts of lichens (Retallack, 1994b; Steiner and Reitner, 2001). A geological view of soils, however, would include rocks and soils altered by hydrolytic weathering, which has been well documented at least as far back as the Archean (2,800 Ma; Rye and Holland, 1998). Hydrolytic weathering has also been proposed for rocks as old as 3,500 Ma (Buick *et al.*, 1995), and meteorites as old as 4,566 Ma (Retallack, 2001a).

The author prefers to follow the US National Aeronautical and Space Administration (NASA) in using the widely understood word soil for nonsedimentary modified surfaces of the Moon and the Mars.

Whether there was or is life on Mars remains uncertain (McSween, 1997). There is no discernible life in lunar or martian soils at the time of this writing, but that may change with future human discoveries and colonization of space. If the Moon and the Mars are considered to have soils, then soil formation goes back to the first alterations of planetismal and planetary surfaces which occurred in place, as opposed to those transported to form sediments, which are distinct and antithetic to soil formation. Thus defined, both soils and sediments are very ancient.

Hydrolytic alteration of mafic minerals (pyroxene and olivine) to clays (iron-rich smectite), oxides (magnetite), carbonates (gypsum, calcite), and salts (kieserite) has been documented in carbonaceous chondritic meteorites (Bunch and Chang, 1980; Volume 1 of this Treatise). Carbonaceous chondrites also show opaque weathering rinds around mafic grains, cross-cutting veins filled with carbonate, clay skins, and distinctive clayey birefringence fabrics (sepic plasmic fabric; Retallack, 2001a). Carbonate veins have been dated radiometrically at no more than 50 Ma younger than enclosing clayey meteorites dated at 4,566 Ma (Birck and Allègre, 1988; Endress *et al.*, 1996). Carbonaceous chondrites are similar to the surface of some asteroids (Veverka *et al.*, 1997). One interpretation of carbonaceous chondrites is

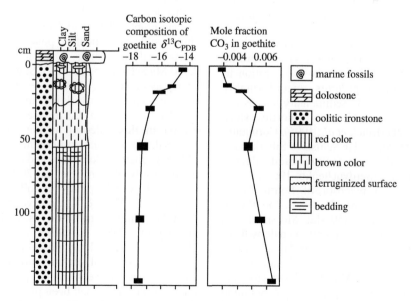

Figure 6 Isotopic composition of carbon in carbonate and mole fraction CO_3 in goethite of a Late Ordovician paleosol from Iron Knob, Wisconsin, showing surprisingly high soil respiration from shallow distance of adjustment of soil to atmospheric values (source Retallack, 2000a; reproduced by permission of Paleontological Society).

as fragments of paleosols from asteroid-sized planetesimals formed early during the formation of the solar system. As primeval soils of the solar system they were similar in their smectites, salts, and carbonates to the soils of Mars, which are probably paleosols relict from a time of free water at the martian surface until at least 2,500 Ma (Retallack, 2001a).

A second interpretation of carbonaceous chondrites is as primary condensates of the solar nebula. By this view, their hydrolytic alteration is due to melting in cometary nuclei during close passes with the Sun, or due to transient heating events by shock waves or collisions (McSween, 1999). Other carbonaceous chondrites show metamorphic alteration with minerals similar to those in Earth formed during deep burial under elevated temperatures and pressures (Brearley, 1999). Like soils and paleosols on Earth and Mars, carbonaceous chondrites demonstrate the great antiquity of hydrolytic weathering in dilute acidic solutions, presumably of carbonic acid derived from water vapor and CO_2. These remain the principal gases released from volcanoes, and soils remain important buffers for this environmental acid.

5.18.3.2 Archean–Paleoproterozoic Greenhouse Paleosols

Despite predictions that Precambrian paleosols would be thin, rocky, and dominated by physical weathering (Schumm, 1956), hundreds of Precambrian paleosols now described have been found to be thick, clayey, deeply weathered, and in some cases with possible traces of life, so that chemical and biological weathering can be traced back almost to the beginning of the suitably preserved sedimentary rock record on Earth (Rye and Holland, 2000). The oldest known profile interpreted to be a paleosol shows alteration to depths of 50 m on granites unconformably underlying the 3,500 Ma sedimentary succession of the Warrawoona Group in northwestern Western Australia (Buick *et al.*, 1995). Corundum ores in the 3,500 Ma Aldan Schists of Siberia have been interpreted as metamorphosed, deeply weathered bauxites (Serdyuchenko, 1968). The Jerico Dam paleosol of South Africa (3,000 Ma; Grandstaff *et al.*, 1986), the Pronto paleosol of Canada (2,450 Ma; Mossman and Farrow, 1992), the Hokkalampi paleosol of Finland (2,200 Ma; Marmo, 1992), a variety of paleosols associated with the Hekpoort Basalt of South Africa (2,100 Ma; Yang and Holland, 2003), and the Sheigra paleosol of Scotland (1,000 Ma; Figures 3 and 4) have been subjected to exceptionally detailed geochemical and petrographic analyses. Along with many other

Precambrian paleosols reviewed by Rye and Holland (1998), these paleosols reveal the antiquity and thoroughness of hydrolytic weathering during the Precambrian. Even then, rock and sediment were under relentless acid attack, which leached base cations (especially Ca^{2+}, Mg^{2+}, and Na^+), and left thick, clayey soil.

It is likely that at least back to 3,500 Ma the principal environmental acid driving this hydrolytic reaction was carbonic acid dissolved in rain water and groundwater (Holland, 1984), as is the case in soils today (Nahon, 1991). Much soil CO_2 may also have come from respiring organisms, which also could have contributed organic acids. Nitric and sulfuric acid may have been locally important in soils developed on particular parent materials, but nitrogen and sulfur salts are so far unreported in Precambrian paleosols, unlike modern soils of mine dumps (Borden, 2001), and hypothesized modern soils on Mars (Bell, 1996; Farquhar *et al.*, 2002), and Venus (Barsukov *et al.*, 1982; Basilevsky *et al.*, 1985).

This view of the likely acids involved in creating Precambrian soils on Earth is supported by the isotopic composition of carbon, nitrogen, and sulfur in sedimentary organic matter, carbonates, sulfates, and sulfides, which are surprisingly similar to their modern counterparts back to 3,500 Ma, and unlike meteoritic or mantle values (Schidlowski *et al.*, 1983; Des Marais, 1997; Canfield and Teske, 1996).

Evidence for life in Precambrian soils comes from isotopic studies of organic carbon within paleosols. Microlaminated chips in the 2,765 Ma Mt. Roe paleosol of Western Australia have extremely depleted carbon isotopic compositions ($\delta^{13}C_{org}$ −40‰). Isotopic fractionation of carbon to this degree is only known in methanogens and methanotrophs (Rye and Holland, 2000). These chips could be fragments of pond scum rather than a true soil microbiota. Organic matter in the 2,560 Ma Schagen paleosols of South Africa is not nearly as depleted (−16‰ to −14‰ $\delta^{13}C_{org}$) as organic matter in overlying marine sediments (−35‰ to −30‰ $\delta^{13}C_{org}$). Interpretation of the carbon in this paleosol as the signature of a hypersaline microbial soil community is compatible with shallow dolocretes and other features of the paleosols (Watanabe *et al.*, 2000).

Normal isotopic values for soil organic matter (−25‰ to −27‰ $\delta^{13}C_{org}$) have been reported from Precambrian paleosols as well (Mossman and Farrow, 1992; Retallack and Mindszenty, 1994). Virtually all Precambrian paleosols have a very low content of organic carbon comparable with to that of well-drained paleosols of the Phanerozoic. If life had been present in the Early Precambrian paleosols, they would have become carbonaceous in the absence of a decomposing microbiota of actinobacteria and of fungi and metazoans during the later Precambrian. Isotopic evidence thus suggests that methanogenic, hypersaline, normal, and decompositional microbes were present in Precambrian paleosols. Other evidence for life in Precambrian paleosols includes microfossils (1,300 Ma; Horodyski and Knauth, 1994), microbial trace fossils (2,200 Ma; Retallack and Krinsley, 1993), chemofossils (2,900 Ma; Prashnowsky and Schidlowski, 1967), plausible megafossils (2,900 Ma; Hallbauer *et al.*, 1977; Retallack, 1994b), and the impressive thickness and soil structure of Precambrian paleosols (3,500 Ma; Retallack, 1986, 2001a; Buick *et al.*, 1995; Gutzmer and Beukes, 1998; Beukes *et al.*, 2002). Life and its by-products such as polysaccharides may have been soil-binders like molasses applied to a cornfield (Foster, 1981), protecting soils from physical weathering so that biochemical weathering could proceed.

The likely existence of microbial mats at the soil surface considerably complicates the use of paleosols as indicators of ancient atmospheres (Ohmoto, 1996). Tropical rainforest soils now have soil CO_2 levels up to 110 times that of the atmosphere, because of high levels of soil respiration by termites and microbes of an abundant supply of soil organic matter that forms in a living membrane separating the subsoil from the atmosphere (Brook *et al.*, 1983; Colin *et al.*, 1992). Nevertheless, modeling by Pinto and Holland (1988) makes it unlikely that microbial scums of the Precambrian were as productive and effective membranes as rainforests.

The observation that Precambrian paleosols were chemically weathered to an extent comparable with rainforest soils today probably indicates much higher levels of CO_2 in the atmosphere at that time (Holland, 1984). The extent of this greenhouse is poorly constrained, but the apparent lack of siderite in paleosols such as the Hekpoort and Mt. Roe paleosols has been used by Rye *et al.* (1995) to argue that CO_2 concentrations could not have been more than ca. 100 times present levels before the rise of oxygen at ca. 2,100 Ma.

Siderite is common in Phanerozoic wetland paleosols (Ludvigsen *et al.*, 1998) in which respired soil CO_2 exceeded this level. Thus, the estimate of Rye *et al.* (1995) of no more than 100 times present levels of soil CO_2 also is a cap on soil respiration and biological productivity during the Precambrian (Sheldon *et al.*, 2001). The contribution of CH_4 to the atmospheric greenhouse effect was probably also much higher than at present, because it was necessary to maintain planetary temperatures above that of the freezing of water at the time of a faint young Sun (Kasting, 1992; Pavlov *et al.*, 2000).

5.18.3.3 Proterozoic Icehouse Paleosols

The oldest known periglacial paleosols are from the 2,300 Ma to 2,400 Ma Ramsay Lake Formation of Ontario, Canada (Young and Long, 1976; Schmidt and Williams, 1999). They have prominent ice wedges, which are strongly tapering cracks filled originally with ice, but now with massive or horizontally layered sand and claystone breccia. Modern ice wedges form in climates with a mean annual temperature of $-4\,°C$ to $-8\,°C$, coldest month temperatures of $-25\,°C$ to $-40\,°C$, warmest month temperatures of $10–20\,°C$, and a mean annual precipitation of $50–500$ mm (Williams, 1986; Bockheim, 1995). Periglacial paleosols of the Late Precambrian (600–1,000 Ma) in Scotland, Norway, and South Australia include sand wedges (Figure 7), which indicate an even drier and more frigid climate: a mean annual temperature of $-12\,°C$ to $-20\,°C$, a mean cold-month temperature of $-35\,°C$, a mean warm-month temperature of $4\,°C$, and mean annual precipitation of 100 mm (Williams, 1986).

Some of the Late Precambrian glaciations were remarkable in extending to very low latitudes, as indicated by the paleomagnetic inclination of glaciogene sediments, and have been dubbed Snowball Earth events (Kirschvink, 1992; Hoffman *et al.*, 1998; Schmidt and Williams, 1999). Between and before these Precambrian episodes of periglacial paleosols and associated glaciogene sediments there is no evidence of frigid conditions, so that the alternation of global

Figure 7 Near-vertical sandstone wedge remaining from fill of ice wedge penetrating the Cattle Grid Breccia (680 Ma), in the Mt. Gunson Mine, South Australia (photo courtesy of G. E. Williams).

icehouse and greenhouse paleoclimates is ancient indeed.

These climatic fluctuations could be attributed to changes in solar luminosity, volcanic degassing, or ocean current reorganization with continental drift (Barley *et al.*, 1997; Dalziel, 1997), but paleosols reveal that these ice ages were also times of change in the atmosphere and life on land. Highly ferruginized pisolitic lateritic paleosols first appear in the geological record at 1,920–2,200 Ma in South Africa (Gutzmer and Beukes, 1998; Beukes *et al.*, 2002). The lateritic paleosols are part of a complex erosional landscape with a variety of paleosols of significantly different geological ages, including mildly oxidized (Retallack, 1986; Maynard, 1992) and chemically reduced paleosols (Rye and Holland, 1998). Opinions differ on the nature and timing of this apparent oxygenation event. Holland (1984), Holland and Beukes (1990), and Yang and Holland (2003) proposed an abrupt rise from less than 0.1% by volume to more than 3% O_2 at ~2,100 Ma.

In contrast, Ohmoto (1996, 1997) and Beukes *et al.* (2002) argue that the Great Oxidation Event interpretation does not take into account the reducing power of biological activity within Precambrian paleosols, and that O_2 levels were close to present levels from 3,000 Ma to 1,800 Ma. An intermediate view of rising, but fluctuating atmospheric oxidation also is compatible with available paleosol data (Retallack, 2001a), and with limited evidence from mass-independent fractionation of sulfur isotopes (Farquhar *et al.*, 2002).

Oxidation of the atmosphere and soils could have come from lichens, possibly actinolichens, considering the small diameter of their filaments, reported from the 2,900 Ma Carbon Leader of South Africa (Hallbauer *et al.*, 1977). Their organic geochemical and isotopic composition gives clear evidence of a photosynthetic component (Prashnowsky and Schidlowski, 1967). The potent greenhouse gas CH_4 was produced by methanogens, detected isotopically in a paleosol dated at 2,765 Ma (Rye and Holland, 2000). Later, plausibly lichenlike and carbon-sequestering organisms are represented by enigmatic, small (1 by 0.5 mm), encrusted, and ellipsoidal objects in the 2,200 Ma Waterval Onder and correlative paleosols (Retallack and Krinsley, 1993; Gutzmer and Beukes, 1998). A later swing to greenhouse conditions could be inferred from molecular sequence data for a Mid- to Late Precambrian (1,458–966 Ma) origin of ascomycete fungi, after the origin of algae and before the origin of metazoans (Heckman *et al.*, 2001). This question is also discussed in Chapter 5.06.

There has long been a debate about plausible permineralized ascomycetes in the 770 Ma Skillogallee Dolomite of South Australia

(Retallack, 1994b). Late Precambrian (600 Ma) enigmatic fossils, widely called "Twitya disks" after their original northwest Canadian discovery site, are probably microbial colonies (Grazdhankin, 2001), and some have been found in ferruginized paleosols (Retallack and Storaasli, 1999). Latest Precambrian (550–540 Ma) interglacial and postglacial circular fossils, widely interpreted as cnidarian medusae, have also been reinterpreted as lichenized microbial colonies and are found in paleosols (Retallack, 1994b; Grazdhankin, 2001; Steiner and Reitner, 2001). The appearance of lichens with their deeply reaching rhizines in a world of cyanobacterial mats could have greatly increased the rate of biochemical weathering, carbon sequestration, oxygenation of the atmosphere, and global cooling (Schwartzmann and Volk, 1991).

5.18.3.4 Cambro-Ordovician Greenhouse Paleosols

The most obvious way in which Ordovician paleosols differ from those of the Precambrian is in the local abundance of animal burrows. Because burrows are known in Late Precambrian marine rocks, the main problem in establishing the presence of animals on land during the Ordovician was to prove that the burrows were formed at the same time as the paleosols, and not during inundation before or after soil formation. This evidence came in part from petrographic studies of soil carbonate in the paleosols, which is cut by some burrows and cuts across other burrows (Figure 8). This carbonate is a largely micritic mixture of calcite and dolomite, as is common in pedogenic carbonates (Retallack, 1985). Compelling evidence also came from the isotopic composition of carbon in this carbonate, which was isotopically too light to have formed in aquatic or marine environments (Retallack, 2001c).

Comparable burrows and tracks of millipede-like creatures have now been reported in several Ordovician paleosol sequences (Johnson *et al.*, 1994; Trewin and McNamara, 1995; Retallack, 2000a), but these were probably only a small part of the overall soil respiration of Ordovician paleosols. Glomalean fungi discovered in Ordovician marine rocks of Wisconsin (Redecker *et al.*, 2000) were also part of an active community of microbial soil respirers. Burrows are not obvious in the Late Ordovician Iron Knob paleosol of Wisconsin, but the short distance of attenuation to atmospheric values of CO_2 mole fraction and $\delta^{13}C$ values of carbon in goethite of that paleosol (Figure 6) indicate soil respiration rates comparable to those of modern savanna grassland soils (Yapp and Poths, 1994). This is remarkable, because there are no clear root traces in

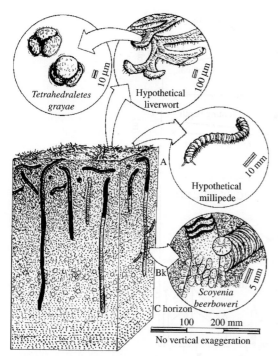

Figure 8 Reconstructed ecosystem of the Late Ordovician Potters Mills paleosol from central Pennsylvania (Retallack, 2000a) (reproduced by permission of Palaeontological Association from Phanerozoic Terrestrial Ecosystems, **2000**, *6*, pp. 21–45).

Ordovician paleosols, and palynological evidence indicates no more than a cover of liverwort-like plants to feed such soil respiration (Strother *et al.*, 1996; Strother, 2000). Primary carbon fixation by these thin thalli with short root hairs could not have created a quantity of biomass or humus comparable to that of modern grasslands.

Furthermore, organic-lean, red Ordovician paleosols contain only sparse reduction spots and soil carbonate nodules (Retallack, 1985; Driese and Foreman, 1992), indicating modest carbon storage in soil organic matter and carbonate compared, e.g., with modern savanna grassland soils (de Wit, 1978). The Ordovician paleosols studied so far show unusually high soil respiration, considering their probable low levels of primary productivity. They also formed at a time estimated from sedimentary mass balance models as the steamiest greenhouse period of all Phanerozoic time, with ~16 times the present atmospheric levels of CO_2 (Berner and Kothavala, 2001). The carbon budget of known Ordovician paleosols would have contributed to this greenhouse.

5.18.3.5 Terminal Ordovician Icehouse Paleosols

Periglacial paleosols, unknown in Cambrian and Early to Middle Ordovician rocks, are found again in latest Ordovician (Hirnantian) rocks. Periglacial paleosols are best documented in South Africa, where patterned ground and sand wedges are common in red beds of the Pakhuis Formation (Daily and Cooper, 1976). The ice sheets extended over much of Africa (Ghienne, 2003).

The causes of this ice age are especially enigmatic, because volcanic activity increased through the Ordovician and the continents were dispersed (Bluth and Kump, 1991), thus working against cold Late Ordovician poles. Mass balance models make the Ordovician ice age seem particularly enigmatic, because they predict atmospheric CO_2 levels 16 times PAL (Berner and Kothavala, 2001). This may be an artifact of the 10 Ma spacing of data points in the model, blurring the less than 10 Ma duration of the ice age that is derived from carbon isotopic data (Brenchley *et al.*, 1994). Studies of carbonate isotopic compositions from paleosols within the glacial interval are needed to re-examine this question.

Also needed is an examination of paleosols within this interval for evidence of fossil mosses, which would have been more deeply rooted than liverworts and so have accelerated weathering and carbon sequestration. Rare Late Ordovician moss-like megafossils (Snigirevskaya *et al.*, 1992) and spores (Nøhr-Hansen and Koppelhus, 1988) support indications from cladistic analysis (Kenrick and Crane, 1997) for a latest Ordovician origin of mosses.

5.18.3.6 Siluro-Devonian Greenhouse Paleosols

Root traces of vascular land plants appear in Silurian paleosols, but until the Early Devonian, root traces are small and shallow within the profiles (Figure 9(b)). The earliest known vascular land plants of the Middle and Late Silurian lacked true roots. Instead, they had stems that ran along the surface and just beneath the surface of the soil as runners and rhizomes furnished with thin unicellular root hairs (Kenrick and Crane, 1997). Plant bioturbation in soils only extended down to a few centimeters, but burrows of millipedes reached more deeply, and in some soils were more abundant than plant traces (Retallack, 1985). In addition to detritivorous and perhaps also herbivorous millipedes (Retallack, 2001c), Late Silurian soil faunas included predatory centipedes and spiderlike trigonotarbids (Jeram *et al.*, 1990). Fungal hyphae and spores in Silurian and Devonian rocks indicate proliferation of chytrids and other fungi (Sherwood-Pike and Gray, 1985; T. N. Taylor and E. L. Taylor, 2000).

Early Devonian paleosols have abundant traces of true roots, including woody tap roots of a variety

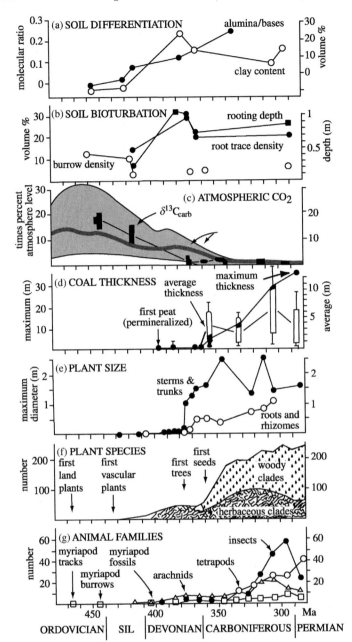

Figure 9 Early Paleozoic changes in (a) soil differentiation as indicated by clay content (volume percent) and alumina/bases (molar ratio) of the most weathered horizon of calcareous red paleosols; (b) soil bioturbation as indicated by proportion of transect in paleosols occupied by roots or burrows (percent) and by measured rooting depth (m); (c) atmospheric CO_2 levels (PAL) calculated from a sedimentary mass balance model; (d) maximum coal seam thickness and average thickness of at least 10 consecutive seams (m); (e) diameter of fossil plant stems and roots (m); (f) diversity of fossil land plants (number of species); (g) diversity of soil animals (number of families) (Retallack, 1997c) (reproduced from *Dinofest*, **1997**, pp. 345–359).

of land plants (Elick *et al.*, 1998). Root traces reached tens of centimeters down into paleosols, extending greatly the depth of the active rhizosphere and its associated mucigel of microbes. Among the numerous roots of Early and Middle Devonian paleosols, the burrows of soil fauna are less prominent (Figure 9(b)). Devonian soils also

have a higher clay content and are more deeply weathered of bases than Silurian or Ordovician soils (Figure 9(a)). They have isotopically lighter pedogenic carbon, closer to the isotopic composition of coexisting organic carbon, than Silurian and Ordovician paleosols (Mora *et al.*, 1996; Mora and Driese, 1999).

Within the parameters of the pedogenic carbonate palaeobarometer of Cerling (1991), these data indicate declining atmospheric levels of CO_2 from the Silurian into the Devonian (Figure 9(c)). Consumption of atmospheric CO_2 by increased hydrolytic weathering, and burial of carbon in limestone and organic matter during the Silurian and Devonian has been widely interpreted as an instance of atmospheric global change induced by the evolution of life (Retallack, 1997b; Berner, 1997; Algeo and Scheckler, 1998).

5.18.3.7 Late Devonian to Permian Icehouse Paleosols

Periglacial paleosols and glaciogene sedimentary facies unknown in Silurian and Early to Middle Devonian appear in the latest Devonian, and remain locally common in Carboniferous and Permian rocks, especially within the Gondwana supercontinent, then positioned near the south pole (Figure 10; Krull, 1999). Unlike periglacial paleosols of the Ordovician and Precambrian however, these Late Paleozoic profiles include root traces of what must have been frost-hardy woody plants. The earliest documented examples of tundra (polar shrubland) vegetation have been found in paleosols with freeze–thaw banding and thufur mounds in Carboniferous glacigene sedimentary rocks near Lochinvar in southeastern Australia (Retallack, 1999a). Taiga (polar forest) paleosols with discontinuous permafrost deformation are found in Early Permian red beds near

Kiama, also in southeastern Australia (Retallack, 1999b).

Milankovitch-scale temporal variation in climate and sea level has long been recognized in cyclothemic sedimentation in North American paleotropical Carboniferous marginal marine sequences, and this in turn has been related to ice-volume fluctuations on the south polar Gondwana supercontinent (Rasbury *et al.*, 1998; Miller and West, 1998). Full glacial coal seams (Histosols) alternating with interglacial marine rocks are a clear indication of these changes. Environmental alternations of full-glacial, dry, calcareous, swelling-clay soils (Vertisols), and interglacial, wet, decalcified, forest soils (Alfisols) indicate a terrestrial contribution to multimillenial-scale change in atmospheric greenhouse gases and paleoclimate (Retallack, 1995).

By Middle Devonian time the evolution of increasingly larger plants culminated in the evolution of trees with trunks up to 1.5 m in diameter, which leave obvious large root traces in paleosols (Driese *et al.*, 1997), as well as abundant permineralized stumps and logs (Meyer-Berthaud *et al.*, 1999). Middle Devonian paleosols are also the oldest known with clay-enriched subsurface horizons (argillic horizons of Soil Survey Staff, 1999). The clay in modern forest soils is partly formed by weathering in place, and is partly washed down root holes, which taper strongly downward in forest trees. Evidence of both neoformation and illuviation of clay can be seen in thin sections of Devonian forested paleosols (Retallack, 1997b).

Figure 10 Deep clastic dike in a coal of the Weller Coal Measures of the Allan Hills, Antarctica, interpreted as infill of periglacial polygonal patterned ground (E. S. Krull and hammer for scale).

Latest Devonian paleosols also include coals from the oldest woody peats. Thin peats of herbaceous plant remains such as the Rhynie Chert of Scotland (Rice *et al.*, 1995) and the Barzass coal of Siberia (Krassilov, 1981) are found in Early Devonian rocks, but by the latest Devonian (Algeo and Scheckler, 1998) and into the Carboniferous, woody coals became widespread and thick (Figure 9(d)). Carbon consumption by accelerated weathering in forest soils and carbon burial in coals are widely acknowledged as the likely cause for mass balance estimates of Late Paleozoic high atmospheric oxygen levels (perhaps 35 vol.%) and near-modern CO_2 levels (Berner *et al.*, 2000). Low Permian atmospheric CO_2 levels are also confirmed by stomatal index studies (Retallack, 2001b). These atmospheric trends and coeval changes in oceanic Mg/Ca ratio could be attributed to changes in volcanic and hydrothermal activity, particularly at mid-ocean ridges (Stanley and Hardie, 1999). However, the abundance of Early Paleozoic pedogenic dolomite, but Late Paleozoic and Neogene pedogenic calcite (Retallack, 1985, 1993), suggests a role for soils in these changes in oceanic ionic chemistry, as well as in changing atmospheric CO_2 levels.

5.18.3.8 Triassic–Jurassic Greenhouse Paleosols

Greenhouse paleoclimates right from the very beginning of the Mesozoic have been revealed by discovery of deeply weathered paleosols in earliest Triassic rocks of Antarctica (Figure 11), which even at that time was at paleolatitudes of 65–77° S (Retallack and Krull, 1999). Comparable modern soils are Ultisols (of Soil Survey Staff, 1999) and Acrisols (of FAO, 1988), which are not found either north of 48° N latitude or south of 40° S, and are rare outside subtropical regions. Greenhouse conditions at this time are also indicated by stomatal index studies of fossil seed ferns (Retallack, 2001b) and by the isotopic composition of carbon and oxygen in marine and nonmarine carbonate and organic matter (Holser and Schönlaub, 1991).

The timing and magnitude of this greenhouse and isotopic excursion immediately at and after the greatest mass extinction of all time has suggested a catastrophic release of methane from permafrost or marine clathrate deposits (Krull and Retallack, 2000; Krull *et al.*, 2000). There is no other source of carbon that is sufficiently large and isotopically depleted to create the observed

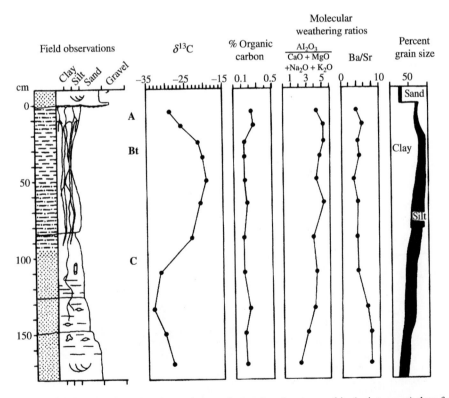

Figure 11 Geochemical (isotopic and major and trace element) and petrographic (point count) data for a deeply weathered paleosol which formed at a latitude of 70° S in the Allan Hills of Antarctica during the Early Triassic. This is the John paleosol, an Ultisol (Sombrihumult), as indicated especially by its high ratios of alumina to bases and barium to strontium, and its strong subsurface enrichment in clay. Extremely light carbon isotopic values deep in the profile imply a role for methanogenic methane in this postapocalyptic greenhouse (Krull and Retallack, 2000) (reproduced by permission of Geological Society of America from *Geol. Soc. Am. Bull.*, **2000**, *112*, 1459–1472).

negative carbon isotopic anomaly. Release mechanisms for methane could have included meteorite impacts, Siberian Traps volcanism, or continental shelf collapse, which also have been invoked as causes for extinctions at this time (Hallam and Wignall, 1997). Really large life crises were also times of transient global greenhouses indicated by stomatal index data (Retallack, 2001b) at the earliest Jurassic (Pliensbachian), Early Jurassic (Toarcian), Mid-Jurassic (Bathonian), Early Cretaceous (Aptian), Mid-Cretaceous (Cenomanian-Turonian), earliest Paleocene (Danian), and earliest Eocene (Ypresian).

During the Early Mesozoic, atmospheric CO_2 minima also were high (at least twice that of the present), and this general and long-term greenhouse calls for a different and noncatastrophic explanation. Paleosols and permineralized wood of forest ecosystems at high latitudes provide evidence for this long-term greenhouse during which no periglacial paleosols are recorded (Ollier and Pain, 1996; Retallack, 2001a). The Triassic appearance of large sauropod dinosaurs such as *Massospondylus* and *Plateosaurus*, together with footprints and other dinoturbation (Lockley, 1991), and of a variety of termite and ant nests in paleosols (Hasiotis and Dubiel, 1995), would have effectively increased the destruction of woody tissues in and on soils (Olsen, 1993). The effect of such evolutionary innovation may have been to decrease carbon sequestration by lignin in swamps, forests, and their soils.

5.18.3.9 Early Cretaceous Icehouse Paleosols

Fossil patterned and hummocky ground reveal permafrost conditions during the Early Cretaceous (Aptian) sediments of southeastern Australia, which at that time was at 66–76° S and attached to the Antarctic portion of the Gondwana supercontinent (Rich and Vickers-Rich, 2000). This ice age does not appear to have been as extensive or severe as the Permo-Carboniferous or modern ice ages. This episode of planetary cooling coincides with a dramatic evolutionary radiation of flowering plants (Retallack and Dilcher, 1986; Truswell, 1987; Crane *et al.*, 1995). The key evolutionary innovation of flowering plants was an abbreviated life cycle, in which pollination, fertilization, and germination followed one another in quick succession (Wing and Boucher, 1998).

Early angiosperms were largely confined to weakly developed soils (Entisols) of disturbed coastal and streamside habitats, which they colonized and weathered more rapidly than associated conifers and cycadlike plants (Retallack and Dilcher, 1981). Angiosperm leaves were less coriaceous and less well defended with resins and other toxins, and so rotted more rapidly to create a richer soil humus than leaves of conifers and cycadlike plants (Knoll and James, 1987). Erosion control and soil humification from newly evolved angiosperms may have played a role in Early Cretaceous chilling.

5.18.3.10 Cretaceous–Paleogene Greenhouse Paleosols

Another long period of generally warmer planetary climates without evidence of polar ice caps or periglacial paleosols lasted from the Mid-Cretaceous to the latest Eocene. Mid-Cretaceous (Cenomanian) tropical paleosols (Ultisols and Oxisols) are known from South Australia, then at 60° S (Firman, 1994), and the US, then at 45° N (Thorp and Reed, 1949; Joeckel, 1987; Mack, 1992). The Mid-Cretaceous greenhouse was unusually long and profound, judging from the stomatal index of fossil ginkgo leaves (Retallack, 2001b). The volcanic activity that created the enormous Ontong-Java Plateau has been cited as a cause for this long-term greenhouse (Larson, 1991), but there is another plausible explanation in the co-evolution with angiosperms of ornithopod dinosaurs such as *Iguanodon*, with their impressive dental batteries for processing large amounts of foliage. The feeding and trampling efficiency of these large newly evolved dinosaurs may have further promoted the spread of early angiosperms with their ability to tolerate higher levels of disturbance than other plants (Bakker, 1985). Newly evolved ornithopod dinosaurs and their trackways are associated with carbonaceous and early successional paleosols (Entisols, Inceptisols, and Histosols), whereas archaic sauropod dinosaurs and their trackways remained associated with less fertile and less carbonaceous paleosols (Aridisols) throughout the Cretaceous (Retallack, 1997c).

Other times of unusually extensive tropical paleosols were the latest Paleocene (55 Ma; Taylor *et al.*, 1992), latest Eocene (35 Ma; Bestland *et al.*, 1996; Retallack *et al.*, 2000), and Middle Miocene (16 Ma; Schwarz, 1997). These events are notable as short-lived (<0.5 Myr) spikelike warmings in both stable isotopic records from the ocean (Veizer *et al.*, 2000; Zachos *et al.*, 2001) and stomatal index studies (Retallack, 2001b). The latest Paleocene warm spike is associated with such profound carbon isotopic lightening that it can only reasonably be attributed to the methane from isotopically light methane clathrates from the ocean floor or permafrost (Koch, 1998). Short-term physical forcings are thus also recorded in the paleosol record of paleoclimate.

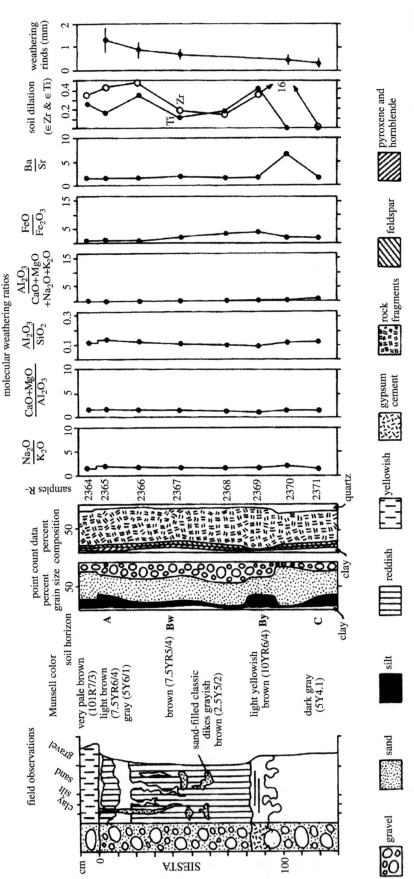

Figure 12 Geochemical data for the Siesta paleosol of Pliocene age (3 Ma) from the Meyer Desert Formation at Oliver Bluffs, central Transantarctic Mountains (Retallack *et al.*, 2001) (reproduced by permission of Geological Society of London from *Geol. Soc. London J.*, **2001**, *158*, 925–935).

5.18.3.11 Neogene Icehouse Paleosols

Periglacial paleosols appear during Late Miocene time (8 Ma) in Antarctica (Sugden *et al.*, 1995; Retallack *et al.*, 2001), where soil development is so slow that some surface soils may be of comparable antiquity (Campbell and Claridge, 1987). Antarctic soil formation is not only promoted by ground ice deformation, but includes the effects of salt accumulation and eolian mass addition in an extremely dry continental frigid climate (Figure 12).

The Late Miocene is best known for the Messinian salinity crisis, when the Mediterranean Sea became a desert (Krijgsman *et al.*, 1999). It was also a significant time for geographic and climatic expansion of grassland biomes and their characteristic soils: Mollisols of Soil Survey Staff (1999) or Chernozems of FAO (1988). Evidence for this transformation in tropical regions comes from the dramatic change to a less depleted (less negative) carbon isotopic composition ($\delta^{13}C$) of pedogenic carbonate and organic matter, and of the apatite of fossil mammalian tooth enamel attributed to the tropical expansion of C_4 grasses (Cerling *et al.*, 1997; MacFadden, 2000). There is also evidence from adaptations to grazing in fossil mammals (Janis *et al.*, 2002), from traces of grassland invertebrates such as dung beetles (Genise *et al.*, 2000), and from increased abundance of silica bodies (phytoliths) and pollen characteristic of grasses (Strömberg, 2002).

Paleosols also demonstrate Late Miocene expansion of grasslands capable of forming sod of the sort that is unrolled to create lawns and golf courses. The dense growth of fine (<2 mm diameter) adventitious roots, together with the slime of abundant earthworms, create a characteristic soil structure consisting of fine crumb peds, which can be preserved in paleosols (Figure 13). Grassland soils are also unusually rich in organic matter, intimately admixed with clay, often with as much as 10 wt.% C down to a meter or more, although this organic matter is not always preserved in paleosols. The soft, low-density upper horizons of grassland soils are also rich in mineral nutrients (Ca^{2+}, Mg^{2+}, Na^+, K^+), and their subsurface horizons commonly include nodules of soil carbonate (usually micritic low magnesium calcite). It has long been known that such pedogenic nodules form at shallow depths within soil profiles in dry climates and deeper within the profile in more humid climates (Jenny, 1941; Retallack, 1994c). Observations of depth to carbonate horizon together with root traces and crumb peds of grassland paleosols can be used to constrain the paleoclimatic range of grasslands (Retallack, 1997d, 2001d).

Observations on hundreds of paleosols in the North American Great Plains, Oregon, Pakistan, and East Africa have revealed a broad schedule of origin and paleoclimatic expansion of bunch and then sod grasslands (Figure 14). The increased organic carbon content, high internal surface area, elevated albedo, and greater water retention capacity of grasslands compared with woodlands of comparable climatic regions would have been a potent force for global cooling as grasslands emerged to occupy almost a quarter of the current land area (Bestland, 2000; Retallack, 2001d). Mountain uplift and ocean currents played a role in Neogene climate change as well (Raymo and Ruddiman, 1992; Ramstein *et al.*, 1997), but there remain problems with the timing and magnitude of carbon sequestration by these physical mechanisms (Retallack, 2001d).

5.18.3.12 Pleistocene Glacial and Interglacial Paleosols

Over the past million years large ice caps have grown to engulf the present-day location of Chicago within more than a kilometer of ice during glacial maxima, then retreated to the current ice caps of Greenland and Alaska during interglacial times at Milankovitch scale frequencies of 100 ka. There have also been less extreme paleoclimatic oscillations on the other Milankovitch frequencies of 42 ka and 23 ka (Hays *et al.*, 1976; Petit *et al.*, 1999). In Illinois, interglacials are defined by paleosols such as the Sangamon paleosol, which is comparable with modern forest soils under oak–hickory forest. The 42 ka and 23 ka interstadials are defined by paleosols such as the Farmdale paleosol, which is comparable with modern boreal forest paleosols under spruce forest

Figure 13 Tall grassland (Mollisol) paleosol with thick, dark brown crumb-textured surface over a deep (79 cm) white nodular calcic horizon, over a thinner short grassland paleosols with carbonate nodules at a depth of 39 cm, in the Late Miocene (7 Ma) Ash Hollow Formation, 13 km north of Ellis, Kansas.

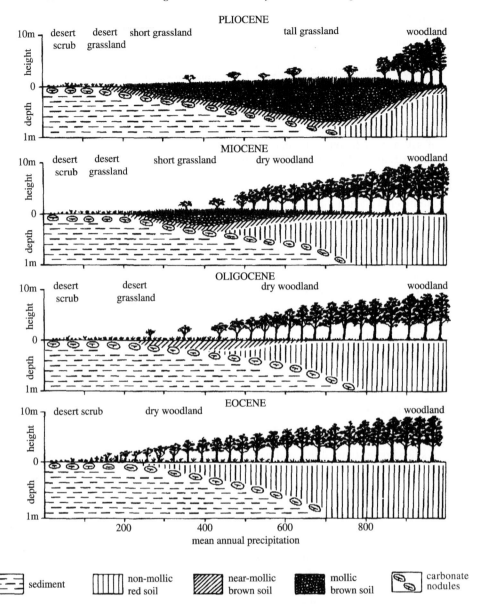

Figure 14   A scenario for climatic and geographic expansion of grasslands and their soils in the Great Plains of North America (Retallack, 1997d) (from *Palaios* **1997**, *12*, pp. 380–390. reproduced by permission of Society for Sedimentary Geology.).

(Follmer *et al.*, 1979). Ice, till, loess, and periglacial soils (Gelisols) alternated with forest soils (Alfisols or Inceptisols) through these paleoclimatic fluctuations.

Oscillations between different ecosystems can be inferred from many paleosol sequences, even beyond the ice margin (Figure 15). In the Palouse loess of Washington, for example, grassland soils (Mollisols) with crumb peds and earthworm castings during interglacials and interstadials alternate with sagebrush soils (Aridisols) with cicada burrows and shallow carbonate horizons during glacials and interstadial minima (Busacca, 1989; O'Geen and Busacca, 2001). Vegetation of the paleosols can be inferred from carbon isotopic values typical for CAM saltbush in the sagebrush paleosols, and for C_3 grasses in the grassland paleosols, as well as from the characteristic phytoliths of these plants (Blinnikov *et al.*, 2002). Comparable alternations of ecosystems with paleoclimatic fluctuation are seen in many Quaternary sequences of paleosols (Paepe and van Overloop, 1990; Feng *et al.*, 1994; Wang *et al.*, 1998). Differences in primary production and carbon sequestration of these alternating ecosystem types on a global basis may have played a role in the relative abundance of greenhouse gases during glacial–interglacial paleoclimatic cycles.

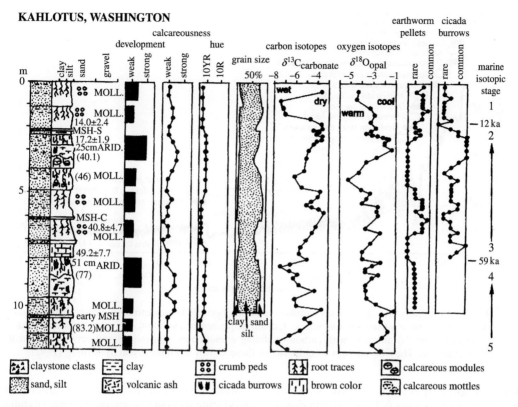

Figure 15 Alternating paleoenvironments of desert CAM shrublands and C₃ grasslands represented by a sequence of Quaternary paleosols (0–100 ka) in the Palouse loess near Kahlotus, Washington, USA; showing (left to right), field section with thermoluminescence dates and paleosol identification (Moll. is Mollisol, Arid. is Aridisol), paleosol position and development (represented by black boxes), paleosol calcareousness (scale based on acid reaction), Munsell hue (measured dry in field), sand-silt-clay proportion, $\delta^{13}C$ of pedogenic carbonate, $\delta^{18}O$ of opal phytoliths, abundance of earthworm pellets, and abundance of cicada burrows (source Retallack, 2001c).

Ice core records show as little as 180 ppmv CO_2 during glacial periods and 280 ppmv during interglacials, in a strongly asymmetric pattern of gradual drawdowns followed by steep rises known as terminations (Figure 2(b)). Even higher CO_2 levels during interglacials are prevented by high plant productivity of forests in humid, previously glaciated terrains and of grasslands in arid rangelands. This slow weathering and biomass building, together with nutrient leakage to the ocean and carbon burial there, could draw down greenhouse gases and bring on cooling. As ice expands and grasslands are converted to deserts, the carbon sequestration capacity of soils and ecosystems is diminished. Large herds of mammals or populations of humans could disturb these impoverished soils into dustbowl conditions and the massive carbon oxidation events of a glacial termination. Such long-term biological trends, metered by steadily declining and then abruptly renewed soil nutrients, could amplify other drivers of climate, which include large ice caps, ocean currents, mountain building, and orbital configuration (Muller and MacDonald, 2000).

5.18.4 SOILS AND GLOBAL CARBON CYCLE CHANGES

Over geological time there have been dramatic changes in soil, life, and air that are well represented in the fossil record of soils. Paleosols are an underexploited record of past environments in land. This review has emphasized mainly the evidence from paleosols for changes in carbon cycling and greenhouse gases (CO_2, CH_4, H_2O) in the atmosphere over geological timescales. It is unremarkable that paleosols would change, particularly at high latitudes, as global climates warmed or cooled with changing atmospheric loads of greenhouse gases. It is notable that sequences of paleosols can, under certain circumstances, be high-resolution records of such paleoclimatic change (Figure 15).

Parallels between biological activity within soils and greenhouse gas composition have been emphasized in this review as fertile ground for future research (Figure 16). Olsen (1993) has suggested that soil producers such as plants cool the planet, but soil consumers such as animals warm it. This idea, which the author has dubbed the Proserpina principle after the ancient Roman

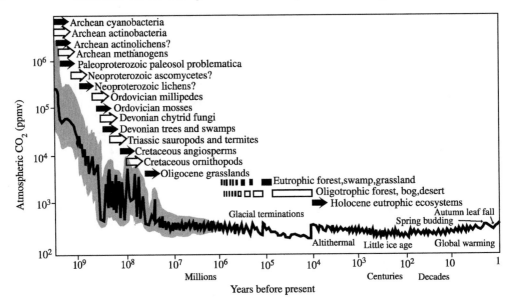

Figure 16 The Proserpina principle relates variation in atmospheric CO_2 concentration with coeval evolutionary and ecological events on a variety of timescales. Carbon sequestering evolutionary innovations and ecological transitions (closed symbols) alternate with carbon oxidizing evolutionary innovations and ecological transitions (open symbols). The CO_2 curve is a composite of those shown in Figure 2 and by Kasting (1992).

goddess of spring (Retallack, 2000b), is undeniable for the annual spring fall and autumn rise of atmospheric CO_2 with northern hemisphere leaf sprouting and shedding (Figure 2(a)). This explanation is especially demonstrated by the muted and out-of-phase annual fluctuation of CO_2 in the southern hemisphere (Mooney et al., 1987), where there is less fertile land, more evergreen plants, and different seasons. The question addressed here is whether the Proserpina principle operates on geologically significant timescales, and so far, such a simple idea does not conflict with the history of life and paleoclimate outlined here.

On evolutionary timescales, it is the biochemical evolution of lignin, pyrethrin, caffeine, and other substances that deter herbivory, digestion, and decay, which affect rates of carbon burial in sediments as the principal long-term control on atmospheric CO_2 levels. The role of trees and their soils in Late Paleozoic carbon sequestration, cooling, and glaciation is widely accepted (Berner, 1997; Algeo and Scheckler, 1998; see Chapter 5.06). The role of humans in global warming is also becoming well known (Vitousek et al., 1997b). According to the Proserpina principle, we may not have been the only organisms to have had significant effects on climate. There remain many other instances of global change less clearly related to changes in life and soils, in part because the numerous paleosols of appropriate age have not yet been studied in detail. Asteroid impacts, volcanic eruptions, and methane clathrate dissociation

events also affect life and the carbon cycle, producing transient greenhouse events (Retallack, 2001b). Ocean currents and mountain building also are likely to play a role in carbon sequestration (Raymo and Ruddiman, 1992; Ramstein et al., 1997). Soils and their ecosystems play an important role in the carbon cycle today, and the history of that role now decipherable from paleosols appears ripe for modeling and other quantitative comparisons with other likely controls on global paleoclimate change.

ACKNOWLEDGMENTS

Nathan Sheldon, Hope Jahren, and Tim White have been sounding boards for the ideas presented here. I also thank J. I. Drever and H. D. Holland for helpful reviews.

REFERENCES

Algeo T. J. and Scheckler S. E. (1998) Terrestrial–marine teleconnections in the Devonian: links between the evolution of land plants, weathering processes and anoxic events. *Roy. Soc. London Phil. Trans.* **B353**, 113–130.

Allen J. R. L. (1974) Geomorphology of Siluro-Devonian alluvial plains. *Nature* **249**, 644–645.

Amundson R., Stern L., Baisden T., and Wang Y. (1998) The isotopic composition of soil and respired CO_2. *Geoderma* **82**, 83–114.

Bakker R. T. (1985) *The Dinosaur Heresies*. William Morrow, New York.

Barley M. E., Pickard A. L., and Sylvester P. J. (1997) Emplacement of a large igneous province as a possible cause of banded iron formation 2.45 billion years ago. *Nature* **385**, 55–58.

Barsukov V. L., Volkhov V. P., and Khodakovsky I. L. (1982) The crust of Venus: theoretical models of chemical and mineral composition. *J. Geophys. Res. Suppl.* **87A**, 3–9.

Basilevsky A. T., Kuzmin R. O., Nikolaeva O. V., Pronin A. A., Ronca A. B., Avdvesky V. S., Uspensky G. R., Cheremukhina Z. P., Semenchenko V. V., and Ladygin V. M. (1985) The surface of Venus as revealed by the Venera landings: Part II. *Geol. Soc. Am. Bull.* **96**, 137–144.

Bell J. F. (1996) Iron sulfate, carbonate and hydrated minerals on Mars. In *Mineral Spectroscopy: A Tribute to Roger G. Burns*, Geochem. Soc. Spec. Publ. (eds. M. D. Dyar, C. MacCammon, and M. W. Schaeffer). Geochemical Society, St. Louis, vol. 5, pp. 359–380.

Berner R. A. (1997) The rise of plants and their effect on weathering and atmospheric CO_2. *Science* **276**, 543–546.

Berner R. A. and Kothavala Z. (2001) GEOCARBIII: a revised model of atmospheric CO_2 over Phanerozoic time. *Am. J. Sci.* **301**, 182–204.

Berner R. A., Petsch S. T., Lake J. A., Beerling D. J., Popp B. N., Lane R. S., Laws E. A., Westley M. B., Cassor N., Woodward F. I., and Quick W. P. (2000) Isotopic fractionation and atmospheric oxygen: implications for Phanerozoic evolution. *Science* **287**, 1630–1633.

Bestland E. A. (2000) Weathering flux and CO_2 consumption determined from paleosol sequences across the Eocene–Oligocene transition. *Palaeogeogr. Palaeoclimat. Palaeoecol.* **156**, 301–326.

Bestland E. A. and Retallack G. J. (1993) Volcanically influenced calcareous paleosols from the Miocene Kiahera Formation, Rusinga Island, Kenya. *Geol. Soc. London J.* **150**, 293–310.

Bestland E. A., Retallack G. J., Rice A. E., and Mindszenty A. (1996) Late Eocene detrital laterites in central Oregon: mass balance geochemistry, depositional setting and landscape evolution. *Geol. Soc. Am. Bull.* **108**, 285–302.

Beukes N. J., Dorland H., Gutzmer J., Nedachi M., and Ohmoto H. (2002) Tropical laterites, life on land, and the history of atmospheric oxygen in the Paleoproterozoic. *Geology* **30**, 491–494.

Birck J.-L. and Allègre C. J. (1988) Manganese–chromium systematics and the development of the early solar system. *Nature* **331**, 571–574.

Bird M. I. and Chivas A. R. (1993) Geomorphic and paleoclimatic implications of an oxygen isotope chronology for Australian deeply weathered profiles. *Austral. J. Earth Sci.* **40**, 345–358.

Blinnikov M., Busacca A., and Whitlock C. (2002) Reconstruction of the late Pleistocene grassland of the Columbia basin, Washington. *Palaeogeogr. Palaeoclimat. Palaeoecol.* **177**, 77–101.

Bluth G. G. S. and Kump L. P. (1991) Phanerozoic paleogeology. *Am. J. Sci.* **291**, 284–308.

Bockheim J. G. (1995) Permafrost distribution in the southern circum-polar region and its relation to the environment: a review and recommendation for further research. *Permafrost Periglac. Process.* **6**, 27–45.

Borden R. (2001) Geochemical evolution of sulphide-bearing waste rock soils at the Bingham Canyon Mine, Utah. In *Evolution and Remediation of Acid–Sulphate Systems at Reclaimed Mine Sites* (eds. J. J. Donovan and A. W. Rose). Geological Society of London, pp. 15–21.

Brearley A. J. (1999) Origin of graphitic carbon and pentlandite in matrix olivines in the Allende meteorite. *Science* **285**, 1380–1382.

Brenchley P. J., Marshall J. B., Carden G. A. F., Robertson D. B. R., Long D. E. F., Meidia T., Hints L., and Anderson T. F. (1994) Bathymetric and isotopic evidence for a short-lived Late Ordovician glaciation in a greenhouse period. *Geology* **22**, 293–298.

Brimhall G. H., Chadwick O. A., Lewis C. J., Compston W., Williams I. S., Danti K. J., Dietrich W. E., Power M. E., Hendricks D., and Bratt J. (1991) Deformational mass transport and invasive processes in soil evolution. *Science* **255**, 695–702.

Brook G. A., Folkoff M. E., and Box E. O. (1983) A world model of soil carbon dioxide. *Earth Surf. Process. Landforms* **8**, 79–88.

Buckland W. (1837) *Geology and Mineralogy Considered with Reference to Natural Theology*. W. Pickering, London.

Buick R., Thronetree J. R., McNaughton N. J., Smith J. B., Barley M. E., and Savage M. (1995) Record of emergent continental crust ~3.5 billion years ago in the Pilbara Craton of Australia. *Nature* **375**, 574–576.

Bunch T. E. and Chang S. (1980) Carbonaceous chondrites: II. Carbonaceous chondrite phyllosilicates and light element geochemistry as indicators of parent body processes and surface conditions. *Geochim. Cosmochim. Acta* **44**, 1543–1577.

Buol S. W., Hole F. D., and McCracken R. W. (1997) *Soil Genesis and Classification*, 4th edn. Iowa State Univ. Press, Ames.

Busacca A. J. (1989) Long Quaternary record in eastern Washington, USA, interpreted from multiple buried paleosols in loess. *Geoderma* **45**, 105–122.

Campbell I. B. and Claridge G. G. C. (1987) *Antarctica: Soils, Weathering Processes and Environment*. Elsevier, Amsterdam.

Canfield D. E. and Teske A. (1996) Late Proterozoic rise in atmospheric oxygen concentration inferred from phylogenetic and sulphur isotope studies. *Nature* **382**, 127–132.

Cerling T. E. (1991) Carbon dioxide in the atmosphere: evidence from Cenozoic and Mesozoic paleosols. *Am. J. Sci.* **291**, 377–400.

Cerling T. E., Harris J. M., MacFadden B. J., Leakey M. G., Quade J., Eisenmann V. V., and Ehleringer J. F. (1997) Global vegetation change through the Miocene/Pliocene boundary. *Nature* **389**, 153–158.

Chadwick O. A., Derry L. A., Vitousek P. M., Huebert B. J., and Hedin L. O. (1999) Changing sources of nutrients during four million years of ecosystem development. *Nature* **397**, 491–497.

Colin F., Brimhall G. H., Nahon D., Lewis C. J., Baronnet A., and Danti K. (1992) Equatorial rain forest lateritic mantles: a geomembrane filter. *Geology* **20**, 523–526.

Crane P. R., Friis E. M., and Pedersen K. J. (1995) The origin and early diversification of angiosperms. *Nature* **374**, 27–33.

Daily B. and Cooper M. R. (1976) Clastic wedges and patterned ground in the Late Ordovician–Early Silurian tillites of South Africa. *Sedimentology* **23**, 271–283.

Dalziel I. W. D. (1997) Neoproterozoic–Paleozoic geography and tectonics: review, hypothesis, environmental speculation. *Geol. Soc. Am. Bull.* **109**, 16–42.

Delvigne J. E. (1998) *Atlas of Micromorphology of Mineral Alteration and Weathering*. Canadian Mineralogist Special Publication, vol. 3, p. 495.

Des Marais D. J. (1997) Isotopic evolution of the biogeochemical carbon cycle during the Proterozoic Eon. *Org. Geochem.* **27**, 185–193.

de Wit H. A. (1978) *Soils and Grassland Types of the Serengeti Plain (Tanzania)*. Medelingen Landbouwhogeschool, Wageningen.

Driese S. G. and Foreman J. L. (1992) Traces and related chemical changes in a Late Ordovician paleosol, *Glossifungites* ichnofacies, southern Appalachians, USA. *Ichnos* **1**, 207–219.

Driese S. G., Mora C. I., and Elick J. M. (1997) Morphology and taphonomy of root and stump casts of the earliest trees (Middle and Late Devonian), Pennsylvania and New York, USA. *Palaios* **12**, 524–537.

Ekart D. P., Cerling T. E., Montañez I. P., and Tabor N. J. (1999) A 400 million year carbon isotope record of pedogenic carbonate: implications for paleoatmospheric carbon dioxide. *Am. J. Sci.* **299**, 805–827.

Elick J. E., Driese S. E., and Mora C. I. (1998) Very large plant root traces from the Early to Middle Devonian: implications for early terrestrial ecosystems and pCO$_2$ estimates. *Geology* **26**, 143–146.

Endress M., Zinner E., and Bischoff A. (1996) Early aqueous activity on primitive meteorite parent bodies. *Nature* **379**, 701–704.

FAO (1988) *Soil Map of the World. Vol. 1.* Revised Legend. UNESCO, Rome.

Farquhar G. D., Lloyd J., Taylor J. A., Flanagan L. B., Syvertsen J. P., Hubick K. T., Chin Wong S., and Ehleringer J. R. (1993) Vegetation effects of the isotopic composition of oxygen in atmospheric CO$_2$. *Nature* **363**, 439–443.

Farquhar J., Wing B. A., McKeegan K. D., Harris J. W., Cartigny P., and Thiemens M. H. (2002) Mass-independent sulfur of inclusions in diamond and sulfur recycling on early Earth. *Science* **298**, 2369–2372.

Feng Z.-D., Johnson W. L., Lu L.-C., and Ward P. A. (1994) Climatic signals from loess-soil sequences in the central Great Plains, USA. *Palaeogeogr. Palaeoclimat. Palaeoecol.* **110**, 345–358.

Firman J. B. (1994) Paleosols in laterite and silcrete profiles: evidence from the southeast margin of the Australian Precambrian Shield. *Earth Sci. Rev.* **36**, 149–179.

Follmer L. R., McKay E. D., Lineback J. A., and Gross D. L. (1979) Wisconsinan, Sangamonian and Illinoian stratigraphy in central Illinois. *Guidebk Illinois, State Geol. Surv.* **13**, 138.

Foster R. C. (1981) Polysaccharides in soil fabrics. *Science* **214**, 665–667.

Genise J. F., Mángano M. G., Buatois L. A., Laza J. H., and Verde M. (2000) Insect trace fossil associations in paleosols: the *Coprinisphaera* ichnofacies. *Palaios* **15**, 49–64.

Ghienne J. F. (2003) Late Ordovician sedimentary environments, glacial cycles and post-glacial transgression in the Taoudeni Basin, West Africa. *Palaeogeogr. Palaeoclimat. Palaeoecol.* **189**, 117–145.

Grandstaff D. E., Edelman M. J., Foster R. W., Zbinden E., and Kimberley M. M. (1986) Chemistry and mineralogy of Precambrian paleosols at the base of the Dominion and Pongola Groups. *Precamb. Res.* **32**, 91–131.

Gray J. (1993) Major Paleozoic land plant evolutionary bioevents. *Palaeogeogr. Palaeoclimat. Palaeoecol.* **104**, 153–169.

Grazdhankin M. (2001) Microbial origin of some of the Ediacaran fossils. *Abstr. Geol. Soc. Am.* **33**(6), A429.

Gutzmer J. and Beukes N. J. (1998) Earliest laterites and possible evidence for terrestrial vegetation in the early Proterozoic. *Geology* **26**, 263–266.

Hallam A. and Wignall P. (1997) *Mass Extinctions and their Aftermath*. Oxford University Press, New York.

Hallbauer D. K., Jahns H. M., and Beltmann H. A. (1977) Morphological and anatomical observations on some Precambrian plants from the Witwatersrand, South Africa. *Geol. Rundsch.* **66**, 477–491.

Hasiotis S. T. and Dubiel D. L. (1995) Termite (Insecta, Isoptera) nest ichnofossils from the Upper Triassic Chinle Formation, Petrified Forest national Monument, Arizona. *Ichnos* **4**, 111–130.

Hays J. D., Imbrie J., and Shackleton N. J. (1976) Variations in the Earth's orbit: pacemaker of the ice ages. *Science* **194**, 1121–1132.

Heckman D. S., Geiser D. M., Eidell B. R., Stauffer R. L., Kardos N. L., and Hedges S. B. (2001) Molecular evidence for early colonization of land by fungi and plants. *Science* **293**, 1129–1133.

Hoffman P. F., Kaufman A. J., Halverson G. P., and Schrag D. P. (1998) A Neoproterozoic snowball Earth. *Science* **281**, 1342–1346.

Holland H. D. (1984) *The Chemical Evolution of the Atmosphere and Oceans*. Princeton University Press, Princeton.

Holland H. D. and Beukes M. J. (1990) A paleoweathering profile from Griqualand West, South Africa: evidence for a sudden rise in atmospheric oxygen between 2.2 and 1.6 bybp. *Am. J. Sci.* **290**, 1–34.

Holser W. T. and Schönlaub H. P. (1991) The Permian–Triassic boundary in the Carbic Alps of Austria (Gartnerkofel region). *Abh. Geol. Bund. Autriche Wien* **45**, 1–232.

Horodyski R. J. and Knauth L. P. (1994) Life on land in the Precambrian. *Science* **263**, 474–498.

Hutton J. (1795) *Theory of the Earth, with Proofs and Illustrations*. J. W. Creech, Edinburgh.

Jahren A. H. and Sternberg L. S. L. (2002) Eocene meridional weather patterns reflected in the oxygen isotopes of Arctic fossil wood. *GSA Today* **12**(1), 4–9.

Jahren A. H., Arens N. C., Sarmiento G., Guerro J., and Armstrong R. (2001) Terrestrial record of methane hydrate degassing in the Early Cretaceous. *Geology* **29**, 159–162.

Janis C. M., Damuth J., and Theodor J. M. (2002) The origins and evolution of the North American grassland biome: the story from hoofed mammals. *Palaeogeogr. Palaeoclimat. Palaeoecol.* **177**, 183–198.

Jenny H. J. (1941) *Factors in Soil Formation*. McGraw-Hill, New York.

Jeram A. J., Selden P. A., and Edwards D. (1990) Land animals in the Silurian: arachnids and myriapods from Shropshire, England. *Science* **250**, 658–661.

Joeckel R. M. (1987) Paleogeomorphic significance of two paleosols in the Dakota Formation (Cretaceous), southeastern Nebraska. *Contrib. Geol. Univ. Wyoming* **25**, 91–102.

Johnson E. W., Briggs D. E. G., Suthren R. J., Wright J. L., and Tunnicliff J. P. (1994) Non-marine arthropod traces from subaerial Ordovician Borrowdale Volcanic Group, English Lake district. *Geol. Mag.* **131**, 395–406.

Kasting J. F. (1992) Proterozoic climates: the effects of changing atmospheric carbon dioxide concentrations. In *The Proterozoic Biosphere: A Multidisciplinary Study* (eds. J. W. Schopf and C. Klein). Cambridge University Press, Cambridge, pp. 165–168.

Keeling C. D., Bacastow R. B., and Whorf T. P. (1982) Measurement of the concentration of carbon dioxide at Mauna Loa Observatory, Hawaii. In *Carbon Dioxide Review 1982* (ed. W. C. Clark). Oxford University Press, New York, pp. 377–385.

Kenrick P. and Crane P. R. (1997) *Early Evolution of Land Plants*. Smithsonian Institution Press, Washington.

Kirschvink J. (1992) Late Proterozoic low latitude global glaciation. In *The Proterozoic Biosphere: A Multidisciplinary Study* (eds. J. W. Schopf and C. Klein). Cambridge University Press, Cambridge, pp. 51–52.

Knoll M. A. and James W. C. (1987) Effect of the advent and diversification of vascular plants on mineral weathering through geologic time. *Geology* **15**, 1099–1102.

Koch P. L. (1998) Isotopic reconstruction of past continental environments. *Ann. Rev. Earth Planet. Sci.* **26**, 573–613.

Krassilov V. A. (1981) *Orestovia* and the origin of vascular plants. *Lethaia* **14**, 235–250.

Krijgsman W., Hilgen F. J., Ruffi I., Sienro F. J., and Wilson R. S. (1999) Chronology, causes and progression of the Messinian Salinity Crisis. *Nature* **400**, 652–655.

Krull E. S. (1999) Permian palsamires as paleoenvironmental proxies. *Palaios* **14**, 530–544.

Krull E. S. and Retallack G. J. (2000) δ^{13}C$_{org}$ depth profiles of paleosols across the Permian–Triassic boundary: evidence for methane release. *Geol. Soc. Am. Bull.* **112**, 1459–1472.

Krull E. S., Retallack G. J., Campbell H. J., and Lyon G. L. (2000) δ^{13}C$_{org}$ chemostratigraphy of the Permian–Triassic boundary in the Maitai Group, New Zealand: evidence for high latitude methane release. *N. Z. J. Geol. Geophys.* **43**, 21–73.

Larson R. L. (1991) Geological consequences of superplumes. *Geology* **19**, 963–966.

Lockley M. (1991) *Tracking Dinosaurs*. Cambridge University Press, Cambridge.

Ludvigsen G. A., González L. A., Metzger R. A., Witzke B. J., Brenner R. L., Murillo A. P., and White T. S. (1998)

Meteoric sphaerosiderite lines and their use in paleohydrology and paleoclimatology. *Geology* **26**, 1039–1042.

MacFadden B. J. (2000) Origin and evolution of the grazing guild in Cenozoic New World mammals. In *Evolution of Herbivory in Terrestrial Vertebrates* (ed. H.-D. Sues). Cambridge University Press, Cambridge, pp. 223–244.

MacFadden B. J., Solounias N., and Cerling T. E. (1999) Ancient diests, ecology and extinction of 5-million-year-old horses from Florida. *Science* **283**, 824–827.

Mack G. H. (1992) Paleosols as an indicator of climate change at the Early–Late Cretaceous boundary, southwestern New Mexico. *J. Sedim. Petrol.* **62**, 484–494.

Marbut C. F. (1935). *Atlas of American Agriculture: Part III. Soils of the United States, US Department of Agriculture Advance Sheets*, Government Printer, Washington, DC, vol. 8.

Marmo J. S. (1992) The lower Proterozoic Hokkalampi paleosol in North Karelia. In *Early Organic Evolution: Implications for Mineral and Energy Resources* (eds. M. Schidlowski, S. Golubic, M. M. Kimberley, D. M. McKirdy, and P. A. Trudinger). Springer, Berlin, pp. 41–66.

Maynard J. B. (1992) Chemistry of modern soils as a guide to interpreting Precambrian paleosols. *J. Geol.* **100**, 279–289.

McSween H. Y. (1997) Evidence for life in a martian meteorite? *GSA Today* **7**(7), 1–7.

McSween H. Y. (1999) *Meteorites and their Parent Planets.* Cambridge University Press, Cambridge.

Merino E., Nahon D., and Wang Y.-F. (1993) Kinetics and mass transfer of pseudomorphic replacement: application to replacement of parent minerals and kaolinite by Al, Fe and Mn oxides during weathering. *Am. J. Sci.* **293**, 135–155.

Meyer-Berthaud B., Scheckler S. E., and Wendt J. (1999) *Archaeopteris* is earliest known tree. *Nature* **398**, 700–701.

Miller K. B. and West R. R. (1998) Identification of sequence boundaries within cyclic strata of the Lower Permian of Kansas, USA: problems and alternatives. *J. Geol.* **106**, 119–132.

Mooney H. A., Vitousek P. M., and Matson P. A. (1987) Exchange of materials between terrestrial ecosystems and the atmosphere. *Science* **238**, 926–932.

Mora C. I. and Driese S. G. (1999) Palaeoenvironment, palaeoclimate and stable carbon isotopes of Palaeozoic red-bed palaeosols, Appalachian Basin, USA and Canada. In *Palaeoweathering, Palaeosurfaces and Continental Deposits.* Int. Assoc. Sedimentology Spec. Publ. (eds. M. Thiry and R. Simon-Coinçon). Blackwell, Oxford, vol. 27, pp. 61–84.

Mora C. I., Driese S. G., and Colarusso L. A. (1996) Middle to Late Paleozoic atmospheric CO_2 from soil carbonate and organic matter. *Science* **271**, 1105–1107.

Mora C. I., Sheldon B. T., Elliott N. C., and Driese S. G. (1998) An oxygen isotope study of illite and calcite in three Appalachian vertic paleosols. *J. Sedim. Res.* **A68**, 456–464.

Mossman D. J. and Farrow C. E. G. (1992) Paleosol and ore-forming processes in the Elliot Lake district of Canada. In *Early Organic Evolution: Implications for Energy and Mineral Resources* (eds. M. Schidlowski, S. Golubic, M. M. Kimberley, D. M. McKirdy, and P. A. Trudinger). Springer, Berlin, pp. 67–76.

Muller R. A. and MacDonald G. J. (2000) *Ice Ages and Astronomical Causes: Data, Spectral Analyses and Mechanisms.* Springer, Berlin.

Nahon D. B. (1991) *Introduction to the Petrology of Soils and Chemical Weathering.* Wiley, New York.

Nøhr-Hansen H. and Koppelhus E. B. (1988) Ordovician spores with trilete rays from Washington Land, North Greenland. *Rev. Palaeobot. Palynol.* **56**, 305–311.

O'Geen A. T. and Busacca A. J. (2001) Faunal burrows as indicators of paleovegetation in eastern Washington. *Palaeogeogr. Palaeoclimat. Palaeoecol.* **169**, 23–37.

Ohmoto H. (1996) Evidence in pre-2.2 Ga paleosols for the early evolution of atmospheric oxygen and terrestrial biota. *Geology* **24**, 1135–1138.

Ohmoto H. (1997) Evidence in pre-2.2 Ga paleosols for the early evolution of atmospheric oxygen and terrestrial biota: reply. *Geology* **25**, 857–858.

Ollier C. and Pain C. (1996) *Regolith, Soils and Landforms.* Wiley, Chichester.

Olsen P. E. (1993) The terrestrial plant and herbivore arms race: a major control of Phanerozoic CO_2? *Abstr. Prog. Geol. Soc. Am.* **25**(3), 71.

Paepe R. and van Overloop E. (1990) River and soil cyclicities interfering with sea level changes. In *Greenhouse Effect, Sea Level Change and Drought* (eds. R. Paepe, R. Fairbridge, and S. Jelgersma). Kluwer Academic, Dordrecht, pp. 253–280.

Pavlov A. A., Kasting J. F., Brown L. L., Rage K. A., and Freedman R. (2000) Greenhouse warming by CH_4 in the atmosphere of early Earth and Mars. *J. Geophys. Res.* **105**, 11981–11990.

Petit J. R., Jouzel J., Raynaud D., Barkov N. I., Barnola J.-M., Basile I., Bender M., Chapellaz J., Davis M., Pelaygue G., Delmotte M., Kotylakov V. M., Legrand M., Lipankov V. Y., Lorius C., Pepin L., Ritz C., Saltzman E., and Stlevenard M. (1999) Climate and atmospheric history of the past 420,000 years from the Vostok ice core, Antarctica. *Nature* **399**, 429–436.

Pinto J. P. and Holland H. D. (1988) Paleosols and the evolution of the atmosphere: Part II. In *Paleosols and Weathering through Geologic Time*, Spec. Pap. Geol. Soc. Am. (eds. J. Reinhardt and W. Sigleo). Geological Society of America, Boulder, vol. 216, pp. 21–34.

Prashnowsky A. A. and Schidlowski M. (1967) Investigation of a Precambrian thucolite. *Nature* **216**, 560–563.

Ramstein G., Fluteau F., Besse J., and Joussame S. (1997) Effects of orogeny, plate motion and land-sea distribution on European climate change over the past 30 million years. *Nature* **386**, 788–795.

Rasbury E. T., Hanson G. N., Meyers W. J., Goldstein R. H., and Saller A. H. (1998) U–Pb dates of paleosols: constraints on Late Paleozoic cycle durations and boundary ages. *Geology* **26**, 403–406.

Raymo M. E. and Ruddiman W. F. (1992) Tectonic forcing of Late Cenozoic climate. *Nature* **359**, 117–122.

Redecker D., Kodner R., and Graham L. E. (2000) Glomalean fungi from the Ordovician. *Science* **289**, 1920–1921.

Retallack G. J. (1976) Triassic palaeosols in the upper Narrabeen Group of New South Wales: Part I. Features of the palaeosols. *Geol. Soc. Australia J.* **23**, 383–399.

Retallack G. J. (1983) Late Eocene and Oligocene paleosols from Badlands National Park, South Dakota. *Geol. Soc. Am. Spec. Pap.* **193**, 82.

Retallack G. J. (1985) Fossil soils as grounds for interpreting the advent of large plants and animals on land. *Roy. Soc. London Phil. Trans.* **B309**, 105–142.

Retallack G. J. (1986) Reappraisal of a 2,200-Ma-old paleosol from near Waterval Onder, South Africa. *Precamb. Res.* **32**, 195–232.

Retallack G. J. (1991) Untangling the effects of burial alteration and ancient soil formation. *Ann. Rev. Earth Planet. Sci.* **19**, 183–206.

Retallack G. J. (1993) Late Ordovician paleosols of the Juniata Formation near Potters Mills. In *Paleosols, Paleoclimate and Paleoatmospheric CO_2: Paleozoic Paleosols of Central Pennsylvania* (ed. S. G. Driese). Univ. Tennessee Dept. Geol. Sci. Stud. Geol., Knoxville, vol. 22, pp. 33–50.

Retallack G. J. (1994a) A pedotype approach to Latest Cretaceous and Early Paleocene paleosols in eastern Montana. *Geol. Soc. Am. Bull.* **106**, 1377–1397.

Retallack G. J. (1994b) Were the Ediacaran fossils lichens? *Paleobiology* **20**, 523–544.

Retallack G. J. (1994c) The environmental factor approach to the interpretation of paleosols. In *Factors in Soil Formation: A Fiftieth Anniversary Retrospective*, Soil Sci. Soc. Am. Spec. Publ. (eds. R. Amundson, J. Harden, and

M. Singer). Soil Society of America, Madison, vol. 33, pp. 31–64.

Retallack G. J. (1995) Pennsylvanian vegetation and soils. In *Predictive Stratigraphic Analysis* (eds. B. Cecil and T. Edgar). US Geol. Surv. Bull., Washington, DC, vol. 2110, pp. 13–19.

Retallack G. J. (1997a) *A Colour Guide to Paleosols*. Wiley, Chichester.

Retallack G. J. (1997b) Early forest soils and their role in Devonian global change. *Science* **276**, 583–585.

Retallack G. J. (1997c) Dinosaurs and dirt. In *Dinofest* (eds. D. Wolberg and E. Stump). Academy of Natural Sciences, Philadelphia, pp. 345–359.

Retallack G. J. (1997d) Neogene expansion of the North American prairie. *Palaios* **12**, 380–390.

Retallack G. J. (1999a) Carboniferous fossil plants and soils of an early tundra ecosystem. *Palaios* **14**, 324–336.

Retallack G. J. (1999b) Permafrost palaeoclimate of Permian palaeosols in the Gerringong volcanics of New South Wales. *Austral. J. Earth Sci.* **46**, 11–22.

Retallack G. J. (2000a) Ordovician life on land and Early Paleozoic global change. In *Phanerozoic Terrestrial Ecosystems*, Paleont. Soc. Short Course Notes (eds. R. A. Gastaldo and W. A. DiMichele). Carnegie Museum, Pittsburg, vol. 6, pp. 21–45.

Retallack G. J. (2000b) The Proserpina principle: a role for soil communities in regulating atmospheric composition on time scales ranging from ecological to geological. *Abstr. Geol. Soc. Am.* **32**(7), A486.

Retallack G. J. (2001a) *Soils of the Past*, 2nd edn. Blackwell, Oxford.

Retallack G. J. (2001b) A 300 million year record of atmospheric CO_2 from fossil plant cuticles. *Nature* **411**, 287–290.

Retallack G. J. (2001c) *Scoyenia* burrows from Ordovician paleosols of the Juniata Formation in Pennsylvania. *Palaeontology* **44**, 209–235.

Retallack G. J. (2001d) Cenozoic expansion of grasslands and global cooling. *J. Geol.* **109**, 407–426.

Retallack G. J. (2002) Carbon dioxide and climate over the past 300 Myr. *Roy. Soc. London Phil. Trans.* **A360**, 659–674.

Retallack G. J. and Dilcher D. L. (1981) A coastal hypothesis for the dispersal and rise to dominance of flowering plants. In *Paleobotany, Paleoecology and Evolution* (ed. K. J. Niklas). Praeger, New York, vol. 2, pp. 27–77.

Retallack G. J. and Dilcher D. L. (1986) Cretaceous angiosperm invasion of North America. *Cretaceous Res.* **7**, 227–252.

Retallack G. J. and Krinsley D. H. (1993) Metamorphic alteration of a Precambrian (2.2 Ga) paleosol from South Africa revealed by back-scatter imaging. *Precamb. Res.* **63**, 27–41.

Retallack G. J. and Krull E. S. (1999) Ecosystem shift at the Permian–Triassic boundary in Antarctica. *Austral. J. Earth Sci.* **46**, 785–812.

Retallack G. J. and Mindszenty A. (1994) Well preserved Late Precambrian paleosols from northwest Scotland. *J. Sedim. Res.* **A64**, 264–281.

Retallack G. J. and Storaasli M. (1999) Problematic impressions from the Precambrian of Montana. *Abstr. Geol. Soc. Am.* **31**(7), A362.

Retallack G. J., Bestland E. A., and Fremd T. (2000) Eocene and Oligocene paleosols in central Oregon. *Geol. Soc. Am. Spec. Pap.* **344**, 192.

Retallack G. J., Krull E. S., and Bockheim J. G. (2001) New grounds for reassessing the palaeoclimate of the Sirius Group, Antarctica. *Geol. Soc. London J.* **158**, 925–935.

Rice C. M., Ashcroft W. A., Batten D. J., Boyce A. J., Caulfield J. B. D., Fallick A. E., Hole M. J., Jones E., Pearson M. J., Rogers G., Saxton J. M., Stuart F. M., Trewin N. H., and Turner G. (1995) A Devonian auriferous hot spring

system, Rhynie, Scotland. *Geol. Soc. London J.* **152**, 229–250.

Rich T. H. and Vickers-Rich P. T. (2000) *Dinosaurs of Darkness*. University of Indiana Press, Bloomington.

Rye R. and Holland H. D. (1998) Paleosols and the evolution of atmospheric oxygen: a critical review. *Am. J. Sci.* **298**, 621–672.

Rye R. and Holland H. D. (2000) Life associated with a 2.76 Ga ephemeral pond? Evidence from Mount Roe #2 paleosol. *Geology* **28**, 483–486.

Rye R., Kuo P. H., and Holland H. D. (1995) Atmospheric carbon dioxide concentrations before 2.2 billion years ago. *Nature* **378**, 603–605.

Schidlowski M. and Aharon P. (1992) Carbon cycle and carbon isotopic record: geochemical impact of life over 3.8 Ga of Earth history. In *Early Organic Evolution: Implications for Energy and Mineral Resources* (eds. M. Schidlowski, S. Golubic, M. M. Kimberley, D. M. McKirdy, and P. A. Trudinger). Springer, Berlin, pp. 147–175.

Schidlowski M., Hayes J. M., and Kaplan I. R. (1983) Isotopic inferences of ancient biochemistries: carbon, sulfur, hydrogen and nitrogen. In *Earth's Earliest Biosphere: Its Origin and Evolution* (ed. J. W. Schopf). Princeton University Press, Princeton, pp. 149–186.

Schmidt P. W. and Williams G. E. (1999) Palaeomagnetism of the Palaeoproterozoic hematitic breccia and paleosol at Ville-Marie, Quebec: further evidence for the low palaeo-latitude of Huronian glaciation. *Earth Planet. Sci. Lett.* **172**, 273–285.

Schmitt J.-M. (1999) Weathering, rainwater and atmospheric chemistry: an example and modeling of granite weathering in present conditions, in a CO_2 rich and in an anoxic palaeoatmosphere. In *Palaeoweathering, Palaeosurfaces and Continental Deposits*, Int. Assoc. Sedimentology Spec. Publ. (eds. M. Thiry and R. Simon-Coinçon). Blackwell, Oxford, vol. 27, pp. 21–41.

Schumm S. A. (1956) The role of creep and rainwash on the retreat of badland slopes. *Am. J. Sci.* **254**, 693–706.

Schwartzmann D. W. and Volk T. (1991) Biotic enhancement of weathering and surface temperatures of Earth since the origin of life. *Palaeogeogr. Palaeoclimat. Palaeoecol.* **90**, 357–371.

Schwarz T. (1997) Lateritic paleosols in central Germany and implications for Miocene paleoclimate. *Palaeogeogr. Palaeoclimat. Palaeoecol.* **129**, 37–50.

Serdyuchenko D. P. (1968) Metamorphosed weathering crusts of the Precambrian: their metallogenic and petrographic fabric. In *Precambrian Geology*. Proc. 13th Int. Geol. Congr. Prague (ed. B. Hejtman). Academia, Prague, vol. 4, pp. 37–42.

Sheldon N. D., Retallack G. J., and Reed M. H. (2001) Siderite-iron-silicate equilibria in paleosols as an atmospheric CO_2 paleobarometer or paleoproductivity index? *Abstr. Geol. Soc. Am. Geol. Soc. London Global Meet. Edinburgh*, 42.

Sheldon N. D., Retallack G. J., and Tanaka S. (2002) Geochemical climofunctions from North American soils and application to paleosols across the Eocene–Oligocene boundary in Oregon. *J. Geol.* **110**, 687–696.

Sherwood-Pike M. A. and Gray J. (1985) Silurian fungal remains: probable records of Ascomycetes. *Lethaia* **18**, 1–20.

Siegenthaler U. and Sarmiento J. L. (1993) Atmospheric carbon dioxide and the ocean. *Nature* **365**, 119–125.

Snigirevskaya N. S., Popov L. E., and Zdebsak D. (1992) Novie nakhodki ostatkov drevnishchikh vishchikh rastenii v srednem ordovike yuzhnogo kazachstana (New findings of the oldest higher plant remains in the Middle Ordovician of South Kazachstan). *Bot. Zh.* **77**(4), 1–8.

Soil Survey Staff (1999) *Keys to Soil Taxonomy*. Pocahontas Press, Blacksburg, Virginia.

Stallard R. F. (1998) Terrestrial sedimentation and the carbon cycle: coupling weathering and erosion to carbon burial. *Global Biogeochem. Cycles* **12**, 231–257.

Stanley S. M. and Hardie L. A. (1999) Hypercalcification: paleontology links plate tectonics and geochemistry to sedimentology. *GSA Today* **9**(2), 1–7.

Steiner M. and Reitner J. (2001) Evidence of organic structures in Ediacara-type fossils and associated microbial mats. *Geology* **29**, 1119–1122.

Strömberg C. A. E. (2002) The origin and spread of grass-dominated ecosystems in the Late Tertiary of North America: preliminary results concerning the evolution of hypsodonty. *Palaeogeogr. Palaeoclimat. Palaeoecol.* **177**, 59–75.

Strother P. (2000) Cryptospores: the origin and evolution of the terrestrial flora. In *Phanerozoic Terrestrial Ecosystems*, Paleont, Soc. Short Course Notes (eds. R. A. Gastaldo and W. A. DiMichele). vol. 6, pp. 3–20.

Strother P. K., Al-Hatri S., and Traverse A. (1996) New evidence for land plants from the lower Middle Ordovician of Saudi Arabia. *Geology* **24**, 55–58.

Sugden D. E., Marchant D. R., Potter N., Souchez R. A., Denton G. H., Swisher C. C., and Tison J. L. (1995) Preservation of Miocene glacier ice in East Antarctica. *Nature* **376**, 412–414.

Taylor F., Eggleton R. A., Holzhauer C. C., Maconachie L. A., Gordon M., Brown M. C., and McQueen K. G. (1992) Cool climate lateritic and bauxitic weathering. *J. Geol.* **100**, 669–677.

Taylor T. N. and Taylor E. L. (2000) The Rhynie Chert ecosystem: a model for understanding fungal interactions. In *Microbial Endophytes* (eds. C. W. Bacon and J. F. White). Dekker, New York, pp. 31–47.

Thaer A. D. (1857) *The Principles of Practical Agriculture* (translated by W. Shaw and C. W. Johnson). Saxton, New York.

Thorp J. and Reed E. C. (1949) Is there laterite in rocks of the Dakota Group? *Science* **109**, 69.

Trewin N. H. and McNamara K. J. (1995) Arthropods invade the land: trace fossils and palaeoenvironments of the Tumblagooda Sandstone (?Late Silurian) of Kalbarri, Western Australia. *Roy. Soc. Edinburgh, Earth Sci. Trans.* **85**, 117–210.

Truswell E. M. (1987) The initial radiation and rise to dominance of angiosperms. In *Rates of Evolution* (eds. K. S. W. Campbell and M. F. Day). Allen and Unwin, London, pp. 101–128.

Veizer J., Godderis Y., and François L. M. (2000) Evidence for decoupling of atmospheric CO_2 and global climate during the Phanerozoic. *Nature* **408**, 698–701.

Veverka J., Thomas P., Harch A., Clark B., Bell J. F., Carcich B., Joseph J., Chapman C., Merline W., Robinson M., Malin M., McFaddem L. A., Murchie S., Hawkins S. E., Farquahar R., Isenberg N., and Cheng A. (1997) NEAR's flyby of 253 Mathilde: images of a C asteroid. *Science* **278**, 2109–2114.

Vitousek P. M., Chadwick O. A., Crews T. E., Fownes J. H., Hendricks D. M., and Herbert D. (1997a) Soil and ecosystem development across the Hawaiian Islands. *GSA Today* **7**(9), 1–8.

Vitousek P. M., Mooney H. A., Lubchenko J., and Melillo J. M. (1997b) Human domination of Earth's ecosystems. *Science* **277**, 494–499.

Wang H., Liu C. L., and Follmer L. R. (1998) Climatic trend and habitat variation based on oxygen and carbon isotopes in paleosols from Liujiapo, Shaanxi, China. *Quat. Int.* **51/52**, 52–54.

Watanabe Y., Martini J. E. J., and Ohmoto H. (2000) Geochemical evidence for terrestrial ecosystems 2.6 billion years ago. *Nature* **408**, 574–578.

Webster T. (1826) Observations on the Purbeck and Portland Beds. *Geol. Soc. London Trans.* **2**, 37–44.

Williams G. E. (1986) Precambrian permafrost horizons as indicators of paleoclimate. *Precamb. Res.* **32**, 233–242.

Williams G. E. and Schmidt P. W. (1997) Palaeomagnetic dating of the sub-Torridonian weathering profiles, NW Scotland: verification of Neoproterozoic palaeosols. *Geol. Soc. London J.* **154**, 987–997.

Wing S. L. and Boucher L. D. (1998) Ecological aspects of the Cretaceous flowering plant radiation. *Ann. Rev. Earth Planet. Sci.* **26**, 379–421.

Yang W. and Holland H. D. (2003) The Hekpoort paleosol at Strata 1 Gaborone, Botswana: soil formation during the Great Oxidation Event. *Am. J. Sci.* **303**, pp. 187–220.

Yapp C. J. and Poths H. (1994) Productivity of pre-vascular biota inferred from $Fe(CO_3)OH$ content of goethite. *Nature* **368**, 49–51.

Young G. M. and Long D. G. F. (1976) Ice wedge casts from the Huronian, Ramsay Lake Formation (2300 m.y. old), near Espanola, northern Canada. *Palaeogeogr. Palaeoclimat. Palaeoecol.* **19**, 191–200.

Zachos J., Pagani M., Sloan L., Thomas E., and Billups K. (2001) Trends, rhythms and aberrations in global climate 65 Ma to present. *Science* **292**, 689–693.

Volume Subject Index

The index is in letter-by-letter order, whereby hyphens and spaces within index headings are ignored in the alphabetization (e.g. Arabian–Nubian Shield precedes Arabian Sea). Terms in parentheses are excluded from the initial alphabetization. In line with normal materials science practice, compound names are not inverted but are filed under substituent prefixes.

The index is arranged in set-out style, with a maximum of three levels of heading. Location references refer to the page number. Major discussion of a subject is indicated by bold page numbers. Page numbers suffixed by *f* or *t* refer to figures or tables.

Printed and bound by CPI Group (UK) Ltd, Croydon, CR0 4YY

08/05/2025

01864784-0001